Andesites

Andesites

Orogenic Andesites and Related Rocks

edited by

R. S. Thorpe
*Department of Earth Sciences,
The Open University,
Milton Keynes*

A Wiley–Interscience Publication

JOHN WILEY & SONS
Chichester · New York · Brisbane · Toronto · Singapore

Copyright © 1982 by John Wiley & Sons Ltd.,

All rights reserved.

No part of this book may be reproduced by any means, nor transmitted, nor translated into a machine language without the written permission of the publisher.

British Library Cataloguing in Publication Data:

Andesites.
  1. Andesite
  I. Thorpe, R. S
552′.2   QE462.A5   80-042307

ISBN 0 471 28034 8

Type-set in Great Britain by Composition House, Salisbury, Wiltshire.

Printed in Great Britain by Page Bros. (Norwich) Limited.

Contents

List of Contributors vii
Preface xi
Acknowledgements xiii

I **Introduction** .. .. R. S. Thorpe 1

II **Classification, Petrology, and Mineralogy of Orogenic Volcanic Rocks**
Evolution and classification of orogenic volcanic rocks .. .. .. P. E. Baker 11
The mineralogy and petrology of Tertiary–Recent orogenic volcanic rocks: with special reference to the andesite–basaltic compositional range .. .. .. A. Ewart 25

III **Regional Distribution and Character of Active Andesite Volcanism**
The Aleutians .. .. B. D. Marsh 99
The Cascade Province .. .. A. R. McBirney and C. M. White 115
Mexico .. .. C. Robin 137
Central America .. .. .. M. J. Carr, W. I. Rose Jr, and R. E. Stoiber 149
The Lesser Antilles .. .. W. J. Rea 167
The Andes .. R. S. Thorpe, P. W. Francis, M. Hammill, and M. C. W. Baker 187
Indonesia .. .. C. S. Hutchison 207
Papua New Guinea .. .. R. W. Johnson 225
Tonga–Kermadec–New Zealand .. .. J. W. Cole 245
Japan .. .. S. Aramaki and T. Ui 259
Mariana–Volcano Islands .. .. A. Meijer 293
Mediterranean island arcs .. .. J. Keller 307
Anatolia and north-western Iran   F. Innocenti, P. Manetti, R. Mazzuoli, G. Pasquaré, and L. Villari 327

IV **Evolution of Andesite Volcanic Provinces**
Cainozoic andesitic rocks of Sardinia (Italy) .. J. Dostal, C. Coulon, and C. Dupuy 353
Volcanic evolution of the northern Antarctic Peninsula and the Scotia arc   J. Tarney, S. D. Weaver, A. D. Saunders, R. J. Pankhurst, and P. F. Barker 371

V **Andesitic Volcanoes and Their Products**
Eruptions of andesitic volcanoes .. .. G. P. L. Walker 403
Andesitic pyroclastic flows .. .. A. L. Smith and M. J. Roobol 415

VI **The Relationship of Andesitic Volcanism to Intrusive Activity**
Calc-alkaline intrusive rocks: their diversity, evolution, and relation to volcanic arcs
G. C. Brown 437

VII Experimental Studies Relevant to Petrogenesis of Andesites
Anatexis of mafic crust and high pressure crystallization of andesite . . *T. H. Green* 465
The role of mantle anatexis . . . . *B. O. Mysen* 489

VIII Chemical and Isotope Characteristics of Destructive Margin Magmas
Trace element characteristics of lavas from destructive plate boundaries *J. A. Pearce* 525
Isotope characteristics of magmas erupted along destructive plate margins . . . .
C. J. Hawkesworth 549

IX Andesite Volcanism Throughout Geological Time
Archaean andesites . . . . *K. C. Condie* 575
Upper Proterozoic (Pan-African) calc-alkaline magmatism in north-eastern Africa and Arabia . . . . *I. G. Gass* 591
Volcanism in the Caledonian orogenic belt of Britain *J. G. Fitton, M. F. Thirlwall, and D. J. Hughes* 611

X Implications of Andesite Volcanism
Andesitic magmatism and continental growth . . *B. L. Weaver and J. Tarney* 639
Andesites–product of geosphere mixing . . . . *W. S. Fyfe* 663
Eruptive gas compositions and fluxes of explosive volcanoes: budget of S and Cl emitted from Fuego volcano, Guatemala *W. I. Rose Jr, R. E. Stoiber, and L. L. Malinconico* 669
Mineral deposits associated with calc-alkaline rocks . . . . *A. H. G. Mitchell and R. D. Beckinsale* 677

Geographical Index 697

Subject Index 713

Contributors

S. Aramaki	Earthquake Research Institute, University of Tokyo, Bunkyo-Ku, Tokyo 113, Japan
P. E. Baker	Department of Geology, University of Nottingham, University Park, Nottingham NG7 2RD, UK
M. C. W. Baker	Department of Earth Sciences, The Open University, Milton Keynes MK7 6AA, UK
P. F. Barker	Department of Geological Sciences, University of Birmingham, PO Box 363, Birmingham B15 2TT, UK
R. D. Beckinsale	Geochemical Division, Institute of Geological Sciences, 64–78 Grays Inn Road, London WC1X 8NG, UK
G. C. Brown	Department of Earth Sciences, The Open University, Milton Keynes MK7 6AA, UK
M. J. Carr	Department of Geological Sciences, Rutgers University, New Brunswick, New Jersey 08903, USA
J. W. Cole	Department of Geology, Victoria University of Wellington, Private Bag, Wellington, New Zealand
K. C. Condie	Department of Geoscience, New Mexico Institute of Mining and Technology, Socorro, New Mexico 87801, USA
C. Coulon	Laboratoire de Pétrologie, Université Saint-Jérôme, 13397 Marseille Cedex, France
J. Dostal	Department of Geology, Saint Mary's University, Halifax, Nova Scotia, Canada
C. Dupuy	Centre Géologique et Géophysique, USTL, 34060 Montpellier Cedex, France
A. Ewart	Department of Geology and Mineralogy, University of Queensland, Brisbane, Queensland, Australia
J. G. Fitton	Grant Institute of Geology, University of Edinburgh, West Mains Road, Edinburgh EH9 3JW, UK
P. W. Francis	Department of Earth Sciences, The Open University, Milton Keynes MK7 6AA, UK
W. S. Fyfe	Department of Geology, University of Western Ontario, London, Ontario, Canada N6A 5B7
I. G. Gass	Department of Earch Sciences, The Open University, Milton Keynes MK7 6AA, UK
T. H. Green	School of Earth Sciences, Macquarie University, North Ryde, New South Wales 2113, Australia
M. Hammill	Department of Earth Sciences, The Open University, Milton Keynes MK7 6AA, UK
C. J. Hawkesworth	Department of Earth Sciences, The Open University, Milton Keynes MK7 6AA, UK
C. S. Hutchison	Department of Geology, University of Malaya, Kuala Lumpur 22–11, Malaysia

D. J. Hughes	Department of Geology, Portsmouth Polytechnic, Burnaby Road, Portsmouth PO1 3QL, UK
F. Innocenti	Istituto di Mineralogia e Petrografia, Via S. Maria 53, Pisa, Italy
R. W. Johnson	Bureau of Mineral Resources, P.O. Box 378, Canberra City, ACT 2601, Australia
J. Keller	Mineralogisches Institut, Universität Freiburg, D-7800 Freiburg, German Federal Republic
L. L. Malinconico	Department of Earth Sciences, Dartmouth College, Hanover, New Hampshire 03755, USA
P. Manetti	Istituto di Mineralogia, Petrografia e Geochimica, Via Lamarmora, 4 Firenze, Italy
B. D. Marsh	Department of Earth and Planetary Sciences, The Johns Hopkins University, Baltimore, Maryland 21218, USA
R. Mazzuoli	Dipartimento di Scienze della Terra, Universitá di Calabria, Cosenza, Italy
A. R. McBirney	Center for Volcanology, University of Oregon, Eugene, Oregon 97403, USA
A. Meijer	Department of Geosciences, University of Arizona, Tucson, Arizona 85721, USA
A. H. G. Mitchell	United Nations Development Project, 7285 A.DC, Mia Road, Pasay City, Metro Manilla, Philippines
B. O. Mysen	Carnegie Institution of Washington, Geophysical Laboratory, 2801 Upton Street NW, Washington, District of Columbia 20008, USA
R. J. Pankhurst	British Antarctic Survey, Natural Environmental Research Council, c/o Institute of Geological Sciences, 64–78 Grays Inn Road, London WC1X 8NG, UK
G. Pasquaré	Istituto di Geologia, Piazzale Gorini 15, Milan, Italy
J. A. Pearce	Department of Earth Sciences, The Open University, Milton Keynes MK7 6AA, UK
W. J. Rea	Department of Geology and Physical Sciences, Oxford Polytechnic, Headington, Oxford OX3 OBP, UK
C. Robin	Laboratoire de Géodynamique Sous-Marine, Université Pierre et Marie Curie, 06230 Villefranche-sur-Mer, France
M. J. Roobol	Institute of Applied Geology, Jeddah, Saudi Arabia
W. I. Rose Jr	Department of Geology and Geological Engineering, Michigan Technological University, Houghton, Michigan 49931, USA
A. D. Saunders	Department of Geology, Bedford College, Regent's Park, London NW1 4NS, UK
A. L. Smith	Department of Geology, University of Puerto Rico, Mayaguez, Puerto Rico 00708, USA
R. E. Stoiber	Department of Earth Sciences, Dartmouth College, Hanover, New Hampshire 03755, USA
J. Tarney	Department of Geology, University of Leicester, University Road, Leicester LE1 7RH, UK
M. F. Thirwall	Department of Earth Science, University of Leeds, Leeds LS2 9JT, UK
R. S. Thorpe	Department of Earth Sciences, The Open University, Milton Keynes MK7 6AA, UK
T. Ui	Department of Earth Sciences, Faculty of Science, Kobe University, Nada, Kobe, 657, Japan

L. VILLARI	*Istituto Internazionale di Vulcanologia, V. la Regina Margherita 6, Catania, Italy*
G. P. L. WALKER	*Hawaii Institute of Geophysics, 2525 Correa Road, Honolulu, Hawaii 96822, USA*
B. L. WEAVER	*Department of Geology, University of Leicester, University Road, Leicester LE1 7RH, UK*
S. D. WEAVER	*Department of Geology, University of Canterbury, Christchurch, New Zealand*
C. M. WHITE	*Center for Volcanology, University of Oregon, Eugene, Oregon 97403, USA*

Preface

The orogenic andesite association is the basalt–andesite–dacite–rhyolite association characteristic of island arcs, active continental margins, and continental collision zones. This book is intended as a reference text for undergraduate, postgraduate, and research workers who wish to gain an insight into the orogenic andesite association. It contains invited contributions covering the geology, volcanology, petrology, and geochemistry of orogenic andesites and related rocks, the role of such rocks through geological time, and the wider implications of andesite volcanism such as continental growth and the formation of economic mineral deposits. There were several reasons for producing such a volume. These include the abundance and importance of orogenic andesite volcanism in continental areas, and the important advances in our understanding of such volcanism achieved since the development of plate tectonics in the late 1960s. Noting the existence of several volumes devoted to the rarer alkaline rocks, it was felt that it would be valuable to produce a volume devoted entirely to the orogenic andesite association.

The volume is divided into 10 sections, chosen to cover all aspects of the association. Section I is a brief introduction to the setting of orogenic andesite volcanism and how the problems posed by these volcanic rocks have evolved during the last few decades. Section II is a general account of the terminology and classification of orogenic volcanic rocks, with a review of the petrology and mineralogy. In Section III, the characteristics of each of the major orogenic volcanic provinces are reviewed and Section IV outlines the evolution of volcanism through time in relation to plate tectonic activity in contrasted provinces. The later sections of the book outline the characteristics of andesite volcanoes and their products (Section V), the relationship of andesitic volcanism to intrusive activity (Section VI), and experimental, chemical, and isotopic characteristics of andesites and related orogenic volcanic rocks (Sections VII and VIII). Some characteristics of andesite volcanism throughout geological time are explained by reference to Archaean volcanic rocks and to Upper Proterozoic volcanism in northeastern Africa and Arabia and Lower Palaeozoic volcanism in the Caledonian orogenic belt of Britain (Section IX). In conclusion, Section X emphasizes the importance of andesitic volcanism by examining the role of such volcanism in continental growth, in large-scale geochemical cycles, in the eruption of volcanic gases, and in the formation of mineral deposits.

Acknowledgements

The decision to prepare the book has stemmed from field work in the Andes, with the Open University Andean Volcanoes Project, from time spent searching for references to different orogenic andesite provinces, and from discussions with colleagues about the potential value of a book devoted to such andesites. The field work in the Andes has been funded by the Natural Environmental Research Council by grants to the Open University made between 1972 and 1979. I am particularly grateful to G. M. Brown, P. E. Baker, P. W. Francis, I. G. Gass, H. A. Jones (John Wiley & Sons), and A. R. McBirney for their helpful ideas and encouragement during all stages of the project. I would also like to thank S. A. Drury, G. C. Brown, P. W. Francis, I. G. Gass, C. J. Hawkesworth, A. T. Huntingdon, J. Tarney, and J. B. Wright for comment and reviews of individual contributions. I am grateful to John Taylor and Jenny Hill for drawing and correcting diagrams, to Marilyn Leggett for cheerfully typing and retyping parts of the manuscript, and to Gwyneth Williams for preparing the geographical index. Finally, I thank my wife, Olwen, for encouragement, help and patience during preparation of the manuscript.

I Introduction

The Earth's crust has formed from the mantle as a result of magmatic activity over 3700 Ma. At present, such activity occurs dominantly at lithospheric plate boundaries, although minor magmatic activity also occurs within plates (Fig. 1). Thus, volcanism can be classified according to its plate tectonic setting as summarized in Table 1. Volcanism at constructive margins (oceanic ridges) is dominantly submarine and basaltic, and the products form the oceanic crust. Destructive plate boundaries form island arcs and active continental margins, where the volcanic products range from basalt through andesite and dacite to rhyolite in composition. This is the orogenic andesite association. A similar spectrum of volcanic rocks occurs in areas of island arc and continental collision. These destructive margins and collision belts are orogenic zones.

Although the characteristic occurrence of the basalt-andesite-dacite-rhyolite association within orogenic regions has been long recognized, the advent of plate tectonics during the 1960s has lead to a re-appraisal of the relationship between volcanism and orogeny. Prior to the plate tectonic era, orogenic activity was considered to conform to a broad pattern or cycle. The earliest stage of the cycle was the eruption of basaltic lavas ('spilites') within a flysch-type sedimentary sequence deposited in a linear belt along a continental margin, or 'geosyncline'. Then, during the early stages of deformation of the sedimentary-volcanic succession, ultrabasic and basic intrusions were emplaced. The third stage, during and following the main episode of deformation, was the intrusion of batholiths of diorite-granodiorite-granite composition. The final stage, following elevation of the folded sedimentary-volcanic pile, was subaerial eruption of volcanic rocks of the orogenic andesitic association. Within the framework of the orogenic cycle, the petrogenesis of the volcanic rocks was generally considered in terms of fractional crystallization of basaltic magma, allied with melting and assimilation of continental crustal rocks. In view of the large role commonly assigned to crustal fusion, there was little discussion of how such orogenic volcanic rocks might contribute to continental crustal growth, and how such volcanic rocks might form part of large scale geochemical cycles, involving the crust and upper mantle.

Prior to plate tectonics, the major problem was to explain the orogenic cycle outlined above, and this was achieved by considering that the geosynclinal belts evolved in ensialic depressions, possibly located over sinking mantle. Deformation could then occur when the continental crust below the depression melted and became less rigid, allowing the sedimentary-volcanic pile to be compressed and deformed. When sinking ceased, uplift of the deformed orogenic belt occurred and melting in the mantle or at the base of the thickened pile would lead to formation of magmas that would be intruded or erupted at the surface. Plate tectonics provides an alternative explanation of the orogenic cycle. The next sections describe active orogenic belts and the setting of active orogenic volcanism, and examine the major problems posed by such activity.

Island arcs, active continental margins, and collision zones have been abundant throughout much of geological time. At present, most of these features are located around the Pacific (Fig. 1). The western Pacific has major arcs extending from the New Zealand-Tonga system through New Britain-Papua New Guinea and the Mariana-Volcano islands to the Japanese-Kurile-Kamchatka system. The Mariana-Volcano system and the Japanese arc are separated from the continent by 'marginal' basins formed by sea-floor spreading in an analogous way to larger ocean basins. With the exception of

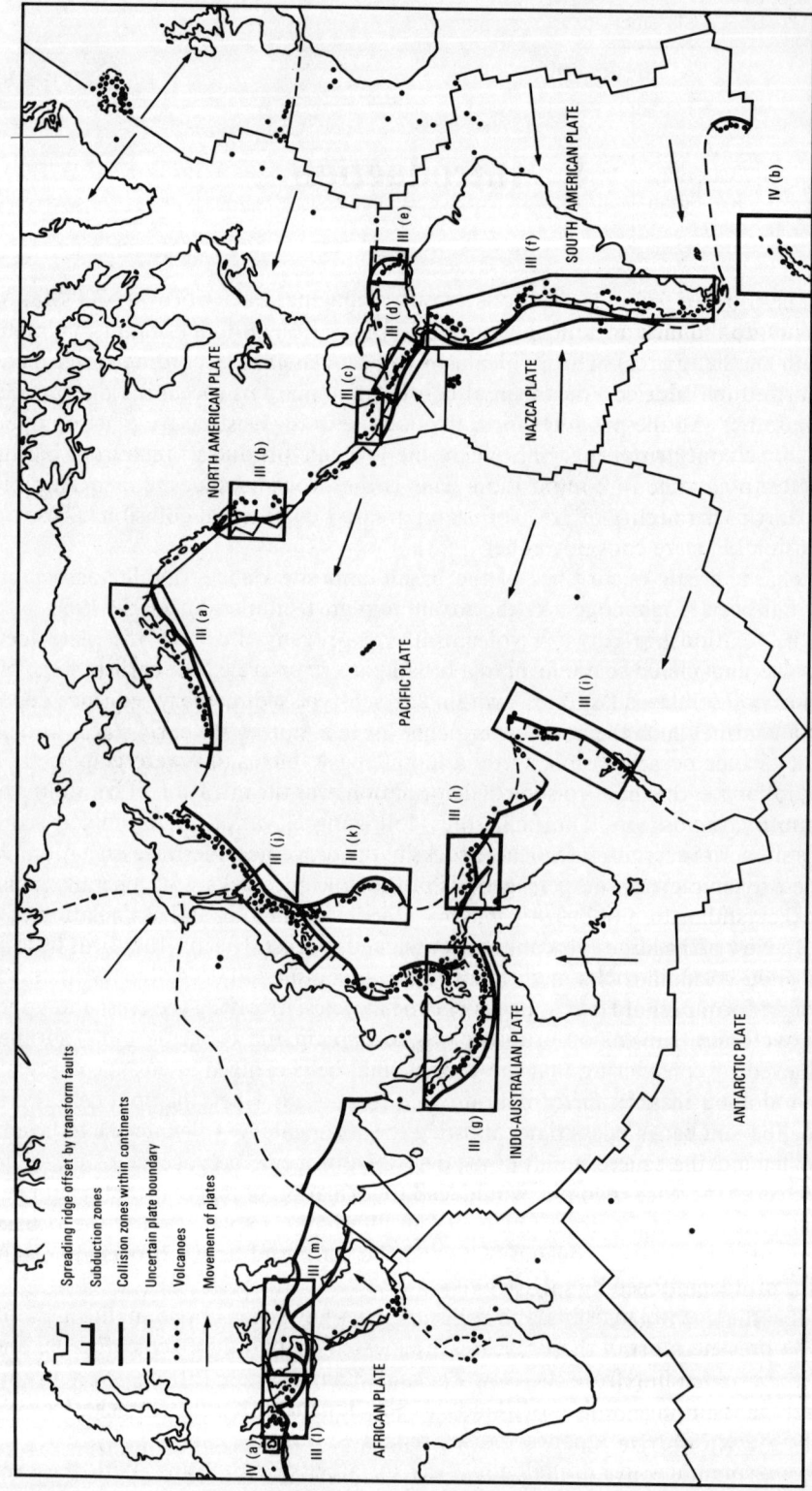

Fig. 1 Map showing lithospheric plate boundaries. The outlined areas are described in Sections III and IX

TABLE 1 Classification of volcanic activity in relation to plate tectonic setting

PLATE MARGIN				WITHIN-PLATE	
Constructive		Destructive	Collision zone	Oceanic	Continental ($\pm$ rifting)
Marginal basin (back-arc spreading centre)	Ocean ridge (large ocean basin)	Volcanic arc (gradational to) active continental margin	Continent and island arc collision		

Japan, these arcs appear to lack ancient continental basement. By contrast to the western Pacific, the eastern Pacific has no island arcs, but has volcanism on an active continental margin extending from the western USA, through Mexico and Central America, to the western margin of South America. In the Atlantic, the Lesser Antilles and the South Sandwich islands are young arcs associated with marginal basins; the characteristics of the basins behind the Mediterranean Aeolian and Aegean arcs are, however, more ambiguous. Finally, volcanism occurs within the active continental collision zone extending from the Alps through Turkey and Iran to the Himalayas.

The volcanic arcs and active continental margins have the following major characteristics:

(1) Arcuate distribution of islands, or a linear belt of volcanism, with a length of the order of several thousands of kilometres and a relatively narrow width (200–300 km).

(2) A deep trench, often 6000–11 000 m, on the oceanic side, and (in island arcs) a relatively shallow (generally less than 3000 m) tray-shaped marginal basin sea on the continental side.

(3) Active volcanism, in which there is an abrupt oceanward boundary of the volcanic zone, parallel to and c. 200 km from the oceanic trench. This is known as the volcanic front. The concentration of volcanoes is greatest at the volcanic front and decreases with distance away from the trench.

(4) Active seismicity, including shallow, intermediate, and deep earthquakes, extending as a well defined plane from below the trench towards the marginal basin or continental side. The mantle below the marginal basin is characterized by low seismic velocities and high attenuation (Q) compared with that below the island arc.

(5) A marked gravity anomaly belt, with a negative anomaly (up to -100 mgal) associated with the trench, and positive anomalies on the arc or continental margin.

(6) A marked heat flow anomaly belt in which the trench area has relatively low heat flow (generally less than 40 mW m^{-2}) and the island arc or continental margin has a higher heat flow (greater than 40 mW m^{-2}).

(7) In some island arcs, notably Japan, there is a distinct zonal arrangement in the compositions of volcanic rocks. The volcanoes near the volcanic front erupt basalts that are tholeiitic, while more distant volcanoes behind the volcanic front erupt basalts of more alkaline composition. A similar zonation expressed as an increase in K content at a given silica percentage occurs in many (but not all) island arcs and active continental margins.

The characteristics of island arcs summarized above may be explained in terms of plate tectonics. The seismic plane or Benioff zone marks the plane of descent of oceanic lithosphere into the mantle. This descent is from the oceanic trench and involves sinking of cold, dense lithosphere into the mantle, as inferred from the negative gravity anomalies and low heat flow associated with oceanic trenches. At a particular depth, the descending slab melts or dehydrates, releasing volatiles into the mantle wedge, thus causing melting, rise and intrusion of magma, and surface volcanism. More extensive upwelling initiates mantle convection and thus may lead to more extensive melting and formation

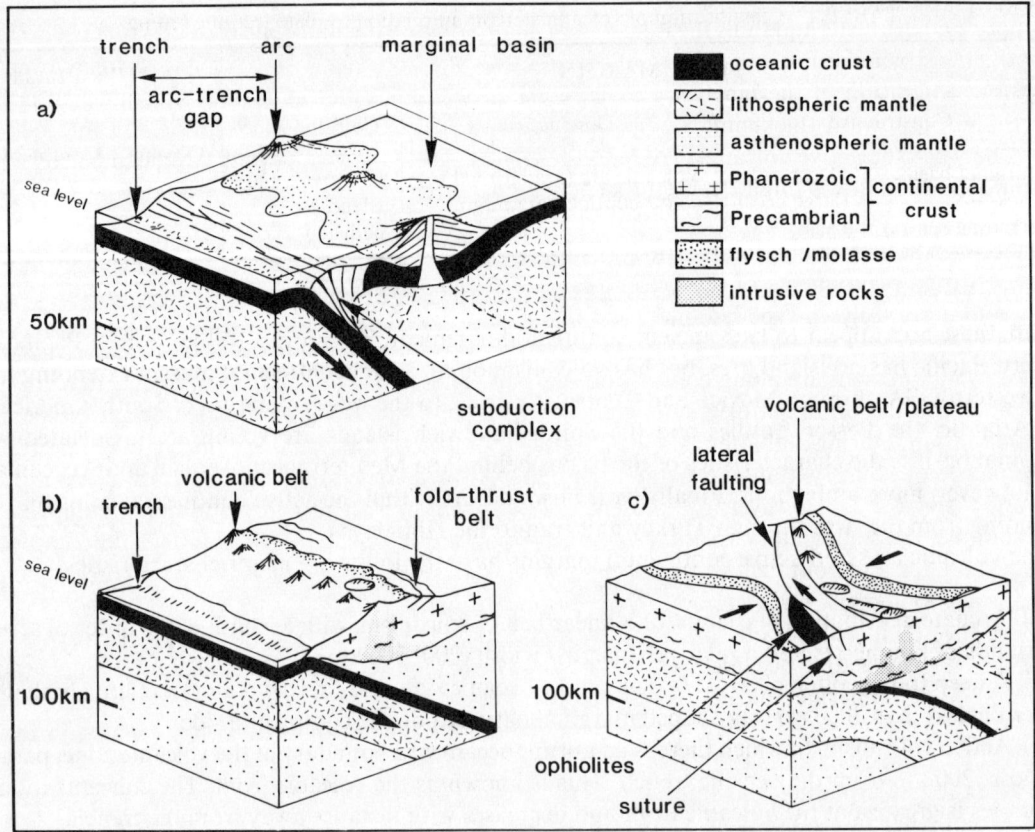

Fig. 2 Schematic sections showing setting of igneous activity in (a) an island arc, (b) an active continental margin or Cordilleran belt, and (c) a continent–continent collision or Alpine–Himalayas belt. Note that the scale of (a) is twice that of (b) and (c). See text for further discussion

of a marginal basin. The formation of island arcs as outlined above has been termed *Pacific-type* orogeny.

The sequence of events in the formation of a single volcanic arc might be as follows (*see* Fig. 2(a)). Development of a trench would be associated with initiation of plate descent. This stage could be accretionary wedge beneath and behind the trench. Such a wedge is therefore a complex mixture, or oceanic sediments (including cherts, argillites, and carbonates) could be swept into the trench by submarine gravity slides. Such slides can carry blocks of basic and ultrabasic rocks derived from disrupted faulted and thrust blocks. This slide deposit will be carried into a deformed sedimentary accretionary wedge beneath and behind the trench. Such a wedge is therefore a complex mixture, or *mélange* of varied sedimentary and igneous rocks scraped from the descending plate, and may experience strong deformation and metamorphism within the blueschist facies (low T/high P, in the low heat flow regime of the trench). Such subduction complexes occur in many western Pacific volcanic arcs.

When the descending plate, capped with hydrated oceanic crust and/or with some oceanic sediment, reaches a depth of *c.* 100 km, dehydration releases hydrous fluids into the overlying mantle wedge. This initiates melting to form the parent magmas for the volcanic arc igneous rocks. These rocks are of basic and intermediate composition; gabbro and diorite intrusions are emplaced into the crust and a volcanic pile of basalt and basaltic andesite begins to accumulate at the surface. Where continued magmatic activity heats and thins the lithosphere, high T/high P metamorphism of the crust occurs. The crust thickens by intrusion and growth of the volcanic pile, together with continued deformation and accretion. Migration of the igneous axis may occur, yielding inner and outer arcs. During crustal growth, an oceanward thickening wedge of flysch-type sediment derived from the volcanic belt develops between the active volcanic front and the trench. At this stage a volcanic arc with paired metamorphic zonation (low T/high P near the trench, high T/high P near the active axis) may be developed. From the description above it is clear that oceanic trenches and their associated sedimentary wedges are likely to be extremely complicated terraines, with complex geometric and time relationships between varied components of the oceanic crust and upper mantle, *mélange*, intrusive and volcanic rocks, and derived flysch-type sediments.

By contrast to the Pacific-type orogeny described above, the orogenic belts of the eastern Pacific are not island arcs but are along continental margins. These are *Cordilleran-type* orogenic belts (Fig. 2(b)). Such a situation might originate where a trench develops along an inactive Atlantic-type continental margin. During and after the formation of a destructive margin near the continental margin, the sequence of events within and behind the newly formed trench may be similar to that described above for island arcs. However, the presence of continental crust provides additional complexities. For example, the sedimentary rocks formed before initiation of the trench may include continental shelf carbonates and continental rise lutites and clastic rocks. The sequence of events may include formation of blueschist *mélanges* and flysch accumulations.

When the subducted oceanic lithosphere reaches depths of c. 100 km, dehydration and possibly melting of the descending oceanic crust will occur. The magmas formed rise so that intrusions may be emplaced, and submarine volcanic rocks may be erupted near the volcanic front. As in island arcs, these igneous rocks are relatively basic in composition and include diorites with basalts and basaltic andesite. The igneous axis may migrate towards the continent and then enroach upon older crust. This leads to underplating and intrusion and therefore thickening of the continental crust at this point. Uplift will cause subsequent volcanism to be subaerial. At this stage the intrusive rocks belong to the diorite–granodiorite–granite association and erupted rocks are andesite and dacite lavas and pyroclastic rocks (air-fall and pyroclastic flows, including ignimbrites) of dacite and rhyolite composition. Formation of these rocks might involve melting of old continental basement. The intrusive and extrusive rocks may show a compositional polarity analogous to that described for island arcs. The subaerial volcanic belt forms an axis of sedimentary polarity and subsequent evolution may be dominated by thrusting towards the continental side of the magmatic axis, involving the flysch-type sediments and sometimes the continental basement.

Active orogenic belts shows a range of characteristics between Pacific-type and Cordilleran-type orogenic belts. More complex orogenic belts may result from collision of an island arc with either an Atlantic-type continental margin or a Cordilleran-type orogenic belt. In addition, multiple collisions of island arcs may lead to formation of larger continental masses with complex structural characteristics.

The most complex situation develops where two continental blocks approach and collide by subduction of the intervening oceanic crust. Such active collision is believed to be responsible for formation of the Alpine–Himalayan continental collision zone (Fig. 2(c)). In the simplest case, we may consider the approach of an Atlantic-type continental margin with another continent bounded by a Cordilleran-type orogenic belt. The structures formed as the Atlantic-type margin meets the trench may involve deformation and thrusting of the continental basements, possibly culminating in the formation of nappes. Oceanic crust, upper mantle, oceanic sediments, and flysch-type deposits are deformed and thrust over earlier thrust sheets. When the buoyancy of the collided continental masses arrests the collision, the collision zone becomes an area of thickened crust characterized by complex faulting (including prominent lateral faulting) and uplift. The Himalayas are the type example of this kind of orogenic belt.

The structures developed during continent–continent collision depend upon the nature of the sedimentary sequence and the basement on the Atlantic-type margin, and the shape of the colliding margins. Where margins are irregular the first continental segments to collide will experience the earliest and most intense deformation. Such segments might be partially subducted, or split so that the upper part becomes 'flaked' on to the continent. Alternatively, the segments might be extended by lateral faulting to generate a more even collision. Against such regions the trench zone becomes a narrow suture from which fragments of oceanic crust, mantle oceanic sediments, and flysch-type sediments are transported. Such regions are likely to have extensive continental molasse-type deposits in external troughs. In contrast to collided segments, incidental areas of the continental margins may never collide and may thus preserve relatively undeformed areas of oceanic crust below flysch-type deposits, within the orogenic belt.

From the discussion above it is clear that continent–continent collision zones are liable to be extremely complex. Prior to collision the ocean might include small island arcs and small continental masses. At the initiation of collision, the margins can become split into a complex pattern of small continental fragments which eventually become sutured along the collision zone. The magmatic activity associated with collision zones is equally complex. Before collision, volcanism on the active continental margin might resemble that described earlier for Cordilleran orogenic belts. After collision, a complex pattern of magmatism may become established. Volcanic rocks of andesite, dacite, and rhyolite composition can be erupted, although these are relatively rare in the Alpine belt and are more abundant in Turkey–Iran and in the Himalayas. However, a range of alkaline volcanic rocks is also erupted, particularly in areas of crustal extension or lateral faulting. Because of the association of continental basement with crustal thickening in collision zones, crustal fusion is likely and some volcanic and intrusive rocks might be largely products of such fusion.

From this brief account, it is clear that intrusive rocks of the gabbro–diorite–granodiorite–granite association and their volcanic equivalents, basalt–andesite–dacite–rhyolite, are ubiquitous components of orogenic belts. The occurrence of the volcanic assocation in oceanic island arcs, the compositional zonation in arcs and Cordilleran belts, and the crustal thickening characteristic of the latter, all indicate that these rocks may be derived from mafic or ultramafic sources within the mantle. The isotopic and trace element characteristics of many of these volcanic rocks further indicate a mantle origin. However, in areas of thickened ancient crust such as Cordilleran belts and continental collision zones, fusion of the lower continental crust might occur, so that the erupted volcanic rocks might be formed by crustal fusion or might be mixtures of lower crustal and mantle-derived components. However, it is clear that, in those settings, intrusive and volcanic rocks might both contain a substantial component derived recently from the mantle.

In conclusion, the formation and eruption of orogenic volcanic rocks are now seen in terms of the plate tectonic cycle. This implies continuous intrusive and volcanic activity during evolution of island arcs, active continental margins, and continental zones. The problems now recognized in the study of the orogenic andesite association are therefore different to those discussed before the advent of plate tectonics. First, the setting of orogenic volcanism should be explained in relation to the pattern of active or past consumptions of oceanic lithosphere. An important problem in this respect is to account for the position of the volcanic front and the distribution of volcanism on its continental side, and to account for variations in volcanism along volcanic belts, as indicated by segmentation. Second, in relation to the petrogenesis of orogenic volcanic rocks, the relative roles of subducted oceanic crust, overlying mantle wedge, and continental crust are hotly debated. Third, the nature of the geological and petrological relationship between intrusive and extrusive igneous activity is not well established. Fourth, there is the possible identification of orogenic volcanic rocks throughout geological time, particularly in the Archaean, where identification of orogenic volcanic rocks analogous to those erupted at present might indicate the occurrence of plate tectonic processes similar to those presently active. Fifth, and based on the problem noted above, is the identification of the extent to which orogenic intrusive and volcanic activity have contributed to recycling of elements through the plate tectonic cycle and have hence contributed to continental growth. Finally, and of greatest practical relevance, is the problem of the relationship of orogenic igneous activity to the formation of economic mineral deposits, and the problem of predicting destructive pyroclastic eruptions in densely populated, orogenic areas.

Andesites
Edited by R. S. Thorpe
© 1982 John Wiley & Sons

II Classification, petrology, and mineralogy of orogenic volcanic rocks

Young mountain belts and island arcs are associated with highly seismic and volcanic belts adjacent to destructive plate margins. This association occurs at intra-oceanic island arcs (e.g. Tonga), at continental margins (e.g. the Andes), and where continental collision is impending (e.g. the Mediterranean) or has occurred (e.g. the Himalayas). Such orogenic areas are characterized by volcanic rocks ranging in composition from basalt through andesite to rhyolite. Although a spectrum of igneous rocks therefore occurs in such regions, the widespread occurrence and local dominance of andesitic rocks is widely regarded as the major characteristic of orogenic volcanism. This section reviews the geological features, the petrology, and the mineralogy of andesite and associated volcanic rocks from active orogenic areas.

In the first contribution, P. E. Baker reviews the classification and terminology of orogenic volcanic rocks. He distinguishes the island-arc, low K tholeiitic series which occurs in primitive island arcs and is dominated by basalts associated with smaller volumes of Fe-rich basaltic andesites and andesites. These contrast with the island arc calc-alkaline association in which andesite is dominant, and the calc-alkaline associations of continental margins and collision zones which are characterized by the occurrence of andesites, dacites, and rhyolites. In addition to these associations, more potash-rich shoshonitic volcanic rocks occur over deeper subduction zones, at a greater distance from the trench than tholeiitic and calc-alkaline volcanic rocks, and true alkaline associations may occur in a back-arc setting.

Using the framework outlined above, A. Ewart presents a review of the mineralogical and petrological characteristics of orogenic volcanic rocks from basalt to andesite (less than 60 per cent SiO_2) in composition from active orogenic regions. This review complements and extends an earlier review of the 'mineralogy and chemistry of Tertiary–Recent dacitic, latitic, rhyolitic, and related salic volcanic rocks' (Ewart, 1979) and, taken together, these reviews are a comprehensive source of data on orogenic volcanic rocks. Ewart confirms that the chemical data show marked regional differences, particularly in the proportions of high K and shoshonitic volcanic associations, which are highest in continental provinces and more dominant among dacite and rhyolite compositions. Orogenic volcanic rocks characteristically have high phenocryst contents (commonly exceeding 20 per cent by volume), and phenocryst mineralogy dominated by plagioclase. The frequent persistence of magnesian olivine in basaltic andesites and andesites, and the significant correlations of phenocryst modes and bulk chemistry, are believed to indicate 'fractional crystallization and/or flow differentiation processes involving both phenocryst accumulation and extraction'. Such processes are examined in greater detail in Sections VII and VIII.

REFERENCE

Ewart, A. (1979). A review of the mineralogy and chemistry of Tertiary–Recent dacitic, latitic, rhyolitic, and related salic volcanic rocks. In *Trondhjemites, Dacites and Related Rocks* (F. Barker, ed.), Elsevier, Amsterdam, pp. 13–121.

Andesites
Edited by R. S. Thorpe
© 1982 John Wiley & Sons

Evolution and classification of orogenic volcanic rocks

P. E. Baker

Department of Geology,
University of Nottingham, University Park, Nottingham NG7 2RD, UK

ABSTRACT

Although orogenic regions were once regarded as being characterized exclusively by calc-alkaline rocks, a considerable diversity of volcanic rock suites has now been recognized. In some instances progressive compositional changes occur in either a lateral or vertical sense, which has led to concepts of island arc evolution. These compositional changes are thought to be depth-related and are currently interpreted in terms of partial melting and fractional crystallization associated with subduction processes: crustal contamination on any significant scale is now less commonly invoked.

The island arc low K tholeiitic series is dominated by basaltic lavas, which are associated with Fe-rich basaltic andesites and andesites. Incompatible element concentrations tend to be low, REE patterns are rather flat, and the magmas are probably derived by fractional crystallization of olivine from a primitive olivine tholeiite magma originating at relatively shallow levels (80–120 km) above the subducted lithosphere. The calc-alkaline series of island arcs, in which andesite is dominant, tends to be more enriched in K and other incompatible elements and is also light REE enriched. On the continental margins dacites and rhyolites become more abundant in the calc-alkaline series and K and related elements are usually higher. Calc-alkaline magmas have probably been derived from the deeper parts of the subduction zone with some partial melting of the descending slab and subsequent reaction of these siliceous magmas with the overlying mantle, followed by fractional crystallization at shallower levels.

K and incompatible elements reach even higher concentrations in the shoshonitic rocks, some of which are nepheline normative. Shoshonitic lavas have been attributed to low degrees of melting of mantle peridotite at greater depths in the vicinity of the subduction zone. Alkaline basalts are sometimes closely associated with calc-alkaline lavas, but more frequently they form a separate zone of activity, for example to the east of the Andes. In this latter instance they have probably evolved in a type of back-arc situation, only indirectly connected with subduction. The basalts of more developed marginal basins have affinities with ocean-floor basalts rather than with orogenic rocks. Boninites may belong to an early stage in subduction zone development.

Introduction

The purpose of this introductory section is to review the development of ideas on orogenic volcanic rock series, to summarize the characteristics of the major subgroups now recognized, and to evaluate possible petrogenetic relations between them.

The term 'Andesite'

According to Johannsen (1937, p. 160) the name andesite was originally given by von Buch in 1836 to 'so-called trachytes' from the Andes: they were lavas containing plagioclase and hornblende. The nature of the ferromagnesian components was at one time used in distinguishing andesites from basalts, particularly on the basis of hornblende in the former and olivine in the latter. Subsequently, emphasis came to be placed on the composition of the plagioclase, with that in the andesites falling within the range An_{50}–An_{30}. However, in a modal mineralogical scheme, as advocated by Johannsen, this poses difficulties in its application since the groundmass feldspars are not readily determined by normal optical methods. In many andesites, zoned plagioclase phenocrysts, which are the only feldspars usually amenable to determination, have relatively calcic composition, ranging up to bytownite, and even anorthite in the cores.

This problem led MacGregor (1938, pp. 48–49) to refer to the labradorite-bearing 'andesitic' rocks of Montserrat as bandaites, a term used by Iddings (1913) and derived from the volcano Bandai-san in Japan. Plagioclase phenocrysts in the Montserrat bandaites are mostly zoned in the bytownite–labradorite range and the normative feldspar is often labradorite, even in rocks with c. 60 per cent SiO_2.

In reality the term andesite has been applied at different times to various volcanic rocks of broadly intermediate composition. If a definition of the term andesite is based largely on the composition of either the modal or normative plagioclase, it will inevitably embrace an assortment of lavas which may differ from each other quite significantly in several other parameters and have diverse petrogenetic affinities. For instance, none of the chemical analyses of andesites listed by Johannsen (1937, p. 168, Table 64) corresponds to an intermediate calc-alkaline rock. The problem was highlighted by Macdonald (1960) who recommended use of the names hawaiite and mugearite for intermediate members of the oceanic alkaline suite to distinguish them from andesites of the island arc and continental orogenic regions. Chayes (1969) took a thoroughly objective look at the chemical composition of Cainozoic andesite, making a survey of 1775 published analyses of rocks 'called andesites by professional petrographers', rejecting only those that were nepheline normative. Chayes' average turned out to be remarkably similar to the well known average andesite of Daly (1933): it is reproduced in Table 2. Current usage tends to restrict the term 'andesite' to intermediate rocks of calc-alkaline affinity, i.e. silica-oversaturated rocks with relatively high Al_2O_3 (c. 15–19 per cent), and moderate total alkalis (see, for example, Jakes and White, 1972, pp. 30–31). Andesites are usually distinguished from other members of the calc-alkaline association on the basis of silica percentage, although the precise boundaries adopted vary slightly from one author to another and tend to be influenced by the incidence of mineralogical changes in a particular suite. For instance, Carmichael et al. (1974, p. 557) show andesite as falling within the range 55–63 per cent SiO_2 and Peccerillo and Taylor (1976) use the range 56–63 per cent.

Historical review

The concept of petrographical provinces was developed by Judd (1886) and Harker (1909). It was Harker's Pacific branch of igneous rocks, richer in CaO and MgO relative to alkalis, that subsequently became the calc-alkaline series of Peacock (1931). Peacock's definition was based on the so-called alkali-lime index, i.e. the silica value at which CaO and total alkalis curves intercepted: in reality, many suites generally referred to as calc-alkaline nowadays would have been calcic on Peacock's criteria. The calc-alkaline series was firmly associated with the circum-Pacific and related volcanoes in contrast to the olivine basalt–trachyte/phonolite suites of oceanic islands.

The 'Andesite line' which marked the boundary between the two great 'provinces' was also regarded as marking the oceanward limit of sialic continental crust (Holmes, 1965, p. 1013). The calc-alkaline suite, dominated by andesite and associated with lesser proportions of basalt, dacite, and rhyolite, was certainly regarded as the principal, if not the exclusive, volcanic suite of orogenic zones. Petrogenetic questions centred on the origin of andesite either by fractional crystallization of basalt, and hence the problem of respective volumes, or by various means of hybridization or contamination by sialic crust.

The situation to the middle of the century was reviewed by Tilley (1950), who, although subscribing to the involvement of sial, stressed the importance of basalt in andesite genesis. At this time views on andesite genesis seem to have been essentially polarized between fractional crystallization of basalt and some form of sialic contamination. Of particular importance was Tilley's observation that a variety of basalt types existed in the orogenic areas, such as porphyritic high alumina basalts, aphyric high alumina types, and 'normal tholeiites' with low alumina.

At about the same time, in his study of the petrology of Hakone volcano, part of the Huzi volcanic zone of Japan, Kuno (1950) recognized two rock suites which he designated the pigeonitic series (characterized by a single monoclinic pyroxene in the groundmass) and the hypersthenic series (characterized by groundmass hypersthene together with augite). The pigeonitic series has tholeiitic affinities and the hypersthenic series broadly corresponds to the calc-alkaline series. Kuno considered that the former was derived from a parental olivine basalt by fractional crystallization, whilst generation of the hypersthenic series involved assimilation of granitic rocks. From then on a steadily increasing diversity of volcanic rock associations gradually came to be recognized in the orogenic zones. Tomkeieff (1949) had already noted the occurrence of alkaline lavas behind several Pacific arcs and Rittmann (1953) distinguished alkaline lavas in the Indonesian arc complex. Much later, Arculus and Curran (1972) and Arculus (1976) described an alkaline series closely associated with calc-alkaline rocks on Grenada, situated on the main axis of the Lesser Antilles volcanic arc. A more potassic, shoshonitic series was recognized in several orogenic areas, e.g. Fiji (Gill, 1970). More recently marginal basin basalts have emerged as a separate group of volcanic rocks whose origins are linked in some way with the subduction process (e.g. Hawkins, 1977; Tarney *et al.*, 1977). Brown and Schairer (1968) and Yoder (1969) presented cases for distinguishing between the calc-alkaline series of circum-oceanic arcs and those of the continental margins. An important point stressed by several authors (e.g. Jakeš and White, 1972) is the continuum of geochemical variation from the island arc tholeiitic series, through the low and high K calc-alkaline suites to the shoshonitic association and even in some instances to leucite tephrites (Barberi *et al.*, 1974).

Alongside the gradual recognition of different rock suites there developed the realization that they tended to be systematically related in both spatial and sometimes stratigraphic senses. Kuno (1959, 1966) discerned a pattern of petrological variation across the Japanese islands ranging from tholeiitic lavas on the Pacific side through calc-alkaline types further inland to more alkaline varieties near the Japan Sea. Jakeš and White (1969) described similar variations across the Melanesian arcs in the vicinity of New Guinea and New Britain, where again tholeiitic lavas on the oceanic side passed laterally into calc-alkaline types and eventually into alkaline or shoshonitic types in the interior.

Kuno (1966) related these progressive changes to increasing depth of earthquake foci from the ocean towards the continent. Similarly, Dickinson and Hatherton (1967) demonstrated a correlation between potash content of the erupted lavas and depth to the inclined Benioff zone in circum-Pacific arcs. These systematic compositional changes were ascribed to differences in the source material with depth (Kuno, 1966) or to differences in the depth of magma segregation or fractionation (Green *et al.*, 1967).

Recognition of these geographical and apparently depth-related variations led to the suggestion that there may also be vertical or stratigraphic changes in orogenic volcanic successions. Early stages in island-arc evolution were thought to be characterized by tholeiitic rocks, more mature stages by the calc-alkaline series, and the late stage by shoshonitic rocks (e.g. Donnelly, 1968; Baker, 1968*a*; Gill, 1970). However, as pointed out by Ringwood (1977), there are many exceptions to an evolutionary scheme such as this which can at best be regarded as expressing 'a broad and general trend rather than a rigorous sequential development'.

Ideas on the petrogenesis of orogenic volcanic rocks were for long dominated by the dual considerations of fractional crystallization and some form of sialic contamination (e.g. Tilley, 1950; Kuno, 1950). Recognition that many island arcs were not associated with sialic crust combined with the evidence of Sr isotope ratios eventually eliminated the latter possibility in many instances, although continental crust may provide a source for at least some rhyolites and ignimbrites on continental margins, e.g. the Andes (Pichler and Zeil, 1971). However, Thorpe and Francis (1979) and Thorpe et al. (1979) argue that andesites and ignimbrites of the central Andes are fractionated products of mantle-derived magmas and that contamination by continental crust is relatively insignificant. With the advent of plate tectonics, attention was directed at ways in which orogenic magmatism might be associated with subduction processes. Possible sources are primarily the subducted oceanic crust or the overlying mantle wedge: oceanic sediments are not generally thought to be subducted in sufficient quantity to play a significant role (Karig and Sharman, 1975).

During the past decade or so a substantial volume of experimental work has been aimed at evaluating the possible means whereby orogenic lavas may have been derived from within or above subducted lithosphere (e.g. Green and Ringwood, 1968; Nicholls and Ringwood, 1972, 1973; see also this volume, Section VIII). The general question of petrogenesis in island arcs has been comprehensively reviewed by Ringwood (1977). It now seems clear that the role of water, derived from the amphibolite and serpentinite in the subducted slab, is of paramount importance in the genesis of orogenic magmas (e.g. Kushiro and Yoder, 1969; Nicholls, 1974). The island arc tholeiitic series is thought to be generated by hydrous melting within the mantle overlying the subducted lithosphere at 80–100 km, followed by fractionation, especially of olivine at shallower depths. Within the depth range 100–150 km, quartz ecolgite of the slab partially melts to form acidic magmas which induce formation of mantle diapirs that ascend, fractionate, and eventually produce calc-alkaline magmas (Ringwood, 1977). Shoshonitic and related alkaline lavas are probably formed by low degrees of partial melting of peridotite mantle under relatively low P_{H_2O} (e.g. Barberi et al., 1974) and their origin may be only indirectly connected with subduction processes (e.g. Thorpe and Francis, 1979, p. 65).

The release of H_2O-rich fluids from the subducted slab may impart distinctive geochemical characteristics to the resultant magmas, causing enrichment in K, Rb, and Sr relative to REE, and displacing $^{87}Sr/^{86}Sr$ to more radiogenic values since Sr added from the slab will have been affected by isotopic ratios in seawater (Hawkesworth et al., 1979).

A detailed review of the mineralogy of orogenic rock series is given by Ewart (this volume, Section II).

Island Arc Tholeiitic Series

Although not described as such, rocks conforming to this suite had been recognized in the Huzi zone of Japan by Tomita in 1931 and by Kuno (1950), who referred to them as the pigeonitic series. However, it was Gill (1970) and Jakeš and Gill (1970) who first formally distinguished an island arc tholeiitic series from the calc-alkaline rocks of island arcs. The series has also subsequently been referred to as the low K tholeiitic series (Pearce and Cann, 1973). It is characteristic of the more isolated 'oceanic' island arcs such as the South Sandwich Islands (Baker, 1968a, 1978) and Tonga (Ewart and Bryan, 1973). However, it also occurs on the oceanic side of more mature arcs such as the Japanese islands, notably in the Izu Peninsula and Izu Islands (Kuno, 1950, 1966). Brown et al. (1977) have shown that there is a distinct variation along the axis of the Lesser Antilles island arc with a tholeiitic suite occurring on St Kitts in the north. In contrast, calc-alkaline lavas prevail in the centre of the arc and some of the southern islands, particularly Grenada, are alkaline. There is also a resemblance between the modern low K tholeiitic series and volcanic rocks of Archaean greenstone belts (e.g. Hart et al., 1970), although

Condie (this volume, Section IX) points out some differences. Gunn (1976) demonstrates that the average intermediate volcanic rock from the greenstone belts is a low K, low Sr andesite of island arc type. In Canadian greenstone belts there is a progressive shift from tholeiitic to calc-alkaline eruptives with height in the stratigraphic pile (Windley, 1977, p. 35; see also Condie, this volume, Section IX), which is also a feature of some more evolved modern arcs such as Fiji (Gill, 1970).

Rock types range from tholeiitic basalt through basaltic andesites and Fe-enriched andesites to dacites and, very rarely, rhyolite. Basalts and basaltic andesites tend to be the most abundant lavas in the island arc tholeiite association (Jakeš and White, 1971; Baker, 1968a) and are often represented by a spilitic submarine assemblage (e.g. Gill, 1970; Donnelly et al., 1971). Pyroclastic rocks are much less abundant than in the calc-alkaline series and pyroclastic flows are extremely rare. One of the few examples of the latter is the unusual pyroclastic flow of basaltic andesite composition which has been described from the island of Tofua, Tonga (Bauer, 1970; Baker et al., 1971).

Although porphyritic rocks are by no means uncommon, the incidence of aphyric types is certainly greater than in the calc-alkaline association. The difference becomes particularly marked in the more siliceous rocks with aphyric Fe-enriched icelandites and dacites of the low K tholeiitic series contrasting with the strongly porphyritic calc-alkaline rocks. Orthopyroxene occurs in the low K tholeiitic series but is less frequent than in the calc-alkaline group. Subcalcic augite or pigeonite are the typical groundmass pyroxenes of the tholeiitic basalts and basaltic andesites. Hydrous minerals, amphibole and biotite, which are common in the more differentiated rocks of the calc-alkaline series, are almost entirely absent from fresh, low K tholeiitic rocks.

Chemically, the island arc tholeiitic series is characterized by a lower SiO_2 mode of c. 53 per cent as compared with 59 per cent for the calc-alkaline series (Jakeš and Gill, 1970), rather lower Al_2O_3, and a marked tendency towards Fe enrichment in intermediate rock types (Table 1). They have distinctly low concentrations of K and associated incompatible elements such as Ba, Sr, Rb, Cs, Zr, Pb, Th, and U, and tend to have lower Mg, Ni, and Cr. The Na/K and K/Rb ratios are both higher in the island arc tholeiitic series than in calc-alkaline rocks. In their plots of Ti–Zr, Ti–Zr–Y, and Ti–Zr–Sr, Pearce and Cann (1973) distinguished a low K tholeiitic field from that of calc-alkaline rocks.

REE abundances usually show a flat, relatively unfractionated chondritic pattern similar to that of ocean-floor basalts, although the concentrations are between four and 20 times the chondritic values in the case of the South Sandwich Islands (Hawkesworth et al., 1977). Some of these lavas show light REE depletion and both positive and negative Eu anomalies which were attributed by Hawkesworth et al. to an original source anomaly or to low pressure plagioclase fractionation. The pattern contrasts with those of calc-alkaline rocks which are generally light REE enriched (Jakeš and Gill, 1970). However, $^{87}Sr/^{86}Sr$ ratios are generally c. 0.704, which is about the same as in many calc-alkaline suites and quite distinct from those of ocean-floor basalts, which average c. 0.7027 (Hart, 1971).

In island arc regions producing low K tholeiites, the depth to the Benioff zone is relatively shallow (80–120 km; Nicholls and Ringwood, 1972), and because of their oceanic situation the influence of continental crust in their generation can usually be discounted. Nicholls and Ringwood (1972) suggest that the basaltic rocks of the island arc tholeiitic association are the hydrous and olivine depleted equivalents of the ocean ridge tholeiites. The geochemical characteristics are consistent with an origin involving olivine fractionation from a primitive olivine-poor tholeiitic magma, indicated by their depletion in Ni and Mg and enrichment in SiO_2, compared with ocean-floor basalts. Experimental work by Nicholls and Ringwood (1973) and Nicholls (1974) demonstrates the expansion of the stability field of olivine by the addition of water to olivine tholeiite and SiO_2-saturated tholeiite com-

TABLE 1 Island arc tholeiitic series

	1	2	3	4	5	6	7	8
SiO_2	51.04	60.01	52.80	50.03	53.09	51.57	51.98	57.40
TiO_2	0.87	1.13	1.40	0.84	1.17	0.80	0.86	1.25
Al_2O_3	17.12	14.87	14.60	15.71	15.44	15.91	17.53	15.60
Fe_2O_3	3.90	1.95	4.30	2.92	4.02	2.74	5.56	3.48
FeO	6.97	9.01	9.64	8.83	9.01	7.04	4.96	5.01
MnO	0.19	0.20	0.22	0.27	0.23	0.17	0.24	—
MgO	5.91	2.47	4.16	7.73	4.66	6.73	5.03	3.38
CaO	11.76	6.03	8.75	11.95	9.68	11.74	9.50	6.14
Na_2O	1.96	3.39	2.28	1.47	2.12	2.41	2.43	4.20
K_2O	0.37	1.18	0.47	0.24	0.45	0.44	0.19	0.43
H_2O^+	0.03	0.02	0.60	—	—	0.35	} 1.47	—
H_2O^-	0.03	0.02	0.60	—	—	0.10		—
P_2O_5	0.07	0.21	0.08	0.09	0.12	0.11	0.09	0.44
Total	100.50	100.48	99.81	100.31	99.99	100.11	100.16	97.33
p.p.m.								
Ba	75	225				150	38	100
Co	47	24				—	28	—
Cr	65	24				125	52	15
Cu	175	125				175	92	—
Ni	27	1				100	27	20
Rb	12	30				5	1.3	6
Sr	135	150				355	115	220
V	290	100				275	290	175
Zr	100	140				30	43	70
$^{87}Sr/^{86}Sr$	0.7038	0.7043						

(1) Basalt (SSW.1). Irving Point, Visokoi, South Sandwich Islands (Baker, 1978, Table VII, No. 8 and Table VIII; Gledhill and Baker, 1973, p. 371). (2) Andesite/icelandite (SSK. 1.1). Glassy dyke, Reef Point, Cook Island, South Sandwich Islands (Baker, 1978, Table VII, No. 28 and Table VIII; Gledhill and Baker, 1973, p. 371). (3) Aphyric basalt (HK 31082204b). Hata basalt group, Izu Peninsula (Kuno, 1950, Table 14, No. 20, p. 1006). (4) Average of five basalts of the Izu–Hakone region (Oki *et al.*, 1978, Table 6–1, No. 1, p. 61). (5) Average of 53 basaltic andesites of the Izu–Hakone region (Oki *et al.*, 1978, Table 6–1, No. 2, p. 61). (6) Basalt, Talasea, New Britain (Lowder and Carmichael, 1970). (7) Average of four tholeiites and basaltic andesites from Eua, Tonga (Ewart and Bryan, 1973, Table 1, No. 1, p. 507). (8) Andesite/icelandite of tholeiitic series. Representative analysis quoted by Jakeš and White (1971, Table 2, No. 2, p. 226).

positions. Such magmas will fractionate olivine to produce SiO_2-saturated basalts and basaltic andesites. At lower pressures, fractionation of amphibole, pyroxene, and plagioclase will drive the residual melt to more evolved compositions in the island arc tholeiitic series. It has been suggested (Nicholls and Ringwood, 1973) that the essential water is derived by dehydration of amphibole in the subducted crust. As discussed by Hawkesworth *et al.* (1979), the release of H_2O-rich fluids from the subducted slab may enrich the resultant magmas in K, Rb, and Sr and may displace the $^{87}Sr/^{86}Sr$ ratios to more radiogenic Sr compositions.

Calc-alkaline Series of Island Arcs

Most of the volcanic rocks of the circum-Pacific arcs, the Lesser Antilles, and Indonesia probably fall within this category, although calc-alkaline eruptives are by no means the exclusive representatives of orogenic volcanism in these areas. The principal rock type is a two-pyroxene andesite and the series as a whole has a silica mode of about 59 per cent (Jakeš and Gill, 1970). Eruptions tend to be much more explosive than in the island arc tholeiitic association and both pyroclastic fall and pyroclastic flow deposits, typically *nuées ardentes*, are common.

The majority of the rocks are porphyritic with calcic plagioclase (often anorthite) the most abundant phenocryst phase: plagioclase phenocrysts often show complex oscillatory zoning. Orthopyroxene (usually hypersthene) phenocrysts are common in the andesites and quartz, hornblende, and biotite characterize the dacites. Garnets have been reported in some of the more siliceous rocks of the series but they are relatively rare (Green and Ringwood, 1968, p. 108). Coarse grained, ejected blocks consisting principally of the assemblage anorthite – olivine – amphibole – pyroxene – magnetite have been described from some provinces such as the Lesser Antilles (e.g., St Vincent; Lewis, 1973)

In contrast to the low K tholeiitic rocks, the calc-alkaline series shows no enrichment in Fe relative to Mg and alkalis and on the AFM diagram corresponds to Daly's (1933) average basalt – andesite – dacite – rhyolite. Al_2O_3 and Fe_2O_3/FeO ratios are higher and TiO_2 marginally lower than in the island arc tholeiitic series, whereas K and associated incompatible elements such as Rb, Sr, Ba, Zr, and U are all more abundant (Table 2) K/Rb ratios are, however, substantially lower (Jakeš and White, 1972). In contrast to the chondritic REE patterns exhibited by the island arc tholeiitic series, lavas of the calc-alkaline suite have patterns showing strong relative enrichment in the light REE and have very high (*c*. 10) La/Yb ratios (e.g. Jakeš and Gill, 1970; Ringwood, 1977). The fractionation of phases such as garnet and clinopyroxene, which discriminate in favour of the heavier REE, would seem to have played a part in the evolution of calc-alkaline magmas.

There is no clear distinction between $^{87}Sr/^{86}Sr$ ratios of calc-alkaline as opposed to island arc tholeiitic lavas. The ratios are often very similar, fluctuating around 0.704, although occasionally significantly higher ratios have been reported in island arc calc-alkaline rocks. For example, in the West Indies most ratios fall within the range 0.7035–0.7054 (Tomblin, 1975, p. 484); however, Pushkar *et al.* (1973) reported ratios of between 0.7073 and 0.7092 from some of the more siliceous rocks of St Lucia and Dominica.

In the case of most island arcs it is reasonable to discount any substantial involvement of continental crust in the generation of calc-alkaline magmas, on the basis of geophysical evidence as well as Sr isotopes. Ringwood (1977) suggests a model in which the magmas are derived ultimately from two sources. Dehydration of serpentinite within the subducted lithosphere in the range 100–150 km leads to partial melting of quartz eclogite, producing dacitic magma. This siliceous melt reacts with overlying pyrolite to produce pyroxenite diapirs from which calc-alkaline magmas subsequently segregate and fractionate.

Calc-alkaline Series of Continental Margins

Attempts have been made to distinguish the calc-alkaline rocks of island arcs from those of continental margins (e.g. Brown and Schairer, 1968; Baker, 1973). Jakeš and White (1972) refer to the latter as the 'Andean' type, although this is not particularly appropriate in view of the considerable diversity of andesites occurring within the Andes (Thorpe and Francis, 1979). Although most 'Andean' andesites are high K calc-alkaline rocks, such types are not completely diagnostic of the continental margin environment. However, in general terms the lavas of this group grade on the one hand into lower K rocks of island arcs and on the other into extreme shoshonitic varieties.

One of the more conspicuous differences between calc-alkaline volcanic rocks of island arcs and those of continental margins is in the greater abundance of more siliceous members in the latter. Pyroxene and hornblende andesites remain the most common rock type but dacites and rhyolites are significantly more abundant than in island arcs. Most of this additional volume of acid rocks is represented

TABLE 2 Chemical analyses of calc-alkaline lavas

	1	2	3	4	5	6	7	8
SiO_2	53.3	59.64	60.24	58.17	60.30	64.2	66.80	70.7
TiO_2	0.89	0.76	0.75	0.80	0.80	0.5	0.23	0.3
Al_2O_3	19.0	17.38	17.42	17.26	16.41	16.7	18.24	13.4
Fe_2O_3	3.98	2.54	5.89	3.07	1.86	0.8	1.25	1.2
FeO	5.48	2.72		4.17	3.35	3.1	1.02	2.5
MnO	0.20	0.09	0.10	—	0.09	0.1	0.06	0.3
MgO	4.12	3.95	2.13	3.23	3.56	1.7	1.50	0.0
CaO	9.00	5.92	5.03	6.93	6.26	6.4	3.17	2.8
Na_2O	3.70	4.40	3.17	3.21	3.95	3.1	4.97	4.9
K_2O	0.46	2.04	3.11	1.61	1.83	1.5	1.92	2.0
H_2O^+	0.23	1.08	1.84	1.24	1.24	1.3	0.26	1.9
H_2O^-	0.17					0.3		0.5
P_2O_5	0.10	0.28	0.20	0.20	0.18	0.1	0.09	0.0
Total	100.63	100.80	99.89	99.87	99.82	99.8	99.51	100.5
p.p.m.								
Ba	170	270	494		542	243	520	
Co	22	—	24		—	5	—	
Cr	24	56	8		—	—	13	
Cu	42	—	19		—	5	—	
Nb	—	—	33		—	—	—	
Ni	4	—	9		—	—	5	
Rb	—	30	94		39	—	45	
Sr	275	385	445		562	277	460	
V	220	175	63		—	—	68	
Y	—	—	60		—	—	—	
Zr	80	110	299		159	88	100	

(1) Basalt, Black Rocks, St Kitts, West Indies (Baker, 1968b, pp. 136, 144). (2) Typical island arc andesite cited by Jakeš and White (1971, p. 226, Table 2). (3) Andesite lava flow from Volcan San Francisco, Andes, 27°S. (spec. 17592). (4) Average of 1775 Cainozoic andesites (Chayes, 1969, p. 2, Table 1). (5) Andesite, Brokeoff Volcano, California (Fountain, 1979, p. 295, Table 1, B). (6) Average dacite from St Lucia, West Indies (Tomblin, 1975, p. 478). (7) Typical island arc dacite (Jakeš and White, 1971, p. 226, Table 2). (8) Rhyolite, St Lucia, West Indies (Tomblin, 1975, p. 478).

by ignimbrites, for example in the central Andes and northern Central America, where they seem to be associated with thicker continental crust. Welded eutaxitic ignimbrites are exceedingly rare in island arcs, where their closest counterparts are the non-welded *nuées ardentes* and pumice flows. Rhyolite flows and domes are also more common than in island arcs. As a consequence, the SiO_2 mode for the calc-alkaline suite of continental margins is slightly higher, probably at *c.* 61–62 per cent compared with 59 per cent for the island arc series.

Chemically the most distinctive features of the continental margin suites are higher concentrations of K and the related elements Rb, Sr, Ba, Zr, Th, U, etc. The K_2O/Na_2O ratio also tends to be higher and is close to unity in the more siliceous rocks (Table 2). K/Rb ratios are generally lower than in the island arc lavas (Jakeš and White, 1972). Forbes *et al.* (1969) show that continental andesites in Alaska contain more Fe and Ca than those of the Aleutian island arc. Also, Fe/Mg ratios are generally higher in continental margin calc-alkaline rocks (Jakeš and White, 1972). $^{87}Sr/^{86}Sr$ ratios suggest a greater range in values for the calc-alkaline rocks of continental margins: for example, Francis *et al.* (1977) show that a range of samples from volcanoes in Ecuador have $^{87}Sr/^{86}Sr$ ratios of *c.* 0.704, which is similar to that found in many island arc calc-alkaline volcanic rocks. However, in northern Chile the range is between 0.7058 and 0.7072, suggesting a significant Sr isotope contribution from the continental crust (Thorpe et al., 1979; Hawkesworth, this volume, Section VIII).

Shoshonite Association

Shoshonites, latites, trachytes, and associated rocks sometimes form a small proportion (c. 2.5 per cent) of the total volume of volcanic rocks to be found in relatively mature orogenic environments. The term shoshonite was given by Iddings (1895) to a mafic variety of olivine trachydolerite in Yellowstone Park, and the term is derived from a river of that name in this area. The term was revived for late Cainozoic lavas in north-western Viti Levu, Fiji, by Dickinson et al. (1968) and was also used by Gill (1970) for lavas of this province. Within the shoshonitic association (Joplin, 1965, 1968), it has been estimated (Jakeš and White, 1971) that basaltic representatives make up c. 50 per cent of the total volume, with intermediates 40 per cent and dacites only 10 per cent. Jakeš and White (1972, p. 33) suggest that when shoshonitic rocks occur, for example in Japan or New Guinea, the age of the basement rocks and associated sediments is usually pre-Mezozoic: they are thought to be characteristic of a late stage in development, as demonstrated on Fiji (Gill, 1970).

Compared with low K tholeiitic and calc-alkaline series, the shoshonitic rocks have higher contents of K, Ba, Rb, Sr, Zr, Th, and U for a given SiO_2 value (see Table 3). They have lower K/Rb ratios and the K_2O/Na_2O ratio is usually close to or in excess of 1: there is virtually no enrichment of Fe relative to Mg and alkalis. They are often, but not invariably, nepheline normative (see Jakeš and White, 1972; Gill, 1970). Sr isotope ratios are similar to those of the calc-alkaline series, as are the REE patterns, which also show slight enrichment in the lighter elements.

Apart from the well described shoshonites of the south-western Pacific, there are sho- shonitic lavas in southern Peru (Lefèvre, 1973) and north-western Argentina (Hörmann et al., 1973). A progression from calc-alkaline lavas to shoshonitic lavas has also been recognized in the Aeolian arc in the southern Tyrhenian Sea by Barberi et al. (1974). The shoshonitic suite from the Aeolian arc has a wide silica range (50–71 per cent) and is similar to the calc-alkaline series in respect of most major oxides, except K_2O, which is higher, and Al_2O_3, which tends to be lower. In the Aeolian arc the shoshonitic rocks are associated with high K andesites and also with leucite tephrites where K_2O is considerably in excess of Na_2O. The chemical composition, especially the Sr isotope ratios (0.7041–0.7061), points to a mantle origin for these rocks. Barberi et al. (1974) suggest that they have been derived by low degrees of partial melting during the closing stages of island arc evolution over a rapidly deepening Benioff zone.

Alkaline Basalts

Although not a very common occurrence, the close association of alkaline rocks with a calc-alkaline suite has been recorded in New Mexico (Baker and Ridley, 1970), Mexico (Robin, this volume, Section III), and on the island of Grenada in the Lesser Antilles (Arculus and Curran, 1972; Arculus, 1976). On Grenada, silica-undersaturated alkaline basalts and basanitoids are associated with calc-alkaline andesites and dacites and the contrasting magmas have sometimes been erupted from the same volcanic centres. The alkaline rocks are low in SiO_2 (44–47 per cent) with relatively high total alkalis, especially Na_2O. Compared with most West Indian calc-alkaline basalts these alkaline rocks also have higher contents of MgO, Ba, Nb, Zr, Sr, Rb, Ni, and Cr. Compared with shoshonitic basalts they have lower SiO_2, K, Rb, and Ba but decidedly higher MgO. However, Arculus (1976) points out that strongly variable trace element abundances characterize individual centres in Grenada. $^{87}Sr/^{86}Sr$ ratios are in the range 0.7043–0.7050 and do not discriminate between the alkaline and calc-alkaline

TABLE 3 Analysis of shoshonitic, alkaline and marginal basin volcanic rocks

	1	2	3	4	5	6	7	8
SiO_2	53.74	50.64	55.04	44.70	47.11	48.8	50.40	55.9
TiO_2	1.05	1.01	0.80	1.16	0.79	1.0	1.42	0.21
Al_2O_3	15.84	16.28	16.30	14.76	16.53	16.2	14.62	10.8
Fe_2O_3	3.25	3.92	3.69	⎱10.36	⎱9.22	1.6	⎱8.81	2.9
FeO	4.85	4.57	3.74	⎰	⎰	7.2	⎰	6.0
MnO	0.11	0.14	0.14	0.24	0.21	0.2	0.17	0.18
MgO	6.36	5.85	4.33	11.81	12.06	9.3	8.13	11.8
CaO	7.90	10.55	7.67	11.99	10.72	12.8	11.10	6.0
Na_2O	2.38	2.84	3.33	3.40	2.85	2.2	3.11	1.8
K_2O	2.57	2.74	3.77	1.27	0.37	0.12	0.36	0.7
H_2O^+ / H_2O^-	1.09	0.58	0.61	—	—	—	—	4.2
P_2O_5	0.54	0.68	0.39	0.32	0.15	0.07	0.15	0.03
Total	99.68	99.80	99.81	100.01	100.01	99.49	98.27	100.52
p.p.m.								
Ba	1000			640	135	<2		51
Co	—			—	—	67		—
Cr	30			—	—	459		263
Cu	—			128	96	101		76
Nb	—			25	7	—		4
Ni	20			292	371	199		72
Rb	75			40	7	<1		5
Sr	700			973	402	97		195
V	200			—	—	224		—
Y	—			120	20	—		28
Zr	40			176	65	—		109
$^{87}Sr/^{86}Sr$						0.7028		

(1) Representative shoshonite (Jakeš and White, 1971, p. 226, Table 2). (2) Average shoshonitic basalt, Eolian Islands (Barberi *et al.*, 1974, p. 271, Table 1, No. 7). (3) Average shoshonite latite (Barberi *et al.*, 1974, p. 271, Table 1, No. 8). (4) Basanitoid, Mount Maitland, Grenada, West Indies (Arculus, 1976, p. 616, Table 1A No. 10). (5) Alkaline basalt, Mt St Catherine, Grenada, West Indies, (Arculus, 1976, p. 616, Table 1A, No. 14). (6) Average of six basalts, Lau Basin (Hawkins, 1977, p. 358, Table 1, No. 1). (7) Marginal basin basalt from South Sandwich spreading axis (Tarney *et al.*, 1977, p. 371, Table 1, No. 1). (8) Boninite, inner wall Mariana trench (Cameron *et al.*, 1979, p. 551, Table 1, No. 4).

rocks. Arculus (1976) interprets the geochemical composition and REE patterns in terms of variable degrees of partial melting of an upper mantle peridotite source. Arculus and Curran (1972) suggested that the island arc calc-alkaline suite may have been derived from an alkaline parent magma. Brown *et al.* (1977) suggest that this compositional progression may have been achieved by successive fractionation of olivine, calcic augite, and spinel followed by amphibole and plagioclase.

Alkaline lavas also occur fairly commonly to the east of the active volcanic chain of the Andes (Thorpe and Francis, 1979) and form the dominant rock type of the Patagonian basalt plateau, where they are considered to have developed in an extensional regime associated with incipient back-arc spreading (Baker and Rea, 1978).

Marginal Basin Basalts

Volcanism associated with the development of back-arc or marginal basins (Karig, 1971) can be considered with justification as part of the spectrum of orogenic activity. Although connected in some way with the subduction process, the precise nature of such a link remains uncertain. One suggestion, put forward by Töksoz and Bird (1977), is that extension in marginal basins is a consequence of induced convective circulation in the mantle wedge overlying the subducted lithosphere. Similarly, Hawkesworth et al. (1977) conclude that subduction provides merely a physical mechanism for magma genesis in the back-arc situation since they detect no chemical contribution from the subducted slab. Another possibility, at least in the early stages of marginal basin development, is of mantle diapirs rising above the subducted slab to split the volcanic arc (Karig, 1974), in which case the eruptives might be expected to bear a resemblance to calc-alkaline or island arc tholeiitic suites, but there is as yet no evidence to substantiate such a transition.

Most authors (e.g. Hawkins, 1977) are agreed that the affinities of back-arc basalts and associated differentiates are very largely with mid-ocean ridge basalts, although with some deviations. Evidence from marginal basins in the Scotia arc (Tarney et al., 1977) suggests that in more mature and well established basins the basalts are essentially similar to mid-ocean ridge basalts, whereas those erupted in the earlier stages of development are slightly different and in particular are large ion lithophile-enriched. In the case of the Sarmiento back-arc basalts (Saunders et al., 1979) there is enrichment in K, Rb, Ba, Fe/Mg, and Ba/Sr, whereas K/Rb ratios are lower than in mid-ocean ridge basalts. In general $^{87}Sr/^{86}Sr$ ratios for marginal basin basalts appear to fall within a range between those of ocean ridge basalts and those of island arcs: this is well illustrated by the basalts of the Scotia Sea, which are in the range 0.7028–0.7034 (Hawkesworth et al., 1977), and Deception Island, where the mean value is c. 0.7034 (Weaver et al., 1979).

Boninites

Similar in some respects to magnesian andesites, boninites are regarded by Cameron et al. (1979) as the nearest Phanerozoic equivalents to the basaltic komatiites of the Archaean. They point to Green's (1976) experimental work on water-saturated partial melting of pyrolite which at 1200 °C and 10 kb yielded a liquid with distinct boninite affinities. Cameron et al. (1979) review the characteristics and distribution of boninites which are mainly associated with ophiolite assemblages, although often in an arc situation. They suggest that boninites may be erupted in the initial stages of breaking of an oceanic plate by wet partial melting of peridotite over the subducting slab. There is a possibility that some pyroxene andesites and dacites are the products of evolved boninitic liquids. Boninites are discussed by Meijer later in this volume (Section II).

REFERENCES

Arculus, R. J. (1976). Geology and geochemistry of the alkali basalt–andesite association of Grenada, Lesser Antilles island arc. *Geol. Soc. Am. Bull.* **87**, 612–624.

Arculus, R. J. and Curran, E. B. (1972). The genesis of the calc-alkaline rock suite. *Earth planet. Sci. Letters* **15**, 255–262.

Baker, I. and Ridley, W. I. (1970). Field evidence and K, Rb, Sr data bearing on the origin of the Mt Taylor volcanic field, New Mexico, USA. *Earth planet. Sci. Letters* **10**, 106–114.

Baker, P. E. (1968a). Comparative volcanology and petrology of the Atlantic island-arcs. *Bull. Volc.* **32**, 189–206.

Baker, P. E. (1968b). Petrology of Mt Misery volcano, St Kitts, West Indies. *Lithos* **1**, 124–150.

Baker, P. E. (1973). Volcanism at destructive plate margins. *J. Earth Sci., Leeds* **8**, 183–195.

Baker, P. E. (1978). The South Sandwich Islands. III. Petrology of the volcanic rocks. *Br. Antarct. Surv. Sci. Rep.* no. 93, 1–34.

Baker, P. E., Harris, P. G., and Reay, A. (1971). The geology of Tofua Island, Tonga. Cook Bicentenary Expedition in the South-west Pacific. *Bull. R. Soc. N.Z.* **8**, 67–79.

Baker, P. E., and Rea, W. J. (1978). Compositional variation in the Patagonian plateau basalts. In *International Geodynamics Conference, Tokyo, Abstracts*, pp. 206–207.

Barberi, F., Innocenti, F., Ferrara, G., Keller, J., and Villari, L. (1974). Evolution of Eolian arc volcanism, southern Tyrrhenian Sea. *Earth planet. Sci. Letters* **21**, 269–276.

Bauer, G. R. (1970). The geology of Tofua Island, Tonga. *Pacific Sci.* **24**, 333–350.

Brown, G. M., Holland, J. G., Sigurdsson, H., Tomblin,

J. F., and Arculus, R. J. (1977). Geochemistry of the Lesser Antilles volcanic island arc. *Geochim. cosmochim. Acta* **41**, 785–801.

Brown, G. M. and Schairer, J. F. (1968). Melting relations of some calc-alkaline volcanic rocks. *Carnegie Instn Wash. Yb.* **66**, 470–477.

Cameron, W. E., Nisbet, E. G., and Dietrich, V. J. (1979). Boninites, komatiites and ophiolitic basalts. *Nature, Lond.* **280**, 550–553.

Carmichael, I. S. E., Turner, F. J., and Verhoogen, J. (1974). *Igneous Petrology*, McGraw-Hill Book Co., New York.

Chayes, F. (1969). The chemical composition of Cenozoic andesite. *Bull. Oregon St. Dept. Geol. Min. Industries* **65**, 1–11.

Daly, R. A. (1933). *Igneous Rocks and the Depths of the Earth*, McGraw-Hill, New York.

Dickinson, W. R. and Hatherton, T. (1967). Andesitic volcanism and seismicity around the Pacific. *Science, N.Y.* **157**, 801–803.

Dickinson, W. R., Rickard, M. J., Coulson, F. I., Smith, J. G., and Lawrence, R. L. (1968). Late Cenozoic shoshonitic lavas in northwestern Viti Levu, Fiji. *Nature, Lond.* **219**, 148.

Donnelly, T. W. (1968). Caribbean island arcs in light of the sea-floor spreading hypothesis. *Trans. N.Y. Acad. Sci.* **30**, 745–750.

Donnelly, T. W., Rogers, J. J. W., Pushkar, P., and Armstrong, R. L. (1971). Chemical evolution of the igneous rocks of the eastern West Indies: an investigation of thorium, uranium and potassium distributions, and lead and strontium isotopic ratios. *Geol. Soc. Am. Mem.* no. 130, 181–224.

Ewart, A. and Bryan, W. B. (1973). The petrology and geochemistry of the Tongan Islands. In *Western Pacific: Island Arcs, Marginal Seas, Geochemistry* (P. J. Coleman, ed.), University of Western Australia Press, Perth, pp. 503–522.

Forbes, R. B., Ray, D. K., Katsura, T., Natsumoto, H., Haramura, H., and Furst, M. J. (1969). The comparative chemical composition of continental vs. island arc andesites in Alaska. *Bull. Oregon St. Dept. Geol. Min. Industries* **65**, 111–120.

Fountain, J. C. (1979). Geochemistry of Brokeoff Volcano, California. *Geol. Soc. Am. Bull.* **90**, 294–300.

Francis, P. W., Moorbath, S., and Thorpe, R. S. (1977). Strontium isotope data for recent andesites in Ecuador and north Chile. *Earth Planet. Sci. Letters*, **37**, 197–202.

Gill, J. B. (1970). Geochemistry of Viti Levu, Fiji, and its evolution as an island arc. *Contr. Mineral. Petrol.* **44**, 179–203.

Gledhill, A. and Baker, P. E. (1973). Strontium isotope ratios in volcanic rocks from the South Sandwich Islands. *Earth planet. Sci. Letters* **19**, 369–372.

Green, D. H. (1976). Experimental testing of 'equilibrium' partial melting of peridotite under water-saturated, high pressure conditions. *Can. Mineral.* **14**, 255–268.

Green, T. H. and Ringwood, A. E. (1968). Genesis of the calc-alkaline igneous rock suite. *Contr. Mineral. Petrol.* **18**, 105–162.

Green, T. H., Green, D. H., and Ringwood, A. E. (1967). The origin of high-alumina basalts and their relationships to quartz, tholeiites and alkali basalts. *Earth Planet. Sci. Letters* **2**, 41–51.

Gunn, B. M. (1976). A comparison of modern and Archaean oceanic crust and island arc petrochemistry. In *The Early History of the Earth* (B. F. Windley, ed.), John Wiley & Sons, London, pp. 389–403.

Harker, A. (1909). *The Natural History of Igneous Rocks*, Methuen and Co., London.

Hart, S. R. (1971). K, Rb, Cs, Sr, and Ba contents and Sr isotope ratios of ocean floor basalts. *Phil. Trans. R. Soc. A* **268**, 573–587.

Hart, S. R., Brooks, C., Krogh, T. E., Davis, G. L., and Nava, D. (1970). Ancient and modern volcanic rocks: a trace element model. *Earth planet. Sci. Letters* **10**, 17–28.

Hawkesworth, C. J., O'Nions, R. K., Pankhurst, R. J., Hamilton, P. J., and Evensen, N. M. (1977). A geochemical study of island-arc and back-arc tholeiites from the Scotia Sea. *Earth planet. Sci. Letters* **36**, 253–262.

Hawkesworth, C. J., Norry, M. J., Roddick, J. C., Baker, P. E., Francis, P. W., and Thorpe, R. S. (1979). $^{143}Nd/^{144}Nd$, $^{87}Sr/^{86}Sr$, and incompatible element variations in calc-alkaline andesites and plateau lavas from South America. *Earth planet. Sci. Letters* **42**, 45–47.

Hawkins, J. W., Jr (1977). Petrologic and geochemical characteristics of marginal basin basalts. In *Island Arcs, Deep Sea Trenches and Back-arc Basins* (M. Talwani and W. C. Pitman III, eds), Maurice Ewing Series no. 1, American Geophysical Union, Washington, DC, pp. 385–365.

Holmes, A. (1965). *Principles of Physical Geology*, Thomas Nelson & Sons, London.

Hörmann, P. K., Pichler, H., and Zeil, W. (1973). New data on the young volcanics in the Puna of NW Argentina. *Geol. Rundschau* **62**, 397–418.

Iddings, J. P. (1895). Absarokite–shoshonite–banakite series. *J. Geol.* **3**, 935–959.

Iddings, J. P. (1913). *Igneous Rocks*, Vol. II, John Wiley & Sons, New York.

Jakeš, P. and Gill, J. (1970). Rare earth elements and the island-arc tholeiitic series. *Earth planet. Sci. Letters* **9**, 17–28.

Jakeš, P. and White, A. J. R. (1969). Structure of the Melanesian arcs and correlation with distribution of magma types. *Tectonophysics* **8**, 223–236.

Jakeš, P. and White, A. J. R. (1971). Composition of island arcs and continental growth. *Earth planet. Sci. Letters* **12**, 224–230.

Jakeš, P. and White, A. J. R. (1972). Major and trace element abundance in volcanic rocks of orogenic areas. *Geol. Soc. Am. Bull.* **83**, 29–40.

Johannsen, A. J. (1937). *A Descriptive Petrography of the Igneous Rocks*, Vol. III, *The Intermediate Rocks*, University of Chicago Press, Chicago.

Joplin, G. A. (1965). The problem of the potash-rich basaltic rocks. *Mineral. Mag.* **34**, 266–275.

Joplin, G. A. (1968). The shoshonite association: a review. *J. geol. Soc. Aust.* **15**, 275–294.

Judd, J. W. (1886). On the gabbros, dolerites and basalts of Tertiary age in Scotland and Ireland. *Q. Jl. geol. Soc. Lond.* **42**, 49–97.

Karig, D. E. (1971). Origin and development of marginal basins in the Western Pacific. *J. geophys. Res.* **76**, 2542–2561.

Karig, D. E. (1974). Evolution of arc systems in the western Pacific. *A. Rev. Earth planet. Sci.* **2**, 50–75.

Karig, D. W. and Sharman, G. F. (1975). Subduction and accretion in trenches. *Geol. Soc. Am. Bull.* **86**, 377–392.

Kuno, H. (1950). Petrology of Hakone Volcano and the adjacent areas, Japan *Geol. Soc. Am. Bull.* **61**, 957–1020.

Kuno, H. (1959). Origin of Cenozoic petrographic provinces of Japan and surrounding areas. *Bull. Volc.* **20**, 37–76.

Kuno, H. (1966). Lateral variation of basalt magma across continental margins and island-arcs. *Geol. Surv. Pap. Can.* no. 66-15, 317–336.

Kushiro, I. and Yoder, H. S. (1969). Melting of forsterite and enstatite at high pressure under hydrous conditions. *Carnegie Instn. Wash. Yb.* **67**, 153–158.

Lefèvre, C. (1973). Les caractères magmatiques dur volcanisme plioquaternaire des Andes dans le Sud de Perou. *Contr. Mineral. Petrol.* **41**, 259–272.

Lewis, J. F. (1973). Petrology of the ejected plutonic blocks of the Soufriere Volcano, St Vincent, West Indies. *J. Petrol.* **14**, 81–112.

Lowder, G. C. and Carmichael, I. S. E. (1970). The volcanoes and caldera of Talasea, New Britain. *Geol. Soc. Am. Bull.* **81**, 17–38.

Macdonald, G. A. (1960). Dissimilarity of continental and oceanic rock types. *J. Petrol.* **1**, 172–177.

MacGregor, A. G. (1938). The Royal Society Expedition to Montserrat, B. W. I. The volcanic history and petrology of Montserrat, with observations on Mt Pelé, in Martinique. *Phil. Trans. R. Soc. B* **229**, 1–90.

Nicholls, I. A. (1974). Liquids in equilibrium with peridotitic mineral assemblages at high water pressures. *Contr. Mineral. Petrol.* **45**, 289–316.

Nicholls, I. A. and Ringwood, A. E. (1972). Production of silica-saturated tholeiitic magmas in island arcs. *Earth planet. Sci. Letters* **17**, 243–246.

Nicholls, I. A. and Ringwood, A. E. (1973). Effect of water on olivine stability in tholeiites and the production of silica-saturated magmas in the island-arc environment. *J. Geol.* **81**, 285–300.

Oki, Y., Aramaki, S., Nakamura, K., and Hakamata, K. (1978). Volcanoes of Hakone, Izu and Oshima. *Bull. Hot Spring Res. Inst. Kanagawa Prefecture*, **9**, (5), 1–88.

Peacock, M. A. (1931). Classification of igneous rocks. *J. Geol.* **39**, 54–67.

Pearce, J. A. and Cann, J. R. (1973). Tectonic setting of basic volcanic rocks determined using trace element analyses. *Earth planet. Sci. Letters* **19**, 290–300.

Peccerillo, A. and Taylor, S. R. (1976). Geochemistry of Eocene calc-alkaline volcanic rocks from the Kastamonu area, northern Turkey. *Contr. Mineral. Petrol.* **58**, 63–81.

Pichler, II. and Zeil, W. (1971). The Cenozoic rhyolite-andesite association of the Chilean Andes. *Bull. Volc.* **35**, 424–452.

Pushkar, P., Steuber, A. M., Tomblin, J. F., and Julian, G. M. (1973). Strontium isotopic ratios in volcanic rocks from St Vincent and St Lucia, Lesser Antilles. *J. geophys. Res.* **78**, 1279–1287.

Ringwood, A. E. (1977). Petrogenesis in island-arc systems. In *Island Arcs, Deep Sea Trenches and Back-arc Basins* (M. Talwani and W. C. Pitman III, eds), Maurice Ewing Series no. 1, American Geophysical Union, Washington, DC, 355–365.

Rittmann, A. (1953). Magmatic character and tectonic position of the Indonesian volcanoes. *Bull. Volc.* **14**, 45–58.

Saunders, A. D., Tarney, J., Stern, C. R., and Dalziel, I. W. D. (1979). Geochemistry of Mesozoic marginal basin floor igneous rocks from southern Chile. *Geol. Soc. Am. Bull.* **90**, 237–238.

Tarney, J., Saunders, A. D., and Weaver, S. D. (1977). Geochemistry of volcanic rocks from the island-arcs and marginal basins of the Scotia arc region. In *Island Arcs, Deep Sea Trenches and Back-arc Basins* (M. Talwani and W. C. Pitman III, eds.), Maurice Ewing Series no. 1, American Geophysical Union, Washington, DC, pp. 367–377.

Thorpe, R. S. and Francis, P. W. (1979). Variations in Andean andesite compositions and their petrogenetic significance. *Tectonophysics* **57**, 53–70.

Thorpe, R. S., Francis, P. W., and Moorbath, S. (1979). Rare earth and strontium isotope evidence concerning the petrogenesis of north Chilean ignimbrites. *Earth. planet. Sci. Letters* **42**, 359–367.

Tilley, C. E. (1950). Some aspects of magmatic evolution. *Q. Jl. geol. Soc. Lond.* **106**, 37–62.

Toksöz, M. N. and Bird, P. (1977). Formation and evolution of marginal basins and continental plateaus. In *Island Arcs, Deep Sea Trenches and Back-arc Basins* (M. Talwani and W. C. Pitman III, eds), Maurice Ewing Series no. 1, American Geophysical Union, Washington, DC, pp. 379–393.

Tomblin, J. F. (1975). The Lesser Antilles and Aves Ridge. In *The Ocean Basins and Margins*, Vol. 3, *The Gulf of Mexico and the Caribbean* (A. E. M. Nairn and F. G. Stehli, eds), Plenum Press, New York, pp. 467–500.

Tomita, T. (1931). Geological and petrological study of Dogo, Oki Islands. *J. geol. Soc. Tokyo* **38**, 461–479.

Tomkeieff, S. I. (1949). The volcanoes of Kamchatka. *Bull. Volc.* **8**, 87–113.

Weaver, S. D., Saunders, A. D., Pankhurst, R. J., and Tarney, J. (1979). A geochemical study of magmatism associated with the initial stages of back-arc spreading. *Contr. Mineral. Petrol.* **68**, 151–169.

Windley, B. F. (1977). *The Evolving Continents*, John Wiley & Sons, Chichester.

Yoder, H. S., Jr (1969). Calc-alkali andesites: experimental data bearing on the origin of their assumed characteristics. *Bull. Oregon St. Dept. Geol. Min. Industries* **65**, 77–89.

The mineralogy and petrology of Tertiary–Recent orogenic volcanic rocks: with special reference to the andesitic–basaltic compositional range

A. EWART

Department of Geology and Mineralogy,
University of Queensland, Brisbane, Queensland, Australia

ABSTRACT

A review is presented of the extrusive rocks, specifically from basalt to andesite, from the orogenic regions covering western (Andean) South America, western USA, Cascades–Aleutian Islands, the north-western and south-western Pacific, and Mediterranean subregions. Comparative data are also included from the K-rich series of the Roman Province, the South Shetland Islands, and the Tertiary anorogenic basalt–tholeiitic andesite association of southern Queensland. The review is based on a compilation of major elements, trace elements (18), plus modal phenocryst and phenocryst occurrence data; some 3500 sets of rock data are included. The data are tabulated into seven chemical groups (less than 48, 48–50, 50–52, 52–54, 54–56, 56–58, 58–60 per cent SiO_2); further breakdown of the data is made on the basis of (1) geographic/tectonic subregions, and (2) division into low K, calc-alkaline, high K, and shoshonitic series. Basalts, basaltic andesites, and andesites are defined here on the basis of SiO_2 percentages of less than 52, 52–56, and 56–63 respectively, and a set of statistics is compiled for these rock types.

The chemical data reveal pronounced regional differences, most strikingly between the active continental orogenic regions or margins (western USA, excluding Cascades; western South America; Mediterranean), and the island arcs; the differences are most evident in terms of the relative proportions, amongst the erupted magmas, of high K and shoshonitic types, which are highest in the continental orogenic regions. The relative proportions of these K-enriched magmas increases even further in the dacitic to rhyolitic compositions. Analysis of data of the south-western Pacific subregion reveal very strong correlations between K, P, Rb, Ba, Sr, Zr, and Pb, and to a less extent light REE, and it is these elements which are found relatively concentrated in the high K to shoshonitic extrusives. In contrast, the low K magmas of the western Pacific exhibit very pronounced depletions of these elements.

Phenocryst mineralogy of 'typical' (i.e. calc-alkaline) andesites and basaltic andesites comprises: calcic plagioclase, augite, orthopyroxene, titanomagnetite, $\pm$ Mg-olivine, with hornblende frequency increasing in andesites. The high K and shoshonitic andesites differ in containing higher frequency occurrences of hornblende and biotite, and more rarely sanidine. Biotite is absent, and hornblende occurs less frequently in the low K intermediate magmas. The ferromagnesian silicate phenocrysts in all the orogenic mafic to intermediate magmas are relatively Mg-rich, while total phenocryst contents in all the basaltic to andesite eruptive groups tend to be quite high (commonly greater than 20 per cent by volume).

Frequency occurrence data illustrate the common occurrence of Mg-olivine in basaltic andesites and andesites (interpreted as a 'relict' phase), and also illustrate that the appearance and levels of occurrence of phenocryst orthopyroxene are related to the level of K enrichment of the magmas; for example, it is shown that orthopyroxene appearance occurs progressively earlier (i.e. at lower SiO_2 magma compositions) in the progressively less K-rich magmas, although regional variations are apparent. Statistical studies of the south-western Pacific data reveal highly significant correlations between the following phenocryst modes and total rock chemical composition: plagioclase with Al_2O_3; olivine with MgO, Ni, Co, and Cr; augite with MgO. These correlations, together with additional evidence such as the phenocryst-rich nature of most basaltic to andesitic magmas, and the occurrence of 'relict' Mg-olivine, are believed to reflect the 'blanketing' effect of low pressure fractional crystallization processes, involving both phenocryst accumulation and extraction, in defining the final petrological characteristics of modern orogenic mafic and intermediate magmas.

Introduction

This paper essentially constitutes the second part of a review of the mineralogy, petrochemistry, and trace element compositions of Tertiary–Recent orogenic magmas. The first paper (Ewart 1979) was primarily concerned with a synthesis of available data on those orogenic eruptives for which SiO_2 is greater than 60 per cent (as recalculated on an anhydrous basis), thus including silicic andesites, dacites, and rhyolites. This paper is concerned with a complementary synthesis of data for those orogenic eruptives for which SiO_2 is less than 60 per cent, i.e. basalts, basaltic andesites, and andesites. Nevertheless, an attempt is made in this paper to overlap data compiled in these two reviews in order to provide sufficient continuity and thus give a reasonably complete overview of certain aspects of modern orogenic volcanism.

The necessary data for these two reviews is based on an accumulated computerized data bank, obtained by the author, of some 6500 sets of rock data. Information stored includes major element analyses and also, where available, accompanying trace element data (18 elements), plus modal and occurrence data for phenocryst phases. The primary aim of this paper is to present a comparative study of the chemical composition and mineralogy of the magmas from the rather diverse geographic and tectonic regions in which modern orogenic volcanism is concentrated. Clearly, the accumulated data will enable the application of a variety of statistical techniques to the complex interrelationships between chemical composition, mineralogy, tectonic environment, etc., to be made. Although beyond the scope of this paper, such statistical analyses will obviously constitute a follow-up study.

Classification

The classification scheme used follows that in Ewart (1979), being based, with minor modification, on the chemical criteria utilized by Peccerillo and Taylor (1976), namely the graph of K_2O against SiO_2 (Fig. 1). Thus, throughout this paper, the following SiO_2-based rock boundaries are used, with compositions recalculated to 100 per cent on an anhydrous basis: basalts below 52 per cent; basaltic andesites 52–56 per cent; andesites 56–63 per cent; dacites 63–69 per cent; rhyolites above 69 per cent SiO_2.

Further subdivisions of the data are made by utilizing the well known variation of K_2O in orogenic magmas, which not only exhibits the normal positive correlation with SiO_2, but also exhibits superimposed and systematically varying concentration levels which relate to the tectonic and/or crustal environment, depth to underlying Benioff zone, or even stratigraphic position within a particular volcanic province or sequence (e.g. Dickinson and Hatherton, 1967; Jakeš and Gill, 1970; Jakeš and White, 1969). Thus, following Peccerillo and Taylor (1976), four volcanic groupings, or series, are recognized (Fig. 1): the shoshonitic series (only distinguished in the compositions with less than 63 per cent SiO_2); the high K series; the calc-alkaline series; and the low K series.

Methods of approach to review

The following procedures have been adopted in data selection and presentation:

Definition of subregions

The data are considered in terms of the following geographic and/or tectonic regional groupings: in the following Table:

Subregion Number (Reference Figs 1, 3, 5, 6, 18, and 19)	Description of Subregion
1	Western (Andean) South America.
2	Western USA—eastern zone; defined on basis of the inferred eastern subduction zone of Lipman *et al.* (1972, their Fig. 9 on p. 235). Specifically excluded are the flood basalts of the Columbia River Plateau and the Snake River Plain.
3	Western USA—western zone; defined as in 2 above, but specifically excluding the Cascades of the north-western USA, and the Columbia River Plateau basalts.
4	North-eastern Pacific comprising the volcanic belt(s) of the Cascades (including Lassen, Medicine Lake, and Newberry volcanic centres), extending northwards to Alaska, and westwards to, and including, the Aleutian Islands. This subregion thus groups together a rather complex continental margin volcanic association with that of the Aleutian arc.
5	North-western Pacific, including the Kuriles, Kamchatka, Japan, Izu–Bonin Islands, Marianas, Taiwan, and the Phillippines. It is obvious that this again constitutes a broad grouping of very diverse and tectonically complex regions, including several arcs, and active continental and sub-continental margin environments.
6	South-western Pacific, including Papua New Guinea, Solomon Islands, New Hebrides, Fiji, Tonga–Kermadec Islands, and New Zealand. This grouping constitutes perhaps the most diverse of all the groupings in this paper, and includes volcanic associations erupted at plate boundaries showing convergent and divergent characteristics (e.g. Johnson *et al.*, 1978), as well as widely varying crustal types, thicknesses, and ages.
7	Mediterranean region, comprising the inferred Aeolian (Calabrian) and Aegean (Hellenic) arcs (e.g. Ninkovich and Hays, 1972; Barberi *et al.*, 1974), and including regions west to Almeria in south-eastern Spain, and eastwards to Anatolia. Specifically excludes the K-rich series of the Roman Volcanic Province (*see* 8, below).
8	The K-rich series of the Roman Volcanic Province. Although these rather extreme magmas can hardly be considered to be typical of orogenic regions, they are erupted within an area which may be affected by a subduction process. Moreover, they provide interesting comparative data on a rather extreme geochemical 'end member' of the orogenic magma association.
9	South Shetland Islands, dominantly Deception Island.
10	Anorogenic basalt–tholeiitic andesite Tertiary association of southern Queensland, Australia. These data are included in order to provide petrological comparisons between the orogenic and a well defined anorogenic bimodal association. This particular region is chosen due to availability of extensive mineralogical and chemical data to the author.

It is noted that no detailed data are included from the Middle Americas or the Indonesian orogenic regions (although included in the previous review; Ewart, 1979) due to limitations on time and space in this work. Similarly, no attempt is made to include data from the Carribean arc, which has been the subject of a comprehensive chemical study by Brown et al. (1977). (The characteristics of volcanism in areas within most of these subregions are described in Section III of this book.)

Sample selection

Within the subregions described above, all available data have been included in the data compilation except rocks specifically identified as sea-floor lavas; thus, a coverage is provided of the rather diverse magma compositions occurring within orogenic regions. No data for pre-Tertiary volcanics are included due to the increasing problems of secondary alteration together with increasing ambiguity as to their tectonic environments.

During data selection, emphasis has been placed on obtaining both chemical and phenocryst data for individual samples. A list of data sources is provided in Ewart (1979) and in the references to this paper. The trace and minor element data utilized in the compilations are based on all available analyses (for which accompanying major element analyses are also presented), irrespective of the analytical technique employed. This has the obvious disadvantage that data of variable precision are mixed (e.g. optical spectrographic and isotope dilution determinations). In the interests, however, of obtaining as broad a coverage as possible, this 'blanket' approach has been adopted, as in the earlier review. (Particular problems may arise, for example, for La, Ce, and Ni in the data for the western USA, judging by comparative data presented for these elements by Lipman (1968) and Zielinski and Lipman (1976) on the same samples, involving spectrographic, X-ray, and activation methods.)

Petrographic data on phenocryst phases are based on both modal abundances (not frequently published) and the percentage frequency occurrence of each phenocryst phase within a given rock series or grouping, which is obtained initially from the stated presence or absence of each phase in a given rock. It must again be noted, however, that many papers which provide excellent chemical data all too often provide rather vague accompanying petrographic information on the rocks analysed. This is especially a problem with respect to the Fe–Ti oxides, and the data for the occurrence of the oxides are, at best, only approximate, with the exception of the data from the south-western Pacific subregion, where ample mineralogical data are available.

Data reduction and presentation

For reduction and presentation of basic statistics, etc., the data are initially subdivided according to two criteria, namely (1) geographic/tectonic subregions (*see* section on Definition of subregions, above); and (2) percentage SiO_2 (calculated on anhydrous basis). The data are grouped according to the following SiO_2 intervals (*see* Table 1): less than 48; 48–50; 50–52; 52–54; 54–56; 56–58; 58–60 per cent. These data groups can be compared and extended directly to the percentage SiO_2 groups presented in Ewart (1979) of 60–63; 63–66; 66–69; 69–73; greater than 73 per cent. This procedure partly overcomes ambiguities caused by wider, arbitrary classification divisions, and, more importantly, allows detailed serial (sequential) changes of both chemical composition and mineralogy to be deduced through the various rock series or associations. In Table 1, the arithmetic means of all the data compiled are presented, subdivided according to the above criteria.

For some aspects of data treatment and interpretation, the data have been further subdivided into the four volcanic series previously noted (shoshonitic, high K, calc-alkaline, low K). Some representative averaged data for these four series are presented in Appendices 1–4 (pp. 88–95).

The majority of chemical analyses incorporated into this survey have included specific FeO and Fe_2O_3 determinations; where these were not given, however, they were calculated according to the method of Le Maitre (1976).

Major Element Compositions

Active continental margins and island arcs

In Table 1, sequential arithmetic means of the total chemical and mineralogical data are presented for each geographic subregion being considered. In Fig. 1, the means of K_2O against SiO_2 are plotted from Table 1, and are compared with the compositional fields of Peccerillo and Taylor (1976). It is clear that the erupted magmas from the western USA (excluding Cascades), Mediterranean, and Andean (South America) regions are, for compositions with greater than 51 per cent SiO_2, consistently enriched in K_2O, on average, by comparison with the orogenic magmas of the western Pacific island arcs and the Cascades–Alaska–Aleutian subregion. These differences are less pronounced in the basaltic compositions. Figure 1 also emphasizes the remarkably K-rich character of the K-rich series of the Roman Volcanic Province.

Table 2 presents a summary of the percentage frequency occurrence of the designated shoshonitic, high K, calc-alkaline, and low K series from each orogenic subregion, through the compositional ranges from basalt to rhyolite, and also the percentage of analysed samples for each of the latter compositional types. These figures simply reinforce Fig. 1, and show that the western USA contains the highest frequency of relatively K-enriched magmas, and moreover, that the frequency of shoshonitic and high K types increases further in the eastern zone compared to the western zone. In contrast, the island-arc systems of the western Pacific, and the Aleutian–Cascade subregions, are characterized by predominantly 'calc-alkaline' volcanism; the rather greater chemical diversity of the south-western Pacific magmas would seem to reflect the greater tectonic complexities encountered within areas of this subregion.

The Fe–Mg–alkali ratios of the four volcanic series are compared in Fig. 2, again based on sequential means for the SiO_2 intervals as previously described, and also including the dacitic and rhyolitic data from Ewart (1979). The characteristic absence of strong Fe enrichment in these orogenic magmas is clearly shown, and the compositional trends of the calc-alkaline, high K, and shoshonitic series, in fact, overlap extensively. In contrast, the low K series from the western Pacific subregions show a distinct relative Fe enrichment (the island arc tholeiite trend of Jakeš and Gill, 1970), but which is not shown by the corresponding low K magmas of the other orogenic zones. The trends in Fig. 2 indicate that only rather small changes occur in $Mg/(Mg + \Sigma Fe(atomic))$ ratios through the mafic to intermediate orogenic magmas: these ratios vary from c 0.5–0.7 in the calc-alkaline to shoshonitic basalts to c 0.4–0.6 for the corresponding andesites and basaltic andesites. The low K magmas of the western Pacific have ratios of c 0.46–0.56 (basalts) to 0.51–0.41 (andesites and basaltic andesites).

The question of silica saturation in orogenic magmas is examined in Fig. 3, again based on sequential mean data for the four volcanic series. The basalts exhibit a complete range from nepheline to quartz normative; the high K and shoshonitic basalts are typically nepheline normative, extending to olivine normative for the more SiO_2-rich types, whereas the western Pacific low K basalts are quartz normative, or nearly so. The basaltic andesites are dominantly quartz normative, although some of the K-rich compositions from the South American and Mediterranean subregions are silica-undersaturated. The levels of silica saturation increase through all the andesitic compositions. Figure 3(d) plots the normative compositions of the weighted means of the basalts, basaltic andesites, and andesites of each subregion (Tables 3–5), which illustrates that the basalts are generally dominated by olivine-normative compositions, the averaged compositions from the Mediterranean and South American subregions lying

TABLE 1 Compilation of total major and trace element data, and phenocryst data, for mafic to intermediate eruptive rocks from orogenic regions, grouped according to geographic and/or tectonic region. Major element analyses are recalculated to 100 per cent on an anhydrous basis

Region:	Western (Andean) South America					Western USA—Eastern Zone					
SiO$_2$ Interval:	<52%	52–54%	54–56%	56–58%	58–60%	<48%	48–50%	50–52%	52–54%	54–56%	56–58%	58–60%
SiO$_2$	51.05	53.10	54.84	56.97	59.19	46.59	48.94	50.85	52.98	55.16	57.20	59.12
TiO$_2$	1.14	1.20	1.35	1.12	1.01	2.43	2.08	1.66	1.31	1.28	1.19	1.09
Al$_2$O$_3$	18.57	18.11	16.78	17.05	17.18	15.38	16.10	15.85	16.25	16.84	16.94	17.07
Fe$_2$O$_3$	3.42	2.87	3.43	3.33	3.32	3.82	4.22	4.15	4.68	4.47	3.79	3.95
FeO	5.48	5.87	4.82	4.29	3.13	8.08	7.11	6.57	4.66	3.81	3.67	2.84
MnO	0.16	0.16	0.14	0.14	0.12	0.18	0.18	0.15	0.15	0.13	0.13	0.11
MgO	5.54	5.46	5.22	4.29	3.41	8.43	7.32	6.83	5.46	4.27	3.54	2.73
CaO	8.87	7.97	7.34	6.47	5.91	9.74	8.91	8.60	7.70	6.76	6.01	5.49
Na$_2$O	3.98	3.62	3.72	3.97	3.99	3.40	3.35	3.34	3.42	3.83	3.83	3.95
K$_2$O	1.42	1.34	1.94	1.99	2.41	1.29	1.24	1.46	2.75	2.89	3.21	3.22
P$_2$O$_5$	0.38	0.29	0.41	0.37	0.33	0.66	0.56	0.53	0.63	0.56	0.47	0.43
n =	27	40	34	30	45	25	35	48	52	55	41	40

Trace elements (p.p.m.)

Rb	49.9 (11)*	40.3 (19)	49.4 (24)	47.1 (17)	67.7 (32)	55.5 (2)	33.3 (6)	30.3 (3)	80.4 (5)	53.6 (9)	84.9 (8)	71.6 (5)
Ba	345 (9)	655 (11)	690 (16)	956 (9)	922 (24)	782 (11)	549 (10)	813 (17)	1015 (13)	1360 (9)	812 (10)	1540 (10)
Sr	608 (10)	662 (19)	629 (24)	679 (17)	681 (32)	957 (15)	675 (16)	726 (19)	1164 (16)	913 (15)	779 (17)	1080 (15)
Zr	162 (8)	199 (10)	166 (15)	190 (3)	200 (11)	184 (9)	169 (8)	153 (15)	182 (12)	181 (6)	183 (10)	296 (9)
Zn	106 (3)	106 (8)	97.6 (8)	92 (2)	110 (2)	n.d.	n.d.	n.d.	92.5 (2)	113 (2)	n.d.	n.d.
La	16.3 (8)	26.4 (9)	22.2 (7)	56.4 (4)	34.1 (4)	89.8 (4)	n.d.	71.6 (5)	80.0 (8)	53.0 (4)	71.8 (8)	120 (6)
Ce	41.6 (8)	55.9 (9)	46.2 (8)	91.6 (6)	53.9 (9)	500 (1)	n.d.	124 (3)	93.3 (6)	89.5 (4)	126 (7)	175 (4)
Yb	2.29 (8)	2.16 (9)	2.49 (8)	2.49 (6)	1.65 (9)	4.22 (9)	3.38 (8)	3.28 (12)	3.90 (10)	2.95 (4)	3.05 (10)	2.71 (7)
Y	31.0 (1)	30.3 (7)	16.8 (4)	22.0 (2)	14.9 (7)	37.9 (9)	33.8 (8)	37.1 (14)	38.8 (12)	24.0 (5)	35.5 (10)	28.9 (9)
Cu	30.0 (3)	56.4 (10)	43.5 (11)	52.4 (5)	43.8 (9)	77.4 (9)	64.3 (11)	84.8 (15)	73.7 (12)	117 (9)	51.4 (14)	33.5 (10)
Ni	57.9 (10)	66.1 (17)	68.7 (18)	48.8 (8)	40.0 (18)	89.6 (9)	89.4 (11)	132 (5)	105 (11)	60.1 (9)	105 (13)	49.7 (11)
Co	29.6 (11)	31.2 (17)	29.8 (15)	21.5 (6)	20.7 (13)	49.0 (9)	35.7 (11)	36.5 (15)	38.7 (11)	25.4 (7)	26.6 (14)	21.5 (11)
Cr	67.9 (8)	204 (11)	198 (7)	74.5 (2)	54.8 (10)	230 (9)	119 (8)	223 (15)	137 (12)	131 (6)	147 (9)	42.1 (9)
V	187 (3)	252 (8)	168 (5)	130 (1)	139 (8)	303 (9)	233 (11)	183 (14)	211 (12)	226 (9)	203 (14)	142 (10)
Nb	n.d.	12.5 (2)	n.d.	n.d.	n.d.	33.9 (8)	34.0 (6)	21.3 (6)	17.0 (6)	12.5 (2)	14.4 (7)	19.7 (6)
Li	n.d.	n.d.	11.4 (5)	10.3 (7)	12.1 (14)	n.d.	8.33 (3)	n.d.	8.0 (1)	10.3 (3)	10.8 (4)	9.5 (2)
Pb	n.d.	n.d.	n.d.	n.d.	n.d.	3.0 (1)	n.d.	8.33 (3)	18.0 (5)	9.67 (3)	12.6 (8)	34.0 (6)
Hf	2.88 (5)	3.80 (7)	3.43 (4)	10.7 (1)	3.35 (2)	n.d.	n.d.	n.d.	n.d.	n.d.	n.d.	n.d.

Phenocryst modes (vol. %)												
Sanidine	0	0	0	0	0.008	0	0	0	0	0	0.04	0.64
Plagioclase	32.2	32.3	8.40	16.6	13.6	5.73	0.42	3.30	5.69	10.4	16.8	13.4
Quartz	0	0	0	0	0.16	7.10	0	3.06	0.02	0.03	0.14	0.04
Olivine	6.27	7.60	1.75	1.99	0.16	7.10	1.34	3.06	3.16	1.35	0.97	0.12
Augite	0.33	4.21	1.31	4.53	2.53	2.13	0	1.71	3.11	3.44	3.65	2.85
Orthopyroxene	0	0	0.14	3.46	2.09	0	0	0	0.38	0.55	1.26	0.70
Hornblende	0	0	0	0.32	3.23	0	0	0	0	2.66	0.13	2.58
Biotite	0	0.43	0	0	0.15	0	0	0	0	0.005	0.04	0.39
Fe–Ti oxides	0	0.11	0.21	0.66	1.10	0.83	0	0.18	0.43	0.83	1.08	1.10
Pigeonite	0			0	0	0	0	0	0	0	0	0
TPC	38.8	44.6	11.8	27.6	23.0	15.8	1.76	8.24	12.8	19.3	24.1	21.8
n =	11	7	8	9	13	6	8	18	14	19	27	19

Phenocryst occurrence (frequency %)												
Hauyne	4.8	0	3.3	0	0	0	0	0	0	0	0	0
Sanidine	0	0	0	0	2.3	0	0	4.3	3.9	3.8	12.2	10.3
Plagioclase	76.2	77.1	40.0	80.0	100	41.7	48.5	48.9	52.9	81.1	85.4	84.6
Quartz	0	0	3.3	0	2.3	0	3.0	0	5.9	1.9	1.5	7.7
Olivine	95.2	94.3	80.0	76.7	22.7	95.8	87.9	87.2	88.2	67.9	46.3	20.5
Augite	66.7	82.9	83.3	90.0	90.9	54.2	45.5	42.6	72.5	79.2	85.4	82.1
Orthopyroxene	14.3	5.7	40.0	73.3	97.7	0	0	2.1	3.9	18.9	46.3	48.7
Hornblende	4.8	2.9	0	33.3	50.0	4.2	0	2.1	7.8	17.0	17.1	41.0
Biotite	4.8	0	20.0	16.7	34.1	4.2	0	0	7.8	5.7	14.6	43.6
Fe–Ti oxides	4.8	20.0	20.0	53.3	79.5	12.5	6.1	10.6	19.6	49.1	46.3	66.7
Pigeonite	0	5.7	0	0	0	0	0	0	0	1.9	0	0
n =	21	35	30	30	44	24	33	47	51	53	41	39

TABLE 1 (*continued*)

Region:	Western USA—Western Zone							Cascades (Western USA)–Alaska–Aleutian Islands						
SiO$_2$ Interval:	<48%	48-50%	50-52%	52-54%	54-56%	56-58%	58-60%	<50%	50-52%	52-54%	54-56%	56-58%	58-60%	
SiO$_2$	47.52	49.29	51.14	52.95	54.95	57.23	59.28	48.79	51.15	53.19	55.11	57.01	59.00	
TiO$_2$	2.21	1.60	1.67	1.41	1.21	1.15	0.99	1.29	1.16	1.29	1.13	0.99	0.93	
Al$_2$O$_3$	16.25	16.77	16.29	16.82	16.86	17.04	17.01	17.71	18.05	17.64	17.60	17.57	17.44	
Fe$_2$O$_3$	3.24	3.79	4.49	3.77	3.83	3.53	3.37	3.23	2.99	3.51	2.89	2.83	2.83	
FeO	7.52	5.71	5.15	4.81	4.13	3.84	3.30	6.76	6.34	5.62	5.22	4.54	4.02	
MnO	0.18	0.14	0.15	0.14	0.13	0.12	0.13	0.16	0.17	0.16	0.14	0.14	0.12	
MgO	8.01	7.76	6.87	5.98	5.08	3.91	3.28	7.57	6.11	5.08	4.81	4.13	3.22	
CaO	9.04	9.77	8.78	8.32	7.48	6.51	5.93	10.69	9.74	8.81	8.09	7.45	6.83	
Na$_2$O	3.69	3.32	3.41	3.51	3.57	3.60	3.64	2.79	3.22	3.53	3.64	3.71	3.90	
K$_2$O	1.69	1.35	1.44	1.77	2.30	2.65	2.68	0.77	0.83	0.90	1.12	1.38	1.47	
P$_2$O$_5$	0.65	0.50	0.61	0.52	0.46	0.43	0.38	0.23	0.25	0.28	0.25	0.24	0.24	
n =	10	24	31	31	27	40	40	40	52	74	97	88	98	

Trace elements (p.p.m.)

Rb	43.3 (6)	36.7 (3)	21.0 (6)	30.8 (6)	61.3 (4)	70.0 (4)	43.3 (3)	42.6 (7)	17.3 (4)	19.9 (14)	31.4 (27)	34.6 (22)	34.6 (20)	
Ba	557 (10)	940 (13)	1085 (15)	1093 (11)	1240 (11)	1274 (11)	1529 (12)	477 (16)	377 (16)	398 (26)	469 (26)	543 (31)	614 (28)	
Sr	941 (10)	1112 (13)	649 (15)	854 (11)	1224 (11)	1264 (11)	1052 (12)	706 (17)	585 (17)	659 (29)	618 (40)	685 (35)	719 (32)	
Zr	212 (10)	192 (13)	215 (14)	183 (11)	200 (11)	225 (11)	215 (11)	72.9 (16)	81.5 (16)	104 (26)	140 (37)	149 (31)	146 (29)	
Zn	94.0 (5)	110 (2)	104 (5)	72 (3)	153 (3)	60.0 (1)	116 (4)	73.7 (3)	n.d.	73.5 (2)	87 (1)	71.2 (6)	69 (3)	
La	68.1 (8)	65.9 (11)	89 (9)	42.0 (5)	65.6 (8)	56.5 (10)	85.6 (9)	8.51 (5)	7.85 (2)	7.65 (4)	9.25 (2)	14.9 (7)	24 (4)	
Ce	90.0 (6)	105 (6)	102 (6)	121 (4)	83.3 (3)	131 (5)	146 (5)	30.8 (4)	n.d.	25.3 (4)	28.0 (1)	48.5 (8)	47 (2)	
Yb	3.0 (2)	3.13 (8)	4.26 (7)	3.33 (3)	3.33 (6)	2.67 (3)	2.80 (5)	2.28 (3)	2.78 (6)	2.87 (6)	2.47 (6)	2.22 (9)	2.93 (12)	
Y	23.3 (9)	32.2 (12)	36.3 (14)	25.4 (11)	31.4 (11)	32.3 (11)	31.7 (12)	28.5 (8)	29.4 (7)	42.4 (14)	23.4 (28)	23.7 (23)	33.3 (22)	
Cu	52.3 (9)	64.1 (12)	58.0 (15)	40.1 (9)	36.9 (8)	54.0 (8)	39.2 (12)	103 (13)	94.2 (15)	107 (23)	68.1 (17)	53.9 (20)	44.5 (23)	
Ni	143 (10)	183 (13)	120 (15)	108 (11)	75.5 (11)	50.5 (11)	48.9 (11)	77.8 (15)	91.8 (16)	47.2 (23)	47.8 (33)	30.2 (29)	20.4 (27)	
Co	43.6 (10)	40.4 (12)	60.7 (11)	30.6 (8)	28.5 (10)	22.5 (10)	21.1 (9)	38.2 (15)	36.3 (15)	27.6 (20)	25.9 (18)	27.5 (22)	20.3 (27)	
Cr	196 (10)	341 (13)	273 (15)	249 (11)	173 (11)	128 (11)	86.4 (11)	151 (14)	194 (13)	99.0 (20)	118 (15)	81.6 (19)	41.8 (27)	
V	217 (10)	276 (13)	187 (11)	167 (11)	192 (11)	161 (11)	140 (12)	283 (14)	285 (14)	260 (24)	203 (36)	169 (30)	156 (29)	
Nb	12.5 (2)	11.1 (7)	26.4 (7)	18.8 (4)	15.0 (7)	11.6 (5)	17.4 (30)	3.27 (3)	n.d.	5.95 (2)	9.30 (1)	3.94 (5)	13.6 (3)	
Li	25.0 (1)	20.0 (3)	20.0 (1)	17.5 (2)	18.3 (3)	25.0 (3)	n.d.	8.67 (3)	2.0 (1)	12.0 (2)	40.0 (3)	17.5 (2)	19.1 (12)	
Pb	14.0 (3)	13.3 (4)	15.0 (6)	14.3 (7)	18.5 (10)	18.0 (10)	23.2 (11)	6.23 (3)	n.d.	6.40 (2)	9.40 (2)	9.32 (5)	14.8 (2)	
Hf	n.d.	n.d.	n.d.	n.d.	n.d.	n.d.	n.d.	n.d.	n.d.	n.d.	n.d.	n.d.	n.d.	

Ewart: Tertiary–Recent orogenic volcanic rocks

Phenocryst modes (vol. %)													
Sanidine	0	0	0	0	0	0	0	0	0	0	0	0	0
Plagioclase	1.69	3.17	9.18	6.22	12.8	11.2	13.8	13.9	15.8	11.6	16.3	19.9	19.0
Quartz	0	0	0.05	0.30	0.10	0.26	0.01	0	0	0	0	0.07	0.21
Olivine	7.91	5.37	4.11	3.19	3.68	1.59	0.11	4.54	3.44	1.27	1.50	1.20	0.21
Augite	1.84	1.70	1.85	0.83	4.98	1.77	1.55	6.29	3.02	3.22	3.31	4.07	2.58
Orthopyroxene	0	0	0.01	0.07	0.35	0.36	0.65	0	0.07	0.70	1.59	2.14	2.20
Hornblende	0	0	0	0.06	0	1.71	4.14	3.17	0.73	0	1.56	0.15	1.21
Biotite	0	0	0	0	0	0.56	0.29	0	0	0	0	0	0.01
Fe–Ti oxides	0	0	0	0.04	0.17	0.33	0.45	1.07	0.25	0.81	0.85	0.98	0.99
Pigeonite	0	0	0	0	0	0	0	0	0	0.04	0	0	0
TPC	11.4	10.2	15.2	10.7	22.1	17.8	21.0	29.0	23.3	17.6	25.1	28.5	26.4
$n =$	8	9	16	12	8	18	17	15	18	29	20	19	22
Phenocryst occurrence (frequency %)													
Sanidine	0	0	0	0	0	0	0	0	0	0	0	0	0
Plagioclase	60.0	85.7	59.3	77.8	83.3	86.1	97.3	79.5	98.0	97.3	97.8	98.7	100.0
Quartz	0	0	3.7	18.5	4.2	8.3	5.4	0	0	0	2.2	7.5	3.2
Olivine	100.0	90.5	100.0	96.3	75.0	66.7	24.3	82.1	92.2	93.2	85.9	67.5	25.5
Augite	70.0	52.4	66.7	74.1	91.7	75.0	86.5	51.3	52.9	65.8	91.3	93.8	95.7
Orthopyroxene	0	0	11.1	11.1	25.0	36.1	67.6	2.6	7.8	32.9	53.3	91.3	91.5
Hornblende	0	4.8	0	7.4	20.8	13.9	62.2	12.8	2.0	2.7	5.4	15.0	16.0
Biotite	0	4.8	0	3.7	16.7	8.3	18.9	0	0	0	0	0	1.1
Fe–Ti oxide	0	0	11.1	11.1	33.3	38.9	62.2	17.9	25.5	30.1	47.8	77.5	87.2
Pigeonite	0	0	3.7	0	0	2.8	0	0	0	2.7	0	0	0
$n =$	10	21	27	27	24	36	37	39	51	73	92	80	94

TABLE 1 (continued)

Region:	North-West Pacific							South-West Pacific					
SiO_2 Interval	<48%	48–50%	50–52%	52–54%	54–56%	56–58%	58–60%	<48%	48–50%	50–52%	52–54%	54–56%	56–58%	58–60%
SiO_2	47.10	49.16	50.99	53.04	54.99	57.01	58.93	47.35	49.37	51.25	53.08	55.27	57.09	59.31
TiO_2	1.61	1.24	1.06	1.00	0.90	0.80	0.80	0.82	0.85	0.86	0.86	0.80	0.77	0.72
Al_2O_3	17.56	17.72	18.09	17.82	17.95	17.70	17.28	12.90	16.17	17.04	17.12	17.03	17.06	16.93
Fe_2O_3	3.79	3.42	3.57	3.40	3.19	3.26	3.02	3.78	2.90	3.36	3.32	3.18	3.05	2.79
FeO	7.55	7.32	7.08	6.54	5.85	5.05	4.53	6.93	7.34	6.24	5.97	5.40	4.66	3.98
MnO	0.18	0.18	0.20	0.17	0.17	0.15	0.15	0.19	0.18	0.17	0.17	0.16	0.15	0.13
MgO	7.99	6.85	5.33	4.94	4.36	3.95	3.47	13.51	7.97	6.42	5.53	4.96	4.31	3.88
CaO	10.27	10.56	10.16	9.24	8.35	7.83	7.01	10.85	11.40	10.36	9.50	8.67	7.83	6.96
Na_2O	2.65	2.53	2.56	2.77	2.99	3.00	3.27	2.16	2.40	2.68	2.82	3.02	3.28	3.41
K_2O	0.99	0.81	0.76	0.89	1.07	1.10	1.36	1.19	1.15	1.33	1.35	1.26	1.55	1.67
P_2O_5	0.30	0.21	0.19	0.19	0.19	0.16	0.17	0.30	0.27	0.29	0.29	0.24	0.25	0.22
n =	21	63	193	227	181	190	188	39	133	169	199	204	171	137

Trace elements (p.p.m.)

Rb	16 (1)	19.6 (8)	25.6 (12)	27.8 (10)	42.5 (11)	22.2 (18)	25.6 (8)	21.8 (21)	28.5 (99)	31.3 (101)	32.7 (94)	27.8 (91)	33.2 (81)	41.7 (70)
Ba	171 (1)	293 (11)	232 (10)	276 (19)	369 (12)	254 (10)	371 (10)	340 (19)	347 (83)	383 (95)	399 (98)	405 (90)	430 (79)	498 (73)
Sr	463 (2)	469 (14)	346 (13)	339 (22)	350 (22)	263 (27)	282 (18)	577 (22)	605 (101)	661 (106)	592 (103)	527 (95)	551 (83)	534 (73)
Zr	88 (1)	76.2 (9)	97.2 (9)	107 (13)	121 (10)	147 (5)	197 (6)	74 (20)	60.6 (95)	77.9 (96)	94.6 (90)	114 (96)	132 (90)	143 (73)
Zn	82 (1)	68.5 (4)	78 (4)	86.1 (7)	75.5 (4)	77.5 (14)	84.0 (9)	81 (7)	70.2 (27)	76.8 (34)	86.0 (33)	77.8 (38)	76.6 (25)	72.3 (22)
La	4.49 (2)	6.47 (3)	7.03 (7)	4.78 (8)	10.6 (8)	9.33 (9)	7.94 (6)	8.3 (4)	8.72 (31)	13.9 (45)	14.9 (43)	24.9 (48)	30.3 (41)	26.4 (33)
Ce	14.6 (2)	31.4 (5)	19.4 (7)	23.4 (9)	25.1 (8)	22.8 (9)	23.3 (6)	24.0 (5)	23.1 (33)	27.9 (49)	29.3 (52)	43.5 (52)	50.0 (39)	45.7 (34)
Yb	2.04 (2)	2.18 (6)	2.16 (7)	2.13 (8)	2.17 (8)	2.35 (9)	3.14 (6)	0.90 (2)	1.68 (10)	1.47 (9)	1.66 (7)	1.52 (12)	2.03 (11)	1.65 (6)
Y	27.5 (2)	24 (7)	28 (4)	20.9 (14)	22.1 (10)	20.5 (10)	18.0 (9)	20.5 (17)	18.7 (91)	20.4 (96)	22.5 (99)	24.2 (83)	26.0 (75)	22.6 (52)
Cu	41.5 (2)	55.5 (8)	67 (6)	54.1 (16)	54.9 (11)	56.2 (21)	36.3 (14)	102 (22)	112 (79)	134 (90)	124 (94)	86.9 (95)	63.3 (89)	43.2 (72)
Ni	187 (1)	75.6 (7)	67 (6)	41.4 (15)	45.9 (9)	43.3 (19)	36.7 (11)	332 (24)	94.8 (87)	56.7 (97)	44.3 (98)	45.5 (95)	37.8 (86)	40.3 (67)
Co	31.8 (2)	31.5 (7)	39.7 (7)	32.5 (15)	34.6 (12)	37.7 (20)	32.8 (13)	71 (19)	41.4 (63)	34.6 (52)	30.8 (51)	28.4 (42)	24.8 (35)	17.9 (14)
Cr	8.70 (1)	250 (3)	114 (4)	60.2 (13)	66.8 (10)	92.5 (19)	61.9 (12)	869 (22)	275 (84)	130 (93)	93.5 (90)	126 (92)	105 (86)	90.1 (76)
V	370 (1)	267 (3)	190 (1)	257 (12)	221 (9)	194 (10)	145 (9)	299 (19)	314 (82)	289 (94)	261 (95)	209 (96)	179 (90)	142 (77)
Nb	n.d.	n.d.	n.d.	2.10 (1)	3.40 (2)	2.67 (3)	0.20 (1)	4.3 (5)	5.91 (25)	5.00 (36)	6.81 (33)	6.25 (41)	6.46 (26)	6.4 (23)
Li	n.d.	n.d.	8.1 (1)	5.47 (6)	5.94 (5)	6.53 (16)	5.51 (9)	n.d.	10 (1)	5.7 (2)	7.0 (1)	11.5 (2)	11.6 (3)	24 (1)
Pb	2.30 (1)	3.21 (7)	4.61 (5)	4.45 (14)	6.51 (9)	9.14 (7)	11.9 (6)	5.4 (12)	5.85 (52)	8.73 (60)	7.96 (54)	8.05 (66)	8.43 (67)	11.5 (53)
Hf	1.60 (1)	0.90 (2)	2.43 (3)	0.88 (5)	2.06 (7)	2.58 (5)	2.67 (3)	0.67 (2)	1.55 (9)	1.24 (7)	1.84 (5)	1.70 (9)	2.48 (10)	2.13 (4)

Phenocryst modes (vol. %)

	1	2	3	4	5	6	7	8	9	10	11	12	13	14
Hauyne	0	0	0	0	0	0	0	0.10	0	0.006	0	0	0.03	0
Sanidine	0	0	0	0	0	0	0	0	0	0.05	0	0	0.00	0
Plagioclase	3.22	10.6	11.0	12.5	15.4	10.7	10.1	8.27	11.0	17.7	18.3	19.1	18.7	20.8
Quartz	0	0	0	0	0	0	0.003	0	0	0	0	0	0	0.02
Olivine	4.90	4.54	1.67	1.18	1.07	0.38	0.99	17.1	5.28	3.55	2.10	0.93	0.79	0.30
Augite	2.69	2.53	1.41	0.77	2.38	1.39	0.97	15.2	7.85	8.61	6.19	5.50	4.57	3.75
Orthopyroxene	0	0.17	0.35	1.26	2.41	1.79	0.99	0	0.06	0.20	0.63	2.59	2.74	2.53
Hornblende	0.07	0	0	0.35	0.15	0.34	1.03	0.03	0.26	0.07	0.29	0.89	1.50	3.57
Biotite	0	0	0	0	0	0	0.03	0.013	0.03	0.04	0.03	0.03	0.10	0.27
Fe–Ti Oxides	0.01	0.07	0.19	0.11	0.59	0.55	0.39	0.75	0.42	0.55	0.50	0.89	1.14	1.25
Pigeonite	0	0	0	0	0	0	0	0	0	0	0	0.05	0	0
TPC	10.9	17.9	14.6	16.2	22.0	15.2	14.5	41.4	24.9	30.8	28.0	30.0	29.6	32.5
$n =$	10	16	49	37	29	24	30	27	98	117	145	157	129	100

Phenocryst occurrence (frequency %)

	1	2	3	4	5	6	7	8	9	10	11	12	13	14
Hauyne	0	0	0	0	0	0	0	5.1	0	0.6	0	0.5	1.2	0
Sanidine	0	0	0	0.4	0	0	0	0	0	1.8	0	0	0.6	0
Plagioclase	61.9	84.1	88.6	92.1	96.1	96.3	96.3	41.0	67.7	89.2	94.8	95.6	95.3	94.1
Quartz	0	0	0.5	0.9	2.2	2.6	5.3	0	0	0	0	0	0	1.5
Olivine	95.2	92.1	85.5	72.2	55.2	36.3	41.5	92.3	82.3	80.7	75.8	55.2	40.2	30.4
Augite	66.7	68.3	73.1	83.3	88.4	93.7	91.5	84.6	68.5	89.2	94.3	95.1	95.3	89.6
Orthopyroxene	0	15.9	38.9	68.7	81.8	92.1	88.3	5.1	3.1	16.3	47.4	68.0	80.5	81.5
Hornblende	4.8	3.2	0.5	3.5	9.9	17.9	21.3	2.6	2.3	3.0	9.3	21.7	31.4	34.1
Biotite	0	0	0	0.4	0.6	4.2	2.7	1.5	33.8	3.6	2.1	3.0	5.3	8.9
Fe–Ti Oxides	9.5	19.0	26.4	40.1	66.9	87.9	87.2	38.5	0	45.8	54.6	66.0	78.1	85.9
Pigeonite	0	0	7.3	1.8	2.8	1.6	0.5	0	0	0.6	0.50	2.0	1.8	0
$n =$	21	63	193	227	181	190	188	39	130	166	194	203	169	135

TABLE 1 (continued)

Region:	Mediterranean (Excluding K-Rich Series of Roman Province)						Potassium-Rich Series of Roman Volcanic Province						
SiO$_2$ Interval:	<50%	50-52%	52-54%	54-56%	56-58%	58-60%	<48%	48-50%	50-52%	52-54%	54-56%	56-58%	58-60%
SiO$_2$	48.47	50.99	53.18	54.98	57.08	59.07	46.61	48.78	51.15	52.76	55.35	56.90	59.33
TiO$_2$	1.44	0.98	0.89	0.86	0.80	0.77	0.93	0.85	0.82	0.78	0.63	0.57	0.52
Al$_2$O$_3$	17.49	17.68	17.58	17.33	17.59	17.32	16.54	17.10	18.96	18.28	19.08	19.60	19.82
F$_2$O$_3$	3.91	3.58	3.44	3.59	3.82	2.93	4.53	3.59	3.54	2.66	2.94	2.62	2.32
FeO	6.24	5.56	4.97	4.10	3.47	3.34	4.40	4.44	3.61	4.16	2.58	1.88	1.32
MnO	0.16	0.16	0.16	0.14	0.15	0.13	0.16	0.14	0.13	0.13	0.13	0.13	0.13
MgO	6.50	6.05	5.44	4.96	3.80	3.31	5.42	4.92	3.37	3.95	2.50	1.73	1.05
CaO	10.57	10.53	9.53	8.56	7.74	6.92	10.93	10.17	7.81	7.42	5.23	4.28	3.16
Na$_2$O	3.49	2.91	2.89	2.94	3.28	3.33	2.17	2.30	2.89	2.46	2.76	3.48	3.89
K$_2$O	1.30	1.23	1.64	2.27	2.04	2.65	7.74	7.11	7.12	6.88	8.43	8.54	8.27
P$_2$O$_5$	0.43	0.34	0.26	0.27	0.22	0.24	0.57	0.60	0.60	0.52	0.36	0.27	0.18
n =	50	42	53	78	63	93	38	37	26	26	32	24	29

Trace elements (p.p.m.)

Rb	44.1 (18)	34.1 (21)	45.2 (25)	64.2 (35)	58.6 (35)	66.6 (31)	395 (8)	310 (25)	322 (16)	448 (21)	542 (23)	478 (15)	580 (18)
Ba	404 (14)	359 (14)	367 (19)	415 (30)	392 (26)	496 (26)	1877 (8)	2077 (25)	1894 (5)	1836 (20)	1809 (17)	1295 (9)	409 (6)
Sr	680 (18)	561 (21)	476 (25)	434 (35)	373 (35)	384 (31)	1242 (8)	1145 (25)	1050 (16)	1099 (21)	1456 (23)	1463 (15)	1353 (19)
Zr	84.0 (2)	45.2 (5)	122 (7)	130 (8)	135 (8)	151 (11)	297 (8)	308 (25)	298 (16)	358 (21)	494 (23)	583 (15)	687 (19)
Zn	98.7 (14)	85.2 (14)	81.1 (17)	82.4 (23)	77.5 (21)	67.0 (23)	74.6 (8)	75.4 (24)	74.8 (13)	76.9 (8)	60.3 (3)	62.8 (5)	73.8 (5)
La	27.0 (1)	24.4 (5)	28.7 (4)	25.7 (4)	11.0 (2)	26.2 (4)	n.d.	n.d.	n.d.	n.d.	n.d.	n.d.	n.d.
Ce	53.0 (1)	47.2 (5)	59.5 (4)	58.1 (4)	26.6 (5)	46.3 (8)	160 (1)	n.d.	183 (1)	n.d.	210 (1)	208 (1)	298 (1)
Yb	n.d.	n.d.	1.86 (2)	1.76 (2)	1.65 (2)	2.28 (3)	n.d.	n.d.	n.d.	n.d.	n.d.	n.d.	n.d.
Y	24.0 (2)	15.0 (5)	24.8 (6)	21.3 (6)	24.0 (8)	25.8 (11)	36.8 (4)	30.8 (6)	41.0 (2)	n.d.	41.5 (2)	35.2 (5)	70.0 (1)
Cu	86.5 (13)	90.4 (14)	73.9 (19)	63.8 (25)	46.2 (23)	41.2 (25)	126 (6)	119 (24)	85.8 (12)	85.6 (8)	30.0 (2)	7.75 (4)	41.3 (4)
Ni	30.4 (13)	31.7 (15)	39.9 (19)	34.4 (29)	21.5 (23)	22.2 (27)	43.3 (3)	37.8 (6)	11.0 (1)	n.d.	11.0 (1)	7.0 (1)	n.d.
Co	33.2 (13)	28.1 (15)	27.8 (18)	26.1 (28)	22.9 (23)	19.4 (27)	33.3 (3)	30.8 (6)	21.0 (1)	n.d.	7.5 (4)	7.5 (4)	n.d.
Cr	47.0 (9)	82.3 (13)	93.7 (19)	85.6 (29)	35.2 (20)	50.5 (25)	157 (3)	207 (6)	86 (1)	n.d.	47.0 (1)	58.3 (4)	n.d.
V	363 (8)	270 (10)	215 (19)	201 (24)	179 (22)	159 (21)	n.d.	n.d.	n.d.	n.d.	n.d.	n.d.	n.d.
Nb	n.d.	n.d.	7.3 (2)	7.5 (2)	6.0 (5)	7.56 (9)	n.d.	13.6 (11)	24.3 (7)	33.3 (6)	60.0 (1)	n.d.	n.d.
Li	11.5 (10)	12.8 (9)	11.8 (15)	14.8 (22)	15.8 (23)	14.8 (19)	n.d.	n.d.	n.d.	n.d.	n.d.	n.d.	48.8 (4)
Pb	n.d.	n.d.	12.7 (2)	7.70 (3)	3.18 (5)	4.80 (9)	n.d.	n.d.	n.d.	n.d.	n.d.	n.d.	n.d.
Hf	n.d.	n.d.	2.51 (2)	2.36 (2)	2.96 (5)	3.10 (4)	n.d.	n.d.	n.d.	n.d.	n.d.	n.d.	n.d.

Phenocryst modes (vol. %)

Hauyne	0	0	0	0	0	0	0	0	0	0	1.83
Leucite	0	0	0	0.11	1.41	9.27	14.8	15.3	19.5	18.6	0.42
Sanidine	0	0.18	0.70	0	0	0.18	0.11	2.80	0.47	1.88	12.6	7.83
Plagioclase	43.4	20.5	24.8	30.5	27.0	0	1.34	5.55	5.89	3.76	4.00	2.25
Quartz	0	0	0.005	0	0	0	0	0	0	0	3.08	0
Olivine	6.88	3.18	2.46	0.41	0.26	2.09	0.67	0.70	0.97	0.61	0.93	0
Augite	6.81	7.23	6.54	5.37	3.96	12.0	13.0	5.75	12.3	6.69	5.64	2.42
Orthopyroxene	0	0.37	2.22	3.16	2.64	0	0	0	0	0	0	0
Hornblende	0	0	0.63	0.55	2.60	0.45	0	0.20	0	0	0	0
Biotite	2.03	1.64	0.53	0.07	1.45	0.18	0	0.50	0.03	0.54	0.79	0.42
Fe–Ti oxides	0	0.41	1.29	1.53	1.10	0	0.18	0.38	0.37	0.93	0.71	0.33
Pigeonite	0	0	0	0	0	0	0	0	0	0	0.007	0.19
TPC	59.1	33.5	39.2	41.7	40.4	24.2	30.1	32.0	39.5	33.0	27.9	15.7
n =	9	11	19	24	26	11	9	10	15	23	14	6

Phenocryst occurrence (frequency %)

Hauyne	0	0	0	0	0	0	0	8.3	0	0	6.3	27.3
Leucite	0	2.5	1.5	4.9	14.6	74.1	86.5	91.7	76.0	92.0	81.3	18.2
Sanidine	0	4.4	9.1	96.7	100	0	5.4	29.2	16.0	52.0	87.5	100.0
Plagioclase	73.3	93.3	95.5	6.6	1.2	7.4	13.5	41.7	52.0	92.0	87.5	36.4
Quartz	0	0	1.5	42.6	28.0	0	0	0	0	0	0	0
Olivine	93.3	100.0	56.1	93.4	90.2	37.0	18.9	20.8	44.0	16.0	18.8	9.1
Augite	77.8	82.5	97.0	70.5	72.0	96.3	94.6	91.7	100	96.0	93.8	90.9
Orthopyroxene	2.2	17.5	63.6	11.5	31.7	7.4	0	4.2	0	0	0	0
Hornblende	28.9	0	9.1	11.5	36.6	3.7	2.7	12.5	4.0	36.0	43.8	45.5
Biotite	0	37.5	10.6	72.1	68.3	18.5	64.9	62.5	72.0	64.0	56.3	45.5
Fe–Ti oxides	51.1	7.5	53.0	0	0	0	0	0	0	0	6.3	36.4
Pigeonite	0											
n =	45	40	66	61	82	27	37	24	25	25	16	11

TABLE 1 (continued)

Region:	South Shetland Islands					Anorogenic Basalt–Tholeiitic Andesite Series of Southern Queensland, Australia			
SiO$_2$ Interval:	<52%	52–54%	54–56%	56–58%	58–60%	<50%	50–52%	52–56%	56–60%
SiO$_2$	50.54	53.15	55.00	57.19	59.07	48.77	50.68	53.41	57.71
TiO$_2$	1.53	1.85	1.93	1.65	1.77	2.56	2.44	2.16	1.69
Al$_2$O$_3$	17.76	16.68	16.35	16.36	16.03	16.40	16.07	15.84	15.19
Fe$_2$O$_3$	3.02	2.92	3.31	2.47	2.57	4.67	3.99	3.68	4.21
FeO	6.27	6.86	6.41	5.90	5.17	7.77	7.53	6.91	5.44
MnO	0.16	0.17	0.17	0.14	0.18	0.16	0.16	0.15	0.20
MgO	6.27	4.68	3.66	3.00	2.51	6.16	5.82	4.41	2.10
CaO	10.10	8.60	7.36	6.37	5.46	7.74	7.34	7.34	5.53
Na$_2$O	3.56	4.03	4.73	5.53	5.81	3.74	3.83	3.84	4.07
K$_2$O	0.51	0.72	0.79	1.10	1.05	1.37	1.48	1.59	3.10
P$_2$O$_5$	0.28	0.33	0.30	0.29	0.39	0.65	0.67	0.66	0.76
n =	25	17	18	9	11	61	44	36	4

Trace elements (p.p.m.)

Rb	10.1 (7)	11.6 (7)	10.8 (5)	20 (1)	30 (1)	23.6 (52)	24.8 (35)	27.4 (26)	42.1 (2)
Ba	118 (4)	148 (3)	168 (3)	180 (1)	195 (1)	268 (21)	441 (31)	385 (21)	520 (1)
Sr	411 (7)	374 (7)	353 (5)	338 (2)	313 (2)	631 (52)	671 (35)	478 (26)	535 (2)
Zr	158 (5)	158 (6)	187 (3)	308 (2)	305 (2)	232 (53)	249 (35)	257 (26)	437 (2)
Zn	n.d.	n.d.	n.d.	n.d.	n.d.	125 (53)	138 (35)	133 (26)	170 (2)
La	n.d.	n.d.	n.d.	n.d.	n.d.	22.8 (21)	25.5 (28)	24.1 (17)	24.0 (1)
Ce	n.d.	n.d.	n.d.	n.d.	n.d.	50.6 (21)	55.6 (28)	54.6 (17)	54.0 (1)
Yb	n.d.	n.d.	n.d.	n.d.	n.d.	n.d.	1.90 (1)	1.67 (1)	n.d.
Y	25.2 (5)	25 (6)	30(3)	39.5 (2)	41 (2)	30.4 (52)	31.7 (35)	31.7 (26)	56.8 (2)
Cu	n.d.	n.d.	n.d.	n.d.	n.d.	54.3 (21)	33.1 (31)	30.6 (21)	26 (1)
Ni	24.7 (3)	25.7 (3)	19.3 (3)	12 (2)	5.5 (2)	92.4 (21)	70.2 (31)	67.4 (21)	46 (1)
Co	24.4 (5)	34.6 (5)	34 (4)	26 (1)	29 (1)	40.2 (6)	34.5 (11)	31.5 (13)	31 (1)
Cr	75 (2)	50 (1)	35 (2)	25 (1)	25 (1)	120 (19)	87.3 (27)	116 (20)	52 (1)
V	179 (4)	175 (3)	228 (3)	195 (1)	57 (1)	191 (21)	162 (31)	144 (21)	120 (1)
Nb	n.d.	n.d.	n.d.	n.d.	n.d.	36.5 (53)	29.4 (32)	25.1 (22)	40 (2)
Li	14.0 (4)	13.5 (2)	16 (2)	20 (2)	30 (1)	n.d.	n.d.	n.d.	n.d.
Pb	n.d.	n.d.	n.d.	n.d.	n.d.	3.01 (50)	3.31 (34)	3.68 (23)	7.5 (2)
Hf	n.d.	n.d.	n.d.	n.d.	n.d.	n.d.	3.96 (1)	3.40 (1)	n.d.

Phenocryst modes (vol. %)

Sanidine	0	0	0	0	n.d.	0	0.008†	0.50†	0
Plagioclase	15.5	19.0	10.8	14.3	n.d.	2.52	1.20	6.43	7.55
Quartz	0	0	0	0	n.d.	0	0	0	0
Olivine	5.65	0.68	0.35	0	n.d.	4.18	2.49	1.73	0.10
Augite	6.46	7.03	2.35	2.13	n.d.	0.58	0.04	0.23	0.05
Orthopyroxene	0	0.10	0	0.78	n.d.	0	0.003	0.16	0.15
Hornblende	0	0	0	0	n.d.	0	0	0	0
Biotite	0	0	0	0	n.d.	0	0	0	0
Fe–Ti oxides	1.03	1.53	0	0.95	n.d.	0.17	0.04	0.06	0.50
Pigeonite	0	0	0	0	n.d.	0	0	0.003	0
TPC	28.6	28.3	13.5	18.2	n.d.	7.45	3.78	9.11	8.35
$n =$	10	4	2	4	n.d.	55	38	30	2

Phenocryst occurrence (frequency %)

Sanidine	0	0	0	0	0	0	2.4†	6.1†	0
Plagioclase	96.0	100	93.8	88.9	100	78.9	70.7	87.9	100
Quartz	0	0	0	0	0	0	0	0	0
Olivine	92.0	71.4	43.8	33.3	36.4	94.7	92.7	69.7	33.3
Augite	80.0	92.9	81.3	88.9	81.8	19.3	19.5	51.5	66.7
Orthopyroxene	0	14.3	12.5	22.2	18.2	0	2.4	12.1	33.3
Hornblende	0	0	0	0	0	0	0	0	0
Biotite	0	0	0	0	0	0	0	0	0
Fe–Ti oxides	8.0	7.1	0	33.3	54.5	40.4	19.5	36.4	33.3
Pigeonite	24.0	21.4	37.5	0	9.1	0	0	3.0	0
$n =$	25	14	16	9	11	57	41	33	3

* Numbers in parentheses refer to number of data used to calculate mean for each trace element.
† Xenocrysts.
n = number of data used to calculate mean for each element.
n.d. = insufficient data.
TPC = total phenocryst content.

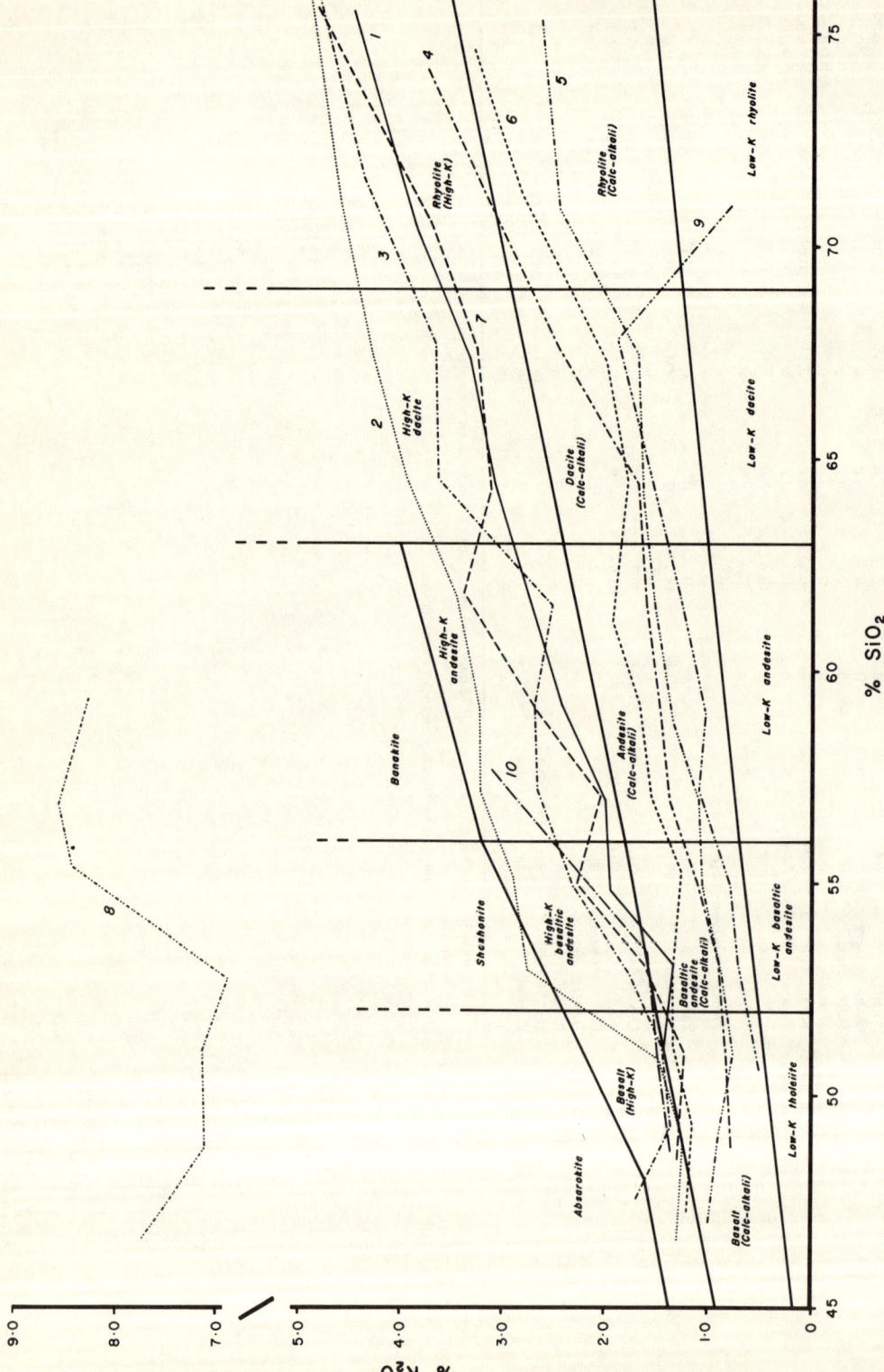

Fig. 1 A graph of K_2O against SiO_2 (in per cent by weight) of the sequentially averaged data from Table 1, together with similar data (for SiO_2 compositions greater than 60 per cent) from the same regional subgroupings given in Ewart (1979). The rock boundaries and nomenclature are slightly modified from Peccerillo and Taylor (1976). The numbers refer to the various subregions, represented by each trend line, as listed in the Introduction

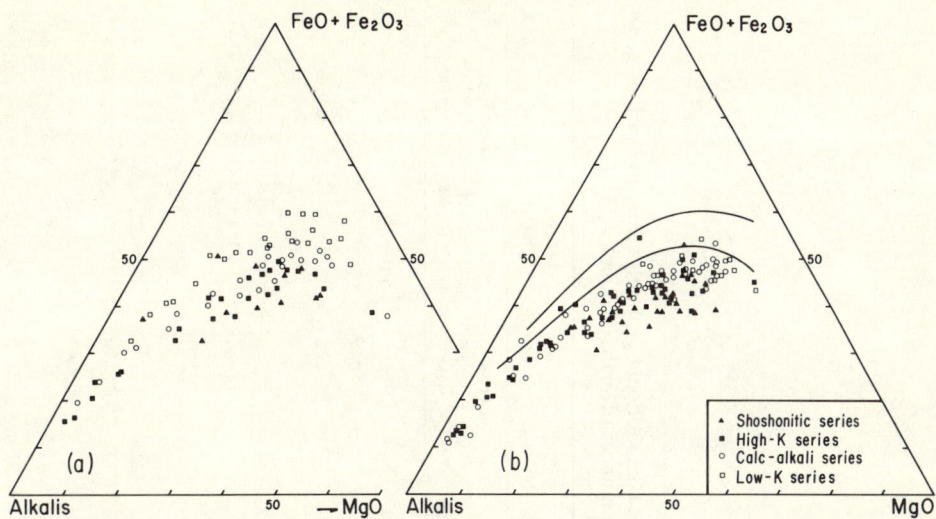

Fig. 2 AFM plot of sequentially averaged SiO_2 data groups for the four volcanic series occurring within the orogenic subregions listed 1–7 (*see* Introduction). The following sequential percentage SiO_2 groups used are: < 48, 48–50, 50–52, 52–54, 54–56, 56–58, 58–60, 60–63, 63–66, 66–69, 69–73, > 73 per cent SiO_2. Data groups with SiO_2 greater than 60 per cent are taken from Ewart (1979). Figure 2(a) is based on data from the two western Pacific subregions (numbers 5 and 6), while Fig. 2(b) is based on the data representing the northern and eastern Pacific subregions, and the Mediterranean subregion (numbers 1–4, and 7). The trend lines defined in Fig. 2(b) enclose the field of the low K compositions plotted in Fig. 2(a)

within the alkaline field of Chayes (1966). The Japanese basalts in contrast are clearly dominantly quartz normative. Figure 3(d) does suggest that there is a higher proportion of quartz-normative or near quartz-normative basalts in the island-arc regions (including the Cascades) than in the active continental margins.

South Shetland Islands

A series of lavas and pyroclastics from basalt through to rhyolite has been erupted, dominantly from Deception Island. These are characterized by rather low K_2O, but abnormally enriched in Na_2O (Table 1). The normative mineralogy (Fig. 3(d)) of the basaltic to andesitic magmas, however, is in no other way distinct from equivalent magmas from the other orogenic regions. The environment in which these magmas are erupted is possibly one in which subduction has ceased in the geologically recent past (e.g. Baker *et al.*, 1969, 1975; Hawkes, 1961; González-Ferrán and Katsui, 1970).

Anorogenic basalt-tholeiitic andesite association (Southern Queensland)

The lavas erupted in this Tertiary province constitute a bimodal association, dominated by mafic magmas, but also having produced significant volumes of silicic magmas of quartz-trachyte, rhyolite, peralkaline trachyte, and comendite compositions. The mafic magmas are dominated by olivine basalts which grade into less abundant, more siliceous types which are generally classified as tholeiitic andesites (e.g. Wilkinson and Binns, 1977). Compared to orogenic magmas, these anorogenic mafic magmas are equivalent to the high K type (Fig. 1), are significantly higher in Ti, and are dominantly olivine normative (Fig. 3(d)), but with individual samples extending to nepheline-normative compositions; the tholeiitic andesites are typically quartz normative. A further chemical characteristic of the anorogenic basalts is their relatively low normative diopside proportions, as compared to orogenic basalts.

TABLE 2 Frequency distribution of low K, calc-alkaline, high K, and shoshonitic series within the compositional groupings from the various geographic/tectonic orogenic subregions considered within this compilation, based on number of analysed samples

Subregion and type of K group	Region no.*	Total number of analyses for each subregion	Basalts (<52% SiO$_2$)			Basaltic andesites (52–56% SiO$_2$)			Andesites (56–63% SiO$_2$)			Dacites (63–69% SiO$_2$)			Rhyolites (>69% SiO$_2$)		
			Frequency % of each K$_2$O group	n†	% of total ‡	Frequency % of each K$_2$O group	n†	% of total ‡	Frequency % of each K$_2$O group	n†	% of total ‡	Frequency % of each K$_2$O group	n†	% of total ‡	Frequency % of each K$_2$O group	n†	% of total ‡
Western (Andean) South America	1	397		27	6.8		74	18.6		144	36.3		103	25.9		49	12.3
Shoshonitic			25.9			17.6			4.9			76.7			93.9		
High K			—			21.6			61.8			23.3			6.1		
Calc-alkaline			74.1			50.0			32.6			—			—		
Low K			—			10.8			0.7								
Western USA— western zone	2	542		65	12.0		58	10.7		127	23.4		92	17.0		200	36.9
Shoshonitic			23.1			15.5			12.6			76.1			95.5		
High K			43.1			56.9			57.5			22.8			4.5		
Calc-alkaline			30.8			25.9			29.9			1.1			—		
Low K			3.1			1.7			—								
Western USA— eastern zone	3	634		108	17.0		107	16.9		143	22.6		133	21.0		143	22.6
Shoshonitic			25.0			47.7			28.7			95.5			98.6		
High K			33.3			40.2			67.1			4.5			1.4		
Calc-alkaline			38.9			10.3			4.2			—			—		
Low K			2.8			1.9			—								
Cascades– Alaska– Aleutian Islands	4	695		92	13.2		171	24.6		260	37.4		84	12.1		88	12.7
Shoshonitic			7.6			—			—			14.3			58.0		
High K			3.3			1.2			8.8			83.3			42.0		
Calc-alkaline			75.0			94.7			87.7			2.4			—		
Low K			14.1			4.1			3.5								

		n	%	n	%	n	%	n	%	n	%	n	%
North-western Pacific	5	1678		277	16.5	408	24.3	578	34.4	241	14.4	174	10.4
Shoshonitic		5.1		1.2		1.6		⎫ 11.6		⎫ 22.4			
High K		10.8		11.5		12.6		⎭		⎭			
Calc-alkaline		52.0		57.1		62.8		54.8		58.6			
Low K		32.1		30.1		23.0		33.6		19.0			
South-western Pacific	5	1484		341	23.0	403	27.2	453	30.5	143	9.6	144	9.7
Shoshonitic		17.6		6.9		2.0		⎫ 15.4		⎫ 45.8			
High K		19.1		21.8		25.6		⎭		⎭			
Calc-alkaline		50.7		43.7		57.8		51.0		43.8			
Low K		12.6		27.5		14.6		33.6		10.4			
Mediterranean	7	656		92	14.0	131	20.0	238	36.3	124	18.9	71	10.8
Shoshonitic		25.0		26.0		21.4		⎫ 61.3		⎫ 74.6			
High K		14.1		16.8		37.8		⎭		⎭			
Calc-alkaline		54.3		49.6		38.7		38.7		25.4			
Low K		6.5		7.6		2.1		—		—			
Middle Americas§		431		80	18.6	97	22.5	150	34.8	69	16.0	35	8.1
Shoshonitic		10.0		—		1.3		⎫ 26.1		⎫ 82.9			
High K		23.8		27.8		26.0		⎭		⎭			
Calc-alkaline		58.8		64.9		70.7		73.9		17.1			
Low K		7.5		7.2		2.0		—		—			
Total for above subregions		6517		1082	16.6	1449	22.2	2093	32.1	989	15.2	904	13.9

* See Introduction

† *n* = number of analyses in each SiO$_2$ division, i.e. basalts, basaltic andesites, etc.

‡ % of total = percentage of analyses falling within basaltic, basaltic andesite, etc., groupings, as percentage of total number of analyses in each subregion.

§ Comprising Mexico, Nicaragua, Costa Rica, Guatemala, El Salvador, Honduras. Frequency distributions based on analyses with more than 60 per cent SiO$_2$ are from compilations in Ewart (1979).

TABLE 3 Compilation of the weighted arithmetic mean compositions of basalts (less than 52 per cent SiO_2) from data in Table 1

	South-western Pacific	North-western Pacific	Cascades–Alaska–Aleutians	Western USA western zone	Western USA eastern zone	South America	South Shetland Islands	Mediterranean	K-rich Roman Province	Anorogenic–southern Queensland
SiO_2	50.07	50.28	50.12	49.90	49.24	51.05	50.54	49.62	48.57	49.57
TiO_2	0.85	1.14	1.22	1.73	1.97	1.14	1.53	1.23	0.87	2.51
Al_2O_3	16.23	17.97	17.90	16.46	15.82	18.57	17.76	17.58	17.37	16.26
Fe_2O_3	3.23	3.55	3.09	4.04	4.10	3.42	3.02	3.76	3.93	4.39
FeO	6.75	7.17	6.52	5.72	7.09	5.48	6.27	5.93	4.21	7.67
MnO	0.18	0.19	0.17	0.15	0.17	0.16	0.16	0.16	0.14	0.16
MgO	7.84	5.88	6.74	7.37	7.36	5.54	6.27	6.29	4.71	6.02
CaO	10.82	10.26	10.15	9.19	8.96	8.87	10.10	10.55	9.85	7.57
Na_2O	2.51	2.56	3.03	3.42	3.36	3.98	3.56	3.23	2.40	3.78
K_2O	1.24	0.79	0.80	1.45	1.35	1.42	0.51	1.27	7.35	1.42
P_2O_5	0.28	0.20	0.24	0.58	0.57	0.38	0.28	0.39	0.59	0.66
$n =$	341	277	92	65	108	27	25	92	101	105
Trace elements (p.p.m.)										
Rb	29.1 (221)*	22.9 (21)	33.4 (11)	33.1 (15)	36.5 (11)	49.9 (11)	10.1 (7)	38.7 (39)	328 (49)	24.1 (87)
Ba	364 (197)	260 (22)	427 (32)	896 (38)	734 (38)	345 (9)	118 (4)	382 (28)	1986 (48)	371 (52)
Sr	628 (229)	413 (29)	646 (34)	884 (38)	772 (48)	608 (10)	411 (7)	616 (39)	1130 (49)	647 (87)
Zr	69.7 (211)	86.8 (19)	77.2 (32)	206 (37)	166 (32)	162(8)	158 (5)	56.3 (7)	303 (49)	239 (88)
Zn	74.6 (68)	74.2 (9)	73.7 (3)	101 (12)	n.d.	106 (3)	n.d.	92.0 (28)	75.1 (45)	130 (88)
La	11.6 (80)	6.47 (12)	8.32 (7)	74.0 (28)	79.7 (9)	16.3 (8)	n.d.	24.8 (6)	n.d.	24.3 (49)
Ce	25.9 (87)	23.0 (14)	30.8 (4)	99.0 (18)	218 (4)	41.6 (8)	n.d.	48.2 (6)	172 (2)	53.5 (49)
Yb	1.54 (22)	2.15 (15)	2.61 (9)	3.58 (17)	3.60 (29)	2.29 (8)	n.d.	n.d.	n.d.	1.9 (1)
Y	19.7 (204)	25.8 (13)	28.9 (16)	31.6 (35)	36.5 (31)	31.0 (1)	25.2 (5)	17.6 (7)	34.5 (12)	30.9 (87)
Cu	121 (191)	58.1 (16)	93.3 (28)	58.6 (36)	76.5 (35)	30.0 (3)	n.d.	88.5 (27)	111 (42)	41.7 (52)
Ni	104 (208)	79.9 (14)	85.0 (31)	148 (38)	108 (35)	57.9 (10)	24.7 (3)	31.1 (28)	36.8 (10)	79.2 (52)
Co	43.0 (134)	35.1 (16)	37.3 (30)	48.1 (33)	39.5 (35)	29.6 (11)	24.4 (5)	30.5 (28)	30.6 (10)	36.5 (17)
Cr	273 (199)	152 (8)	172 (27)	276 (38)	199 (32)	67.9 (8)	75 (2)	67.9 (22)	180 (10)	101 (46)
V	300 (195)	272 (5)	284 (28)	225 (38)	231 (34)	187 (3)	179 (4)	311 (19)	n.d.	174 (52)
Nb	5.3 (66)	n.d.	3.27 (3)	18.0 (16)	30.2 (20)	n.d.	n.d.	n.d.	17.8 (18)	33.8 (85)
Li	7.1 (3)	8.1 (1)	7.00 (4)	21.0 (5)	8.3 (3)	n.d.	14.0 (4)	12.1 (19)	n.d.	n.d.
Pb	7.2 (124)	3.68 (13)	6.23 (3)	14.2 (13)	7.0 (4)	n.d.	n.d.	n.d.	n.d.	3.13 (84)
Hf	1.3 (18)	1.78 (6)	n.d.	n.d.	n.d.	2.9 (5)	n.d.	n.d.	n.d.	3.96 (1)

Phenocryst mineralogy modes (vol. %)

	C1	C2	C3	C4	C5	C6	C7	C8	C9
Hauyne	0	0	0	0	0	0	0	0	0
Leucite	0	0	0	0	0	0	0	0	0
Sanidine	0.02	9.88	0	0	0	0	0	0	0.003†
Plagioclase	13.9		14.9	5.73	3.04	32.2	15.5	43.6	1.98
Quartz				0.02					
Olivine	5.76	2.71	3.94	5.37	3.39	6.27	5.65	6.07	3.49
Augite	9.04	1.82	4.51	1.81	1.36	0.33	6.46	5.81	0.36
Orthopyroxene	0.12	0.25	0.04	0.005					0.001
Hornblende	0.14	0.009	1.84						0
Biotite	0.03	0	0	0	0	0	0	p	0
Fe–Ti oxides	0.52	0.14	0.62		0.26		1.03	1.66	0.12
Pigeonite	0	0	0	0	0	0	0	0	0
TPC	29.6	14.8	25.9	12.9	8.0	38.8	28.6	57.1	5.95
n =	242	75	33	33	32	11	10	14	93

Frequency occurrence (%)

	C1	C2	C3	C4	C5	C6	C7	C8	C9
Hauyne	0.9	0	0	0	0	4.8	0	0	0
Leucite	0	0	0	0	0	0	0	1.2	0
Sanidine	0.9	85.6	0	0	1.9	0	0	0	1.0†
Plagioclase	75.2	0.3	90.0	69.0	47.1	76.2	96.0	76.5	75.5
Quartz				1.7	1.0			0	0
Olivine	82.7	87.7	87.8	96.6	89.4	95.2	92.0	96.5	93.9
Augite	80.6	71.5	52.2	62.1	46.2	66.7	80.0	80.0	79.4
Orthopyroxene	9.3	30.7	5.5	5.2	0.9	14.3	0	9.4	1.0
Hornblende	3.0	1.4	6.7	1.7	1.9	4.8	0	15.3	0
Biotite	2.7	0	0	1.7	1.0	4.8	0	0	0
Fe–Ti oxides	40.3	23.4	22.2	0	9.6	4.8	8.0	44.7	31.7
Pigeonite	0.3	5.1	0	1.7	0		24.0	3.5	0
n =	335	277	90	58	104	21	25	85	98

* Numbers in parentheses refer to number of data used to calculate mean for each trace element.
† Xenocrysts.
n.d. insufficient data.
n = number of data used to calculate means for major elements and phenocryst figures.
TPC = total phenocryst content.

TABLE 4 Compilation of the weighted arithmetic mean compositions of basaltic andesites (52–56 per cent SiO_2), from data in Table 1

	South-western Pacific	North-western Pacific	Cascades–Alaska–Aleutians	Western USA—western zone	Western USA—eastern zone	South America	South Shetland Islands	Mediterranean	K-rich Roman Province	Anorogenic–southern Queensland
SiO_2	54.19	53.91	54.28	53.90	54.10	53.90	54.10	54.25	54.19	53.41
TiO_2	0.83	0.96	1.20	1.31	1.29	1.27	1.89	0.87	0.70	2.16
Al_2O_3	17.07	17.88	17.62	16.84	16.55	17.50	16.51	17.43	18.27	15.84
Fe_2O_3	3.25	3.31	3.16	3.80	4.57	3.13	3.12	3.53	2.81	3.68
FeO	5.68	6.23	5.39	4.49	4.22	5.39	6.63	4.45	3.29	6.91
MnO	0.16	0.17	0.15	0.14	0.14	0.15	0.17	0.15	0.13	0.15
MgO	5.24	4.68	4.93	5.55	4.85	5.35	4.16	5.15	3.15	4.41
CaO	9.08	8.85	8.40	7.92	7.22	7.68	7.96	8.95	6.21	7.34
Na_2O	2.92	2.87	3.59	3.54	3.63	3.67	4.39	2.92	2.63	3.84
K_2O	1.30	0.97	1.02	2.02	2.82	1.62	0.76	2.02	7.74	1.59
P_2O_5	0.26	0.19	0.26	0.49	0.59	0.35	0.31	0.27	0.43	0.66
$n =$	403	408	171	58	107	74	35	131	58	36

Trace elements (p.p.m.)

Rb	30.3 (185)*	35.5 (21)	27.5 (41)	43 (10)	63.2 (14)	45.4 (43)	11.3 (12)	56.3 (60)	497 (44)	27.4 (26)
Ba	402 (188)	312 (31)	440 (63)	1167 (22)	1156 (22)	676 (27)	158 (6)	396 (49)	1824 (37)	385 (21)
Sr	561 (198)	345 (44)	635 (69)	1039 (22)	1042 (31)	644 (43)	365 (43)	452 (60)	1286 (44)	478 (26)
Zr	105 (186)	113 (23)	125 (63)	192 (22)	182 (18)	179 (25)	168 (9)	126 (15)	429 (44)	257 (26)
Zn	81.6 (71)	82.2 (11)	78.0 (3)	113 (6)	103 (4)	102 (16)	n.d.	81.8 (40)	72.4 (11)	133 (26)
La	20.2 (91)	7.69 (16)	8.18 (6)	56.5 (13)	71 (12)	24.6 (16)	n.d.	27.2 (8)	n.d.	24.1 (17)
Ce	36.4 (104)	24.2 (17)	25.8 (5)	105 (17)	91.8 (10)	51.3 (17)	n.d.	58.8 (8)	210 (1)	54.6 (17)
Yb	1.57 (19)	2.15 (16)	2.67 (12)	3.3 (9)	3.63 (14)	2.32 (17)	n.d.	1.84 (4)	n.d.	1.67 (1)
Y	23.3 (182)	21.4 (24)	29.7 (42)	28.4 (2)	34.4 (17)	25.4 (11)	26.7 (?)	23.1 (12)	41.5 (2)	31.7 (26)
Cu	105 (189)	54.4 (27)	90.5 (40)	38.6 (17)	92.3 (21)	49.6 (21)	22.5 (6)	68.2 (44)	74.5 (10)	30.6 (21)
Ni	44.9 (193)	43.1 (24)	47.6 (56)	91.8 (22)	84.8 (20)	67.4 (35)	34.3 (9)	36.6 (48)	n.d.	67.4 (21)
Co	29.7 (93)	33.4 (27)	26.8 (38)	29.4 (18)	33.5 (18)	30.5 (32)	40 (3)	26.8 (46)	11.0 (1)	31.5 (13)
Cr	110 (182)	63.1 (23)	107 (35)	211 (22)	135 (18)	202 (18)	n.d.	88.8 (48)	47.0 (1)	116 (20)
V	235 (191)	242 (21)	226 (60)	180 (22)	217 (21)	220 (13)	202 (6)	207 (43)	n.d.	144 (2)
Nb	6.5 (74)	2.97 (3)	7.07 (3)	16.4 (11)	15.9 (8)	12.5 (2)	n.d.	7.4 (4)	37.1 (7)	25.1 (22)
Li	10 (3)	5.68 (11)	28.8 (5)	18.0 (5)	9.7 (4)	11.4 (5)	14.8 (4)	13.6 (37)	n.d.	n.d.
Pb	8.0 (120)	5.26 (23)	8.4 (3)	16.8 (17)	14.9 (8)	n.d.	n.d.	10.2 (4)	n.d.	3.68 (23)
Hf	1.75 (14)	1.57 (12)	n.d.	n.d.	n.d.	3.67 (11)	n.d.	2.44 (4)	n.d.	3.4 (1)

Phenocryst mineralogy modes (vol. %)

Hauyne	0	0	0	0	0	0	0
Leucite	0	0	0	0	0	0	0
Sanidine	0	0	0	0	0	0	C.50†
Plagioclase	18.7	13.8	13.5	8.85	19.6	16.3	23.2	19.0	6.43
Quartz	0	0	0	0.22	0	0	0.03	1.32	C
Olivine	1.49	1.13	1.36	3.39	4.48	0.57	2.72	4.60	1.73
Augite	5.83	1.48	3.26	2.49	2.66	5.47	6.79	0.75	C.23
Orthopyroxene	1.65	1.77	1.06	0.04	0.07	0.07	1.54	8.90	C.16
Hornblende	0.60	0.26	0.64	0.18	0	0	1.00	0	C
Biotite	0.03	0	0	1.69	0	0	0.33	0.34	C
Fe–Ti oxides	0.70	0.32	0.83	0.003	0.31	1.02	0.97	0.71	0.06
Pigeonite	0.03	0	0.02	0.66	0.05	0	0	0	0.003
TPC	29.0	18.7	20.7	0	27.2	23.4	37.1	35.6	9.11
n =	302	66	49	16.5	15	6	30	38	30
				33					

Frequency occurrence (%)

Hauyne	0.3	0	0	0	1.5	0	0	0	0
Leucite	0	0	0	0	0	0	1.8	84.0	0
Sanidine	0	0.2	0	3.8	0	0	7.2	34.0	6.1
Plagioclase	95.2	93.9	97.6	67.3	60.0	96.7	94.6	72.0	87.9
Quartz	0	1.5	1.2	3.9	1.5	0	0.9	0	0
Olivine	65.3	64.7	89.1	77.9	87.7	56.7	61.3	30.0	69.7
Augite	94.7	85.6	80.0	82.4	83.1	86.7	94.6	98.0	51.5
Orthopyroxene	57.9	74.5	44.3	17.6	21.5	13.3	50.4	0	12.1
Hornblende	15.6	6.3	4.2	13.7	1.6	0	9.9	0	?
Biotite	2.6	0.5	0	9.8	9.2	0	7.2	20.0	?
Fe–Ti oxides	60.4	52.0	40.0	21.5	20.0	3.3	48.6	68.0	35.4
Pigeonite	1.3	2.2	1.2	0	3.1	30.0	0	0	3.0
n =	397	408	165	51	65	30	111	50	33
			104						

* Numbers in brackets refer to number of data used to calculate mean for each trace element.
† Xenocrysts.
n.d. = insufficient data.
n = number of data used to calculate means for major elements and phenocryst figures.
TPC = total phenocryst content.

TABLE 5 Compilation of the weighted arithmetic mean compositions of andesites (56–63 per cent SiO$_2$), from data in Table 1 and from Ewart (1979)

	South-western Pacific	North-western Pacific	Cascades–Alaska–Aleutians	Western USA–Western Zone	Western USA–Eastern Zone	South America	South Shetland Islands	Mediterranean	K-rich Roman Province	Anorogenic–Southern Queensland
SiO$_2$	59.09	59.22	59.07	59.49	59.72	59.89	58.82	59.50	58.23	57.51
TiO$_2$	0.73	0.77	0.91	0.98	1.03	0.95	1.65	0.75	0.54	1.69
Al$_2$O$_3$	16.83	17.27	17.43	16.98	16.84	17.07	16.19	17.33	19.72	15.79
Fe$_2$O$_3$	2.82	3.00	2.65	3.32	3.80	3.31	2.31	3.13	2.46	4.21
FeO	4.16	4.46	4.05	3.23	2.71	3.00	5.52	2.97	1.57	5.44
MnO	0.13	0.14	0.12	0.12	0.11	0.12	0.16	0.12	0.13	0.20
MgO	3.83	3.42	3.35	3.27	2.78	3.25	2.55	3.13	1.36	2.10
CaO	7.05	7.00	6.77	6.01	5.32	5.67	5.56	6.72	3.67	5.53
Na$_2$O	3.41	3.22	3.93	3.65	3.95	3.95	5.80	3.36	3.70	4.07
K$_2$O	1.70	1.33	1.48	2.60	3.31	2.47	1.11	2.73	8.39	3.10
P$_2$O$_5$	0.23	0.17	0.24	0.35	0.42	0.31	0.34	0.24	0.22	0.76
$n =$	453	578	260	127	143	144	24	238	53	4
Trace elements (p.p.m.)										
Rb	41.2 (241)*	23.8 (41)	36.2 (67)	79.1 (10)	86.8 (19)	75.4 (95)	21.4 (5)	64.1 (90)	534 (33)	42.1 (2)
Ba	479 (241)	500 (33)	618 (84)	1528 (39)	1366 (32)	886 (75)	241 (4)	569 (66)	941 (15)	520 (1)
Sr	516 (254)	272 (68)	673 (106)	1118 (39)	949 (50)	648 (95)	276 (7)	407 (90)	1402 (34)	535 (2)
Zr	138 (258)	166 (13)	151 (79)	222 (38)	277 (33)	195 (40)	326 (7)	158 (27)	641 (34)	437 (2)
Zn	75.4 (71)	80 (37)	77.3 (17)	91.4 (7)	95 (1)	97.8 (8)	n.d.	71.5 (52)	68.3 (10)	170 (2)
La	25.4 (118)	11.2 (25)	18.9 (20)	78.2 (33)	100 (24)	38.0 (17)	n.d.	21.5 (10)	n.d.	24.0 (1)
Ce	44.0 (117)	28.7 (25)	54.4 (21)	143 (24)	174 (19)	66.8 (27)	n.d.	42.1 (17)	253 (2)	54 (1)
Yb	1.94 (30)	2.69 (25)	2.30 (35)	2.46 (17)	2.68 (27)	1.94 (27)	n.d.	2.17 (8)	n.d.	n.d.
Y	24.7 (195)	19.2 (29)	36.5 (60)	28.3 (39)	29.5 (31)	12.2 (17)	44.0 (7)	24.4 (28)	41.2 (6)	56.8 (2)
Cu	51.8 (247)	44.4 (56)	42.8 (67)	40.4 (35)	37.0 (38)	40.0 (27)	n.d.	38.7 (65)	24.5 (2)	26 (1)
Ni	34.4 (236)	41.3 (47)	25.9 (75)	42.2 (38)	65.9 (35)	38.6 (51)	9.0 (5)	20.5 (61)	7.0 (1)	46 (1)
Co	21.3 (81)	30.1 (59)	21.1 (72)	21.1 (33)	20.7 (37)	18.6 (42)	23.8 (4)	19.6 (61)	7.5 (4)	31 (1)
Cr	87.4 (250)	69.0 (55)	51.6 (70)	102 (38)	79.9 (32)	48.4 (29)	25.0 (3)	40.9 (49)	58.3 (4)	52 (1)
V	154 (257)	157 (29)	152 (77)	134 (39)	145 (38)	125 (29)	102 (3)	159 (51)	n.d.	120 (1)
Nb	6.3 (73)	2.84 (6)	6.29 (14)	16.6 (28)	17.3 (23)	n.d.	n.d.	10.0 (17)	41.2 (6)	40 (2)
Li	15.1 (6)	6.88 (41)	20.9 (16)	25.4 (5)	10.4 (6)	12.9 (6)	28.8 (5)	15.2 (49)	48.8 (4)	n.d.
Pb	9.9 (179)	9.92 (20)	7.59 (18)	24.2 (36)	22.8 (25)	n.d.	n.d.	4.66 (18)	n.d.	7.5 (2)
Hf	2.7 (24)	2.98 (17)	n.d.	n.d.	n.d.	5.46 (5)	n.d.	3.64 (12)	n.d.	n.d.

Phenocryst mineralogy modes (vol. %)

	1	2	3	4	5	6	7	8	9
Hauyne	0	0	0	0	0	0	0	0.65	0
Leucite	0	0	0	0	0	0	0	8.95	0
Sanidine	0.004	0	0	0	0.08	14.3	1.37	5.15	7.55
Plagioclase	19.0	10.8	19.7	15.6	16.6	0	27.1	2.83	0
Quartz	0.01	0.001	0.11	0.16	0.07	0	0.001	0	0.10
Olivine	0.42	0.53	0.52	0.61	0.65	0	0.24	0.65	0.05
Augite	3.83	1.19	3.08	1.86	2.62	2.13	4.22	4.67	0.15
Orthopyroxene	2.55	1.56	2.28	0.98	2.33	0.78	2.47	0	0
Hornblende	2.18	0.75	0.98	3.06	3.93	0	2.07	0	0
Biotite	0.22	0.02	0.004	1.01	0.17	0	1.61	0.68	0.50
Fe–Ti oxides	1.07	0.45	0.95	0.58	1.19	0.95	1.12	0.60	0
Pigeonite	0	0	0	0	0	0	0	0.06	8.35
TPC	29.3	15.3	27.6	24.3	27.6	18.2	4.02	24.2	
$n =$	333	74	53	55	35	4	71	20	2

Frequency occurrence (%)

	1	2	3	4	5	6	7	8	9
Sodalite + hauyne	0.5	0	0	0	0	0	0	14.9	0
Leucite	0	0	0	0	0	0	0	55.6	0
Sanidine	0.4	0	1.7	13.4	2.2	0	16.4	92.6	100
Plagioclase	95.7	99.6	95.0	86.6	94.1	95.8	99.1	66.7	0
Quartz	1.8	6.8	8.4	10.6	7.4	0	8.5	0	
Olivine	30.8	36.3	29.4	21.8	28.7	41.7	24.4	14.8	33
Augite	88.4	92.4	84.0	83.8	84.5	79.2	88.9	92.6	67
Orthopyroxene	77.2	92.0	61.4	48.6	88.9	29.2	66.3	0	33
Hornblende	34.0	19.0	43.7	40.2	60.3	0	38.2	0	0
Biotite	9.7	0.9	26.9	35.2	36.1	0	40.0	44.5	33
Fe–Ti Oxides	80.5	84.0	67.2	63.4	72.8	45.8	66.7	51.9	0
Pigeonite	0.7	0	0.8	0	0	4.2	0	18.6	
$n =$	442	237	119	142	136	24	225	27	3

* Numbers in brackets refer to number of data used to calculate mean for each trace element.
n.d. = insufficient data.
n = number of data used to calculate means for major elements and phenocryst figures.
TPC = total phenocryst content.

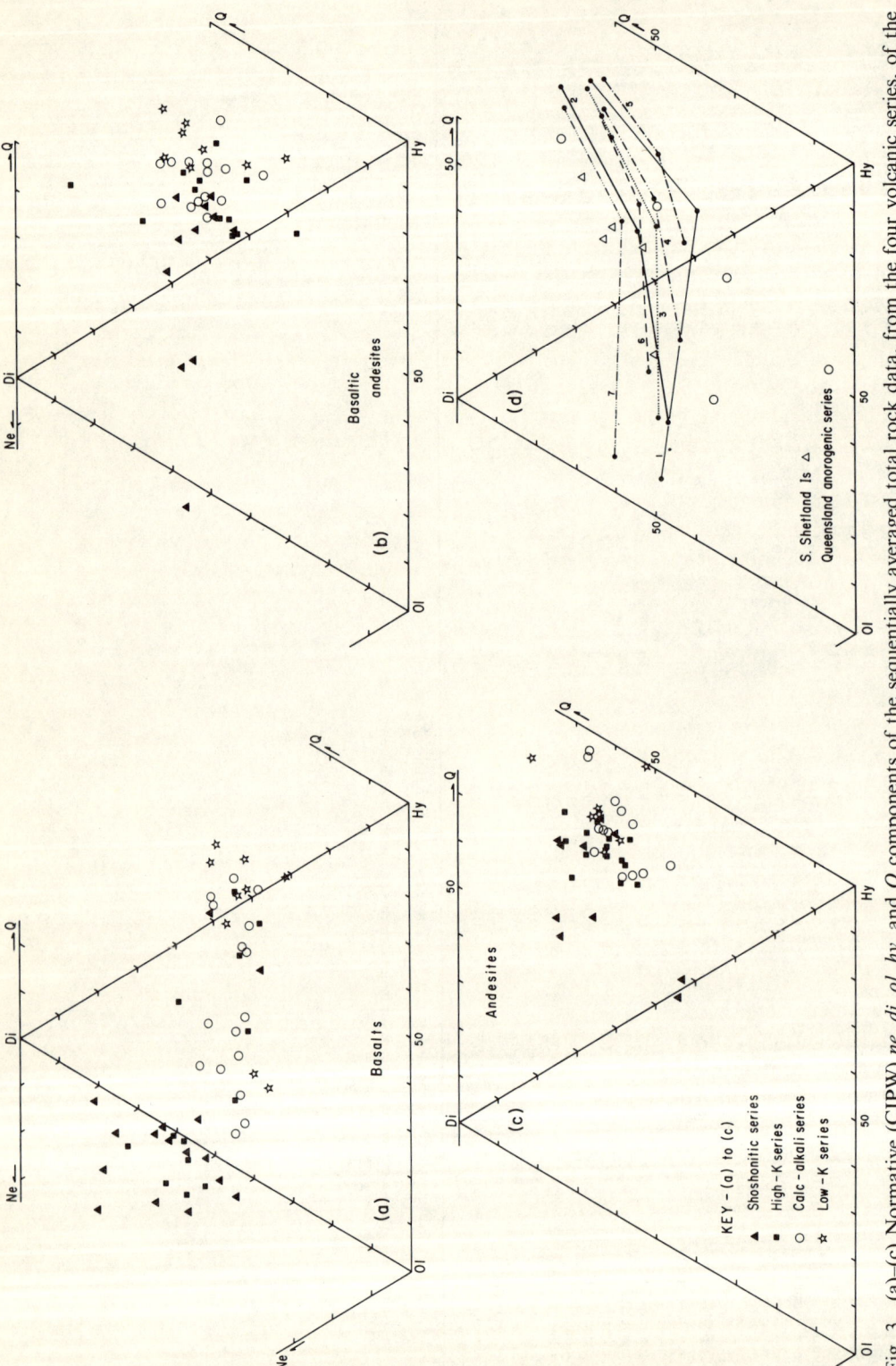

Fig. 3 (a)–(c) Normative (CIPW) *ne, di, ol, hy,* and *Q* components of the sequentially averaged total rock data, from the four volcanic series, of the orogenic subregions numbers 1–7; only data groups for SiO₂ less than or equal to 60 per cent are included, which are as listed in caption in Fig. 2. Fig. 3 (d) Normative components of the weighted average basaltic, basaltic andesite, and andesitic compositions from the various orogenic subregions, based on the analyses presented in Tables 3–5. Plots for each subregion are joined by tie-lines, and the numbers correspond to the appropriate subregion as outlined in the Introduction. Comparative data (Table 1) from the South Shetland Islands, and the anorogenic basalt–tholeiitic andesite series of southern Queensland are also plotted

Mineralogy

Total phenocryst contents

A striking feature of many orogenic volcanic rocks is their crystal-rich nature. Data on the phenocryst contents of orogenic eruptives are compiled in a series of histograms shown in Fig. 4, and these confirm that c. 65 per cent of the rocks contain phenocryst contents in excess of 20 per cent (by volume), quite frequently extending up to 60 per cent or more. Reference to Table 1 indicates that the averaged phenocryst contents of the various SiO_2 divisions within the different subregion groupings are generally relatively high (commonly greater than 20 per cent), and show little evidence of significant systematic differences between equivalent magmas from the various subregions, with the possible exception of certain of the basaltic groupings from the western USA, which contain persistently low phenocryst contents.

Figure 4 suggests, however, that the various volcanic series may exhibit different distributions of crystallinity. For example, the low K and calc-alkaline basaltic andesites and andesites exhibit bimodal phenocryst distribution, which was noted by Ewart (1976a) and explained in terms of fractional crystallization and/or a flow differentiation process acting within feeder conduits, or shallow magma chambers. The bimodal distributions evidently do not extend into the low K and calc-alkaline basalts, nor do the high K and shoshonitic eruptives show such phenocryst distributions; the latter eruptive groups simply have continuous phenocryst contents extending from zero to greater than 50 per cent. The comparable compilation of phenocryst contents in orogenic dacites and rhyolites (Ewart, 1979) shows a definite trend of decreasing crystallinity with increasing SiO_2 of the eruptives.

One aspect of the rather high degree of crystallinity of many orogenic magmas which it is important to note results from application of the data of Shaw et al. (1968), which show that the effective viscosity of crystal-rich magmas is likely to increase drastically due to the almost exponential effect of increasing crystals on viscosity (inducing pseudoplastic behaviour). This would seem to imply that rather crystalline magmas were necessarily erupted from near-surface magma chambers. The observation that the low K to calc-alkaline basaltic andesites and andesites exhibit different phenocryst distributions compared to the more potash-enriched equivalents suggests the possibility of differing ascent and crystallization environments (e.g. ascent rates, geometry, and size of magma reservoirs or feeder channels) between the different volcanic series.

Anorogenic association

Comparative data are also presented in Fig. 4 for the basalts and tholeiitic andesites of the southern Queensland province. It is clear that these exhibit, overall, very much lower phenocryst contents than the orogenic magmas. This is consistent with the observation that the silicic members of the bimodal associations are also characterized by very low phenocryst contents, again consistent with their relatively high equilibration temperatures (Ewart, 1979).

Phenocryst assemblages and phenocryst occurrences

Plots of the sequential changes of specific phenocryst phases with increasing host rock SiO_2 are illustrated in Figs 5–7. The following are some generalized comments on phenocryst assemblages found in the basalts and andesites of the four volcanic series of the orogenic regions; it should be noted, however, that certain differences are evident between the western Pacific magmas and those of the other subregions. Dacitic and rhyolitic phenocryst assemblages are detailed in Ewart (1979).

Low K series

Common assemblages in the basalts are pl + ol + aug ± Fe–Ti oxide, with opx and/or pig becoming important phases in the more SiO_2-rich basalts of the western Pacific; hb occurs

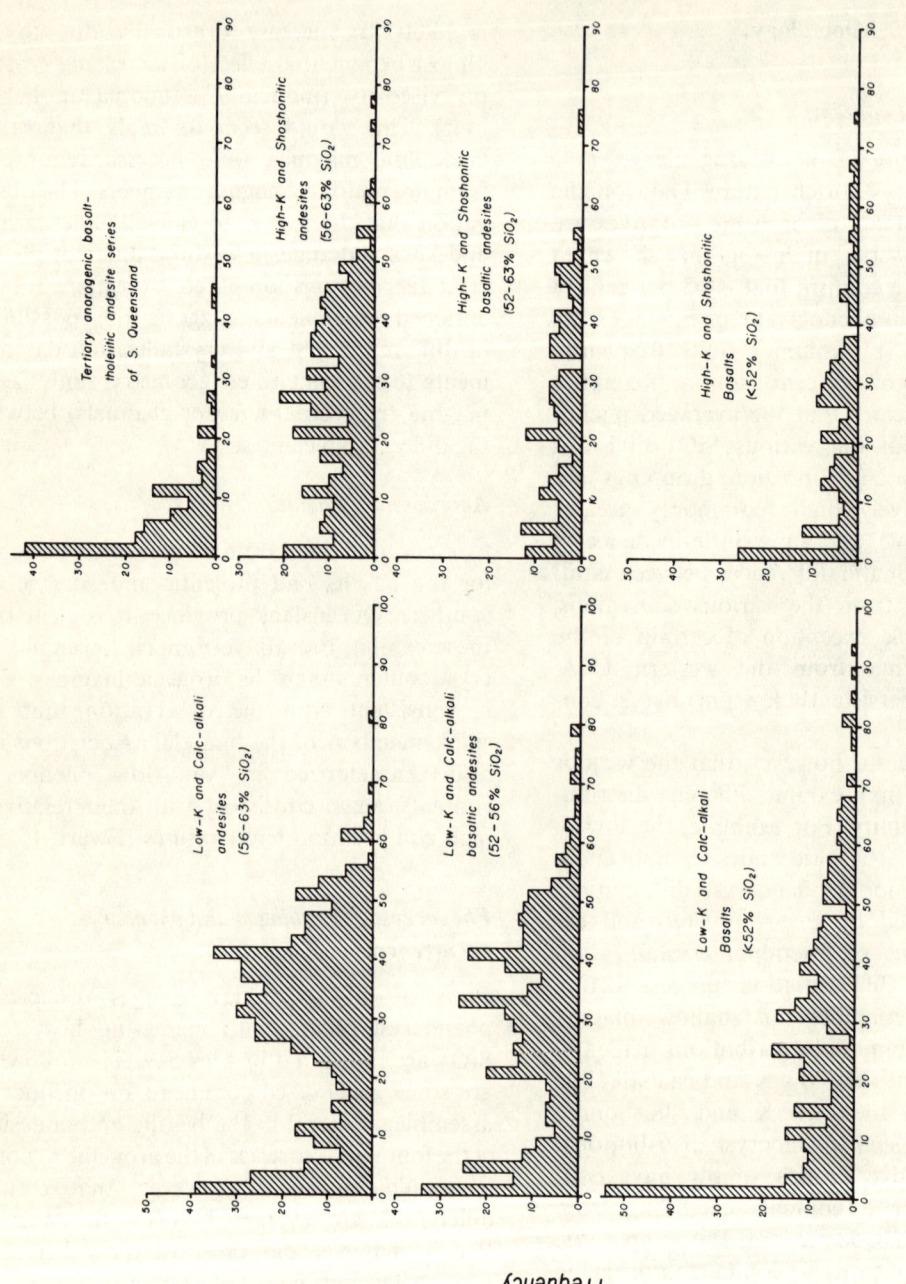

Fig. 4 Histograms showing the distributions of total phenocryst contents of basalts, basaltic andesites, and andesites from Tertiary–Recent orogenic eruptives. The compilations are subdivided into two broad groupings based on the relative volcanic affinities of the host rocks. Comparative data are included from the anorogenic series from southern Queensland

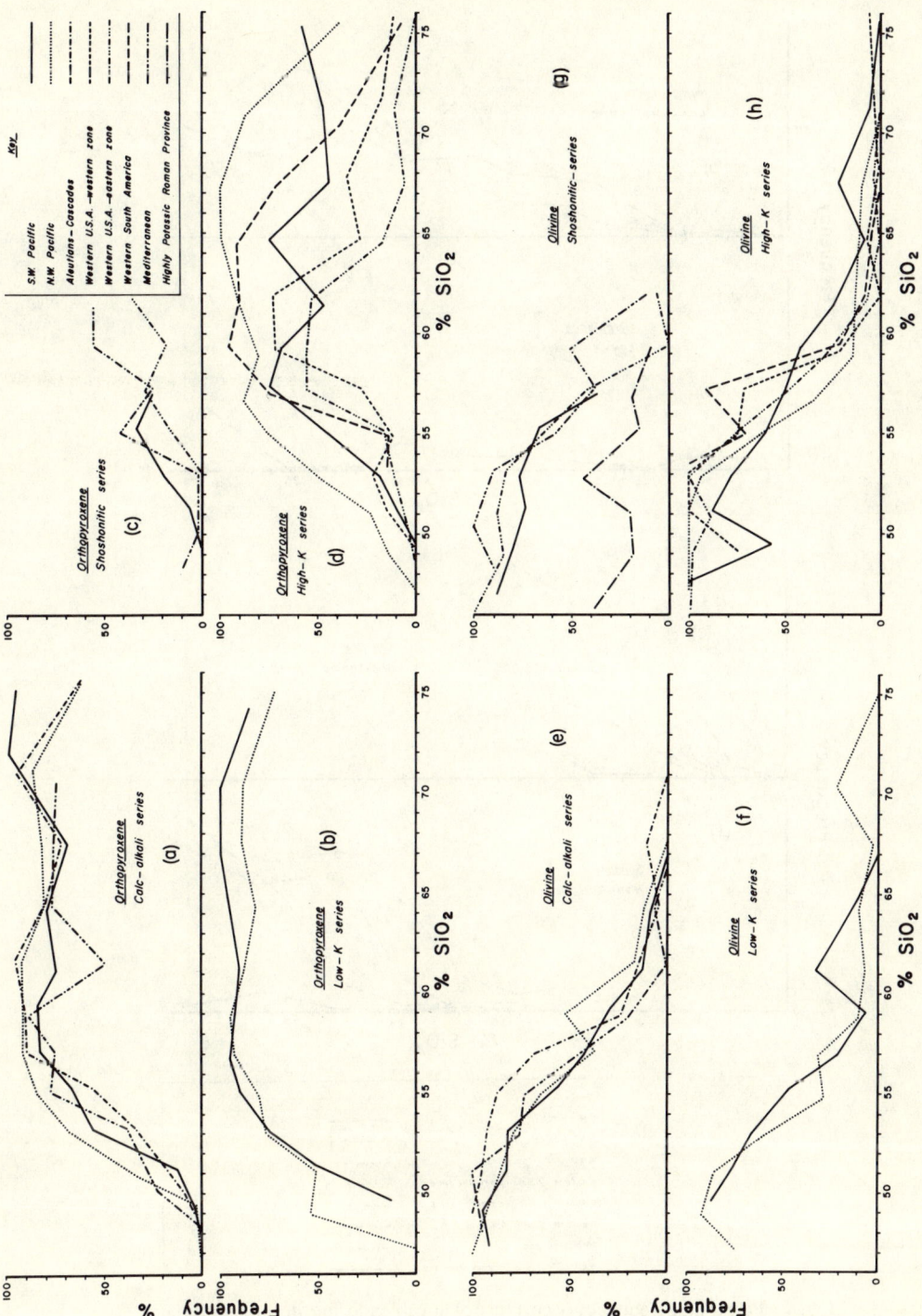

Fig. 5 Percentage occurrence of Mg-olivine and orthopyroxene phenocrysts, plotted as a function of averaged total rock compositions (expressed as SiO_2) from eruptives of selected orogenic subregions, and including the K-rich series of the Roman Province. The data plots are subdivided on the basis of the four volcanic series, and are based on sequentially averaged data, as outlined in the caption to Fig. 2

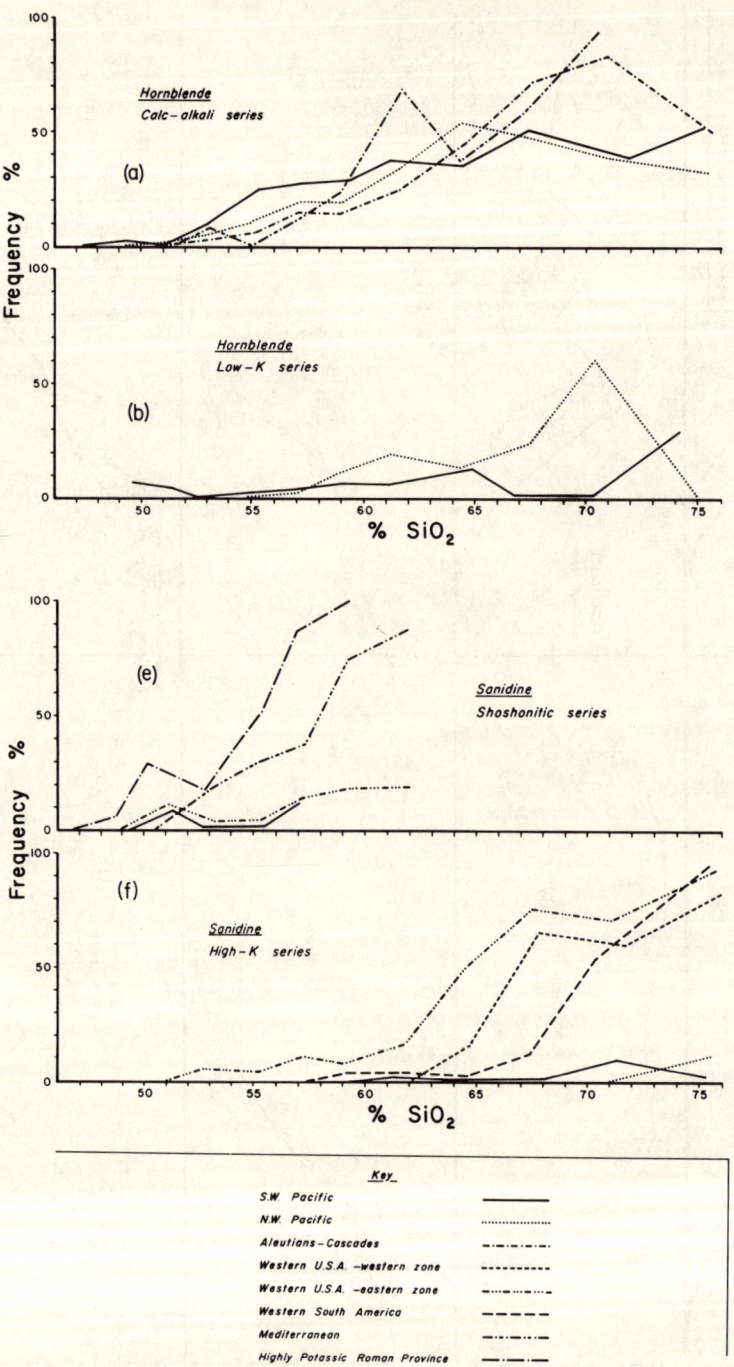

Fig. 6 Percentage frequency occurrence of hornblende, biotite, and sanidine phenocrysts, plotted against averaged total rock compositions (expressed as per cent SiO_2), from eruptives of selected orogenic subregions, and

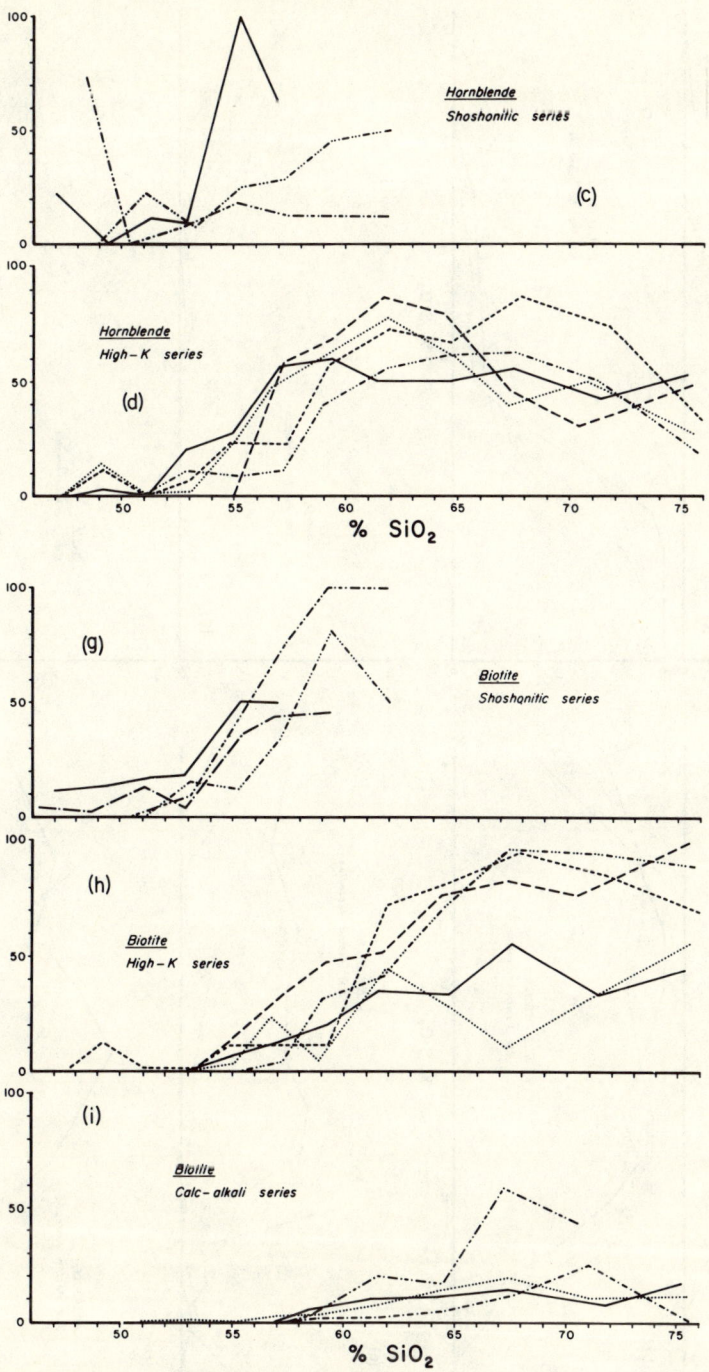

including the K-rich series of the Roman Province. The plots are subdivided on the basis of the four volcanic series, and are based on sequentially averaged data as outlined in the caption to Fig. 2

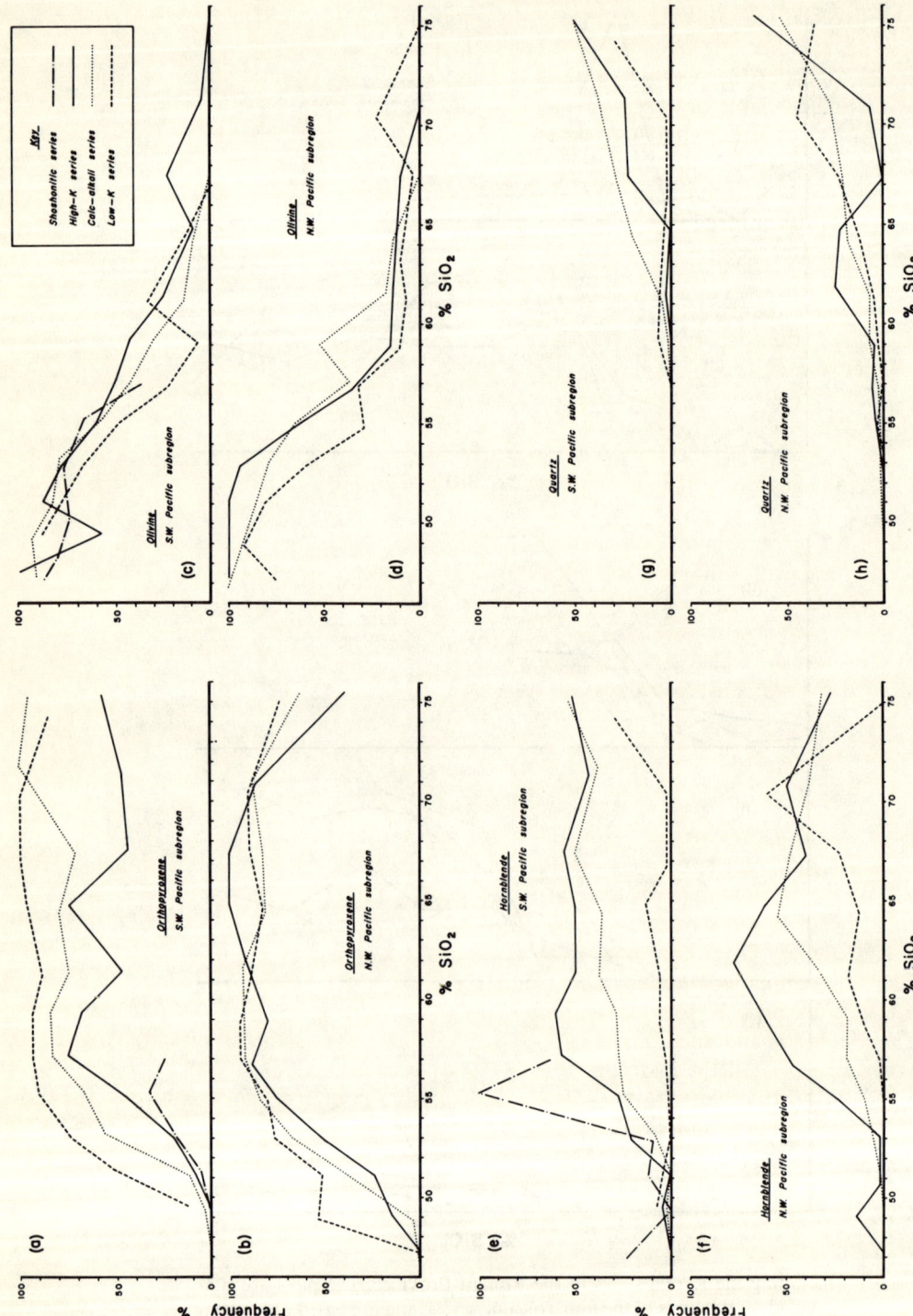

Fig. 7 Comparison of percentage frequency occurrence of Mg-olivine, orthopyroxene, hornblende and quartz in the four volcanic series from the north-western and south-western Pacific subregions. The frequency data are plotted against averaged total rock compositions (as per cent SiO_2), and are based on sequentially averaged data as outlined in the caption to Fig. 2

rather rarely in certain south-western Pacific basalts, while pl is rather less frequent in the low K basalts outside the western Pacific subregion. (pl = plagioclase; ol = Mg-olivine; aug = augite; opx = orthopyroxene; pig = pigeonite; hb = hornblende; Fe–Ti oxide = titanomagnetite; biot = biotite.)

The most typical assemblage in the basaltic andesites and andesites is pl + aug + opx + Fe–Ti oxide ± ol; hb occurs in only low frequency, as do pig (frequently overlooked?) and also quartz, the latter occurring only in the more silicic andesites.

Calc-alkaline series

Phenocryst assemblages are rather similar to those listed for the low K series, the main differences being the increase in the frequency of hb occurrence (especially in the andesites, Fig. 6), the appearance of occasional biotite in andesites, a possible increase in the frequency of Fe–Ti oxide occurrence, and a small decrease in pl occurrence in the most mafic basaltic compositions.

High K series

Assemblages become more diverse with increasing K enrichment. High K basalts contain ol + aug as the two dominant phenocryst phases, with pl occurring less frequently than in the low K magmas. Opx, hb, and Fe–Ti oxide become more common co-existing phases in the more siliceous high K basalts. Within the basaltic andesites and andesites, pl + aug + opx + Fe–Ti oxide ± ol is still the most common assemblage, but both hb + biot become significant phases, especially in the andesites (Fig. 6). Sanidine also appears in the andesites of the western USA (eastern zone) and Mediterranean subregions (Fig. 6).

Shoshonitic series

In the basalts, ol + aug + Fe–Ti oxide is the predominant assemblage, accompanied at lower frequencies by pl, hb, biot, and opx. Sanidine is sporadically reported, while feldspathoids (of the hauyne or sodalite type) occur, for example, in certain shoshonitic lavas of Papua New Guinea (Johnson et al., 1976) and Ecuador (Colony and Sinclair, 1928), and leucite in certain Mediterranean basalts (including some not belonging to the K-rich series of the Roman Province). Within the basaltic andesites, common phenocryst phases include pl, aug, Fe–Ti oxide, biot, hb, ± ol, ±opx, and less commonly sanidine.

Anorogenic basalt – tholeiitic andesite series (southern Queensland)

The most common assemblage is ol + pl ± titanomagnetite ± ilmenite, with augite notably reduced in occurrence compared to equivalent orogenic magmas. Xenocrysts of sanidine-anorthoclase and aluminous pyroxenes are found in the basalts.

Frequency distribution of individual phenocryst phases

Mg-olivine Reference to Figs 5 and 7 shows that olivine is virtually a ubiquitous phase in orogenic basalts, but more significant is the persistence of olivine into the more silicic magmas (normally surrounded by reaction rims). The frequency occurrence curves obviously show a decay in olivine occurrence with increasing SiO_2, and the non-linear form of these curves suggests that the abundance pattern of olivine in the magma compositions extending from basaltic andesite to dacite is not simply due to a mixing (i.e. diluent) process. The most obvious explanation is that olivine is a relict (non-equilibrium) phase in the more silicic magmas, and this receives support from the olivine compositions in, for example, andesites (*see* later).

A further significant aspect of the olivine occurrence is that not only do the eruptives from each region show similar 'olivine decay curves' (Fig. 5(e)–(h)), but that at least in the western Pacific subregions (where most data are available), these 'decay curves' are similar between the four volcanic series (Fig. 7(c)–(d)). The only systematic difference may be the small but consistently lower relative abundance of olivine, for given SiO_2 intervals, in the low K series.

Orthopyroxene Reference to Fig. 5(a)–(d) illustrates that marked differences are apparent in (1) the rate at which orthopyroxene appears in magmas of basalt to basaltic andesite compositions; (2) the overall level of orthopyroxene occurrence in the andesite to dacite compositions; and (3) the frequency at which orthopyroxene persists into rhyolite compositions.

Thus, the frequency occurrence curves show that orthopyroxene appears earlier (i.e. at lower SiO_2 levels) in low K magmas relative to the more K-rich series; but even within the latter there are marked regional differences in the frequency abundance levels of orthopyroxene in the magmas of intermediate compositions. Furthermore, the curves show that orthopyroxene persists into the low K and calc-alkaline rhyolites at higher frequencies than found in the high K rhyolites.

Figure 7(a), (b) compares the orthopyroxene curves from the south-western and north-western Pacific subregions, for the four volcanic series (shoshonitic to low K). These clearly indicate that the appearance of orthopyroxene within the basalt–basaltic andesite compositions is in the order low K (earliest, i.e. at lowest SiO_2 compositions) → calc-alkaline → high K → shoshonite (most delayed precipitation). This observation is understood in terms of the reaction

$$Mg_2SiO_4 + SiO_2 = 2MgSiO_3$$
$$\text{Forsterite} \quad \text{Liquid} \quad \text{Enstatite}$$

Clearly, orthopyroxene stability will depend on $a^{liq}_{SiO_2}$, and this implies that the more potash-rich (i.e. alkaline) magmas have, for given SiO_2 compositions (anhydrous), lower $a^{liq}_{SiO_2}$.

That this is, in fact, the most likely situation can be illustrated in a rather more generalized way by reference to Fig. 3, showing that shoshonitic and high K basalts and basaltic andesites are more silica-undersaturated than their calc-alkaline and low K equivalents. It is also possible, as subsequently discussed, that the more K-rich series are also more hydrous, which will presumably also effectively reduce $a^{liq}_{SiO_2}$, and perhaps further account for some of the regional differences in orthopyroxene occurrence evident in Fig. 5(a)–(d).

The different frequency occurrence levels of orthopyroxene in the different volcanic series, together with the rather variable persistence of orthopyroxene into the rhyolites, may be related to the inverse correlation between orthopyroxene and both hornblende and especially biotite occurrences in these intermediate to acid magmas; this suggests a reaction relationship between orthopyroxene–amphibole–biotite involving the interaction of variables such as $f_{O_2}, f_{H_2O}, a^{liq}_{K_2O}, P_{total}$.

Hornblende and biotite Both hornblende and biotite increase more rapidly in the intermediate compositions of the more K-rich series compared with less potassic magmas (Fig. 6); this is further illustrated by the detailed comparison of the occurrence of hornblende within the four volcanic series given in Fig. 7(e)–(f) for the western Pacific subregions.

Although the correlation between increasing biotite occurrence and increasing K_2O in the host magmas is essentially to be anticipated, a similar explanation is not so obvious for the increasing hornblende occurrence in these same magmas. An alternative explanation, relevant for both phenocryst phases, is to postulate increasing f_{H_2O} within the more potassic magmas, compared especially with the low K magma series. This receives some further support from subsequent data presented.

Comparison of modal phenocryst ratios

The averaged modal data presented in Tables 1 and 3–5, and in Appendices 1–4 (pp. 88–95), emphasizes the overall dominance of phenocryst plagioclase within orogenic magmas. In

Fig. 8, modal percentage plagioclase is plotted against total Fe–Mg silicates (olivine, pyroxenes, hornblende, biotite), for the basalt, basaltic andesite, and andesite compositions. It is clear that the calc-alkaline and low K basalts and basaltic andesites generally display higher modal plagioclase/$\sum$ Fe–Mg silicate ratios than do the corresponding high K and shoshonitic eruptives (including also the K-rich series of the Roman Province). As is to be expected, the numerical values of these ratios change between the basalts and basaltic andesites, due to decreasing Fe and Mg abundances with increasing SiO_2. Within the andesites, however, there is no such clear division of the modal proportions of the phases, although the low K eruptives still exhibit the highest plagioclase/$\sum$ Fe–Mg silicate ratios.

A possible explanation for the variation of these plagioclase–ferromagnesian silicate ratios is found in experimental studies of, for example, Eggler (1972) and Eggler and Burnham (1973) on the phase relations of water-saturated and water-undersaturated melting of andesite compositions (*see also* Green, this volume, Section VII). The important aspect of these experiments with respect to the observed modal data referred to above is the very pronounced lowering of plagioclase stability relative to the pyroxene phases, with increasing $X_{H_2O}^{fluid}$ and P_{H_2O}, together with a corresponding increase of the amphibole stability field. Thus, under anhydrous conditions, plagioclase is the liquidus phase. Under increasing water pressures, however, there will be a tendency for the interval between plagioclase and pyroxene (or amphibole) precipitation to be reduced, thereby favouring decreasing plagioclase/ferromagnesian silicate ratios.

Phenocryst compositions

Plagioclase

Published plagioclase compositions, based on the stated ranges of compositions in individual rocks (microprobe and optically determined), are presented in histograms in Fig. 9. One feature that is common to nearly all orogenic eruptives is the presence of well developed oscillatory normal zoning in plagioclase phenocrysts, and compositional ranges of 20–30 per cent are quite common. This is reflected in the wide compositional ranges in the histograms.

Considering the plagioclases of the low K and calc-alkaline series first, the histograms show that labradorite–bytownite is the dominant composition of basalts and basaltic andesites (the compositions being skewed towards bytownite in the basalts), while labradorite is characteristic of andesites. It is significant to note that throughout the calc-alkaline and low K basalt to andesite range, plagioclase compositions are frequently reported that extend into anorthite. The most sodic compositions reported for the plagioclase represent narrow outermost, and often compositionally rather abrupt rims of phenocrysts, probably produced during quenching accompanying magma eruption.

Comparison of the plagioclase phenocryst compositions of the high K and shoshonitic basalts and andesites reveals rather more sodic compositional ranges; these are dominantly labradorite in the basalt and basaltic andesites, extending to andesine–labradorite in the andesites.

Anorogenic basalt–tholeiitic andesite association (*southern Queensland*) Although phenocryst plagioclase is modally less abundant in this association, the compositions are also plotted in Fig. 9 for comparison, from which it is clear that andesine–sodic labradorite is the dominant compositional range. This is more sodic than found in the equivalent orogenic magmas, and it is also noteworthy that extremely calcic compositions are not encountered in these anorogenic lavas. The anorogenic plagioclases also exhibit rather sodic rim compositions, frequently anorthoclase.

Olivine

Available compositional data (both optically and microprobe determined) for olivines in

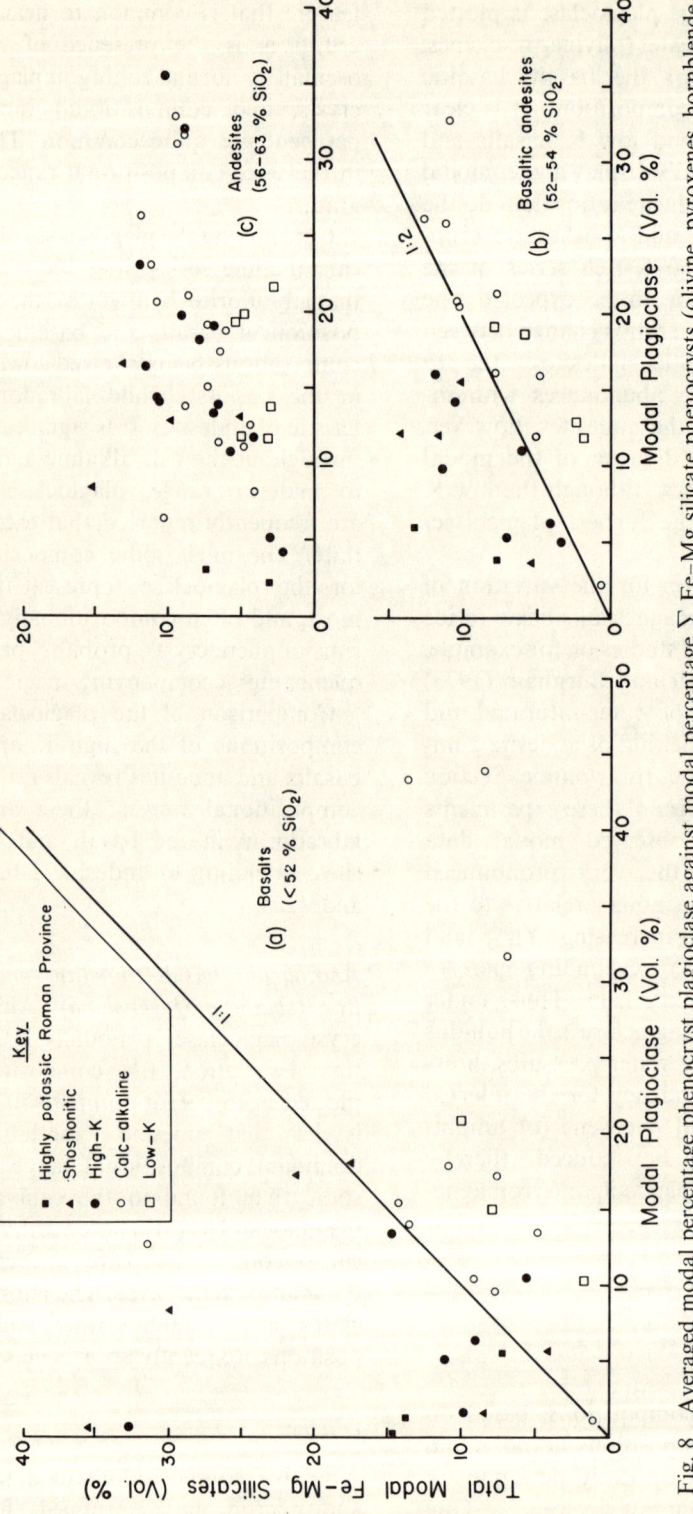

Fig. 8 Averaged modal percentage phenocryst plagioclase against modal percentage $\sum$ Fe–Mg silicate phenocrysts (olivine, pyroxenes, hornblende, biotite) in basalt, basaltic andesites, and andesites, from the orogenic subregions (1–7, *see* Introduction) and including the K-rich series of the Roman Province. Plots are subdivided according to the four volcanic series, and are based on sequentially averaged modal data, divided as indicated in the caption to Fig. 2, for rock compositions up to 63 per cent SiO_2

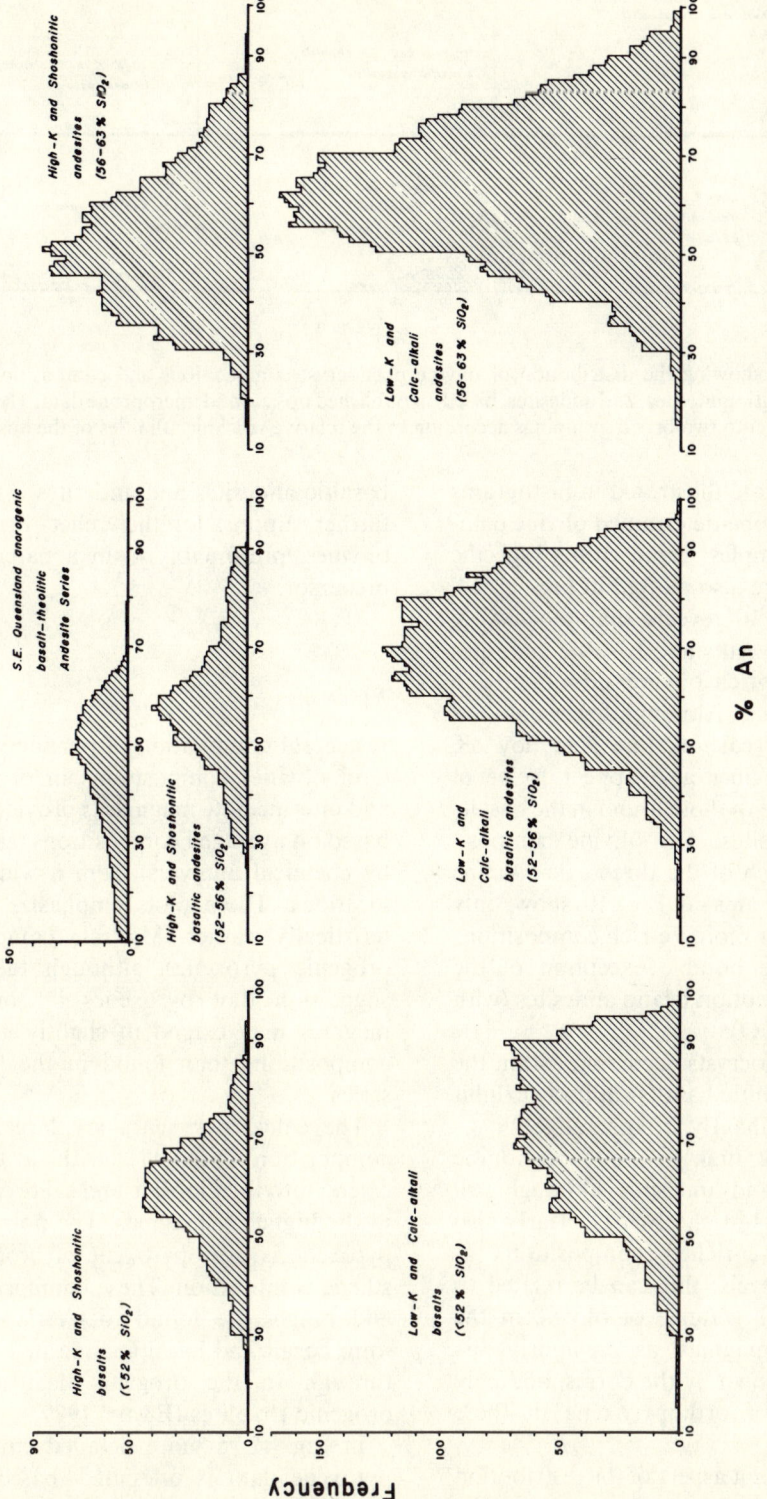

Fig. 9 Histograms showing the distribution of plagioclase phenocryst compositions, and compositional ranges, in orogenic basalts, basaltic andesites, and andesites, based on published optical and probe data. The compilations are subdivided into two broad groupings according to the relative volcanic affinities of the host rocks. Comparative data are also included from the anorogenic series of southern Queensland

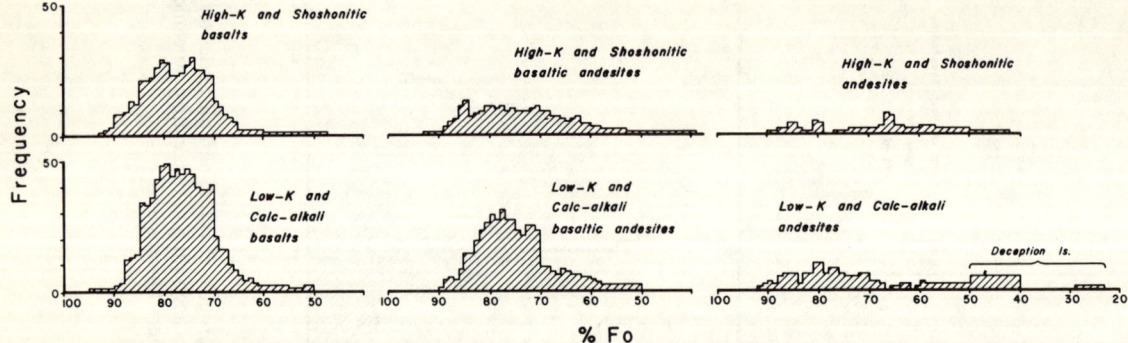

Fig. 10 Histograms showing the distribution of olivine phenocryst compositions and compositional ranges, in orogenic basalts, basaltic andesites, and andesites, based on published optical and microprobe data. The compilations are subdivided into two broad groupings according to the relative volcanic affinities of the host rocks

orogenic eruptives are illustrated in histograms in Fig. 10. Microprobe-determined olivine compositions, for samples representing specific volcanic regions, are also plotted in Figs 11–14. Reference to Fig. 10 reveals that the low K and calc-alkaline basalts and basaltic andesites contain rather Mg-rich olivine, dominantly in the range Fo_{70}–Fo_{85}. Although fewer data are available for the calc-alkaline and low K andesites, their olivines also appear to be of similar composition to those found in the basalts and basaltic andesites. The olivine compositions plotted in each of the three calc-alkaline and low K histograms in Fig. 10 show only limited tendency for more Fe-rich compositions to occur, with the notable exception of the rather unusual Deception Island andesites (with high Na_2O–low K_2O).

The olivine phenocrysts occurring within the high K and shoshonitic basalts similarly exhibit compositions dominantly in the range Fo_{70}–Fo_{85}, whereas the high K and shoshonitic basaltic andesites and andesites, although still containing rather Mg-rich olivines, clearly also contain more Fe-enriched compositions at higher frequency levels; this can be related to the extended stability range of olivine in the more K-enriched magmas, as previously discussed, and also shown by the correspondingly reduced stability of orthopyroxene in these magmas.

The most significant aspect of the distribution of the Mg-olivine compositions within the basaltic andesites and andesites is in providing further support for the 'relict' origin of these olivines, presumably from a parental basaltic precursor.

Pyroxenes

A general compilation of phenocryst pyroxene (and olivine) compositions in orogenic mafic and intermediate magmas is provided in Fig. 11, based on averaged compositions (as determined by chemical analyses) from a wide variety of locations. These plots emphasize the characteristically rather Mg-rich compositions of orogenic pyroxenes, although there is some suggestion that pyroxenes in some high K magmas may extend to slightly more Fe-rich compositions than found in the less potassic series.

The calcic pyroxenes are largely of augite composition (Fig. 11), with a tendency to extend into the diopside and salite compositions in the higher K magmas. The normal Ca-poor pyroxene is orthopyroxene of bronzite–hypersthene composition. These compositional fields and ranges are found to overlap extensively from basalt and basaltic andesite compositions through to the orogenic dacites and even orogenic rhyolites (Ewart, 1979).

In Fig. 12, a more detailed compilation of pyroxene data is presented, based on microprobe determinations on both individual pheno-

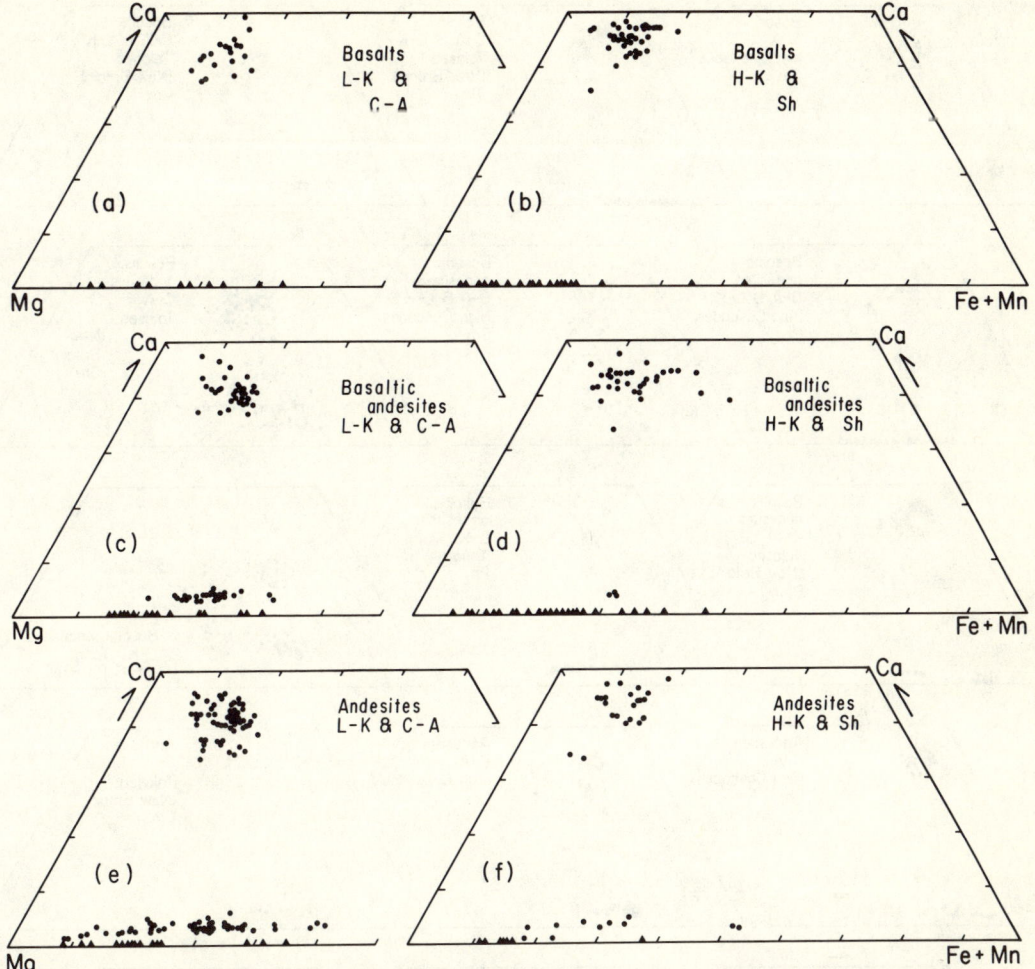

Fig. 11 Averaged compositions of phenocryst pyroxenes (●) and olivines (▲) in basalts, basaltic andesites, and andesites from various orogenic regions, plotted in terms of Mg, $\sum$Fe + Mn, and Ca (atomic per cent). The plots are subdivided into two groupings based on the relative volcanic affinities of the host rocks. Data taken from volcanic centres in western USA, Cascades, Aleutian Islands, Japan, Indonesia, Papua New Guinea, New Zealand, and the Mediterranean region

cryst and groundmass phases, from representative samples of well studied calc-alkaline and low K volcanic provinces from the southwestern Pacific and the Cascades of the northwestern USA. Figure 12 thus gives a rather more complete picture of individual grain variability and crystallization trends within individual volcanic samples. The phenocryst pyroxenes again exhibit rather restricted compositional changes, with trends towards slight Ca depletion in several sets of the augite data.

In contrast, the groundmass pyroxenes of the basaltic andesites especially show very extensive variations in solid solution, and range across the subcalcic pyroxene and pigeonitic compositional fields; those from the Tongan low K basaltic andesites (which include phenocryst rim compositions) also exhibit considerable Fe enrichment of some groundmass crystals. Such extensive subcalcic groundmass pyroxenes appear to be less common in the basalts, and are not present in the analysed Cascade andesites,

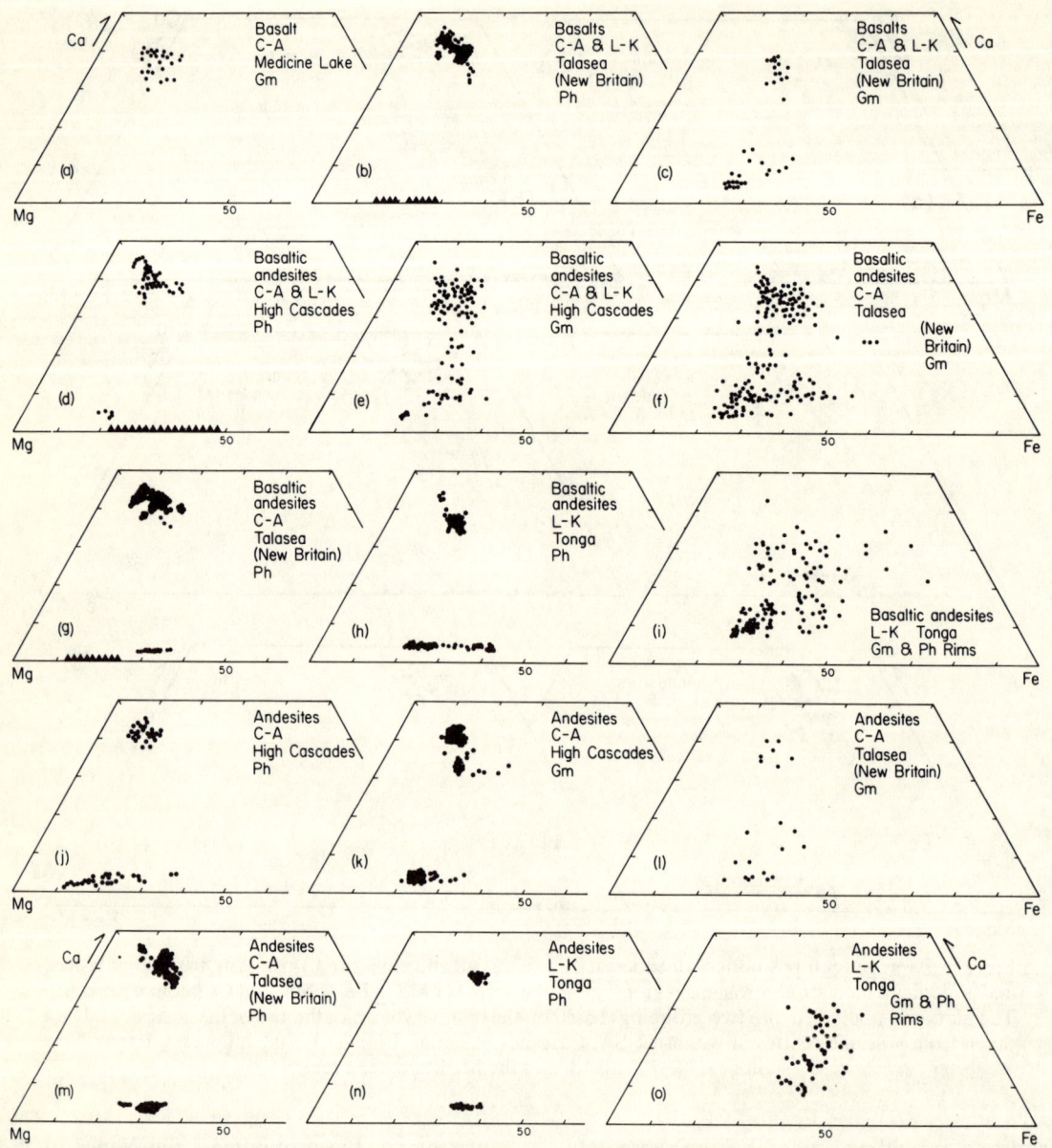

Fig. 12 Detailed microprobe analyses of individual pyroxene (●) and olivine (▲) phases, plotted in terms of Mg, $\sum$Fe, and Ca (atomic per cent), from various low K and calc-alkaline basalts, basaltic andesites, and andesites from volcanic centres in the Cascades (Smith and Carmichael, 1968), Talasea in New Britain (Lowder, 1970), and Tonga (Ewart, 1976b). Ph refers to phenocrysts; Gm refers to groundmass phases

but do occur in the andesites of both Talasea and Tonga.

Figure 13 is a comparable plot of pyroxene and olivine data, but based on detailed probe studies of calc-alkaline to high K volcanic centres in south-western Utah (Lowder, 1973; Hausel and Nash, 1977; from subregion 3 of this paper), and the high K to shoshonitic lavas of Wyoming (Nicholls and Carmichael, 1969; from subregion 2 of this paper), all in the western USA. The analysed samples from the Utah region incorporate volcanic rocks

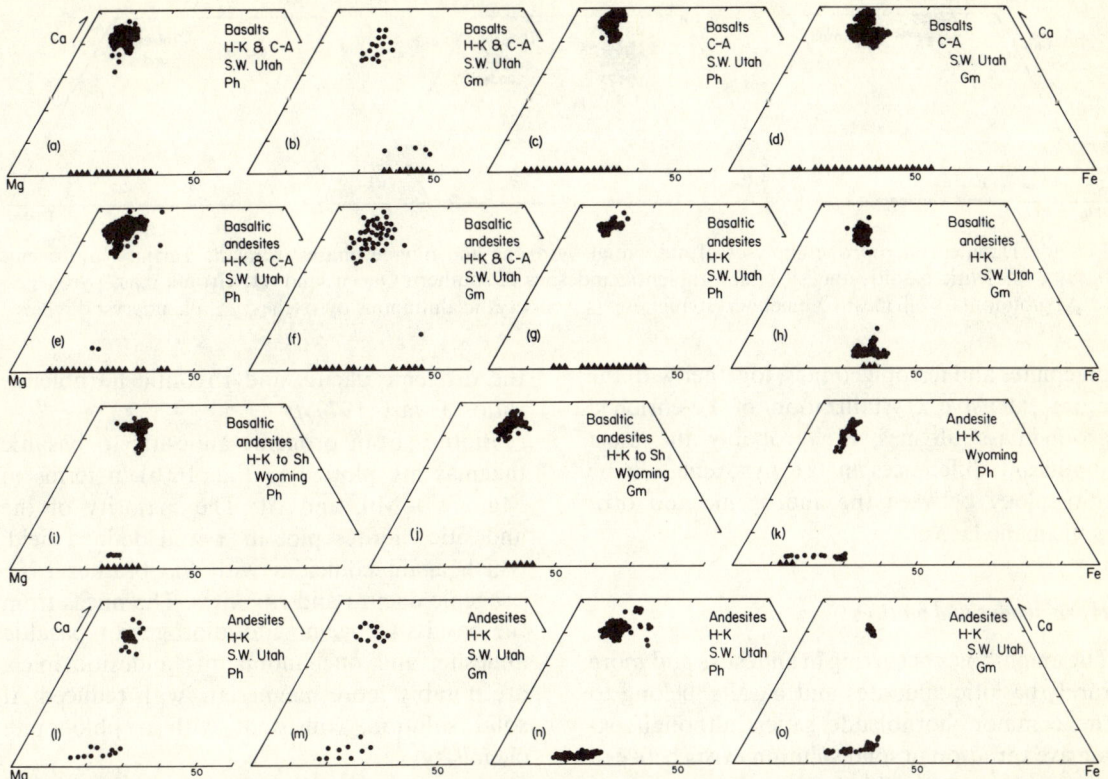

Fig. 13 Detailed microprobe analyses of individual pyroxene (●) and olivine (▲) phases, plotted in terms of Mg, $\sum$Fe, and Ca (atomic per cent), from basalts, basaltic andesites, and andesites of the western USA. Data represent calc-alkaline to high K eruptives from various volcanic centres in south-western Utah (Lowder, 1973, plots labelled c, d, g, h, n, and o; Hausel and Nash, 1977, plots labelled a, b, e, f, l, and m); and high K to shoshonitic eruptives from Wyoming (Nicholls and Carmichael, 1969, plots labelled i, j, and k). Ph refers to phenocryst; Gm refers to groundmass phases

erupted from a region of transitional and rather complex tectonic history. Phenocryst compositions within these plots again exhibit very limited variations, the augites exhibiting some tendency towards Ca depletion. Little variation exists, in terms of Mg, Fe, and Ca, between the augites from the basalts or basaltic andesites; the reduced stability of orthopyroxene is also noteworthy in many samples from these volcanic centres. Finally, attention is drawn to the absence of any tendency for the groundmass pyroxenes to develop subcalcic or pigeonitic compositions.

Anorogenic basalt–tholeiitic andesite association (southern Queensland) Figure 14 plots phenocryst, xenocrystic, and groundmass pyroxenes (and olivines) from this association. Phenocryst pyroxenes, as previously noted, are not so common in these anorogenic basalts and basaltic andesites, and, where present, are of augite to salite compositions. Of particular interest, however, are the occurrences of aluminous augite and aluminous orthopyroxene xenocrysts, which are surrounded by granular olivine and low Al augite reaction rims. Such xenocrysts are not characteristically found in orogenic magmas.

Groundmass pyroxene compositions are somewhat similar to those found in corresponding orogenic lavas, with the 'tholeiitic basaltic andesites' showing some tendency to develop subcalcic and especially pigeonitic pyroxenes. In fact, the occurrence of groundmass

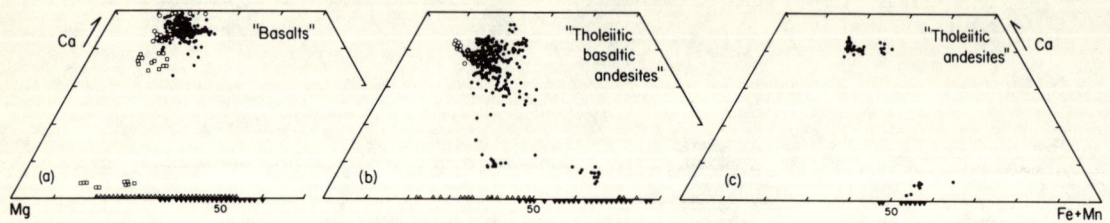

Fig. 14 Detailed microprobe analyses of individual pyroxene and olivine phases from the Tertiary anorogenic basalts, 'tholeiitic basaltic andesites' and tholeiitic andesites of southern Queensland. ●, Groundmass pyroxenes; ▲, groundmass olivines; ○, phenocryst augites; □, xenocrystic aluminous pyroxenes; △, phenocryst olivines

pigeonites and ferropigeonites, together with the more extensive crystallization of Fe-enriched groundmass olivines, are probably the most significant differences in the pyroxene–olivine mineralogy between the anorogenic and orogenic mafic lavas.

Hornblendes and biotites

The amphiboles occurring in andesites and more rarely basaltic andesites and basalts, belong to the common hornblende series, although extensive variation in solid solution exists between the major cations. This is illustrated in Fig. 15 in terms of Si, Ca + Na + K, $\sum$ Al, and Mg + $\sum$ Fe + Mn (O = 23). Superimposed on these plots are the generalized compositional fields of the dacitic and rhyolitic calcic amphiboles (Ewart, 1979). It was demonstrated that the hornblendes in these silicic orogenic magmas show systematic compositional changes with decreasing host rock SiO_2, namely decreasing Si and Mg + $\sum$ Fe + Mn, and increasing Ca + Na + K and $\sum$ Al. Reference to Fig. 15 shows that these solid solution trends are essentially continued with further decrease in host rock SiO_2 from andesite through to basalt. Thus, the hornblendes occurring in the basaltic andesites and basalts are approaching the ideal pargasite end-member ($NaCa_2Mg_4AlSi_6Al_2O_{22}(OH)_2$). No systematic differences are found between hornblendes from the different volcanic series (i.e. low K to shoshonitic). The hornblendes are replotted in Fig. 16(a) in terms of Mg : Fe : Ca ratios, and clearly show overlapping Mg : Fe ratios regardless of co-existing magma compositions (and which also overlap the orogenic dacitic and rhyolitic hornblende ratios, Ewart (1979)).

Biotites from orogenic andesitic to basaltic magmas are plotted in Fig. 16(b) in terms of Mg, Fe + Mn, and Al. The majority of the andesitic biotites plot in a well defined field, which again coincides with the biotites from orogenic dacites and rhyolites. The micas from the basalts (orogenic and anorogenic), basaltic andesite, and one anomalous andesitic mica, are notably more magnesian, with reduced Al solid solution, consistent with a phlogopite chemistry.

Fe–Ti oxides

Microprobe studies confirm that the characteristic oxide phase, both phenocryst and groundmass, of orogenic basalts, basaltic andesites, and andesites is titanomagnetite. Ulvöspinel solid solution tends to fall into the same limits for both phenocrysts and groundmass titanomagnetites, being most commonly Usp_{10}–Usp_{75}. The average values are Usp_{38} (phenocrysts) and Usp_{44} (groundmass). Where co-existing phenocryst and groundmass titanomagnetites are analysed, the groundmass phase is generally enriched in Ti, which Ewart (1976b) has suggested reflects an exchange equilibrium of the type

Fe_2TiO_4 + $CaAl_2Si_2O_8$ $\rightleftharpoons$
Ulvöspinel Anorthite

$2FeSiO_3$ + $CaTiAl_2O_6$
Pyroxene Titanpyroxene

The normal absence of ilmenite co-existing with titanomagnetite in orogenic basaltic to

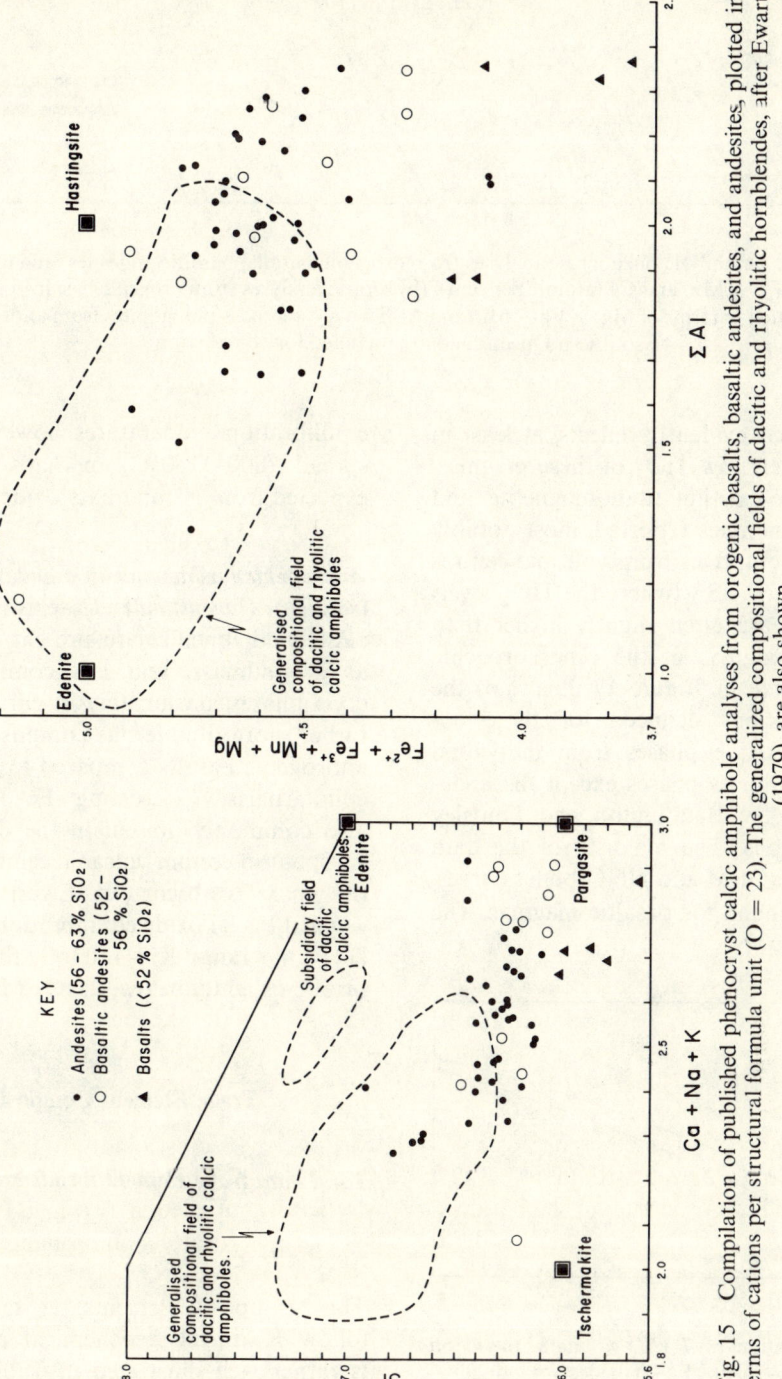

Fig. 15 Compilation of published phenocryst calcic amphibole analyses from orogenic basalts, basaltic andesites, and andesites, plotted in terms of cations per structural formula unit (O = 23). The generalized compositional fields of dacitic and rhyolitic hornblendes, after Ewart (1979), are also shown

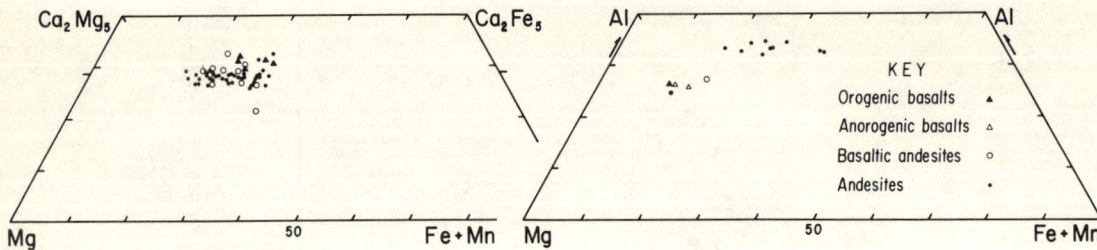

Fig. 16 (a) Calcic amphibole phenocryst analyses from orogenic basalts, basaltic andesites, and andesites, plotted in terms of Mg, $\sum$Fe + Mn, and Ca (atomic per cent). (b) Biotite analyses from orogenic basalts, basaltic andesites, and andesites, plotted in terms of Mg, $\sum$Fe + Mn, and Al. Two groundmass phlogopites from anorogenic basalts of southern Queensland are included for comparison

andesitic magmas evidently reflects, at least in part, the relatively low TiO_2 of these magmas. Nevertheless, co-existing titanomagnetite and ilmenite are sometimes reported, most notably from the mafic eruptives from volcanic centres within the western USA (where the TiO_2 levels in the magmas are often slightly higher than in equivalent magmas of the other orogenic regions; see Table 1). Figure 17 illustrates the T (°C)–f_{O_2} relations deduced for these co-existing Fe–Ti oxide phases from individual lavas (all groundmass phases except the andesite), based on the Buddington and Lindsley (1964) calibrations. The majority of the data plot close to the QFM and NNO buffer curves, as is normally found for basaltic magmas. The equilibration temperatures, however, are widely spread (800–1150°C), perhaps again to be expected from groundmass oxides.

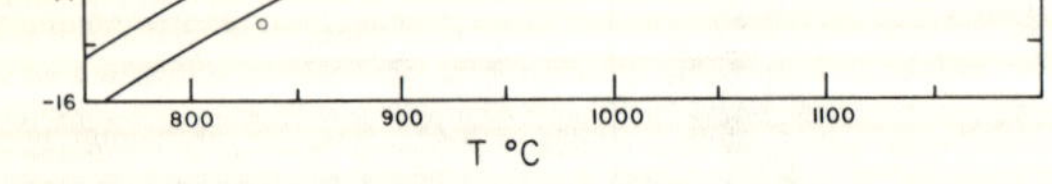

Anorogenic basalt–tholeiitic andesite association (southern Queensland) Co-existing titanomagnetite and ilmenite are present in the basalts as groundmass, and less commonly phenocryst phases, presumably in part reflecting the rather more titaniferous compositions of these anorogenic basalts compared to their orogenic counterparts. Co-existing Fe–Ti oxides are also commonly present in the tholeiitic andesites, but in certain volcanic centres the magnetite phase is absent (e.g. Ewart et al., 1977b). Typical Fe–Ti oxide equilibration temperatures lie in the range 850–1100°C; the data lie on, or often slightly below, the QFM buffer curve.

Trace Element Composition

Total data trace element trends

Rb, Ba, Sr, Zr, Pb

This group of elements, as is well known, follow K during geochemical processes; this is rather well illustrated in Table 6, in which inter-element correlations at significance levels greater than 99.9 per cent are listed, as calculated for the total data set (less than 60 per cent SiO_2) of the south-western Pacific subregion. (This subregion is specifically used in certain

Fig. 17 Compilation of T (°C)–f_{O_2} data, based on equilibration of co-existing Fe–Ti oxides, for individual samples of basalts, basaltic andesites, and andesites from the western USA. Only the andesite point represents phenocryst phases, the remaining points representing groundmass oxides. Data from Hausel and Nash (1977), Lowder (1973), Smith and Carmichael (1968), and Stormer (1972)

TABLE 6 Values of r (simple correlation coefficient) of the major and trace element variables showing correlations at values greater than 0.315, for the total data of the south-western Pacific subregion, for volcanic compositions with SiO₂ less than or equal to 60 per cent

	TiO₂	FeO₁	MgO	MnO	Na₂O	K₂O	P₂O₅	Rb	Ba	Sr	Zr	Pb	La	Ce	Ni	Co	Cr	V
TiO₂																		
FeO₁																		
MgO	0.627																	
MnO	0.354																	
Na₂O																		
K₂O	0.495																	
P₂O₅						0.370												
Rb (557)*						0.705	0.488											
Ba (537)						0.638	0.428	0.583										
Sr (583)						0.533	0.490	0.603	0.734									
Zr (560)	0.418					0.539	0.382	0.381	0.571	0.328								
Pb (364)						0.325	0.316	0.545	0.601	0.512	0.482							
La (245)						0.387			0.492	0.324	0.571	0.327						
Ce (264)									0.540	0.390	0.651	0.415	0.866					
Ni (554)			0.650															
Co (276)			0.545													0.702		
Cr (543)			0.663													0.814	0.674	
V (553)		0.544		0.399														

* Figures in parentheses indicate number of analyses available for statistical calculations on each trace element. Correlations between major and trace elements based on 1052 analyses.

of the following statistical treatments, simply due to the relatively large number of trace element and modal mineralogical determinations available.)

The behaviour of Ba, Sr, and Zr from the various subregions (based on arithmetic means of total data, Table 1), is illustrated in Fig. 18, which also includes plots of these elements from the K-rich series of the Roman Province. Ba shows, in all subregions, an increase from the basaltic to the andesitic–dacitic compositional range; however, in the rhyolitic compositions

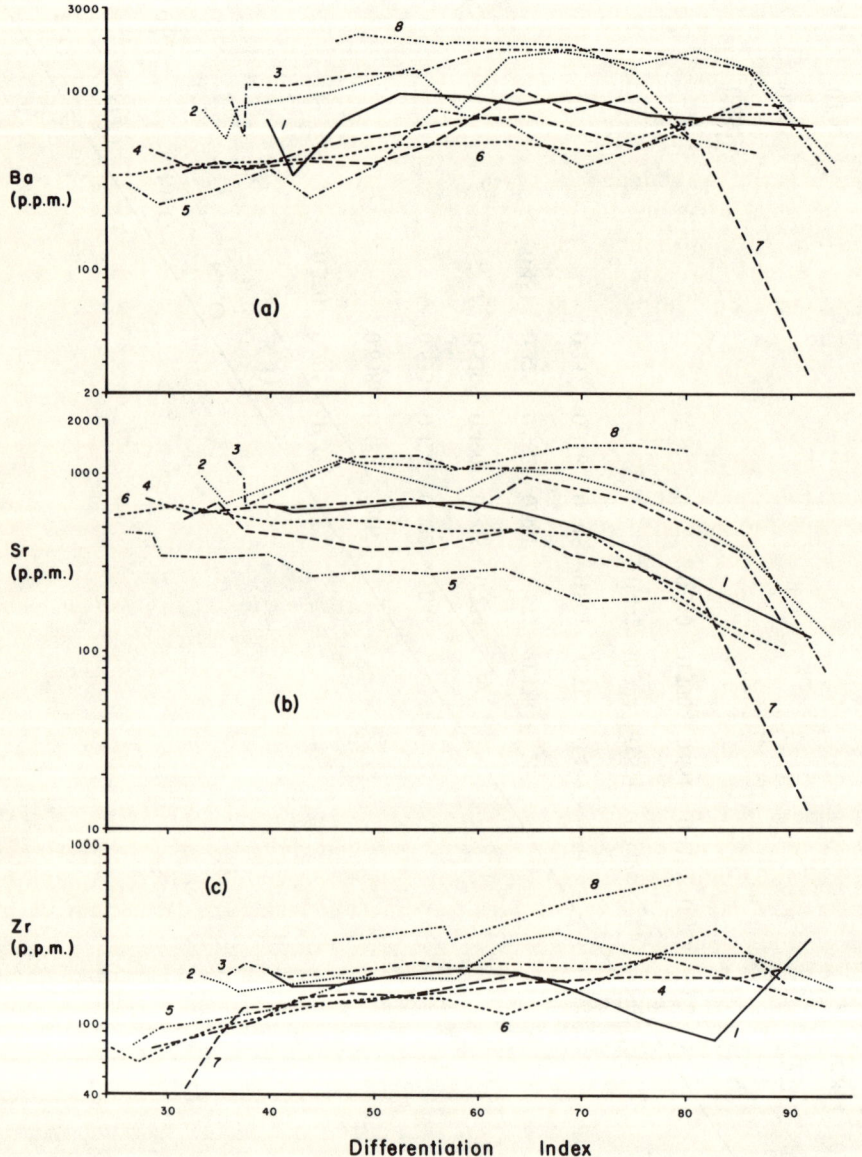

Fig. 18 The averaged trends of Ba, Sr, and Zr abundances (p.p.m.) against differentiation index (*DI*) for the various orogenic subregions (numbers 1–7, as listed in Introduction), and the K-rich series of the Roman Province. The data are plotted from the sequentially averaged total data presented in Table 1, together with equivalent data for compositions of greater than 60 per cent SiO_2 from Ewart (1979)

there are marked depletions of Ba in those rhyolitic magmas from the western USA and especially the Mediterranean subregions. A similar depletion is also shown by the trachytic compositions of the Roman Province. The most obvious explanation for these 'late stage' Ba depletions is by fractional crystallization involving removal of feldspar, but this also implies that the rhyolites from the other subregions have not undergone extensive fractionation (*see* Discussion in Ewart, 1979). Another noteworthy feature shown by the Ba curves in Fig. 18(a) is the rather higher overall Ba abundances of the western USA magmas (excepting the rhyolites).

The curves for Sr (Fig. 18(b)) show many of the features shown by Ba. Sr exhibits a general decrease from the mafic through the intermediate compositions, with the abundances decreasing more rapidly at the higher differentiation index compositions, again most pronounced in the western USA and Mediterranean rhyolites. The western USA mafic to intermediate magmas are again characterized by the highest overall Sr abundances, with, significantly, the volcanic centres of the Cascade-Aleutian subregion also exhibiting rather higher average Sr levels than observed in the other subregions.

Zr abundance patterns (Fig. 18(c)) tend to show irregularly increasing Zr with increasing differentiation index. Marked regional variations are again apparent, which become more divergent at higher differentiation indices. The K-rich series of the Roman Province have highest Zr abundances, a feature found in other highly alkaline magmas (e.g. peralkaline types). With respect to the general geochemical behaviour of Zr, attention is drawn to the very strong correlations of Zr with Ti, Na, La, and Ce, as well as the other 'K-related' elements, as noted previously (Table 6).

Cr, Ni, Co, V

The overall behaviour of these elements is similar within the various orogenic subregions, each becoming depleted, albeit erratically, with increasing differentiation index, as illustrated for Cr, Ni, and V in Fig. 19. Very high levels of correlation are found between Cr, Ni, Co, and MgO (Table 6), whereas V is highly correlated with total Fe and MnO. Again it is apparent from Fig. 19 that the Cr and Ni abundances are rather higher in the mafic to intermediate eruptives of the western USA than in equivalent eruptives from the other subregions; the western USA trends for V are also more erratic (Fig. 19(c)), perhaps again emphasizing the complexity of the magma associations in this region.

Anorogenic basalt–tholeiitic andesite association (southern Queensland)

The most significant trace element characteristics of this association are the tendency for higher abundances (compared to equivalent orogenic eruptives, with the possible exception of some within the western USA subregions) of the elements Zn, Zr, Y, and Nb (also noting the relatively high Ti of these anorogenic lavas); with increasing differentiation index, especially within associated trachytes, these four trace elements reach rather high concentrations. In contrast, V is abnormally low in the anorogenic basalts, relative to their orogenic equivalents.

Trace element abundance patterns in the low K, calc-alkaline, high K, and shoshonitic series

From the very well defined correlations of P, Rb, Ba, Sr, Zr, and Pb with K_2O (Table 6), it follows that the high K, and especially the shoshonitic, series of the orogenic regions will exhibit enrichments of these trace and minor elements, as has been pointed out, for example, by Jakeš and White (1969, 1972) and Jakeš and Gill (1970).

Rb behaviour in the four volcanic series is illustrated in Fig. 20, based on arithmetic means of K/Rb and K_2O for the sequential SiO_2 compositional intervals of the different orogenic subregions (but restricted to compositions from basalt to andesite). The low K series clearly exhibit Rb depletion, as do the soda-rich

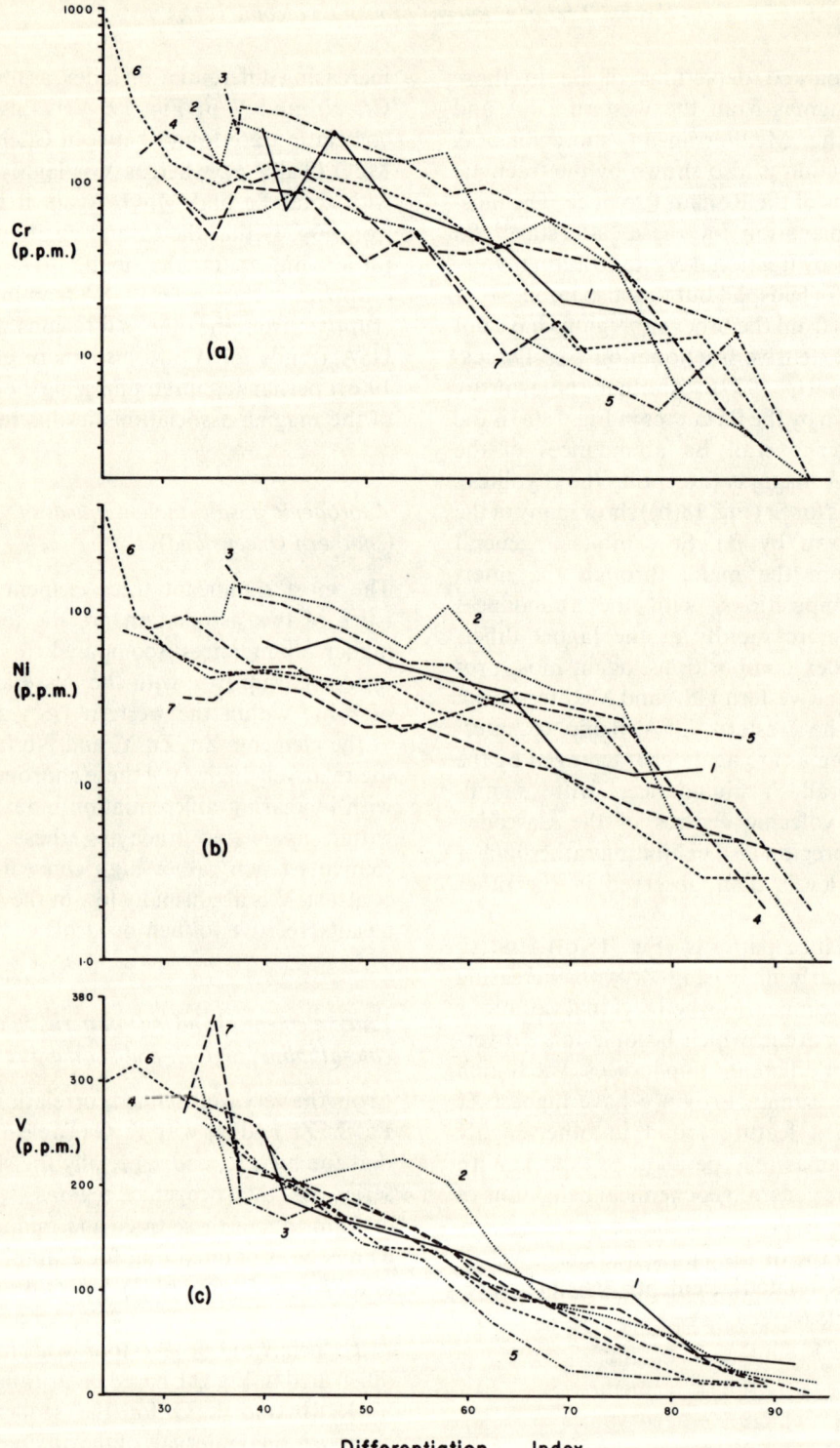

Fig. 19 The averaged trends for Cr, Ni, and V abundances (p.p.m.) against differentiation index, for the various orogenic subregions (numbers 1–7, as listed in Introduction). Data sources as in Fig. 18

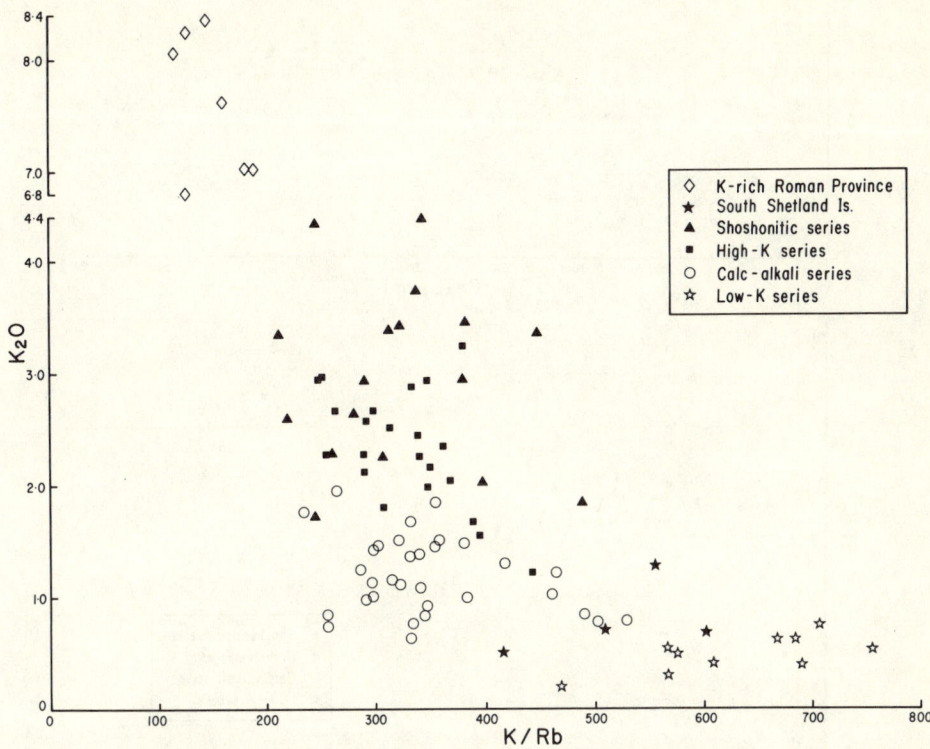

Fig. 20 Comparison of sequentially averaged K_2O against K/Rb ratios of the low K to shoshonitic eruptives, for compositions of less than 60 per cent SiO_2, from the various orogenic subregions (numbers 1–7, as listed in Introduction), and including the South Shetland Islands and the K-rich series of the Roman Province. Data compiled as listed in caption to Fig. 2

Deception Island eruptives, although to a lesser degree. The K/Rb ratios show a further progressive decrease through the calc-alkaline series, to ratios of 200–400, which are evidently characteristic of the high K, shoshonitic, and many of the eruptives of the calc-alkaline series. The K-rich series of the Roman Province exhibit relative Rb enrichment, with K/Rb ratios of less than 200.

The relative depletion of Rb that is seen in the low K series is also found for other elements such as Ba, Zr, light REE, Nb, Pb, and even Ni and Cr, at least in the western Pacific regions. The behaviour of two of these elements is illustrated in Fig. 21, in which averaged values of Ce are plotted against Ce/Y and Ni against MgO for data from the four volcanic series from the south-western Pacific subregion (for compositions extending from basalt to rhyolite). The very low abundances of Ce, and the corresponding low Ce/Y ratios, are notable in the low K series, but increase systematically towards the more K-enriched series. Thus, even the averaged data for the most silicic low K magmas have lower Ce and Ce/Y ratios than even the basalts of the calc-alkaline to shoshonitic series; conversely, Fig. 21 also indicates that only minor relative variation of the heavy REE (approximated by Y) occurs through the four volcanic series. The plot of Ni against MgO (Fig. 21(b)) illustrates further that the low K series contain not only lower Ni abundances for given MgO than the more K-enriched series, but also lower absolute abundances of Ni and MgO. Similar concentration patterns are found for Cr. Although these relatively low Ni and Cr abundances in the low K series could, in principle, be explained by olivine–pyroxene fractionation, such

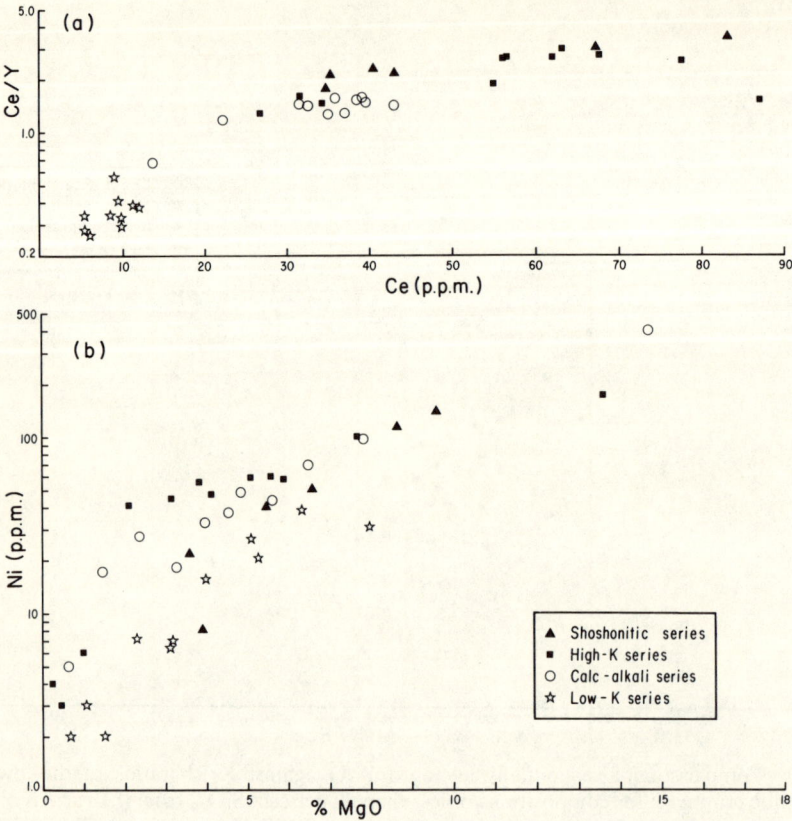

Fig. 21 Sequentially averaged Ce against Ce/Y and Ni against MgO for the four volcanic series from the south-western Pacific subregion. Data from rock composition spanning the compositional range basalt to rhyolite. Data compiled as listed in caption to Fig. 2

fractionation certainly cannot account for the element depletions discussed previously.

Correlations between Modal Mineralogy and Chemical Composition of the Magma

Correlation coefficients between modal phenocryst phases, and total rock major and trace elements of the south-western Pacific data group (less than 60 per cent SiO_2), reveal that four phenocryst mineral phases exhibit well defined correlations with total rock chemical composition. The minerals are plagioclase, olivine, augite, and biotite (Table 7). The biotite data exhibits a correlation with total rock K_2O, which decreases in significance towards the rhyolitic compositions, as might be anticipated. This correlation is, of course, also reflected in the biotite frequency occurrence graphs (Fig. 6), which simply show that biotite becomes a more frequent phenocryst phase in the more K-rich magmas.

Perhaps the more intriguing correlations petrologically are those between (1) total rock Al_2O_3 and modal phenocryst plagioclase; (2) total rock MgO and modal phenocryst olivine; and (3) total rock MgO and modal phenocryst augite (Table 7(a)). Data presented in Table 7(b) show that the plagioclase and augite correlations extend through the whole compositional spectrum from basalts to rhyolites (with the notable exception of augite in the less than 48 per cent SiO_2 basaltic division), whereas the olivine correlations weaken towards the more siliceous total

TABLE 7 Values of r (simple correlation coefficient) between selected modal phenocryst and total rock major and trace element abundances for the south-western Pacific subregion

(a) Modal phenocryst versus major and trace element correlations for total data of south-western Pacific subregion, for compositions with SiO_2 less than or equal to 60 per cent.

Variables	n^*	r	Variables	n^*	r	Variables	n^*	r
Plagioclase vs Al_2O_3	773	0.550	Mg-olivine vs MgO	773	0.751	Augite vs MgO	773	0.349
Plagioclase vs CaO	773	−0.043	Mg olivine vs Ni	554	0.689	Augite vs CaO	773	0.236
Plagioclase vs Sr	583	−0.108	Mg-olivine vs Co	276	0.567	Augite vs Cr	543	0.202
Plagioclase vs Ba	537	−0.182	Mg-olivine vs Cr	543	0.688	Biotite vs K_2O	773	0.343

(b) Modal phenocryst versus total rock major oxide correlations for individual SiO_2 divisions

SiO_2 divisions (per cent): n^*:	<48 27	48–50 98	50–52 117	52–54 145	54–56 157	56–58 129	58–60 100	60–63 104	63–66 62	66–69 35	69–73 32	>73 77
Plagioclase vs Al_2O_3	0.759	0.544	0.543	0.575	0.585	0.498	0.518	0.484	0.555	0.635	0.386	0.327
Mg-olivine vs MgO	0.945	0.726	0.675	0.654	0.397	0.481	0.405	0.196	0.273	—	—	—
Biotite vs K_2O	0.651	0.397	0.501	0.573	0.236	0.516	0.217	0.212	0.408	0.049	0.221	0.029
Augite vs MgO	−0.351	0.410	0.378	0.584	0.515	0.440	0.569	0.403	0.620	0.221	0.626	0.521

n^* = number of data for each SiO_2 division, and set of variables, for which r is calculated
Correlations for SiO_2 divisions greater than 60 per cent are based on data compiled in Ewart (1979).

rock compositions. Two explanations for these relationships seem most probable:

(1) The correlations reflect a strong tendency for the plagioclase, olivine, and augite phenocrysts, by some gravitative or flow differentiation process, to be cumulative phases in the crystal-rich (i.e. most commonly erupted) magmas, which further seems to imply that the crystal-poor magmas have undergone preferential extraction of these same mineral phases.

(2) The erupted lavas and pyroclastics, which are strongly porphyritic in plagioclase, olivine, and/or augite, crystallized from magmas primarily high in Al and Mg. The data, however, indicate that such magmas must contain consistently high modal proportions of these phases, with little representation amongst the phenocryst-poor lavas and pyroclastic rocks.

Further data correlations which are relevant to this problem are the complete absence of correlations between phenocryst plagioclase modes and total rock Sr and Ba; such correlations, if they do exist, must be masked by the strong variation of these two trace elements with K_2O. Moreover, there is only a relatively weak correlation between phenocryst plagioclase and total rock CaO (which simple arithmetic calculations show will be much less modified by addition of calcic plagioclase than is Al_2O_3). In contrast, phenocryst olivine modes are very strongly correlated with total rock Ni, Co, and Cr (Table 7(a)), while modal phenocryst augite is rather more weakly correlated with CaO and Cr (although still significant at the greater than 95 per cent confidence level). These correlations are all readily explained in terms of a phenocryst concentration process, with the possible exception of olivine–Cr. It should be noted that the phenocryst olivine–total rock MgO correlation extends well into the andesite compositions (Table 7(b)), and that evidence previously presented supports the 'relict' (xenocryst) origin for the persistence of olivine into these silicic magma compositions. Thus, it appears that these 'relict' olivines exert a strong control on the Mg. Ni, Co, and Cr (?) abundances in andesites.

A summary of the mutual interrelationships between phenocryst mineralogy, major element, and trace element variables in the south-western Pacific subregion (for those compositions with less than or equal to 60 per cent SiO_2) is presented in Fig. 22, based on the technique of principal components analysis, and showing a plot of component 1 against component 2. Component 1, which explains c. 17 per cent of the variance of the sample population, defines essentially the interaction between the K-related elements and the ferromagnesian and related elements, which reflect the contrasting geochemical characteristics and behaviour of the elements within the shoshonitic to low K series within this subregion. Component 2, which accounts for a further c. 12 per cent of the variance, emphasizes the control that phenocrystic olivine, plagioclase, and to a less extent augite, exert on the behaviour of MgO, Ni, Co, Cr, and Al_2O_3, as previously discussed.

Discussion

Mineralogical aspects

The phenocryst mineralogy of 'typical' (i.e. calc-alkaline) andesites and basaltic andesites is as follows: calcic plagioclase, augite, orthopyroxene, titanomagnetite, ± olivine, with hornblende increasing in frequency in the andesites. The most conspicuous differences in phenocryst mineralogy of the mafic and intermediate magmas belonging to the four volcanic series are, in fact, in the frequency of occurrence of hornblende and biotite, both of which increase with increasing K enrichment in the host magmas (Figs 5–7). Compositional characteristics of these phenocryst phases include the rather calcic compositions of the plagioclases, and the relatively Mg-rich compositions of the ferromagnesian silicates, which show no tendency to become Fe enriched even within orogenic dacites and rhyolites. Further significant aspects

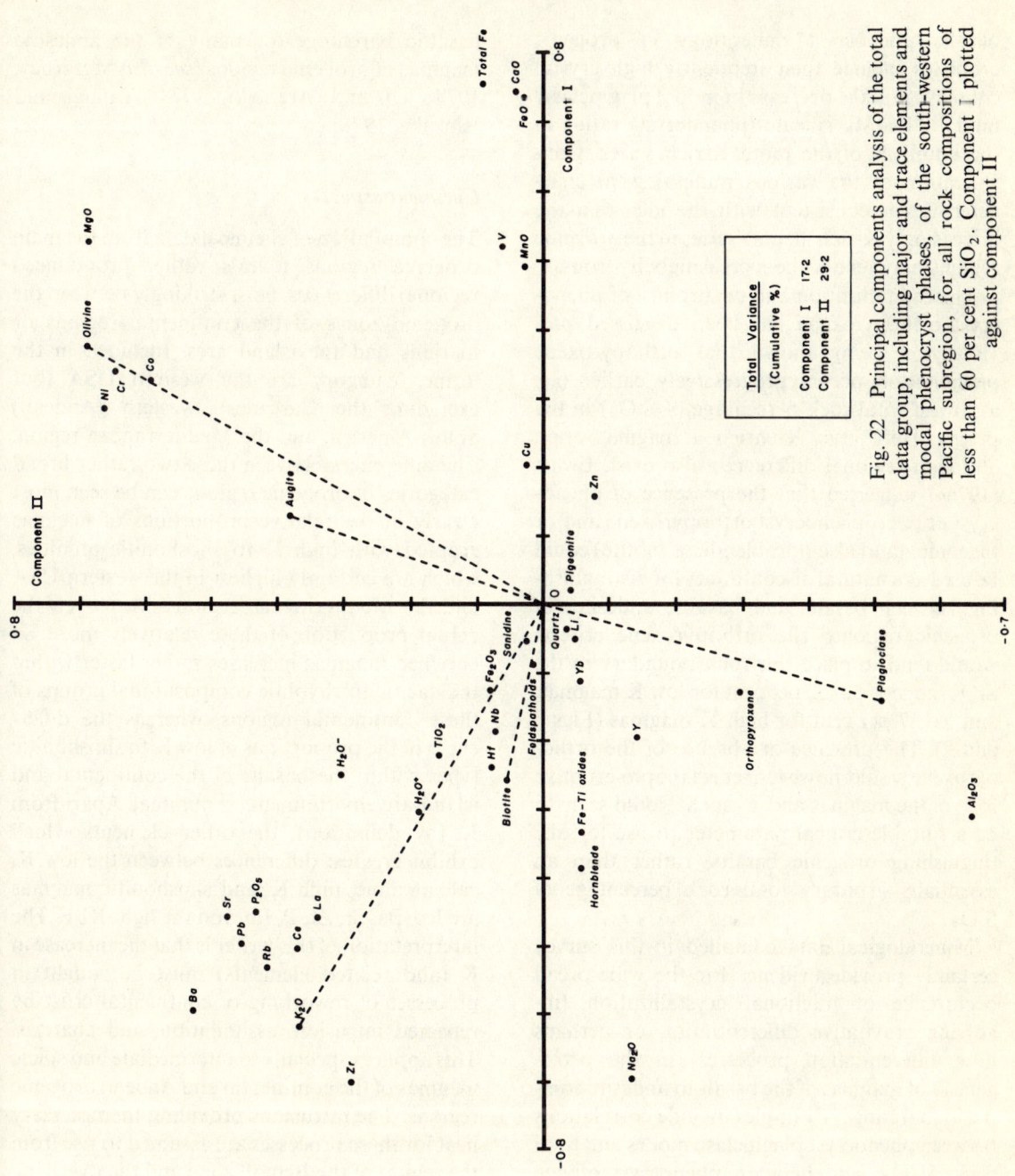

Fig. 22 Principal components analysis of the total data group, including major and trace elements and modal phenocryst phases, of the south-western Pacific subregion, for all rock compositions of less than 60 per cent SiO_2. Component I plotted against component II

of the phenocryst mineralogy of orogenic eruptives include their frequently high crystal contents, and the decrease in modal plagioclase/modal $\sum$Fe–Mg silicate (phenocryst) ratios in the eruptives of the more K-rich series. Consideration of the various mineralogical characteristics is consistent with the idea that the increasingly K-rich magma series in the orogenic regions may tend to be increasingly hydrous.

The crystallization and occurrence of phenocryst orthopyroxene has been discussed previously, it being shown that orthopyroxene precipitation occurs progressively earlier (i.e. at lowest total rock percentage of SiO_2) in the progressively least K-enriched magma series, although regional differences also exist. Ewart (1976a) suggested that the presence of phenocryst or microphenocryst orthopyroxene and/or pigeonite (and also hornblende or biotite) could be used as a natural discontinuity for distinguishing between basalts and basaltic andesites of orogenic regions. The orthopyroxene criteria would tend to place this rock boundary in the SiO_2 range of 50–52 per cent for low K magmas, and 53–57 per cent for high K magmas (Figs 5 and 7). The presence or absence of the orthopyroxene would, however, reflect the pre-eruptive $a_{SiO_2}^{liq}$ of the magmas and, as such, would seem to be a suitable critical parameter to use for distinguishing orogenic basalts, rather than an essentially arbitrary total rock percentage of SiO_2.

Mineralogical data compiled in this survey certainly provide evidence for the widespread occurrence of fractional crystallization (involving gravitative differentiation or perhaps flow differentiation processes) in the petrogenesis of magmas of the basalt to andesite compositional range. Examples include correlations between phenocryst plagioclase modes and total rock Al_2O_3, and between phenocryst olivine and augite modes and total rock MgO; the rather phenocryst-rich character of the majority of orogenic extrusives; and the occurrence of 'relict' Mg-olivines in basaltic andesites and andesites. This latter observation is considered highly significant and seems to suggest a basaltic parentage for many of the andesitic magmas of orogenic regions (*see also* McBirney, 1978; Ui and Aramaki, 1978; Yanagi and Ishizaka, 1978).

Chemical aspects

The compilation of chemical data from the main orogenic regions reveals rather pronounced regional differences, most strikingly between the orogenic zones of the continental regions or margins and the island arcs. Included in the former category are the western USA (but excluding the Cascades), western (Andean) South America, and the Mediterranean region. The differences between these two rather broad categories of orogenic regions can be seen most clearly in the relative proportions of magmas erupted with high K to shoshonitic affinities, which are certainly highest in the western USA and the Mediterranean regions. Moreover, the actual proportion of these relatively more K-enriched magmas increases rather faster within the dacitic to rhyolitic compositional groups of these continental regions, whereas the differences in the proportions of low K to shoshonitic types within the basalts of the continental and island-arc environments is minimal. Apart from K (by definition), the other elements which exhibit greatest differences between the low K, calc-alkaline, high K, and shoshonitic magmas are Rb, Ba, Sr, Zr, P, Pb, and the light REE. The interpretation of this writer is that the increase in K (and related elements) must be sought in processes of reworking of continental crust by repeated intrusion, assimilation, and anatexis. This applies especially to intermediate and silicic magmas of the continental and Andean orogenic regions. The intrusions providing the necessary heat for these processes are assumed to rise from the region of the Benioff zone and the overlying peridotite wedge into a crust that may itself have been produced in a geologically short time span by an earlier episode of orogenic magmatism.

The problem is, however, extremely complex. For example, the statistics summarized in Table 2 emphasize the common occurrence of basalts, some of which are rather alkaline, in orogenic regions (e.g. Fig. 3(a)). Two examples of their occurrence in island arcs (without continental crust) are the shoshonitic hornblende basalts of Bogoslof Island in the Aleutian arc (Arculus et al., 1977; see also Appendix 4), and the alkaline basalts of Grenada in the Lesser Antilles (e.g. Brown et al., 1977), which demonstrate the primary character of at least some orogenic alkali-rich basalts. The most extreme examples of alkali enrichment are the highly K-enriched series of the Roman Province (Table 1), the parental magmas of which were possibly derived from an enriched mantle source (Appleton, 1972; Cox et al., 1976), although Turi and Taylor (1976) interpret their oxygen isotopic data for these lavas in terms of interaction of the parental magmas with crustal rocks. Although the correlation between K_2O and depth to Benioff zone is widely accepted to apply in orogenic zones, Arculus and Johnson (1978) have recently drawn attention to important exceptions to this relationship, as well as to other widely accepted evolutionary sequences assumed to apply in such regions.

One localized example illustrating a correlation between K enrichment of the magmas, crustal type, and the relative abundance of silicic magma compositions is provided by the Tonga–Kermadec–New Zealand island-arc system within the south-western Pacific. The Tonga–Kermadec segment, situated on a narrow zone of 'intra-oceanic' crust (with a thickness intermediate between normal oceanic and continental crusts), is dominated by low K basalts and basaltic andesites, with only localized dacites. The extension of this arc southwards into the rifted Taupo Volcanic Zone of New Zealand, characterized by a continental crust, results in the eruption of predominantly calc-alkaline magmas, volumetrically dominated by rhyolites (Ewart et al., 1977a).

In the case of the western USA (excluding the Cascades), the patterns of volcanicity are often complicated by the occurrence of both subduction and extensional tectonics, with the development of voluminous and predominantly intermediate and silicic eruptives (all relatively K-rich), followed by bimodal suites consisting of silica-rich rhyolites and mafic lavas (e.g. Lipman et al., 1978). The latter association exhibits very similar chemical and mineralogical characteristics shown to exist in the anorogenic Queensland bimodal association (Table 1). The large volumes of erupted magmas belonging to the earlier intermediate-silicic, volcanic phases, especially when considered in terms of inferred associated batholith occurrences (and especially within subregion 2), suggest that certain zones of the crust must have been almost completely molten, at least in the lower and intermediate crustal depths. Given this situation of such voluminous magma within the crust, it seems inconceivable that contemporaneous mantle-derived magmas (which must constitute the necessary heat transfer mechanisms) could be erupted without gross modification either by crystal fractionation, mixing, equilibration, and/or fusion of enclosing country rocks. Indeed, this seems to be implicit in the results and conclusions of Lipman et al. (1978), who have interpreted the Pb and Sr isotopic data for the San Juan volcanic field in terms of lower crustal equilibration of mantle-derived magmas for the earlier intermediate lavas, with shallower crustal equilibration (including partial melting?) and crystal fractionation influencing the following silicic magmas. This author, therefore, would attribute the relative K enrichment of these western USA magmas (excepting again the Cascades) to crustal equilibration (and presumably also fusion), involving significant components of Proterozoic and Palaeozoic cratonic platform-type crustal basement assemblages, although excluding the uppermost crustal materials. By analogy, similar crustal involvement may explain the generally K-enriched characteristics of the intermediate and especially the silicic Andean and Mediterranean magmas, bearing in mind that the isotopic and chemical characteristics of 'lower', 'intermediate', and 'upper' crust will be very different, particularly if the lower crust has been in part formed by accretion of mafic, relatively younger, underplated materials.

It can be argued, however, that similar processes of crustal fusion and equilibration have occurred in the development of the western Pacific orogenic magmas, especially the more silicic magma types, but these do not exhibit the overall level of K enrichment seen in so many of the Andean and western USA regions. A possible explanation, albeit rather generalized, is that the western Pacific orogenic regions are developed 'offshore' from the adjacent continental cratons, with intervening marginal basins, and the crust constituting these orogenic regions may therefore contain relatively higher proportions of orogenic intrusive, volcanic, and volcaniclastic greywacke sequences, predominantly of Paleozoic and younger age, which may be more K-deficient, on average, than the more crystalline basement platform environments of the stabilized cratons. For example, Ewart and Stipp (1968), on the basis of Sr isotope evidence, suggested that the voluminous rhyolites of the Taupo Volcanic Zone of New Zealand were derived by partial fusion of Mesozoic volcanogenic greywacke sequences. Nevertheless, high K and shoshonitic magmas occur locally in, for example, the south-western Pacific, perhaps most notably in eastern Papua and the New Guinea Highlands; these areas, however, are adjacent to the Australian craton, part of which evidently underlies the New Guinea Highlands (Johnson *et al.*, 1978), and craton-derived younger sediments may thus constitute important crustal components in the Papua New Guinea region.

A different problem concerns the petrogenesis of the low K eruptives of especially the western Pacific region, which constitute an interesting group, as evidenced by their extremely depleted abundances of K, Rb, Ba, Sr, Zr, Nb, P, and light REE, even within the dacitic and rhyolitic low K magmas (Ewart, 1979). The low K series is apparently characteristic of the earlier stages of island arc evolution, and the more mafic low K magmas must therefore be considered as potential candidates as 'parental'

orogenic magmas. Nevertheless, the behaviour of the K-related elements through the compositional spectrum from low K basalt to rhyolite indicates that the low K series is unlikely to have given rise to the more K-enriched magmas of the orogenic regions by any simple fractionation process; this is illustrated, for example, in Fig. 21(a), with specific reference to Ce. The origin of the very depleted geochemistry of the low K magmas is problematical. It seems necessary to appeal to partial melting from a source extremely depleted in the K-related elements (which is not necessarily reflected in Sr and Pb isotopic compositions) and also to the presence, during magma generation, of a residual phase which would need to concentrate preferentially such elements as Ti, Zr, P, light REE, etc. Marsh (1976) has suggested rutile, while Beswick and Carmichael (1978) have suggested residual fluorapatite in controlling P and REE abundances in basalts. Another, as yet unexplored, possible residual phase is loveringite (approximately [Ca, REE, Y, Th, U, Pb][Ti, Fe, V, Al, Mg, Zr, Mn, Cr]$_{21}$O$_{38}$, or related minerals; Gatehouse *et al.*, 1978), which seems to have the appropriate chemical composition. The only other possible alternative mechanism to explain the primary element depletions in the low K magmas is by removal of these elements after magma formation; this would perhaps require an unusual type of fractional crystallization or a reaction mechanism of the magma with mantle peridotite. In either case, it would be necessary to invoke fractionation of, or a reaction zone containing, a mineral assemblage consisting of pyroxene–amphibole–phlogopite–apatite, etc., acting as a 'sink' for K, Rb, Ba, Sr, Zr, Nb, Ti, P, light REE (and perhaps H$_2$O).

Anorogenic basalt–tholeiitic andesite association (southern Queensland)

This association, which was erupted in a 'rifted tectonic environment', is bimodal in terms of the distribution of magma compositions; intermediate magmas in the range 55–65 per cent SiO$_2$ are relatively uncommon, and this is in clear contrast with the orogenic association. The anorogenic mafic magmas also exhibit additional chemical and mineralogical differences from equivalent orogenic lavas, for example the higher abundances of Ti, Nb, Y, Zr, Zn; the very low phenocryst contents; the reduced stability of phenocryst augite and, to a less extent, phenocryst plagioclase; relatively Fe-enriched groundmass olivine and pigeonitic phases; and the presence of aluminous pyroxene xenocrysts. The shape and nature of the feeder conduits (Marsh, 1978) may control some of these differences between the anorogenic and orogenic magmas, bearing in mind the contrasting tectonic environments.

Acknowledgements

The author wishes to thank Messrs B. Fordham and F. Audsley, University of Queensland, for much assistance in the computer processing of the data used in the compilations. Dr R. W. Johnson, Dr A. Cundari, Dr I. E. M. Smith, and Mr D. E. Mackenzie very kindly allowed the author access to unpublished data and thin sections of material from Papua New Guinea and Italy. Professor W. Elston offered very helpful comments on aspects of the manuscript. The author would also like to thank those editors of journals who have allowed authors to include basic data within their published papers!

REFERENCES

This list contains references cited in the text together with data sources that were not listed in Ewart (1979).

Anderson, A. T. and Gottfried, D. (1971). Contrasting behaviour of P, Ti, and Nb in a differentiated high-alumina olivine tholeiite and a calc-alkaline andesitic suite. *Geol. Soc. Am. Bull.* **82**, 1929–1942.

Anderson, C. A. (1936). Volcanic history of the Clear Lake area, California. *Geol. Soc. Am. Bull.* **47**, 629–644.

Anderson, C. A. (1940). Hat Creek lava flow. *Am. J. Sci.* **238**, 477–492.

Appleton, J. D. (1972). Petrogenesis of potassium-rich lavas from the Roccamonfina volcano, Roman Region, Italy. *J. Petrol.* **13**, 425–456.

Arculus, R. J. and Johnson, R. W. (1978). Criticism of generalized models for the magmatic evolution of arc-trench systems. *Earth planet. Sci. Letters* **39**, 118–126.

Arculus, R. J., Delong, S. E., Kay, R. W., Brooks, C., and Sun, S. S. (1977). The alkalic rock suite of Bogoslof Island, eastern Aleutian Arc, Alaska. *J. Geol.* **85**, 177–186.

Baker, P. E. (1972). Recent volcanism and magmatic variation in the Scotia Arc. In *Antarctic Geology and Geophysics* (R. J. Adie, ed.), International Union of Geological Science, Series B, No. 1, Universitets Forlaget, Oslo, pp. 57–60.

Baker, P. E. (1976). Volcanism and plate tectonics in the Antarctic Peninsula and Scotia Arc. In *Proceedings of the Symposium on 'Andean and Antarctic Volcanology Problems', Santiago, Chile, Sept 1974* (O. Gonzáles-Ferrán, ed.), International Association for Volcanology and the Chemistry of the Earth's Interior, Special Series, pp. 347–356.

Baker, P. E. and McReath, I. (1971). 1970 volcanic eruption at Deception Island. *Nature Phys. Sci.* **231**, 5–9.

Baker, I. and Ridley, W. I. (1970). Field evidence and K, Rb, Sr data bearing on the origin of the Mt Taylor volcanic field, New Mexico, USA. *Earth planet. Sci. Letters* **10**, 106–114.

Baker, P. E., Davies, T. G., and Roobol, M. J. (1969). Volcanic activity at Deception Island in 1967 and 1969. *Nature, Lond.* **244**, 553–560.

Baker, P. E., Harris, P. G., and Reay, A. (1971). The geology of Tofua Island, Tonga. *Bull. R. Soc. N.Z.* **8**, 67–79.

Baker, P. E., McReath, I., Harvey, M. R., Roobol, M. J., and Davies, T. G. (1975). The geology of the South Shetland Islands. V. Volcanic evolution of Deception Island. *Br. antarct. Surv. scient. Rep.* no. 78, 1–81.

Barberi, F., Borsi, S., Ferrara, G., and Innocenti, F. (1969). Strontium isotopic composition of some Recent basic volcanites of the Southern Tyrrhenian Sea and Sicily Channel. *Contr. Mineral. Petrol.* **23**, 157–172.

Barberi, F., Innocenti, F., Ferrara, G., Keller, J., and Villari, L. (1974). Evolution of Aeolian arc volcanism (southern Tyrrhenian Sea). *Earth planet. Sci. Letters* **21**, 269–276.

Beswick, A. E. and Carmichael, I. S. E. (1978). Constraints on mantle source compositions imposed by phosphorus and the rare-earth elements. *Contr. Mineral. Petrol.* **67**, 317–330.

Blake, D. H. (1976). Madilogo, a late Quaternary volcano near Port Moresby, Papua New Guinea. In *Volcanism in Australasia* (R. W. Johnson, ed.), Elsevier, Amsterdam, pp. 253–258.

Bogue, R. and Hodge, E. T. (1940). Cascade andesites of Oregon. *Am. Mineral.* **25**, 627–665.

Boutwell, J. M. (1912). Geology and ore deposits of the Park City District, Utah. *US geol. Surv. prof. Pap.* no. 77, 1–231.

Brew, D. A., Muffler, L. J. P., and Loney, R. A. (1969). Reconnaissance geology of the Mount Edgecumbe volcanic field, Kruzof Island, south-eastern Alaska. *US geol. Surv. prof. Pap.* no. 650-D, 1–18.

Brothers, R. N. (1967). Andesite from Rumble III volcano, Kermadec Ridge, southwest Pacific. *Bull. Volc.* **31**, 17–19.

Brown, G. M., Holland, J. G., Sigurdsson, H., Tomblin, J. F., and Arculus, R. J. (1977). Geochemistry of the lesser Antilles volcanic island arc. *Geochim. cosmochim. Acta* **41**, 785–801.

Buddington, A. F. and Lindsley, D. H. (1964). Iron-titanium oxide minerals and synthetic equivalents. *J. Petrol.* **5**, 310–357.

Bultitude, R. J. (1976). Eruptive history of Bagana volcano, Papua New Guinea, between 1882 and 1975. In *Volcanism in Australia* (R. W. Johnson, ed.), Elsevier, Amsterdam, pp. 317–336.

Burri, C. and Sonder, R. A. (1936). Beiträge zur Geologie and Petrographie des Jungtertiären und Rezenten vulkanismus in Nicaragua. *Z. Vulk.* **17**, 34–92.

Callaghan, E. (1933). Some features of the volcanic sequence in the Cascade Range in Oregon. *Trans. Am. geophys. Un.* **14**, 243–249.

Chayes, F. (1966). Alkaline and subalkaline basalts. *Am. Mineral.* **264**, 128–145.

Coats, R. R. (1953). Geology of Buldir Island, Aleutian Islands, Alaska. *US geol. Surv. Bull.* no. 989–A, 1–26.

Coats, R. R., Nelson, W. H., Lewis, R. Q., and Powers, H. A. (1961). Geologic reconnaissance of Kiska Island, Aleutian Islands, Alaska. *US geol. Surv. Bull.* no. 1028–R, 563–581.

Cole, J. W. (1970). Petrology of the basic rocks of the Tarawera volcanic complex. *N.Z. Jl Geol. Geophys.* **13**, 925–936.

Cole, J. W. (1973). High-alumina basalts of Tanpo Volcanic Zone, New Zealand. *Lithos* **6**, 53–64.

Cole, J. W. and Teoh, L. H. (1975). Petrography, mineralogy and chemistry of Pureora andesite volcano, North Island, New Zealand. *N.Z. Jl Geol. Geophys.* **18**, 259–272.

Colley, H. and Warden, A. J. (1974). Petrology of the New Hebrides. *Geol. Soc. Am. Bull.* **85**, 1635–1646.

Colony, R. J. and Sinclair, J. H. (1928). The lavas of the volcano Sumaco, eastern Ecuador, South America. *Am. J. Sci.* **16**, 299–312.

Cooke, R. J. S., McKee, C. O., Dent, V. F., and Wallace, D. A. (1976). Striking sequence of volcanic eruptions in the Bismark volcanic arc, Papua New Guinea, in 1972–75. In *Volcanism in Australasia* (R. W. Johnson ed.), Elsevier, Amsterdam, pp. 149–172.

Cornwall, H. R. (1962). Calderas and associated volcanic rocks near Beatty, Nye County, Nevada. In *Petrologic Studies: A Volume in Honour of A. F. Buddington* (A. E. J. Engel, H. L. James, and B. F. Leonard, ed.), The Geological Society of America, pp. 357–371.

Cornwall, H. R. and Kleinhampl, F. J. (1964). Geology of Bullfrog Quadrangle and ore deposits related to Bullfrog Hills caldera, Nye County, Nevada and Inyo County, California. *US geol. Surv. prof. Pap.* no. 454–J, 1–25.

Coulon, C. (1977). Le volcanisme calco-alcalin Cénozoique de Sardaigne (Italie) petrographie, geochimie et genèse de laves andesitiques et des ignimbrites—signification geodynamique. Thèse (Docteur ès—Sciences Naturelles), Université de Droit, d'Economie et des Sciences d'Aix, Marseille III, two volumes.

Cox, K. G., Hawkesworth, C. J., O'Nions, R. K., and Appleton, J. D. (1976). Isotopic evidence for the derivation of some Roman Region Volcanics from anomalously enriched mantle. *Contr. Mineral. Petrol.* **56**, 173–180.

Creasey, S. C. and Krieger, M. H. (1978). Galiuro volcanics, Pinal, Graham, and Cochise Counties, Arizona. *J. Res. US geol. Surv.* **6**, 115–131.

Cristofolini, R. (1972). I basalti a tendenza tholeiitica dell'Etna. *Per di Mineralogia* **41**, 167–200.

Cristofolini, R. (1973). Recent trends in the study of Etna. *Phil. Trans. R. Soc. A* **274**, 17–35.

Crowe, B. M. (1978). Cenozoic volcanic geology and probable age of inception of basin-range faulting in the southeastermost Chocolate Mountains, California. *Geol. Soc. Am. Bull.* **89**, 251–264.

Cundari, A. and Mattias, P. P. (1974). Evolution of the Vico lavas, Roman volcanic Region, Italy. *Bull. Volc.* **38**, 98–114.

Dallwitz, W. B., Green, D. H., and Thompson, J. E. (1966). Clinoenstatite in a volcanic rock from the Cape Vogel area, Papua. *J. Petrol.* **7**, 375–403.

Déruelle, B. and Déruelle, J. (1974). Los volcanes cuaternarios de los Nevados de Chillán (Chile central) y reseña sobre el volcanismo cuaternario de los Andes Chilenos. *Estudios Geologicos* **30**, 91–108.

Dickinson, W. R. and Hatherton, T. (1967). Andesitic volcanism and seismicity around the Pacific. *Science, N.Y.* **157**, 801–803.

Discala, L. and Viramonte, J. G. (1969). Preliminary report on the 1968 eruption of the Cerro Negro volcano, Nicaragua. Smithsonian Institution Center for Short-lived Phenomena, 20th January 1969.

Dostal, J., Dupuy, C., and Lefèvre, D. (1977). Rare earth element distribution in Plio-Quaternary volcanic rocks from southern Peru. *Lithos* **10**, 173–183.

Dostal, J., Zentilli, M., Caelles, J. C., and Clark, A. H. (1977). Geochemistry and origin of volcanic rocks of the Andes (26–28°S). *Contr. Mineral. Petrol.* **63**, 113–128.

Drewes, H. (1963). Geology of the Funeral Peak Quadrangle, California, on the east flank of Death Valley. *US geol. Surv. prof. Pap.* no. 413, 1–78.

Drewes, H., Fraser, G. D., Snyder, G. L., and Barnett, H. F. (1961). Geology of Unalaska Island and adjacent insular shelf, Aleutian Islands. *US geol. Surv. Bull.* no. 1028–S, 583–676.

Duggen, M. B. and Wilkinson, J. F. G. (1973). Tholeiitic andesite of high-pressure origin from the Tweed Shield Volcano, northeastern New South Wales. *Contr. Mineral. Petrol.* **39**, 267–276.

Dupuy, C. and Lefèvre, C. (1974). Fractionnement des éléments en trace Li, Rb, Ba, Sr dans les séries andesitiques et shoshonitiques du Pérou. Comparaison avec d'autres zones orogéniques. *Contr. Mineral. Petrol.* **46**, 147–157.

Eggler, D. H. (1972). Water-saturated and undersaturated melting relations in a Paricutin andesite and an estimate of water content in a natural magma. *Contr. Mineral. Petrol.* **34**, 261–271.

Eggler, D. M. and Burnham, C. W. (1973). Crystallization and fractionation trends in the system andesite–H_2O–CO_2–O_2 at pressures to 10 kb. *Geol. Soc. Am. Bull.* **84**, 2517–2532.

Eichelberger, J. C. and McGetchin, T. R. (1976). Petrogenesis of the 1973 Pacaya Lavas, Guatemala. In Proceedings of the Symposium on '*Andean and Antarctic Volcanology Problems*', *Santiago, Chile, Sept 1974* (O. Gonzáles-Ferrán, ed.), International Association for

Volcanology and the Chemistry of the Earth's Interior, Special Series, pp. 435–449.

Ewart, A. (1976a). Mineralogy and chemistry of modern orogenic lavas—some statistics and implications. *Earth planet. Sci. Letters* **31**, 417–432.

Ewart, A. (1976b). A petrological study of the younger Tongan andesites and dacites, and the olivine tholeiites of Niua Fo'ou Island, south-western Pacific. *Contr. Mineral. Petrol.* **58**, 1–21.

Ewart, A. (1979). A review of the mineralogy and chemistry of Tertiary–Recent dacitic, rhyolitic, and related salic volcanic rocks. In *Trondhjemites, Dacites and Related Rocks* (F. Barker, ed.), Elsevier, Amsterdam, pp. 13–121.

Ewart, A. and Bryan, W. B. (1972). Petrography and geochemistry of the igenous rocks from Eua, Tongan Islands. *Geol. Soc. Am. Bull.* **83**, 3281–3298.

Ewart, A. and Stipp, J. J. (1968). Petrogenesis of the volcanic rocks of the central North Island, New Zealand, as indicated by a study of $^{87}Sr/^{86}Sr$ ratios and Sr, Rb, K, U, and Th abundances. *Geochim. cosmochim. Acta* **32**, 699–735.

Ewart, A., Brothers, R. N., and Mateen, A. (1977a). An outline of the geology and geochemistry, and the possible petrogenetic evolution of the volcanic rocks of the Tonga–Kermadec–New Zealand island arc. *J. Volc. geothermal Res.* **2**, 205–250.

Ewart, A., Oversby, V. M., and Mateen, A. (1977b). Petrology and isotope geochemistry of Tertiary lavas from the northern flank of the Tweed Volcano, south-eastern Queensland. *J. Petrol.* **18**, 73–113.

Fodor, R. V. (1971). Chemistry, mineralogy, and petrology of the mafic and intermediate lavas of the Black Range, New Mexico. Ph.D. thesis, University of New Mexico, Albuquerque.

Ford, J. H. (1976). A geochemical and stable isotope study of the Panguna porphyry copper deposit, Bougainville. Ph.D. thesis, University of Queensland, Brisbane, two volumes.

Fornaseri, M. and Scherillo, A. (1963). Petrografia dei Colli Albani. In *La Regione Vulcanica dei Colli Albani* (M. Fornaseri, A. Scherillo, and U. Ventriglia, eds), Consiglio Nazionale delle Ricerche, Centro di Mineralogia e Petrografia presso l'Universita di Roma, pp. 339–550.

Francis, P. W., Moorbath, S., and Thorpe, R. S. (1977). Strontium isotope data for Recent andesites in Ecuador and North Chile. *Earth planet. Sci. Letters* **37**, 197–202.

Fukuyama, K. (1961a). Drusy pigeonite in the olivine basalt from the Hirado–Shiratake Hill, Nagasaki Prefecture, Japan (I). *Kumamoto J. Sci. B(1)* **4**, 1–18.

Fukuyama, K. (1961b). Drusy pigeonite in the olivine basalt from the Senryu–Shiratake Hill, Nagasaki Prefecture, Japan (II). *Kumamoto J. Sci. B(1)* **4**, 19–24.

Fukuyama, K. (1961c). Magnetite and hypersthene pseudomorph after olivine in the olivine basalt from the Senryu-Shiratake, Nagasaki Prefecture, Japan. *Kumamoto J. Sci. B(1)* **4**, 25–36.

Fukuyama, K. (1961d). Geological and petrological significance of the olivine basalt from Kohochi, northern Kyushu, Japan. *Kumamoto J. Sci. B(1)* **4**, 67–78.

Fukuyama, K. (1961e). The petrochemistry of the Yushima Island, olivine basalt, Kumamoto Prefecture, Japan. *Kumamoto J. Sci. B(1)* **4**, 79–84.

Fuster, J. M., Ibbarrola, E., and Martin, J. (1967). Las andesitas piroxénicas de la Mesa de Roldán (Almeria, SE de España). *Estudios Geologicos, Inst. 'Lucas Mallada'* **23**, 1–13.

Gatehouse, B. M., Grey, I. E., Campbell, I. H., and Kelly, P. (1978). The crystal structure of loveringite—a new member of the crichtonite group. *Am. Mineral.* **63**, 28–36.

Ghiara, M. R. and Lirer, L. (1976–77). Mineralogy and geochemistry of the 'Low Potassium' series of the Roccamonfina volcanic suite. *Bull. Volc.* **40**, 39–56.

Gill, J. B. (1970). Geochemistry of Viti Levu, Fiji, and its evolution as an island arc. *Contr. Mineral. Petrol.* **27**, 179–203.

Gilluly, J. (1946). The Ajo Mining District, Arizona. *US geol. Surv. prof. Pap.* no. 209, 1–112.

González-Ferrán, O. and Halpern, M. (1974). Edad del volcan andrus y relacion inicial Sr 87/Sr 86 de rocas volcanicas de Tierra Maria Byrd e Islas Shetland del sur, Antarctica. *Revista Geográfica de Chile 'Terra Australis'* **22–23**, 193–200.

González-Ferrán, O. and Katsui, Y. (1970). Estudio integral del volcanismo Cenozoico superior de las Islas Shetland del sur Antarctica. *Instituto Antarctico Chileno, Serie Cientifica* **1**, 125–174.

González-Ferrán, O., Munizaga, F., and Moreno, H. (1971). 1970 eruption at Deception Island: distribution and chemical features of ejected materials. *Antarctic J. US* **6**, 87–89.

Greeley, R. and Hyde, J. H. (1972). Lava tubes of the Cave Basalt, Mount St Helens, Washington. *Geol. Soc. Am. Bull.* **83**, 2397–2418.

Green, D. C. (1969). Transitional basalts from the Eastern Australia Tertiary Province. *Bull. Volc.* **33**, 930–941.

Hamilton, W. B. and Neuerburg, G. J. (1956). Olivine-sanidine trachybasalt from the Sierra Nevada, California. *Am. Mineral.* **41**, 851–873.

Hausel, W. D. and Nash, W. P. (1977). Petrology of Tertiary and Quarternary volcanic rocks, Washington County, Southwestern Utah. *Geol. Soc. Am. Bull.* **88**, 1831–1842.

Hawkes, D. D. (1961). The geology of the South Shetland Islands. II. The geology and petrology of Deception Island. *Falk. Isl. Dep. Surv. scient. Rep.* no. 27, 1–43.

Hawkins, J. W. (1970). Petrology and possible tectonic significance of late Cenozoic volcanic rocks, southern California and Baja California. *Geol. Soc. Am. Bull.* **81**, 3323–3338.

Hayatsu, K. (1977). Geologic study on the Myoko volcanoes, central Japan. Part 2. Petrography. *Mem. Fac. Sci., Kyoto Univ., Ser. Geol. Mineral.* **43**, 1–48.

Hernandez-Pacheco, A. and Ibbarrola, E. (1970). Neuvos datos sobre la petrologiá y geoquimica de las rocas volcánicas de la isla de Alborán (Mediterraneo occidental Almería). *Estudios Geologicos, Inst. 'Lucas Mallada'* **26**, 93–103.

Hoffer, J. M. (1971). Mineralogy and petrology of the Santo Tomas–Black Mountain basalt field, Potrillo Volcanics, south-central New Mexico. *Geol. Soc. Am. Bull.* **82**, 603–612.

Hotz, P. E. and Willden, R. (1964). Geology and mineral deposits of the Osgood Mountains Quadrangle Humboldt County, Nevada. *US geol. Surv. prof. Pap.* no. 431, 1–128.

Huber, N. K. and Rinehart, C. D. (1967). Cenozoic volcanic rocks of the Devils Postpile Quadrangle, eastern Sierra Nevada, California. *US geol. Surv. prof. Pap.* no. 554-D, 1-21.

Hunt, C. B. (1938). Igneous geology and structure of the Mount Taylor Volcanic field New Mexico. *US geol. Surv. prof. Pap.* no. 189-B, 51-80.

Iddings, J. P. (1899). Absarokite-shoshonite-banakite series. In *Geology of the Yellowstone National Park*. Part II. *Descriptive Geology, Petrography, and Paleontology*. US Geological Survey Monograph no. 32, US Geological Survey, Washington, DC, pp. 326-355.

Imbo, G. (1965). *Catalogue of the Active Volcanoes and Solfatra Fields of Italy*, Catalogue of the Active Volcanoes of the World, Part 18, International Association of Volcanologists, Naples, pp. 1-72.

Ishizaka, K., Yanagi, T., and Hayatsu, K. (1977). A strontium isotopic study of the volcanic rocks of the Myoko Volcano Group, central Japan. *Contr. Mineral. Petrol.* **63**, 295-307.

Jakeš, P. and Gill, J. (1970). Rare earth elements and the island arc tholeiite series. *Earth planet. Sci. Letters* **9**, 17-28.

Jakeš, P. and White, A. J. R. (1969). Structure of the Melanesian arcs and correlation with distribution of magma types. *Tectonophysics* **8**, 223-236.

Jakeš, P. and White, A. J. R. (1972). Major and trace element abundances in volcanic rocks of orogenic areas. *Geol. Soc. Am. Bull.* **83**, 29-40.

Jaques, A. L. (1976). High-K_2O island-arc volcanic rocks from the Finisterre and Adelbert Ranges, northern Papua New Guinea. *Geol. Soc. Am. Bull.* **87**, 861-867.

Johnson, R. W., Davies, R. A., and White, A. J. R. (1972). Ulawun volcano, New Britain. *Bur. Min. Res., Geol., Geophys. (Canberra) Bull.* no. 142, 1-42.

Johnson, R. W., Wallace, D. A., and Ellis, D. J. (1976). Feldspathoid-bearing potassic rocks and associated types from volcanic islands off the coast of New Ireland, Papua New Guinea: a preliminary account of geology and petrology. In *Volcanism in Australasia* (R. W. Johnson, ed.), Elsevier, Amsterdam, pp. 297-316.

Johnson, R. W., Mackenzie, D. E., and Smith, I. E. M. (1978). Delayed melting of subduction-modified mantle in Papua New Guinea. *Tectonophysics* **46**, 197-216.

Kawano, Y., and Aoki, Ken-Ichiro (1960). Some anorthite bearing volcanic rocks in Japan. *Sci. Rep. Tohoku Univ., Ser. III* **6**, 431-437.

Klerkx, J., Deutsch, S., Pichler, H., and Zeil, W. (1977). Strontium isotopic composition and trace element data bearing on the origin of cenozoic volcanic rocks of the central and southern Andes. *J. Volc. geothermal Res.* **2**, 49-71.

Krieger, M. H., Creasey, S. C., and Marvin, R. F. (1971). Ages of some Tertiary andesitic and latitic volcanic rocks in the Prescott-Jerome area, north-central Arizona. *US geol. Surv. prof. Pap.* no. 750-B, 157-160.

Krushensky, R. D. and Escalante, G. (1967). Activity of Irazu and Poas volcanoes, Costa Rica, November 1964-July 1965. *Bull. Volc.* **31**, 75-84.

Kuno, H. (1965). Fractionation trends of basaltic magmas in lava flows. *J. Petrol.* **6**, 302-321.

Lambert, M. B. (1974). The Bennett Lake cauldron subsidence complex, British Columbia and Yukon Territory. *Geol. Surv. Can. Bull.* no. 227, 1-213.

Laughlin, A. W., Brookins, D. G., and Causey, J. D. (1972). Late Cenozoic basalts from the Bandera lava field, Valencia County, New Mexico. *Geol. Soc. Am. Bull.* **83**, 1543-1552.

Leeman, W. P. and Rogers, J. J. W. (1970). Late Cenozoic alkali-olivine basalts of the Basin-Range Province, USA. *Contr. Mineral Petrol.* **25**, 1-24.

Lefèvre, C. (1973). Les caractères magmatiques du volcanisme plio-quaternaire des Andes dans le sud du Perou. *Contr. Mineral. Petrol.* **41**, 259-272.

Le Maitre, R. W. (1976). Some problems of projecting chemical data into mineralogical classifications. *Contr. Mineral. Petrol.* **56**, 181-189.

Lewis, R. Q., Nelson, W. H., and Powers, H. A. (1960). Geology of Rat Island, Aleutian Islands, Alaska. *US geol. Surv. Bull.* no. 1028-Q, 555-562.

Lipman, P. W. (1968). Geology of the Summer Coon volcanic center, eastern San Juan Mountains, Colorado. *Colorado School Mines* **63**, 211-236.

Lipman, P. W. (1969). Alkalic and tholeiitic basaltic volcanism related to the Rio Grande Depression, southern Colorado and northern New Mexico. *Geol. Soc. Am. Bull.* **80**, 1343-1354.

Lipman, P. W., Doe, B. R., Hedge, C. E., and Stevens, T. A. (1978). Petrologic evolution of the San Juan volcanic field, southwestern Colorado: Pb and Sr isotope evidence. *Geol. Soc. Am. Bull.* **89**, 59-82.

Lipman, P. W. and Moench, R. H. (1972). Basalts of the Mount Taylor volcanic field, New Mexico. *Geol. Soc. Am. Bull.* **83**, 1335-1344.

Lipman, P. W., Prostka, H. J., and Christiansen, R. L. (1972). Cenozoic volcanism and plate-tectonic evolution of the Western United States. I. Early and Middle Cenozoic. *Phil. Trans. R. Soc. A* **271**, 217-248.

Lloyd, E. F. (1972). Geology and hot springs of Orakeikorako. *N.Z. geol. Surv. Bull.* no. 85, 1-164.

Locardi, E. and Mittempergher, M. (1968). Relationship between some trace elements and magmatic processes. *Geol. Rundschau* **57**, 313-334.

Lopez-Escobar, L., Frey, F. A., and Vergara, M. (1976). Andesites from central-south Chile: trace element abundances and petrogenesis. In *Proceedings of the Symposium on 'Andean and Antarctic Volcanology Problems'*, Santiago, Chile, Sept 1974 (O. Gonzáles-Ferrán, ed.), International Association for Volcanology and the Chemistry of the Earth's Interior, Special Series, pp. 725-761.

Lopez-Escobar, L., Frey, F. A., and Vergara, M., 1977. Andesites and high-alumina basalts from the central-south Chile High Andes: Geochemical evidence bearing on their petrogenesis. *Contr. Mineral. Petrol.* **63**, 199-228.

Lowder, G. G. (1970). The volcanoes and caldera of Talasea, New Britain: mineralogy. *Contr. Mineral. Petrol.* **26**, 324-340.

Lowder, G. G. (1973). Late Cenozoic transitional alkali olivine tholeiitic basalt and andesite from the margin of the Great Basin, south-west Utah. *Geol. Soc. Am. Bull.* **84**, 2993-3012.

Luft, S. J. (1964). Mafic lavas of Dome Mountain, Timber

Mountain caldera, southern Nevada. *US geol. Surv. prof. Pap.* no. 501–D, 14–21.

Marsh, B. D. (1976). Some Aleutian andesites: their nature and source. *J. Geol.* **84**, 27–45.

Marsh, B. D. (1978). On the cooling of ascending andesitic magma. *Phil. Trans. R. Soc. A* **288**, 611–625.

Mathews, W. H. (1958). Geology of the Mount Garibaldi map-area, southwestern British Columbia, Canada. *Geol. Soc. Am. Bull.* **69**, 179–198.

Matsumoto, H. (1968). Petrological study on rock from Oninomi-yama volcano, Beppu City, Oita Prefecture. *Kumamoto J. Sci. B(1)* **7**, 91–94.

Matsumoto, H. (1976). Petrological study on some rocks from Rabaul volcano. *Kumamoto J. Sci. (Geol.)* **10**, 19–26.

McBirney, A. R. (1978). Volcanic evolution of the Cascade Range. *A. Rev. Earth planet. Sci.* **6**, 437–456.

McDougall, I. and Wilkinson, J. F. G. (1967). Potassium–argon dates on some Cainozoic volcanic rocks from northwestern New South Wales. *J. geol. Soc. Aust.* **14**, 225–234.

McKee, C. O., Cooke, R. J. S., and Wallace, D. A. (1976). 1974–75 eruptions of Karkar volcano, Papua New Guinea. In *Volcanism in Australasia* (R. W. Johnson, ed.), Elsevier, Amsterdam, pp. 173–190.

Melson, W. G. and Rodrigo, S. R. (1968). The 1968 eruption of volcan Arenal, Costa Rica: preliminary summary of field and laboratory studies. Smithsonian Institution Center for Short-lived Phenomena, 7 November 1968.

Melson, W. G. and Saenz, R. (1973). Volume, energy, and cyclicity of eruptions of Arenal volcano, Costa Rica. *Bull. Volc.* **37**, 416–437.

Mertzman, S. A. (1977a). Recent volcanism at Schonchin and Cinder Buttes, northern California. *Contr. Mineral. Petrol.* **61**, 231–243.

Mertzman, S. A. (1977b). The petrology and geochemistry of the Medicine Lake volcano, California. *Contr. Mineral. Petrol.* **62**, 221–247.

Moore, J. G. and Melson, W. G. (1969). *Nuées ardentes* of the 1968 eruption of Mayon Volcano, Philippines. *Bull. Volc.* **33**, 600–620.

Moreno, R. H. (1976). The Upper Cenozoic volcanism in the Andes of southern Chile (From 40°00′ to 41°30′ S.L.). In *Proceedings of the Symposium on 'Andean and Antarctic Volcanology Problems'*, Santiago, Chile, Sept 1974 (O. Gonzáles-Ferrán, ed.), International Association for Volcanology and Chemistry of the Earth's Interior, Special Series, pp. 143–171.

Murata, K. J., Dondoli, C., and Saenz, R. (1966). The 1963–65 eruption of Irazú volcano, Costa Rica (the period of March 1963 to October 1964). *Bull. Volc.* **29**, 765–796.

Nairn, I. A., Hewson, C. A. Y., Latter, J. H., and Wood, C. P. (1976). Pyroclastic eruptions of Ngauruhoe volcano, central North Island, New Zealand, 1974 January and March. In *Volcanism in Australasia* (R. W. Johnson, ed.), Elsevier, Amsterdam, pp. 385–405.

Nelson, W. H. and Pierce, W. G. (1968). Wapiti Formation and Trout Peak trachyandesite northwestern Wyoming. *US geol. Surv. Bull.* no. 1254-H, 1–11.

Nicholls, J. and Carmichael, I. S. E. (1969). A commentary on the absarokite–shoshonite–banakite series of Wyoming, USA. *Schweiz. Mineral. Petrog. Mitt.* **49**, 47–64.

Ninkovich, D. and Hays, J. D. (1972). Mediterranean island arcs and origin of high potash volcanoes. *Earth planet. Sci. Letters* **16**, 331–345.

Noble, D. C., Bowman, H. R., Hebert, A. J., Silberman, M. L., Heropoulos, C. E., Fabbi, B. P., and Hedge, C. E. (1975). Chemical and isotopic constraints on the origin of low-silica latite and andesite from the Andes of central Peru. *Geology* **3**, 501–504.

Nockolds, S. R. and Allen, R. (1954). The geochemistry of some igneous rock series: Part II. *Geochim. cosmochim. Acta* **5**, 245–285.

Oba, Y. (1972). Petrology of the Late Pliocene basalts of the western part of Hokkaido. *J. Fac. Sci. Hokkaido Univ., Ser. IV* **15**, 11–25.

Oba, Y. (1975). Late Neogene basaltic rocks from the Kitami–Monbetsu District, Northeast Hokkaido, *J. Fac. Sci. Hokkaido Univ., Ser. IV* **16**, 501–510.

Onuma, K. (1962). Petrography and petrochemistry of the rocks from Iwate volcano, northeastern Japan. *J. Jap. Assoc. Mineral. Petrol. econ. Geol.* **47**, 192–204.

Orheim, O. (1972). Volcanic activity on Deception Island, South Shetland Islands. In *Antarctic Geology and Geophysics* (R. J. Adie, ed.), International Union for Geological Science, Series B, No. 1, Universitetsforlaget, Oslo, pp. 117–120.

Parsons, W. H. (1939). Volcanic centers of the Sunlight area, Park County, Wyoming. *J. Geol.* **47**, 1–26.

Peccerillo, A. and Taylor, S. R. (1976). Geochemistry of Eocene calc-alkaline volcanic rocks from the Kastamonu area, northern Turkey. *Contr. Mineral. Petrol.* **58**, 63–81.

Peck, D. L. (1964). Geologic reconaissance of the Antelope–Ashwood area, north-central Oregon. *US geol. Surv. Bull.* no. 1161-D, 1–26.

Pichler, H., Hörmann, P. K., and Braun, A. F. (1976). First petrologic data on lavas of the volcano El Reventador (Eastern Ecuador). *Münster Forsch. Geol. Paläont* **38/39**, 129–141.

Pichler, H. and Weyl, R. (1976). Quaternary alkaline volcanic rocks in eastern Mexico and central America. *Münster Forsch. Geol. Paläont.* **38/39**, 159–178.

Pichler, H. and Zeil, W. (1972). The Cenozoic rhyolite–andesite association of the Chilean Andes. *Bull. Volc.* **35**, 424–452.

Powers, H. A., Coats, R. R., and Nelson, W. H. (1960). Geology and submarine physiography of Amchitka Island Alaska. *US geol. Surv. Bull.* no. 1028-P, 521–554.

Proffett, J. M. and Proffett, B. H. (1976). Stratigraphy of the Tertiary ash-flow tuffs in the Yerington District, Nevada. *Nevada Bur. Mines Geol. Rep.* no. 27, 1–28.

Puchelt, H. and Hoefs, J. (1974). Preliminary geochemical and strontium isotope investigations on Santorini rocks. In *Acta of the First International Scientific Congress on the Volcano of Thera* (A. Kaloyeropoulou, ed.), T.A.P. Service, Athens, pp. 318–327.

Puchelt, H., Murad, E., and Hubberten, H. W. (1977). Geochemical and petrological studies of lavas, pyroclastica and associated xenoliths from the Christiana Islands, Aegean Sea. *Neues Jb. Miner. Abh.* **131**, 140–155.

Remy, J.-M. (1963). L'eruption volcanique de 1960 au Lopévi (Nouvelles-Hebrides). *Bull. Soc. geol. Fr.* **V**, 188–197.

Rhodes, R. C. (1970). Volcanic rocks associated with the western part of the Mogollon Plateau volcano–tectonic complex, southwestern New Mexico. Ph.D. thesis, University of New Mexico, Albuquerque.

Riehle, J. R., McKee, E. H., and Speed, R. C. (1972). Tertiary volcanic center, west-central Nevada. *Geol. Soc. Am. Bull.* **83**, 1383–1396.

Ross, D. C. 1970. Pegmatitic trachyandesite plugs and associated volcanic rocks in the Saline Range–Inyo Mountains region, California. *US geol. Surv. prof. Pap.* no. 614–D, 1–29.

Savelli, C. (1967). The problem of rock assimilation by Somma–Vesuvius magma. I. Composition of Somma and Vesuvius lavas. *Contr. Mineral. Petrol.* **16**, 328–353.

Schneider, H. (1965). Petrographie des Lateravulkans und die magmenentwicklung der Monti Volsini. *Schweiz Mineral. Petrogr. Mitt.* **45**, 331–455.

Schultz, C. H. (1976). Petrology and petrochemistry of Deception Island, Antarctica. In *Proceedings of the Symposium on 'Andean and Antarctic Volcanology Problems', Santiago, Chile, Sept 1974* (O. Gonzáles-Ferrán, ed.), International Association for Volcanology and the Chemistry of the Earth's Interior, Special Series, pp. 498–517.

Schwab, K. (1971). Beobachtungen an jungen Vulkanitvorkommen der argentinischen Puna. *Münster Forsch. Geol. Paläont.* **20/21**, 251–274.

Shaw, H. R., Peck, D. L., Wright, T. L., and Okamura, R. (1968). The viscosity of basaltic magma: an analysis of field measurements in Makaopuhi lava lake, Hawaii. *Am. J. Sci.* **266**, 225–264.

Shepherd, E. S. (1938). The gases in rocks and some related problems. *Am. J. Sci.* **35A**, 311–351.

Sheppard, R. A. (1967). Petrology of a late Quaternary potassium-rich andesite flow from Mount Adams, Washington. *US geol. Surv. prof. Pap.* no. 575–C, 55–59.

Sheridan, D. M., Maxwell, C. H., and Albee, A. L. (1967). Geology and uranium deposits of the Ralston Buttes District Jefferson County, Colorado. *US Geol. Surv. prof. Pap.* no. 520, 1–121.

Siems, P. L. (1968). Volcanic geology of the Rosita Hills and Silver Cliff District, Custer County, Colorado. *Q. Colorado School Mines* **63**, 89–124.

Simons, F. S. and Mathewson, D. E. (1955). Geology of Great Sitkin Island, Alaska. *US geol. Surv. Bull.* no. 1028–B, 21–43.

Smith, A. L. and Carmichael, I. S. E. (1968). Quaternary lavas from the southern Cascades, western USA. *Contr. Mineral. Petrol.* **19**, 212–238.

Smith, A. L. and Carmichael, I. S. E. (1969). Quaternary trachybasalts from southeastern California. *Am. Mineral.* **54**, 909–923.

Sparks, R. S. J. (1975). Stratigraphy and geology of the ignimbrites of Vulsini volcano, central Italy. *Geol. Rundschau* **64**, 497–523.

Staatz, M. and Carr, W. J. (1964). Geology and mineral deposits of the Thomas and Dugway Ranges Juab and Tooele Counties Utah. *US geol. Surv. prof. Pap.* no. 415, 1–188.

Stanton, R. L. and Bell, J. D. (1969). Volcanic and associated rocks of the New Georgia Group, British Solomon Islands Protectorate. *Overseas Geol. Mineral Resources* **10**, 113–145.

Stern, R. J. (1979). On the origin of andesite in the northern Mariana Island arc: implications from Agrigan. *Contr. Mineral. Petrol.* **68**, 207–219.

Stevens, N. C. (1962). The petrology of the Mt Alford ring-complex, S.E. Queensland. *Geol. Mag.* **99**, 501–515.

Stevens, N. C. (1965). The volcanic rocks of the southern part of the Main Range, south-east Queensland. *Proc. R. Soc. Queensland* **77**, 37–52.

Stevens, N. C. (1969). The Tertiary volcanic rocks of Toowoomba and Cooby Creek, south-east Queensland. *Proc. R. Soc. Queensland* **80**, 85–96.

Stevens, N. C., Oba, Y., and Katsui, Y. (1978). Some trace-element data on Pliocene basaltic rocks from two districts in Hokkaido. *J. Fac. Sci. Hokkaido Univ.*, Ser. IV **18**, 485–489.

Stormer, J. C., 1972. Mineralogy and petrology of the Raton–Clayton volcanic field, northeastern New Mexico. *Geol. Soc. Am. Bull.* **83**, 3299–3322.

Sturiale, C. (1968). A subterminal radial fissure eruption on Mt Etna. *Geol. Rundschau* **57**, 765–773.

Tatsumoto, M. and Knight, R. J. (1969). Isotopic composition of lead in volcanic rocks from central Honshu—with regard to basalt genesis. *Geochem. J.* **3**, 53–86.

Tournon, J. (1972). Présence de Basaltes alcalins Récents au Costa Rica (Amerique Centrale). *Bull. Volc.* **36**, 140–147.

Turi, B. and Taylor, H. P. (1976). Oxygen isotopic studies of potassic volcanic rocks of the Roman Province, Central Italy. *Contr. Mineral. Petrol.* **55**, 1–31.

Ui, T. and Aramaki, S. (1978). Relationship between chemical composition of Japanese island-arc volcanic rocks and gravimetric data. *Tectonophysics* **45**, 249–259.

Varnes, D. J. (1963). Geology and ore deposits of the south Silverton mining area, San Juan County, Colorado. *US geol. Surv. prof. Pap.* no. 378–A, 1–56.

Vergara, M. and Katsui, Y. (1969). Contribucion a la geologia y petrologia del volcan Antuco, Cordillera de los Andes, Chile central. *Univ. de Chile, Facultad de Ciencias, Fisicas y Matematicas, Dept Geologia, Publ.* no. 35, 25–47.

Vitaliano, C. J. and Vitaliano, D. B. (1972). Cenozoic volcanic rocks in the southern Shoshone Mountains and Paradise Range, Nevada. *Geol. Soc. Am. Bull.* **83**, 3269–3280.

Warden, A. J. (1967). The 1963–65 eruption of Lopevi volcano (New Hebrides). *Bull. Volc.* **30**, 277–318.

Washington, H. S. (1906). The Roman comagmatic Province. *Carnegie Instn. Wash. Publ.* no. 57, 1–199.

Weyl, R. (1957). Beiträge zur Geologie de Cordillera de Talamanca Costa Ricas (Mittelamerika). *Neues Jb. Geol. Paläontol. Abh.* **105**, 123–204.

Weyl, R. (1969). Magmatische Forderphasen und Gesteinschemismus in Costa Rica (Mittelamerika). *Neues Jb. Geol. Paläeontol. Monatshefte* 423–446.

Whitebread, D. H. (1976). Alteration and geochemistry of Tertiary volcanic rocks in parts of the Virginia City Quadrangle, Nevada. *US geol. Surv. prof. Pap.* no. 936, 1–43.

Wilkinson, J. F. G. (1968). The magmatic affinities of some volcanic rocks from the Tweed Shield Volcano. S.E. Queensland–N.E. New South Wales. *Geol. Mag.* **105**, 275–289.

Wilkinson, J. F. G. and Binns, R. A. (1977). Relatively

iron-rich lherzolite xenoliths of the Cr-diopside suite: a guide to the primary nature of anorogenic tholeiitic andesite magmas. *Contr. Mineral. Petrol.* **65**, 199–212.

Williams, H. (1952). The great eruption of Coseguina, Nicaragua, in 1835. *Univ. Calif. Publ. Bull. Dept Geol. Sci.* **29**, 21–45.

Wilshire, H. G. (1957). Propylitization of Tertiary volcanic rocks near Ebbetts Pass, Alpine County, California. *Univ. Calif. Publ. Geol. Sci.* **32**, 243–271.

Witkind, I. J. (1969). Geology of the Tepee Creek Quadrangle Montana–Wyoming. *US geol. Surv. prof. Pap.* no. 609, 1–101.

Yagi, K., Takeshita, H., and Oba, Y. (1972). Petrological study of the 1970 eruption of Akita–Komagatake volcano, Japan. *J. Fac. Sci. Hokkaido Univ., Ser. IV* **15**, 109–138.

Yamaguchi, M. (1958). Petrography of the Otozan flow on Shodo-Shima Island, Seto-uchi Inland Sea, Japan. *Mem. Fac. Sci. Kyushu Univ. D* **6**, 217–238.

Yanagi, T. and Ishizaka, K. (1978). Batch fractionation model for the evolution of volcanic rocks in an island arc: an example from central Japan. *Earth planet. Sci. Letters* **40**, 252–262.

Zielinski, R. A. and Lipman, P. W. (1976). Trace element variations at Summer Coon volcano, San Juan Mountains, Colorado, and the origin of continental-interior andesites. *Geol. Soc. Am. Bull.* **87**, 1477–1485.

APPENDIX 1 Averaged chemical compositions and phenocryst mineralogy of selected low K volcanic rocks of orogenic regions

	Basalts (<52% SiO_2)			Basaltic Andesites (52-56% SiO_2)		Andesites (56-63% SiO_2)	
	South-Western Pacific	North-Western Pacific	Cascades–Aleutian Islands	South-Western Pacific	North-Western Pacific	South-Western Pacific	North-Western Pacific
SiO_2	50.73	50.63	50.08	54.05	53.64	58.68	58.92
TiO_2	0.83	0.86	1.03	0.72	0.95	0.63	0.73
Al_2O_3	17.38	18.16	18.30	17.36	17.39	16.80	17.04
Fe_2O_3	3.01	3.29	2.10	3.34	3.28	3.00	3.02
FeO	6.96	7.79	6.71	6.35	7.53	5.35	5.01
MnO	0.19	0.19	0.16	0.17	0.18	0.15	0.16
MgO	6.97	5.61	7.99	5.21	4.47	3.57	3.55
CaO	11.51	11.15	10.51	9.95	9.59	8.31	7.78
Na_2O	2.06	1.96	2.74	2.37	2.43	2.78	3.09
K_2O	0.26	0.27	0.24	0.38	0.39	0.62	0.54
P_2O_5	0.09	0.10	0.16	0.09	0.14	0.13	0.17
$n =$	43	89	13	111	123	66	133

Trace elements (p.p.m.)

Rb	4.1 (22)*	2.4 (2)	3.75 (2)	4.9 (46)	9.9 (2)	8.1 (24)	6.8 (21)
Ba	90.2 (21)	195 (8)	110 (4)	116 (46)	181 (12)	176 (24)	139 (5)
Sr	224 (22)	247 (3)	465 (5)	241 (46)	244 (18)	287 (24)	159 (33)
Zr	30.8 (21)	313 (3)	87.5 (4)	37.0 (39)	68.0 (7)	60.5 (23)	121 (4)
Zn	78.2 (17)	n.d.	n.d.	83.6 (32)	94.1 (7)	93.7 (12)	84 (30)
La	2.95 (18)	2.40 (7)	3.7 (1)	2.15 (25)	3.88 (9)	4.03 (12)	3.66 (6)
Ce	7.77 (18)	7.34 (7)	n.d.	5.61 (25)	12.1 (9)	9.48 (11)	9.90 (6)
Yb	2.26 (4)	1.81 (7)	2.77 (3)	1.28 (3)	2.29 (8)	1.57	2.30 (6)
Y	15.8 (21)	17 (1)	26.7 (3)	21.0 (38)	20.0 (14)	27.5 (21)	19.6 (14)
Cu	98.4 (20)	40 (1)	52.3 (3)	123 (40)	52.0 (15)	138 (23)	48.9 (33)
Ni	35.5 (20)	38 (1)	165 (4)	23.3 (40)	36.8 (12)	10.9 (21)	46.6 (32)
Co	35.7 (17)	40.8 (1)	50.3 (4)	31.3 (37)	34.7 (15)	21.9 (23)	31.2 (33)
Cr	82.4 (20)	30.6 (1)	342 (3)	57.3 (40)	45.5 (15)	27.9 (24)	94.6 (32)
V	286 (20)	240 (1)	358 (3)	290 (40)	269 (14)	205 (24)	167 (14)
Nb	1.7 (5)	n.d.	n.d.	0.56 (3)	2.1 (1)	2.55 (2)	0.31 (2)
Li	n.d.	n.d.	5.0 (1)	n.d.	4.85 (8)	8.7 (1)	5.2 (30)
Pb	4.0 (18)	5.28 (1)	n.d.	2.34 (31)	6.0 (10)	2.3 (17)	4.0 (4)
Hf	0.86 (4)	0.90 (1)	n.d.	0.63 (3)	0.92 (6)	1.44 (3)	n.d.

Phenocryst mineralogy modes (vol. %)

Plagioclase	18.7	12.8	6.67	18.9	12.6	17.2	13.8
Quartz	0	0	0	0	0	0.07	0
Olivine	2.94	0.76	2.34	0.89	0.20	0.17	0.11
Augite	4.98	0.35	0.57	3.74	0.51	2.44	0.98
Orthopyroxene	0.66	0.76	0	2.17	1.51	2.26	1.88
Hornblende	0.59	0	0	0.04	0	0.18	0.01
Biotite	0	0	0	0	0	0	0
Fe–Ti oxides	0.39	0	0	0.23	0.13	0.54	0.64
Pigeonite	0	0	0	0.02	0	0	0
TPC	28.3	14.7	9.6	26.0	15.0	22.9	17.4
$n =$	31	21	7	99	25	56	23

Occurrence (frequency %)

Plagioclase	95.1	88.8	75.0	99.1	91.8	95.5	97.7
Quartz	0	0	0	0	0.8	3.1	2.3
Olivine	80.5	82.1	83.3	57.0	48.0	21.2	17.4
Augite	70.8	61.8	8.3	98.1	75.6	90.9	88.7
Orthopyroxene	36.6	49.4	0	82.3	78.0	92.5	92.4
Hornblende	4.9	0	0	0.9	0	4.6	9.9
Biotite	0	0	0	0	0	0	0
Fe–Ti oxides	43.9	7.8	0	42.0	26.0	72.7	89.4
Pigeonite	0	9.0	0	1.9	4.9	0	0.8
$n =$	41	89	12	107	123	66	132

n.d. = insufficient data; TPC = total phenocryst content.
* Figures in parentheses indicate number of data used to calculate each trace element mean.

APPENDIX 2 Averaged chemical compositions and phenocryst mineralogy of selected calc-alkaline volcanic rocks of orogenic regions

	Basalts ($<52\%$ SiO_2)				Basaltic Andesites ($52–56\%$ SiO_2)				Andesites ($56–63\%$ SiO_2)			
	South-Western Pacific	North-Western Pacific	Cascades, Alaska, Aleutians	Mediterranean	South-Western Pacific	North-Western Pacific	Cascades, Alaska, Aleutians	Mediterranean	South-Western Pacific	North-Western Pacific	Cascades, Alaska, Aleutians	Mediterranean
SiO_2	49.91	50.26	50.32	49.89	54.49	53.98	54.29	54.14	59.16	59.21	58.99	58.82
TiO_2	0.81	1.16	1.22	1.05	0.76	0.91	1.21	0.88	0.69	0.77	0.91	0.75
Al_2O_3	16.37	17.97	17.81	18.16	17.00	18.14	17.62	17.60	16.91	17.32	17.45	17.81
Fe_2O_3	3.17	3.59	3.04	3.92	3.19	3.33	3.18	3.22	2.82	3.01	2.64	2.95
FeO	6.92	7.02	6.62	5.79	5.76	5.72	5.40	5.07	4.08	4.42	4.07	3.69
MnO	0.18	0.20	0.17	0.18	0.16	0.17	0.15	0.16	0.13	0.14	0.12	0.14
MgO	8.22	5.90	6.75	6.35	5.18	4.88	4.86	5.28	3.95	3.50	3.41	3.28
CaO	11.00	10.18	10.01	10.77	9.10	8.63	8.38	9.42	7.14	7.00	6.81	7.35
Na_2O	2.39	2.69	3.06	2.84	2.98	3.00	3.61	2.90	3.40	3.14	3.94	3.50
K_2O	0.82	0.83	0.77	0.81	1.15	1.06	1.05	1.15	1.49	1.33	1.42	1.54
P_2O_5	0.20	0.20	0.24	0.24	0.22	0.19	0.26	0.19	0.20	0.16	0.23	0.18
$n =$	173	144	69	50	176	233	162	65	262	363	228	92

Trace elements (p.p.m.)

	South-Western Pacific	North-Western Pacific	Cascades, Alaska, Aleutians	Mediterranean	South-Western Pacific	North-Western Pacific	Cascades, Alaska, Aleutians	Mediterranean	South-Western Pacific	North-Western Pacific	Cascades, Alaska, Aleutians	Mediterranean
Rb	13.4 (97)*	23.9 (18)	17.3 (4)	21.8 (29)	21.2 (68)	27.3 (15)	26.7 (38)	29.8 (38)	33.6 (143)	37.0 (18)	34.4 (61)	40.5 (45)
Ba	237 (79)	226 (10)	395 (21)	240 (20)	324 (62)	320 (19)	439 (60)	282 (28)	391 (144)	432 (25)	531 (69)	430 (26)
Sr	485 (102)	386 (22)	656 (22)	525 (29)	476 (71)	351 (20)	632 (65)	312 (38)	425 (147)	355 (31)	676 (89)	340 (45)
Zr	68.4 (98)	94.6 (11)	63.5 (21)	23.3 (3)	103 (75)	133 (16)	124 (60)	113 (8)	122 (146)	186 (9)	140 (66)	157 (24)
Zn	74.7 (31)	74.4 (8)	n.d.	89.8	75.7 (17)	65.3 (3)	77.0 (2)	82.3 (22)	69.6 (28)	64.3 (6)	77.1 (15)	70.3 (23)
La	10.2 (29)	10.3 (4)	7.9 (2)	16.7 (3)	18.0 (26)	11.1 (6)	9.8 (5)	15.4 (3)	17.8 (54)	12.5 (18)	18.2 (15)	18.1 (7)
Ce	27.0 (31)	32.0 (5)	n.d.	33.7 (3)	38.6 (29)	24.3 (6)	26.3 (3)	30.4 (3)	34.6 (51)	32.5 (18)	56.6 (18)	34.2 (14)
Yb	1.6 (11)	2.5 (5)	2.7 (5)	n.d.	1.5 (10)	1.9 (6)	2.6 (11)	1.92 (2)	1.8 (17)	2.8 (18)	2.4 (28)	2.3 (7)
Y	20.4 (91)	27.0 (10)	30.9 (7)	16.7 (3)	25.2 (64)	24.6 (9)	30.1 (39)	21.0 (5)	24.2 (95)	19.6 (13)	29.7 (50)	24.8 (25)
Cu	121 (82)	61.1 (13)	106 (20)	91.4 (19)	94.5 (69)	55.3 (10)	92.3 (38)	52.4 (26)	45.3 (142)	37.7 (20)	44.1 (54)	30.2 (31)
Ni	146 (88)	77.8 (11)	89.3 (21)	31.1 (20)	47.1 (68)	53.7 (10)	39.5 (53)	41.0 (30)	29.0 (132)	29.3 (14)	26.6 (67)	18.1 (31)
Co	48.2 (60)	35.7 (13)	37.8 (20)	32.4 (20)	28.6 (35)	24.3 (10)	26.3 (35)	25.8 (28)	21.1 (44)	26.7 (23)	21.9 (59)	20.2 (31)
Cr	373 (88)	167 (6)	192 (18)	65.0 (19)	125 (67)	107 (7)	89.3 (32)	114 (31)	86.5 (143)	36.6 (21)	57.0 (57)	52.7 (31)
V	284 (79)	310 (3)	268 (19)	318 (17)	216 (69)	200 (6)	225 (57)	209 (26)	151 (148)	155 (13)	156 (65)	172 (26)
Nb	5.6 (26)	n.d.	n.d.	n.d.	5.1 (24)	3.4 (2)	7.1 (2)	9.0 (2)	4.9 (34)	4.1 (4)	4.4 (12)	10.0 (16)
Li	7.1 (3)	8.1 (1)	1.5 (2)	11.6 (16)	10.0 (3)	6.2 (2)	26.0 (4)	10.8 (23)	16.4 (5)	10.9 (10)	20.9 (16)	10.4 (14)
Pb	4.2 (49)	3.3 (10)	n.d.	n.d.	6.3 (40)	4.6 (12)	8.4 (3)	3.2 (2)	7.8 (113)	11.6 (14)	7.1 (15)	4.7 (17)
Hf	1.6 (10)	1.9 (4)	n.d.	n.d.	1.8 (8)	2.1 (5)	n.d.	2.5 (2)	2.2 (14)	2.9 (16)	n.d.	3.7 (11)

Phenocryst mineralogy modes (vol. %)

Sanidine	0	0	0	0	0	0	0	0	0	0	0	0
Plagioclase	15.9	11.7	16.7	43.6	20.8	18.7	13.9	29.8	21.6	11.3	18.7	32.0
Quartz	0	0	0	0	0	0	0	<0.01	<0.01	<0.01	0.12	<0.01
Olivine	6.8	3.8	5.7	6.1	1.2	0.93	1.1	3.9	0.39	0.81	0.58	0.12
Augite	7.6	2.3	4.1	5.8	6.5	2.3	3.5	4.6	4.4	1.5	2.8	3.1
Orthopyroxene	0.06	0.12	0.06	0	2.0	2.6	1.2	2.2	3.1	1.9	2.4	3.1
Hornblende	0.03	0	0	0	0.66	0.57	0.69	0.38	2.1	0.93	0.77	2.8
Biotite	0	0	0	0	0	0	0	0	0.09	0.04	<0.01	0
Fe–Ti oxides	0.37	0.26	0.47	1.7	0.94	0.49	0.90	1.3	1.2	0.41	0.90	1.3
Pigeonite	0	0	0	0	0.02	0	<0.01	0	0	0	0	0
TPC	30.8	18.2	27.0	57.2	32.1	25.6	21.3	42.2	32.9	16.9	26.3	42.4
$n =$	129	36	20	14	136	30	45	16	195	32	47	35

Occurrence (frequency %)

Sanidine	0	0	0	0	0	0.4	0	0	0	0	0	0
Plagioclase	82.0	86.1	91.2	95.8	97.7	96.1	98.1	98.1	98.0	97.5	99.5	98.8
Quartz	0	0.7	0	0	0	1.7	1.2	1.9	2.0	5.8	7.0	9.6
Olivine	88.3	88.9	97.0	97.9	66.1	72.9	89.8	75.5	29.3	35.8	37.5	24.1
Augite	80.8	75.7	54.4	85.4	93.7	91.0	79.6	98.1	88.6	92.0	92.5	83.1
Orthopyroxene	6.9	24.3	7.4	12.5	62.6	76.0	45.2	58.5	81.0	92.8	92.5	74.7
Hornblende	0.6	0.7	0	0	18.4	7.7	4.4	3.8	30.9	24.2	16.9	30.1
Biotite	0	0	0	0	0	0.4	0	0	0	4.5	0.9	6.0
Fe–Ti oxide	34.3	27.1	17.7	47.9	70.1	61.8	41.4	58.5	86.6	87.8	83.1	62.7
Pigeonite	0	3.5	0	6.2	0.5	1.3	0.6	0	1.2	0.6	0	0
$n =$	172	144	68	48	174	233	157	53	253	360	213	83

n.d. = insufficient data; TPC = total phenocryst content.
* Figures in parentheses indicate number of data used to calculate each trace element average.

APPENDIX 3 Averaged chemical compositions and phenocryst mineralogy of selected high K volcanic rocks of orogenic regions

	Basalts (<52% SiO_2)				Basaltic Andesites (52–54% SiO_2)				Andesites (56–63% SiO_2)			
	Western USA—Eastern Zone	Western USA—Western Zone	North-Western Pacific	South-Western Pacific	Western USA—Eastern Zone	Western USA—Western Zone	Mediter-ranean	South-Western Pacific	Western USA—Eastern Zone	Western USA—Western Zone	Mediter-ranean	South-Western Pacific
SiO_2	49.33	50.11	49.84	49.94	54.24	54.06	54.49	53.94	59.74	59.58	59.80	59.30
TiO_2	2.20	1.74	1.42	0.91	1.43	1.43	1.03	1.05	1.05	0.99	0.73	0.86
Al_2O_3	16.00	16.81	17.70	15.59	16.57	16.66	17.05	16.93	16.86	16.83	17.14	16.62
Fe_2O_3	4.30	4.52	3.85	3.09	4.25	3.63	4.67	3.11	3.81	3.22	3.55	2.72
FeO	6.85	5.05	6.37	7.21	4.67	4.70	3.65	5.16	2.78	3.36	2.57	3.73
MnO	0.16	0.15	0.18	0.18	0.15	0.13	0.14	0.15	0.10	0.12	0.12	0.13
MgO	6.92	6.78	6.40	7.51	4.81	5.42	4.47	5.40	2.86	3.25	3.14	3.73
CaO	8.47	9.10	9.07	10.73	7.32	7.64	8.57	8.29	5.49	5.86	6.57	6.22
Na_2O	3.62	3.43	3.21	2.84	3.77	3.65	3.36	3.37	3.91	3.61	3.41	3.74
K_2O	1.49	1.64	1.59	1.59	2.22	2.15	2.25	2.11	2.99	2.78	2.71	2.59
P_2O_5	0.66	0.66	0.38	0.40	0.58	0.54	0.32	0.48	0.39	0.40	0.26	0.38
n =	36	28	30	65	43	33	22	88	96	73	90	117
Trace elements (p.p.m.)												
Rb	29.8 (4)*	26.7 (3)	44 (1)	33.0 (55)	52.0 (12)	46.7 (6)	61.6 (9)	48.5 (51)	72.6 (15)	82.7 (6)	84.8 (40)	64.9 (71)
Ba	835 (17)	1280 (13)	476 (4)	451 (48)	1190 (14)	1194 (13)	560 (8)	578 (58)	1200 (27)	1640 (25)	664 (36)	760 (70)
Sr	811 (20)	986 (13)	691 (4)	729 (56)	1067 (23)	1015 (13)	636 (9)	742 (60)	961 (40)	1140 (25)	458 (40)	728 (80)
Zr	196 (16)	249 (12)	110 (1)	78.0 (50)	187 (10)	189 (13)	99 (2)	145 (58)	250 (27)	200 (24)	100 (1)	185 (87)
Zn	n.d.	96.3 (4)	73 (1)	69.0 (7)	113 (2)	113 (6)	87 (5)	84.1 (19)	95 (1)	104 (5)	72.8 (26)	73.6 (30)
La	64.5 (6)	91.8 (11)	19.8 (1)	15.2 (12)	77.4 (7)	54.5 (10)	14.7 (2)	31.9 (34)	87.1 (19)	87.7 (23)	21 (1)	38.6 (50)
Ce	86.5 (2)	103 (5)	55.2 (2)	30.5 (15)	93 (6)	107 (5)	30.9 (2)	48.3 (43)	167 (15)	157 (17)	68 (1)	59.4 (54)
Yb	3.5 (13)	3.5 (7)	2.4 (3)	1.35 (2)	3.6 (8)	3.5 (4)	1.7 (2)	1.8 (5)	2.8 (22)	2.2 (9)	n.d.	2.3 (8)
Y	38.0 (15)	33.2 (13)	24 (2)	20.8 (46)	35.7 (9)	30.2 (13)	20.5 (2)	23.0 (64)	30.1 (26)	28.8 (25)	23 (1)	25.0 (77)
Cu	67.0 (17)	55.2 (13)	47.5 (2)	122 (44)	88.8 (14)	34.2 (11)	62.7 (7)	97.3 (64)	38.7 (32)	36.4 (23)	43.3 (30)	53.9 (80)
Ni	104 (17)	140 (13)	57 (2)	86.5 (51)	61.5 (12)	83.2 (13)	42.1 (7)	59.2 (64)	67.5 (29)	42.9 (24)	24.1 (26)	49.2 (81)
Co	32.7 (17)	34.8 (12)	28.3 (2)	43.9 (35)	29.0 (12)	29.2 (9)	33.3 (7)	36.6 (10)	21.8 (30)	21.2 (21)	19.4 (26)	20.8 (13)
Cr	146 (16)	241 (13)	180 (1)	246 (46)	106 (10)	189 (13)	57.7 (6)	134 (60)	83.3 (26)	121 (24)	18.3 (15)	108 (81)
V	216 (17)	235 (13)	190 (1)	375 (47)	181 (14)	182 (13)	220 (6)	213 (61)	157 (32)	129 (25)	148 (22)	143 (82)
Nb	35.9 (12)	15.0 (4)	n.d.	5.3 (12)	17.4 (5)	17.6 (8)	5.8 (2)	6.8 (39)	16.2 (19)	18.6 (19)	n.d.	7.8 (36)
Li	8 (1)	20 (2)	n.d.	n.d.	9.7 (4)	22.5 (6)	15.2 (6)	n.d.	9.4 (5)	25.7 (3)	16.8 (34)	n.d.
Pb	11 (2)	15.5 (4)	4 (1)	8.7 (21)	12.8 (5)	16.0 (10)	17.3 (2)	11.0 (38)	23.7 (20)	26.7 (23)	n.d.	17.0 (47)
Hf	n.d.	n.d.	2.4 (1)	1.4 (2)	n.d.	n.d.	2.4 (2)	1.8 (2)	n.d.	n.d.	n.d.	4.0 (5)

Phenocryst mineralogy modes (vol. %)												
Sanidine	0	0	0	0	0	0	0	0	0.41	0	0	0
Plagioclase	3.8	8.4	2.9	7.4	8.7	6.9	26.4	15.3	16.8	15.0	32.9	14.9
Quartz	0	0	0	0	0.01	0.24	0	0	0.12	0.20	0	0
Olivine	5.6	4.5	4.6	4.7	2.2	3.7	0	2.5	0.42	0.24	0.14	0.61
Augite	2.2	1.8	3.3	9.1	2.3	2.5	9.2	7.1	2.8	1.9	5.3	3.5
Orthopyroxene	0	0.02	0	0.03	0.38	0.05	0.64	0.65	1.3	0.90	2.4	1.5
Hornblende	0	0	0	0.18	0.37	0.21	4.8	0.96	1.8	3.3	1.2	3.4
Biotite	0	0	0	0	0	0	0	0.03	0.63	0.51	0.33	0.60
Fe–Ti oxides	0.21	0	0.04	0.29	0.54	0.11	0.80	0.85	1.31	0.66	1.8	1.2
Pigeonite	0	0	0	0	0	0	0	0	0	0	0	0
TPC	11.8	14.7	10.8	21.7	14.5	13.7	41.8	27.4	25.6	22.7	44.1	25.7
$n =$	15	13	9	48	18	16	5	51	62	36	15	78
Occurrence (frequency %)												
Sanidine	0	0	0	0	4.8	0	3.8	0	12.6	0	2.4	0.9
Plagioclase	44.1	75.0	76.6	52.4	71.4	77.4	95.3	92.1	87.4	94.0	98.9	91.3
Quartz	2.9	0	0	2.4	2.4	16.2	0	0	11.6	10.4	10.2	0.9
Olivine	94.1	91.7	100	73.0	85.7	87.1	38.1	70.5	24.2	26.9	23.9	39.1
Augite	35.3	66.7	90.0	69.8	57.1	74.2	85.7	94.3	83.1	83.6	92.0	86.1
Orthopyroxene	3.0	8.3	16.7	3.2	14.3	16.1	42.9	30.7	54.7	61.2	57.9	63.5
Hornblende	0	4.2	3.3	1.6	9.5	16.1	23.8	23.8	39.0	53.7	60.2	55.7
Biotite	0	0	0	0	0	6.5	0	3.4	28.5	31.3	43.2	22.6
Fe–Ti oxides	5.9	0	43.3	30.2	26.2	16.1	47.6	64.8	66.3	68.7	83.0	72.2
Pigeonite	0	4.2	3.3	0	2.4	0	0	0	0	1.5	0	0
$n =$	34	24	30	63	42	31	21	88	95	67	88	115

n.d. = insufficient data; TPC = total phenocryst content.
* Figures in parentheses indicate number of data used to calculate each trace element average.

APPENDIX 4 Averaged chemical composition and phenocryst mineralogy of selected shoshonitic volcanic rocks of orogenic regions

	Basalts (<52% SiO_2)					Basaltic Andesites (52–56% SiO_2)				Andesites (56–63% SiO_2)	
Regions:	Western South America	Western USA— Eastern Zone	Aleutian Islands	Mediterra- nean	South- Western Pacific	Western South America	Western USA— Eastern Zone	Mediterra- nean	South- Western Pacific	Western USA— Eastern Zone	Mediterra- nean
SiO_2	50.98	48.63	48.14	49.21	50.18	54.37	54.18	54.35	53.42	59.64	60.24
TiO_2	1.44	2.05	1.60	1.42	0.91	1.60	1.12	0.87	1.00	1.01	0.79
Al_2O_3	16.17	15.22	17.79	17.06	15.67	16.24	16.66	16.85	17.03	16.63	16.62
Fe_2O_3	4.50	5.23	5.55	3.60	3.75	3.66	4.99	3.50	3.67	3.82	2.78
FeO	4.07	5.99	5.14	5.34	5.58	3.92	3.41	3.73	4.17	2.49	2.34
MnO	0.15	0.16	0.18	0.14	0.17	0.13	0.14	0.14	0.15	0.12	0.11
MgO	7.03	7.61	5.12	5.99	7.69	5.38	4.67	5.50	5.17	2.65	2.92
CaO	8.50	8.67	11.21	10.24	9.87	6.71	6.81	7.72	7.86	4.87	5.60
Na_2O	3.47	3.52	3.01	3.93	2.82	3.78	3.57	2.91	3.42	4.06	3.07
K_2O	2.97	2.17	1.83	2.42	2.80	3.45	3.75	4.02	3.47	4.25	5.15
P_2O_5	0.73	0.76	0.42	0.65	0.55	0.74	0.68	0.42	0.62	0.46	0.37
$n =$	7	27	7	23	60	13	51	34	28	41	51
Trace elements (p.p.m.)											
Rb	84.0 (5)*	47.4 (5)	58.2 (5)	85.5 (6)	68.8 (47)	89.4 (8)	130 (2)	140 (12)	72.8 (20)	139 (4)	179 (3)
Ba	556 (2)	1149 (3)	741 (6)	827 (6)	599 (49)	1853 (5)	1593 (4)	573 (12)	753 (22)	2017 (6)	768 (3)
Sr	788 (5)	1057 (8)	796 (6)	1028 (6)	995 (49)	1071 (8)	1043 (4)	773 (12)	1030 (21)	902 (10)	752 (3)
Zr	222 (5)	225 (2)	121 (6)	81.0 (4)	82.5 (42)	328 (4)	226 (4)	159 (5)	137 (14)	394 (6)	277 (1)
Zn	106 (3)	n.d.	73.7 (3)	81.2 (6)	73.0 (13)	101 (4)	92.5 (2)	80.4 (12)	77.3 (3)	n.d.	70.7 (3)
La	26.1 (2)	115 (2)	9.7 (4)	33.0 (3)	18.8 (21)	88.8 (2)	70 (3)	47.3 (3)	38.0 (6)	150 (5)	58 (1)
Ce	61.4 (2)	200 (1)	30.8 (4)	62.7 (3)	35.4 (23)	187 (2)	80 (2)	106 (3)	63.7 (7)	200 (4)	143 (1)
Yb	2.26 (2)	3.0 (2)	1.5 (1)	n.d.	1.6 (5)	2.3 (2)	3 (2)	n.d.	2.1 (1)	2.4 (5)	n.d.
Y	31.0 (1)	35.5 (2)	27.6 (5)	18.3 (4)	18.7 (46)	44.0 (2)	26.3 (4)	26.2 (5)	22.0 (16)	26.0 (5)	25 (1)
Cu	30.0 (3)	44.5 (4)	95.0 (4)	108 (5)	130 (45)	31.5 (4)	152 (3)	110 (10)	139 (16)	27.2 (6)	81 (3)
Ni	92 (5)	80.3 (4)	16.8 (5)	30.8 (5)	76.2 (49)	75.3 (4)	63.3 (4)	17.0 (10)	35.3 (21)	59.2 (6)	14.3 (3)
Co	29.4 (5)	56.3 (4)	28.4 (5)	21.8 (5)	32.8 (22)	29.3 (4)	28.0 (2)	24.9 (10)	21.7 (11)	15.4 (7)	16.7 (3)
Cr	63.0 (2)	82.5 (2)	30.8 (5)	86.0 (3)	189 (45)	172 (1)	196 (4)	26.7 (10)	84.9 (15)	65.9 (15)	20 (2)
V	208 (2)	199 (4)	296 (5)	200 (1)	261 (49)	195 (1)	210 (3)	197 (10)	258 (21)	80.2 (6)	144 (2)
Nb	n.d.	32.5 (2)	3.3 (3)	n.d.	5.7 (23)	15 (1)	15 (1)	n.d.	11.8 (8)	16.3 (4)	n.d.
Li	n.d.	8.5 (2)	20 (1)	n.d.	n.d.	11 (4)	n.d.	22.0 (7)	n.d.	15 (1)	26 (1)
Pb	n.d.	n.d.	6.2 (3)	n.d.	12.1 (36)	n.d.	n.d.	n.d.	20.0 (11)	19.4 (5)	n.d.
Hf	n.d.	n.d.	n.d.	n.d.	1.0 (2)	6.3 (1)	n.d.	n.d.	4.8 (1)	n.d.	n.d.

Phenocryst mineralogy modes (vol. %)

	1	2	3	4	5	6	7	8	9	10	11
Hauyne	n.d.	0	0	n.d.	0	n.d.	0	0	0.12	0	0
Leucite	n.d.	0	0	n.d.	0	n.d.	0	0	0	0	0
Sanidine	n.d.	0	0	n.d.	0	n.d.	0	1.7	0	1.1	5.1
Plagioclase	n.d.	0.45	18.9	n.d.	0.16	n.d.	5.1	9.7	15.1	14.5	12.8
Quartz	n.d.	0	0	n.d.	0	n.d.	0.04	0	0	0.04	0
Olivine	n.d.	3.9	0	n.d.	5.9	n.d.	1.9	2.2	3.0	0.31	0.46
Augite	n.d.	0	10.4	n.d.	17.3	n.d.	4.6	9.3	6.0	3.8	5.9
Orthopyroxene	n.d.	0	0	n.d.	0.03	n.d.	0.46	0.96	0.38	1.2	1.7
Hornblende	n.d.	0	9.9	n.d.	0.09	n.d.	0.58	0	0.92	1.2	0
Biotite	n.d.	0	0	n.d.	0.20	n.d.	0.01	1.1	0.32	1.7	5.7
Fe–Ti oxides	n.d.	0	1.8	n.d.	1.5	n.d.	0.90	0.44	0.61	0.90	0
TPC	n.d.	4.4	41.0	n.d.	37.8	n.d.	13.6	25.4	26.5	24.8	31.7
$n =$	—	2	6	—	34	—	11	9	16	21	19

Occurrence (frequency %)

	1	2	3	4	5	6	7	8	9	10	11
Hauyne	20	0	0	15.4	0	0	0	0	3.6	0	0
Leucite	0	0	0	0	5.3	0	0	7.2	0	0	0
Sanidine	0	4	0	0	0	0	4.0	25.0	0	17.1	75.5
Plagioclase	0	36	100	23.1	42.1	100	66.0	85.7	75.0	82.9	100
Quartz	0	0	0	0	0	0	2.0	0	0	7.3	4.1
Olivine	80	92	0	92.3	94.7	0	74.0	71.4	75.0	17.1	28.6
Augite	100	64	100	100	78.9	100	96.0	92.9	89.3	85.4	95.9
Orthopyroxene	20	0	0	38.5	5.3	0	8.0	25.0	21.4	29.3	51.0
Hornblende	8	0	85.7	7.7	42.1	85.7	16.0	14.3	28.6	41.5	12.2
Biotite	20	0	0	38.5	0	0	14.0	28.6	25.0	53.7	95.9
Fe–Ti oxides	0	16	85.7	15.4	31.6	85.7	46.0	46.4	57.1	53.7	46.9
$n =$	5	25	7	13	19	7	50	28	28	41	49

* Figures in parentheses indicate number of data used to calculate each trace element average.
n.d. = insufficient data; TPC = total phenocryst content.

III Regional distribution and character of active andesite volcanism

Orogenic volcanic associations occur in a variety of island arc and continental margin settings, in which the rocks have distinctive geological, petrological, and chemical characteristics as described in Section II. In general terms, intra-oceanic island arcs are dominated by basalts and basaltic andesites, while continental margins and continental collision zones are less basaltic and more andesitic and have rhyolitic members, which form the calc-alkaline rock association. Some continental margins and collision zone provinces have shoshonitic rocks and may be associated with both sodic- and potassic-alkaline volcanic rocks.

In this section, different contributors review the characteristics of active calc-alkaline volcanism in provinces representative of the range of settings outlined above. The provinces described by individual authors are shown in Fig. 1 on p. 2. The aim of this section is therefore to provide a comprehensive data source for young calc-alkaline rocks from presently active provinces. For each province, a contributor reviews the distribution and relationships of active volcanism to regional and plate tectonics, and to the thickness, structure, and age of the underlying crust. In some cases this includes discussion of earlier volcanic and intrusive episodes to place the active volcanism in spatial and temporal perspective. The volcanic forms and products are described and, from consideration of chemical characteristics, most contributors conclude with a discussion of the petrogenesis of the volcanic rocks in each province. This section therefore provides basic data for the evaluation of the distribution and interrelationships of andesites from currently active provinces, and for older andesite terraines in terms of modern counterparts.

Andesites
Edited by R. S. Thorpe
© 1982 John Wiley & Sons

The Aleutians

B. D. Marsh

Department of Earth and Planetary Sciences,
The Johns Hopkins University, Baltimore, Maryland 21218, USA

ABSTRACT

The Aleutian Island arc stretches some 2500 km westward from near Anchorage, Alaska; lying upon continental crust in the east and oceanic crust in the west. Trench formation and subduction, which eventually gave rise to the Aleutian Islands, began about 70 Ma ago. The land-mass of the arc today represents the superpositioning of many periods of volcanism and plutonism. This has been periodic (every 2.5 Ma) for at least the last 10 Ma. The present volcanism occurs along a narrow, segmented volcanic front containing c. 80 volcanoes which form fairly regularly spaced volcanic centres. Two small volcanic centres, Amak and Bogoslof, occur 50 km north of the front itself. The relative plate motion is nearly normal in the east and strike-slip at the western end of the arc. Westward of Buldir volcano, there is no subduction. The flux of lava increases from west to east in direct proportion to the rate of subduction normal to the arc. At least 5000 km^3 of material has been erupted over the last few million years. The principal lava is basaltic and contains c. 50 per cent silica, but it also has the andesitic characteristics of low titania and magnesia, and high alumina and potash; it is called andesitic basalt. These lavas contain between 10 and 60 per cent phenocrysts; during crystallization plagioclase appears first, olivine and magnetite next, and lastly clinopyroxene. Aphyric lavas are rare. Hydrous phases generally only appear in domes and plugs. There are apparently no geographical correlations relating the positions of the volcanic centres to the near surface regional structure, or between the chemical composition of the lava and the nature of the underlying crust. The bulk composition of the andesitic basalt is matched exceedingly well by the composition of melt from a quartz eclogite at 30 kb, and it shows a poor agreement with that of melt from undepleted peridotite.

Introduction

The Aleutian island arc is a sharp, segmented volcanic front containing c. 80 major volcanoes over a distance of c. 2500 km (Fig. 1; Coats, 1950). To the east, volcanism first appears near Anchorage, Alaska (150°W), continues along the Alaska Peninsula, into the Aleutians proper, and ends at tiny Buldir Island (176°E) in the west.

Crust

The arc stretches from a purely oceanic environment to a purely continental one. Progressing from the central Aleutians to the east the crust becomes increasingly thick and continental; the oceanic–continental transition itself probably occurs beneath Umnak, Unalaska, and Unimak Islands, at the intersection of the arc with the Bering Shelf. The crustal structure in the central Aleutians is known from seismic and gravity studies (Shor, 1964; Murdock, 1969; Stone, 1968; Grow, 1973; Helmberger, 1977). Just behind the arc there is a major velocity transition at a depth of 8 km from 3 km s^{-1} to 6.5 km s^{-1}, the Moho lies at 16 km, and the uppermost mantle has a velocity of 8.2 km s^{-1}. Immediately beneath the arc the crust is thicker

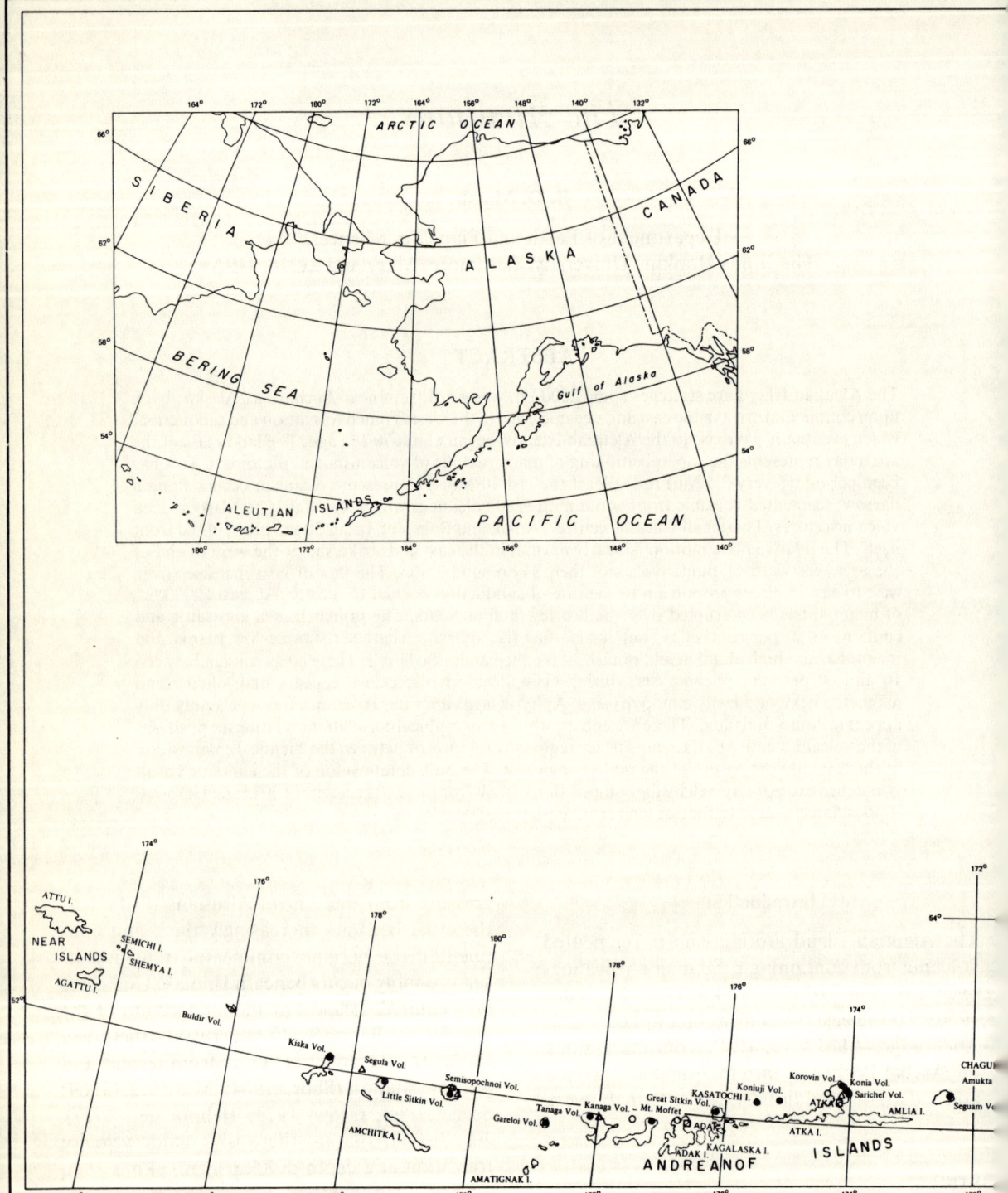

Fig. 1 Volcanoes of the Aleutian Island Arc. Revised from Coats (1950)

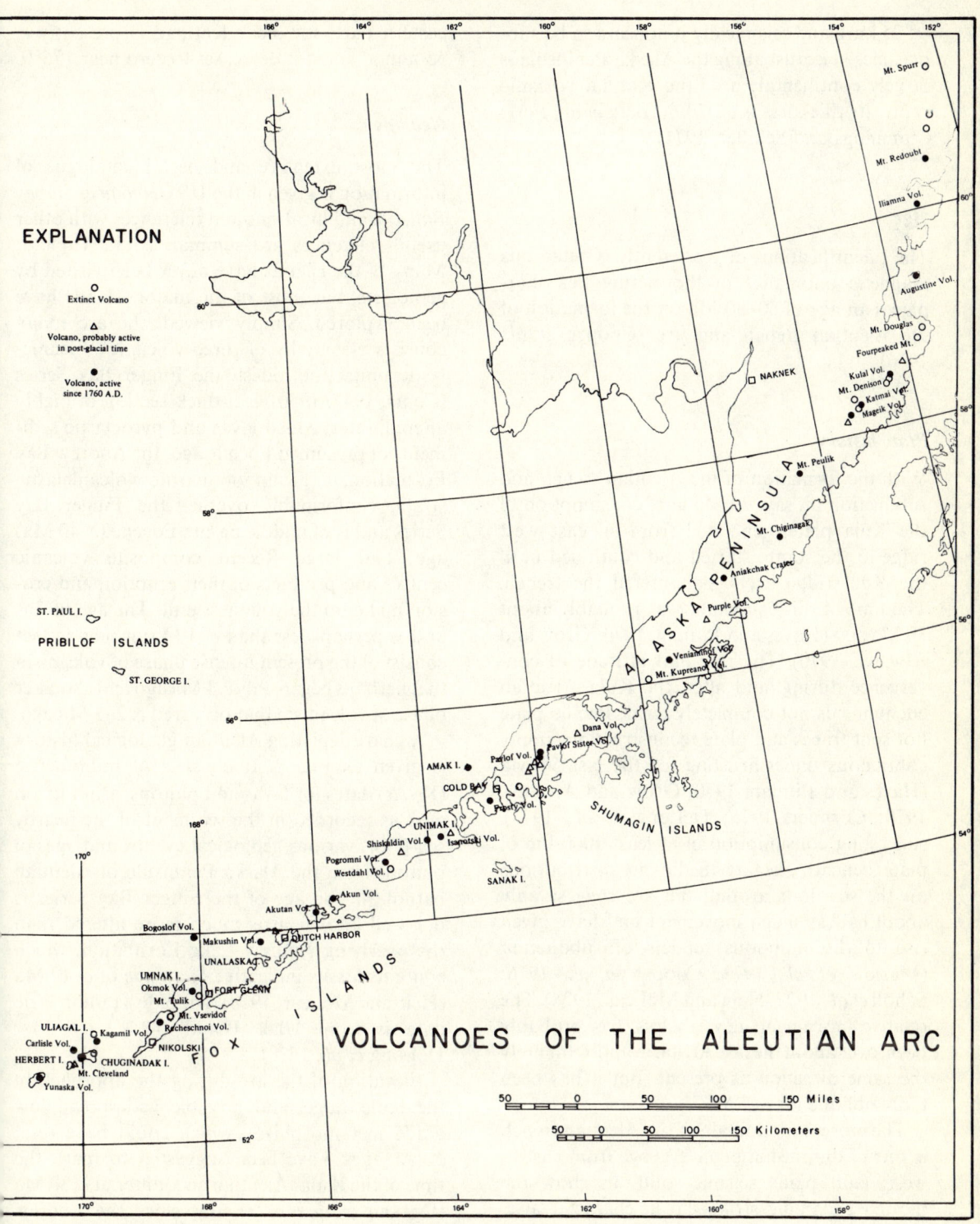

(c. 25 km), and seismically it appears to be suboceanic. The crust along the Alaska Peninsula is largely continental; near the Katmai volcanic centre its thickness is c. 35 km (Berg et al., 1967; Cummings and Schiller, 1971).

Age

The identification of apparently Cretaceous magnetic anomalies in the Bering Sea floor places an age of 60–80 Ma on the formation of the Aleutian trench and arc (Cooper et al., 1976).

Plate history

With the formation of the Aleutian trench and subduction at, say, 70 Ma ago, consumption of the Kula plate, generated from an east–west ridge to the south, started and continued until the Kula ridge itself encountered the trench. The time of this encounter was probably about 30 Ma ago (Hayes and Pitman, 1970; Grow and Atwater, 1970). The rate and attitude of convergence during, and after, the Kula–Aleutian encounter is not completely clear. Pacific plate hot spot traces and plate reconstructions imply continuous underthrusting for the last 50 Ma (Hayes and Pitman, 1970; Grow and Atwater, 1970; Cormier, 1975; DeLong et al., 1978), suggesting consumption of at least 1000 km of plate beneath the arc. Sediment distributions on the sea-floor around the arc suggest only about half as much movement and have given rise to discontinuous models of subduction (Marlow et al., 1973; Cooper et al., 1976; Scholl et al., 1977; Hein and McLean, 1978). The relative movement, nevertheless, has probably been, over about the last 30 Ma, in approximately the same direction as present. But it has been more oblique for c. 7 Ma.

The present motion along the Aleutian trench is one of diminishing convergence from east to west; fault plane seismic solutions show the motion becoming strike-slip in character at c. 176°E (Stauder and Bollinger, 1964, 1966; Cormier, 1975). The convergence rate in the eastern part, say, near Katmai, is presently c. 55 mm a^{-1} and it decreases to zero near 176°E.

Geology

The most extensive and useful catalogue of information is given in the *US Geological Survey Bulletin*, and locations and references with other useful references are summarized in Table 1. Many of the islands have never been visited by geologists, but most of the major islands have been explored. Simply viewed, the arc today consists essentially of three volcanic–volcaniclastic units: the oldest, the Finger Bay Series (Coats, 1947, 1956), is a thick section of highly altered, interbedded lavas and pyroclastic sediments of presumed Eocene age; the Andrew Bay Formation, a group of marine volcaniclastic strata, conformably overlies the Finger Bay Series and is of middle or late Eocene (c. 40 Ma) age; and large Recent composite volcanic centres and products of their eruption and erosion make up the youngest unit. The age of this unit is perhaps less than c. 4 Ma; it may in fact consist of the present intense phase of volcanism that perhaps began only c. 1 Ma ago and a weaker phase of volcanism that occurred c. 2–3 Ma ago.

A chart depicting Aleutian geological history is given as Fig. 2. It consists of radiometric (K–Ar) dates for lavas and plutons, ash horizon ages as recorded in the sediment of the nearby sea-floor, various geological events, and ages of plutons from the Alaska Peninsula or Aleutian batholith. The age of the Finger Bay Series is not well known; it is much more altered than the overlying Andrew Lake Formation, which contains fossils indicative of an age of c. 40 Ma (Hein and McLean, 1978). The oldest radiometric date is from Ulak Island (42.3 ± 4.6 Ma; DeLong et al., 1978).

Elevation of the arc during the approach of the Kula ridge, and possible widespread low grade metamorphism, which could have reset K–Ar ages, have been suggested to mark the time of the Kula–Aleutian encounter at c. 30 Ma (DeLong et al., 1978). But since the Andrew Lake Formation is hardly altered relative to the Finger Bay Series (Hein and McLean, 1978),

TABLE 1 General references

US Geological Survey Bulletin series (areas from east to west)

Volcanic activity of arc	*Bull.* 974-B
Katmai area	*Bull.* 1058-G
Pavlof volcano	*Bull.* 1028-A
Frosty Peak-Cold Bay	*Bull.* 1028-T
Unalaska Island	*Bull.* 1028-S
Umnak Island and Bogoslof Island	*Bull.* 1028-L
Great Sitkin Island	*Bull.* 1028-B
Western Islands (general)	*Bull.* 1028-E
Adak Island	
Northern (volcanoes)	*Bull.* 1028-C
Southern and Kagalaska Island	*Bull.* 1028-M
Kanaga Island (northern)	*Bull.* 1028-D
Delarof and westernmost	
Andreanof Islands	*Bull.* 1028-I
Kanaga (southern)	
Tanaga (southern)	
Amatignak	
Ulak	
Kavalga	
Unalga	
Ilak	
Skagul	
Ogliuga	
Gareloi Island	*Bull.* 1028-J
Semisopochnoi Island	*Bull.* 1028-O
Amchitka	*Bull.* 1028-P
Rat Island	*Bull.* 1028-Q
Little Sitkin Island	*Bull.* 1028-H
Segula, Davidof, Khvostof Island	*Bull.* 1028-K
Kiska Island	*Bull.* 1028-R
Buldir Island	*Bull.* 989-A
Near Islands	*Bull.* 1028-U
Attu	
Agattu	
Shemya	

Others
    Amchitka Island (geology, hydrology, climate, history, etc.)
        M. L. Merritt and R. G. Fuller (1977). *The Environment of Amchitka Island, Alaska*, Technical Information Center (ERDA), US Department of Commerce, Springfield, Virginia 22161.
    Physiographic divisions of Alaska (maps of volcanoes)
        C. Wahrhaftig (1965). *US geol. Surv. prof. Pap.* 682.
    Research in Alaska (abstracts of yearly research)
        Current Research Profile for Alaska, Information Services, Arctic Environmental Information and Data Center, University of Alaska, Anchorage.

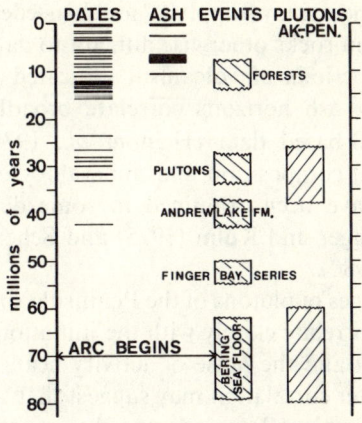

Fig. 2 Aleutian geological history; columns from left to right: radiometric ages of lavas and plutons; major ash horizons recorded on surrounding sea-floor (Hein *et al.*, 1978); geological events; radiometric ages of plutons of the Alaska–Aleutian batholith (Reed and Lanphere, 1973)

either no widespread metamorphism accompanied this interaction, or the Kula ridge arrived before *c*. 40 Ma ago.

From the distribution of dates, there was strong and almost continuous volcanism throughout the arc from *c*. 16 Ma until *c*. 7 Ma ago. After this time volcanism became restricted only to those islands below which subduction is active. Almost all of the present volcanic centres, however, rest on islands which existed during the Tertiary. The present distribution of islands, then, represents the superposition of a Tertiary volcanic arc and a much younger arc (less than *c*. 7 Ma). Active volcanism generally occurs on the north side of each Tertiary island.

A portion of an apparently petrified forest on Atka (*c*. 173°W) (R. R. Coats, personal communication; B. D. Marsh, unpublished maps) is cut perhaps by an andesitic plug dated at 6.6 Ma (B. D. Marsh, R. E. Drake, and G. H. Curtis, unpublished K–Ar dates). Similar logs are found in the Chitka Point Formation on Amchitka Island, which is dated at 12–14 Ma (Gard, 1977). Since trees, old or otherwise, are generally unknown to the Aleutians, these occurrences may supply a much needed time marker in rocks otherwise difficult to date.

The periods of volcanism indicated by the sea-floor ash horizons correlate broadly with the land-based data (Hein et al., 1978) The chemical composition and mineralogy of these ashes have been examined in some detail by Scheidegger and Kulm (1975) and Scheidegger et al. (1980).

The ages of plutons of the Peninsula shown in Fig. 2 correlate closely with the initiation of the arc itself and the pulse of activity near 30 Ma. This latter correlation may suggest that some of these Aleutian dates are not due to resetting during metamorphism.

Distribution of Volcanism and Tectonic Relations

As noted already, plate subduction ceases west of c. 176°E, where the relative plate motion becomes strike-slip in character. Where subduction ceases, so too does volcanism. The westernmost volcanic centre is tiny Buldir Island. This correlation, perhaps better than elsewhere, indicates the intimate connection between subduction and arc volcanism.

The attitude of the subducting plate is quite well known (Jacob, 1972; Abe, 1972; Engdahl, 1973, 1977; Van Wormer et al., 1974). As noted earlier by Coats (1962, p. 102), the volcanic front generally lies almost exactly 110 km above the inclined seismic, or Benioff, zone. The dip of the deeper (i.e. greater than c. 100 km) part of the plate varies somewhat but it is generally c. 45°. The distance between the trench and the volcanic front increases systematically from west to east (Jacob et al., 1977), indicating the decrease in dip of the plate nearest the trench with approach to the continent, perhaps in response to its heavy loading with sediment. The subducting plate is also segmented or fractured into staves (Jacob et al., 1977; Spence, 1977; Holden and Kienle, 1977).

The volcanic front is much sharper today than appears from the distribution of land. For, as mentioned already, the land represents the superpositioning of the Tertiary and present arc, which do not everywhere coincide. Where they do coincide, the volcanism lies on the northern edge of the older islands, and over the last c. 7 Ma it has migrated c. 5 km northward in the central Aleutians and c. 8 km on the Peninsula (Cold Bay).

The front itself consists of a series of large volcanic centres (i.e. clusters or nodes of volcanoes) regularly spaced at 60–70 km intervals along the entire arc (Marsh and Carmichael, 1974). The front of active, or potentially active volcanoes is clearly segmented throughout the arc (Fig. 3; Marsh, 1979, 1979a) and discontinuities in the front correlate closely with breaks or fractures in the subducting Pacific plate (Van Wormer et al., 1974; Holden and Kienle, 1977; Spence, 1977). This correlation seems to point to a second clear connection between the presence of the downgoing plate and the position and nature of the volcanic front. There are, however, two young volcanic centres that each occur almost exactly 50 km north of the front: Amak Island (163°W) and Bogoslof Island (168°W). Whereas the present volcanism of the front may be several million years old, Amak is certainly less than about 10 ka old (Marsh and Leitz, 1979) and Bogoslof arose from the sea in 1796. The implications of the spacing of volcanic centres, the segmentation of the front, and the development of the weak secondary front in Amak and Bogoslof are considered in Marsh (1979, 1980). Briefly, these features are consistent with magma ascent governed by a gravitational instability from a locally (i.e. 100–500 km) continuous, ribbon-like region inclined at the attitude of the local Benioff zone.

The total volume of volcanic material erupted in the most recent (i.e. less than c. 3 Ma) period of volcanism has been found by the author to be c. 4700 km^3. A chart showing this volume as a function of longitude along the arc and as a function of spacing between volcanic centres is given in Fig. 4. The diminution of volume with reduction in convergence rate is clear. The secondary centres, Amak and Bogoslof, occur behind the most voluminous parts of the 'oceanic' front: the segments respectively from Unimak to Pavlof volcano (for Amak) and

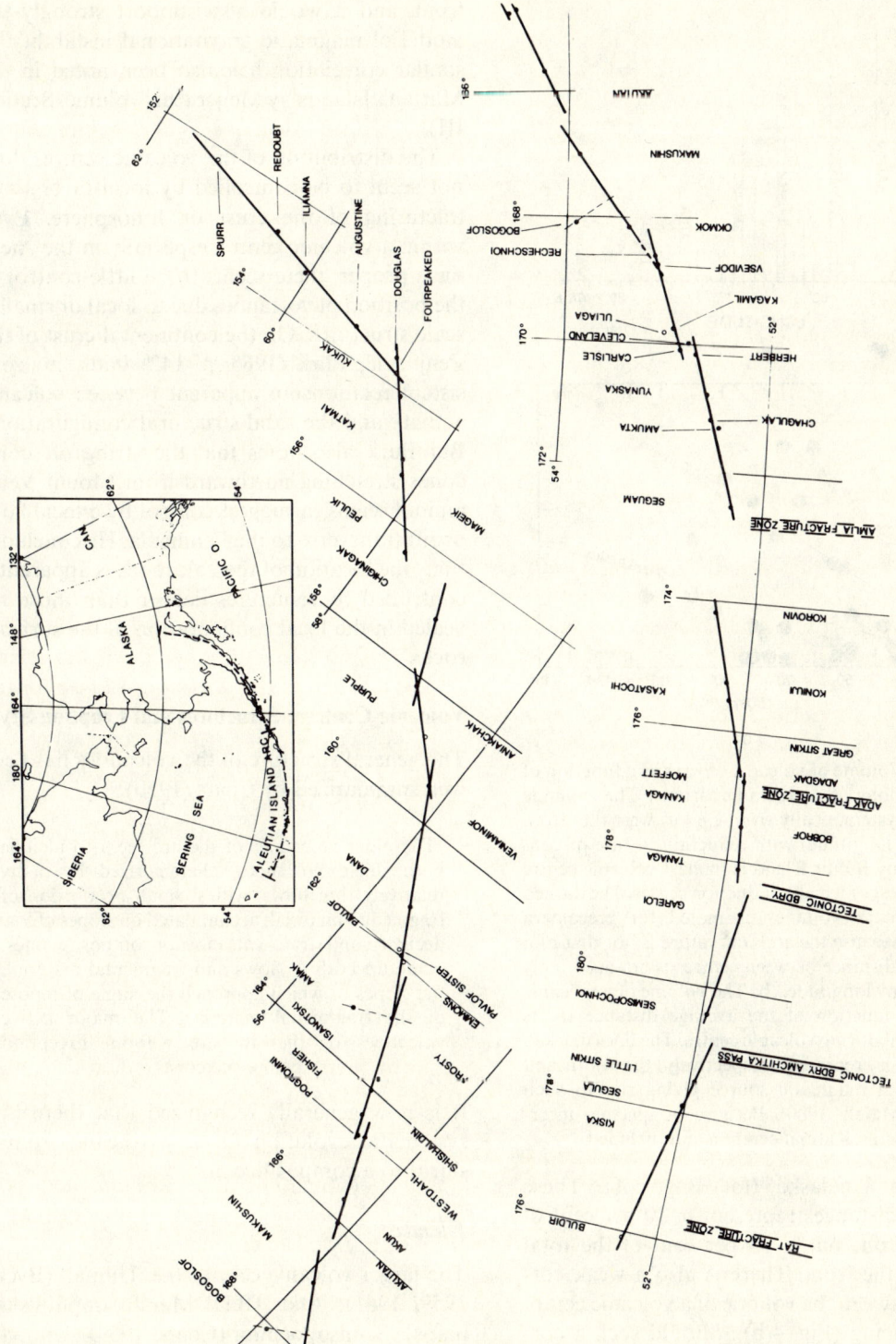

Fig. 3 Segmentation of the volcanic front. Straight lines drawn through as many active or potentially active volcano summits as possible give rise to this segmented pattern. Over 90 per cent of these volcanoes fall on these lines. Major fractures in the Pacific plate correlate with breaks in the volcanic front. Two volcanoes lie 50 km north of the front: Amak (c. 163°W) and Bogoslof (c. 168°W); see also Marsh (1979, 1980)

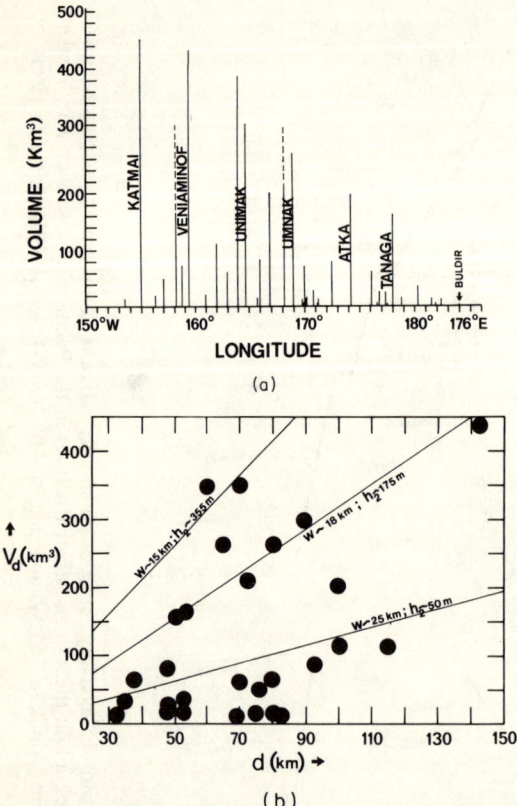

Fig. 4 (a) Volume of volcanic centres as a function of longitude along the volcanic front. The volume diminishes systematically from east to west (i.e. from west to east longitude) with reduction in rate of convergence. Tiny Buldir Island is the last volcanic centre (c. 176°E), west of which subduction ceases. The dashed extensions mark volumes estimated for precaldera formation. Because the arc's curvature is not that of a parallel, the distance between centres is not accurately represented by longitude. (b) The volume of a volcanic centre as a function of the average distance to its nearest neighbouring volcanic centre. The lines marked with W and h_2 represent, respectively, the width and thickness of a magmatic source giving rise to such variations (Marsh, 1980). Because of spacing uncertainty the Katmai centre is not included

Umnak to Unalaska (for Bogoslof). These combined distances represent c. 20 per cent of the main front but c. 35 per cent of the total volume of the front. There is also a weak correlation between the volume of a volcanic centre and its spacing (Fig. 4(b)). Should such a correlation exist, it would imply a source region continuous for great distances beneath the front, and it would also support strongly the model of magmatic gravitational instability. A similar correlation has also been noted in the Mariana Islands by Meijer (this volume, Section III).

The distribution of the volcanic centres does not seem to be controlled by local or regional fracturing of the crust or lithosphere. Even within a volcanic centre, especially in the Aleutians proper, there seems to be little control of the position of volcanoes due to local or smaller scale structures. On the continental crust of the Peninsula, Burk (1965, p. 142) finds 'no consistent relationship apparent between volcanic activity and the local structural configuration'. But Burk also notes that the string of scoria cones stretching northward from Mount Veniaminof seems to suggest control by a local fault or rift transverse to the Peninsula. He concludes that 'the location of the volcanoes is apparently controlled by structures deeper than those revealed in the local configuration of the surficial rocks'.

Volcanic Centres: Structures and Eruption Style

The general structure of the volcanoes has been well summarized by Coats (1950):

> 'The older volcanoes of the arc seem to include both shield volcanoes, characterized by many relatively thin flows, with a small proportion of fragmental material, accumulated on slopes of low declivity, and strato-volcanoes or composite cones, made up both of flows and fragmental material, the slopes of which approach the angle of repose of the fragmental material. The major active volcanoes of the arc are without exception composite cones.'

It is now generally recognized that there is a systematic evolution for any volcano from a shield to a composite cone.

Islands

The larger volcanic centres (i.e. Umnak (Byers, 1959, 1961); Atka (B.D. Marsh, unpublished maps); Semisopochnoi (Coats, 1959); Unimak, Seguam, and Tanaga are yet to be mapped) show a similar pattern of development. A large central composite cone develops, generally with some parasitic vents, and is ringed by smaller satellite or subsidiary cones. The central cone is characterized by summit eruptions, the composition of which finally become substantially more silicic than the bulk of the lavas (see later). The satellite cones are contemporaneously active with one another and also with the central cone, but generally erupt more basic lava throughout their existence. Once the system is well established, the large central cone commonly undergoes caldera collapse. This marks a sharp decline in the activity of whole system, although one or more of the satellite vents may remain active. During caldera formation a dacitic ash flow (Semisopochnoi) or dome (Atka) or a more rhyolitic ash flow (Umnak) may be extruded from near the caldera rim itself.

After caldera formation, new volcanoes may form on the caldera floor (north-eastern Umnak) or on the caldera rim (Atka). Caldera formation, nevertheless, generally signals the decline and extinction of the whole central cone–satellite cone system. Later volcanism generally appears outside, but adjacent to, this system and may be only a single moderately sized cone (c. 2000 m in height). Within any area of volcanism, individual conical volcanoes may be built and become extinct in this fashion but always remain within a relatively small area (15–20 km in diameter). The smaller volcanic centres seem to proceed only partially toward the central cone–satellite cone development, and some (e.g. Great Sitkin) seem to maintain a single major vent.

On the central cone, flank domes commonly develop long after a volcano is extinct, and these are usually about halfway up the volcano; the lava they erupt is generally of relatively basic composition (c. 52–55 per cent SiO_2); summit domes also commonly occur and these are generally dacitic.

Peninsula

Here the overall development is more in the nature of the smaller centres in the islands; generally a large central vent is maintained with only minor extra activity. In the Katmai region, however, several moderate-sized vents occur, but large single centres are more common. Evidence of ash flows is much more common, although this evidence is more difficult to preserve on islands, and caldera formation is common.

Petrography and Petrology

Rock type

A histogram of the silica content of Aleutian lavas and an estimate of volume plotted against silica content is given in Fig. 5. It is clear that the dominant rock type is one containing c. 50 per cent silica. Yet the wellknown chemical characteristics of andesites, namely, low TiO_2 (less than or equal to c. 1 per cent), low MgO (3–6 per cent), high Al_2O_3 (17–21 per cent), and high K_2O (c. 1 per cent), persist throughout the rock suite. The persistence of these characteristics, in even the most basic lavas, sets these rocks distinctly apart from tholeiitic basalts and alkaline basalts. These rocks seem in all regards to have characteristics of primitive magma in

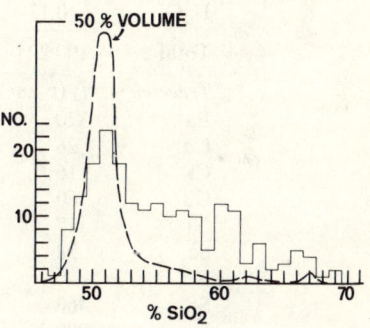

Fig. 5 Histogram of chemical analyses of Aleutian lavas, blocky line; and an estimate of the actual volume of lava as a function of silica content. The latter is based on the author's unpublished work on the Atka volcanic centre. The dominant rock is one with c. 50 per cent silica. Much silicic ash, however, may have been lost to erosion and explosive eruption

their own right and here this magma is called andesitic basalt. With increasing silica the names basaltic andesite (c. 53 per cent), andesite (c. 57 per cent), and dacite (c. 63 per cent SiO_2) are employed.

The trace element content reflects the major element characteristics: base metals are always low (e.g. Ni is less than c. 30 p.p.m.) and the REE content is fairly low and slightly enriched in the lighter REE in the basic lavas, becoming increasingly fractionated and enriched in the lighter REE with increasing silica content (Kay, 1977; B. D. Marsh, unpublished data). Representative whole-rock chemical analyses and trace element data are given in Table 2.

A plot of potash against silica (Fig. 6) reveals two distinct trends of potash enrichment: a sharply non-linear trend leading to a very high (c. 4 per cent) potash content, and a linear trend leading to much lower (c. 2 per cent) potash in dacites. These trends do not represent changes in depth to the underlying Benioff zone and are generally not found at a single volcanic centre (except weakly on Atka), but only appear on an intercentre or island comparison. The low K trend is strongest in the western islands.

TABLE 2 Representative Aleutian chemical data (Atka volcanic centre)

	Andesitic Basalt	Basaltic Andesite	Andesite		Dacite
SiO_2	49.80	53.29	58.24	62.49	66.88
TiO_2	0.96	0.95	0.69	0.62	0.46
Al_2O_3	20.12	18.31	17.33	16.37	15.57
Fe_2O_3	4.36	3.83	3.91	1.99	1.09
FeO	5.51	4.71	3.25	3.95	2.74
MnO	0.18	0.20	0.16	0.15	0.11
MgO	4.72	3.92	3.24	1.69	0.60
CaO	9.66	8.82	7.10	4.23	2.16
Na_2O	3.71	3.29	3.74	4.53	5.55
K_2O	0.78	1.11	2.00	3.00	4.09
P_2O_5	0.23	0.26	0.19	0.19	0.12
H_2O^+	0.06	0.73	0.34	0.14	0.74
H_2O^-	0.13	0.53	0.10	0.07	0.20
Total	100.22	99.95	100.29	99.42	100.31
Trace elements (p.p.m.)					
Ba	320	530	790	900	910
Co	26	29	24	10	38
Cr	16	29	50	<2	18
Cu	60	105	52	34	29
Ni	17	7	11	<2	<3
Sc	28	—	—	20	—
Rb	—†	21	40	—	116
Sr	460	575	487	150	155
V	290	—	—	110	—
Y	30	23	22	48	45
Zr	70	99	128	190	342
Ga	17	—	—	18	—
Zn	—	82	59	—	78
Pb	—	4	15	—	36
Nb	—	7	7	—	14
Th	—	—	—	—	—
Yb	2	—	—	4	—

† Undetermined

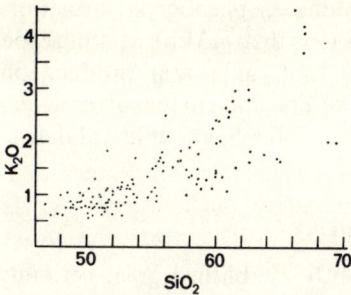

Fig. 6 Potash against silica for Aleutian lavas, showing the two potash trends. Most of these data are from the central and western Aleutians. These differences are not attributable to variations in the depth to the Benioff zone, but may instead represent two styles (open and closed) of near surface magmatic behaviour

These trends may reflect near surface differentiation from closed (high K) and open (low K) magmatic resevoirs.

Petrography and mineralogy

The andesitic basalts are porphyritic, weakly vesiculated, dense rocks. They contain plagioclase, olivine, magnetite, and clinopyroxene, and a groundmass of these minerals. Hydrous minerals appear to be absent; the author has never seen a hydrous phase in hundreds of flows of this type. These lavas generally contain 20–50 per cent phenocrysts (by volume), and, by point counting nearly 100 lavas of similar composition (SiO_2 c. 49 ± 2 per cent), the sequence of crystallization can be established with great certainty: plagioclase is the liquidus phase, olivine and magnetite appear after c. 15 per cent plagioclase is present, and clinopyroxene appears after c. 35 per cent crystallization. A typical mode for, say, 40 per cent crystallization is: plagioclase c. 35 per cent, olivine c. 4.0 per cent, magnetite c. 0.5 per cent, clinopyroxene c. 0.25 per cent, and c. 0.25 per cent vesicles. Only rarely are phenocrysts of magnetite absent.

With increasing silica content, olivine content diminishes in favour of orthopyroxene and clinopyroxene crystallizes earlier, while plagioclase and magnetite remain as phenocryst phases throughout the suite. In the andesites, and probably also the dacites, the sequence of crystallization is: plagioclase, clinopyroxene and magnetite, and lastly orthopyroxene; the crystallinity (i.e. total amount of phenocrysts whereupon each phase appears) is yet to be established.

In the andesitic basalts, the plagioclase composition is generally in the range An_{70}–An_{80}, the olivine is primarily Fo_{65}–Fo_{80}, and when orthopyroxene appears in the basaltic andesite it is commonly En_{60}–En_{70}; a more detailed description and microprobe data are given by Coats (1952), Byers (1959, 1961), and Marsh (1976). Plagioclase is relatively unzoned in the basic lavas, but zoning increases with overall crystallinity; the most calcic part of the crystal is generally not the core. Overall, the groundmass compositions are in all respects, except plagioclase, of a composition similar to that of the phenocryst phases. Ilmenite is yet to be found in the lavas, although it does occur in the very siliceous ash flows of the Peninsula.

Isotopic data

A compilation of all Pb and Sr data is given by Kay et al. (1978). The nature of the underlying crust, whether continental or oceanic, shows no effect in the initial Sr ratios (0.7028–0.7037, oceanic; 0.7030–0.7034, continental). There is also no correlation of initial Sr ratio and $^{87}Rb/^{86}Sr$. The Pb isotopes lie between the values for oceanic ridge rocks and those for continentally derived detritus (see Hawkesworth, this volume, Section VIII).

Geographical variations

There are no apparent correlations between Sr and Pb isotope content and crustal type. But little is known about many areas, particularly the very siliceous ash flows of the Peninsula which might represent crustal melts. It is the author's experience that, whereas andesitic basalt is the lava of the oceanic part of the arc, andesite becomes more common on the Peninsula; this may be due purely to hydrostatic reasons.

The usual sharp volcanic front is broken in two places by the appearance of Amak Island (163°W) and Bogoslof Island (168°W), and these two small volcanic centres sample magmatic variations transverse to the front. The rocks of Amak are in all respects, except potash content, nearly identical to the rocks of the nearest centre on the volcanic front (Marsh and Leitz, 1979). The rocks from Amak are consistently richer in potash by just the amount given by the Dickinson–Hatherton correlations.

The rocks from Bogoslof are also richer in potash than those on the front but in excess of that predicted by the K–h correlations, and at least one rock is nepheline normative. Whereas the lavas from Amak are andesites, Bogoslof's domes range from basaltic to andesitic with little correlation with time (Marsh and Leitz, 1979; Arculus et al., 1977). The variable composition of domes from Bogoslof, which also show differences in initial $^{87}Sr/^{86}Sr$ ratio, may reflect various degrees of interaction with near surface peridotitic wall rock. Such interaction is to be expected in the initial diapirism leading to a new volcano like Bogoslof.

Petrogenesis

The great dominance of andesitic basalt shows it clearly to be the magma the origin of which demands explanation.

Temperature and oxygen fugacity

Simultaneous solution of the f_{O_2}–T equations describing the ferric–ferrous ratio in the melt, as determined in glassy lavas, and the equilibrium olivine = magnetite + enstatite, generally constrains the temperature to the range 1100–1200 °C; $\log f_{O_2}$ is generally -7 to -8. The ubiquitous appearance of magnetite phenocrysts reflects these relatively high oxygen fugacities.

Total pressure

The small compositional contrast between corresponding phenocryst and groundmass phases clearly indicates phenocryst formation at low pressures (less than c. 3 kb). Melting experiments on these lavas at 1 atm produce phases of identical composition to that of the actual phenocrysts (B. D. Marsh, unpublished data).

H_2O content

The recently established great certainty in the sequence of crystallization (see above), the 1100–1200 °C temperature range of crystallization, and the absence of hydrous phases even in rocks having undergone 60 per cent crystallization all point to a low (less than c. 1 per cent by weight) pre-eruption water content. Even in siliceous lavas that represent perhaps 90 per cent crystal fractionation, where large amounts of water could accumulate, hydrous phases are absent. (Recall that, once saturated, it is virtually impossible to undersaturate a magma with regard to water.) There is no sign whatsoever of an unusual importance of water in the genesis and evolution of these magmas.

Source

Stemming from the ideas of Coats (1962), there are two possible source environments for these andesitic basalts: partial melting of the peridotitic mantle above the descending plate, and direct partial melting of the subducted oceanic crust. Although genesis could involve any degree of combination of these sources, this is not the point. The matter of first importance is to determine the region of partial melting which precipitates the whole process of magmatism. It is conceivable, but not likely, that genesis occurs simultaneously in more than a single source rock and leads to the same end product.

Peridotite

Although there is some disagreement, as a result of experiments there have been suggestions (e.g. Mysen et al., 1974; Mysen, this volume, Section VII) that hydrous melting of peridotite at depth may yield andesitic arc rocks. It is fairly clear that at any pressure, with addition of water there is a shift to higher activities of silica in the melt, as indicated by increased olivine stability. (Yet the melt compositions determined by Mysen and Kushiro (1977) actually show a decrease in silica content upon addition of water to peridotite.) Nevertheless, the dominant Aleutian magma has $c.$ 50 per cent SiO_2 and often even less. Thus, whatever the source, the effect of water must be unimportant otherwise the arc magmas would overall be more siliceous than observed (Fig. 5). This reasoning, coupled with the observation that these andesitic basalts are distinct from basalts known to have been produced from partial melting of peridotite (i.e. alkaline and tholeiitic basalts), both in chemical and isotopic composition, seems to preclude peridotite entirely as the host of the primary genesis, although a very subordinate role may be possible (Kay, 1977).

Subducted oceanic crust

Shown in Fig. 7 is the variation in melt composition with degree of partial melting at 30 kb quartz-eclogite containing 5 percent water as estimated from the data of Stern and Wyllie (1978) and Green and Ringwood (1968). Also shown is the same variation in melt composition with melting for a peridotite at 20 kb; no data for 30 kb presently exist (Mysen and Kushiro, 1977). Because relative to oceanridge tholeiite the starting composition of Stern and Wyllie is high in alumina and lime and low in potash and soda, these starting concentrations have been adjusted to match those of tholeiite at 100 per cent melt (see caption). The olivine tholeiite of Green and Ringwood (1968) is close to ocean-ridge tholeiite and these data agree with the adjustments to Stern and Wyllie's data. The effect of 5 per cent water on melt composition is seen to be small by comparing these results with those of Green and Ringwood for dry melting at 27 kb (see caption). This water lowers the solidus temperature, but only affects the melt composition near the solidus; beyond about 25 degrees of the solidus the melt is already undersaturated with water. The titania variation has been calculated assuming rutile as a liquidus phase and garnet and clinopyroxene to be the major contributors to the melt below the liquidus. There is good agreement throughout except for alumina where the results of Stern and Wyllie are more complete and probably more accurate.

Horizontal broken lines across these diagrams mark the composition of Aleutian andesitic basalt. The lined areas represent regions of agreement between each source rock and the arc lava. It is clear that the quartz-eclogite source rock can give rise directly to the typical Aleutian magma after $c.$ 60 per cent melting. This degree of melting is also necessary for magma extraction (B. D. Marsh, manuscript in preparation; Arndt, 1977). There is a poor agreement between the arc lava and the peridotite. This strong disagreement is not a function of pressure, for a comparison of the early 1 atm work by Reay and Harris (1963) and the 20 kb work of Mysen and Kushiro (1977) shows melt composition to be rather insensitive to pressure changes. Even with fractional crystallization it is virtually impossible to match the Aleutian magma with liquids yielded by peridotite.

Upon closer inspection (Marsh, 1979), the belief that the subducted oceanic crust is too cold to melt at the proper depth has been found untenable. The thermal regime in the uppermost regions of the plate at depths of 75–150 km changes on such a small scale that no thermal model yet developed can be used as an accurate measure of the temperature there. Hence present thermal models can not be used to establish a petrologic model. In sum, it is most likely that partial melting of subducted oceanic crust is the ultimate source of Aleutian volcanism. The question of the origin of the more siliceous magma is still largely unanswered. Certainly, near surface fractional crystallization, and possibly even remelting, plays an important role.

Acknowledgements

This work is supported by National Science Foundation Grant EAR-80051095-17617 to The Johns Hopkins University.

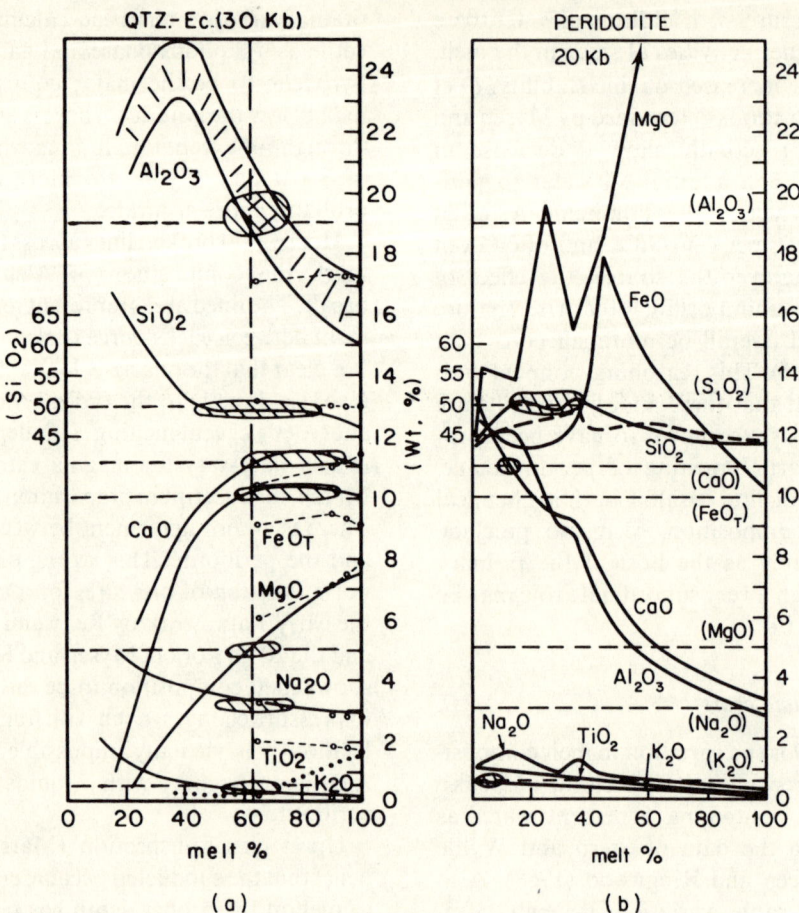

Fig. 7 (a) The variation in melt composition during partial melting of a quartz-eclogite at 30 kb, intended to represent subducted oceanic crust at high pressure, as estimated from the work of Stern and Wyllie (1978) (solid lines), and Green and Ringwood (1968) (dots and dashes). The double line for alumina shows the effect of different source rock composition; the range indicated is intended to cover that actually observed for the oceanic crust. The lined areas on each oxide line mark the areas of agreement of melt composition and the composition of the Aleutian andesitic basalt. The match between partial melt and lava composition is generally excellent, even for titania, whose intersection for clarity is not marked, for $c.$ 60 per cent melt (marked by the vertical broken line), which is also approximate the amount necessary for magma extraction (Marsh, 1981)

Fig. 7 (b) The variation in melt composition during partial melting of an undepleted garnet peridotite at 20 kb (Mysen and Kushiro, 1977). The scales are the same as those for the similar quartz-eclogite diagram. The additional broken line for silica represents this oxide when 1.9 per cent (mass) of water was added to the system. This water seems to have little on melt composition. The horizontal broken lines represent the composition of the Aleutian andesitic basalt and the lined areas mark the intersections or agreements of lava and partial melt composition. It is clear that the agreement is particularly poor: large amounts of magnesia must be removed from the partial melt and lime and alumina must be increased considerably, all without changing the relative agreement with silica

REFERENCES

Abe, K. (1972). Seismological evidence for a lithospheric tearing beneath the Aleutian arc. *Earth Planet. Sci. Letters* **14**, 428–432.

Arculus, R. J., Delong, S. E., Kay, R. W., Brooks, C., and Sun, S. S. (1977). The alkalic rock suite of Bogoslof Island, eastern Aleutian arc, Alaska. *J. Geol.* **85**, 177–186.

Arndt, N. T. (1977). Ultrabasic magmas and high-degree melting of the mantle. *Contr. Mineral. Petrol.* **64**, 205.

Berg, E., Susuma, K., and Kienle, J. (1967). Preliminary determination of crustal structure in the Katmai National Monument, Alaska. *Bull. seismol. Soc. Am.* **57**, 1367–1392.

Burk, C. A. (1965). Geology of the Alaska Peninsula—island arc and continental margin (Part 1). *Geol. Soc. Am. Mem.* no. 99.

Byers, F. M., Jr (1959). Geology of Umnak and Bogoslof Islands, Aleutian Islands, Alaska. *US geol. Surv. Bull.* no. 1028-L, 267–369.

Byers, F. M. Jr (1961). Petrology of three volcanic suites, Umnak and Bogoslof Islands, Aleutian Islands, Alaska. *Geol. Soc. Am. Bull.* **72**, 93–128.

Coats, R. R. (1947). Reconnaissance geology of some western Aleutian islands. In *Alaskan Volcano Investigations*, Reprt no. 2, Progress of Investigations in 1946, US Geological Survey, Washington, DC, pp. 95–105.

Coats, R. R., (1950). Volcanic activity in the Aleutian arc. *US geol. Surv. Bull.* no. 974-B.

Coats, R. R. (1952). Magmatic differentiation in Tertiary and Quaternary volcanic rocks from Adak and Kanaga Islands, Aleutian Islands, Alaska. *Geol. Soc. Am. Bull.* **63**, 485–514.

Coats, R. R. (1956). Geology of northern Adak Island, Alaska. *US geol. Surv. Bull.* no. 1028-C, 45–67.

Coats, R. R. (1959). Geologic reconnaissance of Semisopochnoi Island, western Aleutian Islands, Alaska. *US geol. Surv. Bull.* 1028-O, 477–519.

Coats, R. R. (1962). Magma type and crustal structure in the Aleutian arc. In *The Crust of the Pacific Basin*, American Geophysical Union Monograph no. 6, American Geophysical Union, Washington, DC, pp. 92–109.

Cooper, A. K., Scholl, D. W., and Marlow, M. S. (1976). Plate tectonic model for the evolution of the eastern Bering Sea basin. *Geol. Soc. Am. Bull.* **87**, 1119–1126.

Cormier, V. F. (1975). Tectonics near the junction of the Aleutian and Kuril–Kamchatka arcs and a mechanism for Middle Tertiary magmatism in the Kamchatka Basin, *Geol. Soc. Am. Bull.* **86**, 443–453.

Cummings, D. and Schiller, G. I. (1971). Isopach map of the Earth's crust. *Earth Sci. Rev.* **7**, 97–125.

Delong, S. E., Fox, P. J., and McDowell, F. W. (1978). Subduction of the Kula ridge at the Aleutian trench. *Geol. Soc. Am. Bull.* **89**, 83–95.

Engdahl, E. R. (1973). Relocation of intermediate depth earthquakes in the central Aleutians by seismic ray tracing. *Nature phys. Sci.* **245**, 23–25.

Engdahl, E. R. (1977). Seismicity and plate subduction in the central Aleutians. In *Island Arcs, Deep Sea Trenches and Back-arc Basins* (M. Talwani and W. C. Pitman III, eds), American Geophysical Union, Washington, DC, pp. 259–272.

Gard, L. M., Jr (1977). Geologic history In *The Environment of Amchitka Island, Alaska* (M. L. Merritt and R. G. Fuller, eds), ERDA, pp. 13–34, (*see* Table 1).

Green, T. H., and Ringwood, A. E. (1968). Genesis of the calc-alkaline igneous rock suite. *Contr. Mineral. Petrol.* **18**, 105–162.

Grow, J. A. (1973). Crustal and upper mantle structure of the central Aleutian arc. *Geol. Soc. Am. Bull.* **84**, 2169–2192.

Grow, J. A. and Atwater, T. (1970). Mid-Tertiary tectonic transition in the Aleutian arc. *Geol. Soc. Am. Bull.* **81**, 3715–3722.

Hayes, D. E. and Pitman, W. C., III (1970). Magnetic lineations in the North Pacific. *Geol. Soc. Am. Mem.* no. 126, 291–314.

Hein, J. R. and McLean, H. (1978). Paleogene sedimentary and volcanogenic rocks from Adak Island, central Aleutian Islands, Alaska, *US geol. Surv. prof. Pap.* in press.

Hein, J. R., Scholl, D. W., and Miller, J. (1978). Episodes of Aleutian ridge explosive volcanism. *Science, N.Y.* **199**, 137–141.

Helmberger, D. V. (1977). Fine structure of an Aleutian crustal section. *Geophys. Jl R. astr. Soc.* **48**, 81–90.

Holden, J. C. and Kienle, J. (1977). Geometry of a subducted plate and anti-arc segmentation of Aleutian volcanic chain. *Trans. Am. geophys. Union* **58**, 168.

Jacob, K. H. (1972). Global tectonic implications of anomalous seismic P travel times from the nuclear explosion Longshot. *J. geophys. Res.* **77**, 2556–2572.

Jacob, K. H., Nakamura, K., and Davies, J. N. (1977). Trench–volcano gap along the Alaska–Aleutian arc: facts and speculations on the role of terrigenous sediments for subduction. In *Island Arcs, Deep Sea Trenches and Back-arc Basins* (M. Talwani and W. C. Pitman, III, eds), American Geophysical Union, Washington, DC, pp. 243–258.

Kay, R. W. (1977). Geochemical constraints on the origin of Aleutian magmas. In *Island Arcs, Deep Sea Trenches and Back-arc Basins* (M. Talwani and W. C. Pitman, III, eds), American Geophysical Union, Washington, DC, pp. 229–242.

Kay, R. W., Sun, S. S., and Lee-Hu, C.-N. (1978). Pb and Sr isotopes in volcanic rocks from the Aleutian Islands and Pribilof Islands, Alaska. *Geochim. cosmochim. Acta* **42**, 263–274.

Marlow, M. S., Scholl, D. W., Buffinton, E. C., and Alpha, T. R. (1973). Tectonic history of the central Aleutian arc. *Geol. Soc. Am. Bull.* **84**, 1555–1574.

Marsh, B. D. (1976). Some Aleutian andesites: their nature and source. *J. Geol.* **84**, 27–45.

Marsh, B. D. (1979). Island arc volcanism. *Am. Scient.* **67**, 161–172.

Marsh, B. D. (1979a). Island arc development: some observations, experiments, and speculations. *J. Geol.* **87**, 687–713.

Marsh, B. D. (1981). On the crystallinity of lava and the sampling and rheology of magma. *Contrib. Min. Petrology* (in press).

Marsh, B. D. and Carmichael, I. S. E. (1974). Benioff zone magmatism. *J. geophys. Res.* **79**, 1196–1206.

Marsh, B. D. and Leitz, R. E. (1979). Geology of Amak Island, Aleutian Islands, Alaska. *J. Geol.* **87**, 715–723.

Murdock, J. N. (1969). Crust–mantle system in the central Aleutian region—a hypotheses. *Bull. seismol. Soc. Am.* **59**, 1543–1558.

Mysen, B. O. and Kushiro, I. (1977). Compositional variations of coexisting phases with degree of melting of peridotite in the upper mantle. *Am. Mineral.* **62**, 843–865.

Mysen, B. O., Kushiro, I., Nicholls, I. and Ringwood, A. E. (1974). A possible mantle origin for andesitic magmas; discussion of a paper by Nicholls and Ringwood. *Earth planet. Sci. Letters* **21**, 221–229.

Reay, A. and Harris, P. G. (1963). The partial fusion of peridotite. *Bull. Volc.* **26**, 115–127.

Reed, B. L. and Lanphere, M. A. (1973). Alaska–Aleutian range batholith: geochronology, chemistry, and relation to circum-Pacific plutonism. *Geol. Soc. Am. Bull.* **84**, 2583–2610.

Scheidegger, K. F. and Kulm, L. D. (1975). Late Cenozoic volcanism in the Aleutian arc: information from ash layers in the northeastern Gulf of Alaska. *Geol. Soc. Am. Bull.* **86**, 1407–1412.

Scheidegger, K. F., Carloss, J. B., Jezek, P. A., and Ninkovich, D. (1980). Composition of deep-sea ash layers derived from North Pacific volcanic arcs: Variations in time and space. *J. Volcanol. geothermal Res.* **7**, 107–138.

Scholl, D. W., Hein, J. R., Marlow, M., and Buffington, E. C. (1977). Meiji sediment tongue: North Pacific evidence for limited movement between the Pacific and North American plates. *Geol. Soc. Am. Bull.* **88**, 1567–1576.

Shor, G. G., Jr (1964). Structure of the Bering Sea and the Aleutian ridge, *Marine Geol.* **1**, 213–219.

Spence, W. (1977). The Aleutian arc: tectonic blocks, episodic subduction, strain diffusion, and magma generation. *J. geophys. Res.* **82**, 213–230.

Stauder, W. and Bollinger, G. A. (1964). The S-wave project for focal mechanism studies of earthquakes of 1962. *Bull. seismol. Soc. Am.* **54**, 2199–2208.

Stauder, W. and Bollinger, G. A. (1966). The focal mechanism of the Alaskan earthquake of March 28, 1964, and of its aftershock sequence. *J. geophys. Res.* **73**, 5283–5296.

Stern, C. R. and Wyllie, P. J. (1978). Phase compositions through crystallization intervals in basalt–andesite–H_2O at 30 kbar with implications for subduction zone magmas. *Am. Mineral.* **63**, 641–663.

Stone, D. B. (1968). Geophysics in the Bering Sea and surrounding areas, *Tectonophysics* **6**, 433–460.

Van Wormer, J. D. Davies, J., and Gedney, L. (1974). Seismicity and plate tectonics in south central Alaska. *Bull. seismol. Soc. Am.* **64**, 1467–1475.

The Cascade Province

A. R. McBirney and C. M. White

Center of Volcanology,
University of Oregon, Eugene, Oregon 97403, USA

ABSTRACT

The Cascade Range of north-western North America is a classic example of an orogenic volcanic system of the continental margin type. It has evolved near the leading edge of the American plate in a region where the continental crust varies widely in age, composition, and thickness. Although it has no trench or Benioff zone and has had little historic volcanic activity, the Cascade Range has most of the features associated with volcanic regions near convergent plate boundaries.

The large composite volcanoes of the High Cascades are composed mainly of basaltic andesite and andesite with subordinate amounts of more siliceous calc-alkaline rocks. Most are of the divergent type, having late-stage eruptions of rhyolite or rhyodacite closely associated in time and space with basalt or andesitic basalt. In the central part of the range, where Quaternary activity has been strongest, they are built upon broad overlapping shield volcanoes of high alumina basalt. Although the large cones are topographically imposing, their volumes are small compared to those of the less conspicuous flat-lying basaltic lavas on which they stand.

The volcanoes of the High Cascades have few lavas with reversed magnetic polarities and appear to have been formed entirely within the last million years. They were preceded by a series of earlier volcanic episodes, each of which was characterized by a distinctive spatial distribution, compositional characteristics, and volumetric proportions of rock types. With time, the rocks have declined in volume, average silica content, and concentrations of incompatible trace elements. These changes seem to be related to progressive depletion of lithophile components from the crustal section through which successive batches of magma have risen. No unequivocal link to subduction of oceanic lithosphere has yet been identified.

Introduction

The Cascade Range of western North America was the first andesitic province to be studied in detail, and is still the only orogenic volcanic system for which there is extensive correlated geological and geochemical data. Even here, however, the data are fragmentary and far from adequate. The earliest studies were devoted almost entirely to large composite cones. The classic work of Howel Williams, T. P. Thayer, C. A. Anderson, and others defined the salient geological and petrographic features of the principal volcanoes and provided what still remain some of the most comprehensive accounts of individual eruptive centres. More recent studies have tended to focus on geochemical features of both the Quaternary and Tertiary rocks and have given a clearer picture of the magmatic and tectonic evolution of the system through Cainozoic time.

Although the province is considered a prime example of an active continental margin, it lacks some of the features commonly considered typical of convergent plate boundaries. Most notably, it has neither a trench nor a Benioff zone. In fact, the central part of the province, where Cainozoic volcanism has been most intense, has the lowest seismicity in the western United States. It is also characterized by very sparse historic volcanism. Only four volcanoes (Baker, Saint Helens, Hood, and Lassen) have had recorded activity since the region was first settled more than a century ago.

Taken as a whole, however, the record of Cainozoic volcanism is remarkably complete. Thanks to well preserved sections in the Western Cascades and central Oregon and Washington, it is possible to obtain an unparalleled record of the evolution of the province from the earliest Eocene activity through successive episodes to the most recent post-glacial eruptions in the High Cascades. The summary that follows draws heavily on a recent review of current knowledge (McBirney, 1978), and work in progress in both the Quaternary and Tertiary sequences.

Evolution of the Cascade System

The modern volcanic chain that extends from British Columbia to northern California is the most recent of several igneous belts or zones that have followed the Pacific margin of North America since late Paleozoic time (Fig. 1). There is no visible record of Precambrian igneous activity and little evidence for volcanism prior to the last part of the Paleozoic era, but it is clear that there was an important volcanic episode that began during the Permian period and continued well into early Triassic time. Submarine lavas of this age are exposed in northern California, north-eastern Oregon, and northern Washington, and Gilluly (1963) may well be correct in stating that this outpouring of basalt exceeded that of any other period before or after. The nature of the activity was largely oceanic, however, and it is uncertain whether it had much in common with modern orogenic volcanism.

The earliest igneous activity that was clearly orogenic in nature dates from the second half of the Mesozoic era. It was long thought that the plutonic rocks that were emplaced in such large volumes during this episode had little associated volcanism, but detailed studies of contemporaneous sedimentary deposits, notably by Dickinson and his students (Dickinson, 1962, 1970) have shown quite clearly that the batholithic rocks that are so conspicuous today are in fact the roots of a deeply eroded volcanic belt that may not have been very different from the modern High Cascades.

Segments of uplifted plutonic and weakly metamorphosed sedimentary and volcanic units of Mesozoic age are exposed from British Columbia diagonally across the state of Washington into north-eastern Oregon and in isolated windows that extend across central Oregon to connect with the major axis of the Sierra Nevada system in north-western California. This large sigmoidal belt and the embayment it forms near what is now the central part of the High Cascade Range forms one of the major structural features of the crust in the Pacific North-west Hamilton (1969) has proposed that during the period of strongest activity the western Cordilleran axis may have resembled the modern Andes, especially if one reverses post-Cretaceous deformation that is postulated to have accentuated the kink and segmented it into discontinuous blocks. Unfortunately, little is known about the geochemical or geological relations of the volcanic rocks, because most of them have been removed by erosion or are buried beneath a Cainozic cover.

Recent paleomagnetic studies of Eocene rocks of the Oregon Coast Range (Simpson and Cox, 1977) show evidence of up to 60° clockwise rotation and lend support to the postulated oroclinal deformation of the pre-Cascade basement. Various interpretations have been offered for the paleomagnetic data, including large scale motion of plates from various parts of the eastern Pacific margins; however, in at least one instance where detailed magnetic measurements have been made in an area that has been mapped in detail (R. Wells, work in progress), the rotation appears to be confined to small blocks between north-west-trending transcurrent faults with dextral displacement.

Eocene volcanic episodes

The Cainozoic igneous history of western North America has recently been reviewed and summarized by Armstrong (1978), who shows that it

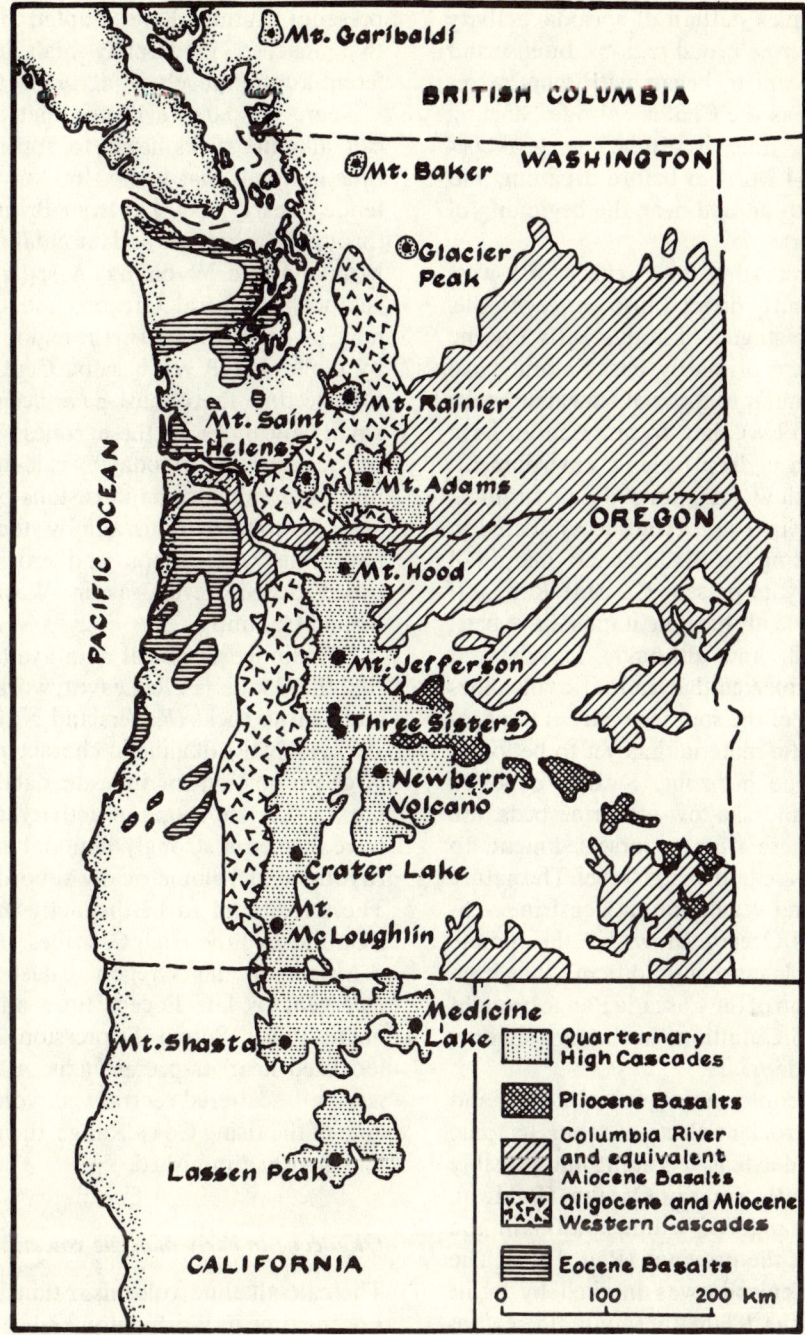

Fig. 1 Generalized map of the Cascade Range and adjacent parts of the northwestern United States

followed a complex pattern of episodic activity that migrated across broad regions. Intense and widespread volcanism began with what Armstrong refers to as the Challis episode. Starting in early Eocene time, it reached its peak between 54 and 44 Ma ago before declining and finally coming to an end near the beginning of the Oligocene era.

The Eocene record of the Pacific North-west can be conveniently divided into early, Middle, and late Eocene stages. Although early Eocene igneous rocks are primarily basaltic and have compositions similar to those of Hawaiian rocks (Snavely et al., 1968), they do not seem to have been erupted in a deep oceanic environment. Lavas and shallow intrusions occur within a thick eugeosynclinal series that accumulated in a shallow subsiding basin, the eastern margin of which was near the present Cascade Range. If there was a trench at this time it must have been far to the west, and although Snavely and Wagner (1963) inferred that andesitic volcanoes were active east of the southern part of the geocyncline, andesitic material has yet to be found in Lower Eocene horizons. Swamp deposits, including coal and shallow estuarine beds, are common, but there are no coarse sediments to indicate a nearby region of high relief. The nature of weathering and vegetation in lacustrine sediments of central Oregon show that the climate of that region was more humid than it has been since the elevation of the Cascade Range brought about more arid conditions in the rain shadow of its leeward side.

A marked unconformity between Lower and Middle Eocene rocks reflects a strong tectonic event that caused extensive folding and possibly thrusting in south-western Oregon (Baldwin, 1965) and a change of sedimentation in the northern part of the province (Rau, 1966). The middle Eocene episode was marked by uplift and erosion of the Klamath Mountain region, together with volcanism, mainly of basaltic character, in localized centres along an axis extending from the partly emergent Coast Range of Oregon along the shallow basin of the Willamette–Puget Sound Depression. At least two volcanic complexes north and south of the present Columbia River erupted tholeiitic rocks that reached moderately high levels of differentiation (Snavely et al., 1965, 1968).

There is sparse evidence that differentiated calc-alkaline rocks began to appear about this time in a belt that was c. 160 km wide and extended nearly 1600 km from British Columbia through Washington and into Idaho and western Montana and Wyoming. A separate field developed in central Oregon, and other centres may have been located in the region now covered by Columbia River basalts. Certainly by late Eocene time there must have been a swarm of small volcanoes in these zones. The eruptive centres are marked today by calc-alkaline lavas, tuffs, and subvolcanic intrusions of the Challis and Clarno formations, and by stocks and small batholiths where uplift and erosion have exposed deeper levels, as in Washington and British Columbia.

The limited chemical data available indicate that the Challis (F. R. Leavitt, work in progress) and Clarno rocks (Rogers and Novitsky-Evans, 1977) are calc-alkaline in character and include large proportions of andesite, dacite, and rhyolite. In the late stages of activity, they seem to have formed a strongly bimodal association of rhyolite and volumetrically subordinate basalt. The series tend to be distinctly more potassic than those of the High Cascades.

Much of the Oregon Coast Range was emergent by late Eocene time, and the Willamette–Puget Sound Depression had become localized near its present axis. Although there were still scattered central-vent volcanoes in the area of the rising Coast Range, their importance had greatly diminished.

Oligocene to early miocene volcanism

The calc-alkaline volcanism that began in late Eocene time in Washington and central Oregon increased in intensity through the Oligocene epoch and spread westward and southward until it covered a broad zone across most of western Oregon and Washington. It is still uncertain when activity began along the Cascade axis. The oldest andesitic rocks from this region that have been dated by radiometric methods are less than 40 Ma old. (Table 1). Andesitic lavas of the Colestin, Fisher, and equivalent formations in southern Oregon were once thought to have come from late Eocene volcanoes, but dating has shown that they are somewhat younger.

TABLE 1 Stratigraphic column for Quaternary and Tertiary rocks of the Cascade region of Central Oregon

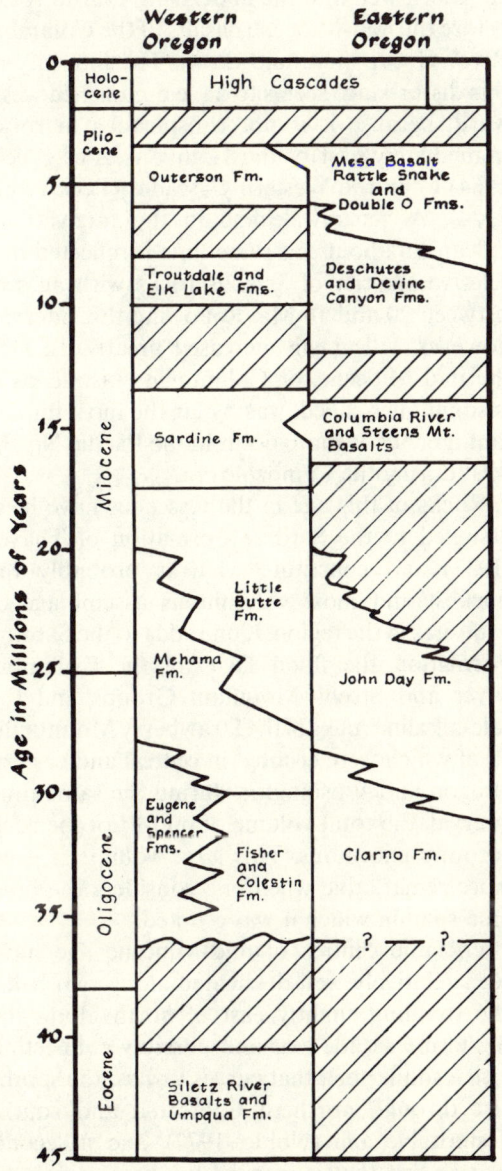

Instead, there is a very marked angular nonconformity separating the Eocene formations from Oligocene calc-alkaline rocks, and the deformation that occurred at that time seems to have coincided with the onset of orogenic igneous activity in the Cascade region.

Certainly by Oligocene time there were many eruptions of andesite and more siliceous rocks from centres in the area of the Western Cascades. Andesitic flows and siliceous tuffs interfinger with Oligocene sediments along the east side and southern end of the Willamette Valley and are abundant in the Oligocene sections of the Clarno and John Day formations east of the Cascades (Hay, 1963; Fisher and Rensberger, 1972). Within the Western Cascade Range, rocks of this age range include the thick Mehama and Little Butte Formations and numerous intrusive bodies that probably represent the eroded remnants of volcanic centres.

The distribution of these centres has not been well defined, but it appears that by the end of Oligocene time small volcanoes were scattered over a broad zone along much of the Pacific continental margin (Fig. 2). In Washington, the locus of activity has been identified in shallow subvolcanic intrusions and associated pyroclastic units near the present Cascade Range, but the full width of the zone cannot be determined there owing to the extensive cover of younger rocks east of the Cascades. In Oregon, Oligocene and early Miocene centres extend from the Coast Range well into central Oregon. In the Coast Range they are marked by subvolcanic intrusions of gabbro and nepheline syenite, while in the Western Cascades and central Oregon, flows, tuffs, and volcanic sediments are associated with vent complexes and shallow stocks, dykes, and sills.

The spatial distribution of the Oligocene rocks is unlike that of modern volcanic belts in that the most alkaline compositions are closest to the ocean. Most of the Oligocene rocks of the Coast Range are alkaline dolerites, gabbros, and nepheline syenites, whereas the main volcanic series of the Western Cascades and central Oregon are strongly subalkaline, and although a few alkaline rocks have been reported from the

John Day formation in central Oregon, they are very subordinate in volume.

As volcanism spread during the Oligocene epoch it tended to become less basaltic and more differentiated with time. The andesites and basalts that dominate the lower parts of the sections in the Western Cascades and make up most of the Clarno formation in the east give way upward to siliceous pyroclastic rocks which, by the end of the episode, reached enormous volumes.

Despite the large volumes of eruptive material, there seem to have been few large volcanoes during this period but rather a scattering of small basaltic and andesitic cones, and a few low-rimmed calderas or broad volcano-tectonic depressions (Walker, 1970). Some of the most voluminous eruptions seem to have come from fissures or from unroofing of shallow intrusions that stoped their way toward the surface. The Coast Range at this time must have formed a peninsula or low island chain, and behind it the Willamette–Puget Sound Depression formed a shallow arm of the sea. There seems to have been no pronounced topographical barrier near the present Cascade axis. Instead, great volumes of tuffaceous debris were deposited in a system of estuaries and shallow basins to form the Eugene formation and equivalent units of the Willamette Valley and the lacustrine beds of the John Day formation in central and eastern Oregon. Coarse detritus and deep erosional channels are notably rare, and the relief on the volcanic landscape could not have been great.

The broad distribution of volcanism during this episode provides an excellent opportunity to compare the compositions of rocks erupted at differing distances inland along a section normal to the continental margin. The total volume of erupted material and the proportions of siliceous pyroclastic rocks increase markedly inland; in the same direction, rocks at the same level of differentiation become richer in K, Rb, Zr, Sr, Ba, and La/Sm and poorer in Ni and Cr. The lavas erupted along the western margin are unusual in that they contain few phenocrysts and seem to have been very hot and fluid when poured out on the surface. They are somewhat more Fe-rich and tholeiitic than the younger calc-alkaline suites (Table 2).

Mid-miocene (columbian) volcanism

An episode of faulting, uplift, and erosion occurred during the later part of early Miocene time throughout much of central Oregon. There is a marked erosional non-conformity at the top of the John Day formation and in places erosion cut down well into the underlying Clarno rocks before the mid-Miocene basalts of the Columbia River Group were laid down. The intensity of this disturbance seems to have diminished westward, because it is not conspicuous in rocks immediately east of the High Cascades (Peck, 1964) or in the Western Cascades (Peck et al., 1964). A general decline in the intensity of volcanism about this same time is reflected in a relative scarcity of igneous rocks with ages of between 20 and 16 Ma. Following this interval, however, volcanism increased greatly. In fact, the mid-Miocene, or Columbian episode, as it is sometimes called, was by far the most important igneous event to occur in the Pacific Northwest during the Cainozoic era.

Rocks of this age in the Cascades have been assigned to the Sardine Formation of Thayer (1937) and constitute what is probably the thickest and most voluminous assemblage of andesites in the region. If one adds to the Sardine Formation the flood lavas of the Columbia River and Steens Mountain Groups and the calc-alkaline rocks of the Strawberry Mountains, all of which were erupted in central and eastern Oregon and Washington during the same time interval, the total volume of mid-Miocene rock becomes enormous. This large volume is even more remarkable when one considers the brief time-span in which it was erupted.

Volcanic centres of mid-Miocene age have been relatively well delineated along two belts, one trending slightly east of north along the Western Cascade axis and possibly connecting with a similar belt that curves toward the southeast through northern California and southwestern Nevada (Noble, 1972), and a second shorter belt that is marked by three middle to

TABLE 2 Weighted average compositions of Cainozoic volcanic rocks of the Oregon Cascades and northern California

	Quaternary rocks	Quaternary rocks	Pliocene rocks	Middle and Upper Miocene rocks	Oligocene–Miocene Rocks
	Northern California	Central Oregon			
Average composition (in wt %)					
SiO_2	52.2	52.7	52.7	57.3	62.4
TiO_2	1.2	1.4	1.3	1.1	0.9
Al_2O_3	18.0	17.4	17.3	16.6	15.7
ΣFeO	9.6	9.1	8.6	7.4	6.1
MnO	0.1	0.1	0.1	0.1	0.1
MgO	5.5	5.3	5.6	3.7	2.3
CaO	9.1	8.4	8.7	6.7	5.3
Na_2O	3.2	3.8	3.5	3.5	3.4
K_2O	0.9	0.9	0.9	1.3	1.7
P_2O_3	0.2	0.3	0.2	0.3	0.2
Total	100.0	99.4	98.9	98.0	98.1
No. of analyses and percentage of rock types by volume (in parentheses)					
Basalt	7 (69)	33 (85)	17 (90)	20 (39)	6 (10)
Andesite	22 (29)	112 (13)	76 (9)	99 (41)	25 (45)
Dacite–rhyolite	35 (2)	31 (2)	6 (1)	17 (20)	10 (45)
Total volume of rocks (in km^3)	3095	4600	2150	24 850	>10 000

late Miocene volcanic centres trending northeast through the Strawberry Mountain complex and other igneous centres in east-central Oregon (Robyn, 1977) (Fig. 2). Of these two chains, that of the Western Cascades was the most extensive and produced the largest volume of calc-alkaline rocks. Its axis is marked by stocks, mainly of quartz diorite, and by broad aureoles of hydrothermal alteration and mineralization. The forms of the cones can be inferred from the lithologic zones that reflect the conditions under which lavas and pyroclastic rocks accumulated on the slopes and lower flanks of large composite volcanoes. These centres seem to have been the first to have the form and alignment that is conventionally associated with andesitic belts. They probably resembled volcanoes of the modern High Cascades, except that the presence of shallow water sediments between them indicates that the belt was a chain of volcanic islands or a broad shelf area, at least during its early stages of development.

There is surprisingly little correlation of mid-Miocene units on opposite sides of the High Cascades. Despite their great thickness in the Western Cascades, andesitic rocks are scarce, or absent, along the eastern base of the modern range. A few thin flows of basalt within the Sardine Formation have been correlated with the Columbia River Group to the east (Peck et al., 1964; White and McBirney, 1978) and indicate that tongues of flood lavas flowed between the andesitic cones, but it is difficult to visualize the topographic configuration that could account for the limited interfingering of two adjacent units of such great thicknesses. The problem is made more difficult by the shallow level of erosion and the extensive cover of younger rocks that limit exposures on the east side.

Most of the products of mid-Miocene volcanism in the Western Cascades were andesitic. Siliceous pyroclastic rocks are much less important than they were in the preceding episode, and basalts were still quite subordinate (Table 2). The relationship, if any, between this andesitic volcanism and the great outpouring of flood lavas of the Columbia River Group has never been explained. The same brief episode during which these large volumes of volcanic rocks were erupted was also marked by strong activity in Central America, the south-western Pacific, and other parts of the circum-Pacific system (McBirney et al., 1974, Kennett et al., 1977).

Late miocene (andean) volcanism

The Columbian episode declined sharply c. 13–14 Ma ago and was followed by moderately strong deformation throughout much of the Pacific North-west. Strong faulting and tilting occurred east of the Cascade Range where an important angular unconformity separates Pliocene rocks from the underlying older units. In the Cascade region, broad folds developed along axes that closely parallel the trend of the earlier mid-Miocene volcanoes.

A brief late Miocene pulse of activity, dated c. 9–10 Ma ago, has recently been recognized in the Western Cascades (McBirney et al., 1974) and appears to have been synchronous with strong volcanism elsewhere in the circum-Pacific region, especially the Andes. The rocks have only been separated from the older Sardine Formation in a few areas where detailed studies have been carried out. They appear to have been erupted from small cones around the lower flanks of the large eroded remnants of mid-Miocene volcanoes and from scattered vents east of the Cascades. Known centres in the Cascades are too few to permit an interpretation of the distribution of vents or their relationship to regional conditions. In central Oregon, however, there appears to be a systematic migration of late Miocene rhyolitic eruptions toward the Cascades (G. W. Walker and N. S. MacLeod, personal communication). In addition, rhyolitic and dacitic ignimbrites and scattered basaltic lavas were erupted throughout much of central and eastern Oregon (Walker, 1970). Taken as a whole, the episode was characterized by andesitic and basaltic lavas in the Cascades and dominantly silicic ignimbrites and domes toward the east.

Pliocene (fijian) volcanism

Kennett et al. (1977) have applied the name Fijian to the important volcanic episode that occurred between c. 3 and 6 Ma ago. The name was taken from the islands in the south-western Pacific where the episode was first established by systematic dating, but volcanism was widespread at this time throughout much of the circum-Pacific region. Rocks of this episode are widespread in the Cascade Range, central Oregon, northern California, and the Basin and Range Province, but it is often difficult to distinguish them from Pleistocene and Holocene units, because they are only moderately affected by weathering and erosion (Fig. 1).

Activity in Oregon during this period produced mainly basaltic lavas and rhyolitic domes and ignimbrites (Fig. 2). Small monogenetic cones and thin but extensive flows of basalt broke out over a broad region extending from the western side of the Cascades across central Oregon almost to the Idaho border. Rhyolitic ignimbrites were also erupted from centres east of the Cascades, and the chain of rhyolitic domes that began to develop in late Miocene time continued its westward migration toward the Cascade Range. Another region of basaltic and rhyolitic activity developed in southern Idaho and migrated eastward along the Snake River Plain toward Yellowstone.

Andesites seem to have been subordinate to basalt and rhyolite, and there were no conspicuously large composite cones near the axis of the Cascades. By this time most of the large Miocene volcanoes had probably been levelled by erosion. Coarse andesitic debris is widespread in alluvial and lacustrine deposits along the eastern side of the Cascades where it was deposited in pediments and a number of subsiding basins.

Quaternary (cascadian) volcanism

The recent episode of volcanism that has been responsible for the familiar volcanoes of the modern High Cascades has been studied in greater detail than any other Cainozoic period, but even today much remains to be learned about the volcanoes and their relations to the system as a whole. This last period of activity, from which the Cascadian episode takes its name, followed closely and, in places, merged with the preceding Pliocene episode. Its main feature was a marked narrowing of the focus of volcanism to form a well defined chain of large composite cones extending from British Columbia to northern California (Figs. 1 and 2).

The earliest Pleistocene activity resembled that of the preceeding Pliocene episode in that it was characterized by basaltic cones, flows, and low overlapping shields. With time, activity became more localized in persistent centres from which progressively more differentiated magmas were discharged. Most of the large andesitic cones that form the crest of the High Cascades began to rise during Pleistocene time c. 1 Ma ago and reached their present elevations by rapid growth during a brief period of intense activity. The fact that few of the lavas have reversed magnetic polarities, even in the lowest levels of deeply glaciated cones, indicates that by far the greatest volumes must have been discharged since the present period of normal magnetic polarity began about 670 ka ago.

Block-faulting occurred concurrently with volcanism in the central Cascade Range and resulted in uplift and westward tilting of the Western Cascades. At the same time, the basement below the active volcanoes of the High Cascades began to subside to form a shallow graben, much of which has been filled by the products of Quaternary volcanoes (Fig. 3). Depression of the Cascade graben has been most pronounced in the central Cascades where volcanism has been strongest, it dies out toward the north and south where individual volcanoes are large but widely spaced, and the total volume of Quaternary volcanic rocks is small.

The topographically imposing volcanoes of the High Cascades give the impression that andesite is the dominant rock type in the modern range. In places this is probably true, but if one considers the total volume of rocks produced in the system as a whole, andesitic cones are seen to account for a very subordinate amount of the erupted volumes. The proportion of andesite is high only in those parts of the chain where the total volume of Quaternary rocks is small, namely in Washington and northern California. In one part of the central Oregon Cascades where absolute volumes have been estimated (McBirney et al., 1974), it has been found that basaltic lavas beneath and between large andesitic cones account for c. 85 per cent of the total volume of Quaternary rocks. Glaciated shield lavas were found to total c. 1282 km^3, while large composite cones in the same area account for c. 189 km^3, and the very recent scoria cones and lava flows amount to c. 55 km^3. Unfortunately, most geological and petrological studies have been concentrated on the high cones, and the great volume of underlying rocks and smaller volcanoes have been largely ignored.

The Quaternary rocks of the High Cascades have most of the petrographic and petrochemical features considered typical of the calc-alkaline rocks of modern continental margins. Almost all rocks, with the exception of the most basic basalts and the most siliceous rhyolitic obsidians, are strongly porphyritic and rich in plagioclase. Basalts commonly contain up to c. 20 per cent of olivine phenocrysts. The olivine may have minute inclusions of reddish brown spinel and rarely shows a reaction relationship to the groundmass. Olivine is not uncommon as phenocrysts in basaltic andesites and andesites, but it contains few, if any, inclusions of spinel and is normally corroded and rimmed with pyroxene. Titaniferous magnetite, although abundant in the groundmass of almost all rocks, is seldom important as phenocrysts. Most andesites contain two pyroxenes, augite and

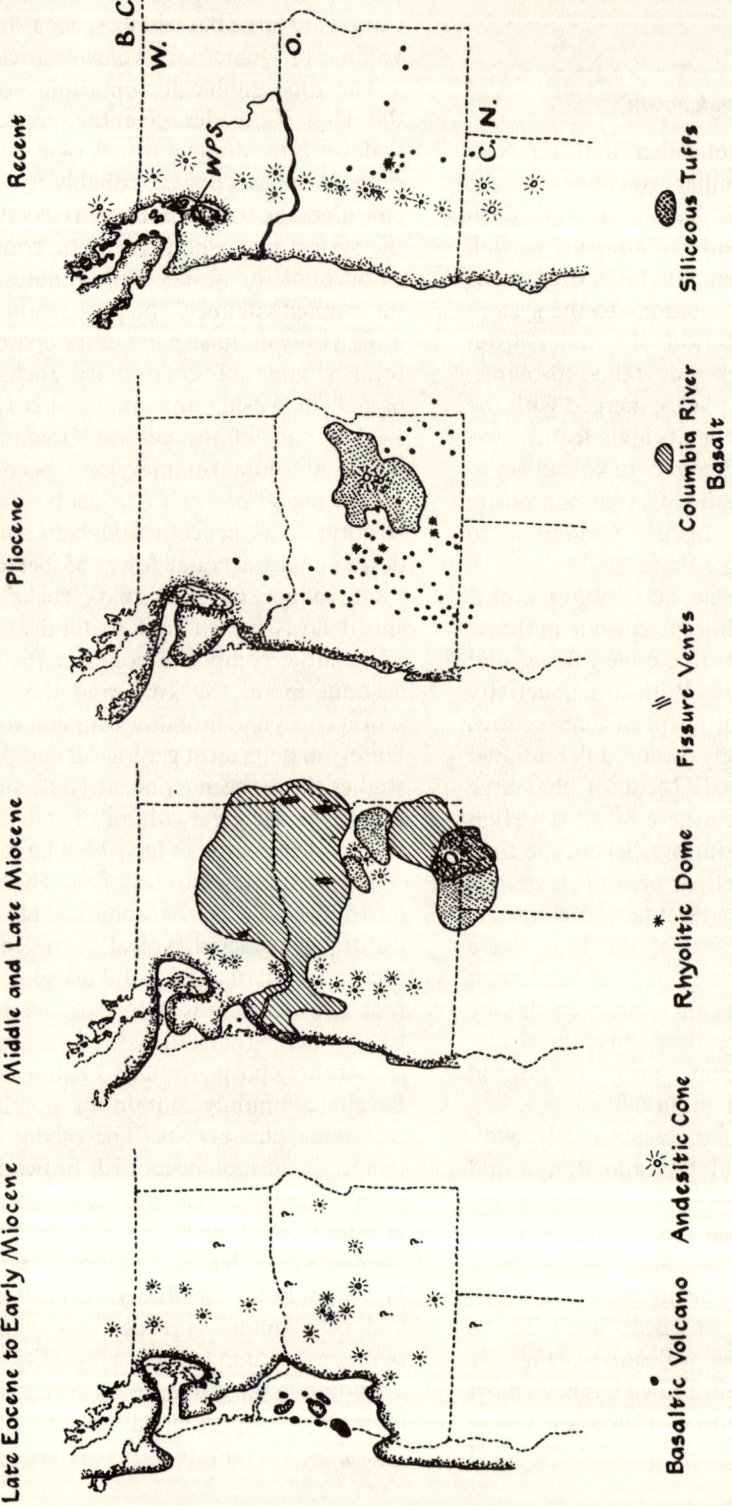

Fig. 2 Paleogeographic reconstruction of the Pacific North-west during Cainozoic time, compiled from various sources, including Snavely et al. (1968) and Armstrong (1978) (from McBirney, 1978). National and state boundaries are indicated by broken lines as indicated on the 'Recent' reconstruction with BC = British Columbia (Canada), W = Washington, O = Oregon, C = California, and N = Nevada; WPS = Willamette–Puget Sound

TABLE 3 Average compositions of Quaternary volcanic rocks of the central Oregon Cascades

	Basalt, 43 Analyses ($<53.5\%$ SiO_2)		Basaltic andesite, 57 Analyses (53.5–57% SiO_2)		Andesite, 56 Analyses (57–63% SiO_2)		Dacite, 16 Analyses (63–68% SiO_2)		Rhyolite, 15 Analyses ($>68\%$ SiO_2)	
	(wt%)	(S.D.)	(wt%)	(S.D.)	(wt%)	(S.D.)	(wt%)	(S.D.)	(wt%)	(S.D.)
SiO_2	51.1	1.84	55.4	0.91	60.0	1.72	64.9	1.78	71.6	2.47
TiO_2	1.4	0.33	1.0	0.17	0.9	0.16	0.7	0.22	0.3	0.13
Al_2O_3	17.3	0.99	17.9	0.72	17.4	0.64	16.2	0.42	13.9	0.65
ΣFeO	9.4	2.40	7.6	0.79	6.4	0.79	4.7	0.83	2.4	0.57
MnO	0.2	0.02	0.1	0.1	0.1	0.03	0.1	0.03	0.1	0.02
MgO	6.1	2.03	4.6	0.92	2.8	0.73	17.	0.56	0.5	0.34
CaO	8.9	0.78	7.5	0.78	6.1	0.75	4.4	0.99	1.7	0.36
Na_2O	3.6	0.56	3.9	0.29	4.3	0.34	4.5	0.56	4.5	0.43
K_2O	0.8	0.24	0.9	0.22	1.2	0.33	1.6	0.33	3.0	0.35
P_2O_3	0.3	0.12	0.2	0.07	0.2	0.06	0.2	0.08	0.1	0.05

hypersthene, which vary little in composition throughout the series. Hornblende is much less common. It occurs as oxidized relicts in a few andesites but is a common phase only in dacites and rhyolites. Plagioclase is by far the most conspicuous phase among the phenocrysts throughout the series. It normally has very complex zoning that may differ from grain to grain, even in a single thin section. Its composition is also varied but is normally in the range of sodic labradorite to calcic andesine. The groundmass plagioclase shows a somewhat wider range of composition between rocks of differing silica contents but is rarely more sodic than oligoclase.

The chemical compositions of the principal Quaternary rock types in the central Oregon Cascades are given in Table 3.

In recent years, several Quaternary High Cascade volcanoes have been examined in considerable detail, but much of this work is still incomplete, and there are few places where the entire volcanic and petrological development of a large cone can be traced. The most complete data are probably those for Mount Jefferson (Thayer, 1937; Walker et al., 1966; Greene, 1968; Condie and Swenson, 1973; Sutton, 1974; White and McBirney, 1978). Four stages of activity have been recognized (Fig. 4(a)). The earliest Pleistocene eruptions formed a broad base of basaltic shield lavas on which the main cone was then built during two separate stages of andesitic activity. Finally, in very recent time, small flows of basalt were discharged from satellite vents on the lower flanks of the main cone. There was a general increase in the silica content through the three main stages of growth and a steady decline in the volumes of erupted rocks, but the flank eruptions of the last stage reverted to a more basic composition similar to that of the Pleistocene shield lavas. The magma of each of the four stages had its own distinctive

Fig. 3 Schematic cross-section through the central Oregon Cascades showing known and inferred structural and stratigraphic relations. Ages of units are designated as follows: Te = Eocene, To = Oligocene, Tm = Miocene, Tmp = Mio-Pliocene, Tp = Pliocene, Qv = Quaternary. Tcr indicates Columbia River Basalt, and Tmi refers to subvolcanic stocks of Miocene age

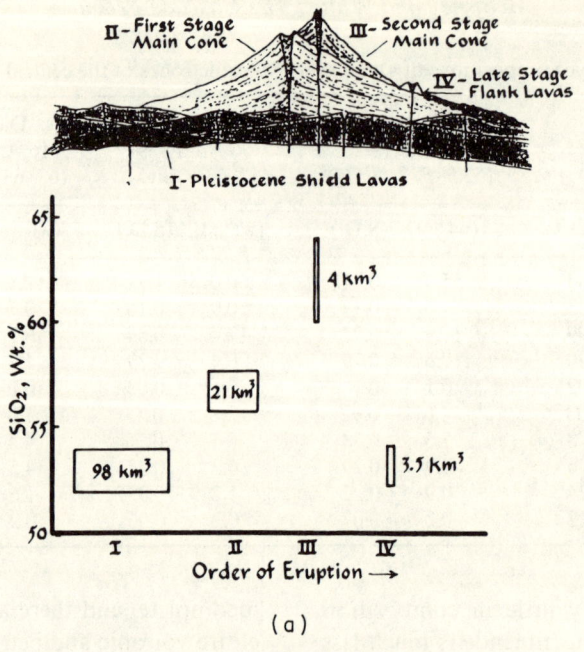

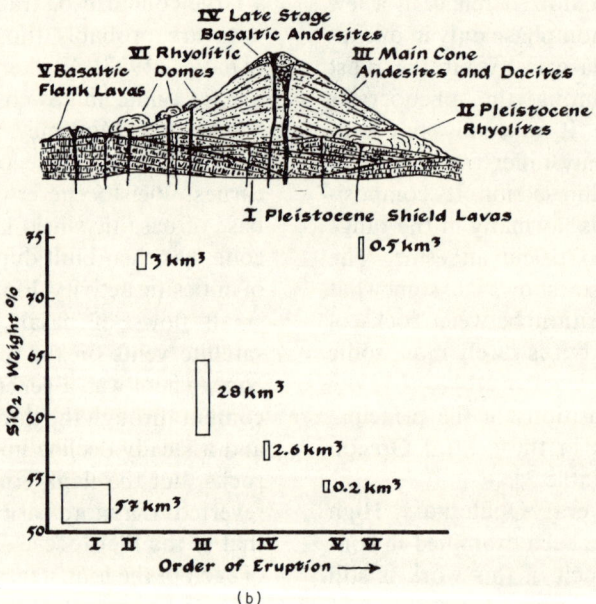

Fig. 4 (a) The Quaternary evolution of Mount Jefferson included four main stages, each of which was characterized by distinctive rocks. Volumes of rocks in each stage are shown by the relative areas of rectangles in the lower diagrams. The mid-point on the vertical dimension of the rectangle is placed at the mean value of SiO_2 for the rocks of that stage, and the vertical length of the edge indicates one standard deviation for the silica value. (b) The Quaternary evolution of the Three Sisters complex resembles that of Mount Jefferson but is characterized by more siliceous rocks, especially in the final stage when the magma became strongly divergent geochemical character and does not seem to have had a direct genetic relation to the others.

Similar patterns of development have been followed by most of the other large Quaternary cones, but in many places late-stage eruptions have produced not only basaltic rocks but rhyolitic or dacitic domes and pumice as well. This pattern in which compositions evolve through andesite and then diverge into more or less contemporaneous basalt and rhyolite in the latest stages of activity is most pronounced in the southern and northern parts of the chain.

At the Three Sisters (see Fig. 1), for example, the main andesitic cones are ringed with domes of rhyolitic obsidian and, at lower elevations, by contemporaneous basaltic scoria cones (Fig. 4(b)). Studies by J. Clark (work in progress) have shown that the divergent late-stage magmas were derived from the same parent, possibly by gravitational stratification of a shallow body of magma. A similar pattern is seen at Mount Mazama (Crater Lake) where siliceous domes and pumice have erupted from vents on the upper flanks of the main andesitic cone while basalt and andesitic basalt were discharged at lower elevations. Clear evidence of a graded magma is seen in the products of the climatic eruption that led to formation of the caldera. Ritchey (1980) has shown that the upper siliceous magma responsible for the large volumes of rhyolitic pumice discharged during the main stages of the Plinean outpourings overlay a crystal-rich zone of more mafic composition that was tapped at the close of the eruption. Geochemical and petrological evidence indicates that the two compositions evolved by gravitational stratification of an initially homogeneous intrusion.

Summary of geological development of the Cascade system

The foregoing descriptions have been very brief and are far from complete, but it is apparent even from this short summary that the Cascade system has evolved to its present form through a varied succession of igneous and tectonic events and that the setting of volcanism today is by no means characteristic of that in earlier periods.

Calc-alkaline volcanism first appeared around the close of the Eocene epoch east of the present Cascade axis and gradually spread across a broad region during Oligocene and early Miocene time. A well defined line of composite volcanoes did not develop until the mid-Miocene Columbian event, and following that episode there was no new chain of large andesitic volcanoes until the modern High Cascades began to rise c. 1 Ma ago.

Tectonic disturbances occurred at several times and in different regions. The strongest deformation seems to have occurred in the southern Coast Ranges around the end of middle Eocene time, but conspicuous faulting and uplift also took place east of the Cascades shortly before the mid-Miocene volcanic episode and again shortly before the late Miocene and Pliocene episodes. Basin and range faulting extended into central Oregon toward the end of Pliocene time and has continued down to the recent past.

Trends of Magmatic Evolution

Several aspects of the Cainozoic activity stand out when the sequence of igneous episodes is viewed as a whole. Although the volumes of volcanic and intrusive rocks produced during the early episodes are difficult to estimate, they were certainly much greater than those of more recent times. There has been a somewhat irregular decline of volcanism with each successive episode since the mid-Tertiary pulses, which were by far the most intense and widespread to occur anywhere near the present Cascade axis. The time span during which most of the Oligocene and early Miocene rocks were erupted seems to have been of the order of 10 Ma, but the mid-Miocene episode was much shorter, possibly only 2 or 3 Ma. Hiatuses in which there was little or no volcanism seem to have separated each of the subsequent volcanic episodes down to the present.

The proportion of basaltic rocks has increased steadily with time (Table 2). The Oligocene–early Miocene episode produced the largest volume of siliceous rocks, mainly rhyolite and dacite; andesite was the dominant rock type produced in the Cascades during the mid-Miocene (Columbian) episode, and since that time basalt has outweighed all other rock types combined. In the modern High Cascade chain, andesite is important only in the southern and northern parts of the chain where the intensity of Cainozoic volcanism has been relatively mild. Elsewhere, there are large composite cones composed largely of andesite, but they constitute a relatively small part of the total erupted volume.

Knowing the relative proportions of the different rock types in each age group and the average compositions of the individual members of each series, it is a simple matter to calculate the average compositions of volcanic rocks produced during successive igneous episodes. This has been done for the central Oregon Cascades by White and McBirney (1978) who obtained the results shown in Table 2. For purposes of comparison, data are also shown for Quaternary rocks of northern California.

The most notable feature of these averages is the decline of silica and potash with time. The Na content is remarkably uniform, mainly because the concentration of that element in basalts has increased by an amount that balances the increased proportions of mafic rocks. Similarly, the average of Quaternary rocks in the central Oregon Cascades does not differ markedly from that of northern California, even though dacites and rhyolites are much more abundant in the southern part of the chain. The reason for this apparent inconsistency lies in the fact that the basalts of northern California tend to be more basic than those of central Oregon.

As the proportions of rock types changed with time, the nature of the individual members of the suites also seem to have evolved in a systematic way (Fig. 5). If basalts of each of the major eruptive episodes are compared, they are seen to become progressively more sodic with time. The younger rocks have lower Fe/Mg ratios, and they are progressively depleted in certain trace elements, notably Zr, Rb, and REE. The rate of decline of these incompatible elements in each successive suite can be correlated directly with the amount of volcanism that has occurred in a given part of the Cascade chain (White and McBirney, 1978). The relationship is shown most clearly in Rb; in those regions where there has been a large amount of Tertiary igneous activity, the modern rocks are more depleted in this element than are those erupted in regions where the magnitude of earlier volcanism was less. At the same time, the isotopic ratio of Sr in basaltic rocks of the central Oregon Cascades has declined from $c.$ 0.7035 to $c.$ 0.7030.

Relations such as these indicate that the source of Cascade magmas has been somewhere in the mantle, possibly between the postulated subduction zone and the overlying continental lithosphere, but that a multistage process is required to explain all the temporal and compositional variations in the system as a whole. The decline of Fe contents and the increase of Na with time can be explained by postulating that the depth at which the basalts last equilibrated with the mantle has increased with time. Experimental studies of the compositions of melts in equilibrium with crystalline phases at various pressures (Kushiro, 1973; Mysen, 1974; Osborn and Watson, 1977) have shown that mantle liquids probably become richer in Na and poorer in Fe with increasing depth and pressure. This increase could come about as a result of progressive thickening of the lithosphere, both by accumulation of eruptive rocks at the surface and by underplating of the brittle layer with a solid residue from which the liquid fraction of each batch of magma was separated as it rose toward the surface.

The progressive depletion of lithophile elements with time may take place at the source or in the lithosphere through which the magma must pass *en route* to the surface. Progressive melting of a single source in the mantle could account for the observed decline of volumes, Fe/Mg ratios, silica, K, and other lithophile components, but similar variations might also result if successive batches of magma pass through the same section overlying the source and deplete the rocks of components that are selectively fractionated into the liquid. The first liquid to pass through a given section would scavenge more of these elements than successive ones following the same path.

The only way to discriminate between the individual effects of these two processes is to compare rocks from regions having different crustal structures and magmatic histories. Table 4 and Fig. 6 provide comparative data for Cainozoic lavas of several different regions and eruptive episodes. The most notable relations are, first, a marked increase in the concentrations and ranges of abundances of lithophile elements in rocks of the same differentiation index erupted at increasing distances toward the continental interior and, second, the decline in the concentrations of the same elements in rocks erupted in the same area over the course of time.

Similar increases of the concentrations of lithophile elements observed toward the interior of other provinces have been attributed to either an increasing depth of an inferred subducted slab or, alternatively, to the increasing thickness of crustal rocks from which rising magma can scavenge elements that are strongly partitioned into the liquid phase. The fact that the range of values increases along with the average abundance is in accord with the direct correlation found between these factors and crustal thicknesses (Condie and Potts, 1969; McBirney, 1976). If the high concentrations were more uniform in rocks of the same differentiation index, they could be explained as the result of equilibrium fractionation during melting or crystallization, but the fact that they vary so widely implies that the enrichment process is due, in large part, to the vagaries of contamination and assimilation.

This same conclusion could be reached from a comparison of rocks erupted in the same area over a period of two or more magmatic episodes. The fact that the abundances decline steadily with time shows that they cannot be attributed to a steady state process of subduction but are

TABLE 4 Selected representative data on orogenic igneous rocks of the Cascade Province compiled from various sources, including Condie and Swenson (1973), White and McBirney (1978), Rogers and Novitsky-Evans (1977), Rogers and Ragland (1980, and personal communication), and unpublished data from work in progress. Data 1A come from Oligocene to early Miocene rocks of the Calapooya Valley on the western edge of the Western Cascades. 2A, 2B, and 2C come from Oligocene–early Miocene, middle to late Miocene, and Quaternary rocks respectively in the area of Mount Jefferson in the central Oregon Cascades. 3C comes from the area around Mount Rainier in Washington, and 4A is from the early Tertiary Clarno Formation of central Oregon. DI = Si/3 − Mg − Ca + K

	1A	2A	2B	2C	3C	4A
Volume (km^3) in 2000 km^2	>6000	>10 000	27 000	4600	1380	
Percentages of rock types						
Basalt	32	10	39	85	33	
Andesite	59	45	41	13	60	
Dacite–rhyolite	39	45	20	82	7	
Weighted average SiO$_2$	56.1	62.4	57.3	52.7	58.1	
Lithophile elements at DI = 5						
K$_2$O (wt%)	1.2 ± 0.5	2.2	2.1	1.4	1.7	1.2 ± 0.6
Rb (p.p.m)	23 ± 10	46 ± 16	36 ± 10	16 ± 5	46	50 ± 21
La/Sm	3.7	2.9	3.45	3.92	3.95	—
Zr (p.p.m.)	163 ± 20	263 ± 35	205 ± 20	120 ± 20	174	235 ± 80
Ba (p.p.m.)	225	462	405	360	338	465 ± 120
Included elements at DI = 1.0						
Ni (p.p.m.)	17	22	70	65	55	66 ± 24
Cr (p.p.m.)	18	42	35	85	74	390 ± 10
^{87}Sr/^{86}Sr	0.7035	0.7035	0.7032	0.7030	0.7038	

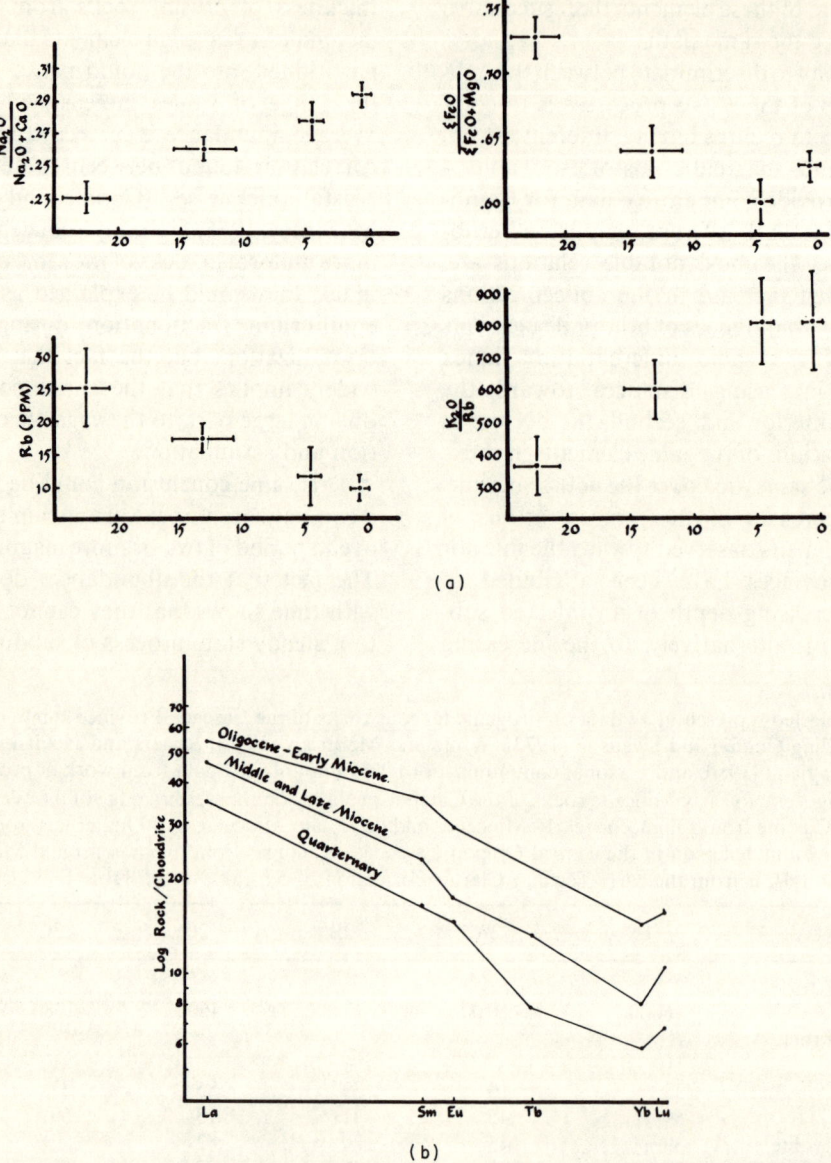

Fig. 5 (a) Variations of $Na_2O/(Na_2O + CaO)$, $\sum FeO/\sum(FeO + MgO)$, Rb, and K_2O/Rb with time in basaltic rocks of the central Oregon Cascade Range. Error bars show one standard deviation in chemical data and the range of radiometric dates for the same rocks. (b) Mean abundances of REE in radiometrically dated samples from the central Oregon Cascades. Data are normalized to the mean silica content of the middle and late Miocene group (58 per cent). (From White and McBirney, 1978)

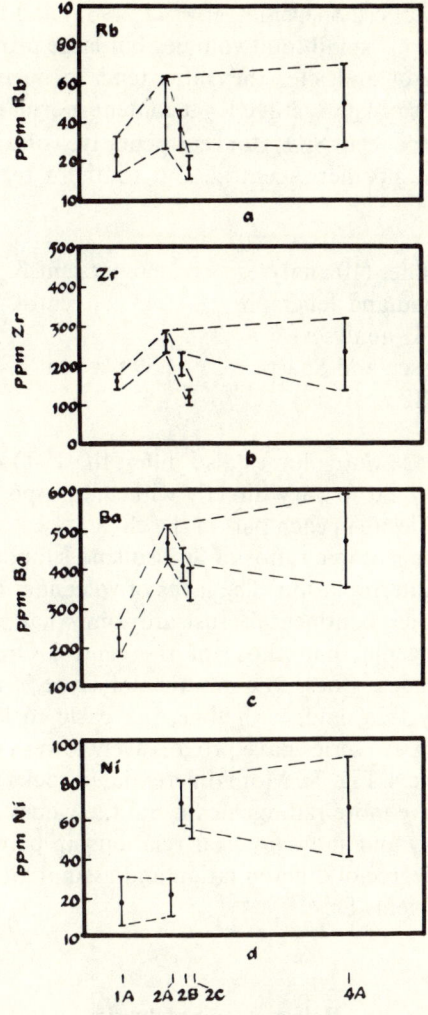

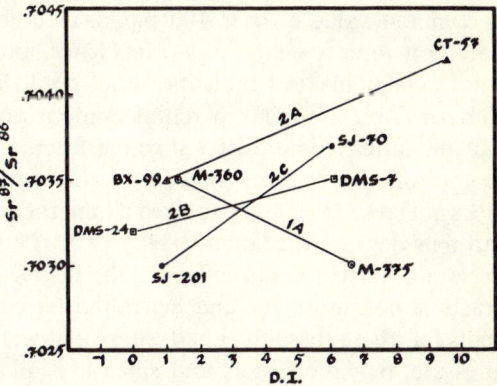

Fig. 7 Isotopic ratios of Sr in rocks of differing ages and differentiation indices. 1A, 2A, 2B, and 2C correspond to the groups of the same designation in Table 4. Determinations by R. L. Armstrong. *See* text for discussion

Fig. 6 Abundances of Rb, Zr, Ba, and Ni in rocks of differing geological settings and ages in central Oregon. Horizontal scale is proportional to the east–west distance from the continental margin; numbers and letters on the scale identify the groups according to the key in Table 4. *See* text for discussion

more probably the result of progressive depletion of mobile elements in the crustal rocks through which the magmas must rise. If the declines were caused by depletion of a fixed source in the mantle, the incompatible elements would be so strongly fractionated into the first liquid that an abrupt rather than gradual decline would be observed in successive melts from the same source.

A further insight into these processes can be gained from comparisons of the abundances of a strongly compatible element, such as Ni, in the same groups of rocks just considered. Because Ni has a high distribution coefficient in olivine, its concentration in liquids is very sensitive to fractionation of that mineral and reflects the conditions under which the magma last equilibrated before erupting.

The patterns of variations of Ni in space and time are quite marked (Fig. 6(d)). The spatial relations show that the abundance of this element in early Cainozoic basalts increases sharply toward the interior of the continent. With time, it also increases by approximately the same amount where successive igneous episodes have taken place in the same area. Unlike the lithophile elements, however, Ni does not have a continuous range in rocks at the same stage of differentiation (DI = 1.0). It tends to have one of two separate average abundances, and we find little or no overlap between the two groups. Tholeiitic basalts contain $c.\ 20 \pm 10$ p.p.m. Ni, whereas calc-alkaline basalts, regardless of their age, contain $c.\ 65 \pm 25$ p.p.m. The limited data we have on very primitive basalts of both types of suites indicate that Ni contents converge on a common value of over 100 p.p.m. at differentiation indices below c. −1.0. Hence, both the calc-alkaline and tholeiitic series could be derived from a similar parental composition, but the latter would evolve through fractionation of olivine, whereas the calc-alkaline series does not seem to have crystallized a mineral that strongly depletes the liquid in Ni.

Several factors could influence the degree of fractionation of olivine and hence the Ni contents of these basalts. Load pressure, water pressure, oxidation state, and silica and alkali contents of the liquid have all been shown to affect the stability field of olivine. The explanation that seems most consistent with the relations we observe is one based on a difference in water content and the relative roles of olivine and amphibole in early stages of differentiation. The amount of Ni fractionated from primitive basaltic magmas would be greatly reduced if amphibole replaced olivine as a liquidus phase. Numerous earlier workers have advocated amphibole as a key mineral in the evolution of calc-alkaline suites, and the increase in Ni contents which we find between the tholeiitic and calc-alkaline rocks of the Cascade system is consistent with this hypothesis.

When considered in terms of the geological setting of the various igneous centres and the patterns of variations of lithophile elements outlined in the preceding pages, the abrupt transition from Ni-poor tholeiitic to Ni-rich calc-alkaline basalts in space and time can be directly related to the nature and thickness of the crust and to the conditions under which olivine gives way to amphibole as an important liquidus phase. The boundary between the stability regimes of these two minerals must become deeper as the amphibolitic continental lithosphere thickens with time and distance from the continental margin.

There is ample evidence that the nature of the crust through which the Quaternary lavas of the High Cascades have risen had an important influence on the volumes and compositions of erupted rocks. Volcanoes at the southern and northern ends of the chain stand on thick continental crust (Dehlinger et al., 1965) and have relatively small total volumes but large proportions of andesite; they also tend to be more potassic. The average K_2O content (normalized to 60 per cent SiO_2) for representative volcanoes in the northern, central, and southern regions are

Rainier (10 analyses) 1.66 per cent K_2O
Hood and Jefferson   1.43 per cent K_2O
 (41 analyses)
Lassen and Shasta   1.59 per cent K_2O
 (10 analyses)

Because the volumes also differ, the K_2O contents tend to vary directly with the proportion of andesite in each part of the chain.

The isotopic ratios of Sr also lend support to this interpretation. The lavas of volcanoes built on thick continental crust are somewhat more radiogenic than those of the central Oregon Cascades, where the continental crust is relatively thin, and, with time, the basic rocks of successive series have progressively lower ratios (Table 4, Fig. 7). More differentiated rocks tend to have more radiogenic Sr, but the values vary widely and show no clear relationship between the degree of differentiation and assimilation of radiogenic Sr.

Relations to Subduction

Remarkably little evidence has been found to relate the Cascade magmas to oceanic lithosphere or sediments. Nearly all workers who have so far examined the petrological relations, trace elements, or isotopic compositions of the rocks have concluded that they reflect a primary mantle origin with little if any contribution from subducted crustal rocks (e.g. Smith and Carmichael, 1968; Peterman et al., 1970; Church and Tilton, 1973; Condie and Swenson, 1973; Church, 1976; White and McBirney, 1978).

No correlation has been found between the rates of production of igneous rocks in the Cascades and subduction along the adjacent plate boundary. The wide variation in the tempo of Cainozoic volcanism, not only in the Cascades but in the circum-Pacific as a whole, is in marked contrast to the nearly constant rates of sea-floor spreading deduced from the spacing of magnetic anomalies on the sea-floor (Kennett et al., 1977). The same appears to be true of the intensity of recent volcanic activity, which varies widely from place to place with no detectable relationship to calculated subduction rates or the amount of crustal material that could be consumed in trenches (McBirney, 1971). The lack of such relations seems to argue against generation of calc-alkaline magmas by flux-melting when water is released by dehydration of subducted hydrous phases and rises into the overlying mantle. It has been pointed out (Fyfe and McBirney, 1975) that there must be a direct relation between the amount of water introduced into the mantle and that of phlogopite reaching depths of 100–150 km in the down-going slab. Because the stability of phlogopite is directly dependent on the amount of K in the rocks, there should be a good correlation between the amount of magma generated and the rate at which the system recycles K (as well as other components that have high solubilities at elevated temperatures and pressures).

One can compare the amount of K entering the oceans from the continents with that being returned to the continents in calc-alkaline magmas and demonstrate a crude balance, at least in orders of magnitude over a period of 20 Ma (McBirney, 1976), but any such calculation is no better than the assumptions on which it rests, and in this case it is little more than an attempt to fit inadequate data to an unproven hypothesis.

Conclusions

It is still too early to offer a comprehensive synthesis of the complex igneous history of the Cascade region, and it would be even more premature to propose that the causes of orogenic volcanism can be discerned in the dim record of events as they are now seen. At best, one can only note a few salient features that have emerged from recent studies.

Igneous activity has been strongly episodic with distinct pulses separated by periods in which there was little volcanism. Some of the episodes seem to have occurred in unison in widely separated parts of the circum-Pacific region. The volumes and compositions of rocks produced during successive periods have varied widely, but there appears to have been an overall decline in the magnitude of volcanism and a gradual change toward more basic compositions. Most petrological evidence points toward generation of magmas in the mantle wedge overlying the zone of subduction and contamination with lithophile components as the magma rises toward the surface.

Perhaps the most important conclusion that can be drawn from the evidence now available is that there is little direct relation, other than a spatial one, between volcanism in the Pacific North-west and the subduction that is commonly thought to be associated with it. Even the spatial relation is somewhat ambiguous because, as the Tertiary record shows, andesitic volcanism has seldom been concentrated in long linear belts near the continental margin as it is today. Even during Quaternary time, andesitic volcanoes are by no means confined to island arcs and continental margins; many are found well within the continental interior.

The ultimate cause of andesitic volcanism in the Cascade province is still unknown.

Acknowledgements

This paper is essentially an updated version of a review of Cascade volcanism that was originally published in *Annual Reviews of Earth and Planetary Sciences* and is used here through the kind permission of the editors of that publication. Unpublished data on the rocks of the Three Sisters have been furnished by James Clark of the Center for Volcanology. Support for this work was furnished by the National Science Foundation, Grant No. GA-35129.

REFERENCES

Armstrong, R. L. (1978). Cenozoic igneous history of the U.S. Cordillera from 42° to 49°N latitude. *Geol. Soc. Am. Mem.* no. 152, 265–282.

Baldwin, E. M. (1965). Geology of the south end of the Oregon Coast Range Tertiary basin. *Northwest Sci.* **39**, 93–103.

Church, S. E. (1976). The Cascade Mountains revisited: a re-evaluation in light of new lead isotopic data. *Earth planet. Sci. Letters* **29**, 175–178.

Church, S. E. and Tilton, G. R. (1973). Lead and strontium isotopic studies in the Cascade Mountains: bearing on andesite genesis. *Geol. Soc. Am. Bull.* **84**, 431–454.

Condie, K. C. and Potts, M. J. (1969). Calc-alkaline volcanism and the thickness of the early Precambrian crust in North America. *Can. J. Earth Sci.* **6**, 1179–1184.

Condie, K. C. and Swenson, D. H. (1973). Compositional variation in three Cascade stratovolcanoes: Jefferson, Rainier, and Shasta. *Bull. Volc.* **37**, 205–230.

Dehlinger, P., Chiburis, E. F., and Collver, M. M. (1965). Local travel-time curves and their geologic implications for the Pacific Northwest states. *Seismol. Bull. Soc. Am.* **55**, 587–608.

Dickinson, W. R. (1962). Petrogenetic significance of geosynclinal andesite volcanism along the Pacific margin of North America. *Geol. Soc. Am. Bull.* **73**, 1241–56.

Dickinson, W. R. (1970). Relations of andesites, granites, and derivative sandstones to arc-trench tectonics. *Rev. Geophys. Space Phys.* **8**, 813–860.

Fisher, R. V. and Rensberger, J. M. (1972). Physical stratigraphy of the John Day formation, central Oregon. *Univ. Calif. Publ. geol. Sci.* no. 1.

Fyfe, W. S. and McBirney, A. R. (1975). Subduction and the structure of andesitic volcanic belts. *Am. J. Sci.* **275A**, 285–297.

Gilluly, J. (1963). The tectonic evolution of the western United States. *Q. J. geol. Soc. Lond.* **119**, 133–174.

Greene, R. C. (1968). Petrography and petrology of volcanic rocks in the Mount Jefferson area, High Cascade Range, Oregon. *US geol. Surv. Bull.* no. 1251–G.

Hamilton, W. (1969). The volcanic central Andes—a modern model for the Cretaceous batholiths and tectonics of western North America. *Oregon Dept geol. Min. Industries Bull.* no. 65, 175–184.

Hay, R. L. (1963). Stratigraphy and zeolitic diagenesis of the John Day Formation of Oregon. *Univ. Calif. Publ. geol. Sci.* **42**, 119–262.

Kennett, J. P., McBirney, A. R., and Thunell, R. C. (1977). Episodes of Cenozoic volcanism in the circum-Pacific region. *J. Volc. geothermal. Res.* **2**, 145–163.

Kushiro, I. (1973). Partial melting of garnet lherzolites from kimberlite at high pressures. In *Lesotho Kimberlites* (P. H. Nixon, ed.), Lesotho National Development Corp., Maseru, Lesotho, pp. 294–299.

McBirney, A. R. (1971). Thoughts on some current concepts of orogeny and volcanism. *Comments on Earth Sciences. Geophysics* **2**, 69–76.

McBirney, A. R. (1976). Some geologic constraints on models for magma generation in orogenic environments. *Can. Mineral.* **14**, 245–254.

McBirney, A. R. (1978). Volcanic evolution of the Cascade Range. *A. Rev. Earth planet. Sci.* **6**, 437–456.

McBirney, A. R. (1980). A model for assessing the genetic characteristics of orogenic igneous rocks. *Geol. Soc. Am. Mem.* in press.

McBirney, A. R., Sutter, J. F., Naslund, H. R., Sutton, K. G., and White, C. M. (1974). Episodic volcanism in the central Oregon Cascade Range. *Geology* **2**, 585–589.

Mysen, B. O. (1974). The oxygen fugacity (f_{O_2}) as a variable during partial melting of peridotite in the upper mantle. *Carnegie Instn. Wash. Yb.* **73**, 237–240.

Noble, D. C. (1972). Some observations on the Cenozoic volcano–tectonic evolution of the Great Basin, Western United States. *Earth planet. Sci. Letters* **17**, 142–150.

Osborn, E. F. and Watson, E. B. (1977). Studies of phase relations in subalkaline volcanic rock series. *Carnegie Instn. Wash. Yb.* **76**, in press.

Peck, D. L. (1964). Geologic reconnaissance of the Antelope–Ashwood area, north-central Oregon. *US geol. Surv. Bull.* no. 1161–D.

Peck, D. L., Griggs, A. B., Schlicker, H. G., Wells, F. G., and Dole, H. M. (1964). Geology of the central and northern parts of the Western Cascade Range in Oregon. *US geol. Surv. prof. Pap.* no. 449.

Peterman, Z. E., Carmichael, I. S. E., and Smith, A. L. (1970). $^{87}Sr/^{86}Sr$ ratios of Quaternary lavas of the Cascade Range, northern California. *Geol. Soc. Am. Bull.* **81**, 311–318.

Rau, W. W. (1966). Stratigraphy and foraminifera of the Satsop River area, southern Olympic Peninsula, Washington. *Wash. Dept. Conserv., Div. Mines Geol. Bull.* no. 53.

Ritchey, J. L. (1980). Divergent magmas at Crater Lake, Oregon: products of fractional crystallization and vertical zoning in a shallow, water-saturated chamber. *J. Volc. geothermal. Res.* **7**, 373–386.

Robyn, T. L. (1977). Geology and Petrology of the Strawberry Mountain volcanic series, central Oregon. Ph.D. dissertation, University of Oregon, Eugene.

Rogers, J. J. W. and Ragland, P. C. (1980). Trace elements in continental margin magmatism: Part I. Trace elements in the Clarno Formation of central Oregon and the nature of the continental margin on which eruptions occurred: summary. *Geol. Soc. Am. Bull.* **91**, 196–198.

Rogers, J. J. W. and Novitsky-Evans, J. M. (1977). The Clarno formation of central Oregon, USA: volcanism on a thin continental margin. *Earth planet. Sci. Letters* **34**, 56–66.

Simpson, R. W. and Cox, A. (1977). Paleomagnetic evidence for tectonic rotation of the Oregon Coast Range. *Geology* **5**, 585–589.

Smith, A. L. and Carmichael, I. S. E. (1968). Quaternary lavas from the southern Cascades. *Contr. Mineral. Petrol.* **19**, 212–238.

Snavely, P. D. and Wagner, H. C. (1963). Tertiary history of western Oregon and Washington. *Wash. Dept Conserv., Div. Mines Geol., Rep. Invest.* no. 22.

Snavely, P. D., Wagner, H. C., and MacLeod, N. S. (1965). Preliminary data on compositional data variations of Tertiary rocks in the central part of the Oregon Coast Range. *The Ore Bin* **27**, 101–117.

Snavely, P. D., Wagner, H. C., and MacLeod, N. S. (1968). Tholeiitic and alkalic basalts of the Eocene Siletz River volcanics, Oregon Coast Range. *Am. J. Sci.* **266**, 454–481.

Sutton, K. G. (1974). Geology of Mount Jefferson. M.Sc. thesis, University of Oregon, Eugene.

Thayer, T. P. (1937). Petrology of later Tertiary and Quaternary rocks of the north-central Cascade Mountains in Oregon. *Geol. Soc. Am. Bull.* **48**, 1611–1652.

Walker, G. W. (1970). Cenozoic ash-flow tuffs of Oregon. *The Ore Bin* **32**, 97–115.

Walker, G. W., Greene, R. C., and Pattee, E. C. (1966). Mineral resources of the Mt Jefferson Primitive Area, Oregon. *US geol. Surv. Bull.* no. 1230-D.

White, C. M. and McBirney, A. R. (1978). Some quantitative aspects of orogenic volcanism in the Oregon Cascades. *Geol. Soc. Am. Mem.* no. 152, 369–388.

Wise, W. S. (1969). Geology and petrology of the Mt Hood area: a study of High Cascade volcanism. *Geol. Soc. Am. Bull.* **80**, 969–1006.

Mexico

C. Robin

Laboratoire de Géodynamique Sous-Marine,
Université Pierre et Marie Curie, 06230 Villefranche-sur-Mer, France

ABSTRACT

Mexico is characterized by two large magmatic provinces: the Mexican volcanic belt (MVB), composed mainly of andesitic rocks erupted during the Miocene and during the Pliocene–Quaternary, and the eastern alkaline province of over- and undersaturated basaltic rocks of Oligocene to Quaternary age. These provinces are difficult to reconcile with the classical concept of a circum-Pacific volcanic zone. Mexico can be understood as a transition zone between areas affected by subduction of the Cocos plate, and areas affected by oblique extension as a result of north–south rifting in the eastern alkaline province and the intracontinental faults of the Mexican volcanic belt. Simple models relating magmatism to a subduction zone are therefore unlikely to be valid.

The MVB can be divided into two sections: a northern area (2.5–1.5 Ma old) and the neo-volcanic chain (less than 1.6 Ma old), each with two major types of associations: (1) *Major basalt–andesite–dacite suite*. These have characteristics of an intermediate (subalkaline to alkaline) magma type: high ($K_2O + Na_2O$) and low Al_2O_3 contents, and the occurrence of titanaugite. The $^{87}Sr/^{86}Sr$ ratios, Ni, Cr, and Mg/Fe contents, and REE and large ion lithophile element concentrations suggest formation as a result of a low degree of partial melting of mantle peridotite. For these magmas the subducted lithospheric plate has not contributed directly to the composition of primary magmas, but has been indirectly involved by the release of hydrous fluids which transported volatiles and incompatible elements into the mantle. Such influence varies with the distance to the trench. (2) *Basalts, basic andesites, and andesites*. These have high Sr, Ni, Cr, Co, and low K and Ba concentrations and high Fe/Mg and K/Rb ratios in relation to SiO_2, so that these lavas have much in common with island arc andesites. Very marked enrichment of U and Th (and also K, Rb in some cases) and increase of La/Tb may result either from contamination of the mantle by liquids or fluids derived from the oceanic plate, or from the presence of garnet-rich bodies in the source as proposed for well known models on andesite magmatism.

Introduction

The Mexican volcanic belt (MVB) crosses the land mass over almost 1000 km between 19 and 21°S in an east–west direction slightly oblique to the Middle America Trench: the distance between the MVB and the trench ranges from 180–200 km at its western end to 350–450 km in the east (Fig. 1). The MVB is on average 100 km in width and is cut by a sequence of grabens that are oblique in relation to the general trend. The oblique trend of the MVB and the occurrence of grabens indicate a zone of crustal weakening traversing the continental crust, which is c. 40 km in thickness (Mooser, 1969; Cummings and Schiller, 1971). According to Mooser (1972), this zone of weakness might represent the rejuvenation of a former geosuture, and the

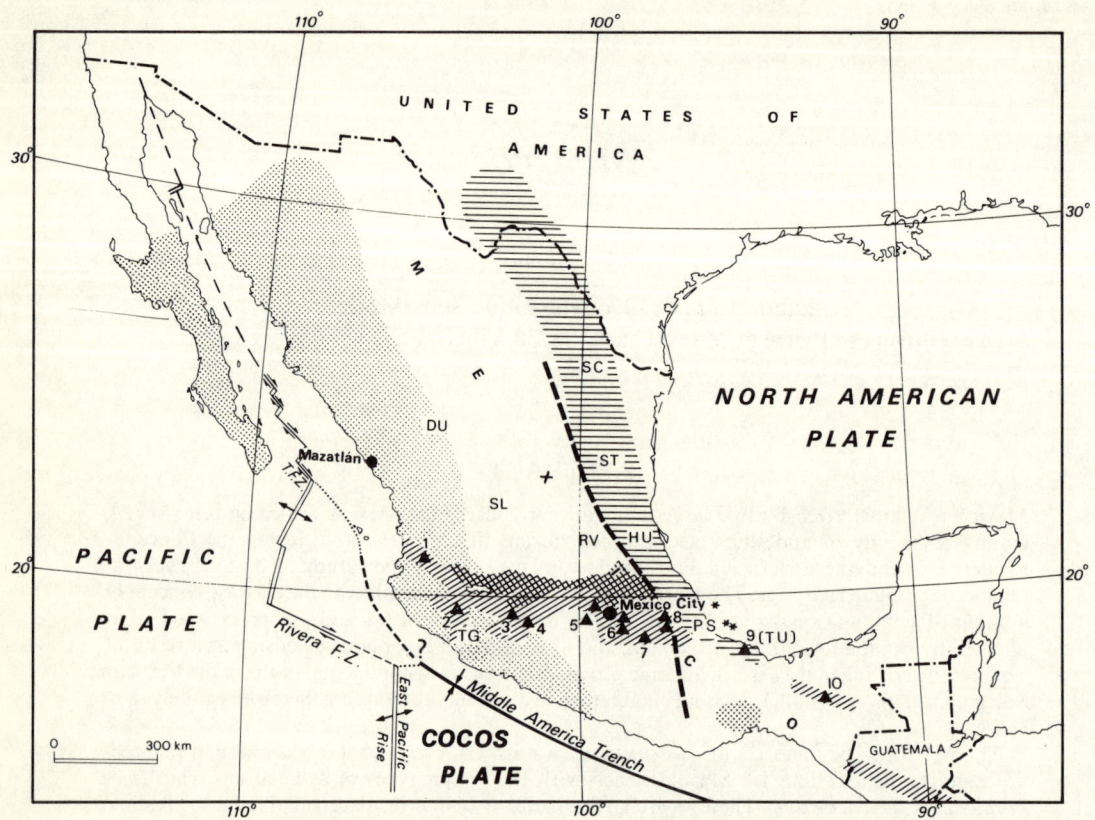

Fig. 1 Distribution of Cainozoic–Recent volcanic rocks in Mexico in relation to plate tectonics and continental structure. Plate boundaries in the western Pacific are from Atwater (1970) and Larsen (1972). TFZ is the Tamayo Fracture Zone. The fine stipple is the Oligocene–Miocene Cordilleran Province, the diagonal lines correspond to the Mexican Volcanic Belt (MVB) and the horizontal lines indicate the distribution of alkaline rocks. The MVB is divided into two parts (*see* text). Recent and active volcanoes of the MVB (▲) are numbered as follows: 1, Ceboruco; 2, Colima; 3, Paricutin; 4, Jorullo; 5, Nevado de Toluca and Ajusco–Xitle complex; 6, Popocatepetl–Iztacchautl; 7, Malinche; 8, Sierra Negra–Orizaba–Cofre de Perote; 9, San Martin; 10, Chichon (related to the Guatemalan Cordillera). The asterisks north-east of San Martin indicate the positions of Pliocene–Recent volcanoes within the Gulf of Mexico. The volcanic centres within the alkaline provinces are labelled as follows: RV, Rio Verde; SL, San Luis; and DU, Durango, are alkaline centres of the Altiplano; SC, Sierra de San Carlos (Oligocene–Miocene); ST, Sierra de Tamaulipas (Miocene–Quaternary); HU, Huasteca plain (Miocene–Pliocene); PS, Palma Sola (Pliocene–Quaternary); TU, San Martin Tuxtla (Quaternary) represent the major centres of the Oriental Province. The fault zone at the altiplano border is indicated by the broken line from SC to south of the MVB. TG is the Tepic Graben. This fault emphasizes a structural break between the Mexican land mass and the Gulf of Mexico–Caribbean region

zigzag arrangement of grabens might reflect parting of blocks away from the southern continental margin. At its western end, the MVB bends towards the Gulf of California; for this reason, and noting the extensional tectonic characteristics of the province and the oblique position of the MVB in relation to the trench, some authors have proposed a relationship between the MVB and the Gulf (Mooser and Maldonado-Koerdell, 1961; Guenther, 1972; Negendank, 1972). However, only the Tepic graben, from its trend and the direction of extension, would seem to be related to the opening of the Gulf of California during the last 5 Ma. Seismic data (Molnar and Sykes, 1969; Dean and Drake, 1978) also indicate a relationship between the MVB and subduction of the Cocos plate in the Middle America Trench.

Boundaries of Current Andesitic Belt within Tectonic Framework

The active volcanoes of the MVB are built upon a largely continuous Miocene andesitic belt that runs parallel with the Pacific coast for over c. 4000 km from California to Guatemala. The northern outcrops of this province are in Lower California (Gastil et al., 1975; Demant and Robin, 1975). However, the present andesitic belt no longer reaches the Gulf and the eastern area is characterized by north–south rifting which orthogonally crossed the earlier province during late Miocene and Pliocene. The rifted area has thinned the continental crust (25 km; Hales, 1971) and is characterized by alkaline and hyperalkaline magmatic suites that resemble those of the African rift system. This alkaline magmatism was initiated during the Eocene in Texas (the Trans-Pecos Province; Parker and McDowell, 1973; Barker, 1977), during the Oligocene and Miocene in the north–east of Mexico (Bloomfield and Cepeda, 1973; Cantagrel and Robin, 1979), during the Upper Miocene at Palma Sola (PS, Fig. 1), and has only recently reached San Andres Tuxtla (TU, Fig. 1). The intersection of this area of rifting with the MVB is linked with 400 km southward extension, between 8 and 3 Ma, of the faults that separated the Altiplano from the coastal plains (Fig. 1). Dyke swarms occur along the fracture zone. So, in the east, the present MVB terminates at Pico de Orizaba, whereas Palma Sola and San Andres volcanoes belong to the eastern alkaline province (Robin, 1976; Robin and Tournon, 1978).

Thus, the southern part of the North American continent is the junction of major structural features related to plate tectonics and the superposition of various tectonic styles (Robin, 1976; Thorpe, 1977). The south-western area is characterized by the collision of the Cocos Plate and the East Pacific Rise with the Mexican land mass, and the accretion and transformation zone of the Gulf of California (Atwater, 1970; Malfait and Dinkelman, 1972; Lynn and Lewis, 1976). The eastern area corresponds to the continental rift that merges in the south with the tensional margin of the Gulf of Mexico. These two structures are separated by the Trans-Mexican tectonic province, an area that is also characterized by extensional tectonics. Simple models relating magmatism with a subduction zone can consequently be excluded, especially if we accept Menard's hypothesis (1978), which involves pivoting subduction of the Cocos Plate. As the rate of convergence varies according to the distance from the pivot, it may determine various aspects of volcanism of the MVB and account for the obliquity of MVB in relation to the trench.

MVB Evolution from Miocene to Quaternary

Early phases of activity

Because of its east–west trend, the Plio-Quaternary province can be distinguished from Miocene andesitic volcanism. The Miocene axis is poorly known; during the Upper Miocene (ages ranging from 9 to 6 Ma; Cantagrel and Robin, 1979), the southern section of the andesitic cordillera was displaced towards the east. This arrangement heralded the future Plio-Quaternary volcanic range. The latter was initiated after a pause in calc-alkaline volcanism between 6 and 2.5 Ma ago, which corresponded to an eastward advance of alkaline volcanism and extensional tectonics in the west, particularly the east and continental margins, and to the opening of the Gulf of California to the west. The geochronological data are consistent with the proposals of Kennett et al. (1977) who suggested that a new episode of andesitic volcanism in Central America was initiated c. 2 Ma ago, in contrast to the period from 10 to 2 Ma ago, when volcanism was scarcely represented.

Recent studies (Cantagrel and Robin, 1979; C. Robin, manuscript in preparation) divide the Plio-Quaternary province into two sections. The northern area consists of basalts and relatively undifferentiated lavas (plateau basalts), the petrological characters of which are intermediate between alkaline and calc-alkaline, and mainly formed between 2.5 and 1.5 Ma ago. The type series can be found near Pachuca (Hidalgo) or in the north of Guanajuanto and Jalisco states (see Fig. 1). In the southern area the Plio-Quaternary rocks overlap in area with the volcanic rocks of the Altiplano province. The oldest volcanic rocks in the southern area are 1.6 Ma in age and include andesites, and basalts with characteristics like those of the northern area.

The active MVB consists of 10 major central volcanoes or volcanic systems, and of several thousand simple monogenetic cones. These cones include basalt to acidic andesite compositions (over 60 per cent SiO_2) but are generally of basaltic andesite (52–55 per cent SiO_2) composition. Most of the activity is less than 700 ka in age (Mooser et al., 1974; Cantagrel and Robin, 1979). In the western section of the MVB, Sanganguey, Ceboruco, Tequila, and Colima volcanoes are medium-sized; reaching 1000–2500 m above the basement. The central area contains Nevado de Toluca (4650 m). In the east are the largest two volcanic systems: that of Popocatepetl (5450 m) accompanied by the eroded volcano Iztaccihuatl, and Pico de Orizaba (5675 m) which, together with Sierra Negra and Cofre de Perote, forms a small north–south range c. 80 km in length. Malinche is located between Pico de Orizaba and the Popocatepetl–Iztaccihuatl system. Immediately east of Orizaba, the eastern province exhibits an east–west geochemical break characterized by the occurrence of lavas containing feldspathoids. Within the MVB there is no gradual chemical zonation within calc-alkaline suites, and genuine shoshonites do not exist.

Volcanic forms and products of volcanic activity

The Upper Pliocene–Lower Quaternary belt (mainly 2.5–1.6 Ma old) has relatively undifferentiated plateau lavas associated in the most recent suites with youthful cones. Magmatic evolution in the high volcano complexes of the active range is controlled by differentiation in crustal magma chambers. After the establishment of an andesitic basement, a major stage in volcanic evolution was the eruption of dacitic pyroclastic flows and the formation of calderas, some of which include domes. Peléan-type eruptions, which in many cases contain the most differentiated products (amphibole dacites), resulted from partial destruction of these domes. In addition there have been St Vincent-type hot avalanches, nuées ardentes, and Plinian eruptions (dacitic or rhyodacitic pumice falls) prominent at Nevado de Toluca and Pico de Orizaba, as well as dacitic or rhyolitic eccentric domes. Large recent calderas, 10–15 km in diameter, with associated ignimbrites, occur in the northern part of the MVB, in the Guadalajara area, and at the north-western side of the Pico de Orizaba–Cofre de Perote andesitic chain (Teziutlan area). The primavera complex, near Guadalajara, was active c. 100 ka ago (Mahood, 1977) and the age of the Los Humeros caldera (Teziutlan area) seems likewise recent.

Petrology, Geochemistry, and Petrogenesis

The chemical composition of recent volcanic rocks from the eastern part of the MVB are summarized in a plot of $Na_2O + K_2O$ against SiO_2 in Figure 2.

Pliocene and Lower Quaternary suites (2.5–1.6) Ma)

In normative terms, the basic lavas lie between the critical planes of silica over- and undersaturation. Normative compositions of most of the basalts have either a little quartz (less than 3 per cent) or hypersthene + olivine and c. 12 per cent have nepheline (less than 3 per cent). The mineral paragenesis in these microlitic lavas is ol (Fo_{80}–Fo_{83})–pl–cpx–Fe–Ti oxides; pigeonite or orthopyroxene are absent. In the less aluminous basalts high Ti (1.7–2.4 per cent TiO_2) is associated with the presence of titanaugite ($Ca_{38}Mg_{44}Fe_{18}$ with 2–2.5 per cent TiO_2). The chemical characteristics of the differentiated lavas lie between alkaline and subalkaline domains; both K-rich and Na-rich trachy-

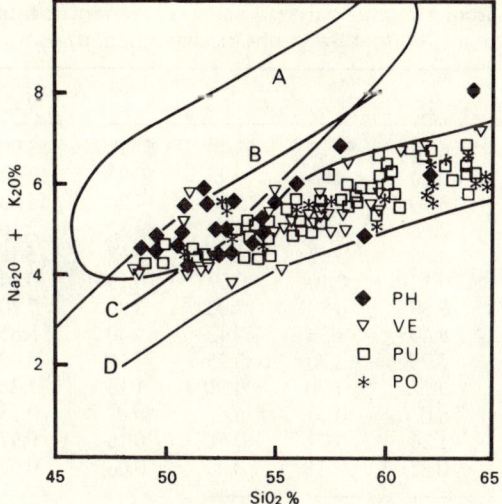

Fig. 2 Plot of $Na_2O + K_2O$ against SiO_2 for Recent andesitic rocks from the eastern part of the MVB (from Toluca to the Gulf of Mexico). A, Alkaline volcanic rocks and intrusions ('traps') from the border zone of the altiplano (cf. Fig. 1); B, limit of alkaline (above) and subalkaline (below) associations after Irvine and Baragar (1971); C, limit of alkaline (above) and high alumina basalt series of Kuno (1959); D, limit of high alumina basalt series (above) and tholeiitic series (below) of Kuno (1959). PH, north part of the MVB; VE, PU, and PO are from the southern part of the MVB, with VE north-east of Orizaba, PU, south-west of Orizaba, and PO, Popocatepetl. (cf. Robin and Tournon, 1978)

basaltic types are observed. In the south, comparable lower Quaternary units near the Pico de Orizaba or within the Valley of Mexico just to the north of Mexico City mark the southern limit of this volcanic formation. The petrography and chemical characteristics of basalts from the southern area are identical to those in the north. However, early precipitation of Fe–Ti oxides with olivine causes a silica increase that is sufficient to allow appearance of opx in the mesostasis of the intermediate lavas (54–56 per cent SiO_2).

In the north of the province, at the centre of Altiplano (RV and SL, Fig. 1), plateau basalts overlie Pliocene sediments. They differ from the contemporaneous lavas at Atotonilco in the north–east of Mexico) by their higher alkali contents (average 1.40 per cent K_2O for 15 basalts analysed), very high TiO_2 (1.8–3.2 per cent), and lower Al_2O_3 and SiO_2 concentrations. These alkali basalts are always nepheline normative and, assuming a contemporaneous age, their presence suggests a certain north–south chemical zonation during the early Quaternary (C. Robin, unpublished data).

$^{87}Sr/^{86}Sr$ ratios, ranging between 0.7041 and 0.7044 (Cantagrel and Robin, 1978), high Ni, Cr contents, and Mg/Fe ratios of basalts, REE, and large ion lithophile element concentrations (Table 1) suggest an origin from low degrees of partial melting of garnet peridotites and this suggestion is consistent with light REE/heavy REE ratios. On the other hand, the D_{Ni}^{ol-liq}, D_{Fe-Mg}^{ol-liq}, and D_{Ni-Mg}^{ol-liq} partition coefficients (C. Robin, manuscript in preparation) suggest temperatures of equilibration between 1150 and 1220 °C, and pressures greater than or equal to 10 kb that indicate, according to the Ni values in whole rocks, some differentiation at depth. The low K, Rb contents of these basalts in relation to the basalts of the MVB (*see* later) and the comparatively low $^{87}Sr/^{86}Sr$ ratios tend to minimize any contribution from fluids released by the subducted plate, as suggested by the hypotheses of Ninkovich and Hays (1972) or Best (1975).

Active andesitic province (*younger than 1.5 Ma*)

Two types of associations are known:

(1) *Basalt–andesite–dacite suites.* Basalts are sometimes absent, e.g. at Ceboruco (Thorpe and Francis, 1976) and at Nevado de Toluca where the compositional range covers the interval 57–66 per cent SiO_2. In the east of Mexico, especially in the Pico de Orizaba area, where volcanic suites seem to have a wider compositional range, cogenetic relationships between basalts and andesites have been proposed (Robin and Nicolas, 1978) and a continuous suite of lavas from basalts (49 per cent SiO_2) to rhyodacites (67–68 per cent SiO_2) is present.

(2) *Basalt–andesitic basalts (51–53 per cent SiO_2), basic andesites (53–56 per cent SiO_2) and more seldom andesites, from monogenetic cones.*

TABLE 1 Chemical analyses of representative volcanic rocks from Mexico. Analyses 1 and 2 (PH series) are from the northern section, analyses 3-13 are from the southern section (PU, south-west of Orizaba; VE, north-east of

	1	2	3	4	5	6	7	8	9	10
SiO_2	49.00	56.90	50.60	58.6	61.20	56.10	55.10	55.40	53.90	53.00
TiO_2	1.75	1.45	1.20	0.70	0.65	0.70	0.90	0.95	0.95	0.90
Al_2O_3	16.40	14.50	15.30	15.30	17.10	13.90	15.10	15.10	14.90	15.60
Fe_2O_3	6.04	3.21	5.12	4.03	2.22	2.85	3.09	3.53	3.88	2.16
FeO	3.57	6.93	3.86	1.82	2.33	3.65	3.62	2.87	3.62	5.18
MnO	0.17	0.23	0.15	0.09	0.08	0.12	0.10	0.09	0.13	0.14
MgO	7.55	1.90	8.70	4.55	2.60	6.50	6.80	8.00	8.80	7.70
CaO	8.90	4.50	8.80	6.10	5.20	8.40	8.30	7.40	9.40	8.65
Na_2O	3.55	4.00	3.40	4.10	4.65	3.25	3.20	3.75	3.10	3.00
K_2O	0.90	3.00	1.05	1.60	2.00	1.55	1.70	1.70	1.15	1.35
P_2O_5	n.d.	0.60	n.d.	n.d.	n.d.	n.d.	0.20	n.d.	n.d.	0.15
H_2O^+	0.73	1.41	0.98	1.19	0.87	1.26	0.79	0.61	0.16	0.97
H_2O^-	0.24	0.47	0.39	0.77	0.00	0.23	0.41	0.00	0.08	0.08
Total	98.80	99.10	99.65	98.85	98.90	98.51	99.31	99.40	100.07	98.78
Rb	13.5	74	20	33	44	22	25	25		25
Sr	612	391	399	507	501	561	915	775		557
Ba	400	750	355	470	640	350	535	405		375
Cs	0.07	0.73	0.27	0.89	1.34	1.41		n.d.		1.32
Co	30	13	32	11	5	17	30	40		32
Cu	27	<5	55	24	16	36	35	20		
Ni	126	<5	132	26	14	53	50	180		60
Cr	278	50	428	59	27	218	220	220		260
U	0.5	1.77	0.61	1.30	1.51	2.07				1.77
Th	1.84	6.98	1.98	3.67	4.07	6.73				6.52
Zr	251	818	146	151	162	121				126
Hf	4.82	16.7	3.19	3.93	4.01	2.91				3.44
Ta	0.95	2.58	0.33	0.33	0.47	0.19				0.33
La	21	61.1	13.8	16	17.8	14.1				19.6
Eu	2.16	5.08	1.57	1.28	1.17	1.21				1.42
Tb	0.75	2.01	0.58	0.45	0.44	0.43				0.52
K/Rb	553	332	435	400	377	580	564	564		298
Ba/Sr	0.65	1.91	0.84	0.92	0.62	0.62	0.58	0.52		0.67
Rb/Sr	0.022	0.189	0.05	0.064	0.087	0.039	0.027	0.032		0.044
Th/Ta	1.93	2.70	6.0	11.1	8.6	35				19
La/Tb	28	30	23.5	35	40.5	33				37.6

(1) Transitional basalt, Atotonilco Province; (2) trachybasalt, Atotonilco Province; (3) calc-alkaline basalt; (4) andesite; (5) andesite (3–5; Pico de Orizaba); (6) basaltic andesite, monogenetic cone; (7) monogenetic cone, (6 and 7; Orizaba); (8) basaltic andesite, monogenetic cone, Valle de Puebla; (9) basaltic andesite, monogenetic cone, Orizaba; (10) basalt; (11) andesite; (12) tholeiitic basalt (10–12; Pico de Orizaba); (13) transitional basalt, Popocatepetl; (14) andesite, Nevado de Toluca; (15)

These cannot be chemically correlated with the preceding suites. They include some lavas described in the western part (for example Paricutin; Wilcox, 1954) or in the Navado de Toluca area (Bloomfield, 1975). In the eastern area, from Mexico City to the Gulf of Mexico, the Ni contents (55–185 p.p.m.), Cr contents (85–285 p.p.m.), and MgO/FeO ratios (c. 1.25 with MgO contents up to 9 per cent) for basic andesites containing 54–56 per cent SiO_2 suggest that these may be primary or relatively undifferentiated lavas. Their eruption was associa-

Orizaba; PO, Popocatepetl), analyses 14 and 15 are from the central section (Toluca), and 16–18 are from the western MVB (16, Colima). Analysis 19 is from the Atliplano border in the east

	11	12	13	14	15	16	17	18	19
SiO_2	58.60	48.30	50.80	57.30	64.30	57.63	47.50	51.00	47.45
TiO_2	0.90	1.65	1.60	0.90	0.55	0.90	1.40	1.70	2.22
Al_2O_3	16.50	15.70	15.80	18.00	15.90	18.04	17.00	17.30	15.60
Fe_2O_3	3.42	2.50	2.40	6.90	1.85	7.11	6.12	3.51	2.57
FeO	1.96	7.66	7.30		1.75		4.85	6.38	9.14
MnO	0.07	0.18	0.16	0.10	0.07	0.12	0.20	0.17	0.17
MgO	4.30	8.50	8.50	3.90	1.80	3.63	8.90	5.70	8.65
CaO	6.30	9.60	8.20	6.70	3.95	6.80	10.30	8.10	9.70
Na_2O	4.00	3.40	3.65	4.00	4.50	4.54	3.00	3.65	3.90
K_2O	1.70	0.70	0.90	1.80	1.95	1.29	0.40	1.10	0.80
P_2O_5	n.d.	n.d.	0.50	n.d.	n.d.	0.18	0.25	n.d.	0.45
H_2O^+	0.83	0.80	0.00	0.34	2.32		0.08	0.35	0.05
H_2O^-	0.20	0.00	0.07	0.56	0.28		0.05	0.61	0.00
	98.78	98.99	99.88	100.50	99.22	100.25	100.05	99.57	100.70
Rb	25	11.5	21			15	5	35	17
Sr	905	416	413			656	335	450	782
Ba	385	146	260				220	380	311
Cs		0.26	0.51					n.d.	0.34
Co	35	36	50				65	50	48
Cu	15		20				70	70	
Ni	40	123	130			15	175	100	93
Cr	20	237	225			13	230	145	210
U		0.42	0.60						0.68
Th		1.53	2.14						3.02
Zr		156	206			121			228
Hf		3.34	4.63						4.9
Ta		0.60	1.17						1.97
La		11.9	18.5						31.3
Eu		1.4	1.99						2.4
Tb		0.64	0.81						0.88
K/Rb	564	505	357			713	664	304	390
Ba/Sr	0.42	0.35	0.62				0.65	0.84	0.39
Rb/Sr	0.027	0.027	0.050			0.022	0.014	0.077	0.021
Th/Ta		2.5	1.8						4.44
La/Tb		18	22						35.6

andesite, Nevado de Toluca; (16) andesite, Colima (R. S. Thorpe, personal communication); (17) basalt; (18) basalt (17, 18; near Guadalajara); (19) transitional basalt, eastern Altiplano border (related to alkaline province).

Major elements and Rb, Sr, Ba, Co, Cu, Ni, Cr analyses determined at the University of Clermont-Ferrand (Laboratory for Volcanology) and U. Th, Zr, Hf, Ta, La, Eu, Tb at CEA SACLAY.

ted with local extension affecting the Valle de Mexico or Valle de Puebla.

Basalt-andesite dacite suites

In the first association, there is no primary opx in basalts; opx microlites are restricted to dacitic residual glass. The early precipitation of Fe–Ti oxide + spinels + ol is followed by that of cpx (augites: around $Ca_{38}Mg_{50}Fe_{13}$ with 4.5 per cent Al_2O_3), opx appears between 53 and 55 per cent SiO_2. Most of the composite volcano lavas contain two pyroxenes + plagioclase.

Rocks in which SiO$_2$ exceeds 60 per cent contain amphibole; they are mainly represented in post-caldera lavas: dacites of *nuées ardentes* often have very much more amphibole than opx + cpx. Biotite-bearing rocks are scarce. Basic andesites and andesites of the second group have the assemblage ol + cpx + opx. Plagioclase is scarce or absent in these porphyric glassy lavas.

East of Mexico City, where the province reaches its broadest width, basalts in the south of the range are richer in K and Rb than those in the north. This might be explained by a supply of volatile products released by the oceanic plate and by scavenging of these elements inside the mantle. As the andesitic cordillera is distant from the trench, such a source may be more important as the flow of volatile products is larger nearer to the trench, in contrast to many provinces in the world and to the well known K–h relationship (Hatherton and Dickinson, 1969); $^{87}Sr/^{86}Sr$ ratios (Moorbath et al., 1978; Whitford and Bloomfield, 1976; Cantagrel and Robin, 1978) lead to the same conclusion. These are all consistent with an origin of magmas in the mantle and suggest that any contribution from the continental crust is unimportant, in contrast to andesites in margins with thick continental crust (James et al., 1976; Briqueu and Lancelot, 1978). Furthermore, K and Rb are less abundant than in other continental margin provinces; conversely, Ni, Cr, Co concentrations are very high: in the andesites of the east, 80–120 p.p.m. Ni and 100–200 p.p.m. Cr are common values in andesites containing 55–62 per cent SiO$_2$.

Correlations between Rb, Sr, Ba, and K can be used to comment on various petrogenetic hypotheses; there is a good K–Rb correlation in most of the series; within near-primary magmas, deviations from this correlation correspond to lavas with low K/Rb ratios (Rb increases sharply in relation to K) and with high Ba concentrations. Other lavas, basalts, or basic andesites (high Ni, Cr concentrations, Mg/Fe ratios), have high K/Rb ratios, with high Sr and low Ba (Sr very much higher than Ba). The K/Rb, Ba/Sr, and Rb/Sr ratios cannot be accounted for either by contamination or by fractional crystallization processes (C. Robin, manuscript in preparation). From the stability conditions of the minerals that control Rb, Sr, K, and Ba in the mantle (amphiboles, micas, and clinopyroxenes; Kushiro et al., 1967; Modreski and Boettcher, 1972; Forbes and Flower, 1974; Mysen and Boettcher, 1975; Beswick, 1976) and from published $D^{sol/liq}$ (Philpotts and Schnetzler, 1970; Hart and Brooks, 1974; Shimizu, 1974; Arth, 1976) estimated for basalts and primary andesites (c. Robin, manuscript in preparation), the following origins are possible.

(1) *Basalts with low K/Rb and high Ba*: low degrees of partial melting of amphibole-bearing peridotites in the upper mantle at less than 75 km, or at depths exceeding 75 km within stability field of phlogopite alone. Under the latter conditions the low K/Rb and high Ba imply a very low degree of melting and total disappearance of mica. The last requirement can be easily satisfied as the melting temperature of mica is lowered (1190 ± 30 °C at 35 kb; Modreski and Boettcher, 1972) when the latter co-exists with enstatite and vapour, as might occur in the presence of fluids released by the oceanic plate. For some basalts in the VE area (north-east of Orizaba; Fig. 1) low La/Tb ratios indicate an origin by partial melting of amphibole- or spinel-peridotite.

(2) *Basalts with 'normal' K/Rb, Sr, and Ba*: most of the suites in eastern Mexico (*see*, for example, PU 17). Variable degrees of melting of spinel- or garnet peridotites, with no influence of the subducted oceanic plate other than fluid supply, with K, Rb, light REE (and possibly Th), causing the melting.

Basalt-andesitic basalt-basic andesite suites

Basalts, basic andesites, and andesites of the second association have high K/Rb, high Sr, and low Ba (high Ni, Cr, Co, and Mg/Fe ratios in relation to SiO$_2$); the high K/Rb ratios result from the very low Rb values, but K is also low and these lavas have much in common with calc-alkaline andesites of island arcs. As Rb and Ba do not increase in the same way as Sr, contamination by continental crust may be excluded and this is consistent with the $^{87}Sr/^{86}Sr$ evidence (Moorbath et al., 1978; Cantagrel and Robin, 1978). It seems that micas are the only minerals that can control high Sr and very low Ba/Sr during melting of the mantle or subducted oceanic lithosphere. SiO_2 contents are also compatible with the hypothesis of reaction (and/or fractional crystallization) of andesitic or dacitic fluids from the oceanic plate with the mantle (Green and Ringwood, 1968; Ringwood, 1974; Stern, 1974; Fyfe and McBirney, 1976; Stern and Wyllie, 1978). Partial melting of new bodies of 'enriched mantle' would produce the andesite primary magmas (Ringwood, 1974; Thorpe et al., 1976). Kay (1978) explained the petrogenesis of high Sr magnesian andesites using a similar model. In this model, dacitic liquid from the subducted plate reacts with mantle peridotite to produce an andesitic melt. A very marked enrichment of U and Th and increase in La/Tb (Table 1) characterizes primary basic andesites. This may result either from contamination within the mantle or from the presence of garnet-rich bodies in the sources.

Discussion

The characteristics of the volcanic suites of Mexico demonstrate the varied influence of subduction over magma composition. Subduction has no effect on alkaline and transitional lavas along the side of plateaux emplaced during rest or slowing periods of subduction, very small effects on Plio-Quaternary suites in the north of the province, and small effects (supply of fluids) on most of the calc-alkaline volcanic series of the MVB. Only a small part of the primary magmas would have been directly affected by fluids from the oceanic plate; this is consistent with absence of gradual variations in the chemical composition from west to east. In the western part of the belt, very low Rb-high Sr andesites (10–20 p.p.m. Rb for SiO_2 57–61 per cent) such as those from the volcano Colima (R. S. Thorpe, personal communication) cannot be considered as differentiated products from basalts (see 16, Table 1, for example). As in the eastern part, the western MVB has a variety of primary magmas: basalts and basic andesites (and andesites?) related to the varied role of fluids expelled from the slab

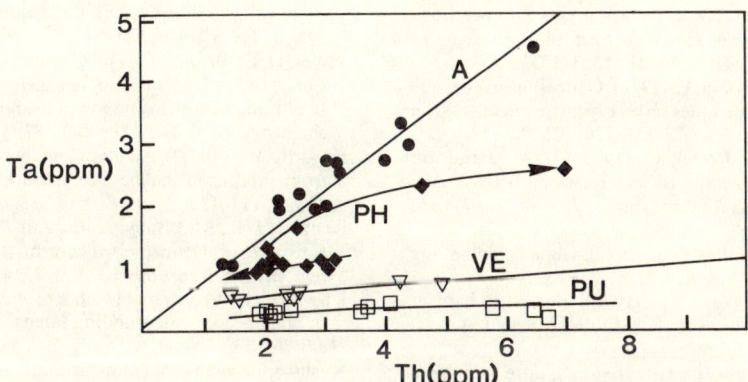

Fig. 3 Plot of Ta against Th for recent volcanic rocks from the eastern part of the MVB (cf. Fig. 2) and the alkaline province A, alkaline volcanic rocks and intrusions ('traps') from the border of the altiplano (cf. Fig. 1). PH, VE, and PU are as in Fig. 2. (cf. Robin and Tournon, 1978.) For the PH series two curves are shown; the lower curve is for basalts and shows the behaviour of Th and Ta during partial melting; the upper curve is for more evolved lavas and shows the behaviour of Th and Ta during fractional crystallization. The arrows indicate increasing degrees of fractional crystallization and partial melting.(cf. Treuil, 1973)

during subduction. The role of such fluids may increase towards the west, where the MVB is nearer to the trench, and this might account for the abundance of monogenetic cones with primary or little differentiated lavas in the west (type (2)), in comparison with the dominance of basalt–andesite–dacite associations (type (1)) in the east. Also, in eastern Mexico there is a distinction between magmatic suites related to the action of subduction and from transitional basalts to andesites of the MVB (Fig. 3). The Ta/Th variations suggest Th enrichment in the areas where the release of fluids is abundant, or variations of $D_{Ta}^{sol-liq}$ during the melting process. In the second case, the variations of D would be a consequence of mineralogical changes between the sources of transitional basalts ('normal' mantle) and those of andesitic suites (more 'enriched' mantle).

REFERENCES

Arth, J. G. (1976). Behaviour of trace elements during magmatic processes. A summary of theoretical models and their applications. *J. Res. US geol. Surv.* **4**, 41–47.

Atwater, T. (1970). Implications of plate tectonics for the cenozoic tectonic evolution of western North America. *Geol. Soc. Am. Bull.* **81**, 3513–3536.

Barker, D. S. (1977). Northern Trans-Pecos magmatic Province: introduction and comparison with the Kenya rift. *Geol. Soc. Am. Bull.* **88**, 1421–1427.

Best, M. G. (1975). Migration of hydrous fluids in the upper mantle and potassium variations in calc-alkalic rocks. *Geology* **3**, 429–432.

Beswick, A. E. (1976). K and Rb relations in basalts and other mantle derived materials. Is phlogopite the key? *Geochim. cosmochim. Acta* **40**, 1167–1183.

Bloomfield, D. K. and Cepeda, L. (1973). Oligocene alkaline igneous activity in NE Mexico. *Geol. Mag.* **110**, 551–555.

Bloomfield, K. (1975). A late-Quaternary monogenetic volcano field in Central Mexico. *Geol. Rundr.* **64**, 476–497.

Briqueu, L. and Lancelot, J. (1978). Nouvelles données analytiques et essai d'interprétation des compositions en strontium des laves calco-alcalines plio-quaternaires du Pérou. *Bull. Soc. Géol. Fr.* **19**, 1223–1232.

Cantagrel, J. M. and Robin, C. (1978). Géochimie isotopique du strontium dans quelques séries types du volcanisme de l'Est mexicain. *Bull. Soc. Géol. Fr.* **20**, 935–939.

Cantagrel, J. M. and Robin, C. (1979). K–Ar dating on eastern Mexican volcanic rocks—relations between the andesitic and the alkaline provinces. *J. Volc. geothermal Res.* **5**, 99–114.

Cummings, D. and Schiller, G. I. (1971). Isopach map of the Earth's crust. *Earth Sci. Rev.* **7**, 77–125.

Dean, B. W. and Drake, C. L. (1978). Focal mechanism solutions and tectonics of the Middle America Arc. *J. geol.* **86**, 111–128.

Demant, A. and Robin, C. (1975). Les quatre provinces volcaniques du Mexique. Relations avec l'Evolution géodynamique depuis le Crétacé: les deux provinces occidentales. *C.r. hebd. Séanc. Acad. Sci. Paris D* **280**, 1437–1440.

Forbes, W. C. and Flower, M. J. F. (1974). Phase relations of titan-phlogopite, $K_2Mg_4TiAl_2O_{20}$: a refractory phase in the upper mantle? *Earth planet. Sci. Letters* **22**, 60–66.

Fyfe, W. S. and McBirney, A. R. (1976). Subduction and the structure of andesitic volcanic belts. *Am. J. Sci.* **275A**, 285–297.

Gastil, R. C., Phillips, R. P., and Allison, E. C. (1975). Reconnaissance geology of the state of Baja California. *Geol. Soc. Am. Mem.* no. 140.

Green, T. H. and Ringwood, A. E. (1968). Genesis of the calc-alkaline igneous rocks suite. *Contr. Mineral. Petrol.* **18**, 105–162.

Guenther, E. W. (1972). Vulkanismus und Tektonik in Mexiko. *Schr. naturwissench. Verk. Schleswieg-Holstein Dt.* **42**, 21–34.

Hales, A. L. (1971). Crustal and upper mantle structure in the region of the Gulf of Mexico. *Bol. Soc. geol. Mex.* **32**, 63–70.

Hart, S. R. and Brooks, C. (1974). Clinopyroxene–matrix partitionning of K, Rb, Cs, Sr and Ba. *Geochim. cosmochim. Acta* **38**, 1799–1806.

Hatherton, T. and Dickinson, W. R. (1969). The relationship between andesitic volcanism and seismicity in Indonesia, the Lesser Antilles, and other island arcs. *J. Geophys. Res.* **74**, 5301–5310.

Irvine, T. N. and Baragar, W. R. A. (1971). A guide to the chemical classification of the common volcanic rocks. *Can. J. Earth Sci.* **8**, 523–548.

James, D. E., Brooks, C., and Cuyubama, A. (1976). Andean Cenozoic volcanism magma genesis in the light of strontium isotopic composition and trace element geochemistry. *Geol. Soc. Am. Bull.* **87**, 592–600.

Kay, R. W. (1978). Aleutian magnesian andesites: melts from subducted Pacific Ocean crust. *J. Volc. geothermal Res.* **4**, 117–132.

Kennett, J. P., McBirney, A. R., and Thunell, R. C. (1977). Episodes of Cenozoic volcanism in the circum-Pacific region. *J. Volc. geothermal Res.* **2**, 145–163.

Kuno, H. (1959). Origin of Cenozoic petrographic provinces of Japan and surrounding areas. *Bull. Volcanol.* **20**, 37–76.

Kushiro, I., Syono, Y., and Akimoto, S. (1967). Stability of phlogopite at high pressures and possible presence of phlogopite in the Earth's upper mantle. *Earth planet. Sci. Letters* **3**, 197–203.

Larsen, R. L. (1972). Bathymetry, magnetic anomalies and plate tectonic history of the mouth of the Gulf of California. *Geol. Soc. Am. Bull.* **83**, 3345–3360.

Lynn, W. S. and Lewis, B. I. R. (1976). Tectonic evolution of the Northern Cocos plate. *Geology* **4**, 718–722.

Mahood, G. A. (1977). A preliminary report on the comenditic dome and ash flow complex of Sierra La Primavera,

Jalisco. *Univ. Nal. Auton. Mexico, Inst. Geologia, Revista* **1**, 177–190.

Malfait, B. T. and Dinkelman, M. G. (1972). Circum-Caribbean tectonic and igneous activity and the evolution of the Caribbean plate. *Geol. Soc. Am. Bull.* **83**, 251–272.

Menard, H. W. (1978). Fragmentation of the Farallon plate by pivoting subduction. *J. Geol.* **86**, 99–110.

Modreski, P. J. and Boettcher, A. L. (1972). The stability of phlogopite and enstatite at high pressures: a model for micas in the interior of the Earth. *Am. J. Sci.* **272**, 852–869.

Molnar, P. and Sykes, L. R. (1969). Tectonics of the Caribbean and Middle America region from focal mechanism and seismicity. *Geol. Soc. Am. Bull.* **80**, 1639–1684.

Moorbath, S., Thorpe, R. S., and Gibson, I. L. (1978). Strontium isotope evidence for petrogenesis of Mexican andesites. *Nature, Lond.* **271**, 437–439.

Mooser, F. (1969). The Mexican volcanic belt. Structure and development—formation of fractures by differential crustal heating. Paper presented at the Pan-American Symposium on the Upper Mantle, Mexico, 1968, Group II. *Inst. Geophys. UNAM, Mexico* **22B**, 15–22.

Mooser, F. (1972). El Eje volcanico Mexicano, debilidad cortical prepaleozoica reactivada en el terciaro. *Soc. geol. Mex. Mem.* no. II, 186–188.

Mooser, F. and Maldonado-Koerdell, M. (1961). Tectonica pene contemporanea a lo largo de la costa Mexicana del Oceano Pacifico. *Geofisica Int.* **1**, 3–20.

Mooser, F., Nairn, A. E. M., and Negendank, J. F. W. (1974). Palaeomagnetic investigations on the Tertiary and Quaternary igneous rocks. VIII. A palaeomagnetic and petrologic study of volcanics of the valley of Mexico. *Geol. Rundschau* **63**, 451–483.

Mysen, B. O. and Boettcher, A. L. (1975). Melting of a hydrous mantle. II. Geochemistry of crystals and liquids formed by anatexis of mantle peridotite at high pressures and high temperatures as a function of controlled activities of water, hydrogen and carbon dioxide. *J. Petrol.* **16**, 549–593.

Negendank, J. F. W. (1972). Volcanics of the valley of Mexico. *Neues Jb. Mineral. Abh.* **116**, 308–320.

Ninkovich, D. and Hays, J. D. (1972). Mediterranean island arcs and origin of high potash volcanoes. *Earth planet. Sci. Letters* **16**, 331–345.

Parker, D. F. and McDowell, F. W. (1973). K–Ar geochronology and eruptive history of Oligocene volcanic rocks, Davis Mountains, Trans-Pecos, Texas. *Geol. Soc. Am. Abstr. Programs* **5**, 764–765.

Philpotts, J. A. and Schnetzler, C. C. (1970). Phenocrysts–matrix partition coefficients for K, Rb, Sr and Ba, with applications to anorthosite and basalt genesis. *Geochim. cosmochim. Acta* **34**, 307–322.

Ringwood, A. E. (1974). Petrological evolution of island arc systems. *J. geol. Soc. Lond.* **130**, 183–204.

Robin, C. (1976). Présence silmutanée de magmatismes de signification tectonique opposée dans l'Est du Mexique. *Bull. Soc. Géol. Fr.* **18**, 1617–1625.

Robin, C. (1981). Les relations entre magmatismes alcalin et calco-alcalin: l'exemple mexicain. Thèse Doctorat d'Etat, Université de Clermont-Ferrand, France.

Robin, C. and Nicolas, E. (1978). Particularités géochimiques des suites andésitiques de la zone orientale de l'axe trans-mexicain, dans leur contexte tectonique. *Bull. Soc. Géol. Fr.* **20**, 193–202.

Robin, C. and Tournon, J. (1978). Spatial relations of andesitic and alkaline provinces in Mexico and Central America. *Can. J. Earth. Sci.* **15**, 1633–1641.

Shimizu, N. (1974). An experimental study of the partitioning of K, Rb, Cs, Sr and Ba between clinopyroxene and liquid at high pressure. *Geochim. cosmochim. Acta* **38**, 1789–1798.

Stern, C. R. (1974). Melting products of olivine tholeiite basalt in subduction zone. *Geology* **2**, 227–230.

Stern, C. R. and Wyllie, P. J. (1978). Phase compositions through crystallization intervals in basalt–andesite–H_2O at 30 kb with implications for subduction zone magmas. *Am. Mineral.* **63**, 641–663.

Thorpe, R. S. (1977). Tectonic significance of alkaline volcanism in eastern Mexico. *Tectonophysics* **40**, 19–26. a major composite volcano of the Mexican volcanic belt.

Thorpe, R. S. and Francis, P. W. (1976). Volcan Ceboruco: *Bull. Volc.* **34**, 1–13.

Thorpe, R. S., Francis, P. W., and Potts, P. J. (1976). Rare earth data and petrogenesis of andesites from the N. Chilean andes. *Contr. Mineral. Petrol.* **54**, 65–78.

Treuil, M. (1973). Critères pétrologiques, géochimiques et structuraux de la genése et de la différenciation des magmas basaltiques: exemple de l'Afar. Thése Doctorat d'Etat, Université de Paris-Sud (Orsay).

Whitford, D. J. and Bloomfield, K. (1976). Geochemistry of late Cenozoic volcanic rocks from the Nevado de Toluca area, Mexico. *Carnegie Instn. Wash. Yb.* **75**, 203–213.

Wilcox, R. E. (1954). Petrology of Paricutin, Mexico. *US geol. Surv. Bull.* no. 965C, 281–353.

Central America

Michael J. Carr, William I. Rose, and Richard E. Stoiber

Department of Geological Sciences,
Rutgers University, New Brunswick, New Jersey 08903, USA,

Department of Geology and Geological Engineering,
Michigan Technological University, Houghton, Michigan 49931, USA

and

Department of Earth Sciences,
Dartmouth College, Hanover, New Hampshire 03755, USA

ABSTRACT

The Central American volcanic chain is the result of convergence between the Cocos and Caribbean plates. The volcanic front is divided into eight segments, separated by zones of transverse faulting and offsets and changes in strike and dip of the seismic zone. The active volcanoes, most of which are probably younger than 100 ka, were preceded by an earlier Quaternary volcanic front, and by several pulses of volcanic activity during the Tertiary.

The active volcanoes are clearly related to contemporaneous seismic activity. Great shallow thrust earthquakes are preceded by lulls in volcanic activity and followed by periods of intense volcanic activity. The most active volcanoes overlie portions of the inclined seismic zone that are largely aseismic, but there are concentrations of seismic activity at intermediate depths located just updip from these aseismic regions. This suggests strongly that magma is produced near the top of the descending slab.

The Quaternary lavas and pyroclastic rocks have mineralogical and chemical characteristics typical of calc-alkaline rocks. At most volcanoes there are continuous fractionation trends from the abundant, high-alumina basalts and basaltic andesites to rare dacites. Very large volumes of contemporaneous rhyolitic pumice in Guatemala appear to have originated by a different mechanism or from a different source. There is relatively little regional variation along the volcanic front. Na_2O contents of basaltic lavas decrease gradually from Guatemala to Nicaragua. K_2O contents are the same in various segments of the arc but tend to be lower at the segment boundaries.

Geological Setting of Active Volcanoes

The Quaternary volcanic chain in Central America extends from the Guatemala–Mexico border to central Costa Rica, a distance of 1100 km. This volcanic chain is clearly the result of plate convergence between the Cocos and Caribbean plates (Molnar and Sykes, 1969; Dengo et al., 1970). The south-eastern end of the chain in central Costa Rica coincides approximately with a proposed fracture zone marking the edge of the Cocos plate (Van Andel et al., 1971). In central and western Guatemala the Polochíc, Motagua, and Jocotán fault system marks the Caribbean–North American plate boundary (Molnar and Sykes, 1969). Although it is not clear how this plate boundary extends to a triple junction with the Cocos plate, it is clear that the Cocos–Caribbean plate interaction produces a distinctly different pattern and rate of volcanic activity in Central America from that produced by the Cocos–North America plate convergence in Mexico (Stoiber and Carr, 1973; Robin, this volume, Section III). Therefore, the Central American volcanic belt is the result of plate convergence bounded to the north-west by a transform zone cutting the overriding plate and bounded to the south-east by a transform zone cutting the underthrusting plate.

The active volcanoes in Central America form eight distinct lineaments which are parallel to the trench, do not overlap, and are separated by changes in strike or offsets of the volcanic front (Fig. 1). The breaks in the volcanic chain outline the Quaternary structure of Central America. Several other geological and geophysical data (Table 1) indicate transverse-

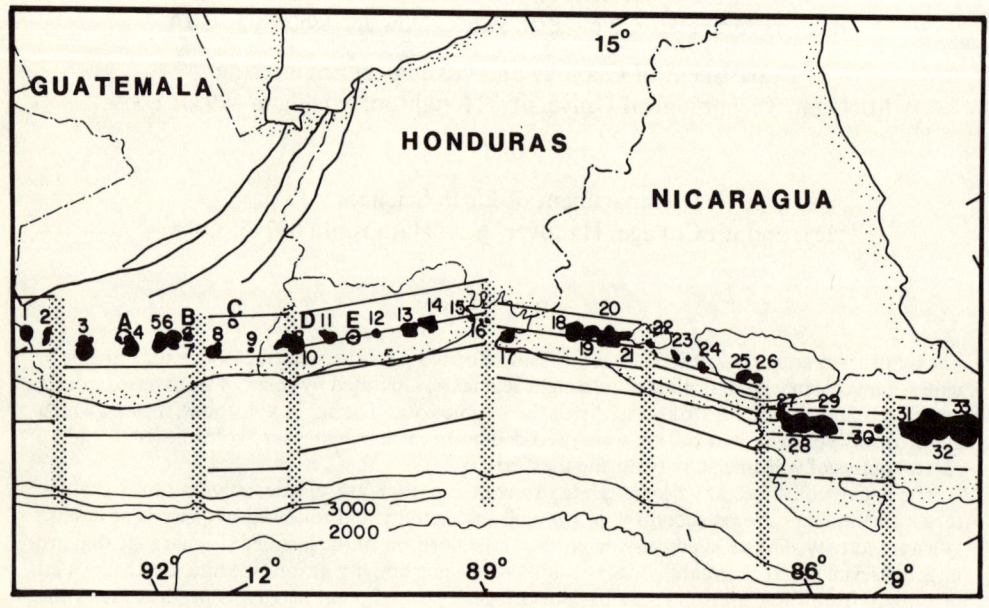

Fig. 1 Segments of the volcanic front in Central America. Black areas are active volcanic centres. Stippled bars mark transverse breaks in the arc. Thin parallel lines are isobaths to the inclined seismic zone. Contour nearest the trench is 50 km. Contour interval is 50 km

Basalt–andesite–dacite volcanic centres

1. Tacaná
2. Tajamulco
3. Santa Mariá–Santiaguito
4. Atitlán
5. Fuego
6. Agua
7. Pacaya
8. Tecuamburro
9. Moyuta
10. Santa Ana–Izalco
11. Boqueron

12. San Vicente
13. Tecapa
14. San Miguel
15. Conchagua
16. Fonseca Islands
17. Cosiguina
18. San Cristobal
19. Telica
20. Cerro Negro
21. Momotombo
22. Nejapa

23. Masaya
24. Apoyo–Mombacho
25. Concepción
26. Madera
27. Orosí
28. Rincón de la Vieja
29. Miravalles
30. Arenal
31. Poás
32. Barba
33. Irázu

Rhyolite centres

A. Atitlán
B. Amatitlan
C. Ayarza
D. Coatepeque
E. Ilopango

TABLE 1 Evidence for transverse breaks that divide Central America into segments

	References
1. Offsets and changes in strike of volcanic lineaments	Stoiber and Carr, 1973
2. Contrasting volcano morphology between segments	Stoiber and Carr, 1973
3. Contrasting eruption styles between segments	Stoiber and Carr, 1973
4. Krakatoan-style eruptions common near breaks	Stoiber and Carr, 1973
5. Clusters of small monogenetic basaltic volcanoes common behind the volcanic front near breaks	Stoiber and Carr, 1973
Seismological evidence	
1. Concentrations of shallow earthquakes near breaks	Stoiber and Carr, 1973
2. Clusters of large, shallow earthquakes at breaks	Carr and Stoiber, 1977
3. Breaks form lateral margins of focal areas of great shallow earthquakes	Carr and Stoiber, 1977
4. Changes in the strike and dip of the inclined seismic zone between segments	Carr, 1976; Carr et al., 1979
5. Changes in the dip of shallow thrust focal mechanisms between segments	Dean and Drake, 1978
6. Transversely oriented strike-slip focal mechanisms in the underthrust zone at breaks	Dean and Drake, 1978
Structural evidence for transverse structures	
1. Offsets or changes in strike of longitudinal depressions	Stoiber and Carr, 1973
2. Large scale topographic changes	Stoiber and Carr, 1973
3. Transverse fault zones	Carr, 1976
4. Offsets of gravity and magnetic anomalies	Brown et al., 1973; Ladd et al., 1978

striking breaks which divide the Central American convergent plate margin into discrete segments. The combined volcanological, seismological, and structural data indicate that every 100–300 km along the plate convergence zone there are structures that strike transverse to the plate margin and cut both the underthrust and overriding lithospheres. Below the seismic zone the segments of underthrust slab can descend into the asthenosphere with different strike and dip directions. The narrow linear segments of the volcanic chain trend parallel to the strikes of the segments of underthrust lithosphere, which suggests that melting occurs at a common depth on each segment. The segment concept is shown schematically in Fig. 2.

There are petrologically distinct groups of recent volcanoes in Central America which occur in three distinct structural settings (Carr et al., 1979). The first structural setting, the volcanic front, is the narrow zone of volcanism that comprises the eight volcanic segments. Since the volcanic front is parallel to the inclined seismic zone and discontinuous at the same breaks, it is clearly and intimately related to this seismic zone. The calc-alkaline volcanic centres of the front are primarily basaltic to andesitic composite cones with associated domes and parasitic cinder cones. A few large centres in this zone have produced voluminous pumiceous rhyolitic ignimbrite and air-fall deposits.

The second structural setting for volcanoes in Central America is the region immediately landward of the volcanic front and adjacent to the transverse breaks. Here, quartz- and olivine-normative basalts and basaltic andesites are common and nepheline-normative basalts are present. Several contemporaneous rhyolite obsidian domes make the distribution of lavas distinctly bimodal. Most of these volcanoes are monogenetic scoria cones and small shield volcanoes forming clusters located 25–100 km behind the volcanic front. These volcanoes are concentrated in zones of normal faulting near several of the transverse breaks and appear to be related to zones of crustal thinning which are common near the transverse breaks. None of these volcanoes has been active in historic time.

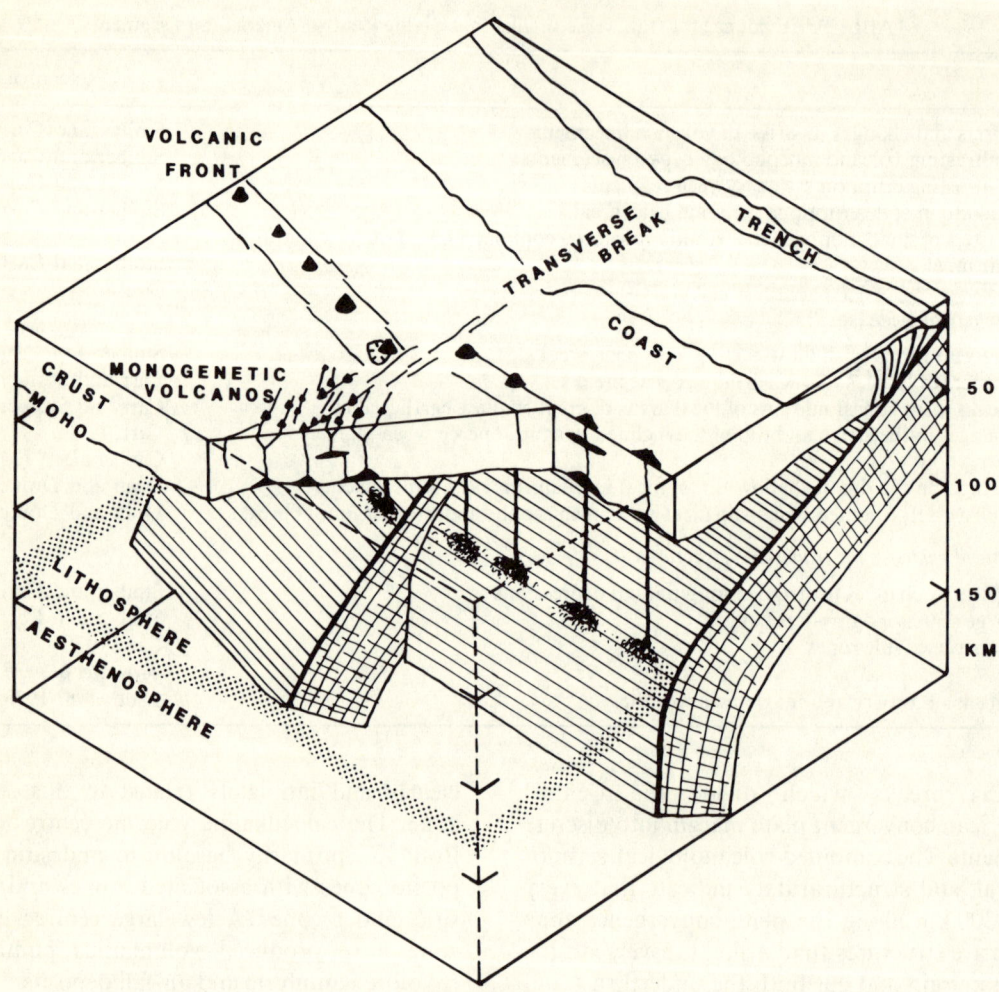

Fig. 2 Block diagram illustrating principle features of a transverse break in Central America. The lined slabs are segments of underthrust lithosphere separated by a break. At the break there are discontinuities in the trench, coastline, and volcanic front. A Krakatoan caldera (hatched circle) is near the break. A cluster of small monogenetic volcanoes occurs behind the volcanic front along the transverse break. The stippled zone marks the transition from lithosphere to astherosphere. Magma for the volcanic front rises from the narrow zone where the lithosphere, the underthrust slab, and the asthenosphere meet. Magma for the monogenetic volcanoes comes from a domal intrusion of asthenosphere into the base of the lithosphere at the transverse break

The third structural setting is far behind the volcanic front where strongly undersaturated lavas have been erupted from small scoria cones that have no apparent spatial relation to the inclines seismic zone (Pichler and Weyl, 1977).

The greatest volume of volcanic products, and all historic eruptions, have come from the volcanic front, which is clearly the focus of volcanic activity in Central America. The following sections will, therefore, be primarily concerned with the geology, recent eruptive history, and petrochemistry of the volcanic front, especially with the relatively well studied sections of the volcanic front in Guatemala, El Salvador, and Nicaragua.

Cainozoic Volcanic History

Until recently Cainozoic volcanic rocks in Central America were arbitrarily divided into Quaternary or Tertiary, based solely on whether the rocks were part of a constructional or erosional landform. This crude criterion is fairly successful because the current pulse of volcanic activity was apparently preceded by a substantial period of low or no volcanic activity which allowed erosion to reduce effectively most Tertiary volcanoes. However, even the Quaternary volcanic centres need further subdivision. Most Quaternary volcanic centres have recorded historic eruptions or fumarolic activity or a very juvenile shape and low degree of erosion. Several volcanic centres appear to be early Quaternary in age, now deeply eroded and probably extinct, but still recognizable as constructional forms. Many of these centres are off the narrow lineaments that comprise the presently active volcanic front. In Guatemala volcanoes of this age are stratigraphically below the voluminous air-fall and ignimbrite layers (H-tephra or Los Chocoyos ash) that cover much of central Guatemala. This marker horizon, which erupted from the Atitlán basin before the formation of the currently active volcanoes, has been correlated with deep sea ash layers dated at c. 84 ka ago (Drexler et al., 1981). Therefore, the current front is very young. The ages of young volcanoes along the front have been estimated by various methods. Based on paleomagnetic data Rose et al. (1977) suggested an age of 30–40 ka for Santa Mariá. Extrapolation of present (1900–79) eruption rates of Fuego volcano suggests an age of only 10 ka (Martin, 1979). Correlation of ^{14}C-dated ashes from Ayarza, Guatemala, with fine rhyolite ashes interbedded with scoria on the slopes of Agua volcano (Peterson and Rose, 1979) shows that Agua's activity began over 23 ka ago.

Recent geochronological (McBirney et al., 1974) and stratigraphic studies (Wiesemann, 1975; Reynolds, 1977) have begun to clarify the Tertiary volcanic history. K–Ar dating has revealed pulses of volcanic activity with a 5 Ma periodicity (McBirney et al., 1974). These pulses of volcanic activity have apparently been synchronous throughout much of the circum-Pacific region. In Central America two rather clear pulses occurred at 0–1 Ma (Quaternary), 3–6 Ma (Pliocene), and a long period of intense activity, 9–17 Ma, was centred at about 14 Ma (middle Miocene). This middle Miocene volcanism was the most voluminous and widespread. Existing data do not reveal anything systematic about earlier periods.

Lithostratigraphic variations, described first in El Salvador (Wiesemann, 1975), have been correlated throughout northern Central America (Reynolds, 1977). Three formations can be defined from middle Miocene to late Pliocene. The more poorly exposed early Tertiary volcanics have not been adequately subdivided.

The earliest widely recognizable lithostratigraphic unit is the Chalatenango formation. This unit of silicic volcanics (Middle to Upper Miocene) lies in the northern or landward portion of the Tertiary volcanic belt. The second unit, the Balsamo Formation (Upper Miocene to Pliocene) comprises andesitic lavas which erupted on the south side of the Tertiary volcanic belt. The third unit, the Cuscatlán Formation (late Pliocene) is comprised of rhyolites and basalts. This unit has been found only in El Salvador and in eastern Guatemala. In eastern El Salvador, the Cuscatlán rocks are on the southern side of the Tertiary belt whereas, in western El Salvador and in eastern Guatemala, Cuscatlán rocks are found on the northern side of the Tertiary volcanic belt.

In summary there have been large changes in the chemical composition of volcanic rocks over time, and a migration of the volcanic belt. The shift in the volcanic belt began with an orderly progression toward the Pacific from mid-Miocene to Pliocene (Chalatenango to Balsamo). This orderly progression changed during the late Pliocene, when the volcanic belt (Cuscatlán) moved slightly landward from the preceding volcanism in eastern Salvador. The volcanic belt moved much farther seaward in eastern Guatemala and western Salvador, and was unrepresented in western Guatemala. The Quaternary belt is more seaward than the late

Pliocene volcanism, coincident with or just landward of the Upper Miocene (Balsamo) volcanics.

Spatial-temporal Variations in Seismic and Volcanic Activity

Earthquakes in Central America occur in two zones, an inclined Benioff zone, and a narrow zone of shallow earthquakes (depth less than 20 km) that coincides with the volcanic front. The shallow earthquakes are commonly associated with volcanic eruptions. Swarms of shallow earthquakes or locally destructive earthquakes have preceded or accompanied volcanic eruptions and are sometimes related to solid earth tides. Perhaps the clearest examples are the destructive earthquakes in San Salvador, El Salvador, in 1658 and 1917 that preceded large flank flows of Boqueron volcano by a few hours (Mooser *et al.*, 1958). The 1879–80 eruption of Islas Quemadas in El Salvador and the swarm of earthquakes that preceded it were affected by semidiurnal and fortnightly earth tides (Golombek and Carr, 1978). Earth tides have also affected recent volcanic activity at Fuego (Martin, 1979).

There is only a short history of seismic instrumentation of Central American volcanoes. The background level of shallow micro-earthquake activity is low (a few events per day) but actively erupting volcanoes such as San Cristobal, in Nicaragua, have several thousand events per day (Wood, 1973). Continuous monitoring is now being carried out at a few Central American volcanoes and the 1973 eruption of Fuego was observed to be preceded by a sharp increase in microseismic activity that began 6 days before the eruption (Ward, 1973).

There are several indications that volcanic activity is closely related to earthquakes along the inclined Benioff zone. First, the strike of the segments of inclined seismic zone closely parallels the strike of the overlying volcanic linea-

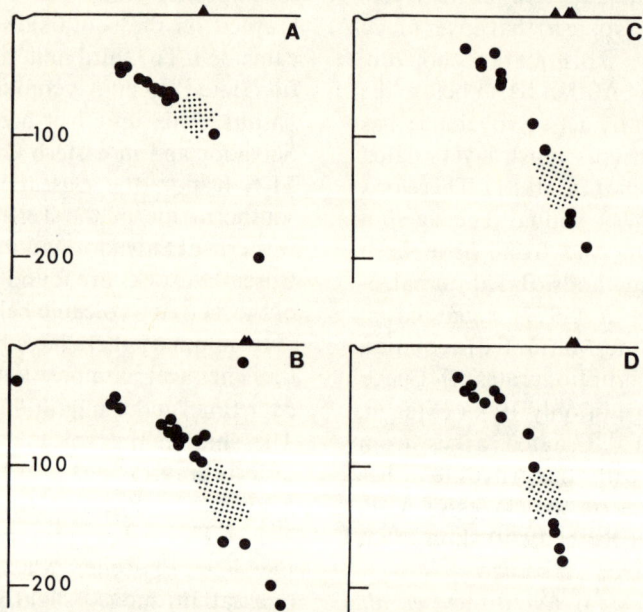

Fig. 3 Cross-sections of segments of the inclined seismic zone. Solid circles are relocated earthquake foci. Triangles are active volcanoes. Stippled areas are gaps in seismicity approximately beneath the volcanoes. A, Central Guatemala; B, El Salvador; C, western Nicaragua; D, eastern Nicaragua

ments (Fig. 1). Second, there is an increase in the dip of the inclined seismic zone approximately beneath the volcanic belt (Fig. 3). Third, there are apparent gaps in the Benioff zone approximately beneath the volcanic front (Fig. 3). These features suggest that where the descending lithosphere loses contact with the overriding lithosphere and sinks into the asthenosphere, there is a change in the angle of descent and a zone of partial melt which reduces seismic activity and is probably the source region for the magmas reaching the volcanic front (Stoiber and Carr, 1976).

There are two prominent spatial-temporal relations between volcanic eruptions and Benioff zone seismic activity. The first is that during the last decade there have been concentrations of small and moderate-size intermediate depth earthquakes whose epicentres are located a few tens of kilometres seaward from the currently active volcanoes (Carr and Stoiber, 1973). Most earthquakes in a cluster just seaward of Fuego and Pacaya volcanos in Guatemala precede or follow eruptions by a few months, but there is no discernable relation between individual earthquakes and specific eruptions of other Central American volcanoes (Carr, 1981).

The second temporal relation is a pattern of volcanic activity associated with great shallow thrust earthquakes (Carr, 1977). One part of the pattern is a period of quiescence or low volcanic activity that begins a few years to a few decades before a great shallow thrust earthquake and ends near the time of the earthquake (Fig. 4). Periods of volcanic quiescence precede most large Central American earthquakes. Quiescence was especially pronounced before two very large earthquakes, the 1902 Guatemala earthquake and the 1850 Nicaragua earthquake. In Central America one of the clearest associations of seismic and volcanic activity is the tendency for volcanoes to be very active for a

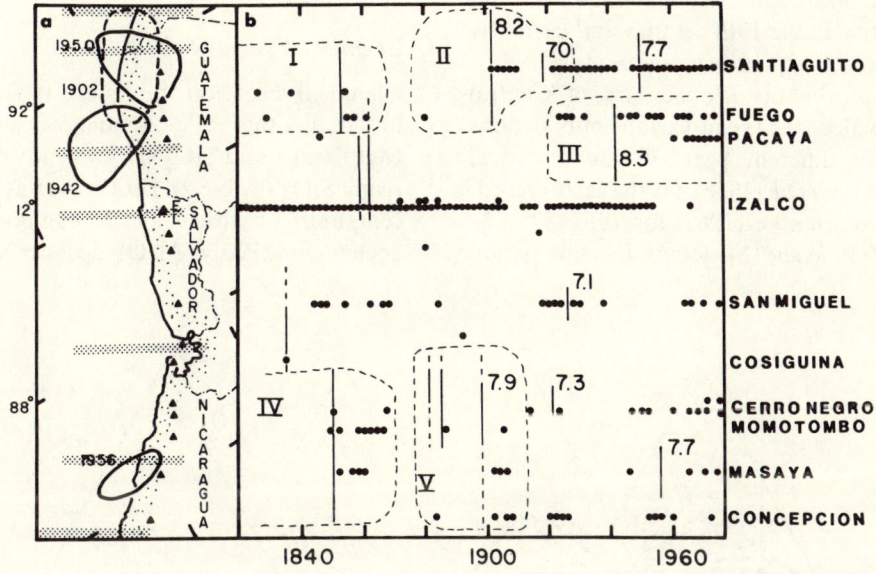

Fig. 4 (a) Map of Central American volcanoes active since 1820 (solid triangles) and rupture zones and times of some recent great earthquakes with areas encircled by solid lines defined by after-shocks. Area encircled by broken lines estimated from damage reports. (b) Space–time plot of volcanic eruptions (solid circles) and great earthquakes (vertical lines dashed where uncertain). The magnitudes of earthquakes are indicated if known. The vertical axis is the position along the arc projected from the adjacent map. Broken lines enclose regions where volcanic activity and great earthquakes are associated

decade or more after a nearby great shallow thrust earthquake. The largest historic eruption in Central America, Santa Mariá in 1902, followed a nearby great shallow thrust earthquake by only a few months.

The distribution of volcanic activity in Central America is apparently influenced by the segmented structure (Stoiber, 1976). Volcanoes that have erupted more or less continuously for several years (Santiaguito, Pacaya, Izalco, Arenal, and Irazú) are on the segment boundaries (Fig. 4). Large and violently explosive eruptions (Santa Mariá in 1902 and Cosigüina in 1835) also occurred on segment boundaries. Within each segment there is usually one volcano which is frequently active (such as Fuego in central Guatemala or San Miguel in El Salvador) while the rest of the volcanoes are infrequently active. The frequently active volcano in each segment must periodically change since all the volcanoes in a segment are roughly the same size. For example, during the 16th and 17th centuries San Cristobal was the frequently active volcano in western Nicaragua. Momotombo was frequently active in the 19th century and currently Cerro Negro is frequently active.

Details of the historic activity of Central American volcanoes are in various publications which cover different parts of the historical record: Mooser *et al.* (1958), for the early record; the report of Hantke (1962) for 1957–59; *The Bulletin of Volcanic Eruptions* for the period 1961–75 (Volcanological Society of Japan, 1961–77); and the *Natural Science Event Bulletin* for the period 1975 to date (Smithsonian Institution, Scientific Event Alert Network, 1975–79). Detailed patterns of recent activity have been described at Arenal (Melson and Saenz, 1973), Santa Mariá–Santiaguito (Rose, 1973; Rose *et al.*, 1977), Fuego (Martin, 1979), and Izalco (Rose and Stoiber, 1969).

Petrology of Volcanic Front Rocks

The Quaternary volcanic rocks of the volcanic front of Central America have a clear calc-alkaline affinity (Figs. 5, 6). Table 2 shows representative chemical composition analyses. Lavas and ashes range from basalt to rhyolite with the distribution peaking in the basalt to andesite range. Basalts are volumetrically more important than previously thought. Rhyolites in the form of voluminous ash flow and tephra deposits comprise a second peak in the distribution of SiO_2 contents.

Chemical Composition

Chemical variation trends are typical of those found in other calc-alkaline volcanic regions. MgO, CaO, and TiO_2 shows steady decline with rising SiO_2 (Table 2). FeO* remains at *c.* 10 per cent until basaltic andesite composition, then declines markedly. Al_2O_3 declines gently from

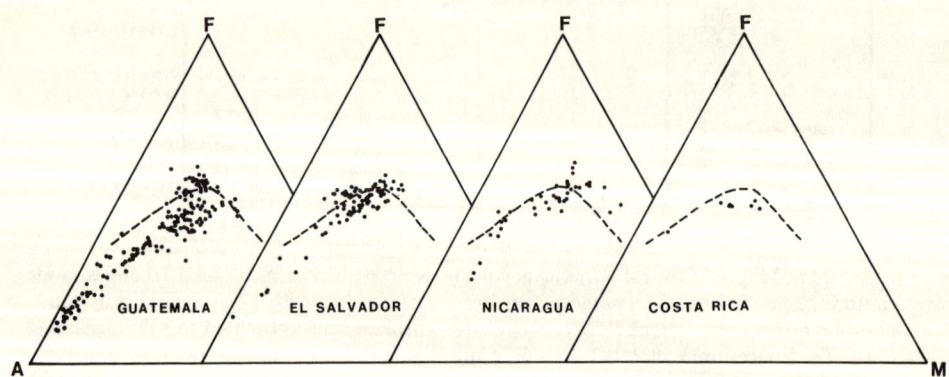

Fig. 5 AFM diagram for Central American volcanic rocks. Broken line separates fields of tholeiitic and calc-alkaline rocks (Irvine and Baragar, 1971)

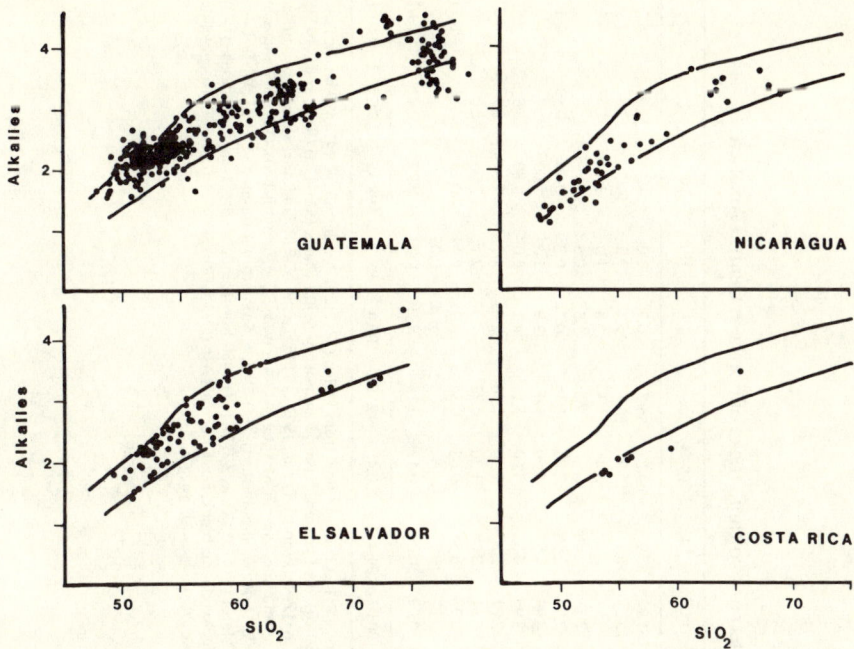

Fig. 6 Alkalis–silica diagram for Central American volcanic rocks. Solid lines outline the alkali basalt (top), high alumina basalt (middle), and tholeiitic basalt (bottom) series defined by Kuno (1966). Most rocks plot in the high alumina basalt series region. Alkali contents steadily decrease from Guatemala to Nicaragua

c. 20 per cent to 13 per cent. Na_2O and P_2O_5 are nearly constant, while K_2O increases markedly. Trace element and Sr isotope data for representative rocks are given in Table 3. Ni, Co, Cr, and V are rapidly depleted as SiO_2 rises, whereas Rb, Ba, Zr, Th, and REE are enriched and Sr decreases slowly.

Sr isotopic data (Pushkar, 1968; Rose, 1972; Rose et al., 1977; Thorpe et al., 1979) on the basalts, andesites, and dacites have given remarkably uniform $^{87}Sr/^{86}Sr$ ratios of c. 0.7040 (Fig. 7). The young rhyolites have nearly identical Sr isotopic ratios (Rose et al., 1979a). New $^{87}Sr/^{86}Sr$ determinations on Guatemalan basalts (Fultz, 1979) also fall in a narrow range, 0.703 82–0.704 21. These data portray two 200 Ma 'pseudo-isochron' arrays. One of these arrays consists entirely of K-rich basalts extruded along the ring fracture of the Atitlán cauldron. Major and minor element data on the K-rich basalts are consistent with contamination of K-poor basalt with c. 15 per cent of rhyolite of the cauldron.

Recent basalts from Fuego volcano (Rose et al., 1978) have rather flat REE abundance curves with La/Lu = 2.0 and show Eu/Eu* = 1 and an enrichment of heavy REE of c. 10 times

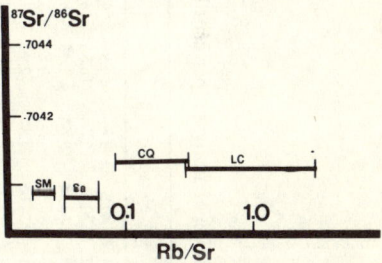

Fig. 7 Mean $^{87}Sr/^{86}Sr$ ratio (±twice the standard error of the mean) and Rb/Sr ratio range for different Quaternary volcanic rock groups from the area near Quezaltenango, Guatemala. SM = Santa Maria (three samples), Sa = Santiaguito (14 samples), CQ = Cerro Quemado (eight samples), LC = Los Chocoyos Ash (nine samples). (From N. K. Grant, unpublished)

TABLE 2 Chemical analyses (per cent by weight) of representative Quaternary volcanic rocks of northern Central America

	1	2	3	4	5	6	7	8	9	10	11	12	13	14	15	16	17	18	19	20	21
SiO_2	50.5	50.9	51.7	50.1	49.5	52.8	54.9	54.0	54.9	58.7	58.0	60.7	61.9	63.2	63.5	65.7	70.3	72.0	68.5	73.9	75.7
Al_2O_3	20.4	19.1	19.1	18.7	20.6	16.8	16.8	18.2	19.18	17.92	16.8	15.5	15.5	16.1	15.9	16.7	15.1	13.0	14.3	12.8	11.9
Fe_2O_3	10.1*	10.6*	1.8	3.9	10.5*	9.8*	2.98	8.26	8.60*	6.82	7.0*	0.34	7.12*	4.86*	6.4*	1.4	0.8	1.0	3.44*	1.54*	0.72
FeO			6.9	6.3			6.86					8.29				2.5	1.3	0.6			
MgO	3.9	4.1	4.6	4.1	4.66	4.8	2.92	3.79	4.29	3.99	3.0	1.55	2.74	2.30	2.88	1.35	0.7	0.2	0.90	0.50	0.1
CaO	9.9	8.5	9.1	10.4	11.3	8.8	8.05	8.61	8.19	7.24	5.99	4.49	4.94	4.91	5.50	4.0	2.9	1.0	3.00	1.7	0.6
Na_2O	3.5	3.90	3.4	2.9	2.1	3.5	3.74	3.34	3.87	3.48	3.75	4.17	3.91	4.33	4.00	5.1	4.1	4.0	3.80	3.6	3.5
K_2O	0.67	0.95	0.92	0.82	0.46	0.91	1.52	1.08	1.10	1.58	2.02	2.71	2.01	2.85	1.90	1.76	2.25	4.7	2.00	2.7	4.4
H_2O			0.10	0.3		0.90	0.23	0.24		0.85	1.2	0.30	0.91	0.05	1.30	0.72	2.3	2.1	2.00	2.0	3.0
TiO_2	0.88	1.13	0.81	1.06	0.63	1.14	2.30	0.74	1.01	0.79	0.65	0.93	0.75	0.56	0.62	0.35	0.25	0.14	0.33	0.21	0.11
P_2O_5		0.23	n.d.		0.21	0.17	0.25		0.22	0.22	0.17	0.16				n.d.	0.05			
MnO	0.13	0.18	0.16	0.16	0.13	0.21	0.21	0.11		0.10	0.21	0.09	0.09	0.12	0.12	0.10	0.03	0.09	0.07	0.07
Total	99.98	99.18	98.8	98.9	99.75	99.58	100.38	98.42	101.39	101.48	98.7	99.41	100.04	99.4	100.4	99.70	100.4	98.7	98.35	99.02	100.01

(1) 1974 olivine basalt of Fuego (F74-34) (Rose et al., 1978). (2) 1961 olivine basalt of Pacaya (E-1) (Eggers, 1971). (3) Basalt flow from Izalco (IZ-112) (Pontier, 1979). (4) 1699 basalt flow from San Miguel (SM-5) (Mayfield, 1978). (5) 1971 olivine basalt of Cerro Negro (SN-1F) (W. I. Rose, unpublished). (6) Pyroxene basaltic andesite of Santa Maria (SM-123) (Rose et al., 1977). (7) Basaltic andesite of Boqueron (B-18) (Fairbrothers et al., 1978). (8) Pyroxene basaltic andesite, summit of Izalco (IZ-1) (Woodruff et al., 1979). (9) Pyroxene basaltic andesite, summit of Atitlán (AT-1) (Woodruff et al., 1979). (10) Pyroxene andesite, flank of Tolimán (L1.0b) (Rose et al., 1981). (11) Pyroxene andesite, summit of Agua (AG-4) (W. I. Rose, unpublished). (12) 1917 andesite from Boqueroncito (B-3) (Fairbrothers et al., 1978). (13) Pyroxene andesite, Summit of Acatenango (ACT-1) (W. I. Rose, unpublished). (14) Dacite, Summit of Tajumulco (TJ-3) (W. I. Rose, unpublished). (15) 1818 dacite, Cerro Quemado (JCQ-15) (Johns, 1975). (16) 1902 hornblende dacite, Santa Mariá (S1115) (Rose, 1972). (17) 1880 hornblende dacite, Islas Quemadas (IL-2) (Mayfield, 1978). (18) Hornblende rhyolite, Cerro Pacho, Lake Coatepeque (SA301) (Carr, 1974). (19) Hornblende rhyodacite, block, Los Chocoyos Ash (OL-5) (Rose et al., 1979a). (20) Hornblende rhyolite block, Los Chocoyos Ash (OL-4) (Rose et al., 1979a). (21) Biotite rhyolite block, Los Chocoyos Ash (SC-7) (Rose et al., 1979a).

* Total Fe as Fe_2O_3 or FeO.
n.d., not determined.

TABLE 3 Selected trace element data (parts per million) on Central American volcanic rocks. Numbers and sources same as Table 2

	1	3	4	6	7	8	9	10	14	16	17	18	19	21
Cs	0.43	n.d.	n.d.	<0.3	n.d.	1.2	1.2	1.1	n.d.	n.d.	n.d.	n.d.	2.0	6.6
Rb	9	17	8	12	30	24	27	45	88	39	40	139	~	117
Sr	580	547	502	633	442	471	504	498	461	420	309	55	404	97
Ba	401	404	492	580	648	402	475	560	1290	930	1045	963	790	1100
Zr	80	67	n.d.	110	n.d.	134	151	177	192	172	n.d.	147	160	109
La	6.88	n.d.	n.d.	9.08	n.d.	14.5	10.7	13.5	n.d.	14.8	n.d.	n.d.	33.3	20.4
Ce	14.3	n.d.	n.d.	17	n.d.	17.9	22.8	22.6	n.d.	27.1	n.d.	n.d.	48.8	31.0
Sm	2.5	n.d.	n.d.	3.25	n.d.	2.75	3.34	3.93	n.d.	3.32	n.d.	n.d.	2.79	2.11
Eu	0.98	n.d.	n.d.	1.10	n.d.	1.06	1.26	1.00	n.d.	1.04	n.d.	n.d.	0.83	0.34
Lu	0.28	n.d.	n.d.	0.29	n.d.	0.30	0.30	0.32	n.d.	0.32	n.d.	n.d.	0.18	0.18
Th	0.58	n.d.	n.d.	0.59	n.d.	1.53	1.63	3.15	n.d.	1.73	n.d.	n.d.	8.48	11.19
Hf	1.71	n.d.	n.d.	2.64	n.d.	1.71	3.29	3.65	n.d.	3.76	n.d.	n.d.	4.82	2.49
Ni	30	16	11	22	11	15.0	17.5	7.0	3	7.0	<1	1	n.d.	n.d.
Co	32.8	n.d.	n.d.	26	n.d.	25.4	24.9	25.0	18	28	n.d.	n.d.	5.43	0.36
Cr	17	13	29	14	11	8	13	n.d.	6.5	3	4	5	3.0	1.0
V	n.d.	253	322	242	189	258	212	155	89	60	16	97	n.d.	n.d.
Sc	23.3	n.d.	n.d.	33.8	n.d.	29.3	24.4	25.7	n.d.	8.93	n.d.	n.d.	6.9	2.8

n.d., not determined.

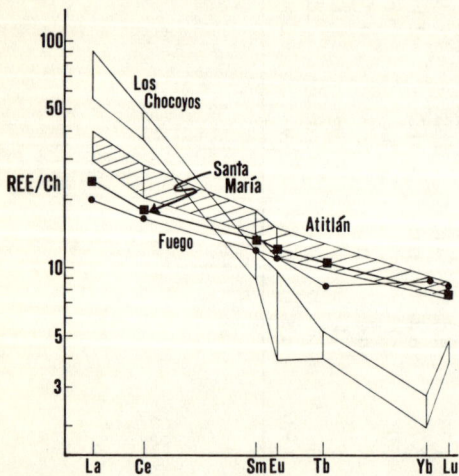

Fig. 8 Chondrite-normalized (Leedey-L-6) REE abundance patterns for selected groups of Guatemalan Quaternary volcanic rocks (from Rose et al., 1981)

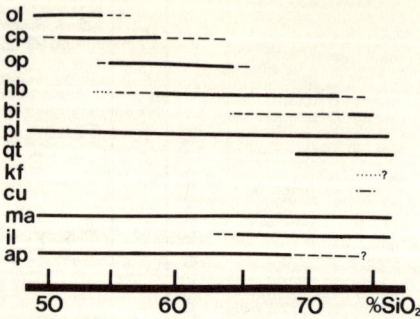

Fig. 9 Graphical summary of phenocryst mineralogy of Central American volcanic rocks. Solid lines indicate usually present, dashed lines indicate sometimes present. Symbols: ol = olivine, cp = clinopyroxene, op = orthopyroxene, hb = hornblende, bi = biotite, pl = plagioclase, qt = quartz, kf = potash feldspar, cu = cummingtonite, ma = magnetite, il = ilmenite, ap = apatite

the chondritic abundance (Fig. 8). Guatemalan andesites and dacites show progressively steeper REE curves. Unlike the more mafic rocks, rhyolites show marked depletion of heavy REE (Rose et al., 1979a), suggesting a role for garnet (or some other mineral which captures heavy REE).

Mineralogy

The mineralogy of phenocrysts in the volcanic rocks, which are generally porphyritic, is presented graphically in Fig. 9. Data on the chemical compositions of these minerals is given in Table 4. Plagioclase is the most important phase, found in virtually all samples. It is zoned, more strongly in mafic rocks, and ranges from anorthite to oligoclase. Olivines are restricted to basalts and basaltic andesites and are forsteritic. Clinopyroxene is found in rocks of all SiO_2 contents, but is most abundant in basaltic andesites. It has a normal augite composition except in rhyolites where it is subcalcic. Augites in basaltic rocks in El Salvador are sometimes rimmed with pigeonite. Hypersthene is found in basaltic andesites, andesites, and dacites. In the dacites it often exceeds clinopyroxene in abundance. Hypersthene is apparently more abundant in Guatemalan lavas than those further south. In Guatemala paragasitic amphibole is found in rocks of all compositions, except the most potassic rhyolites. Its composition varies little with the whole rock composition. Amphiboles are found as inclusions inside phenocrysts in basalts, as badly corroded phenocrysts in andesites and as progressively better preserved phenocrysts in more silicic rocks. In El Salvador amphiboles are present only in dacites and rhyolites. Colour and presumably the degree of dehydration and oxidation of these amphiboles are highly variable. Titaniferous magnetite (Al- and Mg-rich in mafic rocks) and apatite are ubiquitous accessory minerals. The apatite is often included in the phenocryst minerals. Quartz, biotite, and ilmenite are restricted to dacites and rhyolites, while cummingtonite and allanite are rare, occurring only in rhyolites. Alkali feldspar is unknown in any rocks so far examined.

Changes in magma composition with time

Several studies of sequentially erupted lavas from individual vents in northern Central America have emphasized the hypothesis of shallow fractional crystallization. Fuego's 1974

TABLE 4 Microprobe compositions of minerals in Central American volcanic rocks

Source, Rock Type:	1, B	2, A	1, B	1, B	2, A	1, D	3, R	1, D	4, B	5, A	1, D
Mineral:	OL	OL	MAG	CPX	CPX	CPX	CPX	OPX	AMPH	AMPH	AMPH
Volcano:	Fuego	Boqueron	Fuego	Fuego	Boqueron	Santiaguito	Los Chocoyos	Santiaguito	Fuego	Atitlán	Santiaguito
SiO_2	38.1	35.5	0.3	49.0	51.9	50.9	49.4	52.8	42.4	43.1	42.1
Al_2O_3		0.3	10.3	4.4	2.4	1.6	7.0	0.8	13.5	11.7	11.6
FeO*	23.2	43.6	77.9	8.5	11.8	9.2	12.8	19.1	10.4	12.1	12.7
MgO	39.0	21.2	4.9	14.7	14.8	15.1	17.0	25.3	15.1	14.2	14.8
CaO	0.1	0.4	0.1	20.7	19.3	19.4	10.1	0.8	10.4	10.6	10.6
Na_2O				0.5	0.5	0.6	1.5		2.3		2.5
K_2O					0.5		0.1	0.4	0.4		0.4
TiO_2		0.06	4.4	0.6		0.2	1.2	0.1	2.7	2.8	2.3
MnO		1.2				0.5	0.5	1.4			
NiO		0.1	0.2								
Cr_2O_3		0.02									
Total	100.7	102.4	98.5	98.2	101.1	97.5	99.6	100.7	97.2	94.5	97.0

Source, Rock Type:	3, R	1, B	2, BA	3, R	1, B	2, BA	1, D	3, R	3, R	3, R	3, R
Mineral:	AMPH	MAG	MAG	MAG	PL	PL	PL	PL	ILM	CUMM	BIOT
Volcano:	Los Chocoyos	Fuego	Boqueron	Los Chocoyos	Fuego	Boqueron	Santiaguito	Los Chosoyos	Los Chocoyos	Los Chocoyos	Los Chocoyos
SiO_2	43.9	0.3	0.3	0.1	47.7	52.8	56.0	63.6	0.1	54.7	37.8
Al_2O_3	11.4	10.3	4.8	1.6	33.5	27.1	27.7	23.1	0.5	1.8	13.7
FeO*	13.0	77.9	80.0	86.1	0.4	2.2	0.3		53.5	17.5	15.3
MgO	15.0	4.9	3.4	1.1		0.1	0.1		1.7	22.6	13.5
CaO	11.1	0.1			17.2	11.8	9.7	5.0		1.5	0.4
Na_2O	2.5				1.4	6.9	5.5	7.5		0.6	0.3
K_2O	0.4					0.4	0.2	0.6			7.2
TiO_2	2.2	4.4	11.1	4.1		0.04			39.5	0.2	4.0
MnO	0.1		0.5	0.9					1.7	1.1	
NiO											
Cr_2O_3		0.2									
Total	99.6	99.9	98.5	99.9	100.2	100.04	99.5	99.8	99.7*	100.0	92.2

Sources: 1, Rose et al. (1980a); 2, Fairbrothers et al. (1978); 3, Rose et al. (1978); 4, Rose et al. (1981).
Rock types: B, basalt; BA, basaltic andesite; A, andesite; D, dacite; R, rhyolite.
Minerals: OL, olivine; CPX, clinopyroxene; OPX, orthopyroxene; AMPH, amphibole; MAG, magnetite; PL, plagioclase; ILM, ilmenite; CUMM, cummingtonite; BIOT, biotite.
* Total Fe as FeO.

volcanic eruption produced a sequence of high Al basalts which differed markedly in proportions and sizes of phenocrysts. Plagioclase was enriched in the early phases of eruption and olivine was enriched in the later phases which suggested that the shallow magma storage chamber had differentiated markedly (Rose et al., 1978). At Santa Mariá, Rose et al. (1977) interpreted a progressively more silicic sequence of basaltic andesites as representing increasing fractionation during progressively longer repose periods as the composite cone increased in size. Shorter sequences of basaltic andesites at Atitlán and Izalco (Woodruff et al., 1979) show compositional changes which can be interpreted as representing cycles of shallow fractional crystallization. Fairbrothers et al. (1978) found a cyclical variation in SiO_2 content at Boqueron volcano which they attributed to periodic influxes of basaltic magma into a continuously fractionating magma chamber. At the end of one of these cycles the chemical variation pattern abruptly changed from a typical high Al_2O_3 and low FeO calc-alkaline trend to a high FeO and low Al_2O_3 trend that is similar in some respects to a tholeiitic variation pattern. The change was attributed to a decrease in H_2O pressure and a consequent increase in the proportion of plagioclase fractionation. At Izalco volcano, which first erupted in 1770, SiO_2 contents of lavas have slowly and erratically decreased with time (Pontier, 1979). This corresponds most closely with the increased frequency of large lava eruptions. Therefore, the SiO_2 content reflects the length of the preceding repose or fractionation period. Since 1932, Fuego's lavas have also shown an erratic and slow trend to more mafic composition, which also seems to be related to shorter repose periods (Martin, 1979).

Spatial variation of volcanic rocks

Several volcanic centres in Central America have vents separated by a few to 10 km that have erupted systematically different volcanic products. The centres examined so far, such as Santiaguito–Santa Mariá–Cerro Quemado, Atitlán–Tolimán, Fuego–Acatenango in Guatemala and Izalco–Santa Ana in El Salvador, have the following general characteristics. The vent closest to the coast (e.g. Atitlán, Fuego, Izalco) has steep slopes and a high proportion of pyroclastic products. The more inland vent has a lower profile and more lavas and domes. In terms of chemical composition, the most mafic basaltic rocks from each vent are similar, but the vent nearest the coast has distinctly lower incompatible element contents at higher SiO_2 contents.

Fractionation of a higher proportion of more SiO_2-poor phases (olivine, magnetite) at the vent nearest the coast and a higher proportion of plagioclase at the vent farther from the coast could explain these differences. Rose et al. (1977, 1981) have suggested that differences in the H_2O content within shallow magma chambers could cause the apparent differences in proportions of fractionating phases with a systematic distribution of water-rich magma chambers on the coastward side of the volcanic front. The cause of these H_2O variations is unknown. Oxygen isotopic data on Fuego plagioclases (Rose et al., 1980b) show that the water in this H_2O-rich magma chamber (3 per cent H_2O by weight) is of mantle origin.

Although the chemistry of volcanic rocks along the volcanic front can vary greatly between adjacent vents, there are few obvious regional geochemical patterns (Pichler and Weyl, 1973). This is surprising because there are large differences in the shape of the underthrust seismic zone (Fig. 1) and large differences in regional elevation and Bouguer gravity (Fig. 10) which are most probably related to different crustal thicknesses throughout Central America. The crust under the volcanic belt in Guatemala and El Salvador is thought to be a continental block comprising a crystalline Paleozoic basement complex overlain by thick sequences of Paleozoic and Mesozoic sedimentary and igneous rocks (Dengo, 1968). The Quaternary volcanoes are little influenced by this change in crustal structure. One possible effect is that the proportion of silicic rocks is higher in Guatemala and El Salvador, possibly because there is

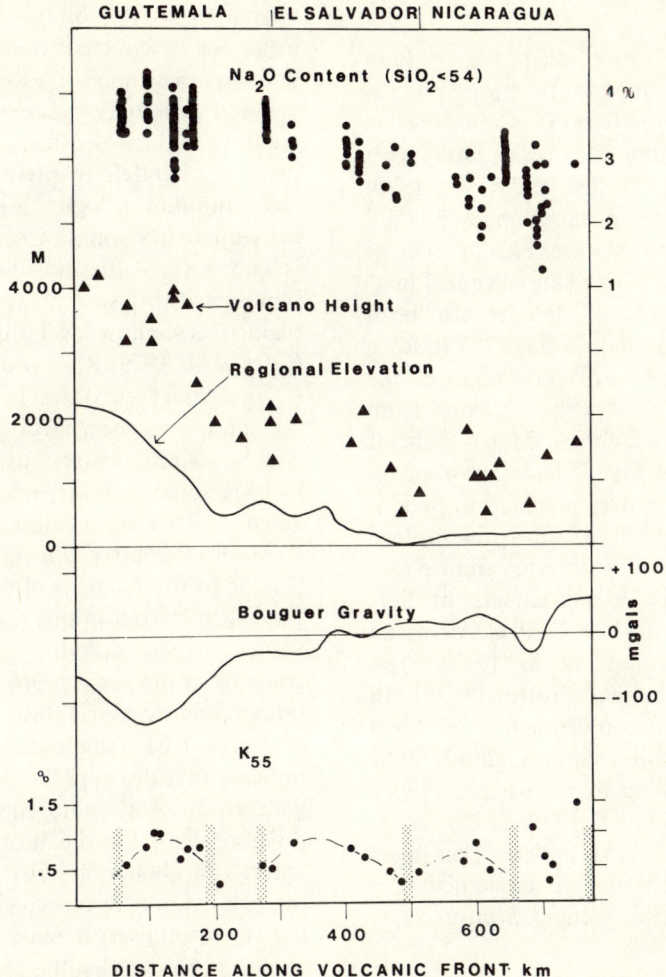

Fig. 10 Regional variation in selected geophysical and geochemical properties. Stippled bars mark transverse breaks in the arc. *See* text for discussion

increased fractional crystallization during passage through thicker crust.

McBirney (1969) demonstrated that Nicaraguan basalts have lower alkali contents than those of other Central American basalts and suggested that this was caused by mantle depletion of alkalis because of the more voluminous Tertiary volcanism in Nicaragua. Carr et al. (1979) demonstrated that the K_2O content of Central American basalts is essentially uniform along the length of the arc and is unrelated to depth to the seismic zone. They suggested that volcanoes with low K contents are associated with transverse breaks in the arc (Fig. 10). The regional variation in alkalis noted by McBirney (1969) is caused by Na_2O. The Na_2O contents of basaltic rocks gradually decreases from Guatemala (3.5 per cent) to Nicaragua (2.0 per cent) (Fig. 10). This variation corresponds with decreasing regional elevation, increasing Bouguer gravity, and presumably with decreasing crustal thickness. However, there is no obvious reason for crustal thickness and Na_2O content to be related.

Volcanic gases

Volcanic gases have been studied extensively in Central America, most frequently during periods of low or moderate activity. Gas condensates from high temperature fumaroles have been extensively analysed (Stoiber and Rose, 1970) and the data show temporal changes of Cl/S related to activity. Incrustations rich in metallic elements from Central American volcanic fumaroles (Stoiber and Rose, 1974) have also been well described. Taylor and Stoiber (1973), Rose et al. (1973), and Rose (1977) have reported the chemical composition of ash leachates from major eruptions for which gas data is difficult to obtain. Rose et al. (1978) have reported on the compositions of glasses included in phenocrysts, which may reflect the pre-eruption magmatic gas contents. A series of remote correlation spectrometry measurements of SO_2 flux has been performed at Central American volcanoes by Stoiber and Jepson (1973), Crafford (1976), and Stoiber and Bratton (1978). All of the above has been complemented by direct sampling within small eruption clouds. The results of direct sampling are reported in a series of publications (Cadle et al. 1979; Lazrus et al., 1979; Rose et al., 1979b, 1980a). The interrelationship of the gas data is explained in a paper by Rose et al. (this volume, Section X).

Origin of magmas

Few data have been obtained to constrain the source materials for Central American basalts. Sr isotopic data, which are essentially uniform at 0.7040, and independent of crustal thickness, preclude much radiogenic crustal material in any of the rocks. Simple derivation of basalt from the subducted oceanic crustal slab is unlikely because Ni, Cr, Rb, and Sr abundances demand strikingly different degrees of partial melting (Gill, 1974; Rose et al., 1977). Complex two-stage models involving partial melting of slab granulite, eclogite, and/or peridotite, and subsequent fractionation of olivine and pyroxene is necessary. Andesitic (common) and dacitic (volumetrically scarce) rocks have been explained as shallow level differentiates of basalts (Carr et al., 1979; Rose et al., 1981).

Quaternary rhyolites in Guatemala and El Salvador have been erupted from a series of widely spaced centres (average interval of c. 90 km), some of which are slightly north of the volcanic front. The volume of rocks produced from these centres during the Quaternary is similar to the volumes of rocks produced from the volcanic front in this region. The Sr isotopic values suggest that the rhyolites could have come from the same source as the mafic rocks, but volumetric considerations and the depletion of heavy REE (suggesting a role for garnet) indicate that the hypothesis used to explain the generation of rhyolite must be substantially different than for the more mafic rocks. The scarcity of Quaternary rhyolites south-east of the El Salvador–Nicaragua border suggests that the continental crust plays a physical, if not a chemical, role in rhyolite genesis.

Acknowledgements

This research was supported by the Earth Sciences section of the National Science Foundation, Grants EAR75-21185, GA 26211, GA 38435, EAR74-19025, and EAR 78-01190.

REFERENCES

Brown, R. D., Ward, P. L., and Plafker, G. (1973). Geologic and seismologic aspects of the Managua Nicaragua earthquakes of Dec. 23, 1972. US geol. Surv. prof. Pap. no. 838.

Cadle, R. D., Lazrus, A. L., Huebert, B. J., Heidt, L. E., Rose, W. I., Woods, D. C., Chuan, R. L., Stoiber, R. E., Smith, D. B., and Zielinski, R. A. (1979). Atmospheric implications of studies of Central American volcanic eruption clouds. J. geophys. Res. in press.

Carr, M. J. (1974). Tectonics of the Pacific margin of northern Central America. Ph.D. thesis, Dartmouth College, Hanover, N.H.

Carr, M. J. (1976). Underthrusting and Quaternary faulting in Central America. Geol. Soc. Am. Bull. **87**, 825–829.

Carr, M. J. (1977). Volcanic acitivity and great earthquakes at convergent plate margins. Science, N.Y. **197**, 655–657.

Carr, M. J. (1981). Active volcanos and intermediate depth

earthquakes. In *Volcanos and Volcanology* (J. Green, ed.), Dowden, Hutchinson and Ross, New York, in press.

Carr, M. J. and Stoiber, R. E. (1973). Intermediate depth earthquakes and volcanic eruptions in Central America, 1961–1972. *Bull. Volc.* **37**, 326–337.

Carr, M. J. and Stoiber, R. E. (1977). Geologic setting of some destructive earthquakes in Central America. *Geol. Soc. Am. Bull.* **88**, 151–156.

Carr, M. J., Rose, W. I., and Mayfield, D. G. (1979). Potassium content of lavas and depth to the seismic zone in Central America. *J. Volc. geothermal Res.* **5**, 387–401.

Crafford, T. (1976). SO_2 emission of the 1974 eruption of Volcan Fuego, Guatemala. *Bull. Volc.* **39**, 1–21.

Dean, B. W. and Drake, C. L. (1978). Focal mechanism solutions and tectonics of the Middle America arc. *J. Geol.* **86**, 111–128.

Dengo, G. (1968). *Estructura Geologica, Historica Tectonica y Morfologia de America Central*, Centro Regional de Ayude Tecnica, Mexico.

Dengo, G., Bohnenberger, O. H., and Bonis, S. (1970). Tectonics and volcanism along the Pacific marginal zone of Central America. *Geol. Rundshau* **50**, 1215–1232.

Drexler, J. W., Rose, W. I., Jr, Sparks, R. S. J., and Ledbetter, M. T. (1981). The Los Chocoyos ash: Quaternary stratigraphic marker in middle America and three ocean basins. *Quaternary Res.* in press.

Eggers, A. A. (1971). Geology of the Amatitlán quadrangle, Guatemala. Ph.D. thesis, Dartmouth College, Hanover, N. H.

Fairbrothers, G. E., Carr, M. J., and Mayfield, D. G. (1978). Temporal magmatic variation at Boqueron volcano, El Salvador. *Contr. Mineral. Petrol.* **67**, 1–9.

Fultz, L. A. (1979). Sr isotopic determinations of Guatemalan basalts. M.S. thesis, Michigan Technological University, Houghton, Mich.

Gill, J. B. (1974). Role of underthrust oceanic crust in the genesis of a Fijian calc-alkaline suite. *Contr. Mineral. Petrol.* **42**, 29–45.

Golombek, M. P. and Carr, M. J. (1978). Tidal triggering of seismic and volcanic phenomena during the 1879–1880 eruption of Islas Quemadas Volcano in El Salvador. *J. Volc. geothermal. Res.* **3**, 299–307.

Hantke, G. (1962). Summary of volcanic activity, 1957–1959. *Bull. Volc.* **24**, 1–2.

Irvine, T. N. and Baragar, W. R. A. (1971). A guide to the chemical classification of the common volcanic rocks. *Can. J. Earth Sci.* **8**, 523–548.

Johns, G. W. (1975). Geology of the Cerro Quemado Dome Complex, Guatemala, M.S. thesis, Michigan Technological University, Houghton, Mich.

Kuno, H. (1966). Lateral variations of basaltic magma across continental margins and island arcs: *Can. geol. surv. Pap.* no. 66-15, 317–335.

Ladd, J. W., Ibrahim, A. K., McMillen, K. J., Latham, G. V., and Worzel, J. L. (1978). Structures and tectonics associated with the Middle America trench offshore Guatemala. *Geol. Soc. Am. Abstr. Programs.* **10**, 439.

Lazrus, A. L., Cadle, R. D., Gangrud, B. W., Greenberg, J. P., Huebert, B. J., and Rose, W. I., Jr (1979). Trace chemistry of the stratosphere and of volcanic eruption plumes. *J. geophys. Res.* in press.

Martin, D. P. (1979). The historic activity of Fuego Volcano, Guatemala: constraints in the subsurface magma bodies and processes therein. M.S. thesis, Michigan Technological University, Houghton, Mich.

Mayfield, D. G. (1978). Magmatic variations in the El Salvador segment of the Middle America arc. M.S. thesis, Rutgers University, New Brunswick, N.J.

Melson, W. G. and Saenz, R. (1973). Volume, energy, and cyclicity of eruptions of Arenal volcano, Costa Rica. *Bull. Volc.* **37**, 416–437.

McBirney, A. R. (1969). Compositional variations in Cenozoic calc-alkaline suites of Central America. *Int. Upper Mantle Project sci. Rep.* no. 15, 185–187.

McBirney, A. R., Sutter, J. F., Naslund, H. R., Sutton, K. G., and White, C. M. (1974). Episodic volcanism in the central Oregon Cascade Range. *Geology* **2**, 585–589.

Molnar, P. and Sykes, L. R. (1969). Tectonics of the Caribbean and Middle America regions from focal mechanisms and seismicity. *Geol. Soc. Am. Bull.* **89**, 1639–1684.

Mooser, F., Meyer-Abich, H., and McBirney, A. R. (1958). *Catalogue of Active Volcanoes of the World*, Part VI, *Central America*, International Association for Volcanology, Naples.

Peterson, P. S. and Rose, W. I., Jr (1979). Dating of caldera forming eruptions at Ayarza, Guatemala and correlation of its Plinian tephra units (abstract). Pacific North-west Section, American Geophysical Union, Bend, Oregon.

Pichler, H. and Weyl, R. (1973). Petrochemical aspects of Central American magmatism. *Geol. Rundshau* **62**, 357–396.

Pichler, H. and Weyl, R. (1977). Quaternary alkaline volcanic rocks in eastern Mexico and Central America. *Münster. Forsch. Geol. Paläont.* **38/39**, 159–178.

Pontier, N. K. (1979). Magmatic evolution of Izalco volcano and relation to the Santa Ana complex, El Salvador. M.S. thesis, Rutgers University, New Brunswick, N.J.

Pushkar, P. (1968). Strontium isotope ratios in volcanic rocks of three island arc areas. *J. geophys. Res.* **73**, 2701–2714.

Reynolds, J. J., III (1977). Tertiary volcanic stratigraphy of northern Central America. M.S. thesis, Dartmouth College, Hanover, N.H.

Rose, W. I., Jr (1972). Santiaguito volcanic dome, Guatemala. *Geol. Soc. Am. Bull.* **83**, 1413–1434.

Rose, W. I., Jr (1973). Pattern and mechanism of volcanic activity at Santiaguito volcano, Guatemala. *Bull. Volc.* **37**, 73–94.

Rose, W. I., Jr (1977). Scavenging of volcanic aerosol by ash: atmospheric and volcanic implications. *Geology* **5**, 621–624.

Rose, W. I., Jr and Stoiber, R. E. (1969). The 1966 eruption of Izalco volcano, El Salvador. *J. geophys. Res.* **74**, 3119–3130.

Rose, W. I., Jr, Bonis, S. B., Stoiber, R. E., Keller, M., and Bickford, T. (1973). Studies of volcanic ash from two recent Central American eruptions. *Bull. Volc.* **37**, 338–364.

Rose, W. I., Jr, Grant, N. K., Hahn, G. A., Lange, I. M., Powell, J. L., Caster, J., and Degraff, J. M. (1977). The evolution of Santa Mariá Volcano, Guatemala. *J. Geol.* **85**, 63–87.

Rose, W. I., Jr, Anderson, A. T., Bonis, S., and Woodruff, L. G. (1978). The October 1974 basaltic tephra, Fuego volcano, Guatemala, description and history of the magma body. *J. Volc. geothermal Res.* **4**, 3–53.

Rose, W. I., Grant, N. K., and Easter, J. (1979a). Geochemistry of the Los Chocoyos ash, Quezaltenango Valley, Guatemala. *Geol. Soc. Am. spec. Pap.* no. 180, in press.

Rose, W. I., Jr, Cadle, R. D., Heidt, L., Lazrus, A. L., and Huebert, B. J. (1979b). Gas and H isotope determinations of volcanic eruption clouds. *J. Volc. Geothermal Res.* **5**, in press.

Rose, W. I., Jr, Chuan, R. C., Cadle, R. D., and Woods, D. C. (1980a). Small particles in volcanic eruption clouds. *Am. J. Sci.* **280**, 671–696.

Rose, W. I., Jr, I. Friedman, and Woodruff, L. G. (1980). Oxygen isotope ratios of successively empted plagioclases from Fuego Volcano, Guatemala. *Bull. Volc.* **42**, in press.

Rose, W. I., Penfield, G. T., Drexler, J. W., and Larson, P. B., (1981). Geochemistry of the andesitic flank lavas of three composite cones within the Atitlan Caldron, Guatemala. *J. Volc. Geothermal Res.*, in press.

Smithsonian Institution, Scientific Event Alert Network (1975–79). *Natural Science Event Bulletin* **1–4**.

Stoiber, R. E. (1976). Frequency of eruption of Central American volcanos in relation to a tectonic model of the Central American Quaternary (abstract). *EOS* **57**, 345.

Stoiber, R. E. and Carr, M. J. (1973). Quaternary volcanic and tectonic segmentation of Central America. *Bull. Volc.* **37**, 304–325.

Stoiber, R. E. and Carr, M. J. (1976). Depths of magma formation and seismic gaps below volcanos. *Geol. Soc. Am. Abstr. Programs* **8**, 1125–1126.

Stoiber, R. E. and Bratton, G. (1978). Airborne correlation spectrometer measurements of SO_2 in eruption clouds (abstract). *EOS*, **59**, 1222.

Stoiber, R. E. and Jepson, A. (1973). Sulfur dioxide contributions to the atmosphere by volcanos. *Science, N.Y.* **182**, 577–578.

Stoiber, R. E. and Rose, W. I. (1970). The geochemistry of Central American volcanic gas condensates. *Geol. Soc. Am. Bull.* **81**, 2891–2912.

Stoiber, R. E. and Rose, W. I. (1974). Fumarolic incrustations at active Central American volcanoes. *Geochim. cosmochim. Acta* **38**, 495–516.

Taylor, P. S. and Stoiber, R. E. (1973). Soluble material on ash from active Central American Volcanos. *Geol. Soc. Am. Bull.* **34**, 1031–1042.

Thorpe, R. S., Francis, P. W., and Moorbath, S. (1979). Strontium isotope evidence for petrogenesis of Central American andesites. *Nature, Lond.* **277**, 44–45.

Van Andel, T. H., Heath, G. R., Malfait, B. T., Heinricus, D. F., and Ewing, J. I. (1971). Tectonics of the Panama basin, eastern equatorial Pacific. *Geol. Soc. Am. Bull.* **82**, 1489–1508.

Volcanological Society of Japan, International Association of Volcanology (1961–77). *Bull. Volc. Eruptions* **1–15**.

Ward, P. L. (1973). A prototype global volcano surveillance system and identification of a volcano-tectonic fault in Central America. *Bull. Volc.* **37**, 438–442.

Wiesmann, G. (1975). Remarks on the geologic structure of the Republic of El Salvador, Central America. *Mitt. Geol. Paläont. Inst. Univ. Hamburg* **44**, 557–574.

Wood, R. (1973). Microearthquakes at Central American volcanos. *Bull. seismol. Soc. Am.* **64**, 275–277.

Woodruff, L. G., Rose, W. I., Jr, and Rigot, W. (1979). Contrasting fractionation patterns for sequential magmas from two Central American volcanos. *J. Volc. geothermal. Res.* **5**, in press.

The Lesser Antilles

W. J. Rea

Department of Geology and Physical Sciences,
Oxford Polytechnic, Headington, Oxford OX3 0BP, UK

ABSTRACT

The Lesser Antilles island arc marks the site of underthrusting of the Caribbean plate by Atlantic ocean floor. From Dominica southwards there is a single line of islands with Oligocene volcanic rocks capped by Miocene limestones and Miocene to Recent volcanic rocks, but north of Dominica there is an eastern arc composed of Oligocene calc-alkaline rocks overlain by Miocene limestones and a western arc composed of predominantly Pleistocene to Recent volcanic rocks. Valley features and offsets of the positive gravity anomaly suggest the existence of faults between some of the islands.

Geophysically, the Lesser Antilles is characterized by paired positive–negative gravity anomaly belts and seismic refraction and reflection studies strongly suggest that the arc is intra-oceanic. The site of subduction in the eastern Caribbean may have varied through time and present subduction along the arc may vary with regard to the dip of the subduction zone and the mechanism and intensity of seismicity.

To a crude approximation, the Lesser Antillean volcanoes can be assigned to two categories—basalt–andesite volcanoes (e.g. Soufrière, St Vincent) and andesite–dacite volcanoes (e.g. Mt Pelée, Martinique). The former group is characterized by the eruption of lava flows and pyroclastic falls, a compositional continuum from basalt to andesite, and the occurrence of ultrabasic cumulate blocks, consisting of variable amounts of sodic anorthite, hastingsitic amphibole, titaniferous magnetite, olivine, and rarer pyroxenes. The more common andesite–dacite volcanoes erupt mainly pyroclastic flows and dome lavas. Basalts and ultrabasic cumulates are absent. Most of the lavas in the Lesser Antilles belong to the calc-alkaline suite but those from Grenada include basanites and alkaline-basalts, and rocks from the northern islands have some affinities with island arc tholeiites. Chemical variation is considerable. Lavas from the northern islands are generally poor in incompatible elements, Ni, Cr, and radiogenic Sr, whereas rocks from the central part of the arc are richer in these elements and also include a higher proportion of dacite. In contrast to the rest of the arc, the southern islands are predominantly basaltic and lavas from Grenada show enrichment in incompatible elements and include undersaturated compositions.

It is likely that many different processes have contributed to petrogenesis in the Lesser Antilles. Partial melting of garnet lherzolite mantle will account satisfactorily for many of the chemical characteristics of the Grenada basanitoids and alkaline basalts. Dehydration of subducted oceanic crust provides a mechanism for the enrichment of elements such as K and Rb in the Grenada magmas. Elsewhere, partial melting of subducted oceanic crust may contribute to petrogenesis and this process could account for the large volumes of andesite and dacite, low in Ni and Cr, which characterize the northern and central Lesser Antilles. Certain chemical properties of Lesser Antillean lavas are compatible with fractionation of SiO_2-poor phases such as olivine, spinel, magnetite, and amphibole. In Grenada, fractionation of olivine–calcic augite–spinel, followed by amphibole, may be responsible for the sequence from alkali-basalt through saturated basalt to andesite. There is strong direct evidence that andesites at the basalt–andesite volcanoes have been produced from basaltic magma by fractionation of material represented in the ultrabasic cumulates. There is also considerable direct evidence for magma mixing and a limited amount of contamination by oceanic sediment and crustal material may also have occurred. It is considered that the 'mix' of petrogenetic processes varies, beneath different parts of the arc, to produce the observed variation in the composition of erupted lavas.

It is possible that the Lesser Antilles has evolved as a series of fault-bounded blocks within which petrogenetic conditions are dependent on the nature of subduction and the composition of mantle and crust above the subduction zone. In the south, partial melting of fertile mantle may be important whereas in the north a more refractory mantle may contribute little to magmagenesis. Partial melting of subducted oceanic tholeiite may be more important beneath the northern and central Lesser Antilles and the contribution of subducted oceanic sediment may be important only beneath the central islands. Processes of fractionation, magma mixing, and contamination are superimposed on these deep level processes and are themselves variable, because of factors such as the variable crustal composition below the Lesser Antilles.

Introduction

History of research

The catastrophic eruptions of 1902 on St Vincent and Martinique led to the first detailed geological research on the Lesser Antilles (Anderson and Flett, 1903; Lacroix, 1904; Flett, 1908). These papers include the first descriptions in the geological literature of *nuée ardente* or pyroclastic flow eruptions and the first geochemical data for the Lesser Antilles. Important regional studies were made subsequently on Montserrat (MacGregor, 1938) and Saba and St Eustatius (Westerman and Kiel, 1961). Great impetus to petrological studies was provided by a research programme initiated at the University of Oxford in the late 1950s and later continued at the University of Durham. A summary of work carried out under this programme is given in Brown et al. (1977). Vital contributions to our present understanding of the Lesser Antilles have also come from geophysical investigations, including gravity determinations (e.g. Andrew et al., 1970), seismic velocity studies (e.g. Officer et al., 1957, 1959; Bunce et al., 1970; Edgar et al., 1971; Westbrook, 1975), and work on the seismicity of the region (Sykes and Ewing, 1965; Molnar and Sykes, 1969; and much work by the Seismic Research Unit in Trinidad, e.g. Tomblin, 1972, 1975). A recently published study of the K–Ar geochronology and palaeomagnetism of the Lesser Antilles (Briden et al., 1979) provides a much improved chronological framework for the area.

General location, physiography, and bathymetry

The islands of the Lesser Antilles form an arc running along the eastern margin of the Caribbean Sea (Fig. 1). The arc marks the eastern boundary of the Caribbean plate which is underthrust by the Atlantic ocean floor. (For a detailed discussion of plate tectonic framework see pp. 171–172). The Lesser Antilles arc lies between the Greater Antilles (Puerto Rico to Cuba) and northern Venezuela, both of which consist essentially of deformed and metamorphosed sedimentary, volcanic, and plutonic rocks of Jurassic to Eocene age. The Lesser Antilles stretch about 700 km from Grenada in the south to Sombrero in the north. North of Dominica the arc is a double arc with the outer islands (Marie Galante to Sombrero, sometimes referred to as the 'Limestone Caribbees') composed of Oligocene calc-alkaline volcanic and plutonic rocks overlain by Miocene limestones. West of these lies a group of young volcanic islands (western Guadeloupe to Saba, 'The Volcanic Caribbees') composed predominantly of Pleistocene volcanic rocks, although a maximum age of 7.7 Ma has been obtained on St Kitts (Baker, 1969). From Dominica southwards there is a single line of islands with a general stratigraphy, suggesting a superimposition of the two northern arcs, that is Oligocene volcanic rocks (although these are very poorly preserved), capped by Miocene limestones, in turn overlain by Miocene to Recent volcanic rocks.

The distribution of the islands described above is reflected in the bathymetry of the area (Fig. 1). The southern end of the arc is marked by a strong crustal upwarp, giving a relatively narrow ridge, but from Guadeloupe northwards the islands represent the highest points on a broad nearly flat-topped platform. There are valley features between some of the islands in the middle of the arc, e.g. between St Lucia and Martinique and between Martinique and

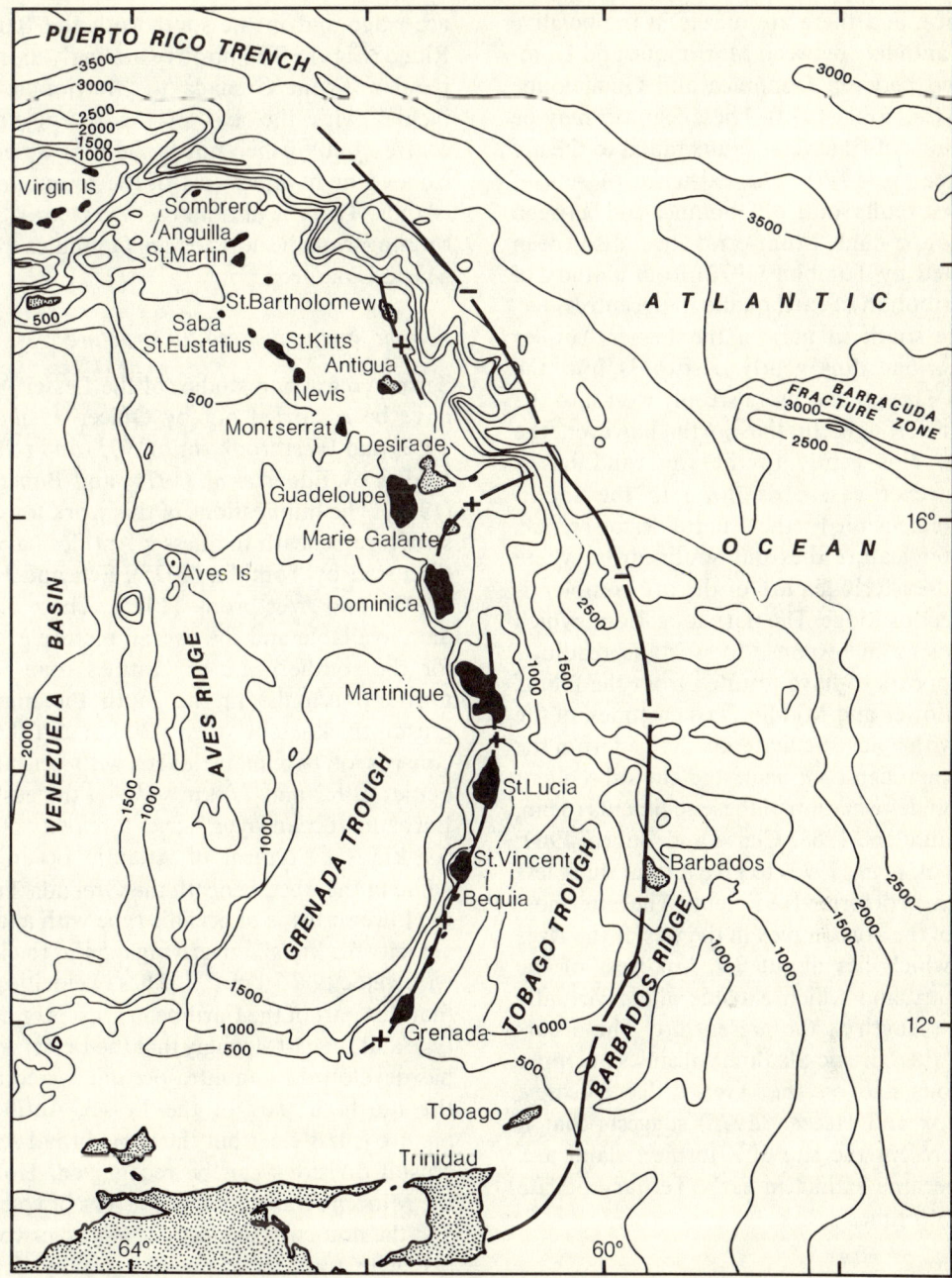

Fig. 1 Map of the eastern Caribbean, showing major physiographic features, bathymetry (1000 fathom intervals) and axes of negative (heavy line with minus signs) and positive (heavy line with plus signs) gravity anomaly. (Based on Tomblin, 1975, Fig. 1.)

Dominica, and there are offsets in the positive gravity anomaly between Martinique and Dominica and between Dominica and Guadeloupe (Fig. 1) (*see* Fink, 1972). These features may be expressions of transverse faults radial to the arc (Westercamp, 1977). The existence of sets of transverse faults south of Dominica and between Guadeloupe and Montserrat has also been postulated by Tomblin (1972) from a study of the distribution of earthquake hypocentres.

In the southern part of the Lesser Antilles the arc ridge dips gently eastwards into the Tobago Trough and more steeply west into the Grenada Trough. In the north, however, the platform dips gently to the west and has a much steeper eastwards dip into the southeastern extension to the Puerto Rico Trench. The latter feature dies out southwards, where the most easterly feature of the arc complex is the Barbados Ridge. The Barbados Ridge, which is still rising in response to isostatic readjustment, appears to have resulted from the filling, during Lower and Middle Tertiary times, of the trench with epiclastic material, derived from the South American continent, and the subsequent uplift and deformation of these sediments (Senn, 1940; Saunders, 1968; Chase and Bunce, 1969; Westbrook *et al.*, 1973) to give a total thickness of *c.* 20 km of deformed sediments. The outermost feature of the arc complex in the west is the Aves Ridge, which lies about 300 km west of the present arc and which extends along virtually the same length as the present arc. The occurrence of altered, calc-alkaline volcanics of Upper Cretaceous age on the Aves Ridge (Kearey, 1974; Fox and Heezen, 1975) suggests that it may represent the site of a former island arc, which became extinct in early Tertiary or late Cretaceous times.

Geophysical data

Gravity

A large negative gravity anomaly belt lies on the eastern side of the Lesser Antilles (Andrew *et al.*, 1970). In the north the negative anomaly coincides with the foot of the eastern flank of the arc ridge and in the south with the Barbados Ridge (Fig. 1). The positive anomaly axis is less regular. From Grenada to Martinique it coincides with the axis of the recent volcanic centres but further north there is a series of isolated high values east of the recent volcanic islands. There is also an east–west gravity high running from the north end of Dominica through Marie Galante.

Seismic studies and crustal structure

Seismic refraction studies of the Lesser Antilles have been carried out by Officer *et al.* (1957, 1959) and Westbrook *et al.* (1973) and reflection studies by Edgar *et al.* (1971) and Bunce *et al.* (1970). The implications of this work for crustal structure beneath the Lesser Antilles have been discussed by Tomblin (1975), Fox and Heezen (1975), and Westbrook (1975). There is more data available and the overall picture is clearer for the southern Lesser Antilles than for the northern islands. In the south the minimum crustal thickness is at least 30 km and the crust consists of two main layers with seismic velocities of 6.2 and 7.0 km s^{-1}. To the east, near Barbados, ocean layer 3 gives a velocity of about 6.7 km s^{-1}, typical of Atlantic ocean crust, while in the west, beneath the Grenada Trough, the lower crust is of oceanic type with a layer of velocity 6.2 km s^{-1} overlying a 5 km thick layer with velocity 7.3 km s^{-1}. These velocities show that this part of the Caribbean crust is of oceanic type and suggest strongly that the Lesser Antilles has developed as an intra-oceanic island arc. In the northern part of the Lesser Antilles the picture is less clear but the same broad seismic/crustal divisions can be recognized. However, the Miocene translation of the axis of volcanism and the non-coincidence of the present volcanic belt with the axis of positive gravity anomaly imply some difference in the history of crustal development and in crustal composition in this area.

Seismicity and nature of subduction

The nature of seismicity in the Eastern Caribbean has been elucidated by various authors, especially Tomblin (1972, 1975). The site of subduction may have varied through time (p. 172) but the present site probably lies c. 150 km east of the active volcanic arc (Tomblin, 1975). An early interpretation of the 1964-70 earthquake data suggested that the earthquake zone dipped westwards at 30° to a depth of 100 km below Grenada and more steeply at 50° to a depth of 160 km below St Kitts (Sigurdsson et al., 1973). Subsequent revision of the data (Tomblin, 1975) suggested an overall conformity in the seismic zone along the arc with a general westwards dip of 30-40° and a depth to the top of the zone of 100 km below the southern and northern islands and 120 km below the central islands. However, Tomblin (1975, Fig. 8) shows considerable variation in detail in the nature of the seismic zone beneath different parts of the arc with the northern part of the zone having a very low dip (17° increasing westwards) compared with the central (20-50° increasing westward) and southern (30°, steepening eastward) parts. This variation, together with the thickness (30-50 km) of the seismic zone means that there must be some doubt concerning the assumption, made by some authors (e.g. Brown et al., 1977), that subduction is essentially uniform along the entire arc.

Evidence of differences in the mechanisms of earthquakes and in the intensity of seismicity along the arc also argues against uniformity in the subduction process. Thus, although the major cause of earthquakes is the westward underthrusting of the Caribbean plate by the Atlantic plate, earthquake mechanisms are also related to transverse faulting and to hinge faulting, especially at the southern end of the arc, and probably also at the northern end, where the subduction zone becomes extremely oblique and eventually passes into a transform along the Puerto Rico Trench. Also, as pointed out by Briden et al. (1979, p.525), the slip vector in the middle of the Lesser Antilles (St Lucia and Martinique) is normal to the subduction zone whereas further south there is southward oblique slip and to the north, northward oblique slip.

It appears also that the northern Lesser Antilles may be more active seismically than the southern part of the arc, where some authors have suggested a gap in seismicity, placed variously at between 14.5° and 11°N (Mauk, 1977) or south of 16°N (McCann and Murphy, 1977). The latter authors have suggested that the reduction in the size and number of large earthquakes and in the definition of the seismic zone may be related to the sedimentary infill of the trench in this part of the arc system, which might inhibit short term seismicity and even subduction itself through a buoyant braking mechanism.

Plate tectonic setting and evolution

Present situation

The Lesser Antilles island arc is situated at the site of underthrusting of the small Caribbean plate by the Atlantic plate. Various factors complicate the basic picture, especially at the northern and southern margins of the arc. Thus at the northern margin of the arc the subduction zone passes into the Puerto Rico Trench, which is an unusual feature described as a 'transform trench' by Schell and Tarr (1978). The southern margin is also complicated, with earthquake mechanism solutions suggesting hinge faults involving rotation on an east-west, near vertical plane (Tomblin, 1975). The situation is complicated further by the fact that the junction between the North and South American plates must extend westwards from the Mid-Atlantic Ridge towards somewhere near the Lesser Antilles. The location of the margin between these plates is unclear. Jordan (1975) and Saunders (1977) have suggested that it may run eastwards from the southern margin of the Lesser Antilles, but Le Pichon et al. (1973) suggest a location to the north, perhaps on the line of the Barracuda Fracture Zone. More recently Briden et al. (1979) have suggested a possible location at about 14°N in the St Vincent-St Lucia-Martinique segment of the Lesser Antilles. This general area coincides with a change in the character of seismicity (Mauk, 1977) and a change in the composition of material erupted from predominantly basaltic to the south (St Vincent and Grenada) to predominantly andesitic in the north. The change from a single arc to a double arc also occurs in this general area.

Evolution of the subduction zone

It seems likely that the site of the subduction zone in the eastern Caribbean has varied through time. Thus in Upper Cretaceous times the subduction zone may have been near the Aves Ridge, as much as 300 km to the west of the present position. The position is, however, far from clear. Mattson (1979) presents evidence for two major periods of subduction in the Northern Caribbean—a Jurassic one related to northward-dipping subduction and a period of southward-dipping subduction ending in the late Cretaceous. The second of these would coincide with the Aves Ridge event. In addition, the evidence from Barbados and the Barbados Ridge implies the existence of a trench in Lower to Middle Tertiary times, in a very similar position to that of the present subduction zone. Further evidence of the site of past subduction may also be provided by a possible ophiolitic remnant of ocean crust on Desirade (Fink, 1970; Mattison *et al.*, 1973). The age of this fragment was originally determined at *c.* 145–150 Ma, an age supported by recent determinations on zircons (Mattison *et al.*, 1980) but at variance with K–Ar ages of 90–80 Ma presented by Dinkelman and Brown (1977) and Briden *et al.* (1979). In either case, the determined age is more likely to give the age of formation than an age of obduction, so that the significance with regard to the history of subduction in the eastern Caribbean is uncertain.

In more recent times, the westward shift in the axis of volcanism in the northern Lesser Antilles at *c.* 9 Ma might be related to a change in the site of subduction. Briden *et al.* (1979, p.525), however, favour an explanation for translation involving either a change in dip of the subduction zone or a change in the depth of magma generation, both of which might be related to the change in relative spreading rate in the North Atlantic at 9 Ma (Pitman and Talwani, 1972). An earlier interruption of spreading in the North Atlantic at 38 Ma (Pitman and Talwani, 1972) might be related in some way to the beginning of volcanism in the Lesser Antilles, which Briden *et al.* (1979) date at 38 Ma.

Volcanic History and Nature of Volcanism

Stratigraphical and palaeontological evidence, summarized by Tomblin (1975), suggest the occurrence of calc-alkaline volcanism of Eocene (?) to Oligocene age along the Grenada–Guadeloupe–Anguilla chain, with the volcanism being mainly subaerial in the south and submarine in the north. General submergence in the Lower Miocene led to the formation of reef limestones in many places, followed by uplift during the end-Miocene 'Andean orogeny'. The more recent geochronological work of Briden *et al.* (1979) and Nagle *et al.* (1976) confines the earlier period of volcanism to between 37 Ma (Lower Oligocene) and 10 Ma (Upper Miocene) and also suggests that the record of this period is much poorer than was thought previously. Briden *et al.* (1979) also show that ages in the range 9–6 Ma are common only on Martinique, where there is a progression of age from the oldest volcanics in the south-east to the active centre of Mt Pelée in the north-west (Westercamp, 1976). The oldest age determined in the northern 'Volcanic Caribbees" is a doubtful age of 7.7 ± 2 Ma from St Kitts (Baker, 1969) and the vast majority of ages in the 'Volcanic Caribbees' and in the southern islands belong to the Pliocene to Recent period, with Pleistocene to Recent ages predominant.

The general nature of the volcanic activity in the Lesser Antilles appears to have been relatively constant throughout the Oligocene to Recent period. For example, the peculiar alkaline character found in younger lavas from Grenada (*see* pp. 175, 177) is also a feature of the older lavas on the island (Arculus, 1973). A change in the location of volcanism occurred in the northern Lesser Antilles in the late Miocene, with the westward translation of the volcanic arc, but further south new volcanic centres formed ajacent to the older ones. Since the Pliocene, the much more complete record suggests that there has been very little change in the nature of the volcanism, although there has been migration of the centre of activity within the islands. It appears, therefore, that age is not a major factor in controlling the nature of volcanism within the Lesser Antilles, whereas geographical location does appear to be important.

Recent volcanic activity in the Lesser Antilles has been described by Robson and Tomblin (1966). Since 1966, eruptions have occurred on St Vincent in 1971–72 and 1978–79 (Shepherd et al., 1979). A phreatic eruption occurred on Guadeloupe in 1976 (Tazieff, 1977) and there was also a seismic crisis (increased seismic and sulphur spring activity, related to magma movement below a volcano but not followed by an actual eruption) on Montserrat in 1966–67 (Shepherd et al., 1971). A summary of the most recent activity in the Lesser Antilles is given in Table 1. Among the eruptions listed are some of the most famous of all known eruptions, including the 1902 eruptions of Mt Pelée on Martinique and Soufrière on St Vincent. Some of the earlier eruptions have been extremely

TABLE 1 Age and nature of volcanic activity in the Lesser Antilles

Island	Radiometric Age Range for Whole Island (Ma)	Recent Centres	Dates of Most Recent Major Eruptions (E), Minor Eruptions (M), Seismic Crises (S)	Composition of Magmas at the Named Centres	Major Products of Eruptions at the Named Centres	Sources
Saba	—	The Mountain	Late pre-Columbian (E)	Andesite	Pyroclastic flows, dome lavas, lava flows	1, 3, 4
St Eustatius	—	The Quill	Late pre-Columbian; volcano formed within last 21 ka? (E)	Andesite	Pyroclastic flows, including pumice flows	1, 3, 5
St Kitts	0–7.7	Mt Misery	2158 ± 94 BC (E) 1843 (M)	Basalt–andesite	Pyroclastic falls, lava flows, later lava domes	1, 6, 7
Nevis	—	Nevis Peak	Late pre-Columbian (E)	Andesite	Pyroclastic flows, dome lavas	1, 8
Montserrat	0–4.4	Soufrière Hills	1646 ± 54 AD (?E) 16416 ± 360 BC (E) 1966–67 AD (S)	Andesite	Pyroclastic flows, pumice flows, dome lavas	1, 2, 9
Guadeloupe	0–2.5 (Basse Terre); ?–11 (Grand Terre)	La Soufrière	1400 ± 150 AD (E) 1976 (M)	Andesite	Pyroclastic flows, dome lavas	1, 2, 10, 11
Dominica	0–1.8	Valley of Desolation (Including Morne Microtrin)	26 900 ± 900 BC (E) 1880 AD (M)	Andesite–dacite	Pyroclastic flows, pumice flows, dome lavas	1, 2, 12
Martinique	0–36	Mt Pelée	1902 AD (E) 1929 AD (E)	Andesite–dacite	Pyroclastic flows, pumice flows, dome lavas	1, 2, 13
St Lucia	0–18	Qualibou	37 500 ± 1500 BC (E)	Andesite–dacite	Pyroclastic flows, pumice flows, dome lavas	1, 2, 14
St Vincent	0–2.7	Soufrière	1902 AD (E) 1979 AD (E) 1971–72 AD (M)	Basalt–andesite	Lava flows, pyroclastic falls, pyroclastic flows (St Vincent type), rare dome lavas	1, 2, 15, 16, 17
Grenadines	0–11.2	Kick-em-Jenny (submarine volcano north of Grenada)	1974 AD (E)	Olivine–basalt	Lava flows, pyroclastics	1, 2, 18
Grenada	0–21	Mt St Catherine	? 0.9 Ma (E)	Basanite–alkali basalt —subalkaline basalt —andesite	Lava flows, pyroclastic falls, dome lavas	1, 2, 19

Notes: (a) —, information not available. (b) Errors on dates of eruptions are on radiocarbon dates. (c) Sources of information: Most information from (1) Robson and Tomblin (1966) and (2) Briden et al. (1979); Other sources are (3) Westerman and Kiel (1961), (4) Baker et al. (1980), (5) Gunn et al. (1976), (6) Baker (1968), (7) Baker (1969), (8) Hutton and Nockolds (1978), (9) Rea (1974), (10) Tazieff (1977), (11) Westercamp and Mervoyer (1976), (12) Wills (1974). (13) Andreieff et al. (1976), (14) Tomblin (1964), (15) Rowley (1978), (16) Shepherd et al. (1979), (17) Aspinall et al. (1973), (18) Sigurdsson and Shepherd (1974), (19) Arculus (1976).

large, e.g. the eruption on St Vincent 14 ka (?) ago (Hay, 1959), which may have been an order of magnitude greater than the largest historic eruptions. Calculation of the size of eruptions (and also interpretation of the geological history) is, however, greatly complicated by factors such as the deposition of much material under the sea, away from the volcano concerned, and the very rapid reworking and stripping of unconsolidated material from volcano flanks. Some indication of the latter effect is given by Shepherd et al. (1979) who showed that a very large proportion of material erupted in the 1979 St Vincent eruption had been removed from the flanks of the volcano within a period of 2 weeks.

The violence of, and the damage and large number of fatalities caused by the West Indian volcanoes (over 30 000 people killed this century) has led to an understandable interest in volcano prediction and disaster planning in the area. So far, however, achievement in these fields has been rather limited. The evacuation of population on St Vincent in 1971–72 and on Guadeloupe in 1976 was followed by quiet dome building and phreatic activity respectively (Tazieff, 1977). Evacuation of population, based on predictions made on St Vincent in 1979, was, however, fully justified by the violent activity of April 1979. Volcano prediction in the West Indies is not made any easier by the common occurrence of seismic crises and phreatic eruptions, which do not lead to violent explosive activity, but prediction aimed at forecasting catacylsmic activity rather than the outbreak of activity and based on geological, geophysical (especially earthquake focal depth migration), and geochemical (e.g. temperature-dependent change in gas composition) criteria may be more successful in the future (Tazieff, 1979).

A study of the eruptive history of the Lesser Antillean volcanoes shows that there is a strong compositional control on eruptive style. Thus basalt magma is erupted as relatively mobile lava and as vertically ejected pyroclasts giving rise to lava flows and pyroclastic fall deposits. The more viscous andesite to dacite magma gives accumulations of lava in the form of domes and pyroclastic activity is often violent in the form of pyroclastic flow eruptions. The latter can be subdivided into St Vincent-type *nuée ardente* eruptions, Peléan-type *nuée ardente* eruptions, ash–pumice flow eruptions, and the crystal–pumice ground surge deposits, produced by the collapse of the eruptive column in Plinian eruptions (Roobol and Smith, 1976, 1980; Smith and Roobol, this volume, Section V).

The abundance of basalt on Grenada and St Vincent is reflected by the large proportions of lava flows and pyroclastic fall deposits on these islands. Elsewhere there is considerable variation in the relative proportions of basalt and andesite and in the relative importance of lava flow/pyroclastic falls and dome lava/pyroclastic flows. This variation occurs between different islands and also between volcanic centres within individual islands. Study of the variation between different volcanic centres suggests that as a crude approximation the West Indian volcanoes belong to two major categories as follows:

(1) *Basalt–andesite volcanoes* (e.g. Soufrière, St Vincent; Mt Misery, St Kitts; South Soufrière Hill, Montserrat). These volcanoes have erupted mainly lava flows and pyroclastic falls which have accumulated to form cone-shaped composite volcanoes, usually with complete central craters. A compositional gradation from basalt through basaltic andesite to andesite is common, although pyroclastic fall successions such as the Mansion Series, St Kitts (Baker and Holland, 1973), and the White River Pyroclast Fall Series, Monserrat (Rea and Baker, 1981), show numerous compositional cycles, each becoming more basic upwards and also show evidence of magma mixing. Ultrabasic cumulate blocks (p. 175) are found at these volcanoes and are particularly abundant at Soufrière, St Vincent (Wager, 1962; Lewis 1973).

(2) *Andesite–dacite volcanoes* (e.g. Mt Pelée, Martinique; Soufrière, Guadeloupe). These volcanoes are much more common in the Lesser Antilles than are the basalt–andesite volcanoes. They have erupted mainly pyroclastic flows and dome lavas with eruptive cycles typically beginning with the former and ending with the latter.

Petrology

Petrography

The majority of the volcanic rocks from the Lesser Antilles belong to the calc-alkaline volcanic suite. They are generally strongly porphyritic, with plagioclase the predominant phenocryst mineral. The plagioclases generally have calcic cores (bytownite to sodic anorthite) and are characterized by normal oscillatory zoning. Resorption often accompanies the larger oscillations. Other minerals present as phenocrysts include olivine and augite in more basic rocks, augite and hypersthene in intermediate rocks, and hypersthene, hornblende, quartz, and rarely biotite in more acidic rocks. Titaniferous magnetite is present throughout the compositional range. Many of the phenocrysts show disequilibrium features such as zoning and resorption. Possible causes of these features include loss of water during eruption (hornblende and plagioclase resorption), reduction of pressure during magma ascent (quartz resorption), and constitutional supercooling (fine scale oscillatory zoning in plagioclase—see Sibley et al., 1976). Mixing of magma may also be the cause of certain unusual petrographic features, e.g. the reversed zoning in plagioclase and the persistence of olivine in acid lavas on Saba (Baker et al., 1980).

In addition to the calc-alkaline rocks, lavas with alkaline affinities occur on Grenada (Arculus, 1976). Some of the basic rocks here are sufficiently rich in olivine to be classed as basanites. Rocks from some of the northern islands have been considered to belong to an island-arc tholeiite suite (Brown et al., 1977) but this is undoubtedly something of an oversimplification (p. 177). Coarse grained cumulates of overall ultrabasic composition occur at basalt–andesite volcanoes (p. 174) throughout the arc. These consist of widely variable amounts of plagioclase (sodic anorthite), amphibole (with high Al/Si ratios), titaniferous magnetite, olivine (generally Fo_{90}–Fo_{80}), and rarer augitic clinopyroxene and orthopyroxene.

Chemical compositions

A very large volume of geochemical data for the entire Lesser Antilles arc has been presented and assessed in an important paper by Brown et al. (1977). Rea and Baker (1980) have reviewed chemical data for the northern Lesser Antilles and there is now a considerable body of data available for the individual islands. A comparison of andesite compositions in the different islands is given in Table 2 and selected geochemical characteristics for different islands in Table 3.

In general the rocks of the Lesser Antilles show a wide range of SiO_2 (typically 48–65 per cent by weight). Apart from the southern part of the arc (Grenada, Grenadines, St Vincent), where basalt (47–54 per cent SiO_2) predominates, andesite (54–63 per cent SiO_2) is by far the most abundant composition represented. Dacites (63–70 per cent SiO_2) occur mainly in the central portion of the arc and rhyolites (SiO_2 greater than 70 per cent) are extremely rare. Other general characteristics of the calc-alkaline rocks in the arc are relatively high contents of Al_2O_3 and CaO. Contents of MgO and TiO_2 are low except for the mafic-rich basic rocks from Grenada. In general the rocks show an increase in silica accompanied by a moderate to low increase in alkalis and virtually no Fe enrichment.

It is, however, difficult to generalize about the chemical composition of the Lesser Antilles arc since one of the most important chemical features is the variation that occurs between different islands. As a first approximation, however, it is possible to consider chemical variation within the arc by reference to the following groups of islands.

(1) *Northern group (Saba, St Eustatius, St Kitts, Nevis, Montserrat)*. These islands are characterized by relatively low volumes of basalt and low contents of K_2O, incompatible trace elements, $^{87}Sr/^{86}Sr$, Ni, and Cr (Table 3).

TABLE 2 Chemical compositions of andesites from the Lesser Antilles

	1	2	3	4	5	6	7	8	9	10	11
SiO_2	59.4	60.0	59.6	60.7	60.0	59.7	60.6	59.5	63.4	54.2	59.2
TiO_2	0.6	0.6	0.6	0.5	0.5	0.8	0.6	0.7	0.6	0.9	0.7
Al_2O_3	17.6	17.5	18.5	17.0	18.5	17.4	17.6	17.8	17.1	18.4	17.7
Fe_2O_3	3.2	3.6	3.8	4.5	3.3	2.0	1.9	2.0	1.4	2.3	1.7
FeO	2.9	2.7	3.4	2.9	3.2	5.5	5.2	5.4	4.0	6.2	4.7
MgO	3.7	2.8	2.4	2.6	2.5	2.3	2.4	2.5	2.2	4.5	3.0
CaO	7.5	7.4	6.9	6.8	7.4	7.8	7.4	7.7	6.9	9.4	7.4
Na_2O	3.6	4.4	3.9	3.3	3.5	3.3	2.9	3.0	2.7	3.3	3.9
K_2O	1.1	0.7	0.6	1.2	0.8	0.9	0.6	1.2	1.5	0.5	1.5
MnO	0.2	0.1	0.1	0.2	0.2	0.2	0.2	0.2	0.1	0.2	0.1
P_2O_5	0.1	0.2	0.2	0.3	0.1	0.1	0.1	0.1	0.1	0.1	0.2
Trace elements (p.p.m.)											
Ba	400	170	260	270	170	165	226	220	459	116	546
Zr	90	90	90	90	110	101	99	102	136	93	169
Sr	350	250	290	300	230	300	293	287	335	236	877
Rb	23	14	—	—	17	23	46	46	70	12	43
Ni	32	4	2	5	6	39	41	38	7	90	66
Cr	80	30	7	3	6	—	—	133	32	265	152
V	—	—	100	150	130	—	—	—	64	209	140

(1) Average andesite, Saba (Baker *et al.*, 1980). (2) Average andesite, Statia (Lacroix, 1926; P. E. Baker, unpublished). (3) Average andesite, Mt Misery, St Kitts, (Baker, 1968). (4) Average andesite, Nevis (Hutton and Nockolds, 1978). (5) Average andesite, Montserrat (Rea, 1970). (6) Average rock, Guadeloupe (Brown *et al.*, 1977, Table 1). (7) Average rock, Martinique (Brown *et al.*, 1977, Table 1). (8) Average andesite, Dominica (Brown *et al.*, 1977, Table 6). (9) Average rock, St Lucia (Brown *et al.*, 1977, Table 1). (10) Average rock, St Vincent (Brown *et al.*, 1977, Table 1). (11) Average andesite, Grenada (Brown *et al.*, 1977, Table 6).
Note: Analyses recalculated to 100 per cent, water-free.

TABLE 3 Selected geochemical characteristics of volcanic rocks from the Lesser Antilles

	1	2	3	4	5	6	7	8	9
SiO_2	58.9	56.6	58.0	60.8	63.0	53.9	53.3	57.4	59.0
K_2O	1.0	0.6	0.8	1.2	1.5	0.5	1.1	0.95	1.0
Rb	21	—	15	42	70	12	31	27	36
Ba	380	220	150	215	459	116	435	247	245
Sr	340	280	300	284	335	236	790	387	340
Ni	39	7	8	37	7	90	160	75	50
Cr	93	19	10	126	32	265	363	—	—
$^{87}Sr/^{86}Sr$ range	—	0.7036–0.7047	0.7024–0.7049	0.7042–0.7073	0.7035–0.7092	0.7038–0.7043	0.7043–0.7050	0.7035–0.7092	—

(1) Average rock analysis, Saba (Baker *et al.*, 1980). (2) Average rock analysis, Mt Misery, St Kitts (Baker, 1968; Sr isotope analyses from Hedge and Lewis, 1971). (3) Average rock analysis, Montserrat (Rea, 1970). (4) Average rock analysis, Dominica (Brown *et al.*, 1977; Sr isotope analyses from Donnelly *et al.*, 1971). (5) Average rock analysis, St Lucia (Brown *et al.*, 1977; Sr isotope analyses from Pushkar *et al.*, 1973). (6) Average rock analysis, St Vincent (Brown *et al.*, 1977; Sr isotope analysis from Hedge and Lewis, 1971, and Pushkar *et al.*, 1973). (7) Average rock analysis, Grenada (Brown *et al.*, 1977; Sr isotope analyses from Arculus, 1976). (8) Unweighted average rock analysis, Lesser Antilles (Brown *et al.*, 1977). (9) Average rock analysis, Lesser Antilles, weighted by area of outcrop on each island (Brown *et al.*, 1977).

It is clear from data presented in Table 2 and also in Rea and Baker (1980) that the northern islands do not form an altogether chemically homogeneous group. For example, rocks from Saba (Baker et al., 1980) contain much higher contents of Ni, Cr, and Ba than do rocks from other islands in the group. It is also clear that among the northern islands there is no consistent compositional trend from north to south, which contrasts with the situation at the southern end of the arc (Brown et al., 1977, p. 791). The chemical composition of the northern islands also suggests that it is an oversimplification to assign the rocks from these islands to an island arc tholeiitic suite. Even on St Kitts, where the rocks are closest compositionally to island arc tholeiites, the rocks have higher average SiO_2, Sr, Ba, lower K/Rb and show less Fe enrichment than the island arc tholeiitic series as defined by Jakeš and White (1972) and Jakeš and Gill (1970). Elsewhere, e.g. Montserrat, the rocks show chemical characteristics such as SiO_2 and Sr contents, Na_2O/K_2O, K/Rb, and La/Yb ratios, and Fe enrichment trends which are closely similar to calc-alkaline volcanic rocks developed in other island arcs.

(2) *Central Group (Guadeloupe, Dominica, Martinique, St Lucia)*. This group includes the largest islands in the Lesser Antilles and the total volume of volcanics above recognizable basement is greater here than in other parts of the Lesser Antilles (Wills, 1974). There is a higher proportion of acid rocks than elsewhere in the arc and values of incompatible trace elements are intermediate between Grenada on the one hand and St Vincent and the northern islands on the other. Contents of Ni and Cr are higher than for the northern islands but lower than for Grenada and St Vincent. There is very little variation in Ni and Cr contents at varying contents of SiO_2. Some higher values of $^{87}Sr/^{86}Sr$ (0.7073–0.7092) have been recorded in acid lavas from St Lucia and Dominica but the average Sr isotope ratios are similar to those found in the northern islands (0.7038) and lower than those determined in the southern islands (0.7048). (For discussion and data sources *see* Brown et al., 1977, p. 795.) As is the case with the northern islands, there is considerable inter- and intra-island variation in chemical composition. Intra-island variation is particularly noticeable on Dominica (Wills, 1974) and Martinique (Gunn et al., 1974).

(3) *St Vincent and Bequia*. Rocks on St Vincent are predominantly basaltic. Contents of incompatible trace elements are generally low, being nearer to those in rocks from St Kitts than those in rocks from Grenada. Contents of Ni and Cr are, however, much higher than those found in the northern islands. Brown et al. (1977) consider that average compositions of Bequia and St Vincent are transitional between those of Grenada to the south and the larger islands to the north. They also suggest that southwards from St Vincent through the Grenadines to Grenada there is a continuous gradation towards more alkaline compositions and higher contents of Zr, Nb, K, Rb, Ni, and Cr.

(4) *Grenada and the southern Grenadines*. Grenada is predominantly characterized by the presence of basalt. Basanitoids and alkaline basalts, which are strongly undersaturated in silica, have, at certain volcanic centres, been erupted with subalkaline basalt, andesite, and dacite (Arculus, 1976). Overall the Grenada rocks are characterized by high contents of Ni, Cr, and large ion lithophile elements, but there is considerable compositional variation between volcanic centres and also within individual volcanic centres. Shimizu and Arculus (1975) have shown that the basanitoids and alkaline olivine basalts are relatively enriched, and have variable contents of light REE, in relation to the heavy REE which show little variation.

Petrogenesis

Various processes, including partial melting, fractional crystallization, magma mixing, and contamination have been considered important in the genesis of volcanic rocks in island arcs. The contribution of these processes to petrogenesis in the Lesser Antilles is discussed below.

Partial melting of mantle peridotite

Arculus (1976) and Brown et al. (1977) have invoked partial melting of upper mantle peridotite to account for the generation of the basanitoids and alkaline basalts on Grenada. REE distributions in these rocks (Shimizu and Arculus, 1975) can be interpreted in terms of 2–17 per cent partial melting of a garnet lherzolite source, and the undersaturated nature of the rocks might imply hydrous melting, perhaps via release of water from underthrust Atlantic ocean crust (Sigurdsson et al., 1973). Brown et al. (1977) also considered that the depth of the subduction zone is approximately constant beneath the Lesser Antilles (see discussion on p. 171) and that, as a result, petrogenetic conditions may also be similar along the length of the arc. The implication is that mantle partial melting is an important petrogenetic process throughout the Lesser Antilles. There is no doubt, however, that the Grenada basalts differ from basalts from elsewhere in the arc and the differences with those from the northern Lesser Antilles are particularly marked. Thus the Grenada basalts are unsaturated and contain much higher contents of large ion lithophile elements, Ni and Cr. The last two elements also show strong depletion with increase in SiO_2, in contrast to the low and constant (with varying SiO_2) contents found in the northern islands. It seems likely that these geochemical differences indicate differences in the conditions of, and processes involved in, magma genesis along the arc. For example, partial melting within the upper mantle may be less important below the northern islands than it is below Grenada, or the effects of subsequent processes such as fractional crystallization may be quite different to those in Grenada, where fractional crystallization of olivine-Ca augite and spinel, followed by amphibole, is believed to be responsible for the transition from alkaline basalt through saturated basalts to andesites (Arculus, 1976). Certainly the very low contents of Cr and Ni in basalts and andesites from the middle and especially the northern islands in the Lesser Antilles cannot be reconciled with an origin involving only partial melting in the upper mantle. Mysen and Kushiro (1978), from a study of Ni partition between forsterite and aluminous melt, suggest Ni contents of between 400 and 500 p.p.m. in andesites generated by equilibrium partial melting of the mantle. Therefore, if magmas in the northern Lesser Antilles originate in the upper mantle, subsequent processes, such as olivine fractionation or separation of an immiscible Ni sulphide liquid, must have occurred. Similarly, the very low contents of Cr would imply efficient fractional crystallization of a Cr-bearing phase such as Cr-spinel. Further evidence, suggesting a limited role for mantle partial melting in the evolution of magmas in the northern and central Lesser Antilles, lies in the low volume of basalt of subalkaline character, erupted in these parts of the arc. The saturated nature of the basalts requires very large amounts of water to be released from the subducted slab since partial melting of peridotite under conditions of $P_{H_2O} < 0.5 P_{tot}$ will give rise to undersaturated melts (Mysen and Boettcher, 1975). Finally, the very large volumes of andesite in the north and central Lesser Antilles are difficult to reconcile with partial melting of a peridotite source, which is capable of yielding relatively small volumes of magma of this composition.

In view of the discussion above it seems that partial melting of peridotite in the upper mantle may be less important in the northern and central parts of the Lesser Antilles than it is in Grenada. Even on Grenada, the petrogenetic picture may be more complicated than previously thought since recent geochemical evidence requires multiple source models, possibly involving subducted lithosphere and sediment (Graham, 1980).

Partial melting of oceanic tholeiite

Various authors (e.g. Delany and Helgeson, 1978) have presented evidence to suggest that the temperatures generated within the upper parts of subducted oceanic crust are unlikely to be high enough for melting to occur. On the other hand, Marsh (1979) has proposed a model which suggests that melting will occur towards the top of the slab, near its contact with the asthenosphere/lithosphere boundary of the overlying plate. Even if no actual melting occurs, the subducted slab may make an important contribution to magma genesis, through the supply of the water necessary for melting to occur above the subduction zone. In addition, certain elements such as K and Rb might be incorporated into this aqueous phase, leading eventually to the enrichment of these elements in melts generated above the slab (see, for example, Hawkesworth et al., 1979a).

However, if as suggested by Marsh (1979) melting within the slab does occur, then oceanic tholeiite or its metamorphosed equivalent is a source capable of giving rise to the large volumes of andesite and dacite observed in the northern and central Lesser Antilles. This source could also account satisfactorily for the low contents of Ni and Cr found in these rocks. Fairly complete melting of the subducted oceanic tholeiite would be required to produce the basalts in the northern and central Lesser Antilles and the chemistry of these basalts compared to that of the subducted basalts would require that elements such as Ti, Ni, and Cr were concentrated in minor refractory phases or were subsequently depleted by fractionation.

Fractional crystallization

Since one of the major chemical characteristics of the Lesser Antilles lavas is strong silica enrichment accompanied by a relatively small degree of Fe enrichment, fractional crystallization of silica-poor phases such as olivine, spinel, magnetite, and amphibole has often been considered important in their genesis. Fractional crystallization of olivine would certainly account for the low contents of Mg and Ni in lavas from the centre and north of the Lesser Antilles, but the experimental evidence (e.g. Nicholls and Ringwood, 1973) suggests that olivine will be a liquidus phase only at low pressures, especially in melts of andesitic composition. Separation of chrome spinel would lead to silica depletion and might also explain the very low Cr contents found in lavas from the northern Lesser Antilles.

The importance of magnetite fractionation has been suggested from the experimental work of Osborn (1959). However, there are a number of objections to this concept (see summary in Cawthorn and O'Hara, 1976), and Taylor et al. (1969) discounted magnetite fractional crystallization on the grounds that high Al basalts and andesites have very similar contents of V, which should be depleted by magnetite fractional crystallization. In certain parts of the Lesser Antilles, especially St Kitts, the difference in V contents between basalts (300 p.p.m.) and andesites (40 p.p.m.) and the close correlation of V and Ti suggest that the possibility of fractional crystallization of titaniferous magnetite cannot be excluded on geochemical grounds. Experimental evidence (e.g. Cawthorn and O'Hara, 1976) has shown the importance of amphibole as a liquidus phase in wet basaltic systems and the chemical characteristics (low SiO_2, high Fe/Mg at least in comparison with pyroxene and olivine) of the amphibole are such that its fractional crystallization will produce a calc-alkaline trend. Even though amphibole is not on the low pressure liquidus of certain calc-alkaline compositions (Ritchey and Eggler, 1978), there is evidence for amphibole fractional crystallization at certain volcanoes in the Lesser Antilles, as discussed below.

The ultrabasic cumulates found at basalt–andesite volcanoes throughout the Lesser Antilles provide direct evidence of the fractional crystallization of basaltic magma. The cumulate blocks contain some or all of the phases anorthite (An_{89}–An_{96}), hastingsitic amphibole, olivine (Fo_{68}–Fo_{88}), titanomagnetite (Ulv_{12}–Ulv_{30}), Ca-rich clinopyroxene (Di_{68}–Di_{79}), and orthopyroxene (En_{63}–En_{72}), often together with a little vesicular glass of high alumina basalt composition. Orthopyroxene is absent from the cumulates on St Vincent, Grenada, and the Grenadines. Anorthite, amphibole, and magnetite are the major cumulus phases on Montserrat and St Kitts, but in some of the Dominican nodules amphibole appears to overgrow and sometimes to replace clinopyroxene, suggesting that it is an intercumulus phase. A detailed description of the mineralogy and mineral chemistry of cumulates from the Lesser Antilles has been given by Wills (1974). A recent investigation of crystallization conditions in the nodules (Powell, 1977) suggested that, assuming a crystallization temperature range of 1100–1200 °C, the pressure of crystallization decreases from north to south along the arc, from c. 9–10 kb in St Eustatius to 3 kb in Grenada. Oxygen activity is uniformly high and P_{H_2O} is generally less than 1 kb, except for Grenada where it is between 1 and 3 kb.

The association of these ultrabasic cumulates with volcanoes showing a compositional continuum from basalt through basaltic andesite to andesite suggests a genetic link, whereby the andesites are derived from the basalts by fractional crystallization of material represented in the cumulates. At certain volcanoes such as Mt Misery, St Kitts, and South Soufrière Hill, Montserrat, the field evidence for this relationship is particularly good. Both these volcanoes have pyroclastic fall successions showing compositional cycles with andesitic pyroclastics, containing cumulates at the base, grading upwards into basaltic compositions. In addition computations of mineral abstracts required to give the andesites by removal from the basalts are very close to mineral modes determined on the cumulates, and the chemical fit for both major and trace elements is also very good (Rea, 1974).

It appears, therefore, that andesites have been produced by the fractional crystallization of basaltic magma at basalt–andesite volcanoes in the Lesser Antilles. However, the complete absence of basalt and ultrabasic cumulates at the much more common andesite–dacite volcanoes makes it unlikely that andesites at these volcanoes have been generated in this way. The only direct evidence of fractional crystallization at these volcanoes is the presence of cognate xenoliths, which appear to represent high level (1–3 km) accumulations of phases which occur mainly as microphenocrysts in the host andesite lavas. Fractional crystallization of this material may have produced limited silica enrichment in andesitic magmas at the andesite–dacite volcanoes.

Magma mixing

There is direct evidence for magma mixing at both types of volcanoes in the Lesser Antilles. At the basalt–andesite volcanoes banded pumices of bimodal composition occur within the pyroclastic fall succession of Mt Misery, St Kitts, and South Soufrière Hill, Montserrat, and Shepherd et al. (1979) noted mixtures of basaltic andesite (SiO_2 57 per cent) and dacite (SiO_2 66 per cent) among the products of the 1979 eruption of Soufrière, St Vincent. In the last case, magma mixing may have been responsible for initiation of the pyroclastic activity, as suggested elsewhere by Sparks et al. (1977). At the andesite–dacite volcanoes banded blocks of lava and pumice have been reported at Soufrière Hills, Montserrat (Rea, 1974), Basse Terre, Guadeloupe (Smith et al., 1975), Mt Pelée, Martinique (Roobol and Smith, 1976), and St Eustatius (reported in Roobol and Smith, 1976). These blocks are generally mixtures of more basic andesite (c. 57 per cent SiO_2) and more acid andesite (c. 62 per cent SiO_2) but chemical data (e.g. Rea, 1974) suggest that the chemical effects of this mixing are limited. Sometimes, however, magma mixing may produce significant compositional effects, as for example on Saba, where Baker et al. (1980) have attributed the many unusual mineralogical and geochemical features to the mixing of two genetically distinct magmas.

Contamination

Possible sources of contamination of magmas in the Lesser Antilles include subducted oceanic sediments and the mantle–crust columns through which magmas rise from their point of origin to the surface. Oceanic sediment has been suggested by Donnelly et al. (1971) as a possible source for the enrichment of radiogenic Pb in lavas from the Lesser Antilles and Graham (1980) has invoked a contribution from oceanic sediment to account for certain incompatible element ratios of the Grenada rocks. There is, however, argument concerning the amount of sediment subducted and Donnelly et al. (1971) have suggested that more sediment may be subducted beneath the central part of the arc where

TABLE 4 Variation in lava composition and possible petrogenetic controls within the Lesser Antilles

	Part of Arc		
	Northern Islands (Saba to Montserrat)	Central Islands (Guadeloupe to St Lucia)	Southern Islands (St Vincent to Grenada)
Nature of magmas Relative volumes	Predominant andesite, some basalt and dacite	Predominant andesite, some dacite and basalt, rare rhyolite	Predominant basalt, including basanites and alkaline basalts on Grenada, some andesite, rare dacite
Magma types	Calc-alkaline with some affinities with island arc tholeiites	Calc-alkaline	Calc-alkaline and alkaline
Chemical characteristics	Low incompatible elements, Cr, Ni, and radiogenic Sr	Higher incompatible elements and some higher radiogenic Sr	High incompatible elements, especially on Grenada
Nature of subduction zone Dip	17° increasing to west	20–50° increasing to west	30° increasing to east
seismic, Activity	High	High	Low
Mechanisms	Underthrusting with northward oblique slip, hinge faulting (?)	Underthrusting with slip vector normal to arc	Underthrusting with southward oblique slip, hinge faulting (?)
Subducted material	No oceanic sediment (?)	Includes some oceanic sediment (?)	No oceanic sediment (?)
Nature of mantle	Depleted by previous subduction (?)	Depleted (?)	Fertile
Nature of crust	Comparatively primitive due to westward translation of arc (?)	Comparatively evolved	Comparatively evolved
		(higher positive gravity anomalies compared with further north)	

the slip vector is normal to the subduction zone (p. 171).

There is very little petrographic evidence for contamination in lavas from the Lesser Antilles. Siliceous metasedimentary xenoliths occur within certain basalts but show no sign of reaction and the rounded quartz crystals in andesites are probably original high pressure phenocrysts, subsequently resorbed at shallow depth (Nicholls et al., 1971). Certain unusual petrographic features of Saba andesites such as the persistence of olivine phenocrysts into relatively siliceous andesites and the frequent occurrence of amphibole in basaltic andesites may, however, indicate that contamination was involved in the origin of these rocks (Baker et al., 1980). Chemical data for the Lesser Antilles, for example the low values of K, Rb, Ba, and $^{87}Sr/^{86}Sr$, especially in the northern islands, rule out large scale contamination by old continental crust in the origin of lavas from the Lesser Antilles.

Composition and petrogenetic variation in the Lesser Antilles

As noted above (pp. 175–177) there is considerable variation in the composition of lavas erupted along the Lesser Antilles. Similar compositional variation occurs along other destructive plate margins such as Japan (Carr et al., 1973), Central America (Stoiber and Carr, 1973), and the Cascades (Hughes et al., 1980) and has been ascribed to variable petrogenetic conditions within arc segments bounded by faults running normal to the volcanic belt. It seems possible that the Lesser Antilles may also be considered as a series of fault-bounded blocks (see discussion on pp. 170 and 171) and the observed compositional variation may reflect variation in petrogenetic conditions and in the mixture of petrogenetic processes operating within these blocks. Factors which may influence petrogenesis and which may vary in different parts of the Lesser Antilles include the nature of the subduction process and the composition of mantle and crust above the subduction zones. A summary of the possible variation in these factors is given in Table 4.

From the information given in Table 4 it appears that the mantle beneath the northern Lesser Antilles may be refractory and contribute relatively little to magmagenesis. Partial melting of subducted oceanic tholeiite will account satisfactorily for most of the chemical characteristics of the lavas in the northern Lesser Antilles and is the favoured primary petrogenetic process in this part of the arc. It is thought that partial melting of the subducted ocean crust may also be more important than mantle melting in the genesis of magmas in the central part of the Lesser Antilles. In this case, however, there may be greater contribution to magma genesis from subducted oceanic sediment and from crustal contamination than in the case beneath the northern islands. Processes directly related to subduction are probably less important beneath the southern end of the arc where short term seismicity and even subduction itself may be inhibited by the sedimentary infill of the trench. The primary process here appears to have been partial melting within relatively fertile garnet lherzolite mantle, although the chemistry of the Grenada lavas probably requires some contribution from subducted ocean crust (Hawkesworth et al., 1979b; Graham, 1980). Processes of fractionation, magma mixing, and contamination are superimposed on the deeper level processes and are themselves variable, because of factors such as the variable crustal composition below the Lesser Antilles.

REFERENCES

Anderson, T. and Flett, J. S. (1903). Reports on the eruption of the Soufrière in St Vincent in 1902 and on a visit to Montagne Pelée in Martinique, Part I. Phil. Trans. R. Soc. A **200**, 353–553.

Andreieff, P., Bellon, H., and Westercamp, D. (1976). Chronometrie et stratigraphie comparée des edifices volcaniques et formations sedimentaires de la Martinique (Antilles Francaises). Bull. B.R.G.M. (2) **4**, 335–346.

Andrew, E. M., Masson-Smith, D., and Robson, G. R. (1970). Gravity anomalies in the Lesser Antilles. Inst. geol. Sci., Geophys. Pap. no. 5.

Arculus, R. J. (1973). The alkali basalt, andesite association of Grenada, Lesser Antilles, Ph.D. thesis, University of Durham, England.

Arculus, R. J. (1976). Geology and geochemistry of the alkali basalt–andesite association of Grenada, Lesser Antilles island arc. *Geol. Soc. Am. Bull.* **87**, 612–624.

Aspinall, W. P., Sigurdsson, H., and Shepherd, J. B. (1973). Eruptions of Soufrière Volcano on St Vincent Island, 1971–72. *Science N.Y.* **181**, 117–124.

Baker, P. E. (1968). Petrology of Mt Misery, St Kitts, West Indies. *Lithos* **1**, 124–150.

Baker, P. E. (1969). The geological history of Mt Misery volcano, St Kitts, West Indies. *Overseas Geol. Mineral. Resources* **10**, 207–230.

Baker, P. E. and Holland, J. G. (1973). Geochemical variations in a pyroclastic succession on St Kitts, West Indies. *Bull. Volc.* **37**, 472–490.

Baker, P. E., Buckley, F., and Padfield, T. (1980). Petrology of the volcanic rocks of Saba, West Indies. *Bull. Volc.* **43**, 337–346.

Briden, J. C., Rex, D. C., Faller, A. M., and Tomblin, J. F. (1979). K–Ar geochronology and palaeomagnetism of volcanic rocks in the Lesser Antilles island arc. *Phil. Trans. R. Soc. A* **291**, 485–528.

Brown, G. M., Holland, J. G., Sigurdsson, H., Tomblin, J. F., and Arculus, R. J. (1977). Geochemistry of the Lesser Antilles volcanic island arc. *Geochim. cosmochim. Acta* **41**, 785–801.

Bunce, E. T., Phillips, J. D., Chase, R. L., and Bowin, C. O. (1970). The Lesser Antilles arc and the eastern margin of the Caribbean Sea. In *The Sea*, Vol. 4 (A. E. Maxwell, ed.), Wiley-Interscience, New York, pp. 359–385.

Carr, M. J., Stoiber, R. E., and Drake, C. L. (1973). Discontinuities in deep seismic zones under the Japanese arcs. *Geol. Soc. Am. Bull.* **84**, 2917–2930.

Cawthorn, R. G., and O'Hara, M. J. (1976). Amphibole fractionation in calc-alkaline magma genesis. *Am. J. Sci.* **276**, 349–329.

Chase, R. L. and Bunce, E. T. (1969). Underthrusting of the eastern margin of the Antilles by the floor of the western North Atlantic Ocean, and the origin of the Barbados Ridge. *J. geophys. Res.* **74**, 1413–1420.

Delany, J. M. and Helgeson, H. C. (1978). Calculation of the thermodynamic consequences of dehydration in subducting oceanic crust to 100 kb and 800°C. *Am. J. Sci.* **278**, 638–686.

Dinkelman, M. G. and Brown, J. F. (1977). K–Ar geochronology and its significance to the geological setting of La Desirade, Lesser Antilles. *8th Caribbean Geology Conference, Curaçao, July 1977. Abstracts* pp. 38–39.

Donnelly, T. W., Rogers, J. J. W., Pushkar, P., and Armstrong, R. L. (1971). Chemical evolution of the igneous rocks of the eastern West Indies: an investigation of thorium, uranium and potassium distributions and lead and strontium isotopic ratios. *Geol. Soc. Am. Mem.* no. 130, 181–224.

Edgar, N. T., Ewing, J. I., and Hennion, J. (1971). Seismic refraction and reflection in the Caribbean Sea. *Am. Assoc. Petr. Geol. Bull.* **55**, 833–870.

Fink, L. K., Jr (1970). Evidence for the antiquity of the Lesser Antilles island arc (abstract). *Trans. Am. Geophys. Un.* **s1**, 326.

Fink, L. K., Jr (1972). Bathymetric and geologic studies of the Guadeloupe region, Lesser Antilles island arc. *Marine Geol.* **12**, 267–288.

Flett, J. S. (1908). Petrographic notes on the products of the eruptions of May 1902 at the Soufrière in St Vincent. *Phil. Trans. R. Soc. A* **208**, 305–332.

Fox, P. J. and Heezen, B. C. (1975). Geology of the Caribbean crust. In *The Ocean Basins and Margins*, Vol. 3 (A. E. M. Nairn and F. G. Stehli, eds.), Plenum, New York, Chap. 10.

Graham, A. M. (1980). Genesis of the igneous rock suites of Grenada, Lesser Antilles, Ph.D. thesis, University of Edinburgh, UK.

Gunn, B. M., Roobol, M. J., and Smith, A. L. (1974). Petrochemistry of Pelean-type volcanoes of Martinique. *Geol. Soc. Am. Bull.* **85**, 1023–1030.

Gunn, B. M., Roobol, M. J., and Smith, A. L. (1976). The petrogenetic implications of a basaltic andesite–soda rhyolite suite on St Eustatius, Lesser Antilles. *Geol. Soc. Am. Abstr. Programs* **8**, 896–897.

Hawksworth, C. J., Norry, M. J., Roddick, J. C., Baker, P. E., Francis, P. W., and Thorpe, R. S. (1979a). $^{143}Nd/^{144}Nd$, $^{87}Sr/^{86}Sr$ and incompatible element variations in calc-alkaline andesites and plateau basalts from South America. *Earth planet. Sci. Letters* **38**, 95–116.

Hawkesworth, C. J., O'Nions, R. K., and Arculus, R. J. (1979b). Nd and Sr isotope geochemistry of island arc volcanics, Grenada, Lesser Antilles. *Earth planet. Sci. Letters* **45**, 237–248.

Hay, R. L. (1959). Formation of the crystal-rich glowing avalanche deposits of St Vincent, BWI. *J. Geol.* **67**, 540–562.

Hedge, C. E., and Lewis, J. F. (1971). Isotopic composition of strontium in three basalt–andesite centers along the Lesser Antilles arc. *Contr. Mineral. Petrol.* **32**, 39–47.

Hughes, J. M., Stoiber, R. E., and Carr, M. J. (1980). Segmentation of the Cascade volcanic chain. *Geology* **8**, 15–17.

Hutton, C. O. and Nockolds, S. R. (1978). The petrology of Nevis, Leeward Islands, West Indies. *Overseas Geol. Mineral Resources Bull.* no. 52.

Jakeš, P. and Gill, J. (1970). Rare earth elements and the island arc tholeiite series. *Earth planet. Sci. Lett.* **9**, 17–28.

Jakeš, P. and White, A. J. R. (1972). Major and trace element abundances in volcanic rocks of orogenic areas. *Geol. Soc. Am. Bull.* **83**, 29–40.

Jordan, T. H. (1975). The present-day motions of the Caribbean Plate. *J. geophys. Res.* **80**, 4433–4439.

Keary, P. (1974). Gravity and seismic reflection investigations into the crustal structure of the Aves Ridge, eastern Caribbean. *Geophys. Jl R. astr. Soc.* **38**, 435–448.

Lacroix, A. (1904). *La montagne Pelée et ses éruptions*, Masson et Cie, Paris.

Lacroix, A. (1926). Les characteristiques lithologiques des Petites Antilles. *Soc. geol. Belg., Livre Jubilaire* pp. 387–405.

Le Pichon, D., Francheteau, J., and Bonnin, J. (1973). Plate tectonics. *Devlmts Geotectonics* **6**, 300 pp.

Lewis, J. F. (1973). Petrology of ejected plutonic blocks of the Soufrière volcano, St Vincent, West Indies. *J. Petrol.* **14**, 81–112.

MacGregor, A. G. (1938). The volcanic history and petrology of Montserrat. *Phil. Trans. R. Soc. B* **229**, 1–90.

Marsh, B. D. (1979). Island-arc volcanism. *Am. Scient.* **67**, 161–172.

Mattison, J. H., Fink, L. K., and Hopson, C. A. (1973). Age and origin of ophiolitic rocks on La Desirade, Lesser Antilles. *Carnegie Instr. Wash. Yb.* **72**, 616–623.

Mattison, J. M., Fink, L. K., Jr, and Hopson, C. A. (1980). Geochronological and isotopic study of La Desirade island basement complex. Jurassic ocean crust in the Lesser Antilles? *Contr. Mineral. Petrol.* **71**, 237–245.

Mattson, P. H. (1979). Subduction, buoyant braking, flipping and strike-slip faulting in the northern Caribbean. *J. Geol.* **87**, 293–304.

Mauk, F. J. (1977). The seismicity and focal mechanisms of the Trinidad inclined seismic zone: a key to contemporary Caribbean tectonics. *8th Caribbean Geology Conference, Curaçao, July 1977. Abstracts* p. 114.

McCann, W. and Murphy, A. (1977). Seismicity and tectonics of the Caribbean. *8th Caribbean Geology Conference, Curaçao, July 1977. Abstracts*, p. 117.

Molnar, P. and Sykes, L. R. (1969). Tectonics of the Caribbean and Middle America regions from focal mechanisms and seismicity. *Geol. Soc. Am. Bull.* **80**, 1639–1684.

Mysen, B. O. and Boettcher, A. L. (1975). Melting of a hydrous mantle. I. Phase relations of natural peridotite at high pressures and temperatures with controlled activities of water carbon dioxide and hydrogen. *J. Petrol.* **16**, 520–548.

Mysen, B. O. and Kushiro, I. (1978). The effect of pressure on the partitioning of nickel between olivine and aluminous silicate melt. *Carnegie Instn. Wash. Yb.* **77**, 706–709.

Nagle, F., Stipp, J. J., and Fisher, D. E. (1976). K–Ar geochronology of the limestone caribbees and Martinique, Lesser Antilles, West Indies. *Earth planet. Sci. Letters* **29**, 401–412.

Nicholls, I. A. and Ringwood, A. E. (1973). Effect of water on olivine stability in tholeiites and the production of silica-saturated magmas in the island-arc environment. *J. Geol.* **81**, 285–300.

Nicholls, J., Carmichael, I. S. E., and Stormer, J. C., Jr (1971). Silica activity and P_{total} in igneous rocks. *Contr. Mineral. Petrol.* **33**, 1–20.

Officer, C. B., Ewing, J. I., Richards, R. S., and Johnson, H. R. (1957). Geophysical investigations in the eastern Caribbean: Venezuelan basin, Antilles island arc and Puerto Rico trench. *Geol. Soc. Am. Bull.* **68**, 359–378.

Officer, C. B., Ewing, J. I., Hennion, J. F., Harkrider, D. G., and Miller, D. E. (1959). Geophysical investigations in the Eastern Caribbean: summary of 1955 and 1956 cruises. In *Physics and Chemistry of the Earth*, Vol. 3 (L. H. Ahrens, F. Press, K. Rankama, and S. K. Runcorn, eds), Pergamon, London, pp. 17–109.

Osborn, E. F. (1959). Role of oxygen pressure in the crystallisation and differentiation of basaltic magmas. *Am. J. Sci.* **259**, 609–647.

Pitman, W. C., III and Talwani, M. (1972). Sea floor spreading in the North Atlantic. *Geol. Soc. Am. Bull.* **83**, 619–646.

Powell, M. (1977). Crystallisation conditions of low-pressure cumulate nodules from the Lesser Antilles island arc. *Earth planet. Sci. Letters* **39**, 162–172.

Pushkar, P., Steuber, A. M., Tomblin, J. F., and Julian, G. M. (1973). Strontium isotopic ratios in volcanic rocks from St Vincent and St Lucia, Lesser Antilles. *J. geophys. Res.* **78**, 1279–1287.

Rea, W. J. (1970). The geology of Montserrat, British West Indies. Ph.D. thesis, University of Oxford, England.

Rea, W. J. (1974). The volcanic geology and petrology of Montserrat, West Indies, *J. geol. Soc. Lond.* **130**, 341–366.

Rea, W. J. and Baker, P. E. (1980). The geochemical characteristics and conditions of petrogenesis of the volcanic rocks of the northern Lesser Antilles—a review. *Bull. Volc.* **43**, 325–336.

Rea, W. J. and Baker, P. E. (1981). A geochemical investigation of the White River Pyroclast Fall Series, Montserrat, West Indies. In preparation.

Ritchey, J. L. and Eggler, D. H. (1978). Amphibole stability in a differentiated calc-alkaline magma chamber. An experimental investigation. *Carnegie Instn. Wash. Yb.* **77**, 790–793.

Robson, G. R. and Tomblin, J. F. (1966). *Catalogue of Active Volcanoes of the World Including Solfatara Fields*, Part XX, *West Indies*, International Association for Volcanology, Rome.

Roobol, M. J. and Smith, A. L. (1976). Mt Pelée, Martinique: a pattern of alternating eruptive styles. *Geology* **4**, 521–524.

Roobol, M. J. and Smith, A. L. (1980). Pumice eruptions in the Lesser Antilles. *Bull. Volc.* **43**, 277–286.

Rowley, K. (1978). Late Pleistocene pyroclastic deposits of Soufrière Volcano, St Vincent, West Indies. *Geol. Soc. Am. Bull.* **89**, 825–835.

Saunders, J. B. (1968). Field trip guide: Barbados. In *Caribbean Geological Conference IV, Trinidad and Tobago, 1965*, Trans-Caribbean Printers, Trinidad, pp. 443–449.

Saunders, J. B. (1977). A review of models for the geological development of the South East corner of the Caribbean region. *8th Caribbean Geological Conference, Curaçao, July 1977. Abstracts*, p. 170.

Schell, B. A. and Tarr, A. C. (1978). Plate tectonics of the northeastern Caribbean sea region. *Geol. en Mijnbouw* **57**, 319–324.

Senn, A. (1940). Paleogene of Barbados and its bearing on history and structure of the Antillean Caribbean region. *Am. Assoc. Petr. Geol. Bull.* **24**, 1548–1610.

Shepherd, J. B., Tomblin, J. F., and Woo, D. F. (1971). Volcano-seismic crisis in Montserrat, West Indies, 1966–67. *Bull. Volc.* **35**, 143–163.

Shepherd, J. B., Aspinall, W. P., Rowley, K. C., Pereira, J., Sigurdsson, H. Fiske, R. S., and Tomblin, J. F. (1979). The eruption of Soufrière volcano, St Vincent, April–June 1979. *Nature, Lond.* **282**, 24–28.

Shimizu, N. and Arculus, R. J. (1975). Rare earth element concentrations in a suite of basanitoids and alkali olivine basalts from Grenada, Lesser Antilles. *Contr. Mineral. Petrol.* **50**, 231–240.

Sibley, D. F., Vogel, T. A., Walker, B. M., and Byerly, G. (1976). The origin of oscillatory zoning in plagioclase: a diffusion and growth controlled model. *Am. J. Sci.* **276**, 275–284.

Sigurdsson, H. and Shepherd, J. B. (1974). Amphibole-bearing basalts from the submarine volcano Kick 'em Jenny in the Lesser Antilles island arc. *Bull. Volc.* **38**, 891–910.

Sigurdsson, H., Tomblin, J. F., Brown, G. M., Holland, J. G., and Arculus. R. J. (1973). Strongly undersaturated magmas in the Lesser Antilles island arc. *Earth planet. Sci. Letters* **18**, 285–295.

Smith, A. L., Gunn, B. M., and Roobol, M. J. (1975). Guadeloupe and island arc andesites (abstract). *EOS* **56**, 474.

Sparks, S. R. J., Sigurdsson, H., and Wilson, L. (1977).

Magma mixing: a mechanism for triggering acid explosive eruptions. *Nature, Lond.* **267**, 315–318.

Stoiber, R. E. and Carr, M. J. (1973). Quaternary volcanic and tectonic segmentation of Central America. *Bull. Volc.* **37**, 304–325.

Sykes, L. R. and Ewing, M. (1965). The seismicity of the Caribbean region. *J. geophys. Res.* **70**, 5065–5074.

Taylor, S. R., Kaye, M., White, A. J. R., Duncan, A. R., and Ewart, A. (1969). Genetic significance of Co, Cr, Ni, Sc, and V content of andesites. *Geochim. cosmochim. Acta* **33**, 275–286.

Tazieff, H. (1977). La Soufrière, volcanology and forecasting. *Nature, Lond.* **269**, 96–97.

Tazieff, H. (1979). What is to be forecast: outbreak of eruption or possible paroxysm? The example of the Guadeloupe Soufrière. *J. geol. Soc. Lond.* **136**, 327–329.

Tomblin, J. F. (1964). The volcanic history and petrology of the Soufrière region, St Lucia. Ph.D. thesis, University of Oxford, England.

Tomblin, J. F. (1972). Seismicity and plate tectonics of the eastern Caribbean. In *Caribbean Geological Conference VI, Isla de Margarita, 1971. Transactions*, Impreso por Cromotip, Caracas, pp. 277–282.

Tomblin, J. F. (1975). The Lesser Antilles and Aves Ridge. In *The Ocean Basins and Margins*, Vol. 3 (A. E. M. Nairn and F. G. Stehli, eds.), Plenum, New York, Chap. 11.

Wager, L. R. (1962). Igneous cumulates from the 1902 eruption of Soufrière, St Vincent. *Bull. Volc.* **24**, 93–99.

Westbrook, G. K. (1975). The structure of the crust and upper mantle in the region of Barbados and the Lesser Antilles. *Geophys. Jl R. astr. Soc.* **43**, 201–242.

Westbrook, G. K., Bott, M. H. P., and Peacock, J. H. (1973). The nature of the Lesser Antilles subduction zone in the region of Barbados. *Nature phys. Sci.* **244**, 118–120.

Westercamp, D. (1976). Stratigraphie des roches volcaniques du massif du Carbet (Martinique). *Transactions de VIIIe Conference Geologique des Caraibes, Saint François, Guadeloupe 1974*, pp. 427–435.

Westercamp, D. (1977). Evolution des series volcaniques de Martinique (FWI) et des arcs insulaires des Petites Antilles dans leur contexte structural. *8th Caribbean Geological Conference, Curaçao, July 1977. Abstracts*, pp. 227–228.

Westercamp, D. and Mervoyer, B. (1976). Les series de la Martinique et de la Guadeloupe (FWI). *Bull. B.R.G.M.* (2), **4**, 229–242.

Westerman, J. H. and Kiel, H. (1961). The geology of Saba and St Eustatius. *Uitvoerige Natuurw. Werkgp. Ned. Antillen* no. 24, 1–175.

Wills, J. K. (1974). The geological history of southern Dominica and plutonic nodules from the Lesser Antilles. Ph.D. thesis, University of Durham, England.

The Andes

R. S. Thorpe, P. W. Francis, M. Hammill, and M. C. W. Baker
Department of Earth Sciences,
The Open University, Milton Keynes MK7 6AA, UK

ABSTRACT

Active andesite volcanism occurs in three areas of the South American Andes—a northern zone (5°N–2°S) in Colombia and Ecuador, a central zone (16°S–28°S) largely in southern Peru and northern Chile, and a southern zone (31°S–52°S) in southern Chile. Each zone has a well defined tectonic setting and range of volcanic products. The northern zone is characterized by basaltic andesites, the central zone by andesite–dacite lavas and dacite–rhyolite ignimbrites, and the southern zone by high alumina basalts, basaltic andesites, and andesites. In the central zone there is a clear K-h relationship which culminates in the occurrence of a 'shoshonitic' association in the eastern part of the volcanic chain. Alkaline volcanic rocks, commonly belonging to the alkaline basalt–trachyte association, occur at scattered localities to the east of the active volcanic chain.

The northern and central active volcanic zones lie $c.$ 140 km above an eastward-dipping Benioff zone, while the southern zone is only 90 km above the Benioff zone. In each area volcanism has occurred episodically since the Mesozoic. In the northern and southern zones the crust is less than 40 km in thickness, is largely younger than Mesozoic in age, and includes an 'oceanic' crustal component. In contrast the central volcanic zone is built on thick (40–70 km) continental crust and includes an area (between 16°S and 20°S) in which Precambrian crystalline basement underlies the volcanic belt. The correlation between the character of volcanism and tectonic setting in each of the three zones is emphasized using trace element and isotope data. In particular it is noted that the central zone andesites, dactites, and ignimbrites are more silicic and have higher K, Rb, Sr, and Ba and higher $^{87}Sr/^{86}Sr$ and lower $^{143}Nd/^{144}Nd$ ratios than lavas from the northern or southern volcanic zones. In addition, the volcanic rocks of the central zone show a well defined eastward increase is K, Rb, Ba and a decrease in Sr.

Andean magmas result from a complex interplay of partial melting and fractional crystallization processes within the mantle, and contamination and fractional crystallization within the crust. Variations in andesite composition *across* the central Andean volcanic chain reflect a diminishing degree of partial melting or an increase in fractional crystallization or crustal 'contamination' in passing eastwards. Variations *along* the Andean chain suggest a significant crustal contribution and a greater degree of fractional crystallization for andesites of the central volcanic zone in comparison with the volcanic rocks of the northern and southern zones. The high alumina basalts and basaltic andesites of the southern zone formed from a shallower mantle source region than the volcanic rocks from the other zones. The dacite–rhyolite ignimbrites of the central zone share a common source with the related andesites and result from fractional crystallization of andesite magma during relatively slow uprise through thick continental crust. The occurrence of alkaline volcanic rocks east of each of the active volcanic zones is attributed to partial melting of mantle peridotite distant from the subduction zone.

Introduction

The descent of the oceanic Nazca plate below western South America is accompanied by the seismicity, tectonism, and magmatism considered to be characteristic of 'Andean' continental plate margins. However, the Andean plate margin itself exhibits pronounced geological and geophysical variations along its length and the volcanic rocks show important petrological and geochemical variations both along and across the Andean chain. There is therefore no single type of 'Andean volcanism'. Here, the features of the volcanic rocks are reviewed and their variations in relation to subduction style and crustal structure are assessed.

Plate Tectonic Setting

There are several hundred active volcanoes along the length of the Andean Cordillera. Only 45 of these are considered to be 'active' by Macdonald (1972; see Fig. 1), although other workers suggest that many more should be considered active (e.g. Gonzáles-Ferrán, 1976) and it seems likely that many have experienced recent volcanic activity for which no records exist. Three linear zones of recent activity are clearly defined in Fig. 1; a northern zone between latitudes 5°N and 2°S (in Ecuador and Colombia), a contral zone between 16°S and 28°S (in Peru, Chile, Bolivia, and Argentina), and a southern zone between 31°S and 52°S (in Chile and Argentina).

Between Colombia and southern Chile the Nazca plate is being subducted below South America. The subduction zone terminates at c. 45°S, where the Chile Rise runs into the Chilean coast and south of this point the boundary between the South American and Antarctic plate is characterized by lateral faulting rather than subduction. Although the southern volcanic zone appears to terminate at this triple junction, the active volcanoes Burney and Lautaro both lie south of 45°S (see Fig. 1).

The detailed studies of Baranzangi and Isacks (1976, 1979) make it possible to establish a relationship between the distribution of active volcanism and the depth to the Benioff zone. From accurately determined earthquake foci, Baranzangi and Isacks have inferred that the Benioff zone below western South America has five well defined segments. Below the three volcanically active zones the Benioff zone has a steeper dip (25–30°) than below the intervening inactive zones where the dip is 10–15°. The volcanoes of the northern and central zones are both therefore located about 140 km above the Benioff zone while the volcanoes of the southern zone lie only 90 km above it because they are nearer to the trench. South of 45°S seismicity is diffuse and is consistent with occurrence of lateral faulting rather than subduction.

The work of Baranzangi and Isacks therefore appears to establish that the active volcanic areas overlie a 'mantle wedge of asthenospheric material' above oceanic lithosphere being subducted at an angle of 25–30°, whereas the volcanically inactive areas overlie thinner lithosphere below which the angle of subduction is 10–15°. However, the seismicity below the volcanically inactive area of central Peru shows that the 'shallow subduction' model of Baranzangi and Isacks is based on clustered seismic data, rather than on data which indicate a well defined inclined seismic plane (see Fig. 1 (b)). James (1978) has therefore suggested that the subduction below Peru is also at 25–30°, and that the earthquakes used by Baranzangi and Isacks to infer shallower subduction occur as a cluster *within* thick continental lithosphere below central Peru.

Distribution and Setting of Volcanism

The volcanoes of the northern zone (Fig. 2) are the least well known despite the fact that some of them are the most active in the Andes, notably Sangay, Reventador, and Cotopaxi in Ecuador (Hantke, 1966; Pichler et al., 1976). Recent eruptions from Colombian volcanoes have been summarized by Ramirez (1968). In Colombia and Ecuador the active volcanoes are built on a range of Mesozoic and Cainozoic rocks. In Ecuador, Upper Cretaceous–Eocene ocean floor and island-arc volcanic rocks form the basement to the Western Cordillera (Fig. 2; see also Henderson, 1979). In contrast, the Eastern Cordillera are underlain by metamorphic rocks of unknown age, while the easternmost Ecuadorian volcano, Sumaco, is built on sedimentary Jurassic to Cretaceous age that comprise the 'Sub-Andean Cordilleras', which presumably overlie older Precambrian rocks of the Brazilian

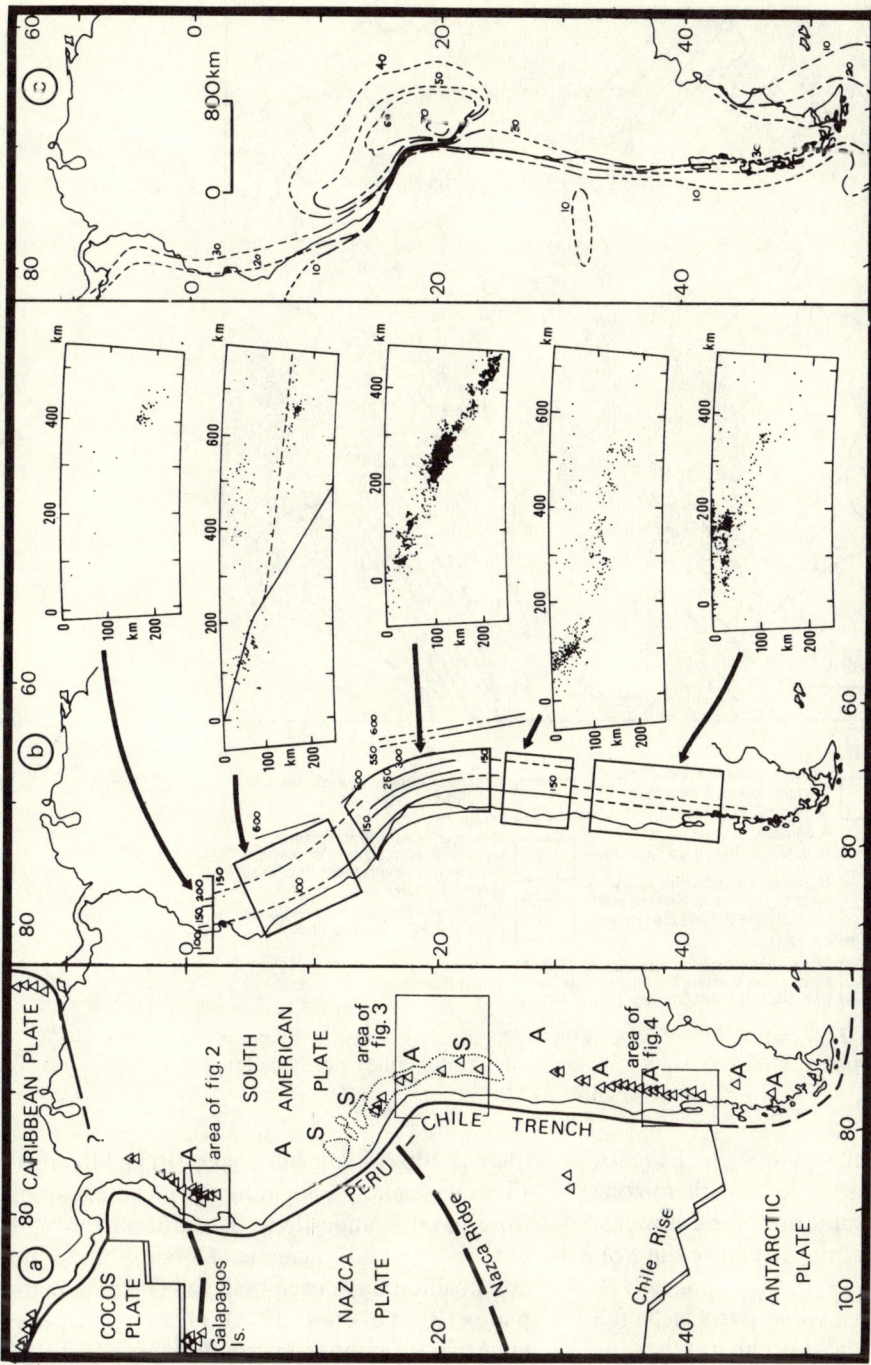

Fig. 1 The distribution of active volcanoes and their relationship to plate tectonics and crustal structure in South America. (a) Active volcanoes (open triangles) in relation to plate tectonics (active volcanoes taken from Macdonald, 1972, with the addition of Volcan Hudson in southern Chile). Solid lines are destructive plate boundaries; thin paired lines are constructive boundaries; thick broken lines are oceanic ridges and/or rises in the Nazca plate; and the thin broken line (south of 42°S) is the boundary of the Antarctic and South American plates. The dotted line outlines the ignimbrite province of the central Andes; S = shoshonitic and A = alkaline volcanic rocks. The areas of Figs 2, 3, and 4 are indicated. (b) Seismicity in South America. Profiles showing the distribution of seismicity at different depths and distances from the oceanic trench (0 km) are shown for the areas distinguished by Baranzangi and Isacks (1976, Figs 2–4) and the solid line is the Benioff zone inferred by James (1978). (c) Crustal thickness in South America. Thicknesses (in kilometres) are from Cummings and Schiller (1971)

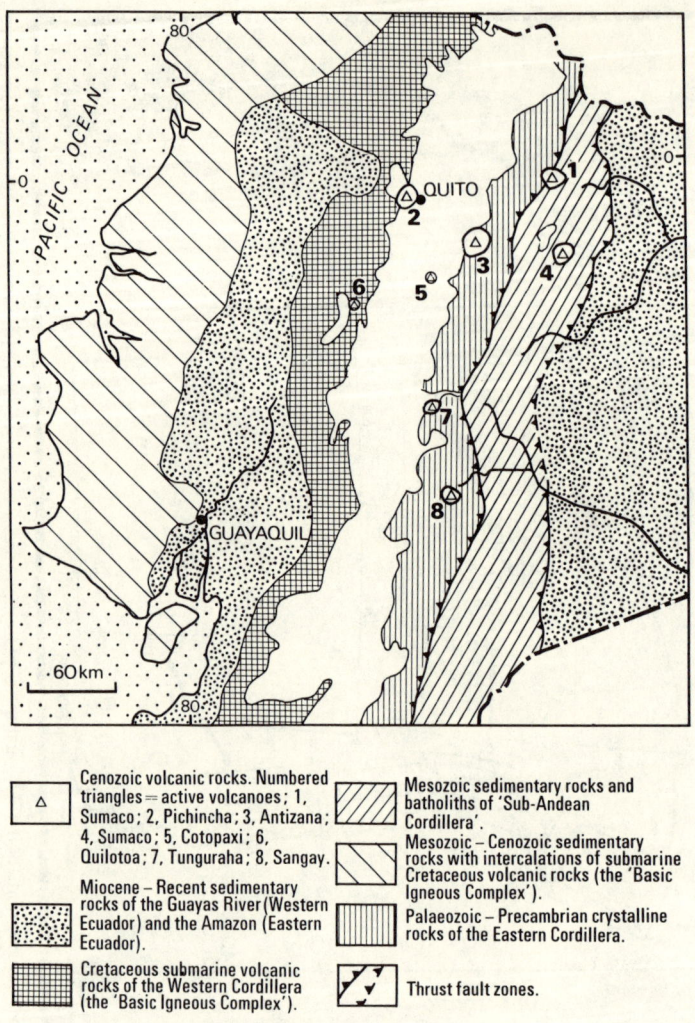

Fig. 2 Relationship of volcanism to regional geology in Ecuador (simplified after the Mapa Geologico de la Republica del Ecuador, Quito, 1969)

Shield (Fig. 2). Recent studies in Ecuador suggest that the volcanoes of the northern zone are characterized by basaltic andesite eruptions, leading to the construction of composite volcanoes (Pichler *et al.*, 1976; Francis *et al.*, 1977). More silicic rocks are known as pyroclastic fall and flow deposits but these occur in relatively small amounts.

Although there is probably a greater density of volcanoes in the central volcanic zone (Fig. 3) than in any other continental area in the world, the volcanic activity is poorly described. Three main groups of volcanic rocks can be identified. First, volcanic rocks ranging in composition from basaltic andesite to dacite occur throughout the zone, and volcanoes of broadly 'andesitic' composition have been built to 6000 m in many places (Roobol *et al.*, 1976). These major peaks are generally confined to the Western Cordillera of the Andes at the margin of the Bolivian altiplano (Fig. 3), but some individual cones occur on the altiplano itself. Second, large volume (up to 100 km^3) ignimbrite sheets are common in the Western Cordillera and also

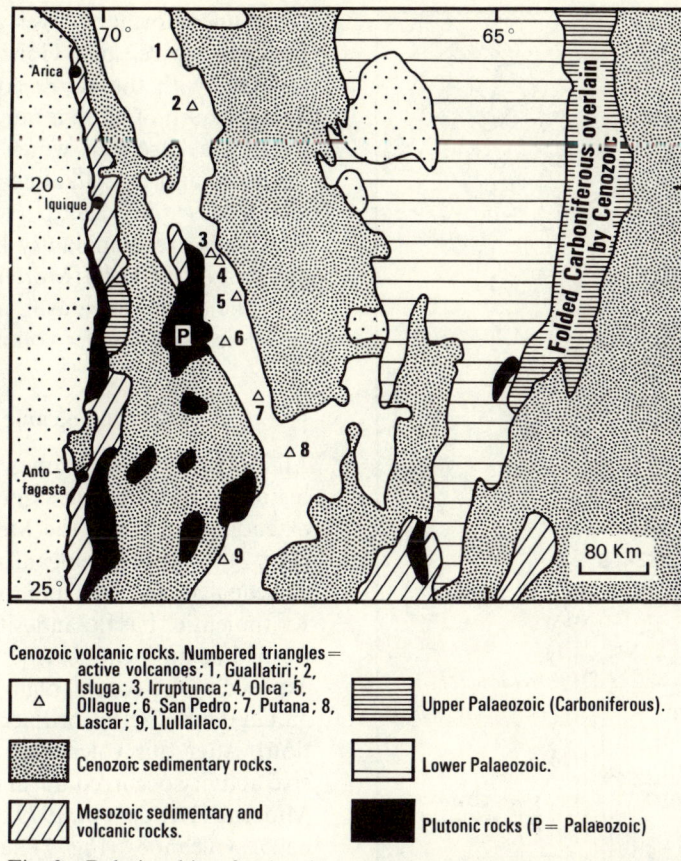

Fig. 3 Relationship of volcanism to regional geology in north Chile (simplified after the Geological Map of South America, 1964)

outcrop in the Eastern Cordillera of Bolivia (Frailes Formation). These ignimbrites range from dacite to rhyolite in composition (Pichler and Zeil, 1972). Francis and Rundle (1976) estimated that, in one representative part of this area, the ratio of volumes of lava to ignimbrite is $c.$ 1.3 : 1. Third, locally there is a volumetrically insignificant but geochemically distinct suite of young basaltic extrusives which have formed throughout the volcanic history so that older examples have been eroded and buried while younger examples are morphologically youthful.

The volcanoes of the central volcanic zone are built upon older Cainozoic and Mesozoic igneous rocks and Palaeozoic crystalline rocks (Fig. 3). However, in contrast to the northern volcanic zone, the occurrence of Precambrian metamorphic rocks on the Peruvian coast (the 'Arequipa Massif'; Cobbing et al., 1977, Shackleton et al., 1979) and below the Bolivian altiplano (Lehmann, 1978) is widely taken to indicate the presence of a Precambrian metamorphic basement below the central Andes. Such a basement might form an extension of the 'Brazilian Shield' below the volcanic belt (e.g. Cobbing and Pitcher, 1972; Pitcher, 1978) and might account partially for the greater thickness of crust below the central volcanic zone as compared with the northern and southern volcanic zones (q.v.) where the basements consist of younger continental crust (Fig. 1(c)).

The southern volcanic zone (Fig. 4) exhibits a higher level of volcanic activity than the central zone. Historic activity has been summarized by Casertano (1962) and Moreno (1974, 1976). Although the relief of the southern zone is much

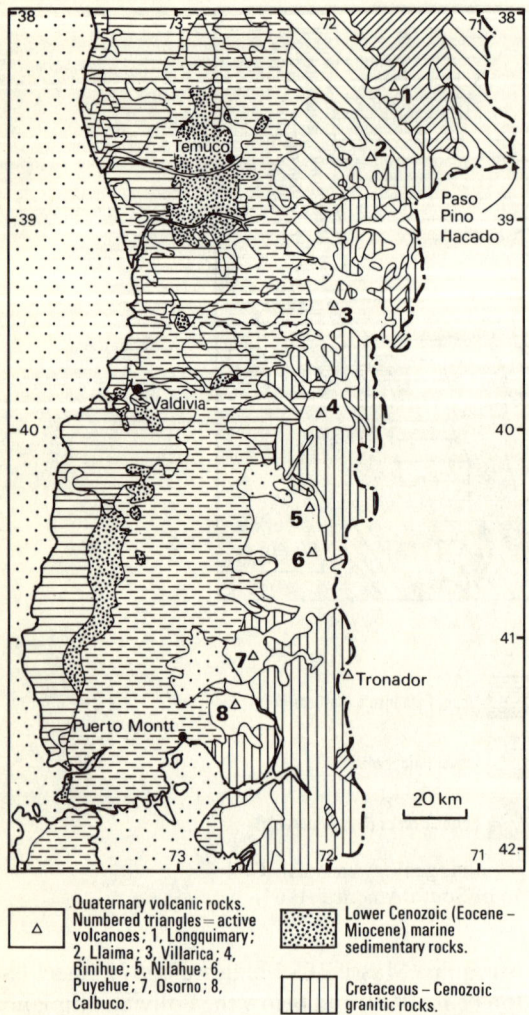

Fig. 4 Relationship of volcanism to regional geology in southern Chile (simplified after the Mapa Geologico de Chile, 1968)

Key:
- Quaternary volcanic rocks. Numbered triangles = active volcanoes; 1, Longquimary; 2, Llaima; 3, Villarica; 4, Rinihue; 5, Nilahue; 6, Puyehue; 7, Osorno; 8, Calbuco.
- Upper Cenozoic – Quaternary sedimentary rocks of the 'Longitudinal Depression'.
- Middle Cenozoic (Miocene – Pliocene) volcanic rocks of the 'Andesite Plateau'.
- Lower Cenozoic (Eocene – Miocene) marine sedimentary rocks.
- Cretaceous – Cenozoic granitic rocks.
- Mesozoic sedimentary and volcanic rocks.
- Palaeozoic – Precambrian crystalline rocks.

lower than that of the central and northern zones, the sizes of the largest volcanoes are similar. The setting of active volcanism in southern Chile is, however, distinct from that further north. Within the Andean Cordillera, two volcanic units occur—an 'andesite plateau' group of Miocene to Pliocene age composed of subhorizontal andesite flows and pyroclastic rocks unconformably overlying Mesozoic basement (Vergara and González-Ferrán, 1972), together with the active composite volcanoes. To the south of 40°S, a range of volcanic rocks overlies a basement which includes Mesozoic tholeiitic and calc-alkaline volcanic and intrusive rocks (see later). However, the active volcanoes are dominantly high alumina basalt and basaltic andesite (Moreno, 1976). Associated with the active volcanism, more silicic volcanic rocks occur in smaller volumes.

Recent Volcanic History

All three volcanic zones have intermittant histories of volcanic and intrusive activity extending back into the Mesozoic. In Ecuador (Fig. 2) the earliest volcanism is of Cretaceous–Eocene age. The volcanic rocks comprise a low K tholeiitic basalt–andesite group probably formed as an intra-oceanic island arc and now exposed in the coastal plain (Piñón Formation) and in the Western Cordillera (Macuchi Formation). After this volcanism, calc-alkaline intrusive activity occurred during the Oligocene and Miocene (Henderson, 1979). The distribution of active volcanoes (Fig. 2) suggests an eastward migration of volcanic activity in that the most active volcanoes are located in the Eastern Cordillera, while those in the Western Cordillera seem to be inactive.

The timing of volcanism in the central zone is better documented. Mesozoic–Eocene intrusive and extrusive activity has been described from the area around the central volcanic zone (e.g. Cobbing and Pitcher, 1972; Pitcher, 1978; Noble et al., 1974; McNutt et al., 1975). The Jurassic–Eocene volcanic rocks, now exposed on the coastal side of the active volcanic chain (Fig. 3), resemble island-arc volcanic products (e.g. McNutt et al., 1975; Dostal et al., 1977b). Some of these rocks might therefore have formed in island arcs which became accreted to the central Andean margin; 'ophiolitic' rocks from associated ocean basins have not been identified in the central Andes (cf. southern Chile).

There is widespread evidence that a major 'pulse' of Cainozoic volcanic activity was initiated during the Miocene. Noble et al. (1974), who compiled radiometric data for volcanic rocks in southern Peru, suggested that a major pulse of Cainozoic volcanism was initiated at c. 25 Ma, and peaked at c. 12 Ma. Detailed geochronological studies by Baker (1977a, b) and Baker and Francis (1978) also indicate initiation of the current episode of volcanism at c. 23 Ma ago. The pattern from this time to the present is, however, complex—Baker and Francis (1978) show that within the area 19°30'–21°30'S there were significant variations in the date of initiation and subsequent intensity of volcanic activity. In the southern half of this area most volcanism is younger than 6 Ma in age, while in the northern half there were significant peaks at 12–9 Ma and 6–3 Ma. The data indicate that andesites have erupted since c. 20 Ma ago, but that widespread ignimbrite volcanism was initiated at times between 21 and 10 Ma in different parts of the Western Cordillera. Finally, ignimbrites less than 1 Ma in age are also known from the Western Cordillera, indicating that this area can be considered 'active' from the point of view of both lava and pyroclastic eruptions.

Morphological data—especially that obtained from Landsat photography—indicate that the oldest composite cones are located in the westernmost parts of the Cordillera, and subsequently migrated eastwards into the Bolivian altiplano and the Argentinian Puna (Baker, 1977b). Although this eastward migration of volcanic activity within the central volcanic zone seems quite clear, and parallels the eastward migration of intrusive centres described by Farrar et al. (1970), it also seems that a *reversal* in this trend has taken place in the recent geological past, since all the currently active volcanoes occur along the western edge of the Western Cordillera.

A detailed geochronological study of part of the southern active zone has been published by Drake (1976). He demonstrated the occurrence of several episodes of volcanism from c. 30 Ma onwards, separated by periods of uplift and folding, and that volcanism has been almost continuous over the last 24 Ma. Further, Drake (1976) demonstrated that a marked westwards migration of volcanic centres has taken place over the last 1–2 Ma, but that no discernable migration of late Cainozoic intrusive centres has occurred. Volcanic activity south of 44°S has been well documented (Dalziel et al., 1974; Suarez and Pettigrew, 1976; Bruhn et al., 1978). To the south of 40°S widespread volcanism, associated with extensional faulting, was initiated during the Jurassic (from c. 170 Ma; Bruhn et al., 1978) and was also associated with the initiation of calc-alkaline plutonism along the site of the present continental margin. This activity ceased during the latest Jurassic when formation of a down-faulted 'back-arc' basin split the calc-alkaline belt from 'continental' South America. During the early Cretaceous, extension of this basin separated an *active* calc-alkaline volcanic arc (built on older pre-Jurassic continental crust) from the inactive continental margin to the east. Closure of this basin, by subduction below the volcanic arc, occurred during the Cretaceous, between c. 140 and 80 Ma ago (Dalziel et al., 1974).

Morphological Characteristics of the Central Andean Volcanic Province

The central Andes is one of the world's most extensive and diverse volcanic zones, and has been studied in more detail than the northern and southern volcanic zones. Some 500 major volcanic structures are recognizable, together with many minor or parasitic structures (Fig. 5). These volcanic structures have all formed since c. 23 Ma ago, and the relative ages of volcanoes within part of the province are indicated on Figs 5 and 6. The distribution of these volcanoes is characterized by a sharply defined 'volcanic front' in the west, with volcanoes scattered irregularly to the east (Fig. 5). Several different classes of volcanic structure, defined by morphology and composition, have been recognized.

Composite cones

These are mainly the classic andesite volcanoes. Some 116 have been recognized within the area

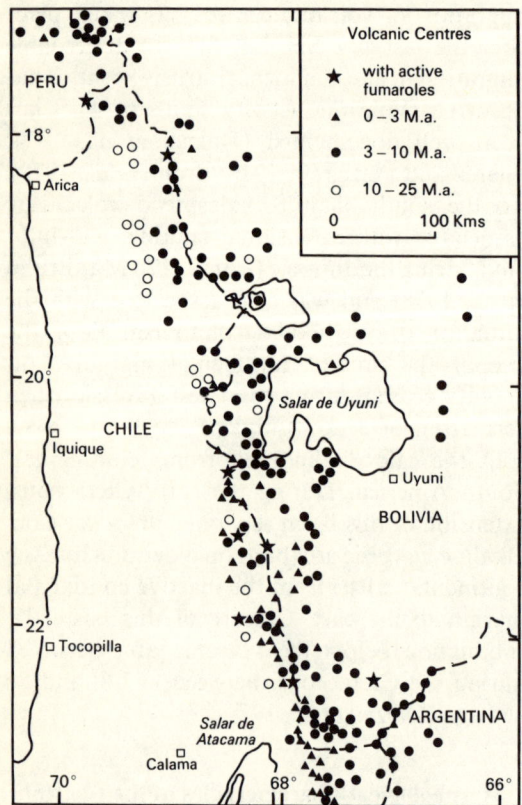

Fig. 5 Distribution and ages of Miocene–Recent volcanic centres in the central Andean volcanic province (from Baker, 1977b)

of Fig. 5. Such cones have symmetrical profiles and rise consistently to maximum heights of c. 6000 m. The steepest slopes on these cones never exceed 35°.

Monogenetic lava extrusions

These are individual lava extrusions which are sometimes but not invariably associated with a larger volcanic structure. The most common type includes dacite–rhyolite domes with aspect ratios (height/thickness) of 0.3–0.4. These domes, of which c. 80 examples are known, have steep, high flow fronts and in at least one case (the Chanka domes of northern Chile) collapse of the flow front has produced small aprons of hot avalanche debris (Francis et al., 1974). A less common variety includes flat-topped dacitic domes ('tortas') of which c. 20 examples are known, mostly associated with ignimbrite source areas. These disc-shaped extrusions may reach up to 500 m thick, and have aspect ratios of 0.2 or less. Apart from domes, there are many other lava extrusions which are more difficult to categorize. These range from the Chao lava, a massive dacite extrusion with a volume of 24 km^3 (Guest and Sanchez, 1969) to tiny extrusions of andesite and dacite.

Compound volcanoes

This term is introduced to describe those volcanoes in which two or more morphological units can be recognized, and which lack radial symmetry. They are the most abundant type of central Andean composite volcano and include Nevado Ojos del Salado, 6800 m, the Earth's highest volcano. In some cases, compound volcanoes can be observed to have developed from simpler composite cones by the eruption of more silicic lavas, following an hiatus. An example is the San Pedro volcano of northern Chile, where an ancestral basaltic andesite volcano experienced substantial erosion before the growth of a new pile of hornblende andesite lavas on its western flanks (Francis et al., 1974). Other compound volcanoes have discrete, young, dacite domes situated either on their summits or lower flanks. In no case do basaltic andesite lavas appear to have been erupted from a compound volcano *after* more silicic lavas. A secondary characteristic of compound volcanoes is that the eruption of silicic lavas near the summit, and the extrusion of domes there, means that the volcanoes acquire distinctive 'over-steepened' profiles, and their upper slopes are often between 35 and 40°. These steep profiles are accentuated by the lack of pyroclastic materials of more silicic compositions, so the slopes are not smoothed out, as they are in simple composite cones.

Ignimbrite centres

Ignimbrite sheets cover some 200 000 km^2 in the central Andes. Because they occupy low topographic levels, it has often been erroneously

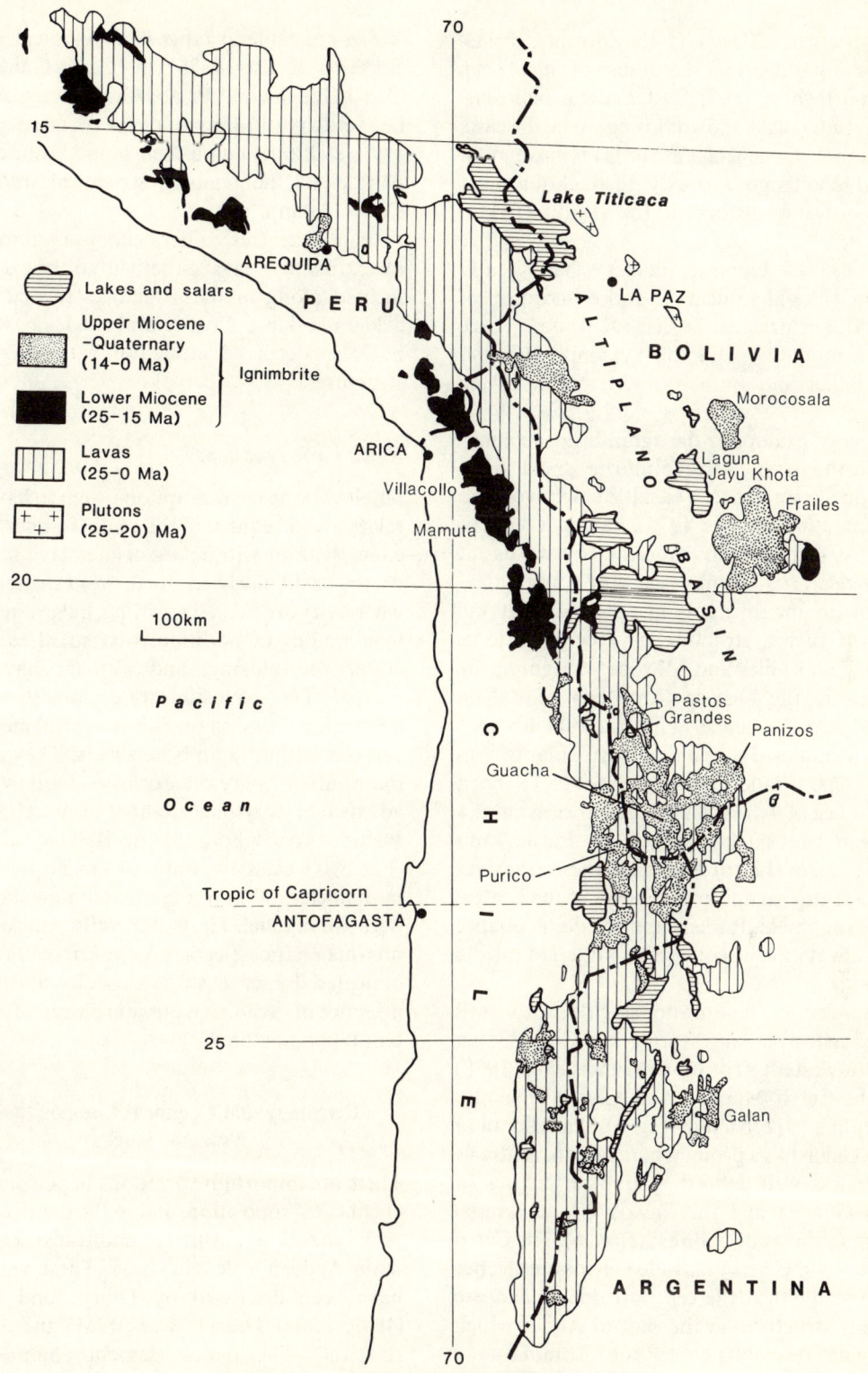

Fig. 6 Volcanic geology of the Central Andes

supposed that collectively they form a stratigraphic unit older than the other volcanic rocks (Zeil and Pichler, 1967). Systematic geochronological studies have shown this not to be the case, and ignimbrites and andesitic lavas have been erupted contemporaneously throughout most of the volcanic history of the region (Baker, 1977a, b).

The use of Landsat imagery has greatly facilitated the identification and description of ignimbrite centres in the central Andes. There are now more than 20 known examples (Guest, 1969; Baker and Francis, 1978; Noble et al., 1979).

In many examples, the ignimbrite accumulated in the form of an ignimbrite shield, morphologically similar to a basaltic shield volcano (Vincent, 1963; Sparks, 1975), having a lenticular cross-section and tapering outwards at low angles (1–3°) from a central source area. The maximum thickness of ignimbrite at the centre of such a structure may be as little as 100 m. The Frailes and Morocosala ignimbrite plateaux in the Eastern Cordillera of Bolivia (Fig. 6) are low angle ignimbrite shields that have accumulated over a highly irregular topography. The Frailes plateau appears to be a coalescence of two such shields. Other examples of ignimbrite shields are Cerro Purico and Cerro Panizos (Fig. 6). In none of these examples was there any significant collapse of the central part of the shield. It seems that caldera collapse is not always a consequence of large ignimbrite eruptions.

Examples of ignimbrite shields with well defined central collapse features are Cerro Galan in north-western Argentina (Francis et al., 1978) and the Cerro Mamuta and Cerro Villacollo centres in northern Chile (Fig. 6). The caldera of Cerro Galan has a prominent resurgent centre as do the less well defined ignimbrite centres of Cerro Guacha and the Pastos Grandes area, both in south-western Bolivia (Fig. 6). The Cerro Mamuta and Cerro Villacollo shields are steeper than average near the central calderas and are the only structures in the central Andes which bear any resemblance to the 'somma'-type calderas common in some other ignimbrite areas.

Sizes of calderas range from a typical 5 km × 5 km to 40 km × 24 km (Cerro Galan); the Cerro Guacha and Pastos Grandes centres may be even larger. Many calderas are circular but a few are elliptical with their long axis orientated parallel to the regional structural trend (e.g. Cerro Galan).

In most centres, caldera collapse was followed by extrusion of lava, either within the caldera or located along arcuate fractures related to the caldera collapse. Post-ignimbrite lava extrusion has also occurred in centres without collapse structures.

Small basic eruptions

Small volume basic eruptions seem to have been relatively infrequent in the central Andes. Scoria cones, with or without associated lava flows, do occur, particularly in north-western Argentina where they are related to rifting, but are generally uncommon. In addition, two small explosion craters, a tuff-ring, and a maar have been located. The explosion craters are steep-sided depressions less than 500 m in diameter developed within ignimbrite sheets. They may be the results of phreatic explosions following the advance of a pyroclastic flow over wet ground. Laguna Jaya Khota on the Bolivian altiplano (Fig. 6) is a classic example of a maar. It is 1.1 km in diameter, 30 m in depth, and almost entirely negative in relief. The crater walls expose mostly non-magmatic ejecta. A volcanic origin is indicated by base surge deposits and by the presence of juvenile magmatic blocks of alkaline basalt composition.

Petrology and Chemical Composition of Volcanic Rocks

There are important variations in petrology and chemical composition, and in the distribution of rock types both parallel to and transverse to the main Andean volcanic chain. These variations have been discussed by Thorpe and Francis (1979a,b and Thorpe et al., 1981) and are now reviewed. The major element compositional characteristics of Andean andesites are summarized in Fig. 7 (cf. Fig. 1 in Thorpe and Francis, 1979b).

In Ecuador (northern volcanic zone) the lavas are chiefly olivine- and two pyroxene-bearing basaltic andesite and andesite, with $SiO_2 = 53-61$ per cent (Pichler et al., 1976; P. W. Francis and R. S. Thorpe, unpublished). Some have phenocrysts of corroded amphibole. The mineralogical and major element characteristics of Ecuadorian lavas (Fig. 7) resemble those of island arc andesites (e.g. Pichler et al., 1976; Ewart, 1976, and this volume, Section II).

By contrast, lavas from the central volcanic zone (16–28°S; Peru, northern Chile, and Bolivia) are more acidic in composition, largely falling within the range 56–66 per cent SiO_2 (Fig. 7). Basalts are almost entirely absent and the commonest lavas are pyroxene andesites (occasionally olivine-, hornblende-biotite-bearing andesites); generally with over 60 per cent SiO_2) and dacites (SiO_2 over 63 per cent), which may have quartz phenocrysts (Lefévre, 1973; Francis et al., 1974: Roobol et al., 1976; Dostal et al., 1977a, b). The central volcanic zone also has extensive ignimbrite sheets—these are commonly of dacite to rhyolite composition ($SiO_2 = 64-75$ per cent), and show continuous mineralogical and chemical variation from andesite and dacite lavas (Francis et al., 1974; Zeil and Pichler, 1967). Dacite and rhyolite pumice clasts from ignimbrites range from nearly aphyric types, through plagioclase-phyric, to rhyolitic pumice containing euhedral phenocrysts of hornblende, biotite, quartz, and plagioclase.

In the southern volcanic zone (31–42°S; southern Chile), the volcanic products resemble those of the northern, rather than the central zone. The commonest lavas are high alumina basalts (olivine, two pyroxenes, and labradorite), and basaltic andesites (with plagioclase of andesine–labradorite composition), containing $SiO_2 = 50-60$ per cent. Andesites and dacites (Fe-rich, occasionally with fayalite) are less abundant (Vergara, 1972; Katsui, 1972; Moreno, 1976; Lopez-Escobar et al., 1976, 1977). The northern part of the southern zone (33–34°S) appears to be characterized by the occurrence of amphibole-bearing andesites, while the southern part (37–41°S) is characterized by a bimodal high alumina basalt–dacite association with minor andesite (Moreno, 1976; Lopez-Escobar et al., 1976, 1977). However, the inland volcano Tronador, at c. 41°S, has amphibole- and orthopyroxene-bearing andesites and appears to be an exception to the pattern noted above (Moreno, 1976).

The petrological variations described above are matched by major and trace element, mineralogical, and isotopic variations (Table 1, Fig. 7). The lavas from the central zone are generally more silica-rich and show less relative Fe enrichment than those from the northern and southern volcanic zones (Fig. 7). For a given SiO_2 content the central zone volcanic rocks tend to have higher concentrations of K, Rb, Sr, and Ba, higher $^{87}Sr/^{86}Sr$ ratios and $\delta/^8O$ values, and lower $^{143}Nd/^{144}Nd$ ratios than rocks from either the northern or southern volcanic zones (see Table 1; Janes et al., 1976; Francis et al., 1977; Klerkx et al., 1977; Hawkesworth et al., 1979; Thorpe and Francis, 1979a, b, Thorpe et al., 1981). In addition there are differences in the REE patterns for volcanic rocks from these three zones. Basaltic andesites and andesites from the northern and southern zones and central zone are similar, with Ce/Yb greater than 20, while the high alumina basalts from the southern zone have Ce/Yb below 20 (Thorpe et al., 1976; Dostal et al., 1977a, b; Lopez-Escobar et al., 1976, 1977; Hawkesworth et al., 1979).

In addition to the variations described above, there are important petrological and geochemical variations across the Andean volcanic chain (reviewed in Thorpe and Francis, 1979a). These have been well described within the central volcanic zone, in southern Peru (Lefévre, 1973; Dupuy and Lefévre, 1974) and in northern Chile–south-western Bolivia and north-western Argentina (Roobol et al., 1976; Dostal et al., 1977a; Deruélle, 1978). The most characteristic variation described by these workers is the increase in K_2O relative to SiO_2 in passing from west to east across the volcanic chain (Fig. 8). This variation is associated with an increase in Rb and a decrease in Sr at a given SiO_2 content,

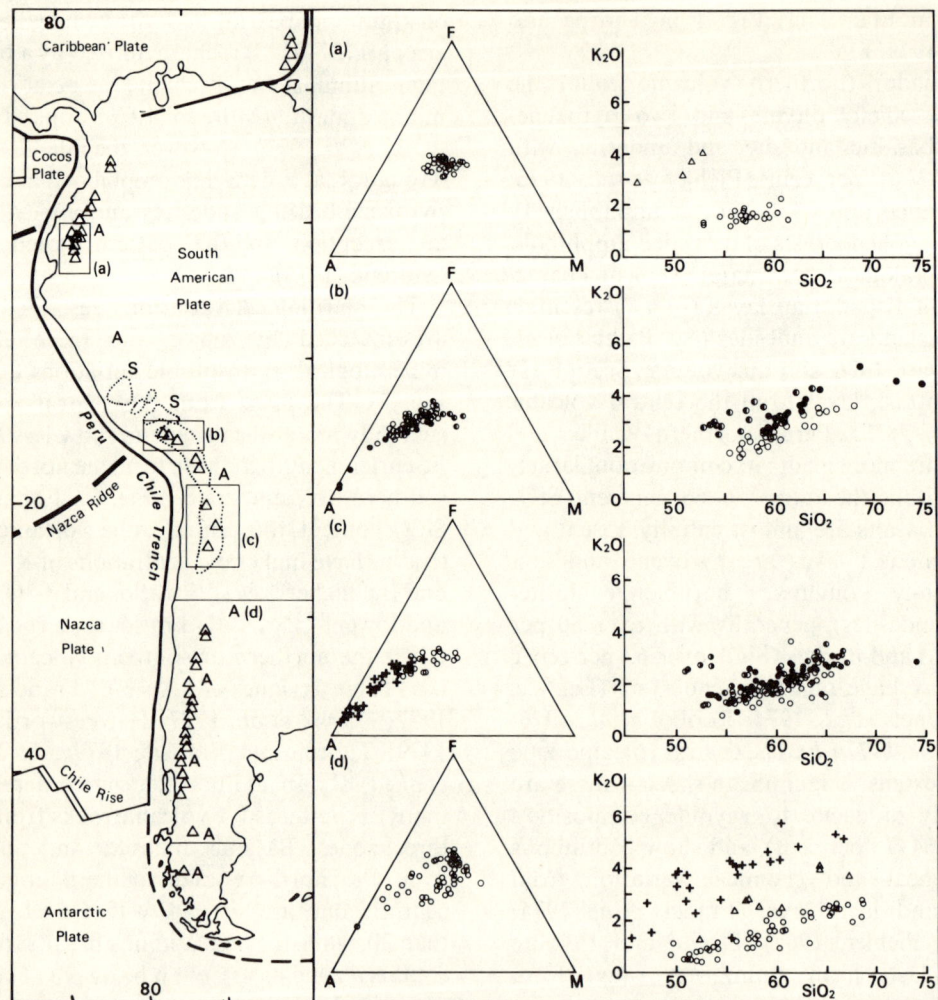

Fig. 7 The distribution and compositional characteristics of active volcanoes, and their relationships to plate tectonics in the South American Andes.

The AFM diagram and plots of K_2O against SiO_2 ((a)–(d)) are for rocks from the areas indicated on the map and are as follows: (a) Ecuador, calc-alkaline (open circles; P. W. Francis and R. S. Thorpe, unpublished; Pichler *et al.*, 1976) and alkaline lavas from Sumaco (open triangles; Colony and Sinclair, 1928). (b) Southern Peru, calc-alkaline and shoshonitic lavas with the open circles = A_1 series, solid circles = A_2 series, and half-filled circles = B series (cf. Lefévre, 1973, Fig. 4). (c) Northern Chile, calc-alkaline and shoshonitic lavas. In the AFM diagram the open circles are lavas (Francis *et al.*, 1974) and crosses are ignimbrites (Francis *et al.*, 1974; Thorpe *et al.*, 1979), and in the K_2O–SiO_2 plot (cf. Roobol *et al.*, 1976, Fig. 2) open circles = Western Chain lavas; filled circles = Eastern Chain lavas; right-filled circles = Argentinian lavas; and left-filled circles = Bolivian lavas. (d) Southern Chile, calc-alkaline and alkaline lavas. The AFM diagram is for calc-alkaline lavas from Moreno (1976), and Lopez-Escobar *et al.*, (1976, 1977) and in the K_2O–SiO_2 plot the open circles are calc-alkaline lavas (Deruélle, 1978; Lopez-Escobar *et al.*, 1976, 1977), the open triangles are for alkaline lavas from Pino Hachado (Vergara, 1972; Lopez-Escobar *et al.*, 1976) and Volcan Hudson (Ponce, 1976), and the crosses are for alkaline lavas from Pocho and Payun Matru in north-western Argentina (Vergara, 1972)

TABLE 1 Chemical analyses of representative Andean lavas. (Major elements in weight per cent, trace elements in parts per million, n.d. = not determined.)

	1	2	3	4	5	6	7	8
SiO_2	62.50	61.9	65.21	53.4	55.72	52.70	52.88	59.20
TiO_2	0.63	0.8	0.98	1.45	0.89	0.89	0.68	1.06
Al_2O_3	15.90	16.6	16.01	15.5	16.89	20.81	18.96	17.08
Fe_2O_3	1.29	1.2	2.16	n.d.	n.d.	n.d.	2.92	2.41
FeO	3.31	4.0	2.03	7.4	7.85	6.11	2.88	3.34
MnO	0.07	0.07	0.07	0.16	0.10	0.17	0.13	0.15
MgO	3.53	2.9	1.56	5.91	5.12	4.80	2.22	1.48
CaO	4.57	5.2	3.65	8.02	7.51	10.50	6.40	2.44
Na_2O	4.19	3.7	2.33	3.8	3.86	3.32	5.09	6.00
K_2O	2.81	3.2	4.29	2.86	1.14	0.44	4.05	3.71
P_2O_5	0.17	0.2	0.28	n.d.	0.23	0.21	0.42	0.42
Others	1.79	n.d.	1.57	n.d.	n.d.	n.d.	2.61	2.40
Total	100.76	99.7	100.14	98.5	99.31	99.95	99.24	99.69
Rb	84	85	194	52	21	18	n.d.	100
Sr	555	510	375	2220	638	374	n.d.	470
$^{87}Sr/^{86}Sr$	0.7063	n.d.	0.7133	0.7042	0.7044	0.7039	0.7040	n.d.
Ce/Yb	30.45	n.d.	n.d.	79.7	27.0	9.9	n.d.	21.9

(1) Andesite, San Pedro volcano, northern Chile. Westernmost chain of active volcanoes (No. 185, Francis et al., 1974, 1977; Thorpe et al., 1976). (2) Andesite, Ollague volcano, northern Chile (30 km east of analyses 1) (No. 414, unpublished data.) (3) 'Rhyodacite', Uturuncu volcano, south-western Bolivia (No. 1/5, Fernandez et al., 1973; Klerkx et al., 1977). (4) 'Low Si latite' (shoshonite), Ayacucho, central Peru (No. AYA-1A, Noble et al., 1975). (5) Basaltic andesite, Cotopaxi volcano (No. EF7, Francis et al., 1977, and unpublished data). (6) High alumina basalt, Villarica volcano, southern Chile (No. 802, Lopez-Escobar et al., 1977; Sr isotope ratio from 'olivine andesite' No. Vi24 in Klerkx et al., 1977). (7) Hauyne-bearing 'andesite tephrite', Sumaco volcano (No. 1, Colony and Sinclair, 1928; Sr isotope ratio for sample of hauyne tephrite AW 387 determined by Dr S. Moorbath). (8) 'Trachyandesite', Pino Hachado, southern Chile (No. TH-34, Lopez-Escobar et al., 1976).

in the same direction (Roobol et al., 1976). As a culmination of the eastward increase in K_2O, some of the easternmost volcanic rocks are distinguished as shoshonites—these include lavas in southern Peru (Lefévre, 1973), northern Chile (Deruélle, 1978), and north-western Argentina (Hörmann et al., 1973; Deruélle, 1978), which have similar Ti and lower Sr to associated calc-alkaline lavas.

A characteristic feature of the central volcanic zone is the occurrence of voluminous ignimbrite sheets of andesite–dacite composition. The ignimbrite magmas have SiO_2 = 64–75 per cent and show continuous major element chemical variation with andesite–dacite lavas (Pichler and Zeil, 1972; Fernandez et al., 1973). However, the ignimbrites have low Sr and high Rb in comparison from andesite–dacite lavas (El-Hinnawi et al., 1969; Thorpe et al., 1979). In addition the $^{87}Sr/^{86}Sr$ ratios and REE abundances of ignimbrite in northern Chile (0.705–0.710; Klerkx et al., 1977; Thorpe et al., 1979) overlap those of northern Chilean andesite and dacite lavas ($^{87}Sr/^{86}Sr$ = 0.705–0.707; Klerkx et al., 1977; Francis et al., 1977; Thorpe and Francis, 1979a, b; Thorpe et al., 1979).

In adddition to the calc-alkaline associations described above, alkaline lavas are known from scattered localities to the east of the Andean volcanic chain in Ecuador, Peru, Bolivia, Chile, and Argentina (cf. Figs 1 and 6). In Ecuador, Sumaco volcano is composed of hauyne tephrite (Colony and Sinclair, 1928). Pliocene–Recent alkaline igneous activity in eastern Peru has been summarized by Stewart (1971). Further south, lavas belonging to a continental alkaline

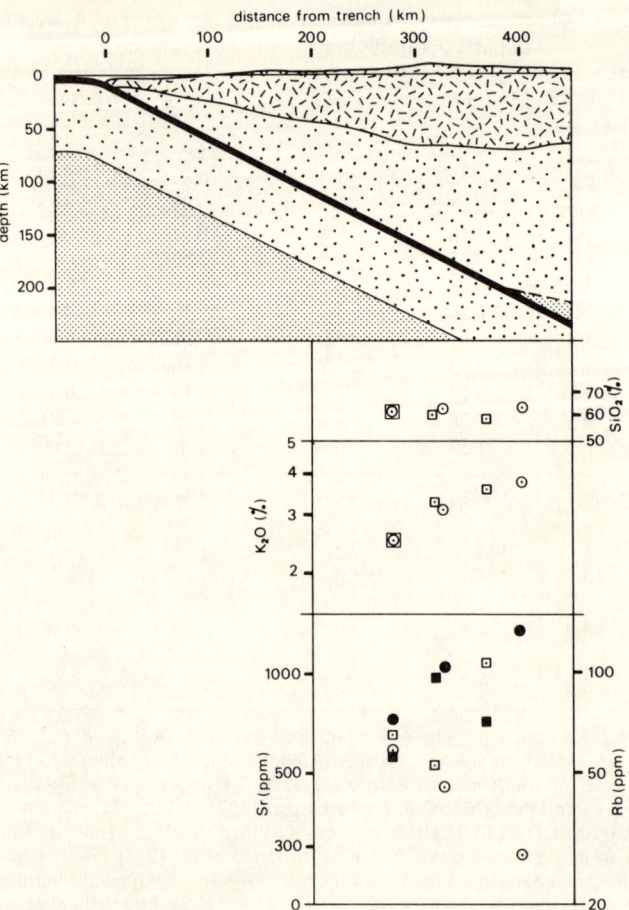

Fig. 8 Relationships between crustal and mantle structure and chemical composition of lavas in a section across the central Andes. Ornament is as follows: ticks = continental crust, dots = lithosphere, black = oceanic crust, and fine stipple = asthenosphere. The schematic cross-section is at 21–22°S, based on data in James (1971a, b), Stauder (1973), and Baranzangi and Isacks (1976). Variations in SiO_2, K_2O, Sr, and Rb for averages of groups of volcanic rocks at different distances from the oceanic trench are shown in the correct position relative to the cross-section. The circles are data from Roobol *et al.* (1976, Table 2, analyses C and D), and Fernandez *et al.* (1973, Tables 2 and 3). The squares are data from Dupuy and Lefévre (1974, Table 2, Groups A_1, A_2, and B). In the lower graph the open symbols refer to Sr and the filled symbols refer to Rb (from Thorpe and Francis, 1979a. Reproduced by permission of Elsevier Scientific Publishing Co.)

basalt–trachyte association occur in south-western Bolivia (Thorpe and Francis, 1979a, and unpublished data), southern Chile (Pino Hachado: Vergara, 1972; Lopez-Escobar et al., 1976; Volcan Hudson: Ponce, 1976), and Argentina (references in Vergara, 1972). In southern Argentina (Patagonia) an alkaline basalt association has been described by Skewes and Stern (1979) and Hawkesworth et al. (1979). These lavas form a varied association and include basaltic rocks ranging from quartz- and olivine tholeiite to alkaline basalt and basanite in composition. In addition, highly potash-rich leucite basanites occur. Although these rocks are products of volcanism spanning the entire Cainozoic era, the youngest products (less than 1 Ma in age) include basanites, alkaline basalts and hawaiites, and highly potassic leucite basanites (Skewes and Stern, 1979). Patagonian lavas have primitive chemical compositions, and samples have $^{87}Sr/^{86}Sr = 0.7033$–0.7052 and $^{143}Nd/^{144}Nd = 0.5126$–0.5130 (Hawkesworth et al., 1979).

Petrogenesis of Volcanic Rocks

Here we review evidence bearing on the petrogenesis of the major volcanic groups described from the Andean Cordillera. The most important and characteristic volcanic group is the high Al basalt–andesite–dacite association, which is responsible for building the major composite volcanoes of the Cordillera. Locally, members of this association grade into shoshonites and ignimbrites of dacite to rhyolite composition. There is also a distinct alkaline basalt–trachyte association. Within the tectonic framework described, a range of different processes has been proposed to account for the petrogenesis of the high Al basalt–andesite–dacite association (this volume, Section VIII). The occurrence of such active Andean volcanism above relatively thick wedges of asthenospheric mantle, (Baranzangi and Isacks, 1976, 1979), the similarity of major and trace elements and Sr–Nd isotope ratios between Andean and intra-oceanic volcanic rocks (e.g. Hawkesworth et al., 1979), and the local massive Cainozoic growth of the Andean crust all indicate a mantle origin for andesitic magmas. Sources might therefore include subducted oceanic crust and the overlying asthenospheric mantle wedge.

Partial melting of subducted oceanic crust produces melts with intermediate SiO_2 contents (Green and Ringwood, 1968), but these must be modified at shallower depths by fractional crystallization or reaction with the surrounding mantle in order to reach the crust as andesites (Stern, 1974; Fyfe and McBirney, 1975; Stern and Wyllie, 1978). The origin for the andesite association must therefore involve processes taking place *within* the mantle wedge overlying the subduction zone (cf. Nicholls and Ringwood, 1973; Ringwood, 1974). For example, in the model of Ringwood, partial melting of subducted oceanic crust produces melts which rise into and *react with* the overlying mantle wedge. Reactions between SiO_2-rich melts with peridotite would produce pyroxene and garnet-rich peridotite ('enriched mantle'). Similarly, models based on Sr–Nd isotope relationships involve enrichment of the mantle wedge in radiogenic Sr (plus other mobile elements?) released during dehydration of the descending oceanic slab. Partial melting of such 'enriched' peridotite might produce parental magmas of the andesites (cf. Ringwood, 1974; Thorpe et al., 1976; Dostal et al., 1977a, b). In this context the origin of andesites might involve some or all of the following processes: (1) partial melting of enriched mantle above the underlying zone; (2) fractional crystallization of the magmas at mantle depths; (3) interaction with the lower continental crust where it exceeds c. 45 km thickness, involving 'scavenging' of Sr (and probably some other elements); and (4) fractional crystallization of olivine, pyroxene, and (over 35 km depth) plagioclase *within* the continental crust.

According to this model, the sharply defined western boundary of the volcanic chain (Fig. 5) is critically controlled by pressure/temperature conditions at the subduction zone. In view of the earlier discussion of migration of the volcanic chain such migration might reflect changes in the depth of the subduction zone through time. We use the processes (1)–(4) above to comment on the petrogenesis of the high Al basalt–andesite–dacite suite described above.

For the high Al basalts of southern Chile, the high Al_2O_3 (19–22 per cent) allied with low SiO_2 (49–54 per cent), low MgO/FeO_t (0.5–0.8), and low Cr, Ni, and Co are generally taken to indicate an origin by partial melting of mantle peridotite followed by considerable fractional crystallization of olivine and pyroxenes (this volume, Section VIII). This model has been developed for southern Chile by Lopez-Escobar et al. (1977) who argue that the parent magma of the high Al basalt was formed by partial melting of garnet-free plagioclase—or spinel-bearing peridotite. This is consistent with the relatively unfractionated REE patterns (Lopez-Escobar et al., 1976, 1977; Thorpe and Francis, 1979a). However, such parental melts would have higher MgO/FeO_t and Cr, Ni, Co than the high Al basalts. Lopez-Escobar et al. (1977) therefore suggest that the southern Chilean high Al basalts were formed by c. 12 per cent partial melting of garnet-free peridotite followed by c. 20 per cent fractional crystallization of a mixture of olivine (80 per cent) and clinopyroxene (20 per cent). The associated basaltic andesites, andesites, and dacites, characteristic of the southern Chilean Andes, might then be derived by further fractional crystallization of such high Al basalt magmas (Moreno, 1976).

By contrast to the high Al basalt–andesite association some andesite–dacite lavas of southern Chile (e.g. Tupungato and Marmolejo) and the abundant andesite–dacite lavas of northern Chile–southern Peru (where basalts are absent) are derived by different processes to those described above. The andesites of the central volcanic zone have a similar major element chemical composition to those from southern Chile, including low Cr, Ni, and Co, but have higher contents of K and related trace elements, more fractionated REE patterns (some with Eu anomalies), with high $^{87}Sr/^{86}Sr$ and low $^{143}Nd/^{144}Nd$ in comparison with the calc-alkaline volcanic rocks of Ecuador and southern Chile. These chemical characteristics may be taken to indicate an origin by partial melting of garnetiferous peridotite, followed by fractional crystallization of olivine, pyroxenes, and plagioclase and (in northern Chile) contamination by continental crustal material. Such crustal contamination may therefore become important where the crust reaches the thickness of that in northern Chile (50–70 km) as compared with that of southern Chile (less than 40 km; Francis et al., 1977; Thorpe and Francis, 1979a, b; see also Zentilli and Dostal, 1977).

For andesite at Tupungato, Lopez-Escobar et al. (1977) suggested a model involving 3 per cent partial melting of garnetiferous peridotite (containing 6 per cent garnet), followed by 20 per cent fractional crystallization of olivine (80 per cent) and clinopyroxene (20 per cent), and similar models can be applied to andesite lavas in northern Chile. However, the higher content of incompatible elements in such andesites, compared with the concentrations inferred for mantle peridotite, suggests that the mantle source of such andesite might be enriched in some incompatible elements (light REE, K, Rb, Ba, Sr; Thorpe et al., 1976; Dostal et al., 1977a, b, Lopez-Escobar et al., 1976, 1977). Following formation of parental andesite magmas from such enriched garnetiferous peridotite, these magmas undergo fractional crystallization of olivine and pyroxenes (removing Cr, Ni, and Co) and plagioclase (removing Eu).

The processes described above can be used to explain the $K-h$ relationships described from the central volcanic zone, and the occurrence of the shoshonitic lavas. These variations can be explained as a result of an increase in amount of fractional crystallization, or a decrease in proportion of partial melting in passing from west to east across the Andean volcanic chain. In addition, these variations can be accounted for by increasing 'contamination' or by zone refining as the andesite parent magma passes through a greater thickness of mantle (Thorpe and Francis, 1979a).

The shoshonitic lavas which occur on the eastern side of the central volcanic zone are chemically variable, as described earlier. The intermediate shoshonites form a continuation of the calc-alkaline chemical trends (see Fig. 8) and might therefore result from a culmination of the processes which account for the K–h variation. However, some basic shoshonites have lower SiO_2, with higher Ti, K, Rb, Ba, Sr, and REE and higher Ce_N/Yb_N than associated calc-alkaline rocks (cf. Hörmann et al., 1973, Noble et al., 1975; Dostal et al., 1977a, b). The basic shoshonites described by Noble et al. (1975) also have lower $^{87}Sr/^{86}Sr$ (c. 0.704) than for calc-alkaline lavas from northern Chile (c. 0.705–0.707; see earlier). These data indicate that some basic shoshonitic lavas might be genetically unrelated to associated calc-alkaline lavas. Dostal et al. (1977b) argue that such rocks might be derived by small degrees (c. 5 per cent) of partial melting of garnetiferous peridotite and might therefore be transitional in origin towards the alkaline volcanic associations which occur on or to the east of the Andean volcanic chain (see below).

The volcanic rocks of the central volcanic zone include the widespread ignimbrite sheets of dacite–rhyolite composition. Several authors (Zeil and Pichler, 1967; Pichler and Zeil, 1972; Fernandez et al., 1973; Klerkx et al., 1977) emphasize the importance of crustal fusion in the petrogenesis of these volcanic rocks. However, the ignimbrites overlap the andesite lavas in both space and time, and both groups show intergradation of mineralogical and petrological features (Fig. 7(c)). The overlapping $^{87}Sr/^{86}Sr$ ratios and REE patterns also provide a link between lavas and ignimbrites, and Rb, Sr, and REE data are consistent with a model in which the ignimbrites result from plagioclase-dominated fractional crystallization of andesite magma (Thorpe et al., 1979). We therefore suggest that the dacite–rhyolite ignimbrites have a similar source to andesite magmas and that contamination by continental crust plays a minor role (Thorpe and Francis, 1979a, b; Thorpe et al., 1979). In this case the restriction of the ignimbrite province to the central volcanic zone is considered to reflect extensive fractional crystallization of andesite magma during its slow rise through thick continental crust.

Finally, we consider the petrogenesis of the alkaline volcanic rocks which occur as products of isolated volcanoes and 'basalt plateaux' areas which form an irregular eastern margin to the Andean Cordillera. Since these groups resemble volcanic associations formed within continental plates, and appear to show little apparent influence of a subduction zone in their chemical and isotopic composition (Hawkesworth et al., 1979), their origin can be attributed to partial melting of mantle peridotite distant from the subduction zone (and fractional crystallization of the derived magmas). The varied compositional characters of the alkaline volcanic provinces described suggest variation in conditions of partial melting (cf. Kay and Gast, 1973) and/or source composition (see Hawkesworth et al., 1979, for Patagonia).

REFERENCES

Baker, M. C. W. (1977a). Geochronology of Upper Tertiary volcanic activity in the Andes of North Chile. Geol. Rundschau 66, 455–465.

Baker, M. C. W. (1977b). Geochronology and volcanology of Upper Cenozoic volcanic activity in north Chile and southwest Bolivia. Ph.D. thesis, Open University, Milton Keynes.

Baker, M. C. W. and Francis, P. W. (1978). Upper Cenozoic volcanism in the central Andes—ages and volumes. Earth planet. Sci. Letters 41, 175–187.

Baranzangi, M. and Isacks, B. L. (1976). Spatial distribution of earthquakes and subduction of the Nazca plate below South America. Geology 4, 686–692.

Baranzangi, M. and Isacks, B. L. (1979). Subduction of the Nazca plate beneath Peru: evidence from spatial distribution of earthquakes. Geophys. Jl R. astr. Soc. 57, 537–555.

Bruhn, R. L., Stern, C. R., and de Wit, M. J. (1978). Field and geochemical data bearing on the development of a Mesozoic volcano-tectonic rift zone and back-arc basin in southernmost South America. Earth planet. Sci. Letters 41, 32–46.

Casertano, L. (1962). General characteristics of active Andean volcanoes and a summary of their activities during recent centuries. Bull. seismol. Soc. Am. 53, 1415–1433.

Cobbing, E. J. and Pitcher, W. S. (1972). Plate tectonics and the Peruvian Andes, Nature, Lond. 240, 51–53.

Cobbing, E. J., Ozard, J. M., and Snelling, N. J. (1977). Reconnaissance geochronology of the crystalline basement rocks of the Coastal Cordillera of southern Peru. Geol. Soc. Am. Bull. 88, 241–246.

Colony, R. J. and Sinclair, J. H. (1928). The lavas of the volcano Sumaco, Eastern Ecuador, South America. Am.

J. Sci. **16**, 299–312.
Cummings, D. and Schiller, G. I. (1971). Isopach map of the Earth's crust. *Earth Sci. Rev.* **7**, 97–125.
Dalziel, I. W. D., de Wit, M. J., and Palmer, K. F. (1974). Fossil marginal basin in the Southern Andes. *Nature, Lond.* **250**, 291–294.
Deruélle, B. (1978). Calc-alkaline and shoshonitic lavas from five Andean volcanoes (between latitudes 21°45′ and 24°30′S) and the distribution of plio-Quaternary volcanism of the south-central and southern Andes *J. Volc. geothermal Res.* **5**, 281–298.
Dostal, J., Dupuy, C., and Lefévre, C. (1977a). Rare earth distribution in Plio-Quaternary volcanic rocks from Southern Peru. *Lithos* **10**, 173–183.
Dostal, J., Zentilli, M., Caelles, J. C., and Clark, A. H. (1977b). Geochemistry and origin of volcanic rocks from the Andes (26°–28°S). *Contr. Mineral Petrol.* **63**, 113–128.
Drake, R. E. (1976). The chronology of Cenozoic igneous and tectonic events in the central Chilean Andes. In *Proceedings of the Symposium on Andean and Antarctic Volcanology Problems, Santiago, Chile, 1974*, (O. Gonzáles-Ferrán, ed.), pp. 670–697.
Dupuy, C. and Lefévre, C. (1974). Fraccionnement des elements en trace Li, Rb, Ba et Sr dans les series andesitiques et shoshonitiques du Perou—Comparison avec d'autres zones orogeniques. *Contr. Mineral. Petrol.* **46**, 147–157.
El-Hinnawi, E. E., Pichler, H., and Zeil, W. (1969). Trace element distribution in Chilean ignimbrites. *Contr. Mineral. Petrol.* **24**, 50–62.
Ewart, A. (1976). Mineralogy and chemistry of modern orogenic lavas—some statistics and implications. *Earth planet. Sci. Letters* **31**, 417–432.
Farrar, E., Clark, A. H., Haynes, S. J., Quirt, G. S., Conn, H., and Zentilli, M. (1970). K-Ar evidence for the post-Paleozoic migration of granitic intrusion foci in the Andes of northern Chile. *Earth planet. Sci. Letters* **10**, 60–66.
Fernandez, A., Hörmann, P. K., Kussmaul, S., Meave, J., Pichler, H., and Subieta, T. (1973). First petrologic data on young volcanic rocks of S. W. Bolivia. *Tschermaks Mineral. Petrol. Mitt.* **19**, 149–172.
Francis, P. W. and Rundle, C. (1976). Rates of production of the main magma types in the central Andes. *Geol. Soc. Am. Bull.* **87**, 474–480.
Francis, P. W., Roobol, M. J., Walker, G. P. L., Cobbold, P. R., and Coward, M. P. (1974). The San Pedro and San Pablo volcanoes of northern Chile and their hot avalanche deposits, *Geol. Rundschau* **63**, 357–388.
Francis, P. W., Moorbath, S., and Thorpe, R. S. (1977). Strontium isotope data for recent andesites in Ecuador and North Chile. *Earth Planet. Sci. Letters* **37**, 197–202.
Francis, P. W., Hammill, M., Kretzschmar, G., and Thorpe, R. S. (1978). The Cerro Galan Caldera, Northwest Argentina and its tectonic setting. *Nature, Lond.* **274**, 749–751.
Fyfe, W. S. and McBirney, A. R. (1975). Subduction and the structure of andesitic volcanic belts. *Am. J. Sci.* **275A**, 285–297.
González-Ferrán, O. (1972). Volcanoes activos de Chile (map), Instituto Geografico Milatar de Chile.
Green, T. H. and Ringwood, A. E. (1968). Genesis of the calc-alkaline igneous rock suite. *Contr. Mineral. Petrol.* **18**, 105–162.

Guest, J. E. (1969). Upper Tertiary ignimbrites in the Andean cordillera of part of Antofagasta province, Northern Chile. *Geol. Soc. Am. Bull.* **80**, 337–362.
Guest, J. E. and Sanchez, J. (1969). A large dacitic lava flow in northern Chile. *Bull. Volc.* **33**, 778–790.
Hantke, G. (1966). The volcanoes of Ecuador. In *Catalogue of Active Volcanoes of the World*, Part 19, International Association for Volcanology and Chemistry of the Earth's Interior, Naples, pp. 26–61.
Hawkesworth, C. J., Norry, M. J., Roddick, J. C., Baker, P. E., Francis, P. W., and Thorpe, R. S. (1979). $^{143}Nd/^{144}Nd$ and $^{87}Sr/^{86}Sr$ variations in calc-alkaline andesites and plateau lavas from South America. *Earth planet. Sci. Letters* **42**, 45–57.
Henderson, W. G. (1979). Cretaceous to Eocene volcanic arc activity in the Andes of Northern Ecuador. *J. geol. Soc. Lond.* **136**, 367–378.
Hörmann, P. K., Pichler, H., and Zeil, W. (1973). New data on the young volcanism in the Puna of N.W. Argentina. *Geol. Rundschau* **62**, 397–418.
James, D. E. (1971a). Plate tectonic model for the evolution of the central Andes. *Geol. Soc. Am. Bull.* **82**, 3325–3346.
James, D. E. (1971b). Andean crustal and upper mantle structure. *J. geophys. Res.* **76**, 3246–3271.
James, D. E. (1978). Subduction of the Nazca plate beneath central Peru. *Geology* **6**, 174–178.
James, D. E., Brooks, C., and Cayambamba, A. (1976). Andean Cenozoic volcanism magma genesis in the light of strontium isotopic composition and trace element geochemistry. *Geol. Soc. Am. Bull.* **87**, 592–600.
Katsui, Y. (1972). Late Cenozoic volcanism and petrographic provinces in the Andes and Antarctica. *J. Fac. Sci. Hokkaido Univ.* **15**, 27–40.
Kay, R. W. and Gast, P. W. (1973). The rare earth content and origin of alkali-rich basalts. *J. Geol.* **81**, 653–682.
Klerkx, J., Deutsch, S., Pichler, H., and Zeil, W. (1977). Strontium isotope composition and trace element data bearing on the origin of Cenozoic volcanic rocks of the central and southern Andes. *J. Volc. geothermal Res.* **2**, 48–71.
Kussmaul, S., Hörmann, P. K., Ploskonka, E., and Subieta, T. (1977). Volcanism and structure of southwest Bolivia. *J. Volc. geothermal Res.* **2**, 73–111.
Lefévre, C. (1973). Les caracteres magmatiques du volcanitisme plioquaternaire des Andes dans le Sud de Perou. *Contr. Mineral. Petrol.* **41**, 259–272.
Lehmann, B. (1978). A Precambrian core sample from the Altiplano/Bolivia *Geol. Rundschau* **67**, 270–278.
Lopez-Escobar, L., Frey, F. A., and Vergara, M. (1976). Andesites from central-south Chile: trace element abundances and petrogenesis. In *Proceedings of the Symposium on Andean and Antarctic Volcanology Problems, Santiago, Chile, 1974* (O. Gonzáles-Ferrán, ed.), pp. 725–761.
Lopez-Escobar, L., Frey, F. A., and Vergara, M. (1977). Andesites and high-alumina basalts from the central-south Chile High Andes: geochemical evidence bearing on their petrogenesis. *Contr. Mineral. Petrol.* **63**, 199–228.
MacDonald, G. A. (1972). *Volcanoes*, Prentice-Hall, Englewood Cliffs, N.J.
McNutt, R. H., Crocket, J. H., Clark, A. H., Caelles, J. C., Farrar, E., Haynes, S. J., and Zentilli, M. (1975). Initial $^{87}Sr/^{86}Sr$ ratios of plutonic and volcanic rocks of the

central Andes between latitudes 26° and 29° south. *Earth planet. Sci. Letters* **27**, 305–313.

Moreno, R. H., 1974. Airplane flight over active volcanoes of central-south Chile. In *Symposium on Andean and Antarctic Volcanology Problems, Santiago, Chile, 1974*, Guide Book D 3, Department of Geology, University of Santiago, Chile, p. 56.

Moreno, R. H. (1976). The Upper Cenozoic volcanism in the Andes of southern Chile (from 40°00′ to 41°30′ S.L.). In *Proceedings of the Symposium on Andean and Antarctic Volcanaology Problems, Santiago, Chile, 1974*, (O. Gonzáles-Ferrán, ed.), pp. 143–171.

Nicholls, I. A. and Ringwood, A. E. (1973). Effect of water on olivine stability in tholeiites and the production of silica-saturated magmas in the island arc environment. *J. Geol.* **81**, 285–300.

Noble, D. C., McKee, E. H., Farrar, E., and Peterman, U. (1974). Episodic Cenozoic volcanism and tectonism in the Andes of Peru. *Earth planet. Sci. Letters* **21**, 213–220.

Noble, D. C., Bowman, H. R., Herbert, A. J., Silberman, M. L. Heropoulos, C. E., Fabbi, B. P., and Hedge, C. E. (1975). Chemical and isotopic constraints on the origin of low-silica latite and andesite from the Andes of central Peru. *Geology* **3**, 501–504.

Noble, D. C., Farrar, E., and Cobbing, E. J. (1979). The Nazca Group of south-central Peru: age, source and regional volcanic and tectonic significance. *Earth planet. Sci. Letters* **45**, 80–86.

Pichler, H. and Zeil, W. (1972). The Cenozoic rhyolite-andesite associations of the Chilean Andes. *Bull. Volc.* **35**, 424–452.

Pichler, H., Hörmann, P. K., and Braun, A. F. (1976). First petrologic data on lavas of the volcano El Reventador (eastern Ecuador). *Münster. Forsch. Geol. Paläont.* **38/39**, 129–141.

Pitcher, W. S. (1978). The anatomy of a batholith. *J. geol. Soc. Lond.* **135**, 157–182.

Ponce, R. F. (1976). El volcan Hudson. In *Proceedings of the Symposium on Andean and Antarctic Volcanology Problems, Santiago, Chile, 1974*, (O. Gonzáles-Ferrán, ed.), pp. 80–87.

Ramirez, J. E. (1968). Los Volcans de Colombia. *Rev. Acad. Colomb. Ciencos Exactos* **13**, 227–235.

Ringwood, A. E. (1974). Petrological evolution of island arc systems. *J. geol. Soc. Lond.* **130**, 183–204.

Roobol, M. J., Francis, P. W., Ridley, W. I., Rhodes, M., and Walker, G. P. L. (1976). Physico-chemical characters of the Andean volcanic chain between 21° and 22° south. In *Proceedings of the Symposium on Andean and Antarctic Volcanology Problems, Santiago, Chile, 1974* (O. Gonzáles-Ferrán, ed.), pp. 450–464.

Shackleton, R. M., Ries, A. C., Coward, M. P., and Cobbold, P. R. (1979). Structure, metamorphism and geochronology of the Arequipa Massic of coastal Peru. *J. geol. Soc. Lond.* **136**, 195–214.

Skewes, M. A. and Stern, C. R. (1979). Petrology and geochemistry of alkali basalts and ultramafic inclusions from the Palei–Aike volcanic field in southern Chile and the origin of the Patagonian plateau Davas. *J. Volc. geothermal Res.* **6**, 3–25.

Sparks, R. S. J. (1975). Stratigraphy and geology of the ignimbrites of Vulsini volcano, central Italy. *Geol. Rundschau* **64**, 497–523.

Stauder, W. M. (1973). Mechanism and spatial distribution of Chilean earthquakes with relation to subduction of the oceanic plate. *J. geophys. Res.* **78**, 5033–5061.

Stern, C. R. (1974). Melting products of olivine tholeiite basalt in subduction zones. *Geology* **2**, 227–230.

Stern, C. R. and Wyllie, P. J. (1978). Phase compositions through crystallization intervals in basalt–andesite–H_2O at 30 kb with implications for subduction zone magmas. *Am. Mineral.* **63**, 641–663.

Stewart, J. W. (1971). Neogene peralkaline igneous activity in eastern Peru. *Geol. Soc. Am. Bull.* **82**, 2307–2312.

Suarez, M. and Pettigrew, T. H. (1976). An Upper Mesozoic island-arc–back-arc system in the southern Andes and South Georgia. *Geol. Mag.* **113**, 305–400.

Thorpe, R. S. and Francis, P. W. (1979a). Variations in Andean andesite compositions and their petrogenetic significance. *Tectonophysics* **57**, 53–70.

Thorpe, R. S. and Francis, P. W. (1979b). Petrogenetic relationships of volcanic and intrusive rocks of the Andes. In *Origin of Granite Batholiths, Geochemical Evidence* (M. P. Atherton and J. Tarney, eds), Shiva Press, Orpington, pp. 65–75.

Thorpe, R. S., Francis, P. W., and Potts, P. J. (1976). Rare earth data and petrogenesis of andesites from the N. Chilean Andes. *Contr. Mineral. Petrol.* **54**, 65–78.

Thorpe, R. S., Francis, P. W., and Moorbath, S. (1979). Rare-earth and strontium isotope evidence concerning the petrogenesis of north Chilean ignimbrites. *Earth planet. Sci. Letters* **42**, 359–367.

Thorpe, R. S., Francis, P. W., and Harmon, R. S. (1981). Andean andesites and continental growth. *Phil. Trans. R. Soc. Lond.* A **301**, 305–320.

Vergara, M. (1972). Note on the zonation of the Upper Cenozoic volcanism of the Andean area of central-south Chile and Argentina. In *Symposium on the Results of Upper Mantle Investigations with Emphasis on Latin America*, International Upper Mantle Project, Buenos Aires, pp. 381–397.

Vergara, M., and Gonzáles-Ferrán, (1972). Structural and petrological characteristics of the late Cenozoic volcanism from Chilean Andean region and West Antarctica. *Krystalinikum* **9**, 157–184.

Vincent, P. M. (1963). Le volcanisme ignimbritique du Tibesti (Sahara Tchadian). Essai d'interprétation dynamique. *Bull. Volc.* **26**, 259–272.

Zeil, W. and Pichler, H. (1967). Die Känozoische Rhyolith-Formations in mitttleren Abschnitt der Andes. *Geol. Rundschau* **57**, 48–81.

Zentilli, M. and Dostal, J. (1977). Uranium in volcanic rocks from the central Andes. *J. Volc. geothermal Res.* **2**, 251–258.

Indonesia

C. S. Hutchison

Department of Geology,
University of Malaya, Kuala Lumpur 22-11, Malaysia

ABSTRACT

The volcanic arc extends some 6000 km from northern Sumatera to the Molucca Sea. From west to east it may be subdivided into the Sumatera cordilleran sector, the predominantly ensimatic Sunda island arc from Java to Flores, the convolute Banda arc, and the Molucca Sea collision complex. The Indian Ocean crust subducts with perpendicular incidence beneath the Sunda arc and with high obliquity beneath Sumatera. An eastwards decrease of $^{87}Sr/^{86}Sr$ ratios of the volcanic rocks from Java to Bali suggests a transition from continental Sunda Shelf to oceanic basement. The Banda arc is located on the oceanic Banda Sea, separated by trenches on the south and north from the converging Australian Platform. The Molucca Sea region represents a collision of two opposed arc–trench systems—Sulawesi–Sangihe and Halmahera. The trenches have coalesced into a single complex filled with 8–10 km of sialic material, yet the two volcanic arcs continue active. The Sumatera to Java and the now extinct arc of western Sulawesi have a history of volcanic activity dating back to the Paleocene. The Banda sector dates back only to the Pliocene.

Tambora (1815) and Krakatau (1883) rank first and fourth in the world's greatest historic eruptions. Tambora spread thick layers of andesitic ash widely in a southerly direction. The now extinct Sumateran Lake Toba ignimbritic mega-eruptive centre, which dates back 1.9 Ma, spread rhyolitic ash widely over the Indian Ocean 75 ka ago and over peninsular Malaysia 30 ka ago. Rhyolitic ash layers in the Indian Ocean must also be related to the Krakatau eruption.

Throughout Indonesia, the volcanism is predominantly calc-alkaline to high K calc-alkaline, with minor tholeiite and shoshonite. The lavas are predominantly andesitic, but range from basalt to dacite in composition. Most volcanoes are fairly uniform in composition, but some, like Krakatau, are of bimodal basalt–dacite composition. The rocks are strongly porphyritic, and are composed of two pyroxenes, plagioclase, magnetite, and ilmenite in a glassy groundmass. The tholeiitic series may contain pigeonite. Higher K contents of the high K calc-alkaline series are accommodated in hornblende, biotite, the plagioclase rims, and the glassy groundmass. Olivine and quartz are sparse. High alumina basalt is known only from Bali.

There is a generally good positive correlation between depth of the underlying Benioff zone with the contents of K_2O, selected trace elements, and the Sr isotope ratios, indicating that the magmas are of mantle wedge origin, are subduction-related, and that there is a decreasing volume of partial melting with increasing depth.

The volcanoes which lie over deep seismic contours are shoshonitic, with the exception of Gunung Api in the Banda Sea. They occur only along the northern edge of the Sunda arc and in the extinct western arc of Sulawesi, but there is also a zone on Lombok and Sumbawa which overlies anomalously shallow seismic contours. The leucite-bearing rocks do not fit on the same K–Si variation diagrams as the calc-alkaline series. Their Sr isotope ratios are generally lower than the high K calc-alkaline series and their strong enrichment in the light REE suggests an origin by a low degree of mantle partial melting. Pliocene cordierite dacites and granites of Ambon have extremely high Sr isotope ratios consistent with an origin by partial melting of continental crust.

The andesites of the Banda arc appear from their mineralogy, Sr and O isotope ratios, and trace element contents, to have resulted from partial melting of the mantle, but with sialic contamination. Slices of the leading edge of the Australian Platform must have been involved in subduction beneath the oceanic Banda Sea. Extinction of the arc south of Seram and north of Timor followed encroachment of the thicker Australian Platform.

Tectonic Setting

At its north-western end, the Indonesian volcanic arc is separated from the extinct Burmese volcanic arc by the active Andaman Sea marginal basin, and at its north-eastern end it continues into the Philippine volcanic arc.

The Indonesian arc to the south-east of the Andaman Sea may be conveniently subdivided into four sectors from west to east: the cordillera of Sumatera, the Sunda island arc from Java to Flores, the convolute Banda arc, and finally the complex Molucca Sea collision zone.

Andaman Sea

To the north of Sumatera, the Andaman Sea consists of short, NE–SW-trending, spreading axes which are offset by *en echelon* northwards continuations of the Semangko Fault (Fig. 1), continuing into Burma as the Sagaing Fault (Curray *et al.*, 1979). Extensional tectonics (Eguchi *et al.*, 1979) imply that before its north–south opening, the Sumateran and the now extinct Burmese volcanic arcs must once have been continuous (Hutchison, 1978).

Sumatera

Sumatera is composed of an ensialic basement of Paleozoic and Mesozoic rocks (Bemmelen, 1970). Indian Ocean lithosphere subducts obliquely beneath it. There is no seismicity deeper than *c.* 200 km. This may be related to the large transform component of the subduction, or to significant wrench displacement along the complex Semangko Fault system (Holcombe, 1977). The volcanoes are located on approximately east–west lineaments over strong aeromagnetic anomalies which are assumed to indicate the presence of subvolcanic plutons. Active volcanoes occur at intersections of the lineations with the NW–SE-trending Semangko Fault system (Posavec *et al.*, 1973).

Sunda arc (Java to Flores)

The Indian Ocean–Australian plate converges northwards on South-east Asia (Sclater and Fisher, 1974), as indicated by the magnetic anomalies (Fig. 1). Oceanic lithosphere is being subducted normal to the predominantly ensimatic island arc of Java. The very active Benioff zone was contoured by Hatherton and Dickinson (1969) and updated by Hamilton (1978). Seismicity in the Java sector extends to a maximum depth of about 600 km in the Java Sea (Fig. 1). The major cross-cutting Sumba Fracture separates Sumbawa from Flores (Audley-Charles, 1975).

Banda arc

The Banda arc is an eastwards continuation of the Sunda arc (Cardwell and Isacks, 1978), complicated by the change in character of the converging Indian Ocean–Australian plate, which is oceanic west of Timor and to the east is of 40 km thick continental crust, depressed by 3 km at the axis of the Timor Trough (Bowin *et al.*, 1980). The depression may be interpreted as attempted subduction beneath the oceanic southern Banda basin. The result of the failed subduction has been the extinction and uplift of the volcanic arc sector north of Timor (Alor to Romang, 508–603 on Fig. 1), in which volcanism ceased *c.* 3 Ma ago (Abbott and Chamalaun, 1978). The subducting Banda arc slab is contorted at its eastern end (Fig. 1), where the trench and the line of the active volcanoes curve to the north-east (Cardwell and Isacks, 1978). The volcanic arc from Damar to Banda (605–610, Fig. 1) is considered to be as young as Pliocene to Recent (Bowin *et al.*, 1980).

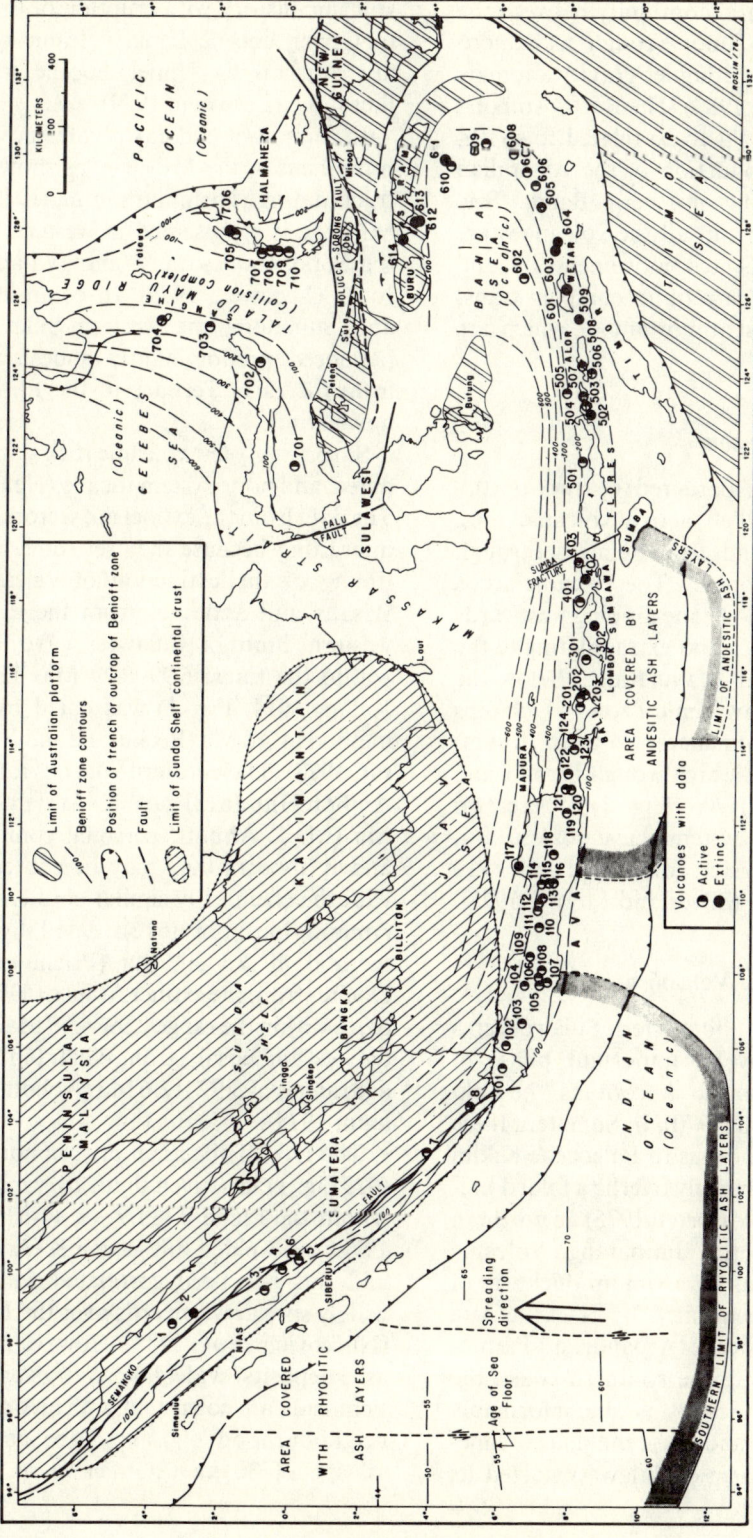

Fig. 1 Outline map of the Indonesian volcanic arc showing volcanoes for which data are available. Numbers refer to the volcanoes listed in Tables 1 and 2. Benioff zone contours are from Hamilton (1978), modified by Cardwell and Isacks (1978). The outline of the Sunda Shelf (Sundaland) is from Hutchison (1982) and of the Australian Platform from Bowin et al., (1980)

There is no structural continuity between the Seram and the Aru–Timor Trough, as demonstrated by the free air Bouguer gravity anomaly pattern (Bowin et al., 1980). The Buru–Ambon–Seram sector is thought to be related to southwards attempted subduction of the Australian Platform beneath the northern Banda Sea. Seismicity is shallow and all the volcanoes are extinct. Such complicated movements in and around the Banda Sea require complex transform faults, the most important of which are shown on Fig. 1.

Molucca sea collision zone

The Molucca Sea is bordered by two north–south-trending, parallel, active volcanic arcs. The western one extends from Sulawesi through Sangihe to the Philippines. The volcanic arc is related to a deep Benioff zone dipping westwards under the Suluwesi Sea and extrapolating to the surface at the Talaud–Mayu Ridge. Below the same ridge is a shallow Benioff zone which dips eastwards under Halmahera, with an accompanying active volcanic arc through Ternate and Halmahera (707 and 705, Fig. 1). These two opposed subduction systems have coalesced in the Molucca Sea collision zone, filled with 8–10 km of sialic material (Silver and Moore, 1978).

Age of Volcanism

The Sumatera, Java, and western Sulawesi arcs are characterized by an important phase of early Tertiary volcanism known as the 'old andesites' (Bemmelen, 1970). In Sumatera these include the late Cretaceous to Paleocene Kikim Tuffs which unconformably overlie a folded pre-Tertiary basement (De Coster, 1975). In northern Java they include the Jatibarang Volcanic Formation, which has a maximum thickness of 1200 m, and has been dated by K–Ar as late Eocene to early Oligocene (Arpandi and Patmosukismo, 1976). Along the southern coast, the Paleocene 'Old Andesite' is unconformably overlain by early to mid-Miocene marine limestone. The western arm of Sulawesi started its volcanic history with eruption of the Paleocene to Upper Eocene Langi volcanic rocks, which are overlain by Upper Eocene to Oligocene limestone (Leeuwen, 1979).

In Sumatera and Java, volcanic activity was prominent in the Miocene, as shown by fission track dating (Nishimura et al., 1978a), and the Miocene is represented in western Sulawesi by the Sopo, Pamesurang, and Walanae volcanic rocks (Leeuwen, 1979). In common with Java and Sumatera, the western Sulawesi arc experienced predominantly andesitic volcanism from the early Tertiary to the Pliocene, but is now extinct.

Surprisingly few volcanic rock suites have been dated, and none systematically (Hehuwat, 1976). The data from the extinct arc sectors are the most interesting because they set some limits on the timing of the extinction of volcanic activity. Basalts and andesites from the extinct southwestern limb of Sulawesi gave K–Ar dates within the range 6.99–9.29 Ma. The basalt on Kelang (614, Fig. 1) was dated by the Rb–Sr method at 7.6 Ma (Beckinsale and Nakapadungrat, 1978). The cordierite dacites (ambonites) of Ambon Hitu (613) gave K–Ar dates of 3.4–4.35 Ma (M. J. Abbott, personal communication), and it appears they may be contemporaneous with the chemically similar cordierite granite of Ambon Laitimor (612), dated by the Rb–Sr method at 3.3–3.8 Ma (Priem et al., 1978). Dacites from the extinct Atauro (509) gave K–Ar dates within the range 2.99–4.49 Ma and diorites, dacites, and basalts from Wetar (601) gave K–Ar age ranges of 3.97–12.6 Ma (Abbott and Chamalaun, 1978).

The Lake Toba ignimbrite of north Sumatera (02, Fig. 1) has given dates by various methods within the range 0.1–1.9 Ma (Nishimura et al., 1978b). A welded tuff gave a K–Ar age of 75 ka and this has been correlated with ash layers widely spread westwards over the Indian Ocean (Ninkovich et al., 1978). However, the rhyolitic ash deposits widely spread over the Malay Peninsula are consistent with an origin from the neighbourhood of Lake Toba and these give an age of 30 ka (Stauffer et al., 1980). This appears to have resulted from the last major eruption.

Eruptive Activity

There are more than 500 young volcanoes, of which 78 have had eruptions in historic times, and 50 are in the solfataric and fumarole stage (Neumann van Padang, 1951).

Types of eruption

(1) Solfatara and fumaroles are characterized by quiet gas emission and S deposition.

(2) Phreatic explosions are caused by gas expansion above the magma chamber. Mud eruptions result from exploding gas. The greatest ever recorded was the 1933 eruption of Suoh, Sumatera, when 0.21 Km^3 of mud were erupted, triggered by a great earthquake (Bemmelen, 1970). The word *lahar* is used in Indonesia to describe a mud flow containing angular volcanic blocks. Cold lahars result from heavy rainfall. Hot lahars are caused when a crater lake empties due to crater wall collapse, or to explosion.

(3) Normal volcanic activity is characterized by paroxysmal gas outbursts, followed by viscous lava slowly squeezed from the vent. Magmas poor in gases erupt slowly. Magmas rich in gases cause *nuées ardentes*.

(4) Plinian outbursts are spectacular paroxysmal eruptions. The Sunda arc is well represented in the world's greatest historic eruptions (Yokoyama, 1957). The greatest was the 1815 eruption of Tambora (401), with a total energy release of 8.4×10^{26} ergs. Krakatau (101) came fourth with 1×10^{25} ergs, during its well documented 1883 eruption (Bemmelen, 1970).

The Tambora eruption has been variously estimated to have produced between 100 and 318 km^3 of volcanic ejectamenta. The volume of the cone which was blown away was itself 30 km^3. The ash layer on Sumbawa island at a distance of 70 km is of 60 cm thickness. Andesitic ash layers have been found in the Indian Ocean up to a distance of 600 km towards the south, but none have been found in the Flores Sea to the north (Ninkovich and Donn, 1977). This spread of andesitic ash (Fig. 1) must be related to the great Tambora eruption. No thicknesses of the layers are available, and the piston cores are limited to the Pleistocene. The lava was of glassy leucite basanite and tephrite, containing phenocrysts of plagioclase, augite, olivine, biotite, and Fe–Ti oxides. Leucite is confined to the groundmass (Petroeschevsky, 1949). The ash, however, contains only glass, plagioclase, augite, olivine, and biotite (Bemmelen, 1970). Since the volcano and the ash layers are both andesitic to basaltic, the enormity of the eruption must be attributed to spectacular gas build-up (Yoder, 1976). Older flows around Tambora consist of olivine basalt, olivine andesite, and biotite-bearing andesite. There is no record of any acidic volcanism.

The well documented Krakatau eruption (Symons, 1888) ejected 18 km^3 of pumice and ash. The volcano had a previous history of bimodal basaltic–dacitic volcanism and the paroxysmal eruption was of dacitic composition c. 66 per cent SiO_2). The spread of rhyolitic ash over the Indian Ocean adjacent to western Java and southern Sumatera must be predominantly from the Krakatau eruption. However, drill cores have penetrated rhyolitic ash layers dating back to the Pliocene and Miocene (Ninkovich and Donn, 1977). Piston cores in the Java Sea have encountered no ash. The new Anak Krakatau ash cone has been built up since 1927 in the centre of the caldera.

(5) *Nuées ardentes* eruptions may result in ignimbrites and tuff-lavas. The most spectacular centre of eruption is the extinct Lake Toba in northern Sumatera (2, Fig. 1). There is a total volume of ignimbritic tuffs of 1500–2000 km^3. During its long history of eruptions, rhyolitic ash was widely strewn according to the prevailing winds. The eruption 75 ka ago resulted in a wide field of ash over the Indian Ocean as far west as Sri Lanka (Ninkovich and Donn, 1977; Ninkovich et al., 1978). An eruption 30 ka ago resulted in rhyolitic ash being strewn over peninsular Malaysia. The distributions and

thicknesses of the ash are discussed by Stauffer et al. (1980).

Petrography

Although there is no universally accepted classification of volcanic rocks, a framework must be provided for comparison with other publications. The author feels that the K_2O–SiO_2 classification of Peccerillo and Taylor (1976) is eminently suitable for description of the Indonesian arc. It has therefore been adopted in the text and in Figures 2–6. The sloping solid lines divide the fields into a low K or tholeiitic series, a calc-alkaline series, a high K calc-alkaline series, and a shoshonitic series (Fig. 2). The vertical broken lines separate the rocks into basalt (less than 52 per cent SiO_2), basaltic andesite (52–56 per cent SiO_2), andesite (56–63 per cent SiO_2), dacite (63–70 per cent SiO_2), and rhyolite (over 70 per cent SiO_2). This classification obviously oversimplifies the complex nature of the lavas, and the dividing lines arbitrarily split a continuous spectrum (e.g. Fig. 3).

The Indonesian calc-alkaline basaltic to andesitic rocks are generally glassy and contain abundant phenocrysts of strongly zoned plagioclase (An_{80}–An_{50}), augite, hypersthene, and smaller crystals of ilmenite and magnetite. Most volcanoes exhibit a good equilibrium partitioning of Fe–Mg in their pyroxenes, with the outstanding exception of Serua (608), which has extreme disequilibrium partition coefficients

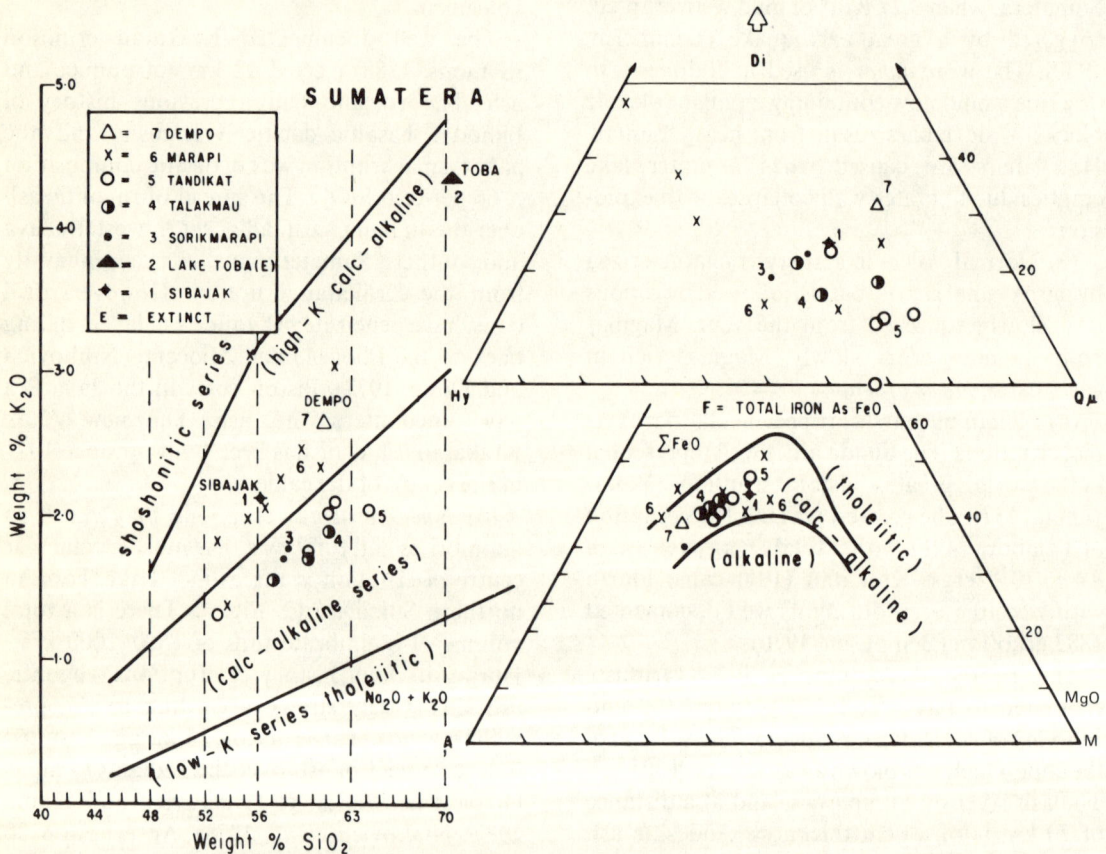

Fig. 2 Volcanoes of Sumatera. Graph of K_2O against SiO_2 in weight per cent, normative diopside: hypersthene: quartz diagram, and A ($K_2O + Na_2O$): F (total iron as FeO): M (MgO) diagram. Norms calculated by the programme of Hutchison (1975b).

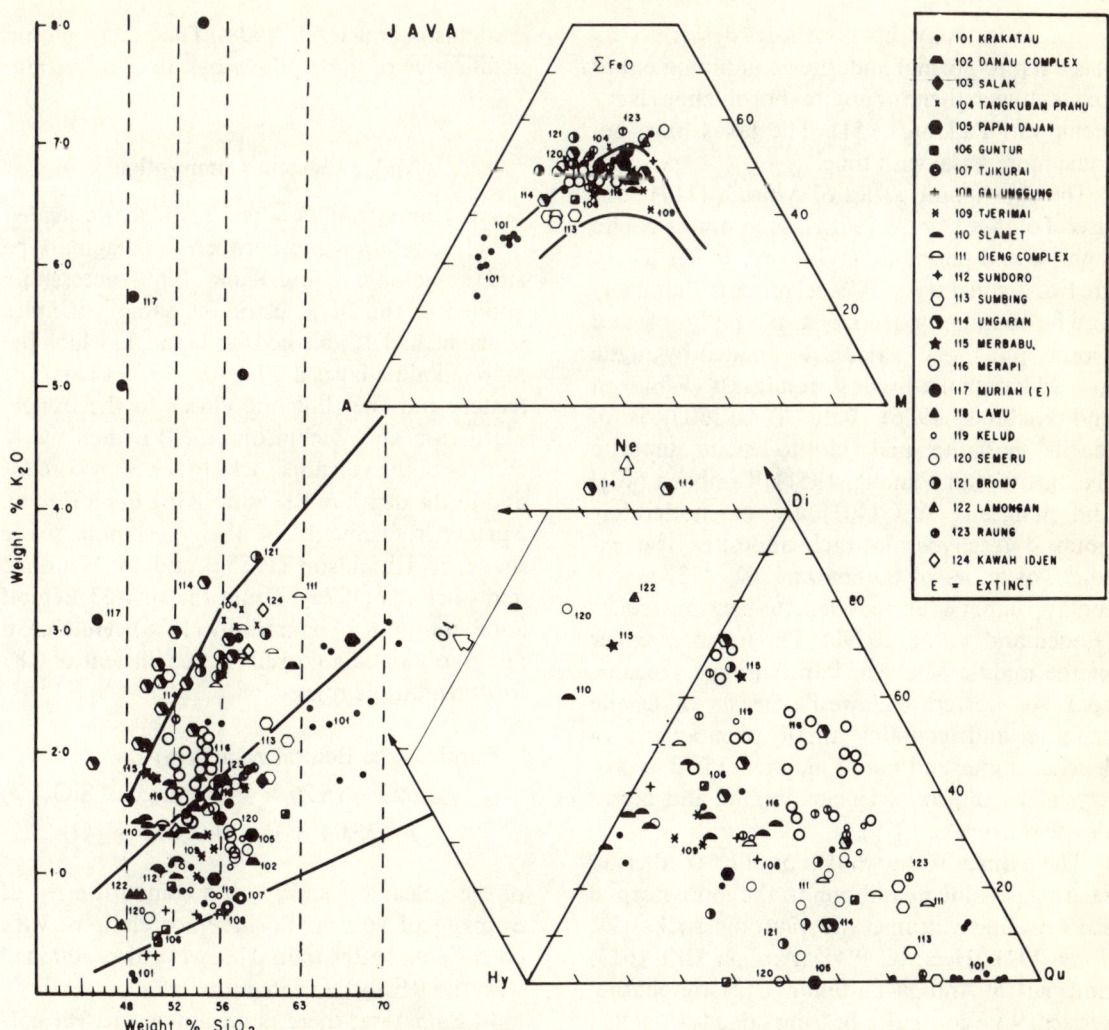

Fig. 3 Volcanoes of Java. Diagrams similar to Fig. 2. Normative proportions extended to include olivine and nepheline. CIPW normative computer program based on Hutchison (1975b)

$K_D(Fe^{opx-cpx})$ ranging from 0.8 to 2.5 (Hutchison and Jezek, 1978), indicating a mixed origin for the magma.

The more tholeiitic rocks may contain minor olivine (Fo_{70}–Fo_{80}) and small crystals of pigeonite. Olivine-rich ankaramite occurs on Lombok (Foden and Varne, 1979a). The high K calc-alkaline series contains hornblende or hornblende and biotite in addition to plagioclase, two pyroxenes, and Fe-Ti oxides. Microprobe analyses (Jezek and Hutchison, 1978) show that the higher K contents are held in the biotite, hornblende, the outer rims of the plagioclase phenocrysts (up to 2 mol per cent Or), and also in the groundmass glass.

Bimodal volcanoes have extruded dacite in addition to basalt, and they appear to be restricted to the tholeiitic and calc-alkaline series. The dacites are extremely fine grained or glassy. They contain phenocrysts of plagioclase (An_{50}–An_{25}), biotite, augite, and frequently small crystals of pigeonite and quartz (Hutchison and Jezek, 1978).

Rare high alumina basalt (Batur, 201, Fig. 4) is known to occur only within the tholeiitic and calc-alkaline series.

The lavas may be concisely described as glass-rich to normal andesites containing either augite–hypersthene or augite–hornblende (Neumann van Padang, 1951). The lavas have become more basic with time.

The shoshonitic series of Muriah (117) consists of olivine leucitite, with less common leucite tephrite and phonolite (Whitford, 1975). There are two distinct types. A 'wet' series is characterized by abundant phenocrysts of amphibole and biotite, and a 'dry' series is dominated by augite in a feldspathoidal-rich groundmass (Whitford and Nicholls, 1976). Batu Tara (505) is of leucite basanite and biotite-leucite tephrite (Neumann van Panang, 1951). Tambora (401) and Sangeang Api (403) are of moderately potassic trachybasalt–trachyandesite. The extinct volcanoes of Sumbawa (302, 402) are of highly undersaturated leucite-bearing rocks (Foden and Varne, 1979b). The upper member of the middle Miocene Pamesurange volcanic rocks of western Sulawesi consists of leucite tephrite and contains small phenocrysts of leucite, augite, and minor magnetite in a microcrystalline matrix of leucite, augite, and labradorite (Leeuwen, 1979).

The extinct Wetar (601) contains cordierite-bearing rhyolite in addition to the more normal calc-alkaline volcanic and plutonic rocks (De Jong, 1941; Heering, 1941). Ambon Hitu (613) and part of Ambon Laitimor (612) are characterized by cordierite-bearing dacites, called ambonites (Kuenen, 1949). The phenocrysts are of Fe-rich cordierite, biotite, labradorite–bytownite, orthopyroxene, and almandine in a high silica, high potash glass. The cordierite, which may contain needles of sillimanite and graphite, is thought to have resulted from the digestion of pelitic xenoliths (Jezek and Hutchison, 1978). The chemically similar granite on Ambon Laitimor contains abundant pelitic xenoliths and cordierite phenocrysts (Priem et al., 1978).

The Pliocene tholeiitic pillow basalts of Keleng island (614) may be correlated with those of south-western Ambon Laitimor (612). The fresh glassy rims of the pillows contain unzoned plagioclase (An_{77}) and olivine (Fo_{86}) (Hutchison and Jezek, 1978). The exact tectonic significance of these pillow basalts is uncertain.

Major Element Composition

Several important studies have demonstrated the close relationship between the magma type and the tectonic setting. Kuno (1966) successfully applied to the Java sector the same principles which he had established for Japan, in which the more alkaline lavas are located farther from the trench, and the tholeiitic closer to the trench. Hatherton and Dickinson (1969) related the K content of the volcanic rocks to the SiO_2 content and to the depth of the underlying Benioff zone. Further refinements of their technique were made by Hutchison (1975a) and by Whitford and Nicholls (1976). Using the revised Benioff zone contours of Hamilton (1978), Hutchison (1976) obtained a correlation coefficient of 0.87 for the multiple regression

$$\text{Depth of the Benioff zone (km)} = 397 - (5.26 \times \text{Percentage of } SiO_2) + (35.04 \times \text{Percentage of } K_2O)$$

of the volcanic rocks, with a standard error of estimate of 26 km. Positive correlations, with coefficients better than 0.80, were also obtained with the Rb and Sr contents.

In Sumatera, there is an almost perfect correlation ($r = 0.95$) between the K content adjusted for SiO_2 and Benioff zone depth (Hutchison, 1975a). All the analyses are quartz, hypersthene, and diopside normative (Fig. 2).

The Java sector is predominantly calc-alkaline to high K calc-alkaline, but some are tholeiitic and some shoshonitic (Fig. 3). Krakatau is distinctly bimodal. Muriah is distinctly shoshonitic and only its 'wet' series shows the same K_2O relationship to the Benioff zone as the calc-alkaline series. The 'dry' series shows no overall trend of K_2O with SiO_2 (Whitford and Nicholls, 1976). Most lavas are quartz normative (Fig. 3).

The Bali volcano of Batur (201) belongs to the high alumina series, with Al_2O_3 average values of c. 20 per cent, in contrast to most other

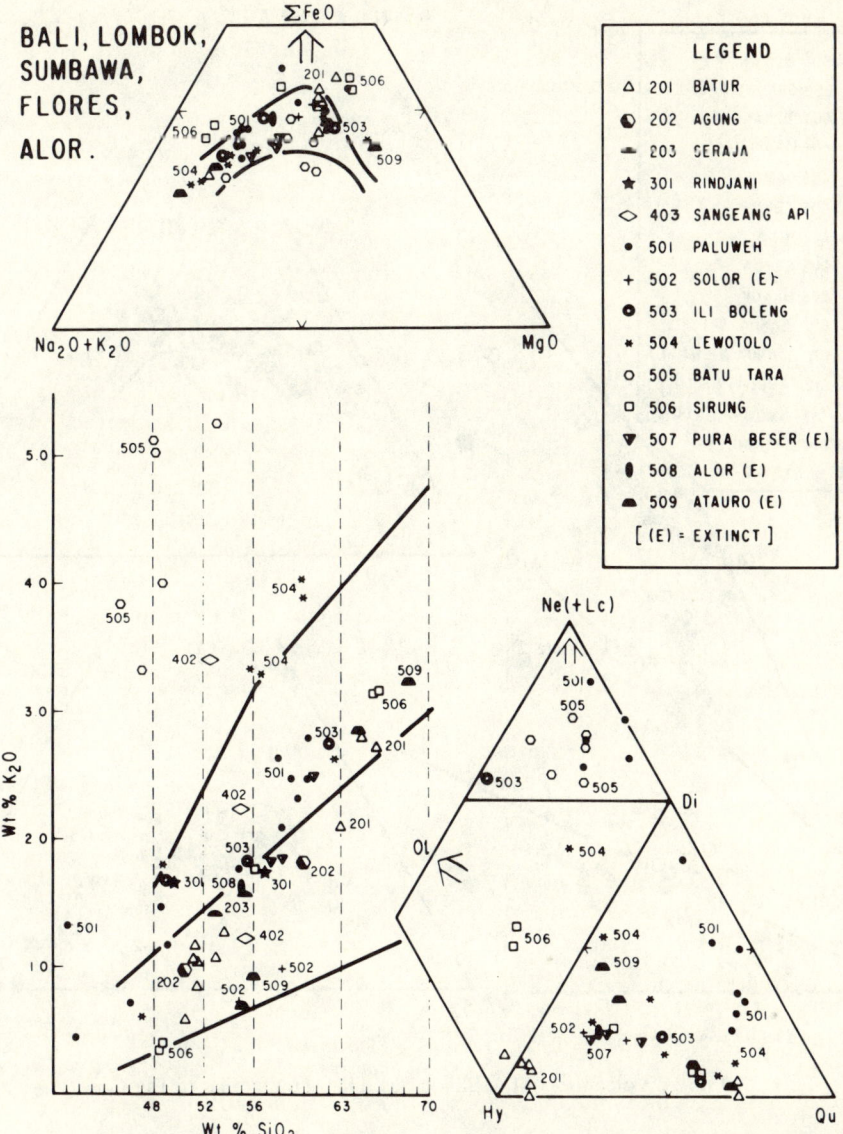

Fig. 4 Volcanoes of Bali to Alor. Diagrams similar to Fig. 2. Normative amounts extended to include olivine and nepheline + leucite

Indonesian volcanoes, where Al_2O_3 is commonly between 15 and 17 per cent. Unfortunately no petrographic descriptions are available to allow further comment. Sirung (506) is strongly enriched in Fe (Fig. 4). The K_2O–Benioff zone depth relationship holds for the calc-alkaline suite of Lombok and Sumbawa (301, 302). However, the more potassic rocks (302, 401, 402, 403) show no obvious relationship between K_2O content and seismic depth (Foden and Varne, 1979b).

The volcanic rocks of the Banda arc are entirely quartz normative (Fig. 5). Api north of Wetar (602) is completely out of character with its underlying Benioff zone depth of 405 km (Fig. 1). Its whole rock chemical composition and mineralogy are similar to Damar (605) and bear no similarity with Batu Tara (505), as

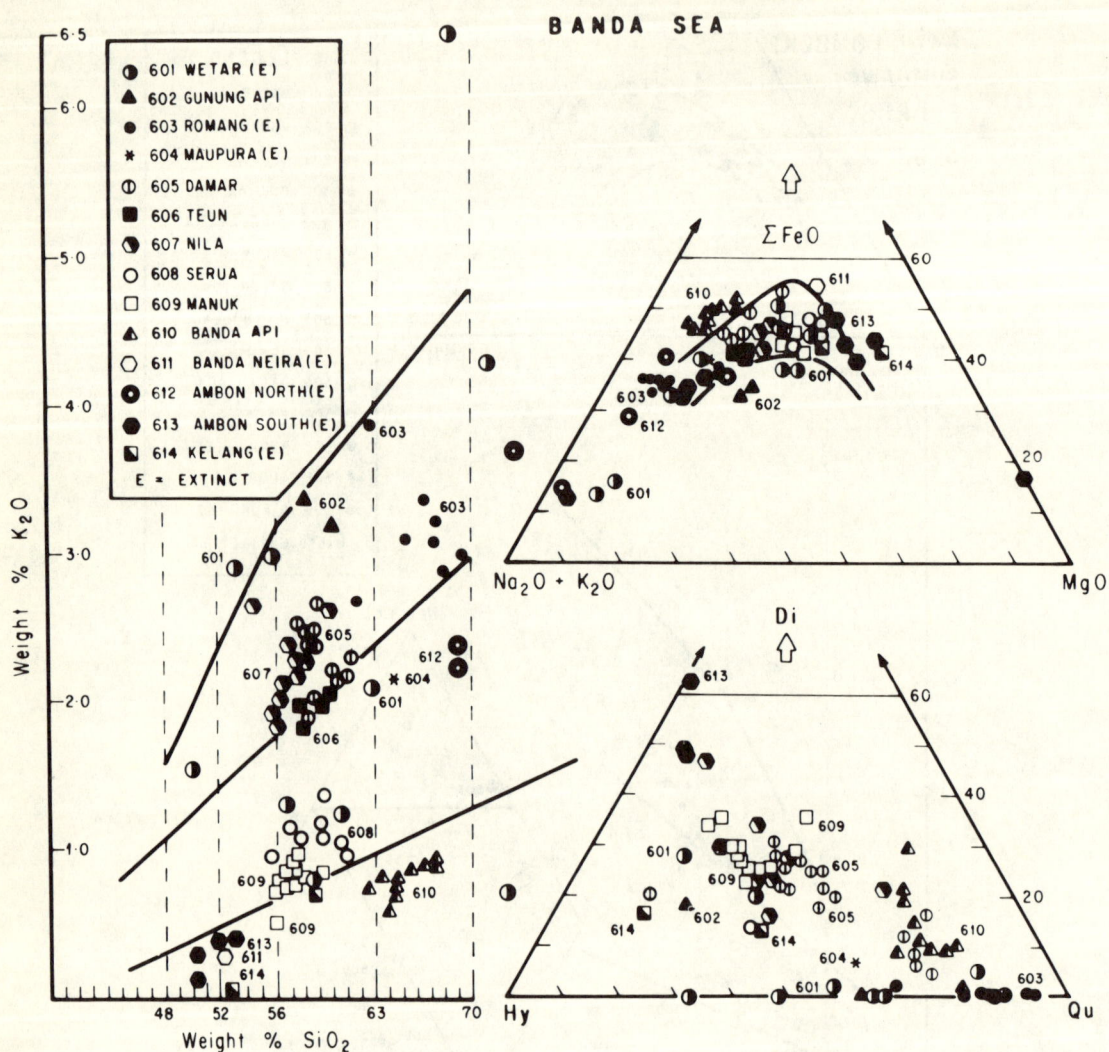

Fig. 5 Volcanoes of the Banda Sea. Diagrams similar to Fig. 2

might be expected from the tectonic setting. The two extinct and uplifted ends of the Banda arc (601, 613) are characterized by high silica dacite and rhyolite (Fig. 5). Banda Neira (610) and Banda Api (611) form a tholeiitic basalt–dacite volcanic pair. The excellent clustering of the data from the Banda arc may reflect the young age of the volcanism in which the magma is produced from a source as yet uncomplicated from a changing tectonic pattern.

The Sulawesi-Moluccas sector has been studied only superficially. The small amount of available evidence (Neumann van Padang, 1951) shows some interesting features. Both Lokon Empung (702, Fig. 6) and Dukono (705) are distinctly bimodal, characterized by basalt containing less than 1 per cent K_2O and dacite with more than 3 per cent K_2O. Although the K_2O–SiO_2 relationships suggest a spectrum from calc-alkaline to high K calc-alkaline, the AFM relationships (Fig. 6) suggest that some of the volcanoes (702, 704, 706, 707) have tholeiitic affinities. The K_2O contents are generally higher than may be expected from the shallow Benioff

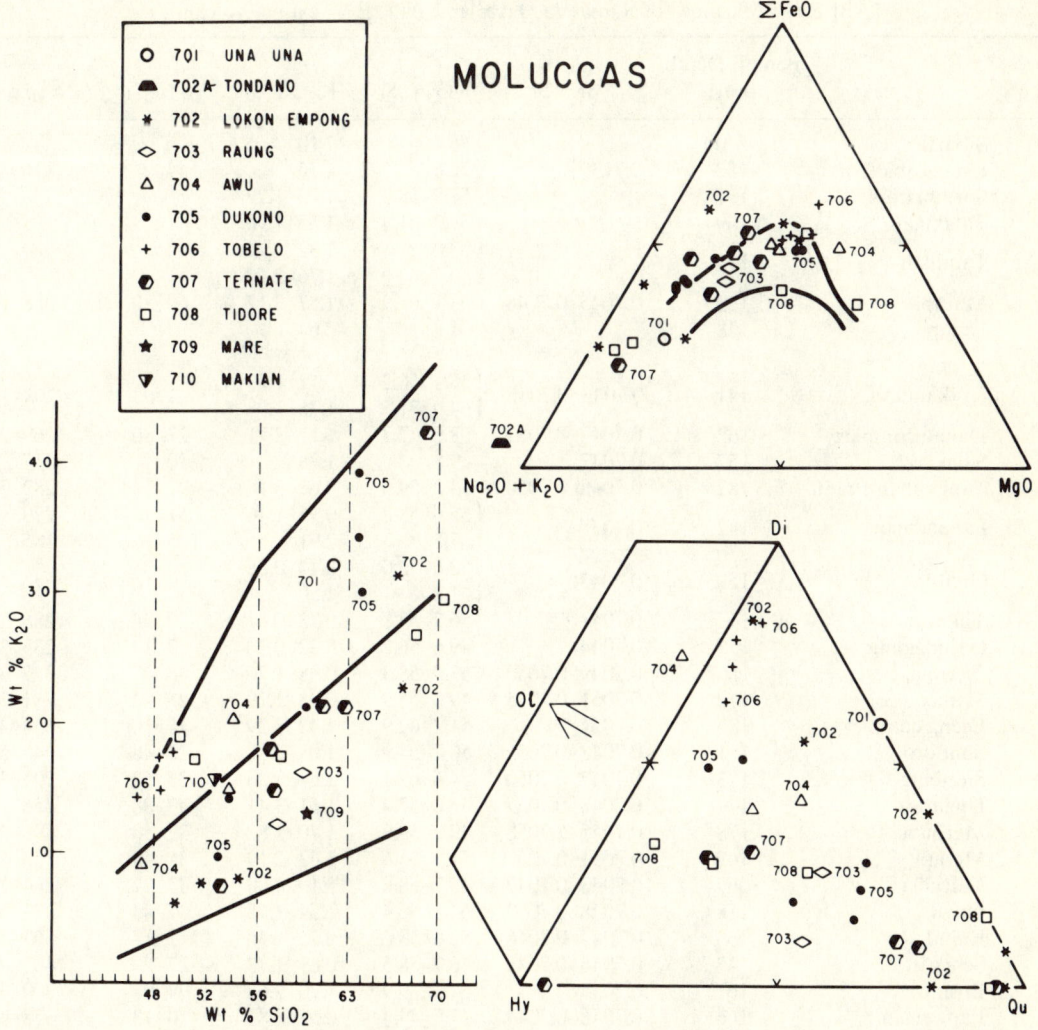

Fig. 6 Volcanoes of the Moluccas. Diagrams similar to Fig. 2. Normative amounts extended to include olivine

zone contours of the region. Some of these discrepancies may be solved by work now in progress (Jezek and Gill, 1979). The significance of andesites on Talaud and other central islands is uncertain.

Isotope Data

The data of Tables 1 and 2 have been summarized from Whitford (1975), Whitford et al. (1977), Whitford and Jezek (1979), and Magaritz et al. (1978). The following general observations may be made:

(1) Rocks of the tholeiite suite have $^{87}Sr/^{86}Sr$ ratios within the range 0.7040–0.7045, lower than the adjacent calc-alkaline rocks.

(2) The calc-alkaline series has ratios variable from 0.7040 to 0.7055, with a regular decrease from western Java to Bali, and some evidence for increasing ratio with increasing Benioff zone depth (Whitford, 1975). The regular decrease from western to eastern Java may be

TABLE 1 Volcanoes of Sumatera, Java, and Bali (E = extinct volcano)

Number	Name	Benioff Depth (km)	$^{87}Sr/^{86}Sr$	SiO_2 (%)	K_2O (%)	Rb (p.p.m.)	Sr (p.p.m.)
1	Sibajak	170		56.4	2.10		
2	Lake Toba (E)	157	0.7139	70.6	4.36	144	116
3	Sorikmarapi	107		57.9–58.3	1.72–1.79		
4	Talakmau	109		57.2–61.4	1.55–1.90		
5	Tandikat	115		{ 53.0 59.9–64.2	1.32 1.76–2.10		
6	Marapi	129	0.7045–0.7048	53.1–61.9	1.37–3.12	60–80	358–390
7	Dempo	128		61.2	2.64		
101	Krakatau	141	0.7043–0.7046	{ 48.4–56.2 64.0–71.5	0.12–1.73 1.78–3.08	3–63	208–373
102	Danau complex	162	0.7047–0.7051	58.9–62.3	1.11–2.71	27–80	269–339
103	Salak	157	0.7047	59.8	1.78	70	217
104	Tangkuban Prahu	182	0.7048–0.7050	54.2–59.3	0.63–3.18	97–144	287–302
105	Papandajan	142	0.7057	{ 55.6–58.8 67.5	0.95–1.46 2.91	51 145	279 185
106	Guntur	159	0.7043	{ 50.6–52.00 61.8	0.44–0.88 1.50	9–10	240–296
107	Tjikurai	142	0.7041	56.7–57.8	0.73–0.81	20–24	286–291
108	Galunggung	155	0.7044	49.6–56.2	0.32–0.70	7–19	233–342
109	Tjerimai	186	0.7049–0.7050	53.9–58.3	0.98–1.70	44–47	327–363
110	Slamet	172	0.7051–0.7055	49.0–55.9	0.98–1.79	28–44	281–304
111	Dieng complex	185	0.7052	51.2–62.9	1.42–3.29	78–100	334–405
112	Sundoro	181	0.7048–0.7052	50.7–54.9	1.08–2.00	26–48	443–578
113	Sumbing	173	0.7047–0.7050	54.6–62.1	1.54–2.26	35–53	295–344
114	Ungaran	179	0.7048–0.7053	45.0–57.4	1.61–3.41	43–88	456–629
115	Merbabu	178	0.7055–0.7058	49.9–59.9	1.70–1.82	37–50	496–497
116	Merapi	169	0.7051–0.7059	51.9–55.6	1.43–2.19	36–49	558–570
117	Muriah (E)	369	0.7043–0.7047	45.2–58.6	3.10–8.00	132–378	264–2100
118	Lawu	118	0.7048–0.7049	54.1–56.4	1.54–1.75	37–43	471–500
119	Kelud	153	0.7044–0.7046	52.6–58.0	0.75–1.30	12–17	507–587
120	Semeru	143	0.7046–0.7048	46.7–58.5	0.63–1.73	27–29	403–414
121	Bromo	161	0.7044	54.3–59.9	1.70–3.62	94–97	300–390
122	Lamongan	166	0.7042–0.7044	47.5–49.4	0.57–0.81	10–13	378–390
123	Raung	123	0.7041–0.7044	52.4–58.9	1.50–2.93	43–100	449–585
124	Kawah Idjen	169	0.7041–0.7045	52.2–59.7	1.36–3.18	40–103	347–584
201	Batur	165	0.7040	{ 50.5–53.7 62.8–65.9	0.58–1.27 2.08–2.79	20 65	453 300
202	Agung	160	0.7040	{ 50.4 60.0	0.98 1.81	17 45	483 414
203	Seraja	153	0.7039	52.8–55.2	1.42–1.56	30–33	462–497

taken to suggest a gradual transition from underlying continental to entirely ensimatic crust.

(3) The island of Serua (608) has anomalously high ratios of 0.7075–0.7095 for the calc-alkaline suite.

(4) The 'dry' shoshonitic series of Muriah (117), predominantly of olivine leucitite but also including more silicic leucite tephrite and phonolite, has low Sr isotope ratios in the range 0.7043–0.7047, in contrast to the calc-alkaline series of Java which ranges up to 0.7055.

TABLE 2 Volcanoes from Lombok, through Banda Sea, to Moluccas (E = extinct volcano)

Number	Name	Benioff Depth (km)	$^{87}Sr/^{86}Sr$	SiO_2 (%)	K_2O (%)	Rb (p.p.m.)	Sr (p.p.m.)	Cs (p.p.m.)	Ba (p.p.m.)
301	Rindjani	164	0.7039–0.7049	48.0–65.6	0.99–3.67	18–91	297–611		
302	Sangenges (E)	150	—	44.7–60.5	2.21–4.42	61–319	724–837		
401	Tambora	182	0.7039	51.5–56.1	2.89–5.69	86–142	923–981		
402	Soromundi (E)	173	0.7053	45.5–47.5	0.93–0.97	114–168	959–1511		
403	Sangeang Api	192	0.7042–0.7049	47.8–55.3	1.23–3.56	20–112	408–1010		
501	Paluweh	192	—	41.0–48.6	0.44–1.48	—	—		
				54.9–60.1	1.76–2.81				
502	Solor (E)	120	0.7057–0.7058	54.8–58.4	0.73–0.98	17–28	350–355		
503	Ili Boleng	159	0.7065	49.0–62.0	1.68–2.72	41	836		
504	Lewotolo	163	—	47.2–48.9	0.61–1.80	—	—		
			0.7058	55.6–62.6	2.61–4.06	75	578		
505	Batu Tara	248	—	45.4–53.1	3.23–5.25	—	—		
506	Sirung	100	0.7055–0.7059	48.6–56.4	0.40–177	9–68	365–397		
			0.7057–0.7058	65.7–66.1	3.15–3.16	116–134	276–280		
507	Pura Beser (E)	120	0.7054–0.7056	57.4–60.7	1.84–2.49	51–63	720–760		
508	Alor (E)	110	0.7077	55.0	1.61	62	426		
509	Atauro (E)	115	0.7072–0.7082	54.9–56.0	0.73–0.94	22–30	224–296		
			0.7066–0.7068	64.5–68.4	2.84–3.24	93–113	279–350		
601	Wetar (E)	120	—	50.1–62.9	0.80–3.00	50–150	50–190		
			—	68.8–72.5	0.73–6.51	50–460	80–160		
602	Gunung Api	405	—	58.1–60.0	3.19–3.38	—	—		
603	Romang (E)	150	—	61.7–69.2	2.71–3.89	140–430	500–1080		
604	Maupura (E)	135	0.7091	64.4	2.18	74	226		
605	Damar	155	0.7065–0.7066	56.0–61.25	1.90–2.67	87–93	501–575	2.8–7.2	1010–1070
606	Teun	130	0.7075–0.7077	58.2–60.1	1.89–2.12	67–73	480–500	2.2–4.9	691–752
607	Nila	125	0.7076–0.7078	54.3–59.7	1.84–2.66	83–90	388–421	2.8–8.0	680–824
608	Serua	115	0.7075–0.7095	55.9–61.2	0.93–1.38	33–53	249–204	1.2–3.1	238–301
609	Manuk	110	0.7051–0.7058	56.1–59.4	0.51–0.97	31–36	198–213	1.5–2.5	234–301
610	Banda Api	130	0.7046–0.7048	62.4–67.8	0.59–0.89	18–25	163–180	0.4–1.6	232–262
611	Banda Neira (E)	130	0.7045	52.2	0.31	9	277	0.5	106
612	Ambon Laitimor	90	0.7042–0.7042	50.4–53.1	0.15–0.42	5–9	365–492	0.2–0.3	81–92
613	Ambon Hitu	86	0.7158–0.7175	69.0–76.0	2.26–7.5	102–298	118–146	18–32	424–470
614	Kelang (E)	50		52.9–58.5	0.02–0.71				
701	Una Una	185		61.7	3.20				
702	Lokon Empung	159		49.8–54.5	0.62–0.80				
				66.7–74.9	2.27–4.16				
703	Raung	130		57.7–59.5	1.20–1.62				
704	Awu	153		47.1–53.9	0.90–2.03				
705	Dukono	165		52.9–53.7	0.97–1.42				
				59.6–64.1	2.13–3.93				
706	Tobelo	170		46.6–49.5	1.42–1.77				
707	Ternate	125		53.2–57.2	0.74–1.80				
				61.0–69.0	2.11–2.24				
708	Tidore	130		49.9–57.7	1.90–1.75				
				68.4–70.4	2.95–2.69				
709	Mare	130	0.7045	59.8	1.31				
710	Makian	130	0.7039	52.7	1.57				

(5) The high K calc-alkaline series of Damar, Teun, and Nila (605, 606, 607 respectively) have ratios in the range 0.7065–0.7078 (Whitford and Jezek, 1979).

(6) The $^{87}Sr/^{86}Sr$ ratios for the extinct portion of the arc from Alor to Wetar (508–601) are unusually high, 0.7066–0.7082, in comparison with the active arc (Whitford et al., 1977).

(7) The Lake Toba ignimbrite (2, Table 1) has an extremely high ratio of 0.7139. The cordierite-bearing ambonites of Ambon Hitu (613) also have high ratios, 0.7158–0.7175 (Whitford and Jezek, 1979).

The Banda volcanic rocks from Ambon, Banda, Manuk, Serua, Teun, and Damar show a clear positive correlation between the $^{87}Sr/^{86}Sr$ and the $\delta^{18}O$ values. The andesites have $\delta^{18}O$ values ranging from 5.6 to 9.2 per mil. Over that range in $\delta^{18}O$, the $^{87}Sr/^{86}Sr$ increases from 0.7044 to 0.7095 (Magaritz et al., 1978). The cordierite dacite (ambonite) from Hitu has $\delta^{18}O$ values of c. 15 per mil, corresponding with Sr ratios of 0.717.

The similarity between the $\delta^{18}O$ values and the $^{87}Sr/^{86}Sr$ ratios in total rocks and in the separated plagioclase phenocrysts indicate that the measured isotope ratios are primary and have not been affected by post-eruptive alteration.

Trace Element Composition

The REE data are shown in Fig. 7, extracted from Nishimura and Ikeda (1978) and Nicholls and Whitford (1976). The Lake Toba ignimbrite (2) is enriched in the light REE, suggesting either derivation from continental crust or a low degree of partial melting of the mantle. The alkaline volcanic rocks of Muriah (117) are extremely enriched in the light REE, which may be interpreted as implying that the parent magma resulted from a low degree of partial melting (Yoder, 1976). The generally high relative proportions of the heavy REE argue against the derivation of the lavas by partial melting of eclogite in subducted basaltic crust (Nicholls and Whitford, 1976). The low Ni contents of the basalts suggests that most of the lavas have experienced some fractional crystallization (Whitford et al., 1979).

The Rb, Sr, Cs, and Ba contents have been analyzed by Whitford (1975), Whitford et al. (1977), Whitford and Jezek (1979), and Hutchison (1977). A summary is given in Tables 1 and 2. Both Rb and Sr are strongly enriched in the Muriah alkaline rocks. Several extinct volcanoes of the Banda arc are also abnormally rich in Rb and Sr. The high Cs contents of the ambonites of Ambon Hitu (613) may be related to contamination of the magma by pelitic xenoliths. There is no obvious explanation for the high Ba content of the Damar (605) lavas. There is a very good positive correlation between both Rb and Sr contents and depth to the Benioff zone for rocks of the Java sector (Hutchison, 1976). The incompatible elements such as K, Rb, Cs, Sr, Ba, the light REE, U, and Th show an increase in abundance of almost an order of magnitude with

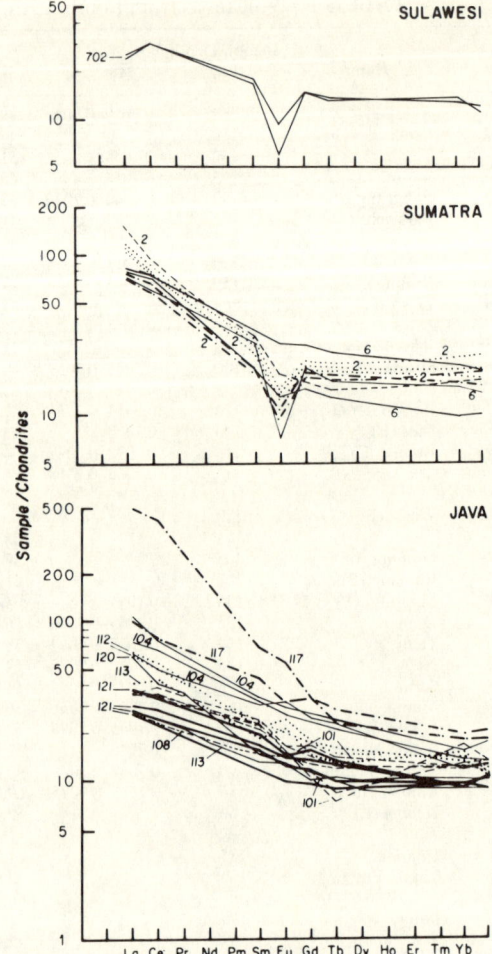

Fig. 7 REE data for Volcanoes of Sulawesi, Sumatera, and Java

increasing Benioff zone depth (Whitford et al., 1979). The rocks are characteristically depleted in compatible elements such as Ni, Co, Cr, and Sc and these show no correlation with the depth of the seismic zone.

Mineralization

The major porphyry copper deposits are located in Sulawesi at Tapadaa and Sasak. This arc appears to continue through Sangihe to the Philippines, where the giant porphyry copper deposits are located. By contrast, the Indian Ocean sector from Sumatera to Java contains no known economic copper deposits, although present work may lead to some discoveries in Sumatera. The Sumateran arc has produced major Au and Ag epithermal deposits at Mangani, Salida, and Benkulu, and in western Java at Bajah dome (Hutchison and Taylor, 1978).

Petrogenesis

The arc includes a wide range of volcanic rocks which can be demonstrated to have been derived by processes which vary along its length, in keeping with the varying tectonic situation. The data can be used to distinguish origins from the oceanic lithosphere and overlying mantle wedge, from the continental crust, and from mixtures of these components.

Oceanic lithosphere and mantle wedge

The rocks of the Sumatera to Sumbawa sector have a spectrum from tholeiitic through calc-alkaline to high K calc-alkaline series. There is an apparent eastwards decrease of $^{87}Sr/^{86}Sr$ from 0.7059 to 0.7039 from western Java to Bali (Whitford, 1975), which suggests progressive eastwards thinning of the underlying Sunda Shelf continental crust. There is also some evidence that the ratio increases with Benioff zone depth from about 0.7040 at 130 km to 0.7050 at 200 km, over the range tholeiitic to high K calc-alkaline series. However, the 'dry' shoshonitic series of Muriah does not continue this pattern, and its ratio drops to c. 0.7045.

The major magma genesis is thought to be in the mantle wedge overlying the Benioff zone, by 20–25 per cent melting at 30–40 km depth for primary tholeiites, by 5–15 per cent melting at 40–60 km for primary high K calc-alkaline magmas, and by 5 per cent melting at 80–90 km depth for shoshonitic magmas (Whitford *et al.*, 1979). However, the increase in Sr isotope ratio concomitant with increase in K_2O and incompatible trace element contents with Benioff zone depth would imply a contribution to the magma resulting from partial melting of subducted oceanic lithosphere at the Benioff zone.

Of the shoshonite series volcanoes, only the extinct Muriah (117), Sangenges (302), and Soromundi (402) have been adequately studied. The relatively low $^{87}Sr/^{86}Sr$ ratio of 0.7043–0.7047 precludes any contribution from the underlying crust, rules out a hypothesis of limestone contamination (cf. Bemmelen, 1970), and is consistent with a mantle origin. The strong enrichment in K and the light REE favours a very low degree of partial melting. The shoshonitic volcanoes overlie deep seismic contours at Muriah (370 km) and Batu Tara (250 km); however, those on Sumbawa overlie shallow contours of 150 and 173 km. There is a complete absence of evidence to suggest that this series is part of a general spectrum from the high K calc-alkaline series. On the contrary, the high K series contains amphibole and biotite and fits on the K_2O–SiO_2 variation diagrams as the tholeiitic and calc-alkaline series. But the 'dry series' feldspathoidal rocks show no relationship between K_2O and SiO_2 and are distinctly different from the calc-alkaline series. Their origin is therefore a separate problem, which is yet unresolved, but the suggested model is that the magma results from a very low degree of partial melting of the mantle wedge which has already been enriched in incompatible elements which have risen from the Benioff zone (Whitford *et al.*, 1979; Foden and Varne, 1979*b*).

If the Benioff zone depth is the sole controlling factor of magma type, then Api north of Wetar (405 km depth) should be shoshonitic. However, it belongs to the high K series and contains biotite and hornblende but no leucite or nepheline. This is a good example to illustrate the observation of Arculus and Johnson (1978) that there are important exceptions to the relationships between magma type and tectonic setting, and indicates that the rules of magma generation are not understood—all we have are imperfect models. It seems necessary to conclude that the mantle wedge is heterogeneous and has become locally enriched in incompatible elements rising from the Benioff zone (Whitford *et al.*, 1979). In the case of the shoshonites of Sumbawa, which overlie the 150–173 km seismic contours, Foden and Varne (1979b) suggest that K and other incompatible elements have migrated up the cross-cutting Sumba Fracture Zone.

Continental crust

The high $^{87}Sr/^{86}Sr$ ratio of 0.7139 for the Lake Toba ignimbrite, extruded between 0.1 and 1.9 Ma ago, together with its enrichment in light REE, imply a continental crustal anatectic origin, although petrographic evidence such as fused xenoliths is lacking. The cause of this megaeruptive centre is obscure (Nishimura et al., 1978b).

The Pliocene cordierite-bearing dacitic ambonites of Ambon Hitu, where the crust is thick and apparently continental, show clear evidence of a crustal origin. Cordierite, almandine, high concentrations of K, Rb, Cr, and high $^{87}Sr/^{86}Sr$ ratios of c. 0.715 suggest that they have been formed by melting of pelitic material with the continental crust (Whitford and Jezek, 1979). The source of the ambonites is therefore likely to be the margin of the Australian continent, which has subducted beneath the Seram accretionary wedge along a Benioff zone which crops out at the Seram Trough (Fig. 1). The cordierite granites of Ambon Laitimor, with an $^{87}Sr/^{86}Sr$ ratio of 0.7221, and an age of 3.3 Ma (Priem et al., 1978), are probably related to the cordierite dacite and of similar origin.

Mixed oceanic lithospheric mantle wedge and continental crustal origin

The volcanic rocks of the Banda Sea have trace element contents and Sr and O isotopic ratios which indicate that a mantle-derived magma has been contaminated by sialic material. Since the oceanic Banda Sea is surrounded on three sides by the Australian Platform, and Seram and Timor are characterized by thrust tectonics, it is considered that slices of continental crust were subducted at the Seram, Aru, and Timor troughs along with predominantly oceanic material. The island of Serua particularly supports this mixing hypothesis. Its Sr isotope ratios range widely from 0.7074 to 0.7095, and there is an exceptionally strong disequilibrium between the phenocryst compositions.

REFERENCES

Abbott, M. J. and Chamalaun, F. H. (1978). New K/Ar age data for Banda arc volcanics. Institute for Australasian Geodynamics, Flinders University of South Australia, no. 78/5.

Arculus, R. J. and Johnson, R. W. (1978). Criticism of generalized models for the magmatic evolution of arc-trench systems. Earth planet. Sci. Letters 39, 118–126.

Arpandi, D. and Patmosukismo, S. (1976). The Cibulakan Formation as one of the most prospective stratigraphic units in the north-west Java basinal area. In Proceedings of the Indonesian Petroleum Association, June 1975, Fourth Annual Convention, Indonesian Petroleum Association, Jakarta, Vol. 1, pp. 181–210.

Audley-Charles, M. G. (1975). The Sumba fracture: a major discontinuity between eastern and western Indonesia. Tectonophysics 26, 213–228.

Beckinsale, R. D. and Nakapadungrat, S. (1978). A late Miocene K–Ar age for the lavas of Pulau Kelang, Seram, Indonesia. J. Phys. Earth, Tokyo, 26, Suppl., S199–S201.

Bemmelen, R. W. van (1970). The Geology of Indonesia, Vol. 1A: General Geology of Indonesia, Martinus Nijhoff, The Hague.

Bowin, C., Purdy, G. M., Johnston, C., Shor, G., Lawver, L., Hartono, H. M. S., and Jezek, P. (1980). Arc-continent collision in the Banda Sea region. Bull. Am. Assoc. Petrol. Geol. 64, 868–915.

Cardwell, R. K. and Isacks, B. L. (1978). Geometry of the subducted lithosphere beneath the Banda Sea in eastern Indonesian from seismicity and fault plane solutions. J. geophys. Res. 83, 2825–2838.

Curray, J. R., Moore, D. G., Lawver, L. A., Emmel, F. J., Raitt, R. W., Henry, M., and Kieckhefer, R. (1979). Tectonics of the Andaman Sea and Burma. In Geological and geophysical investigations of continental margins. (J. Watkins, L. Montardent and P. Dickinson, eds.), Mem. Am. Assoc. Petrol. Geol., 29, 189–198.

De Coster, G. L. (1975). The geology of the central and south Sumatra basins. In Proceedings of the Indonesian Petroleum Association Third Annual Convention, June 1974, Indonesian Petroleum Association, Jakarta, pp. 77–110.

De Jong, J. D. (1941). Geological Investigations in West Wetar, Lirang and Solor, N.V. Noord-Hollandsche Uitgevers Maatschappij, Amsterdam.

Eguchi, T, Uyeda, S., and Maki, T. (1979). Seismotectonics and tectonic history of the Andaman Sea. Tectonophysics 57, 35–51.

Foden, J. D. and Varne, R. (1979a). The geochemistry and petrology of the basalt–andesite–dacite suite from Rindjani volcano, Lombok. Implications for the petrogenesis of island arc calc-alkaline magmas. Paper presented to the ESCAP CCOP-IOC SEATAR ad hoc Working

Group Meeting, Bandung, Indonesia, 9–14 July 1979.
Foden, J. D. and Varne, R. (1979b). Petrogenetic and tectonic implications of near coeval calc-alkaline to highly alkaline volcanism on Lombok and Sumbawa islands in the eastern Sunda arc. Paper presented to the ESCAP CCOP-IOC SEATAR ad hoc Working Group Meeting, Bandung, Indonesia, 9–14 July, 1979.
Hamilton, W. (1978). Tectonic map of the Indonesian region 1:5,000,000. Map 1-875-D, US Geological Survey.
Hatherton, T. and Dickinson, W. R. (1969). The relationship between andesitic volcanism and seismicity in Indonesia, the Lesser Antilles, and other island arcs. *J. geophys. Res.* **74**, 5301–5310.
Heering, J. (1941). *Geological Investigations in East Wetar, Alor and Poera Besar.* N.V. Noord-Hollandsche Uitgevers Maatschappij, Amsterdam.
Hehuwat, F. (1976). Isotopic age determinations in Indonesia: the state of the art. In *Proceedings of the Seminar on Isotopic Dating, Bangkok, May 1975*, CCOP, UNDP, Thailand, pp. 135–157.
Holcombe, C. J. (1977). Earthquake foci distribution in the Sunda Arc and the rotation of the back arc area. *Tectonophysics* **43**, 169–180.
Hutchison, C. S. (1975a). Correlation of Indonesian active volcanic geochemistry with Benioff zone depth. *Geologie en Mijnbouw* **54**, 157–168.
Hutchison, C. S. (1975b). The norm, its variations, their calculation and relationships. *Schweiz. mineral. Petrogr. Mitt.* **55**, 243–256.
Hutchison, C. S. (1976). Indonesian active volcanic arc: K, Sr, and Rb variation with depth to the Benioff zone. *Geology* **4**, 407–408.
Hutchison, C. S. (1977). Banda Sea volcanic arc: some comments on the Rb, Sr and cordierite contents. *Geol. Soc. Malaysia Newsletter* **3**, 27–35.
Hutchison, C. S. (1978). Southeast Asian tin granitoids of contrasting tectonic setting. *J. Phys. Earth, Tokyo* **26**, Suppl., S221–S232.
Hutchison, C. S. (1982). Southeast Asia, In *The Indian Ocean Basin and Margins* (A. E. M. Nairn, F. G. Stehli, and M. Churkin, eds), Plenum Press, New York, in press.
Hutchison, C. S. and Jezek, P. A. (1978). Banda Arc of eastern Indonesia: petrography, mineralogy and chemistry of the volcanic rocks. In *Proceedings of the Third Regional Conference on the Geology and Mineral Resources of South-East Asia, Bangkok, 14–17 November 1978* (Prinya Nutalaya, ed.), Asian Institute of Technology, Bangkok, pp. 607–619.
Hutchison, C. S. and Taylor, D. (1978). Metallogenesis in S.E. Asia. *J. geol. Soc. Lond.* **135**, 407–428.
Jezek, P. A. and Gill, J. B. (1979). Mineralogy and geochemistry of igneous rocks from the Banda and Molucca Sea island arcs, eastern Indonesia. Abstract presented at the ESCAP CCOP-IOC SEATAR ad hoc Working Group Meeting, Bandung, Indonesia, 9–14 July, 1979.
Jezek, P. A. and Hutchison, C. S. (1978). Banda Arc of eastern Indonesia: petrography and geochemistry of the volcanic rocks. *Bull. Volc.* **41**, 586–608.
Kuenen, P. H. (1949). Ambon and Haroekoe. *Verh. K. Ned. Geol. Mijnb., Gen. Geol. Ser.* **15**, 44–62.
Kuno, H. (1966). Lateral variation of the basalt magma type across continental margins and island arcs. *Bull. Volc.* **29**, 195–222.
Leeuwen, Th. M. van (1979). The geology of southwest Sulawesi with special reference to the Biru area. In *Proceedings of a Workshop on Eastern Indonesia, Bandung, July 1979*.
Magaritz, M., Whitford, D. J., and James, D. E. (1978). Oxygen isotopes and the origin of high $^{87}Sr/^{86}Sr$ andesites. *Earth planet. Sci. Letters* **40**, 220–230.
Neumann van Padang, M. (1951). *Catalogue of the Active Volcanoes of the World Including Solfatara Fields*, Part 1, Indonesia, International Volcanology Association, Naples.
Nicholls, I. A. and Whitford, D. J. (1976). Primary magmas associated with Quaternary volcanism in the western Sunda arc, Indonesia. In *Volcanism in Australasia* (R. W. Johnson, ed.), Elsevier, Amsterdam, pp. 77–90.
Ninkovich, D. and Donn, W. L. (1977). Cenozoic explosive volcanism related to east and Souteast Asian Arcs. *Island Arcs, Deep Sea Trenches, and Back Arc Basins* (M. Talwani and W. C. Pitman, eds), American Geophysical Union, Maurice Ewing Series, no. 1, pp. 337–347.
Ninkovich, D., Shackleton, N. J., Abel-Monem, A. A., Obradovich, J. D., and Izett, G. (1978). K–Ar age of the late Pleistocene eruption of Toba, north Sumatra. *Nature, Lond.* **276**, 574–577.
Nishimura, S. and Ikeda, T. (1978). Geochemical studies of volcanic rocks of Sunda Island Arc, Indonesia. In *Studies of Physical Geology on the Sunda Island Arc* (S. Sasajima, ed.), Kyoto University Press, Kyoto, pp. 63–83.
Nishimura, S., Sasajima, S., Hirooka, K., Thio, K. H., and Hehuwat, F. (1978a). Radiometric ages of volcanic products in Sunda Arc. In *Studies of Physical Geology on the Sunda Island Arc* (S. Sasajima, ed.), Kyoto University Press, Kyoto, pp. 34–37.
Nishimura, S., Abe, E., Yokoyama, T., Sugiyarta, and Dharma, A. (1978b). Danau Toba—the outline of Lake Toba, North Sumatera, Indonesia. In *Studies of Physical Geology on the Sunda Island Arc* (S. Sasajima, ed.), Kyoto University Press, Kyoto, pp. 54–62.
Peccerillo, A. and Taylor, S. R. (1976). Geochemistry of Eocene clac-alkaline volcanic rocks from the Kastamonu area, Northern Turkey. *Contr. Mineral. Petrol.* **58**, 63–81.
Petroeschevsky, W. A. (1949). A contribution to the knowledge of the Gunung Tambora (Sumbawa). *Tijdschr. K. Ned. Aardr. Gen.* **66**, 688–703.
Posavec, M., Taylor, D., Leeuwen, Th. van, and Spector, A. (1973). Tectonic controls of volcanism and complex movements along the Sumateran Fault System. *Geol. Soc. Malayasia Bull.* **6**, 43–60.
Priem, H. N. A., Andriessen, P. A. M., Boelrijk, N. A. I. M., Hebeda, E. H., Hutchison, C. S., Verdurmen, E. A. Th., and Verschure, R. H. (1978). Isotopic evidence for a Middle to Late Pliocene age of the cordierite granite on Ambon, Indonesia. *Geologie en Mijnbouw* **57**, 441–443.
Sclater, J. G. and Fisher, R. L. (1974). Evolution of the east central Indian Ocean, with emphasis on the tectonic setting of the Ninetyeast Ridge. *Geol. Soc. Am. Bull.* **85**, 683–702.
Silver, E. A. and Moore, J. C. (1978). The Molucca Sea collision zone. Indonesia. *J. geophys. Res.* **83**, 1681–1691.
Stauffer, P. H., Nishimura, S., and Batchelor, B. C. (1980). Volcanic ash in Malaya from a catastrophic eruption of Toba, Sumatra 30 000 years ago. In *Physical Geology of Indonesian Island Arcs* (S. Nishimura, ed.). Kyoto

University Press, Kyoto, pp. 156–164.

Symons, G. J. (ed.) (1888). *The Eruption of Krakatoa and Subsequent Phenomena*, Royal Society, London.

Whitford, D. J. (1975). Strontium isotopic studies of the volcanic rocks of the Sunda arc, Indonesia, and their petrogenetic implications. *Geochim. cosmochim. Acta* **39**, 1287–1302.

Whitford, D. J. and Jezek, P. A. (1979). Origin of late Cenozoic lavas from the Banda Sea, Indonesia: trace element and Sr isotope evidence. *Contr. Mineral. Petrol.* **68**, 141–150.

Whitford, D. J. and Nicholls, I. A. (1976). Potassium variation in lavas across the Sunda arc in Java and Bali. In *Volcanism in Australasia* (R. W. Johnson, ed.), Elsevier, Amsterdam, pp. 63–75.

Whitford, D. J., Compston, W., Nicholls, I. A., and Abbott, M. J. (1977). Geochemistry of late Cenozoic lavas from eastern Indonesia: role of subducted sediments in petrogenesis. *Geology* **5**, 571–575.

Whitford, D. J., Nicholls, I. A., and Taylor, S. R. (1979). Spatial variations in the geochemistry of Quaternary lavas across the Sunda arc in Java and Bali. *Contr. Mineral. Petrol.* **70**, 341–356.

Yoder, H. S., Jr (1976). *Generation of Basaltic Magma* National Academy of Sciences, Washington, DC.

Yokoyama, I. (1957). Energetics in active volcanoes. 2nd paper. *Tokyo Univ. Earthquake Res. Inst. Bull.* **35**, 75–97.

Andesites
Edited by R. S. Thorpe
© 1982 John Wiley & Sons

Papua New Guinea

R. W. Johnson

Bureau of Mineral Resources,
PO Box 378, Canberra City, ACT 2601, Australia

ABSTRACT

Andesite is present on most of the 100 or more late Cainozoic volcanoes in Papua New Guinea, and has been produced historically by eight—and possibly as many as 10—of the 14 active eruptive centres. All the arc–trench-type volcanoes are grouped roughly into seven provinces, six of which have andesites (the exception is the alkaline Tabar-to-Feni province). Papua New Guinea andesites may be divided into two broad types, one having generally greater contents of incompatible elements, higher $^{87}Sr/^{86}Sr$ values, and more fractionated REE patterns, than the other. The origin of the andesites is related to convergent plate boundary tectonics, but there is an incomplete correlation between andesitic volcanism and intermediate and deep focus earthquake activity, particularly in the Highlands and Eastern Papua provinces.

Introduction

The Cainozoic volcanic geology of the subaerial parts of Papua New Guinea seems to be characterized by andesites—that is, mainly silica-oversaturated rocks containing 53–63 per cent SiO_2 (see below for a more precise definition of the chemical composition). Andesite appears to be the main constituent of the volcanic components of Tertiary island arcs, and is found on most of the 100 or more late Cainozoic volcanoes scattered throughout the region (Fig. 1). Andesite production is apparently related, in a general sense, to convergence of the major Indo-Australian (or Indian) and Pacific plates, and to the descent of lithospheric slabs. However, tectonic relationships are much more complex, owing to the existence of at least two minor plates (Fig. 2), and to the overprinting of tectonic events throughout the Cainozoic. At least six, and possibly as many as 10 plate boundaries may be present in Papua New Guinea.

There are 14 active volcanoes in Papua New Guinea, of which eight to 10 have erupted andesite during the past 100 a. Three or four of these eruptions caused losses of human life. The best known one was the 1951 eruption of Mount Lamington which killed almost 3000 people (Taylor, 1958), and the most recent one was in March 1979 when an andesitic explosive eruption caused the death of two volcanologists on Karkar volcano.

Papua New Guinea may be divided into three broad tectonic provinces (Dow, 1977). A zone of island arcs occupies all but the south-western third of the region, and is mainly covered by shallow seas (Fig. 3). The crust is 30–45 km thick beneath the New Britain and New Ireland island arcs (Finlayson and Cull, 1973), and the crust beneath the Solomon Sea has a similar thickness to oceanic crust elsewhere (Furumoto et al., 1970). The south-western corner of Papua New Guinea is the northern edge of the Australian continent which makes up a relatively

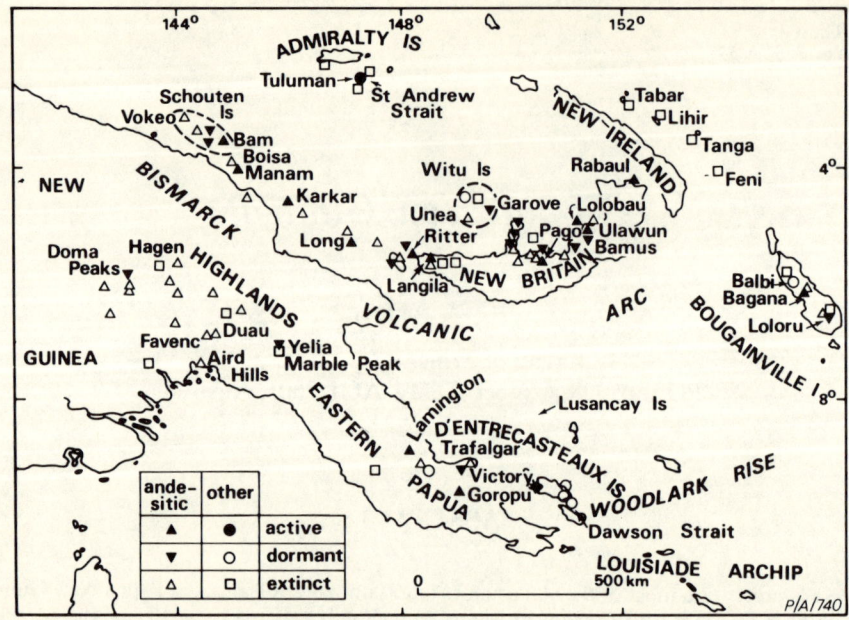

Fig. 1 Subaerial volcano distribution in Papua New Guinea. Volcanoes are identified as andesitic only where andesite has been identified by chemical analysis (a few of the other volcanoes may have andesite). Classification of an andesitic volcano as active does not necessarily mean that andesite was produced by eruptions recorded historically (compare with Table 1)

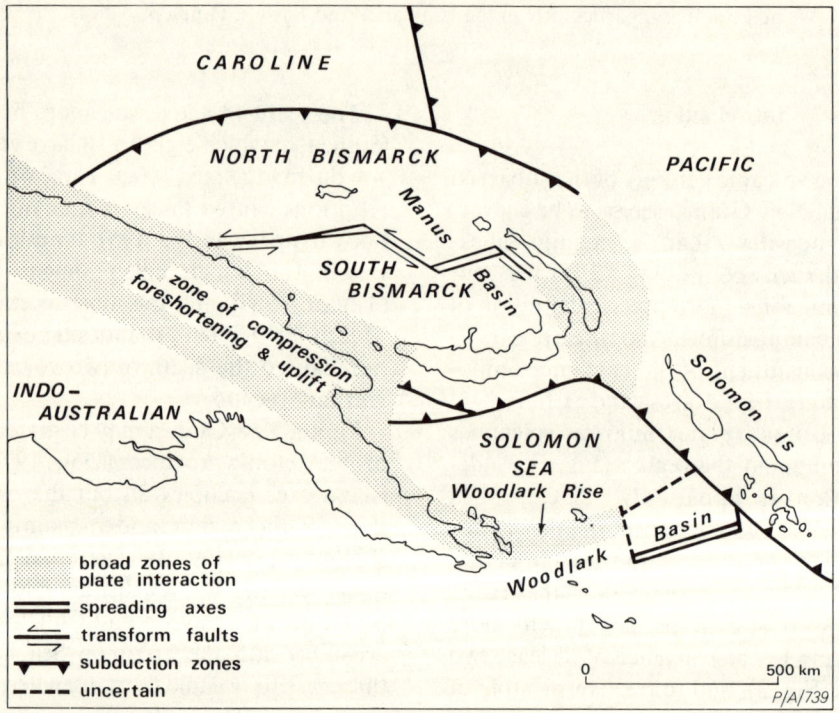

Fig. 2 Schematic plate-boundary distribution in Papua New Guinea and surrounding areas (synthesized from several published sources). Teeth on subduction-zone symbol point in direction of presumed underthrusting

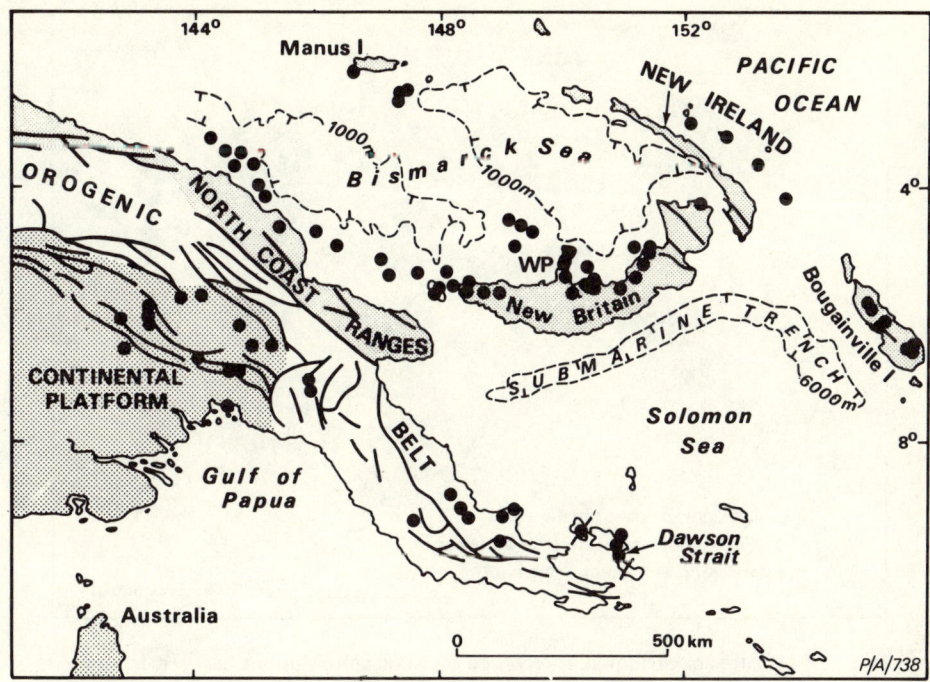

Fig. 3 Simplified geology of Papua New Guinea showing three tectonic provinces. Main late Cainozoic volcanic centres shown as solid circles, and main faults represented by thick lines. Broken lines represent isobaths that outline the Manus Basin (Fig. 2) and New Britain Trench. WP, Willaumez Peninsula

stable, and generally low lying platform (Fig. 3). Entry of this continental mass into the region appears to have had an important influence on the late Cainozoic tectonic history of Papua New Guinea.

The continent is thought to have collided with the zone of island arcs in the mid-Tertiary, blocking a subduction zone, and causing major uplift of New Guinea island, the eastern half of which is the largest land mass in Papua New Guinea (Jaques and Robinson, 1977). The collision zone between the continental platform and the arcs is an orogenic metamorphic belt consisting of deformed continental margin sediments, high angle faults (of which at least some may be strike-slip faults), ophiolite sequences, and blueschist terrains. This zone forms the central cordillera of mainland Papua New Guinea. It reaches a maximum height of 4509 m (Mount Wilhelm), and has Pleistocene glacial landforms (Löffler, 1972).

The late Cainozoic volcanoes of Papua New Guinea are found in all three tectonic provinces (Fig. 3). However, most of the active volcanoes, and most of the andesitic ones, are in the zone of island arcs and shallow seas, particularly along the southern margin of the Bismarck Sea (including New Britain) and on Bougainville Island.

Plate Boundaries

Two minor plates in Papua New Guinea are reasonably well defined, although their senses of relative motion are not yet clearly determined. These are the South Bismarck and Solomon Sea plates (Fig. 2). The boundary between them is marked by the axis of the New Britain Trench (Fig. 3) where the Solomon Sea plate is thought to underthrust the South Bismarck plate northwards at rates possibly as high as 12.4 cm a^{-1} (Krause, 1973), although Johnson and Molnar

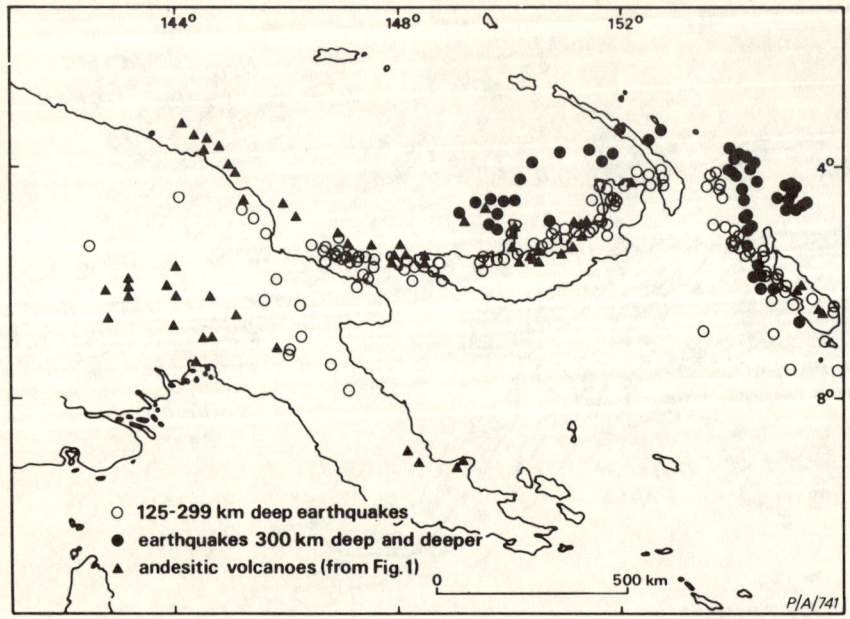

Fig. 4 Epicentres of earthquakes recorded by 15 or more stations between 1 January 1964 and 31 December 1977, and whose depths are 125 km or greater (BMR earthquakes data file), illustrating the imperfect correlation between deep seismicity and late Cainozoic andesitic volcanism

(1972) and Curtis (1973) estimated somewhat lower rates of 9.2 and 12.0 cm a^{-1}, respectively. A Benioff zone dips northwards beneath New Britain at c. 70°, reaching a depth of almost 600 km (Fig. 4). The western end of the trench splits into two well defined valleys (Fig. 2), and the eastern end bends abruptly through almost 90°, extending down the south-western side of Bougainville Island (Fig. 3). There, the Solomon Sea plate dips steeply north-eastwards beneath the Pacific plate; rates of subduction are estimated to be between 6.7 cm a^{-1} (Krause, 1973) and 10.9 cm a^{-1} (Curtis, 1973). The northern margin of the South Bismarck plate is believed to be a series of transform faults and spreading axes (Fig. 2). Maximum sea-floor spreading rates (13.2 cm a^{-1}) are in the east where the Manus Basin formed during the last 3.5 Ma (Taylor, 1979).

The Indo-Australian/South Bismarck boundary is a broad zone of plate interaction, and is not adequately represented by a single line (Fig. 2). This boundary coincides with the Tertiary continent–arc collision zone, and currently is almost certainly under compression, although one plate does not appear to underthrust the other. South-westward subduction of the South Bismarck plate below Papua New Guinea has been proposed in some studies (e.g. Hamilton, 1979), but there is no recognizable submarine trench off the New Guinea north coast. Rather, a narrow chain of active andesitic volcanoes overlies a nearly vertical Benioff zone in the east.

Earthquake epicentres in south-eastern New Guinea define a zone which is narrower than that of the Indo-Australian/South Bismarck boundary. The sense of relative motion is unclear, but is possibly a combination of compressive and strike-slip components. However, vertical movements are suggested by high-angle faults and raised coral reefs in the region. The earthquake zone trends along the Woodlark Rise (Fig. 2) where, again, velocity vectors are uncertain, although there is a consensus that rifting may be taking place. Late Cainozoic sea-floor spreading is more clearly identified in the eastern Woodlark Basin (Luyendyk et al., 1973), and this spreading may connect with the rest of the Solomon Sea/Indo-Australian boundary farther west by a transform fault. The spreading axis is being subducted beneath the Solomon Islands where there is a marked shoaling of the New Britain Trench (Fig. 2).

Brooks (1969) determined a depth of c. 125 km to the low velocity zone beneath the New Guinea continental platform. Epicentres of earthquakes equal to or deeper than this are plotted in Fig. 4, together with the positions of late Cainozoic, andesitic volcanic centres. There is a reasonably close correlation between andesitic volcanism and the deeper seismicity that may be taken as potential evidence for the existence of subducted lithosphere. The correlation is also good in Bougainville Island, New Britain, and beneath the eastern end of the volcanic island chain off the north coast of New Guinea. Elsewhere the correlation is poor, particularly in the continental platform region of eastern New Guinea where no deeper earthquakes have been recorded beneath the andesitic volcanoes of the central cordillera.

Trench-like submarine troughs in northern Papua New Guinea may be late Cainozoic subduction zones that define the margins of two additional minor plates—the Caroline and North Bismarck plates (Fig. 2). These plate boundaries are very poorly defined by seismicity, and their age of inception and the extent of the presumed subduction are not known. However, they appear to be young features, and subduction may be slow and infrequent. Curtis (1973) and Krause (1973) ignored the North Bismarck and Caroline plates in their vector calculations, whereas Johnson and Molnar (1972) and Weissel and Anderson (1978) assigned them greater importance. No subaerial andesitic volcanoes are known to be related to these submarine troughs in Papua New Guinea.

Volcanic Provinces

Andesitic provinces

The late Cainozoic volcanoes of Papua New Guinea are spread throughout areas of sharply contrasting tectonic character, and clearly they are unlikely to be related to the kind of simple tectonic pattern proposed, for example, by Jakeš and White (1969). Johnson et al. (1978a) recognized seven distinct arc–trench-type volcanic provinces, each characterized by different tectonic settings and a distinctive spatial pattern of chemical variation. Six of these provinces contain andesite, and the seventh—the Tabarto-Feni Islands (north and east of New Ireland, Fig. 1)—is an alkaline province in which the only quartz-normative rocks are trachyte (see below).

Three of the six andesitic provinces constitute the 1000 km long chain of volcanoes that border the southern margin of the South Bismarck plate (Figs. 2 and 3). This chain has 10 of the 14 active volcanoes in Papua New Guinea, and is conveniently called the Bismarck volcanic arc (Carey, 1938). However, the volcanoes correspond to two (possibly three) different plate boundaries, and therefore in a strict sense two volcanic arcs should be recognized (Johnson, 1977). A western arc extends eastwards from Vokeo Island to Langila in western New Britain, and may also include the two extinct centres a little farther to the east (Fig. 1). This arc is related to tectonism at the Indo-Australian/South Bismarck plate boundary. An eastern arc is made up of the volcanoes on the central-north coast of New Britain and the Witu Islands to the northwest. These volcanoes overlie the New Britain Benioff zone which marks the descent of the Solomon Sea plate. The Rabaul caldera complex at the north-eastern tip of New Britain also overlies the New Britain Benioff zone, but it also coincides with the broad zone of interaction between the South Bismarck, Pacific, and Solomon Sea plates, and is separated from the eastern arc by an area of complex faulting. It is considered to be a separate province.

The three other andesitic provinces are: in the Highlands region of the central cordillera of New Guinea; at the south-eastern end of New Guinea, extending to the D'Entrecasteaux Islands, Woodlark Rise, and Louisiade Archipelago (Eastern Papua province); and on Bougainville Island (Fig. 1).

Non-andesitic provinces

The Tabar, Lihir, Tanga, and Feni Islands form a chain of equally spaced volcanic groups that parallels the Tertiary island arc of New Ireland, and which is in line with the andesitic volcanic chain on Bougainville Island (Fig. 1). Late Cainozoic volcanic rocks are found on all four island groups, and Tertiary ones are known in the Tabar Islands. The alkaline rocks of the Tabar-to-Feni province are mainly alkaline basalt, basanite, tephrite, phonolitic tephrite, and trachybasalt, but transitional basalt, olivine nephelinite, trachyandesite, tephritic phonolite, and clinopyroxene-rich cumulates are also represented (Johnson et al., 1976; Perfit et al., 1978; Heming, 1980). Mafic rocks are dominated by calcic augite, plagioclase, olivine, Fe–Ti oxide, and leucite, and phenocrysts of hornblende (pargasite), haüyne, phlogopite, and apatite are common in more differentiated samples. Anorthoclase, sodalite, haüyne, leucite, and analcite are present in the groundmass. K_2O/Na_2O values are mainly 0.5–1.1, and low TiO_2 contents (mainly less than 1 per cent), low Th/U, and high Zr/Nb ratios are unusual features that may represent an island arc signature. This suite of rocks is not represented elsewhere in Papua New Guinea and the tectonic significance of the volcanism is still obscure.

There are at least two other volcanic provinces in Papua New Guinea, although neither is considered to have been produced in an arc–trench-type environment. The first is a group of volcanoes in the Admiralty Islands, north of the northern margin of the South Bismarck plate (Figs. 1 and 2), and at the north-western end of a broad submarine rise that crosses the Bismarck Sea (Fig. 3). Only rocks from the St Andrew Strait area have been studied in detail, and these are dominated by alkaline rhyolite which makes up the active volcano Tuluman (Johnson and Smith, 1974; Reynolds and Best, 1976; Johnson et al., 1978b). Other rocks are transitional basalt, quartz-tholeiitic basalt, and dacite high in Na_2O/K_2O. This volcanism appears to be intraplate, and is possibly related to a mantle hot spot.

The other non-arc province is in the Dawson Strait area where comendite (peralkaline rhyolite) makes up parts of the D'Entrecasteaux Islands, including three well defined, dormant volcanoes (Fig. 1). There are also lesser amounts of basalt, benmoreite, and trachyte, and all the rocks constitute an association typical of those from oceanic islands, continental and oceanic rift systems, and other continental environments (Smith, 1976a). This volcanism appears to have taken place close to the hinge-point of opening of the Woodlark Basin (Smith et al., 1977; Fig. 2).

Present-day sea-floor spreading in the Woodlark Basin and in the ridge-transform plate boundary crossing the Bismarck Sea, is probably being accompanied by active volcanism. The character of this volcanism is unknown, however, although Luyendyk et al. (1973) reported one microprobe analysis of a low K_2O tholeiitic basalt glass from the Woodlark Basin. The rocks of these episodes of sea-floor spreading presumably constitute provinces separate from those mentioned above.

Active Andesitic Volcanism

A volcano may be defined as active if there are written reports of direct observations of an eruption. The 14 active volcanoes in Papua New Guinea have all erupted this century, and most of them during the last 25 years (Cooke and Johnson, 1980). Eight of the 14 have produced andesite historically, and Ritter and Ulawun may have too, although there are no chemical analyses of material definitely known to have been produced by recorded eruptions (Table 1).

Other volcanoes may be termed dormant. Twenty-two subaerial volcanoes mapped as such are shown in Fig. 1 because they have particularly youthful morphology or active (or recently active) fumaroles or solfataras, or there are indigenous stories (or legends) of eruptions, or unsubstantiated reports of events which may have been an eruption. Sixteen of these dormant volcanoes are andesitic.

Five of the 10, active, andesitic volcanoes—including Ritter (Table 1)—are in the western part of the Bismarck volcanic arc (Fig. 1). Four

TABLE 1 Ten Papua New Guinea volcanoes that have produced andesite historically

Volcano co-ordinates, and province	Morphology and geology	Recorded activity and principal references†
Bam 3.60°S, 144.85° E Western arc of southern Bismarck Sea	Composite, conical, island volcano c. 2200 m above sea-floor. Active summit crater. Remnant of older crater in west.	A few brief reports of activity in 19th and early 20th centuries. Mild explosive eruptions 1954–1960. Cooke and Johnson (1981).
Manam 4.10°S, 145.06°E Western arc of southern Bismarck Sea	Composite, conical, island volcano c. 3100 m above sea-floor. Two active summit craters. Four conspicuous radial valleys.	Frequent, dating back to 1616. Incandescent lava fragments and ash. Intermittent *nuées ardentes* and lava flows. Palfreyman and Cooke (1976), Cooke et al. (1976).
Karkar 4.65°S, 145.96°E Western arc of southern Bismarck Sea	Composite, conical, island volcano c. 3000 m above sea-floor. Two eccentric summit calderas. Active cone, Bagiai, on floor of inner caldera.	1643, 1895, and possibly 1885 and 1962. Major eruptive period, including lava flow, 1974–1975. New period of activity 1978. Two volcanologists killed March 1979. McKee et al. (1976).
Ritter‡ 5.52°S, 148.12°E Western arc of southern Bismarck Sea	Composite, conical, island volcano before 1888. Now a mainly submarine caldera. Remnant of older cone preserved as arcuate island and escarpment.	Several reports of activity 1700–1888. Caldera formation and explosive activity 1888. Tsunami caused heavy casualties. Submarine explosions 1972, 1974. Cooke et al. (1976).
Langila 5.52°S, 148.42°E Western arc of southern Bismarck Sea	Cluster of composite cones and craters on northern flank of extinct Talawe volcano. Two active craters.	Frequent activity since 1954. Incandescent lava fragments and ash. Infrequent lava flows from both craters. Cooke et al. (1976).
Ulawun§ 5.04°S, 151.34°E Eastern arc of southern Bismarck Sea	Composite cone 2350 m above sea level. Several active summit cones and craters. Remnant of older cone in south.	Six eruptions 1960–1978. Several known before 1960. Incandescent lava fragments and ash. Intermittent *nuées ardentes* and lava flows. Johnson et al. (1972), Melson et al. (1972), Cooke et al. (1976).
Rabaul 4.24°S, 152.21°E Rabaul province	Composite volcanic complex. Two overlapping calderas breached by sea. Peripheral extinct volcanoes. Four intra-caldera active and dormant centres.	Eruptions in 18th and 19th centuries. Tavurvur and Vulcan both active May 1937. About 500 people killed. Tavurvur activity early 1940s. Fisher (1939, 1976).
Bagana 6.14°S, 155.19°E Bougainville Island	Lava cone c. 800 m high. Active summit crater at times occupied by cumulodome.	Frequent, dating back to 1880s. Long periods of slow lava effusion. Infrequent explosive eruptions and *nuées ardentes*. Bultitude (1976, 1979), Bultitude et al. (1978).
Lamington 8.94°S, 148.17°E Eastern Papua	Composite cone c. 1600 m above base. Summit crater occupied by cumulodome.	Peléen eruption, 1951. Cumulodome and *nuées ardentes*. Almost 3000 people killed. 180 km² totally devastated. Taylor (1958).
Goropu (Waiowa) 9.57°S, 149.07°E Eastern Papua	Small cone with crater lake	Volcano formed by explosions and probable *nuées ardentes* 1943–1944. Baker (1946), Smith and Davies (1976).

† In addition to Fisher (1957) and Cooke and Johnson (1980).

‡ Ritter Island has both andesite and basalt, but products of its historical eruptions have not been identified unequivocally, and their andesitic character is therefore uncertain.

§ Ulawun is included in this table solely on the basis of one chemical analysis of an andesite bomb (53.47 per cent SiO_2) which was collected in 1969 from the surface of the vegetation-free slopes of the volcano. The bomb is obviously young, but the date of its eruption is unknown. All analysed products of Ulawun eruptions since 1970 are basalt.

of them (including Ritter) were in eruption in 1972–75, and within a single 8 month period from February to October 1974. Details of this remarkable space–time clustering of mainly andesitic volcanic activity were given by Cooke et al. (1976). Ulawun volcano was also active in 1973, and is the only active centre in the eastern arc of the southern Bismarck Sea that may have produced andesite historically (see Table 1). Pago and Lolobau produced dacite.

Four separate, active, or potentially active centres are found inside the caldera at Rabaul, but in Table 1 are grouped together as a single eruptive complex. One of them, Tavurvur, is an andesitic cone that last erupted in 1943 (Fisher, 1976). The Rabaul area is monitored by a sophisticated seismological network from which signals are transmitted to a volcanological observatory overlooking Rabaul harbour (Myers, 1976). This observatory is staffed by the Volcanological Section of the Geological Survey of Papua New Guinea, and is responsible for volcanic surveillance throughout Papua New Guinea.

Bagana is the only active andesitic volcano on Bougainville Island, and has probably been the most persistently active volcano in Papua New Guinea in modern times. Two active andesitic volcanoes are present in the Eastern Papua province—Lamington, and Goropu (or Waiowa) which was formed by activity in 1943–44. Volcanoes in the Highlands province are extinct, except for the andesitic Doma Peaks and Yelia volcanoes which retain solfataric activity (Fig. 1).

Andesitic volcanic activity in Papua New Guinea is mainly explosive—except at Bagana where eruptions are characterized by the slow effusion of pasty lava that may flow for several years, although *nuées ardentes* have also been discharged at times. Lava flows were also a major feature of the 1974–75 Karkar eruption. Lava extrusion and *nuées ardentes* have both been produced at Manam, Ulawun, and Lamington, and small *nuées* may also have been expelled from Goropu. Lava flows have been erupted at Langila.

Petrology

Andesite is here defined chemically as those rock compositions which plot to the right of the line W-Z-35-0 in Fig. 5, and which contain 53 per cent or more, and less than 63 per cent silica (Johnson *et al.*, 1978a). These rocks are hypersthene normative, and virtually all of them are also quartz normative. Andesites are part of a compositional continuum between basalt (less than 53 per cent SiO_2), dacite, and rhyolite (63 per cent SiO_2 and greater), and other

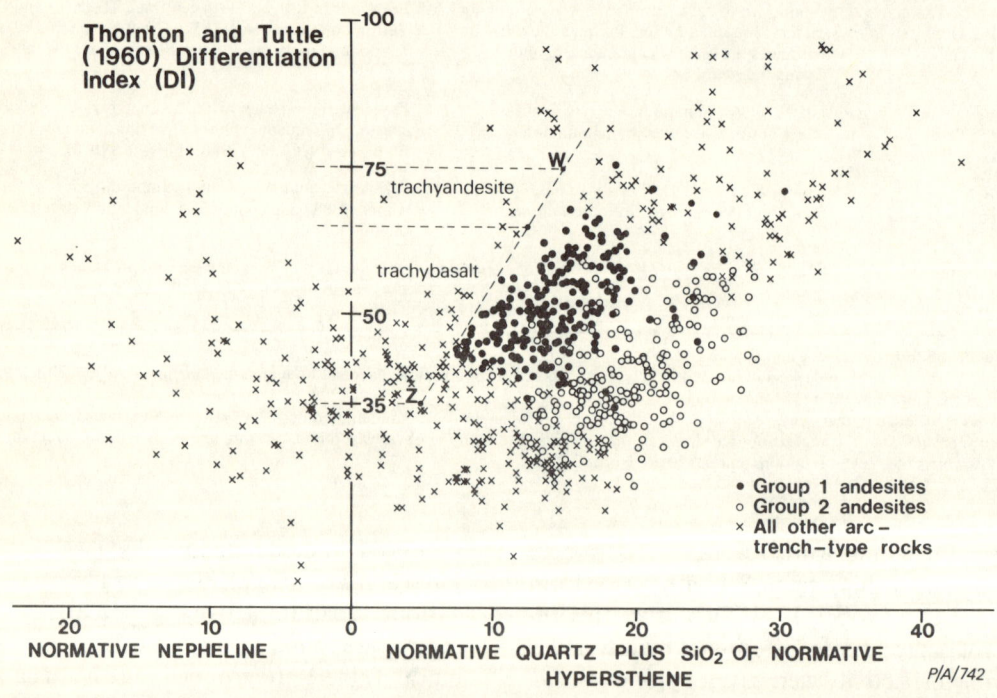

Fig. 5 Graph of differentiation index (DI) against normative-mineral abundances, adapted from Figs. 2 and 8 of Johnson *et al.* (1978a), illustrating two groups of andesite (Table 3) relative to other arc-trench-type rocks in Papua New Guinea

hypersthene-normative intermediate rocks which, because of their higher alkali contents and commonly olivine-normative character, are more appropriately termed trachybasalt and trachyandesite (Fig. 5). The compositional boundaries between these rock types and andesite are arbitrary.

Bultitude et al. (1978) proposed that the average of 31 Bagana andesite analyses (Table 2, column 1) is a suitable reference for the purposes of comparing Papua New Guinea andesites, which have a particularly wide range of compositions. The major element values of this reference composition are close to the average values for 868 arc–trench-type rocks given by Johnson et al. (1978a). The reference composition has the familiar features of typical 'calc-alkaline' andesite. Its TiO_2 content is less than 1 per cent, the Al_2O_3 content is high, there is no marked enrichment of Fe relative to Mg, the Mg number is much lower than those normally considered applicable to primary, mantle-derived magmas, CaO is quite high for intermediate rocks, and the Na_2O content exceeds that for K_2O, although K_2O is not as low as in many so-called 'tholeiitic series' rocks. These major element characteristics are also shown by the average of 453 Papua New Guinea andesites given in Table 3. K and other large-ion trace elements are also moderately high in the Bagana reference composition, and light REE are moderately enriched relative to heavy REE (Fig. 6). Ni and Cr are characteristically low (less than 10 p.p.m.), and much lower than those likely to be appropriate for primary magmas derived from the mantle.

Two broad groups of andesite may be distinguished in Papua New Guinea (Fig. 5; Table 3). These correspond to two of three volcanic rock associations that have been identified for arc–trench-type rocks in the region (Johnson et al., 1978a). Group 1 andesites are found in an association made up of the Highlands, Eastern Papua, Rabaul, and Bougainville provinces. Group 2 andesites are those in the western and eastern arcs at the southern margin of the Bismarck Sea (that is, the Bismarck volcanic arc, excluding Rabaul province). The third association is the andesite-free, alkaline province of the Tabar-to-Feni Islands.

There is considerable overlap between the two groups of andesite, although on some variation

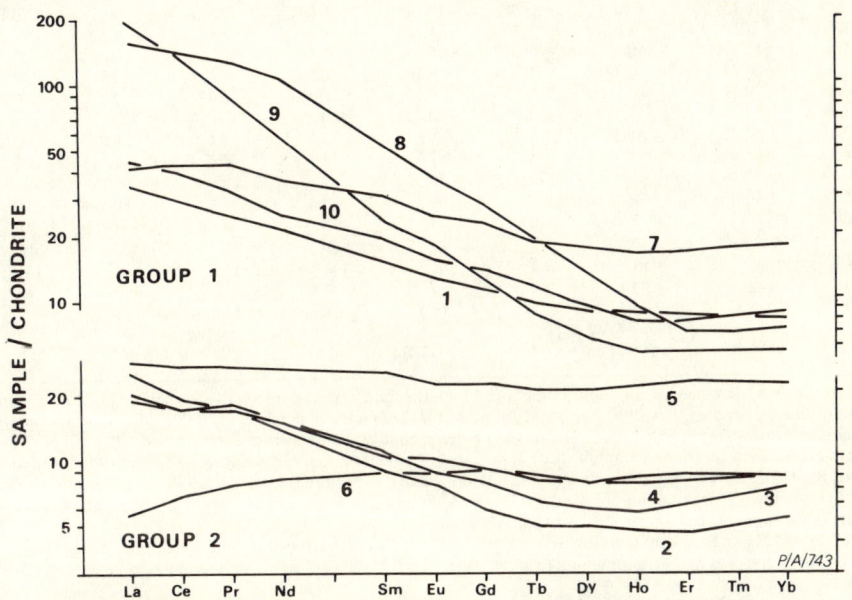

Fig. 6 Chondrite-normalized REE patterns of 10 andesites whose analyses are listed in Table 2. Normalizing values are those given by Taylor and Gorton (1977)

III Regional distribution and characteristics

TABLE 2 Chemical analyses of 10 Papua New Guinea andesites

	1	2	3	4	5	6	7	8	9	10
SiO_2	55.5	54.9	61.7	53.8	57.0	55.4	61.95	55.68	59.6	5.55
TiO_2	0.81	0.38	0.27	0.56	0.62	1.38	0.83	0.84	0.65	0.82
Al_2O_3	17.7	16.5	17.2	19.6	19.8	16.4	16.31	14.29	17.3	16.8
Fe_2O_3	4.38	4.80	3.30	2.45	1.78	3.00	1.91	2.58	3.45	2.15
FeO	4.00	4.25	2.45	6.30	5.30	6.10	4.28	3.86	1.74	4.67
MnO	0.18	0.16	0.13	0.16	0.13	0.20	0.17	0.10	0.10	0.13
MgO	3.38	5.40	3.20	3.15	2.40	4.35	2.20	6.97	2.75	6.22
CaO	8.01	9.40	6.80	10.3	8.90	8.10	5.24	8.15	5.69	6.40
Na_2O	3.95	2.25	3.35	2.50	2.70	3.80	4.39	3.37	3.65	3.57
K_2O	1.61	1.36	1.20	0.88	0.45	0.53	2.24	2.54	3.25	1.39
P_2O_5	0.35	0.19	0.16	0.16	0.09	0.17	0.31	0.49	0.39	0.30
H_2O^+	—	0.36	0.15	<0.01	0.56	<0.01	0.19 ⎫		1.10	2.00
H_2O^-	—	0.10	0.05	0.06	0.25	0.16	0.07 ⎬ loss = 0.31		—	—
CO_2	—	0.04	0.09	0.05	<0.05	0.10	0.12 ⎭		0.02	—
Total	—	100.09	100.05	99.97	99.98	99.69	100.21	99.18	99.69	99.95
Mg no†	47	57	55	44	42	51	44	70	54	66
Q†	3.01	6.75	16.35	6.63	13.85	5.93	12.11	0.97	9.14	4.50
Cs	0.26	0.51	0.34	1.8	0.18	0.11	1.0	1.5	5.4	0.97
Rb‡	25	18.6	15.8	16.6	7.4	7.6	32.0	—	92	30.0
Ba‡	250	145	355	980	85	95	370	1150	915	500
Pb	5‡	4.7	6.8	13	3.1	2.0	6.7	24	24	12
Sr‡	810	720	740	475	239	210	381	980	917	610
La	11	8‡	6.4	6.1	1.8	9‡	13	50	61	14
Ce	24	16‡	14	14	5.6	22	35	(120)	110	32
Pr	2.9	2.0	2.1	2.0	0.90	3.2	5.0	15	10	3.8
Nd	13	8.3	9.1	9.0	5.0	16	22	63	34	15
Sm	2.9	1.7	2.1	2.0	1.7	4.9	5.9	10	4.5	3.8
Eu	0.93	0.55	0.63	0.74	0.63	1.6	1.8	2.6	1.3	1.1
Gd	2.9	1.5	2.0	(2.4)	2.3	5.7	5.8	7.1	(3.4)	3.6
Tb	0.47	0.24	(0.31)	0.39	0.40	1.0	0.89	(1.0)	0.42	0.57
Dy	2.9	1.6	(1.9)	2.5	2.5	6.7	5.6	(5.0)	2.2	3.0
Ho	0.64	0.34	0.41	0.56	0.60	(1.6)	1.2	0.68	0.42	0.57
Er	1.8	0.97	1.3	1.7	1.8	4.8	3.6	1.5	1.2	1.7
Tm	0.25	(0.15)	(0.20)	(0.24)	(0.24)	(0.66)	(0.52)	0.21	(0.17)	(0.25)
Yb	1.7	1.1	1.5	1.7	1.7	4.5	3.7	1.5	1.2	1.8
Y‡	17	11	10	15	15	37	32	19§	22	19
Th	1.3	0.76	0.87	1.3	0.15	0.80	1.4	12	22	3.6
U	0.55	0.47	0.70	0.52	0.17	0.39	0.91	2.8	6.0	1.2
Zr‡	81	30	37	34	50	130	133	220	245	130
Hf	2.2	0.72	1.2	1.2	1.3	3.6	3.4	6.8	5.9	3.7
Nb‡	4	<1	<1	1	<1	4	2	5.5§	35.5	5.4§
Zn‡	74	63	48	77	59	87	83	—	—	—
Cu‡	46	40	99	169	100	25	53	80	60	33
Ni‡	6	14	9	10	4	18	2	54	19	119
Sc‡	16	31	16	25	22	24	16	24	15	18
V‡	178	269	161	263	161	194	110	86	131	169
Cr‡	9	35	22	14	9	22	—	300	60	237
Ga‡	19.5	13.5	13.5	17.0	15.5	18.5	17.0	—	—	—
$^{87}Sr/^{86}Sr$‖	—	0.7034	—	—	0.7034	—	—	—	0.7045	0.7044

(1) Bagana reference andesite composition (Bultitude et al., 1978). (2) Bam andesite, 18NG1010. Analysis previously unpublished, except for $^{87}Sr/^{86}Sr$ value. (3) Boisa andesite, 19NG0959 (Gust and Johnson, 1981). (4) Karkar andesite, 74710001. Major element analysis from McKee et al. (1976). Other data previously unpublished. (5) Bamus andesite, 51NG0195X. Data from Johnson and Chappell (1979), except for spark source mass spectrographic data (previously unpublished) and $^{87}Sr/^{86}Sr$ value. (6) Garove andesite, 48NG0559 (Johnson and Arculus, 1978). (7) Rabaul andesite, RABAUL 5. Analysis previously unpublished. (8) Victory andesite, 6514. Data from Jakeš and Smith (1970), except for spark source mass spectrography data (previously unpublished). (9) Doma Peaks andesite, H0212. Analysis previously unpublished. (10) Hagen andesite, H1607. Analysis previously unpublished, except for $^{87}Sr/^{86}Sr$ value.

† Mg number (100 Mg/[Mg + Fe^2]) and Q (normative quartz) calculated using FeO = 0.85 total Fe as FeO.

‡ Determined by X-ray fluorescence spectrometry. All other trace element data by spark source mass spectrography. Interpolated values shown in parenthesis.

§ Determined by spark source mass spectrography.

‖ Data from Page and Johnson (1974) and D. E. Mackenzie (personal communication).

TABLE 3 Average major element analyses of Papua New Guinea andesites

	Group 1					Group 2			Both Groups
	Highlands	East Papua	Rabaul	Bougainville	All	West	East	Both	
SiO_2	56.70	57.72	59.80	55.89	57.09	56.03	56.81	56.44	56.79
TiO_2	0.86	1.01	0.95	0.78	0.90	0.53	0.60	0.57	0.75
Al_2O_3	17.22	16.12	15.66	17.45	16.78	16.56	16.83	16.70	16.74
Fe_2O_3	3.00	2.73	2.48	4.38	3.10	3.50	3.17	3.32	3.20
FeO	3.90	3.19	4.05	3.64	3.60	5.22	5.11	5.16	4.33
MnO	0.14	0.11	0.19	0.17	0.13	0.16	0.15	0.16	0.14
MgO	4.02	4.91	2.31	3.33	4.16	4.50	4.20	4.34	4.25
CaO	6.90	6.33	5.16	7.84	6.76	8.76	8.47	8.61	7.62
Na_2O	3.52	3.83	3.87	3.95	3.72	2.61	2.77	2.69	3.24
K_2O	2.09	2.26	2.19	1.62	2.08	1.26	0.76	1.00	1.58
P_2O_5	0.43	0.41	0.27	0.34	0.40	0.20	0.12	0.16	0.29
n	102	90	11	39	242	100	111	211	453

n = number of chemical analyses.
Data are the same as used by Johnson et al. (1978a), except for the Bougainville average in which analysis 2 of Blake and Miezitis (1967) is eliminated (late Cainozoic age of rock now considered doubtful), and in which analysis 16 of Blake and Miezitis (1967) is replaced by the new one given by Bultitude et al. (1978).

diagrams (e.g. Fig. 5) they are reasonably well polarized. The two groups do not correspond closely with the classification into tholeiitic, calc-alkaline, and shoshonitic associations used widely in other regions. Tholeiitic-type andesites are found in group 2, and shoshonitic-type andesites are found in group 1, but calc-alkaline andesites are present in both groups. The concept of these three world-wide associations is not readily applicable to Papua New Guinea.

A major difference between the two groups of andesite is that group 1 generally has greater abundances of alkaline and incompatible elements. This may be seen in Table 3, where the average of 242 group 1 andesite analyses has higher TiO_2, Na_2O, K_2O, and P_2O_5 than does the average for 211 group 2 andesites. SiO_2, Al_2O_3, and MgO are similar comparing both averages, but total Fe oxides and CaO values are both lower in the group 1 andesites. For the trace element values listed in Table 2 for five selected andesites from each group, there is a bias towards higher incompatible element abundances in the group 1 rocks. Strongly fractionated, light REE-enriched patterns characterize the group 1 andesites, whereas many group 2 rocks are less light REE enriched and some are light REE depleted (Fig. 6). Group 1 andesites also have generally higher initial $^{87}Sr/^{86}Sr$ values—mainly in the range 0.7039–0.7049, in contrast to group 2 values which are mainly 0.7034–0.7037 (Fig. 7).

The phenocryst and groundmass mineralogy of Papua New Guinea andesites is dominated by plagioclase, clinopyroxene, orthopyroxene, and Fe–Ti oxides. Some olivine phenocrysts are present in mafic andesites, and may be rimmed by low Ca pyroxene, particularly in samples where the groundmass is well crystallized. Hornblende phenocrysts are found in many group 1 andesites, but are surprisingly uncommon in those of group 2. Rare biotite is known in some group 1 andesites, but is more common in trachybasalts and trachyandesites. High temperature polymorphs of silica are common in the groundmass of more coarsely crystalline rocks. Alkaline feldspar is also present in group 1 andesites—in some, as rims on plagioclase phenocrysts although, like biotite, it is more commonly seen in rocks richer in alkalis. Apatite is a common accessory phase, particularly in group 1 andesites.

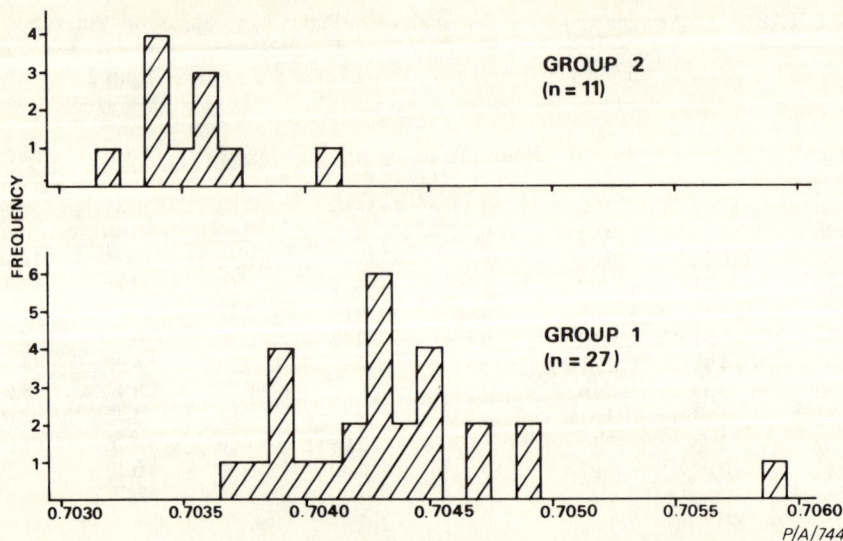

Fig. 7 Histograms of $^{87}Sr/^{86}Sr$ values for the two groups of Papua New Guinea andesites. Data from Peterman et al. (1970), Page and Johnson (1974), Peterman and Heming (1974), and DePaolo and Johnson (1979), plus unpublished data of D. E. Mackenzie and I. E. M. Smith (personal communications)

The character of the volcanism in each of the six andesitic provinces in Papua New Guinea is summarized below, and broad scale petrological and geological features are stressed. However, regional considerations alone do not take into account important aspects of andesitic volcanism which are observed only on the scale of individual volcanoes. Ten volcanoes have therefore been selected to illustrate some particular aspect of andesite geology. Any one of these examples is not necessarily representative of all the volcanoes in each province, but all 10 together illustrate the wide range of compositions, structure, stratigraphy, and morphology of andesitic volcanoes throughout Papua New Guinea. A chemical analysis of an andesite from each of the volcanoes is given in Table 2.

Western Arc of the South Bismarck Sea

The narrow chain of andesitic volcanoes off the north coast of New Guinea marks the northern edge of the broad zone of convergence between the Indo-Australian and South Bismarck plates (Figs. 1 and 2) that formed as a result of the collision of the Australian continent with the Tertiary island arc now represented by the north coast ranges (Fig. 3). A north-eastward-dipping subduction zone is believed to have existed on the south-western side of the island arc, but this was destroyed by the collision, and the underlying slab of subducted lithosphere is thought to be hanging almost vertically beneath the north coast ranges at the present day (Johnson, 1977). This slab is believed to be providing the requisite conditions for andesite formation in an arc that lacks a submarine trench.

Andesite is known on all the volcanoes of the chain mapped in Fig. 1. Tholeiitic basalt is also common, but dacite is rare, and rhyolite is unknown; basalt has yet to be found on any of the five volcanoes in the Schouten Islands (Vokeo to Bam) at the north-western end of the chain (Fig. 1). A striking petrological feature is the change in chemical compositions of the andesites along the chain from one group of volcanoes to the next, comparing rocks with similar silica contents. Alkali contents, for example, increase from the Vokeo–Bam group in the north-west to maximum values in Long Island andesites, and decrease farther eastwards (Johnson, 1977).

Another feature is that the volume of volcanic rocks appears to increase generally south-eastwards (note, for example, the closer grouping of volcanic centres south-east of Long Island in Fig. 1), although two of the largest volcanoes—Manam and Karkar—are in the north-west. The most easterly volcanoes overlie intermediate focus earthquakes defining a nearly vertical Benioff zone, but to the north-west the Benioff zone is absent (Fig. 4).

These features may be accounted for by a tectonic model in which sea-floor between the continent and arc in the Pliocene was triangular in plan, widening south-eastwards. This area may have closed during the late Cainozoic in scissor-like fashion about a point near the western corner of the triangle, and subduction may have taken place north-eastwards beneath the arc at rates greater in the south-east than in the north-west (Johnson, 1977; Jaques and Robinson, 1977; Johnson and Jaques, 1980). Thus, earthquakes defining the Benioff zone are found in the south-east, but not in the north-west (Fig. 4) where rates of subduction were so low that the downgoing slab heated up to temperatures above those that would normally favour the production of intermediate depth earthquakes. Volcanic rock compositions are different along the arc because thermal regimes are different beneath different parts of the arc, and volumes of erupted magmas appear to be greater in the south-east possibly because the downgoing slab there has had greater opportunity to introduce greater amounts of slab-derived fluids to the overlying mantle.

Bam is a simple, composite cone about 3 km wide at sea level. It is the westernmost of three active, andesitic volcanoes (Manam and Karkar are the other two) which are associated with neither a Benioff zone (Fig. 4) nor a submarine trench, although the existence of an underlying slab of lithosphere is implied if the tectonic model given above is accepted. Another feature of interest is the contrast in geology and petrology between Bam and other volcanoes of the Schouten Islands. The islands are offset northwards from the main trend of the western arc, and appear to be cut by the western part of the ridge-transform plate boundary that crosses the Bismarck Sea (Fig. 2). Bam and two other volcanoes south-east of this plate boundary are younger, and their andesites have different compositions (e.g. higher K_2O and P_2O_5), compared to the volcanoes to the north-west (Johnson, 1977). The tectonic significance of these differences is still obscure, although evidently the volcanism is controlled in some way by the ridge-transform tectonics of the northern margin of the South Bismarck plate. Compared to the Bagana reference composition, the Bam andesite (Table 2, column 2) is generally poorer in incompatiable elements, and has a low $^{87}Sr/^{86}Sr$ ratio.

Boisa is another small (1.7 km wide at sea level), inactive volcano. It is overshadowed by Manam (Fig. 1), a much larger and frequently active, andesitic volcano (Table 1), but is of interest here on account of its amphibole-bearing inclusions, and because of a well defined temporal change in andesite compositions: low silica andesites are overlain by high silica andesites. Mineral compositions determined by electron microprobe were used in least-squares mixing calculations to examine the compositional relationships between the older and younger (Table 2, column 3) andesites and the cumulates on Boisa (Gust and Johnson, 1981). Closer matches between the two lava types were obtained using both phenocryst compositions and the compositions of minerals in the amphibole-bearing inclusions.

Karkar is one of the largest andesitic volcanoes in the Bismarck volcanic arc. The island has a maximum diameter of *c.* 25 km, and is a composite cone truncated by a double caldera 5.5 km wide. An inner caldera, 3.2 km in diameter, contains the active volcano Bagiai which produced a major andesitic eruption in 1974–75, and which started a new active phase in July 1978. About 1.7×10^8 m^3 of low silica andesite (Table 2, column 4) was extruded from Bagiai in 1974–75 producing a compound sheet up to *c.* 50 m thick on the floor of the inner caldera (McKee *et al.*, 1976). A systematic change in andesite compositions took place during the eruption: plagioclase-rich lava was extruded during the first few months of 1974, but thereafter lavas had fewer plagioclase phenocrysts.

Eastern Arc of the South Bismarck Sea

Andesite genesis in the eastern arc is apparently related to subduction of the Solomon Sea plate northwards beneath the South Bismarck plate (Fig. 2), but the complex distribution pattern of volcanic centres shown in Fig. 1 is suggestive that the tectonic configuration is more complicated than simply one plate underthrusting another (Johnson, 1977). Willaumez Peninsula in particular is a curious feature (Fig. 3), consisting of a 60 km long chain of andesitic volcanoes extending at right angles from the northern coast of New Britain (Lowder and Carmichael, 1970). Volcanoes are numerous east of the southern end of the peninsula, but are strikingly absent west of it—at least, as far west as the extinct volcanoes at the western end of New Britain (Fig. 1). The Witu Islands are an isolated group of volcanoes north of western New Britain. They rest on the south-western end of the submarine rise on the Bismarck Sea floor (Fig. 3), and overlie the deepest part (deeper than 300 km) of the New Britain Benioff zone, but there are no equivalent volcanoes to the east in the southern Manus Basin (Figs. 1 and 2). This anomalous distribution of volcanoes may be caused by a tectonic pattern related to the splitting of the New Britain trench south of western New Britain (a thrust slice may exist under western, but not eastern New Britain; Fig. 2), and to the formation of the submarine rise on the Bismarck Sea floor (Johnson, 1977; Johnson and Arculus, 1978; Johnson et al., 1979).

Andesite is at least as common in the eastern arc as in the western arc, but unlike the western arc, dacite and rhyolite are well represented and basalt is apparently less common (Johnson and Chappell, 1979). The average major element compositions of andesites in the western and eastern arcs are similar, except that K_2O and P_2O_5 are lower in the eastern arc average (Table 3). Differences in andesite composition in the eastern arc volcanoes correspond to different depths, h, to the Benioff zone. In general, alkaline and incompatible elements are more abundant in andesites over the deeper parts of the Benioff zone, whereas CaO, MgO, and Al_2O_3 values, for example, are lower (comparing rocks with similar SiO_2 contents). There are some notable exceptions, however—including K_2O which, in rocks containing more than c. 60 per cent SiO_2, has progressively *lower* values as h increases to depths greater than c. 200 km, and which does not in general reach the high values found in andesites over the deep parts of other Benioff zones (Johnson, 1976). There are no detectable differences in $^{87}Sr/^{86}Sr$ and $^{143}Nd/^{144}Nd$ across the New Britain Benioff zone (Page and Johnson, 1974; DePaolo and Johnson, 1979).

Bamus and Ulawun are two andesitic centres that tower above other volcanoes along the central-north coast of New Britain, reaching over 2200 m above sea level (Fig. 1). Ulawun is the better known one because of its frequent volcanic activity (Table 1), but Bamus is potentially active, and may have erupted in the late 19th century. Like Boisa, the younger andesites of Bamus are more SiO_2-rich than the older ones, but the main feature of interest is the contrast in rock compositions between Bamus and neighbouring Ulawun. Most Bamus andesites are a little more felsic than the rocks of Ulawun, but this slight difference in degree of fractionation is inadequate to explain, for example, Zr being twice as abundant in Bamus andesites. Both volcanoes have grown side by side throughout the Quaternary; magma conduits were apparently unconnected, and andesitic magmas were derived independently at each volcano. Bamus andesites are characterized by incompatible element contents lower than those of the Bagana reference composition, but appropriate for a volcano over the shallower part of a Benioff zone (Table 2, column 5; Fig. 6).

Garove is the largest of the Witu Islands (Fig. 1). Its rocks range from basalt through andesite and dacite to rhyolite. The andesites do not seem to be typical of those found in island arcs—or at least, those expected above the deeper parts of Benioff zones. For example, in the Garove andesite whose analysis is given in Table 2, column 6, TiO_2 is high for a normal island-arc rock, most incompatible element contents are low, the REE pattern is almost chondritic (Fig. 6), and there is no enrichment of elements such as Ba relative to K and light REE (as is commonly observed in other andesites). Other rocks in the Witu Islands appear to have a stronger island arc signature (for example the andesites of Unea), but some basalts more closely resemble those of back-arc basins. The problem of determining an island arc origin for the andesites and related rocks of the Witu Islands is dealt with elsewhere (Johnson and Arculus, 1978).

Rabaul

The Rabaul volcanic complex consists of two overlapping calderas, breached by the sea in the south-east, rimmed by pre-caldera volcanoes on its north-eastern side, and characterized by a dissected mantle of pyroclastic flow and air-fall deposits to the west and south (Fisher, 1939; Heming and Carmichael, 1973; Heming, 1974). Heming (1974) reported two periods of cauldron subsidence on the basis of ^{14}C dates (c. 1400 and 3500 a ago). Two main centres of post-caldera volcanism are Tavurvur and Vulcan (Table 1), and two outlying andesitic centres are Varzin to the south-south-west and Watom Island to the north-west. Epicentres of local earthquakes plot in an elliptical zone (Myers, 1976) that may outline the boundaries of a still active cauldron block.

The caldera complex at Rabaul is one of five major calderas in Papua New Guinea (Cooke and Johnson, 1980), and is an excellent example of the geological complexity to be found at larger andesitic centres in Papua New Guinea, compared to relatively simple volcanoes such as Bam and Boisa. Rocks range in composition from basalt through to rhyolite, and there are apparently no simple stratigraphic changes in composition. Indeed, according to the results of Heming (1974), both andesite and dacite were produced virtually simultaneously in May 1937 when Tavurvur (andesitic) and Vulcan (dacitic) were in eruption (Table 1).

The chemical compositions of andesites from Rabaul (Table 2, column 7) more closely resemble those of the Highlands, Eastern Papua, and Bougainville provinces than do the group 2 andesites of the southern Bismarck Sea, particularly with regard to alkaline and incompatible element abundances (Tables 2 and 3) and $^{87}Sr/^{86}Sr$ (Fig. 7). There are some similarities with rocks from the Witu Islands in the eastern arc (Johnson, 1977), but the Benioff zone beneath Rabaul is less than 300 km deep, whereas it is deeper than 300 km beneath the Witu Islands, and the tectonic setting of both volcanic areas is quite different. The average SiO_2 content of Rabaul andesites given in Table 3 is the highest of all those for the six provinces, and higher than that of the Bagana reference andesite. However, this may be due to disproportionate sampling of the relatively small number (11) of analysed samples from Rabaul.

Bougainville Island

Two groups of late Cainozoic volcanoes are present on Bougainville Island, both of which are considered to be related to present-day subduction of the Solomon Sea plate north-eastwards beneath the Pacific plate (Figs. 1 and 2). A more numerous, northerly group of eruptive centres is separated from a southern group by an area of mid-Tertiary rocks intruded by high level dioritic bodies, one of which—at Panguna Mine—has rich, porphyry-copper-type mineralization (e.g. Blake and Miezitis, 1967; Baldwin et al., 1978). Reversal of arc polarity may have taken place beneath the Solomon Islands (including Bougainville) in the Tertiary (e.g. Kroenke, 1972), and there is the possibility that the deeper earthquakes beneath Bougainville represent a detached piece of lithosphere that formerly was part of a south-west-dipping slab (Halunen and Von Herzen, 1973). The mid-Tertiary arc–trench-type rocks may relate to this earlier subduction system whose polarity was opposite to that of the present-day island arc (Mitchell and Garson, 1972). The northern group of late Cainozoic volcanoes includes Bagana, the active volcano (Table 1), and Balbi, which has an extensive thermal area (Fisher, 1957). The southern group consists of two, north-west-trending, parallel chains of eruptive centres, including Loloru, which is thought to be dormant (Fig. 1).

Chemical analyses of Bougainville volcanic rocks are relatively few, and the only published data for andesites, other than from Bagana, are those for one andesite from each of four volcanoes (Blake and Miezitis, 1967; Taylor et al., 1969). Average major element values for 39 Bougainville andesites in Table 3 are similar in most respects to the average of all group 1 andesites, although, in particular, TiO_2 and K_2O values are a little low, and CaO and total Fe values are quite high. Other analysed late Cainozoic volcanic rocks from Bougainville are basalt, dacite, trachybasalt, and trachyandesite.

Bagana is the most fully documented of all the Bougainville volcanoes (Bultitude, 1976, 1979; Bultitude et al., 1978). It is an 800 m high andesitic cone that consists of thick, blocky lava flows and minor volcaniclastic debris. Volcanic activity has been persistent during the past 30 years or so, but activity before 1947 is poorly known. Many of the lava flows are well dated, and they can be divided into three stratigraphic groups that may be distinguished chemically. Lavas of a pre-1943 group are the most fractionated; those of a 1943–53 group are the most mafic; and a compositionally intermediate, but younger, 1959–75 group of lavas have K, Rb, and Ba contents which are similar to, or lower than, those of the 1943–53 group. The three groups appear to represent distinct batches of andesite magma that were erupted successively from Bagana, possibly from a high level reservoir that was periodically emptied and refilled (Bultitude et al., 1978).

Eastern Papua

Late Cainozoic volcanic rocks are widespread throughout eastern Papua and the D'Entrecasteaux Islands, and they constitute a province that may also include islands on the Woodlark Rise and in the Louisiade Archipelago, although the rocks there are probably Tertiary (Fig. 1; Smith, 1976b). There are only four major eruptive centres. All four are andesitic, and they include Lamington, the active centre (Table 1), and Victory, a dormant one (Figs. 1 and 4). Other eruptive centres in the province are minor cones, most of them probably less than 100 m high—for example the active volcano Goropu (Table 1).

Two belts of volcanism may be distinguished in this province (Johnson et al., 1978c). Andesite is common in a northern one that includes most of the eruptive centres shown in Fig. 1. A southern belt is still poorly studied, but appears to be dominantly basaltic. Rocks of the entire province are alkaline basalt and trachybasalt, together with tholeiitic basalt, andesite, dacite, and rhyolite, which are characterized by greater abundances of alkaline and incompatible elements, and higher $^{87}Sr/^{86}Sr$ values, than are the basalt-to-rhyolite rocks of the south Bismarck Sea (Tables 2 and 3; Fig. 7). Trachytes characterized by extreme REE fractionation are found in the Lusancay Islands (Fig. 1), but their relationship to the andesitic volcanism farther south is uncertain (Smith et al., 1979).

The province as a whole coincides roughly with the Indo-Australian/Solomon Sea plate boundary (Fig. 2), and the late Cainozoic history of that part of the boundary represented in south-eastern New Guinea appears to have been dominated by vertical movements. Rare, intermediate focus earthquakes are known in the area, but there is little evidence that the andesitic volcanoes are underlaid by a lithospheric slab (see, however, Hamilton, 1979). A speculative interpretation is that, like the Highlands province (see below), the late Cainozoic andesites may have been derived from primary magmas formed by partial melting of mantle sources that were chemically modified by subduction processes much earlier than the late Cainozoic.

Victory and Trafalgar are two, major, coalesced andesitic volcanoes that form a peninsula extending into the Solomon Sea (Figs. 1–3). Trafalgar is extinct, but Victory appears to have been in eruption in the late 19th century (Fisher, 1957). Twin andesitic centres are known elsewhere in Papua New Guinea—for example the double caldera at Rabaul (*see* above), and Duau/Favenc and Yelia/Marble Peak in the Highlands (Fig. 1)—but the significance (if any) of the pairing is unclear. The analysis of a Victory andesite in Table 2, column 8, is an example of the strong enrichment in incompatible elements found in eastern Papua andesites (compared with the Bagana reference andesite), and of the high Mg numbers, and high Ni and Cr contents, that appear to characterize many of them (Jakeš and Smith, 1970; Smith, 1976*b*).

Highlands

Andesite is found on most, and possibly on all of the 19 major volcanic centres in the Highlands province, but its overall volume is probably less than that of basalt (D. E. Mackenzie, personal communication). Seventeen volcanoes overlie the northern edge of the Australian continent in a region of late Pliocene uplift and intense faulting, whereas the two most easterly volcanoes—Yelia and Marble Peak—overlie the orogenic belt (Figs. 1 and 3; Mackenzie, 1976; Johnson *et al.*, 1978*c*). Intermediate depth earthquakes (deeper than 125 km) have not been recorded beneath the continental part of the Highlands province, but some beneath the two eastern volcanoes (Fig. 4) define a Benioff zone that appears to dip south-westwards from the north-western part of the Solomon Sea (Dent, 1976).

Highlands rocks are alkaline basalt, trachybasalt, trachyandesite, tholeiitic basalt, andesite, and rare dacite. Dacite found in the Aird Hills (Fig. 1) has extreme REE fractionation similar to that of the Lusancay Islands trachytes (Smith *et al.*, 1979). Like other group 1 andesites, those of the Highlands province are characterized by alkaline and incompatible element enrichment, and by higher $^{87}Sr/^{86}Sr$ values than those of group 2 (Tables 2 and 3). Rocks nearer the centre of maximum Pliocene uplift are especially enriched—for example those of Hagen (Table 2, column 10)—but so too are some in the west— including Doma Peaks (Table 2, column 9).

Andesite formation beneath the continental part of the Highlands province does not appear to be related to a contemporaneous downgoing slab of lithosphere. Rather, volcanism may have been triggered by the Pliocene uplift, which caused partial melting of mantle sources that may have been modified chemically by Cretaceous subduction (Johnson *et al.*, 1978*c*). Cretaceous arc–trench-type rocks are known in the orogenic belt to the north, and are thought to be related to a subduction system that dipped southwards beneath the continental margin. A Cretaceous pseudo-isochron has been identified in the most mafic rocks of the late Cainozoic Highlands province (Johnson *et al.*, 1978*c*). This interpretation does not necessarily apply to Yelia and Marble Peak in the east, which overlie a Benioff zone, but it may apply to the andesitic volcanism of the Eastern Papua province, although evidence for an earlier period of subduction is not as clear there as in the Highlands.

Hagen is one of the largest of the Highlands volcanic centres. It is a complex of three coalesced cones forming a range that rises *c*. 2100 m above the surrounding plains (Mackenzie, 1973). The western and northern sides of the complex are much more deeply dissected than elsewhere, owing at least in part to higher rainfall there (Ollier and Mackenzie, 1974), and a Pleistocene ice cap covered about 21 km² of the summit, forming cirques and moraines (Löffler, 1972). D. E. Mackenzie (personal communication) reported that, in common with other volcanoes in the Highlands province, andesites generally overlie more mafic rocks. However, on Hagen the andesites have *lower* contents of several incompatible elements than the older basalts, and cannot have been derived from them by removal of the phenocrysts found in the basalts.

Doma Peaks is one of only two dormant volcanoes in the Highlands province (Fig. 1). It retains cold solfataric activity, and there are stories among the local people of eruption-like phenomena some generations ago, although Blong (1979) believed these relate to ash-fall from Long Island in the Bismarck volcanic arc. Doma Peaks volcano is strongly asymmetric, and is dominated by westward-facing escarpments overlooking a rugged, dissected area that slopes abruptly to an almost flat apron farther west. Much of the apron may be made up of volcaniclastic materials scoured out from the central area of Doma Peaks, including those of lahars. The andesites of Doma Peaks are characterized by extreme enrichment in incompatible elements compared to others in Papua New Guinea (Table 2). The REE pattern of sample 9 (Table 2) is strongly fractionated, and crosses over the pattern for the Bagana reference andesite, reaching heavy REE abundances only five to six times those of chondrites (Fig. 6).

Conclusion

Late Cainozoic andesitic volcanism in Papua New Guinea has taken place in six tectonically independent provinces—and possibly in more if outlying centres of volcanism such as Yelia and Marble Peak (Highlands) are regarded as separate, minor provinces. Conditions of andesite magma genesis are probably different in each province, but some generalizations can be made.

Papua New Guinea andesites are unlikely to represent primary magmas from the upper mantle, as their Mg numbers and Ni and Cr contents are in general too low to have been produced in equilibrium with Mg-rich mantle olivine. Evidently they represent fractionated magmas, and have lost olivine in particular. The primary magmas may have been basaltic. Basalt is represented in all six andesitic provinces (although apparently it is rare on Bougainville Island), and in several volcanoes there is a temporal relationship, andesitic rocks overlying more mafic ones which are closer in composition to those of primary magmas. Exceptions to these generalizations were noted above: some andesites in the Eastern Papua province have higher Mg numbers and higher Ni and Cr contents, and their basaltic parentage is therefore open to question; on Hagen the younger andesites do not appear to be related by fractionation to the older basalts.

Andesite production is possibly related, in one way or another, to influxes of water and alkaline and incompatible elements from downgoing lithospheric slabs. Nevertheless, there is an incomplete correlation between late Cainozoic volcanism and Benioff zones and submarine trenches. The correlation is good for New Britain and Bougainville Island, but a submarine trench is apparently missing in the western chain of the Bismarck volcanic arc, and a Benioff zone is developed only beneath the eastern half of the chain. Most andesitic volcanism in the Highlands and Eastern Papua provinces does not seem to relate to contemporaneous subduction. If true, this conclusion is important, for andesite in any one area can no longer be taken as geological evidence that the area was necessarily underlain by a subduction zone at the time of andesite eruption.

Acknowledgements

D. E. Mackenzie, I. E. M. Smith, and the late R. J. S. Cooke generously gave free access to unpublished information. Some of the trace element data are from studies currently under way in conjunction with B. W. Chappell and S. R. Taylor. This review is published with the permission of the Director of the Bureau of Mineral Resources.

REFERENCES

Baker, G. (1946). Preliminary note on volcanic eruptions in the Goropu Mountains, southeastern Papua, during the period December, 1943, to August, 1944. *J. Geol.* **54**, 19–31.

Baldwin, J. T., Swain, H. D., and Clark, G. H. (1978). Geology and grade distribution of the Panguna porphyry copper deposit, Bougainville, Papua New Guinea. *Econ. Geol.* **73**, 690–702.

Blake, D. H. and Miezitis, Y. (1967). Geology of Bougainville and Buka Islands, New Guinea. *Bur. Mineral. Resour. Aust. Bull.* no. 93.

Blong, R. J. (1979). Huli legends and volcanic eruptions,

Papua New Guinea. *Search* **10**, 93–94.

Brooks, J. A. (1969). Rayleigh waves in southern New Guinea. II. A shear velocity profile. *Bull. seismol. Soc. Am.* **59**, 2017–2038.

Bultitude, R. J. (1976). Eruptive history of Bagana volcano, Papua New Guinea, between 1882 and 1975. In *Volcanism in Australasia* (R. W. Johnson, ed.), Elsevier, Amsterdam, pp. 317–336.

Bultitude, R. J. (1979). Bagana volcano, Bougainville Island: geology, petrology, and summary of eruptive history between 1875 and 1975. *Geol. Surv. Papua New Guinea Mem. no.* 6.

Bultitude, R. J., Johnson, R. W., and Chappell, B. W. (1978). Andesites of Bagana volcano, Papua New Guinea: chemical stratigraphy and a reference andesite composition. *Bur. Mineral. Resour. Aust. J.* **3**, 281–295.

Carey, S. W. (1938). The morphology of New Guinea. *Aust. Geogr.* **3**, 3–31.

Cooke, R. J. S. and Johnson, R. W. (1980). Papua New Guinea. In *Encyclopedia of Volcanoes and Volcanology* (J. Green, ed.), Dowden, Hutchinson, and Ross, Stroudsberg, in press.

Cooke, R. J. S. and Johnson, R. W. (1981). Bam volcano: morphology, geology, and reported eruptive history. In *Cooke-Ravian Volume of Volcanological Papers* (R. W. Johnson, ed.), *Geol. Surv.* Papua New Guinea Mem. no. 10, in press.

Cooke, R. J. S., McKee, C. O., Dent, V. F., and Wallace, D. A. (1976). Striking sequence of volcanic eruptions in the Bismarck volcanic arc, Papua New Guinea, in 1972–75. In *Volcanism in Australasia* (R. W. Johnson, ed.), Elsevier, Amsterdam, pp. 149–172.

Curtis, J. W. (1973). Plate tectonics of the Papua–New Guinea–Solomon Islands region. *J. geol. Soc. Aust.* **20**, 21–36.

Dent, V. F. (1976). The seismicity pattern near Mount Yelia, Papua New Guinea. *Geol. Surv. Papua New Guinea Rep. no.* 76/22.

De Paolo, D. J., and Johnson, R. W. (1979). Magma genesis in the New Britain island arc: constraints from Nd and Sr isotopes and trace-element patterns. *Contr. Mineral. Petrol.* **70**, 367–379.

Dow, D. B. (1977). A geological synthesis of Papua New Guinea. *Bur. Mineral. Resour. Aust. Bull.* no. 201.

Finlayson, D. M. and Cull, J. P. (1973). Structural profiles in the New Britain–New Ireland region. *J. geol. Soc. Aust.* **20**, 37–48.

Fisher, N. H. (1939). Geology and vulcanology of Blanche Bay and the surrounding area, New Britain. *Terr. New Guinea geol. Bull.* no. 1.

Fisher, N. H. (1957). *Catalogue of the Active Volcanoes of the World, Including Solfatara Fields*, Part 5, *Melanesia*, International Volcanological Association, Naples.

Fisher, N. H. (1976). 1941–42 eruption of Tavurvur volcano, Rabaul, Papua New Guinea. In *Volcanism in Australasia* (R. W. Johnson, ed.), Elsevier, Amsterdam, pp. 201–210.

Furumoto, A. S., Hussong, D. M., Campbell, J. F., Sutton, G. H., Malahoff, A., Rose, J. C., and Woollard, G. P. (1970). Crustal and upper mantle structure of the Solomon Islands as revealed by seismic refraction survey of November–December 1966. *Pacific Sci.* **24**, 315–332.

Gust, D. and Johnson, R. W. (1981). Amphibole-bearing inclusions from Boisa Island, Papua New Guinea: evaluation of the role of fractional crystallization in an andesitic volcano. *J. Geol.* **89**, 219–232.

Halunen, A. J., Jr and Von Herzen, R. P. (1973). Heat flow in the western equatorial Pacific. *J. geophys. Res.* **78**, 5195–5208.

Hamilton, W. (1979). Tectonics of the Indonesian region. *US geol. Surv. prof. Paper* no. 1078.

Heming, R. F. (1974). Geology and petrology of Rabaul caldera, Papua New Guinea. *Geol. Soc. Am. Bull.* **85**, 1253–1264.

Heming, R. F. (1980). Undersaturated lavas from Ambittle Island, Papua New Guinea. *Lithos* **12**, 173–186.

Heming, R. F. and Carmichael, I. S. E. (1973). High-temperature pumice flows from the Rabaul caldera Papua, New Guinea. *Contr. Mineral. Petrol.* **38**, 1–20.

Jakeš, P., and Smith, I. E. (1970). High potassium calc-alkaline rocks from Cape Nelson, eastern Papua. *Contr. Mineral. Petrol.* **28**, 259–271.

Jakeš, P. and White, A. J. R. (1969). Structure of the Melanesian arcs and correlation with distribution of magma types. *Tectonophysics* **8**, 223–236.

Jaques, A. L. and Robinson, G. P. (1977). The continent/island-arc collision in northern Papua New Guinea. *Bur. Mineral. Resour. Aust. J.* **2**, 289–303.

Johnson, R. W. (1976). Potassium variation across the New Britain volcanic arc. *Earth planet. Sci. Letters* **31**, 184–191.

Johnson, R. W. (1977). Distribution and major-element chemistry of late Cainozoic volcanoes at the southern margin of the Bismarck Sea, Papua New Guinea. *Bur. Mineral. Resour. Aust.* Rep. no. 188.

Johnson, R. W. and Arculus, R. J. (1978). Volcanic rocks of the Witu Islands, Papua New Guinea: the origin of magmas above the deepest part of the New Britain Benioff zone. *Bull. Volc.* **41**, 609–655.

Johnson, R. W. and Chappell, B. W. (1979). Chemical analyses of rocks from the late Cainozoic volcanoes of north-central New Britain and the Witu Islands, Papua New Guinea. *Bur. Mineral. Resour. Aust.* Rep. no. 209.

Johnson, R. W. and Jaques, A. L. (1980). Continent–arc collision and reversal of arc polarity: new interpretations from a critical area. *Tectonophysics* **63**, 111–124.

Johnson, R. W. and Smith, I. E. (1974). Volcanoes and rocks of St Andrew Strait, Papua New Guinea. *J. geol. Soc. Aust.* **21**, 333–351.

Johnson, R. W., Davies, R. A., and White, A. J. R. (1972). Ulawun volcano, New Britain. *Bur. Mineral. Resour. Aust. Bull.* no. 142.

Johnson, R. W., Wallace, D. A., and Ellis, D. J. (1976). Feldspathoid-bearing potassic rocks and associated types from volcanic islands off the coast of New Ireland, Papua New Guinea: a preliminary account of geology and petrology. In *Volcanism in Australasia* (R. W. Johnson, ed.), Elsevier, Amsterdam, pp. 297–316.

Johnson, R. W., Mackenzie, D. E., and Smith, I. E. M. (1978a). Volcanic rock associations at convergent plate boundaries: re-appraisal of the concept using case histories from Papua New Guinea. *Geol. Soc. Am. Bull.* **89**, 96–106.

Johnson, R. W., Smith, I. E. M., and Taylor, S. R. (1978b). Hot-spot volcanism in St Andrew Strait, Papua New Guinea: geochemistry of a bimodal rock suite. *Bur. Mineral. Resour. Aust. J.* **3**, 55–69.

Johnson, R. W., Mackenzie, D. E., and Smith, I. E. M. (1978c). Delayed partial melting of subduction-modified mantle in Papua New Guinea. *Tectonophysics* **46**, 197–216.

Johnson, R. W., Mutter, J. C., and Arculus, R. J. (1979). Origin of the Willaumez–Manus Rise, Papua New Guinea. *Earth planet. Sci. Letters* **44**, 247–260.

Johnson, T. and Molnar, P. (1972). Focal mechanisms and plate tectonics of the southwest Pacific. *J. geophys. Res.* **77**, 5000–5032.

Krause, D. C. (1973). Crustal plates of the Bismarck and Solomon Seas. In *Oceanography of the South Pacific 1972* (R. Fraser, ed.), New Zealand Nat. Comm. UNESCO, Wellington, pp. 271–280.

Kroenke, L. W. (1972). Geology of the Ontong Java Plateau. *Hawaii Inst. Geophys. Rep.* no. HIG-72-5.

Löffler, E. (1972). Pleistocene glaciation in Papua New Guinea. *Z. Geomorph. N.F.* **13**, 32–58.

Lowder, G. G., and Carmichael, I. S. E. (1970). The volcanoes and caldera of Talasea, New Britain: geology and petrology. *Geol. Soc. Am. Bull.* **81**, 17–38.

Luyendyk, B. P., Macdonald, K. C., and Bryan, W. B. (1973). Rifting history of the Woodlark Basin in the southwest Pacific. *Geol. Soc. Am. Bull.* **84**, 1125–1134.

Mackenzie, D. E. (1973). Quaternary volcanoes of the central and southern Highlands of Papua New Guinea. *Bur. Mineral. Resour. Aust. Rec.* no. 1973/89.

Mackenzie, D. E. (1976). Nature and origin of late Cainozoic volcanoes in western Papua New Guinea. In *Volcanism in Australasia* (R. W. Johnson, ed.), Elsevier, Amsterdam, pp. 221–238.

McKee, C. O., Cooke, R. J. S., and Wallace, D. A. (1976). 1974–75 eruptions of Karkar volcano, Papua New Guinea. In *Volcanism in Australasia* (R. W. Johnson, ed.), Elsevier, Amsterdam, pp. 173–190.

Melson, W. G., Jarosevich, E., Switzer, G., and Thompson, G. (1972). Basaltic *nuées ardentes* of the 1970 eruption of Ulawun volcano, New Britain. *Smithson. Contr. Earth Sci.* **9**, 15–32.

Mitchell, A. H. G. and Garson, M. S. (1972). Relationship of porphyry copper and circum-Pacific tin deposits to palaeo-Benioff zones. *Trans./Sect. B Inst. Min. Metall.* **81**, 10–25.

Myers, N. O. (1976). Seismic surveillance of volcanoes in Papua New Guinea. In *Volcanism in Australasia* (R. W. Johnson, ed.), Elsevier, Amsterdam, pp. 91–99.

Ollier, C. D. and Mackenzie, D. E. (1974). Subaerial erosion of volcanic cones in the tropics. *J. trop. Geogr.* **39**, 63–71.

Page, R. W. and Johnson, R. W. (1974). Strontium isotope ratios of Quaternary volcanic rocks from Papua New Guinea. *Lithos* **7**, 91–100.

Palfreyman, W. D. and Cooke, R. J. S. (1976). Eruptive history of Manam volcano, Papua New Guinea. In *Volcanism in Australasia* (R. W. Johnson, ed.), Elsevier, Amsterdam, pp. 117–131.

Perfit, M. R., Arculus, R. J., Johnson, R. W., and Chappell, B. W. (1978). Mineralogy, major, trace and isotope chemistry of volcanic rocks from the Tabar-to-Feni Islands, Papua New Guinea: an alkalic island arc? *Geol. Soc. Am. Abstr. Programs* **10**, 470.

Peterman, Z. E. and Heming, R. F. (1974). Sr^{87}/Sr^{86} ratios from the Rabaul caldera, Papua New Guinea. *Geol. Soc. Am. Bull.* **85**, 1265–1268.

Peterman, Z. E., Lowder, G. G., and Carmichael, I. S. E. (1970). Sr^{87}/Sr^{86} ratios of the Talasea series, New Britain, Territory of New Guinea. *Geol. Soc. Am. Bull.* **81**, 39–40.

Reynolds, M. A. and Best, J. G. (1976). Summary of the 1953–57 eruption of Tuluman volcano, Papua New Guinea. In *Volcanism in Australasia* (R. W. Johnson, ed.), Elsevier, Amsterdam, pp. 287–296.

Smith, I. E. M. (1976a). Peralkaline rhyolites from the D'Entrecasteaux Islands, Papua New Guinea. In *Volcanism in Australasia* (R. W. Johnson, ed.), Elsevier, Amsterdam, pp. 275–285.

Smith, I. E. M. (1976b). Volcanic rocks from southeastern Papua. The evolution of volcanism at a plate boundary. Ph.D. thesis, Australian National University, Canberra.

Smith, I. E. and Davies, H. L. (1976). The geology of the southeast Papuan mainland. *Bur. Mineral. Resour. Aust. Bull.* no. 165.

Smith, I. E. M., Chappell, B. W., Ward, G. K., and Freeman, R. S. (1977). Peralkaline rhyolites associated with andesitic arcs of the south-west Pacific. *Earth planet. Sci. Letters.* **37**, 230–236.

Smith, I. E. M., Taylor, S. R., and Johnson, R. W. (1979). REE-fractionated trachytes and dacites from Papua New Guinea and their relationship to andesite petrogenesis. *Contr. Mineral. Petrol.* **69**, 227–233.

Taylor, B. (1979). Bismarck Sea: evolution of a back-arc basin. *Geology* **7**, 171–174.

Taylor, G. A. (1958). The 1951 eruption of Mount Lamington, Papua. *Bur. Mineral. Resour. Aust. Bull.* no. 38.

Taylor, S. R. and Gorton, M. P. (1977). Geochemical application of spark source mass spectrography. III. Element sensitivity, precision, and accuracy. *Geochem. cosmochim. Acta* **41**, 1375–1380.

Taylor, S. R., Capp, A. C., Graham, A. L., and Blake, D. H. (1969). Trace element abundances in andesites II. Saipan, Bougainville and Fiji. *Contr. Mineral. Petrol.* **23**, 1–26.

Thornton, C. P., and Tuttle, O. F. (1960). Chemistry of igneous rocks. I. Differentiation index. *Am. J. Sci.* **258**, 664–684.

Weissel, J. K. and Anderson, R. N. (1978). Is there a Caroline Plate? *Earth planet. Sci. Letters* **41**, 143–158.

Andesites
Edited by R. S. Thorpe
© 1982 John Wiley & Sons

Tonga–Kermadec–New Zealand

J. W. Cole

Department of Geology,
Victoria University of Wellington, Wellington, New Zealand

ABSTRACT

The Tonga, Kermadec, and Taupo (New Zealand) arcs are separate units within a single arc system which gets progressively younger towards the south. Recent volcanism in Tonga is confined to the western group of islands which comprise mostly basaltic andesite with minor andesites and dacites. The Kermadec Islands are mainly basalt with some basaltic andesite. Chemically, the lavas of both the Tonga and Kermadec arcs are 'island arc tholeiites' and are probably derived initially from the subduction zone, with magma generated in the early stages of arc development causing partial melting of the overlying mantle wedge to form the range of compositions from basaltic andesite to basalt. Andesitic activity in New Zealand occurs at either end of the Taupo Volcanic Zone. In the north, White Island comprises mainly dacite, but andesite was erupted in 1977. Most of the andesites within the Tongariro Volcanic Centre, New Zealand, are related to an older north-west-trending arc (the Northland arc), and only andesites erupted from NNE-trending vents, such as the olivine-bearing andesites erupted from Tongariro within the last 50 ka, are regarded as relating to the present Taupo arc. Andesites of both New Zealand arcs are calc-alkaline, and their chemical characteristics indicate contamination of the magma by crustal material. Subsequent fractional crystallization must have occurred, however, suggesting that this contamination may have occurred prior to subduction.

Introduction

The Tonga, Kermadec, and Taupo arcs form the south-western end of the circum-Pacific andesite province. The three arcs are regarded as forming parts of a single arc system (Karig, 1970), but it is evident from the bathymetry (Brodie and Hatherton, 1958) and the seismicity (Sykes, 1966; Hamilton and Gale, 1968; Eiby, 1977) that each arc is a separate unit (Fig. 1). The dip of the Benioff zone is generally 40–50° beneath the Tongan Islands (Sykes, 1966), 55–60° beneath the Kermadec Islands (Sykes, 1966), and 50° beneath New Zealand (Adams and Ware, 1977).

The morphology of the Tonga and Kermadec arcs has been described by Karig (1970), who suggested that most features of these arcs continue directly into the North Island, New Zealand. Cole (1978b) considered that the Havre Trough extends south-westwards into the Ngatoro Basin (Fig. 1) intersecting the coast of New Zealand at Tauranga (Fig. 3). The Taupo Volcanic Zone is thus a separate parallel marginal basin of the Taupo arc.

Development of the Arc System

The time at which the subduction system developed is uncertain. Eocene volcanic rocks exposed on the island of Eua (Fig. 2) are interpreted by Ewart and Bryan (1972) as part of an underlying ophiolite complex, and by Kroenke and Tongilava (1975) as representing fracture zone volcanism along a transform fault aligned along the present-day axis of the Tonga Trench.

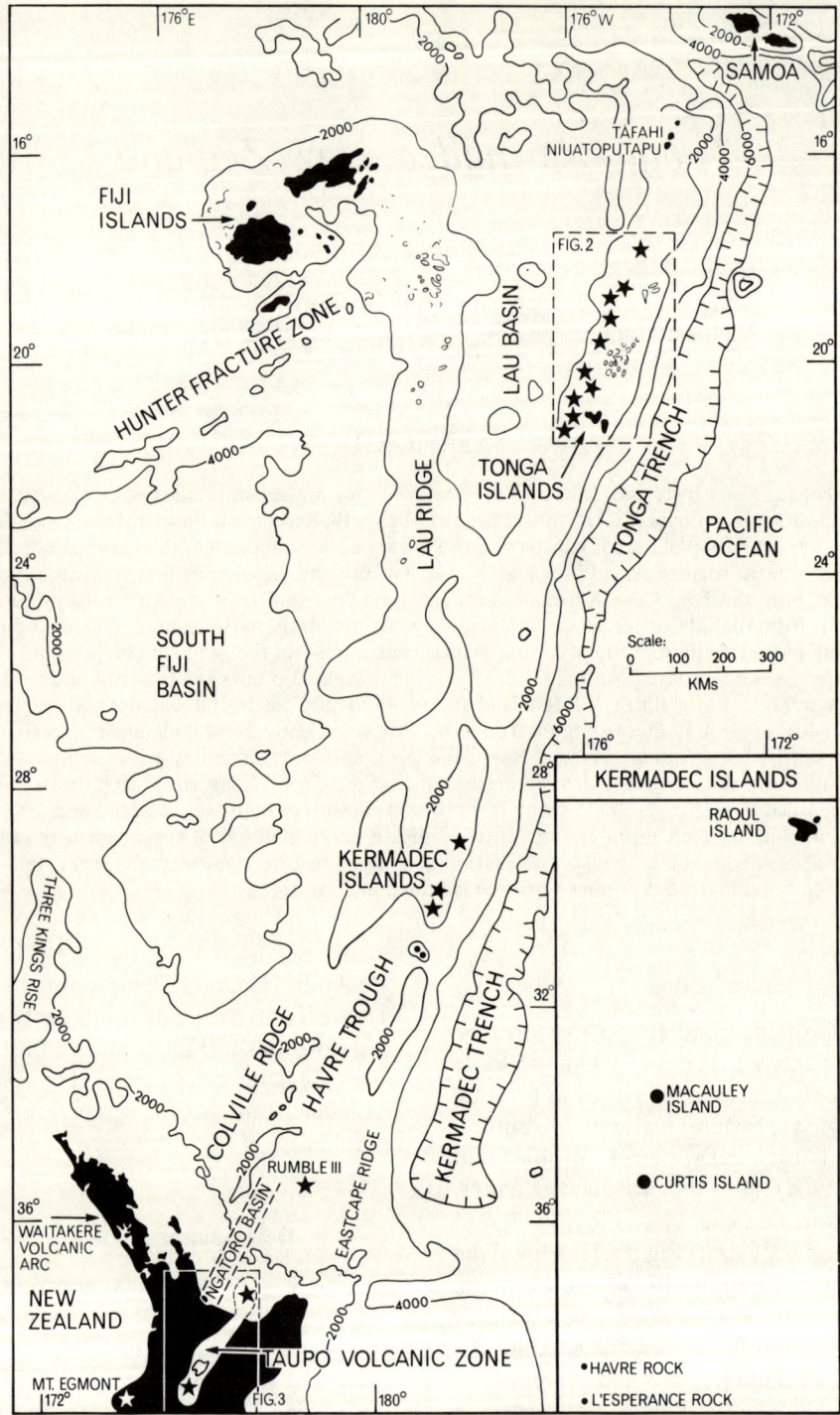

Fig. 1 Distribution of active andesite volcanoes (stars) within the Tonga–Kermadec–Taupo (New Zealand) arc system. Bathymetry shown in metres

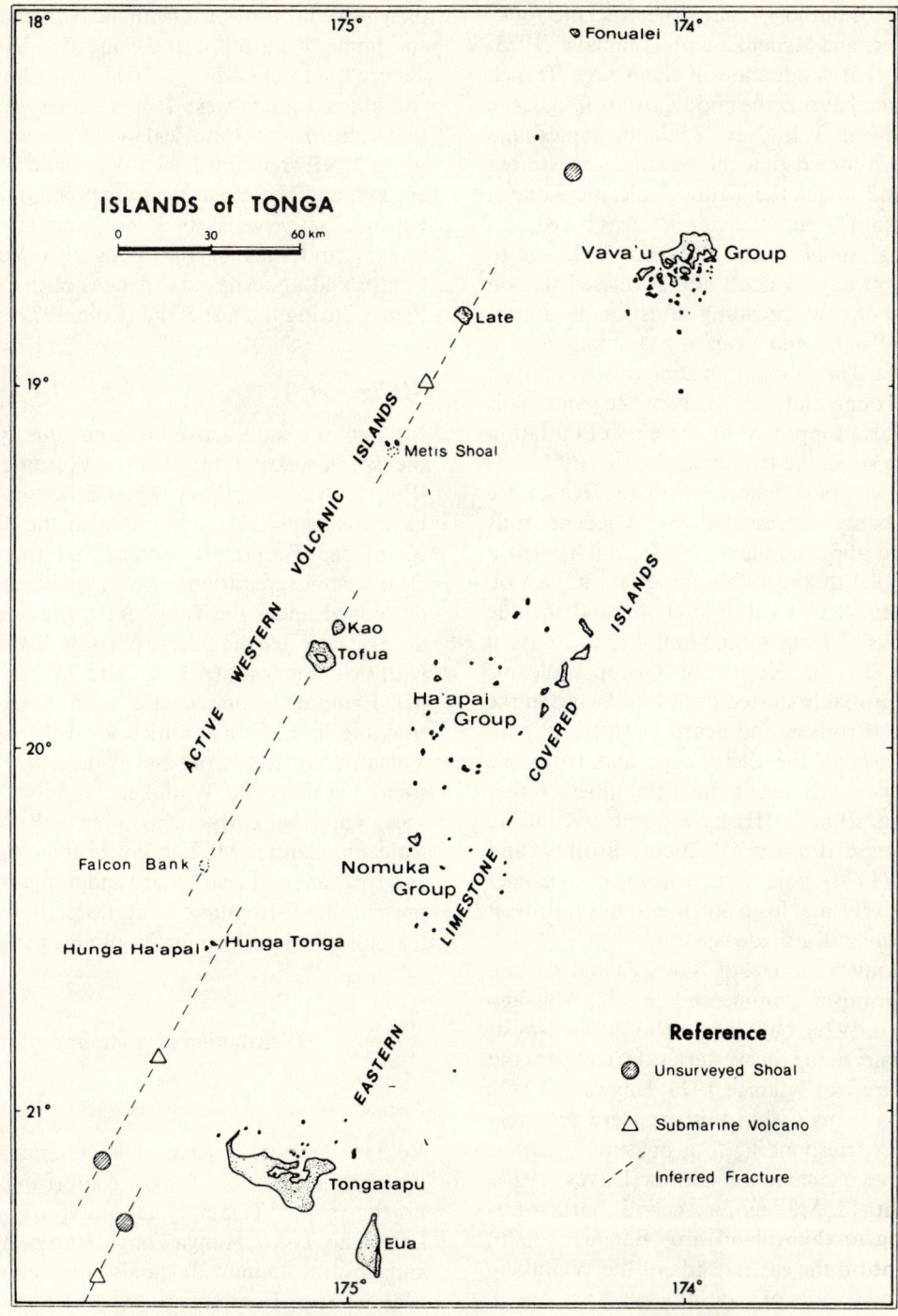

Fig. 2 Location map of the principal Tongan Islands. (Reproduced from Ewart and Bryan (1973) with permission of the University of Western Australia Press)

Both sets of authors regard the volcanic rocks as oceanic, and Kroenke and Tongilava (1975) consider that subduction of the Tonga Trench was initiated during the middle to late Oligocene (c. 30 Ma ago). In New Zealand, Sameshima (1975) considered that the oceanic slab started to descend under the Taupo Volcanic Zone c. 20 Ma ago. The latter calculation was based on a spreading rate of 55 mm a^{-1} from 5 Ma ago to the present day. Walcott (1978) revised this on the basis of new spreading rates for the south-western Pacific and gives a maximum age of 12–15 Ma. The subduction zone thus developed first in Tonga and then in New Zealand. This order is also supported by the ages of initiation of volcanism in the two areas.

Initial stages of volcanism in the Tonga arc are probably represented by Miocene tuffs (Kroenke and Tongilava, 1975) which form a line parallel to, but offset some 50 km east of, the present active volcanic chain, and include the islands of Tongatapu, Nomuka, and Vava'u (Fig. 2). In the Kermadec Group, volcanic activity probably started on Raoul Island in the Pliocene (Brothers and Searle, 1970), and continued through the Pleistocene and Holocene with minor activity in historic times. Other vents (e.g. Rumble III) have developed during historic time. Brothers (1970) and Brothers and Martin (1970) note that volcanism becomes progressively later from north-east to south-west in the Tonga–Kermadec group.

In the northern part of New Zealand, Cainozoic volcanism commenced c. 22 Ma ago (Hayward, 1979), but along a single north-west-trending arc along the western side of Northland (Waitakere arc: Ballance, 1976; Hayward, 1979). The main vents within this arc were probably offshore, corresponding to a present-day series of positive magnetic anomalies (Davey, 1974). At about 18 Ma ago a second north-west-trending arc (Northland arc: Ballance, 1976) developed on the eastern side of the Waitakere arc to form a series of andesite–dacite volcanoes, the southern part of which now forms the Hauraki Volcanic Region (Fig. 3). From 18 to 15 Ma ago a double arc was in existence (Ballance, 1976; Cole, 1979). Volcanic activity then moved southwards from the Northland arc, and finally occurred in the Tongariro Volcanic Centre less than 1 Ma ago. Vents were, however, still aligned north-west. It is only very recently that volcanism in New Zealand has been related to the NNE-trending Tonga–Kermadec–Taupo arc system. The most recent expression of this volcanic activity has been eruption of olivine low Si andesites of the Tongariro Volcanic Centre and andesites and dacites of the Bay of Plenty during the last 50 ka (Cole, 1979).

Mt Egmont

Mt Egmont is an active andesite cone lying to the south-west of the Taupo Volcanic Zone (Fig. 1). Many authors (e.g. Hatherton, 1969) have associated Mt Egmont with the Tonga–Kermadec–Taupo arc system, but there is a clear seismic separation between the Benioff zone developed under the Taupo Volcanic Zone and deep-seated earthquakes possibly associated with Mt Egmont (Adams and Ware, 1977). Mt Egmont is one centre of a north-west-trending line of cones which form the Egmont Volcanic District (Cole and Nairn, 1975). This trend parallels the Waitakere and Northland arcs, and the early lavas of the Tongariro Volcanic Centre. Mt Egmont is thus regarded as a remnant of an earlier arc and not part of the present NNE-trending Taupo arc. It will not therefore be described further in this paper.

Distribution of Andesites

Tongan Islands

Recent andesitic volcanism is confined to the western islands of Tonga comprising, from north to south: Tafahi, Niuatoputapu, Fonualei, Late, Kao, Tofua, Hunga Tonga, Hunga Ha'apai, and several unnamed shoals and submarine islands (Figs. 1 and 2).

Tafahi and Niuatoputapu are small islands which are no longer regarded as active (Ewart et al., 1977). The remainder have all had recent activity. Fonualei comprises a low cone, the top

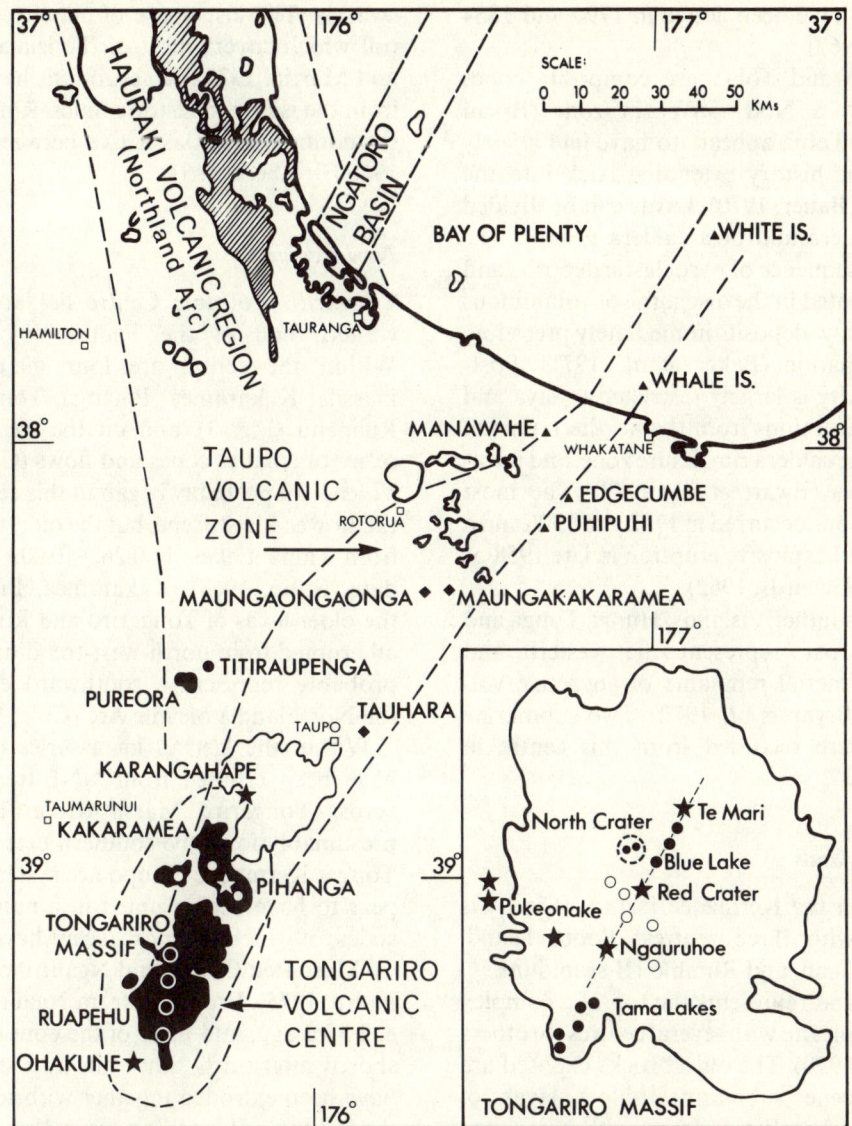

Fig. 3 Andesites and dacites of the Taupo Volcanic Zone, New Zealand: ▲, andesites–dacites of Bay of Plenty; ♦, dacites of Rotorua–Taupo area; ★, olivine low Si andesites of Tongariro Volcanic Centre; ○, older north west-trending andesites of Tongariro Volcanic Centre; ●, other vents of Tongariro Volcanic Centre

of which has collapsed to form a caldera (Bryan et al., 1972). Within this caldera is a pyroclastic cone from which blocky lava flows originated during an eruption in 1846–47 (Richards, 1962). The most recent eruption occurred in 1939 (Richards, 1962; Brodie, 1970), when lava flowed through a breach in the caldera wall and down the south-western slopes of the island. Late is a conical volcano with a well developed central crater, and several scoria cones on the outer slopes (Bryan et al., 1972). One of the most prominent features is a graben striking east-north-east, on the eastern side of the island, within which are two pit craters. One of these craters may have been active in 1790 and 1854 (Richards, 1962).

Both Kao and Tofua are composite cones separated by a NNE–SSW rift zone (Bryan et al., 1972). Tofua appears to have had a fairly long volcanic history extending back into the Pleistocene (Bauer, 1970). Lavas can be divided into pre-caldera and post-caldera groups. The pre-caldera sequence of pyroclastic deposits and flows culminated in the discharge of voluminous pyroclastic flow deposits immediately preceding caldera formation (Baker et al., 1971). Post-caldera activity is largely restricted to lava and pyroclastic eruptions from the northern part of the island, the caldera rim fissure zone, and intra-caldera cones (Ewart et al., 1977). The most recent eruptions occurred in 1792 and 1906 and a possible small explosive eruption in late 1958 or early 1959 (Richards, 1962).

The most southerly islands, Hunga Tonga and Hunga Ha'apai, represent the western and northern subaerial remnants of an active volcanic cone (Bryan et al., 1972). Two submarine eruptions were recorded from this centre in 1912 and 1937.

Kermadec Islands

Volcanism in the Kermadec Islands (Fig. 1) is associated with three centres; Raoul Island, Macauley Island, and Rumble III seamount.

Raoul Island represents the top of a complex composite volcano with several centres (Brothers and Searle, 1970). The oldest rocks exposed are Plio-Pleistocene submarine pillow lavas of tholeiite and basaltic andesite with associated fossiliferous volcanogenic sediments. These must have been subsequently uplifted prior to sub-aerial volcanism which built a composite cone, possibly as high as 1000 m above sea level (Brothers and Searle, 1970). Voluminous pumice eruptions commenced c. 2000 BC (E. F. Lloyd, personal communication), and were accompanied by progressive collapse of a 3.3 km diameter caldera.

Macauley Island also represents the summit of a larger structure formed by repeated eruptions from a single centre and intrusions of dyke swarms. The last phase of activity produced a tuff which covers most of the island (Brothers and Martin, 1970). No eruptions have occurred from the island in historic times. Rumble III is a seamount which was active between 1958 and 1966 (Brothers, 1967).

New Zealand

Tongariro Volcanic Centre lies at the southwestern end of the Taupo Volcanic Zone. Within the centre are four major andesite massifs: Kakaramea, Pihanga, Tongariro, and Ruapehu (Fig. 3), and on the western side a series of smaller cones and flows (Cole, 1978a). Volcanism probably began in this centre during the Lower Pleistocene, but the oldest dated lava, from Tama Lakes, is 0.26 ± 0.003 Ma (K–Ar date; Stipp, 1968). Kakaramea, Pihanga, and the older lavas of Tongariro and Ruapehu were all erupted from north-west-trending vents, and probably represent a southward extension of the Northland Volcanic Arc (Cole, 1979).

Within the last 50 ka a series of andesites have been erupted from NNE-trending vents across Tongariro Massif (Cole, 1978a), and presumably form the southern extension of the Tonga–Kermadec–Taupo arc system. They appear to have been erupted in a number of episodes, of which the most recent have been from Te Mari, Red Crater, and Ngauruhoe (Cole and Nairn, 1975). Eruptions from Ngauruhoe began c. 2.5 ka ago, and most of the cone was formed shortly afterwards. Since this time smaller flows have been extruded together with some tephra; the last major lava flows having been erupted in 1954. Intermittent small pyroclastic eruptions occurred between 1955 and 1974, and then in 1974 and 1975 explosive eruptions formed eruption columns rising to 1300 m and accompanied by pyroclastic avalanches (Nairn et al., 1976; Nairn and Self, 1978). Ejecta from these eruptions was chemically identical to 1954 lava and thus probably represents material derived by disintegration of a solid lava plug rather than new lava (Nairn et al., 1976).

Ruapehu is an andesite composite volcano which has been in existence since early Quaternary times (Fleming and Steiner, 1951). Grindley (1965) suggests it may have had a two-stage development, but it is not yet clear whether this relates to the two arc trends identified on Tongariro massif. The volcano consists of interbedded andesite flows texturally similar to those of Tongariro massif and thick pyroclastic deposits cut in places by vertical andesite dykes.

The broad summit of Ruapehu consists of a main depression occupied by three craters which Cole and Nairn (1975) consider to represent explosive vents which have been later modified by erosion. One crater is occupied by a crater lake and this is the site of most recent activity. The last eruption of lava occurred in 1945 when a tholoid displaced the original crater lake. Activity increased until two explosive eruptions created a new crater within the tholoid. Once activity ceased this filled with water to create a new crater lake. Since 1945, phreatic explosions from the crater lake have been common, particularly in 1966 when lava was extruded on to the floor of the crater lake (Cole and Nairn, 1975). Phreatomagmatic eruptions have also occured in 1969 (Healy et al., 1978), 1971, 1975, and 1977.

White Island

White Island is the emergent summit of a large (16 km × 18 km) submarine volcanic edifice which lies at the north-eastern end of the Taupo Volcanic Zone (Fig. 3). It comprises two overlapping composite cones (Black, 1970; Duncan, 1970), an older cone forming the western side of the island and a younger cone forming the eastern and central part of the island.

Eruptive activity began in the late Pleistocene, and the youngest flow is probably a small 1-2 m thick flow on the summit of Mt Gisborne (Black, 1970). The present crater has apparently been formed by explosion (Duncan, 1970) and breached by marine erosion. The volcano has been in a state of continuous solfataric and fumarolic activity with intermittent steam and tephra eruptions since 1826 when the first recorded landing was made by Europeans (Cole and Nairn, 1975). Many explosion craters have formed within the main crater, but after a while these became obliterated by later activity. The most recent eruption began in December 1976, and in March 1977 fresh lava was recorded for the first time since observations of the island began (Clark et al., 1979).

Petrography

Tongan Islands

Most of the islands are composed of basaltic andesites with phenocrysts of plagioclase (An_{84}-An_{89}), clinopyroxene, and orthopyroxene in a pigeonitic groundmass (Ewart et al., 1973; Ewart and Bryan, 1973). Olivine is rare, appearing only as phenocrysts in basaltic andesites of Tofua and Kao (Bryan et al., 1972). No hydrous phases (e.g. amphibole) or phenocrystic magnetite occur in these lavas. Fonualei is unusual as it is composed predominantly of dacite, although some siliceous andesites occur in the pre-caldera succession (Ewart et al., 1977). Occasional dacites also occur on Tofua. Mineralogy of the dacites is similar to the andesites, with the addition of titanomagnetite (Ewart and Bryan, 1973). The groundmass mineralogy differs, however, in having hypersthene, Fe-Ti oxides, rare apatite, and frequent glass. Fine grained potash feldspar and quartz are identifiable in the more coarsely crystalline dacites (Ewart and Bryan, 1973).

The lava erupted from Metis Shoal in 1967-68 appears to be unique within the Tongan islands as it comprises vesicular rhyolitic glass containing phenocrysts of augite, orthopyroxene, bytownite, titanomagnetite, and corroded olivine (Melson et al., 1970).

Kermadec Islands

Volcanism in the Kermadec Islands (Fig. 1) is primarily basaltic (Brothers, 1970), but basaltic andesites have been erupted from Raoul Island (Brothers and Searle, 1970) and Macauley Island (Brothers and Martin, 1970). Brothers (1967) records basaltic andesites from the Rumble III seamount which was active between 1958 and 1966. However, an analysis of the lava indicates that it is a tholeiitic basalt (Ewart et al., 1977).

New Zealand

Differences occur in the mineralogy of andesites associated with the north-west- and NNE-trending arc structures. Andesite lavas of the Tongariro Volcanic Centre associated with the north-west-trending arc range from aphyric to porphyritic in texture with phenocrysts of labradorite, orthopyroxene, clinopyroxene, and rare hornblende in a pilotaxitic groundmass. Aggregates of plagioclase, orthopyroxene, clinopyroxene, and occasionally olivine are common, particularly in the more strongly porphyritic lavas, and some are extensively oxidized (Cole, 1978a), but comagmatic xenoliths are rare. Andesites associated with the NNE-trending structure are commonly olivine-bearing, with phenocrysts of clinopyroxene, orthopyroxene, labradorite, and fosteritic olivine in a pilotaxitic ground-mass.

Lavas of White Island are predominantly dacitic and consist of flows, breccias, and tuffs with occasional plugs of lava and agglomerate. Texturally the lavas are porphyritic with phenocrysts of plagioclase (An_{40}–An_{65}), augite, hypersthene, and titaniferous magnetite in a pilotaxitic groundmass (Duncan, 1970). Much of the lava is strongly hydrothermally altered. The new lava erupted in March 1977 was an olivine-bearing andesite similar to that erupted from NNE-trending vents of the Tongariro Volcanic Centre (Clark et al., 1979).

Chemical Composition

Major and trace elements

Clear differences exist between the chemical characteristics of lavas from the Tonga, Kermadec, and Taupo arcs. The main difference between Tonga and Kermadec arcs is in SiO_2 content, with most Tongan lavas ranging between 53 and 66 per cent SiO_2, and most Kermadec lavas between 48 and 56 per cent SiO_2 (Fig. 4). Other petrochemical characteristics are very similar. Lavas from both Tonga and Kermadec arcs are typical of the 'island arc

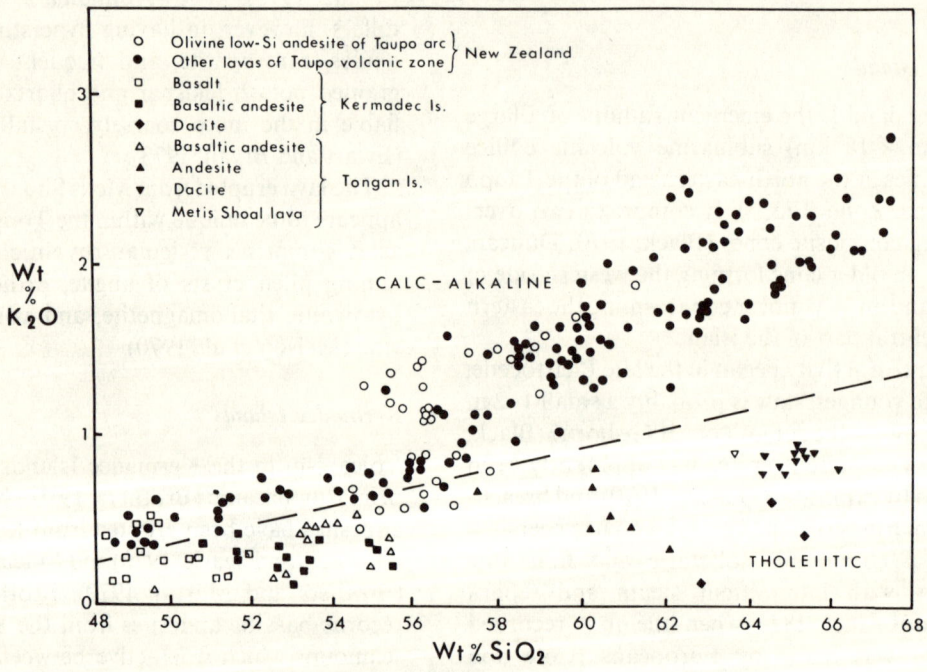

Fig. 4 Plot of K_2O against SiO_2 for basaltic–dacitic lavas of the Tonga, Kermadec, and New Zealand arcs. (Data sources listed in Table 1)

tholeiitic series' defined by Jakeš and Gill (1970). They are characterized by relatively high total Fe ($\sum$ FeO) and CaO, and low alkalis, particularly K_2O (Table 1 and Fig. 4). The andesites also have low Al_2O_3. Trace elements conform with major elements as the lavas have low concentrations of Rb, Sr, Ba, Zr, U, and Th (Table 1). K/Rb ratios are typically high and Rb/Sr ratios low. REE abundances (Ewart and Bryan, 1973; Ewart et al., 1977) are low, with slightly higher $\sum$ REE values in the Kermadec arc (Table 1), and are all subparallel to the chondritic trend.

Lavas of both Northland and Taupo arcs, however, differ markedly in chemical composition from both Tonga and Kermadec arcs, and are clearly 'calc-alkaline' (Fig. 4). They are enriched in the incompatible elements (K, Rb, Ba, Zr, Th, U, light REE) compared with Tonga and Kermadec, and are also higher in MgO, Ni, and Cr, but lower in V and Sc (Table 1). REE (Ewart et al., 1968, 1977) are more abundant in the New Zealand lavas (Table 1) and there is a marked enrichment of light REE (Fig. 5).

Andesites of the Tonga and Kermadec arcs, and olivine low Si andesites of the Taupo arc are plotted on an AFM diagram for comparison in Fig. 6. The Taupo lavas are clearly enriched in alkalis and show less Fe enrichment than lavas of the other arcs.

Isotopic abundances

$^{87}Sr/^{86}Sr$ ratios are very low for lavas of the Tonga and Kermadec arcs ranging from 0.7036 for basaltic andesites to 0.7043 for the dacites (Ewart et al., 1973). Lavas of the Taupo Volcanic Zone (Ewart and Stipp, 1968) are slightly higher, ranging from 0.7042 for high alumina basalts to 0.7061 for acid andesites and dacites from Bay of Plenty, New Zealand.

Pb isotopic compositions vary within each arc. The Tonga arc has a very narrow range ($^{206}Pb/^{204}Pb$ = 18.310–18.727; $^{207}Pb/^{204}Pb$ = 15.505–15.586; $^{208}Pb/^{204}Pb$ = 37.930–38.473), indicating that the lavas were derived from a very homogeneous source region. Kermadec lavas form a linear array, indicating a heterogeneous source similar to that found for mid-

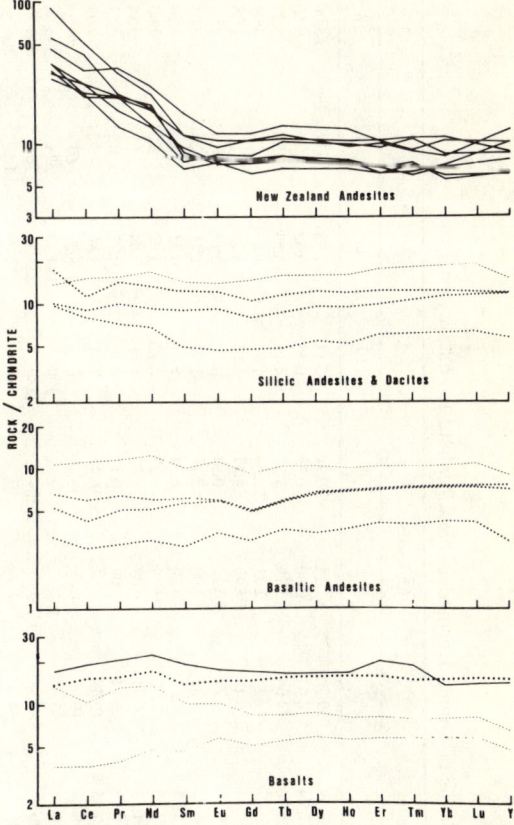

Fig. 5 REE data, normalized to chondritic abundances, of representative volcanic samples from the Tonga, Kermadec, and New Zealand arcs. Solid lines = New Zealand; broken lines = Kermadec; dotted lines = Tonga. (Reproduced from Ewart et al. (1977) with permission of Elsevier Scientific Publishing Co., Amsterdam)

oceanic, mantle-derived lavas (Oversby and Ewart, 1972). New Zealand isotopic ratios from the north-west-trending arc (Armstrong and Cooper, 1971) are substantially more radiogenic than either those of Tonga or Kermadec ($^{206}Pb/^{204}Pb$ = 18.682–18.837; $^{207}Pb/^{204}Pb$ = 15.518–15.683; $^{208}Pb/^{204}Pb$ = 38.382–38.787). Unfortunately no data is yet available for olivine low Si andesites of the NNE-trending arc.

Petrogenesis

The origin of andesite lavas from the Tonga–Kermadec–Taupo arc system is still a matter for debate. Ewart et al. (1977) consider that basaltic

TABLE 1 Basaltic, andesitic, and dacitic lavas of Tonga–Kermadec–New Zealand Island arc system

	Basalts			Basaltic/Low Si Andesites			Andesites			Dacites			
	1	2	3	4	5	6	7	8	9	10	11	12	13
SiO_2	49.95	50.43	52.53	53.40	53.04	56.59	59.70	61.14	59.22	65.11	65.78	63.60	66.08
TiO_2	0.74	1.02	0.69	0.52	0.90	0.64	0.67	0.80	0.72	0.58	0.65	0.62	0.50
Al_2O_3	17.80	17.13	15.21	16.70	17.73	15.36	14.06	15.79	16.83	14.13	15.53	15.55	15.27
Fe_2O_3	2.82	10.35	2.46	6.63	4.09	2.17	4.81	2.31	2.54	1.87	1.63	6.04	3.01
FeO	7.36		6.79	3.61	6.94	5.47	6.15	5.46	4.17	6.13	3.39		1.38
MnO	0.18	0.16	0.16	0.18	0.21	0.13	0.21	0.18	0.13	0.18	0.13	0.11	0.08
MgO	5.91	6.30	7.29	5.12	4.31	6.42	2.83	2.00	3.78	1.52	1.76	2.60	2.04
CaO	12.11	10.99	10.33	11.05	9.35	8.24	7.84	6.10	6.77	5.87	4.71	5.54	4.29
Na_2O	1.94	2.67	2.58	1.61	2.26	2.86	2.50	3.44	3.27	2.97	3.82	3.01	3.60
K_2O	0.28	0.51	0.67	0.16	0.29	1.10	0.71	0.43	1.50	1.11	0.44	1.98	2.34
P_2O_5	0.05	0.25	0.12	0.08	0.09	0.12	0.14	0.09	0.12	0.19	0.13	0.11	0.13
ΣFeO	9.89	9.31	9.00	9.58	10.62	7.45	10.48	7.54	6.45	7.81	4.86	5.21	4.09
Rb	5.2	12.9	29	5	5.1	18	10	6	36	16	6.2	69	78
Sr	215	329	296	198	185	289	260	167	179	300	184	228	236
Ba	100	—	146	98	121	220	180	173	334	280	253	706	600
Zr	31	106	42	22	42	85	34	64	119	47	72	121	206
Cu	133	38	60	131	118	80	131	36	39	27	17	26	5.6
Ni	32	31	54	24	14	65	4	4	22	<2	4	14	4.6
Sc	—	37		38		29	36		21	26		19	11
V	322	230	222	248	325	170	285	165	146	93	59	139	26
Cr	72	68	240	48	27	305	9		69	4	1.2	42	0.9
Th	0.31	0.85		0.29	0.40	2.7	0.62	5.9	3.3	0.70	1.27	6.61	8.5
U	0.15	0.16		0.12	0.19	0.70	0.25	1.24	1.05	0.28	0.49	1.48	1.7
ΣREE	25.28	71.5		17.43	36.52	73.9	26.58	0.62	51.7	41.3	53.76	79.3	—
La/Yb	1.2	3.6		1.2	1.5	2.8	1.8	—	5.9	2.1	1.1	8.57	—
K/Rb	447	328	192	266	472	508	589	595	346	576	589	238	249
Rb/Sr	0.024	0.039	0.098	0.028	0.028	0.062	0.037	0.035	0.201	0.052	0.034	0.303	0.330
$^{87}Sr/^{86}Sr$		0.7043		0.7037		0.7060	0.7042		0.7056	0.7043		0.7061	0.7052

(1) Average of 16 basalts from Raoul Group and Macauley Island, Kermadec Arc (Ewart et al., 1977). (2) Average of 15 (10†) high alumina basalts from the Taupo Volcanic Zone, New Zealand (Cole, 1979). (3) Average of three basalts from Tongariro Volcanic Centre, New Zealand (Cole, 1979). (4) Average of 13 basaltic andesites from Tafahi, Late, and Hunga Ha'apai Islands. Tonga arc (Ewart and Bryan, 1973; Ewart et al., 1973, 1977). (5) Average of 14 basaltic andesites from Raoul Group and L'Esperance Islands, Kermadec arc (Ewart et al., 1977). (6) Average of 38 (12†) olivine low Si andesites from Tongariro Volcanic Centre, Karangahape and White Island, Taupo Volcanic Zone, New Zealand (Cole, 1979; Taylor and White, 1966; Ewart et al. 1968;‡ Ewart and Stipp, 1968§). (7) Average of four andesites from Niuatoputapu, Fonualei, and Late Islands, Tonga (Ewart and Bryan, 1973; Ewart et al., 1973, 1977). (8) Average of three andesites from North and South Meyer Islands, Raoul Group, Kermadec arc (Ewart et al., 1977). (9) Average of 47 (41†) labradorite and labradorite–pyroxene andesites from Tongariro Volcanic Centre, Taupo Volcanic Zone, New Zealand (Cole, 1979; Ewart et al. 1968;‡ Ewart and Stipp, 1968§). (10) Average of 11 dacites from Fonualei, Tonga (Ewart and Bryan, 1973; Ewart et al., 1973, 1977). (11) Average of four dacites from Curtis Island, Kermadec arc (Ewart et al., 1977). (12) Average of 39 (36†) acid andesites and dacites from Bay of Plenty, Taupo Volcanic Zone, New Zealand (Cole, 1979; Ewart et al., 1968;‡ Ewart and Stipp, 1968§). (13) Average of 10 (2†) dacites from Rotorua–Taupo area, New Zealand (Cole, 1979; Ewart and Stipp, 1968§).
† Numbers in brackets refer to number of trace element analyses where different from major element analyses.
‡ REE data.
§ $^{87}Sr/^{86}Sr$ data.

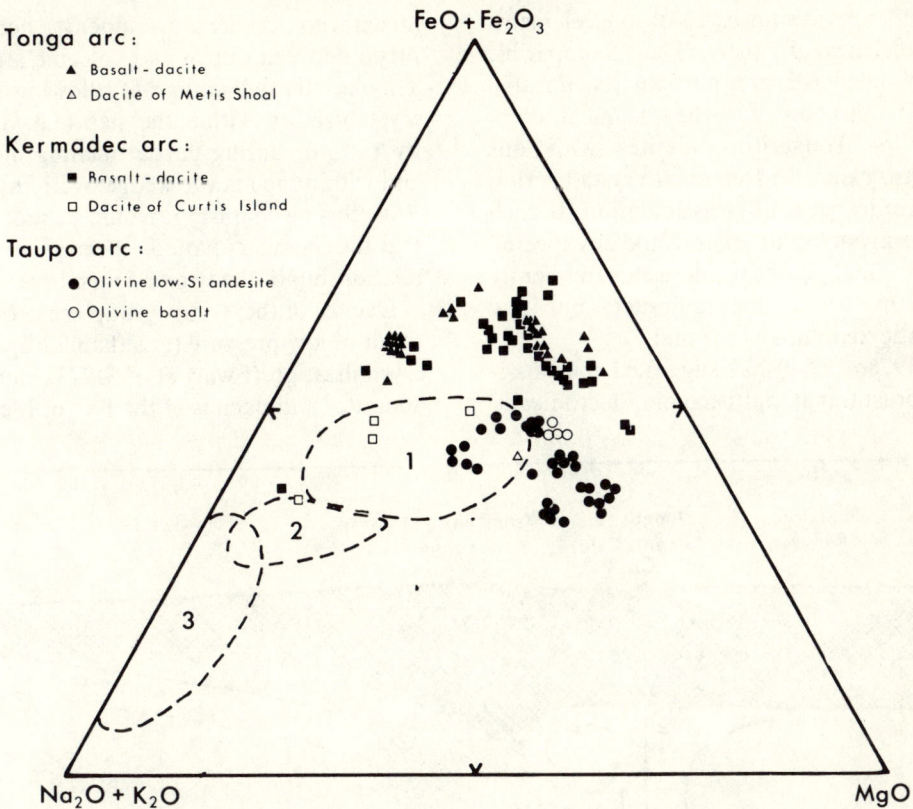

Fig. 6 AFM diagram for lavas of the Tonga, Kermadec, and Taupo arcs. Broken lines enclose fields of lava composition from Taupo Volcanic Zone, New Zealand: 1 = andesites of Tongariro Volcanic Centre and andesites–dacites of Bay of Plenty; 2 = dacites from Rotorua–Taupo area; 3 = rhyolites and ignimbrites of Rotorua–Taupo area

andesite magma is initially generated by lithospheric partial melting under essentially anhydrous conditions and propose a model in which bodies of basaltic andesite rise from the subduction zone during the early stages of arc development, causing partial melting of the peridotite in the overlying mantle wedge and forming a range of compositions from basalt to basaltic andesite. During later stages the bodies will rise in established and relatively stable magma conduits, allowing little interaction with mantle peridotite. This would result in a more homogeneous basaltic andesite assemblage such as that of Tonga.

The chemical and isotopic differences between andesites of the Tonga/Kermadec arcs and the New Zealand arcs indicate that different processes have been involved in their formation. The lavas of both appear to be associated with subduction zones, and the most obvious difference is that the New Zealand arcs are upon a continental margin. Ewart et al. (1977) consider, therefore, that in the Taupo Volcanic Zone the magma is modified by intrusion into the continental crust with resultant mixing and equilibration of the magma. However, the possibility that high alumina basalt of a composition similar to that found north of Lake Taupo had assimilated either a partial or a total melt of greywacke–argillite was discounted by Cole (1978a) on geological and chemical grounds. Mixing of a melt of the composition of Tongan basaltic andesite (Table 1, column 4) and a partial or total melt of greywacke–argillite is also unlikely, as once again the best-fit mixture, using a least squares mixing programme (Bryan *et al.*, 1969), provides an excess of some elements and a deficiency of others. The incompatible elements and REE are particularly unsatisfactory. To account for the chemical compositions of Tongariro andesites with this mechanism, extensive fractional crystallization would have to occur after assimilation. If such a mechanism had occurred, it would be expected that the residual phases would occur frequently as xenoliths within the andesites, but few comagmatic xenoliths are found.

Cole (1978a, 1979) has suggested the possibility of crustal material becoming tectonically mixed with oceanic crust under the accretionary prism between trench and volcanic arc (Fig. 7). This has the advantage of allowing fractional crystallization within the slab (e.g. due to dehydration); during partial melting of the slab, and within the mantle wedge overlying the slab. With this mechanism it would be most unlikely that the chemistry would reflect a simple mixing relationship of the two end members.

Dacites of the Tonga group are probably the result of low pressure (less than 2 kb) fractional crystallization (Ewart *et al.*, 1973), but the acid andesites and dacites of the Bay of Plenty, New

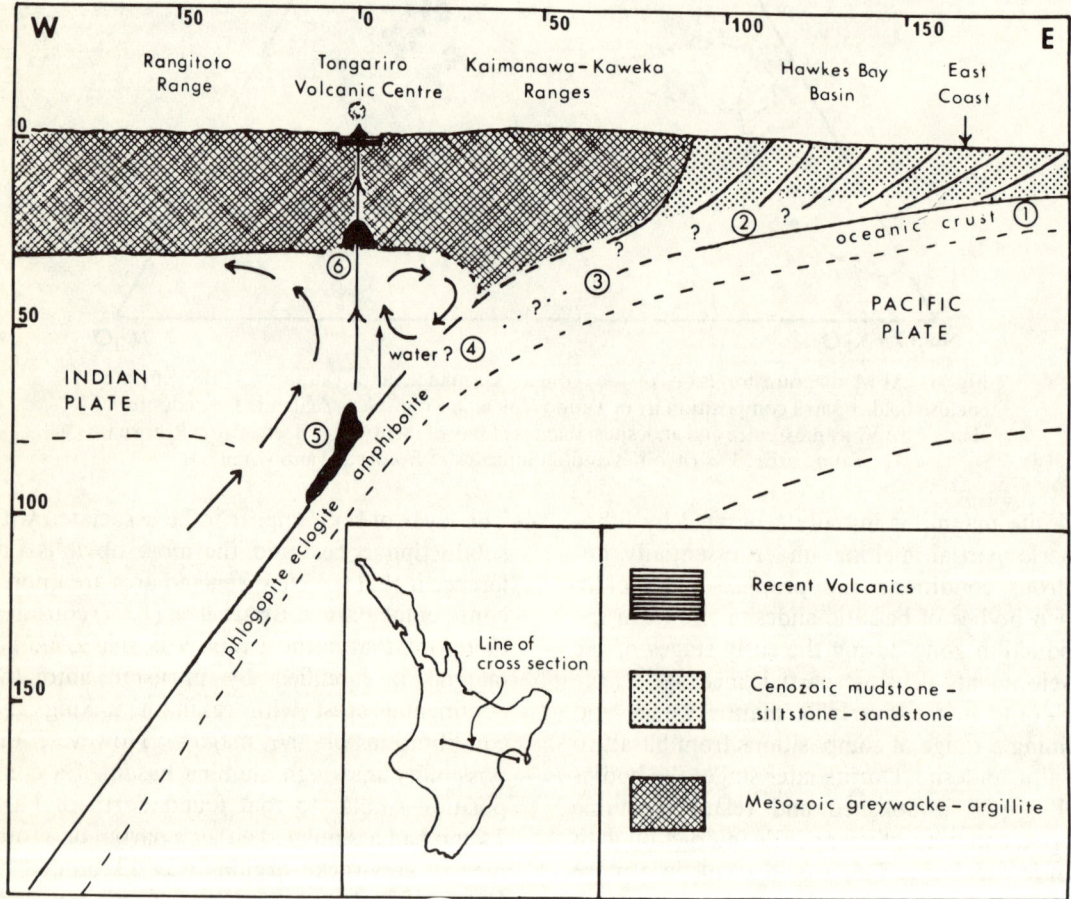

Fig. 7 Schematic diagram illustrating a WNW–ESE cross-section through the central part of the North Island (*see* inset for location). 1 = spilitized oceanic crust; 2 = accretionary prism undergoing thrusting due to plate movement; 3 = mixing of oceanic crust with sediments of the accretionary prism and possibly Mesozoic sediments under the Kaimanawa–Kaweka Ranges; 4 = formation of amphibolite, breaking down between 60 and 90 km, releasing water; 5 = partial melting of amphibolite to produce 'andesitic' magma; 6 = fractionation of magma to produce 'normal' and 'low Si' andesites

Zealand, have a more variable trace element composition (Table 1). It is, however, consistent with assimilation of crustal material or rhyolitic magma by an andesite magma (Cole, 1978a).

Dacites of the Rotorua–Taupo area are more likely to have formed by near complete melting of greywacke–argillite sediments, as proposed by Ewart and Stipp (1968).

REFERENCES

Adams, R. D. and Ware, D. E. (1977). Subcrustal earthquakes beneath New Zealand; locations determined with a laterally inhomogeneous velocity model. *N.Z. J. Geol. Geophys.* **20**, 59–83.

Armstrong, R. L. and Cooper, J. A. (1971). Lead isotopes in island arcs. *Bull. Volc.* **35**, 27–63.

Baker, P. E., Harris, P. G., and Reay, A. (1971). The geology of Tofua Island, Tonga. *R. Soc. N.Z. Bull.* **8**, 67–79.

Ballance, P. F. (1976). Evolution of the Upper Cenozoic magmatic arc and plate boundary in northern New Zealand. *Earth planet. Sci. Letters* **28**, 356–370.

Bauer, G. R. (1970). The geology of Tofua Island, Tonga. *Pacific Sci.* **24**, 333–350.

Black, P. M. (1970). Observations on White Island Volcano, New Zealand. *Bull. Volc.* **34**, 158–167.

Brodie, J. W. (1970). Notes on the volcanic activity at Fonualei, Tonga. *N.Z. J. Geol. Geophys.* **13**, 30–38.

Brodie, J. W. and Hatherton, T. (1958). The morphology of Kermadec and Hikurangi trenches. *Deep. Sea Res.* **5**, 18–28.

Brothers, R. N. (1967). Andesite from Rumble III Volcano, Kermadec Ridge, southwest Pacific. *Bull. Volc.* **31**, 17–19.

Brothers, R. N. (1970). Petrochemical affinities of volcanic rocks from the Tonga–Kermadec island arc, southwest Pacific. *Bull. Volc.* **34**, 308–329.

Brothers, R. N. and Martin, K. R. (1970). The geology of Macauley Island, Kermadec Group, southwest Pacific. *Bull. Volc.* **34**, 330–346.

Brothers, R. N. and Searle, E. J. (1970). The geology of Raoul Island, Kermadec Group, southwest Pacific. *Bull. Volc.* **34**, 7–37.

Bryan, W. B., Finger, L. W., and Chayes, F. (1969). Estimating proportions in petrographic mixing equations by least squares approximation. *Science, N.Y.* **163**, 926–927.

Bryan, W. B., Stice, G. D., and Ewart, A. (1972). Geology, petrology and geochemistry of the volcanic islands of Tonga. *J. geophys. Res.* **77**, 1566–1585.

Clark, R. H., Cole, J. W., Nairn, I. A., and Wood, C. P. (1979). Magmatic eruption of White Island Volcano, New Zealand, December 1976–April, 1977. *N.Z. J. Geol. Geophys.* **22**, 175–190.

Cole, J. W. (1978a). Andesites of the Tongariro Volcanic Centre, North Island, New Zealand. *J. Volc. geothermal Res.* **3**, 121–153.

Cole, J. W. (1978b). Tectonic setting of Mayor Island Volcano (note) *N.Z. J. Geol. Geophys.* **21**, 645–647.

Cole, J. W. (1979). Structure, petrology and genesis of Cenozoic volcanism, Taupo Volcanic Zone, New Zealand—a review'. *N.Z. J. Geol. Geophys.* **22**, 631–657.

Cole, J. W. and Nairn, I. A. (1975). *Catalogue of the Active Volcanoes of the World Including Solfatara Fields*, Part 22, New Zealand, International Association of Volcanology and Chemistry of the Earth's Interior, Naples.

Davey, F. J. (1974). Magnetic anomalies off the West Coast of Northland, New Zealand. *Jl. R. Soc. N.Z.* **4**, 203–216.

Duncan, A. R. (1970). Eastern Bay of Plenty Volcanoes. Ph.D. thesis, Victoria University of Wellington, Wellington.

Eiby, G. A. (1977). The junction of the main New Zealand and Kermadec seismic regions. In *International Symposium on Geodynamics in S.W. Pacific Noumea (New Caledonia)*, Editions Technip, Paris, pp. 167–178.

Ewart, A. and Bryan, W. B. (1972). The petrology and geochemistry of the igneous rocks from Eua, Tongan Islands. *Geol. Soc. Am. Bull.* **83**, 3281–3298.

Ewart, A. and Bryan, W. B. (1973). The petrology and geochemistry of the Tongan Islands. In *The Western Pacific: Island Arcs, Marginal Seas, Geochemistry* (P. J. Coleman, ed.) University of Western Australia Press, Perth, pp. 503–522.

Ewart, A. and Stipp, J. J. (1968). Petrogenesis of the volcanic rocks of the Central North Island, New Zealand, as indicated by a study of Sr^{87}/Sr^{86} ratios, and Sr, Rb, K, U and Th abundances. *Geochim. cosmochim. Acta* **32**, 699–735.

Ewart, A., Taylor, S. R., and Capp, A. C. (1968). Trace and minor element geochemistry of rhyolitic volcanic rocks, Central North Island, New Zealand—total rock and residual liquid data. *Contr. Mineral. Petrol.* **18**, 76–104.

Ewart, A., Bryan, W. B., and Gill, J. B. (1973). Mineralogy and geochemistry of the younger volcanic islands of Tonga, S.W. Pacific. *J. Petrol.* **14**, 429–465.

Ewart, A., Brothers, R. N., and Mateen, A. (1977). An outline of the geology and geochemistry and the possible petrogenetic evolution of the volcanic rocks of the Tonga–Kermadec–New Zealand island arc. *J. Volc. geothermal Res.* **2**, 205–250.

Fleming, C. A. and Steiner, A. (1951). Sediments beneath Ruapehu Volcano. *N.Z. J. Sci. Technol. B* **32**, 31–32.

Grindley, G. W. (1965). Tongariro National Park—stratigraphy and structure. In *New Zealand Volcanology—Central Volcanic Region* (A. Ewart, J. Healy, and B. N. Thompson, eds), N.Z. DSIR Information Series no. 50, pp. 79–86

Hamilton, R. N. and Gale, A. W. (1968). Seismicity and structure of North Island, New Zealand. *J. geophys. Res.* **73**, 3859–3876.

Hatherton, T. (1969). The geophysical significance of calc-alkaline andesites in New Zealand. *N.Z. J. Geol. Geophys.* **12**, 436–459.

Hayward, B. W. (1979). Eruptive history of the early to mid-Miocene Waitakere volcanic arc, and paleogeography of the Waitemata Basin, northern New Zealand. *Jl R. Soc. N.Z.* **9**.

Healy, J., Lloyd, E. F., Rishworth, D. E. H., Wood, C. P., Glover, R. B., and Dibble, R. R. (1978). The eruption of

Ruapehu, New Zealand on June 22, 1969. *N.Z. DSIR Bull.* no. 224.

Jakeš, P. and Gill, J. B. (1970). Rare earth elements and the island arc tholeiitic series. *Earth planet. Sci. Letters* **9**, 17–28.

Karig, D. E. (1970). Ridges and basins of the Tonga–Kermadec island arc system. *J. geophys. Res.* **75**, 239–254.

Kroenke, L. and Tongilava, S. L. (1975). A structural interpretation of two reflection profiles across the Tonga arc. *S. Pacific marine Geol. Notes* **1**, 9–15.

Melson, W. G., Jarosewich, E., and Lundquist, C. A. (1970). Volcanic eruption at Metis Shoal, Tonga, 1967–1968. Description and petrology. *Smithson. Contr. Earth Sci.* no. 4.

Nairn, I. A., Hewson, C. A. Y., Latter, J. H., and Wood, C. P. (1976). Pyroclastic eruptions of Ngauruhoe Volcano, Central North Island, New Zealand, 1974, January and March. In *Volcanism in Australasia* (R. W. Johnson, ed.), Elsevier, Amsterdam, pp. 385–405.

Nairn, I. A. and Self, S. (1978). Explosive eruptions and pyroclastic avalanches from Ngauruhoe in February 1975. *J. Volc. geothermal Res.* **3**, 39–60.

Oversby, V. M. and Ewart, A. (1972). Lead isotopic compositions of Tonga–Kermadec volcanics and their petrogenetic significance. *Contr. Mineral. Petrol.* **37**, 181–210.

Richards, J. J. (1962). *Catalogue of the Active Volcanoes of the World, Including Solfatara Fields*, Part 13, *Kermadec, Tonga and Samoa*, International Association of Volcanology and Chemistry of the Earth's Interior, Naples.

Sameshima, T. (1975). Silica indices of volcanoes in and around New Zealand with reference to volcanic zones in the North Island. *N.Z. J. Geol. Geophys.* **18**, 523–539.

Stipp, J. J. (1968). The geochronology and petrogenesis of the Cenozoic volcanics of the North Island, New Zealand. Ph.D. thesis, Australian National University, Canberra.

Sykes, L. (1966). The seismicity and deep structure of island arcs. *J. geophys. Res.* **66**, 1265–1278.

Taylor, S. R. and White, A. J. R. (1966). Trace element abundances in andesites. *Bull. Volc.* **29**, 177–194.

Walcott, R. W. (1978). Present tectonic and late Cenozoic evolution of New Zealand. *Geophys. J. R. astr. Soc.* **52**, 137–164.

Japan

S. Aramaki and T. Ui

Earthquake Research Institute,
University of Tokyo, Bunkyo-Ku, Tokyo 113, Japan

and

Department of Earth Sciences,
Faculty of Science, Kobe University, Nada, Kobe 657, Japan

ABSTRACT

Japan and the surrounding islands form three chains of island arcs which meet at a triple junction in central Japan. These are the North-east Honshu (or North-east Japan) and Kurile arcs, the South-west Honshu and Ryukyu arcs (or South-west Japan arc) and the Izu–Bonin and Mariana arcs in the south. These arcs have characteristics ranging from youthful single arcs (central Kurile arc) to mature double arc systems (the North-east Honshu arc). The Pacific side of the volcanic belts is sharply delineated by the volcanic front, which is characterized by the densest population of volcanic vents. The volcanic front is parallel with the inclined Wadati–Benioff zone, but the parallelism is broken at the junction of the Izu-Bonin and North-east Honshu arcs, where the front makes an acute angle pointing away from the trench.

Miocene volcanism was widespread within the Japanese arcs and the distribution of submarine, calc-alkaline volcanic products roughly parallels the Quaternary volcanic belts. High Mg bronzite andesites of Miocene age are known from the Bonin Islands and the outer zone of South-west Japan arc, where they occur with high K, calc-alkaline volcanic and intrusive rocks. The Quaternary volcanism is characterized by (1) conical composite volcanoes in which the oldest rocks are mafic andesites and the youngest are more felsic products that were erupted explosively as pyroclastic falls, flows, and avalanches; (2) large scale dacite–rhyolite pyroclastic flows and resultant Crater Lake-type calderas and post-caldera cones and domes; and (3) groups of monogenetic volcanoes such as basaltic scoria cones, lava flows, and domes. Many major eruption episodes follow a pattern starting with Plinian pumice- and ash-fall, progressing through pyroclastic flow generation and ending with lava eruptions.

The bulk composition of the Japanese Quaternary volcanic rocks shows a unimodal distribution for most elements, with a principal mode at 57 per cent SiO_2, when large scale pyroclastic flows are disregarded. Three basalt series (with differentiates) can be distinguished—the low alkali tholeiitic, high alumina and/or high alkali tholeiitic, and the alkaline series, these being arranged parallel to each other at progressively greater distances from the trench. However, the calc-alkaline series (equivalent to Kuno's hypersthenic rock series) is much more abundant than these basaltic series and is characterized by a greater abundance of andesites and dacites in comparison with basalt. Such calc-alkaline rocks occur throughout the areas characterized by the basaltic series.

Throughout the volcanic arcs there is a tendency for K, Ba, Sr, Rb, REE, Hf, Th, U, Pb, La/Sm, and La/Yb to increase away from the volcanic front, although the levels of absolute concentration are locally variable. The southern (oceanic) part of the Izu–Bonin arc is characterized by the abundance of low alkali tholeiite and its differentiates, while in the north part, around Fuji, large volumes of aphyric high alumina basalt were erupted. The South-west Japan arc is characterized by relatively high levels of K and other incompatible elements, possibly suggesting crustal contamination. This is consistent with REE patterns and Sr isotope data. But a major process responsible for variation in calc-alkaline magma is fractional crystallization. Studies of F contents and the order of crystallization of hornblende and biotite indicate that volatile contents increase away from the volcanic front, where volatile concentration and fractional crystallization of calc-alkaline magma takes place in crustal magma chambers.

Present-day Configuration of the Island-arc System and Distribution of Quaternary Volcanoes

Japan and the surrounding islands form three chains of island arcs which meet at a trench–trench–trench-type triple junction located at c. 34°N and 142°E (Fig. 1). The north-eastern chain consists of the North-east Honshu (or North-east Japan) and Kurile arcs. The southern chain consists of the Izu–Bonin and Mariana arcs (or Izu–Mariana arc), and the south-western chain consists of the South-west Honshu and Ryukyu arcs (or South-west Japan arc; Sugimura and Uyeda, 1973).

The North-east Honshu arc is a mature double arc made up of a non-volcanic outer arc on the Pacific side and a volcanic arc on the Japan Sea side. The double arc structure terminates in the Central Kurile islands (Fig. 2 of Gorshkov, 1970) where the outer arc is lacking. However, to the north the arc again widens and becomes double and in Kamchatka the inner arc has a graben structure (Erlich, 1968). The trench running parallel with the Kurile arc is deeper than 7000 m along most of its axis to the junction with the Aleutian trench. The southern part of the Izu–Mariana arc is a very wide complex arc, mostly with a submerged outer arc and a narrow volcanic arc and a deep, well defined trench. The features of South-west Honshu arc are somewhat obscure because it is built upon complex pre-Quaternary structures. The accompanying Nankai trough is shallow and broad, because it is buried by sediments. The Ryukyu arc is also built on older continental basement with well defined volcanic chains on the inner arc. The Ryukyu trench is clearly delineated with a depth of more than 6000 m for most of its length.

Quaternary volcanoes are densely distributed along these arcs. Sugimura (1960) grouped these into two broad volcanic belts: the East Japan Volcanic Belt including the Kurile, North-east Honshu, and Izu–Mariana arcs and the West Japan Volcanic Belt including the South-west Honshu and Ryukyu arcs. The north-eastern extension of the former arc includes Kamchatka. Catalogues of geographic position and various parameters of the Quaternary volcanoes may be found in the several references summarized in Table 1. Although the identification and grouping of the Quaternary volcanoes and the numbering system of individual volcanic centres differ within these references, c. 440 volcanic centres may be recognized in these areas. This is about 10 per cent of the total number of Quaternary volcanic centres in the world. About 140 of these (32 per cent) are considered as active, usually meaning that they have erupted in historic time, although the records of observed eruptions may be based on relatively short periods of up to the last 1300 a in Japan.

The limit of the volcanic belt on the trench side is sharp and runs parallel to and at a distance of 150–300 km away from the trench axis. This line (the thick solid line in Fig. 2) was defined as the 'volcanic front' by Sugimura (1960). The density of distribution of volcanoes and the volume of volcanic products are both at the maximum along the volcanic front, gradually decreasing towards the back-arc side (Fig. 3). The width of the volcanic belt ranges from 270 km (Rishiri volcano near the junction of Kurile and North-east Honshu arcs, Fig. 2) to almost a single chain of volcanic islands in parts of Izu–Mariana and Ryukyu arcs.

Local Grouping and Alignment of Volcanic Centres

The distribution density of volcanoes varies along the volcanic front (Fig. 2). In the North-east Honshu arc, 27 volcanic edifices lie along 880 km of the volcanic front. The distance between the volcanoes ranges from 6 to 40 km with

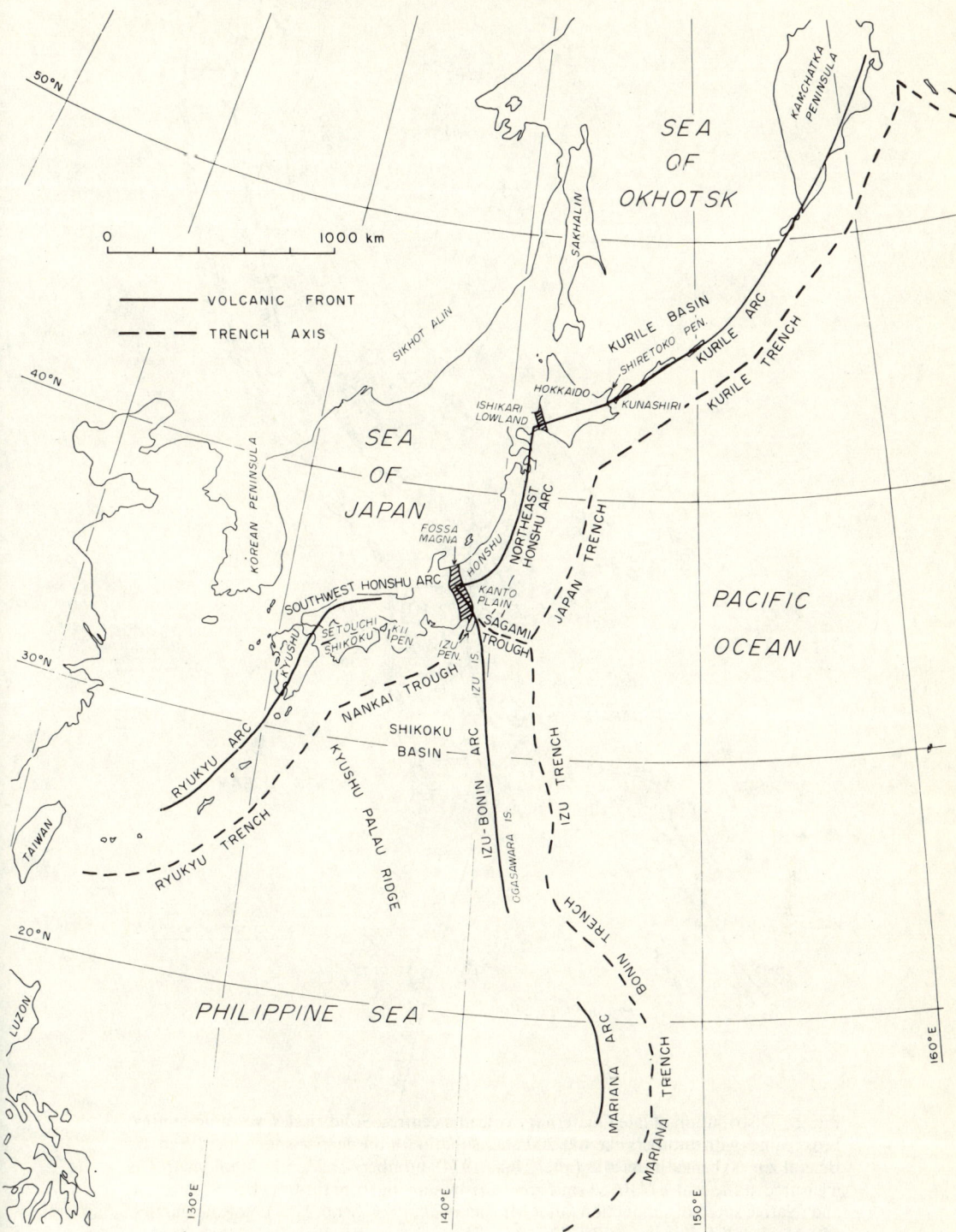

Fig. 1 Index map showing the tectonic setting of Japan and surrounding areas

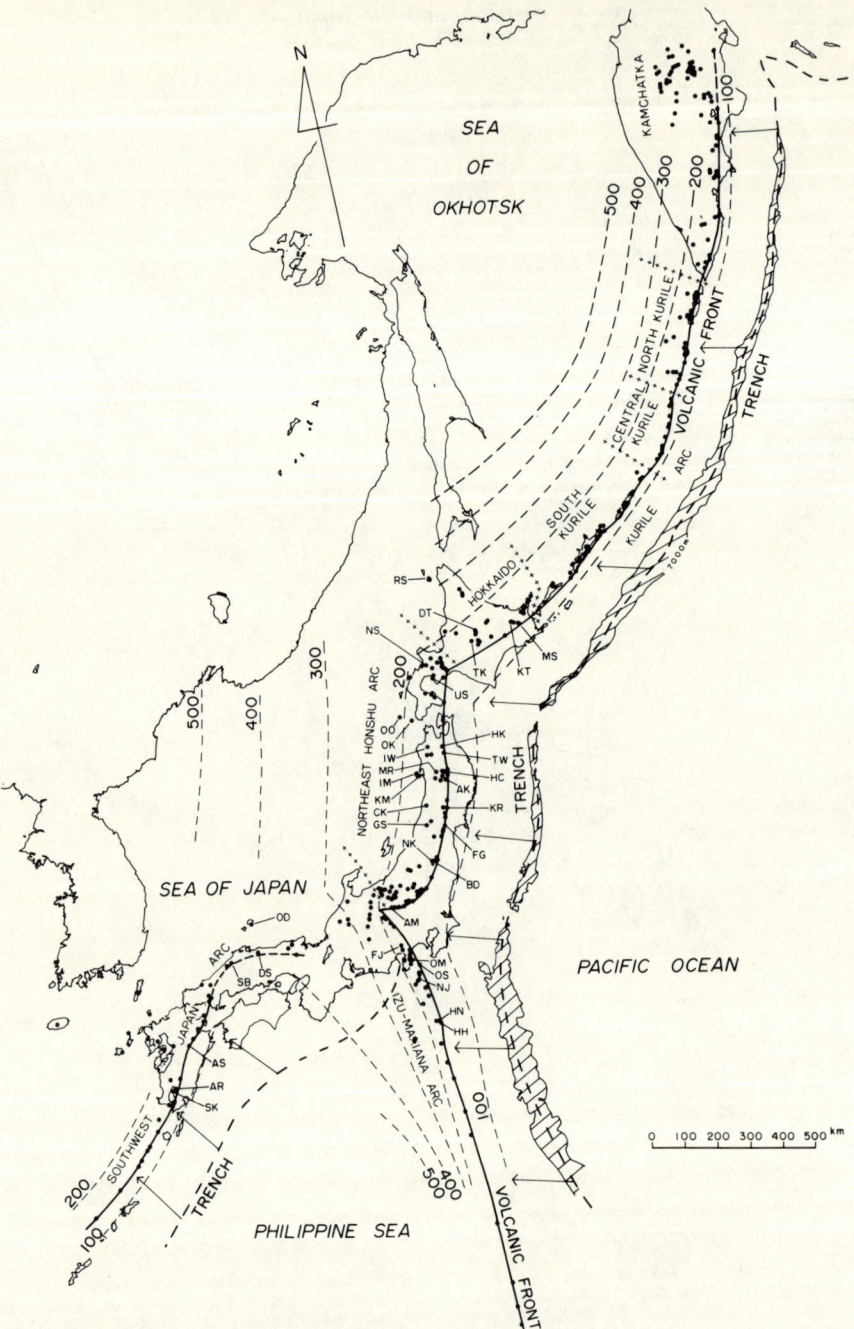

Fig. 2 Distribution of late Quaternary volcanic centres. Solid circles, volcanic centres; broken lines with numbers like 100, 200, etc., isobaths for the deep seismic zones (Wadati–Benioff zones) beneath the arcs (after Utsu, 1974), numbers are depths in kilometres as measured at the centre of the seismic zone; arrow starting from the trench axes, direction and relative speed of subduction (after Minster *et al.*, 1974; Seno, 1977). Volcano names: MS, Mashu; KT, Kutcharo; DT, Daisetsu; TK, Tokachi; RS, Rishiri; NS, Niseko; US, Usu; OO, Oshima–Oshima; OK, Oshima–Kojima; IW, Iwaki; HK, Hakkoda; TW, Towada; HC, Hachimantai; MR, Moriyoshi; KM, Kampu; IM, Ichinomegata; AK, Akita–Komagatake; KR, Kurikoma; CK, Chokai; GS, Gassan; FG, Funagata; BD, Bandai; NK, Nekoma; AM, Asama; FJ, Fuji; OM, Omuroyama; OS, Oshima (Izu–Oshima); NJ, Niijima; HN, Hachijo–Nishiyama; HH, Hachijo–Higashi-yama; DS, Daisen; OD, Oki–Dogo; SB, Sambe; AS, Aso; AR, Aira; SK, Sakurajima

TABLE 1 Itemized list of papers which give systematic references and/or parameters for the volcanoes and volcanic rocks of Japan, Mariana, Kurile, and Kamchatka

Item	Japan	Mariana	Kurile	Kamchatka
Geographic position	1,2,3,4,7,8,11	1,2,3,4,18	1,2,3,5,12	1,2,3,6
Trench–volcanic gap	1,7,19	1	1,19	1
Height of volcano	2,3,4,11	2,3,4,18	2,3,5	2,3,6
Type of volcano	2,3,4,7,15	2,3,4,18	2,3,5,12	2,3,6
Volume of volcanic edifice	1,11,15	1	1	1
Summary of geology	4	4,19	5,12	6
Eruptive activity	2,3,4,7	2,3,4	2,3,5,12	2,3,6
Petrology and chemistry	4,20,21,22,23,24	4,18	12,21	6
Bouguer anomaly	1,9,13			
Depth of Wadati–Benioff zone	1,17		1,17	1,17
Thickness of crust	1,14		1,16	1,16
Heat flow	1,10		1,10	

(1) Aramaki and Ui (1978b); (2) Working Group on The World Volcanological Map, IAVECI (1973); (3) Katsui (1971); (4) Kuno (1962); (5) Gorshkov (1958); (6) Vlodavetz and Piip (1959); (7) Isshiki et al. (1968a); (8) Isshiki et al. (1968b); (9) Tomoda (1973); (10) Uyeda (1972); (11) Sugimura (1965); (12) Gorshkov (1970); (13) Hagiwara (1967); (14) Kanamori (1963); (15) Moriya (1977); (16) Tuyezov (1971); (17) Utsu (1974); (18) Ishikawa and Egawa (1977); (19) Geographical Survey Institute of Japan (1973); (20) Kawano et al. (1961); (21) Katsui (1961); (22) Ono (1962); (23) Shibata (1968); (24) Katsui et al. (1978a).

a mean of 23 km except for five gaps where the distance is 60–80 km. Where the density of volcanoes at the front is high, the density of volcanic centres within the back-arc region is also generally high. Locally, *en echelon* arrangements of short chains of volcanic centres at an acute angle to the volcanic front are conspicuous (for example in eastern Hokkaido and southern Kurile, and in the northern part of Izu–Bonin arc).

On a small scale, preferred orientation is common where lateral (parasitic) craters are abundant on the surface of a major composite volcano, or in a group of monogenetic volcanoes. Nakamura (1977) examined examples of such preferred orientation on 26 of 200 Quaternary volcanoes in the Japanese islands. A distribution map for the Japanese islands indicates a similarity in the direction of alignments between neighbouring volcanoes. This suggests that it is difficult to interpret such alignments in terms of localized 'weak zones', but that they are better interpreted as reflecting the trace of regional σ_{Hmax}, which is the easiest direction to form a tensional fracture in the crust near the surface (Jacob et al., 1977). Thus, considering the regional stress to be compressional, the direction of eruptive fissures and intrusive dykes are aligned parallel to the axis of maximum compression.

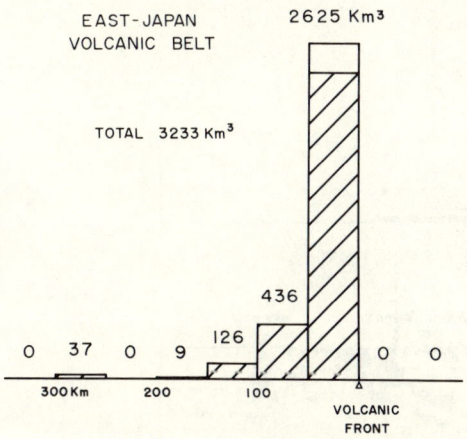

Fig. 3 Variations of the volume of late Quaternary volcanic material across the East Honshu volcanic belt. After Fig. 13 of Sugimura et al. (1963) and Figure 121 of Sugimura and Uyeda (1973) but modified using the data of Aramaki and Ui (1978a). The hatched column indicates the lavas and pyroclastic rocks which form the composite volcanoes and the open part of the column indicates pyroclastic flow and fall deposits associated with large calderas

Structure of Lithosphere and Asthenosphere

Figure 2 shows the contour lines of a well defined Wadati–Benioff zone extending across the Sea of Japan (Japan Sea) as far as 1200 km from the trench axis and more than 600 km deep. This part represents one of the widest horizontal extensions of the inclined deep seismic zone in the world, but the occurrence of active volcanoes is apparently restricted to much narrower zones within this area.

Cross-sections in Fig. 4 illustrate the subsurface structure across the North-east Honshu arc (Yoshii, 1979, partly modified). A double-planed structure of the deep seismic or Wadati–Benioff zone has been found by the Tohoku University seismic network in the traverse

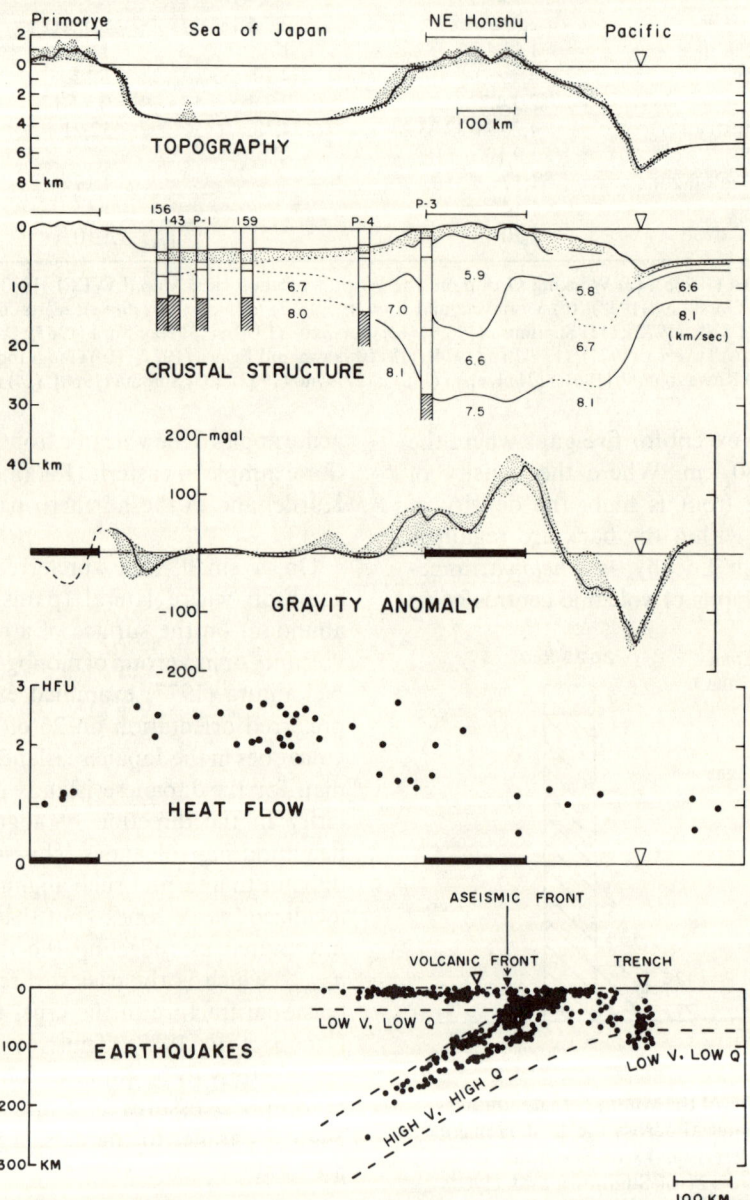

Fig. 4 Topography, crustal structure, gravity anomaly, heat flow, and depth of earthquake foci along the transverse section at 38–40°N latitude across the North-east Honshu arc. From Yoshii (1979) with earthquake foci from Hasegawa et al. (1978)

between 38°N and 41°N latitude (Hasegawa et al., 1978). The two planes are nearly parallel with a vertical separation of 30–40 km and the double-planed structure terminates at a depth of c. 180–200 km. This structure has been determined only for the North-east Honshu arc, probably due to the high resolution of the seismic network in this area.

The relation between the trench–volcano gap and the depth of the Wadati–Benioff zone for individual volcanic centres is shown in Fig. 5. The depth of the Wadati–Benioff zone along the volcanic front ranges from 130 to 200 km. The trench–volcano gap is smaller in the South-west Honshu arc, and wider in the North-east Honshu arc.

A region of very low seismicity occupies most parts of the wedge-shaped section of asthenosphere above the seismic zone extending into the lower half of the crust below the Japanese arcs (Fig. 4). The eastern edge of this region is well defined and is located 50–100 km east of the volcanic front. Yoshii (1975) named the horizontal delineation on the map of this limit the 'aseismic front' which corresponds to the easternmost limit of the relatively low Q zone above

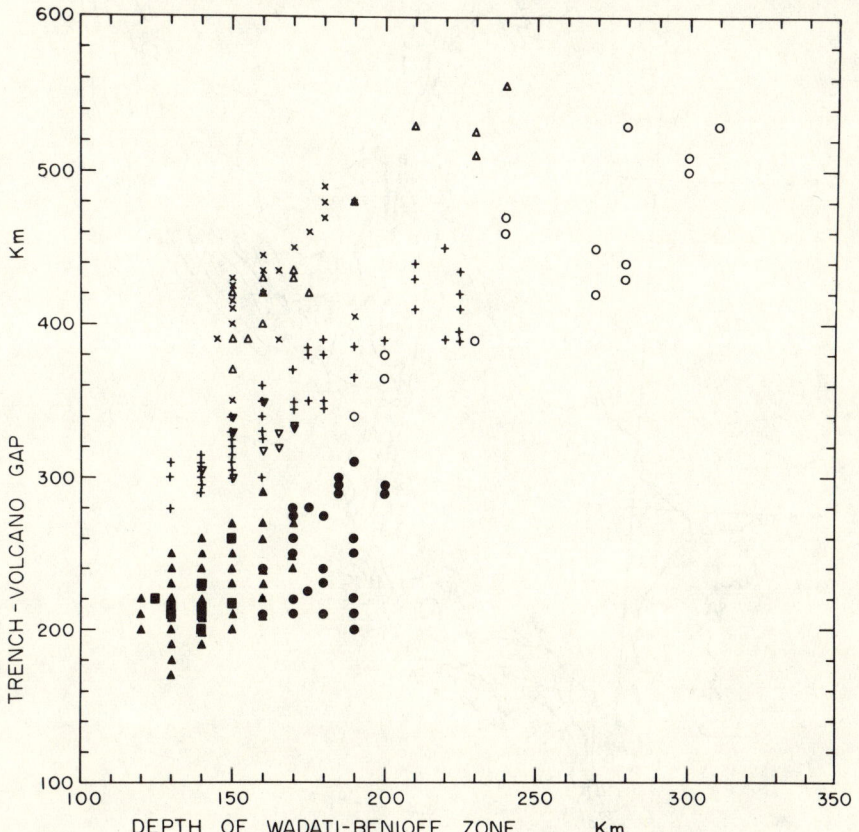

Fig. 5 Relation between depth of Wadati–Benioff zone and trench–volcano gap for individual volcanic centres (Aramaki and Ui, 1978a). Symbols denote different arcs as follows: solid square (■), Ryukyu arc; solid circle (●), Izu–Bonin arc excluding the central Honshu; open circle (○), central Honshu (Izu–Bonin arc); plus (+), North-east Honshu arc excluding western Hokkaido; cross (×), western Hokkaido (North-east Honshu arc); open triangle (△), central Hokkaido (Kurile arc); open inverted triangle (▽), eastern Hokkaido (Kurile arc); solid triangle (▲), Kurile arc excluding Hokkaido. No data for the Wadati–Benioff zone is available for the Mariana arc

the descending slab which is characterized by a high Q value (Utsu, 1971, 1974). The heat flow is low (less than 1.5 heat flow units) to the east of the volcanic front as far as the trench axis, while it is high (more than 1.5 heat flow units) to the west, including most of the Sea of Japan (Uyeda, 1972) (Fig. 4). This contrast is generally true for the whole of the East Japan Volcanic Belt (Uyeda, 1972). The presence of a low V, low Q mantle wedge and higher heat flow above the descending slab of oceanic lithosphere beneath the North-east Honshu arc may reflect the partially molten state of the asthenosphere (Sugimura and Uyeda, 1973).

The regional gravity anomaly is negative (-150 mgal) near the trench axis and positive ($+100 - 150$ mgal) along the axis of the outer arc (Fig. 4). The shallow structure of the '6.6

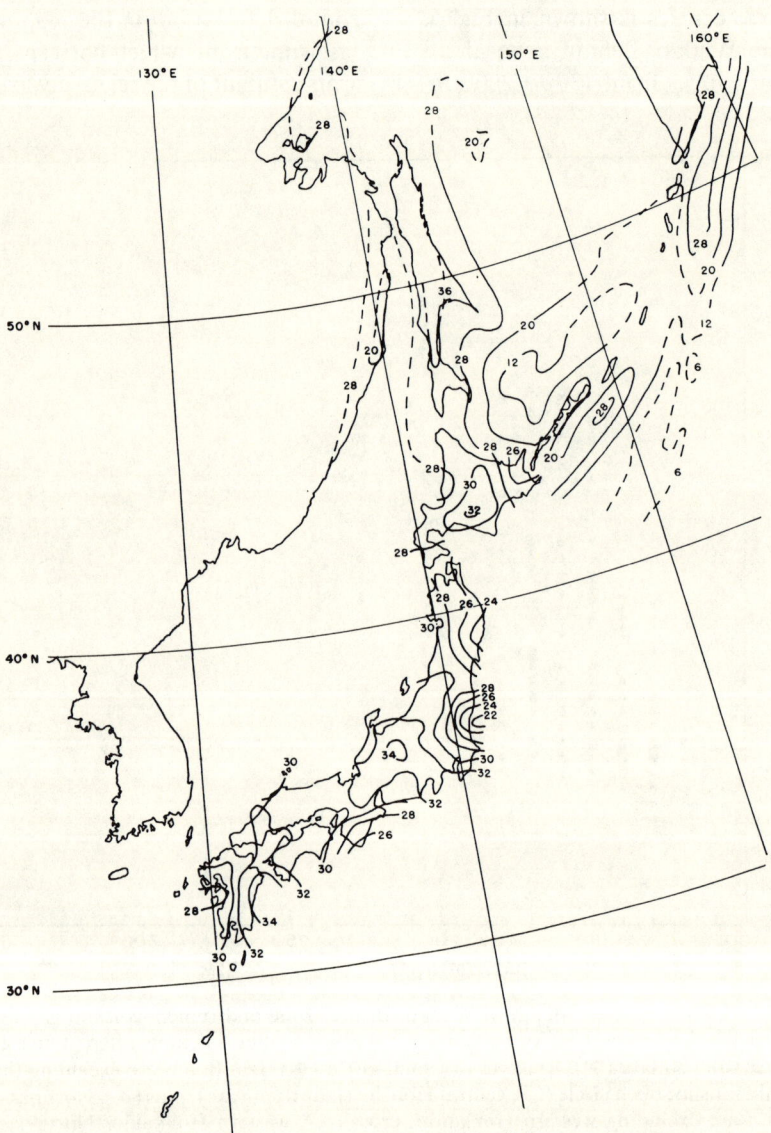

Fig. 6 Contour map of the crustal thickness determined from gravimetric and seismic explosion data (compiled from Kanamori, 1963; Tuyezov, 1971)

km s⁻¹ layer' in the outer arc of north-eastern Honshu (Fig. 4) indicates the presence of 'basement' rocks of probable Precambrian or early Paleozoic age. Regional variations in crustal thickness show that the crust is 24–34 km in thickness below Japan but decreases to c 15 km in the central Kurile islands (Fig. 6). The thickness of the Japanese crust (Kanamori, 1963; Tuyezov, 1971), however, does not show any regular pattern parallel to the axis of the double arc. This may reflect pre-Cainozoic faulting. Two models of the petrological features of the lower crust and uppermost mantle are available (Fig. 7), one at Ichinomegata on the Japan Sea coast of north-eastern Honshu and the other for Oki-Dogo in the Sea of Japan off south-western Honshu (Takahashi, 1978). These models are based upon the study of mafic and ultramafic inclusions in Neogene–Quaternary alkaline and high alumina basalts. The common occurrence of hornblende gabbro and amphibolite inclusions at Ichinomegata indicates the presence of a hydrous lower crust beneath the western part of north-east Japan while the absence of these rock types at Oki-Dogo indicates dry conditions at and below the Moho beneath the northern part of south-western Japan (Takahashi, 1978).

In contrast to north-eastern Honshu, the volcanic front is only poorly defined along the South-west Honshu arc (Fig. 1). This lower activity is reflected in (1) distinctly lower eruption rates and the abundance of monogenetic rather than polygenetic volcanoes (Fig. 8); (2) sporadic occurrence of hot springs in the outer zone south of the volcanic front, in contrast with the East Japan Volcanic Belt where high temperature (over 30 °C) hot springs are characteristically absent in the outer zone; (3) the extension of the anomalous high heat flow zone from the Sea of Japan to the Shikoku basin, crossing the volcanic zone without a negative anomaly as in the outer arc of the North-east Honshu arc (Sugimura, 1978). Finally, seismic activity along the Wadati–Benioff zone terminates beneath Shikoku island and the southern

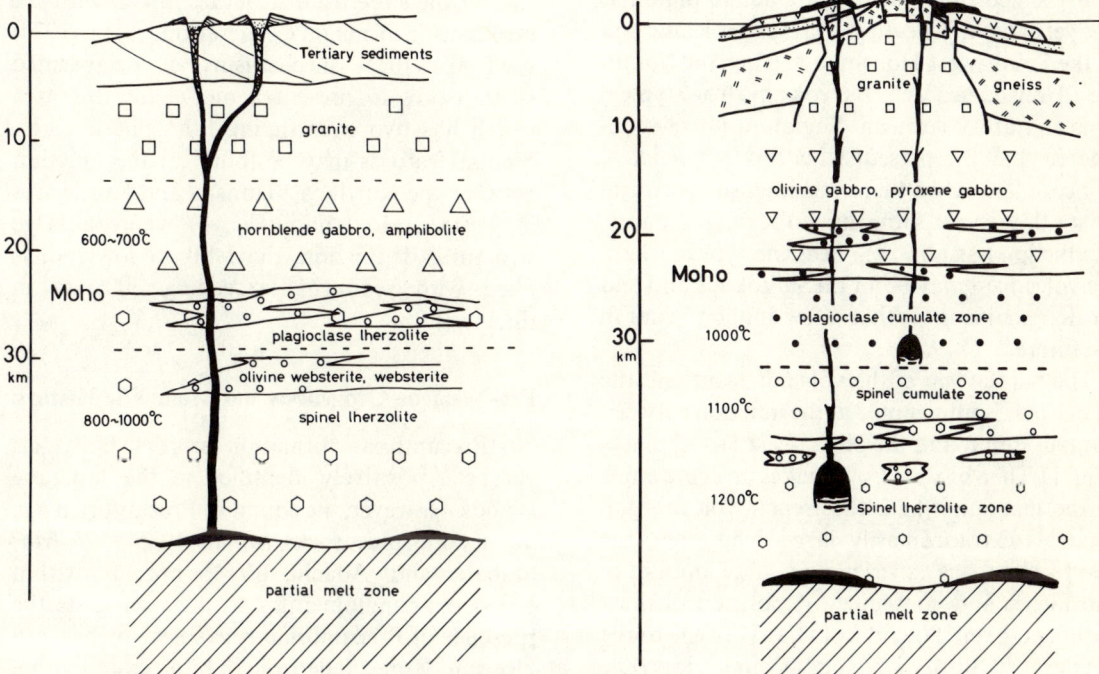

Fig. 7 Petrological model for the crust and the uppermost mantle in Oki–Dogo area, South-west Honshu arc (right) and Ichinomegata area, North-east Honshu arc (left) (Takahashi, 1978)

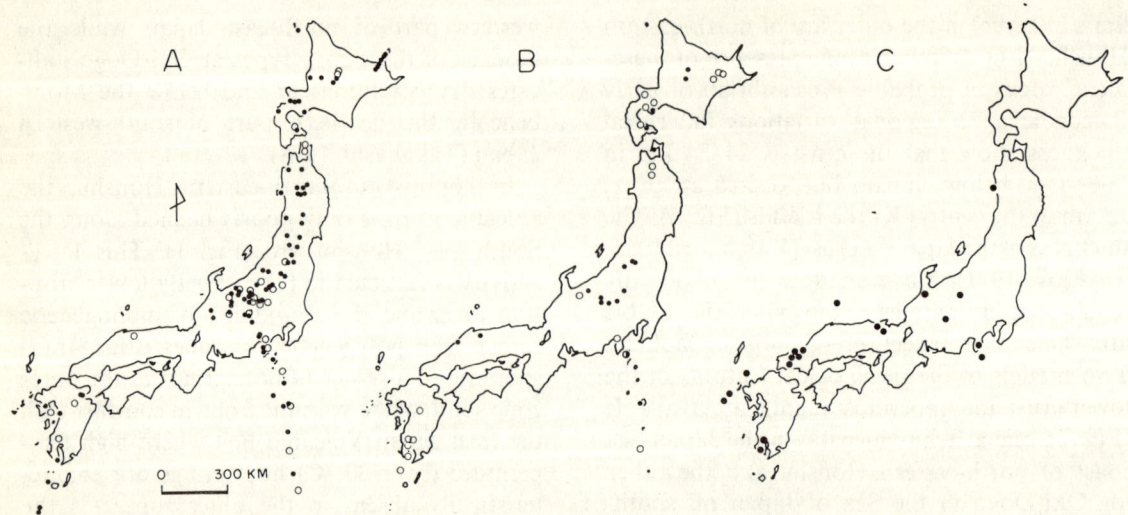

Fig. 8 Distribution of three types of Japanese late Quaternary volcanoes (Table 5). After Moriya (1973, 1977) with modifications. (A) Composite volcanoes; solid circle (●) = simple composite volcanoes, open circle (○) = complex or fully grown composite volcanoes. (B) Collapse calderas formed by pyroclastic flow eruptions. Solid circle (●) = calderas of Haruna type (smaller scale), open circle (○) = calderas of Crater Lake (Krakatau) type, half solid circle (◐) = caldera of Kilauea type (Oshima). (C) Monogenetic volcano groups

part of Kii peninsula, Honshu, at a depth of only 80 km (Kanamori and Tsumura, 1971; Shiono, 1974), suggesting that the subducting plate has not yet arrived beneath the present volcanic belt of the South-west Honshu arc. Thus the Southwest Honshu arc lacks the volcanic front typical of island arc volcanism. Sugimura (1978) considered that the present situation is a relict of Pliocene age, when there was intense volcanism in south-western Honshu. Following a pause, subduction has now resumed, and typical island arc volcanism may occur on Shikoku island and the Kii peninsula within a few million years in the future.

The parallelism of the volcanic front and the trench axis is interrupted at the junction between the Izu–Bonin and the North-east Honshu arcs (Fig. 1). The volcanic front makes an acute angle at the junction where the trench–volcano gap becomes extraordinarily large. This area, the Kanto plain, has experienced a large amount of subsidence and consequent deposition of thick sediment (over 3000 m) from the Miocene to the Quaternary. Since the normal asymmetric profile of the volcanic arc and the low K_2O of the volcanic rocks along the front (Fig. 9), is retained in the section near the junction, the sharp bend in the volcanic zone may be the surface expression of the subcrustal structure linked with the mechanism of magma generation. Aoki (1974) used the focal mechanisms of deep-seated earthquakes to present a model for this area which has two overlapping lithospheric slabs. Similar features may be found at the junction between the North-east Honshu and Kurile arcs in western Hokkaido (Fig. 2) where a large structural depression, the Ishikari lowland, is filled with sediments exceeding 3000 m in thickness.

Pre-Neogene Orogenesis and Magmatic History

No Precambrian formation or rock body has yet been positively identified in the Japanese islands. However, the common Precambrian age of metamorphic rocks (up to 1985 ± 25 Ma; Shibata and Adachi, 1974) present within Paleozoic conglomerates strongly suggests the presence of Precambrian rocks near or beneath Honshu. Upper Paleozoic ophiolitic submarine basalt volcanism in south-western Honshu produced a variety of basalt series including low

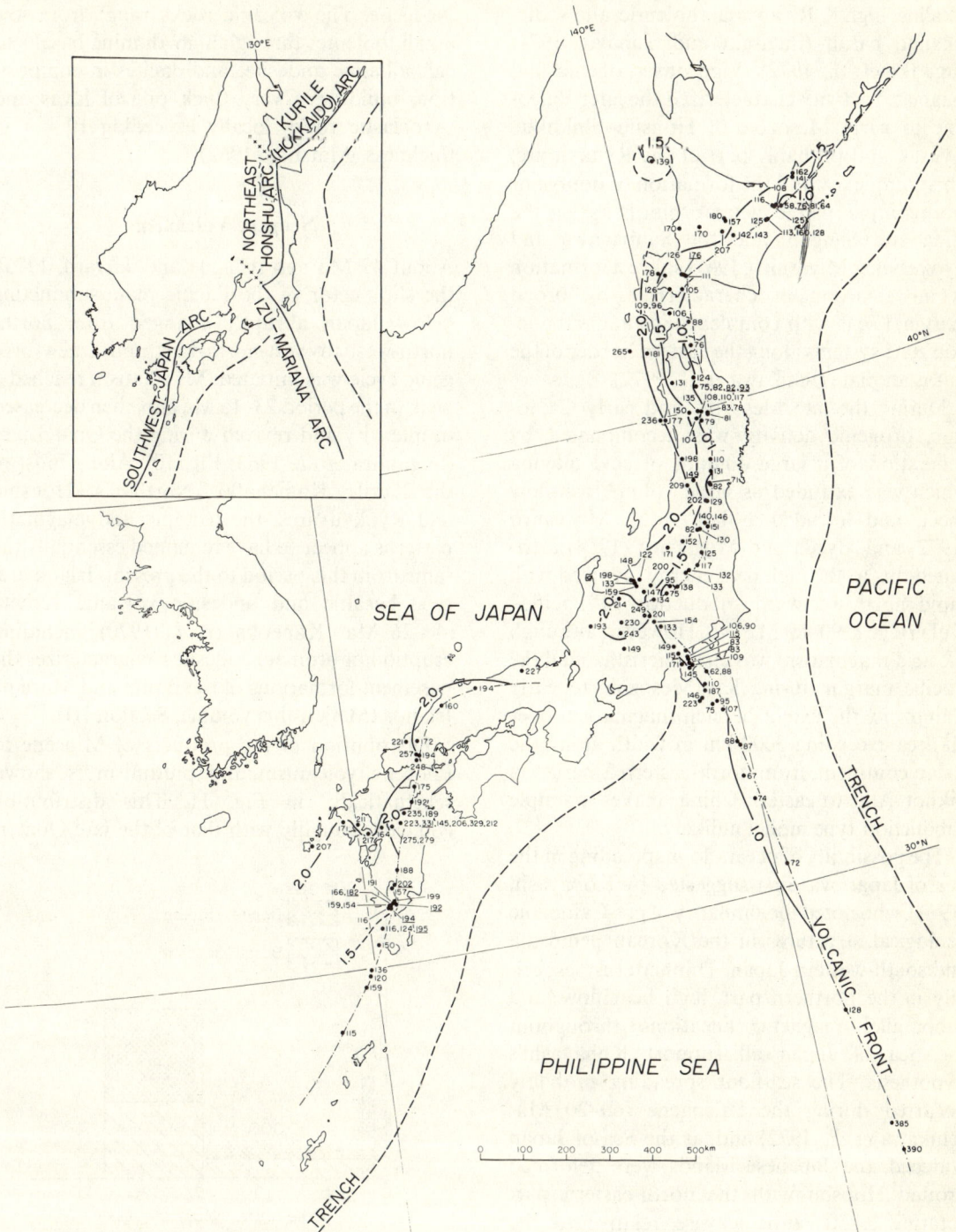

Fig. 9 Map showing lateral variation of K_2O (in per cent by weight) in Japanese Quaternary volcanic rocks (S. Aramaki and T. Ui, unpublished data). The K_2O content is normalized to $SiO_2 = 60$ per cent by weight on an empirically determined differentiation trend. Average values are given for individual centres. Values for large scale pyroclastic flows are underlined. Contours are for $K_2O = 1.0$, 1.5, and 2.0 per cent

alkaline, high K/Rb abyssal tholeiitic, and sodic-alkaline basalt (Sugisaki and Tanaka, 1971; Sugisaki et al., 1972). Vigorous orogenic and magmatic activity characterized the later Paleozoic to early Mesozoic of Honshu–Shikoku–Kyushu and probably part of the Ryukyu arc. This complexity and the formation of double or more complex paired metamorphic belts (e.g. the Hida and Sangun, and the Sanbagawa and Ryoke belts; Miyashiro, 1961) suggest formation in an environment characterized by broad marginal seas with complex ridge and subduction zone systems along the eastern border of the Eurasian plate (Ichikawa et al., 1972).

During the late Mesozoic and early Cainozoic, orogenic activity was accompanied by generation of a large amount of acid magma which was extruded as large volume ash-flow sheets and intruded as batholiths. Miyashiro (1972) and Uyeda and Miyashiro (1974) attributed this to the high heat energy involved with rapid north-westward subduction of Pacific–Kula ridge c. 90–80 Ma ago. However, although the acid magmatism was characteristic of all the Pacific margin during late Mesozoic to early Cainozoic, the extent of such magmatism over an area exceeding 3000 km in width along the Asian continent, from north-eastern Siberia via Sikhot–Alin to eastern China, makes a simple subduction-type model unlikely.

The possibility of ocean-floor spreading in the Sea of Japan was first suggested by Kobayashi (1941), who noted the similarity of pre-Cainozoic geological structures in the Korean peninsula and south-western Japan. Thinner crust, especially in the northern part, high heat flow, and subparallel magnetic lineations throughout the Sea of Japan all support Kobayashi's hypothesis. The sea-floor spreading probably occurred during the Paleogene (60–20 Ma; Ichikawa et al., 1972) and, as the Sea of Japan widened, the Japanese islands were deformed around Honshu with the north-eastern part rotating c. 50° anticlockwise relative to the south-western part (Kawai et al., 1961). The gap which opened between north-eastern and south-western Honshu is a depression called the 'Fossa Magna' where submarine sedimentation and volcanism were widespread during the early Neogene. The volcanic rocks range from low alkali tholeiites through high alumina basalts to calc-alkaline andesites and dacites in composition, building a very thick pile of lavas and pyroclastic rocks locally exceeding 10 km in thickness (Matsuda, 1962).

Neogene Volcanism

About 42 Ma ago (Clague and Jarrard, 1973), the slip vector of the Pacific plate subducting below Japan abruptly changed from north-north-west to west-north-west and a new orogenic cycle was initiated. Magmatism reached a peak in the period 23–15 Ma ago, then decreased in intensity and revived during the Quaternary (Sugimura et al., 1963; Fig. 10). Along most of the Kurile, Kamchatka, North-east Honshu, and Ryukyu arcs, the tectonic and magmatic patterns appear to have remained essentially the same from this period to the present. Island arc-type basaltic and andesitic volcanic activity (40–26 Ma; Kaneoka et al., 1970), including eruption of bronzite andesites, characterizes the basement formations of the Bonin and Mariana Islands (Meijer, this volume, Section III).

Distribution of the products of Miocene to Pliocene volcanism and plutonism is shown schematically in Fig. 11. This distribution coincides roughly with that of the late Quater-

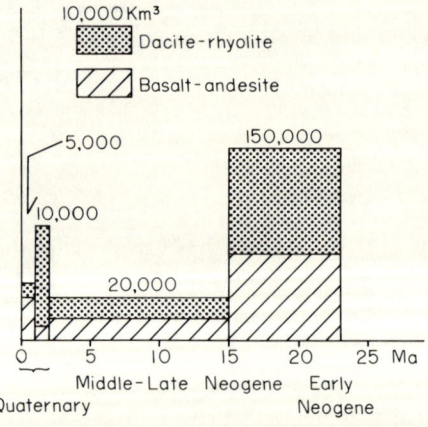

Fig. 10 Variation with time of the volcanic products in the Neogene and Quaternary throughout the Japanese islands (after Sugimura et al. (1963, Fig. 10) partly modified by the present authors)

modal basalt and dacite–rhyolite associations became more abundant (Konda, 1974) and a thick pile of hyaloclastites, breccias, and pillow lavas was formed with abundant intrusive sheets and domes. Hydrothermal alteration is widespread, especially in the lower part of the succession, and the whole Neogene submarine sequence is called the 'green tuff' formation. This formation extends along the inner arc of the Kurile islands as far as Kamchatka. Submarine volcanism, similar to the green tuff in north-eastern Honshu, was especially intense in the southern Fossa Magna (Fig. 11). Here the volcanic strata are intensely deformed by subsequent folding (Matsuda, 1962; Mikami, 1962). Similar volcanism may extend to the submerged ridge of Izu islands forming the immediate basement of the Quaternary volcanoes.

Where the green tuff volcanic formation is thick, there are widespread intrusive masses of soda-rich, potash-poor quartz diorite and granodiorite of Miocene age with variable texture and grain size, often displaying continuous gradation to an extrusive facies (Fig. 11). In contrast to this, more potassic and less calcic magmatism occurred in south-western Japan $c.\ 14 \pm 2$ Ma ago. High level batholiths and their extrusive equivalents in the form of large lava lakes and ash-flow composite volcano complexes associated with Valles-type cauldrons (Aramaki et al., 1977) are distributed along the Pacific coast but displaced distinctly to the east of the Quaternary volcanic front. Intimately associated with these, in the Setouchi area are the high Mg, high K andesites and their derivatives, (Ujike, 1970). The main rock type is bronzite-bearing andesite, similar to boninite from the Eocene volcanics of the Bonin islands (Meijer, this volume, Section III; Shiraki et al., 1977), and bronzite andesite from Papua New Guinea. This is unique in the Japanese islands and its possible role as a parent magma for arc-type calc-alkaline series makes further study desirable.

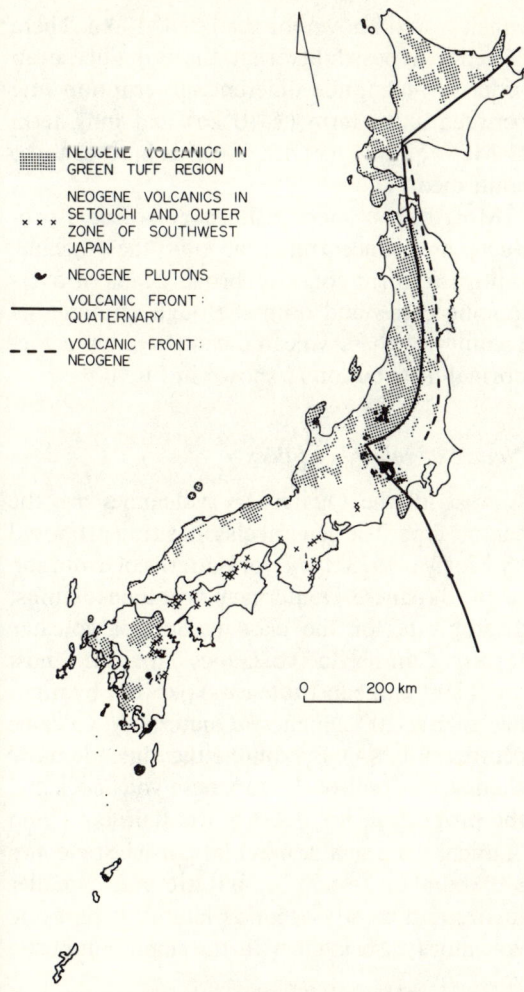

Fig. 11 Map showing the distribution of Neogene volcanic materials and position of volcanic fronts in Miocene (broken line) and Quaternary (solid line) for the East Japan volcanic belt. Data mainly after Isshiki et al. (1968a). In North-east Honshu, the Miocene volcanic front is up to 60 km east of the present-day front, while in south-western Honshu it is displaced to the north (towards the inner side) of the present front. Nakamura (1969) explained the apparent retreat of the front in northern Honshu as due to the slow forward movement of the continental lithosphere beneath the Japanese islands as its front margin is slowly being dragged into the subduction zone

nary (Fig. 2) and similar rocks of similar age and chemical composition have been drilled on the Kyushu–Palau ridge (Kroenke and Scott, 1978).

In the North-east Honshu arc, the most abundant andesitic volcanism was of early Miocene age. During the later Miocene, bi-

Quaternary Volcanism

The last volcanic episode clearly distinguishable from that of the present day in the Japanese islands was the eruption of large scale felsic pyroclastic flows. Moriya (1977) pointed out the presence of more than 10 regions in Hokkaido, Honshu, and Kyushu where deeply dissected pyroclastic plateaus are recognized, indicating widespread occurrence of thick, mostly welded pyroclastic flow deposits. Most of the rocks range in age from 1 to 2 Ma and are characterized by biotite and/or hornblende phenocrysts and locally by abundant bipyramidal quartz phenocrysts. The magmas are dacite and rhyolite with $SiO_2 = 65$–74 per cent and were erupted subaerially from major vent areas which became large scale volcano-tectonic depressions after eruption. During the last 1 Ma smaller eruptions of intermediate magmas have resulted in formation of many volcanic centres. The eruption rate during this period in the terrestrial part of the Izu–Mariana arc, i.e. volcanoes in Izu peninsula to Shirouma-Oike in central Honshu, is 4.2×10^2 km^3 Ma^{-1} per 100 km length of the arc (Table 2), which is much higher than the average for the North-east Japan arc (2.2×10^2 km^3 Ma^{-1} per 100 km), Hokkaido (2.0×10^2 km^3 Ma^{-1} per 100 km), and the land part of south-western Japan (1.2×10^2 km^3 Ma^{-1} per 100 km, Table 2).

The total volume of the volcanic products for the last 1 Ma shown in Table 2 (4500 km^3) is probably a considerable underestimate, as wind-distributed pyroclastic materials are not included. On the other hand the estimate of 2.7×10^3 km^3 Ma^{-1} per 100 km of arc by Nakamura (1974) is primarily based on several examples of Japanese volcanoes the eruptive history of which is well known for the last 1–10 ka. There is hence a possibility that there might be an order of magnitude difference in eruption rate between short term (1–10 ka) and long term (1 Ma) volcanic activity, but this has yet to be confirmed.

Most of the later Quaternary volcanic products are concentrated around the volcanic centres to form cones or broad heaps of overlapping cones and domes. Rough estimates of volumes of these volcanic centres show a log-normal distribution as shown in Fig. 12.

Nature of volcanic edifices

Almost all the Quaternary volcanoes are the central type. The general classification proposed by Moriya (1973, 1977) for patterns of evolution of the Japanese Quaternary volcanoes is most appropriate for the classification of volcanic forms. Composite volcanoes are the most abundant and constitute c. 60 per cent by number of the 200 Japanese Quaternary volcanic centres (Fig. 8A). By volume they include more than 80 per cent of the Japanese volcanoes and the proportion is similar in the Kurile arc and Kamchatka. Lava domes and pyroclastic cones are abundant in number but are much smaller in size and mostly occur as lateral or parasitic volcanoes associated with the main composite volcano.

The early stage of a typical composite volcano is characterized by the construction of the main cone by alternating eruptions of lava flows and

TABLE 2 Volume of volcanic products and eruption rate during the late Quaternary in the Japanese islands. Major data source: Aramaki and Ui (1978a)

Arc	Length Along Volcanic Front (km)	Volume of Products (km^3)	Eruption Rate ($\times 10^2$ km^3/10^6 100 km)
South-western Japan	(1694)†	(797)†	(0.47)†
South-western Japan (land part)	686	787	1.15
Izu–Bonin	(2110)†	(1076)†	(0.5)†
Izu–Bonin (land part)	246	1041	4.23
North-eastern Japan	880	1939	2.20
Kurile–Kamchatka (Hokkaido only)	360	713	1.98

† For the whole length of the arc including the island chains in the ocean. The volume of volcanic islands is estimated only for the emerged part so that total volume of products is grossly underestimated.

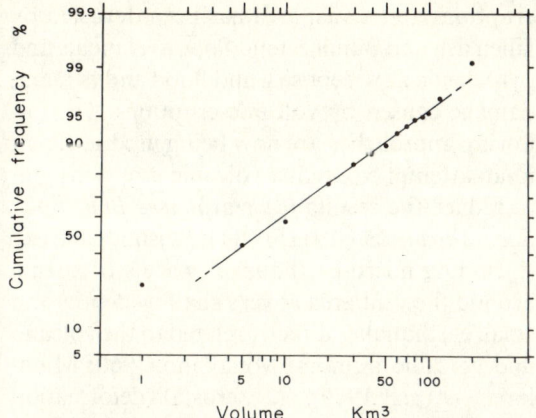

Fig. 12 Cumulative frequency of the sizes of the late Quaternary volcanic centres plotted on a probability paper. Data after Aramaki and Ui (1978b). Volcanic centres larger than 5 km³ in size are well represented by a log-normal distribution. The accuracy of volume estimation of the small volcanoes is very poor

pyroclastic materials. The eruption rate is greatest early in the volcanic history and gradually decreases toward the latter stages of the growth. The early volcanic products are typically andesitic to basaltic andesite but later products are less basic in composition. Near the end of the growth of the main cone, the explosive eruption of ash and pumice and generation of pyroclastic flows become more frequent and the magma becomes more felsic and viscous. The former activity produces extensive ash- and pumice-fall sheets which are distributed mainly to the east of the main cone, as a result of westerly wind dispersal of the eruption column rising to the stratosphere. Formation of lateral (parasitic) lava domes, lava flows, and pyroclastic eruptions (up to 10 km³ of ejecta) may produce collapse and enlargement of the central crater, giving rise to small 'Haruna-type' calderas (Aramaki, 1969).

Close to the end of the growth history, extended intervals between eruptions allow stream erosion to breach the central crater to form cirque-like erosion calderas and large radial canyons. Large scale phreatic explosions are also characteristic in the declining stage, resulting in huge, air-cushioned, high speed avalanches and explosion calderas. The 1888 phreatic explosion of Bandai-san is a good example of such a late stage phreatic eruption (Sekiya and Kikuchi, 1890).

While the magmas of the major composite volcanoes are relatively mafic (andesite to basaltic andesite), eruption of large amounts of dacitic and rhyolitic magma is characteristic of a distinctive type of late Quaternary volcanism. This is characterized by voluminous pyroclastic eruptions leading to formation of large collapse calderas (Fig. 8B). Most of these may be classified as Crater Lake (or Krakatau) type calderas (Williams, 1941, 1942), but many of the Japanese examples lack the early formation of a main cone, like Mt Mazama. Instead the bulk of the magma is driven out of the newly formed vent by extensive underground vesiculation and generation of pyroclastic flows with a huge eruptive column. The main phase typically lasts a very short time (probably a few days). Collapse calderas measure up to 25 km in diameter (e.g. Aso, Kutcharo, Shikotsu, and Aira) but the subvolcanic structure is funnel-shaped so that the horizontal cross-section of the shattered and foundered portion becomes narrower at lower levels. Wide pyroclastic flow fields surround the caldera and the aggregate volume of the deposits often exceeds 100 km³. Post-caldera activity is characterized in many cases by more mafic magmas and the characteristic products are small composite volcanoes, lava domes, lava flows, and pyroclastic cones formed inside or on the rim of the caldera. No Valles-type calderas (Smith and Bailey, 1968) with well marked ring fractures are identified in the late Quaternary, although some are found or suspected in earlier Quaternary and Neogene volcanic structures (Aramaki et al., 1977; Yoshida, 1970). The higher eruption rate during the early Quaternary (Fig. 10) was mainly due to felsic pyroclastic flow eruptions from several large centres throughout the Japanese islands.

The third type of volcanic centre is represented by isolated groups of small monogenetic volcanoes such as pyroclastic cones, lava domes, flows, and maars (Fig. 8C). Most of these occur in the Sea of Japan side of the volcanic belt (south-western Honshu and Ichinomegata in northern Honshu), and are produced by magmas of more alkaline affinity. Nakamura (1977) and Jacob et al. (1977) suggested that such monogenetic volcanoes form in the back-arc region where the regional stress in the crust tends to be extensional.

Mode of volcanic eruption

The activity of a composite volcano is characterized by an alteration of long periods of relative quiescence and short bursts of intense eruption by which the bulk of the volcanic edifice is constructed. Oshima volcano, a basaltic composite volcano, has been active at 100–150 a intervals during the last 10^4 a (Nakamura, 1964; Isshiki, 1964). The last five major eruptions occurred in 1421(?), 1552(?), 1684, 1777, and 1950–1951.

Combined study of historic records, ^{14}C dates, and the volcanic stratigraphy of andesitic composite volcanoes such as Asama and Sakurajima suggests an interval of several hundred to thousand years for very large eruptions and several tens of years for large eruptions. Explosive volcanic activity and sporadic ash emission occur more commonly on these volcanoes, but this kind of activity, although dangerous to the local population, does not contribute much to the growth of the volcano. The typical sequence of events during a large eruption is (1) formation of a vertical eruption column reaching well into the stratosphere and causing ash and pumice deposition over wide areas on the eastern, lee side; (2) eruption of pyroclastic flows; and (3) eruption of lava flows (stage 2 may be absent in some cases; Aramaki and Yamasaki, 1963). The best example is the 1783 activity of Asama (Aramaki, 1956). During this sequence the vesicularity of the erupted essential (magmatic) materials decreased with time.

Prediction of eruption and hazard mitigation

Out of 200 late Quaternary volcanic centres in Japan c. 70 have erupted in historic times or display vigorous solfataric activity (Kuno, 1962). At least 20 000 lives have been lost by historic eruptions and a vast area has been devasted by fallen ash and pumice, mud flow, avalanche, and pyroclastic flow deposits, and flood and tsunami damage caused by volcanic eruptions. The following approaches are now being used in Japan in an attempt to predict volcanic eruptions and to reduce the resultant hazards (see Bull. Volc. Soc. Japan 23, 1–115): (1) Seismic activity (including micro-earthquake swarms occurring around the vent area at very shallow depths and small earthquakes directly related to the volcano and volcanic tremors; Minakami, 1960; Minakami et al., 1969); (2) crustal deformation (Yokoyama, 1974, 1978); (3) changes in geomagnetism and geoelectricity (Yokoyama, 1974, 1978); (4) geothermal anomalies (Yokoyama, 1974, 1978); (5) chemical composition of volcanic gas issuing from active crater and fumaroles, hot (and cold) spring water, etc. (Ossaka et al., 1978); (6) mineralogical and petrological characteristics of magmatic ejecta such as ash, pumice, scoria, and lava (Katsui et al., 1978b); (7) assessment from the past activity—the pattern and characteristics of the past eruptions of the same or similar volcano greatly help the prediction of future eruptions (Katsui et al., 1978b). Much manpower and financial support have promoted research during the last national five year project. More than 16 active volcanoes are routinely monitored (Suwa, 1978) and in the case of increased activity an ad hoc committee will help to coordinate various research groups from universities and governmental organizations such as the Meteorological Agency. The committee serves to communicate the collected data and on the spot assessments of the volcanic activity made by volcanologists to government bodies and to the mass media.

Bulk Chemical Composition of the Volcanic Rocks

Approximately 5000 major element analyses have been published for Neogene to Quaternary volcanic products from Japan, Mariana, Kurile, and Kamchatka (Table 1). The average chemical compositions of the Japanese arcs seen as a whole are given in Table 3 (columns 1–5),

TABLE 3 Mean composition of Japanese Quaternary volcanic rocks

Composition range: Number of analyses:	1	2	3	4	5	6 SI= 39–30	7 SI= 29–20	8 SI= 19–10	9 SI= 9–0	10 SI= 39–30	11 SI= 29–20	12 SI= 19–10	13 SI= 9–0	14 <50% SiO₂	15 50–60% SiO₂	16 60–70% SiO₂	17 >70% SiO₂
	1641	1596	1641	541	415	5	53	4	8	8	13	9	10	119	35	32	8
SiO₂	59.13	57.55	57.19	63.58	54.88	50.03	53.09	58.98	71.37	53.81	55.55	66.55	75.93	48.62	52.19	63.96	73.13
TiO₂	0.81	0.86	0.88	0.64	1.05	0.84	1.17	1.06	0.48	0.95	0.83	0.68	0.18	1.95	2.12	0.67	0.20
Al₂O₃	16.75	16.96	17.25	15.92	17.35	15.71	15.44	14.96	13.59	17.79	17.24	15.35	13.47	16.76	17.84	17.20	13.70
Fe₂O₃	2.79	3.01	3.06	2.23	3.13	2.92	4.02	3.29	1.89	2.44	3.68	2.21	0.77	3.47	3.79	2.50	1.48
FeO	4.37	5.05	4.92	2.69	5.93	8.83	9.01	7.59	2.94	6.60	5.96	3.19	0.91	7.18	6.44	2.35	1.23
MnO	0.14	0.16	0.15	0.11	0.18	0.27	0.23	0.19	0.18	0.19	0.18	0.15	0.10	0.17	0.19	0.17	0.06
MgO	3.16	3.67	3.64	2.21	3.88	7.35	4.66	2.74	0.66	5.87	4.42	1.74	0.41	7.93	3.34	0.56	0.15
CaO	6.68	7.38	7.47	4.67	8.27	11.95	9.68	7.30	3.44	8.79	8.40	5.11	1.90	8.75	6.74	2.10	0.58
Na₂O	3.19	3.10	3.04	3.76	2.94	1.47	2.12	3.03	3.84	2.76	2.92	3.66	3.92	3.18	4.29	5.17	4.31
K₂O	1.55	1.23	1.32	2.50	1.20	0.24	0.45	0.69	1.42	0.62	0.68	1.17	2.31	1.55	2.47	5.11	5.11
H₂O⁺	0.99	0.70	0.66	1.50	0.64	—	—	—	—	—	—	—	—	—	—	—	—
H₂O⁻	0.43	0.38	0.42	0.45	0.42	—	—	—	—	—	—	—	—	—	—	—	—
P₂O₅	0.20	0.20	0.20	0.25	0.23	0.09	0.12	0.15	0.19	0.19	0.13	0.19	0.12	0.44	0.59	0.21	0.05
Total	100.19	100.25	100.20	100.51	100.10	99.70	99.99	99.98	100.00	100.01	99.99	100.00	100.02	100.00	100.00	100.00	100.00
CIPW norms																	
Q	15.15	12.93	12.50	19.03	9.11	3.36	9.79	17.11	34.28	6.02	10.53	26.78	39.52	—	—	9.01	26.81
or	9.16	7.27	7.80	14.77	7.09	1.42	2.66	4.08	8.39	3.66	4.02	6.91	13.65	9.16	14.60	30.20	30.20
ab	26.99	26.23	25.72	31.82	24.88	12.44	17.94	25.64	32.49	23.35	24.71	30.9	33.17	26.91	36.30	43.75	36.47
an	26.81	28.73	29.52	19.18	30.60	35.56	31.28	25.18	15.65	34.32	31.93	22.00	8.64	26.88	22.13	8.63	2.55
C	—	—	—	—	—	—	—	—	—	—	—	—	1.35	—	—	—	0.14
wo	2.10	2.75	2.61	0.98	3.72	9.67	6.66	4.21	0.07	3.36	3.71	0.88	—	5.71	3.12	0.17	—
en	7.87	9.14	9.06	5.51	9.67	18.30	11.60	6.82	1.64	14.62	11.01	4.34	1.02	5.28	6.34	1.39	0.37
fs	4.64	5.66	5.33	2.25	6.90	12.91	11.72	9.82	3.38	8.89	6.87	3.19	0.92	1.98	4.23	1.46	0.82
fo	—	—	—	—	—	—	—	—	—	—	—	—	—	10.14	1.38	—	—
fa	—	—	—	—	—	—	—	—	—	—	—	—	—	4.19	1.02	—	—
mt	4.05	4.36	4.44	3.23	4.54	4.23	5.83	4.77	2.74	3.54	5.34	3.20	1.12	5.03	5.50	3.62	2.15
il	1.54	1.63	1.67	1.22	1.99	1.60	2.22	2.01	0.91	1.80	1.58	1.29	0.34	3.70	4.03	1.27	0.38
ap	0.46	0.46	0.46	0.58	0.53	0.21	0.28	0.35	0.44	0.44	0.30	0.44	0.28	1.02	1.37	0.49	0.12

SI = solidification index.

(1) Volume-weighted average composition, Japanese Quaternary volcanic rocks; caldera-forming pyroclastic flows are included (S. Aramaki and T. Ui, unpublished da a). (2) Same as 1, but caldera-forming pyroclastic flows are excluded (S. Aramaki and T. Ui, unpublished data). (3) Same as 1 except that it is not volume-weighted (S. Aramaki and T. Ui, unpublished data). (4) Volume-weighted average composition of Quaternary volcanic rocks of South-west Japan arc; caldera-forming pyroclastic flows are included (S. Aramaki and T. Ui, unpublished data). (5) Volume-weighted average composition of Quaternary volcanic rocks of Izu–Mariana arc; caldera-forming pyroclastic flows are included (S. Aramaki and T. Ui, unpublished data). (6–9) Average compositions of aphyric rocks of the tholeiite series (pigeonitic rock series) from Izu–Hakone region for different SI (100 × MgO/MgO + FeO*Fe₂O₃ + Na₂O + K₂O) ranges (Kuno, 1968a). (10–13) Average compositions of aphyric rocks of calc-alkaline series (hypersthenic rock series) from Izu–Hakone region for different SI ranges (Kuno, 1968a). (14–17) Average compositions of volcanic rocks of alkaline series in south-western Japan for different SiO₂ ranges (Aoki, 1978).

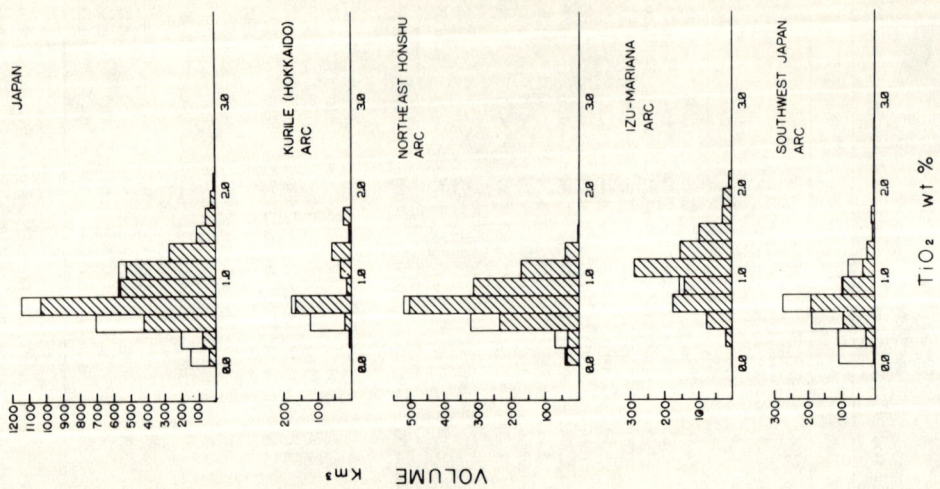
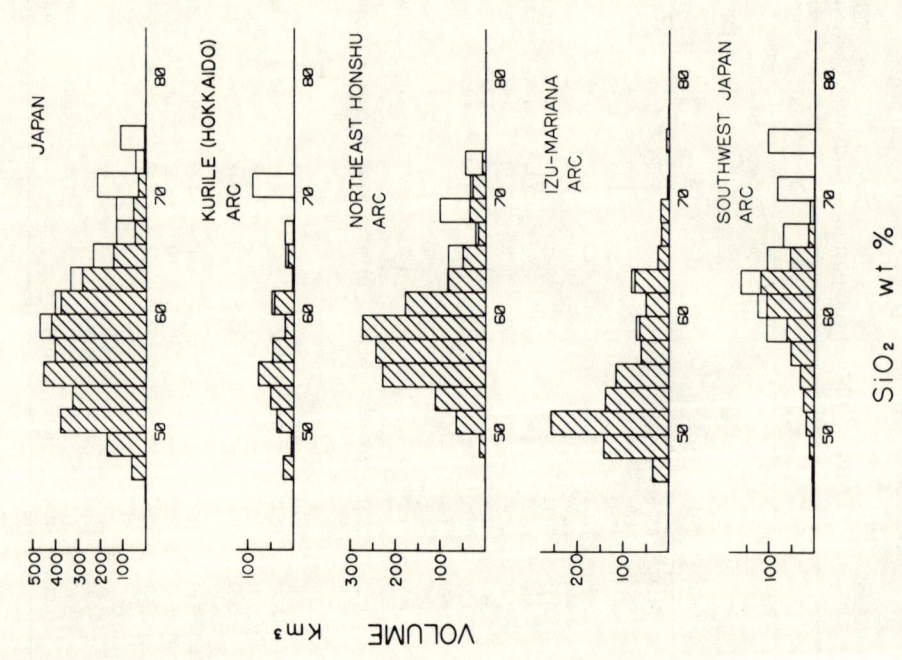

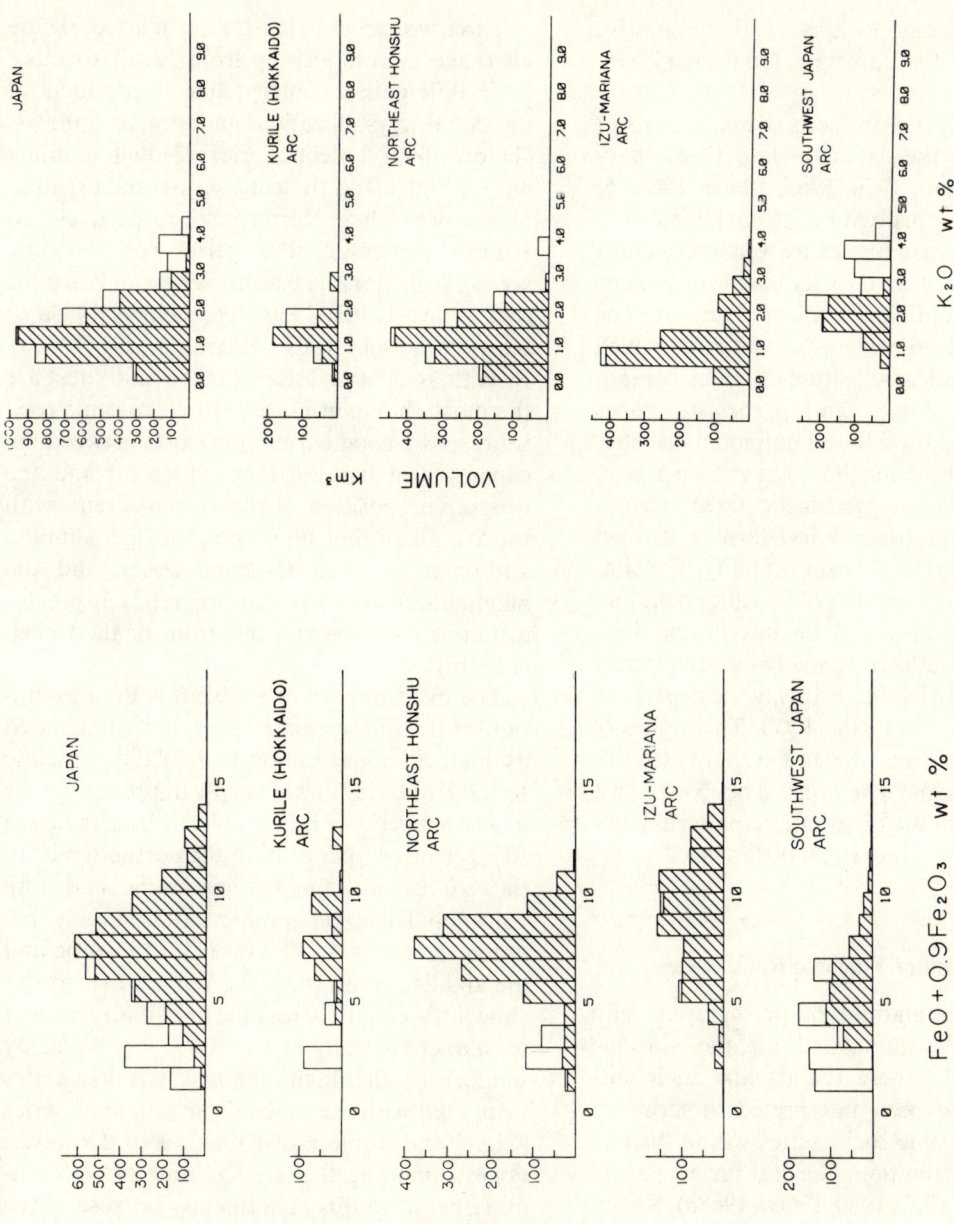

Fig. 13 Bar diagrams showing the frequency distribution of the volume-weighted chemical analyses for the Japanese Quaternary volcanic rocks (c. 1600 analyses; Aramaki and Ui, 1978a). The top diagram for each component represents the overall histogram for the total of the four arcs. The hatched portions of the bars indicate the lavas and pyroclastics which constitute the ordinary volcanic edifices such as composite volcanoes, lava domes, and pyroclastic cones. The open portions of the bars indicate the pyroclastic flows and fall deposits associated with large calderas

together with compositions of representative rock types for the Japanese Quaternary volcanic belts.

The frequency distributions of major element abundances of the Japanese late Quaternary rock series (Kuno, 1950, 1965, 1966a, 1968a,b; Ui (1978a). The histograms of major element abundance shown in Fig. 13 are weighted against the volume of volcanic rocks based on volume estimations of individual volcanic centres. The lava flows and volcaniclastic materials which form positive volcanic features such as composite volcanoes, domes, and pyroclastic cones (hatched bars) show a broad unimodal distribution in SiO_2 with a maximum at c. 57 per cent, whereas large scale pyroclastic flows accompanying calderas (open bars) form a distinct secondary mode at c. 70 per cent SiO_2. The contrasted abundance profiles of basaltic rocks and alkali-rich andesites and dacites in the Izu-Mariana and South-west Japan arcs respectively are clearly shown by the positively and negatively skewed peaks of SiO_2 and K_2O. The modes of TiO_2 and FeO^* (total Fe as FeO) are also different between the two arcs. The North-east Honshu arc shows distribution patterns very similar to those of the whole of Japan.

Recognition of Volcanic Rock Series

The very large amount of petrographic and chemical data accumulated for the volcanic rocks of the Japanese islands and their surroundings have been interpreted in terms of contrasted volcanic rocks series with a distinctive spatial distribution. General review papers include Kuno (1959, 1960, 1966a, 1968a), Katsui (1961), Sugimura (1960), Miyashiro (1974), and Katsui et al., (1978a).

Table 4 shows the generalized relation between the spatial distribution and chemical characteristics of the major volcanic rock series identified in Japanese islands in Neogene to Quaternary times. A similar picture is valid for the Mariana, Kurile, and Kamchatka arcs (Gorshkov, 1970), although all series do not necessarily occur in the same segment of the arc.

Three volcanic series are recognized. These all range in composition from basalt to more felsic differentiates interpreted as products of fractional crystallization and are as follows: (1) low alkali tholeiitic series; (2) high alumina and/or high alkali tholeiitic series; and (3) alkaline series. These correspond respectively to Kuno's pigeonitic rock series (or tholeiitic series), high alumina basaltic series, and alkaline rock series (Kuno, 1950, 1965, 1966a, 1968a,b; Kawano et al., 1961; Katsui et al., 1978a). Basaltic rocks (inclusive of mafic andesites) are the most abundant in these three volcanic rock series and a good correlation exists between the chemistry of basaltic rocks (Fig. 14) and the geographic position of the volcanic vent, with the low alkali tholeiitic series, the high alumina and/or high alkali tholeiitic series, and the alkaline series occurring at progressively greater distances from the volcanic front or the trench (Fig. 15).

The most important problem is the recognition of the intermediate series, here designated as high alumina and/or high alkali tholeiitic series. Kuno (1960) was deeply impressed by the wide occurrence of nearly aphyric basalts rich in Al_2O_3 (over 17 per cent) in the northern part of the Izu islands, the Izu peninsula, and Fuji volcano. Their geographic position falls between the zones of the low alkali tholeiitic and the alkaline series (Fig. 15). While other oxides show little change across the arc, Al_2O_3 content is characteristically higher (by c. 2 per cent by weight) in the high alumina basaltic series compared with the tholeiitic and alkaline series (Fig. 3 and Table 6 of Kuno, 1960). However, notwithstanding the large volume of high alumina basalt in this area (mainly because of the large volume, c. 300 km³, of Fuji volcano), aphyric, high alumina basalts are not known elsewhere in the Japanese islands. The volcanoes indicated as high alumina basaltic series (Fig. 8 of Kuno, 1960; Fig. 1 of Kuno, 1966a; Table 4 of Katsui et al., 1978a) are characterized by porphyritic rocks with bulk compositions falling within the field of high alumina basalt given by Kuno (Fig. 7 of Kuno, 1960). The rocks may have been derived from the high alumina

TABLE 4 Relations between the chemical characteristics and spatial distribution of the four volcanic rock series

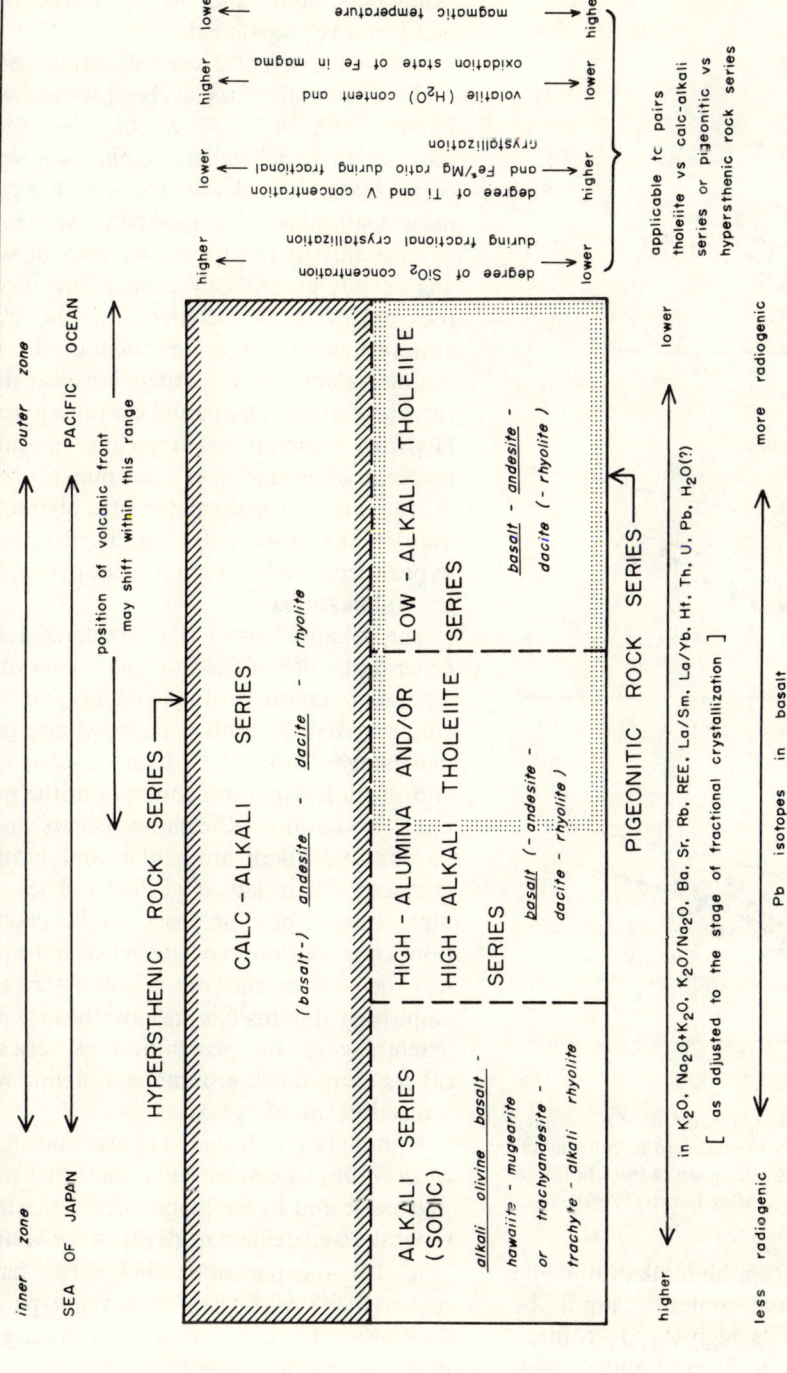

Rocks from each series are genetically connected by fractional crystallization within the crust. The four rock series grade each other as indicated by broken-line boundaries. The horizontal axis corresponds to (1) the spatial (geographic) position of the vent across the arc and (2) chemical parameters of the rocks such as K_2O that change monotonously across the arc. The vertical axis indicates the systematic change of another set of chemical, mineralogical, and physical parameters which are interrelated and give rise to distinction between the calc-alkaline and non-calc-alkaline series (or between hypersthenic and pigeonitic rock series.) The most abundant rocks in each rock series are underlined and rare rocks are in parentheses. Many Japanese petrographers follow Kuno's nomenclature as defined mainly by the colour index of the groundmass (Kuno, 1954): colour index is above 35 in basalt and trachybasalt, 10–35 in andesite and trachyandesite, and below 10 in dacite and rhyolite. In dacite, normative An is more abundant than Or, in rhyolite Or is less abundant than An, and trachyte does not contain silica minerals.

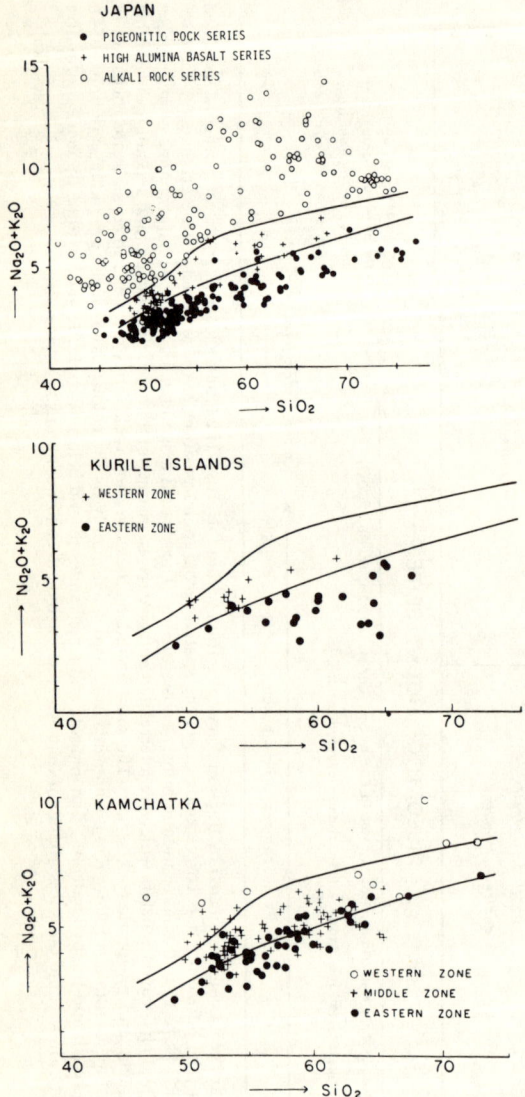

Fig. 14 Total alkali–SiO$_2$ diagrams for Quaternary volcanic rocks from Japan (including the continental part of east Asia) (top), Kurile islands (middle), and Kamchatka (bottom). After Kuno (1966a)

sion of its significance to the whole volcanic belt is not yet warranted.

The mafic mineral crystallization sequence of the three rock series has been summarized by Kuno (1950, 1968b) (Fig. 16). The pigeonitic rock series is defined by Kuno as a series of comagmatic non-alkaline rocks with a groundmass containing a Ca-poor clinopyroxene such as pigeonite or ferropigeonite with or without augite, but no orthopyroxene. The pigeonitic rock series is characterized by the frequent appearance of pigeonite phenocrysts in the middle to late stages and there is a clear reaction relation between olivine and Ca-poor pyroxenes. Hydrous minerals are typically absent. The presence of groundmass silica minerals such as cristobalite and tridymite is also characteristic even in the most mafic basalts. Rocks of the hypersthenic series contain orthopyroxene as a groundmass phase.

The alkaline series is characterized by the general absence of Ca-poor pyroxenes and of a reaction relation between olivine and Ca-rich clinopyroxene in both phenocryst and groundmass stages (Aoki, 1959; Kuno, 1968b). Olivine and alkali feldspar are common in the groundmass of basalts. Kaersutite occurs in some basalts and alkali amphibole and biotite are characteristic in late stage rocks. Rocks of the high alumina basaltic series exhibit crystallization sequences intermediate between the pigeonitic rock series and the alkaline series. Thus some high alumina basalts have the groundmass assemblage of the pigeonitic rock series while others carry much groundmass olivine without a reaction rim of pyroxene.

Kuno (1953), Katsui (1954), and Kawano et al. (1961) demonstrated that rocks of the pigeonitic and hypersthenic rock series fall into separate, well defined fields on an AFM diagram (Fig. 17), the pigeonitic rock series having a higher FeO*/MgO ratio than the hypersthenic rock series. The correlation between the groundmass pyroxene assemblage and the MgO–FeO*–alkali ratio was attributed to a higher H$_2$O content in the liquid of hypersthenic rock series. This enhances oxidation of Fe, leading to the early separation of magnetite and inhibition

liquid, but derivation from high alkali tholeiite with 'normal' alumina content cannot be excluded. Apart from its high Al$_2$O$_3$ content and low TiO$_2$, the composition of Kuno's high alumina basalt falls within the range of the olivine tholeiite such as those from Hawaii. Thus the occurrence of high alumina basaltic liquid in the Fuji–Izu region is unique, but the exten-

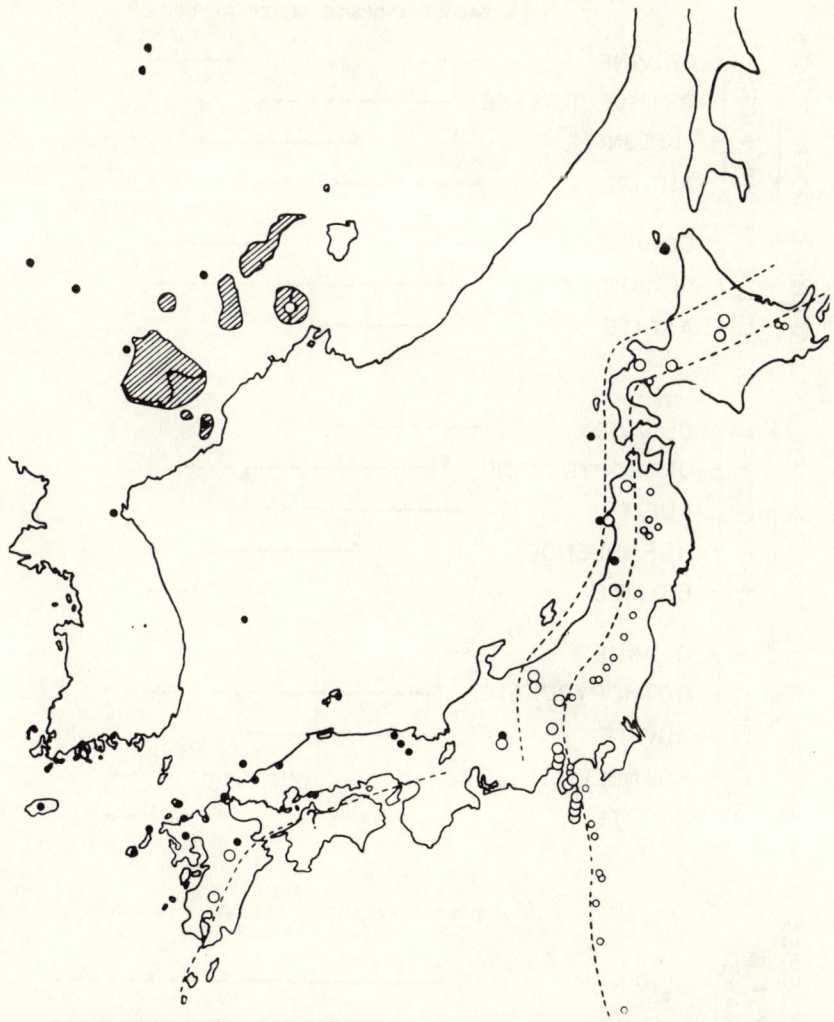

Fig. 15 Map showing zonal distribution of tholeiite (small open circles), high alumina (large open circles) and alkali olivine basalt (solid circle) in Quaternary volcanoes as defined by Kuno (1966a). The hatched areas in northern Korea and north-eastern China are basalt plateau mostly of alkali olivine basalt of younger Tertiary and Quaternary age. Broken lines show the boundaries of the three basalt zones

of Fe concentration in later liquids, and lowers the crystallization temperature, leading to crystallization of hypersthene in place of pigeonite (Kuno, 1966b).

Kuno (1959, 1964, 1965) proposed to restrict the term calc-alkaline rock series to basalt–andesite–dacite–rhyolite of his hypersthenic rock series and to its plutonic equivalent (if any) because the chemical variation trends of this rock series agree with those generally accepted as characteristic of the calc-alkaline series. This definition has been accepted by most Japanese petrographers (e.g. Kawano *et al.*, 1961; Aoki and Oji, 1966; Katsui *et al.*, 1978a) and is adopted in this paper. Thus in Table 4, the range of the hypersthenic rock series is shown to cover the entire field of the calc-alkaline series. The horizontal broken line separating the field of the calc-alkaline series from the fields of two tholeiitic and alkaline series is the boundary

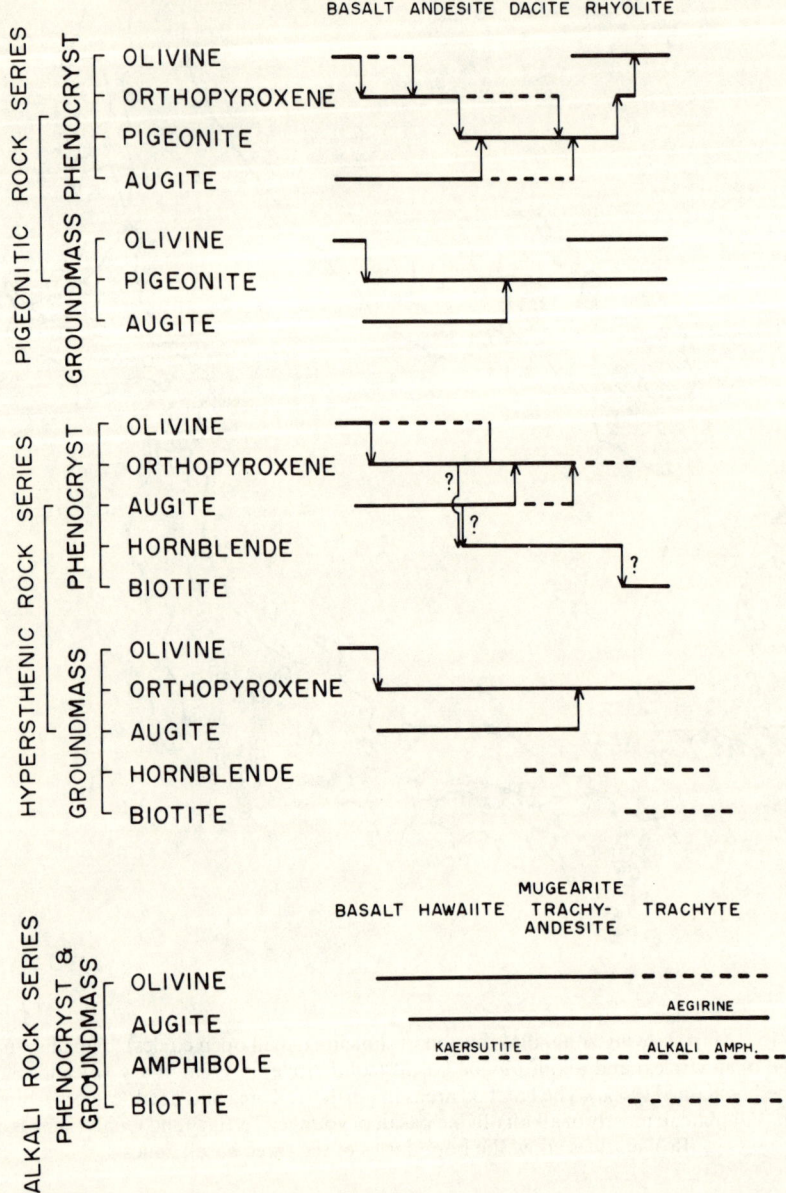

Fig. 16 Mafic mineral crystallization sequences for pigeonitic, hypersthenic, and alkaline rock series. After Kuno (1968b) with minor modifications. Arrows indicate reaction relations

above which orthopyroxene is the characteristic groundmass mineral and below which it is absent. The vertical axis of Table 4 corresponds roughly with the degree of Fe enrichment of an AFM diagram, except in the case of the alkaline series which, in most cases, falls in the same area as that of the hypersthenic rock series on the AFM diagram.

The relative abundances of rocks with various combinations of phenocryst and groundmass mafic minerals are summarized in Table 5 (Kuno, 1950, 1954). Comparing this data with

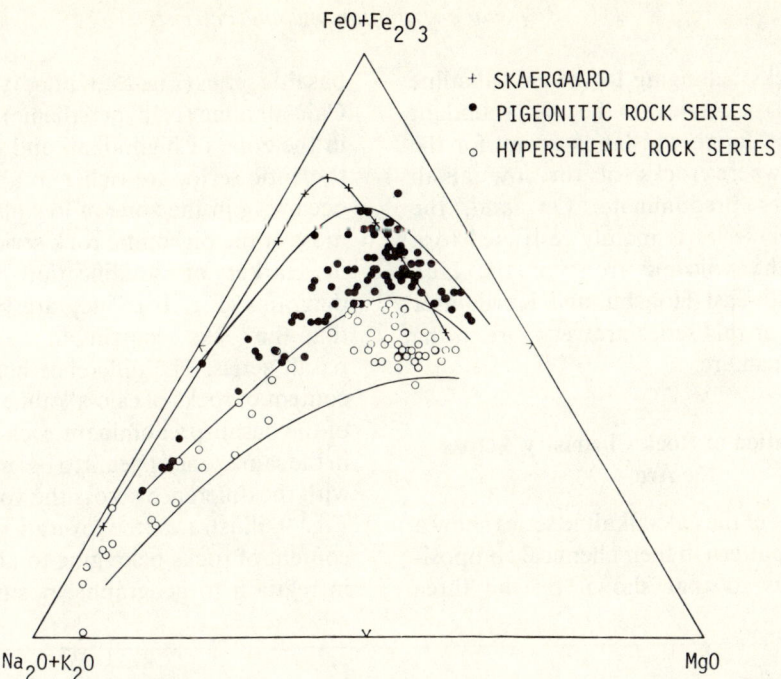

Fig. 17 Alkali–total Fe–MgO (AFM) diagram showing separation of the fields of pigeonitic and hypersthenic rock series for the Quaternary aphyric rocks of Izu–Hakone region (Kuno, 1965). The Skaergaard liquid trend is from Wager (1960)

TABLE 5 Assemblage of phenocryst and groundmass minerals of the Japanese Neogene and Quaternary volcanic rocks. After Kuno (1950, 1954).

Phenocryst	Groundmass	a	a → d	b	b → c	c	d → c	d	e
		OL + CPX + OPX	(OL)OPX + CPX	OL + CPX	(OL)CPX	CPX	(OPX)CPX	OPX + CPX	OPX
I	OPX	+	−	+	−	++	++	++	++
II	OL + OPX	+	+	+	−	++	++	++	+
III	OL	+	++	++	++	++	+	++	−
IV	OL + CPX	+	+	++	++	++	++	++	−
V	CPX + OPX ± OL	+	+	+	+	+++	++	+++	++
VI	CPX + OPX + AM ± OL	−	−	−	−	−	−	++	++
VII	OPX + AM + OL	−	−	−	−	−	−	++	++
VIII	AM ± OL	−	−	−	−	−	−	+	++
IX	CPX + AM ± OL	−	−	−	−	−	−	+	!
X	OPX	+	+	+	+	+	+	+	+

OL, olivine; CPX, clinopyroxene; OPX, orthopyroxene; AM, amphibole.
+++, abundant; ++, common; +, rare; −, not found.

(OL)OPX means that the groundmass olivine is surrounded by orthopyroxene forming a reaction rim. (OL)CPX and (OPX)CPX also indicate similar reaction relations respectively. Symbols such as IIIb and V$d \to c$ as defined in this table are frequently used by Japanese petrographers. V$d \to c$ indicates that the phenocrysts are clinopyroxene and orthopyroxene with or without olivine, and that the groundmass orthopyroxene has a reaction rim of pigeonite indicating a reaction relation. The pigeonitic rock series as defined by Kuno (1950) includes groundmass assemblages of b, $b \to c$, c, and $d \to c$. The hypersynthetic rock series includes assemblages of a, d, and e. When biotite is present as a phenocryst, X is added, or example as XV IIId and XXe.

Fig. 3, the rocks belonging to the calc-alkaline (hypersthenic) rock series are the most abundant throughout the Japanese islands except for the Izu islands, where rocks of the low alkali tholeiitic series predominate. On land, the pigeonitic rock series is mainly restricted to a zone along the volcanic front of the Izu–Mariana, North-east Honshu, and Kurile arcs and members of this series are very rare in the South-west Japan arc.

Zonal Variation of Rock Chemistry Across the Arc

Volcanic rocks of the calc-alkaline series show a distinct zonal pattern in their chemical composition analogous to that shown by the three basaltic series (Fig. 15; Kuno, 1964, 1965, 1968b). Calc-alkaline (or hypersthenic) rocks occurring in the zone of high alkali and/or high alumina tholeiitic series are richer in alkalis than those occurring in the zone of low alkali tholeiite (or most of the pigeonitic rock series) when plotted in Harker or solidification index variation diagrams (Fig. 14). They are poorer in alkalis than the rocks occurring in the zone of alkaline basalt series. The difference between the alkali content of rocks of calc-alkaline series and those of the basalt-predominant rock series occurring in the same zone appears to be small as compared with the difference across the volcanic arc. Thus Fig. 9 illustrates the overall change in K_2O content of rocks belonging to all the rock series in relation to geographic position of the vent.

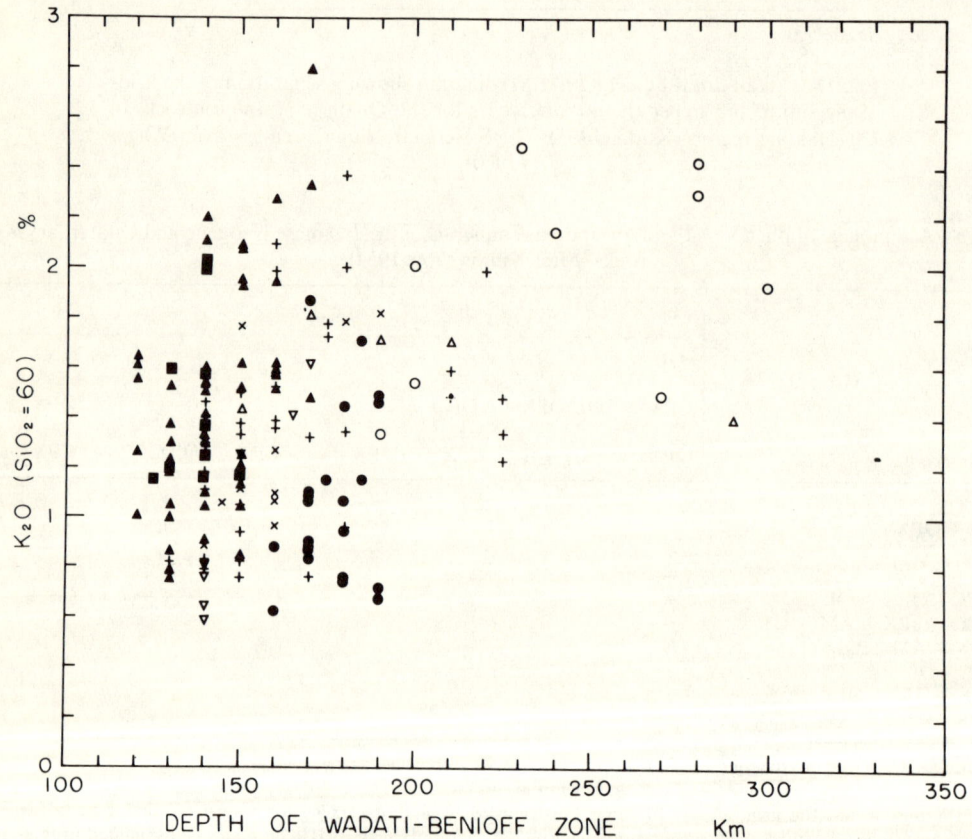

Fig. 18 Relations between K_2O contents and the depth of the Wadati–Benioff zone for individual volcanic centres in the Japanese islands (S. Aramaki and T. Ui, unpublished data). The K_2O values are averaged normalized to K_2O percentages at $SiO_2 = 60$ per cent on a Harker variation diagram and the same as those shown in Fig. 9. The symbols are the same as in Fig. 5

$Na_2O + K_2O$ and K_2O/Na_2O show essentially the same zonal distribution as K_2O, as pointed out by many workers (Yamasaki, 1956; Sugimura, 1960; Katsui, 1961; Kuno, 1966a; Katsui et al., 1978a).

The relationship between the SiO_2-normalized K_2O content and the depth of the Wadati–Benioff zone for individual volcanic centres is shown in Fig. 18. As pointed out by Nielson and Stoiber (1973) and others, the correlation is by no means unique and different regression lines are obtained for different arcs. For the same depth of Wadati–Benioff zone of the arc, the SiO_2-normalized K_2O content for individual arcs increases in the following order: Izu–Mariana arc < North-east Honshu arc < Southwest Honshu arc < Kurile–Kamchatka arc. This order bears no direct relationship to either the speed of subduction or the dip of the Wadati–Benioff zone. Gradual changes in Ba, Sr, Rb, Th, U, Hf, and La (or REE) across the arc in a similar way to K has been confirmed in Hokkaido and north-eastern Honshu (Table 6 and Fig. 19).

Masuda (1966, 1968), Tanaka (1977), and Fujimaki (1977, 1980) demonstrated that REE patterns in Quaternary volcanic rocks vary regularly across the arc. Fujimaki (1977, 1980) showed that the light REE in basaltic rocks gradually increase from the Pacific side to the Japan Sea side of the volcanic arc whereas the heavy REE in the same rocks are relatively uniform, falling within a narrow range (mostly less than 10 times chondrite normalized values; Fig. 20). Distinct depletion in the light REE relative to heavy REE (with distinctively high Ba) is characteristic of low alkali, low Ti, high Mg/Fe* tholeiite such as those from Hachijo–Higashiyama and Izu–Oshima (Fujimaki, 1977) and Funagata and Mashu (Masuda et al., 1975; Masuda and Aoki, 1979). In some volcanic

TABLE 6 Contents of K, Ba, Th, Hf, and La in rocks of some late Quaternary volcanoes in Hokkaido and north-eastern Japan. The contents are normalized against $SiO_2 = 60$ per cent on a Harker diagram

(a) Hokkaido (after Katsui et al., 1978a)

Volcano	Trench–volcano gap (km)	Depth of Wadati–Benioff zone (km)	K_2O (wt%)	Ba	Th	U	Hf	La	La/Sm
				(p.p.m. at $SiO_2 = 60\%$)					
Mashu	305	140	0.4	200	1	—	1.5	4	1.1
Usu	420	160	0.6	300	2	—	2.0	7	2.0
Yotei	445	160	1.1	800	5	—	3.5	17	2.1
Oshima–Oshima	440	190+	2.8	1200	13	—	3.0	30	4.5

(b) North-eastern Japan (after Masuda and Aoki, 1979)

Group	Volcanoes	K_2O (wt%)	Ba	Th	U	Hf	La
			(p.p.m. at $SiO_2 = 60\%$)				
Tholeiitic series	Usu, Towada, Hachimantai, Akita–Komagatake, Funagata, Nekoma	0.69	200	2.18	0.49	2.8	9.3
Calc-alkaline series I	Hakkoda, Hachimantai, Moriyoshi, Kurikoma, Bandai	1.34	356	4.15	1.14	3.9	13.2
Calc-alkaline series II	Niseko, Iwaki, Chokai, Gassan	1.55	593	4.9	1.7	3.0	11.0
Calc-alkaline series III	Oshima–Oshima, Oshima–Kojima, Kampu	2.58	1115	13.7	2.9	3.5	26.2

The calc-alkaline series I, II, III are located sequentially away from the volcanic front toward the Sea of Japan. Average values for each group are given.

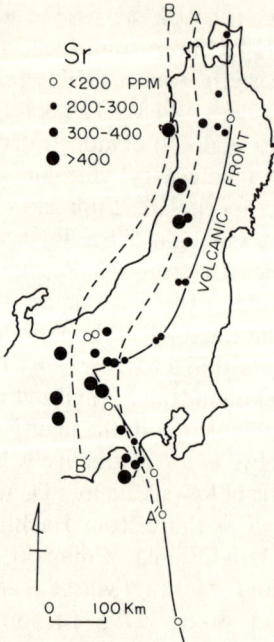

Fig. 19 Zonal variation of Sr in Quaternary volcanic rocks in north-eastern Honshu (data after Fukuyama, 1978). Broken lines A–A' and B–B' are the boundary lines between the tholeiitic and high alumina basaltic series and high alumina basaltic and alkaline series respectively, as given by Kuno (1966a)

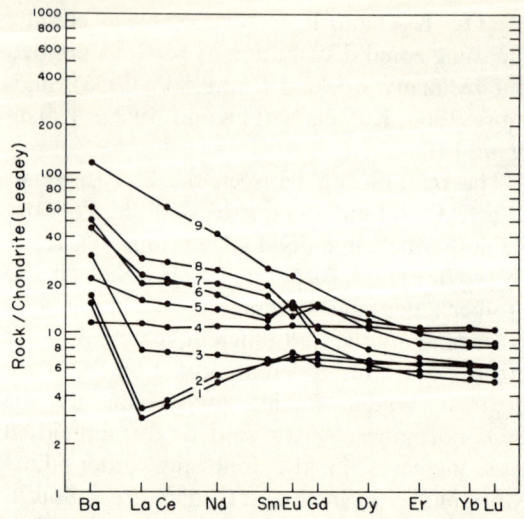

Fig. 20 Chondrite-normalized REE patterns for the basaltic rocks of Japan. (Fujimaki, 1977, 1980). (1) Hachijo–Higashiyama; (2) Hata basalt (Pliocene); (3) Usu; (4) Niijima; (5) Omuroyama; (6) Aso; (7) Fuji; (8) Setouchi; (9) Oki-Dogo

centres (e.g. Usu, Hachijo–Higashiyama, and Fuji), successive increases in REE with subparallel profiles from basalt through andesite to dacite indicates fractional crystallization in shallow reservoirs. However, in other series, crossing of REE patterns with mafic members having higher heavy REE and lower light REE than the felsic members (e.g. Tokachi and Daisetsu; Fujimaki, 1977) and very discordant REE patterns may co-exist in rocks from the same volcanic centre or regions (e.g. Aso Omuroyama, see discussion on p. 289). The distribution of different assemblages of the hydrous minerals hornblende and biotite in volcanic rocks, and their order of crystallization in relation to other minerals, are independent of the nature of the rock series and characterize volcanic associations which have distributions roughly parallel with the volcanic front (Kuno, 1954; Sakuyama, 1977a). In Fig. 21A, the order of appearance of quartz and hornblende phenocrysts is compared (Sakuyama, 1977a, 1979). In volcanoes close to the volcanic front quartz appears earlier than hornblende, while in volcanoes further away from the front the order is reversed, indicating a higher H_2O content in the magma reservoir, which should be located in the crust as the phenocryst assemblage clearly indicates pressures below 10 kb (Sakuyama, 1979).

Figure 21B illustrates a zonal arrangement of phenocryst association in which the order (1) no hornblende or biotite, (2) hornblende but no biotite, (3) biotite, occurs from the volcanic front towards the back-arc side (Sakuyama, 1977a,b). This may reflect the combined effect of H_2O concentration and alkali (especially K) concentration in the magma (compare with Fig. 9). Ishikawa et al (1980) showed that the content of another volatile, F, also increases systematically from tholeiitic basalt to alkaline basalts, across the arc, also suggesting increasing volatile contents away from the front.

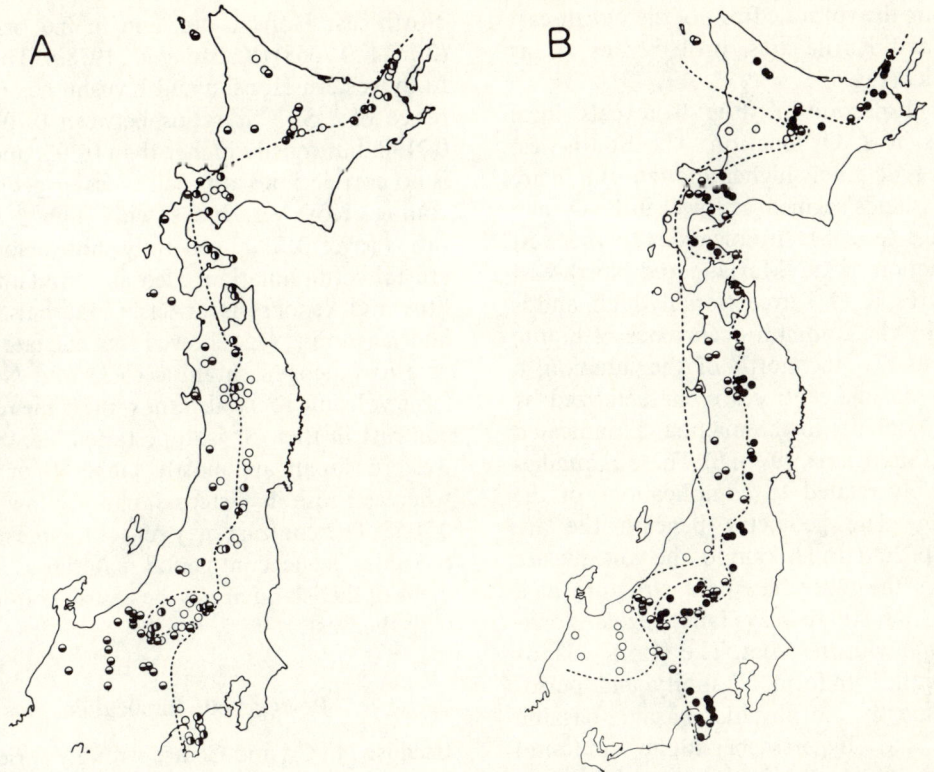

Fig. 21 Diagram showing lateral variation of mafic phenocryst assemblages of late Quaternary volcanic rocks of north-eastern Honshu (Sakuyama, 1977a, b). Symbols in (A): circles with right-hand half solid (◐), volcanoes with rocks containing quartz but no hornblende phenocrysts; circle with bottom half solid (◒), those with rocks containing hornblende but no quartz phenocrysts; circle with diagonal solid quarters (◓), volcanoes in which above two types of rocks are observed; open circles (○), those without rocks containing hornblende and/or quartz, or those in which hornblende and quartz always occur together. Symbols in (B): solid circles (●), volcanoes without biotite and hornblende phenocrysts in any of the lavas or pyroclastic rocks; half solid circles (◒), those with hornblende phenocrysts but without biotite phenocryst at least in one sample; open circles (○), those with biotite phenocrysts at least in one sample

Variation of Rock Association and Chemical Composition along the Arc

The zonal arrangement of volcanic rocks is best developed in the North-east Honshu arc. In the South-west Japan arc, the volcanic front is poorly defined, the zonation is not sharp and the rocks are generally more alkaline in comparison with those of north-eastern Japan. Volcanic rocks of Pliocene to Quaternary age include alkali olivine basalt, trachybasalt, trachyandesite, and melilite olivine nephelinite forming monogenetic volcanoes, as well as hornblende- and biotite-bearing andesites and dacites forming composite cones (Sambe and Daisen volcanoes). In northern Kyushu, hornblende- and sometimes biotite-bearing andesites occur close to the front but in southern Kyushu and the Ryukyu islands pyroxene andesite is the prominent rock type. In the Aso caldera areas, rocks with different levels of K_2O co-exist independent of geographic zonation (Fig. 9). Although rocks of low alkali tholeiitic series do not occur in the South-west Japan arc, they occur along the volcanic front of the North-east Honshu and Kurile arcs, probably as far as Kamchatka (Fig. 14).

Closer inspection of Fig. 9 reveals local anomalies in K_2O zonation. The South-west Japan arc is definitely higher and Izu–Bonin arc in its Izu islands segment is lower in K_2O than the average (p. 278). In volcanoes to the west of the junction of Izu–Mariana and North-east Honshu arcs, K_2O is exceptionally high, and is reflected by the common occurrence of biotite phenocrysts. To the north of the junction, a group of volcanic centres are characterized by low H_2O contents in the magma, as indicated by Fig. 21 (Sakuyama, 1977a,b). These anomalies are probably related to complications of the subducting plate geometry beneath the arc junction (p. 268). In Hokkaido, the volcanic arc cuts across the older basement structure at a high angle, but still displays fairly regular zonation parallel with the front. The Kurile islands show a rather uniform alkalinity and petrography (mostly porphyritic basalt, basaltic andesite, and hypersthene–augite andesite) along the volcanic front, irrespective of the large difference in the nature of the crust (Central Kurile with a thin, suboceanic and South and North Kuriles with thicker, subcontinental crust; Gorshkov, 1970).

Several correlations indicate a link between the petrogenesis of the volcanic rocks and the structure of the continental crust. Ui and Aramaki (1978) pointed out a negative correlation between the K_2O values and Bouguer gravity anomalies throughout the Japanese islands, possibly reflecting the varying role of continental crust during fractional crystallization. The $^{87}Sr/^{86}Sr$ ratios show no distinct zonal variations parallel with the volcanic front. Instead, generally low values (less than 0.7055) in the East Japan Volcanic Belt contrast with widely variable and generally high values (over 0.7055) in south-western Japan (Kurasawa, 1970, 1975; Shuto and Kagami, 1975). The boundary between these provinces is located somewhere near to the western margin of the Fossa Magna. The low value group includes rocks from Hokkaido (0.7028–0.7039), the North-east Honshu arc, and in Izu peninsula (0.7024–0.7058) (Katsui et al., 1978a). The rocks from western Honshu and Kyushu have a wide range of $^{87}Sr/^{86}Sr$ ratios between 0.7035 and 0.7190, but mostly higher than 0.705, and there is no correlation with rock types or geographic zoning. However, rocks with high $^{87}Sr/^{86}Sr$ ratios (over 0.7055) generally show evidence of crustal contamination, such as abundant xenoliths and xenocrysts of felsic materials. Shuto and Kagami (1975) showed that the late Mesozoic to Paleogene granitic rocks and Neogene felsic volcanic rocks also show the same regional contrast in their Sr isotope ratios; i.e. those in western Japan are mostly higher than 0.7055 whereas those in eastern Japan are lower than 0.7055. The contrast may reflect the presence of the older, felsic, continental material as a basement of the island arc in the east in comparison with the west.

Petrogenetic Implications

Because Ni, Cr, and Co are strongly enriched in olivine (in the case of Cr, also in clinopyroxene), they fall rapidly during fractional crystallization (Katsui et al., 1978a; Fukuyama, 1978). Fukuyama (1978) found that in the most mafic (magnesian) basalts in Japan the Ni content is distinctly lower (less than 100 p.p.m.) than that of a basaltic liquid in equilibrium with olivine. Therefore the basalts in the Japanese islands must have undergone considerable fractional crystallization of olivine and clinopyroxene before eruption. Sekine et al. (1979) showed that the pyroxene andesite magmas of Akita–Komagatake, Asama, and Sakurajima were undersaturated with water prior to eruption except for those erupted from a relatively small part of the top of the magma chamber. Taken with the results of Sakuyama (1979), the andesitic magmas erupting close to the volcanic front appear to be mostly water-undersaturated, making a model of andesite magma generation by direct wet melting of the mantle peridotite more difficult.

The tendency of Pb isotopes in Quaternary volcanic rocks to become less radiogenic across

Honshu towards the Sea of Japan was demonstrated by Masuda (1966), Kurasawa (1968), Hedge and Knight (1969), and Tatsumoto and Knight (1969). As $^{238}U/^{204}Pb$ and $^{232}Th/^{204}Pb$ ratios increase in the same direction, giving a negative correlation between parent and daughter isotope ratios, the data cannot be explained by long established horizontal mantle heterogeneity. The possible explanations of the isotope data are (1) crustal contamination, (2) magma generation at different levels within a vertically fractionated mantle, and (3) addition of sedimentary Pb from the subducting oceanic crust (Tatsumoto, 1969; Tatsumoto and Knight, 1969). However, as suggested by Oversby and Ewart (1972) and Matsuhisa (1977), this zonation cannot be uniquely attributed to any one of these models. Sato (1975) pointed out a similar trend of Pb isotope ratios in ore minerals from the Miocene vein-type and Kuroko-type deposits and concluded that the overall general mechanism of magma generation was similar in Neogene to that in Quaternary. Contamination of continental materials is generally precluded by the low Sr isotope ratios (p. 288) except for the South-west Japan arc, where discordant REE patterns, sometimes with very high La/Yb ratios, also suggest substantial contamination by the crustal materials. The low level of heavy REE (many lower than abyssal tholeiite; Fujimaki, 1977, 1980) may be due to the presence of heavy REE-depleted upper mantle beneath Japan or to partial melting in the presence of garnet.

The zonal variation of alkalis and other 'incompatible' elements generally correlating with the depth of the Wadati–Benioff zone argues strongly for a mechanism of magma generation determined by plate subduction. Presumably the parental basaltic magmas are produced either by different degrees of partial melting of the mantle wedge or by different depths of melting or by a combination of both mechanisms. At depths of initiation of melting garnet may have played an important role (see REE patterns), but in the uppermost mantle olivine and pyroxene ($\pm$ plagioclase) were the main crystallizing phases. The effect of volatiles (mainly H_2O?) during the earliest stages of basalt magma generation is still obscure, but in magmas which ascended to the lower crust the volatile content was higher away from the front. Therefore, crustal contamination coupled with fractional crystallization is largely responsible for the predominance of andesite–dacite–rhyolite of the calc-alkaline series over the three basaltic series. The effect is the most pronounced in the South-west Japan arc.

REFERENCES

Aoki, H. (1974). Plate tectonics of arc-junction at central Japan. *J. Phys. Earth* **22**, 141–161.

Aoki, K. (1959). Petrology of alkali rocks of the Iki Islands and Higashi–Matsuura district. Japan, *Sci. Rep., Tohoku Univ., Ser. 3*, **6**, 261–310.

Aoki, K. (1978). Petrological characters of the Quaternary volcanic rocks in Japanese islands. In *Material Science of the Earth*, II, *Igneous Rocks and Their Origin* (I. Kushiro and S. Aramaki, eds), Earth Science Series, Vol. 3, Iwanami-Shoten, Tokyo, pp. 153–170 (in Japanese)

Aoki, K. and Oji, Y. (1966). Calc-alkaline volcanic rock series derived from alkali-olivine basalt magma. *J. geophys. Res.* **71**, 6127–6135.

Aramaki, S. (1956). The 1783 activity of Asama volcano, Part 1. *Japan. J. Geol. Geogr.* **27**, 189–229.

Aramaki, S. (1969). Some problems of the theory of caldera formation. *Bull. Volc. Soc. Japan Ser. 2* **14**, 55–76 (in Japanese).

Aramaki, S. and Ui, T. (1978a). Major element frequency distribution of the Japanese Quaternary volcanic rocks. *Bull. Volc. Ser. 2* **41**, 390–407.

Aramaki, S. and Ui, T. (1978b). List of geodynamic parameters of Quaternary volcanoes of Japan, Mariana, Kurile and Kamchatka (preliminary). *Contr. Geodyn. Project Japan* no. 78–2, 9p.

Aramaki, S. and Yamasaki, M. (1963). Pyroclastic flows in Japan. *Bull. Volc. Ser. 2* **26**, 89–99.

Aramaki, S., Takahashi, M., and Nozawa, T. (1977). Kumano acidic rocks and Okueyama complex; two examples of ten granitic rocks in the outer zone of Southwestern Japan. In *Plutonism in Relation to Volcanism and Metamorphism, Proceedings of the 7th Circum-Pacific Plutonism Project Meeting, IGCP*, pp. 127–147.

Clague, D. A. and Jarrard, R. D. (1973). Tertiary Pacific plate motion deduced from the Hawaii–Emperor chain. *Geol. Soc. Am. Bull.* **84**, 1135–1154.

Erlich, E. N. (1968). Recent movements and Quaternary volcanic activity within the Kamchatka. *Pacific Geol.* **1**, 23–39.

Fujimaki, H. (1977). Studies on rare-earth elements in volcanic rocks from Japan and their petrological implications. Ph.D. thesis, University of Tokyo, 162p.

Fujimaki, H. (1980). Lateral variation of rare earth patterns of basaltic magma across the Japan arc. *Earth planet. Sci. Letters* in press.

Fukuyama, H. (1978). Petrological study of Ni, Rb and Sr in magmas with special reference to the mantle–magma equilibrium. *Ph.D. thesis, University of Tokyo*, Part 2, pp. 59–220.

Geographical Survey Institute of Japan (1973). *Japan and its Environment:* 1:3 000 000 topographic map.

Gorshkov, G. S. (1958). *Catalogue of the Active Volcanoes of the World Including Solfatara Fields*, Part 7, *Kurile Islands*, International Association for Volcanology, Naples.

Gorshkov, G. S. (1970). *Volcanism and the Upper Mantle. Investigations in the Kurile Island Arc*, Plenum Press, New York, 385p.

Hagiwara, Y. (1967). Analysis of gravity values in Japan. *Bull. Earthquake Res. Inst. Univ. Tokyo* 45, 1091–1228.

Hasegawa, A., Umino, N., and Takagi, A. (1978). Double-planed structure of the deep seismic zone in the north-eastern Japan arc. *Tectonophysics* 47, 43–58.

Hedge, C. E. and Knight, R. J. (1969). Lead and strontium isotopes in volcanic rocks from northern Honshu, Japan. *Geochem. J.* 3, 15–24.

Ichikawa, K., Matsumoto, T., and Iwasaki, M. (1972). Development of Japanese islands. *Kagaku* 42, 181–191, (in Japanese).

Ishikawa, M. and Egawa, R. (1977). Volcanic rocks from northern Mariana islands. *Chikyu Kagaku* 31, 55–69 (in Japanese).

Ishikawa, K., Kanisawa, and Aoki, K. (1980). Content and Behaviour of fluorine in Japanese Quaternary volcanic rocks and petrogenetic application, *J. Volc. geothermal Res.* 8, 161–175.

Isshiki, N. (1964). *Oshima Volcano*, Guide-book for Excursion no. 2, Geological Survey of Japan, Kawasaki.

Isshiki, N., Matsui, K., and Ono, K. (1968a). *Volcanoes of Japan:* 1:2 000 000 map series, Geological Survey of Japan, Kawasaki, 78p.

Isshiki, N., Matsui, K., and Ono, K. (1968b). *Selected Bibliography of Japanese Volcanoes*, Geological Survey of Japan, Kawasaki.

Jacob, K. H., Nakamura, K., and Davies, J. N. (1977). Trench–volcano gap along the Alaska–Aleutian arc: facts, and speculations on the role of terrigineous sediments; (3) analysis of the surface wave data. *Bull. Earth-Basins*, M. Ewing Series no. 1, pp. 243–258.

Jakeš, P. and White, A. J. R. (1972). Major and trace element abundances in volcanic rocks of orogenic areas. *Geol. Soc. Am. Bull.* 83, 29–40.

Kanamori, H. (1963). Study on the crust-mantle structure in Japan: (1) analysis of gravity data; (2) interpretation of the results obtained by seismic refraction studies in connection with the study of gravity and laboratory experiments; (3) analysis of the surface wave data. *Bull. Earthquake Res. Inst., Univ. Tokyo* 41, 743–759, 761–779, 801–818.

Kanamori, H. and Tsumura, K. (1971). Spatial distribution of earthquakes in the Kii peninsula, Japan, south of the median tectonic line. *Tectonophysics* 12, 327–342.

Kaneoka, I, Isshiki, N., and Zashu, S. (1970). K–Ar ages of the Izu-Bonin islands. *Geochem. J.* 4, 53–60.

Katsui, Y. (1954). Chemical composition of the lavas of the Chokai volcanic zone, Japan. *J. geol. Soc. Japan* 60, 185–191.

Katsui, Y. (1961). Petrochemistry of the Quaternary volcanic rocks of Hokkaido and surrounding areas. *J. Fac. Sci., Hokkaido Univ.*, Ser. 4, 11, 1–58.

Katsui, Y. (1971). *List of the World's Active Volcanoes* (with Map), Volcanolical Society of Japan/International Association for Volcanology and Chemistry of the Earth's Interior, Naples.

Katsui, Y., Oba, Y., Ando, S. Nishimura, S., Masuda, Y., Kurasawa, H., and Fujimaki, H. (1978a). Petrochemistry of the Quaternary volcanic rocks of Hokkaido, North Japan, *J. Fac. Sci., Hokkaido Univ.*, Ser 4, 18, 449–484.

Katsui, Y., Oba, Y., and Soya, T. (1978b). Records of volcanic eruptions in historic times and estimation of future eruptions. *Bull Volc. Soc. Japan*, Ser. 2, 23, 41–52 (in Japanese).

Kawai, N., Ito, H., and Kume, S. (1961). Deformation of the Japanese islands as inferred from rock magnetism. *Geophys. J. R. Astr. Soc.* 6, 124–130.

Kawano, Y., Yagi, K., and Aoki, K. (1961). Petrography and petrochemistry of the volcanic rocks of Quaternary volcanoes of northeastern Japan. *Sci. Rep., Tohoku Univ.*, Ser. 3, 7, 1–46.

Kobayashi, T. (1941). The Sakawa orogenic cycle and its bearing on the origin of the Japanese islands, *J. Fac. Sci., Imp. Univ. Tokyo, Sec. 2* 5, 219–578.

Konda, T. (1974). Bimodal volcanism in the Northeast Japan arc. *J. geol. Soc. Japan* 80, 81–89 (in Japanese).

Kroenke, L. and Scott, R. (1978). In the Phillippine Sea old questions answered and new ones asked. *Geotimes* 23 (7), 20–23.

Kuno, H. (1950). Petrology of Hakone volcano and adjacent areas, Japan. *Geol. Soc. Am. Bull.* 61, 957–1020.

Kuno, H. (1953). Formation of calderas and magmatic evolution. *Trans. Am. geophys. Union* 34, 267–280.

Kuno, H. (1954). *Volcanoes and Volcanic Rocks*, Iwanami, Tokyo (in Japanese), 255p.

Kuno, H. (1959). Origin of Cenozoic petrographic provinces of Japan and surrounding areas. *Bull. Volc. Ser. 2* 20, 37–76.

Kuno, H. (1960). High-alumina basalt. *J. Petrol.* 1, 121–145.

Kuno, H. (1962). *Catalogue of the Active Volcanoes of the World Including Solfatara Fields*, Part 9, *Japan, Taiwan and Marianas*, International Association for Volcanology and Chemistry of the Earth's Interior, Rome.

Kuno, H. (1964). Igneous rock series. In *Chemistry of the Earth's Crust*, Nauka, Moscow, pp. 107–121 (in Russian).

Kuno, H. (1965). Some problems on calc-alkali rock series. *J. Japan. Assoc. Petrol. Mineral. Econ. Geol.* 53, 131–142 (in Japanese).

Kuno, H. (1966a). Lateral variation of basalt magma type across continental margins and island arcs. *Bull. Volc. Ser. 2*, 29, 195–222.

Kuno, H. (1966b). Review of pyroxene relations in terrestrial rocks in the light of recent experimental works. *Mineral. J.* 5, 21–43.

Kuno, H. (1968a). Origin of andesite and its bearing on the island arc structures. *Bull. Volc. Ser. 2* 32, 141–176.

Kuno, H. (1968b). Differentiation of basaltic magmas. In *Basalts*, Vol. 2 (H. H. Hess, ed.), Interscience, New York, pp. 623–688.

Kurasawa, H. (1968). Isotopic composition of lead and concentrations of uranium, thorium and lead in volcanic rocks from Dogo of the Oki islands, Japan. *Geochem. J.* **2**, 11–28.

Kurasawa, H. (1970). *Isotope Geology*, Lattice, Tokyo (in Japanese), 275p.

Kurasawa, H. (1975). Isotope geology of volcanic rocks. *Bull. Volc. Soc. Japan, Ser. 2* **20**; 307–317 (in Japanese).

Masuda, A. (1966). Lanthanides in basalts of Japan with three distinct types. *Geochem. J.* **1**, 11–26.

Masuda, A. (1968). Geochemistry of lanthanides in basalts of central Japan. *Earth planet. Sci. Letters* **4**, 284–292.

Masuda, Y. and Aoki, K. (1979). Trace element variations in the volcanic rocks from Nasu Zone, northeast Japan. *Earth planet. Sci. Letters* in press.

Masuda, Y., Nishimura, S., Ikeda, T., and Katsui, Y. (1975). Rare-earth and trace elements in the Quaternary volcanic rocks of Hokkaido, Japan, *Chem. Geol.* **15**, 251–271.

Matsuda, T. (1962). Crustal deformation and igneous activity in the South Fossa Magna, Japan. In *The Crust of the Pacific Basin* (L. Knopoff *et al.*, eds), American Geophysical Union Monograph no. 6, pp. 140–150.

Matsuhisa, Y. (1977). Isotope geochemistry of an island arc traverse. *Geochem. J.* **11**, 107–109.

Mikami, K. (1962). Geological and petrographical studies on the Tanzawa mountainland. *Sci. Rep., Yokohama National Univ., Sec. II* **8**, 57–110, **9**, 59–108.

Minakami, T. (1960). Fundamental research for predicting volcanic eruptions, Part 1. *Bull. Earthquake Res. Inst. Univ. Tokyo* **38**, 497–544.

Minakami, T., Hiraga, S., Miyazaki, T., and Uchibori, S. (1969). Fundamental research of predicting volcanic eruptions. Part 2. *Bull. Earthquake Res. Inst. Univ. Tokyo* **47**, 893–949.

Minster, J. B., Jordan, T. H., Molnar, P., and Haines, E. (1974). Numerical modelling of instantaneous plate tectonics, *Geophys. J. R. astr. Soc.* **36**, 541–576.

Miyashiro, A. (1961). Evolution of metamorphic belts, *J. Petrol.* **2**, 277–311.

Miyashiro, A. (1972). Metamorphism and related magmatism in plate tectonics. *Am. J. Sci.* **272**, 629–656.

Miyashiro, A. (1974). Volcanic rock series in island arcs and active continental margins. *Am. J. Sci.* **274**, 321–355.

Moriya, I. (1973). Geomorphological evolutions of the Japanese Quaternary volcanoes, *Hokkaido Chiri* **48**, 1–7 (in Japanese).

Moriya, I. (1977). Geomorphic evolution of Japanese Quaternary volcanoes. *Ph.D. thesis, University of Tokyo.*

Nakamura, K. (1964). Volcano-stratigraphic study of Oshima volcano, Izu. *Bull. Earthquake Res. Inst., Univ. Tokyo* **42**, 649–728.

Nakamura, K. (1969). Island arc tectonics, a hypothesis. In *Symposium on Problems Concerning 'Green Tuffs'*, Annual Meeting of the Geological Society of Japan, pp. 31–38 (in Japanese).

Nakamura, K. (1974). Preliminary estimates of global volcanic production rate. In *The Utilization of Volcanic Energy*, (J. L. Colp and A. S. Furmoto, eds), Hilo, Hawaii, pp. 273–285.

Nakamura, K. (1977). Volcanoes as possible indicators of tectonic stress orientation—principle and proposal. *J. Volc. geothermal Res.* **2**, 1–16.

Nielson, D. R. and Stoiber, R. E. (1973). Relationship of potassium content in andesitic lavas and depth to the seismic zone. *J. geophys. Res.* **78**, 6887–6892.

Ono, K. (1962). *Chemical Composition of Volcanic Rocks in Japan*, Geological Survey of Japan, Kawasaki, 44p.

Ossaka, J., Hirabayashi, J., and Ozawa, T. (1978). Volcanic observations and prediction of volcanic eruption by geochemical methods. *Bull. Volc. Soc. Japan, Ser. 2*, **23**, 33–40.

Oversby, V. M. and Ewart, A. (1972). Lead isotopic compositions of Tonga–Kermadec volcanics and their petrogenetic significance. *Contr. Mineral. Petrol.* **37**, 181–210.

Sakuyama, M. (1977*a*). Lateral variation of phenocryst assemblages in volcanic rocks of the Japanese islands. *Nature, Lond.* **269**, 134.

Sakuyama, M. (1977*b*). Lateral variation of H_2O contents in Quaternary magma of northeastern Japan. *Bull. Volc. Soc. Japan. Ser. 2* **22**, 263–271 (in Japanese).

Sakuyama, M. (1979). Lateral variations of H_2O contents in Quaternary magmas of northeastern Japan. *Earth planet. Sci. Letters* **43**, 103–111.

Sato, K. (1975). Unilateral isotopic variation of Miocene ore leads from Japan. *Econ. Geol.* **70**, 800–805.

Sekine, T., Katsura, T., and Aramaki, S. (1979). Water saturated phase relations of some andesites with application to the estimation of the initial temperature and water pressure at the time of eruption. *Geochim. cosmochim. Acta* **43**, 1367–1376.

Sekiya, S. and Kikuchi, Y. (1890). The eruption of Bandaisan. *J. Coll. Sci., Imp. Univ. Tokyo* **3**, 91–172.

Seno, T. (1977). The instantaneous rotation vector of the Philippine Sea plate relative to the Eurasian plate. *Tectonophysics* **42**, 209–226.

Shibata, K. and Adachi, M. (1974). Rb–Sr whole-rock ages of Precambrian metamorphic rocks in the Kamiaso conglomerate from central Japan. *Earth planet. Sci. Letters* **21**, 277–287.

Shibata, S. (1968). *Nihon Ganseki-shi*, Vol. 3, Asakura Shoten, Tokyo (in Japanese), 389p.

Shiono, K. (1974). Travel time analysis of relatively deep earthquakes in southwest Japan with special reference to the underthrusting of the Philippine sea plate. *J. Geosci., Osaka City Univ.* **18**, 37–59.

Shiraki, K., Kuroda, N., and Urano, H. (1977). Boninite: an evidence for calc-alkalic primary magma. *Bull. Volc. Soc. Japan, Ser. 2* **22**, 257–261.

Shuto, K., and Kagami, H. (1975). Regional variations in the Sr isotopic composition in the Cenozoic tholeiitic-calc-alkaline volcanic rocks from the intra- and circum-Pacific regions. *Chikyu-Kagaku* **29**, 75–86 (in Japanese).

Smith, R. L. and Bailey, R. A. (1968). Resurgent cauldrons, *Geol. Soc. Am. Mem.* no. 116, 613–662.

Sugimura, A. (1960). Zonal arrangement of some geophysical and petrological features in Japan and its environs. *J. Fac. Sci., Univ. Tokyo, Sec. 2*, **12**, 133–153.

Sugimura, A. (1965). Distribution of volcanoes and seismicity of the mantle in Japan. *Bull. Volc. Soc. Japan, Ser. 2* **10**, 37–58 (in Japanese).

Sugimura, A. (1978). Major topography, volcanoes and earthquakes in island arcs. In *Earth Sciences*, Vol. 10 (K. Kasahara and A. Sugimura, eds), Iwanami, Tokyo pp. 159–181 (in Japanese).

Sugimura, A. and Uyeda, S. (1973). *Island Arcs: Japan and Its Environs*, Elsevier, Amsterdam, 247p.

Sugimura, A., Matsuda, T., Chinzei, K., and Nakamura, K. (1963). Quantitative distribution of late Cenozoic volcanic materials in Japan. *Bull. Volc. Ser. 2* **26**, 125–140.

Sugisaki, R. and Tanaka, T. (1971). Magma types of volcanic rocks and crustal history in the Japanese pre-Cenozoic geosynclines. *Tectonophysics* **12**, 393–413.

Sugisaki, R., Mizutani, S., Hattori, H., Adachi, M., and Tanaka, T. (1972). Late Paleozoic geosynclinal basalt and tectonism in the Japanese Islands. *Tectonophysics* **14**, 35–56.

Suwa, A. (1978). The surveillance of volcanic activities in Japan. *Bull. Volc. Soc. Japan, Ser. 2* **23**, 83–90 (in Japanese).

Takahashi, E. (1978). Petrological model of the crust and upper mantle of the Japanese island arcs. *Bull. Volc., Ser. 2* **41**, in press.

Tanaka, T. (1977). Rare earth abundances in Japanese Paleozoic geosynclinal basalts and their geological significance. *Bull. geol. Surv. Japan*, **28**, 529–559.

Tatsumoto, M. (1969). Lead isotopes in volcanic rocks and possible ocean-floor thrusting beneath island arcs. *Earth planet. Sci. Letters* **6**, 369–376.

Tatsumoto, M. and Knight, R. J. (1969). Isotopic composition of lead in volcanic rocks from central Honshu—with regard to basalt genesis. *Geochem. J.* **3**, 53–86.

Tomoda, Y. (1973). *Maps of Free Air and Bouguer Gravity Anomalies in and around Japan*, University of Tokyo Press, Tokyo.

Tuyezov, I. K. (1971). Crustal structure of the Okhotsk and Japanese area from regional seismic prospecting data. In *Island Arc and Marginal Seas* (S. Asano and G. B. Udinstev, eds), Tokai University Press, Tokyo, pp. 121–136 (in Japanese).

Ui, T. and Aramaki, S. (1978). Relationship between chemical composition of Japanese island volcanic rocks and gravimetric data. *Tectonophysics* **45**, 249–259.

Ujike, O. (1970). Petrological study of Tertiary volcanic rocks from northeastern Shikoku. *J. Japan. Assoc. Mineral. Petrol. Econ. Geol.* **63**, 43–62.

Utsu, T. (1971). Seismological evidence for anomalous structure of island arcs with special reference to the Japanese region. *Rev. Geophys. Space Phys.* **9**, 839–890.

Utsu, T. (1974). Epicenter distribution around Japan. *Kagaku* **44**, 739–746 (in Japanese).

Uyeda, S. (1972). Heat flow. In *The Crust and Upper Mantle of the Japanese Area*, Part 1, pp. 97–106.

Uyeda, S. and Miyashiro, A. (1974). Plate tectonics and the Japanese islands: a synthesis. *Geol. Soc. Am. Bull.* **85**, 1159–1170.

Vlodavetz, V. I. and Piip, B. I. (1959). *Catalogue of the Active Volcanoes and Solfataric Fields of the World*, Part 8, *Kamchatka and Continental Areas of Asia*, International Association for Volcanology, Naples.

Wager, L. R. (1960). The major element variation of the layered series of the Skaergaard intrusion and a re-estimation of the average composition of the hidden layered series and of the successive residual magmas. *J. Petrol.* **1**, 364–398.

Williams, H. (1941). Calderas and their origin, *Calif. Univ., Dept. geol. Sci., Bull.* **25**, 239–346.

Williams, H. (1942). The geology of Crater Lake National Park, Oregon, with a reconnaissance of the Cascade Range southward to Mt Shasta. *Carnegie Instn. Wash. Publ.* no. 540, 162p.

Working Group on the World Volcanological Map, IAVCEI (1973). *Data Sheets of the Post-Miocene Volcanoes of the World*, International Association for Volcanology and Chemistry of the Earth's Interior, Rome.

Yamasaki, M. (1956). Petrologic significance of the K_2O/Na_2O ratios of volcanic rocks of the Fuji and Nasu volcanic zones in Japan. *J. geol. Soc. Japan* **62**, 504–514.

Yokoyama, I. (1974). Geomagnetic and gravity anomalies in volcanic areas. In *Physical Volcanology* (L. Civetta et al., eds), Elsevier, Amsterdam, pp. 167–194.

Yokoyama, I. (1978). Predictions of volcanic eruptions by means of geodetic, geomagnetic, geoelectric and geothermic observations. *Bull. Volc. Soc. Japan, Ser. 2*, **23**, 19–32 (in Japanese).

Yoshida, T. (1970). Ishizuchi collapse caldera and Tengu-dake pyroclastic flow, Shikoku Island. *J. Japan. Assoc. Mineral. Petrol. econ. Geol.* **64**, 1–12.

Yoshii, T. (1975). A proposal on 'aseismic front'. *Bull. Seismol. Soc. Japan, Ser. 2* **28**, 365–367 (in Japanese).

Yoshii, T. (1979). A detailed cross-section of the deep-seismic zone beneath northeastern Honshu, Japan. *Tectonophysics* **55**, 349–360.

Andesites
Edited by R. S. Thorpe
© 1982 John Wiley & Sons

Mariana Volcano Islands

A. Meijer

Department of Geosciences,
University of Arizona, Tuscon, Arizona 85721, USA

ABSTRACT

The Mariana and Volcano Islands are island arcs formed since the Eocene upon crust of oceanic character some 17–20 km in thickness. The volcanic evolution might represent the early stages of development of continental crust. The volcanoes of the Mariana arc appear to follow a relatively consistent mode of evolution. During the initial stage of this evolution the eruptive products are dominated by 'thin' lava flows. Subsequently, the volume of lava flows decreases in relation to pyroclastic materials as the cone develops the form of a classic composite volcano. The pyroclastic eruptions generally climax in the formation of a collapse structure, although such features do not occur on all the islands. The formation of the collapse structure is followed either by the eruption of a new intra-crater cone or the extrusion of viscous domes and flows within the crater and from the crater rim. The volcanoes of the Mariana arc are predominantly of olivine–augite basalts and basaltic andesites. Pyroxene andesites occur locally and hornblende andesites occur on Sarigan Island. Igneous differentiation within most of the volcanoes produces substantial Fe-enrichment trends and, although the Mariana arc lavas may be considered to belong to the 'island arc tholeiitic series', the samples analysed from Sarigan Island define 'calc-alkaline differentiation trends'. Older lavas erupted in the Volcano Island arc appear to be similar to the younger lavas erupted in the Izu and Mariana arcs, but younger lavas are generally rich in alkalis (particularly soda) and may be called trachyandesites.

Introduction

Because they are bounded to the east and west by oceanic lithosphere, the northern Mariana and Volcano island chains are intra-oceanic or 'oceanic arcs'. Various geophysical and geochemical characteristics suggest that they represent early stages in the evolution of orogenic magmatism which is thought to ultimately lead to the formation of continental crust. The purpose of this paper is to review the general aspects of volcanism in the Mariana and Volcano arcs and to discuss some preliminary results of recent field studies in the Mariana arc.

Previous Work

The northern Mariana Islands (Fig. 1) have received surprisingly little attention from geologists until recently. Early observations by European and Japanese naturalists (*see* review by Cloud *et al.*, 1956, pp. 9–15) established that the volcanoes of this arc were of orogenic type, but details of the volcanic successions on the islands remained obscure. During the time the Mariana Islands were under Japanese Mandate (1920–44), several geological investigations were carried out. This work, in addition to providing an overview of the general geological characteristics of each of the islands (Tayama, 1936; Tanakadate, 1940), included a relatively detailed investigation of the stratigraphy on the island of Pagan (Tayama, 1936). Geologists of the US Geological Survey visited a number of the islands briefly (Schmidt, 1957; pp. 156–158) during the early 1950s and carried out a detailed investigation of the island of Pagan (Corwin *et al.*, 1957).

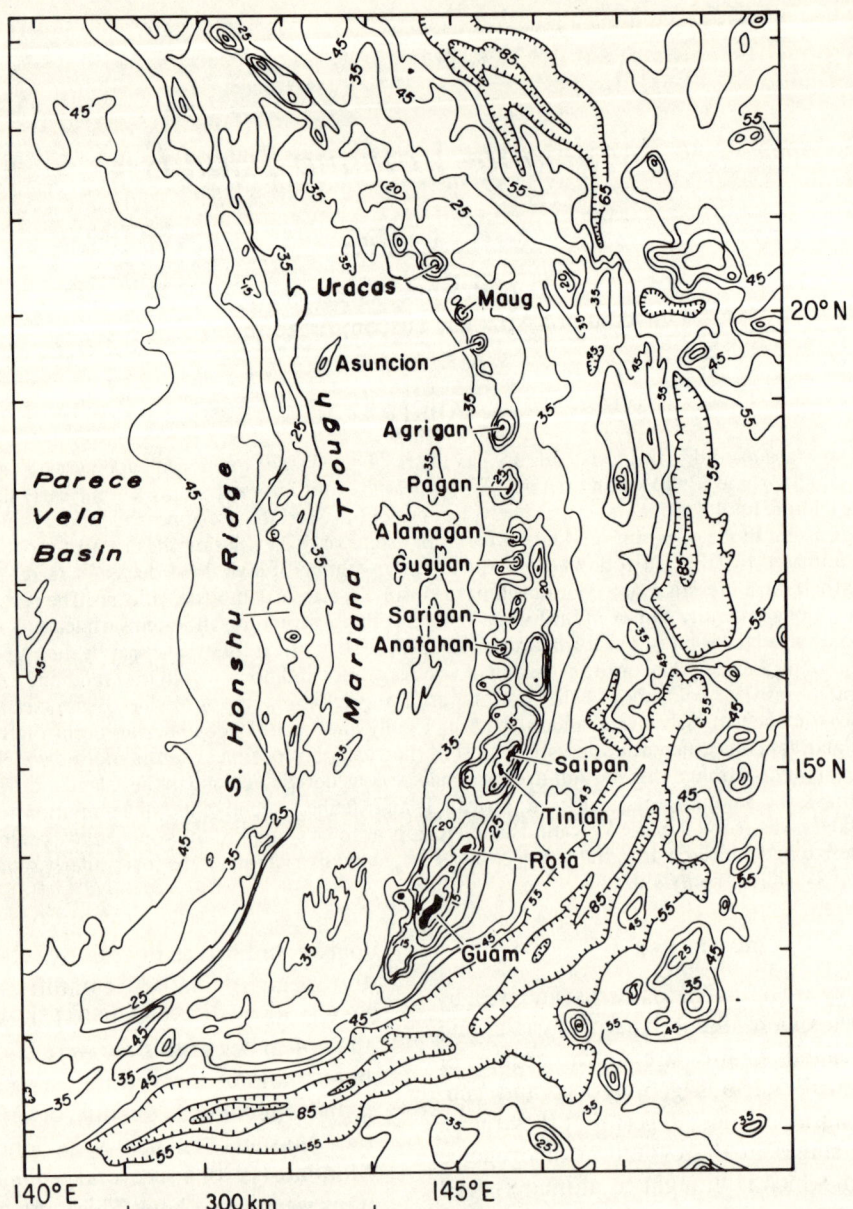

Fig. 1 Generalized bathymetric map of Mariana island-arc system (modified after Karig, 1971). The Mariana Trench is indicated by hatched contours (depths in hundreds of metres)

Schmidt (1957) compiled all available chemical analyses of lavas from the northern Mariana Islands (10 in all) and this compilation was enlarged by Larson et al. (1974). Samples collected by the US Geological Survey have been analysed for Zr (Chao and Fleisher, 1960), U and Th (Gottfried et al., 1964), and Sr isotope ratios (Hedge, 1966; Pushkar, 1968). Recent contributions to the geology and geochemistry of the islands have been presented by Meijer and Glassley (1973), Meijer (1974, 1976a), Meijer et al. (1978), Youngblood and Meijer (1978), and Stern (1978, 1979). Available information on historic eruptions of volcanoes in the northern Mariana Islands has been discussed by Kuno (1962).

Islands in the Volcano Group, although fewer in number, have been more extensively studied. Kuno (1962) has summarized the available data on their geological characteristics and historic activity. Recent contributions to the chemical compositions of lavas from these islands include those of Hedge (1966), Philpotts et al. (1971), and Matsuda et al. (1977).

Regional Setting

The Mariana island arc system (Fig. 1, from Karig, 1971) comprises from east to west: (1) the Mariana Trench; (2) the arc–trench slope; (3) the frontal arc including the islands of Guam, Rota, Tinian, and Saipan; (4) the active arc extending from Anatahan north to Farallon de Pajaros (also called Uracas); (5) an inter-arc basin known as the Mariana Trough; and (6) a 'remnant arc' known as the South Honshu Ridge (also known as the West Mariana Ridge).

The Mariana Trench is among the deepest in the world, a characteristic attributed in part to the lack of significant sediment sources in the upper trench slope regions. Dredges taken at various depths along the inner trench wall have yielded mafic and ultramafic rocks at several points along the arc trend (Bogdanov, 1977; Dietrich et al., 1978; Hawkins et al., 1979). Recent drilling by Leg 60 of International Program of Ocean Drilling (IPOD) at latitude 18°N penetrated olistostrome-like deposits on the lower trench slope (7055 m maximum depth, 20 m total penetration; Shipboard Scientific Staff, 1978). The diversity of fossil assemblages (reworked Cretaceous(?) to Quaternary) and rock types observed in these deposits suggests considerable reworking, although the detailed nature of tectonic processes operative along the lower trench slope remain obscure.

The results of drilling in the upper arc–trench slope at latitude 18°N suggests that this region is floored by arc-related volcanic rocks of Eocene to Oligocene age (Shipboard Scientific Staff, 1978). Surprisingly, volcanic rocks with geochemical characteristics suggestive of arc associations were drilled just 50 km west of the Mariana Trench. These data, combined with available information on the geology of Guam and Saipan (see below), suggest that the Mariana arc is superimposed on crust not older than late middle Eocene in age. This is considerably younger than the Cretaceous to Jurassic age inferred for the westernmost portion of the Pacific plate currently descending beneath the Mariana arc system (Larson and Chase, 1972).

Geological studies of the islands of Guam (Tracey et al., 1963), Tinian (Doan et al., 1960), and Saipan (Cloud et al., 1956) indicate that the frontal arc is comprised of Eocene to Miocene arc volcanic rocks locally interbedded with and overlain by shallow water limestone and other sediments. Although much of the exposed volcanic section consists of Eocene–Oligocene low alkali lavas belonging to the 'island arc tholeiitic series' (Jakeš and White, 1972), pre-late Miocene alkaline lavas are exposed on both Guam (Stark and Tracey, 1963) and Saipan (Schmidt, 1957). These observations, in addition to measurements of the crustal thickness beneath the arc (17–20 km; Murauchi et al., 1968; Bibee, personal communication), imply that the Mariana arc lavas erupted on to crust slightly more evolved (both in thickness and composition) than oceanic crust but not as evolved as the crust in most other circum-Pacific island arcs (e.g. Aleutians; Grow, 1973; see also Marsh, this volume, Section III).

The Mariana Trough is a young (c. 0–6 Ma old; Shipboard Scientific Staff, 1978) basin apparently created by a form of sea-floor spreading behind the Mariana arc (Karig, 1971). The results of drilling by Leg 60 of IPOD suggest a half-spreading rate for the basin of $c.\ 1.5\ \mathrm{cm\ a}^{-1}$. Volcanic rocks recovered from different parts of the basin by dredging and drilling (Leg 60) indicate it is floored by volcanic rocks similar to ocean ridge basalts (Hart et al., 1972; Meijer, 1974, 1976a,b; Pineau et al., 1976; Fryer, 1978). One characteristic feature distinguishing them from most ocean ridge basalts is their high degree of vesicularity. In fact, samples analysed to date

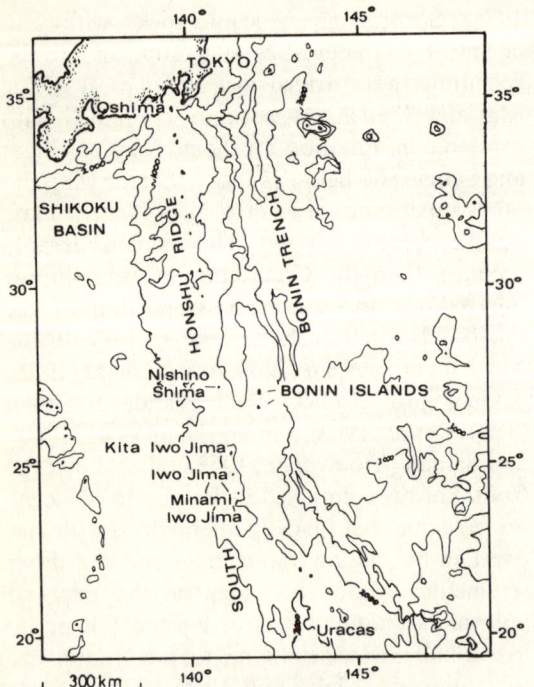

Fig. 2 Generalized bathymetric map of Izu–Bonin–Volcano island-arc system. The Izu arc is a chain of active volcanoes extending from the Izu peninsula in Japan southwards along the Honshu Ridge to the island north of Nishino Shima. The Bonin Islands are comprised largely of Eocene–Oligocene volcanic arc rocks and can be considered as a frontal arc to the active Volcano arc. The latter extends from Nishino Shima to Minami Iwo Jima. Sin Iwo Jima (not shown) is located 5 km north-east of Minami Iwo Jima. Note that islands along the Izu–Volcano arcs define a very consistent linear trend. Bathymetry in fathoms. (After Karig and Ingle, 1975)

suggest that some of the trough magmas were nearly H_2O-saturated at their time of eruption (Muenow et al., 1977).

Less is known about the crust beneath the Volcano Islands (Fig. 2). Volcanic units exposed on the volcanically inactive Bonin Islands directly east of the Volcano Islands show similarities in age (Eocene to Oligocene; Kaneoke et al., 1970) and composition (Kuroda and Shiraki, 1975) to units drilled by Leg 60 in the Mariana fore-arc region. This, in addition to a reported crustal thickness of 17 km beneath the islands (Murauchi et al., 1968), suggests that they may be underlain by volcanic rocks similar to those comprising the Mariana arc crust.

The Benioff zone beneath the Mariana arc system is essentially vertical (Katsumata and Sykes, 1969) and may even be overturned toward the east at the deepest levels (400–600 km; B. Isacks et al., unpublished data). Although there are not sufficient data to document variations in the dip of the Benioff zone directly beneath the arc, on a larger scale the data of Katsumata and Sykes (1969) suggest a general decrease in dip from 90° beneath the central Mariana arc to c. 45° in the northern Izu arc. This suggests that the Benioff zone beneath the Volcano Islands should be less steep than the zone beneath the Mariana Islands, although the pronounced gap in earthquake activity observed within the Volcano Islands (Katsumata and Sykes, 1969) makes definition of this zone difficult.

Distribution of Volcanism

More than in any other volcanic arc, the volcanoes of the Mariana Islands lie with amazing regularity along an arc (see Fig. 1) with a radius of c. 750 km centred at longitude 139°E, latitude 17°20′N. Those volcanoes located more than 5 km from this arc (i.e. Sarigan, Anatahan, and South Volcano on Pagan) generally lack signs of historic activity (Kuno, 1962), suggesting that their deviation is due to a relocation of the locus of activity. The regularity shown by the other c. 12 volcanoes and seamounts is likely to reflect some fundamental characteristic of the magma generation mechanism which remains to be identified.

The spacing of islands and shallow seamounts along the arc ranges from 20 to c. 80 km. In a general way, this spacing is correlated with the size of each volcano, small volcanoes being more closely spaced than large volcanoes (Table 1). It is tempting to conclude from this that distance and size must somehow be a function of the volume of mantle from which a particular volcano derives its magma supply. However, testing of such a correlation would require accurate data on the bathymetry (i.e. the volume) of each volcano, the total interval during which

TABLE 1 Characteristics of active and inactive volcanoes and seamounts forming the Mariana arc

Name	Area (km²)	Maximum Elevation (m)	Spacing (km)	Collapse Structure	Most Recent Activity
Seamount	—	−1756		?	1974+
			103		
Seamount	—	−512		?	?
			16		
Uracas	2.1	353		Yes	1977
			51		
Supply Reef	—	−9.1		?	?
			16		
Maug	2.1	227		Yes	Old
			42		
Asuncion	7.3	8.57		No	1906
			32		
Seamount	—	−106		?	?
			72		
Agrigan	47.4	882		(No)	1917
			72		
Pagan	47.7	570		Yes	1925
			63		
Alamagan	11.4	744		No	Old
			32		
Guguan	4.1	287		Yes	Recent
			49		
Zealandia Bank	—	−2		?	?
			21		
Sarigan	4.9	538		Yes	Old
			42		
Anatahan	32.4	787		?	Recent
			49		
Seamount	—	−874		?	?
			34		
Seamount	—	−1265		?	?
			84		
Seamount	—	−102		?	1975+
			78		
Seamount	—	−728		?	?
			94		
Seamount	—	−1611		?	?

each volcano was active, the initial degree of partial melt in the source volume for each volcano, and other factors. Such data are not currently available.

The time of the inception of volcanism along the active arc trend can only be estimated at present. An upper limit of 6–8 Ma is suggested by the age of initiation of spreading in the Mariana Trough (Shipboard Scientific Staff, 1978). A lower limit could be obtained by radiometric dating of the oldest volcanic rocks exposed on islands in the active arc. Such dating is currently in progress but definitive data have not yet been obtained. Another approach to the problem would involve dating of ash layers in sedimentary sequences deposited in basins to the east and west of the arc. Although this approach has considerable potential, it requires some definitive criterion for distinguishing ash from the Mariana active arc from ash derived from other sources (e.g. South Honshu Ridge).

Kuno (1962) has summarized available information on historic activity in the Mariana active arc. In general, all the northern Mariana Islands except Sarigan and Alamagan have shown some form of volcanic activity during this century. Since the date of Kuno's compilation, documented eruptions have occurred on Uracas and on at least two seamounts within the arc, one located north of Uracas at 20°55.5′N, 143°26.5′E (January 1974) and the other located south-west of Tinian at 15°00′N, 145°15′E (May 1975) (Smithsonian Center for Short Lived Phenomena, Event Cards no. 1794 and no. 2185).

The distribution of volcanoes in the Volcano Islands is very different from that observed in the Marianas. Instead of forming an arc, the locus of activity in the Volcano and Izu Islands follows a linear trend from Oshima Island near Tokyo all the way south to Minami Iwo Jima at its juncture with the Mariana arc system (Fig. 2). If we can assume that arc magmas originate at approximately the same depth beneath each volcano within these particular arcs, as suggested by the consistency in depth to the Benioff zone beneath each volcano (Katsumata and Sykes, 1969), the variations in arc curvature can be easily explained; that is, the curvature of the arc would be a function of the curvature of the upper surface of the subducted plate at the depth of magma generation. In the Marianas, the subducted plate is nearly vertical at depth and therefore would have a convex downward shape beneath the area of the volcanic arc resulting in a volcanic chain with a small radius of curvature. In the Izu–Volcano arc, the plate has a moderate dip at depth ($c.$ 45°) and presumably has a slight convex upward shape beneath the volcanic arc resulting in a linear volcanic chain. A more pronounced convex upward shape of the plate at the depth of magma generation could result in a volcanic arc convex toward the back-arc region. The seamounts and islands located at the juncture of the Mariana and Volcano arcs appear to define this sort of geometry.

The spacing of volcanoes in the Volcano arc is generally more regular than is observed in the Mariana arc (Table 2). However, this regularity refers primarily to the spacing of the large volcanic complexes which in some cases consist of several individual cones spaced 10–20 km apart (Table 2). Nearly all the historic activity within the Volcano Islands has been either submarine or very close to sea level (Kuno, 1962). These eruptions resulted in the formation of two new islands: Sin Iwo Jima at 24°17′N, 141°31′E (5 km north-east of Minami Iwo Jima) and Nishino Jima at 27°14′N, 140°52′E (see Fig. 2). Recent activity at these locations includes an eruption at Sin Iwo Jima in December 1973 and at Nishino Jima in March 1975 (Smithsonian Institution Center for Short Lived Phenomena Event Card no. 2130).

Volcanic Evolution

On the basis of published data and observations, much of the subaerial evolution of volcanoes in the Mariana active arc appears to follow a consistent pattern. A summary of this pattern will be presented here and details of our field studies will be presented elsewhere.

The initial phase of subaerial activity on each of the islands in the arc is dominated by the extrusion of a large number of relatively thin (2–3 m) lava flows of basalt and basaltic andesite. The general lack of intervening erosional effects and soil horizons suggests these flows erupted in rapid succession. Characteristically, they have scoriaceous tops and bottoms which together nearly equal the thickness of their more massive interiors. Where access to the interior structure of the volcanoes is possible (Maug, Sarigan), the ratio of scoria to massive material in the flows increases towards the end of this stage. Pyroclastic materials are generally subordinate in volume at this stage and, when present, generally consist of ash-fall deposits. The young cone on the island of Uracas appears to be in this stage of development at present.

TABLE 2 Characteristics of active and inactive volcanoes and seamounts forming the Volcano arc

Name	Area (km²)	Maximum Elevation (m)	Distance Between Volcanoes (km)	Reported Date of Latest Activity
Seamount	—	−1271		?
			87	
Seamount	—	−1463		?
			51	
Nishino Shima	Several (?)	−50		1975
			56	
Seamount	—	−163		?
			61	
Seamount	—	−139		?
			67	
Kita Iwo Jima	~6	804		Old
			67	
Iwo Jima	~22	165		Recent
			36	
Sin Iwo Jima	—	Close to SL		1973
			21	
Minami Iwo Jima	~3.8	918		Old

Following the initial extrusive stage, the ratio of pyroclastic to flow material increases upward in the section as the cone assumes the classic composite volcano form. Pyroclastic materials erupted during this stage include ash-fall deposits as well as poorly sorted lithic tuff breccias containing older lavas and occasional gabbroic fragments. Lava flows erupted during this stage are generally basaltic andesite or andesite and trend toward greater thicknesses and higher viscosities as their frequency decreases. The end of this stage is generally marked by the formation of some type of collapse structure; usually an enlarged crater or caldera.

After the collapse event, the volcanoes no longer follow a consistent pattern of development. In some cases, a new cone of basalt, basaltic andesite, and/or andesite has formed within the crater boundary as exemplified by the young cones on Pagan, Guguan, and Uracas. Although these cones are not fully developed at present, they may follow a similar evolutionary pattern. The Maug islands should also be included in this group, although the new cone here is small and still below sea level. Most of the original cone on Maug was destroyed by a collapse event which was similar to the 1883 Krakatoa eruption in Indonesia. All that remains are three circumferential island remnants representing the lower slopes of the original cone. The post-collapse development of Sarigan Island is distinct from that of the other volcanoes. In place of a new intra-crater cone, highly viscous flows and domes of pyroxene and hornblende andesite were extruded from the crater rim during the last phases of activity.

Other significant characteristics of the Mariana arc volcanoes include the abundance and orientation of dykes, satellitic vents, and faults and the general asymmetry of each cone. Although dykes have been identified on all of the islands, they are most evident on Maug, Agrigan, and Pagan. Characteristically, they radiate from a central axis and cut vertically through most of the volcanic section. Subaerial satellitic vents are rare on most of the islands, with the exception of Pagan, where many have been identified (Corwin et al., 1957). The island of Alamagan has a single satellitic vent on its north-eastern slope while Agrigan appears to have one at its south-western tip (Stern, 1978). Major intrusions other than dykes and domes have not been found on any of the islands.

Faults mapped on many of the islands (Corwin et al., 1957; Stern, 1978; A. Meijer and E. Youngblood, unpublished data) are generally associated with slump structures resulting from the undercutting of the lower subaerial and submarine slopes by wave action and ocean currents. The wave action also contributes to the pronounced east–west asymmetry of each island. Because ocean currents and storm waves nearly always travel from east to west in this region, the eastern slopes of the islands are more rapidly eroded than those on the west. This results in erosional steepening of the eastern slopes and produces an asymmetric east–west profile. Our field studies suggest that little, if any, of this asymmetry can be ascribed to regional tilting. In particular, elevated limestone reef terraces present on several of the islands show no preferential inclination to the east or west.

The volcanic geology of islands in the Volcano arc has been described by Tsuya (1936, 1937) and summarized by Kuno (1962). In brief, the island of Iwo Jima (8 km × 4 km) is comprised of a platform of breccias, tuffs, and sands surmounted at its south-western end by a steep-sided cone (0.7 km diameter) of trachyandesite (~hawaiite) flows and breccias. Of the four volcanoes in the arc north of Iwo Jima, only Kita Iwo Jima and Nishino Jima are above sea level. Kita Iwo Jima (3 km × 2 km) is reported to be an old, dissected cone of basaltic flows and pyroclastics traversed by numerous basaltic dykes (Tsuya, 1937) while Nishino Jima (less than 1 km^2) is a newly emergent island comprised of basaltic lava flows and scoria cones (Smithsonian Center for Short Lived Phenomena, Annual Report for 1974, pp. 126–129).

Petrology

As pointed out most recently by Larson et al. (1974), lavas erupted in the Mariana arc show rather limited petrological variations. Olivine–augite basalts and hypersthene–augite basaltic andesites (Coats, 1968) appear to dominate the eruptive products while augite–hypersthene andesite, hypersthene–augite andesites, and hornblende andesite are found locally.

Nearly all the lavas are porphyritic, with plagioclase as the dominant phenocryst phase followed in order of abundance by augite, hypersthene, olivine, Fe–Ti oxides, and amphibole. Plagioclase is often highly anorthitic (e.g. An_{95} cores in basaltic andesite on Agrigan: Stern, 1978) and commonly shows the complex zoning and inclusions so characteristic of this mineral in orogenic lavas. Megacrysts as large as 5 cm in length have been found in scoriaceous flows on several of the islands. Augite phenocrysts are characteristically euhedral and extensively zoned. Megacrysts up to 7 cm in length have been found on the island of Agrigan. Olivine, although a common phase, rarely comprises more than 5 per cent of the lavas. Hypersthene-bearing lavas have been found on all of the islands in the active arc. Hypersthene is commonly mantled by reaction rims of clinopyroxene. Microphenocrysts (0.5–1 mm) of Fe–Ti oxide phases are common in basaltic andesites, particularly as components of glomeroporphyritic clots with pyroxene and plagioclase. Such clots may be related to the gabbroic inclusions (5–10 cm diameter) found within flows on most of the islands. Amphibole has been found only in several massive flows of hornblende andesite on the island of Sarigan (Youngblood and Meijer, 1978). It is generally highly oxidized and often entirely replaced by anhydrous phases including plagioclase, pyroxene, and Fe–Ti oxide.

Lavas from islands in the southern portion of the Volcano arc are petrologically distinct from the Mariana arc lavas. The dominant rock type on these islands is olivine–augite trachyandesite with andesine as the dominant feldspar. Anorthoclase, hornblende, and biotite are found as minor phases while Fe–Ti oxide and apatite are common accessory minerals. The variability in lava types on these islands appears to be very restricted (Tsuya, 1936; MacDonald, 1948; Kuno, 1962).

Islands in the northern Volcano arc appear to be comprised of lavas quite distinct from those on the southern islands. Tsuya (1937) has described an augite–olivine basalt from Kita Iwo Jima with anorthic plagioclase (An_{90}–An_{95}) and a chemical composition similar to basalts found on islands in the Izu arc to the north and the Mariana arc to the south. This suggests that the alkaline lavas are restricted to that portion of the Izu–Volcano–Mariana trend (i.e. at the intersection of the Volcano and Mariana arcs) which has the smallest radius of curvature as noted earlier.

The Bonin Islands are the type area for an unusual rock type called boninite (Petersen, 1891). This rock, a plagioclase-free bronzite andesite, has high concentrations of refractory elements such as Mg, Ni, Cr, and very low abundances of large ion lithophile elements (Kuroda and Shiraki, 1975; Sun and Nesbitt, 1978). Their geochemical and petrological characteristics suggest that they represent a very early phase of arc volcanism. Current studies being carried out on these interesting rocks may clarify their significance to the evolution of arc magmatism.

Chemical Composition

As Corwin (1961) and Larson et al. (1974) have pointed out, the Mariana volcanic suite is not easily classified using existing geochemical and petrographic classification schemes. Petrographic characteristics and Fe enrichment trends suggest that the lavas on the island of Pagan, for instance, should be included in Kuno's (1960) pigeonitic series (Fig. 3). However, on a graph of total alkali against SiO_2 (Kuno, 1960), many of these lavas would be included in Kuno's high alumina basaltic series (Fig. 4). Similar contradictions develop when other classification schemes are used (e.g. Irvine and Baragar, 1971). Kuno (1968) anticipated this situation when he noted that trends in Fe enrichment are not always correlated with alkali abundances. This presumably was the reason that he emphasized the graph of total alkalis against SiO_2. Unfortunately, this graph is not useful for most of the volcanic rocks erupted in the Mariana arc

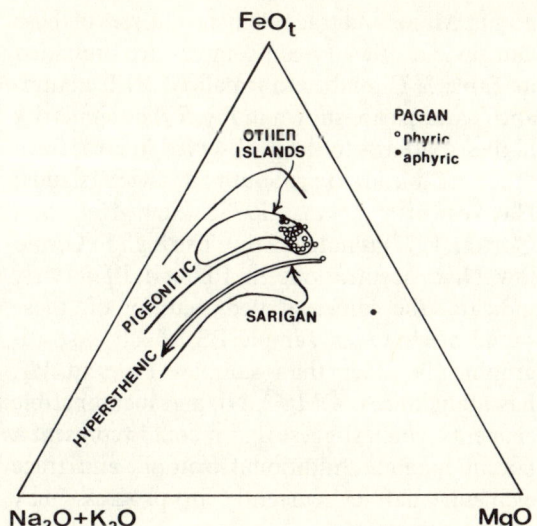

Fig. 3 AFM diagram showing position of individual analyses of samples from Pagan Island, the trend of analyses for Sarigan Island, and a field enclosing available analyses from the other islands in the Mariana active arc. Pigeonitic and hypersthenic series boundaries from Kuno (1968). Note that the Sarigan trend differs substantially from that of the other islands even though alkali abundances are similar in both groups of samples

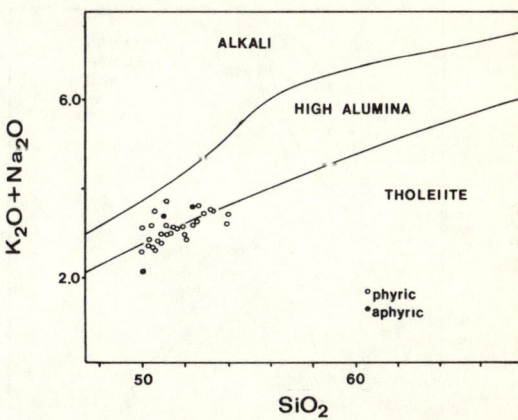

Fig. 4 Graph of total alkalis against silica for samples from Pagan Island. Analyses of samples from other islands in the Mariana arc would plot within this grouping. Curves separating alkaline, high alumina, and tholeiitic fields from Kuno (1968)

because they show little change in SiO_2 upon differentiation.

A more useful diagram for discussion of these lavas is one in which total FeO (FeO_t) is plotted against K_2O. Because most of the mineral phases observed in the Mariana lavas have very low solid/liquid partition coefficients for K, its concentration is an excellent differentiation index. On the other hand, the behaviour of Fe during fractional crystallization is very sensitive to the oxidation state of the magma at any given stage. Therefore, in a cogenetic suite of lavas, the concentrations of FeO_t and K_2O will be very sensitive to changes in the crystallization sequence and oxidation state of the magma. Conversely, if the concentrations of FeO_t and K_2O in a given set of lavas do not define simple trends in a graph of FeO_t against K_2O, they are probably not cogenetic.

Because we wish to determine the differentiation sequence and ultimately the parent magma for a given suite of lavas, only analyses of aphyric rocks and groundmass separates should be plotted. However, because aphyric rocks are rare and data on groundmass separates have not been obtained for lavas from the Marianas, all available analyses will be included. An example of how the trend of liquid compositions can be affected by phenocrystic (xenocrystic?) phases is shown in Fig. 5 for samples from Tonga arc (Ewart et al., 1973).

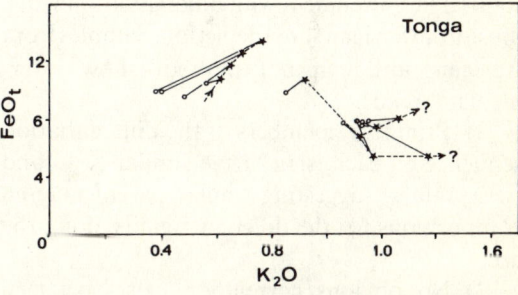

Fig. 5 Graph of total Fe (weight per cent) as FeO against K_2O for lavas from the Tonga arc. Open circles represent whole rock analyses and stars represent groundmass analyses. Data from Ewart et al. (1973). Simple trend of groundmass analyses suggests crystal fractionation control of compositional variations. Note analyses less than 0.8 per cent of K_2O by weight are from Late Island while others are from Fonualei Island

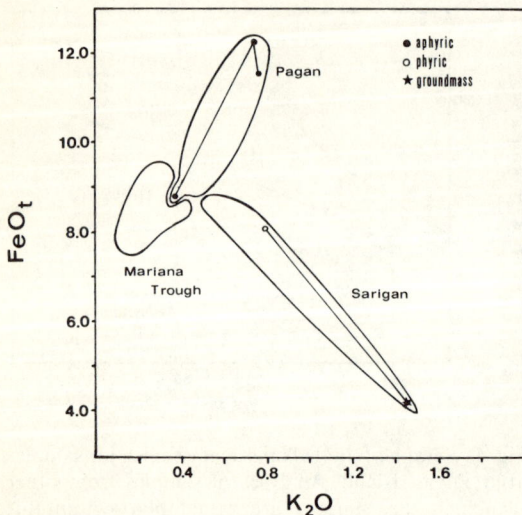

Fig. 6 Graph of total Fe (weight per cent) as FeO against K_2O for lavas from Pagan and Sarigan Islands in the Mariana active arc and for samples dredged from the Mariana Trough (inter-arc basin). The closed curves include all available analyses of phyric samples from each location. Individually plotted are several aphyric samples from Pagan and a groundmass–whole rock pair from Sarigan

Data for samples from the Mariana arc are plotted in Fig. 6. Several important features of the Mariana arc lavas are evident in this figure:

(1) Lavas from different islands in the arc show substantial differences in their Fe enrichment trends. Samples from Pagan show substantial Fe enrichment with differentiation while those from Sarigan show depletion. Samples from Agrigan, although not plotted, show intermediate trends.

(2) 'Primitive' members of the differentiation sequence on each island have similar K_2O and FeO_t values, suggesting similar parent magma compositions for the different islands along the arc.

(3) No obvious correlation exists between degree of Fe enrichment observed on a particular island and position along the arc.

(4) Samples from the Mariana Trough have compositions which are distributed along the lower K_2O end of the active arc cluster.

Several aphyric samples collected on the island of Pagan are plotted in Fig. 6 as separate points. Major and trace element analyses of these and several other Pagan samples are presented in Table 3. Chondrite-normalized REE abundance patterns are shown in Fig. 7. The similarity of these patterns to those observed in lavas from other oceanic arcs (e.g. South Sandwich Islands: Hawkesworth et al., 1978; New Hebrides: Gorton, 1977) is noteworthy. The high FeO_t and low Ni concentrations in PL and PF clearly indicate the differentiated nature of these samples. However, sample PS, which is stratigraphically older than samples PL and PT, has abundances of MgO, Ni, and incompatible elements which suggest that it could represent a parent magma. Additional isotopic and trace element analyses currently in progress may resolve this question.

The wide range in Fe enrichment observed in lavas found within and among islands in the active arc suggest local control of this parameter (e.g. variations in f_{O_2} within shallow magma chambers). It is important to note that hornblende andesites from Sarigan are very similar to andesites erupted along continental margins. However, in conjunction with isotopic results (De Paulo and Wasserberg, 1977), these data suggest such andesites can be produced without the involvement of continental materials.

The similarity of the chemical trends for volcanic rocks from the Mariana Trough and active arc, although suggestive, does not necessarily indicate similar modes of origin for lavas erupted in these areas. In fact, differences in the isotopic compositions of Pb and Sr (Hart et al., 1972; Meijer, 1976b) imply that they are derived from separate source materials.

The trachyandesites of the Volcano arc have unusual compositions relative to alkaline lavas in other arc systems. They have very high Na_2O and K_2O abundances (Table 3) but are clearly not members of the shoshonitic arc series. The unusual chemical compositions of these lavas may in some way be related to the fact that intermediate to deep earthquakes are extremely rare beneath the Volcano arc. Katsumata and Sykes (1969) reported only five events with depths greater than 70 km. DeLong et al. (1975) have suggested that the unusual

TABLE 3 Geochemical data

	Pagan Island					Iwo Jima No. 118‡	Kita Iwo Jima No. 75‡
	PA†	PC†	PF†	PL†	PS†		
SiO_2§	50.97	51.85	51.82	51.21	49.91	59.60	48.49
TiO_2	0.78	0.77	0.86	0.97	0.67	0.86	0.76
Al_2O_3	18.86	17.57	18.15	16.50	15.20	16.81	18.72
FeO	9.82	10.98	10.60	12.30	8.85	6.62	10.67
MnO	0.16	0.19	0.19	0.25	0.18	0.26	0.22
MgO	4.71	5.31	4.48	4.57	11.42	1.34	5.17
CaO	10.85	10.52	10.76	10.29	11.51	3.10	12.47
Na_2O	2.38	2.22	2.36	2.69	1.72	6.11	2.02
K_2O	0.60	0.67	0.73	0.74	0.37	4.17	0.33
H_2O_t	0.45	0.43	0.60	0.59	0.14	0.35	0.98
CO_2	0.44	0.26	0.22	0.24	0.11	—	—
P_2O_5	0.13	0.10	0.18	0.15	0.04	0.50	0.05
Total	100.15	100.87	100.95	100.50	100.12	96.68	99.88
Trace elements‖							
Ba (± 15)	130	134	186	195	100	—	—
Co (± 4)	36	40	38	44	49	—	—
Cu (± 8)	152	131	175	181	111	—	—
Ni (± 4)	34	23	25	7	69	—	—
Rb (± 1.5)	7.5	11.3	10.7	14.5	6.5	—	—
Sr (± 4)	371	324	339	317	261	—	—
Y (± 2)	19.3	23.9	23.5	29.7	13.9	—	—
Zr (± 4)	56	52	72	68	43	—	—

† Samples PA, PC, PF are olivine–augite basalts. Samples PL and PS are aphyric basalts.
‡ From Tsuya (1937). No. 118: olivine–augite–andesine trachyandesite; No. 75: augite–olivine basalt.
§ Major element analyses in weight per cent by X-ray fluorescence.
‖ Trace element analyses (p.p.m.) by X-ray fluorescence methods including primary and secondary absorption, and enhancement corrections.

compositions are produced by extraction of magmas from 'regions beneath (or beside) subducted lithosphere'. Several authors have pointed out the similarity between the Volcano arc trachyandesites and the trachytes of oceanic islands such as Hawaii (Tsuya, 1937; MacDonald, 1948). These observations and the 'oceanic' character of the isotopic composition of Pb and Sr in these rocks (Tatsumoto, 1966; Meijer, 1976a, and unpublished data) suggest they could have been derived directly from the mantle without contributions from the subducted lithosphere. If so, the substantial enrichment in light REE measured by Philpotts *et al.* (1971) in an Iwo Jima trachyandesite (Fig. 7) would suggest that these rocks were produced by small degrees of melting of this mantle. On the basis of the only chemical analysis known to the author (Table 3), it appears that the lavas of the northern Volcano Islands are similar in composition to those found in the Mariana and Izu arcs.

Acknowledgements

The able assistance of Elizabeth Youngblood and Steven Rooke both in the field and the laboratory is gratefully acknowledged. Peter Gromet initiated the author to the intricacies of REE analyses while Professor Leon Silver made available laboratory facilities.

The author wishes to express his appreciation to the people of Agrigan and Alamagan and the Government of the Northern Mariana Islands Commonwealth for support and assistance

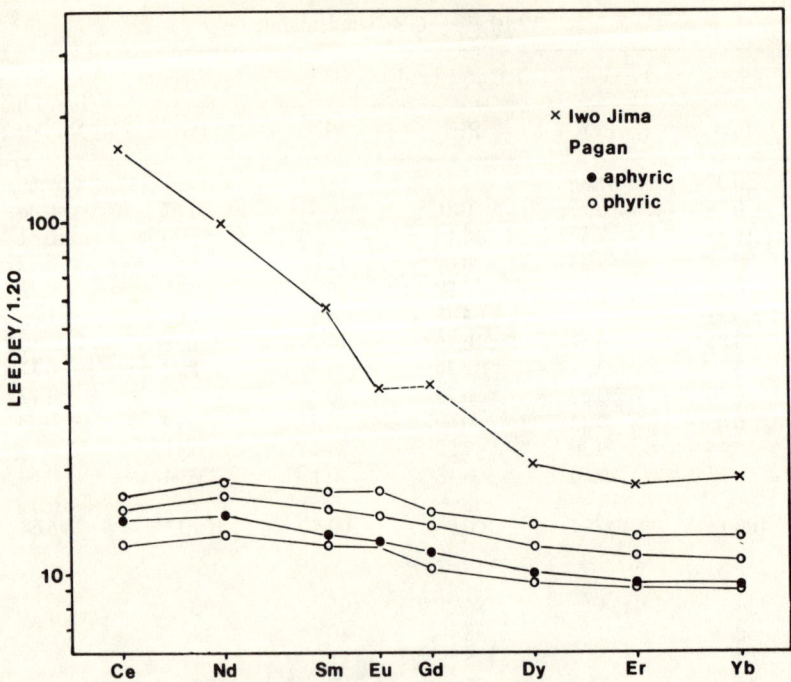

Fig. 7 Normalized REE abundance patterns for samples from Pagan Island in the Mariana arc (Meijer et al., 1978) and Iwo Jima (from Philpotts et al., 1971) in the Volcano arc. The normalization factors are the abundances of REE measured in the Leedey chondrite (Masuda et al., 1973) divided by 1.20. This procedure results in normalization factors similar to those based on average REE abundances in chondrites. All analyses shown are by isotope dilution techniques with 3 per cent maximum absolute errors

during the field work. This work has been funded as part of the SEATAR Program in the Office of International Decade of Ocean Exploration of the National Science Foundation (Grant No. OCE-78-16760). The author is indebted to Joyce Lekawa and Sandra L. Meijer for their assistance in the typing and editing of the manuscript.

REFERENCES

Bogdanov, N. (ed.) (Shipboard Scientific Staff) (1977). Initial report of the Geological Study of the Oceanic Crust of the Philippine Sea Floor—Investigations by the International Working Group on the IGPC Project 'Ophiolites' (R/V *Dmitry Mendeleev*, Cruise 17, June–August, 1976). *Ofioliti* 2, 137–168.

Chao, E. C. T., and Fleisher, M. (1960). Abundance of zirconium in igneous rocks. In *Proceedings of the 21st International Congress, Copenhagen*, Part 1, pp. 106–131.

Cloud, P. E., Jr, Schmidt, R. G., and Burkes, H. W. (1956). Geology of Saipan, Mariana Islands. *US geol. Surv. prof. Paper* no. 280-A, 1–126.

Coats, R. R. (1968). Basaltic andesites. In *Basalts*, Vol. 2 (H. H. Hess and A. Poldervaart, eds), Interscience, New York, pp. 689–736.

Corwin, G., (1961). Volcanic suite of Pagan, Mariana Islands. *US geol. Surv. prof. Paper* no. 204, 204–206.

Corwin, G., Bonham, L. D., Terman, M. J., and Viele, G. W. (1957). *Military Geology of Pagan, Mariana Islands*, Intelligence Division, US Army.

De Long, S. E., Hodges, F. N., and Arculus, R. J. (1975). Ultramafic and mafic inclusions, Kanaga Island, Alaska, and the occurrence of alkaline rocks in island arcs. *J. Geol.* 83, 721–736.

De Paulo, D. J. and Wasserburg, G. J. (1977). The sources of island arcs as indicated by Nd and Sr isotopic studies, *Geophys. Res. Letters* 4, 465–468.

Dietrich, V., Emmermann, R., Oberhänsli, R., and Puchelt, H. (1978). Geochemistry of basaltic and gabbroic rocks from the west Mariana Basin and the Mariana Trench. *Earth planet. Sci. Letters* 39, 127–144.

Doan, D. B., Burke, H. W., May, H. G., and Stensland, C. H. (1960). *Military Geology of Tinian, Mariana Islands*, Intilligence Division, US Army.

Ewart, A., Bryan, W. B., and Gill, J. B. (1973). Mineralogy and geochemistry of the younger islands of Tonga. *J. Petrol* **14**, 429–465.

Fryer, P. (1978). Chemical variations in basalt from marginal basins (abstract). Western Pacific and Magma Genesis, International Geodynamic Conference, Tokyo.

Gorton, M. P. (1977). The geochemistry and origin of Quaternary volcanism in the New Hebrides. *Geochim. cosmochim. Acta* **41**, 1257–1270.

Gottfried, D., Moore, R., and Campbell, E. (1964). Thorium and uranium in some volcanic rocks from the circum-Pacific province. *US geol. Surv. prof. Paper* no. 450-E, 85–89.

Grow, J. A. (1973). Crustal and upper mantle structure of the central Aleutian arc. *Geol. Soc. Am. Bull.* **7**, 2169–2192.

Hart, S. R., Glassley, W. E., and Karig, D. E. (1972). Basalts and sea floor spreading behind the Mariana Island arc. *Earth. planet. Sci. Letters* **15**, 12–18.

Hawkesworth, C. J., O'Nions, R. K., Pankhurst, R. J., Hamilton, P. J., and Evensen, N. M. (1978). A geochemical study of island-arc and back-arc tholeiites from the Scotia Sea. *Earth planet. Sci. Letters*, **36**, 253–262.

Hawkins, J. W., Melchior, J., Bloomer, S., and Evans, C. (1979). Petrology of Mariana trough and Mariana arc basalts. *EOS* **60**, 413.

Hedge, C. E. (1966). Variations in radiogenic strontium found in volcanic rocks. *J. geophys. Res.* **71**, 6119–6126.

Irvine, T. N., and Baragar, W. R. A. (1971). A guide to the chemical classification of the common volcanic rocks. *Can. J. Earth Sci.* **8**, 523–548.

Jakeš, P., and White, A. J. R. (1972). Major and trace element abundances in volcanic rocks of orogenic areas. *Geol. Soc. Am. Bull.* **83**, 29–40.

Kaneoke, I., Isshiki, N., and Zashu, S. (1970). K–Ar ages of the Izu–Bonin Islands. *Geochem. J.* **4**, 53–460.

Karig, D. E. (1971). Structural history of the Mariana Island arc system. *Geol. Soc. Am. Bull.* **82**, 323–344.

Karig, D. E. and Ingle, J. C., Jr (1975). *Initial Reports of the Deep Sea Drilling Project no 31*, US Government Printing Office, Washington, D.C.; see Fig. 1.

Katsumata, M. and Sykes, L. R. (1969). Seismicity and tectonics of the Western Pacific: Izu–Mariana–Caroline and Ryukyu–Taiwan regions. *J. geophys. Res.* **74**, 5923–5948.

Kuno, H. (1960). High alumina basalt. *J. Petrol.* **1**, 121–145.

Kuno, H. (1962). *Catalogue of the Active Volcanoes of the World Including Solfatara Fields*, Part II, *Japan, Taiwan, and Marianas*, International Association for Volcanology, Naples.

Kuno, H. (1968). Differentiation of basalt magmas. In *Basalts*, Vol. 2 (H. H. Hess and A. Poldervaart, eds), Interscience, New York, pp. 623–688.

Kuroda, N., and Shiraki, K. (1975). Boninite and related rocks of Chichi-jima, Bonin Islands, Japan, *Rep. Fac. Sci., Shizuoka Univ.* **10**, 145–155.

Larson, E. E., Reynolds, R. L., Merrill, R., Levi, S., Ozima, M., Aoki, Y., Kinoshita, H., Zasshu, S., Kawai, N., Nakajima, T., and Hirooka, K. (1974). Major-element petrochemistry of some extrusive rocks from the volcanically active Mariana Islands. *Bull. Volc.* **38**, 361–377.

Larson, R. L. and Chase, C. G. (1972). Late Mesozoic evolution of the Western Pacific Ocean. *Geol. Soc. Am. Bull.* **83**, 3627–3644.

MacDonald, G. A. (1948). Petrography of Iwo Jima. *Geol. Soc. Am. Bull.* **59**, 1009–1018.

Masuda, A., Nakamura, N., and Tanaka, T. (1973). Fine structure of mutually normalized rare earth patterns of chondrites. *Geochim. cosmochim. Acta.* **37**, 239–248.

Matsuda, J., Zashu, S., and Ozima, M. (1977). Sr isotopic studies of volcanic rocks from the island arcs in the western Pacific. *Tectonophysics* **37**, 141–151.

Meijer, A. (1974). A study of the geochemistry of the Mariana Island arc system and its bearing on the genesis and evolution of volcanic arc magmas. Ph.D. dissertation, University of California, Santa Barbara.

Meijer, A. (1976a). Pb and Sr isotopic data bearing on the origin of volcanic rocks from the Mariana Island arc system. *Geol. Soc. Am. Bull.* **87**, 1358–1369.

Meijer, A. (1976b). Pb isotopes, major elements and selected trace elements and mineralogy of altered and metamorphosed sea-floor basalts (abstract). *EOS* **57**, 342.

Meijer, A. and Glassley, W. (1973). A survey of trace element abundances in volcanic rocks from the Mariana Island arc (abstract). *EOS* **54**, 1218.

Meijer, A., Rookes, S., Ruth, J., and Matuska, M. (1978). Rare earth elements in volcanic rocks from the Mariana Islands (abstract). *Geol. Soc. Am. Abstr. Programs* **10**.

Muenow, D., Delaney, J. R., Meijer, A., and Liu, N. (1977). Water-pillow basalt rims in the Marianas back arc basin (abstract). *EOS* **58**, 530.

Murauchi, S., Den, N., Asano, S., Hona, H., Yoshii, T., Asanuma, T., Hagiwara, K., Iohikawa, K., Sato, T., Ludwig, W. J., Ewing, J. I., Edgar, N. T., and Houtz, R. E. (1968). Crustal structure of the Philippine Sea. *J. geophys. Res.* **73**, 3143–3172.

Peterson, J. (1891). Beiträge zur petrographie von Sulphur Island, Peel Island, Hachijo and Mijakeshima. *Jb. Hamburg, Wiss. Anst.* **8**, 25.

Philpotts, J. A., Martin, W., and Schnetzler, C. C. (1971). Geochemical aspects of some Japanese lavas. *Earth planet. Sci. Letters* **12**, 89–96.

Pineau, F., Javoy, M., Hawkins, J. W., and Craig, H. (1976). Oxygen isotope variations in marginal basin and ocean-ridge basalts. *Earth planet. Sci. Letters* **26**, 299–307.

Pushkar, P. (1968). Strontium isotopic ratios in volcanic rocks of three island arc areas. *J. geophys. Res.* **73**, 2701–2714.

Schmidt, R. G. (1957). Petrology of the volcanic rocks. In *Geology of Saipan, Mariana Islands*, Part 2, Chap. B, pp. 127–175.

Shipboard Scientific Staff, Leg. 60, IPOD (1978). Leg 60 ends in Guam. *Geotimes* **23**, 19–22.

Stark, J. T. and Tracey, J. I. (1963). Petrology of the volcanic rocks of Guam. *US geol. Surv. prof. Paper* no. 403-C.

Stern, R. J. (1978). Agrigan: an introduction to the geology of an active volcano in the northern Mariana Island arc. *Bull. Volc.* **41**, 1–13.

Stern, R. J. (1979). On the origin of andesite in the northern Mariana island arc: implications from Agrigan. *Contr. Mineral. Petrol.* **68**, 207–219.

Sun, S-S. and Nesbitt, R. W. (1978). Geochemical regularities and genetic significance of ophiolitic basalts. *Geology* **6**, 689–693.

Tanakadate, H. (1940). Volcanoes in the Mariana Islands in the Japanese Mandated South Seas. *Bull. Volc. Ser. 2*, **6**, 199–223.

Tatsumoto, M. (1966). Isotopic composition of lead in volcanic rocks from Hawaii, Iwo Jima, and Japan. *J. geophys. Res.* **71**, 1721–1733.

Tayama, R. (1936). Topography, geology, and coral reefs of the Northern Mariana group. *Tohoku Imp. Univ. Fac. Sci., Inst. Geol. Paleo., Jap. Lang. Contr.* no. 23, 1–88 (in Japanese).

Tracey, J. I., Schlanger, S. O., Stark, J. T., Doan, D. B., and May, H. G. (1963). General geology of Guam. *US geol. Surv. prof. Paper* no. 403-A.

Tsuya, H. (1936). Geology and petrology of Iwo-Jima (Sulphur Island), Volcano Islands Group. *Bull. Volc. Soc. Japan* **3**, 28–52.

Tsuya, H. (1937). On the volcanism of the Huzi volcanic zone with special references to the geology and petrology of Izu and the southern islands. *Bull. Earthquake Res. Inst., Tokyo Univ.* **15**, 215–357.

Yoder, H. S., Jr and Tilley, C. E. (1962). Origin of basalt magmas: an experimental study of natural and synthetic rock systems. *J. Petrol.* **3**, 346–532.

Youngblood, E. and Meijer, A. (1978). Petrology and geochemistry of an island arc volcanic sequence, Sarigan, Mariana Islands (abstract). *EOS* **59**, 1182.

Andesites
Edited by R. S. Thorpe
© 1982 John Wiley & Sons

Mediterranean island arcs

J. Keller

Mineralogisches Institut,
Universität Freiburg, D-7800 Freiburg, German Federal Republic

ABSTRACT

Active island arc volcanism occurs in two areas of the Mediterranean basin: the Aeolian Islands in the southern Tyrrhenian Sea and the arc of volcanic islands within the Aegean Sea. Both areas are distinguished by intermediate or deep-focus seismicity, suggesting active subduction processes. The Tyrrhenian Sea and the central Aegean have certain characteristics of back-arc basins, with thinned crust, anomalous mantle, and high heat flow. The island arc volcanism is young to Recent in age. The exposed part of the Aeolian Islands has probably formed entirely within the last 0.5 Ma. The Aegean volcanic rocks associated with the present setting of the arc have formed within the last 3 Ma. Some older lavas in the Aegean area belong to an earlier Oligocene–Miocene subduction episode. In both provinces the volcanic products belong to a calc-alkaline high alumina basalt–andesite–dacite association.

In the Aeolian Islands the dominant rock is of basaltic composition, but characterized by great variations in K_2O and K_2O/Na_2O. Accordingly, calc-alkaline, high K calc-alkaline, shoshonitic, and leucite-tephritic series are distinguished. A distinct K_2O–time relationship exists, with calc-alkaline lavas characterizing the early evolution and shoshonitic to leucite-tephritic products dominating the recent activity. Interpretation of these compositional variations in terms of a $K–h$ relationship is possible with the assumption that there is a steep to vertical Benioff zone. In the Aegean arc, Santorini has extensive basaltic members, while andesites, dacites, and rhyolites dominate the other islands. The dominant part of the Aegean arc is calc-alkaline and only a few small volcanic occurrences provide a significant cross-section perpendicular to the arc. In these cases, there is a trend of increasing K_2O with distance from the volcanic front.

In both provinces the most primitive parental high alumina basalts have $Mg/Mg + Fe^{2+}$ ratios and Ni and Cr contents too low to be direct equilibrium melts from a peridotitic mantle. But their major and trace element compositions are consistent with formation from a primary mantle melt by a small degree of fractionation crystallization dominated by olivine. Both active Mediterranean island arcs exhibit noticeable differences in important features including age, structural setting, depth of subduction, and chemical composition (particularly large ion lithophile element concentrations) of the overall lavas. However, similarities in the petrological characteristics of the typical calc-alkaline high alumina basalt–andesite–dacite clan indicate petrogenesis by broadly similar processes in both arcs.

Introduction

The features of Alpine geology in the Mediterranean area have resulted from the complex collision between Africa and Europe, involving smaller microplates between the major Eurasian and African plates (Dewey et al., 1973; Biju-Duval et al., 1977). This plate convergence resulted in folding, collision, and subduction throughout the evolution of the Alpine belt. Volcanism related to this plate convergence is typically represented by the calc-alkaline andesite series of trench–island arc or continental margin systems, and was widespread in the peri-Mediterranean foldbelts, especially during the Tertiary. Major examples are found from the

Betic Cordillera in Spain, the Maghrebinian belt of North Africa, Sardinia (Dostal et al., this volume, Section IV) to the Carpathians and Turkey (Innocenti et al., this volume, Section III). In the Alps, the Apennines, and Sicily the occurrence of Tertiary andesite volcanism is mainly indicated by the occurrence of extensive volcanoclastic sedimentary series (Marinelli, 1975). The volcanism of this extensive 'peri-Mediterranean andesite belt' is largely extinct and the aim of petrological analysis is concentrated upon reconstructing the structural setting—former arcs, chains, and Benioff zones—from geochemical parameters.

However, active seismicity distinguishes two domains within the Mediterranean basin, where the occurrence of intermediate to deep focus earthquakes and active volcanicity are consistent with active subduction processes. These two areas are the southern Tyrrhenian Sea with the Aeolian (or Lipari) Islands, and the Aegean Sea with the Aegean or Hellenic arc volcanoes. In both areas active Benioff zones have been proposed (Caputo et al., 1972; Ninkovich and Hays, 1972) and in both areas volcanism has characteristics of island arc and continental arc systems.

Structure and Geophysical Setting of the Southern Tyrrhenian Sea

The Tyrrhenian Sea is an extremely complex area with structural features changing drastically within short distances (Barberi et al., 1973; Morelli, 1970). Calabria and north-eastern Sicily form a crystalline block within the Apennine system, with the northern edge of the African plate passing through northern Sicily. A thick sialic crust reaches 40–45 km underneath Calabria and thins towards the central Tyrrhenian Sea, which has a more oceanic character. The Moho depth is about 20 km underneath the Aeolian Islands and less than 15 km under the Tyrrhenian abyssal plain (Giese and Morelli, 1973). The Tyrrhenian Sea is floored by Upper Miocene and Pliocene tholeiitic basalts with chemical characteristics which indicate an origin by back-arc spreading processes (Dietrich et al., 1977; Barberi et al., 1978). The occurrence of such basalts and the presence of shallow, anomalous mantle and high heat-flow values in the central Tyrrhenian Sea (the Tyrrhenian abyssal plain), suggest that it might be a young marginal sea (Menard, 1967; Boccaletti and Guazzone, 1972).

The south-eastern Tyrrhenian area has earthquake epicentres increasing in depth in a south-easterly to north-easterly direction from Calabria through the volcanic arc to the Tyrrhenian abyssal plain (Fig. 1). This seismicity defines a Benioff plane dipping c. 50–60° north-west to west-north-west. This is generally accepted as representing a subduction zone (Caputo et al., 1972; Barberi et al., 1973, 1974; Ninkovich and Hays, 1972; Keller, 1974; Blot, 1978).

The presence of a buried trench area is supected on the Ionian side of Calabria (Fig. 1). The subduction area has a limited width of only c. 100 km. The distribution of earthquake centres is not regular over the whole depth range but clusters between 200 and 350 km (Fig. 2). Two isolated 450 km deep centres can only tentatively be related to the same Benioff zone. In the depth range 200–50 km no seismicity has been reliably recorded and this discontinuity in Benioff zone seimicity has been explained by a model of a detached slab (Keller, 1974; Biju-Duval et al., 1978).

Geology and Volcanic History of the Aeolian Arc

The Aeolian arc consists of seven major islands (Fig. 1) which are all composed exclusively of volcanic material. Several seamounts extend the arc to either side, to the west of Alicudi and northeast of Stromboli. Two huge seamounts in the Tyrrhenian abyssal plain, Marsili and Palinuro, cannot be unambiguously linked with the Aeolian arc on the basis of existing dredge samples (Del Monte, 1972; Keller and Leiber, 1974).

The visible parts of the islands are of Quaternary age. K–Ar age measurements have been determined for the oldest formations of three islands. These have yielded an age of 0.5 Ma for

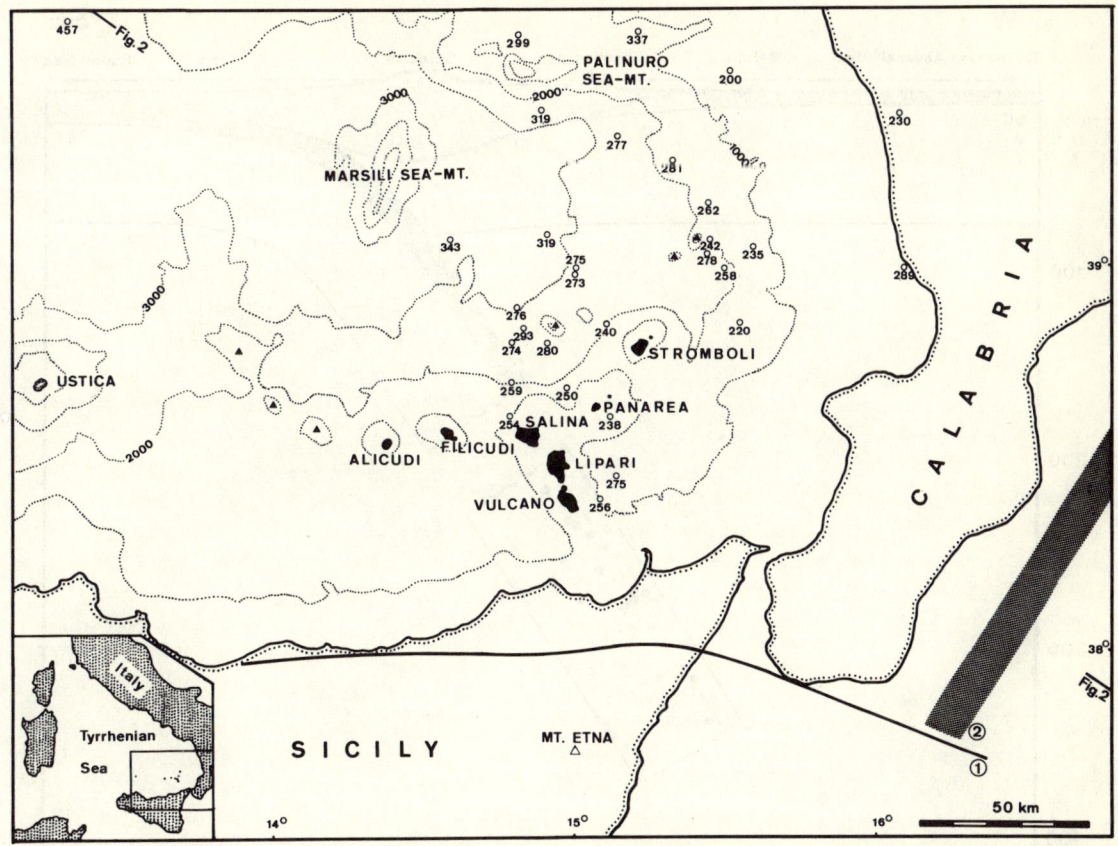

Fig. 1 The Aeolian Islands in relation to submarine topography and seismicity in the Tyrrhenian Sea. Open circles are earthquake epicentres with depths marked in kilometres (Caputo *et al.*, 1972). The depth contours of the Tyrrhenian Sea are in metres. The triangles mark seamounts which are probably related to the Aeolian arc, while Marsili and Palinuro are seamounts of ambiguous tectonic setting and petrogenetic association. 1 = Approximate northern margin of the African plate (Barberi *et al.*, 1973); 2 = suggested area of a buried trench (Ninkovich and Hays, 1972)

Panarea (Civetta, in Barberi *et al.*, 1977) and age limits of less than 0.5 Ma for Salina and less than 1 Ma for Filicudi (Barberi *et al.*, 1974). This is in good agreement with a stratigraphic system based upon marine glacio-eustatic terrace levels (Keller, 1967; Pichler, 1967). In this scheme Panarea is the oldest island with an evolution which started during the Middle Pleistocene.

Within the short span of Aeolian volcanic activity of about 500 ka two clearly separated periods can be distinguished. During a Middle Pleistocene phase the main parts of Lipari, Salina, Alicudi, Filicudi, and Panarea were constructed. Later extensive marine erosion formed terraces and raised beaches, and the volcanic products of a younger phase cover this unconformity. In the post-erosional phase the structures of Salina, Lipari, and Filicudi were completed, and Vulcano and Stromboli were formed. The post-erosion volcanic series formed during the Upper Quaternary and is probably less than 100 ka old. Historic and/or active volcanism is known from Lipari, Vulcano, and Stromboli. The petrological importance of the two phases lies in the fact that all pre-erosion lavas are calc-alkaline, while shoshonitic rocks appear after the erosion hiatus. However, some

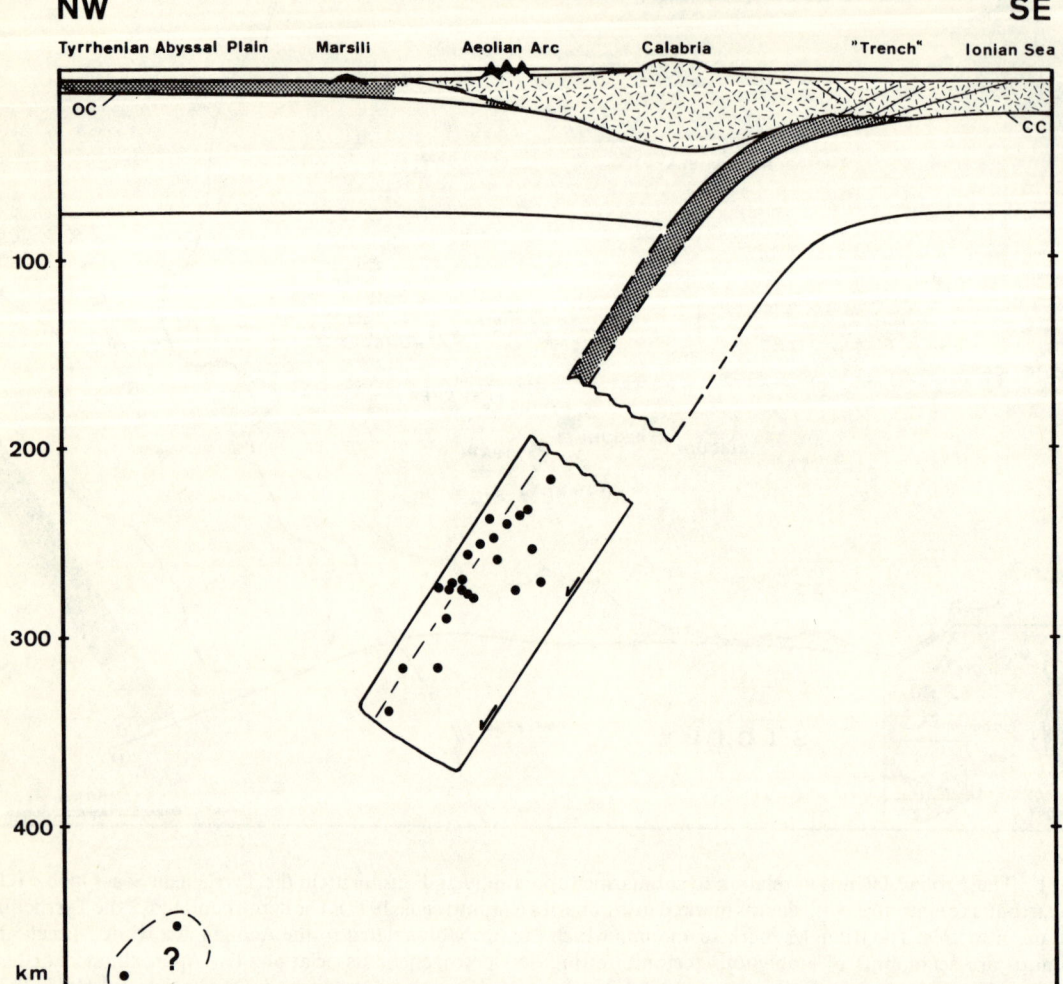

Fig. 2 The Calabria–Aolian arc subduction system. All subcrustal earthquake centres from a 100 km wide strip have been projected on to the north-west–south-east profile line in Fig. 1. The gap in seismic activity with focal depths between 50 and 200 km is interpreted in terms of the existence of a detached slab between 200 and 350 km, characterized by seismic activity. The active volcanism is therefore over a Benioff zone c. 200 km in depth. OC = oceanic crust, CC = continental crust. Data and interpretation are from Caputo et al. (1972), Keller (1974), Görler and Giese (1978), Biju-Duval et al. (1978)

typical calc-alkaline volcanoes were still active contemporaneously with the post-erosional shoshonitic suite (Keller, 1974).

Geological mapping at 1 : 10 000 scale has been completed for all the islands and full descriptions of geology, volcanology, and petrology for every island have been published, or are in press. A summarized outline of the geological features of the individual islands is given here.

Lipari (38 km^2) is the largest island of the group. Pichler (1967, 1980) subdivided its activity in four periods. Low Si andesites (in the classification of Peccerillo and Taylor, 1976, as used in Fig. 3) form the Middle Pleistocene period I, in the western part of the island, in which volcanic forms have been heavily eroded by marine terracing. The terraces are covered by younger pyroxene–andesites of period II. These include garnet–cordierite andesites (with xenoliths of

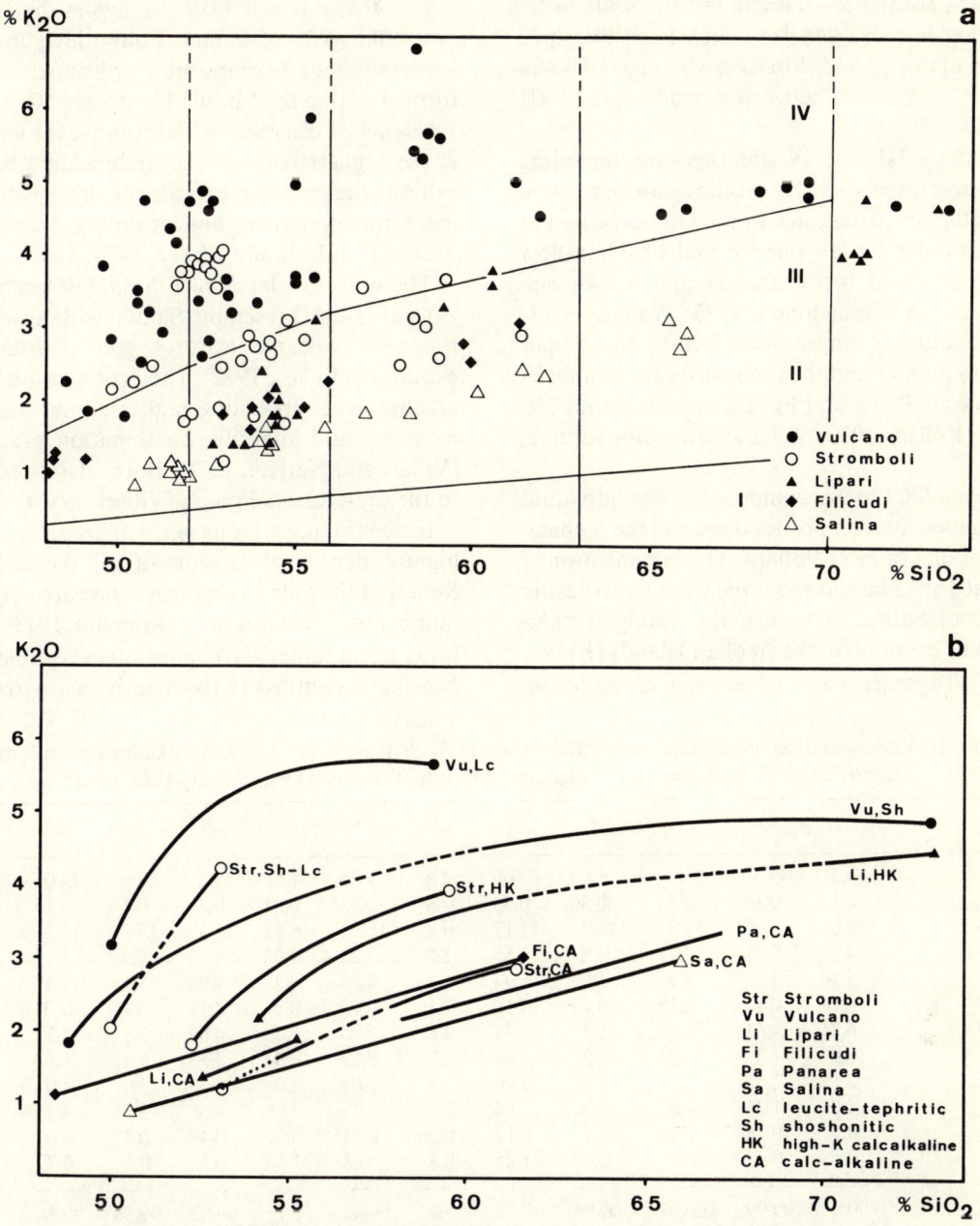

Fig. 3 (a) Graph of K_2O against SiO_2, for Aeolian arc lavas. The field boundaries are from Peccerillo and Taylor (1976) for (I) island-arc tholeiitic series; (II) calc-alkaline series; (III) high K calc-alkaline series; and (IV) shoshonitic series. Basalts have SiO_2 under 52 per cent, low Si andesites 52–56 per cent SiO_2, andesites 56–63 per cent SiO_2, dacites 63–70 per cent SiO_2 and rhyolites over 70 per cent SiO_2. Data are from Villari (1972), Romano (1973), Keller (1974, 1980), Pichler (1980), Rosi (1980). (b) Graph of K_2O against SiO_2 for Aeolian arc lavas summarizing the fractionation trends for the various volcanic series (cf. (a)) for the different islands

garnet-, cordierite-, sillimanite-, and andalusite-bearing hornfels schists) which form late lava flows of this period. A radiocarbon age of 40 ka dates a paleosoil between periods II and III (Pichler, 1980).

Periods III and IV are rhyolitic, including extrusive domes and viscous flows and extensive air-fall pumice deposits. In the Holocene period IV an older air-fall pumice and obsidian-flow cycle occurred between c. 11 and 8.5 ka ago (Bigazzi and Bonadonna, 1973; Wagner et al., 1976) and a younger cycle formed the Upper Pelato pumice and the renowned obsidian flows of Rocche Rosse and Forgia Vecchia about 550 AD (Keller, 1970; Bigazzi and Bonadonna, 1973).

Salina (26 km^2) is composed of six individual volcanoes, five composite cones, and the pumice-explosion crater of Pollara. The last-mentioned is dated at 13 ka ago and is the youngest volcanic event of Salina. It is also the youngest calc-alkaline eruption of the Aeolian Islands (Keller, 1974). Together with the very regular andesitic cone of Monte dei Porri, it forms the 'post-erosional' part of Salina. Four older, in part superimposed, composite volcanoes were formed during the Middle Pleistocene. Of these, Fossa delle Felci reaches 960 m above the sea and is the highest cone of the archipelago. Salina exhibits the most typical calc-alkaline evolution from the dominant high alumina basalts to dacites (Table 1, and Keller, 1974, 1980).

The western islands of *Filicudi* (10 km^2) and *Alicudi* (5 km^2) resemble Salina, with a smaller range in volcanic features and petrological evolution (Villari, 1972). The products are calc-alkaline with prevailing high alumina basaltic members and andesitic fractionation products (Villari and Nathan, 1978). Both islands belong to the pre-erosional phase (Villari, 1980).

In accordance with its age, *Panarea* shows the highest degree of erosion of all the islands. Relicts of the oldest composite cone are cut by a number of volcanic domes (Romano, 1973). The lavas are andesites and dacites, often hornblende-bearing in contrast to the mainly two-pyroxene

TABLE 1 Representative and average chemical analyses of Aeolian arc lavas. 1–5, Calc-alkaline series; 6, high K calc-alkaline; 7–12 shoshonitic association. Mg values computed with $Fe_2O_3/FeO = 0.15$

	1	2	3	4	5	6	7	8	9	10	11	12
SiO_2	51.3	49.5	58.2	65.8	63.60	54.0	51.85	51.65	52.5	58.5	72.0	74.65
TiO_2	0.5	0.86	0.54	0.36	0.58	0.6	0.95	0.83	0.7	0.6	0.15	0.08
Al_2O_3	17.0	18.95	17.3	16.2	15.17	16.8	16.97	16.45	15.5	15.7	13.2	12.86
Fe_2O_3	4.25	5.22	2.7	2.8	3.85	2.5	3.23	4.03	4.10	2.9	1.1	0.6
FeO	5.15	4.88	4.6	1.6	1.94	6.1	4.75	4.7	4.90	3.55	1.45	1.14
MnO	0.18	0.17	0.12	0.12	0.12	0.16	0.16	0.15	0.16	0.16	0.07	tr
MgO	6.75	4.61	3.1	1.7	2.72	4.8	5.05	5.48	4.6	3.3	1.2	tr
CaO	10.8	11.3	7.8	4.2	4.62	8.7	9.62	9.04	8.3	5.1	1.3	0.79
Na_2O	2.15	2.52	3.2	3.95	2.96	2.5	2.64	3.0	3.6	3.9	4.0	3.7
K_2O	1.0	1.25	1.8	2.7	2.84	2.4	3.47	2.95	4.6	5.3	4.7	4.92
P_2O_5	0.17	0.28	0.2	0.17	0.12	0.26	0.61	0.32	0.47	0.37	0.07	0.02
H_2O	0.55	0.45	0.4	0.45	1.25	0.8	0.68	1.13	0.5	0.3	0.7	0.99
Total	99.8	99.99	100.01	100.05	99.77	99.62	99.98	99.73	99.93	99.68	99.94	99.75
$Mg/Mg + Fe^{2+}$	60	49	47	45	50	54	57	57	52	52	50	—
K_2O/Na_2O	0.46	0.5	0.56	0.68	0.96	0.96	1.31	0.98	1.28	1.36	1.18	1.34

(1) High alumina basalts, Salina. Group average of five least fractionated samples (Keller, 1980). (2) High-alumina basalts, Filicudi (Villari, 1972), average of three, normalized for LOI). (3) Andesite, Salina–Fossa delle Felci (Keller, 1980; sample Sa 179). (4) Dacite, Salina–Fossa delle Felci (Keller, 1974; sample Sa 197). (5) Dacite, Panarea (Romano, 1973; sample PR 91). (6) High K andesite, Lipari, Monte Rosa (Pichler, 1980). (7) Shoshonitic basalts, Stromboli. Average of 14 (Rosi, 1980). (8) Shoshonitic basalts, Vulcano. Average of eight (Keller, 1974). (9) Leucite tephrite, Vulcano–Vulcanello (Keller, 1980; sample VO 18). (10) Trachyte, Vulcano–Vulcanello (Keller, 1980; sample VO 6). (11) Rhyolite, Vulcano–Lentia (Keller, 1980; sample VL 6). (12) Rhyolitic obsidian, Lipari, Rocche Rosse flow (J. Keller, unpublished data).

andesites of the other islands. The uninhabited islet of Basiluzzo (c. 4 km north-east of Panarea) is formed by a thick, viscous rhyolite flow, which is possibly much younger than Panarea.

Vulcano belongs entirely to the post erosional Upper Pleistocene group. An early shoshonitic composite cone in Southern Vulcano developed a caldera which was subsequently filled by volcanic products in which shoshonitic basalts and leucite tephrites alternate. Later, extrusion of rhyolites, similar and probably coeval to the Lipari group III rhyolites (Table 1, no. 11) was followed by the formation of a second caldera in which the active cone of the 'Fossa di Vulcano' (361 m) is nested. The Fossa volcano evolved from leucite tephrites through highly potassic trachytes to alkaline rhyolites. The last eruption was between 1888 and 1890. During historical times, from 183 BC to c. AD 1550, the peninsula of Vulcanello was formed by leucite tephritic and subordinate trachytic lava flows (Table 1, nos 9 and 10), and three ash cones. More details are given in Keller (1974, 1980).

Stromboli: Under the regular morphology of a cone rising from 2000 m below the sea to 924 m above sea-level at least two volcanic units can be distinguished, called the 'old cycle' and the 'young cycle' by Rosi (1980). The old cycle forms the eastern part of the cone and includes high alumina basalts and andesites of both calc-alkaline and high K calc-alkaline affinity. The neck of Strombolicchio, c. 1.7 km offshore, is related by Rosi to the old cycle and has the most pronounced calc-alkaline characteristics of Stromboli. The young cycle developed from a collapse structure on the western flank of the cone and comprises the present Strombolian activity of the Sciarra volcano. These products have a constant shoshonitic composition (Table 1, no. 7). In some lavas leucite appears in the groundmass, indicating a gradual passage into leucite tephrites as known from Vulcano and Vulcanello. The entire evolution of Stromboli is attributed to the post-erosional activity of Aeolian volcanism and is thus of Upper Quaternary age. For the oldest parts, and for Strombolicchio, this tentative attribution has still to be confirmed by reliable stratigraphical or radiometric results.

Petrology and Chemical Composition of the Aeolian Arc

Representative and average chemical analyses of Aeolian arc lavas are presented on Table 1. The Aeolian lavas range from basalt to rhyolite in composition, covering an SiO_2 range of 49–75 per cent. Basaltic rocks are the most abundant, followed by andesites. The most striking petrological feature is represented by extreme variation in K_2O between the closely associated series of neighbouring islands (Fig. 3). This has led to the recognition of different evolution series which differ primarily in their K_2O content and K_2O/Na_2O ratio (Barberi et al., 1974; Keller, 1974): (1) calc-alkaline series (Salina, Filicudi, Panarea, Alicudi, Lipari, Stromboli); (2) high K calc-alkaline series (Lipari, Stromboli); (3) shoshonitic series (Vulcano, Stromboli); (4) leucite-tephritic series (Vulcano, Stromboli). The K_2O/Na_2O ratios increase from c. 0.35 in the calc-alkaline suite up to 1.5 in leucite tephrites. No representatives of low K island-arc tholeiites (Jakeš and Gill, 1970) are known. The main series develop independently and are not connected by evolutionary lines. K_2O differences therefore appear to be a primary feature, inherited from the parental magmas or from the source. Of the main series, only the shoshonites and leucite tephrites which occur together on Vulcano and Stromboli show a direct genetic relationship, and a low pressure fractionation line can be drawn from shoshonitic basalts to leucite-bearing tephrites and late stage, high K trachytes (Barberi et al., 1974).

The calc-alkaline series

An example of a typical calc-alkaline suite is the high alumina basalt andesite–dacite series of Salina. Figure 3 (open triangles) shows that the variations in K_2O and other oxides form similar smooth and continuous trends in variation diagrams. The Aeolian calc-alkaline rocks have phenocrysts of plagioclase and two pyroxenes, together with olivine in the basic rocks. Hornblende and biotite are rare (*see* Barberi et al., 1974, Table 2, for modal variations).

Variation within the calc-alkaline series can be explained by fractional crystallization of the phenocrysts phases present. The occurrence of xenoliths of gabbroic cumulates composed of olivine, two pyroxenes, plagioclase, and magnetite (Honnorez and Keller, 1968; Keller, 1980) is consistent with a fractional crystallization model. In the Fe-enrichment plot (Fig. 4(a)) the fractional crystallization is dominantly low pressure/moderate f_{O_2} according to Osborn (1976). The relative volumes, with basalts forming up to 80 per cent of erupted lavas, are consistent with such a fractionation series.

The volumes and petrographic characteristics of the basaltic lavas indicate an upper mantle origin (Keller, 1980). However, the most primitive high alumina basalts have Mg values (Mg/(Mg + Fe^{2+} = 0.59–0.62, with Fe_2O_3/FeO = 0.15) and Ni (less than 30 p.p.m.) and Cr concentrations (less than 60 p.p.m.) too low to be equilibrium partial melts of peridotitic mantle. Crystallization dominated by olivine with some chrome-spinel and subordinate pyroxene must therefore have occurred before eruption of these basalts. This is consistent with the models proposed by Osborn (1976) and Nicholls (1978) for the evolution of the volcanic rocks of Santorini (see later). REE data available for the calc-alkaline lavas (Klein et al., 1975; Klerkx et al., 1976; J. Keller, S. R. Taylor, and P. Muir, unpublished data) resemble data for calc-alkaline series elsewhere with light REE enrichment (40–60 chondritic in basalts) and a flat heavy REE pattern with concentrations of seven to 10 times chondritic. This heavy REE pattern is different from that expected when garnet is residual in melting of a subducted slab (Gill, 1974; Lopez-Escobar et al., 1977).

The shoshonitic series

Despite the difference in K_2O and related trace elements e.g. Rb, Ba, Sr, light REE (Keller, 1974) many other petrological characteristics are similar or show only small differences between the calc-alkaline and the shoshonitic series. Island arc geochemical features such as low TiO_2, high Al_2O_3, and weak Fe enrichment are shown by both associations. The shoshonites are similar in their modal composition with phenocrysts of plagioclase, augite, and olivine but no low Ca pyroxene (pigeonite or hypersthene). Distinctive K-feldspar is confined to the groundmass and leucite appears in the tephrites. Basaltic lavas also predominate in the shoshonitic association. Mg values of 60 and Ni and Cr contents of c. 40 and 150 p.p.m. respectively indicate a similar degree of fractional crystallization for the most abundant basalts in both series.

The trend of shoshonitic variation is shown in the Fe-enrichment plot of Fig. 4(a). In terms of this plot the evolution is very similar to the low P/moderate f_{O_2} calc-alkaline fractionation trend of Osborn (1976). The shoshonitic variation (line (2)) includes shoshonitic basalts and tephrites and terminates with late stage K-trachyte (e.g. Table 1, no. 10). Rhyolitic members of the shoshonitic series (Table 1, no. 11) belong to a quite distinct fractionation path which might be explained by high pressure (more than 8 kb) fractionation processes (Osborn, 1976). Of all Aeolian lavas these rhyolites have the most extreme Eu anomalies, with Eu/Eu* = c. 0.05 (Klerkx et al., 1976; Kiesl et al., 1978), suggesting very extensive plagioclase fractionation. To a smaller degree, Eu anomalies (Eu/Eu* = c. 0.4) also characterize the most potassic trachytes. Other Aeolian rock types are generally characterized by lack of any europium anomalies.

Sr isotope data have a broad range and scatter from 0.7030 to 0.7065 (Table 2). However, for each of the main series and for the major rock types, including rhyolites and tephrites, values between 0.7040 and 0.7045 have been reported. This range is considered significant for the Aeolian arc and the source for all Aeolian magmas seems isotopically of rather constant composition. Some higher values up to 0.7065 may result from crustal contamination.

Island arc structure related to geochemical features

K variation across island arcs and relationships between K content and depth to seismic zone play an important role in magma-genetic discussions for island arcs (Dickinson and Hatherton, 1967). Recent detailed results (e.g. Whitford

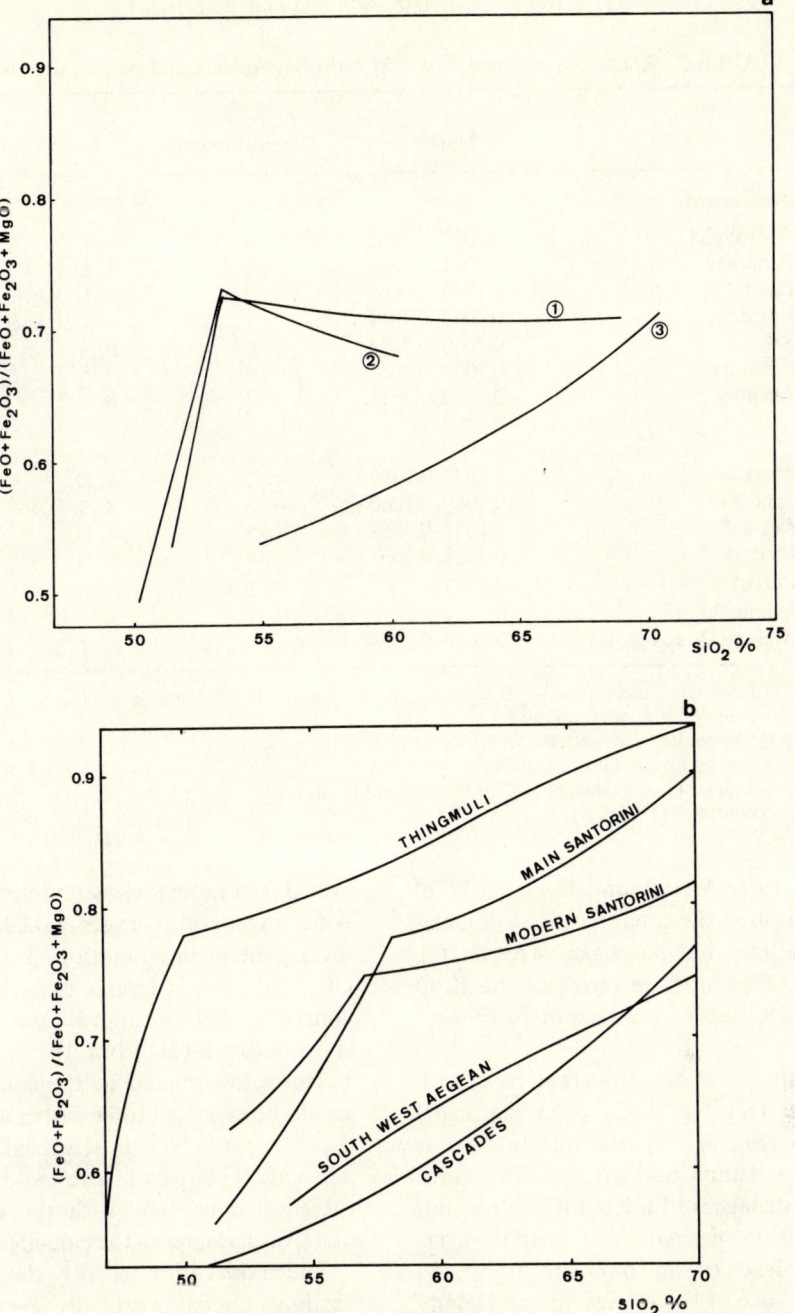

Fig. 4 Plot of Fe enrichment (FeO + Fe_2O_3/FeO + Fe_2O_3 + MgO) against SiO_2 showing fractionation trends for various volcanic suites (cf. Osborn, 1959)

(a) Aeolian arc series: calc-alkaline basalt–andesite–dacite evolution from Salina (1) and shoshonitic basalt–tephrite–trachyte series from Vulcano (2) are consistent with low P/moderate f_{O_2} fractionation (Osborn, 1976) while the rhyolitic trend of the shoshonitic series from Vulcano (3) is consistent with Osborn's high pressure fractionation pattern

(b) Aegean arc trends compared with low P tholeiitic (Thingmuli) and high P calc-alkaline fractionation lines (Cascades; Nicholls, 1971; Osborn, 1976)

TABLE 2 Range of published $^{87}Sr/^{86}Sr$ ratios for Aeolian and Aegean arc lavas

	$^{87}Sr/^{86}Sr$	Number of Determinations	Rock Types†
Aeolian arc			
Stromboli‡	0.7053–0.7065	4	shB, A,
Panarea‡	0.7041, 0.7052	2	A, D,
Lipari‡	0.7045–0.7061	8	A, D, R,
Vulcano‡	0.7041–0.7061	7	shB, shA, R, LT,
Salina‡	0.7040–0.7053	6	B, A,
Filicudi‡	0.7030–0.7054	4	B, A,
Alicudi‡	0.7040, 0.7055	2	B,
Aegean arc			
Nisyros§	0.7037, 0.7050	2	A, D,
Santorini ‖	0.7040–0.7060	8	B, A, D, R,
Aegina¶	0.7041–0.7068	5	A, D,
Methana¶	0.7058–0.7067	4	A, D,
Poros¶	0.7073	1	D,
Isthmus¶	0.7134	1	D,
Atalanti–Volos group¶	0.7086–0.7098	4	A,

† Rock-type abbreviations: B = basalt, A = andesite, D = dacite, R = rhyolite, LT = leucite tephrite; sh = shoshonitic.
‡ Source: Barberi *et al.* (1974); Klerkx *et al.* (1976).
§ Source: Pe and Gledhill (1975).
‖ Source: Pe and Gledhill (1975); Puchelt and Hoefs (1971).
¶ Source: Pe (1975).

and Nicholls, 1976; Arculus and Johnson, 1978) have re-emphasized the consistent K–*h* correlation in many arcs, but have also stressed the individuality of each arc in terms of the K_2O content and the rate of increase of K_2O with SiO_2.

The Aeolian arc is an important case with K_{55} variation (K_2O at $SiO_2 = 55$ per cent) from 1.4 to 6.0 (Fig. 3). A spatial variation is not obvious: calc-alkaline and shoshonitic suites occur within distances of a few kilometres and the interpretation of small scale spatial variations would lead to an opposite trend, as Vulcano ($K_{55} = 6.0$) lies closest to the trench. However, in the Aeolian arc there is a clear temporal sequence from calc-alkaline to high K calc-alkaline to shoshonitic and to tephritic series. Shoshonites and tephrites appear to indicate the final stage of arc evolution (Barberi *et al.*, 1974; Keller, 1974).

Bearing in mind that K_2O concentrations cannot 'measure' the Benioff depth (nor vice versa) it is nevertheless tempting to use worldwide ranges and averages of the K–*h* relationship for a general intepretation. Tyrrhenian seismicity with focal depths exceeding 250 km is consistent with the high K concentrations in the shoshonitic lavas from the active volcanoes. Earthquakes related to the calc-alkaline series would be expected to be rather shallower, in the depth range which is seismically inactive (less 200 km). This gap in present-day seismicity is therefore consistent with the older ages and inactive character of the calc-alkaline volcanism.

The model of a recently detached slab descending almost vertically best explains the seismic evidence (Fig. 2). The depth of the slab will control the concentrations of large ion lithophile element fluids derived from the slab (Nicholls, 1974; Ringwood, 1974) and different travel distances through the overlying mantle to the site of partial melting will further contribute to the fluid by scavenging incompatible elements from the surrounding asthenosphere (Ninkovich and Hays, 1972). In this model all the Aeolian rock series are derived from a similar mantle source. The main variation is explained by adding various amounts of scavenged incompatible large ion lithophile elements, although additional influences of varying degrees of partial melting cannot be ruled out.

Figure 2 shows that subduction has reached at least 350 km (leaving open the significance of two c. 450 km foci below the Tyrrhenian abyssal plain). Using average convergence rates, subduction must have started in the order of 20 Ma ago. The opening of the back-arc basin in the Tyrrhenian Sea started during the Miocene, and basalts resembling those from active spreading centres have been dated at 7 Ma (Barberi et al., 1978; Dietrich et al., 1977). Recorded island arc volcanism covers only the last 0.5 Ma and therefore provides insights into only the very late stages of the presumed arc evolution in the southern Tyrrhenian/Aeolian area.

Structure and Geophysical Setting of the Aegean Area

The Aegean area forms an individual small plate between the Eurasian and the African plate (McKenzie, 1972). The Aegean microplate overrides the northward-moving African block (Fig. 5). The crust is continental on either side of the collision zone and a system of trenches follows the arcuate plate boundary (Fig. 5). Earthquake centres between the trench system and the central Aegean reach depths of 180 km. A seismic plane is not sharply defined, but can be inferred as an amphitheatre-like conical plane dipping c. 35° from the southerly trenches towards the central Aegean (Comninakis and Papazachos, 1972; Ninkovich and Hays, 1972). An example of seismic centre distribution from which the presence of a Benioff zone can be inferred is given in Fig. 6. The arcuate line of recent calc-alkaline volcanoes lies above a Benioff zone 130–150 km in depth. This is generally termed the Aegean or Hellenic volcanic arc.

The Central Aegean area behind that arc has several geophysical features of a back-arc basin or marginal sea. The heat flow is high, a mean value of 2.08 heat flow units is given by Jongsma (1974). A shallow Moho depth and an anomalous mantle characterize the central Aegean (Makris, 1977). The crust–mantle boundary rises from 40 km under the border areas to c. 20 km depth under the central Aegean. However, there are also some important differences with typical active back-arc basins (Berkhemer, 1978). These include the continental character of the crust (Makris, 1977) and the apparent absence of active basaltic back-arc volcanism. Using an average subduction rate for the area, the start of subduction can be estimated to be c. 12 Ma ago (Fytikas et al., 1976).

Geology and Volcanic History of the Aegean Arc

The Aegean volcanic arc extends from the Greek mainland across the Aegean Sea and comprises volcanic formations of Crommyonia (also referred to as the Isthmus of Corinth), Methana, Aegina, Poros, the Milos group, Santorini, Nisyros, Yali, and Kos (Fig. 5). In general the width of the arc does not exceed c. 20 km. The position of the rhyolitic rocks of Antiparos, which occur some 50 km inside the arc, is not clear. Their Pliocene age is consistent with a relationship to the Aegean arc.

According to the stratigraphical record and available radiometric results (Table 3) volcanism became active during the Pliocene and reached its maximum during the Quaternary. Fytikas et al. (1976) and Barberi et al. (1977) suggested that calc-alkaline activity was initiated c. 3 Ma ago in the present arc and this is consistent with the most recent data which show that the oldest volcanic rocks have ages of between 3.15 and 3.4 Ma (Ferrara et al., 1980; Innocenti et al., 1979). Two islands are considered active: Santorini, which has experienced numerous historic eruptions up to 1950, and Nisyros, with strong fumarolic activity and with phreatic eruptions up to the end of the 19th century. Methana last had a lava eruption around 230 BC and Yali has experienced several prehistoric pumice eruptions.

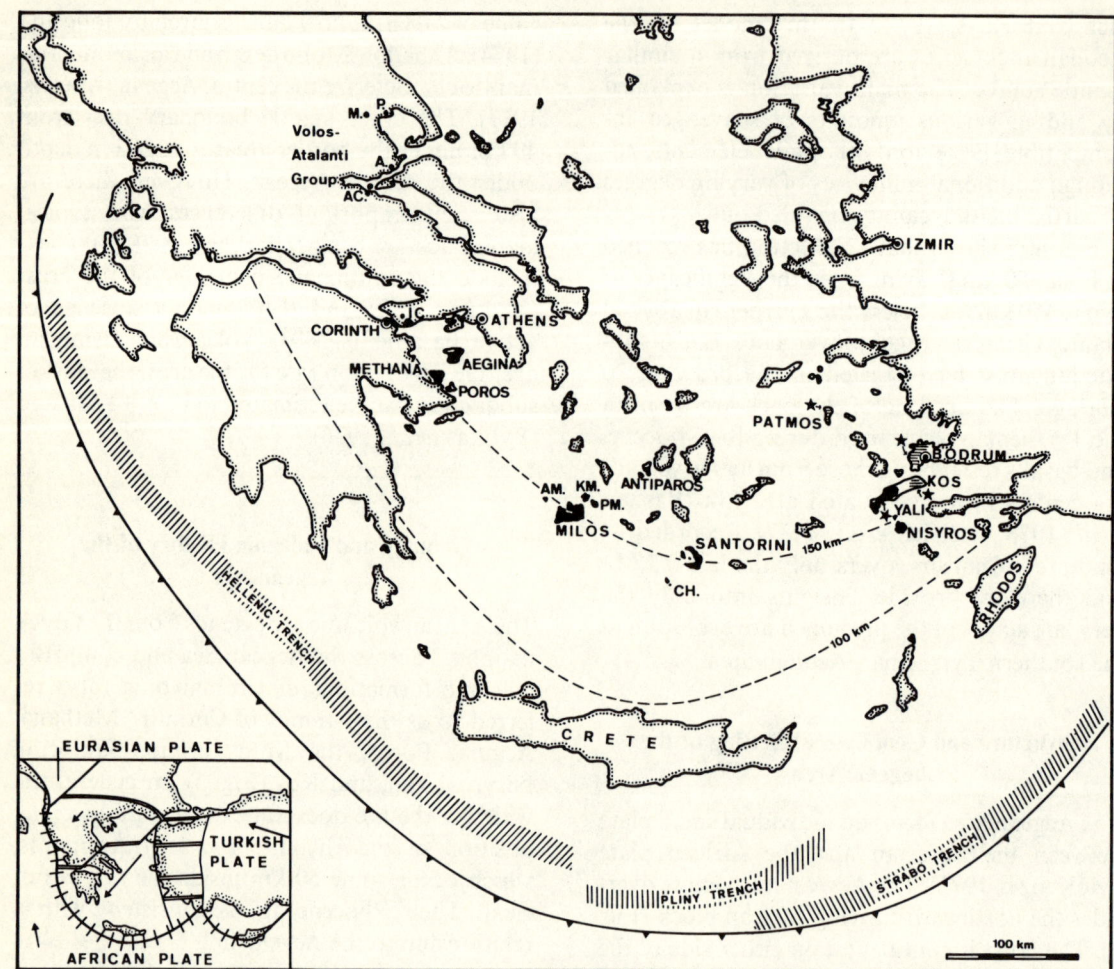

Fig. 5   The Aegean volcanic arc (in black) in relation to topography and seismicity in the eastern Mediterranean. The plate tectonic interpretation of McKenzie (1972) is shown in the inset. In the eastern part of the arc, the horizontal shading and asterisks indicate Miocene high K volcanic rocks of Patmos, Bodrum, and part of Kos. The diagonal shading indicates the Hellenic trenches with the adjacent solid line showing the outer limit of Aegean seismicity. Also shown are 100 km and 150 km isodepth lines for the Benioff zone (based on data from Ninkovich and Hays, 1972; McKenzie, 1972; Comninakis and Papazachos, 1972; Kiskyras, 1978; Richter and Stobach, 1978). *Abbreviations*: Volos–Atalanti group: Achilleion (A), Porphyrion (P), Microthebe (M), Atalanti Channel with Likhades Islands and Kammena Vourla (AC). Corinth: Isthmus–Crommyonia (IC). Milos group: Kimolos (KM), Antimilos (AM), Poliegos (PM). Christiana Islands: CH

In the eastern Aegean, volcanic rocks of different ages and belonging to different subduction episodes occur together. These overlap on the *Island of Kos* which has K-rich alkaline lavas (Keller, 1977) similar to rocks on Patmos and Bodrum (Robert, 1973, 1976; Burri *et al.*, 1967). These have been linked with a Miocene subduction episode (Keller, 1969, 1971) to which also belongs the bulk of the north-western Anatolian andesitic series (Borsi *et al.*, 1972; Innocenti and Mazzuoli, 1971; Barberi *et al.*, 1977). A potassic trachyte (Table 4, no. 13) on Kos has yielded a K–Ar age of 10.4 Ma. (Besang *et al.*, 1977). The potassic suite of Bodrum is of similar age (10.6–10.5 Ma; cf. Innocenti *et al.*, this volume, Section III). Pleistocene volcanic

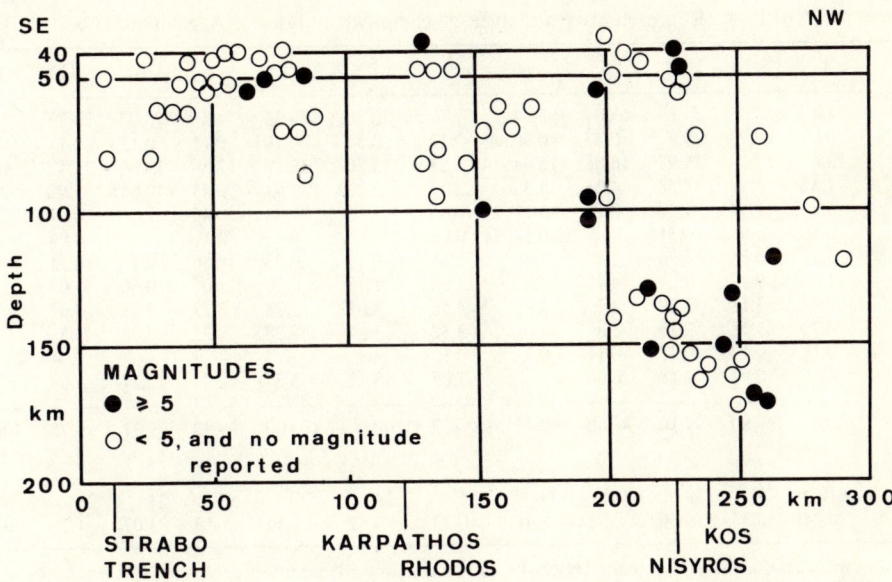

Fig. 6 Example of the seismicity in the Aegean area in a section perpendicular to the arc (from Richter and Strobach, 1978). The diagram includes all earthquake foci in a c. 80 km wide strip from Strabo Trench to Kos projected on to a vertical plane

TABLE 3 Synopsis of radiometric ages and stratigraphical data for Aegean arc volcanism

	Radiometric Ages				Geological Age Range
	Age	Method†	Dated Unit	Reference‡	
Nisyros	0.2 Ma	K–Ar	Pre-caldera dacite	1	Pliocene/Quaternary–Upper Quaternary
Yali	0.024 Ma	FT	Obsidian dome	2	Upper Quaternary–prehistoric
Kos	—		Calc-alkaline series		Lower Quaternary–Upper Quaternary
	10.4 Ma	K–Ar	K-trachytic ignimbrite	3	Miocene
Santorini	1.59–0.63 Ma	K–Ar	Lumaravi and Akrotiri units	4	Plio-Pleistocene
	0.1 Ma	FT	Lower Thera pumice series (BU)	8	Upper Quaternary
	37–18 ka	C-14	Younger Thera volcanics	9	Upper Quaternary
	1450–1480 BC	C-14	Minoan eruption		Prehistoric
	197 BC–1950 AD		Pala and Nea Kaimeni		Historic
Christiana islands	—				
Kimolos	3.15 Ma	K–Ar	Granitic subvolcanites	4	Quaternary, pre-Minoan
Milos	2.5–0.48 Ma	K–Ar	Rhyolites, dacites, Andesites	1, 5	Pliocene/early Quaternary–Recent Quaternary
Antiparos	—				Upper Pliocene–early Quaternary
Methana	0.90–0.32 Ma	K–Ar	Dacites	1	Quaternary–historic (~230 BC)
Aegina	—				Pliocene (?), mainly Lower Quaternary
Poros	—				Plio-Pleistocene
Isthmus	2.7–2.2 Ma	K–Ar	Dacite	1, 7	Pliocene
Atalanti–				1, 6	Plio-Pleistocene
Volos group	3.4–0.5 Ma	K–Ar	Achilleion, Microthebe, Porphyrion, Likhades		

† K–Ar = potassium–argon dating; FT = fission track; C-14 = radiocarbon dating.
‡ References: (1) Fytikas et al. (1976); (2) Wagner et al. (1976); (3) Besang et al. (1977); (4) Ferrara et al. (1980); (5) Angelier et al. (1977); (6) Innocenti et al. (1979); (7) Schroeder (1976); (8) Seward et al. (1980); (9) Friedrich and Pichler (1976).
Reference added in proof: H. Bellon, J. J. Jarrige and D. Sorel (1979) Rev. Geol. Dyn. Geogr. Phys. 21, 41–55.

TABLE 4 Representative and average chemical analyses of Aegean arc lavas

	1	2	3	4	5	6	7	8	9	10	11	12	13
SiO_2	51.4	54.85	57.69	63.6	65.34	70.7	70.76	74.35	74.5	75.02	73.73	54.18	63.1
TiO_2	0.85	0.73	0.5	0.47	0.86	0.47	0.31	0.19	0.11	0.12	0.11	0.71	0.4
Al_2O_3	18.4	17.10	16.97	16.82	15.44	12.23	13.70	12.29	12.58	12.86	12.4	18.3	16.7
Fe_2O_3	1.35	3.98	4.38	4.0	2.34	2.0	2.29†	1.41†	0.43	0.84†	1.02	3.42	2.45
FeO	6.9	1.44	1.36	0.1	3.11	0.65	—	—	0.43	—	0.91	2.84	1.1
MnO	0.1	0.08	0.11	—	0.14	0.06	—	—	0.1	—	0.09	0.12	0.1
MgO	6.3	6.75	3.73	2.57	1.4	1.51	0.33	0.23	0.56	0.1	0.31	4.38	1.4
CaO	11.0	9.53	7.27	4.75	4.0	2.94	1.43	0.90	1.05	0.66	1.0	7.7	3.2
Na_2O	2.9	3.18	3.29	3.7	4.9	4.56	4.64	2.78	3.73	3.78	3.6	3.84	3.6
K_2O	0.7	1.18	2.17	2.8	2.0	3.32	3.34	4.28	4.23	4.06	4.5	2.6	6.8
P_2O_5	0.1	0.17	0.15	0.11	0.2	0.1	—	—	—	—	0.04	0.54	0.16
H_2O	—	0.81	1.48	0.68	—	1.65	3.2‡	3.57‡	2.2	2.27	2.04	1.37	0.92
Total	100.0	99.80	99.10	99.60	99.73	100.19	100.00	100.00	99.92	99.71	99.75	100.00	99.93
$Mg/Mg + Fe^{2+}$	61§	73	59	58	35	55	24	27	58	21	26	60	46
K_2O/Na_2O	0.24	0.37	0.66	0.76	0.41	0.73	0.72	1.54	1.13	1.07	1.25	0.68	1.9

(1) High-alumina basalt average, Santorini (Nicholls, 1978). (2) Low-silica andesite, Nisyros (Di Paola, 1974, no. NY 39). (3) Andesite, Aegina (Pe, 1973, no. 6). (4) Dacite, Kos-Kefalos (J. Keller, unpublished). (5) Dacite, average of modern lavas, Santorini–Nea Kaimeni (Puchelt, 1978). (6) Rhyodacite, Nisyros (Di Paola, 1974; no. NY 11). (7) Minoan Santorini tephra residual glass (Keller, 1980a). (8) Residual glass of caldera-phase pumice, Nisyros (Keller, 1980a). (9) Rhyolites, Milos (Burri and Soptrajanova, 1967; average of four biotite-liparites). (10) Rhyolite, Kos-Zini (J. Keller, unpublished). (11) Rhyolites, Antiparos (Anastopoulos, 1963; average of 12). (12) High K andesite, Volos–Atalanti Group (Innocenti et al., 1979; no. IC 6). (13) Potassic trachyte, Kos-Fokas, Miocene shoshonitic province (J. Keller, unpublished).

† Total Fe as Fe_2O_3.
‡ Volatiles by difference.
§ All Mg values with $Fe_2O_3/FeO = 0.15$.

rocks on Kos which are related to the present arc include calc-alkaline dacites, but are dominantly rhyolitic (Table 4, nos. 4 and 10; Fig. 7).

Nisyros is the only major island of the arc in which no pre-volcanic basement is exposed. Three main stages can be distinguished: (1) an early submarine stage which has been extensively eroded and is now exposed with several raised-beach levels—this is estimated to be of Plio-Quaternary age; (2) a composite volcano stage, culminating in the formation of a central caldera with a diameter of c. 4 km—a pre-caldera dacite has yielded a K–Ar age of 0.2 Ma (Fytikas et al., 1976); (3) a post-caldera stage in which large domes grew inside the caldera. The caldera formation and post-caldera activity are believed to belong to the Upper Quaternary (Keller, 1971). The lavas range from low Si andesites (54 per cent SiO_2) to rhyodacites (70–72 per cent SiO_2 (Davis, 1967; Di Paola, 1974).

Yali is a small islet only 4–5 km off the northern coast of Nisyros. It is exclusively rhyolitic, with obsidian domes in the north-eastern part, and air-fall pumice deposits in the south-western part. A fission track age of 24 ka dates the obsidian domes (Wagner et al., 1976). However, pumice beds of the most recent eruptions cover soils which contain primitive pottery and Neolithic obsidian artefacts, proving that the most recent volcanic activity is younger than Neolithic in age (Keller, 1980a).

The ring-islands of the *Santorini group* were shaped by the 'Minoan' eruption and associated caldera collapse in 1480–1450 BC. The pre-caldera volcanic island was formed of volcanic rocks erupted from different volcanic centres—the Lumaravi–Archangelo, Akrotiri, Thera, Skaros–Therasia, Simandir, and Peristeria volcanoes (Nicholls, 1971; Pichler and Kussmaul, 1972). These pre-caldera volcanic rocks were grouped by Nicholls into three basalt–andesite–dacite series. Within the caldera the new islands of Palaea Kameni and Nea Kameni were formed by lava extrusions between 197 BC and AD 1950.

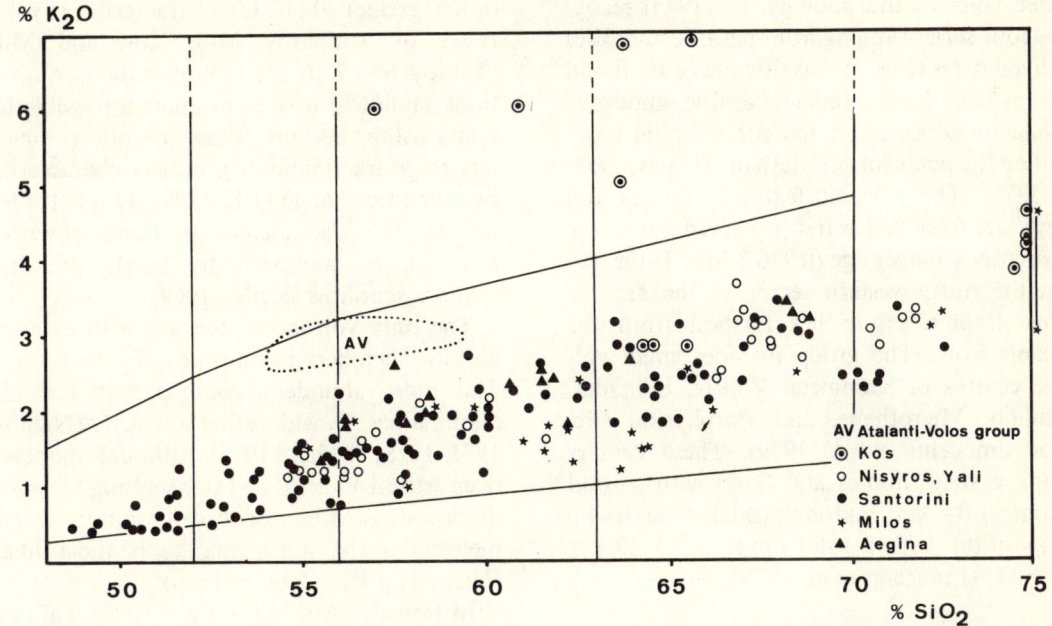

Fig. 7 Graph of K_2O against SiO_2 for the main Aegean arc series, forming a uniform calc-alkaline suite. The field boundaries are from Peccerillo and Taylor (cf. Fig. 3). The Atalanti–Volos group (AV) is a distinct high K association. Examples of potassic lavas from the Miocene of Kos are added to emphasize the overlap of two different magmatic events in this island. *Data sources*: Atalanti–Volos group: Innocenti *et al.* (1979). Kos: J. Keller (unpublished data). Nisyros and Yali: Davis (1967), Di Paola, (1974). Santorini: Nicholls (1971), Pichler and Kussmaul (1972), Puchelt (1978). Milos: Burri and Soptrajanova (1967). Aegina: Davis (1957), Pe (1973)

Here, dacites are by far the dominant rock type, forming the 'Modern Series' of Nicholls (1971). Additional detailed geochemical data for Santorini are found in Puchelt (1978).

The uninhabited islands of the *Christiana group* are situated 25 km south-west of Santorini. The whole surface area of the group is less than 1.5 km² with an elevation of 284 m. Puchelt *et al.*, (1977) describe andesitic lavas, some transitional to dacites, and rhyolitic pumices.

Milos: A metamorphic, and Neogene sedimentary basement is covered by the earliest volcanic rocks, a silicic pyroclastic series of late Pliocene age. Lava domes and less common lava flows followed this explosive activity. These lavas include rhyolites, rhyolitic obsidians, dacites, and andesites which were randomly erupted between 2.5 and 0.95 Ma (Angelier *et al.*, 1977; Fytikas *et al.*, 1976). Two Recent eruptions have led to the formation of the large explosion craters of Trachylas and Phyriplaka–Traphores and their associated perlitic rhyolite flows dated (K–Ar) at 0.48 Ma (Fytikas *et al.*, 1976). The youngest volcanic activity has produced phreatic explosion craters (Fytikas and Marinelli, 1976). The petrography and chemical compositions of lavas from Milos have been described by Burri and Soptrajanova (1967).

The volcanic outcrops near *Crommyonia* together with *Poros, Aegina*, and *Methana* form the Saronic Gulf subprovince of the Aegean arc. The small outcrops of volcanic rocks at Crommyonia are dacites (Pe, 1972; Davis, 1957) dated as Pliocene (Fytikas *et al.*, 1976). Small outcrops of dacite on Poros are tentatively attributed to the same period (Davis, 1957). Volcanic activity on Aegina started during the Pliocene but reached a maximum during the Lower Pleistocene (Davis, 1957). Volcanic forms in Aegina are endogeneous domes and lava flows, with deeply eroded volcanic morphology. Pe (1973) recognizes four series ranging from basaltic andesites to rhyodacites (Fig. 7). Basaltic rocks are found only as xenoliths. Similar basaltic andesite–rhyodacite series cover the pre-volcanic basement on the peninsula of Methana (Davis, 1957; Pe, 1974). The volcanic features—domes and flows—are fresh and better preserved due to the apparently younger age (0.9–0.3 Ma; Table 3).

In the north-western sector of the arc the *Volos–Atalanti* group lies furthest from the volcanic front. The group includes small volcanic centres of Kammena Vourla, Likhades, Achillion, Microthebe, and Porphyrion (Pe, 1975; Innocenti et al., 1979). These centres comprise small domes and flows with a total volume of 0.4 km^3 and are of high K andesitic composition, with a total range of 54–59 per cent SiO$_2$ (Innocenti et al., 1979).

Petrology and Chemical Composition of the Aegean Arc

Representative and average chemical analyses of Aegean arc lavas are shown in Table 4. The volcanic rocks of the Aegean arc belong to a typical calc-alkaline suite. In general, andesites and dacites predominate (on Methana, Aegina, Crommyonia, Poros, and Nisyros) but some islands (Yali, Kos, Milos, Antiparos) are dominantly rhyolitic. Santorini exhibits a high alumina basalt–andesite–dacite–rhyodacite association, with basalts as the most abundant rock type. The graph of K$_2$O against silica (Fig. 7) shows a well defined trend with little scatter consistent with the narrow width of the arc, and only the small centres of the Volos–Atalanti group which lie behind the volcanic front show significantly higher K$_2$O values in relation to SiO$_2$ (Innocenti et al., 1979). On a smaller scale, a section over 40 km perpendicular to the arc is represented by the line Christiana–Santorini–submarine Kolomvos volcano (7 km north-east of Santorini) with K$_{60}$ (K$_2$O at 60 per cent SiO$_2$) increasing in the order of 1.68, 1.84, 2.00 (Puchelt et al., 1977).

High K$_2$O concentrations and K$_2$O/Na$_2$O ratios greater than 1.0 characterize rhyolitic rocks of Antiparos, Yali, Kos, and Milos (Table 4, nos. 9, 10, 11). However, these compositions cannot be used to evaluate a possible K-*h* relationship because these rhyolites include late-stage fractionation products characterized by a drastic change of K$_2$O/Na$_2$O (cf. Table 4, no. 8), and also include products of crustal anatexis, as suggested by partly mobilized granitic xenoliths (Keller, 1969).

The only volcano of the arc with extensive basaltic members is Santorini (Table 4, no. 1). The most abundant compositions are high alumina basalts with rather low K$_2$O (Nicholls, 1971, 1978; Puchelt, 1978). Although models of deep crustal anatexis and slab melting have been discussed, an origin of these basalts by partial melting in the upper mantle is most likely, (Nicholls, 1978; Osborn, 1976).

In terms of Mg/Mg + Fe^{2+} (0.59–0.67) and Ni (30–100 p.p.m.) and Cr contents (20–350 p.p.m.), it is unlikely that even the most primitive of the erupted basalts represent unfractionated equilibrium partial melts from the mantle (Nicholls, 1978). Nicholls presents calculations which indicate that fractionation of 6–13 per cent olivine and small amounts of Cr-rich spinel and diopsidic pyroxene from the primary magma gave rise to basalts with the observed Mg, Ni, and Cr contents. The initial fractional crystallization occurred under high pressure conditions (over 8 kb; Osborn, 1976). The calc-alkaline evolution to andesites and dacites, very similar on Santorini and Nisyros, is a low pressure/moderate f_{O_2} fractionation line in terms of the Fe-enrichment trend (Fig. 4(b)). In this plot lavas of the south-western arc show a different trend line, indicating that differentiation occurred dominantly at high pressure.

Trace element information is not yet available for all parts of the Aegean arc. It is, however, evident from published data (Pe and Gledhill, 1975; Pe, 1975; Puchelt, 1978; Innocenti et al., 1979) that a large scatter in large ion lithophile element concentrations characterizes the arc. As an example the average Sr contents at SiO$_2$ 60 per cent are 150–200 p.p.m. for Santorini, 270 p.p.m. for the Methana, 550 p.p.m. for

Aegina and 380 p.p.m. for Poros. The respective value for the calc-alkaline Aeolian islands is 650 p.p.m. (e.g. Salina; Keller, 1974) and for an average 'island arc type' andesite it is 385 p.p.m. (Taylor, 1969; Jakeš and White, 1972).

The large variation in Sr isotope ratios, with $^{87}Sr/^{86}Sr$ varying between 0.7037 and 0.7134 (Table 2), may reflect the influence of slab-derived fluids in an area of great complexity in structure and composition of subducted crust together with contamination by continental crust. However, the lowest values of 0.7037 (Nisyros), 0.7040 (Santorini), and 0.7041 (Aegina) are consistent with derivation from a mantle-derived parent, and are similar to values from the Aeolian arc.

REFERENCES

Anastopoulos, J. (1963). Geological study of Antiparos island group (in Greek). *I.G.S.R. Athens* 7, (5), pp. 375.

Angelier, J., Cantagrel, J. M., and Vilminot, J. C. (1977). Néotectonique cassante et volcanisme plio-quaternaire dans l'arc égéen interne: l'Ile de Milos (Grèce). *Bull. Soc. géol. Fr.* 19, 119–121.

Arculus, R. J. and Johnson, R. W. (1978). Criticism of generalized models for magmatic evolution of arc trench systems. *Earth planet. Sci. Letters* 39, 118–126.

Barberi, F., Gasparini, P., Innocenti, F., and Villari, L. (1973). Volcanism of the Southern Tyrrhenian Sea and its geodynamic implications. *J. geophys. Res.* 78, 5221–5232.

Barberi, F., Ferrara, G., Keller, J., Innocenti, F. and Villari, L. (1974). Evolution of Aeolian arc volcanism. *Earth Planet. Sci. Letters* 21, 269–276.

Barberi, F., Innocenti, F., Marinelli, C., and Mazzuoli, R. (1977). Vulcanismo e tettonica a placche: esempi nell' area mediterranea. *Mem. Soc. Geol. It.* 13, 327–358.

Barberi, F., Bizouard, H., Capaldi, G., Ferrara, G., Gasparini, P., Innocenti, F., Joron, J. L., Lambret, B., Treuil, M., and Allègre, C. (1978). Age and nature of basalts from the Tyrrhenian abyssal plain. *Init. Rep. DSDP* no. 42, 509–514.

Berkhemer, H. (1978). Some aspects of the evolution of marginal seas deduced from observations in the Aegean region. In *Alps, Apennines, Hellenides* (H. Closs et al., eds), I. C. G. Scient. Rep. no. 38, Stuttgart, pp. 527–529.

Besang, C., Eckhardt, F.-J., Harre, W., Kreuzer, H., and Müller, P. (1977). Radiometrische Altersbestimmungen an neogenen Eruptivgesteinen der Türkei. *Geol. Jb. B* 25, 3–36.

Bigazzi, G. and Bonadonna, F. (1973). Fission track dating of the obsidian of Lipari island (Italy). *Nature, Lond.* 242, 322–323.

Biju-Duval, B., Dercourt, J., and Le Pichon, X. (1977). From the Tethys ocean to the Mediterranean seas: a plate tectonic model of the evolution of the western alpine system. In *Proceedings of the International Symposium on the Structural History of the Mediterranean Basins, Split, 1976* (B. Biju-Duval and L. Montadert, eds), Edition Technip, Paris, pp. 143–164.

Biju-Duval, B., Letouzey, J., and Montadert, L. (1978). Structure and evolution of the Mediterranean basins. *Init. Rep. DSDP* no. 42, 951–986.

Blot, C. (1978). Volcanism and seismicity in Mediterranean island arcs. In *Thera and the Aegean World*, Vol. I (C. Doumas ed.), Athens, pp. 33–44.

Boccaletti, M. and Guazzone, G. (1972). Gli archi appenninici, il Mar Ligure ed il Tirreno nel quadro della tettonica dei bacini marginali retro-arco. *Mem. Soc. Geol. It.* 11, 201–216.

Borsi, S., Ferrara, G., Innocenti, F., and Mazzuoli, R. (1972). Geochronology and petrology of recent volcanics in the Eastern Aegean Sea (West Anatolia and Lesvos Island). *Bull. Volc.* 36, 473–496.

Burri, C. and Soptrajanova, G. (1967). Petrochemie der jungen Vulkanite der Inselgruppe von Milos (Griechenland) und deren Stellung im Rahmen der Kykladenprovinz. *Vierteljahresschr. Naturforsch. ges. Zürich* 112, 1–27.

Burri, C., Tartar, Y., and Weibel, M. (1967). Zur Kenntnis der jungen Vulkanite der Halbinsel Bodrum. *Schweiz. Mineral. Petrol. Mitt.* 47, 833–854.

Caputo, M., Panza, G. F., and Postpischl, D. (1972). New evidences about the deep structure of the Lipari arc. *Tectonophysics* 15, 219–231.

Comninakis, P. E. and Papazachos, B. C. (1972). Seismicity of the Eestern Mediterranean and some tectonic features of the Mediterranean ridge. *Geol. Soc. Am. Bull.* 83, 1093–1102.

Davis, E. (1957). Die jungvulkanischen Gesteine von Aegina, Methana und Poros und deren Stellung im Rahmen der Kykladenprovinz. *Publ. Vulkaninst. Imm. Friedländer* 6, 3–74.

Davis, E. (1967). Zur Geologie und Petrologie der Inseln Nisyros und Jali (Dodecanes). *Praktika Akad. Ath.* 42, 235–252.

Del Monte, N. (1972). Vulcanesimo del Mar Tirreno: nota preliminare sui vulcani Marsili e Palinuro. *Giorn. Geol.* 38, 231–252.

Dewey, J. E., Pittman, W. C., III, Ryan, W. B. F., and Bonnin, J. (1973). Plate tectonics and evolution of the Alpine system. *Geol. Soc. Am. Bull.* 84, 3137–3180.

Dickinson, W. R. and Hatherton, T. (1967). Andesitic volcanism and seismicity around the Pacific. *Science, N.Y.* 157, 801–803.

Dietrich, V., Emmermann, R., Keller, J., and Puchelt, H. (1977). Tholeiitic basalts from the Tyrrhenian sea floor. *Earth planet. Sci. Letters* 36, 285–296.

Di Paola, G. M. (1974). Volcanology and petrology of Nisyros Island (Dodecanese, Greece). *Bull. Volc.* 38, 944–987.

Ferrara, G., Fytikas, M., Giuliani, O., and Marinelli, G. (1980). Age of the formation of the Aegean active volcanic arc. In *Thera and the Aegean World*, Vol. 2, Athens, 37–41.

Friedrich, W. L. and Pichler, H. (1976). Radiocarbon dates of Santorini volcanics. *Nature, Lond.* **262**, 373–374.

Fytikas, M., Giuliani, O., Innocenti, F., Marinelli, G., and Mazzuoli, R. (1976). Geochronological data on recent magmatism of Aegean Sea. *Tectonophysics* **31**, T29–T34.

Fytikas, M. and Marinelli, G. (1976). Geology and geothermics of the island of Milos (Greece). Paper presented to the International Congress on Thermal Waters, Geothermal Energy and Volcanism of the Mediterranean Area, Athens, October 1976.

Giese, P. and Morelli, C. (1973). Main features of crustal and upper amntle velocity distribution in Southern Italy. Paper presented at IASPEI, Lima.

Gill, J. B. (1974). Role of underthrust oceanic crust in the genesis of a Fijian calc-alkaline suite. *Contr. Mineral. Petrol.* **43**, 29–45.

Görler, K. and Giese, P. (1978). Aspects of the evolution of the Calabrian arc. In *Alps, Apennines, Hellenides* (H. Closs *et al.*, eds), ICG Scientific Report no. 38, Stuttgart, pp. 374–388.

Honnorez, J. and Keller, J. (1968). Xenolithe in vulkanischen Gesteinen der Äolischen Inseln. *Geol. Rundschau* **57**, 719–736.

Innocenti, F. and Mazzuoli, R. (1971). Petrology of the Izmir-Karaburun volcanic area (Western Turkey). *Bull. Volc.* **36**, 83–104.

Innocenti, F., Manetti, P., Peccerillo, A., and Poli, G. (1979). Inner arc volcanism in NW Aegean Arc: Geochemical and geochronological data. *N. Jb. Min. Jg.* **1979**, 145–158.

Jakeš, P. and Gill, J. B. (1970). Rare earth elements and the island arc tholeiitic series. *Earth planet. Sci. Letters* **9**, 17–28.

Jakeš, P. and White, A. J. R. (1972). Major and trace element abundances in volcanic rocks of orogenic areas. *Geol. Soc. Am. Bull.* **83**, 29–40.

Jongsma, D. (1974). Heat flow in the Aegean Sea. *Geophys. Jl. R. astr. Soc.* **37**, 337–346.

Keller, J. (1967). Alter und Abfolge der vulkanischen Ereignisse auf den Äolischen Inseln/Sizilien. *Ber. Naturf. Ges. Freiburg i. Br.* **57**, 33–67.

Keller, J. (1969). Origin of rhyolites by anatectic melting of granitic crustal rocks. The example of rhyolitic pumice from the island of Kos (Aegean Sea). *Bull. Volc.* **33**, 942–959.

Keller, J. (1970). Datierung der Obsidiane und Bimstuffe von Lipari. *N. Jb. Geol. Mh.* **1970**, 90–101.

Keller, J. (1971). The major volcanic events in recent Eastern Mediterranean volcanism and their bearing on the problem of Santorini ash-layers. In *Acta of the 1st International Scientific Congress on the Volcano of Thera*, pp. 152–169.

Keller, J. (1974). Petrology of some volcanic rock series of the Aeolian Arc, Southern Tyrrhenian Sea: Calc-alkaline and shoshonitic associations. *Contr. Mineral. Petrol.* **46**, 29–47.

Keller, J. (1980). The Island of Salina. The Island of Vulcano. In *The Aeolian Islands. An Active Island Arc in the Mediterranean Sea*, Rend. Soc. Ital. Min. Petrogr., **36**, 369–414 and 489–524.

Keller, J. (1980a). Prehistoric pumice tephra on Aegean islands. In *Thera and the Aegean World*, Vol. 2, Athens, 49–56.

Keller, J. and Leiber, J. (1974). Sedimente, Tephra-Lagen und Basalte der südtyrrhenischen Tiefsee-Ebene im Bereich des Marsili-Seeberges. *METEOR-Forschungsberichte C* **19**, 62–76.

Kiesl, W., Kluger, F., Weinke, H. H., Scholl, H., and Klein, P. (1978). Untersuchungen an süditalienischen Vulkaniten: Lipari, Vulcano. *Chem. Erde* **37**, 40–49.

Kiskyras, D. (1978). The geotectonic state of the Greek Region: vulcanism, intermediate earthquakes and plate tectonics. In *Thera and the Aegean World*, Vol. 1 (C. Doumas, ed.), Athens, pp. 85–96.

Klein, P., Kluger, F., and Kiesl, W. (1975). Untersuchungen an süditalienischen Vulkaniten: Alicudi, Filicudi. II. Die Seltenen Erden. *Chem. Erde* **34**, 283–292.

Klerkx, J., Deutsch, S., Hertogen, J., Gijbels, R., and Pichler, H. (1976). Strontium isotope and rare earth data relating to the petrogenesis of the Aeolian arc volcanism. In *Proceedings of the International Congress on Thermal Waters, Geothermal Energy and Vulcanism of the Mediterranean Area, Athens, 1976*, Vol. 3, *Vulcanism*, pp. 76–96.

Lopez-Escobar, L., Frey, F. A., and Vergara, M. (1977). Andesites and high-alumina basalts from the central-south Chile High Andes: geochemical evidence bearing on their petrogenesis. *Contr. Mineral. Petrol.* **63**, 199–228.

Makris, J. (1977). Geophysical investigations of the Hellenides. *Hamb. Geophys. Einzelschr.* **34**, 1–124.

Marinelli, G. (1975). Magma evolution in Italy. In *Geology of Italy*, Ed. Earth Sci. Soc., Tripoli, pp. 165–219.

McKenzie, D. (1972). Active tectonics of the Mediterranean region. *Geophys. Jl R. astr. Soc.* **30**, 109–185.

Menard, H. W. (1967). Transitional types of crust under small ocean basins. *J. geophys. Res.* **72**, 3061–3073.

Morelli, C. (1970). Physiography, gravity and magnetism of the Tyrrhenian Sea. *Boll. Geof. Teor. Appl.* **12**. 274–308.

Nicholls, I. A. (1971). Petrology of Santorini volcano, Cyclades, Greece. *J. Petrol.* **12**, 67–119.

Nicholls, I. A. (1974). Liquids in equilibrium with peridotitic mineral assemblages at high water pressures. *Contr. Mineral. Petrol.* **45**, 289–316.

Nicholls, I. A. (1978). Primary basaltic magmas for the precaldera volcanic rocks of Santorini. In *Thera and the Aegean World*, Vol. 1 (C. Doumas, ed.), Athens, pp. 109–120.

Ninkovich, D., and Hays, J. D. (1972). Mediterranean island arcs and origin of high potash volcanoes. *Earth planet. Sci. Letters* **16**, 331–345.

Osborn, E. F. (1959). Role of oxygen pressure in the cristallization and differentiation of basaltic magma. *Am. J. Sci.* **257**, 609–647.

Osborn, E. F. (1976). Origin of calc-alkali magma series of Santorini volcano type in the light of recent experimental phase-equilibrium studies. In *Proceedings of the International Congress on Thermal Waters, Geothermal Energy and Vulcanism of the Mediterranean Area, Athens, 1976*, Vol. 3, *Vulcanism*, pp. 154–167.

Pe, G. G. (1972). Geochemistry and chemical mineralogy of the lavas of Crommyonia. *Annls. Géol. Pays Helléniques* **24**, 257–275.

Pe, G. G. (1973). Petrology and geochemistry of volcanic rocks of Aegina, Greece. *Bull. Volc.* **37**, 491–514.

Pe, G. .G. (1974). Volcanic rocks of Methana (South Aegean Arc, Greece). *Bull. Volc.* **38**, 270–290.

Pe, G. G. (1975). Strontium isotope ratios in volcanic rocks from the northwestern part of the Hellenic Arc. *Chem. Geol.* **15**, 53–60.

Pe, G. G. and Gledhill, A. (1975). Strontium isotope ratios in volcanic rocks from the south-eastern part of the Hellenic arc. *Lithos* **8**, 209–214.

Peccerillo, A. and Taylor, S. R. (1976). Geochemistry of Eocene calc-alkaline volcanic rocks from the Kastamonu Area, Northern Turkey. *Contr. Mineral. Petrol.* **58**, 63–81.

Pichler, H. (1967). Neue Erkenntnisse über Art und Genese des Vulkanismus der Äolischen Inseln (Sizilien). *Geol. Rundschau* **57**, 102–126.

Pichler, H. (1980). The Island of Lipari. In *The Aeolian Islands. An Active Island Arc in the Mediterranean Sea.* Soc. Ital. Min. Petrogr., Rend., **36**, 415–440.

Pichler, H. and Kussmaul, S. (1972). The calc-alkaline volcanic rocks of the Santorini group (Aegean Sea, Greece). *N. Jb. Miner. Abh.* **116**, 268–307.

Puchelt, H. (1978). Evolution of the volcanic rocks of Santorini. In *Thera and the Aegean World*, Vol. 1 (C. Doumas ed.), Athens, pp. 131–146.

Puchelt, H. and Hoefs, J. (1971). Preliminary geochemical and strontium isotope investigations on Santorini rocks. In *Acta of the 1st International Scientific Congress on the Volcano of Thera*, Athens, pp. 318–327.

Puchelt, H., Murad, E., and Hubberten, H. W. (1977). Geochemical and petrological studies of lavas, pyroclastics and associated xenoliths from the Christiana islands, Aegean Sea. *N. Jb. Miner. Abh.* **131**, 140–155.

Richter, I. and Strobach, K., (1978). Benioff zones of the Aegean arc. In *Alps Apennines, Hellenides* (H. Closs, *et al.*, eds), ICG Scientific Report no. 38, Stuttgart, pp. 315–321.

Ringwood, A. E. (1974). The petrological evolution of island arc systems. *J. geol. Soc. Lond.* **130**, 183–204.

Robert, U. (1973). Les roches volcaniques de l'ile de Patmos (Dodecanese, Grèce). Thesis, 3rd Cycle, Université de Paris VI.

Robert, U. (1976). Données nouvelles sur le volcanisme du Sud-Est de la Mer Egée: existence d'une épisode à caractère alcalin. In *Proceedings of the International Congress on Thermal Waters, Geothermal Energy and Vulcanism of the Mediterranean Area, Athens, 1976*, Vol. 3, *Vulcanism*, pp. 211–224.

Romano, R. (1973). Le Isole di Panarea e Basiluzzo. *Riv. Mineraria Sicil.* **139–141**, 3–40.

Rosi, M. (1980). The island of Stromboli. In *The Aeolian Islands. An Active Island Arc in the Mediterranean Sea*, Soc. Ital. Min. Petrogr., Rend., **36**, 345–368.

Seward, D., Wagner, G. A., and Pichler, H. (1978). Fission track ages of Santorini volcanics (Greece). In *Thera and the Aegean World*, Vol. 2, Athens, 101–108.

Schroeder, B. (1976). Volcanism, neotectonics and postvolcanic phenomena east of Corinth/Greece. In *Proceedings of the International Congress on Thermal Waters, Geothermal Energy and Vulcanism of the Mediterranean Area, Athens, 1976*, Vol. 3, *Vulcanism*, pp. 240–248.

Taylor, S. R. (1969). Trace element chemistry of andesites and associated calc-alkaline rocks. *Oregon Dept Geol. Mineral Industries Bull.* no. 65, 43–63.

Villari, L. (1972). L'isola di Filicudi ed il suo significato magmatologico. *Rend. Soc. Ital. Mineral. Petrol.* **28**, 475–506.

Villari, L. (1980). The Island of Filicudi. In *The Aeolian Islands. An Active Island Arc in the Mediterranean Sea*, Soc. Ital. Min. Petr., Rend., **36**, 467–488.

Villari, L. and Nathan, S. (1978). Petrology of Filicudi, Aeolian Archipelago. *Bull. Volc.* **41**, 81–96.

Wagner, G. A., Storzer, D., and Keller, J. (1976). Spaltspurendatierungen quartärer Gesteinsgläser aus dem Mittelmeerraum. *N. Jb. Miner. Mh. Jg.* **1976**, 84–94.

Whitford, D. J., and Nicholls, I. A. (1976). Potassium variation in lavas across the Sunda arc in Java and Bali. In *Volcanism of Australasia* (R. W. Johnson, ed.), Elsevier, Amsterdam, pp. 63–75.

Anatolia and north-western Iran

F. Innocenti, P. Manetti, R. Mazzuoli, G. Pasquaré, and L. Villari

Istituto di Mineralogia e Petrografia,
Via S. Maria 53, Pisa, Italy,

Istituto di Mineralogia, Petrografia e Geochimia,
Via Lamarmora, 4 Firenze, Italy,

Dipartimento di Scienze della Terra,
Università di Calabria, Cosenza, Italy,

Istituto di Geologia,
Piazzale Gorini 15, Milan, Italy,

and

Istituto Internazionale di Vulcanologia,
Via la Regina Margherita 6, Catania, Italy

ABSTRACT

The volcanism which characterizes the inner margin of the Taurus–Central Iran Range produced a large volume of calc-alkaline and high K calc-alkaline products, ranging in age between Upper Oligocene and Quaternary. Shoshonitic volcanic products have been found, belonging to the latest phases of volcanic activity in western and eastern Anatolia. The orogenic volcanic cycle terminated at different times in the different sectors of the belt and ceased by the Upper Miocene in western Anatolia, but continued to the Quaternary in central and eastern Anatolia. In this latter region, the volcanism has a complex pattern characterized by northward migration of the andesitic volcanic front during Mio-Pliocene times, with the almost contemporary appearance of sodic-alkaline to transitional volcanic products in the Lake Van region, lasting to the present. Sodic-alkaline volcanic activity was also locally developed during the late Miocene and Quaternary in western Anatolia.

The calc-alkaline and the high K calc-alkaline series consist of rocks ranging in composition between basalts and rhyolites; but andesites and dacites are dominant. These rocks are similar to continental margin andesite series but are characterized by a relatively high MgO content and hence a high $Mg/Mg + Fe_{tot}$ ratio. Shoshonitic rocks, dominated by latites, amount to 10 per cent of the total volume.

The volcanic evolution reflects the complex convergent interaction between the Eurasian and Afro-Arabian plates, which was dominated by two main tectonic events. The first was diachronous continent–continent collision, marking the progressive cessation of lithosphere subduction processes. The second was the penetration of the Arabian wedge into the Anatolian–Iranian continental mass, giving rise to lithosphere fracturing and eruption of magmas with intraplate characteristics around the site of the suture.

Introduction

One of the most characteristic effects of plate convergence is the development of calc-alkaline volcanic belts. These volcanic rocks are believed to form by partial melting at the subduction zone or, more probably, from the asthenosphere overlying the inclined seismic zone (Marsh and Carmichael, 1974; Ringwood, 1974). Although the cessation of subduction is often clearly related to the latest stages of volcanism, there may be a slight delay between the cessation of subduction and the extinction of the arc's volcanic activity (Snyder et al., 1976). The petrogenetic affinity of the volcanic products in space and time appears to be determined by the plate interaction mechanism and by the nature and thickness of the lithosphere on which the magmatic arc develops. It is therefore difficult to develop a generalized model for the magmatic evolution of a converging system (Arculus and Johnson, 1978).

The present paper is a description of the volcanic cycle that developed within the Anatolian–Iranian sector of the southern margin of the Eurasian plate between the Neogene and the Quaternary. During this time a diachronous continental collision led to the formation of a number of continental microplates, each showing different geotectonic characteristics (McKenzie, 1972; see inset of Fig. 1). The complex history of the convergent interaction between the Eurasian and African plates has controlled the occurrence of volcanic products in space and time. The distribution pattern of volcanic rocks provides an example of a magmatic cycle in an area of active continental collision (McKenzie, 1972; Bird et al., 1975; Nowroozi, 1972).

Structural Outline

The main structural units of the area are represented by two orogenic belts: the Pontus–Minor Caucasus–Alborz to the north, and the Taurus–Central Iran Range to the south (Fig. 1). These two folded ranges surround the block-faulted masses of central and western Anatolia.

The northern orogenic belt marks the continental margin of the Eurasian plate, and was earlier deformed during the Hercynian and the Upper Eocene–Lower Oligocene, during the Pyrenhaic cycle (Brinkmann, 1976).

The southern orogenic belt is a broadly linear articulated orogenic system which is bordered to the south by the Arabian platform, the northern edge of which is deeply folded (Zagros–Border folds–Cyprus Arc–Mediterranean Ridge). Within this sector, between the Afro-Arabian mass and the Pontus orogenic belt, there is a rigid nucleus defined as the Anatolian Median Mass (central and western Anatolia, Fig. 1) in addition to a wide mobile belt—the Taurus Range to the west and the Iranian Range to the east (Pinar-Erdem and Ilhan, 1977). The Taurus–Iranian belt behaved as a single structural element during the Cretaceous and Eocene–Oligocene orogenic phases, but has become segmented since the Oligocene. During the Miocene this orogenic belt became divided into four segments, each characterized by different behaviour:

(1) *Western Taurus*, affected by tangential tectonics and characterized by the emplacement of the Lycian nappes (Brunn et al., 1970).

(2) *Central Taurus*, affected by mainly vertical positive movements (Akarsu, 1960) contrasted with the strong subsidence of the Adana basin (Bizon et al., 1974) in its eastern part.

(3) *Eastern Taurus*, showing an uplifted and deeply folded sector to the south (Altinli, 1966) and a series of intermountain basins to the north (Lange, 1971; Kurtman and Akaus, 1971).

(4) *Central–Western Iran Ranges*, characterized by largely extended and strongly subsiding intermountain basins. The depressions are separated by faulted blocks, deformed during earlier orogenic phases (Stöcklin, 1968; Stöcklin and Nabayi, 1973).

During the Pliocene, strong deformation affected only the eastern Taurus and central-western Iran. The overthrust of the crystalline massif occurred in the eastern Taurus, along the folded margin of the Puturge–Bitlis Arabian platform (Fig. 1: Border Folds; Braud and Ricou,

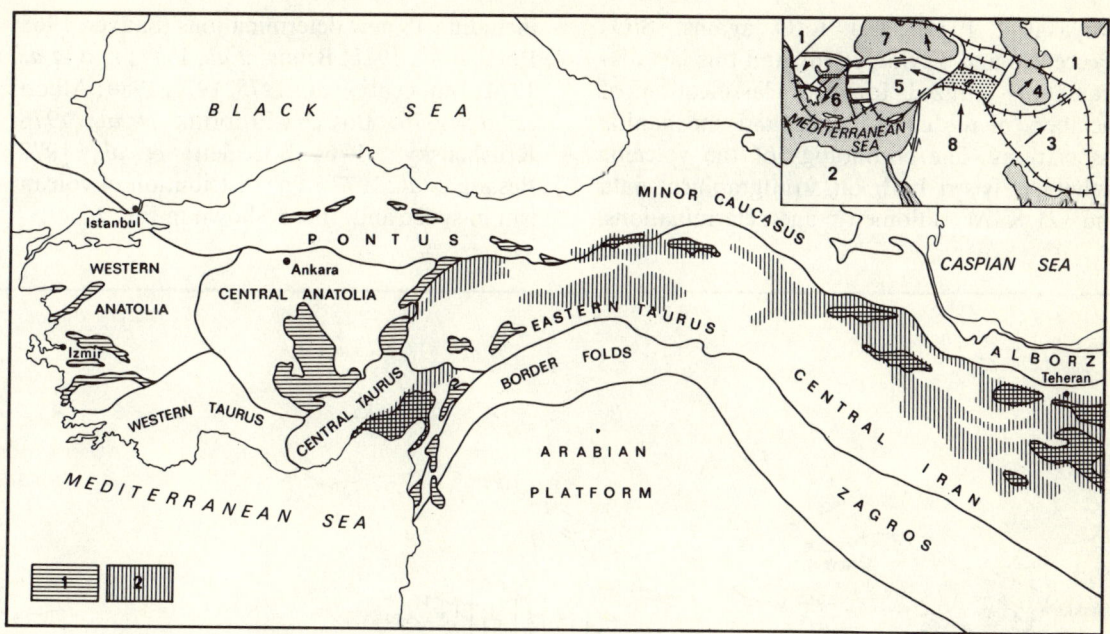

Fig. 1 Structural sketch map of Turkey and north-western Iran. Horizontally hatched areas (1) show Neogene to Quaternary subsiding depressions; vertically hatched areas (2) are Oligo-Miocene molassic basins. *Inset*: plate boundary and motions in Middle East after McKenzie (1977); the dotted area represents the zone of lithosphere deformation of Van (Innocenti *et al.*, 1976); the arrow length is approximately proportional to the velocity relative to Eurasia; plates are labelled as follows: 1, Eurasian; 2, African; 3, Iranian; 4, South Caspian; 5, Turkish; 6, Aegean; 7, Black Sea; 8, Arabian

1975), while transcurrent structures developed, including the North Anatolian Fault and East Anatolian Fault (McKenzie, 1972, 1976; Arpat and Saroglu, 1972; Tchalenko, 1977). In western Iran general contraction occurred, with consequent folding of the intermountain basins and further deformation of the associated uplifted blocks. Transcurrent movements played only a minor role, in spite of their present seismic and tectonic importance (Tchalenko and Braud, 1974).

In such a context the median masses of central Anatolia and western Anatolia (Menderes Massif) acted as rigid cores. Central Anatolia was emergent during most of the Palaeozoic and Mesozoic, showing mainly block-faulting during the Alpine orogenic cycle (Turkish Gulf Oil Co., 1961). The Menderes Massif showed limited mobilization during the initial Alpine phases (Durr *et al.*, 1978), but became welded to the Hercynian massifs to the north during the Pyrenean phase in Eocene time (Bingol, 1971; Fourquin, 1975). During the Neogene, western Anatolia showed rigid behaviour with development of extended coastal grabens, which are related to post-Miocene tensional tectonics in the adjacent Aegean area.

Analytical Data

Chemical and petrographic data on the volcanic products from Anatolia and north-western Iran are irregularly distributed over the area. The following discussion is based on 526 major element analyses (including 129 new analyses mainly from volcanic sequences in eastern Anatolia and north-western Iran). The petrological character of the rocks examined was defined from standard plots including AFM, alkalis against SiO_2 (Irvine and Baragar, 1971), FeO_{tot}, TiO_2, SiO_2, against FeO_{tot}/MgO

(Miyashiro, 1974), and K_2O against SiO_2 (Peccerillo and Taylor, 1976), and this last diagram was utilized for the classification of members of the calc-alkaline and shoshonitic associations. The chronology of the volcanic activity is based both on stratigraphical data and 171 K–Ar radiometric age determinations, including 18 new determinations (Sanver, 1968; Borsi et al., 1972; Benda et al., 1974; Leo et al., 1974; Innocenti et al., 1975, 1976, 1980; Alberti et al., 1976; Bosscha-Erdbrink et al., 1976; Krushensky, 1976; Boccaletti et al., 1977; Besang et al., 1977). The distribution of volcanism in space and time is shown in Fig. 2.

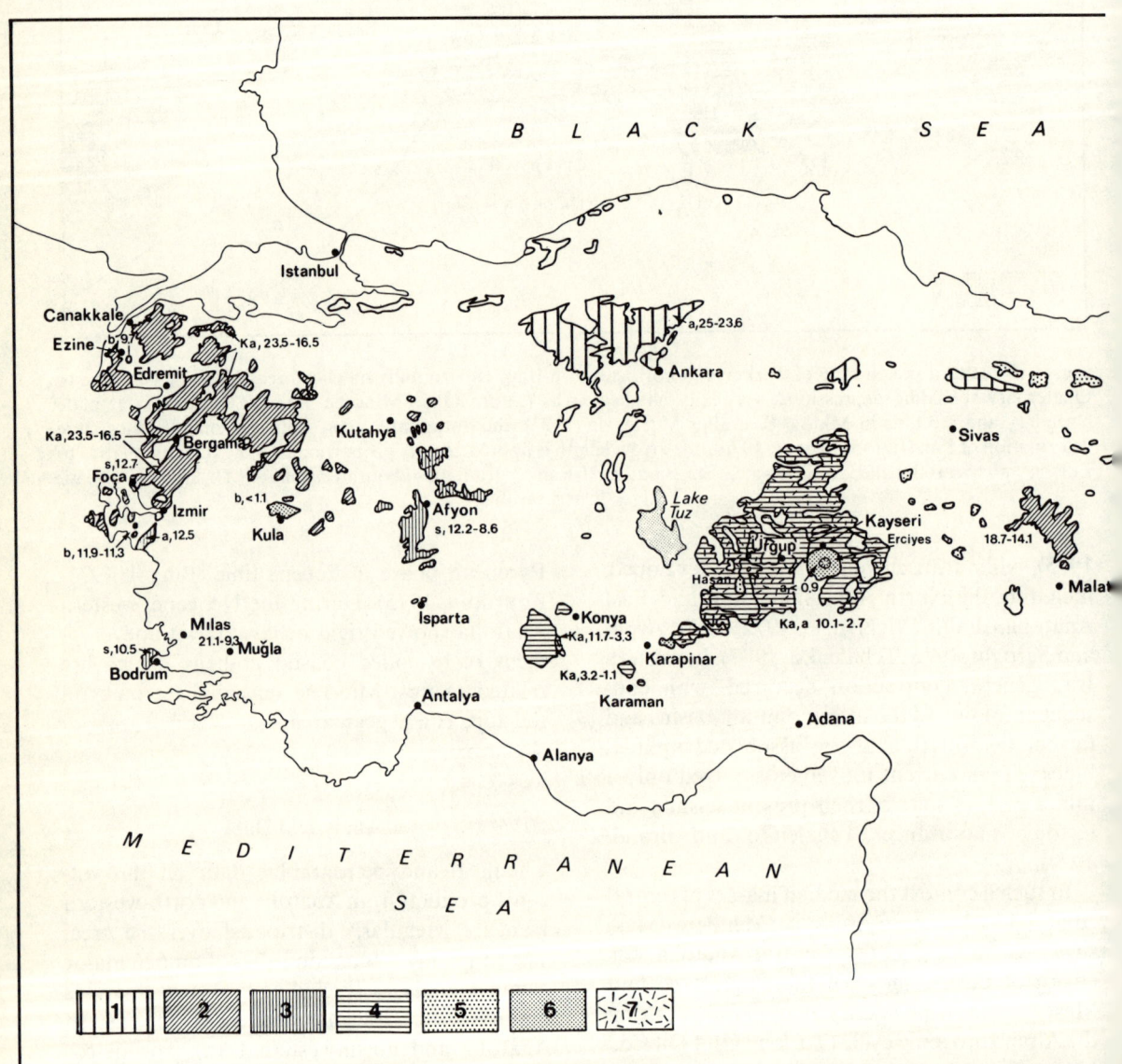

Fig. 2 Neogene and Quaternary volcanism of Anatolia and north-western Iran. *a*, calc-alkaline series; *Ka*, high K calc-alkaline series; *s*, shoshonitic series; *b*, alkaline series; figures represent age intervals by K–Ar dating. *Shading key*: 1, Upper Oligocene–Lower Miocene; 2, Lower–Middle Miocene; 3, Middle–Upper Miocene; 4, Upper Miocene–

Distribution of Neogene and Quaternary Volcanism

The inner margin of the Taurus Zagros Range was affected by widespread volcanic activity which forms a continuous belt consisting mainly of products with calc-alkaline affinity (Fig. 2).

The evolution of volcanism can be related to the structural complexity of the folded belt. Three main sectors can be distinguished in which the volcanic evolution has a relatively uniform character. These are western and central Anatolia and eastern Anatolia–north-western Iran (see Fig. 1). Volcanism is most extensive in the last area (see Fig. 2).

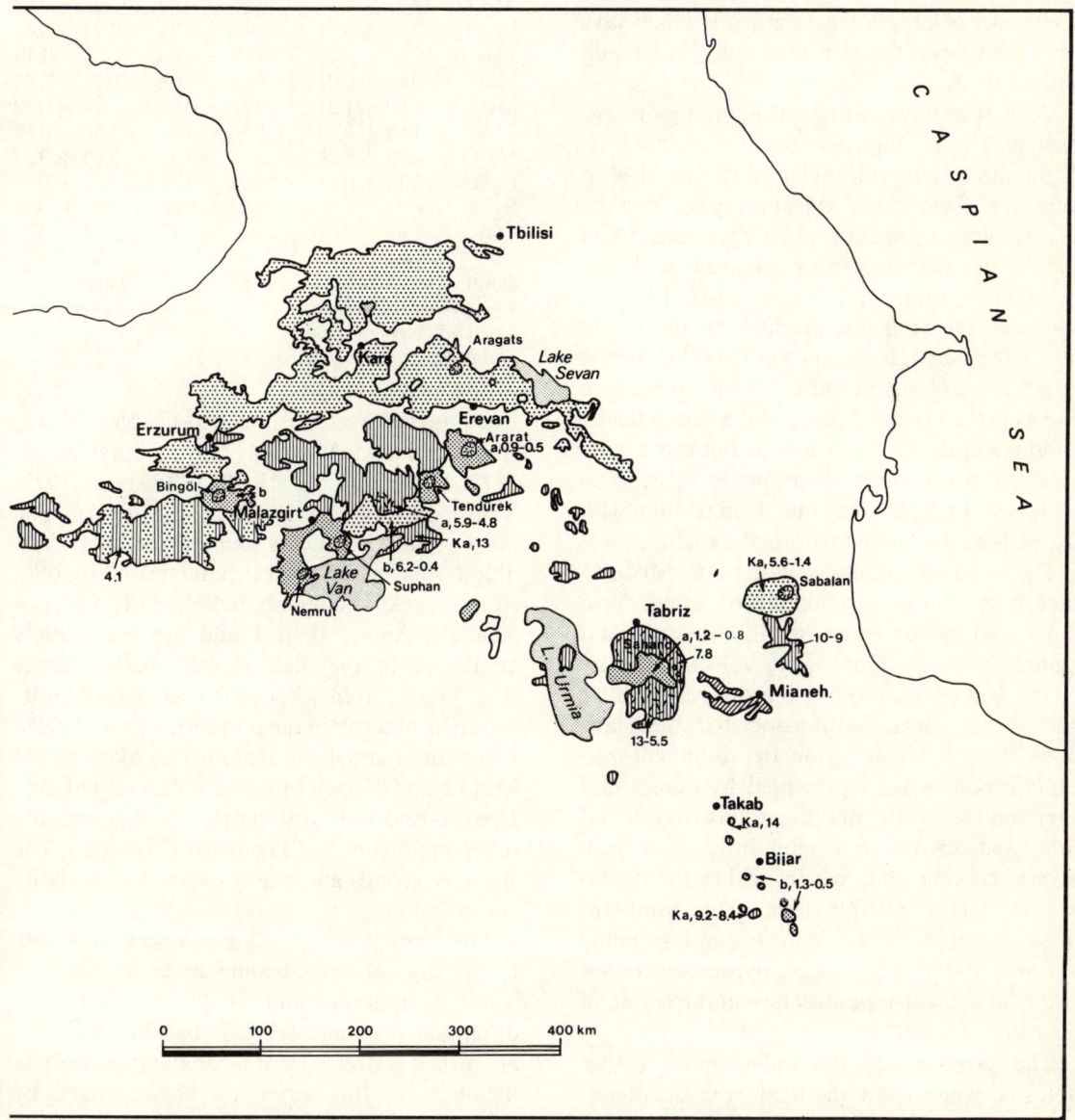

Pliocene; 5, Pilo-Quaternary; 6, Quaternary; 7, dominant pyroclastic products and ignimbrites; ⊕ Quaternary central volcanoes

Western Anatolia

Western Anatolian volcanic rocks are particularly widespread along the Mediterranean coast, forming a belt which extends for c. 200 km inland (Fig. 2). To the east, large volumes of volcanic products outcrop in the Afyon region, where volcanic activity was controlled by north–south regional tectonic trends, marked by the occurrence of large Neogene grabens which have developed since the late Miocene (Keller and Villari, 1972).

Volcanism in western Anatolia ranges in age between Upper Oligocene and Pleistocene (Fig. 2), but most of the volcanic products are of Miocene age. Two main volcanic cycles can be clearly distinguished: the older cycle came to an end in the Middle–Upper Miocene with the eruption of Bodrum volcanics (10.6–10.5 Ma ago) and the youngest products in the Afyon region (8.6 Ma ago; Besang et al., 1977). During this first cycle the volcanic activity reached a climax in the Lower Miocene and was practically absent after the Upper Miocene, but was locally renewed in a younger, less intense Pleistocene cycle (e.g. the Kula area), mainly in relation with the east–west-trending tensional structures.

The volcanic products along the Mediterranean margin consist mainly of domes, lava flows, and agglomerates, with minor tuffs and ignimbrite sheets. Quaternary volcanic activity in the Kula area produced prominent spatter and scoria cones, with associated thin lava flows. In the Afyon region the dominant volcanic products are represented by domes and agglomerates, with subordinate lava flows and tuffs, and an extensive ignimbrite sheet with several cooling units occurs within the northeastern sector of this area. The southernmost outcrops of this north–south-trending volcanic system are domes, pyroclastic rocks, and a large maar-type structure in the region of Isparta.

The products of the older, more active, volcanic phase show the distinctive characteristics of orogenic associations: the lack of any Fe enrichment, a consistently high Al_2O_3 concentration, and relatively low TiO_2 content in basic and intermediate members (Tables 1 and 2) (Innocenti and Mazzuoli, 1972; Borsi et al., 1972; Keller and Villari, 1972; Savascin, 1972; Krushensky, 1976; Burri et al., 1967). The $Na_2O + K_2O$ contents span a wide range. The oldest volcanic rocks are characterized by relatively high K_2O contents and Na_2O/K_2O ratios generally higher than 1 and are accordingly attributed to the high K calc-alkaline series (Fig. 3(a)), even though minor shoshonitic products outcrop in the northern sector (Tuzla-Ezine area) and Foca region. The more recent Middle and Upper Miocene volcanics (Afyon–Isparta–Bodrum) are more alkaline, mainly reflecting higher K_2O contents (Fig. 3(b)). The younger groups are largely of shoshonite–latite composition.

The average composition of rocks belonging to the high K calc-alkaline series is shown in Table 1: andesites and dacites are by far the dominant products, whereas basaltic andesites are rare. It is stressed that basic and intermediate members of this series are characterized by relatively high MgO content and consequently by $Mg/(Mg + Fe_{tot})$ atomic ratios which are distinctly higher than the reported values for

TABLE 1 Western Anatolia (Miocene)—average chemical composition of high K calc-alkaline association

	Basaltic Andesites ($n = 3$)†	Andesites ($n = 44$)	Dacites ($n = 26$)
SiO_2	54.57 ± 0.94‡	60.28 ± 1.57	65.88 ± 1.91
TiO_2	0.84 ± 0.10	0.70 ± 0.13	0.52 ± 0.09
Al_2O_3	15.52 ± 1.36	16.59 ± 0.90	15.61 ± 0.87
Fe_2O_3	2.92 ± 0.81	3.48 ± 1.25	3.23 ± 0.63
FeO	3.60 ± 0.27	1.79 ± 1.00	0.75 ± 0.40
MnO	0.12 ± 0.01	0.09 ± 0.03	0.06 ± 0.02
MgO	7.50 ± 1.27	2.65 ± 0.86	1.14 ± 0.52
CaO	7.88 ± 0.38	5.56 ± 0.83	3.58 ± 0.68
Na_2O	2.89 ± 0.26	3.31 ± 0.33	3.13 + 0.37
K_2O	2.17 ± 0.31	3.01 ± 0.48	3.68 ± 0.56
P_2O_5	0.33 ± 0.08	0.28 ± 0.08	0.24 ± 0.08
LOI	1.66 ± 0.69	1.89 ± 1.22	2.04 ± 0.92
Total	99.99	99.63	99.86

† n is number of samples
‡ Average values ± 1 s.d.

TABLE 2 Western Anatolia (Miocene)—average chemical composition of shoshonitic association

	Shoshonitic basalts ($n = 8$)†	Shoshonites ($n = 13$)	Latites ($n = 31$)	Trachytes ($n = 7$)
SiO_2	50.35 ± 1.99‡	53.68 ± 1.08	59.74 ± 1.79	66.14 ± 1.91
TiO_2	1.15 ± 0.16	1.09 ± 0.30	0.80 ± 0.25	0.57 ± 0.00
Al_2O_3	15.72 ± 1.63	15.59 ± 2.15	15.85 ± 1.03	16.01 ± 0.42
Fe_2O_3	4.72 ± 1.79	4.90 ± 1.31	4.12 ± 1.06	2.32 ± 1.09
FeO	2.26 ± 1.96	1.77 ± 1.27	0.89 ± 0.64	0.99 ± 1.31
MnO	0.14 ± 0.02	0.14 ± 0.05	0.10 ± 0.05	0.05 ± 0.03
MgO	6.45 ± 2.04	4.80 ± 2.37	2.73 ± 1.15	0.97 ± 0.53
CaO	8.52 ± 1.03	6.90 ± 0.85	4.94 ± 0.85	2.55 ± 0.81
Na_2O	3.40 ± 0.59	3.11 ± 0.93	3.52 ± 0.44	3.87 ± 0.82
K_2O	4.05 ± 1.23	4.89 ± 1.27	4.56 ± 0.82	5.28 ± 0.50
P_2O_5	0.60 ± 0.22	0.67 ± 0.19	0.41 ± 0.16	0.24 ± 0.11
LOI	2.42 ± 0.99	2.32 ± 1.31	2.23 ± 0.82	1.20 ± 0.75
Total	99.79	99.86	99.89	100.19

† n is number of samples.
‡ Average values ± 1 S.D.

orogenic associations elsewhere (Ewart, 1976, and this volume, Section II).

The high K calc-alkaline rocks examined are all porphyritic. Basaltic andesites are characterized by a phenocryst association dominated by olivine, clino- and orthopyroxene, and plagioclase. Olivine is absent in andesites, whereas hornblende, with associated biotite, is more common, and these may be the only mafic phenocrysts in the dacites. Sanidine and corroded quartz phenocrysts are sporadically present, mainly in lavas from the northernmost sector (Edremit area). Fe–Ti oxides occur as microphenocrysts in both basaltic andesites and andesites.

More recent volcanic products from the Bodrum area and Afyon–Isparta volcanic system are characterized by K_2O always exceeding 4 per cent and by a Na_2O/K_2O ratio distinctly lower than 1. Their chemical characters indicate that they can be attributed to the shoshonitic association (Table 2; Fig. 3(b)). Latites are dominant, although shoshonites and shoshonitic basalts also occur. The phenocryst association of rocks belonging to this series is characterized by olivine in both shosponitic basalts and shoshonites, and plagioclase and clinopyroxene with associated biotite are present throughout the series. Hornblende and sanidine occur both as phenocrysts, and in the groundmass mainly in latites and trachytes. Leucitites and phonolitic leucitites occur in the Afyon region, forming outcrops of limited extent in the southernmost part of the volcanic complex and also occurring as xenoliths in a thick pyroclastic deposit (Keller and Villari, 1972).

Rhyolites, which are associated in time and space with the shoshonitic volcanic rocks, have been found to the south of Izmir as well as to the north-east of Afyon, where they consist of extensive ignimbrite sheets. Chemical, petrographic, and Sr isotopic data (initial $^{87}Sr/^{86}Sr$ = 0.7121) have been interpreted as evidence for formation of these rocks by crustal anatexis (Keller and Villari, 1972; Borsi et al., 1972). The shoshonitic volcanic rocks, mainly cropping out in Bodrum, Isparta, and Afyon, mark the end of the Tertiary orogenic phase in western Anatolia. Upper Miocene alkaline basalts and hawaiites are locally reported in the area of Urla, southwest of Izmir and Ezine (Table 4).

The second volcanic cycle, with a distinct sodic-alkaline affinity, has developed in association with the tensional tectonics which affected this area from Upper Miocene to Quaternary. The most abundant products of this activity are

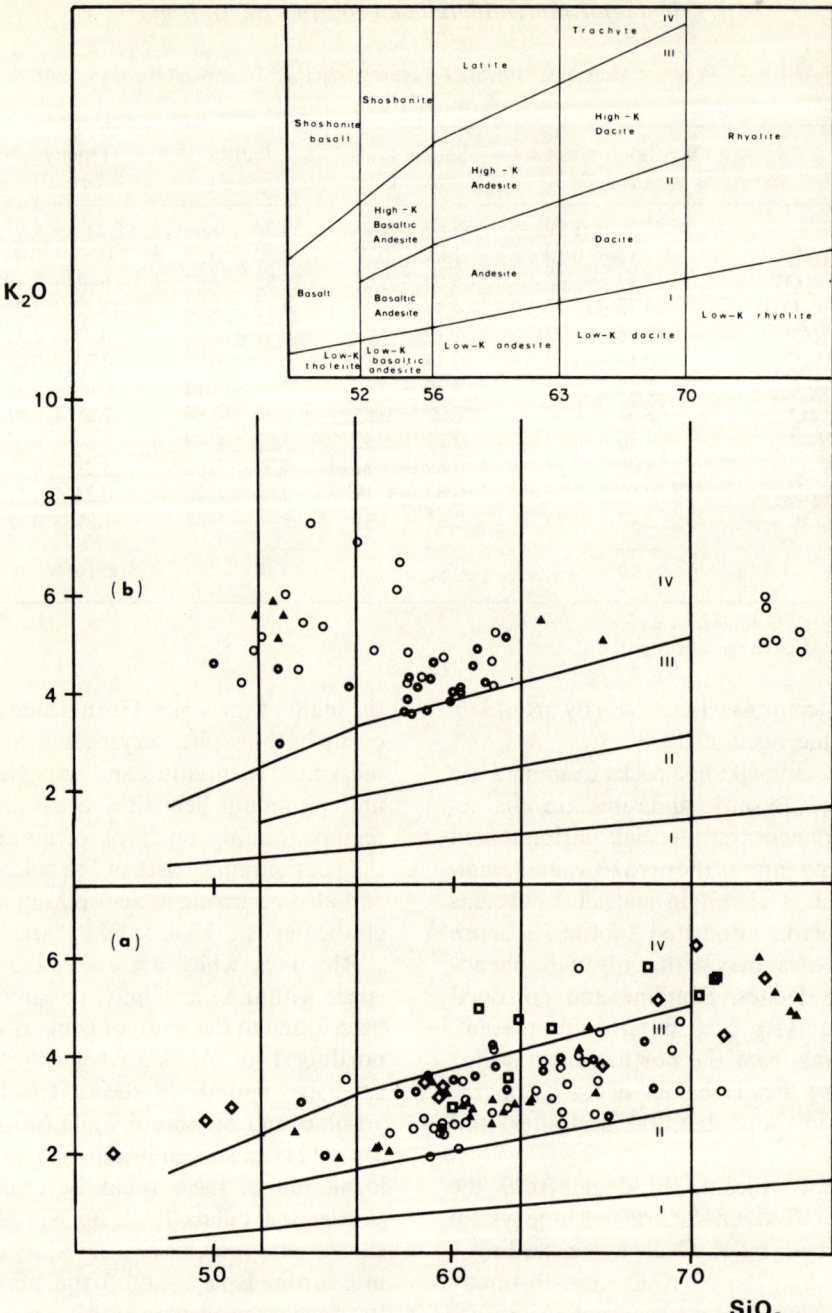

Fig. 3 Graphs of K_2O versus SiO_2 for the western Anatolia Miocene orogenic association. (a) Older products. Data sources: □, Ezine area (Borsi *et al.*, 1972); ○, Edremit–Balikesir area (Krushensky, 1966); **O**, Bergama area (Borsi *et al.*, 1972); ◇, Foça area (Savascin, 1972; authors' unpublished data); ▲, Izmir area (Innocenti and Mazzuoli, 1972). (b) Younger products. Data sources: **O**, Bodrum area (Burri *et al.*, 1967; authors' unpublished data); ○, Afyon area (Keller and Villari, 1972; authors' unpublished data); ▲, Isparta (authors' unpublished data). Inset represents the adopted classification after Peccerillo and Taylor (1976) modified. I, island-arc tholeiitic series; II, calc-alkaline series; III, high K calc-alkaline series; IV, shoshonitic series

the Quaternary volcanic rocks of Kula, which occur along the northern margin of the east–west-trending Salihili Graben affecting the northern edge of the Menderes Massif ('Neogene–Quaternary subsiding depressions' in Fig. 1). Volcanic rocks from Kula are alkaline and undersaturated (Table 4), and are characterized by abundant clinopyroxene, olivine, Ti-magnetite, and kaersutitic amphibole phenocrysts. Labradoritic plagioclase is always restricted to the groundmass, together with nepheline and alkali feldspar.

Central Anatolia

Volcanic products in this region occur mainly along a north-west to south-east trend, running for c. 350 km from Karaman to the north-east of Kayseri. This volcanic belt attains its maximum width, of more than 100 km, in its northernmost part and, with the exception of the Konya volcanic complex west of Karaman, it forms an almost continuous range. The volcanic rocks are mainly domes and lava flows with subordinate tuffs and agglomerates, as well as several ignimbrite units which have their maximum area and volume in the Urgup Basin (Innocenti et al., 1975). In this region large central volcanoes also occur, the most important of which is Erciyes composite volcano. Phreatic and phreato-magmatic eruptions were particularly common in the southernmost part of the belt, producing typical hyaloclastite tuff-rings and maar craters, occasionally with associated pyroclastic surge deposits (Keller, 1974).

The activity of this volcanic belt is well documented as continuous from the Middle Miocene to the Quaternary. During the Pleistocene, volcanism was restricted to the Hasan and Erciyes central volcanoes, and to small monogenetic volcanic centres in the Karapinar area (Innocenti et al., 1975; Besang et al., 1977). The westernmost volcanic products of the Konya Massif range in age from Miocene to Upper Pliocene.

Rocks belonging to this belt form a well defined association, with a marked calc-alkaline affinity (Tables 3 and 4; Figs. 4 and 5). The K_2O

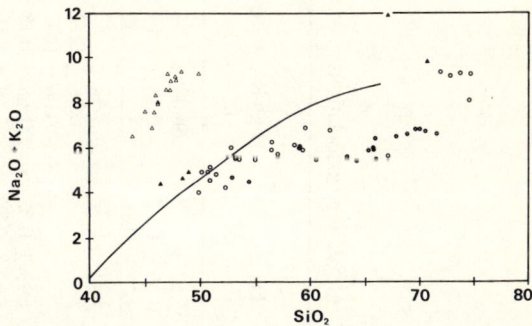

Fig. 4 Graph of $Na_2O + K_2O$ against SiO_2 for Quaternary volcanics in western and central Anatolia: dividing line from Irvine and Baragar (1971). △, Kula area (authors' unpublished data); ○, Karapinar and Acigol areas (Keller, 1974; Jung et al., 1972); ●, Erciyes Dag volcano (Innocenti et al., 1975; Ayranci and Weibel, 1973). The Upper Miocene alkaline volcanics of Urla and Ezine areas are also reported: ▲ (Innocenti and Mazzuoli, 1972)

contents are variable; volcanic rocks erupted from Quaternary central volcanoes (Erciyes Dag, Karapinar, and Acigol, west of Urgup; Fig. 6(b)) plot into the calc-alkaline field together with the rocks of Hasan Dag and most of the lavas from Urgup Basin. All the volcanic products from the Konya and Karadag areas, as well as some of the Urgup volcanics, plot into the high K calc-alkaline field (Fig. 6(a)). The distribution of these volcanic centres suggests a westward increase of K_2O content, almost

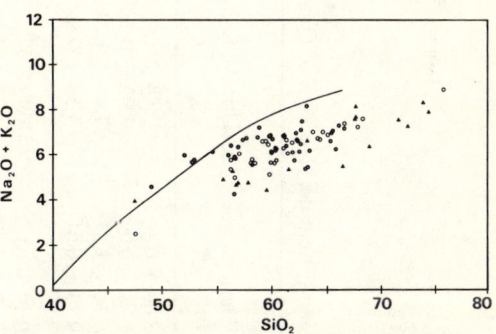

Fig. 5 Graph of $Na_2O + K_2O$ against SiO_2 for the Upper Miocene to Quaternary volcanics of central Anatolia. ●, Hasan Dag–Karadag area (Jung et al., 1972; Schleicher and Schwarz, 1977); ○, Konya area (Jung et al., 1972; Keller et al., 1977); ▲, Urgup Basin (Innocenti et al., 1975; Ayranci and Weibel, 1973)

TABLE 3 Central Anatolia (Neogene to Quaternary)—average chemical composition of calc-alkaline and high K calc-alkaline associations

	Basalts (n = 3)†	Basaltic Andesites (n = 3)	High K Basaltic Andesites (n = 3)	Andesites (n = 9)	High K Andesites (n = 50)	Dacites (n = 7)	High K Dacites (n = 18)
SiO_2	48.00 ± 0.84‡	53.44 ± 1.76	54.34 ± 1.66	58.91 ± 2.61	59.63 ± 2.00	65.52 ± 1.58	66.08 ± 1.84
TiO_2	1.28 ± 0.09	1.02 ± 0.27	0.93 ± 0.15	0.74 ± 0.13	0.70 ± 0.15	0.46 ± 0.07	0.48 ± 0.17
Al_2O_3	16.74 ± 1.22	17.93 ± 1.24	15.79 ± 1.15	17.37 ± 0.87	16.63 ± 0.80	16.23 ± 0.86	15.22 ± 1.08
Fe_2O_3§	9.73 ± 0.61	7.34 ± 0.83	7.37 ± 1.12	3.29 ± 1.06¶	4.07 ± 0.83††	3.98 ± 0.73	2.42 ± 0.92
FeO				2.35 ± 0.85¶	1.71 ± 0.76††	—	0.98 ± 0.52
MnO	0.19 ± 0.06	0.11		0.11 ± 0.02‖	0.09 ± 0.02‡‡		0.06 ± 0.02
MgO	8.57 ± 1.58	5.59 ± 0.76	5.29 ± 1.60	3.93 ± 1.45	3.04 ± 0.97	1.93 ± 0.52	1.68 ± 0.78
CaO	11.28 ± 0.80	8.55 ± 1.32	8.27 ± 1.11	6.71 ± 0.99	6.19 ± 1.24	5.02 ± 0.80	3.50 ± 1.10
Na_2O	2.80 ± 1.41	4.39 ± 0.38	3.79 ± 0.67	3.60 ± 0.58	3.53 ± 0.45	3.99 ± 0.66	3.65 ± 0.61
K_2O	0.89 ± 0.41	1.18 ± 0.17	2.13 ± 0.43	1.57 ± 0.41	2.78 ± 0.39	2.30 ± 0.27	3.34 ± 0.55
P_2O_5	0.25 ± 0.03	0.30 ± 0.06	0.43 ± 0.22	0.19 ± 0.07	0.26 ± 0.09§§	0.16 ± 0.12	0.16 ± 0.07
LOI	0.50 ± 0.47	0.53 ± 0.92	1.56 ± 1.71	1.33 ± 0.38	1.21 ± 0.95	1.22 ± 0.74	2.16 ± 2.09
Total	100.23	100.38	99.90	100.10	99.84	100.81	99.71

† n is number of samples.
‡ Average value ± 1 s.d.
§ Fe_2O_3 is total iron when FeO is not reported.
¶ On eight samples.
‖ On six samples.
†† On 37 samples.
‡‡ On 32 samples.
§§ On 49 samples.

TABLE 4. Miocene and Quaternary alkaline volcanic rocks (western Anatolia) and Quaternary calc-alkaline association (Central Anatolia)

	Western Anatolia—Alkaline Association					Central Anatolia—Calc-Alkaline Association					
	Miocene		Quaternary			Karapinar–Acigol		Quaternary		Ercyes Dag	
	Ezine	Urla	Kula								
	T76	K165	K136	K142	K130	Basalts (n = 6)†	Basaltic Andesites (n = 4)	Andesites (n = 10)	Rhyolites (n = 5)	Basaltic Andesites (n = 3)	Dacites (n = 10)	Rhyolites (n = 3)
SiO_2	46.38	49.03	43.90	45.00	46.93	50.67 ± 0.57‡	53.37 ± 1.16	58.81 ± 1.76	73.48 ± 1.14	53.49 ± 0.80	66.38 ± 1.99	70.76 ± 0.77
TiO_2	2.71	1.63	2.42	2.48	2.03	1.12 ± 0.16	1.07 ± 0.21	0.70 ± 0.05	0.09 ± 0.05	1.16 ± 0.41	0.47 ± 0.16	0.38 ± 0.04
Al_2O_3	12.89	16.44	15.75	15.94	17.79	17.27 ± 1.05	16.71 ± 1.07	15.76 ± 0.46	13.11 ± 0.70	17.23 ± 1.12	15.98 ± 1.20	14.48 ± 0.85
Fe_2O_3	4.86	3.49	4.99	3.90	3.15	8.28 ± 0.89	3.06 ± 1.18	1.99 ± 0.38	1.27 ± 0.33	4.15 ± 0.56	1.68 ± 1.08	0.93 ± 0.37
FeO	6.65	6.36	4.35	5.28	4.80		4.52 ± 0.36	3.54 ± 0.59		3.74 ± 0.73	1.57 ± 0.51	1.49 ± 0.27
MnO	0.19	0.17	0.15	0.58	0.15	0.13 ± 0.01				0.21 ± 0.13	0.07 ± 0.03	0.07 ± 0.01
MgO	8.28	5.40	7.96	77.80	6.06	7.05 ± 1.04	6.05 ± 1.52	5.16 ± 0.91		5.43 ± 1.11	1.82 ± 0.48	1.08 ± 0.13
CaO	10.12	8.71	10.71	9.60	8.39	10.51 ± 0.75	8.80 ± 0.91	7.06 ± 0.78	0.96 ± 0.36	8.06 ± 0.62	4.46 ± 1.05	3.34 ± 0.28
Na_2O	2.86	3.41	4.14	4.99	5.37	3.79 ± 0.41	4.14 ± 0.71	4.11 ± 0.42	4.67 ± 0.51	3.85 ± 0.33	3.77 ± 0.21	3.90 ± 0.10
K_2O	1.58	1.56	2.47	2.70	3.31	0.96 ± 0.09	1.18 ± 0.34	2.02 ± 0.22	4.43 ± 0.43	1.05 ± 0.40	2.29 ± 0.35	2.81 ± 0.09
P_2O_5	0.67	0.30	1.32	1.32	0.83	0.27 ± 0.11	0.48 ± 0.19	0.20 ± 0.19		0.32 ± 0.09	0.12 ± 0.04	0.10 ± 0.02
LOI	2.41	3.18	1.76	0.32	0.55	0.25 ± 0.21	0.35 ± 0.17	0.50 ± 0.29	1.36 ± 1.01	1.27 ± 0.93	1.35 ± 1.25	0.62 ± 0.16
	99.60	99.68	99.92	99.91	99.36	100.30	99.73	99.85	99.37	99.98	99.97	96.96

† n is number of samples.
‡ Average value ± 1 S.D.

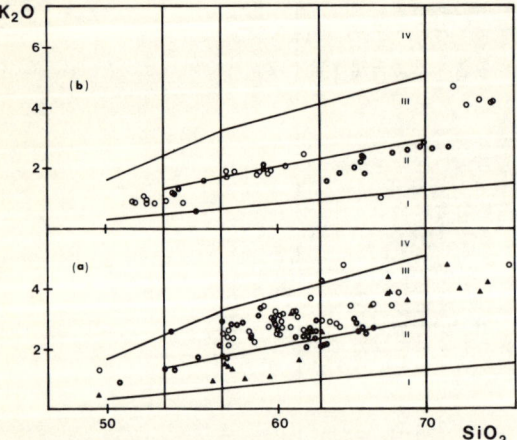

Fig. 6 Graphs of K_2O against SiO_2 for Neogene to Quaternary volcanics of central Anatolia. (a) Upper Miocene to Quaternary; ○, Hasan Dag–Karadag area; ○, Konya areas; ▲, Urgup Basin; for data sources see Fig. 5. (b) Quaternary; ○, Karapinar and Acigol area; ○, Erciyes Dag volcano; for data source see Fig. 4

perpendicular to the trend of the orogenic belt. The average chemical composition of Neogene and Quaternary products is reported in Tables 3 and 4. Members of the calc-alkaline series show a chemical composition similar to the average chemical composition of volcanic products from active continental margins (Ewart, 1976), although (as in western Anatolia) MgO and hence $MgO/(MgO + FeO_{tot})$ are higher than other calc-alkaline volcanic rocks.

In the calc-alkaline series all the rocks are porphyritic in texture. The groundmass, commonly containing glass, typically forms 40–75 per cent of the rocks. Basalts are characterized by phenocrysts of olivine, pyroxene, and plagioclase, and slightly resorbed and pyroxene-rimmed olivine which occasionally occurs in basaltic andesites and more rarely in andesites, where orthopyroxene is always present. Hornblende occurs sporadically in andesites and dacites and is the only mafic mineral in the rhyolites. In the high K calc-alkaline series hornblende phenocrysts occur throughout, even in basaltic andesites where they are associated with clino- and orthopyroxene, rare embayed olivine, plagioclase, and Fe–Ti oxide; however, frequent subordinate biotite also occurs.

Minor amounts of corroded quartz are often reported in dacites.

Eastern Anatolia and North-western Iran

In this sector volcanic products outcrop in a continuous belt extending for c. 900 km from the Sivas–Malatya region, in Turkey, to Bijar in north-western Iran (Fig. 2). The maximum width is attained in the central part, where volcanic rocks extend northward, for more than 350 km, from the Van area to Armenia.

Volcanic landforms are dominated by the occurrence of numerous domes and lava flows locally forming a continuous plateau-like cover, which is widest in the Kars region. In this area the plateau lavas are deeply eroded and form a stratigraphic sequence consisting mainly of volcanic rocks with subordinate sedimentary layers; pillow lavas and hyaloclastites are locally observed. Several ignimbrite units are present at the base of this section. Wide ignimbrite sheets also occur to the north-west of Lake Van and along the border between Turkey and Soviet Armenia. The pyroclastic basement of the Sahand volcanic complex (north-west Iran) is also extensive. This volcanic structure may be regarded as a system of adjoining volcanic domes, rather than a central volcano, as previously reported (Gansser, 1966). Large central volcanoes of Pliocene and Quaternary age occur in the central part of the belt (Fig. 2), often characterized by large summit calderas (Nemrut, Tendurek, and Sabalan). The volcanism of this sector is believed to range in age from Middle Miocene to Quaternary; it is nevertheless possible to distinguish areas with different volcanic evolution. Before examining the volcanic history of these areas, it must be noted that very few geochronological, stratigraphical, and petrographic data are available on both the south-western segment and the region immediately to the east of Erzurum (Sanver, 1968; Pasquarè, 1971; Leo et al., 1974; Ota and Dincel, 1975) and the assumed age of the volcanism is therefore largely tentative.

The oldest exposed volcanic products form a relatively narrow belt, running along the eastern Taurus to central Iran. This volcanism is mainly of Middle–Upper Miocene age; Lower Miocene products are reported in the Mianeh and Sivas areas (Lescuyer et al., 1974), to the south-eastern and western margins respectively. In the Van region the lowermost part of the main volcanic sequence consists of domes and lava flows with subordinate agglomerate, while the upper part is characterized by large ignimbrite units (Innocenti et al., 1980). Upper Miocene ignimbrites and pyroclastic rocks are present in the Sahand area (Bosscha-Erdbrink et al., 1976).

During this volcanic cycle, rocks of mainly calc-alkaline to shoshonitic affinity were erupted, while dacites belonging to the high K calc-alkaline series are dominant (Table 5; Fig. 7(a)). The most common phenocryst paragenesis consists of plagioclase, abundant hornblende and biotite, and rare quartz; anhydrous para-geneses with clino- and orthopyroxene as mafic phenocrysts also occur. Sanidine is always observed in the relatively potassic ignimbrites. Furthermore, on the eastern margin of the belt, to the south of Sabalan volcano, is a small plateau of Upper Miocene sodic-alkaline lavas (Comin-Chiaramonti et al., 1978).

The most widespread volcanic episode in eastern Anatolia–north-western Iran developed between the Pliocene and the Quaternary. To the north of the Taurus Range, along the Pontus–Minor Caucasus chain, an andesite belt extends from Erzurum through the Kars plateau into Soviet Armenia and further on towards the Sabalan volcano. Such a belt is defined, in the following discussion, as the Northern Belt, in order to make a distinction with the Southern Belt (around Lake Van). Several central volcanoes occur in the Northern Belt, e.g. Aragats and Sabalan, and Ararat may belong to this belt. Volcanic products erupted in the Kars–Ararat area belong to a calc-alkaline series, dominated by andesitic members (Table 6; Fig. 7(b)). Rocks of the high K calc-alkaline series occur in the uppermost part of the volcanic sequence and Sabalan consists entirely of high K calc-alkaline rocks. Quaternary shoshonitic volcanism, including leucite-bearing rocks is reported in Soviet Armenia (Aslanian, 1977; Adamia et al., 1977).

Basalts of the calc-alkaline series are characterized by olivine, clinopyroxene, plagioclase, and Fe–Ti oxide phenocrysts. Resorbed olivine is still present in basaltic andesites and, frequently, even in andesites. Pyroxene andesites are peculiar to this calc-alkaline association, while hornblende sporadically occurs, mainly

TABLE 5   Eastern Anatolia and north-western Iran (Miocene)—representative chemical analyses of high K calc-alkaline and shoshonitic associations

	Van Area				Bijiar Area			
	AG 50	AG 63	AG 109	AG 125	IR 139	IR 143	IR 146	IR 149
SiO_2	67.49	66.40	68.45	73.66	62.21	62.61	64.41	65.78
TiO_2	0.52	0.64	0.63	0.29	0.60	0.77	0.73	0.51
Al_2O_3	17.27	15.17	14.41	13.06	15.59	14.65	14.23	15.66
Fe_2O_3	2.48	1.67	1.35	0.78	1.79	3.42	2.32	1.60
FeO	0.44	1.57	1.41	0.56	2.24	0.79	1.06	1.57
MnO	0.02	0.07	0.11	0.05	0.07	0.12	0.11	0.11
MgO	0.67	1.11	0.50	0.23	3.81	3.20	2.82	2.22
CaO	3.03	2.51	0.77	0.63	5.34	4.25	3.29	3.86
Na_2O	3.71	4.12	5.12	3.96	4.26	3.91	3.51	4.18
K_2O	1.86	4.55	4.33	4.50	2.32	4.54	5.05	2.80
P_2O_5	0.12	0.19	0.11	0.04	0.24	0.68	0.78	0.20
LOI	2.38	2.01	2.82	2.24	1.53	1.07	1.49	1.67
Total	99.99	100.01	100.01	100.00	100.00	100.01	99.80	100.16

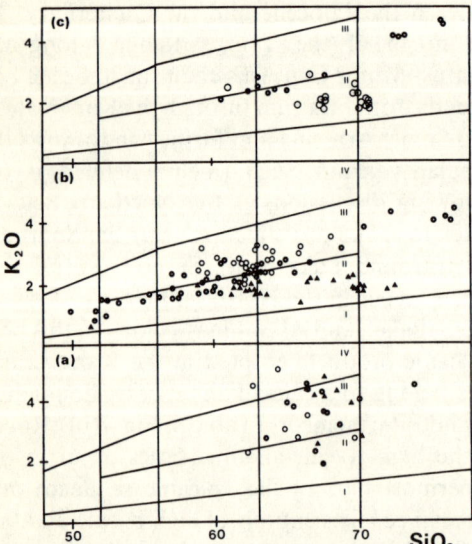

Fig. 7 Graphs of K_2O against SiO_2 for Neogene to Quaternary orogenic associations of eastern Anatolia and north-western Iran. (a) Miocene: ●, Van area (Innocenti *et al.*, 1976, 1980; authors' unpublished data); ○, Bijar area (Boccaletti *et al.*, 1977); ▲, ignimbrite from Van and Sahand areas (Innocenti *et al.*, 1976; authors' unpublished data). (b) Plio-Quaternary: ○, Kars area (authors' unpublished data); ▲, Ararat volcano (Lambert *et al.*, 1974; authors' unpublished data); ○, Sabalan volcano (Alberti *et al.*, 1975; Didon and Gemain, 1976). (c) Quaternary: ●, Suphan Dag volcano (Innocenti *et al.*, 1976; authors' unpublished data); ○, Sahand volcanic complex (authors' unpublished data)

Plio-Quaternary volcanism also developed in the Southern Belt, attaining its widest extension to the north of Lake Van. Volcanic activity began in the Lower Pliocene with the eruption of products showing a well defined sodic-alkaline affinity (ne-normative and hy-normative): small lava plateaux were built up, as well as more recent central volcanoes, i.e. Tendurek and Nemrut. Volcanic products range in composition from alkaline basalts to benmoreites (Table 7; Fig. 8). Despite the persistence of the alkaline volcanic activity, resumption of calc-alkaline volcanism took place during the Quaternary (Suphan Dag and Sahand), together with local eruption of volcanic products showing a transitional or tholeiitic affinity (Malazgirt Graben; Innocenti *et al.*, 1980). Strongly undersaturated, mainly potassic, alkaline products (Table 7) outcrop more to the south-east in the Bijar region (Boccaletti *et al.*, 1977). They may have been erupted through a fault system running parallel to the 'Main Recent Fault'. Both the Suphan and Sahand Quaternary central volcanoes consist of calc-alkaline rocks among which dacites and rhyolites are dominant (Table 8; Fig. 7(c)). These have mineralogy analagous to equivalent rocks from the Northern Belt.

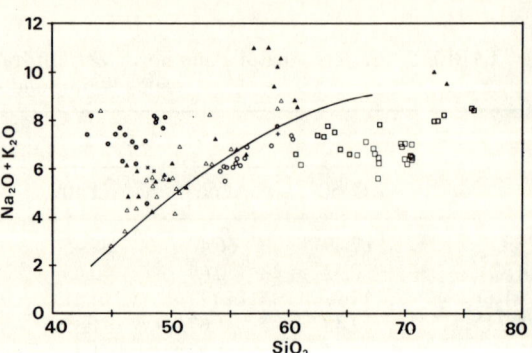

Fig. 8 Graph of $Na_2O + K_2O$ against SiO_2 for Pliocene and Quaternary volcanics from the Southern Belt (eastern Anatolia and north-western Iran). ○, Bijar area; △, Van area (hy-normative alkaline); ▲, Van area (ne-normative alkaline); ○, Malazgirt Graben (Van area); +, Aravil volcano (Ota and Dincel, 1975); ×, Bingol area (authors' unpublished data); ■, Suphan Dag volcano; □, Sahand volcanic complex. Data sources as in Fig. 7, if not otherwise reported

in dacitic rocks. Rhyolites, including both perlite and obsidian lava flows and domes, are common and are characterized by a phenocryst paragenesis consisting of plagioclase and quartz with associated sanidine and, sometimes, augite and/or biotite. Rocks belonging to the high K calc-alkaline series and in particular those from the Sabalan volcano, are characterized by relative abundance of hydrous phases, hornblende, and minor biotite, with associated plagioclase and rare augite. In the Kars plateau as well as in Soviet Armenia, scattered occurrences of recent alkaline basalts (Table 7) are reported, forming to spatter cones and lava flows of limited extent.

TABLE 6 Turkish Armenia and north-western Iran (Pliocene to Quaternary)—average chemical composition of calc-alkaline and High K calc-alkaline associations

			Kars Area					Ararat			Sabalan	
	Basalts ($1 = 3$)†	Basaltic Andesites ($n = 5$)	Andesites ($n = 16$)	High K Andesites ($n = 5$)	Dacites ($n = 5$)	High K Dacites ($n = 6$)	Rhyolites ($n = 6$)	Basalts ($n = 1$)	Andesites ($n = 9$)	Dacites ($n = 14$)	Rhyolites ($n = 5$)	High K Andesites ($n = 14$)	High K Dacites ($n = 5$)
SiO_2	51.28 ± 0.91†	53.87 ± 1.46	58.97 ± 2.41	59.21 ± 1.50	64.44 ± 1.57	64.98 ± 2.10	74.53 ± 2.51	51.56	61.89 ± 1.19	67.36 ± 2.60	71.22 ± 0.91	61.53 ± 1.59	65.21 ± 1.94
TiO_2	1.24 ± 0.23	1.32 ± 0.43	1.02 ± 0.33	1.34 ± 0.28	0.77 ± 0.13	0.77 ± 0.15	0.19 ± 0.10	1.63	1.08 ± 0.40	0.64 ± 0.18	0.42 ± 0.05	0.72 ± 0.41	0.54 ± 0.12
Al_2O_3	17.07 ± 1.10	17.40 ± 0.57	17.00 ± 0.49	16.60 ± 0.42	15.78 ± 0.36	15.28 ± 0.53	13.15 ± 0.44	18.30	16.88 ± 0.08	15.84 ± 0.65	14.54 ± 0.61	17.01 ± 0.41	16.00 ± 0.62
Fe_2O_3	3.07 ± 1.00	4.19 ± 0.74	3.29 ± 1.61	3.11 ± 1.14	2.81 ± 1.01	2.09 ± 0.63	1.05 ± 0.65	9.70	5.59 ± 0.41	3.93 ± 0.97	2.80 ± 0.19	3.38 ± 0.34	1.68 ± 0.50
FeO	5.28 ± 0.51	3.19 ± 0.87	2.92 ± 1.51	3.37 ± 1.38	1.61 ± 0.80	1.61 ± 1.05	0.30 ± 0.17					1.02 ± 0.51	1.38 ± 0.37
MnO	0.16 ± 0.01	0.16 ± 0.03	0.12 ± 0.02	0.15 ± 0.02	0.08 ± 0.01	0.11 ± 0.03	0.11 ± 0.11	0.16	0.10 ± 0.01	0.09 ± 0.02	0.06 ± 0.01	0.05 ± 0.01	0.05 ± 0.01
MgO	5.54 ± 0.72	5.37 ± 1.36	3.36 ± 0.71	2.28 ± 0.41	2.19 ± 0.53	1.84 ± 0.67	0.36 ± 0.27	3.80	2.02 ± 0.67	1.13 ± 0.45	0.86 ± 0.24	2.17 ± 0.39	1.42 ± 0.31
CaO	8.56 ± 1.07	7.88 ± 0.91	6.15 ± 0.70	5.27 ± 0.68	4.55 ± 0.66	3.95 ± 0.96	0.63 ± 0.18	9.50	5.82 ± 0.33	4.46 ± 0.57	3.80 ± 0.56	4.68 ± 0.58	3.40 ± 0.63
Na_2O	3.55 ± 0.39	3.45 ± 0.37	3.69 ± 0.55	3.85 ± 0.15	3.40 ± 0.24	3.80 ± 0.29	3.57 ± 0.87	4.50	4.61 ± 0.28	4.42 ± 0.29	4.18 ± 0.28	5.06 ± 0.15	4.93 ± 0.18
K_2O	1.04 ± 0.19	1.46 ± 0.17	1.88 ± 0.22	2.49 ± 0.28	2.42 ± 0.18	2.76 ± 0.24	4.20 ± 0.21	0.60	1.84 ± 0.11	1.94 ± 0.27	2.00 ± 0.34	2.75 ± 0.30	3.19 ± 0.28
P_2O_5	0.29 ± 0.03	0.41 ± 0.14	0.29 ± 0.14	0.45 ± 0.13	0.20 ± 0.03	0.22 ± 0.06	0.08 ± 0.07	0.24	0.23 ± 0.04	0.18 ± 0.08	0.11 ± 0.04	0.45 ± 0.07	0.33 ± 0.03
LOI	2.91 ± 1.63	1.17 ± 0.34	1.31 ± 0.45	1.87 ± 0.68	1.76 ± 0.41	2.10 ± 0.50	1.93 ± 2.04					0.75 ± 0.39	1.43 ± 0.65
Total	99.99	99.87	100.00	99.99	100.01	100.01	100.10	99.99	100.06	99.99	99.99	59.57	99.56

† n is number of samples.
‡ Average value ± 1 S.D.

TABLE 7 Eastern Anatolia and north-western Iran (Pliocene to Quaternary)—representative chemical analyses of alkaline association

		Van Area						Kars Area		B;jar Area		
	AG 52	AG 2	BIN 72	AG 53	AG 122	AG 40	DY 1	AG 223	AG 237	IR 114	IR 117	IR 124
SiO_2	46.33	45.42	48.48	50.56	53.78	57.19	72.51	46.10	49.97	42.90	44.72	49.45
TiO_2	1.47	2.02	2.20	1.49	2.24	1.30	0.29	2.24	2.48	1.90	2.01	1.72
Al_2O_3	17.94	18.60	17.58	18.06	16.63	17.52	11.75	17.69	18.33	15.80	14.47	14.17
Fe_2O_3	3.24	2.33	2.59	4.86	3.89	3.48	1.92	11.95	2.22	3.79	3.58	3.92
FeO	6.00	8.37	7.82	4.53	5.47	3.66	2.41	0.43	8.33	5.06	5.35	3.64
MnO	0.18	0.18	0.18	0.14	0.19	0.20	0.13	0.22	0.22	0.14	0.14	0.14
MgO	7.24	4.38	5.67	5.45	2.91	0.88	0.01	5.66	4.31	9.09	8.99	7.80
CaO	10.13	9.72	7.88	8.50	5.47	3.46	0.32	8.74	7.16	9.77	9.97	8.79
Na_2O	3.43	5.87	4.41	4.17	5.07	7.23	5.55	3.64	4.16	6.09	4.42	5.08
K_2O	0.87	1.49	1.52	0.87	2.21	3.80	4.53	0.65	1.25	1.40	2.71	3.13
P_2O_5	0.40	0.84	0.67	0.46	0.82	0.22	0.01	0.73	0.65	1.79	1.52	1.64
LOI	2.78	0.77	1.02	0.91	1.32	1.07	0.57	1.95	0.93	2.26	2.11	0.52
Total	100.01	99.99	100.00	100.00	100.00	100.01	100.00	100.00	100.01	99.99	99.99	100.00

TABLE 8 Eastern Anatolia and north-western Iran (Quaternary)—average chemical composition of calc-alkaline volcanoes

	Suphan Dag			Sahand		
	Andesites ($n = 2$)†	Dacites ($n = 6$)	Rhyolites ($n = 5$)	Andesites ($n = 1$)	Dacites ($n = 9$)	Rhyolites ($n = 8$)
SiO_2	61.54 ± 1.30‡	64.60 ± 1.78	74.20 ± 1.55	61.20	67.92 ± 1.26	70.55 ± 0.30
TiO_2	1.11 ± 0.66	0.84 ± 0.14	0.17 ± 0.07	0.87	0.39 ± 0.06	0.26 ± 0.02
Al_2O_3	16.31 ± 0.15	16.04 ± 0.50	13.42 ± 0.32	16.24	15.89 ± 0.56	15.40 ± 0.28
Fe_2O_3	2.15 ± 0.08	1.82 ± 0.68	0.78 ± 0.26	3.61	1.67 ± 0.60	0.88 ± 0.12
FeO	3.87 ± 0.10	3.25 ± 0.75	0.79 ± 0.10	1.53	1.02 ± 0.35	0.87 ± 0.13
MnO	0.13 ± 0.01	0.11 ± 0.01	0.06 ± 0.02	0.12	0.06 ± 0.03	0.04 ± 0.02
MgO	2.08 ± 0.51	1.45 ± 0.20	0.23 ± 0.16	2.72	1.34 ± 0.44	1.13 ± 0.44
CaO	4.31 ± 0.83	3.49 ± 0.47	1.01 ± 0.42	5.64	3.54 ± 0.58	3.10 ± 0.22
Na_2O	4.73 ± 0.45	4.79 ± 0.38	3.95 ± 0.10	3.68	4.46 ± 0.35	4.65 ± 0.19
K_2O	2.27 ± 0.11	2.56 ± 0.24	4.34 ± 0.23	2.52	2.19 ± 0.35	1.94 ± 0.16
P_2O_5	0.28 ± 0.06	0.31 ± 0.11	0.05 ± 0.02	0.45	0.22 ± 0.06	0.12 ± 0.02
LOI	1.26 ± 0.11	0.74 ± 0.22	1.11 ± 0.78	1.42	1.26 ± 0.63	1.06 ± 0.43
Total	100.04	100.00	100.11	100.00	99.96	100.00

† n is number of samples
‡ Average value ± 1 S.D.

Summary and Conclusions

Chemical and petrographical data

The inner margin of the Taurus Range and north-western Iran are characterized by Neogene to Quaternary orogenic volcanism. Chemical variations of the erupted products allow the distinction of calc-alkaline, high K calc-alkaline, and shoshonitic series. The calc-alkaline and the high K calc-alkaline volcanic activity gave rise to similar volumes of products, ranging in composition from basalt to rhyolite. Intermediate members, andesites, and dacites clearly dominate: andesites are the most abundant rock type in both the calc-alkaline and high K calc-alkaline associations (Fig. 9(a)). Shoshonitic rocks are subordinate and have only c. 10 per cent of the combined volume of the calc-alkaline rocks. Latites are the dominant rock type in this series (Fig. 9(b)).

The average chemical composition of members of the three series is shown in Table 9. Intermediate members of the calc-alkaline series show chemical characters which are comparable

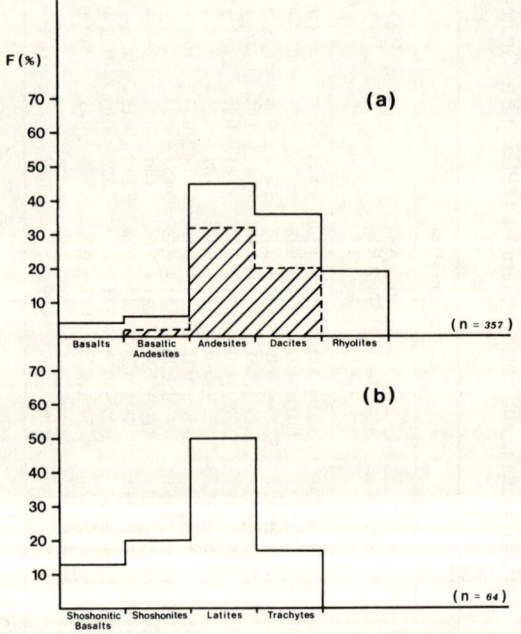

Fig. 9 Percentage distribution of (a) the whole calc-alkaline and high K calc-alkaline samples (dashed area represents the high K calc-alkaline series); (b) shoshonitic series.

TABLE 9 Average chemical composition and phenocryst mineralogy of orogenic associations (Anatolia and north-western Iran)

	Calc-Alkaline Association					High K calc-Alkaline Association			Shoshonitic Association			
	Basalt	Andesite Basalt	Andesite	Dacite	Rhyolite	High K Andesite Basalt	High K Andesite	High K Dacite	Shoshonitic Basalt	Shoshonite	Latite	Trachyte
SiO_2	50.26	53.57	59.69	66.57	72.49	54.46	60.10	66.09	50.35	53.68	58.83	66.46
TiO_2	1.22	1.16	0.91	0.56	0.24	0.89	0.73	0.56	1.15	1.09	0.80	0.60
Al_2O_3	17.18	17.29	16.71	15.92	14.06	15.66	16.66	15.46	15.72	15.59	15.81	15.48
Fe_2O_3	8.88*	7.86*	2.87	1.70†	0.90§	7.15*	3.67††	2.50	4.72	4.90	4.10	2.58
FeO	—	—	2.94	1.82‡	0.77§	—	1.73‡‡	1.06	2.26	1.77	0.89	1.02
MnO	0.15	0.12	0.11	0.08‡	0.07¶	0.12‖	0.09‡‡	0.06	0.14	0.14	0.10	0.06
MgO	6.80	5.61	3.62	1.51	0.72¶	6.40	2.74	1.31	6.45	4.80	2.74	1.08
CaO	10.16	8.30	6.28	4.16	2.06	8.08	5.72	3.21	8.52	6.90	4.92	2.40
Na_2O	3.56	3.90	3.99	4.11	4.21	3.34	3.65	3.80	3.40	3.11	3.53	3.93
K_2O	0.93	1.25	1.88	2.24	3.26	2.15	2.85	3.52	4.05	4.89	4.56	5.11
P_2O_5	0.27	0.39	0.24	0.19	0.09¶	0.38	0.30	0.21	0.60	0.67	0.42	0.27
LOI	0.77	0.84	1.11	1.43	1.30¶	1.61	1.45	2.04	2.42	2.32	2.19	1.44
n =	13	15	48	56	33	6	113	73	9	13	32	11
plag	(+)	+	+	+ +	+	(+)	+ +	+ +	+ +	(+)	+ +	(+)
ol	+ +	+ +	(+)			+ +	(+)		+ +	+ +		
cpx	+ +	+ +	+ +	+ +	(+)	+ +	+ +	(+)	+ +	+ +	(−)	+
opx			+ +	+	(+)	(+)	+ +				+	
hb			(+)	(+)	(+)		(+)	+ +		+ +	+ +	+ +
bi					(+)							
san											(−)	
Q			(+)	(+)	(+)		(+)	(+)			−	−
ore	+	+	(+)	+		+	+	+	+	+		

* Fe_2O_3 is total Fe when FeO is not reported   ¶ On 28 samples   + +, Always present
† On 35 samples   ‖ On 3 samples   +, Frequently present
‡ On 49 samples   †† On 100 samples   (+), Sporadically present
§ On 23 samples   ‡‡ On 112 samples

with those reported for products from continental margins (Ewart, 1976, and this volume, Section II). There are, however, some differences, in particular the higher MgO content and the lower FeO_{tot} (and hence higher $Mg/(Mg + Fe_{tot})$ ratio) seen mainly in basaltic and basaltic andesites. This distinctive character is clearly observed in the AFM plot (Fig. 10), showing a distribution of volcanic trends, mostly plotting near or below the lower limit of orogenic associations (Ringwood, 1974). The Na_2O/K_2O ratio is 1–2 in the calc-alkaline series but is generally below 1 in the shoshonitic series, due to increased K_2O, while Na_2O remains constant.

The phenocryst mineralogy, summarized in Table 9, indicates an anhydrous mineral association in basic and intermediate members of the calc-alkaline series. Hornblende sometimes occurs in dacitic members of the calc-alkaline series, but is remarkably widespread in the intermediate and acid rocks of the high K calc-alkaline series. Biotite dominates the paragenesis of the shoshonitic series, even in the most basic members.

Isotopic and geochemical data concerning the studied orogenic series are sparse and uneven (Borsi et al., 1972; Lambert et al., 1974; Keller et al., 1977; Dostal and Zerbi, 1978; Innocenti et al., 1979a, 1980). $^{87}Sr/^{86}Sr$ ratios are available for some samples of the high K calc-alkaline series from western Anatolia (average value on eight samples: $^{87}Sr/^{86}Sr$ = 0.7078, S.D. = 0.0009) and central Anatolia, Konya region (average value on eight samples: $^{87}Sr/^{86}Sr$ = 0.7068, S.D. = 0.001) (Borsi et al., 1977). The only Sr isotope data for rocks belonging to the calc-alkaline series are some determinations on Quaternary rocks from Mt Ararat (average value on seven determinations: $^{87}Sr/^{86}Sr$ = 0.7050, s = 0.0004) (Lambert et al., 1974). The large ion lithophile element concentration (Rb, Sr, Ba, La, and Ce) in intermediate and basic rocks of the high K calc-alkaline series from western Anatolia and the Konya and Van regions is significantly higher than the reported abundance in orogenic associations (Taylor, 1969). Data for REE distributions are available for Erciyes and Sabalan Plio-Quaternary central volcanoes as well as on samples of Miocene volcanics from western Anatolia (Fig. 11). The high K calc-alkaline and shoshonitic rocks from western Anatolia show the fractionation of both light REE and heavy REE. Similar fractionated patterns are also observed for the calc-alkaline Plio-Quaternary rocks from Sabalan volcano (north-western Iran), while calc-alkaline rocks from Erciyes Quaternary volcano (central Anatolia) show a relative flat heavy REE pattern, associated with a fractionated pattern of less enriched light REE. Significant negative Eu anomalies have only been observed in some trachytes of the western Anatolia shoshonitic association.

The available geochemical and isotopic data are inadequate to draw general conclusions

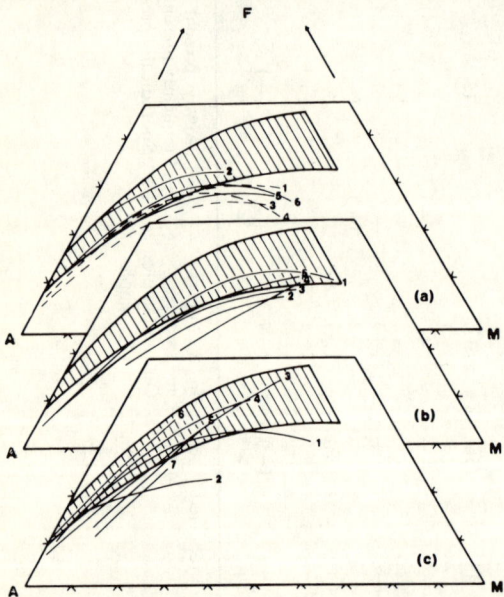

Fig. 10 AFM plot for calc-alkaline, high K calc-alkaline, and shoshonitic (broken lines) volcanic trends from (a), western Anatolia (b), central Anatolia and (c), eastern Anatolia–north-western Iran. Dashed areas represent the field of orogenic suites after Ringwood (1974). Key: (a) 1, Foca area; 2, Edremit–Balikesir area; 3, Afyon–Isparta area; 4, Bodrum area; 5, Bergama area; 6, Izmir area; (b) 1, Konya area; 2, Karadag; 3, Hasan Dag; 4, Karapinar and Acigol areas; 5, Urgup basin. (c) 1, Van area; 2, Bijar area; 3, Kars area; 4, Ararat volcano; 5, Sabalan volcano; 6, Suphan volcano; 7, Sahand volcanic complex

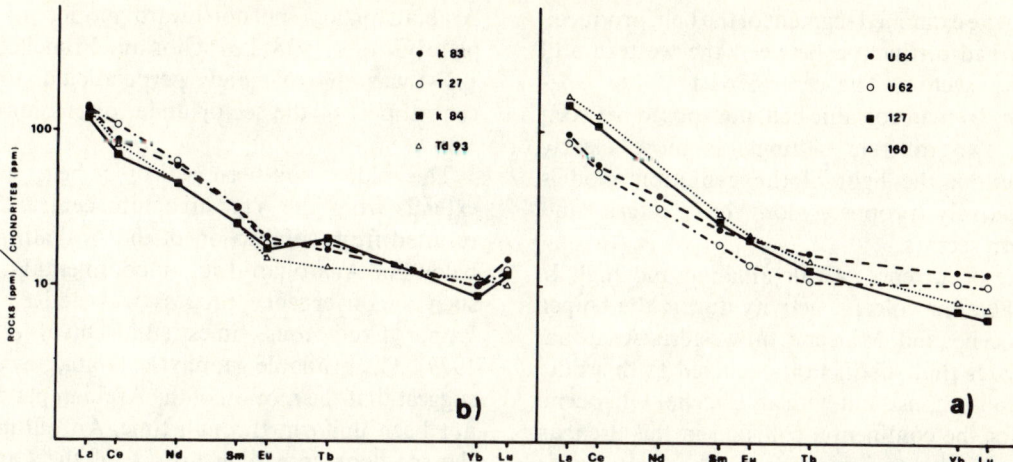

Fig. 11 Chondrite-normalized REE patterns for some representative samples of the Anatolia–north-western Iran orogenic association. (a) Pliocene–Quaternary volcanic rocks from central Anatolia and north-western Iran: basaltic andesite (U84) and andesite (U62) of the calc-alkaline series from Erciyes Dag (Innocenti *et al.*, 1979a); andesites (127 and 160) from Sabalan volcano (Dostal and Zerbi, 1978). (b) Miocene volcanics from western Anatolia: high K basaltic andesite (K 84) from Karaburun (south-west of Izmir); high K andesites (K 83 and Td 93) from Karaburun and Bergama respectively; latite (T 27) from the south of Edremit (Innocenti *et al.*, 1979a).

about the petrogenesis of this orogenic volcanism on the whole. The REE distribution and Sr isotope ratios have, nevertheless, been interpreted as evidence suggesting magma genesis by partial melting of mantle anomalously enriched in large ion lithosphere elements and radiogenic Sr above a subduction plane (Dostal and Zerbi, 1978; Innocenti *et al.*, 1980).

Evolution of volcanism and geodynamic implications

The data enable us to review the major characteristics of the Neogene to Quaternary volcanic history within the Anatolian–Iranian belt. The space–time distribution and magmatic affinity of volcanism in different areas is summarized in Fig. 12. One of the most important features shown by these data is that the high K calc-alkaline volcanism ended during the Middle Miocene in western Anatolia, but continued into the Quaternary in the other sectors of the belt. There are major differences in the volcanological evolution of eastern and central Anatolia.

The variations observed in the distribution of volcanism reflect the complex pattern of past and present interaction between plates, which are now dominated by subduction of the Afro-Arabian plate below the Eurasian plate (Biju-Duval *et al.*, 1977). In such a context, the diachronism of the continental collision events

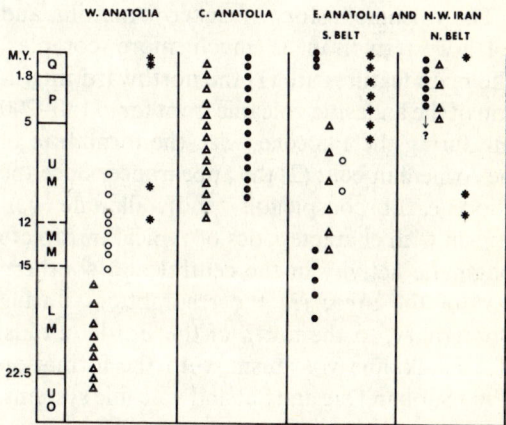

Fig. 12 Time distribution of volcanism. ●, calc-alkaline; △, high K calc-alkaline; ○, shoshonitic; *, sodic-alkaline

along the examined segment of the belt, produced a marked distinction between the western and eastern sectors. The central part of the Anatolian–Iranian volcanic belt, interposed between these two different settings, is more clearly defined in the light of the evolution models respectively proposed for the western and eastern sectors.

The occurrence of calc-alkaline and high K calc-alkaline volcanic activity during the Upper Oligocene and Miocene in western Anatolia, indicates that subduction occurred at this time and this is consistent with an Eocene–Oligocene age for the continental collision in this area, as suggested by geological data (Boccaletti et al., 1974). The sporadic shoshonitic activity confirms that western Anatolia became stable during the Middle–Upper Miocene. A subsequent, important tectonic phase developed in relation to later southward migration of the subduction, which took place c. 12 Ma ago (Fytikas et al., 1976), giving birth, 3 Ma ago, to the active Aegean volcanic arc (Keller, this volume, Section III). The present setting of plate convergence in this area is the result of the south-westward motion of the Aegean plate (McKenzie, 1972). Such movement gave rise to the tensional tectonic regime which has caused the formation of east–west-trending grabens in western Anatolia since the Miocene. The scattered alkaline volcanism reported in this area is related to this tectonic setting.

The volcanic history of eastern Anatolia and north-western Iran is much more complex. The main features are (1) the northward migration of the andesitic volcanic front for c. 150–200 km during the Pliocene, with the formation of the Armenian belt; (2) the appearance, since the Pliocene, of conspicuous sodic-alkaline volcanism with characteristics of typical intraplate magmatic activity in the central and southern part of the area; (3) the resumption, during Quaternary, to the north of the Border Folds, of calc-alkaline volcanism, with the formation of the Suphan Dag and Sahand volcanic systems.

The interpretation of this volcanism in terms of plate interaction is strongly influenced by the position of the area, facing the apex of the Arabian plate. The northward motion of this plate (Girdler, 1978; Le Pichon and Francheteau, 1978) was approximately perpendicular to the central part of the sector under discussion (the Van area).

The older, southern volcanic belt, which extends from the Van area into central Iran, resulted from subduction of the Arabian plate below the Anatolian–Iranian continental mass: such a convergence process was active from Upper Cretaceous times (Biju-Duval et al., 1977). The available geophysical data, however, suggest that the motion of the Arabian plate has not been uniform through time. An outline of the sea-floor spreading history of the Gulf of Aden and the Red Sea indicates an early major spreading phase taking place at least from the Upper Oligocene to the Middle Miocene (Girdler, 1978). It is therefore suggested that after the major event of continental collision between Africa and Europe in Eocene–Oligocene times, an overall decrease of the convergence rate also affected the eastern sector of the Anatolian–Iranian volcanic range. The Arabian plate took shape successively in Oligocene times (Le Pichon and Francheteau, 1978) and its northward velocity increased because of the sea-floor spreading to the south.

A relationship between relative convergence rate and angle of subduction is commonly proposed, with lower angles of subduction being associated with faster convergence (Luyendyk, 1970). It is therefore suggested that as a consequence of the velocity increase of the Arabian plate, at least up to Middle Miocene, a significant decrease of the slab dip took place in the area of maximum convergence velocity, i.e. the northern apex of Arabia. The observed northward migration of the volcanic front, with the formation of the Armenian belt in the Upper Miocene, is inferred to be the result of this variation of subduction geometry. Continental collision was definitely taking place during the Pliocene (Dewey et al., 1973), causing the development of the Bitlis suture and folding of the Arabian platform. It is suggested that, after such an episode of continental collision, the subducted slab detached, and continued to induce magma generation with progressively lower intensity beneath an area (the Armenian belt) located progressively farther from the continental suture.

Active convergence between Arabia and Anatolia–Iran, even after the continental collision, produced intense continental deformation in the area of maximum stress accumulation. The penetration of the Arabian mass into the area in front of its apex caused divergent motion of the Iranian and Anatolian microplates (McKenzie, 1972). The eruption of sodic-alkaline volcanic products in the Van area is regarded as genetically related to such a tensional process (Innocenti et al., 1976).

The local resumption of calc-alkaline volcanism during the Quaternary, to the Bitlis suture, is more difficult to explain. It can be tentatively attributed to the increasing convergence rate, due to the most recent Red Sea floor spreading phase (Le Pichon and Francheteau, 1978). If such northward motion could not be wholly absorbed by continental deformation in the Van area, this could be the cause of subduction at the base of the two colliding continental lithosphere segments. The scattered calc-alkaline volcanic activity (Suphan Dag and Sahand) may be the result of this process. The occurrence of a NE–SW-trending andesite belt in central Anatolia, persisting up to the Quaternary, is explained within this framework. Among features which may be stressed in this area are the lack of evidence for any migration of the volcanic front, and the distribution of volcanic products showing a north-westward increase in K_2O content. The geology of central Anatolia is dominated by vertical movements, without important thrusting events, which characterize both the western and eastern sectors of the orogenic belt. Such behaviour might reflect a north-westward subduction process (Innocenti et al., 1975), persisting throughout the Quaternary within an area of 'tectonic shadow'. This area was situated at the western end of an original embayment of the Anatolian continental margin, between the termination of the western Anatolia continental suture and the Africa–Arabia plate boundary (the Dead Sea Rift System).

REFERENCES

Adamia, Sn. A., Lordkipanidze, M. B., and Zakariadze, G. S. (1977). Evolution of an active continental margin as exemplified by the alpine history of the Caucasus. *Tectonophysics* **40**, 183–199.

Akarsu, I. (1960). Geology of the Mut region. *MTA Bull.* **54**, 38–43.

Alberti, A., Comin-Chiaromonti, P., Di Battistini, G., Sinigoi, S., and Zerbi, M, (1975). On the magmatism of the Savalan volcano (North-West Iran). *Rend. Soc. Ital. Mineral. Petrol.* **31**, 337–350.

Alberti, A., Comin-Chiaromonti, P., Di Battistini, G., Nicoletti, M., Petrucciani, C., and Sinigoi, S. (1976). Geochronology of the Eastern Azerbaijan volcanic plateau (N-W Iran). *Rend. Soc. Ital. Mineral. Petrol.* **32**, 579–589.

Altinli, I. (1966). Geology of the Eastern and Southeastern Anatolia. *MTA Bull.* **66**, 35–76.

Arculus, R. J. and Johnson, R. W. (1978). Criticism of generalized models for magmatic evolution of arc–trench system. *Earth planet. Sci. Letters* **39**, 118–126.

Arpat, E. and Saroglu, F. (1972). The East Anatolian fault system; thoughts on its development. *MTA Bull.* **78**, 33–49.

Aslanian, A. T. (1977). Volcano-tectonic activity in Armenian Highland in Pliocene and Pleistocene. *Bull. Acad. Sci. Armenia, Earth Sci.* **6**, 3–11 (in Russian).

Ayranci, B. L. and Weibel, M. (1973). Zum Chemismus der Ignimbrite des Erciyes Vulkans (Zentral-Anatolien). *Bull. Schweiz. Mineral. Petrol.* **53**, 49–60.

Benda, L., Innocenti, F., Mazzuoli, R., Radicati, F., and Steffens, P. (1974). Stratigraphic and radiometric data of the Neogene in Northwest Turkey. *Z. dt. Geol. Ges.* **125**, 183–193.

Besang, C., Eckhardt, F. J., Harre, W., Kreuzer, G. and Muller, P. (1977). Radiometrische Alterbestimmungen an neogenenen Eruptivgesteinen der Turkei. *Geol. Jh.* **25**, 3–36.

Biju-Duval, B., Dercourt, J., and Le Pichon, X. (1977). From the Tethys ocean to the Mediterranean Seas: a plate tectonic model of the evolution of the Western Alpine System. In *International Symposium on the Structural History of the Mediterranean Basins, Split, 1976* (B. Biju-Duval and L. Montadert, eds), Editions Technip, Paris, pp. 143–164.

Bingol, E. (1971). Essai d'application de mesures géochronologiques au massif de Kazdag, Tuquie. *Bull. Soc. Geol. Tur.* **14**, 1–16.

Bird, P., Toksoz, M. N., and Sleep, N. H. (1975). Thermal and mechanical models of continent–continent convergence zones. *J. geophys. Res.* **80**, 4405–4416.

Bizon, G., Biju-Duval, B., Letouzey, J., Monod, O., Poisson, A., Ozer, B., and Oztumer, E. (1974). Nouvelles précisions stratigraphiques concernant les bassins tertiaires au sud de la Turquie (Antalya, Mut, Adana). *Rev. Inst. Fr. Pétrol.* **29**, 305–318.

Boccaletti, M., Manetti, P., and Peccerillo, A. (1974). The Balkanides as an istance of back-arc thrust belt: possible relation with the Hellenides. *Geol. Soc. Am. Bull.* **85**, 1077–1084.

Boccaletti, M., Innocenti, F., Manetti, P., Mazzuoli, R., Motamed, A., Pasquaré, G., Radicati, F., and Amin Sobhani, E. (1977). Neogene and Quaternary Volcanism of the Bijar area (Western Iran). *Boll. Volc.* **42**, 1–12.

Borsi, S., Ferrara, G., Innocenti, F., and Muzzuoli, R. (1972). Geochronology and petrology of recent volcanics in the Eastern Aegean Sea (West Anatolia and Lesvos Island). *Boll. Volc.* **36**, 473–496.

Bosscha-Erdbrink, D. P., Priem, A. N. A., Hebeda, E. H., Cup, C., Dankers, P., and Cloetingh, S. A. P. L. (1976). The bone bearing beds near Maragheh, N.W. Iran. *Konik. Nederl. Akad. Van Wetenschappen, Amsterdam*, B **79**, 85–112.

Braud, J. and Ricou, L. E. (1975). Elements de continuité entre le Zagros et la Turquie de Sud-Est. *Bull. Soc. Géol. Fr.* **17**, 1015–1023.

Brinkman, R. (1976). *Geology of Turkey*. F. Enke Verlag, Stuttgart, p. 158.

Brunn, J. H., De Graciansky, P., Gutnic, M., Juteau, T., Lefévre, R., Marcoux, J., Monod, O., and Poisson, A. (1970). Structures majeures et corrélations stratigraphiques dans les Taurides occidentales. *Bull. Soc. Géol. Fr.* **7**, 515–556.

Burri, C., Tatar, Y., and Weibel, M. (1967). Zur Kenntnis der jungen Vulkanite des Halbinsel Bodrum (SW-Turkei). *Schweiz. Mineral. Petrogr. Mitt.* **47**, 833–853.

Comin-Chiaramonti, P., Mosca, R., Sinigoi, S., and Di Battistini, G. (1978). Miocene Volcanism in the Nir district (Eastern Azerbaijan, Iran). *Neues Jb. Miner. Abh.* **133**, 23–32.

Dewey, J. F., Pitman, W. C., III, Ryan, W. B. F., and Bonnin, J. (1973). Plate tectonics and the evolution of the Apline system. *Geol. Soc. Am. Bull.* **84**, 3137–3180.

Didon, J. and Gemain, Y. M. (1976). Le Sabalan, volcan plio-quaternaire de l'Azerbaijan oriental (Iran): étude géologique et pétrographique de l'edifice et de son environment. Thèse 3° cycle, pp. 304.

Dostal, J. and Zerbi, M. (1978). Geochemistry of the Savalan volcano (Northwestern Iran). *Chem. Geol.* **22**, 31–42.

Durr, St., Altherr, R., Keller, J., Okrusch, M., and Seidel, E. (1978). The Median Aegean Crystalline Belt: stratigraphy, structure, metamorphism, magmatism. In *Alps, Apennines, Hellenides* (H. Closs et al., eds), IVCG Scientific Report no. 38, Stuttgart, pp. 455–477.

Ewart, A. (1976). Mineralogy and chemistry of modern orogenic lavas. Some statistics and implications. *Earth planet. Sci. Letters* **31**, 417–432.

Farhoudi, G. (1978). A comparison of Zagros geology to island arcs. *J. Geol.* **86**, 323–334.

Fourquin, C. (1975). L'Anatolie de Nord-Quest, marge méridionale du continent européen, histoire paléogéographique, tectonique et magmatique durant le Secondaire et le Tertaire. *Bull. Soc. Géol. Fr.* **17**, 1058–1070.

Fytikas, M., Giuliani, O., Innocenti, F., Marinelli, G., and Mazzuoli, R. (1976). Geochronological data on recent magmatism of the Aegean Sea. *Tectonophysics* **31**, T29–T34.

Gansser, A. (1966). The volcanoes of Iran. In *Catalogue of Active Volcanoes of the World*, Part 17, International Association for Volcanology and Chemistry of the Earth's Interior, Naples, pp. 7–20.

Girdler, R. W. (1978). Comparison of the East African rift system and the Permian Oslo. In *Tectonics and Geophysics of Continental Rifts* (E.-R. Neumann and I. B. Ramberg, eds), D. Reidel, Dordrecht, pp. 329–345.

Innocenti, F. and Mazzuoli, R. (1972). Petrology of Izmir-Karaburum volcanic area (West Turkey). *Bull. Volc.* **36**, 83–104.

Innocenti, F., Mazzuoli, R., Pasquarè, G., Radicati di Brozolo, F., and Villari, L. (1975). The neogene calc-alkaline volcanism of Central Anatolia: geochronological data on Kayseri–Nidge area. *Geol. Mag.* **112**, 349–360.

Innocenti, F., Mazzuoli, R., Pasquarè, G., Radicati di Brozolo, F., and Villari, L. (1976). Evolution of volcanism in the area of interaction between the Arabian, Anatolian and Iranian plates (Lake Van, Eastern Turkey). *J. Volc. geothermal Res.* **1**, 103–112.

Innocenti, F., Manetti, P., Mazzuoli, R., Peccerillo, A., and Poli, G. (1979*a*). REE distribution in tertiary and quaternary volcanic rocks from central and western Anatolia. In *Proceedings of the 6th Colloquium on Geology of the Aegean Region, Izmir*, in press.

Innocenti, F., Mazzuoli, R., Pasquarè, G., Serri, G., and Villari, L. (1980). Geology of the volcanic area north of Lake Van (Turkey). *Geol. Rdsc.*, **69**, 1, 292–322.

Irvine, T. N. and Baragar, W. R. A. (1971). A guide to the chemical classification of the common volcanic rocks. *Can. J. Earth Sci.* **8**, 523–548.

Jung, D., Keller, J., and Eckhardt, F. J. (1972). Der Känozoische Vulkanismus Zentralanatolien. DFG Programm G-a 120/3, Bundesanstalt für Bodenforschung, Internal Rep.

Keller, J. (1974). Quaternary Maar volcanism near Karapinar in Central Anatolia. *Bull. Volc.* **38**, 378–396.

Keller, J. and Villari, L. (1972). Rhyolitic ignimbrites in the region of Afyon (Central Anatolia). *Bull. Volc.* **36**, 342–358.

Keller, J., Jung, D., Burgath, K., and Wolf, F. (1977). Geologie and Petrologie des Neogenen Kalkalkali-Vulkanismus von Konya (Erenler-Dag-Alaca Dag-Massif, Zentral Anatolien). *Geol. Jh.* **25**, 37–117.

Krushensky, R. D. (1976). Neogene calc-alkaline extrusive and intrusive rocks of the Karaler–Yesiller area, northwest Anatolia, Turkey. *Bull. Volc.* **39**, 336–360.

Kurtman, F. and Akaus, M. (1971). Inter-mountain basins in eastern Anatolia and their oil possibilities. *MTA Bull.* **77**, 1–9.

Lambert, R. S. J., Holland, J. G., and Owen, P. F. (1974). Chemical petrology of a suite of calc-alkaline lavas from Mt Ararat, Turkey. *J. Geol.* **82**, 419–438.

Lange, S. P. (1971). The subdivision of the Cenozoic in Eastern Central Anatolia. *Newsletter Stratigraphy* **1**, 37–40.

Leo, G. W., Marvin, R. F., and Mehnert, H. H. (1974). Geologic framework of the Kuluncak–Sofular area, east-central Turkey, and K-Ar ages of igneous rocks. *Geol. Soc. Am. Bull.* **85**, 1785–1788.

Le Pichon, X. and Francheteau, J. (1978). A plate-tectonic analysis of the Red Sea–Gulf of Aden area. *Tectonophysics* **46**, 369–406.

Lescuyer, J. L., Michel, R., Riov, R., and Vivier, G. (1976). Etude géochimique du volcanisme tertiare de la région de Mianeh (Azerbajan, Iran). *Géol. Alpine* **52**, 85–98.

Luyendyk, B. P. (1970). Dips of downgoing lithospheric plates beneath island arcs. *Geol. Soc. Am. Bull.* **88**, 1479–1487.

Marsh, B. D. and Carmichael, I. S. E. (1974). Benioff zone and magmatism. *J. geophys. Res.* **79**, 1196–1206.

McKenzie, D. P. (1972). Active tectonics of the Mediterranean region. *Geophys. Jl R. astr. Soc.* **30**, 109–185.

McKenzie, D. P. (1976). The East Anatolian fault: a major structure in eastern Turkey. *Earth. planet. Sci. Letters* **29**, 189–193.

McKenzie, D. P. (1977). Can plate tectonics describe continental deformation? In *Proceedings of the International Symposium on the Structural History of the Mediterranean Basins, Split, 1976* (B. Biju-Duval and L. Montadert, eds), Editions Technip, Paris, pp. 189–196.

Miyashiro, A. (1974). Volcanic rocks series in island arcs and active continental margins. *Am. J. Sci.* **274**, 321–335.

Nowroozi, A. A. (1972). Focal mechanism of earthquakes in Persia, Turkey, West Pakistan and Afganistan and plate tectonics of the Middle East. *Bull. seismol. Soc. Am.* **61**, 317–341.

Ota, R. and Dincel, A. (1975). Volcanic rocks of Turkey. *Bull. geol. Soc. Japan* **26**, 393–419.

Pasquaré, G. (1971). Cenozoic volcanics of the Erzurum area (Turkish Armenia). *Geol. Rundschau* **60**, 900–911.

Peccerillo, A. and Taylor, S. R. (1976). Geochemistry of Eocene calc-alkaline volcanic rocks from Kastamonu area, Northern Turkey. *Contr. Mineral. Petrol.* **68**, 63–81.

Pinar-Erdem, N. and Ilhan, E. (1977). Outlines of the stratigraphy and tectonics of Turkey, with notes on the geology of Cyprus. In *The Ocean Basins and Margins* (A. E. M. Nairn, W. H. Kanes, and F. G. Stehli, eds), Plenum Press, New York, Vol. 4A, pp. 277–318.

Ringwood, A. E. (1974). The petrological evolution of island arc system. *J. geol. Soc. Lond.* **130**, 183–204.

Sanver, M. (1968). A paleomagnetic study of quaternary volcanic rocks from Turkey. *Phys. Earth planet. Interiors* **1**, 403–421.

Savascin, M. Y. (1972). Beiträge zur Frage der Genese westanatolischer Andesite und Basalte. Thesis, University of Tübingen.

Schleicher, H. and Schwarz, G. (1977). Zur Geologie und Petrographie des Karadag, Zentralanatolien. *Geol. Jh.* **25**, 119–138.

Snyder, W. S., Dickinson, W. R., and Siberman, M. L. (1976). Tectonic implications of space–time patterns of Cenozoic magmatism in the western United States. *Earth planet. Sci. Letters* **32**, 91–106.

Stöcklin, J. (1968). Structural history and tectonics of Iran: a review. *Am. Assoc. Petrol. Geol. Bull.* **52**, 1229–1258.

Stöcklin, J. and Nabayi, N. H. (1973). Tectonic map of Iran. Scale 1:2 500 000. Geological Survey of Iran, Teheran.

Taylor, S. R. (1969). Trace element chemistry of andesites and associated calc-alkaline rocks. *Bull. St. Oregon Dept. Geol. Mineral. Industries* no. 65, 43–63.

Tchalenko, J. S. and Braud, J. (1974). Seismicity and structure of the Zagros (Iran): the main recent fault between 33 and 35 N. *Phil. Trans R. Soc. A* **277**, 1–25.

Tchalenko, J. S. (1977). A reconnaissance of the seismicity and tectonics at the Northern border of the Arabian plate (Lake Van region). *Rev. Géogr. Phys. Géol. Dyn.* **19**, 189–208.

Turkish Gulf Oil Co. (1961). Regional geology and oil possibilities in the Tüz Golu basin of Central Anatolia. *Petrol. Activ., Ankara* **6**, 29–32.

Andesites
Edited by R. S. Thorpe
© 1982 John Wiley & Sons

IV   Evolution of andesite volcanic provinces

The earlier contributions in Section II outlined the characteristics of calc-alkaline rocks erupted in active volcanic provinces. Some of these contributions emphasized important spatial and temporal variations in the volcanic products in relation to tectonic setting. This section therefore has two contributions which present a more detailed account of two contrasted Mesozoic–Cainozoic calc-alkaline volcanic provinces, showing how such spatial and temporal variations of the volcanic rocks are related to regional tectonics, and how such variations provide clues to the petrogenesis of the volcanic rocks.

For the early Cainozoic (c. 30–15 Ma) volcanic province of Sardinia, J. Dostal et al. establish important regional compositional variations in which the volcanic rocks of southern Sardinia have characteristics resembling an island arc tholeiitic suite, while the volcanic rocks of northern Sardinia are more similar to island arc calc-alkaline associations. These regional variations are consistent with formation over a northward-dipping Cainozoic subduction zone below Sardinia, developed during Cainozoic displacement of the Sardinian microplate. In contrast, taking the larger area of the Antarctic Peninsula and the Scotia arc, J. Tarney et al. show how calc-alkaline intrusive and volcanic activity has evolved in response to changing tectonic environment. During the Mesozoic, the Antarctic Peninsula area experienced extensive calc-alkaline magmatic activity related to an eastward-dipping subduction zone consuming Pacific oceanic lithosphere. The opening of the Weddell Sea and later the Drake Passage initiated complex plate tectonic motions in the Scotia Sea which culminated in the present extended loop of the Scotia arc and the subduction of Atlantic oceanic lithosphere below the young South Sandwich island arc. These two contributions therefore provide detailed accounts of calc-alkaline volcanism and its relationship to varied tectonic evolution in two contrasted Mesozoic–Cainozoic provinces.

Cainozoic andesitic rocks of Sardinia (Italy)

J. Dostal, C. Coulon, and C. Dupuy

Department of Geology,
Saint Mary's University, Halifax, Nova Scotia, Canada,

Laboratoire de Pétrologie,
Université Saint-Jérôme, 13397 Marseille Cedex, France,

and

Centre Géologique et Géophysique,
USTL, 34060 Montpellier Cedex, France

ABSTRACT

Cainozoic calc-alkaline suites from Sardinia (Italy) are composed of high Al basalts, andesites, dacites, and subordinate rhyolites. The individual volcanoes have different geochemical characteristics (continental margin or island arc types), and the andesitic rocks show an overall spatial chemical zonation with respect to K and related elements indicating the presence, during the Cainozoic, of a northward-dipping subduction zone under Sardinia. The basaltic rocks were probably formed by partial melting of upper mantle peridotite. The chemical variations within individual volcanic complexes are compatible with a mechanism involving both fractional crystallization and contamination. Contamination is probably related to an interaction with ignimbrites which are spatially and temporally associated with andesitic rocks.

Introduction

Sardinia and Corsica are usually considered to be continental microplates which rotated counterclockwise with respect to Europe (Nairn and Westphal, 1968; De Jong et al., 1969; Coulon et al., 1974a, Westphal et al., 1976). Their displacement, which started c. 30 Ma ago and culminated 15–16 Ma ago (Coulon, 1977; Bellon et al., 1977), was accompanied in Sardinia by extensive andesitic and ignimbritic volcanism. The calc-alkaline volcanic rocks in Sardinia were emplaced on continental crust c. 30 km thick (Morelli et al., 1967). The volcanism was probably related to the subduction of oceanic lithosphere, in a north-north-westerly direction below the European continent, in response to the relative motions of the African and European plates. The particular geotectonic setting of Sardinia, together with the limited amount of geochemical data on the volcanic rocks of the western Mediterranean basin, led us to a detailed geochemical and petrological study of Sardinian calc-alkaline volcanic rocks. The purpose of this paper is to present a synthesis of both previously published and unpublished data on the mineralogy and chemistry of the andesitic rocks of Sardinia, to relate this data to the magmatic evolution of the island, and to compare these rocks with those from island arc and continental margin environments.

Geological Notes

The andesite rocks occur in the form of relatively small isolated massifs outcropping in a graben

354 IV Evolution of andesite volcanic provinces

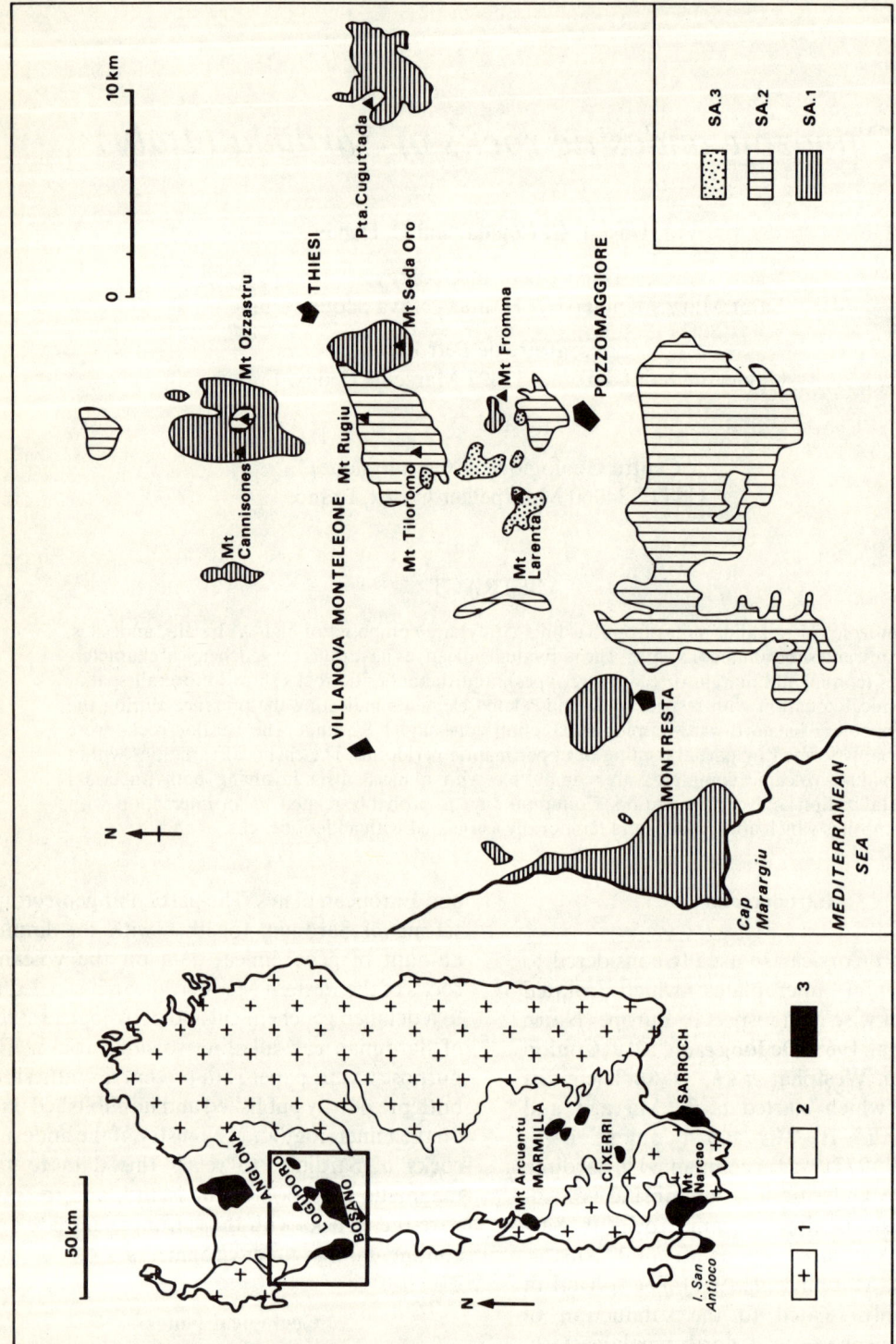

Fig. 1 Generalized geological map of Sardinia (left) and detailed map of the Logudoro–Bosano area (right). 1, Paleozoic metamorphic and granitic rocks; 2, Tertiary and Quaternary sedimentary and volcanic rocks; 3, andesites and associated rocks. SA_1, SA_2 and SA_3 represent three different volcanic episodes of the Logudoro–Bosano area

which forms most of the western part of Sardinia (Fig. 1). The massifs are composed mainly of pyroclastic rocks associated with lava flows and lava domes, and are intruded by dykes. The presence of andesitic pillow lavas in the volcanic complex of Marmilla indicates local submarine volcanic activity.

According to the classification of Taylor (1969), the lavas range from basalts (under 53 per cent SiO_2) through basic andesites (53–56 per cent SiO_2), andesites (56–62 per cent SiO_2), and dacites (62–68 per cent SiO_2) to rhyolites (over 68 per cent SiO_2). Basic andesites and andesites are the dominant rock types (c. 75 per cent) of the calc-alkaline suites in the northern part of the island while basalts and basic andesites are the most abundant types (c. 70 per cent) in southern Sardinia. The studied volcanic rocks are of Late Oligocene to Middle Miocene age with K–Ar dates ranging from 29 to 13 Ma (Coulon et al., 1974b; Bellon, 1976).

In the Bosano–Logudoro regions three successive volcanic episodes have been recognized on the basis of detailed geological mapping (Coulon et al., 1973; Coulon, 1977). The first episode (SA_1), with an age ranging between 24 and 21 Ma, is composed of basalts, basic andesites, and subordinate andesites. The second cycle (SA_2) formed at c. 17 Ma ago includes basalts, basic andesites, andesites, and dacites. A volumetrically minor third episode (SA_3), dated at 13–14 Ma old, comprises basic andesites, dacites, and minor rhyolites.

The andesitic rocks are spatially and temporally associated with ignimbrites. Although it is difficult to estimate the volumetric proportions of these rocks, ignimbrites are c. four times more abundant on the surface. The calc-alkaline volcanics are partly covered by Plio-Quaternary alkaline basalts and related rocks (Demant and Coulon, 1973; Beccaluva et al., 1976).

Mineralogy and Petrography

All the volcanic rocks are porphyritic; phenocrysts of strongly zoned plagioclase and clinopyroxene are present together with olivine in basalts and basic andesites, and orthopyroxene in andesites and dacites. The phenocrysts of amphibole and biotite are sporadic and those of quartz are rather scarce. In addition to glass, the mesostasis contains plagioclase, clinopyroxene, orthopyroxene, and Fe–Ti oxides. In more acid rocks, tridymite and/or cristobalite and, occasionally, K-feldspar and biotite also occur. Some basaltic rocks contain pigeonite in the groundmass.

Feldspar

Plagioclase is the most abundant phenocryst phase in all rock types. It shows a distinct compositional zoning particularly in andesites and dacites. Bulk compositions range from An_{90}–An_{60} in basalts and basic andesites to An_{60}–An_{40} in dacites. In the groundmass, plagioclase has a higher content of Na than in co-existing phenocrysts. K-feldspar is frequently present in the groundmass of the rocks from northern Sardinia.

Pyroxenes

Clinopyroxene occurs in most andesitic rocks. It has a relatively constant composition corresponding to augite and does not show any Fe-enrichment trends during differentiation. The contents of Al, Ti, and Na in clinopyroxene are typically low. Orthopyroxene also has a rather uniform composition (En c. 65). According to Ewart (1976a), the small compositional variations of both pyroxenes seem to imply a rather restricted range of T, P, f_{O_2}, and $a_{SiO_2}^{liquid}$ during crystallization of pyroxene. Orthopyroxene is sometimes resorbed with augite rims, suggesting disequilibrium conditions. Groundmass pyroxenes also have a relatively uniform composition with the exception of some basalts where clinopyroxene tends to evolve towards the subcalcic field. Pigeonite occurs sporadically in the Arcuentu and Seda Oro volcanoes. The presence of either pigeonite or orthopyroxene in similar rocks from the same volcano (Arcuentu) suggests that crystallization took place along the orthopyroxene–clinopyroxene inversion curve (Bowen and Schairer, 1935; Kuno, 1952).

Olivine

The composition of olivine ranges from Fo_{75}–Fo_{50} in phenocrysts to Fo_{40}–Fo_{35} in the groundmass. In some basalts and basic andesites orthopyroxene replaces olivine to a degree that increases with increase of the SiO_2 content of the host rocks.

Titanomagnetite

Fe–Ti oxides occur both as phenocrysts and in the groundmass. The exceptions are the volcanic complexes of Arcuentu and Marmilla where titanomagnetite is present only in the groundmass. As in typical calc-alkaline rocks (Ewart, 1976b), Fe–Ti oxide is relatively low in Ti (Usp_{23}–Usp_{35}).

Other phenocrysts

The rare hornblende is usually replaced by aggregates of clinopyroxene and/or orthopyroxene, plagioclase (An_{80}–An_{70}) and opaque minerals. Biotite occurs only in the most differentiated rocks. The common accessory minerals are apatite, zircon, and sphene.

The mineralogical features of the volcanic complex of Seda Oro from the Logudoro-Bosano area of north western Sardinia are summarized in Table 1 and are comparable to those of other andesitic suites.

Conditions of Crystallization

The order of crystallization deduced from petrographic observations on the whole series from basalt to dacite, together with the chemical composition of the minerals, provides some constraints on the physical conditions of solidification. With respect to the order of crystallization, Fe–Ti oxide is the first mineral to appear, followed by olivine (basalts), orthopyroxene, clinopyroxene (andesites), and plagioclase, although the crystallization periods of the last two minerals overlap. The appearance of Fe–Ti oxide as the earliest phase suggests high f_{O_2} conditions (Osborn, 1959, 1962) while the late crystallization of plagioclase indicates a relatively high water content in the magma (Green, 1972; Eggler, 1972). The simultaneous crystallization of clinopyroxene and plagioclase preceded by orthopyroxene (Eggler, 1972; Green, 1972; Eggler and Burnham, 1973) implies water content in the range 4–5 per cent and P_{tot} = c. 4.5 kb (Coulon, 1977). The abundant phenocrysts of plagioclase probably reflect water-undersaturated conditions (Ewart, 1976b).

Relatively homogeneous distributions of Na in clinopyroxene (Thompson, 1974), of Al_2O_3 in orthopyroxenes (Marsh, 1976; Mertzman, 1977), and of Ca in olivines (Stormer, 1973) suggest low P crystallization, probably less than or equal to 5 kb (Mertzman, 1977). The data obtained for the Sardinian andesites using several geothermometers indicate temperatures similar to those estimated for other andesitic suites. The plagioclase–liquid geothermometer of Kudo and Weill (1970) gives temperature estimates in the range 1100–1200 °C for basalts and basic andesites and 1050–1120 °C for acid andesites and dacites at P_{H_2O} = 1 kb (Coulon, 1977). The temperatures obtained from the orthopyroxene–clinopyroxene pairs are 1000–1030 °C for phenocrysts and 930–1000 °C for groundmasses. The differences in estimated temperatures between plagioclase and pyroxene geothermometers probably resulted from the crystallization of plagioclase under P_{H_2O} of over 1 kb.

Chemical Composition

The average chemical compositions of individual rock-types from several volcanic complexes of Sardinia are given in Table 2. The basalts (under 53 per cent SiO_2) are further subdivided into SiO_2 content intervals of 3 per cent. The major and trace elements were determined by atomic absorption with the exception of REE which were analysed either by instrumental or radiochemical neutron activation. The precision and accuracy of the methods were given by Coulon et al. (1978).

TABLE 1 Mineralogy of volcanic rocks from the Logudoro–Bosano area

		Basalts	Basic	Andesites	Acidic	Dacites
Phenocrysts	Phenocrysts	60–47%	40–30%	55–40%	28–20%	35–25%
	Plagioclase	50–40% An 90–70	An 90–60	An 90–60	An 90–50	Ar 80–40
	Olivine	5–7% Fo 74–53 (zoned)	1%			
	Orthopyroxene		Reaction 1–3% Ca3 Fe32 Mg65	Ca3 Fe34 Mg63 1–4%	Ca3 Fe33 Mg64 2–3%	
	Clinopyroxene	2–5% Ca43 Fe17 Mg40	7–9% Ca43 Fe15 Mg42	Reaction Ca41 Fe16 Mg43 3–4%	Ca41 Fe19 Mg40 1–2%	
	Titanomagnetite	Usp 47–30	Usp 35	Usp 42	Usp 51	
	Biotite					
Mesostasis	Plagioclase	An 80–60	An 65–60	An 65–50	An 50–40	
	Olivine	Fo 38				
	Orthopyroxene		Ca3 Fe35 Mg62	Ca3 Fe39 Mg58	Ca3 Fe38 Mg59	
	Clinopyroxene	{Ca32 Fe27 Mg41 Ca8 Fe47 Mg45 (pigeonite)	Ca44 Fe17 Mg39	Ca43 Fe16 Mg41	Ca42 Fe19 Mg39	
	Biotite					
	K-feldspar	Or59 Ab45 An2	Or64 Ab31 An5	Or68 Ab30 An2	Or70 Ab28 An2	
	Quartz					
	Titanomagnetite	Usp 53	Usp 23	Usp 39	Usp 55	

TABLE 2(a) Average major and trace element compositions of studied volcanic rocks from northwestern Sardinia (Logudoro–Bosano area)

	Cuguttada (SA)		Cap Marargiu (SA$_1$)		Seda Oro–Rugiu–Tiloromo (SA$_1$ + SA$_2$)			
	$n = 3$†	$n = 3$	$n = 3$	$n = 3$	$n = 3$	$n = 4$	$n = 10$	$n =$
Major elements (%)								
SiO_2	44.72 ± 0.28‡	48.77 ± 0.95	51.14 ± 0.47	54.23 ± 0.23	48.64 ± 0.89	54.07 ± 0.92	59.48 ± 1.62	64.43 ±
Al_2O_3	17.69 ± 0.51	18.86 ± 0.38	19.15 ± 0.62	19.78 ± 1.12	18.45 ± 0.35	17.20 ± 0.17	16.91 ± 0.34	15.89 ±
Fe_2O_3 T	12.42 ± 0.26	10.60 ± 0.88	9.71 ± 0.50	7.84 ± 0.82	11.05 ± 0.54	8.71 ± 0.52	6.32 ± 0.85	4.57 ±
MnO	0.18 ± 0.00	0.19 ± 0.00	0.17 ± 0.01	0.15 ± 0.01	0.20 ± 0.00	0.13 ± 0.01	0.10 ± 0.02	0.10 ±
MgO	6.93 ± 0.37	4.85 ± 0.54	4.01 ± 0.45	3.27 ± 0.57	4.41 ± 0.69	3.76 ± 0.28	2.11 ± 0.83	1.15 ±
CaO	11.87 ± 0.11	10.22 ± 0.12	9.05 ± 0.56	8.16 ± 0.26	10.00 ± 0.93	7.62 ± 0.30	5.44 ± 0.64	3.75 ±
Na_2O	2.17 ± 0.18	2.46 ± 0.10	2.83 ± 0.13	2.84 ± 0.00	2.58 ± 0.23	2.75 ± 0.12	3.32 ± 0.35	3.75 ±
K_2O	0.70 ± 0.5	0.97 ± 0.19	1.24 ± 0.22	1.49 ± 0.09	1.08 ± 0.4	2.81 ± 0.30	3.32 ± 0.37	3.25 ±
TiO_2	1.24 ± 0.04	1.02 ± 0.07	0.95 ± 0.00	0.83 ± 0.01	0.98 ± 0.02	0.78 ± 0.02	0.67 ± 0.06	0.58 ±
P_2O_5	0.15 ± 0.01	0.18 ± 0.03	0.22 ± 0.03	0.21 ± 0.01	0.21 ± 0.02	0.25 ± 0.01	0.19 ± 0.03	0.17 ±
H_2O^+	0.25 ± 0.08	0.35 ± 0.08	0.56 ± 0.17	0.40 ± 0.10	0.60 ± 0.17	0.58 ± 0.16	0.39 ± 0.12	0.50 ±
H_2O^-	1.61 ± 0.15	1.50 ± 0.21	0.80 ± 0.24	0.61 ± 0.18	1.70 ± 0.25	1.18 ± 0.35	1.49 ± 0.38	1.64 ±
Trace elements (p.p.m.)								
Li	11 ± 1	10 ± 2	14 ± 2	16 ± 2	11 ± 1	24 ± 13	19 ± 5	17 ±
Rb	17 ± 3	21 ± 3	30 ± 5	37 ± 2	23 ± 3	89 ± 12	109 ± 15	131 ±
Sr	455 ± 33	361 ± 52	348 ± 32	367 ± 22	544 ± 59	438 ± 23	376 ± 20	322 ±
Ba	107 ± 9	188 ± 34	265 ± 42	346 ± 3	194 ± 4	291 ± 25	395 ± 28	504 ±
V	438 ± 7	321 ± 8	250 ± 23	196 ± 43	303 ± 23	216 ± 14	140 ± 43	50 ±
Cr	63 ± 21	19 ± 13	13 ± 8	8 ± 3	18 ± 2	20 ± 7	11 ± 4	5 ±
Co	40 ± 1	29 ± 3	23 ± 3	20 ± 3	37 ± 5	29 ± 5	17 ± 4	12 ±
Ni	41 ± 4	11 ± 1	12 ± 3	10 ± 4	19 ± 1	17 ± 3	11 ± 3	5 ±
Cu	84 ± 5	29 ± 22	51 ± 20	27 ± 8	134 ± 11	90 ± 30	51 ± 20	15 ±
Zn	93 ± 5	92 ± 5	90 ± 5	79 ± 6	99 ± 1	80 ± 6	70 ± 6	75 ±

† n = number of samples.
‡ Average values ± 1 S.D.

Major elements

The SiO_2 contents of volcanic rocks from Sardinia vary from c. 44 to 71 per cent. The range of SiO_2 within the individual volcanoes is, however, variable. In some of them, such as Mt Seda Oro, the SiO_2 ranges in a wide interval from 48 to 66 per cent while in others, such as Mt Fromma, the variations of SiO_2 are rather limited (50–55 per cent).

The volcanic rocks studied are quartz normative, have high contents of CaO and Al_2O_3 and high Fe^{3+}/Fe^{2+} ratios, but they are low in TiO_2 and have low $MgO/(MgO + FeO_{tot})$ ratios. The rocks from the individual volcanoes have variable abundances of alkalis, particularly K_2O, which gradually increase with the increase of SiO_2. On the graph of ($Na_2O + K_2O$) against

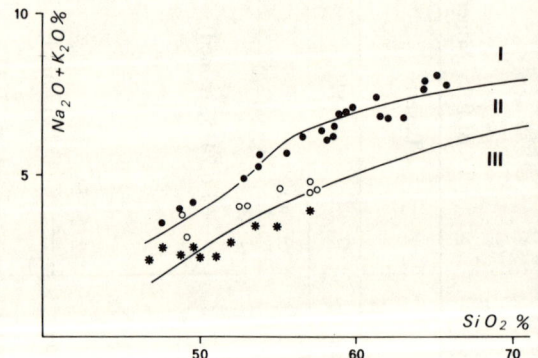

Fig. 2 Graph of ($Na_2O + K_2O$) against SiO_2 for andesitic rocks from Mt Seda Oro (full circles), Mt Cannisones-Ozzastru (empty circles), and Marmilla (stars). The solid lines delineate the fields of alkaline (I), calc-alkaline (II), and tholeiitic (III) rocks of Kuno (1959)

	Cannisones–Ozzastru (SA$_1$ + SA$_2$)			Fromma (SA$_2$)		Larenta (SA$_3$)		
	n = 2	n = 2	n = 2	n = 2	n = 2	n = 3	n = 5	n = 2
Major elements (%)								
)$_2$	49.02 ± 0.11	54.00 ± 1.43	57.11 ± 0.00	50.29 ± 0.15	55.60 ± 0.35	53.06 ± 0.17	64.37 ± 1.42	71.05 ± 0.25
O$_3$	19.00 ± 0.20	18.74 ± 1.39	17.17 ± 0.13	18.40 ± 0.15	17.33 ± 0.00	17.62 ± 0.31	15.94 ± 0.74	14.67 ± 0.02
O$_3$ T	11.10 ± 0.10	8.07 ± 0.32	7.74 ± 0.16	10.25 ± 0.10	7.44 ± 0.29	8.52 ± 0.25	4.66 ± 0.73	2.65 ± 0.06
O	0.20 ± 0.00	0.18 ± 0.01	0.17 ± 0.01	0.20 ± 0.00	0.17 ± 0.01	0.16 ± 0.01	0.10 ± 0.01	0.04 ± 0.06
O	4.58 ± 0.14	2.96 ± 0.86	3.44 ± 0.16	5.20 ± 0.28	3.30 ± 0.46	4.69 ± 0.18	1.32 ± 0.27	0.29 ± 0.02
O	10.15 ± 0.15	8.71 ± 1.14	7.25 ± 0.17	9.64 ± 0.08	7.91 ± 0.36	8.62 ± 0.43	4.50 ± 0.41	2.39 ± 0.00
$_2$O	2.14 ± 0.29	2.74 ± 0.11	2.68 ± 0.11	2.37 ± 0.01	2.84 ± 0.01	2.69 ± 0.14	3.53 ± 0.14	3.43 ± 0.37
O	1.06 ± 0.08	1.33 ± 0.23	1.63 ± 0.04	0.97 ± 0.01	1.72 ± 0.07	1.40 ± 0.52	2.76 ± 0.31	4.11 ± 0.12
)$_2$	0.97 ± 0.01	0.73 ± 0.02	0.68 ± 0.02	0.86 ± 0.01	0.70 ± 0.04	0.69 ± 0.02	0.41 ± 0.06	0.25 ± 0.19
)$_5$	0.15 ± 0.01	0.18 ± 0.01	0.17 ± 0.00	0.23 ± 0.04	0.32 ± 0.01	0.24 ± 0.01	0.21 ± 0.06	0.11 ± 0.00
)$^-$	0.42 ± 0.9	0.70 ± 0.12	0.59 ± 0.17	0.55 ± 0.14	0.70 ± 0.19	0.39 ± 0.54	0.45 ± 0.23	0.20 ± 0.00
)$^+$	1.10 ± 0.36	1.38 ± 0.34	1.20 ± 0.28	0.67 ± 0.25	1.73 ± 0.27	1.60 ± 0.33	1.40 ± 0.26	0.70 ± 0.00
Trace elements (p.p.m.)								
	11 ± 1	14 ± 5	12 ± 4	12 ± 1	13 ± 2	13 ± 1	15 ± 3	25 ± 7
	29 ± 2	39 ± 8	54 ± 2	19 ± 1	50 ± 1	37 ± 10	75 ± 8	137 ± 6
	400 ± 5	358 ± 38	302 ± 3	620 ± 10	539 ± 9	711 ± 45	665 ± 112	422 ± 4
	150 ± 10	245 ± 5	267 ± 12	196 ± 10	297 ± 3	298 ± 49	506 ± 47	614 ± 14
	323 ± 39	193 ± 9	189 ± 10	265 ± 1	165 ± 20	211 ± 10	55 ± 19	34 ± 13
	15 ± 3	8 ± 3	7 ± 1	35 ± 5	24 ± 6	46 ± 35	4 ± 1	4 ± 1
	32 ± 2	21 ± 3	21 ± 1	30 ± 4	22 ± 3	29 ± 6	11 ± 2	4 ± 1
	14 ± 1	9 ± 1	8 ± 1	21 ± 2	16 ± 4	21 ± 7	5 ± 1	6 ± 2
	102 ± 38	31 ± 19	30 ± 3	77 ± 11	46 ± 2	60 ± 6	16 ± 4	21 ± 6
	99 ± 1	75 ± 3	73 = 1	88 ± 4	70 ± 7	88 ± 41	55 ± 19	39 ± 11

SiO$_2$ (Fig. 2) most rocks lie in the hypersthene (calc-alkaline) field of Kuno (1959). Some rocks from southern Sardinia (e.g. Marmilla and Arcuentu volcanoes) fall into the tholeiitic field, while some from northern Sardinia straddle the calc-alkaline–alkaline boundary. Subtle differences between the andesitic suites from southern and northern Sardinia are also shown on the AFM diagram (Fig. 3) and by the variations of TiO$_2$. On the AFM diagram, the rocks from northern Sardinia have typical calc-alkaline linear trends towards the alkaline apex while the rocks from southern Sardinia show small but distinct Fe-enrichment trends, characteristic of tholeiitic series. In northern Sardinia, TiO$_2$ shows a negative correlation with SiO$_2$ but in the south, TiO$_2$ remains nearly constant over the observed intervals of SiO$_2$ content or even

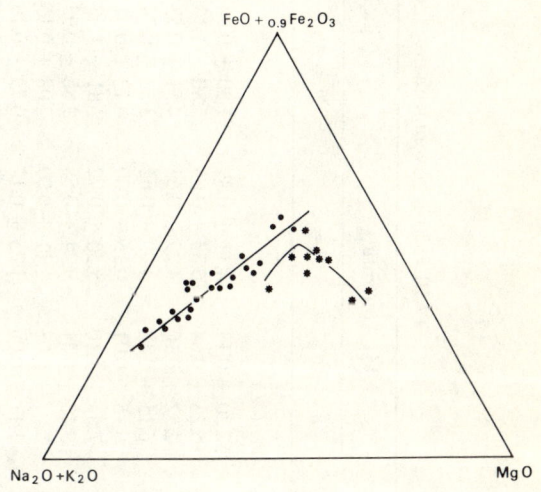

Fig. 3 AFM diagram for andesitic rocks from Mt Seda Oro (full circles) and Marmilla (stars)

TABLE 2(b) Average major and trace element compositions of volcanic rocks from southern Sardinia

	Mt Arcuentu		La Marmilla				Mt Narcao		Sarroch		Cixerri
	$n = 3$†	$n = 5$	$n = 4$	$n = 1$	$n = 1$	$n = 1$	$n = 2$	$n = 4$	$n = 1$	$n = 1$	$n = 5$

Major elements (%)

SiO_2	52.20 ± 0.13‡	54.11 ± 0.39	51.30 ± 0.84	54.76	56.79	51.83	54.54 ± 1.09	57.17 ± 0.67	52.32	53.44	58.00 ± 0.68
Al_2O_3	15.23 ± 0.64	16.33 ± 0.65	15.80 ± 1.13	17.80	17.25	18.00	17.77 ± 0.63	16.75 ± 0.65	17.55	19.20	17.06 ± 0.22
Fe_2O_3	2.52 ± 0.10	4.29 ± 0.63	4.51 ± 0.53	3.54	4.08	3.68	3.56 ± 0.45	3.35 ± 0.64	4.56	3.40	3.76 ± 0.61
FeO	6.15 ± 0.48	4.00 ± 0.62	3.32 ± 0.74	3.50	1.56	5.32	4.10 ± 0.75	3.52 ± 0.48	3.98	5.34	1.66 ± 0.31
MnO	0.16 ± 0.01	0.15 ± 0.00	0.12 ± 0.01	0.13	0.14	0.15	0.15 ± 0.01	0.13 ± 0.00	0.16	0.26	0.13 ± 0.02
MgO	7.59 ± 1.10	5.50 ± 0.72	7.81 ± 1.48	4.18	3.72	4.84	3.81 ± 0.01	4.15 ± 0.71	5.26	3.98	2.52 ± 0.15
CaO	9.13 ± 0.31	9.24 ± 0.68	9.88 ± 0.93	7.90	6.24	9.34	8.24 ± 0.06	7.68 ± 0.31	9.47	8.84	7.00 ± 0.40
Na_2O	2.07 ± 0.03	2.11 ± 0.09	2.08 ± 0.15	2.18	3.12	2.65	2.80 ± 0.10	2.57 ± 0.70	2.41	2.58	3.20 ± 0.14
K_2O	1.11 ± 0.09	1.25 ± 0.17	0.79 ± 0.19	1.15	1.05	0.87	1.35 ± 0.21	1.74 ± 0.14	0.58	0.87	1.73 ± 0.26
TiO_2	0.72 ± 0.06	0.84 ± 0.09	0.57 ± 0.05	1.06	0.45	1.00	0.82 ± 0.03	0.74 ± 0.10	0.56	0.99	0.58 ± 0.05
P_2O_5	0.14	0.22 ± 0.08	0.17 ± 0.02	0.34	0.18	0.22	0.37 ± 0.16	0.25 ± 0.00	0.30	0.21	0.50 ± 0.04
H_2O^+	0.10	0.50 ± 0.24	1.80 ± 0.30	1.10	0.33	0.22	0.86 ± 0.17	0.77 ± 0.35	0.25	0.34	0.35 ± 0.03
H_2O^-	2.80	0.63 ± 0.38	1.76 ± 0.28	1.47	3.84	1.36	0.53 ± 0.32	0.35 ± 0.29	1.20	0.36	2.63 ± 0.45

Trace elements (p.p.m.)

Li	9 ± 4	7 ± 1	9 ± 2	11	14	8	9 ± 1	9 ± 1	4	6	13 ± 3
Rb	31 ± 11	32 ± 10	19 ± 7	35	25	22	42 ± 21	61 ± 10	7	17	37 ± 16
Sr	199 ± 16	238 ± 32	296 ± 160	277	281	290	282 ± 28	318 ± 15	366	240	606 ± 310
Ba	252 ± 60	402 ± 201	202 ± 27	650	n.d.	n.d.	n.d.	462 ± 72	285	310	683 ± 376
V	238 ± 23	209 ± 15	208 ± 15	170	150	209	187 ± 4	179 ± 13	234	170	130 ± 39
Cr	231 ± 181	148 ± 110	292 ± 225	55	20	34	17 ± 9	78 ± 113	93	15	15 ± 14
Co	39 ± 6	27 ± 7	30 ± 7	16	17	19	19 ± 1	23 ± 6	25	18	11 ± 3
Ni	96 ± 68	35 ± 16	89 ± 64	16	12	12	9 ± 3	21 ± 20	17	9	9 ± 4
Cu	84 ± 58	38 ± 25	62 ± 28	14	46	34	42 ± 1	41 ± 21	41	13	13 ± 5
Zn	80 ± 4	87 ± 5	75 ± 10	94	72	70	87 ± 3	81 ± 8	92	93	73 ± 10

† n = number of samples.
‡ Average values ± 1 s.D.
n.d. = not determined.

TABLE 2(c) Average major and trace element compositions of volcanic rocks from northern Sardinia

	$n = 2$†	$n = 7$	$n = 5$
Major elements (%)			
SiO_2	52.02 ± 0.0‡	54.37 ± 0.72	57.94 ± 1.40
Al_2O_3	17.25 ± 0.42	16.02 ± 1.44	17.52 ± 0.95
Fe_2O_3	4.17 ± 0.42	4.21 ± 0.59	3.76 ± 0.61
FeO	4.45 ± 0.04	3.53 ± 1.05	2.92 ± 0.76
MnO	0.14 ± 0.01	0.13 ± 0.02	0.13 ± 0.02
MgO	4.34 ± 0.14	3.50 ± 1.09	2.83 ± 0.62
CaO	8.80 ± 0.38	8.08 ± 0.61	6.93 ± 0.51
Na_2O	2.69 ± 0.01	2.81 ± 0.26	3.09 ± 0.18
K_2O	2.90 ± 0.11	2.69 ± 0.69	2.16 ± 0.40
TiO_2	0.85 ± 0.01	0.77 ± 0.08	0.67 ± 0.03
P_2O_5	0.29 ± 0.04	0.24 ± 0.07	0.21 ± 0.04
H_2O^+	0.67 ± 0.16	1.09 ± 0.59	1.54 ± 0.35
H_2O^-	1.01 ± 0.04	0.72 ± 0.23	0.57 ± 0.18
Trace elements (p.p.m.)			
Li	18 ± 6	19 ± 4	12 ± 3
Rb	113 ± 2	99 ± 35	86 ± 6
Sr	520 ± 62	385 ± 107	300 ± 19
Ba	238 ± 88	299 ± 28	385 ± 65
V	243 ± 18	208 ± 58	164 ± 43
Cr	39 ± 4	24 ± 15	10 ± 8
Co	34 ± 3	25 ± 5	23 ± 4
Ni	24 ± 1	14 ± 4	10 ± 2
Cu	103 ± 38	87 ± 36	53 ± 27
Zn	85 ± 1	79 ± 6	74 ± 4

† n = number of samples.
‡ Average values ± 1 S.D.

increases slightly towards the intermediate rocks. Such trends are similar to those of island arc tholeiites (Ewart *et al.*, 1973).

Trace and minor elements

Li, Rb, Sr, and Ba

Li is low in the Sardinian andesitic rocks and is within the range of values for andesitic rocks (Taylor *et al.*, 1969; Dupuy and Lefévre, 1974). It ranges between 4 and 25 p.p.m., although in more than 75 per cent of the rocks analysed Li varies in a relatively narrow interval from 10 to 15 p.p.m. Li initially shows a positive correlation with increasing SiO_2 content, but it decreases in the more differentiated rocks. The decrease of Li is probably due to the crystallization of plagioclase which, in the more differentiated rocks, has a partition coefficient ($D^{s/l}$) greater than 1 (Dupuy and Coulon, 1973).

Rb and Ba show large variations in the rocks analysed, ranging from 17 to 137 p.p.m. and from 107 to 614 p.p.m., respectively. Both these elements exhibit a systematic increase with increasing fractionation, although in the individual volcanoes the rocks with high Rb and Ba abundances show a distinct scatter. In most volcanic complexes, Sr has a negative correlation with SiO_2 probably due to the crystallization of plagioclase with $D^{s/l} > 1$ (Dupuy and Coulon, 1973). However, in several southern Sardinian volcanoes (Arcuentu, Marmilla, and Narcao), Sr is relatively constant in various rocks, suggesting that plagioclase fractionation played only a limited role similar to that in island arc tholeiitic series (Ewart *et al.*, 1973). In addition to the variations within the individual volcanoes, there are differences in the abundances of these elements (Sr, Rb, Ba) and also of K, among equivalent rocks of the different volcanoes.

Rare earth elements

The abundances of REE in 15 representative samples of andesitic rocks are given in Table 3. The basalts and basic andesites can be divided into two groups on the basis of their geographic distribution and the slope of their chondrite-normalized REE patterns. The rocks from northern and central Sardinia have REE patterns typical of calc-alkaline rocks (Jakeš and Gill, 1970) with an enrichment of light REE and gradual depletion of heavy REE. The REE patterns of the samples from southernmost Sardinia have only small, light REE enrichment and unfractionated heavy REE. Their patterns are similar to continental tholeiites (Herrmann, 1970).

The REE patterns of the various rock types from the two neighbouring volcanoes (Mt Seda Oro and Mt Larenta) are shown in Fig. 4. The REE contents and the La/Yb ratio for the rocks from Mt Seda Oro increase with the increase of SiO_2 while in Mt Larenta, basalt, dacite, and rhyolite have similar light REE abundances. As

TABLE 3 REE abundances (in p.p.m.) in volcanic rocks from Sardinia

	Cuguttada	Bosano	Seda Oro					Larenta			Arcuentu		Marmilla	Narcao	Sarroch
	1531†	1587	1532	1570	1795	1569	906	1528	1526	1566	2059	2351	2325	2348	2343
La	7.5	18.4	15.0	23.9	24.5	28.9	30.3	36.0	37.4	39.9	15.5	14.8	13.9	5.0	12.5
Ce	17.7	38.2	34.5	53.5	53.5	59.7	64.5	71.9	71.0	73.3	34.1	31.7	32.4	12.8	29.8
Nd	12.5	20.0	22.4	27.4	—	28.9	—	31.7	30.5	28.5	—	—	—	—	—
Sm	3.15	4.51	4.80	5.57	5.80	6.02	6.20	5.89	4.86	4.83	3.62	3.45	4.12	2.10	4.50
Eu	1.04	1.25	1.49	1.39	1.30	1.44	1.50	1.59	1.22	1.19	1.02	0.87	1.16	0.57	1.25
Gd	3.71	3.74	3.83	4.31	—	4.76	—	4.23	3.35	3.33	—	—	—	—	—
Tb	0.63	0.63	0.63	0.69	0.81	0.72	1.00	0.63	0.46	0.48	0.62	0.68	0.67	0.38	0.93
Ho	0.74	0.89	0.86	0.95	—	0.96	—	0.88	0.67	0.72	—	—	—	—	—
Yb	1.94	2.14	1.99	2.13	2.60	2.47	3.01	2.00	1.74	1.89	2.10	2.20	2.26	1.65	3.82
Lu	0.29	0.39	0.32	0.33	0.39	0.38	0.46	0.32	0.29	0.32	0.33	0.36	0.35	0.28	0.61

† *Sample numbers*: basalts, 1531, 1532, 2059, 2351; basic andesites, 1587, 1570, 1528, 2325, 2343, 2348; andesites, 1569, 1795; dacites, 906, 1526; rhyolite, 1566.

with K, Rb, Sr, and Ba, equivalent rocks from various volcanoes frequently differ to a large extent in REE abundances while their variations within the individual volcanoes are frequently limited.

Transition elements (V, Cr, Co, Ni, Cu, and Zn)

The average contents of basaltic rocks from the individual volcanic complexes are plotted in Fig. 5. The basalts from Marmilla and Arcuentu have concentrations of transition elements similar to ocean-floor tholeiites with the exception of a slight depletion of Ti in the studied rocks. All other basalts are low in Ni and Cr and have low Ni/Co and high V/Ni ratios, features typical of volcanic rocks of orogenic zones (Taylor *et al.*, 1969; Cole, 1973; Lewis, 1971; Brown *et al.*, 1977). Transition elements show a negative correlation with SiO_2 except for

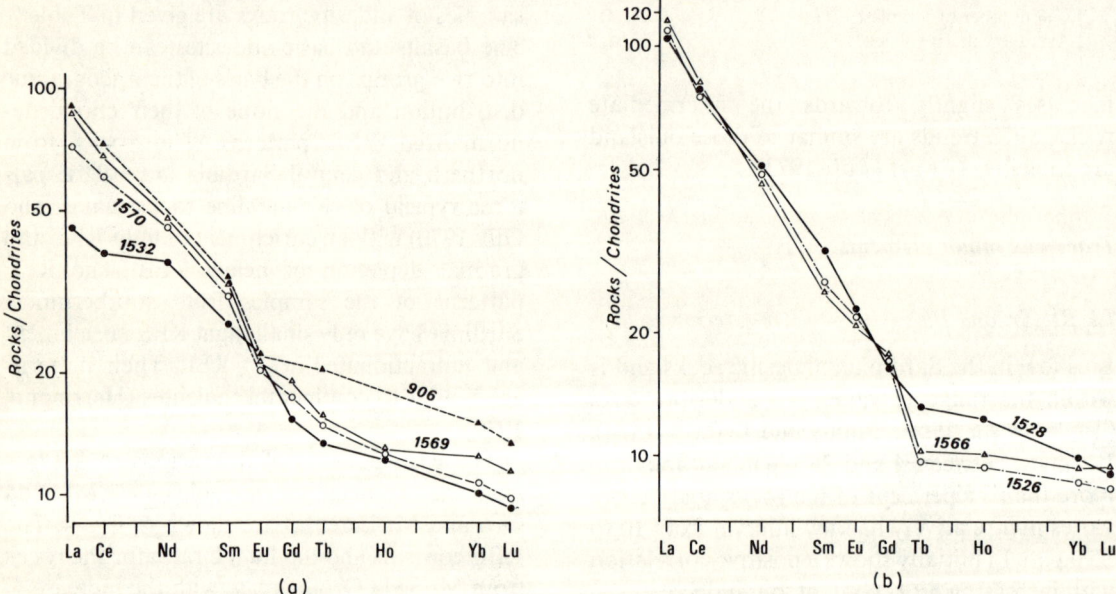

Fig. 4 Chondrite-normalized REE abundances for volcanic rocks from (a) Mt Seda Oro and (b) Mt Larenta. Chondrite-normalizing values from Frey *et al.* (1968)

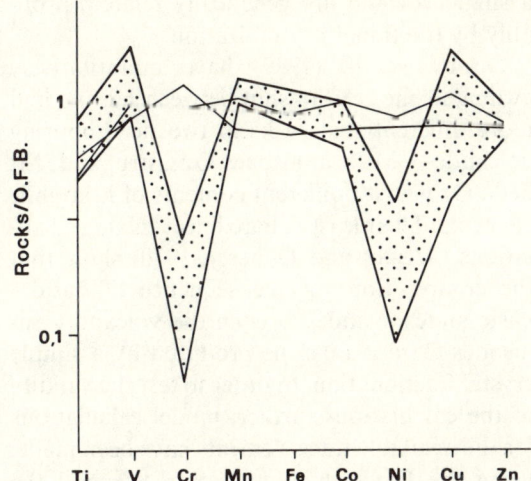

Fig. 5 Transition element abundances of basaltic rocks from Sardinia normalized to ocean-floor basalts (Andriambololona, 1976). Dashed area, field of variation for all high Al basalts except those of Marmilla and Arcuentu, which are represented by continuous lines

V. In the southern Sardinian volcanoes, V remains nearly constant or only slightly decreases with the increase of SiO_2 while in northern Sardinia it decreases rapidly. The differences in the trends of V are probably related to the modal proportions of Fe–Ti oxide microphenocrysts in the rocks. The low amount of Fe–Ti oxide in the rocks from southern Sardinia is consistent with their tholeiitic affinities (Miyashiro and Shido, 1975). As rock-forming minerals of andesitic rocks have $D^{s/l}$ for Cu less than 1, the negative correlation of Cu and SiO_2 implies the crystallization of a sulphide phase (Andriambololona and Dupuy, 1978).

Regional variations

The mineralogical and geochemical features of calc-alkaline volcanic rocks from Sardinia are closely comparable to those from other orogenic areas (Lopez-Escobar et al., 1977; Thorpe et al., 1976; Taylor et al., 1969). Table 2 shows that equivalent rocks from the different volcanoes of Sardinia differ in their abundances of K and other lithophile elements. These differences also appear when two neighbouring volcanoes are compared. The low contents of lithophile elements in the volcanic complex of Mt Cannizones-Ozzastru (Fig. 1) resemble island arc calc-alkaline rocks (Jakeš and White, 1972) while the volcanics of Mt Seda Oro are similar in their abundances to continental margin (Andean) rocks (Dupuy and Lefèvre, 1974; Dostal et al., 1977). It is noteworthy that the content of SiO_2 in the most evolved rocks of Mt Cannizones-Ozzastru is c. 57 per cent while in Mt Seda Oro it reaches 66 per cent. Furthermore, several volcanoes from southern Sardinia display tholeiitic fractionation trends for Fe, Ti, and V (Coulon, 1977), although their relatively high content of lithophile elements is typical of calc-alkaline rocks of island arcs (Jakeš and White, 1972).

The calc-alkaline volcanic rocks of Sardinia also show systematic regional variations for several lithophile elements. In rocks of a given SiO_2 content, the abundances of K, Li, Rb, and Sr and the K/Na, K/Ba, and Rb/Sr ratios increase while the K/Rb ratio decreases from south to north. These variations, together with the occurrence of rocks with tholeiitic fractionation trends in southern Sardinia, led Coulon and Dupuy (1975) to suggest the presence, during the Cainozoic, of a northward-dipping subduction zone under Sardinia. This interpretation is in agreement with the recent geodynamic models for Sardinia (Alvarez, 1972; Auzende et al., 1973).

Petrogenesis

Two aspects of the petrogenesis of the Sardinian calc-alkaline volcanic rocks are considered. The first is the relationship between the various rock-types within the individual volcanoes and the second is the possible origin of the basalts, which represent the most 'primitive' exposed magma.

Relationships between various rock-types

The two types of element variations observed in the Sardinian rocks, variations among various rocks within a single volcano and among the

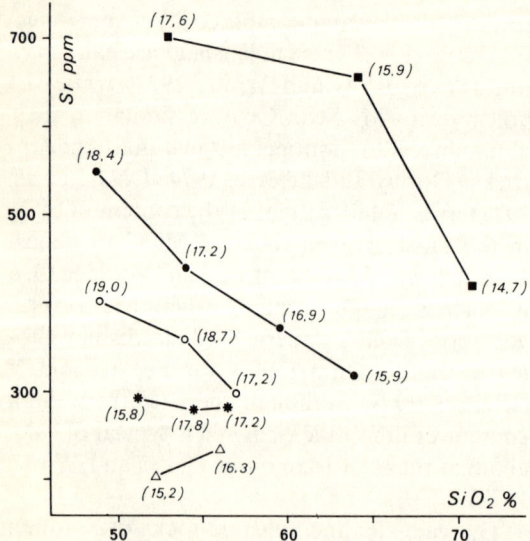

Fig. 6 The variations of Sr against SiO_2 in the individual rock-types of several andesitic complexes from Mt Larenta (squares), Mt Seda Oro (full circles), Mt Cannisones-Ozzastru (empty circles), Marmilla (stars), and Mt Arcuentu (triangles). The numbers in parentheses are the averages of Al_2O_3 contents (Table 2)

equivalent rocks of different volcanoes, are well illustrated by the variations of Sr with respect to SiO_2 (Fig. 6). In individual volcanoes, which are distinctly separated on this plot, Sr either decreases, or increases, or remains nearly constant with increasing SiO_2. The variations of Sr are paralleled by those of Al_2O_3 (Fig. 6). Such variations, together with other mineralogical and chemical data, suggest that the rocks from a single volcano are genetically related, probably by fractional crystallization.

Dupuy et al. (1979) have quantitatively evaluated the relationship between basalts and more differentiated rocks in two neighbouring volcanoes—Mt Cannisones-Ozzastru and Mt Seda Oro—with different contents of lithophile elements. Their least-square linear mixing calculations (Wright and Doherty, 1970) show that the compositions of the sequence of basalt-basic andesite-andesite from the volcano Cannisones-Ozzastru can be produced by a simple crystal fractionation. In order to test the validity of the calculations further, model calculations for the available trace elements have been made, using the Rayleigh fractionation law and the partition coefficients given in Table 4, together with the proportions of phases obtained from the least-square calculations. The results are shown in Table 5 and are consistent with a simple low pressure fractional crystallization model interrelating basalts, basic andesites, and andesites. Such a process was dominated by plagioclase with subordinate clinopyroxene and Fe-Ti oxide, together with olivine in the early stages. However, fractional crystallization alone cannot explain the evolutionary trend of the Mt Seda Oro volcanic complex, which has high contents of lithophile elements comparable to continental margin andesitic series. Thus Dupuy et al. (1979) invoked a mechanism involving both fractional crystallization and crustal contamination, specifically an interaction with ignimbrites

TABLE 4 Partition coefficients (solid/liquid) obtained for minerals from three basalts of Mt Seda Oro

	Plagioclase	Clinopyroxene	Fe–Ti Oxide	Olivine
Rb	0.3 ± 0.1	0.0	0.0	0.0
Sr	2.0 ± 0.2	0.1	0.0	0.0
Ba	0.5 ± 0.1	0.0	0.0	0.0
V	0.0	0.8 ± 0.2	20.0 ± 2.0	0.0
Cr	0.0	2.0 ± 0.3	16.0 ± 3.0	0.5 ± 0.1
Co	0.0	2.0 ± 0.2	5.0 ± 1.0	4.0 ± 1.0
Ni	0.0	1.8 ± 0.3	4.0 ± 1.0	10.0 ± 2.0
Zn	0.0	3.0 ± 1.0	6.0 ± 1.0	3.0 ± 0.5

The measured partition coefficients (D) from basic andesites differ from those of basalts for Cr and Zn of clinopyroxene ($D_{Cr} = 5$; $D_{Zn} = 4$). The values are average values ± 1 S.D.

TABLE 5 Model calculations of fractional crystallization for Mt Cannisones-Ozzastru

		P			Abundances in Residual Liquid							
F	Plg	Cpx	Fe–Ti Oxide	Ol	Rb	Sr	Ba	V	Cr	Co	Ni	Zn
1	0.54	61	15	9	15	48	346	230	183	9.1	26	7.0	75
						(4)	(26)	(18)	(31)	(2.4)	(3)	(1.5)	(9)
2	0.69	78	18	5		50	288	307	182	6.0	24	10.0	74
						(6)	(27)	(25)	(30)	(2.0)	(4)	(1.8)	(10)

1 and 2 represent the calculated abundances (in p.p.m.) in basic andesites and andesites, respectively. Degrees of solidification (F) and mineral proportions (P) used are from the least-square calculations for the major elements given by Dupuy et al. (1979). Figures in parentheses are estimated errors of calculations due to the uncertainty in the values of partition coefficients.

which were generated by crustal anatexis (Coulon et al., 1978) and are spatially and temporally associated with andesitic rocks. Contamination of the rocks of Mt Seda Oro is also indicated by the high and variable Sr isotope data ($^{87}Sr/^{86}Sr$ ratios range from 0.7044 to 0.7081; Dupuy et al., 1974) and is consistent with the abundances of trace elements. The variations of Ba against Rb and Ba against La in the rocks of Mt Seda Oro are plotted in Fig. 7. According to the Rayleigh fractionation law, the contents of these elements during fractional crystallization should fall on a straight line which passes through the origin. The rocks from Mt Seda Oro, however, do not fall on such a line and their variation trends are extended towards the position of ignimbrites. In order to ascertain whether similar variation trends are present in other volcanic series of orogenic zones, the data on the rocks of New Zealand, southern Peru (Arequipa), and the younger Islands of Tonga are also plotted in Fig. 7. The volcanic rocks of Tonga were emplaced in an area devoid of continental crust while the other series were erupted onto continental crust and are closely associated with voluminous ignimbrites. Figure 7 shows that the trends of the volcanic rocks of Tonga, which have the lowest abundances of lithophile elements, are consistent with the fractional crystallization model. On the other hand, the andesitic lavas of the other suites display trends in a direction towards the ignimbrites and the contents of some lithophile elements in the andesitic series appear to be related to their abundances in associated ignimbrites. The high contents of these elements in andesites from southern Peru are accompanied by their high concentrations in ignimbrites, while in New Zealand both andesites and ignimbrites have lower abundances (Fig. 7). This indicates that the rocks emplaced on continental crust were also affected by interaction with the associated ignimbrites. It may also suggest that the differences in the lithophile element abundances between Andean (continental margin) and island arc andesitic rocks are, to a large degree, related to crustal contamination.

Origin of basalts

Two types of basalt can be recognized on the basis of their chemical composition, including REE patterns. The first type, from the Logudoro–Bosano area in northern Sardinia, has typical calc-alkaline features while the second type, from southern Sardinia, shows tholeiitic affinities.

The calc-alkaline basalts from Cuguttada volcano (Table 2) represent the most 'primitive' magma type of the andesitic suites in northern Sardinia. In order to evaluate whether other northern Sardinian basalts can be derived from these rocks, a series of least-squares linear mixing calculations (Wright and Doherty, 1970) have been made for combinations of whole-rock and phenocryst major element compositions. For the sake of simplicity only weight fractions of variables are reported in Table 6. The differences between the calculated and observed compositions are very small and are not significant. The

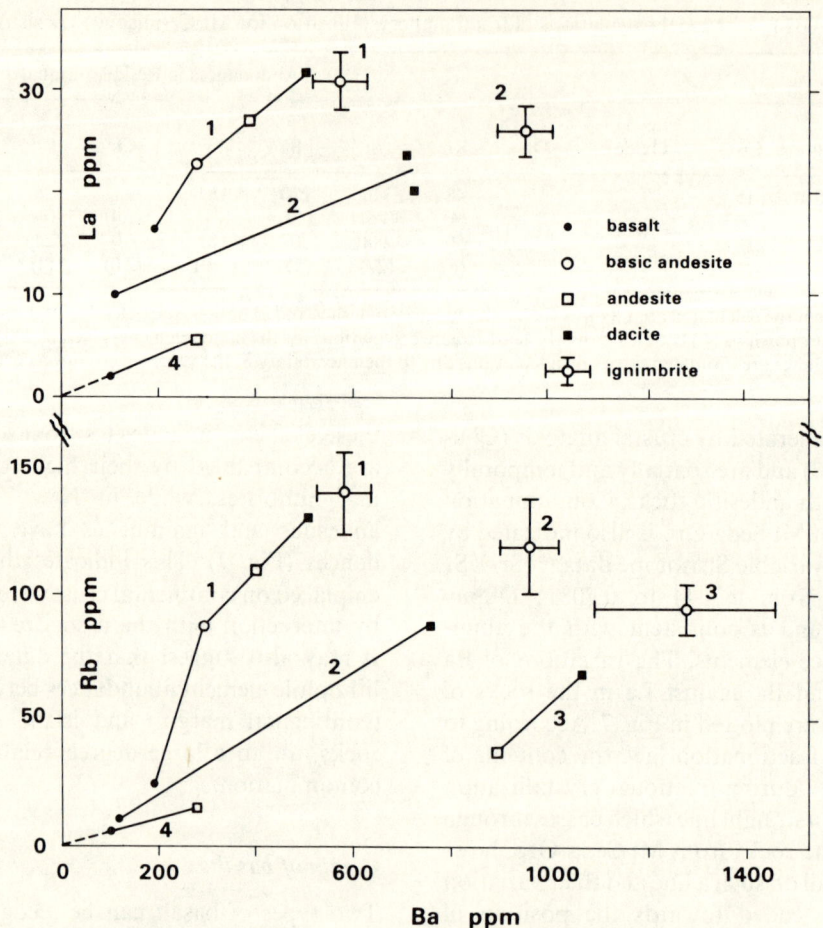

Fig. 7 The relations between Ba and Rb and between Ba and La in andesitic series and associated ignimbrites of Mt Seda Oro (1: Dupuy *et al.*, 1979), New Zealand (2: Ewart *et al.*, 1968; Cole, 1978), southern Peru–Arequipa (3: our unpublished data), and Tonga (4: Ewart *et al.*, 1973). The bars for ignimbrites correspond to 1 S.D. (1: $n = 18$; 2: $n = 18$; 3: $n = 13$)

results suggest that basalts from other volcanic complexes of northern Sardinia could have been derived from a primitive magma with a composition similar to that of basalts from Cuguttada. Such a process probably involved fractional crystallization of clinopyroxene, olivine, titanomagnetite, and plagioclase in the proportions given in Table 6. Table 7 shows the results of the model fractional crystallization calculations for trace elements using the degrees of solidification and mineral proportions given in Table 6 and the partition coefficients given in Table 4. A generally good agreement between calculated and observed trace element abundances is consistent with the 'parental' nature of the basalts from Cuguttada. A discrepancy in Ba and Rb for some basaltic rocks might be the result of crustal contamination.

The composition of the basalts from Cuguttada is similar to those of the New Hebrides (Gorton, 1977), particularly in their REE abundances, and to tholeiitic basalts from the Tyrrhenian Sea (Dietrich *et al.*, 1977). They also have characteristic features of island arc basalts (Ewart, 1976b). The slight deviations in K and Rb might be related either to secondary processes

TABLE 6 Least-squares calculations of fractional crystallization (in per cent by weight) for derivations of basalts from "primitive magma" of Cuguttada

	Basalt	Cpx	Fe–Ti Oxide	Ol	Plg
1. Bosano	50.4	13.8	4.7	7.4	23.7
2. Seda Oro	47.0	13.9	4.5	8.3	26.3
3. Cannizones	48.4	14.9	4.5	7.8	24.4
4. Fromma	43.7	15.0	5.6	7.5	28.2

The compositions of whole rocks are given in Table 2 while those of constituent phases were reported by Coulon (1977).

such as contamination or alteration, or to differences in composition of the source material. Regarding the origin of the Cuguttada basalts, Dostal et al. (1976) have argued that they are of upper mantle origin. However, their relatively low $MgO/(MgO + FeO_{tot})$ ratio and the Ni and Cr contents indicates that the basalts had already undergone low P fractionation, which might have also led to an increase of K, Rb, and Ba. The model calculations for mantle anatexis using the crystal–liquid partition coefficients suggest that the differences in the REE patterns between the basalts from southern and northern Sardinia can be attributed to their derivation from upper mantle peridotite with different mineral assemblages (Dostal et al., 1976). The basalts with calc-alkaline affinities from northern Sardinia could have been derived from garnet peridotite while the basalts with tholeiitic affinities may have been derived from a spinel peridotite source (Dostal et al., 1976). However, equivalent calculations employing the crystal–vapour partition coefficients of Mysen (1978) indicate that the basalts could be produced by partial melting of a single upper mantle peridotite (either garnet- or spinel-bearing) which experienced variable degrees of metasomatic alteration (Mysen, 1978).

Conclusions

The various rock-types in tholeiitic suites occurring in island-arc environments are generally considered to be interrelated by fractional crystallization processes (Ewart et al., 1973; Kay, 1977). On the other hand, simple low pressure crystal fractionation cannot explain the variations observed within the individual series of calc-alkaline volcanic rocks (Kay, 1977; Gill, 1978). The most plausible explanation for the variations in chemical composition of the Mt Seda Oro volcano is a mechanism involving both fractional crystallization and

TABLE 7 Model calculations of fractional crystallization for the derivation of basalts of Mt Seda Oro from basalts of Cuguttada

	Rb	Sr	Ba	V	Cr	Co	Ni	Zn
Calculated	32	492	186	239	14	32	15	86
	(6)	(42)	(23)	(73)	(9)	(4)	(3)	(8)
Observed	23	554	184	303	18	37	19	99
	(3)	(59)	(4)	(23)	(2)	(5)	(1)	(1)

Degrees of solidification and mineral proportions are given in Table 6 and partition coefficients in Table 4. Figures in parentheses for calculated abundances are estimated errors of calculations due to the uncertainty in the values of partition coefficients while those for the observed abundances correspond to 1 S.D. The abundances are in p.p.m.

contamination. Contamination is probably related to ignimbrites which are spatially and temporally associated with andesitic rocks. It seems likely that mantle-derived basaltic magma provided heat for partial melting in the crust (Coulon et al., 1978). The similarities in the depths of fractional crystallization of andesitic magma and of crustal melting (Coulon, 1977) suggest that longer residence times of magma in differentiating chambers facilitated large scale melting of the surrounding crustal rocks. The resulting anatectic dacitic to rhyolitic liquids then interacted with the basic magma. Contamination of basaltic magma probably started deeper in the crust, continued during ascent through the crust and was accompanied by fractional crystallization. The degrees of contamination and homogenization are probably affected by the crustal residence time of the magma. Figure 7 shows that a similar process may have also occurred within other calc-alkaline series such as those of New Zealand and southern Peru where the contents of lithophile elements appear to be related to the composition of associated ignimbrites. The interaction of the two magmas is also consistent with high and variable Sr isotope data (Dupuy et al., 1979).

It seems that continental crust plays an important role during volcanic activity in island-arc environments, although other factors, including mantle heterogeneity, thermal regime, and spatial and temporal evolution of volcanic rocks, may also significantly affect the compositions of rocks from such orogenic zones. Compressional magmatism related to two oceanic plates produces rocks of tholeiitic character, whereas if one of the plates is continental in character, the magmatism is calc-alkaline. These differences are shown in the Tonga–Kermadec–New Zealand island arcs where the primitive calc-alkaline andesitic magma of New Zealand is isotopically comparable to the tholeiitic basalts of Tonga–Kermadec (Ewart et al., 1977). The major differences in the compositions of the rocks of the two calc-alkaline and tholeiitic associations, such as the higher abundances of lithophile elements and the early crystallization of magnetite in the calc-alkaline series can be explained, to a large degree, as an influence of the continental crust.

REFERENCES

Alvarez, W. (1972). Rotation of the Corsica Sardinia microplate. *Nature, Lond.* **235**, 103–105.

Andriambololona, R. (1976). Les éléments de transition dans les suites andesitiques et shoshonitiques du Sud du Pérou. Thesis, 3° cycle, USTL, Montpellier.

Andriambololona, R. and Dupuy, C. (1978). Répartition et comportement des éléments de transition dans les roches volcaniques. I. Cuivre et zinc. *Bull. B.R.G.M.* **2**, 121–138.

Auzende, J. M., Bonnin, J., and Olivet, J. L. (1973). Hypothesis on the origin of the western Mediterranean basin. *J. geol. Soc. Lond.* **129**, 607–620.

Beccaluva, L., Macciotta, G., and Venturelli, G. (1976). Data geochemici e petrografici sulle vulcaniti plio-quaternarie della Sardegna Centro-occidentale. *Bull. Soc. Geol. It.* **84**, 1437–1457.

Bellon, H. (1976). Series magmatiques Neogenes et quaternaires du pourtour mediterraneen Occidental, comparees dans leur cadre geochronometrique—Implications geodynamiques. Thesis, Université d'Orsay, Paris.

Bellon, H., Coulon, C., and Edel, J. B. (1977). Le deplacement de la Sardaigne: synthese des donnees geochronologiques, magmatiques et paleomagnetiques. *Bull. Soc. Geol. Fr.* **19**, 825–831.

Bowen, N. L. and Schairer, J. F. (1935). The syetm MgO–FeO–SiO_2. *Am. J. Sci.* **29**, 151–217.

Brown, G. M., Holland, J. G., Sigurdsson, H., Tomblin, J. F., and Arculus, R. J. (1977). Geochemistry of the Lesser Antilles volcanic island arc. *Geochim. cosmochim. Acta* **41**, 785–801.

Cole, J. W. (1973). High alumina basalt of Taupo volcanic zone, New Zealand. *Lithos* **6**, 53–64.

Cole, J. W. (1978). Andesites of the Tongariro volcanic center, North Island, New Zealand. *J. Volc. geothermal Res.* **3**, 121–153.

Coulon, C. (1977). Le volcanisme calco-alcalin cenozoique de Sardaigne, Italie. Thesis, Université St Jérôme, Marseille.

Coulon, C. and Dupuy, C. (1975). Evolution spatiale des caracteres chimiques du volcanisme andesitique de la Sardaigne, Italie. *Earth planet. Sci. Letters* **5**, 170–176.

Coulon, C., Baque, L., and Dupuy, C. (1973). Les andesites cenozoiques et les laves associées en Sardaigne nord-occidentale (Province du Logudoro et du Bosano)—Caractères mineralogiques et chimiques. *Contr. Mineral. Petrol.* **42**, 125–139.

Coulon, C., Demant, A., and Bobier, C. (1974a). Contribution du paleomagnetisme à l'étude des series volcaniques cenozoiques et quaternaires de Sardaigne nord-Occidentale. *Tectonophysics* **22**, 59–82.

Coulon, C., Demant, A., and Bellon, H. (1974b). Premières datations par la méthode K–Ar de quelques laves cenozoiques et quaternaires de Sardaigne nord-Occidentale. *Tectonophysics* **22**, 41–57.

Coulon, C., Dostal, J., and Dupuy, C. (1978). Petrology and geochemistry of the ignimbrites and associated lava-domes from N.W. Sardinia. *Contr. Mineral. Petrol.* **68**, 89–98.

De Jong, K. A., Manzoni, M., and Zijderveld, J. D. A. (1969). Paleomagnetism of the Alghero trachyandesites. *Nature, Lond.* **224**, 67–69.

Demant, A. and Coulon, C. (1973). Sur la presence de laves à affinites calco-alcalines au sein du volcanisme alcalin plio-quaternaire de la Sardaigne nord occidentale. *C.r. hebd. S anc. Acad. Sci., Paris* **276**, 3073–3076.

Dietrich, V., Emmermann, R., Keller, J., and Puchelt, H. (1977). Tholeiitic basalts from Tyrrhenian sea floor. *Earth planet. Sci. Letters* **36**, 285–296.

Dostal, J., Dupuy, C., and Coulon, C. (1976). Rare-earth elements in high-alumina basaltic rocks from Sardinia. *Chem. Geol.* **18**, 251–262.

Dostal, J., Zentilli, M., Caelles, J., and Clark, A. (1977). Geochemistry and origin of volcanic rocks of the Andes (26–28°). *Contr. Mineral. Petrol.* **63**, 113–128.

Dupuy, C. and Coulon, C. (1973). Li, Rb, Sr, Ba dans les plagioclases de la suite andesitique du Logudoro et du Bosano, Sardaigne nord occidentale. *C.r. hebd. Séanc. Acad. Sci., Paris* **277**, 1593–1596.

Dupuy, C. and Lefèvre, C. (1974). Fractionnement des elements en trace Li, Rb, Ba, Sr dans les series andesitiques et shoshonitiques du Perou. Comparaison avec d'autres zones orogeniques. *Contr. Mineral. Petrol.* **46**, 147–157.

Dupuy, C., McNutt, R. H., and Coulon, C. (1974). Determination $^{87}Sr/^{86}Sr$ dans les andesites cenozoiques et les laves associees de Sardaigne nord-occidentales, Italie. *Geochim. cosmochim. Acta* **38**, 1287–1296.

Dupuy, C., Dostal, J., and Coulon, C. (1979). Geochemistry and origin of andesitic rocks from N.W. Sardinia. *J. Volc. geothermal Res.* **6**, 375–389.

Eggler, D. H. (1972). Water saturated and undersaturated melting relations in Paricutin andesite and an estimate of water content in the natural magma. *Contr. Mineral. Petrol.* **34**, 261–271.

Eggler, D. H. and Burnham, C. W. (1973). Crystallization and fractionation trends in the system andesite–H_2O–CO_2–O_2 at pressures to 10 kb. *Geol. Soc. Am. Bull.* **84**, 2523–2548.

Ewart, A. (1976a). A petrological study of the younger Tongan andesites and dacites and the olivine tholeiites of Niua Fo'ou island, S.W. Pacific. *Contr. Mineral. Petrol.* **58**, 1–21.

Ewart, A. (1976b). Mineralogy and chemistry of modern orogenic lavas. Some statistics and implications. *Earth planet. Sci. Letters* **31**, 417–432.

Ewart, A., Taylor, S. R., and Capp. A. C. (1968). Trace and minor element geochemistry of the rhyolitic volcanic rocks, central North Island, New Zealand. *Contr. Mineral. Petrol.* **18**, 76–104.

Ewart, A., Bryan, W. B., and Gill, J. B. (1973). Mineralogy and geochemistry of the younger volcanic islands of Tonga, S.W. Pacific. *J. Petrol.* **14**, 489–508.

Ewart, A., Brothers, R. N., and Mateen, A. (1977). An outline of the geology and geochemistry and the possible petrogenetic evolution of the volcanic rocks of the Tonga–Kermadec–New Zealand island arc. *J. Volc. geothermal Res.* **2**, 205–250.

Frey, F. A., Haskin, M. A., Poetz, J., and Haskin, L. A. (1968). Rare earth abundances in some basic rocks. *J. geophys. Res.* **70**, 6085–6098.

Gill, J. B. (1978). Role of trace element partition coefficients in models of andesite genesis. *Geochim. cosmochim. Acta* **42**, 709–724.

Gorton, M. P. (1977). The geochemistry and origin of quaternary volcanism in the New Hebrides. *Geochim. cosmochim. Acta* **41**, 1257–1270.

Green, T. H. (1972). Crystallization of calc-alkaline andesite under controlled high-pressure hydrous conditions. *Contr. Mineral. Petrol.* **34**, 150–166.

Herrmann, A. G. (1970). Yttrium and lanthanides. In *Handbook of Geochemistry*, vol. II-2 (K. H. Wedepohl, ed.), Springer, New York, pp. 57–72.

Jakeš, P. and Gill, J. B. (1970). Rare earth elements and the island arc tholeiitic series. *Earth planet. Sci. Letters* **9**, 17–28.

Jakeš, P. and White, A. J. R. (1972). Major and trace element abundances in volcanic rocks of orogenic areas. *Geol. Soc. Am. Bull.* **83**, 29–40.

Kay, R. W. (1977). Geochemical constraints on the origin of Aleutian magmas. In *Island Arcs, Deep Sea Trenches and Back-arc Basins* (M. Talwani and W. C. Pitman III, eds), Maurice Ewing Series no. 1, American Geophysical Union, Washington, DC, pp. 229–242.

Kudo, A. M. and Weill, D. F. (1970). An igneous plagioclase thermometer. *Contr. Mineral. Petrol.* **25**, 52–65.

Kuno, H. (1952). Chemical composition of hypersthene and pigeonite in equilibrium in magma. *Am. Mineral.* **37**, 1000–1006.

Kuno, H. (1959). Origin of Cenozoic petrographic provinces of Japan and surrounding areas. *Bull. Volc.* **20**, 37–76.

Lewis, J. F. (1971). Composition, origin and differentiation of basalt magma in the Lesser Antilles. *Geol. Soc. Am. Mem.* no. 130, 159–179.

Lopez-Escobar, L., Frey, F. A., and Vergara, M. (1977). Andesites and high-alumina basalts from the central-south Chile: geochemical evidence bearing on their petrogenesis. *Contr. Mineral. Petrol.* **63**, 199–228.

Marsh, B. D. (1976). Some Aleutian andesites: their nature and source. *J. Geol.* **84**, 27–45.

Mertzman, S. A. (1977). The petrology and geochemistry of the Medicine Lake volcano, California. *Contr. Mineral. Petrol.* **62**, 221–247.

Miyashiro, A. and Shido, F. (1975). Tholeiitic and calc-alkaline series in relation to the behaviours of titanium, vanadium, chronium and nickel. *Am. J. Sci.* **275**, 265–277.

Morelli, C., Bellemo, S., Finetti, I., and De Visentini, G. (1967). Preliminary depth contour maps for the Conrad and Moho discontinuities in Europe. *Boll. Geofis. teor. appl.* **9**, 142–157.

Mysen, B. O. (1978). Experimental determination of crystal-vapor partition coefficients for rare earth elements to 30 kbar pressure. *Carnegie Instn. Wash. Yb.* **77**, 689–695.

Nairn, A. E. M. and Westphal, M. (1968). Possible implications of the palaeomagnetic study of late palaeozoic igneous rocks of northwestern Corsica. *Palaeogeogr. Palaeoclimat. Palaeocol.* **5**, 179–204.

Osborn, E. F. (1959). Role of oxygen pressure in the crystallization and differentiation of basaltic magma. *Am. J. Sci.* **257**, 609–647.

Osborn, E. F. (1962). Reaction series of sub-alkaline igneous rocks based on different oxygen pressure conditions. *Contr. Mineral. Petrol.* **34**, 261–271.

Stormer, J. C. (1973). Calcium zoning in olivine and its relationship to silica activity and pressure. *Geochim. cosmochim. Acta* **37**, 1815–1821.

Taylor, S. R. (1969). Trace element chemistry of andesites and associated calc-alkaline rocks. *Dept Geol. Min. Res. Oregon Bull.* no. 65, 43–64.

Taylor, S. R., Capp, A. C., Graham, A. L., and Blake, D. H. (1969). Trace element abundances in andesites. II. Saipan, Bougainville and Fiji. *Contr. Mineral. Petrol.* **23**, 1–26.

Thompson, R. N. (1974). Some high pressure pyroxenes. *Mineral. Mag.* **39**, 768–787.

Thorpe, R. S., Potts, P. J., and Francis, P. W. (1976). Rare earth data and petrogenesis of andesites from the North Chilean Andes. *Contr. Mineral. Petrol.* **54**, 65–78.

Westphal, M., Orsini, J., and Vellutini, P. (1976). Le microcontinent corsosarde, sa position initiale: données palaeomagnetiques et raccords géologiques. *Tectonophysics* **30**, 141–157.

Wright, T. L. and Doherty, P. C. (1970). A linear programming and least-squares computer method for solving petrologic mixing problems. *Geol. Soc. Am. Bull.* **81**, 1995–2008.

Volcanic evolution of the northern Antarctic Peninsula and the Scotia arc

J. Tarney, S. D. Weaver, A. D. Saunders, R. J. Pankhurst, and P. F. Barker

Department of Geology,
University of Leicester, University Road, Leicester LE1 7RH, UK,

Department of Geology,
University of Canterbury, Christchurch, New Zealand,

Department of Geology,
Bedford College, Regent's Park, London NW1 4NS, UK,

British Antarctic Survey, Natural Environmental Research Council,
c/o Institute of Geological Sciences, 64–78 Gray's Inn Road, London WC1X 8NG, UK,

and

Department of Geological Sciences,
University of Birmingham, PO Box 363, Birmingham B15 2TT, UK

ABSTRACT

The volcanic evolution of the northern Antarctic Peninsula (Graham Land) and Scotia arc is described in relation to the plate tectonic history of the region. In the Mesozoic, before the breaking up of Gondwanaland and the opening of the Weddell Sea, the tectonic situation of the Antarctic Peninsula was similar to that of the present Pacific margin of South America, and there was extensive subduction-related volcanism and plutonism. The volcanic rocks are mainly basaltic near the trench on the western coast of the Peninsula, but more acid volcanic rocks predominate on the eastern coast. There is also a significant transverse 'K-h' variation in that K, Rb, Th, and Ce/Yb increase from west to east in lavas of equivalent SiO_2 percentage. Other normally incompatible elements such as P, Ti, Sr, Nb, and Zr do not display this transverse variation, nor do they show a progressive increase with fractionation in calc-alkaline rocks. Similar transverse variations in chemical composition are observed in the plutonic rocks.

During the Cainozoic, sections of a spreading ridge (the Aluk Ridge), lying to the west, were progressively consumed beneath the Peninsula, first in the south and later in the north. A consequence of this was that both subduction and calc-alkaline igneous activity ceased progressively in the same direction. Only on the South Shetland Islands, near the northern tip of the Peninsula, did subduction and igneous activity persist until recent times, and the Mesozoic, Tertiary, and Quaternary lavas are mostly low K basalts and basaltic andesites. This apparent lack of significant compositional change with time suggests that the transverse K-h variation is spatially rather than temporally dependent.

The small marginal basin of Bransfield Strait separates the South Shetland Islands from the Antarctic Peninsula and appears to have opened within the last 1–2 Ma as spreading at the ridge in Drake Passage ceased. The recent volcanoes of Deception and Bridgeman Islands lie on the axis of back-arc spreading in Bransfield Strait and their lavas have geochemical characteristics transitional between ocean basalts and calc-alkaline volcanic rocks. The small recent volcano of Penguin

Island, north-west of the axis of back-arc spreading, is mildly alkaline. Much more voluminous alkaline volcanism occurs on the eastern coast of the Peninsula, and has been regarded as the expression of an extensional tectonic regime following the ending of Pacific ocean-floor subduction beneath the Peninsula.

The opening of Drake Passage in the late Oligocene finally separated the Antarctic Peninsula from southern Chile, and resulted in a complex sequence of plate motions in the Scotia Sea. Atlantic Ocean floor is now subducting beneath the younger lithosphere of the Scotia Sea, and the primitive South Sandwich arc has built up during the last $c.$ 5 Ma on lithosphere created at the South Sandwich 'back-arc' spreading centre. However a precursor island arc, also composed mainly of basalts of the island arc tholeiitic series, has been located at the eastern end of the South Scotia Ridge.

Introduction

The Antarctic Peninsula, perhaps the most inaccessible segment of the circum-Pacific belt of andesitic magmatism, has often been considered as a continuation of the Andean magmatic belt of southern Chile. Throughout most of the Mesozoic, and probably for much of the Palaeozoic, the western cost of South America, the Antarctic Peninsula, Marie Byrd Land, and the New Zealand continental block formed a continuous continental margin bordering the Pacific margin of Gondwanaland. However, with the commencement of the break-up of Gondwanaland in the late Jurassic, the region gradually began to evolve on a separate and rather distinctive course, and this is reflected in the magmatic history. First, the opening of the Weddell Sea separated the Peninsula from eastern Antarctica, leaving it as a narrow sliver of continental crust bounded on either side by ocean crust. Second, during the Tertiary, the Aluk Ridge crest began to be progressively consumed beneath the Peninsula. Third, the opening of Drake Passage in the mid-Tertiary not only severed the connection between the Peninsula and southern South America, but also initiated a complex sequence of spreading motions in the Scotia Sea which have resulted in the present extended loop of the Scotia arc and the subduction of Atlantic ocean floor beneath the young South Sandwich Island arc. Fourth, recent cessation of spreading in Drake Passage has been associated with active opening of the Bransfield Strait marginal basin, separating the South Shetland arc from the Peninsula, and possibly also with the recent volcanic activity on James Ross Island in the northern Weddell Sea. Thus during the last 100 Ma there has been a change from a typical Andean-type continental margin, with extensive subduction of Pacific ocean floor beneath the Antarctic Peninsula, to a situation where there is now virtually no subduction of Pacific lithosphere beneath the region, although there is now limited, but active, subduction of Atlantic lithosphere beneath the Scotia Sea. The present plate tectonic configuration of the region is summarized in Fig. 1.

The aim of this contribution is to assemble the available petrological and geochemical data for the volcanic rocks erupted in the northern Antarctic Peninsula–Scotia arc region during the last $c.$ 180 Ma and to attempt to relate the changing character of the volcanism to the complex plate tectonic history (so far as it is understood) and to the processes of magma generation in subduction zones. It is first necessary, however, to outline briefly the tectonic evolution of the region.

Tectonic Evolution of the Antarctic Peninsula and the Scotia Arc

Most of the exposed rocks of the Antarctic Peninsula and offshore islands were produced by subduction-related magmatic activity over the past 180 Ma. The pre-Jurassic record is fragmentary, but there is a consensus that the Peninsula lay at the Pacific margin of Gondwanaland at least as far back as the Carboniferous. The exact relative positions of South America, the Antarctic Peninsula, Marie Byrd Land, and New Zealand within Gondwanaland, and the timing of their fragmentation and

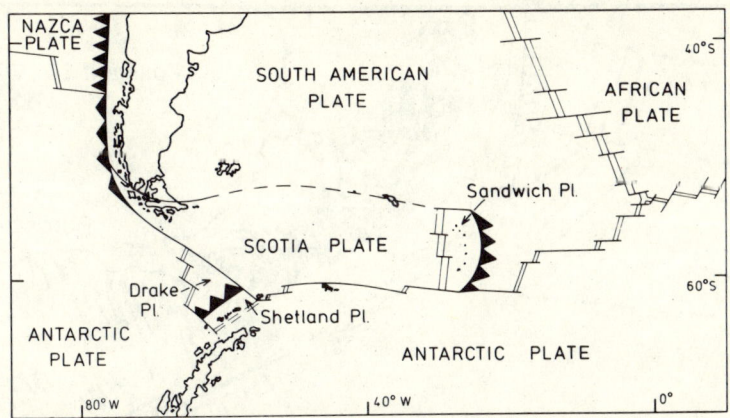

Fig. 1 Simplified present-day plate tectonic features of the Scotia arc region, showing relationship between the larger plates (African, Antarctic, South American, Nazca) and smaller plates (Scotia, Sandwich, Drake, Shetland). Modified from Barker and Dalziel (1980)

dispersal, are still disputed (see, for example, Molnar et al., 1975; Herron and Tucholke, 1976; Barker and Griffiths, 1977; de Wit, 1977; Norton and Sclater, 1980; Dalziel, 1980). A probable Jurassic age for the Weddell Sea (R. A. Jahn, 1978 and personal communication) does, however, offer some support to the persistent speculation (Barker and Griffiths, 1972; Dalziel, 1974; Suarez, 1976) that the Antarctic Peninsula separated from eastern Antarctica by an essentially back-arc extensional process during the very early stages of Gondwanaland break-up. From this time, subduction of Pacific ocean floor in this region occurred beneath a narrow peninsula rather than, as in South America, beneath a major continental landmass.

The present highly asymmetric age pattern of ocean floor in the Pacific indicates that considerable amounts of older Pacific lithosphere have been consumed beneath South America and the Antarctic Peninsula. Most of the ocean floor making up the southern Pacific formed at the Pacific–Antarctic Ridge over the past 80 Ma (Pitman et al., 1974). However, in the southeastern Pacific, adjacent to southern South America and the Antarctic Peninsula respectively, two additional spreading systems existed (Fig. 2; Herron and Tucholke, 1976) at least during the Cainozoic. The spreading centre adjacent to southern South America appears to have been a more southerly extension of the Chile Ridge. Magnetic anomalies young eastwards, showing that the spreading centre migrated into the trench. Moreover, anomalies of a particular age are located progressively to the west in passing northwards from one spreading section to another. Thus, at least over the past 20 Ma, the Chile Ridge appears to have been a transient feature, linking the Pacific–Antarctic Ridge crest to the trench and migrating northwards along the South America margin. With its passing, fast subduction, like that now seen north of 45°S, was replaced by slow subduction of c. 2.0–2.4 cm a^{-1} (Chase, 1978; Minster and Jordan, 1978) which represents the relative motion of the South American and Antarctic plates. Late Cainozoic deformation and the eruption of plateau basalts onshore have both been linked to this northward-migrating reduction in subduction rate (Winslow, 1980), but the main effect has been a marked reduction in the volume of calc-alkaline igneous activity.

A similar ridge system (the Aluk Ridge of Herron and Tucholke, 1976) existed to the west of the Antarctic Peninsula during the early Cainozoic. Three sets of north-east-trending magnetic anomalies occur, which become progressively younger northwards (Fig. 2). The most northerly set, to the north-west of the South Shetland Islands, youngs away from the

374 IV *Evolution of andesite volcanic provinces*

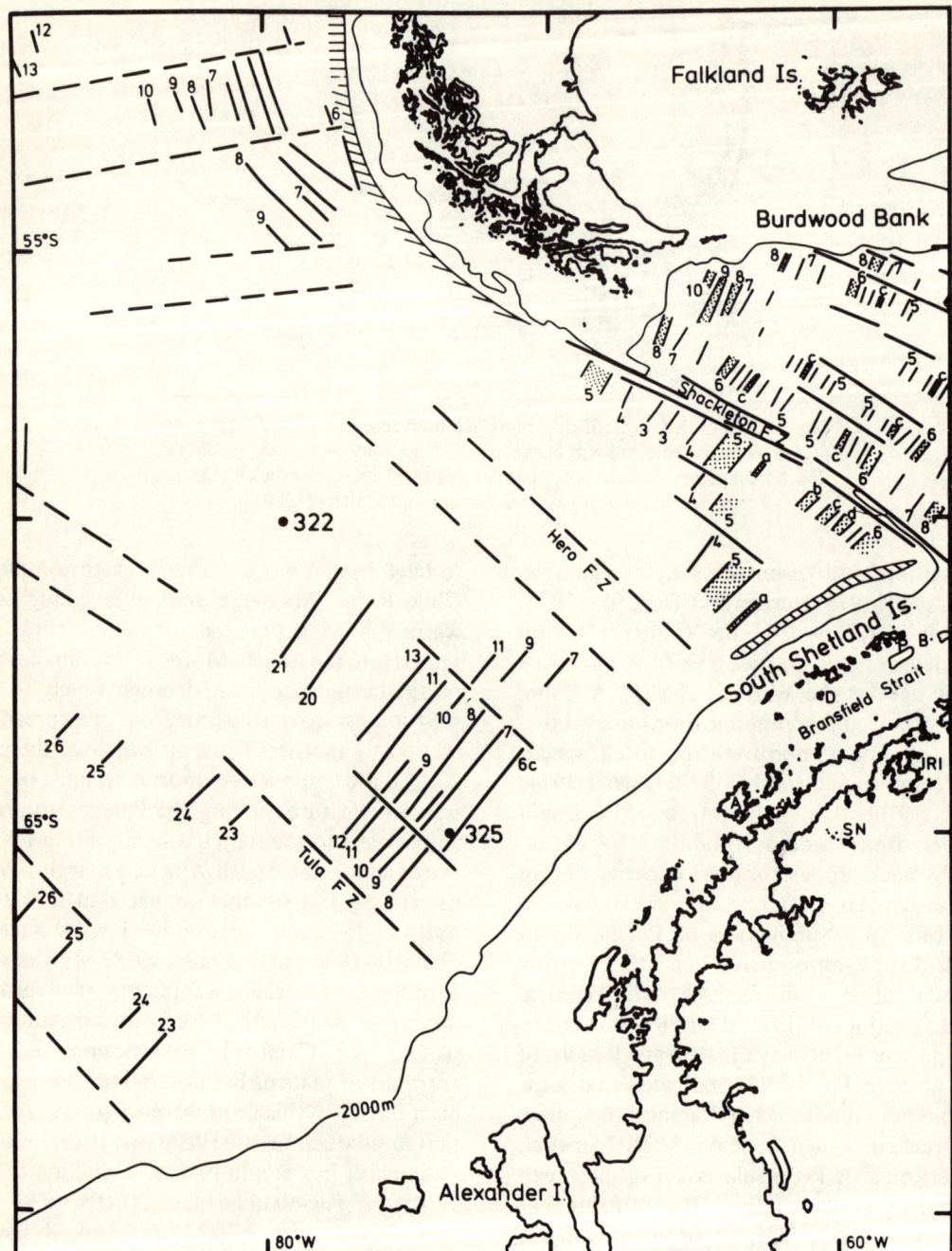

Fig. 2   Magnetic lineation patterns to the west and north of the Antarctic Peninsula in relation to plate tectonics. Magnetic anomaly numbers are conventional and DSDP sites 322 and 325 are shown. Note that off southern Chile, and south-west of the Hero Fracture Zone, magnetic anomalies become younger towards the continental margins, indicating that the ridge crest has been subducted. Slow subduction continues off southern Chile (lined area) but elsewhere has stopped altogether. A = Anvers Island; P = Penguin Island; B = Bridgman Island; JRI = James Ross Island; SN = Seal Nunataks. Redrawn from Barker and Dalziel (1980)

Peninsula and includes both flanks of the recently active Drake Passage spreading centre (Barker, 1970; Barker and Burrell, 1977), but the two southerly sets (the 'Peninsula' and 'Ellsworth' anomalies of Herron and Tucholke, 1976) young towards the Peninsula, indicating consumption of the ridge crests beneath the Peninsula. In contrast to southern Chile, however, where reflection profiles show a typical oceanic layer 2, dipping steeply towards the continental margin (Hayes and Ewing, 1970), reflection profiles crossing the two southerly anomaly sets show oceanic basement lying horizontally at the margin. Thus subduction appears to have stopped completely once the ridge crests met the trench. From the observed anomalies, ages and spreading rates (Herron and Tucholke, 1976; Barker and Burrell, 1977) it is possible to make a rough estimate of the times the ridge crests met the trench: c. 50 Ma ago off southern Alexander Island, c. 33 Ma ago off northern Alexander Island, and 20–23 Ma ago near southern Anvers Island. There has therefore been a northward migration of the cessation of subduction along the Antarctic Peninsula Margin similar to that in southern Chile. This is reflected in the age pattern of calc-alkaline intrusions along the Peninsula (Saunders et al., 1980a; Weaver et al., 1980). There also seem to have been large variations in subduction rate with time, in particular the onset of fast spreading 60–70 Ma ago, indicated by the Ellsworth anomalies.

That portion of the Antarctic Peninsula margin lying opposite the most northerly anomaly set is the only section having a topographic trench, half-filled with sediment, and a clearly defined oceanic layer 2 dipping towards the margin (Barker, 1976). Because the Pacific ocean floor further south became coupled to the Antarctic margin when the ridge crest met the trench some 20 Ma ago, the continued subduction at this South Shetland trench would have exactly complemented the spreading at the Drake Passage ridge crest (Fig. 2). Now that spreading at the ridge has stopped, or slowed markedly 4 Ma ago, it is likely that active subduction beneath the South Shetland Islands has virtually ceased. Thus the volcanic record on the South Shetland Islands is the most complete of any part of the Peninsula, and indeed, because of the additional development and formation of Bransfield Strait, the region is still volcanically active. The reason for the cessation of spreading in Drake Passage is not clear. Forsyth (1975) considers that the small oceanic 'Drake Plate' (Fig. 1) is now under east–west compression, perhaps transmitted westward from events farther east in the Scotia Sea; this may have been sufficient to arrest motion.

The 2 km deep and 65 km wide trough of Bransfield Strait, which separates the South Shetland Islands from the Antarctic Peninsula mainland, matches in lateral extent the subducted margin of the small Drake Plate (Figs. 1, 2, and 3). The crustal structure of the South Shetland Islands is continental and its early geological history is similar to that of the west coast of the Peninsula. The structure of the axial trough of Bransfield Strait, however, more closely resembles that of an oceanic ridge crest, but with a thicker main crustal layer yielding seismic velocities of 6.5–6.9 km s^{-1}, overlying 7.6–7.7 km s^{-1} mantle (Ashcroft, 1972). The trough is asymmetric in section with a steep, normal-faulted, north-western margin bordering the South Shetland Islands. Pliocene to Recent volcanic activity has occurred along this fault zone. The more gradual south-eastern slope of the trough bordering the Antarctic Peninsula is covered by thick sediments which are thought to conceal a similar but broader zone of faulting (Davey, 1972). The active and recently active volcanoes of Deception and Bridgeman Islands, together with a number of other seamounts, lie along an axial line of rough topography within the trough which coincides with a large positive magnetic anomaly. Barker (1976) pointed out that this anomaly could be explained approximately by normal magnetization of the observed topography which, given the present-day volcanic and seismic activity, implies its formation within the past 0.69 Ma. Roach (1978), from a more detailed study of the magnetic anomaly patterns, has identified reversely magnetized material flanking the central zone, and suggests

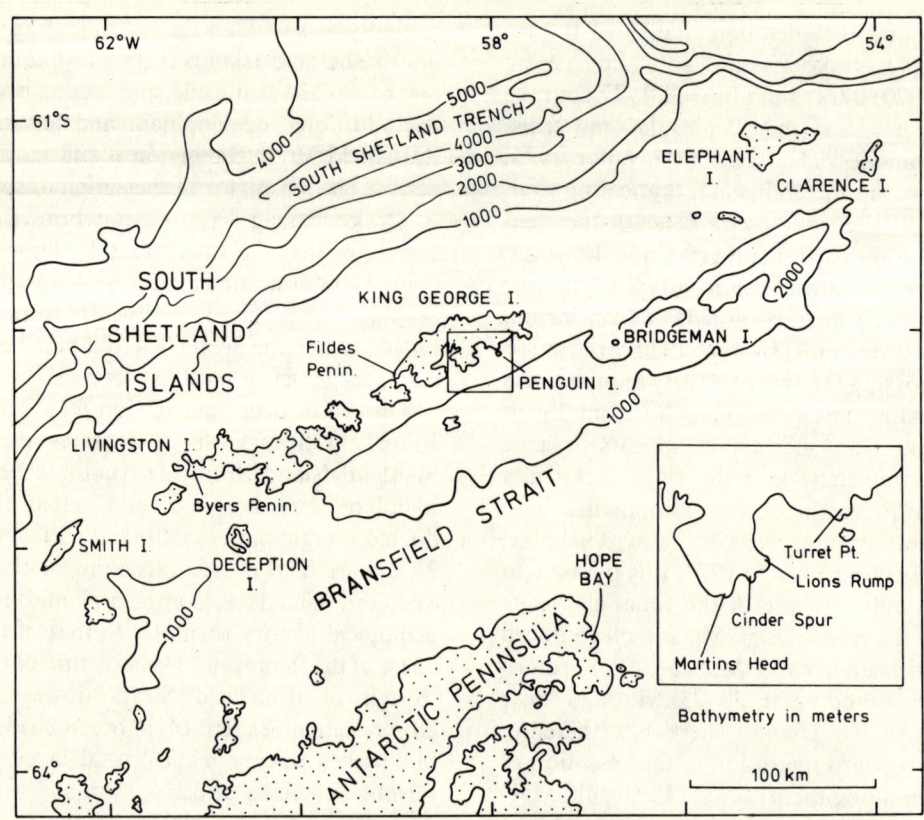

Fig. 3 Map of the South Shetland Islands, separated from the Antarctic Peninsula by the small marginal basin of Bransfield Strait. The volcanoes of Deception and Bridgeman Islands lie close to the axis of back-arc spreading. Penguin Island volcano lies to the north of the axis

that the axial 30 km of the strait has formed by sea-floor spreading over the last 1.3 Ma. Thus Bransfield Strait appears to represent the very early stages of formation of an ensialic back-arc marginal basin which, curiously, seems to be related to the dying stages of subduction beneath the South Shetland Island arc. The Pliocene to Recent alkaline volcanism on the eastern coast of the Peninsula, south-east of Bransfield Strait, may also reflect this change from compressional to extensional tectonics during the last few million years.

To the north of the Antarctic Peninsula lies the Scotia Sea (Fig. 1), which is now almost entirely oceanic and has developed essentially as a series of complications to the South America–Antarctic plate boundary following the separation of the Antarctic Peninsula from South America 30 Ma ago (Barker, 1970; Barker and Burrell, 1977). The extended loop of the Scotia Ridge, which bounds the Scotia Sea on three sides, is composed in large part of strung-out fragments of an originally compact continental connection between South America and the Antarctic Peninsula. Indeed some of these continental fragments contain subduction-related magmatic products which, from their age and tectonic relationships, must have been formed near the Pacific margin (e.g. South Georgia: Suarez and Pettigrew, 1976; Storey et al., 1977). The eastern part of the Scotia Sea has extended itself rapidly into the Atlantic during the last few million years as a result of back-arc spreading (Barker, 1970, 1972; Saunders and Tarney, 1979) connected with subduction of Atlantic lithosphere beneath the younger and more buoyant lithosphere of the Scotia Sea. The present South Sandwich arc appears to be built on this younger lithosphere. The recent discovery of primitive arc tholeiites, similar to the South Sandwich volcanic rocks, but of mid-Miocene age, at the eastern end of the South Scotia Ridge (Barker *et al.*, 1980), indicates that subduction from the Atlantic also occurred much earlier in the evolution of the Scotia arc.

In summary, the subduction history of the Antarctic Peninsula–Scotia arc region has been complex and differs considerably from that of the typical Andean continental margin in Chile and Peru. The last 100 Ma have seen a progressive reduction in the consumption of Pacific lithosphere, increasing 'oceanization' in which back-arc spreading has played a major role, and a switch to consumption of Atlantic lithosphere. We now consider the way in which subduction-related magmatism has responded to this changing situation, and try to elucidate the bearing this may have on the mechanisms of generation of andesitic magmas.

Geological History of Volcanism in the Antarctic Peninsula–Scotia Arc

Mesozoic volcanic rocks

A calc-alkaline basalt–andesite–rhyolite suite of Mesozoic age has long been known to occur throughout the Antarctic Peninsula. These volcanic rocks are predominantly pyroclastic but include numerous lava flows and have a maximum recorded thickness of 3000 m. They have frequently been referred to as the 'Upper Jurassic Volcanic Group' (Adie, 1972), since sediments with plant remains of supposed Middle Jurassic age conformably underlie the volcanics in the Hope Bay area (e.g. Bibby, 1966) and the lavas are cut by late Cretaceous–early Tertiary plutons of the 'Andean Intrusive Suite'. However, fossil control is sparse and Taylor *et al.* (1980), who have challenged the stratigraphic value of the plant remains, have pointed out that on Alexander Island and on the South Shetland Islands ages from late Jurassic to late Cretaceous may be demonstrated for the volcanic rocks (*see also* Pankhurst *et al.*, 1980). Lower Jurassic K–Ar ages of 186 Ma have in fact been reported for lavas from the Jason Peninsula area by Rex (1976). Consequently, Thomson (1980) has proposed the new name 'Antarctic Peninsula Volcanic Group'. However, for the present we will use the informal term 'Mesozoic volcanic rocks'. Such rocks are best described and studied in the South Shetland Islands, where they occur along with Tertiary volcanic sequences.

Stratigraphical and petrological information on the South Shetland Islands may be found in Hawkes (1961*a*,*b*), Barton (1965), Hobbs (1968), and Baker *et al.* (1975). Our studies have concentrated on the Mesozoic lavas of Byers Peninsula, Livingston Island (Fig. 3), which have been described also by Valenzuela and Hervé (1972), Pankhurst *et al.* (1980), and Smellie *et al.* (1980). Basalts, andesites, and rhyolites are interbedded with marine, passing up into non-marine, sediments and are cut by numerous sills, plugs, and dykes. K–Ar ages on lavas from the non-marine sequence range from 125 to 90 Ma (Pankhurst *et al.*, 1980) but older zeolitized flows probably date back to 140 Ma according to faunal evidence. K–Ar ages of *c.* 90 Ma have also been obtained on lavas at Coppermine Cove, Robert Island (R. J. Pankhurst, unpublished data).

Whether volcanism in the Antarctic Peninsula and the South Shetland Islands was more or less continuous throughout the Jurassic and Cretaceous (and even into the Tertiary), or whether it occurred in a number of distinct episodes cannot be deduced from the available data. In view of the scattered exposure and altered nature of many of the older volcanics, it is even doubtful whether intensive K–Ar dating will solve this problem. We have therefore not attempted to subdivide the analysed Mesozoic volcanic rocks.

A petrological and geochemical survey of 200 Mesozoic volcanic rocks from the northern Antarctic Peninsula and the South Shetland Islands has been undertaken by Weaver *et al.*

(1980). This study revealed that there is a marked change in the composition of the erupted lavas across the Antarctic Peninsula (Fig. 4). Thus, those erupted on the South Shetland Islands, closest to the trench, are predominantly basalts and basaltic andesites, those erupted on the west side of the Peninsular mainland include a higher proportion of andesites and dacites, whereas the eastern coast volcanic successions include many rocks of dacitic and rhyolitic composition.

It is worth referring at this stage to the calc-alkaline plutonic rocks of the Peninsula because they dominate the geology and provide a record of the magmatic history. Adie's (1954, 1955)

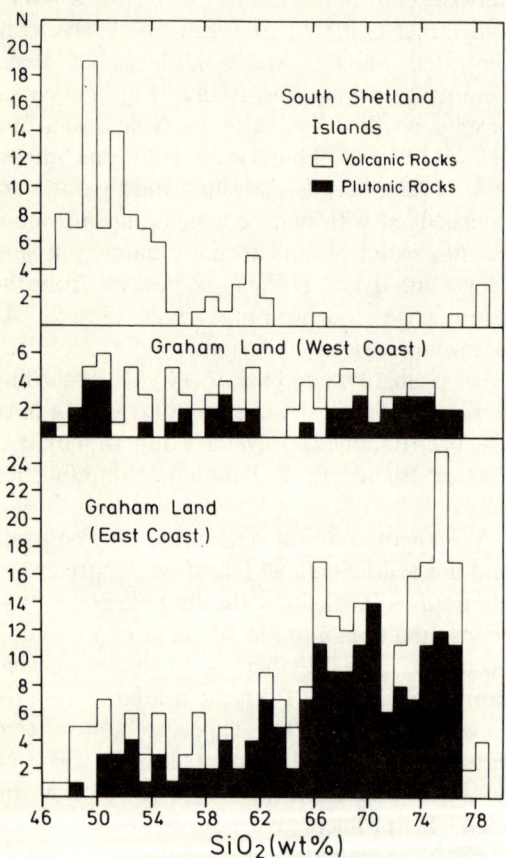

Fig. 4 Histogram showing change in Mesozoic volcanic and plutonic magma compositions (as per cent SiO_2) across the Antarctic Peninsula: South Shetlands; Graham Land, western coast, eastern coast. From Saunders et al. (1920b). Reproduced by permission of Elsevier Scientific Publishing Co.

initial division of the plutons into a 'late Palaeozoic suite' and a later more voluminous 'Andean Intrusive Suite' needs to be modified because, apart from K–Ar mineral ages of 384–351 Ma on one small pluton (Rex, 1976), other supposed Palaeozoic plutons have yielded Mesozoic dates. The largest single body of radiometric data for Antarctic Peninsula plutonic rocks is given by Rex (1976), who recognized four possible intrusive 'events' or groups of K–Ar mineral ages: 180–160 Ma (late Jurassic), 140–130 Ma (Jurassic–Cretaceous), 110–90 Ma (mid-Cretaceous), and 75–45 Ma (late Cretaceous–Tertiary) (see Fig. 5). Some of these ages are supported by Rb–Sr dating (Rex, 1976; Gledhill et al., 1980). Additionally, the granodiorite of Cornwallis Island, east of the South Shetland Islands, has given an age of 9.5 Ma (Rex and Baker, 1973). In reviewing these and other published data, Saunders et al. (1980a) observed that the older ages are restricted to the eastern coast of the Peninsula whereas the younger (70–45 Ma) ages occur along the western coast, at least in the northern part of the Peninsula. Thus the locus of magmatic activity seems to have migrated from east to west (trenchwards) with time in this area. Otherwise, there is a distinct concentration of determined ages in the range 110–90 Ma which appears to represent a widespread mid-Cretaceous culmination of magmatic activity.

Tertiary volcanic rocks

Tertiary igneous activity appears to have been confined mostly to the northern sector of the western coast of the Peninsula and the South Shetland Islands. Plutonic rocks from the Argentine Islands, Port Lockroy (Wiencke Island), and King George Island, for instance, have yielded K–Ar ages between 48 and 57 Ma (Grikurov et al., 1970; Rex, 1976; Watts, 1980), although an older Rb–Sr isochron age of 72 Ma has been reported for a granodiorite from the Argentine Islands (Gledhill et al., 1980). Volcanic rocks from the central and northern parts of Fildes Peninsula, King George Island, have given K–Ar ages between 43 and 60 Ma (un-

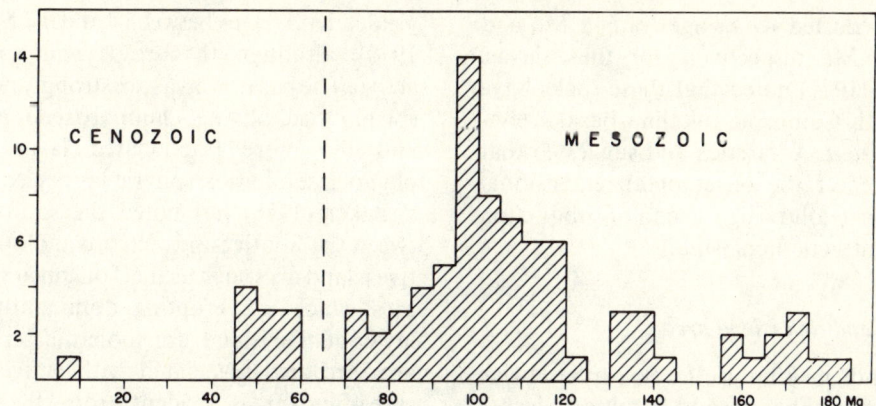

Fig. 5 Histogram of determined ages of igneous activity in the Antarctic Peninsula

published data of the authors; Grikurov et al., 1970; Watts, 1980). Rex (1976) has reported K–Ar ages of 35 Ma for basalts from Two Hummock Island and 54–63 Ma for those from Tower Island, both situated on the southeastern side of Bransfield Strait.

Our studies have been mainly concentrated on King George Island. Fildes Peninsula consists of basaltic and andesitic lavas cut by numerous dykes and several large dolerite plugs. The lava pile represents the products of at least four volcanic centres. Tertiary basaltic and andesite lavas also occur along the southern margin of King George Island. Those at the locality known as Lion's Rump immediately underlie the Pliocene 'Pecten Conglomerate'. The volcanic rocks at Cinder Spur, Martin's Head, and Turret Point contain unconsolidated tephra and may be no more than a few million years old. These deposits are remnants of volcanoes that have largely been downfaulted beneath what is now Bransfield Strait and it seems probable that they heralded the opening of the strait.

Quaternary volcanic rocks

The most recently erupted lavas in the Antarctic Peninsula are closely associated with the extensional feature of Bransfield Strait. Deception Island is an active composite volcano with a diameter of c. 14 km and a prominent caldera, which lies near the axial high in Bransfield Strait. The geology of Deception has been described by Hawkes (1961a), González-Ferran and Katsui (1970), Baker et al. (1975), and Weaver et al. (1979). Deception lavas range from basalt to rhyodacite in composition. Pre-caldera pyroclastic rocks and lavas are predominantly basaltic wheras post-caldera eruptions are more silicic, although basaltic magma appears to have been available throughout the history of the volcano.

Bridgeman Island, lying towards the northeastern extremity of Bransfield Strait, is a small remnant of an eroded volcano made up mainly of basalts and basaltic andesites (González-Ferran and Katsui, 1970; Weaver et al., 1979). Morphological features suggest an age similar to the older post-caldera lavas on Deception Island.

Penguin Island lies north-west of the axial feature in Bransfield Strait, c. 1 km off the southern coast of King George Island. The island, 1.7 km in diameter, is dominated by the regular uneroded slopes of a basalt scoria cone built on a platform of mildly alkaline olivine basalt lavas (González-Ferran and Katsui, 1970; Weaver et al., 1979).

Volcanic rocks of James Ross Island and Seal Nunataks

Alkaline olivine basalts and hawaiites occur on and around James Ross Island (Nelson, 1975) and further south at Seal Nunataks and Jason Peninsula on the Larsen Ice Shelf (Fig. 1). Rex

(1976) has reported K–Ar ages of 6–1 Ma and less than 1 Ma respectively for these lavas. Baker *et al.* (1977) noted that these rocks have affinities with Cainozoic alkaline basalts elsewhere in western Antarctica and suggested that they may reflect the onset of an extensional tectonic regime following the end of subduction along the Antarctic Peninsula.

The South Sandwich island arc

The 11 islands making up the South Sandwich arc are built up of basalts and basaltic andesites with minor andesites and dacites, which have yielded K–Ar ages between 4 and 0.7 Ma (Baker, 1978), although the arc is still volcanically active. The basaltic lavas are strongly porphyritic (plagioclase, olivine, clinopyroxene, magnetite) and the more fractionated lavas generally plagioclase-phyric. Aphyric lavas also occur.

Baker (1978) has noted the similarities between the South Sandwich arc and other primitive island arcs such as the Tongan and Mariana arcs which are erupting dominantly basaltic lavas of the island arc tholeiitic series. These arcs are also associated with active back-arc spreading. It is evident from the magnetic anomaly patterns associated with the South

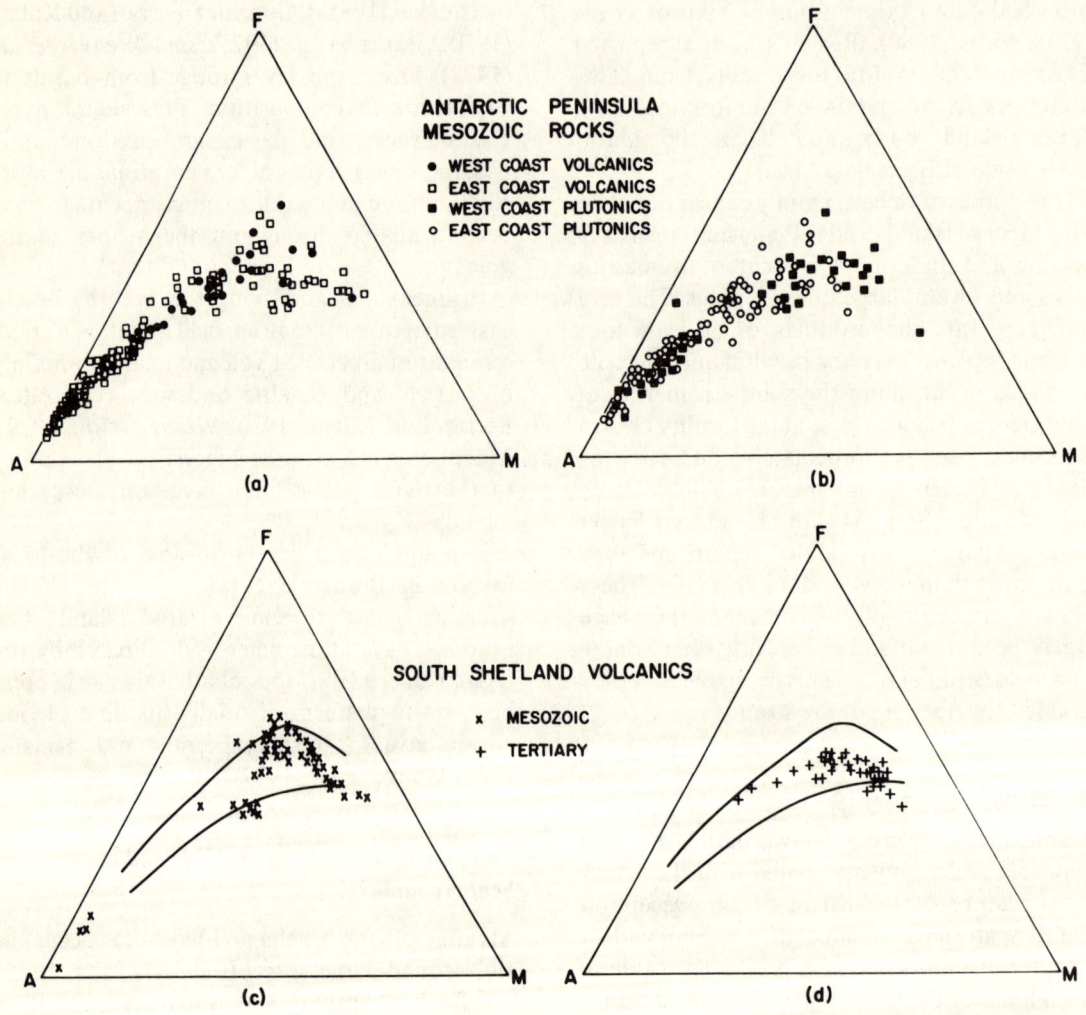

Fig. 6 AFM diagrams for igneous rocks from the Antarctic Peninsula and the South Shetland Islands

Sandwich back-arc spreading centre that the islands of the South Sandwich arc are built on lithosphere created at that spreading centre (Barker, 1972; Saunders and Tarney, 1979); thus it is likely that the age of the arc is no more than 6 Ma.

Evidence for the existence of a precursor volcanic arc near the southern extremity of the present arc has come from dredging operations on the eastern South Scotia Ridge (Barker et al., 1980). Basaltic lavas recovered from this area are petrographically and chemically indistinguishable from those of the South Sandwich arc, but have yielded K–Ar ages between 12 and 20 Ma. Barker et al. have suggested that this older arc ('Discovery arc') may have been associated with earlier back-arc extension in the central Scotia Sea, but that activity may have ceased following subduction of a section of the south-western Atlantic spreading centre.

Chemical Composition

In considering the geochemical data we will deal first with the Mesozoic igneous activity which has the widest geographic distribution throughout the Antarctic Peninsula, and enables the transverse spatial variations to be assessed. Temporal variations are best considered in the South Shetland Islands where the volcanic record is more continuous. The individual tectono-magmatic provinces of Bransfield Strait, James Ross Island, and the South Sandwich Islands will be described separately.

Mesozoic igneous rocks

Antarctic Peninsula igneous rocks are typically calc-alkaline, displaying low Fe-enrichment on AFM diagrams (Fig. 6), high levels of K, Rb, Ba, Sr and Th, high Ba/Sr and low K/Rb ratios, and in fact their geochemical characteristics (Table 1)

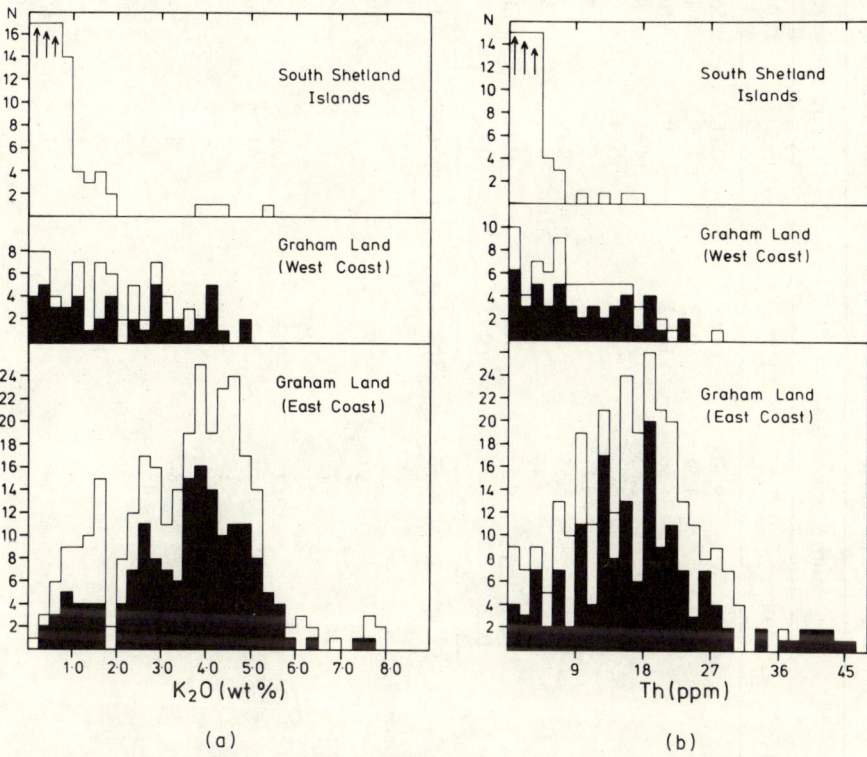

Fig. 7 Histogram of abundance of (a) K_2O (in per cent by weight) and (b) ppm Th (in p.p.m.) in Mesozoic igneous rocks across the Peninsula. See also Fig. 4. From Saunders et al. (1980b). Reproduced by permission of Elsevier Scientific Publishing Co.

TABLE 1 Representative analyses of Mesozoic volcanic rocks from the South Shetland Islands and the Antarctic Peninsula

	South Shetland Islands			Western Coast, Antarctic Peninsula			Eastern Coast, Antarctic Peninsula		
Sample:	P.849.1	P.725.1	P.846.1	O.985.1	O.932.3	O.528.3	TL.554.4	TL.556.7	TL.615.1
SiO_2	47.81	59.50	61.65	50.66	55.34	67.41	50.04	57.93	75.79
TiO_2	0.71	1.34	0.83	1.29	1.16	0.64	1.54	0.77	0.14
Al_2O_3	19.46	15.50	16.02	16.79	16.96	14.19	16.59	16.40	12.83
tFe_2O_3	9.78	8.85	6.43	14.61	9.02	4.07	10.37	7.09	1.40
MnO	0.20	—	0.23	0.24	0.14	0.09	0.16	0.13	0.03
MgO	5.50	2.11	4.10	5.51	4.68	1.04	8.38	6.78	0.02
CaO	12.41	6.55	5.73	8.18	3.92	2.40	6.73	4.04	0.32
Na_2O	2.14	3.86	4.24	2.73	4.60	4.54	3.56	2.62	4.62
K_2O	0.31	0.54	0.86	0.49	1.74	2.97	1.87	2.52	4.56
P_2O_5	0.12	0.32	0.25	0.12	0.23	0.15	0.54	0.19	0.04
Total	98.44	98.57	100.34	100.62	97.79	97.50	99.79	98.47	99.75
Trace elements (p.p.m.)									
Ni	18	3	2	<1	6	<1	28	32	<1
Cr	46	7	9	19	42	3	98	304	1
Rb	2	31	25	8	55	98	102	96	121
Ba	117	352	283	166	505	625	644	583	601
Sr	330	411	363	344	264	225	570	388	84
Th	<1	5	4	<1	6	11	<1	10	20
Pb	5	11	5	7	12	18	8	19	13
La	7	18	14	8	22	26	35	27	69
Ce	16	43	38	17	54	60	57	54	133
Y	17	40	25	17	28	37	28	22	29
Zr	37	192	160	57	218	354	198	198	139
Nb	3	6	6	4	9	14	12	13	12
Zn	—	119	82	97	85	52	91	79	11
Ga	—	—	—	23	18	18	22	12	13

are similar to those of many Andean igneous rocks (Roobol *et al.*, 1976; Pichler and Ziel, 1972; Dostal *et al.*, 1977). This is not surprising since at the time they were generated the tectonic situations in South America and the Antarctic Peninsula were very similar. Figures 4 and 6–8 also demonstrate that there are no significant chemical differences between volcanic and plutonic rocks from the same area. There is, however, considerable real scatter on many of these inter-element plots which points to the complex processes involved in generating calc-alkaline magmas. In spite of this scatter it is possible to discern consistent transverse variations in magma composition across the Peninsula (Saunders *et al.*, 1980a,b).

From the South Shetland Islands, through the western coast to the eastern coast of the Peninsula, there is, as we have already noted, a progressive increase in SiO_2 mode (Fig. 4), with basalts predominating in the west and dacites and rhyolites on the eastern coast. Moreover, lavas in the South Shetlands have lower K, Rb, Th and light REE, and higher K/Rb and Na/K ratios than those on the mainland. Western coast plutonic rocks are trondhjemitic compared

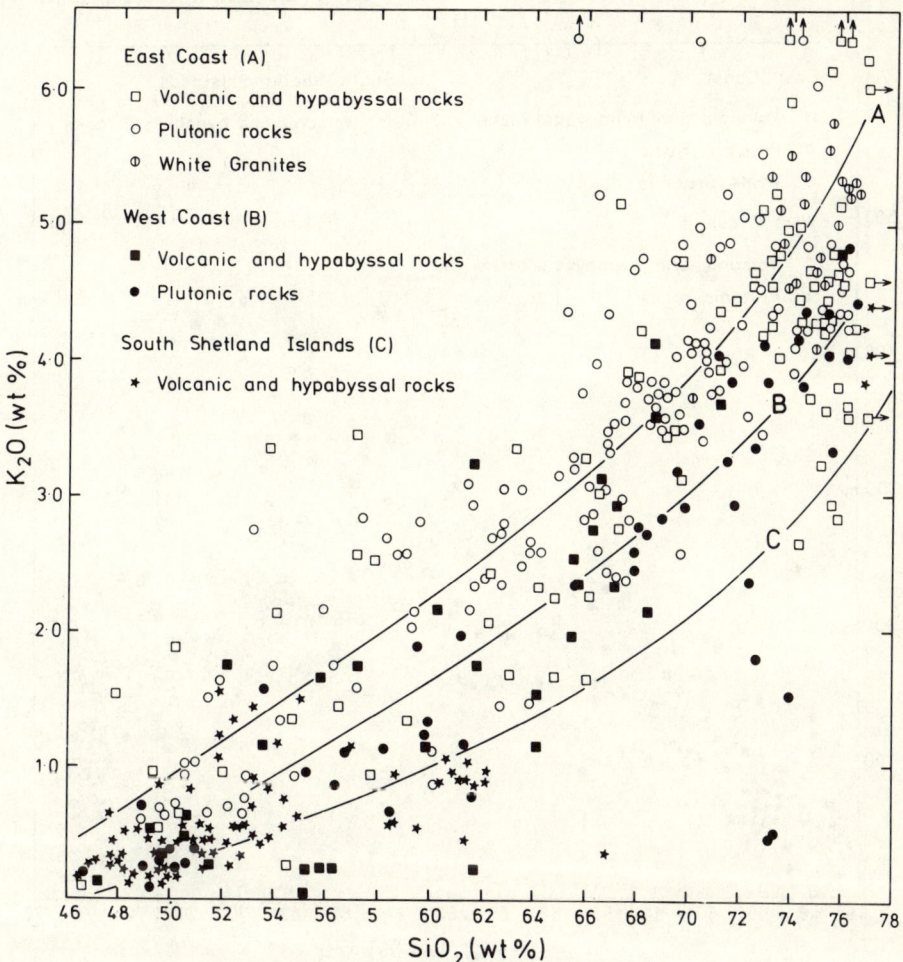

Fig. 8 Plot of K_2O against SiO_2 for Antarctic Peninsula Mesozoic volcanic and plutonic rocks. Curves represent mean K_2O–SiO_2 trend for each region: A, eastern coast; B, western coast; C, South Shetland Islands. From Saunders *et al.* (1980b). Reproduced by permission of Elsevier Scientific Publishing Co.

with the tonalite–granodiorite–granite series of the eastern coast of the Peninsula (Saunders et al., 1980a). There are also significant variations in K_2O mode (Fig. 7(a)), Th (Fig. 7(b)), Rb, Ba, and light REE across the Peninsula. More specifically, at a given SiO_2 content there is a progressive increase in average K_2O (Fig. 8) across the Peninsula, equivalent to the 'K-h' variation in many other arc systems (Dickinson and Hatherton, 1967; Dickinson, 1975; Nielson and Stoiber, 1973). It is worth stressing that this 'K-h' variation applies to the plutonic as well as to the volcanic rocks and that, of the other incompatible elements, it is only the large ion lithophile elements Rb, Th, La, Ce, (Ba), and probably Cs and U which show a similar variation. These are the elements that normally increase with SiO_2 during fractionation. Other smaller high field strength trace elements, such as Ti, P, Zr, Hf, Ta, Nb, which are normally incompatible in basaltic systems (e.g. Weaver et al., 1972; Wood et al., 1979; Tarney et al., 1979) do not show similar transverse variations. Nor do these elements increase progressively with fractionation in individual calc-alkaline suites. Zr begins to decrease at intermediate silica levels (Fig. 9) and Nb, Y, and the heavy REE may show an initial increase, but thereafter

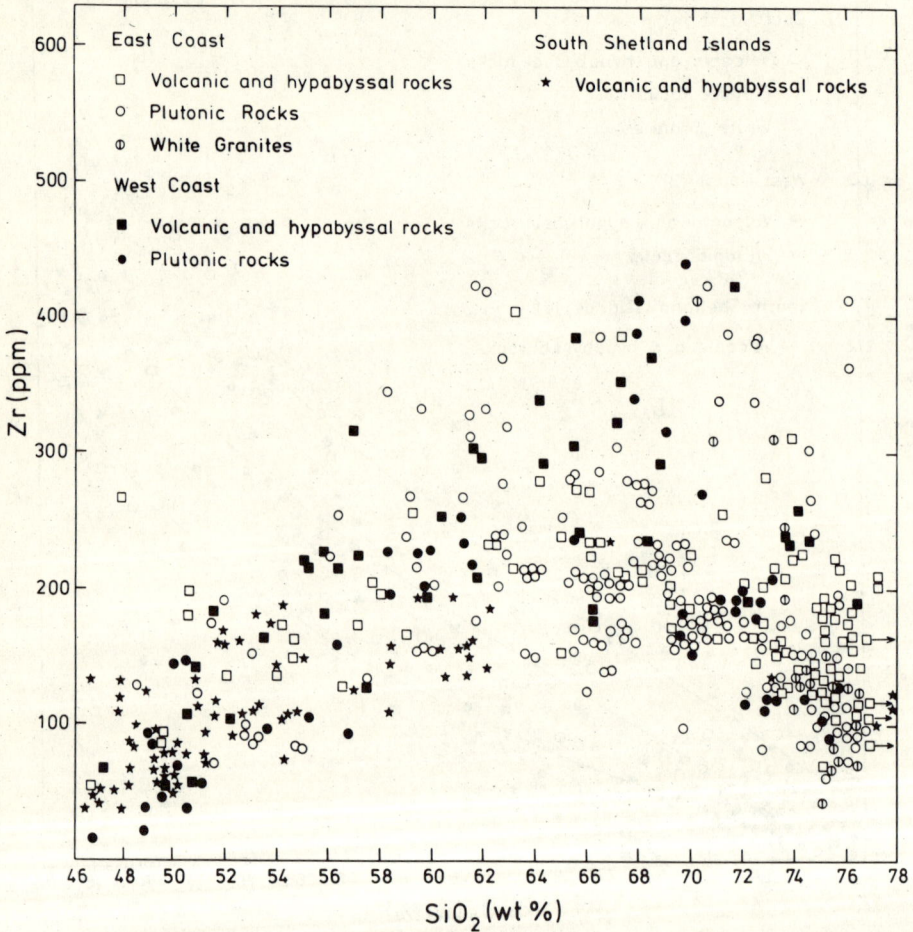

Fig. 9 Plot of Zr against SiO_2 for Antarctic Peninsula Mesozoic volcanic and plutonic rocks. Following an initial increase, Zr begins to decrease in intermediate to acid compositions. There is no significant difference in Zr–SiO_2 correlation across the Peninsula. From Saunders et al. (1980b). Reproduced by permission of Elsevier Scientific Publishing Co.

remain constant or even decrease with increasing SiO$_2$ (Saunders et al., 1980b).

REE data are not yet available for the Mesozoic volcanic rocks other than those from the South Shetland Islands (see below). Chondrite-normalized REE plots for contemporaneous plutonic rocks are shown in Fig. 10 (from Saunders et al., 1980a). The patterns are moderately fractionated with Ce$_N$/Yb$_N$ in the range 2.5–6.1 and the more silicic rocks have strong negative Eu anomalies. The nature of the transverse REE variation in the volcanics can be judged from X-ray fluorescence data for Ce and Y (Fig. 11), using Y as an indicator of heavy REE behaviour. The average Ce$_N$/Y$_N$ ratio increases from near 2.0 on the South Shetland Islands to 5.0 on the eastern coast of the Peninsula. The plutonic rocks show a similar increase. Stern and Stroup (1980) have also noted transverse variations in Ce$_N$/Yb$_N$ ratio in the plutonic rocks of southern Chile.

Mesozoic-Tertiary lavas from the South Shetland Islands

Mesozoic volcanic rocks from Byers Peninsula, Livingston Island are mainly low K basalts and basaltic andesites (Weaver et al., 1980) and follow a trend of moderate Fe-enrichment (Fig. 6(c), and Tables 1 and 2). Eight samples have been analysed for REE by isotope dilution (Table 3) and represent the compositional range from basalt to andesite with K–Ar ages from 125 to 90 Ma. The REE patterns (Fig. 12(a)) are subparallel with Ce$_N$/Yb$_N$ ranging from 1.8 to 3.0. Total REE concentrations vary by a factor of four, and correlate crudely with SiO$_2$ and other major element parameters, strongly with

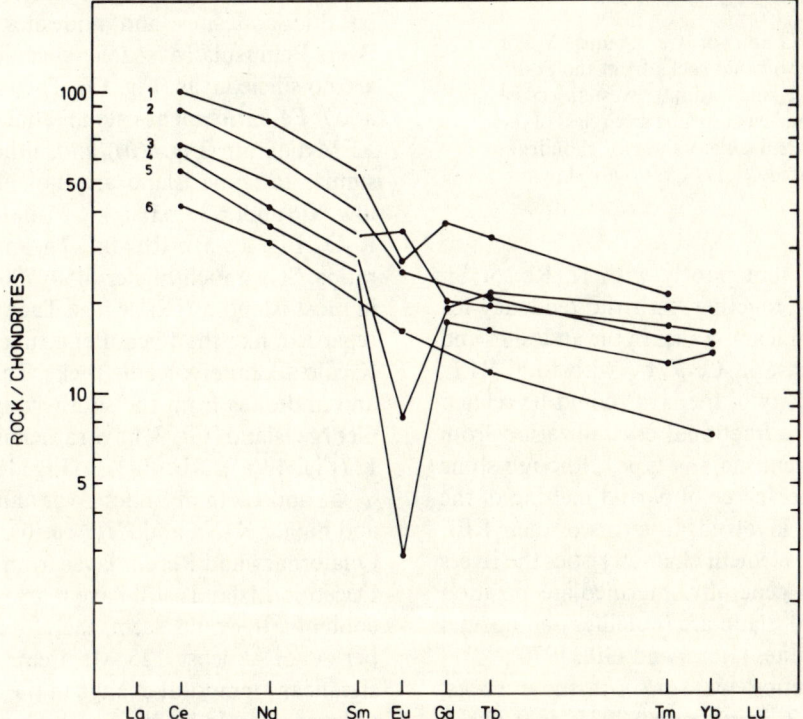

Fig. 10 Chondrite-normalized REE data for Antarctic Peninsula plutonic rocks: 1, granophyre (western coast); 2, tonalite (eastern coast); 3, tonalite (eastern coast); 4, granite (western coast); 5, granite (eastern coast); 6, gabbro (eastern coast). Data points for Sm are interpolated. From Saunders et al. (1980a). Reproduced by permission of the Board of Regents of the University of Wisconsin System

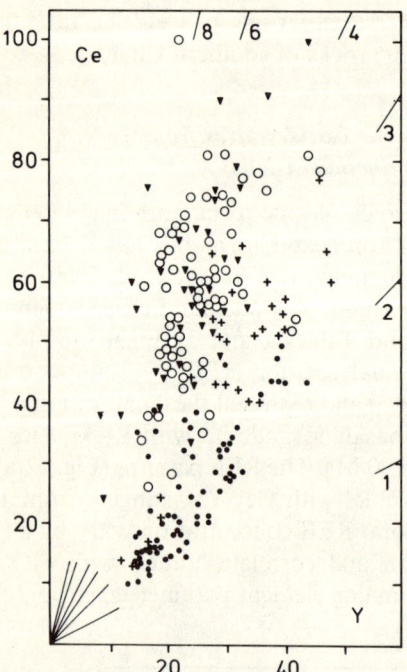

Fig. 11 Graph of Ce against Y for Mesozoic volcanic rocks from the South Shetland Islands (dots), western coast (crosses) and eastern Coast (circles) of the Antarctic Peninsula. Values for chondrite-normalized Ce_N/Yb_N ratio are shown

Zr, Ba, and Y, but poorly with K, Rb, or Sr. These features, together with the tendency for small negative Eu anomalies in the andesites and the slow increase in Ce_N/Yb_N with total REE, suggest that many of the lavas could be related by low pressure fractional crystallization from a common parent magma type, although some variation in the degree of partial melting of the source may be involved. In terms of their REE and other trace element characteristics the Byers lavas fall in a generally intermediate position between typical island arc tholeiites and normal calc-alkaline suites (Jakeš and Gill, 1970).

$^{87}Sr/^{86}Sr$ ratios for some of these rocks, given in Table 3, range from 0.70350 to 0.70432, but show no simple relationship to any other geochemical parameters. The $(^{87}Sr/^{86}Sr)_0$ ratios are, however, significantly higher in the older rocks of Byers Peninsula. This distribution of Sr isotope ratios is different to that seen in many other arc sequences, where the youngest rocks generally have the highest $^{87}Sr/^{86}Sr$ ratios. The initial Sr isotope ratios of the Byers Peninsula rocks (0.7034–0.7044) are typical of primitive calc-alkaline volcanic rocks such as the Andean basalts and andesites from Equador (Francis et al., 1977) and Patagonia (Hawkesworth et al., 1979), although the initial Sr isotope ratios of the Fildes Peninsula rocks are much lower (0.7031–0.7035) than these. Some of the samples have Rb/Sr ratios too low to account for their present Sr isotope ratios, which implies either a depletion event in their source regions or removal of some high Rb/Sr phase, or, perhaps more probably, incorporation of unsupported radiogenic Sr derived from seawater-altered rocks in the subducted oceanic crust.

The early Tertiary lavas of Fildes Peninsula, King George Island, are mainly plagioclase-phyric olivine basalts together with a few basaltic andesites and andesites. Unlike the Byers Peninsula Mesozoic volcanic rocks, there are no silicic lavas (Fig. 13). The samples define a low Fe-enrichment calc-alkaline trend on the AFM diagram (Fig. 6(d)), and, although they are similar to some island arc tholeiites in having low Ni (under 30 p.p.m.), Cr (under 90 p.p.m.), K_2O (Fig. 13(a)), Rb, Ba, Th and high K/Rb ratios, Sr is much higher (450–700 p.p.m.) than in most island arc tholeiites. They are thus best regarded, like the Byers Peninsula lavas, as low K calc-alkaline volcanic rocks. Tertiary basalts and andesites from the southern coast of King George Island (Fig. 3) have rather higher levels of K (Fig. 13(a)), Rb, Ba, Zr (Fig. 13(b)), Zr, Th, Y, Ce and La than Fildes lavas and lower K/Rb and higher K/Na and Ce/Y ratios. However the Quaternary and Recent lavas from Penguin and Deception Islands still have markedly low K_2O contents. It would seem, therefore, that over a period of at least 125 Ma there has been no significant temporal change in the K_2O content of lavas erupted on the South Shetland Islands. This implies that the K–h relationship exhibited by many island arc systems is essentially a spatial feature, although, as in northern Chile, where the locus of magmatic activity migrates

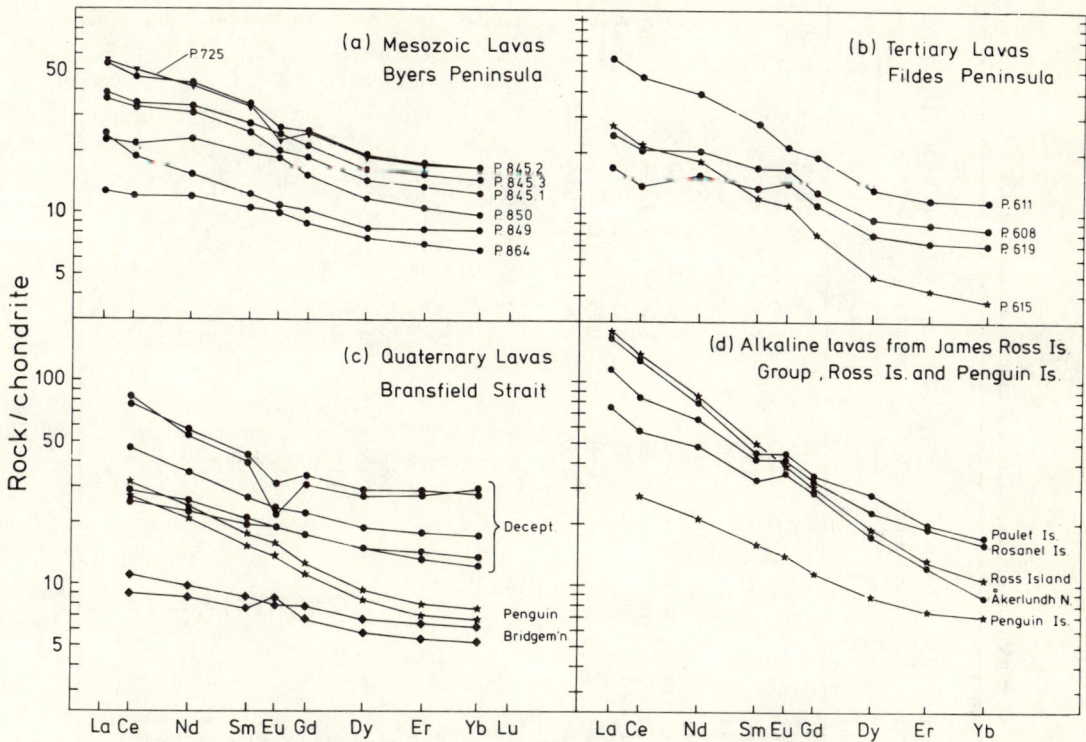

Fig. 12 Chondrite-normalized REE patterns of volcanic rocks from the South Shetland Islands: (a) Mesozoic lavas from Byers Peninsula, Livingston Island; (b) Tertiary lavas from Fildes Peninsula, King George Island; (c) Quaternary lavas from Bransfield Strait. REE patterns of alkaline basalts from three localities in western Antarctica are shown in (d)

away from the trench with time, it may sometimes be expressed as a temporal feature.

REE data for four of the freshest Tertiary lavas (Table 3, Fig. 12(b)) demonstrate that the light REE contents have much the same range as in the earlier Mesozoic lavas, but the tendency toward lower heavy REE concentrations results in higher and more variable Ce_N/Yb_N ratios (2.0–5.8). The highest Ce_N/Yb_N ratio occurs in a basalt with high MgO and CaO, over 700 p.p.m. Sr and 30 p.p.m. Ni. Some of these chemical characteristics are seen in the mildly alkaline basalts of the Recent volcano of Penguin Island on the south coast of King George Island (see below and Weaver et al., 1979) and they suggest that the later Tertiary lavas may include some basalts derived by melting at greater depths, leaving some residual garnet.

The Tertiary REE patterns differ in one important respect from other igneous rocks so far analysed from the South Shetland Islands and the Antarctic Peninsula: they show negative Ce anomalies of up to 15 per cent. Although there is an uncertainty of c. 5 per cent attributable to analytical error (see results for BCR-1 in Table 3), two of the Fildes Peninsula samples are well outside the range of such effects. It is well known that Ce differs from its neighbouring REE in exhibiting a quadrivalent state as well as the normal magmatic trivalent state, but only under extreme oxidizing conditions such as are encountered in the hydrosphere does this occur. Thus Ce^{4+} is preferentially incorporated into deep-sea manganese nodules, and seawater itself has a large negative Ce anomaly (as have some marine clays). Although the abundances of REE are considered to remain relatively constant during the weathering and alteration of igneous rocks, pronounced Ce anomalies (Ce/Ce* down to 0.1) have been recorded in samples

TABLE 2 Representative analyses of Tertiary and Quaternary volcanic rocks from Antarctic Peninsula and Scotia arc

Sample:	South Shetland Islands			Penguin Island	Bridgeman Island	Deception Island		Seal Nunataks	South Sandwich Islands	'Discovery' Arc
	P.619.1	P.608.5a	P.611.1	P.721.3	P.640.16	B.138.2	P.870.1	D.4689.1	SSR1.3	D34.34
SiO_2	49.57	53.53	62.07	49.02	52.88	51.89	68.02	49.87	50.22	49.40
TiO_2	0.70	1.04	0.89	1.08	0.64	1.49	0.55	1.96	0.56	0.59
Al_2O_3	20.47	16.19	16.29	15.81	17.68	16.20	14.99	15.28	19.59	19.64
tFe_2O_3	9.08	10.71	6.15	9.95	7.43	9.46	4.97	11.43	10.60	9.81
MnO	0.18	0.25	0.19	0.18	0.13	0.18	0.18	0.14	0.17	0.16
MgO	5.25	4.48	2.36	8.95	6.14	6.11	0.33	8.84	5.60	4.70
CaO	11.00	8.47	5.01	10.08	10.30	10.07	1.69	8.40	12.53	13.09
Na_2O	3.00	4.16	5.59	3.95	3.54	4.07	7.45	4.21	1.76	2.30
K_2O	0.38	0.41	0.99	0.48	0.47	0.28	1.69	0.99	0.15	0.20
P_2O_5	0.14	0.16	0.34	0.28	0.06	0.21	0.10	0.20	0.05	0.10
Total	99.77	99.24	99.78	99.78	99.27	99.96	99.97	101.12	101.23	99.99
Trace elements (p.p.m.)										
Ni	17	3	2	159	40	35	2	167	11	1
Cr	48	17	9	494	130	141	7	276	65	13
Rb	7	2	46	5	11	3	32	13	1	2
Ba	122	200	516	148	70	88	242	115	46	26
Sr	508	513	446	534	332	340	134	410	143	239
Th	2	<1	3	1	2	2	7	4	—	—
Pb	2	5	11	7	5	4	12	1	—	1
La	6	8	19	10	2	8	28	15	—	7
Ce	12	18	41	23	8	22	71	35	11	12
Y	14	18	29	12	10	26	71	17	23	25
Zr	51	68	185	80	58	144	665	145	2	1
Nb	1	5	6	2	1	2	17	26	—	—
Zn	—	—	—	69	62	76	124	96	—	—
Ga	—	—	—	21	19	16	27	25	13	21

TABLE 3 REE and Rb–Sr data for volcanic rocks from the South Shetland Islands

Sample	Rock Type	Age (Ma)	La	Ce	Nd	Sm	Eu	Gd	Dy	Er	Yb	Ce_N/Yb_N	Eu/Eu^*	Ce/Ce^*	Rb	Sr	$^{87}Sr/^{86}Sr$	$(^{87}Sr/^{86}Sr)_0$
Byers Peninsula																		
P.862.4	Basalt sill	125	7.30	16.5	10.4	2.74	0.92	2.93	3.10	1.95	1.84	2.3	0.99	0.95	6	366	0.704 04	0.703 96
P.845.1b	Basalt lava	110	11.9	28.6	19.4	5.01	1.54	5.21	4.90	2.97	2.73	2.7	0.93	0.96	—	—	—	—
P.845.2c	Basaltic andesite lava	110	17.7	40.3	27.3	6.89	2.01	6.93	6.52	3.96	3.67	2.8	0.89	0.93	25	387	0.704 57	0.704 28
P.845.3a	Basaltic andesite lava	110	12.8	29.7	21.1	5.59	1.88	5.86	5.59	3.43	3.18	2.4	1.01	0.93	—	—	—	—
P.845.9	Dolerite plug	110	—	—	—	—	—	—	—	—	—	—	—	—	3	536	0.703 94	0.703 91
P.848.5	Andesite plug	110	—	—	—	—	—	—	—	—	—	—	—	—	23	409	0.703 68	0.703 43
P.849.1	Basalt lava	(110?)	7.82	16.3	9.58	2.45	0.84	2.80	2.88	1.87	1.84	2.3	0.99	0.92	—	—	—	—
P.850.8	Basalt lava	95	7.58	18.9	14.5	3.92	1.41	4.20	4.00	2.38	2.13	2.3	1.07	0.95	1	535	0.704 05	0.704 04
P.864.4	Dolerite plug	95	4.19	10.4	7.59	2.13	0.77	2.43	2.59	1.58	1.45	1.8	1.04	0.96	—	—	—	—
P.864.1a	Basalt sill	80	—	—	—	—	—	—	—	—	—	—	—	—	9	382	0.704 26	0.704 18
P.725.1	Andesite plug	(80?)	17.9	42.7	26.4	6.67	1.73	6.73	6.46	3.85	3.62	3.0	0.79	0.99	31	411	0.704 60	0.704 35
Fildes Peninsula																		
P.608.5a	Basaltic andesite lava	60	8.14	18.0	13.0	3.45	1.31	3.60	3.28	2.02	1.87	2.4	1.14	0.89	2	513	0.703 52	0.703 51
P.615.1	Basalt lava	60	8.95	18.8	11.5	2.47	0.86	2.23	1.71	0.96	0.82	5.8	1.13	0.91	4	713	0.703 15	0.703 14
P.619.1	Dolerite plug	50	5.60	12.0	9.98	2.75	1.12	3.08	2.73	1.64	1.56	2.0	1.18	0.84	7	508	0.703 50	0.703 47
P.611.1	Andesite plug	45	19.2	40.9	24.6	5.68	1.63	5.29	4.48	2.66	2.54	4.1	0.91	0.93	46	446	0.703 73	0.703 53
Standards																		
BCR-1	Basalt		25.7	53.2	29.1	6.78	1.98	6.75	6.35	3.69	3.39	4.0	0.90	0.94	—	—	—	—
BOB-1	Mid-ocean ridge tholeiite		5.01	13.6	10.8	3.38	1.25	4.32	4.75	2.95	2.76	1.3	1.00	0.99	—	—	—	—

Rare earths determined by mass-spectrometric isotope dilution, Rb and Sr by X-ray fluorescence (all concentrations in p.p.m.). $^{87}Sr/^{86}Sr$ ratios are relative to 0.708 00 for Eimer & Amend $SrCO_3$, errors ± 0.01 per cent (2 s.e.).

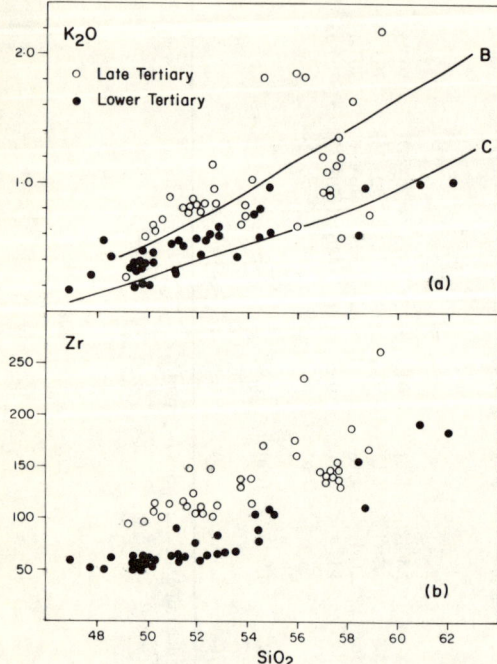

Fig. 13 Graphs of K_2O against SiO_2 and Zr against SiO_2 for early Tertiary and late Tertiary lavas from King George Island

from the zeolitized parts of ophiolite complexes (Robertson and Fleet, 1975; Menzies et al., 1977) and in altered ocean-floor basalts (Ludden and Thompson, 1979), where they are attributed to intense rock-seawater interaction at low to moderate temperatures.

Only recently have similar Ce anomalies been recorded in island-arc lavas. Heming and Rankin (1979) found negative Ce anomalies in three andesites from the Rabaul caldera, New Guinea (Ce/Ce* between 0.6 and 0.8), although other associated lavas ranging from basalt to dacite showed no significant anomalies. They rejected alteration as a possible explanation since the rocks were considered fresh according to petrographic and other geochemical criteria—an argument which applies equally in the present case. Thus it would seem that the Ce anomaly must be a characteristic of the source region or the melting process, either as a result of pre-magmatic alteration or else due to retention of Ce^{4+} during magma separation.

Partial melting of highly altered subducted oceanic crust would seem to be a plausible solution, although it is not possible to exclude the involvement of subducted sediments with negative Ce anomalies. This does not necessarily imply that the magmas were *primarily* derived by melting of the slab, but that light REE as well as large ion lithophile elements may have been transferred into the main magma-producing reservoir in the overlying mantle wedge during dehydration of the slab, as implied by some current models of calc-alkaline and island arc tholeiitic magma generation (Hawkesworth et al., 1977, 1979; Mysen, 1979; Saunders et al., 1980b).

Quaternary lavas of Bransfield Strait

A detailed acount of the chemical compositions of Deception, Bridgeman, and Penguin Island lavas has been published by Weaver et al. (1979), and only a brief summary is presented here.

Like the earlier lavas erupted on the South Shetland Islands, the Bransfield Strait volcanic rocks are low in K_2O (Fig. 14) and all have high Na/K ratios (Table 2). The Na/K ratios of Deception basalts are similar to those in depleted mid-ocean ridge basalts. Deception lavas range from olivine-tholeiite to rhyodacite, although it is apparent from the low MgO (6 per cent), Cr (141 p.p.m.), and Ni (35 p.p.m.) contents of the most primitive lavas that even these are non-primary or evolved mantle melts. The smooth, consistent, chemical trends within the Deception lava suite (Fig. 15) can best be explained by fractional crystallization. There is moderate Fe enrichment until both Ti and Fe decrease sharply with titanomagnetite crystallization. Phosphorous also increases until apatite separates, while Sr consistently decreases because of early and continued plagioclase crystallization. Other incompatible elements such as Zr, Nb, K, Rb, and even Ba and Na_2O, increased progressively with fractional crystallization, Na_2O reaching 7.5 per cent by weight in the rhyodacites. Na_2O/Zr and Ba/Zr ratios do, however, fall slightly, reflecting some entry of

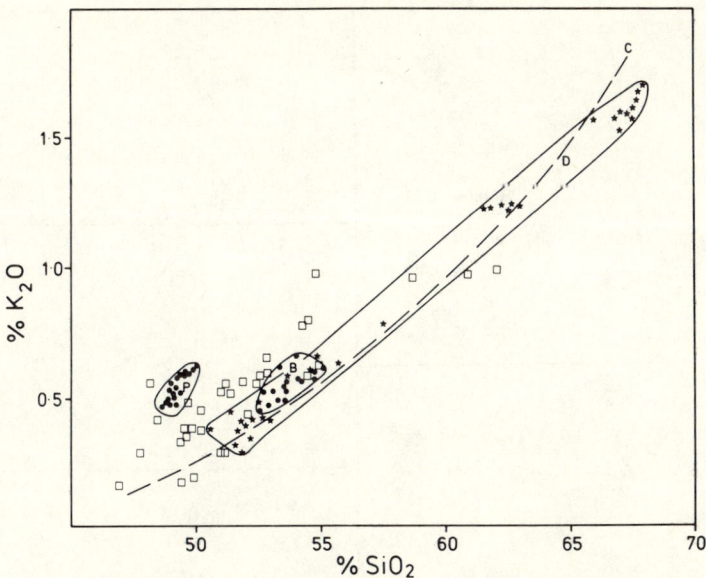

Fig. 14 Graph of K_2O against SiO_2 for Quaternary lavas from Deception (stars; D) Bridgeman (filled circles; B), and Penguin Islands (filled circles; P), Data for Tertiary volcanics (Fildes Peninsula; squares) and the trend for Mesozoic volcanics (Byers Peninsula; broken line; C) are shown for comparison

Na and Ba into plagioclase, but K/Zr and Rb/Zr ratios remain constant, confirming that no potassic phase crystallized. REE patterns of Deception lavas (Fig. 12(c)) have similar Ce_N/Yb_N ratios to other South Shetland calc-alkaline basalts (i.e. c. 2.0) but total REE levels increase with fractional crystallization, with only slight change in the Ce/Yb ratio, and with the eventual development of distinct negative Eu anomalies in the dacites. Ce/Zr and Y/Zr ratios decrease with fractional crystallization, especially in the more silicic compositions, indicating increasing partitioning of REE into the major mineral phases as the magma becomes more silicic. Deception Island REE distributions are very similar to those in the Sarmiento marginal basin ophiolite complex in southern Chile (Saunders et al., 1979), which is also characterized by abundant plagiogranite.

Whereas Deception Island lava compositions are, in a broad sense, calc-alkaline, it is worth pointing out the differences in trace element behaviour between these and other typically calc-alkaline lavas. The high field strength elements, Zr and Nb, increase progressively in Deception lavas with fractional crystallization, the former reaching values near 700 p.p.m. in the rhyodacites, very much higher than in any equivalent calc-alkaline compositions. The REE and Y also attain much higher concentrations. This distribution of the high field strength elements is distinct from that seen in calc-alkaline magmas, where the abundance of the high field strength elements tends to decrease at intermediate to high SiO_2 levels. This implies either that these elements are partitioned more effectively into the separating minerals during fractionation of calc-alkaline magmas, or that they are retained in the source regions of calc-alkaline magmas during magma generation.

Bridgeman Island basaltic andesites (Table 2) have higher alumina contents than those of Deception Island and the range of fractional crystallization, judged from major and trace element abundances (Fig. 15), is very much less. The lavas have lower values for the incompatible

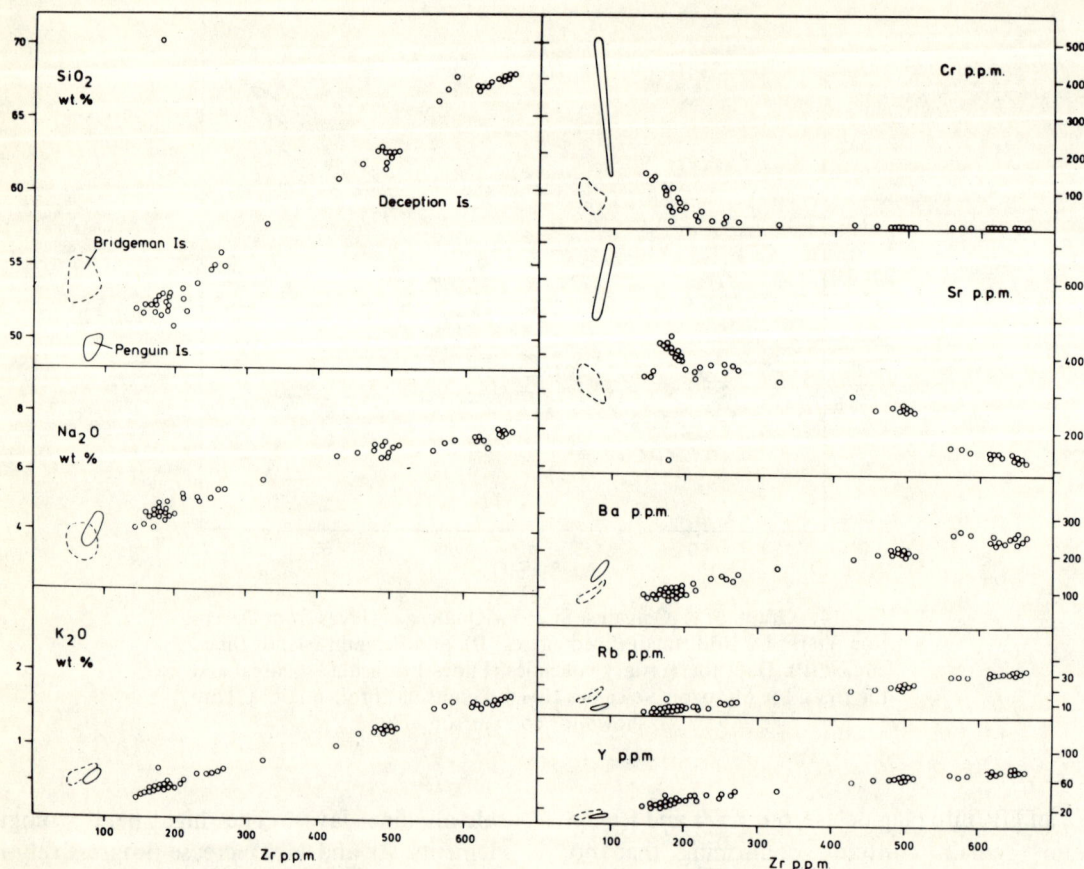

Fig. 15 Trace element relationships in Quaternary lavas from Deception, Bridgeman, and Penguin Islands. Zr is used as an index of fractionation (from Weaver *et al.*, 1979). Reproduced by permission of Springer-Verlag, Heidelberg

elements Zr, Nb, Y and the REE than any of the Deception rocks, but Cr and Ni are lower and K and Rb higher than in equivalent basaltic andesites from Deception. REE patterns for Bridgeman lavas are thus similar in shape to those of Deception basalts (i.e. Ce_N/Yb_N is c. 2) but have less than half the REE abundances (Fig. 12(c)). With these differences it is not possible to relate Bridgeman and Deception lavas by fractional crystallization. Nevertheless, it is possible to relate them by differing degrees and conditions of partial melting of a similar mantle source. Thus the initial Sr isotope ratios of the lavas from the two volcanoes are similar (mostly within the range 0.703 42–0.703 49). Weaver *et al.* (1979) argued that Bridgeman Island lavas were generated under more hydrous conditions, producing a higher degree of melting (hence lower trace element abundances) but more aluminous and silica-rich magmas. The higher K and Rb contents and higher Rb/Sr and lower K/Rb ratios were attributed to addition of K and Rb to their source regions during the initial stages of back-arc spreading, the hydrous fluids probably being derived from the subducting slab. Evidence for more hydrous melting conditions exists in other marginal basins (e.g. Saunders and Tarney, 1979).

Penguin Island, situated on the north-western margin of Bransfield Strait, may be regarded as an off-axis, mildly alkaline volcano. The more primitive magnesian basalts (Table 2) are high in Cr (500 p.p.m.) and Ni (170 p.p.m.) and the rapid depletion of Cr and Ni within the lava suite (Fig. 15) can be accounted for by olivine, clinopyroxene, and/or chromite fractional crystallization. Plagioclase phenocrysts occur only in the more evolved Penguin basalts and could not have segregated from the more Mg-rich compositions. Indeed Sr behaves incompatibly and increases proportionally with Zr and other incompatible elements (Fig. 15). Major element variation between the most Mg-rich and Mg-poor Penguin basalts can be accounted for by separation of 5 per cent olivine (Fo_{86}), 6 per cent clinopyroxene ($Ca_{44}Mg_{46}Fe_{10}$), and 2 per cent plagioclase (An_{71}). The enrichments of Ba, Ce, and Zr predicted by Rayleigh crystallization suggest 14 per cent fractional crystallization, which agrees fairly well with the major element modelling (Weaver et al., 1979).

REE patterns of Penguin lavas (Fig. 12(c)) are more fractionated than those of Deception and Bridgeman lavas and have Ce_N/Yb_N values c. 4.0. Zr, Y, and heavy REE concentrations are similar to those in Bridgeman lavas, but light REE contents are within the range of Deception basalts. Penguin lavas have high K/Rb and low Rb/Sr ratios, similar to Deception, but in fact have higher K/Zr, Rb/Zr, Ba/Zr, and Sr/Zr ratios. These relationships could be accounted for if a Zr-retaining phase such as garnet was stable during the generation of Penguin basalts. Garnet would also retain Y and heavy REE more than Zr and would explain the higher Zr/Y and Ce/Yb ratios in Penguin basalts. On the other hand, $^{87}Sr/^{86}Sr$ ratios of Penguin basalts (0.7039) are significantly higher than the ratios in Deception lavas, which could imply that the source of Penguin basalts had been enriched in incompatible elements at some earlier period. This could also account for many of the trace element relationships. But since the Rb/Sr ratios of Penguin basalts are insufficiently high to account for their present $^{87}Sr/^{86}Sr$ ratios it would still be necessary to invoke a later Rb-depletion event. Weaver et al. (1979) commented that some of these complexities might merely reflect the equally complex nature of the mantle source under a region which had already suffered more than 100 Ma of continual magma generation.

James Ross Island and Seal Nunataks

These volcanics are mildly alkaline olivine basalts and hawaiites. In terms of major elements they are similar to Penguin Island basalts but have higher total alkalis and K/Na ratios. Incompatible elements such as Rb, Zr, Nb, and the light REE are also higher than in Penguin basalts, although Ba is slightly lower (Table 2). Zr/Nb ratios of Seal Nunatak basalts (c. 7.0) are similar to those in basalts from ocean islands and continental rift valleys, whereas Penguin Island basalts have high Zr/Nb ratios (c. 80) more typical of island arc and depleted mid-ocean ridge basaltic magmas (Erlank and Kable, 1976; Tarney et al., 1979). Similar comparisons may also be made with respect to REE patterns (Fig. 12(d)). The James Ross–Seal Nunatak basalts have more strongly fractionated patterns with $Ce_N = 50-150$ and $Ce_N/Yb_N = 7.2-8.2$ (Pankhurst, 1980). Like the Penguin basalts they have small positive Eu anomalies, possibly due to equilibration with clinopyroxene rather than plagioclase accumulation, but otherwise the REE patterns are essentially linear.

All these features, as well as high Ni (over 150 p.p.m.) and Sr (450–1000 p.p.m.), are typical of alkaline basalts from oceanic islands, and suggest small degrees of partial melting of a garnet peridotite source with little subsequent fractional crystallization (indeed many of the lavas carry mantle nodules). The higher Ce_N/Yb_N ratios compared with Penguin basalts and the comparable, or somewhat higher, heavy REE contents (Yb = 8.6–16.7) imply that either smaller degrees of partial melting and/or higher concentrations of light REE (and other incompatible elements) in the sources are the factors responsible, rather than simply an increasing proportion of residual garnet.

$^{87}Sr/^{86}Sr$ ratios of these lavas (Pankhurst, 1980) show considerable variation, but are below the range usually found in island arcs (0.7030–0.7035 for James Ross; 0.7028–0.7030 for Seal Nunataks). Some are unsupported by their Rb/Sr ratios, but they show a strong positive correlation in an Rb/Sr isochron plot which Pankhurst (1977) interpreted as implying that trace element variations in the source regions could have been introduced as a discrete event 500 Ma ago.

The South Sandwich and Discovery Arcs

The data of Baker (1978) and Hawkesworth *et al.* (1977) show that the South Sandwich lavas have typical island arc tholeiite characteristics. The basaltic rocks follow an Fe-enrichment trend and have low concentrations of K, Rb, Ba, Sr, and other incompatible elements (Table 2) compared with the more typically calc-alkaline lavas of the Antarctic Peninsula. REE patterns vary from strongly light REE depleted with low total REE levels (*c.* five times chondrite) and small positive Eu anomalies in the oceanitic basalts to slightly light REE enriched with over 20 times chondritic abundances and negative Eu anomalies in the basaltic andesites and andesites. Hawkesworth *et al.* (1977) considered that these features could be explained by low pressure fractional crystallization involving plagioclase and clinopyroxene and minor olivine. The fact that their $^{87}Sr/^{86}Sr$ ratios (0.7038–0.7039) are high relative to their $^{143}Nd/^{144}Nd$ ratios (0.513 01–0.513 14) and to the $^{87}Sr/^{86}Sr$ ratios of the basalts from the South Sandwich spreading centre (thought to be representative of the mantle source beneath the South Sandwich arc) was taken by Hawkesworth *et al.* (1977) to indicate derivation of some ^{87}Sr from the subducted oceanic crust.

Samples dredged from the Discovery arc (Barker *et al.*, 1980) are all strongly plagioclase-phyric basalts with low contents of K_2O, TiO_2, Zr, Nb, La, Ce, Y, Ba, Cr, and Ni and high Na_2O/K_2O ratios and low Ce/Y ratios (Table 2). These features are almost identical to those in basalts from the active South Sandwich arc. It would appear then that the tectonic situation and conditions of magma generation were essentially similar *c.* 10 Ma earlier—i.e. that the Discovery arc was an intra-oceanic island arc.

Discussion

The processes of magma generation in subduction zones are complex and involve many uncertainties with respect to the chemical and physical nature of the source regions. In addition, the $P-T-X$ conditions of magma generation and any superimposed effects resulting from fractional crystallization, the degree of sediment or continental crust contamination, assimilation, etc., must be accounted for. Some of these factors are very difficult to quantify in a single area, and it was our hope that by looking at magma variations over a wide range of space and time in a single evolving region that we might begin to constrain some of these uncertainties.

Transverse geochemical variations in the Antarctic Peninsula

Transverse variations in magmas related to seafloor subduction can, in theory, be due to a number of factors: (1) changing nature of the melting process with depth in the subduction zone; (2) changing chemical or physical nature of the mantle beneath island arcs, either laterally or vertically; (3) increasing zone refining or fractional crystallization in magmas ascending from greater depth; and (4) increasing crustal contamination (especially in continental based arcs).

In the Antarctic Peninsula the evidence suggests that the transverse geochemical variations are a spatial rather than a temporal feature, and are manifest in the following form:

(1) Mafic lavas predominate near the trench, but become increasingly subordinate to intermediate and silicic lavas away from the trench.

(2) Only the large ion lithophile elements, K, Rb, Th, La, and Ce (and probably Cs and U), show a '$K-h$' type relationship. These are the elements that normally increase in abundance with increasing SiO_2 in calc-alkaline magmas. Small high field strength elements do not show this type of variation.

(3) REE patterns become more fractionated away from the trench (Ce_N/Yb_N ratios range from 2.0 to more than 6.0).

Increasing crustal contamination is a possible explanation for the $K-h$ variation in the Antarctic Peninsula, but not in other, intra-oceanic arcs. Nor is it easy to reconcile with the fact that the K–h variation occurs independently of the abundance of SiO_2 in the rock.

Best (1975) has proposed zone refining. The enrichment of a trace element is proportional to $1/D$ (D being the bulk distribution coefficient of that element). Thus if D is 0.1 then the enrichment cannot exceed 10, and therefore significant enrichment can only be achieved for the most incompatible elements. With calc-alkaline magmas this certainly holds for large ion lithophile elements such as Rb and Th which have very low D values, probably less than 0.01 (see Wood et al., 1979). However, it is difficult to account for the fact that other normally incompatible elements such as Nb and Ta show no enrichment in calc-alkaline magmas. This is a major obstacle to an active zone-refining mechanism.

Evaluating the nature of mantle heterogeneity beneath island arcs to account for the K–h relationship is a difficult exercise because of the need to accommodate the effects of the complex (and uncertain) magma generation processes in and above the subduction zone. At mid-ocean ridges it is rather easier to decipher heterogeneities because the erupted basalts represent relatively high degrees of mantle melting (cf. Tarney et al., 1979). Ideally, some independent evidence for varying mantle compositions beneath active arcs is necessary. Fortunately, in the Antarctic Peninsula region some independent evidence, although still rather tenuous, does suggest some form of lateral transverse variation in mantle composition, and is worth reviewing briefly. This concerns the late alkaline volcanism in the area.

The erupted alkaline basalts, all post-dating the calc-alkaline activity, occur on Penguin Island and probably some of the adjacent centres, in the James Ross Island area on the east coast of the Peninsula, and further south in Marie Byrd Land and Victoria Land (González-Ferran and Vergara, 1972; Katsui, 1972). Baker et al. (1977) have suggested that the occurrence reflects a change in the tectonic regime to an extensional one after the cessation of subduction along the South Shetland trench (and of course further south). In terms of trace element geochemistry Pankhurst (1980) has noted that there is also a transverse geographical progression, as revealed for instance by REE patterns from Penguin Island (Ce_N/Yb_N = 4.0) to James Ross Island and Seal Nunataks (Ce_N/Yb_N = 7.2–8.2) (Fig. 12(d)). No comparative data are yet available from Marie Byrd Land, but the alkaline basalts of Ross Island, Victoria Land, have Ce_N/Yb_N ratios averaging c. 14 (Sun and Hanson, 1975b). Of course there are various ways of interpreting this apparent transverse variation. It could reflect smaller degrees of partial melting associated with an increasing thickness of subcontinental lithosphere and lower geothermal gradient. On the other hand, it could also imply that the mantle source is increasingly enriched in incompatible elements away from the trench.

There is no doubt from the high MgO, Cr, and Ni contents of these alkaline basalts that they represent near-primary mantle melts (Sun and Hanson, 1975b; Pankhurst, 1980; Weaver et al., 1979). Because of the obvious geochemical similarities between the more primitive Penguin, Deception, and Bridgeman Quaternary basalts, Weaver et al. (1979) also considered that Deception and Bridgeman magmas were mantle-derived, but generated under more hydrous melting conditions and, in the case of Bridgeman Island basaltic andesites, with an added component of K and Rb derived from the subducted slab. As we have seen, the negative Ce anomalies exhibited by some Tertiary South Shetland lavas suggest that at least a proportion of their REE content may have been derived from seawater-altered basalts in the subducted slab. There is also an increasing body of isotopic evidence suggesting an added component of seawater Sr in island arc magmas (Meijer, 1976; Hawkesworth et al., 1977, 1979). The most plausible explanation for the genesis of modern calc-alkaline magmas would seem to be that they are essentially mantle-derived magmas, but with a variable component derived from the subducted slab.

Antarctic Peninsula calc-alkaline magmas (indeed all subduction-zone magmas) have higher large ion lithophile to high field strength element ratios than equivalent non-orogenic tholeiitic or alkaline magmas (Saunders et al., 1980b). Clearly, since large ion lithophile elements such as K, Rb, Sr, and Ba are commonly mobile during low temperature alteration of sea-floor basalts, it is feasible that they could be transported into the overlying mantle wedge by hydrous fluids released during dehydration of the slab. However, the high field strength elements (Ti, Nb, Ta, Zr, Hf, P, etc.) are also low in absolute terms in calc-alkaline magmas. Thus suggests that minerals such as ilmenite, sphene, rutile, zircon, and apatite might have enhanced stability under hydrous melting conditions and could retain these elements during partial melting. Of interest in this regard is that the more geochemically 'enriched' mantle nodules from kimberlite pipes— the so called 'marid' nodule suites (Gurney and Harte, 1980)—are rich in mica, amphibole, rutile, ilmenite, diopside, and zircon. Thus there seems no doubt that appropriate minerals capable of retaining high field strength elements during partial melting are present in the subcontinental mantle.

Jordan (1978) has argued that there are significant differences between subcontinental and suboceanic mantle. The subcontinental mantle ('tectosphere') appears to be lighter and more refractory, but, as Jordan points out, is still fertile enough to yield 'enriched' basaltic magmas on demand. This dilemma can be resolved if the upper mantle, and particularly the subcontinental mantle, is veined by incompatible element-enriched liquids (or fluids) approximating to alkaline basalt in composition. Such a veined mantle model has been suggested from studies of mantle nodules (Frey and Green, 1974) and has been invoked to explain trace element relationships in basalts (Frey et al., 1978), North Atlantic ridge basalts (Tarney et al., 1979; Wood et al., 1979) and has indeed been suggested for the source regions of Ross Island alkaline basalts (Sun and Hanson, 1975). Incipient melting in the upper mantle (as opposed to the larger degrees of melting resulting from mantle diapirism) and upward migration of such melts may be the cause of the veining. However, while incipient melting and consequent veining may occur in both suboceanic and subcontinental mantle, the veins are more likely to be preserved beneath stable continents whereas convective recycling in the oceanic mantle will continually disperse any significant build-up of enriched veins.

An increase in the proportion of veins of such material inwards from the continental margin (or with increasing depth below the continental margin) could provide a simple explanation for the $K-h$ variation, since during the time the Mesozoic volcanics were erupted the Antarctic Peninsula bordered the Gondwanaland supercontinent. This model also circumvents the perennial difficulty of trying to account for the high large ion lithophile element abundances in calc-alkaline magmas by fusion of depleted basalt in the slab or depleted lherzolitic mantle above the slab. Instead both alkaline and calc-alkaline magmas can be regarded as being derived from a similar enriched mantle source with alkaline magmas being generated under low P_{H_2O} conditions and calc-alkaline magmas at high P_{H_2O} (with consequent effects on major element compositions and large ion lithophile to high field strength element ratios). Alkaline basalts do in fact occur in many mature arc systems such as Japan, the Aleutians (DeLong et al., 1975), the Lesser Antilles, and of course the South Shetland arc.

Magmatic evolution of the Antarctic Peninsula–Scotia Arc

The above model can be adapted to account for the changing character of volcanicity in the region with time. As we have noted, the last c. 100 Ma has seen a progressive reduction in the amount of Pacific ocean floor being subducted beneath the Antarctic Peninsula and a switch to Atlantic subduction. Initially, subduction occurred beneath a major continental margin. Following the opening of the Weddell Sea, the locus of magmatic activity in the Cretaceous and early Tertiary seems to have moved trenchwards. Progressive oceanization of the region continued in the mid–late Tertiary with the opening of

Drake Passage, the formation of the Scotia Sea, and the opening of Bransfield Strait and the eastern Scotia Sea by back-arc spreading. From the mid-Tertiary to the present the subduction-related volcanism has become distinctly more basaltic and the lavas have generally lower contents of lithophile elements. This is the reverse of the 'normal' evolutionary sequence of an island arc (Jakeš and Gill, 1970). In broad terms it is possible to interpret this as a result of the changing character of the mantle source being sampled by subduction-related magmatism, although in detail many subsidiary factors may be involved.

Molnar and Atwater (1978) have drawn attention to the contrasted magmatic expressions of subduction between the western Pacific, where the Benioff zone dips steeply, and the eastern Pacific, where the slab descends at a much shallower angle beneath the Andes. In the former there is extensive back-arc spreading and low K basaltic to andesitic arc volcanism, whereas along the Andean cordillera there is extensive calc-alkaline plutonism and no back-arc spreading. Molnar and Atwater suggested that the difference in dip might be related to the age of the subducting lithosphere (that in the western Pacific being much older and colder). However, it is worth noting that the South American plate is, in absolute terms, overriding Pacific lithosphere (Chase, 1978; Minster and Jordan, 1978), while buoyant hot oceanic mantle is rising behind the subducting slab in the western Pacific marginal basins. More important is the fact that subcontinental mantle is being involved in magma genesis in the Andes, whereas oceanic mantle is being sampled beneath the western Pacific arcs. During the last 100 Ma the Antarctic Peninsula region has progressed from an Andean-type situation to one where recent subduction beneath the South Shetland and South Sandwich arcs is directly associated with active back-arc spreading in Bransfield Strait (Barker and Griffiths, 1972) and the eastern Scotia Sea (Barker, 1972) respectively.

The advantage of a veined mantle model in calc-alkaline magma genesis is that it is easier to explain the high large ion lithophile element abundances and $^{87}Sr/^{86}Sr$ ratios in continental calc-alkaline magmas without resort to crustal contamination. Depending on the time interval between veining and magma generation the veins would have higher $^{87}Sr/^{86}Sr$ ratios that the host mantle and, since the degree of melting to produce calc-alkaline magmas is relatively small, the veins would contribute significantly to the melt. Moreover, if minor mineral phases such as ilmenite, rutile, and zircon are present in the enriched veins and remain stable under hydrous melting conditions, then veining readily accounts for the low abundances of Ta, Nb, Ti, etc., and high large ion lithophile to high field strength element ratios in the resultant magmas —a feature which will be enhanced if additional large ion lithophile elements are introduced during dewatering of the subducting slab. In this context the $K-h$ variation may be more a function of the increased degree of veining inwards from the continental margin and/or the larger proportion of vein material encountered by magmas rising from greater depths in the Benioff zone. The increasing fragmentation and oceanization of the region since the Mesozoic, resulting from the opening of the Weddell Sea, Scotia Sea, and Bransfield Strait, may have led to the dispersal of enriched, veined subcontinental mantle and its progressive replacement with oceanic mantle having 'mid-ocean ridge basalt'-like characteristics: this process is reflected in the geochemical characteristics of the volcanic rocks.

Acknowledgements

We would like to thank the British Antarctic Survey for providing logistic support on the South Shetland Islands during 1975, and for allowing access to their collections of Antarctic Peninsula rocks. A.D.S. gratefully acknowledges receipt of a Natural Environment Research Council Fellowship; studies in the Scotia arc and Antarctic Peninsula have been supported by NERC (Grants GR3/2993, GR3/3279 and F60/10/1).

REFERENCES

Adie, R. J. (1954). The petrology of Graham Land. I. The basement complex; early Palaeozoic plutonic and volcanic rocks. *Falkland Is. Dep. Surv. scient. Rep.* no. 11.

Adie, R. J. (1955). The petrology of Graham Land. II. The Andean granite-gabbro intrusive suite. *Falkland Is. Dep. Surv. scient. Rep.* no. 12.

Adie, R. J. (1972). Evolution of volcanism in the Antarctic Peninsula. In *Antarctic Geology and Geophysics* (R. J. Adie, ed.), Universitatsforlaget, Oslo, pp. 137–141.

Ashcroft, W. A. (1972). Crustal structure of the South Shetland Islands and Bransfield Strait. *Br. antarct. Surv. scient. Rep.* no. 66.

Baker, P. E. (1978). The South Sandwich Islands. II. Petrology and geochemistry. *Br. antarct. Surv scient. Rep.* no. 93.

Baker, P. E., McReath, I., Harvey, M. R., Roobol, M. J., and Davies, T. G. (1975). The geology of the South Shetland Islands. V. Volcanic evolution of Deception Island. *Br. antarct. Surv. scient. Rep.* no. 78.

Baker, P. E., Buckley, F., and Rex, D. C. (1977). Cenozoic volcanism in the Antarctic. *Phil. Trans. R. Soc. B* **279**, 131–142.

Barker, P. F. (1970). Plate tectonics of the Scotia Sea Region. *Nature, Lond.* **228**, 1293–1297.

Barker, P. F. (1972). A spreading centre in the east Scotia Sea. *Earth planet. Sci. Letters* **15**, 123–132.

Barker, P. F. (1976). The tectonic framework of Cenozoic volcanism in the Scotia Sea region: a review. In *Symposium on Andean and Antarctic Volcanology Problems, Santiago, 1974* (O. González-Ferran, ed.).

Barker, P. F. and Burrell, J. (1977). The opening of Drake Passage. *Marine Geol.* **25**, 15–34.

Barker, P. F. and Dalziel, I. W. D. (1980). Progress in geodynamics in the Scotia arc region. *Geodynamics Project WG2, Final Reports Series*, American Geophysical Union, in press.

Barker, P. F. and Griffiths, D. H. (1972). The evolution of the Scotia Ridge and Scotia Sea. *Phil. Trans. R. Soc. A* **271**, 151–183.

Barker, P. F. and Griffiths, D. H. (1977). Towards a more certain reconstruction of Gondwanaland. *Phil. Trans. R. Soc. B* **279**, 143–159.

Barker, P. F., Hill, I. A., Weaver, S. D., and Pankhurst, R. J. (1980). The origin of the eastern South Scotia Ridge as an intra-oceanic island arc. In *Antarctic Geoscience* (C. Craddock, ed.), University of Wisconsin Press, Madison, in press.

Barton, C. M. (1965). The geology of the South Shetland Islands. III. The stratigraphy of King George Island. *Br. antarct. Surv. scient. Rep.* no. 44.

Bibby, J. S. (1966). The stratigraphy of part of north-east Graham Land and the James Ross Island Group. *Br. antarct. Surv. scient. Rep.* no. 53.

Best, M. G. (1975). Migration of hydrous fluids in the upper mantle and potassium variation in calc-alkali rocks. *Geology* **3**, 429–432.

Chase, C. G. (1978). Plate kinematics: the Americas, East Africa and the rest of the world. *Earth planet. Sci. Letters* **37**, 355–368.

Dalziel, I. W. D. (1974). Evolution of the margins of the Scotia Sea. In *The Geology of Continental Margins* (C. A. Burk and C. L. Drake, eds), Springer-Verlag, New York, pp. 567–579.

Dalziel, I. W. D. (1980). The pre-Jurassic history of the Scotia arc: a review and progress report. In *Antarctic Geoscience* (C. Craddock, ed.), University of Wisconsin Press, Madison, in press.

Davey, F. J. (1972). Marine gravity measurements in Bransfield Strait and adjacent areas. In *Antarctic Geology and Geophysics* (R. J. Adie, ed.), Universitatsforlaget, Oslo, pp. 39–46.

DeLong, S. E., Hodges, F. N., and Arculus, R. J. (1975). Ultramafic and mafic inclusions, Kanaga Island, Alaska, and the occurrence of alkaline rocks in island arcs. *J. Geol.* **83**, 721–736.

de Wit, M. J. (1977). The evolution of the Scotia Arc as the key to the reconstruction of southwest Gondwanaland. *Tectonophysics* **37**, 53–81.

Dickinson, W. R. (1975). Potash–depth ($K-h$) relations in continental margin and intra-oceanic magmatic arcs. *Geology* **3**, 53–56.

Dickinson, W. R. and Hatherton, T. (1967). Andesite volcanism and seismicity around the Pacific. *Science, N.Y.* **157**, 801–803.

Dostal, J., Zentilli, M., Caelles, J. C., and Clark, A. H. (1977). Geochemistry and origin of volcanic rocks from the Andes (26°–28°S), *Contr. Mineral. Petrol.* **63**, 113–128.

Erlank, A. J. and Kable, E. J. D. (1976). The significance of incompatible elements in Mid-Atlantic Ridge basalts from 45°N with particular reference to Zr/Nb. *Contr. Mineral. Petrol.* **54**, 281–291.

Forsyth, D. W. (1975). Fault plane solutions and tectonics of the South Atlantic and Scotia Sea. *J. geophys. Res.* **80**, 1429–1443.

Francis, P. W., Moorbath, S., and Thorpe, R. S. (1977). Strontium isotope data for recent andesites in Ecuador and North Chile. *Earth planet. Sci. Letters* **37**, 197–202.

Frey, F. A. and Green, D. H. (1974). The mineralogy, geochemistry and origin of lherzolite inclusions in Victoria basanites. *Geochim. Cosmochim. Acta* **38**, 1023–1059.

Frey, F. A., Green, D. H., and Roy, S. D. (1978). Integrated models of basalt petrogenesis. *J. Petrol.* **19**, 463–513.

Gledhill, A., Rex, D. C., and Tanner, P. W. G. (1980). K–Ar and Rb–Sr geochronology of igneous and metamorphic rock suites from the Antarctic Peninsula. In *Antarctic Geoscience* (C. Craddock, ed.), University of Wisconsin Press, Madison, in press.

González-Ferran, O. and Katsui, Y. (1970). Estudio integral del volcanismo cenozoico superior de las Islas Shetland de Sur, Antarctica. *Inst. Ant. Ch. Ser. Cient.* **1**, 123–174.

González-Ferran, O. and Vergara, M. (1972). Post-Miocene volcanic petrographic provinces of west Antractica and their relation to the southern Andes of South America. In *Antarctic Geology and Geophysics* (R. J. Adie, ed.), Universitatsforlaget, Oslo, pp. 187–195.

Grikurov, G. E., Krylow, A. Ya., Polyakov, M. M., and Trobun, Ya. N. (1970). Vozrast porod v severnoy chasti Antarkticheskogo poltostrova i na Yudznyth Shetlandskikh osyrovakh (po dannym kaily-argonovogo metoda). *Inf. Byull. sov. antarkt. Eksped.*, no. 80, 30–34.

Gurney, J. J. and Harte, B. (1980). Chemical variations in

upper mantle nodules from southern African Kimberlites. *Phil. Trans. R. Soc. A* in press.
Hawkes, D. D. (1961a). The geology of the South Shetland Islands. II. The geology and petrology of Deception Island. *Falkland Is. Dep. Surv. scient. Rep.* no. 27.
Hawkes, D. D. (1961b). The geology of the South Shetland Islands. I. The petrology of King George Island. *Falkland Is. Dep. Surv. scient. Rep.* no. 26.
Hawkesworth, C. J., O'Nions, R. K., Pankhurst, R. J., Hamilton, P. J., and Evensen, N. M. (1977). A geochemical study of island-arc and back-arc tholeiites from the Scotia Sea. *Earth planet. Sci. Letters* **36**, 253–262.
Hawkesworth, C. J., Norry, M. J., Roddick, J. C., Baker, P. E., Francis, P. W., and Thorpe, R. S. (1979). $^{143}Nd/^{144}Nd$, $^{87}Sr/^{86}Sr$ and incompatible element variations in calc-alkaline andesites and plateau lavas from South America. *Earth planet. Sci. Letters* **42**, 45–57.
Hayes, D. E. and Ewing, M. (1970). Pacific boundary structure. In *The Sea*, Vol. 4, Part II (A. E. Maxwell, ed.), Wiley–Interscience, New York.
Heming, R. F. and Rankin, P. C. (1979). Ce-anomalous lavas from Rabual Caldera, Papua-New Guinea. *Geochim. cosmochim. Acta* **43**, 1351–1355.
Herron, E. M. and Tucholke, B. E. (1976). Sea-floor magnetic patterns and basement structure in the southeastern Pacific. In *Initial Report on Deep Sea Drilling Project* no. 35 (C. D. Hollister *et al.*, eds) US Government Printing Office, Washington, pp. 263–278.
Hobbs, G. J. (1968). The geology of the South Shetland Islands. IV. The geology of Livingston Island. *Br. antarct. Surv. scient. Rep.* no. 47.
Jahn, R. A. (1978). A preliminary interpretation of Weddell Sea magnetic anomalies. Abstr. UKGA2. *Geophys. Jl R. astr. Soc.* **53**, 164 (abstract only).
Jakeš, P. and Gill, J. (1970). Rare earth elements and the island arc tholeiite series. *Geol. Soc. Am. Bull.* **83**, 29–40.
Jordan, T. H. (1978). Composition and development of the continental tectosphere. *Nature, Lond.* **274**, 544–548.
Katsui, Y. (1972). Late Cenozoic petrographic provinces of the volcanic rocks from the Andes to Antarctica. In *Antarctic Geology and Geophysics* (R. J. Adie, ed.), Universitatsforlaget, Oslo, pp. 181–185.
Lloyd, F. and Bailey, D. K. (1975). Light element metasomatism of the continental mantle: the evidence and consequences. *Phys. Chem. Earth* **9**, 389–416.
Ludden, J. N. and Thompson, G. (1979). An evaluation of the behaviour of the rare earth elements during the weathering of sea floor basalt. *Earth planet. Sci. Letters* **43**, 85–92.
Meijer, A. (1976). Pb and Sr isotopic data bearing on the origin of volcanic rocks from the Mariana Island Arc system. *Geol. Soc. Am. Bull.* **87**, 1358–1369.
Menzies, M. A., Blanchard, D., and Jacobs, J. (1977). Rare earth and trace element geochemistry of metabasalts from the Point Sal ophiolite, California. *Earth planet. Sci. Letters* **37**, 203–215.
Minster, J. B. and Jordan, T. H. (1978). Present-day plate motions, *J. geophys. Res.* **83**, 5331–5354.
Molnar, P. and Atwater, T. (1978). Interarc spreading and cordilleran tectonics as alternates related to the age of subducted oceanic lithosphere. *Earth planet. Sci. Letters* **41**, 330–340.
Molnar, P., Atwater, T., Mammerickx, J., and Smith, S. M. (1975). Magnetic anomalies, bathymetry and the tectonic evolution of the South Pacific since the late Cretaceous. *Geophys. Jl R. astr. Soc.* **40**, 383–420.
Mysen, B. O. (1979). Trace element partitioning between garnet peridotite minerals and water-rich vapour: experimental data from 5 to 30 kbar. *Am. Mineral.* **64**, 274–287.
Nelson, P. H. H. (1975). The James Ross Island Volcanic Group of north-east Graham Land. *Br. antarct. Surv. scient. Rep.* no. 54.
Nielson, D. R. and Stoiber, R. E. (1973). Relationship of potassium content in andesitic lavas and depth to the seismic zone. *J. geophys. Res.* **78**, 6887–6982.
Norton, I. and Scalter, J. G. (1980). A model for the evolution of the Indian Ocean and the breakup of Gondwanaland. *J. geophys. Res.* in press.
Pankhurst, R. J. (1977). Strontium isotope evidence for mantle events in the continental lithosphere. *J. geol. Soc. Lond.* **134**, 255–268.
Pankhurst, R. J. (1980). Sr-isotope and trace element geochemistry of Cenozoic volcanics from the Scotia Arc and northern Antarctic Peninsula. In *Antarctic Geoscience* (C. Craddock, ed.) University of Wisconsin Press, Madison, in press.
Pankhurst, R. J., Weaver, S. D., Brook, M., and Saunders, A. D. (1980). K–Ar chronology of Byers Peninsula, Livingston Island, South Shetland Islands. *Br. antarct. Surv. Bull.* in press.
Pichler, H. and Ziel, W. (1972). The Cenozoic rhyolite–andesite association of the Chilean Andes. *Bull. Volc.* **35**, 424–452.
Pitman, W. C., III, Larson, R. L., and Herron, E. M. (1974). The age of the Ocean Basins. *Geol. Soc. Am. Map Chart Series* no. 6.
Rex, D. C. (1976). Geochronology in relation to the stratigraphy of the Antarctic Peninsula. *Br. antarct. Surv. Bull.* **43**, 49–58.
Rex, D. C. and Baker, P. E. (1973). Age and petrology of the Cornwallis Island granodiorite. *Brit. antarct. Surv. Bull.* **32**, 55–61.
Roach, P. J. (1978). Geophysical investigation into the evolution of Bransfield Strait. Ph.D. thesis, University of Birmingham.
Robertson, A. H. F. and Fleet, A. J. (1976). The origins of rare earths in metalliferous sediments of the Troodos massif, Cyprus. *Earth planet. Sci. Letters* **28**, 385–394.
Roobol, M. J., Francis, P. W., Ridley, W. I., Rhodes, M., and Walker, G. P. L. (1976). Physico-chemical characters of the Andean volcanic chain between latitudes 21° and 22° south. In *Symposium on Andean and Antarctic Volcanology Problems, Santiago, 1974* (O González-Ferran, ed.), pp. 450–464.
Saunders, A. D. and Tarney, J. (1979). The geochemistry of basalts from a back-arc spreading centre in the East Scotia Sea. *Geochim. cosmochim. Acta* **43**, 555–572.
Saunders, A. D., Tarney, J., Stern, C. R., and Dalziel, I. W. D. (1979). Geochemistry of Mesozoic marginal basin floor igneous rocks from southern Chile. *Geol. Soc. Am. Bull.* **90**, 237–258.
Saunders, A. D., Weaver, S. D., and Tarney, J. (1980a). The pattern of Antarctic Peninsula plutonism. In *Antarctic Geoscience* (C. Craddock, ed.), University of Wisconsin Press, Madison, in press.

Saunders, A. D., Tarney, J., and Weaver, S. D. (1980b). Transverse geochemical variations across the Antarctic Peninsula: implications for calc-alkaline magma genesis. *Earth planet. Sci. Letters* **46**, 344–360.

Smellie, J. L. S., Davies, R. E. S., and Thomson, M. R. A. (1980). Geology of a Mesozoic intra-arc sequence on Byers Peninsula, Livingston Island, South Shetland Islands. *Br. antarct. Surv. Bull.* in press.

Stern, C. R. and Stroup, J. (1980). The petrochemistry of the Patagonia batholith between 51°S and 52°S latitude. In *Antarctic Geoscience* (C. Craddock, ed.), University of Wisconsin Press, Madison, in press.

Storey, B. C., Mair, B. F., and Bell, C. M. (1977). The occurrence of Mesozoic ocean floor and ancient continental crust on South Georgia. *Geol. Mag.* **114**, 203–208.

Suarez, M. (1976). Plate tectonic model for southern Antarctic Peninsula and its relation to southern Andes. *Geology* **4**, 211–214.

Suarez, M. and Pettigrew, T. H. (1976). An upper Mesozoic island arc–back arc system in the southern Andes and South Georgia. *Geol. Mag.* **113**, 305–328.

Sun, S.-S. and Hanson, G. N. (1975e). Origin of Ross Island basanitoids and limitations upon the heterogeneity of mantle sources for alkali basalts and nephelinites. *Contr. Mineral. Petrol.* **52**, 77–106.

Tarney, J., Wood, D. A., Saunders, A. D., Varet, J. and Caan, J. R. (1979). Nature of mantle heterogeneity in the North Atlantic: evidence from leg 49. In *Results of Deep Sea Drilling in the Atlantic: Ocean Crust*, Maurice Ewing Series 2, American Geophysical Union (M. Talwani, G. C. Harrison and D. E. Hayes, eds.).

Taylor, B. J., Thomson, M. R. A., and Willey, L. E. (1980). The geology of the Ablation Point and Keystone Cliff area, Alexander Island. *Br. antarct. Surv. scient. Rep.* in press.

Thomson, M. R. A. (1980). Mesozoic palaeogeography of western Antarctica. In *Antarctic Geoscience* (C. Craddock, ed.), University of Wisconsin Press, Madison, in press.

Valenzuela, E. and Hervé, F. (1972). Geology of Byers Peninsula, Livingston Island, South Shetland Islands. In *Antarctic Geology and Geophysics* (R. J. Adie, ed.), Universitatsforlaget, Oslo, pp. 83–89.

Watts, D. R. (1980). Potassium–argon palaeomagnetic results from King George Island, South Shetland Islands. In *Antarctic Geoscience* (C. Craddock, ed.), University of Wisconsin Press, Madison, in press.

Weaver, S. D., Sceal, J. S. C., and Gibson, I. L. (1972). Trace-element data relevant to the origin of trachytic and pantelleritic lavas in the East African rift system. *Contr. Mineral. Petrol.* **36**, 181–194.

Weaver, S. D., Saunders, A. D., Pankhurst, R. J., and Tarney, J. (1979). A geochemical study of magmatism associated with the initial stages of back-arc spreading: Quaternary volcanics of Bransfield Strait, South Shetland Islands. *Contr. Mineral. Petrol.* **68**, 151–169.

Weaver, S. D., Saunders, A. D., and Tarney, J. (1980). Mesozoic–Cenozoic volcanism in the South Shetland Islands and the Antarctic Peninsula: geochemical nature and plate tectonic significance. In *Antarctic Geoscience* (C. Craddock, ed.) University of Wisconsin Press, Madison, in press.

Winslow, M. A. (1980). The structural evolution of the Magallanes Basin and neotectonics in the southern most Andes. In *Antarctic Geoscience* (C. Craddock, ed.), Universty of Wisconsin Press, Madison, in press.

Wood, D. A., Tarney, J., Varet, J., Saunders, A. D., Bougault, H., Joron, J. L., Treuil, M., and Cann, J. R. (1979). Geochemistry of basalts drilled in the North Atlantic by IPOD Leg 49: implications for mantle heterogeneity. *Earth planet. Sci. Letters* **42**, 77–97.

V Andesitic volcanoes and their products

Andesitic volcanoes have distinctive forms which reflect their characteristic styles of eruption and products. The study of andesitic volcanoes is therefore incomplete without consideration of characteristic volcanic forms and products. In this section G. P. L. Walker outlines the major controls on eruption of andesite, these including viscosity, rheology, and gas content, and reviews the range of explosive and lava eruptions. For andesitic volcanoes, explosive eruptions are dominant and range (in order of increasing violence) from strombolian, through vulcanian and plinian to ignimbrite and peléan (pyroclastic flow) eruptions. Although eruptions of andesitic volcanoes have caused great damage and loss of life, the average annual loss of life is much less than that from other natural disasters (e.g. earthquakes) and human activities (e.g. wars). However, the greatest hazard arises from peléan explosive eruptions, and the greatest potential danger is of an ignimbrite eruption within a densely populated area. Since pyroclastic flows are a typical component of peléan and ignimbrite eruptions, they are described by A. L. Smith and M. J. Roobol, who delineate the characteristics of such flows and emphasize their complexity and variability, and the wide range of resultant pyroclastic deposits.

Andesites
Edited by R. S. Thorpe
© 1982 John Wiley & Sons

Eruptions of andesitic volcanoes

G. P. L. Walker
Hawaii Institute of Geophysics,
2525 Correa Road, Honolulu, Hawaii 96822

ABSTRACT

The extremely varied styles of eruptive activity of andesitic volcanoes cover almost the entire range known on volcanoes, and the diversity is accounted for mainly by the wide range in magma viscosity, yield strength, and gas content of andesites. Most andesitic eruptions are explosive, and they range from the mildly explosive strombolian style to the exceedingly powerful plinian and exceedingly violent vulcanian styles; *nuées ardentes*, which often develop, are interpreted as a secondary consequence of primary volcanic activity on a high and steep cone. Lava extrusions, often dome-like in form, are also produced in some eruptions; alternatively, the magma may intrude the surface rocks to produce cryptodomes. Any attempt to account in detail for the nature of the volcanic activity must face the fact that remarkably little is yet known about the fundamental controls. The survey ends by drawing attention to certain aspects of the hazard arising from volcanic activity, and particularly the great danger posed by ignimbrite eruptions.

Introduction

Eruptions of andesitic volcanoes embrace practically the entire range of known volcanic styles. These range from the mildly explosive fire-fountaining of strombolian activity at the one extreme to explosions of extreme power and violence and the catastrophic eruption of ignimbrites at the other; from the quiet outflowing of highly fluid lavas to the laboured extrusion of high and steep-sided lava domes. Eruptive style is mainly determined by controls such as magma viscosity, yield strength, and gas content; it is appropriate, therefore, first to look at what is known about the variations in these controls before considering the eruptions themselves.

The Main Controls

Viscosity is an important control and, for andesitic lavas at the time of eruption, is believed to vary over a wide range, from low values typical of basalt to high values typical of rhyolite. Field measurements of viscosity are, however, few. Direct measurements obtained by inserting probes into the lava, as have occasionally been made on basalt, have rarely been attempted on andesite, and the value of $5 \times 10^5 - 10^7$ P (P means poise, a measure of viscosity) obtained at Hekla in 1947 (Einarsson, 1951) was for lava which, although intermediate in composition, was not calc-alkaline. The practical difficulties of safely getting close enough to make such measurements are usually great, except perhaps for a few of the most mafic andesite lavas.

Indirect measurements of viscosity, employing Nichols' formula, which relates the viscosity to the velocity of lava of known thickness flowing down a known slope angle, enable the range of lavas studied to be greatly extended. Values for basaltic andesite include $7 \times 10^5 - 10^{10}$ P for Paricutin (Krauskopf, 1948) and $10^4 - 10^5$ P for Miyakesima in 1940 (Tsuya et al., 1941); for andesite they include $10^6 - 10^9$ P for Sakurajima in 1946 (Nagata *et al.*, 1946); and for dacite they include the $5 \times 10^8 - 10^{10}$ P of Santiaguito (Rose, 1973) and the 10^{10} P of Usu in 1944–45 (Ishikawa, 1950). The drawback of this indirect method is that it assumes the lava to be a Newtonian fluid, and fails to distinguish the effect of viscosity from that of yield strength.

Not only are there few field measurements, but few laboratory determinations have been reported on the viscosities of fused andesites. An andesite from Mt Hood having $SiO_2 = 60.7$ per cent yielded a value of 10^5 P at 1100 °C, near the liquidus temperature, falling to 10^3 P at 1400 °C (Murase and McBirney, 1973), and similar values around the upper end of the temperature scale were obtained for other andesites (SiO_2 55–64 per cent) by Volarovich (1936) and Euler and Winkler (1957), increasing to 10^{11} P at 750 °C (Scarfe, 1977). It is known that at any given temperature there is a general increase in viscosity with an increase in the content of 'glass former' constituents, notably silica and alumina (Scarfe, 1973); the more detailed correlation of viscosity with composition is discussed by Bottinga and Weill (1972) and Shaw (1972).

A second control which is probably just as important as viscosity is the yield strength of the magma. Many andesitic magmas probably have a yield strength when erupted, and although this has never been demonstrated by field measurements there are three reasons for believing that it is so. First, the morphology of andesite lava extrusions is similar to that predicted for lava having a yield strength (Hulme, 1974; Hulme and Fielder, 1977). Second, the highly porphyritic condition of many andesites indicates that they may be well below the liquidus temperature when erupted, and by analogy with basaltic lava (which is known to have a yield strength at a temperature well below the liquidus: Shaw, 1969; Pinkerton and Sparks, 1978) it is likely that andesitic magma likewise has a yield strength. Third, it is known that the purely mechanical effect of crystals is to impose a yield strength on the liquid/crystal mix even if the liquid itself is Newtonian.

It is probable that magmas approximate to Newtonian fluids only at or above the liquidus temperature, and they may be expected to depart increasingly from Newtonian fluids the farther their temperature is depressed below the liquidus (but *see* Scarfe, 1973, for a contrary view). The more salic the magma and the more polymerized its structure, the higher the yield strength. Note that the higher the content of crystals the more salic is the residual liquid.

Viscosity and rheology exercise a control on the form and thickness of extrusive lava bodies; more importantly, they control the ease with which gas bubbles form and grow in or escape from magma and hence they control the explosivity of volcanoes and the violence of their explosive outbursts. They also affect the ability of magma to rise through the Earth's crust, and may play an important part in deciding whether the uprising magma reaches the surface or forms an intrusion instead.

Perhaps equally important is the control exercised by gases, but relatively few direct data exist on the content and composition of magmatic gas in andesitic magmas. A great many gas measurements have, of course, been made, but field measurements should ideally be made by probes inserted into the lava within the eruptive vent, and to date this has seldom been achieved due to the practical difficulties of gaining close access to andesitic vents. The more distant in time or place from the vent the gas is collected and analysed, the greater will be the loss of the more volatile constituents and the chance of contamination of, or reactions among, the remaining gas, and hence the greater the possibility that the analysed gas composition will depart from the true magmatic composition.

The amount of water and other volatiles that can be dissolved in silicate melts has been established by experimental studies (reviewed by Anderson, 1975), but these studies tell nothing about the quantity actually dissolved in natural magmas. Various approaches to this problem have been tried, and yield values ranging from *c.* 1 to 10 per cent. They include field estimation during the course of the eruption (Foshag, 1950; Fries, 1953), phase stability study (Eggler, 1972), calculations on the energetics of explosive eruptions (Markhinin, 1962), and fluid inclusion study (Anderson, 1973).

While there is no doubt about how important is the role played by gases, opinions still differ on the extent to which the participating gases are truly magmatic, or are derived from heated groundwaters within the volcanic edifice (*see*, for example, Schmincke, 1977).

Styles of Explosive Eruptions

Most andesitic eruptions are explosive; thus of the 47 Japanese volcanoes recorded to have erupted in the century from 1858 to 1958 in the *Catalogue of Active Volcanoes* (Kuno, 1962), 80 per cent appear to have been exclusively explosive and did not produce lava extrusions. For the smaller explosive eruptions visual observations are often sufficient to document the activity. The larger that eruptions become, however, and the less common their occurrence, the more difficult and indeed more hazardous it is to make visual observations, and the fewer are the opportunities to make them; the more reliance must then be placed on post-mortem measurements of the erupted products and theoretical or experimental simulation. In the following the most mildly explosive style is described first, and the more powerful or violent styles later.

Mild strombolian

The most weakly explosive activity, characteristic of the most mafic and most fluid of andesitic magmas, is that of mild strombolian type. Most observations and quantitative studies have been made on basaltic eruptions, but basaltic andesite eruptions, involving lava with physical properties similar to those of basalt, are in no way different from the basaltic ones. Eruptions are typically of open vent type, in which the magma column has free access to the surface. The activity is sometimes spasmodic, the outbursts being correlated with the arrival at the surface of trains of bubbles, and is sometimes continuous (Wilson and Head, 1981).

The gas jet typically emerges from the vent with a velocity of 50–200 m s^{-1} (Blackburn *et al.*, 1976; Chouet *et al.*, 1974; McGetchin *et al.*, 1974). The gas disrupts the upper levels of the lava as it bursts through, and escapes carrying lava tatters up with it. The content of ash is characteristically small, and the lava fragments seldom reach a height exceeding several hundred metres, and they then decouple relatively completely from the gases so that the convective gas column rising above the jet contains little ash and is pale-coloured. Because of the low column height, the fall-out thickness–distance curve near the vent is steeper than the *c.* 33° angle of repose of loose pyroclasts. The redistribution of pyroclasts then generates a scoria (cinder) cone having slopes standing at the repose angle. The pyroclastic mantle extends some distance beyond the limits of the cone but typically thins to less than 10 cm within 1–10 km from the vent.

The scoria cone is made mostly of coarse cinders, many having the irregular shape and rough spinose surface caused by the tearing apart of foamy magma, although their outer surfaces are to some degree smoothed by surface tension. Others are partly bounded by fracture surfaces, and are derived from larger lumps which broke up in the air or on impact with the ground or fell apart later on cooling joints.

Bombs having fusiform (i.e. spindle-shaped), subspherical, or other distinctive forms are locally abundant and can exceed 5 m in size; they tend to have a higher density than the bulk of the scoria, and originate from partially degassed fractions of the magma which evidently had a longer than average residence time in the vent. Some bombs have grown by the accretion of layers where lava or other rock fragments fell into the lava pool in the crater and were then tossed out, this cycle often being repeated several times. The larger fragments of lava have flattened under their own weight on impact, and layers of agglutinate are often found on the crater rim where the spatter fragments have welded together. Much more extensive layers of welded tuff are sometimes formed when the accumulation rate of scoria was particularly high. A good example occurs on Tongariro, New Zealand (Healy, 1963).

Violent strombolian

Mild strombolian eruptions generate predominantly coarse scoria, but in many andesitic strombolian eruptions ash predominates among the ejecta and the designation 'violent strombolian' is then appropriate. An example is the Paricutin eruption of 1943–52 (Mexico). From eye-witness accounts (e.g. Foshag and Gonzalez, 1955) the eruptive column rose often to a height of several kilometres and was dark-coloured and heavily laden with ash. Some explosions were heard, and ash fell, as far as 350 km away.

The reasons for the more violent activity can only be surmised. Such activity can arise when a vent becomes partially blocked by back-fallen material or when groundwater enters the vent in large amount, but the violence at Paricutin seems to have been due to a more fundamental cause (but *see* Foshag, 1950); possibly it was a consequence of the lava possessing a yield strength at the time of eruption. The scoria cone of Paricutin differs from a mild strombolian one only in being larger; the pyroclastic mantle around it is, however, much more extensive.

Vulcanian

The vulcanian is perhaps the most common eruptive style shown by andesitic volcanoes and is characterized by violent and short, often more or less instantaneous ('cannon-like'), explosions. The cauliflower-shaped cloud which develops soars rapidly to between 5 and 20 km above the vent, and is dark because it is heavily charged with ash. On a steep-sided cone, back-falling debris commonly generates *nuées ardentes*. Initial ejection velocities of 300–600 m s^{-1} have been calculated from the 4–5 km range of metre-sized ballistic blocks (Minakami, 1950; Fudali and Melson, 1972) and from a supersonic shock wave made visible by condensation of water vapour (Nairn, 1976).

An andesitic volcano may experience a great many vulcanian explosions over a period of years; more than 2000 on Asama volcano from 1934 to 1958 for example (Kuno, 1962). The quantity of ejecta from each explosion is relatively small (commonly of the order of 10^4–10^6 m^3), but the products of many such explosions can produce an impressive accumulation. Typically such products are intercalated on the volcanic cone with strombolian and other pyroclastic deposits and with lava extrusions, while associated mudflows and pyroclastic flow deposits become conspicuous on the lower slopes.

Examination of the products of vulcanian explosions reveals that a high proportion of the ejecta were solid, or nearly solid, at the time of the explosion. The solid ones include vent-wall lithics and solidified juvenile lava. The nearly solid ones include the highly distinctive objects called breadcrust blocks (bombs) made of lava which was sufficiently 'stiff' to fracture as though solid when subjected to the rapidly applied stresses in the explosion, and hence produce more or less angular blocks and flakes bounded by smooth fracture surfaces. However, being still fluid, slow vesiculation took place and inflated them. The outer skin accommodated the increased volume in part by out-bulging, and in part by fracturing and the opening of gaping surface clefts. Breadcrust blocks are too fragile to withstand impact, and clearly became inflated after ejection in the position on the ground or within the pyroclastic deposit in which they are now found.

All gradations exist between strombolian and vulcanian eruptions; any boundary drawn between them must be an arbitrary one and is perhaps best based on the vesicularity of the juvenile clasts, the former being strongly vesiculated and the latter weakly so. The degree of fragmentation is high in vulcanian deposits, and near the vent, where ballistic blocks, fine dust, and all intermediate sizes are mingled together, the sorting is poor. Beyond a few kilometres from the vent the fallout is predominantly fine ash and, being thin (because of the generally wide dispersal and the small total volume), and having a low survival potential, vulcanian deposits tend to be inconspicuous.

Various models have been proposed for vulcanian activity (e.g. Self et al., 1979; Schmincke, 1977), but one feature of which account has not hitherto been taken is the almost invariably highly porphyritic nature of the erupting magma. This suggests that the magma is likely to have a yield strength as well as a high viscosity, and the characteristics of vulcanian explosions may be directly attributed to this.

A mechanism is now proposed in which the volcanic conduit is envisaged as being filled with lava having a high yield strength. The strength impedes the exsolution of magmatic gases: it presumably slows the expansion of the gas spaces (whether they are vesicles or microrifts in the lava) which do form, and it certainly opposes any movement of gas bubbles through the lava (cf. McBirney and Murase, 1971). A gradual expansion of the upper part of the magma column in the conduit therefore takes place. The nearly or quite solidified capping bulges upwards and eventually fails, material is explosively removed from the top, and a fragmentation front rapidly descends through the conduit to reach the level of non-vesiculated fresh magma below. The explosion ceases at this level, and an interval elapses before the next explosion while this fresh magma slowly vesiculates and rises in the conduit to take the place of that just released.

This model can account for the violence of vulcanian explosions, the low mass output rate, and the episodic nature with a period between explosions varying with different magma batches and on different volcanoes from hours to months. The vulcanian craters are often deep, and being deep the collapse of material from the crater walls generates much lithic debris that is ejected together with the juvenile material. With repeated explosions in a crater so deep that a large proportion of ejecta fall back into the crater, the volcano functions as a rock-crushing machine and generates much fine ash and dust which escapes, as happened at Irazu in 1963–65 (Murata et al., 1966).

Plinian

The plinian style is characterized by an exceedingly powerful gas jet and a convective column soaring to a height of 30–60 km, so producing an exceptionally wide dispersal of coarse pumice. In known examples the jet was sustained for a period ranging between less than 1 h and 2 days.

Plinian eruptions are low frequency events, and occur on andesitic volcanoes at the rate of only a few per century, examples being Santa Maria (Guatemala) 1902 and Quizapu (Chile) 1932. They are, however, important because the volume of magma erupted tends to be large: the output of one plinian eruption thus equals that of a great number of vulcanian eruptions. Moreover, a fair proportion of polygenetic andesitic volcanoes experience them: of the volcanoes familiar to the author, more than 40 per cent have generated plinian deposits (although many are rhyolitic rather than andesitic). Being widely dispersed, well beyond the limits of the volcano itself, the deposits make ideal horizons for tephrochronological studies. Extensive disruption is caused by the fallout from a plinian eruption.

Being low frequency events, relatively little is known from direct observations about the characteristics of plinian eruptions. It is probable that only one (Hekla, Iceland, in 1947; a rhyolitic example) has yet been observed by volcanologists (Thorarinsson, 1950, 1954; Einarsson, 1950). Most of what is known has therefore been determined indirectly from measurements made on their pyroclastic deposits. The pumice which is the main constituent often has a low density and has a ragged form produced by the tearing apart of foamy magma. Partial or complete welding of the pumice in some parts of the deposit is commonly found near the vent, and the resulting welded air-fall tuffs are superficially very similar to welded ignimbrites; good criteria are, however, available to distinguish them (Sparks and Wright, 1979).

Plinian deposits tend to be homogeneous through their thickness, indicative of uniform sustained eruption from an open vent. Some stratification may also occur and may reflect fluctuations in eruptive vigour or interruptions in the continuity of the eruption caused by temporary vent blockage, or rainshowers during the eruption which bring fine ash prematurely to earth at the same time that coarse pumice is falling, so generating beds having a bimodal grain-size distribution (Walker et al. 1980b). Some fine beds intercalated with the coarse pumice are also formed from intraplinian pyroclastic flows, generated during the course of the plinian outburst. Besides stratification, many plinian deposits show a general upward coarsening due to a progressive increase in vigour of the outburst with time.

Ignimbrite eruptions

The greatest eruptions are those which generate ignimbrite, and two moderately large ones have occurred on andesitic volcanoes in the past century: Krakatau in 1883 (Williams, 1941; Self and Rampino, 1979) and Katmai in 1912 (Curtis, 1968). A high proportion of andesitic volcanoes are liable to have ignimbrite eruptions: 60 per cent of the polygenetic andesitic volcanoes familiar to the author have had them in the past (although many of the ignimbrites are rhyolitic rather than andesitic). These volcanoes all have calderas, and there is a general supposition that the ignimbrite eruptions were the caldera-forming ones, as was convincingly demonstrated at Crater Lake (Williams, 1942).

Practically nothing is known from direct observation about ignimbrite eruptions. None has been closely observed, and it is rather doubtful if observations sufficiently close to be useful are feasible. An analogy has often been drawn with the *nuée ardente*, but observed *nuées ardentes* are several orders of magnitude smaller, and the *nuée ardente* depends for its mobility on the existence of a steep and high volcanic cone which is not always present in an ignimbrite eruption. Understanding of ignimbrite eruptions must therefore depend heavily on deductions made from features of the ignimbrite themselves and experimental studies.

'Ignimbrite' is used here for a pyroclastic sheet, made predominantly from pumice and pumiceous ash, showing characters indicating that it was emplaced as a highly concentrated particulate flow (concentrated in the sense that the ratio of particulate matter to gas was high). It may or may not be welded. 'Ignimbrite' is thus more or less synonymous with 'ash-flow', and welded ignimbrite is synonymous with 'ash-flow tuff'. In being predominantly made from relatively low density pumice, ignimbrite is distinguished from the products of most *nuées ardentes* in which the juvenile material is either weakly vesiculated dense pumice (e.g. La Soufrière, St Vincent, in 1902), or is largely non-vesiculated (e.g. Mt Pelée, Martinique in 1902).

The biggest ignimbrites, which have individual volumes of $100-1000$ km^3, are rhyolitic. Rhyolitic ignimbrites are also more numerous than those of any other composition, and the Krakatau and Katmai eruptions both involved magmas at the dacite/rhyolite boundary. Andesitic ignimbrites are, however, quite common and include some having individual volumes of $10-100$ km^3.

The best described andesitic ignimbrites occur in Japan. They include the far-reaching Aso and Ito flows in Kyusyu (Matumoto, 1943; Lipman 1967; Yokoyama, 1974), each of which is distributed over a crudely circular area 150 km in diameter, and the smaller Shikotsu flow in Hokkaido (Minato et al., 1972). Some of these ignimbrites are partially welded, and a densely welded facies is also found locally where they are thickest. Compositionally zoned ignimbrites occur at Acatlan (Wright and Walker, 1977) and Acambay in Mexico which have a non-welded rhyolitic lower part overlain by a more densely welded basaltic andesite.

Ignimbrites are emplaced as particulate flows which, it is widely believed, are fluidized, and much of the uncertainty and discussion has centred on the degree of inflation (the volumetric ratio of gas to particulate material) of the flow. Flowage while in a highly expanded condition has been favoured by some workers. The remarkable mobility of flows such as Ito (Yokoyama, 1974), which at 50 km from source crossed mountain passes up to 700 m high to reach valleys on the far-vent side, can be explained by the lapping of a highly expanded 'cloud' hundreds of metres deep over the passes. Subsequently this 'cloud' condensed to form an ignimbrite.

Various features shown by ignimbrites are, however, incompatible with this mechanism, notably the relative homogeneity of grain size and lack of internal stratification, and the conservation of fine ash in the ignimbrite. Moreover, the coarse pumice concentration zone at the top of some ignimbrites (Sparks, 1976), attributed to the buoyant uprise of pumice through the flow, argues that the flows were only slightly expanded compared with the non-welded ignimbrite at rest. The ability to cross mountain passes must then be explained by a high velocity gained during descent of a steep volcanic cone or by collapse from a high eruptive column. A model for the generation of pyroclastic flows by eruptive column collapse has been presented by Sparks et al. (1978).

Certain ignimbrites having an anomalously low aspect ratio have recently been described (Walker et al., 1980a, b; Ui, 1973), and their unusual geometry has been attributed to an exceptionally high velocity resulting from a very high magma discharge rate.

Peléan activity

The *nuée ardente*, one of the most dangerous manifestations of volcanism on andesitic volcanoes, is a gravity-controlled avalanche of incandescent material which travels as a density flow down the side of a volcanic cone. Activity in which *nuées ardentes* are generated is often termed 'peléan', but it is perhaps best not to regard it as a distinct style of volcanism but rather as a secondary consequence of primary activity (e.g. of strombolian or vulcanian styles) operating on a high and steep cone. Even a small primary eruption on such a cone can result in a powerful *nuée ardente* being generated by the topography.

Sometimes the *nuée ardente* originates by the back-fall of pyroclastic ejecta from normal explosive activity; sometimes it is due to the mechanical failure and partial collapse of extrusive lava bodies which, having a high viscosity and yield strength, have a limited ability to flow away from vent; and sometimes it is initiated by laterally directed explosions. Examples generated in these ways are described as of St Vincent, Merapi, and Pelée type respectively (*see* Smith and Roobol, this volume, section V).

Another feature of *nuées ardentes* is that generally the lava involved has a high viscosity and (by implication) a high yield strength. It is magma in this condition that is most likely to give rise to the instantaneous explosion which generates a heavy back-fall, or is most likely to build the large and mechanically unstable lava extrusions (with their limited capacity to flow downslope) from which avalanching occurs.

The products of a typical *nuée ardente* include two kinds of deposits formed more or less simultaneously from two kinds of flows which have responded differently to the topography. One kind (the pyroclastic flow *sensu stricto*) is channelled by the topography and forms relatively thick chaotic and unsorted accumulations in valleys and topographic lows or on low angle slopes. The other kind of flow (the pyroclastic surge, often included under pyroclastic flows *sensu lato*, as by Smith and Roobol, this volume, Section V) forms a thin and fairly well sorted deposit draping the landscape and standing even on quite steep slopes and lacking both the coarsest and finest material. The surge deposit is immediately overlain by a thin airfall ash in which much of the finest materials reside, having settled out after the passage or decay of the turbulent surge.

These two kinds of flows and their deposits are illustrated by the 1902 activity of Mt Pelée. The first kind formed an accumulation in the Rivière Blanche valley up to tens of metres thick. The second kind, the pyroclastic surge, spread widely on either side of the Rivière Blanche valley, and devastated St Pièrre, depositing in the town a layer only a few tens of centimetres thick.

The surge deposit and its associated air-fall ash layer, being thin and easily eroded, have a low preservation potential. The inconspicuous nature of the deposits has long delayed recognition of pyroclastic surge deposits among the products of volcanoes. Particular care must be taken to search for and identify such deposits when compiling volcanic hazard maps of volcanoes.

Phreatic and phreatomagmatic eruptions

The crater of many andesitic volcanoes is occupied by a lake, and the existence of this lake partly determines the incidence of eruptions and can have a significant effect on the style of eruption. Water of external origin when present in large amount may in general play one of three alternative roles: (1) if the rate at which it extracts thermal energy from the magma is sufficiently high it can inhibit or prevent eruptions taking place; (2) even where magmatic eruptions are prevented, the heat exchange may result in phreatic eruptions taking place instead; or (3) in the event that magma erupts, the water can greatly enhance the explosivity and cause the eruption to be phreatomagmatic in type. It is particularly common for eruptions in crater lakes to be of phreatic type, and the sudden explosive expulsion of water from the lake can generate great lahars.

The Extrusion of Lava

Lava extrusions develop in a fair proportion of the eruptions of andesitic volcanoes, particularly when the magma is a relatively mafic one. Sometimes the lava erupts synchronously with the explosive activity, as happened on Lopevi in 1960 (Williams and Curtis, 1965), but more commonly lava emerges at the end of the main explosive phase, as happened at Mt Lamington in 1951 (Taylor, 1958), at Bezymianny in 1956 (Gorshkov, 1959), and at the time of writing apparently at Mt St Helen's. This sequence is attributable to the arrival at the surface of gas-poor magma after the explosive expulsion of the more gas-rich fractions. The 1783 eruption of Asama (Aramaki, 1956, 1957) shows a particularly clear sequence correlated with the appearance of progressively less gas-rich magma fractions: it opened with a plinian phase, continued with the eruption of a pumiceous pyroclastic flow and then a block pyroclastic flow (such as might have arisen by lava collapse), and terminated with the effusion of an andesitic lava flow.

The extrusion of lava is not invariably associated with explosive activity. Thus the 1971–72 lava dome of the Soufrière, St Vincent, grew quietly over 6 months without explosive activity or significant seismic activity (Aspinall et al., 1973).

The time taken for lava extrusions to form varies from a few months to more than a year. Examples are Mt Lamington, the 1951–52 dome of which grew over a period of 13 months, and Akita–Komagatake, in 1970, in which the lava, which had an unusually low viscosity, flowed for c. 2 months (Yagi et al., 1972). Average lava discharge rates vary from 0.3 to 30 $m^3 s^{-1}$, and these low rates favour the formation of lava domes rather than lava flows. Examples are also known where lava flows out more or less continuously for years; one is the Santiaguito dacite lava dome and flow complex, which has been growing more or less steadily and continuously for more than 50 a since its activity began in 1929 (Rose, 1973).

Andesitic lava extrusions vary from steep-sided bun-shaped lava domes often having an aspect ratio of more than 0.2 and a height of up to 1000 m, to lava sheets having an aspect ratio of 0.01 and a thickness of only 10 m, but domes and short stubby flows (coulées) with an aspect ratio near 0.1 predominate. The tendency is for the more mafic andesites to form sheet-like lava flows rather than domes, but exceptions are known: the aspect ratio of a lava extrusion is determined more by the discharge rate than by the lava viscosity or yield strength.

Occasionally the magma intrudes the near-surface rocks to form cryptodomes instead of erupting at the surface as lava. Three times this century (in 1909, 1943–45, and 1977–78) it has happened on Usu Volcano, Hokkaido (Ishikawa, 1950; Minakami et al., 1951; Katsui et al., 1978). Explosive activity accompanied all three events, and an extrusive lava dome broke through the Showa Shinzan cryptodome during the 1943–45 event. Cryptodomes probably having a similar origin are known on many volcanoes, for instance Mt Misery (Baker, 1970) in the West Indies, where marine limestones have been uptilted.

Volcanic Hazard

Volcanic eruptions of andesitic volcanoes can cause great damage and loss of life, and the catastrophes of Krakatau in 1883 and St Pièrre in 1902 at once spring to mind, but it is easy to overestimate the danger. Volcanic activity kills on average a few tens of people per year, and this low background level is punctuated by occasional catastrophes (on average one per decade) in which 10^3–10^5 lives are lost. The average annual loss is much less than that arising from earthquakes, floods, hurricanes, wars, or automobile accidents. Volcanic hazard must, however, be taken seriously because occasional great eruptions occur which may have the potential of causing many thousands of deaths.

The main hazard arises from eruptions of explosive type. The *nuée ardente* holds the record for causing the most destruction and loss of life (e.g. Mt Pelée, 1902; St Vincent, 1902; Mt Lamington, 1951; Mt Agung, 1963), followed by mudflows (e.g. Kelud, 1918; Mt Agung, 1963), tsunamis generated by volcanic action (e.g. Unzen, 1973; Krakatau, 1883), and the collapse of parts of volcanic cones (e.g. Unzen, 1793; Bandai-san, 1888; Mt St Helens, 1980). Lava flows cause much damage but seldom travel so fast that people cannot escape; moreover, they are to some extent controllable.

It is beyond the scope of this chapter to detail the nature of volcanic hazards, but it is worth drawing attention to a few points which may not yet have received the attention they deserve.

The first concerns the *nuée ardente* which is a feature of high and steep volcanoes, and is liable to be generated on a large proportion of such volcanoes. There is sufficient experience of the phenomenon that the area likely to be affected on any particular volcano can be fairly well delineated, and the problem is the agonising choice facing the volcanologist and the administrator he advises on whether to order the evacuation of the local population or not, when an eruption which may generate a dangerous *nuée ardente* appears to be impending. The recent example of Guadeloupe (1976) springs to mind. In such circumstances consideration might usefully be given to the design and construction of volcano shelters as an alternative to evacuation, with all the social and economic problems that evacuation causes.

The second point is that the potential damage and loss by volcanic eruptions is all the time increasing as, with increasing population pressure, ever more people come to live on or near volcanoes and as, with increasing affluence, a society becomes in some ways more vulnerable to volcanic effects. The reality of the vulnerability is shown by the recent activity of Mt St Helens, where a light ash-fall which might barely have been noticed by a primitive community caused great expense, inconvenience, and consternation to a highly civilized one.

The third point is that undoubtedly the greatest volcanic danger of all, that stemming from an ignimbrite eruption, is a much less remote or rare event than hitherto supposed. It is now known that both the Tambora, 1813, and Krakatau, 1883, events were ignimbrite eruptions (Williams, 1942; Self and Rampino, 1979). Including Katmai, 1912, there have thus been at least three fair-sized ignimbrite eruptions in 170 years. The period since 1912 can be characterized as a volcanologically quiet one with no volcanic events of large magnitude, but there seems on present information to be roughly a 1:5 probability that there will be an ignimbrite eruption somewhere before the end of the present century.

There are many places in Indonesia, Italy, Japan, Mexico, New Zealand, Peru, the

Philippines, Turkey, and the western USA where 10^5–10^6 people live today on a single ignimbrite sheet. Modern man has not yet experienced an ignimbrite eruption and is not aware of how catastrophic it might be. Within the area overwhelmed by the ignimbrite itself (up to c. 15 000 km^2 of country) destruction of life and property would be complete. Very severe disruption would also be caused over an area of up to 10^5–10^6 km^2 by the accompanying fall of co-ignimbrite ash. An ignimbrite eruption in a densely populated area has the potential to cause many thousands of deaths and result in disruption over an immense area, in comparison with which the effects of the recent Mt St Helens event would seem quite negligible.

REFERENCES

Anderson, A. T. (1973). The before-eruption water content of some high-alumina magmas. *Bull. Volc.* **37**, 530–552.

Anderson, A. T. (1975). Some basaltic and andesitic gases. *Rev. Geophys. Space Phys.* **13**, 37–55.

Aramaki, S. (1956). The 1783 activity of Asama volcano. Part 1. *Jap. J. Geol. Geogr.* **27**, 189–229.

Aramaki, S. (1957). The 1783 activity of Asama volcano. Part 2. *Jap. J. Geol. Geogr.* **28**, 11–34.

Aspinall, W. P., Sigurdsson, H., and Shepherd, J. B. (1973). Eruption of Soufrière volcano on St Vincent Island, 1971–1972. *Science, N.Y.* **181**, 117–124.

Baker, P. E. (1970). The geological history of Mt Misery volcano, St Kitts, West Indies. *Bull. Overseas Geol. Mineral Resources* **10**, 207–230.

Blackburn, E. A., Wilson, L., and Sparks, R. S. J. (1976). Mechanisms and dynamics of strombolian activity. *J. geol. Soc. Lond.* **132**, 429–440.

Bottinga, Y. and Weill, D. F. (1972). The viscosity of magmatic silicate liquids: a model for calculation. *Am. J. Sci.* **272**, 438–475.

Chouet, B., Hamisevicz, N., and McGetchin, T. R. (1974). Photoballistics of volcanic jet activity at Stromboli, Italy. *J. geophys. Res.* **79**, 4961–4976.

Curtis, G. H. (1968). The stratigraphy of the ejectaments of the 1912 eruption of Mt Katmai and Novarupta, Alaska, *Geol. Soc. Am. Mem.* no. 116, 153–210.

Eggler, D. H. (1972). Water-saturated and undersaturated melting relations in a Paricutin andesite and an estimate of water content in the natural magma. *Contr. Mineral. Petrol.* **34**, 261–271.

Einarsson, T. (1950). The eruption of Hekla 1947–1948. II, 2. A study of the earliest photographs of the eruption. *Vísindafélag Íslendinga*, H. F. Leiftur, Reykjavik.

Einarsson, T. (1951). 'The eruption of Hekla 1947–1948. IV, 3. The flowing lava', *Vísindafélag Íslendinga*, H. F. Leiftur, Reykjavik.

Euler, R. and Winkler, H. G. F. (1957). Über die Viskositäten von Gesteins und Silikatschmelzen. *Glastech. Ber.* **30**, 325–332.

Foshag, W. F. (1950). The aqueous emanation from Paricutin volcano. *Am. Mineral.* **35**, 749–755.

Foshag, W. F. and Gonzalez, R. J. (1955). Birth and development of Paricutin volcano. *US geol. Surv. Bull.* no. 965-D.

Fries, C. (1953). Volumes and weights of pyroclastic material, lava and water erupted by Paricutin volcano, Michoacán, México. *Trans. Am. geophys. Union* **34**, 603–616.

Fudali, R. F. and Melson, W. G. (1972). Ejecta velocities, magma chamber pressure and kinetic energy associated with the 1968 eruption of Arenal volcano. *Bull. Volc.* **35**, 383–401.

Gorshkov, G. S. (1959). Gigantic eruption of the volcano Bezymianny. *Bull. Volc.* **20**, 77–109.

Healy, J. (1963). Welded pyroclastic rock at Tongariro. *N.Z. J. Geogr. Geophys.* **6**, 712–714.

Hulme, G. (1974). The interpretation of lava flow morphology. *Geophys. Jl. R. astr. Soc.* **39**, 361–383.

Hulme, G. and Fielder, G. (1977). Effusion rates and rheology of lunar lavas. *Phil. Trans. R. Soc. A* **285**, 227–234.

Ishikawa, T. (1950). New eruption of Usu volcano, Hokkaido, Japan, during 1943–1945. *J. Fac. Sci. Hokkaido Univ. Ser. 4*, **7**, 237–360.

Katsui, Y. and 21 other authors (1978). Preliminary report of the 1977 eruption of Usu volcano. *J. Fac. Sci. Hokkaido Univ.* Yagi volume, 385–408.

Krauskopf, K. B. (1948). Lava movement at Paricutin volcano, Mexico. *Geol. Soc. Am. Bull.* **59**, 1267–1284.

Kuno, H. (1962). *Catalogue of Active Volcanoes of the World Including Solfatara Fields. Part XI. Japan, Taiwan and Marianas*. International Association for Volcanology, Rome.

Lipman, P. W. (1967). Mineral and chemical variations within an ash-flow sheet from Aso caldera, southwestern Japan. *Contr. Mineral. Petrol.* **16**, 300–327.

Markhinin, E. K. (1962). On the possibility of estimating the amount of juvenile water participating in volcanic explosions. *Bull. Volc.* **24**, 187–191.

Matumoto, T. (1943). The four gigantic caldera volcanoes of Kyusyu. *Jap. J. Geol. Geogr.* **19**, special no., 1–57.

McBirney, A. R. and Murase, T. (1971). Factors governing the formation of pyroclastic rocks. *Bull. Volc.* **34**, 372–384.

McGetchin, T. R., Settle, M., and Chouet, B. A. (1974). Cinder cone growth modeled after North-east Crater, Mount Etna, Sicily. *J. geophys. Res.* **79**, 3257–3272.

Minakami, T. (1950). The explosive activities of volcano Asama in 1935. *Bull. Earthquake Res. Inst. Univ. Tokyo* **13**, 629–644, 790–800.

Minakami, T., Ishikawa, T., and Yagi, K. (1951). The 1944 eruption of volcano Usu in Hokkaido, Japan. *Bull. Volc., Ser. 2* **11**, 45–157.

Minato, M., Hashimoto, S., Fujiwara, Y., Kumano, S., and Okada, S. (1972). Stratigraphy of the Quaternary pumiceous volcanic products, S. W. Hokkaido. *J. Fac. Sci. Hokkaido Univ. Ser. IV* **15**, 679–736.

Murase, T. and McBirney, A. R. (1973). Properties of some common igneous rocks and their melts at high temperatures. *Geol. Soc. Am. Bull.* **84**, 3563–3592.

Murata, K. J., Dondoli, C., and Saenz, R. (1966). The 1963–65 eruption of Irazu volcano, Costa Rica (the period of March 1963 to October 1964). *Bull. Volc.* **29**, 765–796.

Nagata, T., Sakuma, S., and Fukyshima, N. (1946). On the lava flow newly ejected from Sakura-jima volcano. *Bull. Earthquake Res. Inst. Univ. Tokyo* **24**, 161–170.

Nairn, I. A. (1976). Atmospheric shock waves and condensation clouds from Ngauruhoe explosive eruptions. *Nature, Lond.* **259**, 190–191.

Pinkerton, H. and Sparks, R. S. J. (1978). Field measurements of the rheology of lava. *Nature, Lond.* **276**, 383–384.

Rose, W. I. (1973). Pattern and mechanism of volcanic activity at the Santiaguito volcanic dome, Guatemala. *Bull. Volc.* **37**, 73–94.

Scarfe, C. M. (1973). Viscosity of basaltic magmas at varying pressures. *Nature, Lond.* **241**, 101–102.

Scarfe, C. M. (1977). Viscosity of some basaltic glasses at one atmosphere. *Can. Mineral.* **15**, 190–194.

Schmincke, H.-U. (1977). Phreatomagmatische Phasen in quartären Vulkanen der Osteifel. *Geol. Jb. A* **39**, 3–45.

Segerstrom, K. (1950). Erosion studies at Paricutin, State of Michoacan, Mexico. *US geol. Surv. Bull.* no. 965-A.

Self, S. and Rampino, M. R. (1979). Abstract, IUGG Symposium, Canberra, 1979.

Self, S., Wilson, L., and Nairn, I. A. (1979). Vulcanism eruption mechanisms. *Nature, Lond.* **277**, 440–443.

Shaw, H. R. (1969). Rheology of basalt in the melting range. *J. Petrol.* **10**, 510–535.

Shaw, H. R. (1972). Viscosities of magmatic silicate liquids: an empirical method of prediction. *Am. J. Sci.* **272**, 870–893.

Sparks, R. S. J. (1976). Grain size variations in ignimbrites and implications for the transport of pyroclastic flows. *Sedimentology* **23**, 147–188.

Sparks, R. S. J., Wilson, L., and Hulme, G. (1978). Theoretical modeling of the generation, movement, and emplacement of pyroclastic flows by column collapse. *J. geophys. Res.* **83**, 1727–1739.

Sparks, R. S. J. and Wright, J. V. (1979). Welded air-fall tuffs. *Geol. Soc. Am. spec. Pap.* no. 180, 155–166.

Taylor, G. A. (1958). The 1951 eruption of Mount Lamington, Papua. *Aust. Bur. Mineral. Resources Geol. Geophys., Bull.* no. 38.

Thorarinsson, S. (1950). The eruption of Hekla 1947–1948. II, 1. The approach and beginning of the Hekla eruption. Eyewitness accounts. *Visindafélag Íslendinga*, H. F. Leiftur, Reykjavik.

Thorarinsson, S. (1954). The eruption of Hekla 1947–1948. II, 3. The tephrafall from Hekla on March 29th 1947. *Visindafélag Íslendinga*, H. F. Leiftur, Reykjavik.

Tuya, H., Takahasi, R., Hagiwara, T., Nagata, T., Omote, S., and Hirano, K. (1941). The eruption of Miyake-sima, one of the Seven Izu islands, in 1940. *Bull. Earthquake Res. Inst. Univ. Tokyo* **19**, 260–401.

Ui, T. (1973). Exceptionally far-reaching, thin pyroclastic flow in southern Kyusyu, Japan. *Bull. Volc. Soc. Japan* **18**, 153–168.

Volarovich, M. P. (1936). Sur la viscosité des roches fondues. *C.r. hebd. Séanc. Acad. Sci. Paris* **202**, 78–80.

Walker, G. P. L., Heming, R. F., and Wilson, C. J. N. (1980a). Low-aspect ratio ignimbrites. *Nature, Lond.* **283**, 286–287.

Walker, G. P. L., Heming, R. F., Sprod, T. J., and Walker, H. R. (1980b). Latest major eruptions of Rabaul volcano. *Papua New Guinea geol. Surv. Mem.* **10**, in press.

Williams, C. E. and Curtis, R. (1965). The eruption of Lopevi, New Hebrides, July 1960. *Bull. Volc.* **27**, 423–433.

Williams, H. (1941). Calderas and their origin. *Publ. Univ. Calif. geol. Sci. Bull.* **25**, 239–346.

Williams, H. (1942). The geology of Crater Lake National Park, Oregon, with a reconnaissance of the Cascade Range southward to Mount Shasta. *Carnegie Instn. Wash. Publ.* no. 540.

Wilson, L. and Head, J. W. (1981). Ascent and emplacement of basaltic magma on the Earth and Moon. *J. geophys. Res.*, **86**, 2971–3001.

Wright, J. V. and Walker, G. P. L. (1977). The ignimbrite source problem: a co-ignimbrite lag fall deposit from Mexico. *Geology* **5**, 729–732.

Yagi, K., Takeshita, H., and Oba, Y. (1972). Petrological study on the 1970 eruption of Akita-Komagatake. *J. Fac. Sci. Hokkaido Univ. Ser. IV* **15**, 109–138.

Yokoyama, S. (1970). Geomorphology of the Ito pyroclastic flow deposit to the north of the Aira caldera, *Geogr. Rev. Japan* **43**, 464–482.

Yokoyama, S. (1974). Mode of movement and emplacement of Ito pyroclastic flow from Aira caldera, Japan. *Sci. Kyoiku Daigaku C* **12**, 17–62.

Andesitic pyroclastic flows

A. L. Smith and M. J. Roobol

Department of Geology,
University of Puerto Rico, Mayaguez, Puerto Rico 00708, USA

and

Institute of Applied Geology, Jeddah, Saudi Arabia

ABSTRACT

Pyroclastic flows are gravity-controlled, high concentration gas/solid dispersions that are probably the most destructive form of explosive activity of terrestrial volcanoes. They have been described from most volcanic environments and from all geological ages. The flows themselves can travel at high speeds over considerable distances, and are separable into a lower unit, or underflow, which has a high particle concentration and probably moves by either laminar and/or plug flow, and an overriding cloud which has a low particle concentration and moves in a turbulent manner. The former gives rise to true pyroclastic flow deposits, while the latter produce pyroclastic surge and air-fall ash deposits. The deposit produced by a pyroclastic flow is poorly sorted, although coarse tail grading may be common. Two main eruptive mechanisms have been proposed to produce pyroclastic flows: the gravitational or explosive collapse of an active dome or lava flow, and the gravitational collapse of an overloaded vertical eruptive column. The latter type of collapse may be either interrupted or continuous. Pyroclastic flows may also be subdivided on the basis of their volume into small volume flows, characteristic of historic eruptions, that have high yield strength and probably move by grain flow, and large volume flows that have low yield strength and are thought to be partly fluidized.

Introduction

The eruptions of Mt Pelée, Martinique, and Soufrière, St Vincent, and the destruction of the city of St Pierre by the former, brought to worldwide attention the volcanic phenomenon of pyroclastic flows, that to all purposes was previously unknown (Anderson and Flett, 1903; LaCroix, 1904). A small pyroclastic flow eruption on Cotopaxi had been described previously (Wolf, 1878) but its significance was not widely recognized. Since 1902 numerous pyroclastic flow eruptions have been observed. Investigations have also shown that many of the destructive eruptions of early historic times, e.g. Vesuvius in AD 79 and Santorini in 1470 BC produced pyroclastic flows. It is now known that pyroclastic flows have occurred throughout geological time from the Precambrian to the present.

The distribution of pyroclastic flows is worldwide and they have been described from many tectonic settings and are developed in magmas of contrasted series. They are best known and are most widely represented in the calc-alkaline magmas of the continental margins such as the Andes (Zeil and Pichler, 1967), the larger islands of Japan and New Zealand (Aramaki and Yamasaki, 1963; Marshall, 1935), and the island arcs such as the Lesser Antilles.

They are also well described in the alkaline magmas of the continental rifts of East Africa and Ethiopia (Baker et al., 1972; Gibson, 1970) and the alkaline oceanic islands such as the Canary Islands (Schminke and Swanson, 1967). Examples are also known from the dominantly tholeiitic magmas of Iceland (Walker, 1962).

The limitation of this discussion to 'andesitic pyroclastic flows' largely confines one to the examples from the tectonic environment of destructive plate margins. Such margins can be regarded as ranging between two end-member settings—the ensialic continental margin type and the ensimatic island arc type. In the latter, rhyolites are rare, and the most prominent type of pyroclastic flow is of andesitic composition (Baker 1968, and this volume, Section II). In the continental margins and large islands rhyolite–dacite pyroclastic flows dominate (Ross and Smith, 1961; Ewart et al., 1971).

Characteristics of Pyroclastic Flows

A pyroclastic flow can be defined as the lateral movement of clasts as a gravity-controlled, hot gas–solid dispersion. As used here the term includes all hot fragmental flows produced by explosive eruptions, as well as hot rock avalanches formed by the collapse of actively growing unstable lava domes and flows, but excludes volcanic mudflows and cold rock avalanches.

Based on descriptions of historic eruptions, which at the moment we assume to be representative of all types of pyroclastic flows (both small and large volume), a pyroclastic flow can be seen to be composed of two parts: an underflow or, according to common usage, a pyroclastic flow (*sensu stricto*) and an overriding turbulent cloud of gas and dust (Figs 1 and 2) which has been termed the surge component. The underflow, in which the coarsest material is transported, can be classed as a high concentration gas–solid dispersion while the surge represents a highly expanded, low particle concentration medium. Both surges and underflows are types of flow and, although there is a gradation in character between the low concentration, low density surges and the high concentration, high density underflows, the characteristics of their deposits are significantly different. Thus it is appropriate to differentiate flows from surges and to classify their different types of deposits separately.

Once initiated the flow moves away from the eruptive vent either radially or within a definite sector depending on local conditions. The pyroclastic flow (underflow), which is topographically controlled possibly moves as a series of pulses (Davies et al., 1978; A. L. Smith and M. J. Roobol, unpublished data) and when confined whithin a valley often follows a sinuous path (Fig. 3). The overriding gas–ash cloud tends to expand in size and become darker in colour with distance from the vent. This cloud may in some instances overtake the associated underflow and also tends to spread out laterally on either side. The relationship between the area affected by the underflow and that by the cloud can vary from approximately the same dimensions, e.g. Fuego in 1973, to instances when the cloud-affected area can be 100 times greater, e.g. Santiaguito in 1973 (Rose et. al., 1976).

The two-part concept of a pyroclastic flow accounts for the formation of contrasted types of deposit. For example, in the classic account of the 1902–05 eruptions of Mt Pelée by LaCroix (1904, p. 331, 359–60; 1908, p. 79) the main valley infill deposits on the western flanks were noted to contain blocks up to several metres in diameter while those deposits in Morne Rouge and Ajoupa Bouillon on the eastern flanks (high above the main valley infill deposits) contained no large blocks. The main pyroclastic flow deposit which contains the coarsest clasts originates from the underflow. The deposits found on the eastern flanks of the volcano and those outside the area of the main valley infill on the western side, e.g. within the town of St. Pierre, tend to lack large clasts, be enriched in crystals and vitric components, and often show unidirectional sedimentary bed forms (e.g. cross-stratification). It is concluded that these deposits

Fig. 1 *Nuée ardente* from the 16 December 1902 eruption of Mt Pelée, Martinique. The maximum height attained by the cloud was 4000 m. (From LaCroix, 1904, Plate VIII; reproduction by permission of Masson et Cie)

Fig. 2 Scoria flow produced by collapse of dense interior of eruptive column during 1975 eruption of Nguaruhoe, New Zealand. Photograph taken 30 s after initial explosion. (From Nairn and Self, 1978; reproduced by permission of Elsevier Scientific Publishing Co.)

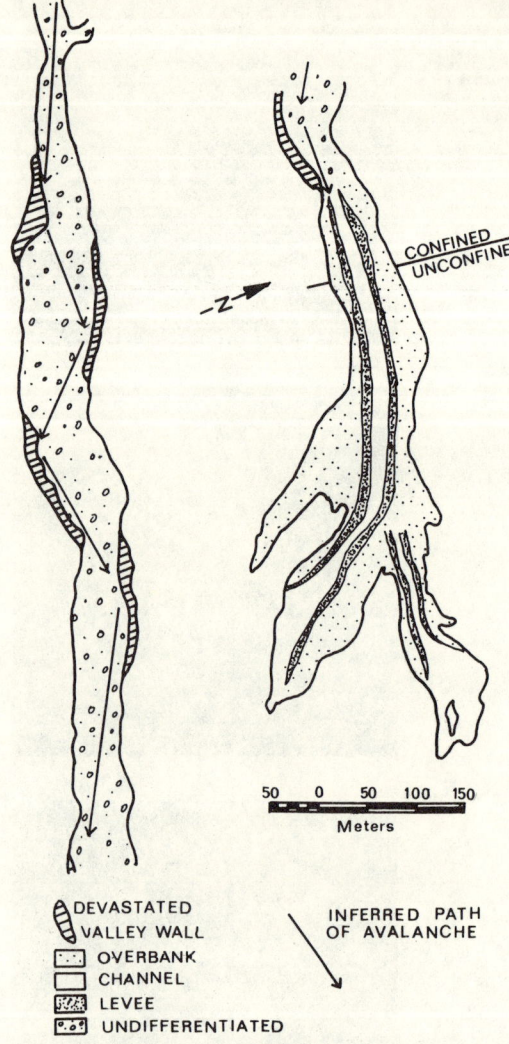

Fig. 3 Map of pyroclastic flow deposits produced during the 1974 eruption of Fuego, showing the inferred path of travel of the flows, areas of heaviest vegetation devastation on the valley sides, and depositional environments. (From Davies et al., 1978; reproduced by permission of the Geological Society of America)

represents the surge component of the eruption. These surge deposits may be of two types:

(1) Groundsurge deposits which represent the products of initial eruptive column collapse or direct eruptions from the crater. They generally underlie associated pyroclastic flow deposits, although they may be produced without the generation of an accompanying pyroclastic flow (Fisher, 1979; A. L. Smith and M. J. Roobol, unpublished data).

(2) Ash-cloud surges which represent the material elutriated from the top of an accompanying pyroclastic flow. These can vary from relatively coarse grained lateral equivalents of the underflow deposits to fine grained vitric enriched types (R. V. Fisher, A. L. Smith, and M. J. Roobol, unpublished data). The finest vitric enriched material of the ash cloud

is dispersed over a wide area as air-fall vitric ashes (Sparks and Walker, 1978).

According to Sparks *et al.* (1978), initially the flow may move as a turbulent suspension, although Fisher (1979) and Wright and Walker (1977) have presented evidence that some flows at least may move in a laminar fashion from their initiation. Depending on groundslope an area of non-deposition may occur close to the vent. In many instances, in this region the flow actually erodes pre-existing material (Perret, 1937; Taylor, 1958; Davies *et al.*, 1978; Nairn and Self, 1978). Further from the vent either laminar and/or plug flow is thought to be the dominant type of movement and probably as the velocity decreases below a certain critical value deposition begins almost instantaneously (Wright, 1979). (Laminar flow occurs when different layers or particles of a fluid slide smoothly past one another, and there are no irregular eddies producing movement from one layer to another. Plug flow is the flow of a high concentration material (Bingham plastic) in which the yield strength is such that the flow can only move on a sheared basal layer, where the critical yield stress has been exceeded. The rest of the flow moves as a rigid plug undergoing no deformation.)

The velocities of observed pyroclastic flows are very variable, both from one flow to another and depending on topography, even within one flow. Estimates of speeds of observed flows have ranged from 8 to 45 m s^{-1} with most having average speeds of *c.* 20 m s^{-1}. On a theoretical basis Sparks *et al.* (1978) have suggested that some of the large volume flows have exceeded velocities of 300 m s^{-1}.

Not only do pyroclastic flows travel at high speeds but they can also extend considerable distances from their source (Table 1). Their mobility is also expressed in their apparent ability to surmount significant topographic barriers (Sparks, 1976; Miller and Smith, 1977;

TABLE 1 Distance travelled by pyroclastic flows from their source

Eruption	Distance (km)	Reference
(a) *Historic eruptions*		
Valley of 10 000 Smokes	29	Curtis (1968)
Bezymianny 1955–56	18	Gorshkov (1959)
Sheveluch 1964	18	Gorshkov and Dubik (1970)
Mt Lamington 1951–56	16	Taylor (1958)
Agung 1963	14	Zen and Hadikusumo (1964)
Merapi 1930	13.5	Bemmelen (1949)
Asama 1783	11	Aramaki (1956)
Komagatake 1929	8	Kozu (1934)
Mt Pelée 1902–05	8	LaCroix (1904)
Soufriére, St Vincent 1902–03	8	Anderson and Flett (1903)
Hibok–Hibok 1948–51	7	MacDonald and Alcaraz (1956)
Mayon 1968	5	Moore and Melson (1969)
Arenal 1968	3.6	Melson and Saenz (1973)
Nguarahoe 1975	2	Nairn and Self (1978)
(b) *Pre-historic eruptions*		
Ash Flow H, Guatemala	*c.* 125	Koch and McLean (1975)
Aso, Japan	*c.* 96	Matumoto (1943)
Crater Lake, Oregon	*c.* 64	Williams (1942)
Aniakchak Caldera, Alaska	*c.* 50	Miller and Smith (1977)
Hakone, Japan	*c.* 24	Kuno (1950)
Fisher Caldera, Alaska	*c.* 20	Miller and Smith (1977)
Mt Pelée, Martinique (1.9 ka ago)	*c.* 20	Roobol and Smith (1980)

Fisher, 1976; Francis and Baker, 1977; Roobol and Smith, 1980). This feature can be adequately explained in terms of their momentum and it is not necessary to invoke a highly expanded flow as suggested by Aramaki and Ui (1966), Yokoyama (1974), and Sheridan and Ragan (1976) to account for the distribution of some pyroclastic flow deposits. Attempts to estimate the mobility of pyroclastic flows (Francis *et al.*, 1974; Sparks, 1976; Nairn and Self, 1978) have indicated that, in spite of their high temperature, they appear to be no more mobile than cold rock avalanches or mudflows. It was originally thought that the larger volume pyroclastic flows were more mobile (Sparks, 1976); however, the recognition that such flows are produced by eruptive column collapse, thus increasing their gravitational head, tends to eliminate this difference (Sparks *et al.*, 1978).

In terms of a real extent and volume, pyroclastic flow deposits can show great variability (Table 2). Deposits from individual flows are known to cover thousands of square kilometres in area and to have volumes to the order of thousands of cubic kilometres, while pyroclastic flow fields can have volumes ex-

TABLE 2 Estimated volumes of pyroclastic flows

Name of Pyroclastic Flow/Volcano	Volume (km^3)	Reference
Fish Canyon Tuff	3000	Steven and Lipman (1976)
Toba Tuff	2000	Bemmelen (1949)
Timber Mountain Tuff (Ranier Canyon Member)	1200	Byers *et al.* (1976)
Timber Mountain Tuff (Ammonia Tanks Member)	900	Byers *et al.* (1976)
Nelson Mountain Tuff	500	Steven and Lipman (1976)
Upper Bandelier Tuff	205	Smith and Bailey (1966)
Aso III Pyroclastic Flow Deposit	175	Lipman (1967)
Bishop Tuff	160	Gilbert (1938)
Ito Pyroclastic Flow Deposit	110	Yokoyama (1974)
Peach Springs Tuff	90	Young and Brennan (1974)
Kutcharo Welded Tuff	90	Murai (1961)
Shikotsu I	80	Aramaki and Yamasaki (1963)
John Day Pyroclastic Flow Deposit	75	Fisher (1966)
Akan Welded Tuff	60	Murai (1961)
Tokachi Welded Tuff	60	Murai (1961)
Ash Flow H	20–50	Koch and McLean (1975)
Crater Lake Pumice Flows	37	Moore (1934)
Towada Pyroclastic Flow I	25	Aramaki and Yamasaki (1963)
Hakone Pyroclastic Flow Deposit	15	Murai (1961)
Valley of 10 000 Smokes	6.1	Curtis (1968)
Krakatoa (1883)	10	Williams (1941)
Mashu Pyroclastic Flow Deposit	5	Aramaki and Yamasaki (1963)
Roseau Pyroclastic Flow	3	Sigurdsson (1972)
Asama Pumice Flow I	2	Aramaki and Yamasaki (1963)
Bezymianny (1955–56)	1.8	Gorshkov (1959)
Nantai Pumice Flow Deposit	0.8	Aramaki and Yamasaki (1963)
Soufriére, St Vincent (1902–03)	0.4–1.4	Robson and Tomblin (1966)
Oiwake, Asama (1281?)	0.6	Aramaki and Yamasaki (1963)
Komagatake (1929)	0.5	Aramaki and Yamasaki (1963)
Asama (1783)	0.11	Aramaki and Yamasaki (1963)
Mt Pelée (1903–03)	0.1	A. L. Smith and M. J. Roobol (unpublished data)
Mt Lamington (1951–56)	0.1	Taylor (1958)
Mt Pelée (1929–32)	0.04	A. L. Smith and M. J. Roobol (unpublished data)
Mayon (1968)	0.015	Moore and Melson (1969)
Sakurajima (1939)	0.001	Aramaki and Yamasaki (1963)

ceeding 10 000 km³. At the other extreme, some historic pyroclastic flow deposits have volumes of less than 10^3 m³ (Nairn and Self, 1978). The estimation of volume of erupted material in a pyroclastic flow deposit presents many difficulties. First, there is the problem of estimating the volume of the deposit, using data which varies from isopach maps to measurements made in a few stratigraphic sections. Second is the problem of different degrees of vesiculation of the clasts. Large volume deposits tend to be highly vesicular while small volume deposits are often poorly vesicular (Aramaki and Yamasaki, 1963). Data are therefore best converted to dense rock equivalent, although this was not done in Table 2, since there was usually a lack of information on density and degree of vesiculation, both of which are needed to convert a given volume estimate to its dense rock equivalent.

As was pointed out by R. L. Smith, most large volume pyroclastic flow eruptions are associated with large calderas or volcano-tectonic depressions (Smith, 1960a, Fig. 3), and there appears to be a linear relationship between the size of the caldera and the volume of the resulting pyroclastic flows. It has been suggested by a number of workers that these large volume deposits were erupted from fissures. This idea is largely based on the assumption that the Valley of Ten Thousand Smokes eruption was a fissure eruption. Recent work by Curtis (1968) has convincingly shown that these pyroclastic flows were erupted from Novarupta and Mt Katmai, and not the supposed fissures in the valley floor. The linear fumaroles, which were thought to represent the fissures, were merely surface reflections of faults along which groundwater moved. Field evidence for prehistoric fissure eruptions is also very poor while studies of some of the larger pyroclastic flows and associated deposits (e.g. Bishop Tuff, Bandelier Tuff) indicate that many of their features can only be explained by eruption from a centralized vent (Wright et al., 1979). It is thus the authors' opinion that most pyroclastic flows are produced by eruptions from centralized vents.

Field observations on historic eruptions have indicated that individual clasts within a pyroclastic flow can vary in temperature from incandescent juvenile clasts (c. 900 °C) to cold, accidental lithic fragments, while mineralogical studies based on Fe-Ti oxides indicate that the quench temperatures for clasts in pumice flows can vary from 640 to 1000 °C (Lipman, 1971; Ewart et al., 1971; Heming and Carmichael, 1973). Estimates of temperatures of pyroclastic flow deposits based on degree of welding (Smith, 1960a, b) seem to range from 580 to 900 °C, indicating that if the temperatures of the essential clasts represent the actual temperature of the pyroclastic flows, then most deposits should show signs of welding. In order to explain this discrepancy, it has been suggested (Smith, 1960a; Sparks et al., 1978) that the emplacement temperature of a pyroclastic flow (as compared with the temperature of the clasts themselves) depends on many factors. These are: (1) the volume of the released gas which determines the particle concentration in the eruptive column—the highest particle concentrations correspond to greatest heat conservation; (2) the initial or muzzle velocity of the rising column will control the height from which collapse occurs—the greater the height of collapse the greater will be the opportunity for cooling; (3) the amount of air that is entrapped during the formation and movement of the flow. Thus variations in emplacement temperatures between flows of similar composition, physical properties, and thickness, e.g. the Rabaul pumice flow deposits (Heming and Carmichael, 1973), can be simply explained by variations in the properties of the eruptive column. Rapid column collapse without the generation of a large vertical column would produce a condition in which heat would be strongly conserved (Sparks et al., 1978). Deposits produced by such an eruption should therefore show features indicative of emplacement at higher temperatures (e.g. welding) than those associated with a vertical eruptive column. Substantiating evidence for this is provided by Sparks and Wilson (1976), who show that welding is more common in flows lacking an underlying plinian air-fall phase than in those in which such a deposit is present.

Some pyroclastic flow sequences have been described that show a systematic change of temperature with time. The best described example of such a feature is the Upper Bandelier Tuff (Smith and Bailey, 1966), which, based on welding characteristics, appears to have been emplaced at temperatures ranging from 550 °C (base) to 800 °C (top). Such an increase in temperature has been attributed to draining of hotter, deeper levels of the magma chamber (Smith and Bailey, 1966). An additional explanation is that it is due to changes in such parameters as vent width and gas content (Sparks and Wilson, 1976; Sparks et al., 1978).

In a number of instances pyroclastic flows from a common source show changes in composition through a vertical sequence. An example of such a change has been described from the Topopah Spring Member of the Paintbrush Tuff, southern Nevada. This widespread ash flow sheet is a multiple flow compound cooling unit which grades upwards from a crystal-poor rhyolite (77 per cent SiO_2) at the base to a crystal-rich quartz latite (69 per cent SiO_2) at the top (Lipman et al., 1966). A somewhat different situation is illustrated by the pumice flow deposits of Crater Lake, Oregon. According to Williams (1942), c. 28 km^3 of the deposit have a silica content of between 64 and 69 per cent SiO_2, while c. 4 km^3 are composed of crystal-rich basaltic andesite (54–57 per cent SiO_2). There are no andesitic ejecta and the transition from dacite to basaltic andesite takes place vertically over a distance of c. 1 m. There appear to be no flow unit boundaries within the deposit, suggesting that it represents the product of a single eruption. The 1902 eruption of Soufrière, St Vincent, also appears to show a change in composition within a single eruption. The first material erupted (7 May 1902) had a silica content of 55.04 per cent SiO_2, while in the last flows (March 1903) it had dropped to 50.55 SiO_2 (Roobol and Smith, 1975). Whether or not material of intermediate composition was produced is not certain. However, from the evidence of 1903 Soufrière ash layers cored from the Atlantic and Caribbean (Carey and Sigurdsson, 1978), in which glass shards of two distinct compositions (56 and 65 per cent SiO_2) were obtained, it seems probable that two distinct magma compositions were erupted here as well.

Classification

In the almost 80 years since they were first studied, various classifications for pyroclastic flows have been proposed. A summary of some of the more important ones, together with the one used here, is presented in Table 3. The classification proposed here is based on the mode of eruption as well as the characteristics of the resultant deposits and can serve independently to describe either flows or their deposits. Specific locality names, e.g. Peléan, have been avoided since in many cases individual volcanoes often have produced more than one type of pyroclastic flow (Roobol and Smith, 1976). The initial subdivision of pyroclastic flows is based on whether they formed (1) by the gravitational or explosive collapse of an active dome or lava flow of (2) by the collapse of an eruptive column.

Pyroclastic flows produced by the lava/dome collapse mechanism are termed *nuées ardentes* (the term being restricted here only to this type of pyroclastic flow). The resultant deposits are characteristically composed of very poorly vesicular, dense lava and are termed 'block and ash'. The historic eruptions of Mt Pelée, Martinique (LaCroix, 1904; Perret, 1937; Roobol and Smith, 1975), can be taken as examples of this type of pyroclastic flow.

Recent work by Fisher, Smith, and Roobol on the early eruptive products of the 1902 eruption of Mt Pelée indicates that pyroclastic flows and surges identical to those produced by dome collapse may be formed by rapid collapse from a short eruptive column produced by a vertical explosion through an actively growing dome. Column collapse on the other hand can, according to Sparks and Wilson (1976) and Nairn and Self (1978), be either

TABLE 3 Classification of pyroclastic flows

LaCroix (1930)	Escher (1933) and Fenner (1937)	MacGregor (1952)	Aramaki (1957)	Williams (1957)
Nuée ardente peléene d'explosion dirigée	Pelée type	Pelée discharge lateral type / Pelée directed lateral type	*Nuée ardente*	Peléan type
Nuée ardente peléene d'avalanche	Merapi type	Merapi lateral disintegration		
Nuée ardente d'explosion vulcanienne	St Vincent type	Pelée vertical type	Intermediate type	
		St Vincent vertical type / Mt Katami tuff flow type		Vertical explosion type (Krakatoan)
Nuée ardente du Massif du Katmai	Katmaian type	Novarupta tuff flow type / Concealed fissure–orifice type	Pumice flow	Fissure eruption type (Valley of 10000 Smokes)

This paper

	Murai (1961)	MacDonald (1971)	Eruptive mechanism	Pyroclastic flow	Underflow deposit	Surge deposit†
Nuée ardente	Pelée type	Pelée type	Lava/dome collapse	*Nuée ardente*	Block and ash	Dense andesite surge
	Merapi type / Sakurajima type	Merapi type				
Intermediate type	St Vincent type	Soufrière type	Eruptive column collapse	Scoria flow	Scoria and ash	Scoriaceous surge
Ash flow	Krakatoan type	Ash flow type		Pumice–ash flow (small volume)	Pumice and ash	Ash pumice surge
	Valley of 10000 Smokes type			Pumice–ash flow (large volume)	Ignimbrite	

† Surge deposits include both groundsurge and ash-cloud surge.

interrupted or continuous. The former give rise to small volume deposits which may be classified on the basis of magma composition and vesiculation into scoria and ash deposits produced by scoria flows (e.g. St Vincent, 1902: Anderson and Flett, 1903; Ngaurahoe, New Zealand, 1975: Nairn and Self, 1978) and small volume ash/pumice flows (e.g. Komagatake, 1929: Kozu, 1934). The latter is thought to produce large volume deposits (ignimbrites), e.g. Valley of Ten Thousand Smokes in 1912 (Fenner, 1920; Curtis, 1968). We prefer to restrict the term ignimbrite to the deposits of the continents and large islands such as New Zealand which are generally of large volume but not to use it for the very small volume deposits of the young ensimatic island arcs.

This fourfold division of pyroclastic flows (Table 3) from *nuées ardentes* and scoria flows to small and large volume pumice/ash flows corresponds generally with an increase in the volumes of the deposits, an increase in the vesicularity of the deposits, and a decrease in maximum clast size. The relationship between volume of deposit and clast vesicularity was first noted by Aramaki and Yamasaki (1963). These authors demonstrated that pyroclastic flow deposits with dense and poorly vesicular clasts usually have volumes of less than 0.2 km^3.

A division of small and large volume ash/pumice flow deposits at c. 5 km^3 may have useful stratigraphic application. The small volume ash/pumice flow deposits are found in both island arc and continental margin environments (Francis *et al.*, 1974) while the larger volume deposits are largely absent from the island arc type.

The deposits of surges were suggested by Sparks and Walker (1973) to represent a third group of pyroclastic rocks (together with flow and fall deposits). Surges associated with pyroclastic flows may be subdivided into ground surges, produced by eruption, column collapse, or directly from the crater, without an accompanying vertical eruption column, and ash cloud surges formed by elutriation from the top of a moving pyroclastic flow (Fisher, 1979).

Ground surge deposits can occur beneath a pyroclastic flow (Sparks *et al.*, 1973; Sparks, 1976) while ash cloud surges occur at the top of a pyroclastic sequence (Fisher, 1979). Surge deposts can also be found around the margins of the underflow (Taylor, 1958; Rose *et al.*, 1976) and in some instances can be seen to pass laterally into a pyroclastic flow (Roobol and Smith, 1976, 1980). Both ash cloud and ground surges may move independently of their associated underflows.

Combining these observations it is possible to conclude that each type of pyroclastic flow deposit can have an associated surge deposit. Individual surge deposits can be identified on the nature and vesicularity of their juvenile clasts. Thus surge deposits are linked with pyroclastic flow deposits in the classification of pyroclastic flows and their deposits proposed here.

Deposits of Pyroclastic Flows

The most distinctive feature of pyroclastic flow deposits is their extremely poor sorting, resulting in the characteristic appearance of large clasts contained in an ashy matrix. Individual deposits lack internal stratification and often occur as valley infills. Two features which the authors have found most useful in distinguishing the flow deposits from other volcaniclastic types (especially mudflow deposits) present on West Indian volcanoes are the presence of ash-poor fossil fumarole pipes and carbonized wood (Walker, 1971, 1972; Roobol and Smith, 1976). Combinations of the above features distinguish flow deposits from pyroclastic air-fall deposits which tend to be well sorted, internally stratified, mantle underlying topography (rather than infill gulleys) and show a detectable decrease in grain size and thickness away from the vent.

Surge deposits are fine grained and better sorted than the flow deposits and were described by Moore and Melson (1969) and Sparks and Walker (1973) as showing variable thickness (pinch and swell structure). They have an internal, low angle stratification, often show cross-bedding and are commonly less than 1 m thick. Sparks and Walker (1973) included the base surge deposits (from phreatomagmatic eruptions usually associated with maars) amongst their surge deposits. Here, however, we are limiting the discussion only to those surge deposits found intimately associated with the pyroclastic flows.

The three main lithological types of pyroclastic flow deposits (block and ash, scoria and ash, pumice and ash) and their associated surges are readily distinguished in the field by the appearance and vesicularity of their main components. Block and ash deposits and their associated surge deposits are characterized by dense, poorly vesicular, angular, andesite clasts (which are cognate lithic fragments of the collapsed dome). Many of the larger blocks, from 0.5 m up to several metres in diameter, are characterized by a radial jointing resulting from post-depositional cooling. Vesicular clasts which represent the juvenile magma are present in the smaller grain sizes and show an inverse abundance relationship with the cognate lithics. The vesicularity of the clasts can vary within one sample from low to high. The density of andesite clasts from Lesser Antillean block and ash deposits varies from 1.2 to 2.4 g cm^{-3}.

The scoria and ash flow deposits and their associated surge deposits are distinct and usually composed of blue-black basaltic andesite. The clasts contain spherical and irregularly shaped vesicles. The largest vesicular clasts are rounded scoriaceous blocks with a ropey surface and a maximum diameter of 1 m. Nairn and Self (1978) report minor amounts of non-vesiculated cognate lithic blocks up to 1 m diameter from some scoria and ash flow deposits. The density range for basaltic andesite clasts from Lesser Antillean scoria and ash deposits is from 1.4 to 1.8 g cm^{-3}.

The pumice/ash flow deposits of various volumes and their associated surge deposits are all distinct because the highly vesiculated andesite (also dacite and rhyolite) clasts have a bright yellow or white colour. The larger clasts are usually rounded and the vesicles are often elongated. The density range for vesiculated andesite pumice clasts from the Lesser Antilles is from 0.7 to 1.1 g cm^{-3}. Ash flow deposits are similar to the ash/pumice types but lack the larger pumice clasts. Deposits of appearance and vesicularity intermediate between those of block and ash and ash/pumice flow types as well as mixed magma pyroclastic flow deposits have been reported (Smith and Roobol, 1979).

Although pyroclastic flows are poorly sorted, certain consistent variations in grain size have been observed. On the basis of these variations it is possible to construct standard pyroclastic flow units. Following the nomenclature of Sparks *et al.* (1973), a typical pyroclastic flow can be subdivided into a fine grained basal layer (layer 2a) and a coarser main unit (layer 2b). Beneath the pyroclastic flows produced by column collapse occur groundsurge deposits (layer 1). This relationship between surge and pyroclastic flow does not seem to apply to eruptions that produce *nuées ardentes*. For ignimbrites (including pumice and ash and scoria and ash deposits) layer 2b commonly shows coarse tail grading. (Coarse-tail grading is the type of grading found in high concentration sedimentary debris flow and pyroclastic flow deposits. In these deposits only the largest clasts of the grain size distribution show vertical variation in size. This is due to the fact that sorting in high concentration flows is controlled by the hindered settling velocities of the clasts. Only the largest clasts have sufficiently high velocities to become vertically sorted, thus producing coarse tail grading.) The pumice clasts are usually reversely graded while the lithic clasts are normally graded. It is thought that the density contrast between the individual clasts and the matrix of the flow (which is ungraded) is responsible for this feature (Sparks, 1976). In contrast, block and ash deposits generally show reversed grading of both their lithic and vesicular clasts, although examples of both ungraded and normally graded deposits have been observed.

Separating the main part of the flow from the ground surface is a fine grained basal layer (layer 2a) which tends to show an enrichment in vesicular fragments in the finer fractions

(A. L. Smith and M. J. Roobol, unpublished data). This layer is similar in its grain size characteristics to the matrix of the main part of the flow. What large clasts are present tend to be reversely graded with the result that the basal layer gradually passes upwards into the main flow unit as the amount of large clasts increases.

In a longitudinal sense pyroclastic flows appear to be a very homogeneous, with little change in the amount, size, and roundness of the vesicular clasts, although lithic clasts do seem to become smaller and less frequent away from the vent (Murai, 1961; Kuno et al., 1964; Sparks, 1976; Davies et al., 1978). Lateral variations in ignimbrite, especially those of large volume, also appear to be minor. However, deposits of nuées ardentes and scoria flows can show quite large variations, especially when not confined within valleys. The more important of these lateral variations appear to be related to the presence of channels and levées (Fig. 3). The former are composed predominantly of fine ash and blocks or bombs, while the latter show a concentration of blocks with very little fine ash and are generally better sorted (Nairn and Self, 1978; Dayies et al., 1978; A. L. Smith and M. J. Roobol, unpublished data).

Studies by a number of workers (Murai, 1961; Walker, 1971; Sheridan, 1971; Sparks, 1976) on non-welded, unlithified deposits have shown that granulometric analysis is a convenient way to describe pyroclastic flows and provides an important confirmation in distinguishing pyroclastic flows from falls (Walker, 1971; Fig. 2). Cumulative curves of grain-size distribution clearly illustrate the lack of sorting shown by pyroclastic flows. On the basis of the data available, Smith (1960a) concluded that most pyroclastic flows had a preponderance of material less than 2 mm. This observation still appears to hold for the large volume flows and the small volume relatively fine grained ash flows (ash hurricanes of Roobol and Smith, 1976); however, most deposits from nuées ardentes, scoria flows, and pumice flows tend to be much more coarsely grained.

Nearly all pyroclastic flow deposits are polymodal, with the number and position of the individual peaks depending on the relative proportions of the different compositional components that make up a pyroclastic flow. These components may be conveniently divided into three groups: (1) Vesicular essential clasts, which represent the juvenile magma—these clasts can show a complete range in degree of vesiculation (poorly vesicular to pumiceous) and in grain size (from bombs, up to a metre or so in diameter, to glass shards and dusts); (2) crystals, which can be regarded as forming a discrete juvenile component with their own behaviour pattern during transportation; (3) lithic clasts, which may be of two kinds, cognate lithics representing non-vesiculated juvenile material and accessory or accidental lithics which represent country rock explosively ejected during the eruption or picked up during transportation.

Representative frequency curves illustrating the distribution of the various components are shown in Fig. 4. From these curves it can be seen that the coarser fractions consist of vesicular essential clasts and lithics, the medium size fractions of crystals, vesicular materials, and lithics, while the finer fractions are composed essentially of crystals and glass shards. Nearly all these deposits exhibit a deficiency around the 1ϕ size range, which is thought to be the result of the weakness of fragments of this size, due to the relative large size of the component crystals compared to the size of the fragments, resulting in their rapid disintegration (Davies et al., 1978).

The medium size fractions, in all types of flow deposits, are dominated by the presence of crystals which usually show a relative enrichment when compared to the original magma due to the separation of the vitric-enriched ash cloud (Hay, 1959; Lipman, 1967; Walker, 1972; Sparks, 1976). In the finest fraction, shards produced by the break-up of the vesicular clasts usually become the dominant component (Sparks, 1976; Rose et al., 1976).

Other deposits which appear to be intimately associated with pyroclastic flows are co-ignimbrite lag-fall deposits and vitric-enriched air-fall ashes. The co-ignimbrite lag-fall deposits (Wright and Walker, 1977) are fall deposits restricted to

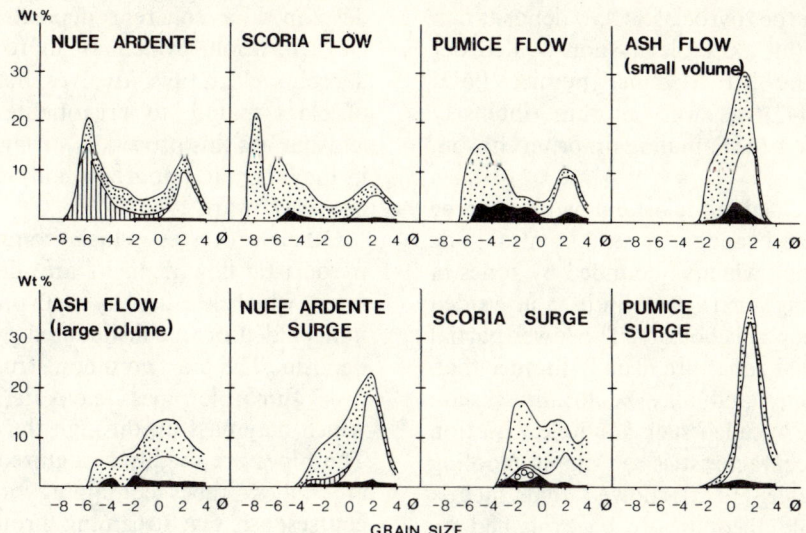

Fig. 4 Frequency curves showing the distribution of the various constituents for the different types of pyroclastic flows and surges. Symbols used are: ▥, cognate lithics; ▨, essential vesicular clasts; ■, accidental and accessory lithics; ☐, crystals; ▩, accretionary lapilli. All data, except ashflow (large volume) from Mt Pelée, Martinique (A. L. Smith and M. J. Roobol, unpublished data). Large volume ashflow represents Plateau Ignimbrite, Dodecanese Islands, Greece (Wright, 1979)

a near vent environment and are found to consist mainly of vesicular clasts and lithic fragments which appear to have been too large or too heavy to be transported by the associated pyroclastic flow. In the example described by Wright and Walker (1977) lithic fragments form the dominant component and these are much better sorted than the smaller pumice clasts.

Fine grained ashes (layer 3 of Sparks et al., 1973) produced by the fall-out from the over-riding cloud are probably associated with all pyroclastic flow eruptions, but are not always preserved (Sparks and Walker, 1978). When present they form thin, well bedded deposits which tend to mantle the topography. Their grain size distribution indicates that most of the clasts are finer than 125 μm and the deposits are generally unimodal, although a minor secondary mode sometimes can be found in the coarser size fractions. Compared to their associated pyroclastic flows, these air-fall ashes are highly enriched in the vitric component and depleted in the crystal and lithic contents (Sparks, 1976; Sparks and Walker, 1978; A. L. Smith and M. J. Roobol, unpublished data).

Post-depositional Processes

After emplacement certain changes may affect a pyroclastic flow deposit as a consequence of its high initial temperature. Some of these changes that may affect a flow during its cooling are compaction, welding, and recrystallization as a result of vapour phase and devitrification processes and chemical alteration. This aspect of pyroclastic flows has been particularly well studied (Smith, 1960a, b; Ross and Smith, 1961; Ragan and Sheridan, 1972; Riehle, 1973; Schmincke, 1974) and is only briefly reviewed here.

The major factors contributing to physical changes in a pyroclastic flow deposit are emplacement temperatures, lithostatic load (thickness), the composition of the residual volatiles, and the physical properties of the glass. Depending on the relationship between these different properties pyroclastic flow deposits can include both thick completely non-welded deposits, e.g. the Mt Mazama pumice flows (Williams, 1942), as well as thin, intensely welded flows, e.g. the Fantale ignimbrite (Gibson, 1970).

A typical welded pyroclastic flow unit can be subdivided into five main zones, a central zone of dense welding which is bounded by zones of partial welding that pass into non-welded zones at the top and bottom. The lower partial and non-welded zones are usually thinner than the upper zones and may be locally absent. Smith (1960a, b) called such a simple variation in welding characteristics a simple cooling unit and Riehle (1973) showed that such a variation could theoretically be explained by the cooling of a sheet emplaced at a uniform temperature. Other deposits which have several zones of dense and partial welding have been termed compound cooling units (Smith, 1960a, b). Such units are thought to represent deposits emplaced at slightly different times so that the underlying units had sufficient time to cool before the emplacement of the subsequent flows.

As a consequence of welding, the juvenile clasts become flattened and the deposit as a whole compacts. The maximum amount of compaction occurs within the zone of dense welding (Ragan and Sheridan, 1972). A measure of this compaction can be obtained by the vertical variations in bulk density and porosity within a flow unit. Within the welded zone the flattened juvenile clasts (fiamme) and glass shards define a planar foliation or eutaxitic texture. In some welded tuffs the fiamme appear to be stretched and sometimes show a well defined lineation indicating secondary mass flowage (rheomorphism). Rheomorphic welded tuffs may in some instances be confused with lava flows.

The percolation of heated groundwater or hot juvenile gases through a pyroclastic deposit leads to the deposition of cristobalite, tridymite, and alkali feldspars as cavity fillings. The result of this vapour phase crystallization is to reduce the pore space and produce, when well developed, a coherent deposit termed sillar. Devitrification, which is more prevalent in densely welded tuffs, involves the crystallization of glass mainly to cristobalite and alkaline feldspar. As this process is strongly exothermic, it may be an important aid in the welding process (Ewart, 1965).

Other processes which commonly modify pyroclastic flow deposits are: degassing structures, which are usually well preserved in the non-welded or unconsolidated pyroclastic flow deposits. The most common structure is that of fossil fumarole pipes—elongate, vertical tubes which may pass up through the entire deposit. The pipes are readily recognized because they are hollow tubes containing only lapilli and coarsest ash. Gas streaming through the deposit carries out the finer ash grade materials and leaves the pipe structures. They range in diameter from less than 1 cm to c. 20 cm. Local concentrations of pipe vesicles occur in areas where the flow was deposited on wet ground and are also found above and leading from carbonized logs within the flow deposit. These pipes can be useful in thick monotonous sequences for separating several flow units of one deposit (all penetrated by the same pipe vesicles) from other deposits composed of one flow unit (pipes truncated by overlying deposit). Surface mounds and ridges related to degassing have been described from the Bishop Tuff, California (Sheridan, 1970), and the Rio Caliente ignimbrite, Mexico (Wright, 1979).

Mechanical reworking of unconsolidated pyroclastic deposits occurs in most environments and is another important modifying process. On land the deposits are particularly prone to fluviatile erosion and reworking. Vertical walled gullies cut back into the youthful deposits and can entirely rework the deposit. The Pleistocene core of Mt Pelée, Martinique, is composed almost entirely of reworked fluviatile block and ash and pumiceous materials (Roobol and Smith, 1976). The stratigraphy of alternating block and ash and pumice deposits is preserved as a series of alternations of dense andesite boulder beds, alternating with yellow, weathered fluviatile ash and pumice beds.

Where pyroclastic materials are deposited in the sea they may be mechanically sorted into conglomerate, pumice-, crystal-, lithic, and mixed tuffs (Roobol, 1976). Alternatively, the pyroclastic material may be ponded along shorelines and slump into deeper water as ash turbidites (Carey and Sigurdsson, 1978).

Mechanisms of Transportation

One mechanism capable of generating pyroclastic flows is the collapse of an active dome or lava flow. Such flows may be generated either by gravitational collapse of an unstable actively growing dome or flow front or as the result of an initial explosion associated with dome growth (Fig. 5(a)). In both cases the dominant driving force of the flow appears to be gravity, although explosively generated flows are thought to have, at least initially, a pronounced lateral component, as was noted by LaCroix (1904) and MacDonald and Alcaraz (1956) (Fig. 6).

A second mechanism is the gravitational collapse of an overloaded eruptive column. This has been observed to generate small pyroclastic flows (Moore and Melson, 1969; Nairn and Self, 1978) and it is thought that this mechanism can also be applied to large volume flows (Smith, 1960a; Sparks and Wilson, 1976; Sparks et al., 1978). The small volume flows can be regarded as the products of discrete explosions (Nairn and Self, 1978) producing intermittent collapse, whereas the larger volume ones are thought to be produced by continuous column collapse (Fig. 5(b), (c)). In both cases the generation of a pyroclastic flow from a vertically rising eruptive column is due to an increase in the effective density of the column so that it exceeds that of the surrounding atmosphere. This critical density depends on a combination of gas content, vent radius, and gas velocity. Once collapse of the column begins, it has been theoretically calculated (Sparks et al., 1978) that reinstatement of vertical activity would require a major change in eruptive conditions, so that once started a pyroclastic flow should continue to form until the eruption has finished. Some eruptions appear to show a sequence of events from an initial plinian phase through a pyroclastic surge and flow phase to a final effusive phase. Such a sequence is thought to represent the tapping of progressively deeper levels of a magma chamber and the subsequent eruption of magma with lower gas contents (Sparks et al., 1973). As a consequence of such a progressive decrease in gas content the initial plinian eruptive column would tend to become overloaded, resulting in collapse and the generation of pyroclastic flows. Erosion and widening of the vent during the eruption would also drive the condition of the column towards collapse (Sparks et. al., 1978).

On the basis of the poor sorting of the deposits it has been suggested (Smith, 1960a; Ross and Smith, 1961; Murai, 1961; Fisher, 1966; Koch and McLean, 1975; Sheridan and Ragan, 1976) that pyroclastic flows are turbulent and in some

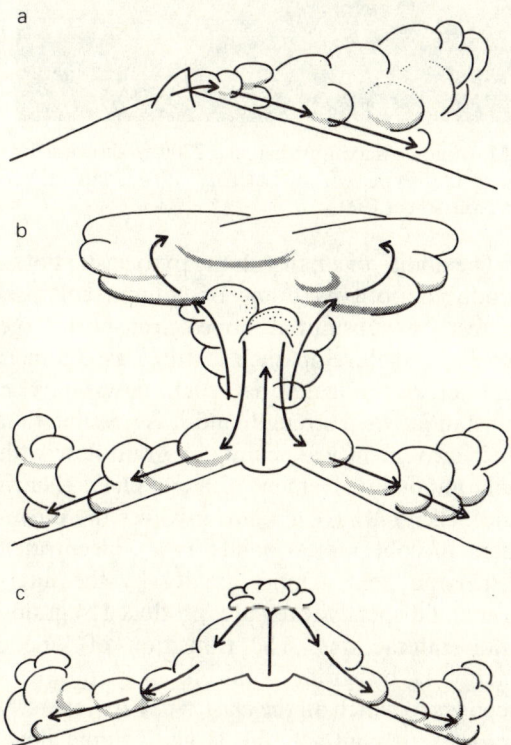

Fig. 5 Diagrammatic representation of the different mechanisms of formation of pyroclastic flows. (a) Dome/lava flow collapse; (b) interrupted column collapse; (c) continuous column collapse

Fig. 6 *Nuée ardente* of 25 January 1903 eruption of Mt Pelée, Martinique (maximum height *c.* 2300 m) showing the pronounced initial lateral component. d represents position of the dome. (From LaCroix, 1904, Plate XIV; reproduced by permission of Masson et Cie)

instances highly expanded (Aramaki and Ui, 1966; Sheridan and Ragan, 1976). As a result of detailed studies by Sparks (1976) and Sparks *et al.* (1978), it appears that, although pyroclastic flows may move as turbulent suspensions initially, they will quickly become segregated into a dense lower zone of high particle concentration (the underflow) and a dilute turbulent upper zone (the associated gas–ash cloud). Once the underflow becomes established it probably travels either by laminar and/or plug flow, i.e. in a similar manner to sedimentary debris flows. The sorting within the underflow can be attributed solely to the particle concentration, since within a high particle dispersion sorting is controlled by hindered settling velocities. In such an environment only the larger particles would have sufficient velocities to move through the flow with the result that the flows develop coarse tail grading.

The ability of small volume pyroclastic flows, produced both by dome or column collapse, to rest on substantial slopes, transport large blocks, and develop such features as channels and *levées*, indicate that such flows have a substantial yield strength and it is possible that their movement was mainly by grain flow with only minor involvement of a gas phase (Nairn and Self, 1978). (*Grain flow* involves the movement of cohesionless solids as a concentrated dispersion held against gravity by the intergranular dispersive pressures produced by grain–grain interactions. The transition of kinetic energy is by grain collision. This produces deposits in which all the clast types are reversely graded.) In contrast, the larger volume flows generally do not show the features mentioned above. Their greater amounts of vesiculated material and the relative movement of their larger clasts suggest instead that they possess low yield strengths of the order of 10^1–10^3 P. The vent conditions necessary to erupt large volumes of magma rapidly can also account for their high velocities, large distances of travel, and ability to surmount substantial topographic obstacles (Sparks et al., 1978).

Many of the features of the large volume flows also seem to suggest that they were partially fluidized (Sparks, 1976; Sparks et al., 1978). (*Fluidization* is defined as condition where a fluid moving vertically through a particle aggregate exerts an upward drag on the individual particles that is comparable with the gravitational forces.) Sparks (1978) suggested that in medium and large volume flows the loss of residual gas from vesicular particles would be sufficient to fluidize the fine and medium ash size particles. The larger clasts are transported by this fluidized matrix.

Summary

The generation of pyroclastic flows represents perhaps the most important form of volcanic activity found in both continental and island arc settings, and can be regarded as the characteristic type of volcanism occurring at destructive plate margins.

Although different types of pyroclastic flows has been discussed in the preceding sections, it should be stressed that there is a complete gradational relationship from cold avalanches to pyroclastic flows, from gravitational dome collapse to explosive dome collapse to small volume, from interrupted column collapse to large volume continuous column collapse. The mechanism of transportation also varies continuously from grain flow associated with dome collapse to partial fluidization in the large volume pumice–ash flows as well as from laminar flow in the underflow to turbulent flow in the ash cloud.

Many of the characteristics of pyroclastic flows and their resulting deposits appear to be controlled by the conditions in the conduit and vent at the time of eruption. It has been suggested, (for example: Murai, 1961; Aramaki and Yamasaki, 1963) that the site of vesiculation plays an important role. Vesiculation within the upper levels of the magma chamber produces large volumes of low density material giving rise to plinian eruptions and the generation of ash–pumice flows. On the other hand, if the main period of vesiculation takes place at the head of the magma column, or after eruption, then the resulting pyroclastic flows would tend to be of small volume with relatively dense fragments. Similarly, the height of the eruptive column and the degree of turbulence in the conduit have been suggested as reasons why ground surge deposits are associated with pyroclastic flows produced by collapse of plinian eruptive columns while they are absent from *nuée ardente* sequences (R. V. Fisher, A. L. Smith, and M. J. Roobol, unpublished data).

In addition to variations in volume, clast vesicularity, and mechanism of formation and transportation, each pyroclastic flow undergoes a mechanical differentiation into an underflow and an accompanying ash cloud. Recent work on the 1902 deposits of Mt Pelée have suggested that the ash cloud may undergo further differentiation to produce a secondary 'underflow' with its accompanying ash cloud.

REFERENCES

Anderson, T. and Flett, J. S. (1903). Report on the eruptions of the Soufriére in St. Vincent, and on a visit to Montagne Pelée, in Martinique. *Phil. Trans. R. Soc. A* **200**, 353–553.

Aramaki, S. (1956). The 1783 activity of Asama Volcano. Part I. *Jap. J. Geol. Geogr.* **27**, 189–229.

Aramaki, S. (1957). The 1783 activity of Asama Volcano. Part II. *Jap. J. Geol. Geogr.* **28**, 11–33.

Aramaki, S. and Ui, T. (1966). The Aira and Ata pyroclastic flows and related caldera depressions in southern Kyushu, Japan. *Bull. Volc.* **29**, 29–47.

Aramaki, S. and Yamasaki, M. (1963). Pyroclastic flows in Japan. *Bull. Volc.* **26**, 89–99.

Baker, B. H., Mohr, P. A., and Williams, L. A. J. (1972). Geology of the Eastern Rift System of Africa. *Geol. Soc. Am. spec. Paper* no. 136.

Baker, P. E. (1968). Comparative volcanology and petrology of the Atlantic island arcs. *Bull. Volc.* **32**, 189–206.

Bemmelen, R. W. van (1949). The Geology of Indonesia. In *General Geology of Indonesia and Adjacent Archipelagoes*, Vol. 1A, US Government Printing Office, The Hague.

Byers, F. M., Jr, Carr, W. J., Orkild, P. P., Quinlivan, W. D., and Sargent, K. A. (1976). Volcanic suites and related cauldrons of Timber Mountain–Oasis Valley caldera complex, southern Nevada. *US geol. Surv. prof. Paper* no. 919.

Carey, S. N. and Sigurdsson, H. (1978). Deep-sea evidence for distribution of tephra from the mixed magma eruption of the Soufrière on St Vincent 1902: ash turbidites and airfall. *Geology* **6**, 271–274.

Curtis, G. H. (1968). The stratigraphy of the ejecta from the 1912 eruption of Mount Katmai and Novarupta, Alaska. *Geol. Soc. Am. Mem.* no. 116, 153–210.

Davies, D. K., Quearry, M. W., and Bonis, S. B. (1978). Glowing avalanches from the 1974 eruption of the volcano Fuego, Guatemala. *Geol. Soc. Am. Bull.* **89**, 369–384.

Escher, B. G. (1933). On a classification of central eruptions according to gas pressure of the magma and viscosity of the lava. *Leidsche Geol. Mededeel.* **6**, 45–49.

Ewart, A. (1965). Mineralogy and petrogenesis of the Whakamaru Ignimbrite in the Maraetai area of the Taupo Volcanic Zone, New Zealand. *N.Z. J. Geol. Geophys.* **8**, 611–677.

Ewart, A., Carmichael, I. S. E., Brown, F. H., and Green, D. C., (1971). Voluminous low temperature rhyolite magmas in New Zealand. *Contr. Mineral. Petrol.* **33**, 128–144.

Fenner, C. N. (1920). The Katmai Region, Alaska, and the Great Eruption of 1912. *J. Geol.* **28**, 556–606.

Fenner, C. N. (1937). Tuffs and other volcanic deposits of Katmai and Yellowstone Park. *Trans. Am. geophys. Union, 18th a. Mtg*, 236–239.

Fisher, R. V. (1966). Geology of a Miocene ignimbrite layer, John Day Formation, eastern Oregon. *Calif. Univ. Publ. Dept. geol. Sci. Bull.* no. 67.

Fisher, R. V. (1976). The mobility of hot pyroclastic flows. *Geol. Soc. Am., Rocky Mtn Section, Abstr. Program* 586.

Fisher, R. V. (1979). Models for pyroclastic surges and pyroclastic flows. *J. Volc. geothermal. Res.* in press.

Francis, P. W. and Baker, M. C. W. (1977). Mobility of pyroclastic flows. *Nature, Lond.* **270**, 164–165.

Francis, P. W., Roobol, M. J., Walker, G. P. L., Cobbold, P. R., and Coward, M. (1974). The San Pedro and San Pablo volcanoes of northern Chile and their hot avalanche deposits. *Geol. Rundschau* **63**, 357–388.

Gibson, I. L. (1970). A pantelleritic welded ash-flow tuff from the Ethiopian Rift Valley. *Contr. Mineral. Petrol.* **28**, 89–111.

Gilbert, C. M. (1938). Welded tuff in Eastern California. *Geol. Soc. Am. Bull.* **49**, 1829–1862.

Gorshkov, G. S. (1959). Gigantic eruption of Bezymianny volcano. *Bull. Volc.* **20**, 77–109.

Gorshkov, G. S. and Dubik, Y. M. (1970). Gigantic directed blast at Shiveluch Volcano (Kamchatka). *Bull. Volc.* **34**, 261–288.

Hay, R. L. (1959). Formation of the crystal rich glowing avalanche deposits of St Vincent, B.W.I. *J. Geol.* **67**, 540–562.

Heming, R. F. and Carmichael, I. S. E. (1973). High temperature pumice flows from the Rabaul Caldera, Papua, New Guinea. *Contr. Mineral. Petrol.* **38**, 1–20.

Koch, A. J. and McLean, H. (1975). Pleistocene tephra and ash-flow deposits in the volcanic highlands of Guatemala. *Geol. Soc. Am. Bull.* **86**, 529–541.

Kozu, S. (1934). The great activity of Komagatake in 1929. *Mineral. Petrol. Mitt.* **45**, 133–174.

Kuno, H. (1950). Geology of Hakone Volcano and adjacent areas, Part I. *Tokyo Univ. Fac. Sci. J.* **7**, 257–279.

Kuno, H., Ishikawa, T., Katsui, Y., Yamasaki, M., and Taneda, S. (1964). Sorting of pumice and lithic fragments as a key to eruptive and emplacement mechanism. *Jap. J. Geol. Geogr.* **35**, 223–238.

LaCroix, A. (1904). *La Montagne Pelée et ses éruptions*, Masson et Cie, Paris.

LaCroix, A. (1908). *La Montagne Pelée apres ses éruptions*. Masson et Cie, Paris.

LaCroix, A. (1930). Remarques sur les matériaux de projection des volcano et sur la genese des roches pyroclastiques qu'ils constituent. In *Livre Jubilaire 1830–1930, Centenaire de la Société Geologique de la France* Vol. II, pp. 431–472.

Lipman, P. W. (1967). Mineral and chemical variation within an ash-flow sheet from Aso Caldera, SW Japan. *Contr. Mineral. Petrol.* **16**, 300–327.

Lipman, P. W. (1971). Iron–titanium oxide phenocrysts in compositionally zoned ash-flow sheets from Southern Nevada. *J. Geol.* **79**, 438–456.

Lipman, P. W., Christiansen, R. L., and O'Conner, J. T. (1966). A compositionally zoned ash-flow sheet in southern Nevada. *US geol. Surv. prof. Pap.* no. 524–F.

MacDonald, G. A. (1971). *Volcanoes*, Prentice-Hall, Englewood Cliffs, N.J.

MacDonald, G. A. and Alcaraz, A. (1956). *Nuée ardentes* of the 1948–1953 eruption of Hibok-Hibok. *Bull. Volc.* **18**, 169–178.

MacGregor, A. G. (1952). Eruptive mechanisms: Mt Pelée, the Soufrière of St Vincent and the Valley of Ten Thousand Smokes. *Bull. Volc.* **12**, 49–74.

Marshall, P. (1935). Acid rocks of the Taupo Rotorua volcanic district. *Trans. R. Soc. N.Z.* **64**, 323–366.

Matumoto, T. (1943). The four gigantic caldera volcanoes of Kyushu. *Jap. J. Geol. Geogr.* **19**, 57p.

Melson, W. G. and Saenz, R. (1973). Volume, energy and cyclicity of eruptions of Arenal Volcano, Costa Rica. *Bull. Volc.* **37**, 416–437.

Miller, T. P. and Smith, R. L. (1977). Spectacular mobility of ash flows around Aniakchak and Fisher calderas, Alaska. *Geology* **5**, 173–176.

Moore, B. N. (1934). Deposits of possible *nuée ardente* origin in the Crater Lake region, Oregon. *J. Geol.* **42**, 358–375.

Moore, J. G. and Melson, W. G. (1969). Nuées ardentes of the 1968 eruption of Mayon Volcano, Phillipines. *Bull. Volc.* **33**, 600–620.

Murai, I. (1961). A study of the textural characteristics of pyroclastic flow deposits in Japan. *Earthquake Research Inst. Bull.* **39**, 133–248.

Nairn, I. A. and Self, S. (1978). Explosive eruptions and pyroclastic avalanches from Ngauruhoe in February 1975. *J. Volc. geothermal Res.* **3**, 39–60.

Perret, F. A. (1937). *The Eruption of Mt Pelée 1929–32*, Carnegie Institution of Washington, Washington, D.C.

Ragan, D. M. and Sheridan, M. F. (1972). Compaction of the Bishop Tuff, California. *Geol. Soc. Am. Bull.* **83**, 95–106.

Riehle, J. R. (1973). Calculated compaction profiles of rhyolitic ash flow tuffs. *Geol. Soc. Am. Bull.* **84**, 2193–2216.

Robson, G. R. and Tomblin, J. F. (1966). *Catalogue of the Active Volcanoes of the World*, Part XX, *West Indies*, International Association for Volcanology, Naples.

Roobol, M. J. (1976). Post-eruptive, mechanical sorting of pyroclastic material an example from Jamaica. *Geol. Mag.* **113**, 429-440.

Roobol, M. J. and Smith, A. L. (1975). A comparison of the recent eruptions of Soufrière, St Vincent and Mt Pelée, Martinique. *Bull. Volc.* **39**, 214-240.

Roobol, M. J. and Smith, A. L. (1976). Mt Pelée, Martinique: a pattern of alternating eruptive styles. *Geology* **4**, 521-524.

Roobol, M. J. and Smith, A. L. (1980). Pumice eruptions of the Lesser Antilles. *Bull. Volc.* **43**, 277-286.

Rose, W. I., Jr, Pearson, T., and Bonis, S. (1976). Nuée ardente eruption from the foot of a dacite lava flow, Santiaguito Volcano, Guatemala. *Bull. Volc.* **40**, 1-16.

Ross, C. S. and Smith, R. L. (1961). Ash-flow tuffs, their origin, geologic relations and identification. *US geol. Surv. Bull.* no. 366.

Schminke, H.-U. (1974). Volcanological aspects of peralkaline silicic welded ash-flow tuffs. *Bull. Volc.* **38**, 594-636.

Schminke, H.-U. and Swanson, D. A. (1967). Laminar viscous flowage structures in ash flow tuffs from Gran Canara, Canary Islands. *J. Geol.* **75**, 641-664.

Sheridan, M. F. (1970). Fumarolic mounds and ridges of the Bishop Tuff, California. *Geol. Soc. Am. Bull.* **81**, 851-868.

Sheridan, M. F. (1971). Particle size characteristics of pyroclastic tuffs. *J. geophys. Res.* **76**, 5627-5634.

Sheridan, M. F. and Ragan, D. M. (1976). Compaction of ash-flow tuffs. In *Compaction of Coarse-grained Sediments, Developments in Sedimentology*, Vol. 18B (G. V. Chiligaran and K. H. Wolf, eds), Elsevier, New York, pp. 677-713.

Sigurdsson, H. (1972). Partly welded pyroclastic flow deposits in Dominica, Lesser Antilles. *Bull. Volc.* **36**, 148-163.

Smith, A. L. and Roobol, M. J. (1979). Characteristics of pyroclastic flows and surges from the Lesser Antilles (abstract). Pacific North-west Meeting of the American Geophysical Union.

Smith, R. L. (1960a). Ash flows. *Geol. Soc. Am. Bull.* **71**, 795-842.

Smith, R. L. (1960b). Zones and zonal variations in welded ash-flows. *US geol. Surv. prof. Pap.* no. 345F, 149-159.

Smith, R. L. and Bailey, R. A. (1966). The Bandelier Tuff, a study of ash-flow eruption cycles from zoned magma chamber. *Bull. Volc.* **29**, 83-104.

Sparks, R. S. J. (1976). Grain size in ignimbrites and implications for the transport of pyroclastic flows. *Sedimentology* **23**, 147-188.

Sparks, R. S. J. (1978). Gas release rates from pyroclastic flows: an assessment of the role of fluidization in their emplacement. *Bull. Volc.*, **41**, 1-9.

Sparks, R. S. J. and Walker, G. P. L. (1973). The ground surge deposit: a third type of pyroclastic rock. *Nature phys. Sci.* **241**, 62-64.

Sparks, R. S. J. and Walker, G. P. L. (1978). The significance of vitric enriched air-fall ashes associated with crystal-enriched ignimbrites. *J. Volc. geothermal. Res.* **2**, 329-341.

Sparks, R. S. J. and Wilson, L. (1976). A model for the formation of ignimbrite by gravitational column collapse. *J. geol. Soc. Lond.* **132**, 441-451.

Sparks, R. S. J., Self, S., and Walker, G. P. L. (1973). Products of ignimbrite eruptions. *Geology* **1**, 115-118.

Sparks, R. S. J., Wilson, L., and Hulme, G. (1978). Theoretical modeling of the generation, movement and emplacement of pyroclastic flows by column collapse. *J. geophys. Res.* **83**, 1727-1739.

Steven, T. A. and Lipman, P. W. (1976). Calderas of the San Juan volcanic field, southwestern Colorado. *US geol. Survey prof. Pap.* no. 958.

Taylor, G. A. (1958). The 1951 eruption of Mt Lamington, Papua. *Aust. Dep. natn. Devlmt. Bull.* no. 38, 117p.

Walker, G. P. L. (1962). Tertiary welded tuffs in eastern Iceland. *Q. Jl geol. Soc. Lond.* **118**, 275-293.

Walker, G. P. L. (1971). Grain size characteristics of pyroclastic deposits. *J. Geol.* **79**, 696-714.

Walker, G. P. L. (1972). Crystal concentration in ignimbrites. *Contr. Mineral. Petrol.* **36**, 136-146.

Williams, H. (1941). Calderas and their origin. *Calif. Univ. Publ. Dept geol. Sci. Bull.* no. 25, 239-346.

Williams, H. (1942). The geology of Crater Lake National Park, Oregon, with a reconnaisance of the Cascade Range southward to Mount Shasta. *Carnegie Instn. Wash. Publ.* no. 549.

Williams, H. (1957). Glowing avalanche deposits of the Sudbury Basin. *Ontario Dept Mines 65th Ann. Dept.* no. 65, 57-89.

Wolf, T. (1878). Der Cotopaxi and seine letzte Eruption an 26 Juni 1877. *Neues Jb. Mineral. Geol. Palanlol.*, 113-167.

Wright, J. V. (1979). Formation, transport and deposition of ignimbrites and welded tuffs. Ph.D. dissertation, University of London.

Wright, J. V., Self, S., and Fisher, R. V. (1979). Eruption sequence of the Bandelier Tuffs, Jemez Volcano, New Mexico (abstract). Pacific North-West Meeting, American Geophysical Union.

Wright, J. V. and Walker, G. P. L. (1977). The ignimbrite source problem: significance of a co-ignimbrite lag-fall deposit. *Geology* **5**, 729-732.

Yokoyama, S. (1974). Flow and emplacement mechanism of the Ito pyroclastic flow in southern Kyushu, Japan. *Sci. Rep. Tokyo Kyoiku C.* **12**, 17-62.

Young, R. A. and Brennan, W. J. (1974). Peach Spring Tuff. Its bearing on structural evolution of the Colorado Plateau and development of Cenozoic drainage in Mohave County, Arizona. *Geol. Soc. Am. Bull.* **85**, 83-90.

Zeil, W. and Pichler, H. (1967). Die känozoische rhyolith formation in mittleren Abschnitt der Anden. *Geol. Rundschau* **57**, 48-81.

Zen, M. T. and Hadikusumo, D. (1964). Preliminary report of the 1963 eruption of Mt. Agung in Bali (Indonesia). *Bull. Volc.* **27**, 269-299.

VI The relationship of andesitic volcanism to intrusive activity

The volcanic andesite–dacite–rhyolite association is similar in mineralogical and chemical composition to the intrusive tonalite–granodiorite–granite ('granitoid') association of orogenic belts. These similarities and the close spatial and temporal links between the volcanic and intrusive rocks in many orogenic belts leaves little doubt that the associations share a common origin. Therefore, although the title of this volume refers to volcanic rocks, it is important to consider the characteristics of granitoids and their relationships to calc-alkaline volcanic rocks. In this contribution, G. C. Brown reviews the occurrence of intrusive rocks in island arcs, continental margins, and continental collision zones. He emphasizes that in island arcs both volcanic and intrusive rocks must be largely mantle-derived but with a significant crustal component in continental margin associations, in contrast with continental collision zone granitoids which appear to result largely or entirely from melting of crustal rocks. It is concluded that calc-alkaline granitoids have formed throughout the Earth's history and, together with the volcanic association, have been the major contributors to continental growth throughout geological time.

Andesites
Edited by R. S. Thorpe
© 1982 John Wiley & Sons

Calc-alkaline intrusive rocks: their diversity, evolution, and relation to volcanic arcs

G. C. Brown

Department of Earth Sciences,
The Open University, Milton Keynes MK7 6AA, UK

ABSTRACT

Calc-alkaline granitoids from magmatic arcs have geochemical characteristics, show compositional trends with time, and have petrogenetic features that vary in relation to arc maturity (as defined by crustal thickness), the presence or absence of ancient sialic basement, and are paralleled by the characteristics of associated volcanic rocks. There are general evolutionary trends with time from the calcic diorite–monzonite stocks of immature island arcs to variable, but generally calc-alkaline tonalite–granodiorite batholiths of mature continental arcs. Accepting that there is a petrogenetic relationship between the intrusive and extrusive associations, it is proposed that the magmas are mainly derived from the subcontinental lithosphere and may either freeze at depth or be erupted; the former is favoured by a large thickness of continental crust, but a given magma may form both intrusive and extrusive products.

In addition to the calc-alkaline granitoids of magmatic arcs, two other types of granitoid, mainly of alkali-calcic composition, also occur in young crustal settings. The first type occurs in compressional arc–continent or continent–continent collision zones, is characterized by silicic compositions and small compositional variation, and is believed to result from crustal melting. The second type occurs in zones of crustal extension, is associated with rocks of basic to acid composition, and is considered to have a deep mantle source.

There are long term changes in the chemical characteristics of calc-alkaline magmas throughout the Earth's history, from early Precambrian arcs dominated by diorite–trondhjemite series to the varied calcic and calc-alkaline associations of modern arcs. These long term changes in chemical characteristics are reminiscent of the short term (c. 10^8 a) trends within modern magmatic arcs. The long term trend might reflect a decreasing involvement of subducted oceanic lithosphere and increasing involvement of the mantle wedge in the melting process. Finally, there is good evidence that the ratio of calc-alkaline (dominant) to alkali-calcic granitoids has decreased during geological history reflecting changes in the relative movements and positions of continental lithospheric fragments, but calc-alkaline batholith formation has been the dominant process in crustal growth.

Introduction

Although orogenic andesites and related volcanic rocks, forming over regions of oceanic plate subduction, are the principal subject of this volume, it is well known that intrusive suites also characterize many Mesozoic and Cainozoic orogenic belts. The cessation of active volcanism, followed by uplift and erosion, exposes intrusive rocks which are assumed, therefore, to underlie active volcanic chains. The volcanic rocks range in composition from basalt through basaltic andesite, andesite, and dacite to rhyolite (*see* Sections II–IV) and this is paralleled by intrusive rocks ranging from gabbro through diorite, tonalite, and granodiorite to true granite in composition. Collectively, the intrusive rocks of intermediate to acidic composition are known as granitoids. At many ocean–continent convergence zones these granitoids form huge linear batholiths with volumes which may exceed 10^6 km^3. These are enormous volumes and in the most extensive batholith developments of the central Andes and California their volume must exceed that of lavas and other eruptive rocks by an order of magnitude (Baker and Francis, 1978). The close spatial association of the intrusive and extrusive igneous suites from modern continental arcs and their similar compositional range leads naturally to the view that they are genetically related (for reviews see Ringwood, 1974; Brown, 1977; Thorpe and Francis, 1979a). Moreover, there is a great difficulty in finding a source for the large volumes of intrusive magmas emplaced within the continental crust unless their source, like that of volcanic rocks, lies dominantly within the subcrustal lithosphere. As it is impossible to divorce the intrusive and volcanic products of continental magmatic arcs, this contribution examines the link between these two groups of rocks and the role of granitoids in continental evolution.

Although studies of plutonic rocks of Mesozoic and Cainozoic arcs have most commonly referred to the western Americas, crystalline plutonic rocks are also exposed in the island arcs of the western Pacific, Indonesia, the Aleutians, and the Caribbean. Unlike large parts of western America, many of these arcs do not contain any ancient (Precambrian) continental basement, but they evolve by the progressive thickening of a volcanic pile founded on oceanic crust until the eruptive pile is thick enough to support plutonism. Eventually such island arc zones may become 'continentalized' and during the Phanerozoic, the New Zealand, Japanese, and Central American (Nicaragua–Panama) areas have all evolved to this intermediate stage between immature island arcs and the relatively mature continental margin arcs characterized by plutonic rock series emplaced within Precambrian continental crust. With increasing maturity, the volume ratio of intrusive to extrusive igneous rocks seems to increase, although the picture is complicated because our view of all these arcs is essentially two-dimensional.

These three types of magmatic arc were termed 'intra-oceanic', 'continental island', and 'western American' arcs by Ewart (1976, and this volume, Section II). On the basis of geochemical and isotopic studies of volcanic and intrusive rocks it is appropriate to subdivide the western American arcs into those with Precambrian basement (western USA, Peru, and northern Chile) and those without (British Columbia, Central America (Nicaragua–Panama), Ecuador, and southern Chile).

At the other end of the evolutionary scale to intra-oceanic island arcs there are continent–continent convergence zones such as the Alps, Himalayas, and south-eastern Asia which contain granitoids, with variable and generally smaller associated volumes of volcanic products. These intra-continental 'arcs' are considered to be the most mature convergent boundary zones of magma generation.

In recent years, much emphasis has been placed on the distinction of I-type and S-type granitoids from different tectonic environments (see Mitchell and Beckinsale, this volume, Section X, for a review of Chappell and White's (1974) definitions). This terminology will not be adopted here as the present author finds the genetic implications over-simplified and restrictive (see Plant et al., 1980, for a full critique).

This contribution describes the geological and chemical characteristics of examples of modern arc-type plutonism and volcanism and reviews granite magmatism and magmagenesis from the geological record (cf. Brown, 1981). The conclusion is essentially uniformitarian: although the rates and style of crustal processes have varied in the past, arc-type magmas have contributed to the evolution and growth of the continental crust throughout the Earth's history. First, however, there follows a discussion of igneous rock nomenclature that is relevant to a review of granitoid evolution.

Petrochemical Variations and the Nomenclature of Magma Series

The characteristic igneous rocks from ocean–continent and ocean–ocean convergence zones are calc-alkaline associations (see Baker, this volume, Section II, for a review of terminology).

The original definition of this term came from Peacock's (1931) alkali-lime index: the percentage of SiO_2 by weight in the chemical analysis of a continuous, related rock series at which the abundances of CaO and $Na_2O + K_2O$ are equal. Calc-alkaline rocks were distinguished as those for which this index (given as percentage of SiO_2 by weight) lies between 56 and 61; suites with an index greater than 61 are calcic, those with an index less than 56 are alkali-calcic (51–56) or alkalic (less than 51). As will be shown later, intrusive rocks from different tectonic settings have characteristic alkali-lime indices.

In contemporary petrogenetic studies, the AFM plot has become widely adopted as a discriminant diagram for igneous rocks. Figure 1 is an AFM plot incorporating data for volcanic rocks from modern magmatic arcs: each lava series plots along its own smooth curve, with progressive enrichment of alkaline elements towards the more silica-rich members. The cause of the variations between curves is generally believed to be the fractional crystallization of different Fe–Mg-rich minerals, such as olivines, pyroxenes, and amphiboles, from parental magmas at the basic end of the series. The most primitive island arc magmas show tholeiitic Fe-enrichment trends (Fig. 1) and are *calcic* in terms of the Peacock index. These lavas generally have higher K_2O and SiO_2 contents and higher Fe/Mg ratios than ocean-floor tholeiites, and

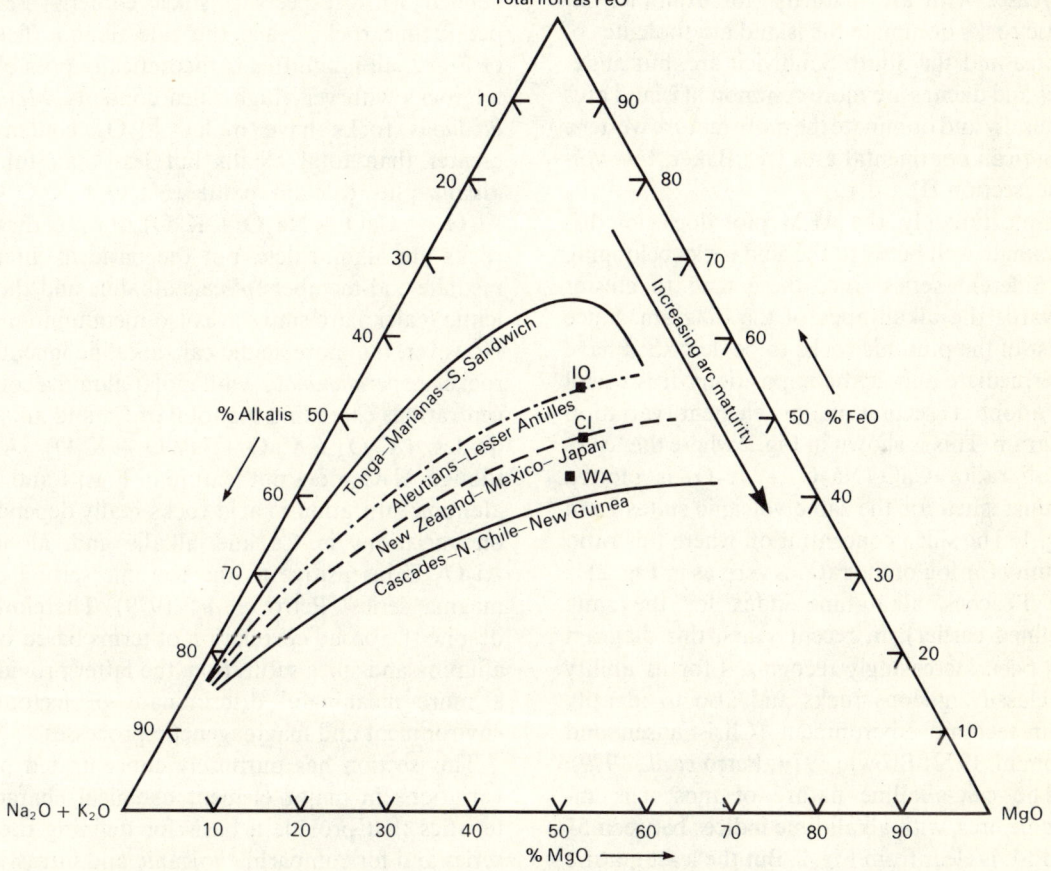

Fig. 1 AFM diagram for volcanic suites from modern magmatic arcs showing the differences due to arc maturity. Also shown are Ewart's (1976) averages for western American volcanic arcs (WA), continental island arcs (CI), and intra-oceanic island arcs (IO). (Other data from Baker, 1968; Carmichael, *et al.*, 1974; and Ringwood, 1974, supplemented from Scheidegger and Kulm, 1975 (Aleutians); Thorpe and Francis, 1975 (Mexico); Stern, 1979 (Marianas); and Thorpe and Francis, 1979b (northern Chile))

are termed 'island arc' tholeiites (cf. Jakeš and Gill, 1970). Other volcanic arc magmas lack strong Fe enrichment on the AFM plot (cf. Larsen, 1948), have calc-alkaline trends, and, for this reason, are termed calc-alkaline, regardless of the original definition of the term. However, some of them are calcic according to Peacock's (1931) index, particularly those close to the boundary with calcic series (tholeiites), and others are alkali-calcic. The AFM diagram is most useful because it provides a crude pictorial definition of arc maturity; there exists a family of curves for arc magmas ranging from the island arc tholeiites to the mature calc-alkaline series of continental margins (Fig. 1 and caption). The volume ratio of basic to intermediate and acidic members of each series also increases with arc maturity: for example, basaltic rocks dominate the island arc tholeiites of Tonga and the South Sandwich arc, but andesites and dacites are more common in island arcs generally and dominate the more mature western American continental arcs (see Baker, this volume, section II).

Unfortunately, the AFM plot does not discriminate well between the acid rocks belonging to different series since these tend to cluster towards the alkali apex of the diagram. Since most of the plutonic rocks to be described have intermediate and acid compositions, it is useful to adopt a second major element variation diagram. This is shown in Fig. 2 where the 'calc-alkali ratio' $CaO/(Na_2O + K_2O)$ is plotted against silica for the same volcanic suites as in Fig. 1. The silica concentration where this ratio is unity (or log of the ratio is zero as in Fig. 2) is the Peacock alkali-lime index for the suite (defined earlier). In recent years, this diagram has been increasingly recognized for its ability to classify igneous rocks and also to identify their tectonic environment (Christiansen and Lipman, 1972; Brown, 1979; Petro et al., 1979).

The calc-alkaline nature of most circum-Pacific arcs, with alkali-lime indices between 57 and 61, is clear from Fig. 2. But the least mature island arcs with tholeiitic products, and the New Zealand Taupo volcanic zone (Ewart and Stipp, 1968), are calcic rocks; Ewart's (1976) average intra-oceanic island arc analysis is also calcic. At the opposite extreme, the volcanic suite of south-eastern New Guinea is an alkali-calcic suite; this is the high K shoshonite group of Jakeš and Smith (1970) and is just one of the many varied magma types from this complex region (Mason and McDonald, 1978; Johnson, this volume, Section III).

Before leaving this topic, chemical nomenclature based on alumina rather than silica saturation should be mentioned because mixed terms frequently occur in the literature on intrusive rock series. First, there are *peralkaline* rocks in which the *molar proportions* of the combined alkali oxides exceed alumina ($Na_2O + K_2O > Al_2O_3$). Most of these rocks fall into the alkaline and alkali-calcic fields already defined with respect to silica content. Few peralkaline rocks reach the calc-alkaline field of Fig. 2, although this is theoretically possible for rocks with very high silica contents. *Metaluminous* rocks have molar Al_2O_3 contents greater than total alkalis but less than total alkalis plus calcium oxide ($Na_2O + K_2O < Al_2O_3 < CaO + Na_2O + K_2O$). Many of these rocks are alkali-calcic, but the basic to intermediate end-members of calc-alkaline and tholeiitic (calcic) arc suites are also metaluminous. However, the more acidic calc-alkaline igneous rocks are *peraluminous*, with molar alumina concentrations exceeding the total of Ca and alkali oxides ($Al_2O_3 > CaO + Na_2O + K_2O$). Although Al_2O_3 does not feature in Figs. 1 and 2, alumina saturation in acid rocks really depends on variations in Ca and alkalis and, alone, Al_2O_3 is insensitive to the tectonic setting of magma series (Petro et al., 1979). Therefore, despite the broad correlation of terms based on alumina and silica saturation, the latter provide a more meaningful discriminant of tectonic environment and magmagenetic processes.

This section has purposely concentrated on variations in major element chemical characteristics that provide a basis for defining rock series and for comparing volcanic and intrusive rocks. However, many trace elements are also sensitive to the nature of such rock series. For example, there are trends towards increasing

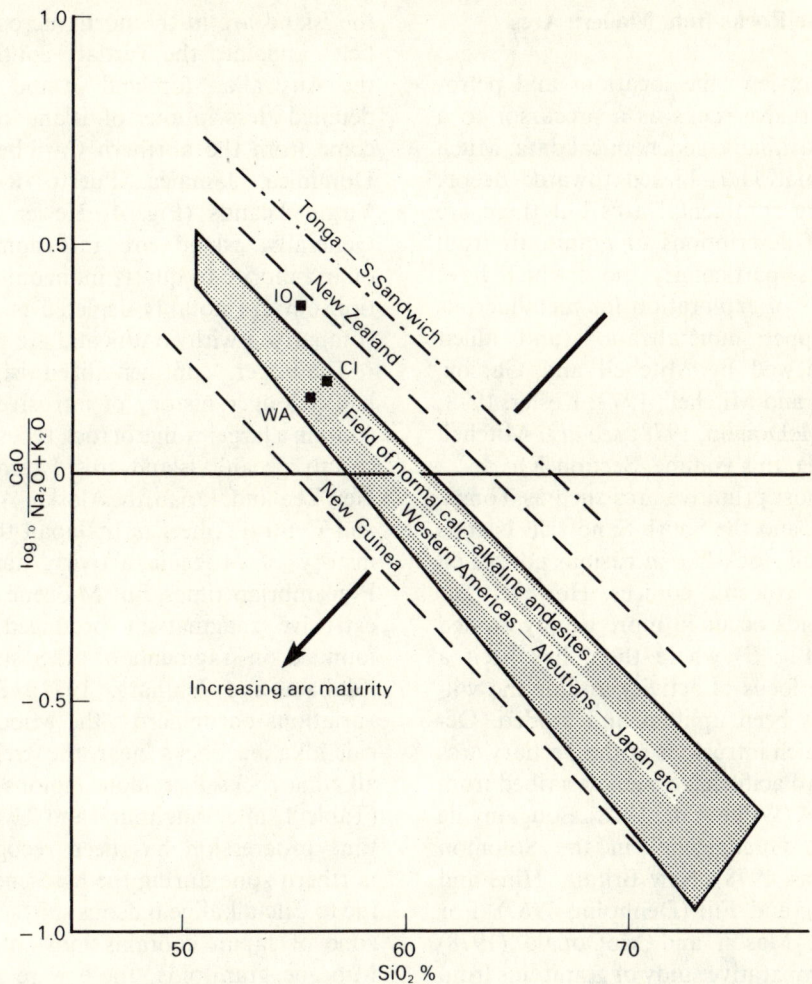

Fig. 2 Calc-alkali ratio–silica trends for volcanic suites from modern magmatic arcs showing the change from calcic to calc-alkaline (abundant) suites with increasing maturity. (Data sources as Fig. 1.) The alkali-lime index (SiO_2 concentration where $CaO = Na_2O + K_2O$) lies between 56 and 61 for calc-alkaline suites (see text)

concentrations of incompatible elements such as U, Th, Rb, Cs, Ba, Sr, Pb, and REE with arc maturity, REE patterns become more strongly fractionated and trace elements such as Cr, Co, and Ni fall (Taylor, 1969; Ringwood, 1974). There is some evidence that high field strength elements (which are not easily leached by weathering and hydrothermal processes), such as Ti, Zr, Y, and Nb, may index arc maturity because they are sensitive to changes in fractionating mafic and minor phases (Pearce and Norry, 1979; Tarney and Saunders, 1979). These elements are well established as discriminators between different types of arc and within-plate magmas (Pearce and Gale, 1977) but futher work is required to clarify their inter-arc variations. Also, unfortunately, less trace element data are available for granitoids than for volcanic rocks and only preliminary trace element comparisons can be made (*see* later). Another difficulty is that, unlike many lavas, it is not easy to determine whether a particular granitoid represents a liquid. Many may be of cumulate origin and this complicates the trace element picture.

Intrusive Rocks from Modern Arcs

This section reviews the locations and petrography of intrusive rocks as a precursor to a discussion of available geochemical data. Much of this information is biased towards deeply eroded mature continental arcs but there are several recent descriptions of granitoids from immature arcs, particularly those which have been the focus of exploration for metalliferous porphyry copper mineralization and allied deposits (reviewed by Mitchell and Garson, 1976; Garson and Mitchell, 1977; Kesler, 1978; Mason and McDonald, 1978; see also Mitchell and Beckinsale, this volume, Section X).

Even the most primitive arcs such as Tonga, the Marianas, and the South Sandwich Islands may have small stock-like intrusions concealed beneath their volcanic edifices. However, exposed granitoids occur in more deeply eroded island arcs (Fig. 3) where there has been a change in the focus of activity and/or the volcanic pile has been uplifted and eroded. Occurrences of such intrusions in the Tertiary arcs of the western Pacific have been described from the Philippines (Wolfe et al., 1978), Bougainville (Ford, 1978), Guadalcanal in the Solomon Islands (Chivas, 1978), New Britain (Hine and Mason, 1978), and Fiji (Denholm, 1967). For New Guinea, Mason and McDonald (1978) reported a comparative study of granitoids from the island arc in the north, across the 'mobile belt', and into the Tertiary continental arc of the Australian foreland in the south. Other detailed descriptions of island arc granitoids come from the northern Caribbean islands of Dominica, Jamaica, Puerto Rico, and the Virgin Islands (Fig. 4; Kesler et al., 1975). Generally, island arc intrusions vary from quartz diorite to quartz monzonite in composition but are notably depleted in K-feldspar in comparison with continental arc plutons.

The larger, 'continentalized' island arcs have had a longer history of intrusive activity and contain a larger range of rock types than occur in strictly oceanic island arcs. Examples occur in New Zealand, Japan, the Alaska–Aleutian range, and Central America. In Japan there is a long history of orogenic activity, dating back to Precambrian times, but Miocene intrusive and extrusive magmatism produced a new arc founded on fragments of older sialic basement (Shibata and Ishihara, 1979). Petrochemical variations occur across the Miocene arc from calc-alkaline rocks near the trench to more alkaline rocks in remote regions to the north (Table 1, after Sugimura and Uyeda, 1973). A time progression has been recognized in the northern zone during the Miocene from tholeiitic to calc-alkaline igneous suites. The southern zone of Japan contains most of the exposed Miocene granitoids; these were intruded into

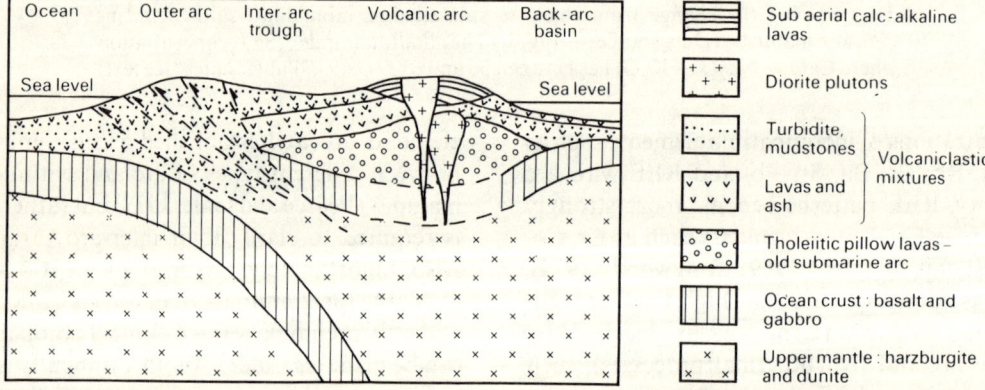

Fig. 3 Cross-section of a young island arc which has evolved from tholeiitic to calc-alkaline magmatism and which contains exposed K-poor diorite intrusions. The continued development of outer arc imbricate zones of volcaniclastic flysch may also lead to intrusive activity in this area (e.g. Japan, Aleutians; see text). (Cross-section adapted from Garson and Mitchell, 1977)

TABLE 1 The Miocene igneous activity of Japan (after Sugimura and Uyeda, 1973)

Age (Ma)	Northern Zone	Median Zone	Southern Zone
14 ± 3	Alkaline rocks	Andesite, dacite, rhyolite, and 'subalkaline' magmas	Granodiorite, granite porphyry, adamellite
21 ± 1	Calc-alkaline rocks	—	Granodiorite and quartz porphyry
25 ± 1	Basalt and andesite—tholeiitic magmas	—	

an imbricate outer arc zone which is more fully developed than indicated in Fig. 3 (Oba, 1977). On average, these Japanese intrusions are richer in quartz and K-feldspar than the diorite stocks of true oceanic island arcs but the outer arc granitoids include a low K gabbro–tonalite suite (Ishizaka and Yanagi, 1977).

The Central American arc (Fig. 4) is rather less mature than Japan in the sense that, like most of the arcs around the Caribbean, it has a Cretaceous basement of ocean crust (Case, 1974) rather than a sialic foundation. The southern limit of the North American craton is thought to be within northern Nicaragua according to Kesler (1978) and this means that the Central American continental crust has developed in the last 60–70 Ma (Kesler et al., 1977). The section from Costa Rica to Colombia is interesting because it is progressively more deeply eroded further south. Figure 4 shows that the

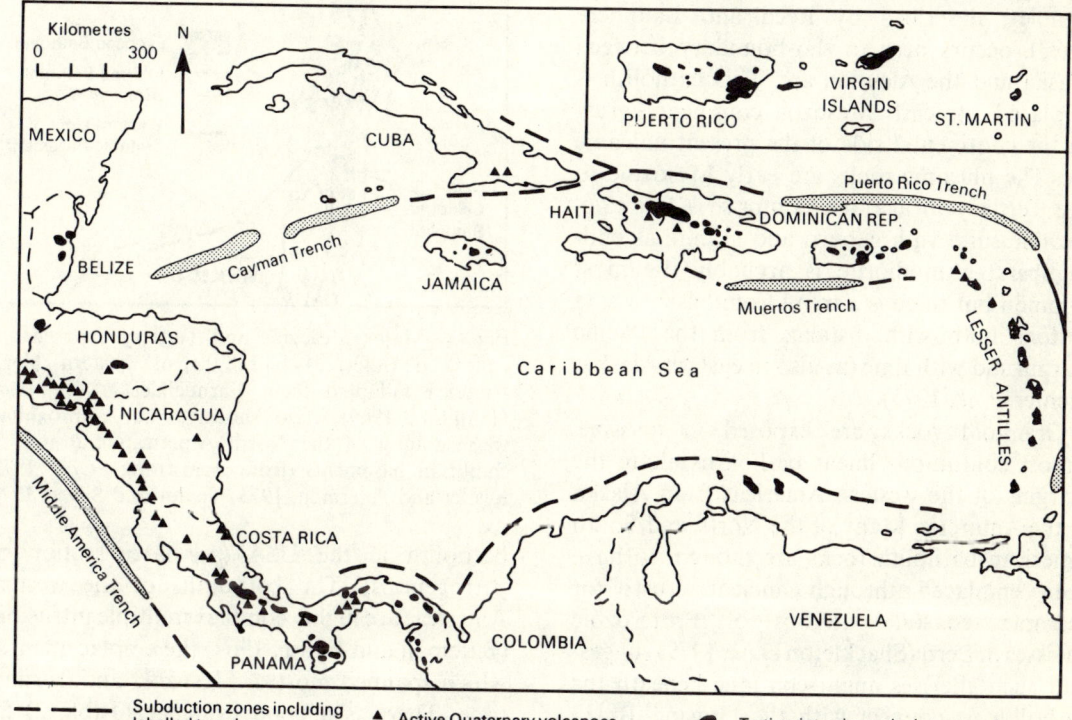

Fig. 4 Map of the Caribbean area to illustrate the relationship between active volcanoes and post-Cretaceous intrusive activity in Central America and the Antilles arcs. Note that the more deeply eroded terrain in Panama and the Greater Antilles exposes intrusive rocks. (Data from Kesler et al., 1975, 1977; Kesler, 1978; Pichler and Weyl, 1975)

line of volcanoes in Nicaragua and Costa Rica gives way southwards to a line of intrusions and occasional volcanoes emplaced through earlier volcaniclastic sequences. This alignment strongly suggests that these arc intrusions are subvolcanic, a conclusion which will be examined again in relation to the Andes. The oldest intrusive rocks of Panama (60–70 Ma, close to the age of initiation of the arc) are calcic quartz diorites and, in the same area, these give way to 50 Ma old calc-alkaline quartz diorites and granodiorites. Younger intrusive events at 35 and 5 Ma are characterized by relatively K-rich granodiorites; this trend towards more siliceous and more alkaline intrusives provides an excellent reflection of the increased growth and maturity of an arc.

The Aleutian arc contains granitoids which, like those of southern Japan, intrude the imbricate 'flysch' belt which has formed an outer arc (cf. Fig. 3) but the main batholithic intrusive complex, described by Reed and Lanphere (1973), occurs near to the boundary between Alaska and the Aleutian arc. This batholith is emplaced into early Mesozoic continental crust on the continental side of the present volcanic arc. The plutonic rocks are early Mesozoic to late Tertiary in age and comprise a full calc-alkaline suite with gabbro and granite as end-members; granodiorite is arguably the most common but there is a trend from basic to acid plutons both with distance from the Pacific margin and with time (as also in eastern Alaska; Richter et al., 1975).

Granitoid rocks are exposed in massive, almost continuous linear batholiths along the margins of the western Americas from Alaska to the Antarctic. Many of the North and South American batholith rocks are thought to have been emplaced through ancient crust; for example, coastal outcrops of Proterozoic gneisses in Peru (Shackleton et al., 1979) suggest that such gneisses might continue beneath the batholith to connect with the Guyana–Brazil craton in the east. The great batholiths of North America are identified in Fig. 5 together with the westernmost limit of the North American craton which underlies the eastern part of the

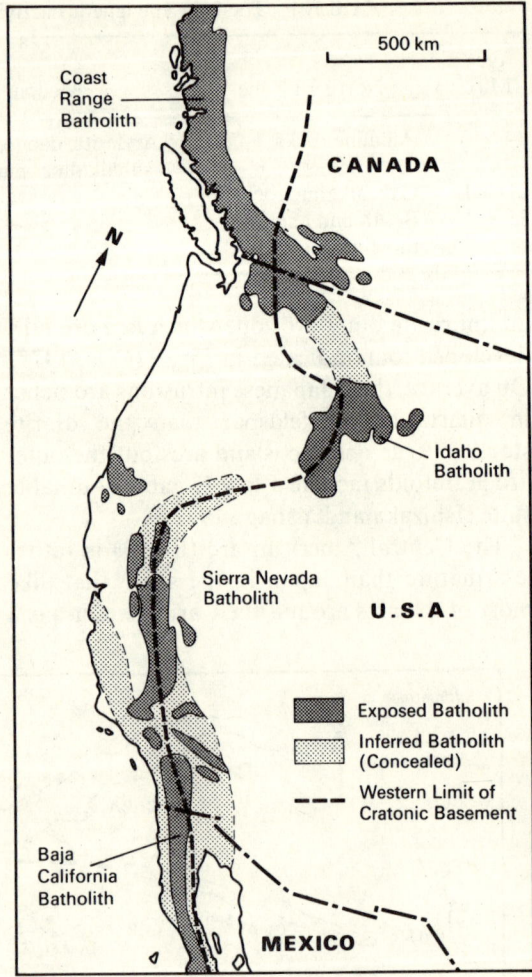

Fig. 5 Major Mesozoic and Tertiary exposed and concealed (inferred) batholiths of western North America (adapted from Carmichael et al., 1974; Hamilton, 1969). Also shown is the approximate western limit of the North American craton (Precambrian basement) (from Armstrong et al., 1977; Kistler and Peterman, 1973; Taylor and Silver, 1978)

batholith in the USA (see later section on petrogenesis). The batholiths of the western Americas are highly complex multiple intrusions (Bateman and Eaton, 1967) the emplacement of which spanned most of Mesozoic and Tertiary time. There is a good evidence (Table 2) for several episodes of magmatism, each lasting 10–25 Ma, with an indication of eastwards migration (away from the trench) during the early Tertiary. Detailed studies in British Col-

TABLE 2 The timing of plutonic episodes in the Western American batholiths (from Rutland, 1973, and related to the major deformation phase, where defined)

	Canada (Ma ago)	USA (Ma ago)	Chile (Ma ago)
Generative phase on east margins	200–175	235	210–195
Earliest 'orogenic' phase, east and west	170–160	180–160	190–180
Middle 'orogenic' phases	143–110	148–132 / 121–104	144
Late main orogenic phases showing eastward migration	105–85	90–79	107–87
Quartz diorites and granodiorites	76–64	75–50	67–56
Post-orogenic phases showing further eastward migration	50–40	40–20	43–29
Quartz monzonites		16–0	10–3

umbia (Preto et al., 1979), Washington–Idaho–Montana (Klepper et al., 1971; Armstrong et al., 1977), the Sierra Nevada (Bateman and Dodge, 1970), and southern California (Baird et al., 1974; Gastil et al., 1975) have identified trends, in calc-alkaline provinces, towards more alkaline and silicic rocks. In some provinces these trends develop with time, and in others they develop with distance from the subduction margin.

The trends with time are nowhere better displayed than in the Peruvian coastal batholith (Cobbing and Pitcher, 1972; Pitcher, 1978), which is part of a segmented chain of batholiths stretching 2500 km from Ecuador to southern Chile. Although these batholiths have composite structures and varied compositions, Pitcher (1978) and his co-workers have estimated the following proportions of rock types in the Peru coastal batholith: gabbros and diorites 7–16 per cent, tonalites 48–60 per cent, granodiorites and monzogranites 20–30 per cent, syenogranites 1–4 per cent (ranges reflect data from several transects). Figure 6 illustrates both the episodic nature of batholith emplacement in this continental arc and the change of composition with time. The bracketed time groups or 'super-units' represent secondary 'differentiation rhythms' superimposed on a primary basic–acidic evolution for the entire batholith (Pitcher, 1978). These trends are typical for western American batholiths which show a remarkable consistency both in the timing of their activity (Table 2) and their products (e.g. Fig. 6).

The western American batholiths were emplaced within, and are flanked by, volcanic rocks (particularly well developed to the west; Armstrong et al., 1977) the ages of which bracket the local age span of the batholith rocks. However, because of the particular level of erosion, the Peru coastal batholith is a good location for demonstrating the possible links between intrusive and extrusive suites; indeed, several papers have been devoted specifically to this subject.

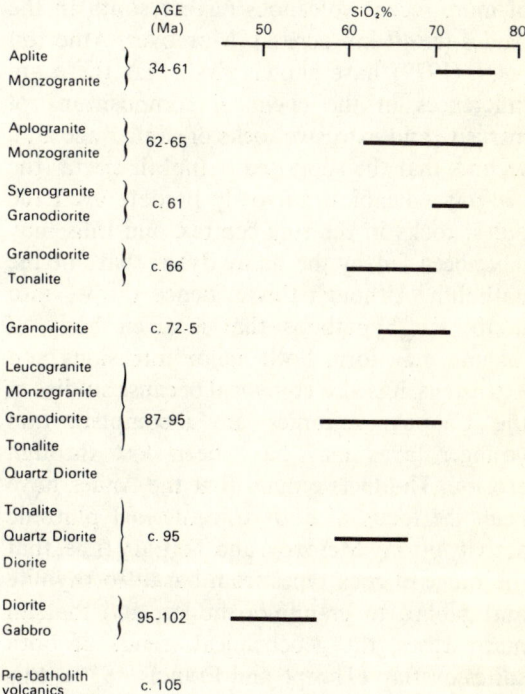

Fig. 6 The evolution of rock types and their silica contents with time (absolute age determinations) in the Lima segment of the Peru coastal batholith. (Data from Atherton et al., 1979; Pitcher, 1978; Wilson, 1975)

The field evidence is illustrated in Fig. 7 and the case for intrusive rocks being emplaced into their extrusive cover (Hamilton and Myers, 1967; Myers 1975) is succinctly summarized in the abstract of Bussell et al. (1976): 'A special feature of the composite Coastal Batholith of Peru is the presence of ring complexes, which suggests a very high-level crustal environment and a direct connection between the magmas of the constituent plutons and caldera-centred volcanicity'. The ring complexes, thought to have been subvolcanic, form the youngest un-roofed intrusions in the coastal batholith (Fig. 7) and they contain many stoped volcanic remnants from the roof. Despite this circumstantial evidence, there is as yet no documented example of an intrusive centre feeding a lava flow, and so there is still an element of doubt about the field connection between intrusive and volcanic rocks. Recently, Thorpe and Francis (1979a) have noted that the ring complexes (10–20 km across, see Fig. 7) are smaller than the calderas of more recent volcanoes further south in the Andes (c. 40 km across). Moreover, Atherton et al. (1979) have shown, first, that there are differences in the chemical compositions of intrusive and extrusive rocks of similar age and, second, that the supposed batholith ejecta (the Calipuy volcanics) narrowly predate even the oldest rocks in the ring centres, and thus may have been fed by the many dykes that cut the batholith. Although this evidence throws into doubt the hypothesis that a given body of magma may form both major intrusions and extrusions, it is also equivocal because studies of the Calipuy volcanics are incomplete and younger lavas may have been lost through erosion. The facts remain that the Andes have been the focus of both volcanic *and* plutonic activity during Mesozoic and Tertiary time, that the range of rock types from basalt to rhyolite and gabbro to granite is similar, and that, in many cases, the geochemical trends of both suites overlap (Thorpe and Francis, 1979a, this volume). These authors envisage an alternative explanation of the fate of an igneous magma whereby its water content determines the particular level at which it will crystallize and

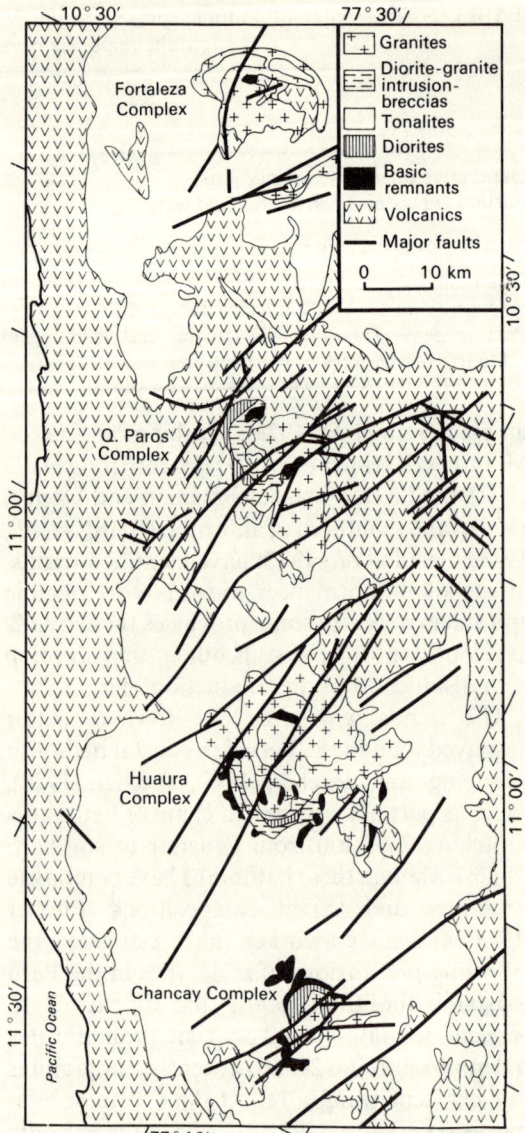

Fig. 7 The relationship of four major ring complexes (comprising mainly 35–70 Ma old batholith units; compare Fig. 6) to the volumetrically predominant 80–100 Ma tonalites and the pre-batholith volcanics of the Lima segment, Peru coastal batholith—see text for further discussion. (Redrawn from Bussell et al., 1976, with permission)

whether it will be erupted. Whilst endorsing this view in general terms the present author would add that magma temperature in relation to the relevant solidus (which varies with water content *and* composition) and the rate of liquid ascent are equally important factors (*see* discussions by Brown and Fyfe, 1970; Brown and Hennessy, 1978).

The last groups of granitoids to be considered are those of continent–continent and continent–island arc convergence zones that have similar magmatic products, and this is probably because both are characterized by sialic overthrusting. These arcs are produced by the continuation of subduction beneath a volcanic arc until the arc is underthrust by a continental mass so that the outer arc, and often the magmatic arc (Fig. 3), are thrust on to the continental foreland (Oxburgh, 1972; Garson and Mitchell, 1977). These plutons are often described as 'late orogenic' and the best known examples occur in the Himalayas, Malaysia, and Thailand. They show little variation from predominantly silica and alkali-rich leucogranites with tourmaline and muscovite; frequently they carry economic concentrations of Sn, W, and Nb. Many of the Eocene–Miocene granites of the Higher Himalayas (Tibet, Nepal, northern India, and Pakistan) are of this type (Le Fort, 1975; Hamet and Allègre, 1976; Desio, 1979) and were emplaced into metamorphic rocks of the continental forelands during the collision of India with Asia. Similar alkali-rich granitoids are found all along the late Cainozoic–Mesozoic collision zone from the Alps and eastern Europe down to the Sunda arc of the north-eastern Indian Ocean and into Indonesia (*see also* Mitchell and Beckinsale, this volume, Section X, Figs. 7 and 8). In the hinterland behind the Sunda arc, earlier (Permo-Triassic) arc–continent collisions took place with uplift and granitoid development in the Triassic–Cretaceous time interval, particularly along the Main Range of Peninsular Malaya (Hutchison, 1977) and to the north in western Thailand (Beckinsale *et al.*, 1979). These are extremely homogeneous, K-feldspar-rich, two-mica alkaline granites; many are greisenized, Sn-bearing, and intruded into a thick tightly folded succession of mainly Palaeozoic metasediments.

Other Sn-bearing alkali-rich granites are found in ocean–continent arcs where they occur behind the main magmatic arc (possibly analogous to oceanic back-arc zones) and are associated with shoshonitic lavas. The best known examples occur in the Eastern Cordillera of Bolivia where there is a 500 km long magmatic arc which is parallel to and 700 km west of the trench (Evernden *et al.*, 1977). An extensive K-rich diorite–syenite intrusive suite with shoshonites (Fig. 2), but without economic Sn, also occurs remote from the subduction margin in south-eastern Papua New Guinea (Smith, 1972).

To summarize, the most primitive island arc intrusions comprise diorite and monzonite stocks which have tholeiitic affinities and are depleted in K-feldspar. Intrusive rocks from the thicker more mature island arcs are characterized by a temporal evolution from diorites towards normal calc-alkaline tonalites and granodiorites as the arc itself matures; often, they also show spatial variations across the arc towards increasingly alkali-rich rocks remote from the ocean trench. There seems to be little *petrographic* difference between the granitoids of arcs built upon old sialic crust or ocean crust. Calc-alkaline intrusive rocks form, apparently without erupted equivalents, in the outer arc flysch of mature island and some continental arcs and, on the remote side of the main arc, alkali-rich syenite–shoshonite lava associations are sometimes observed. Uniformly alkali-rich, silicic granitoids are typical of intracontinental arcs resulting from continent–continent and continent–island arc collisions. As arc maturity increases so the volume ratio of intrusive to extrusive rocks and the silica mode also increase.

Major Element Geochemistry

Geochemical data for the many intrusive rock series of different arc settings are summarized, with references to the source literature, in Figs. 8–10. The AFM diagram (Fig. 8) shows that the vast majority of Mesozoic–Cainozoic arc intrusives follow the established calc-alkaline trend (Larsen, 1948), lying between the broken lines which indicate the extremes of trends for volcanic magmas in AFM diagrams. However, the majority of the exposed intrusive rocks are more evolved than the volcanics and plot to the left of

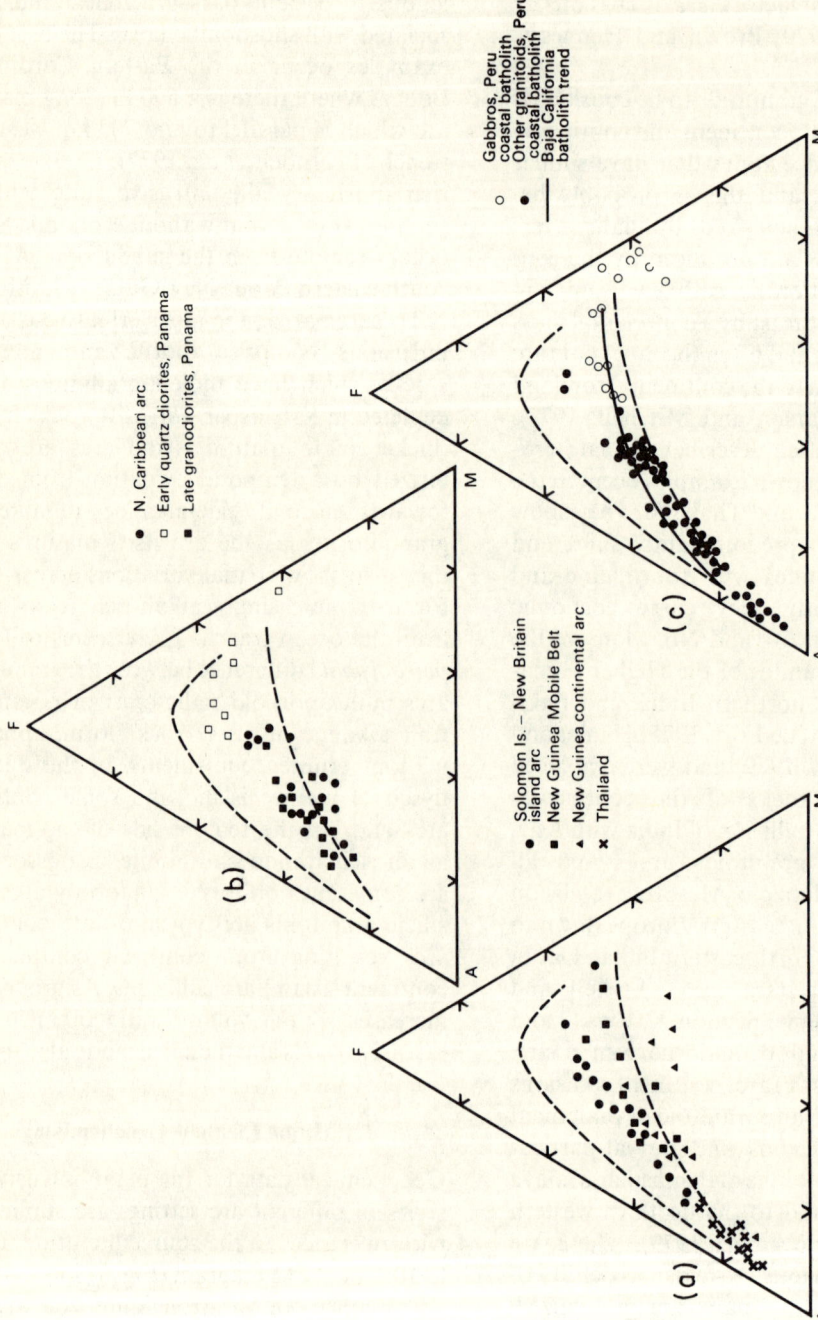

Fig. 8 Comparative AFM diagrams for intrusive suites from Mesozoic and Tertiary magmatic arcs: (a) south-eastern Asia and the western Pacific; (b) Central America and the Caribbean; (c) California and Peru. The broken lines are the limits of AFM variations among volcanic rocks identified in Figure 1. (Data sources: (a) Beckinsale et al., 1979; Hine and Mason, 1978; Mason and McDonald, 1978; Smith 1972; (b) Kesler et al., 1975, 1977; (c) Larsen, 1948; W. McCourt, in Pitcher, 1978; Atherton et al., 1979)

the volcanic averages, shown in Fig. 2. Figure 8(a) clearly illustrates the increasing maturity of the western Pacific arc granitoids adjacent to the Australian craton; the Solomon Islands–New Britain diorite–monzonite rocks are mostly tholeiitic in character whilst, at the other extreme, the New Guinea continental arc rocks plot on the Fe-poor side of the normal calc-alkaline trend. These extreme differences between adjacent immature intra-oceanic island arc and mature continental arc plutons are emphasized by a plot of total alkalis against silica (Fig. 9). This diagram resolves the difference between the alkali contents of silicic magmas more clearly than the AFM plot; for example, the alkali-rich Thai granites (Fig. 8(a)) are seen to be comparable with mature continental arc magmas in Fig. 9.

Both the northern Caribbean and Central American granitoids (Fig. 8(b)) matured from an early subtholeiitic trend, displayed rather well by the early Panamanian data, to a later calc-alkaline evolutionary sequence. The slope of the alkali-enrichment trend for the Panamanian rocks also changes with time and increasing arc maturity (Fig. 9). This raises an important point: *the spatial petrochemical trends normal to a modern arc at a given time (Fig. 8(a)) are comparable with the temporal trends in a given location (Fig. 8(b))*. Chemical data for granitoids from the Peru coastal batholith lie close to the Baja California batholith calc-alkaline trend (Fig. 8(c)). Also shown are data for the early gabbroic rocks of the Peru coastal batholith which, according to McCourt and Taylor (1980), define an almost tholeiitic trend at low total Fe contents (but this may be criticized because at least some of these rocks are cumulates). The Peruvian granitoids from the different rhythms identified in Fig. 6 are not separated by the AFM plot and it seems that the processes causing primary and secondary evolution trends may be identical. Pitcher (1978) and McCourt and Taylor (1980) attributed normal calc-alkaline evolution trends of this kind to hornblende and/or pyroxene-dominated fractionation (but *see* below). The positions of two trend lines for western American batholiths shown in the alkali–silica diagram (Fig. 9) are consistent with the high maturity of these arcs (the Sierra Nevada more than the Alaskan range). The only granitoids which are more alkaline, at a given silica content, are those from behind continental arcs (e.g. New Guinea) and in collision zone locations (e.g. Thailand) Bateman and Dodge (1970) also found that the K_2O/Na_2O ratio of intrusive rocks (at constant silica) increases across the Sierra Nevada away from the trench and this is borne out by data from other regions (the K-rich 'back-arc' intrusive rocks of Bolivia, Japan, etc).

The AFM data for intrusive arc rocks show closely similar ranges, diversity, and trends with arc maturity to those of extrusive rocks from the same arcs (compare Figs. 1 and 8). This strengthens the case for a petrogenetic link between intrusive and extrusive suites, a link which is endorsed by Fig. 10 (for comparison with Fig. 2). The New Britain–Solomon Islands arc, the northern Caribbean arc, and the early Panamanian granitoids are clearly calcic (*see also* discussion by Kesler *et al.*, 1977); the more

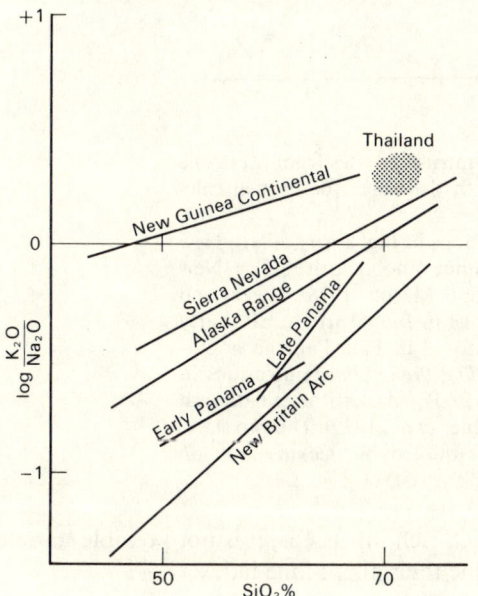

Fig. 9 Alkali ratio–silica trends for representative intrusive suites from Mesozoic and Tertiary magmatic arcs. (Data sources as Fig. 8 but also including Bateman and Dodge, 1970; Reed and Lanphere, 1973)

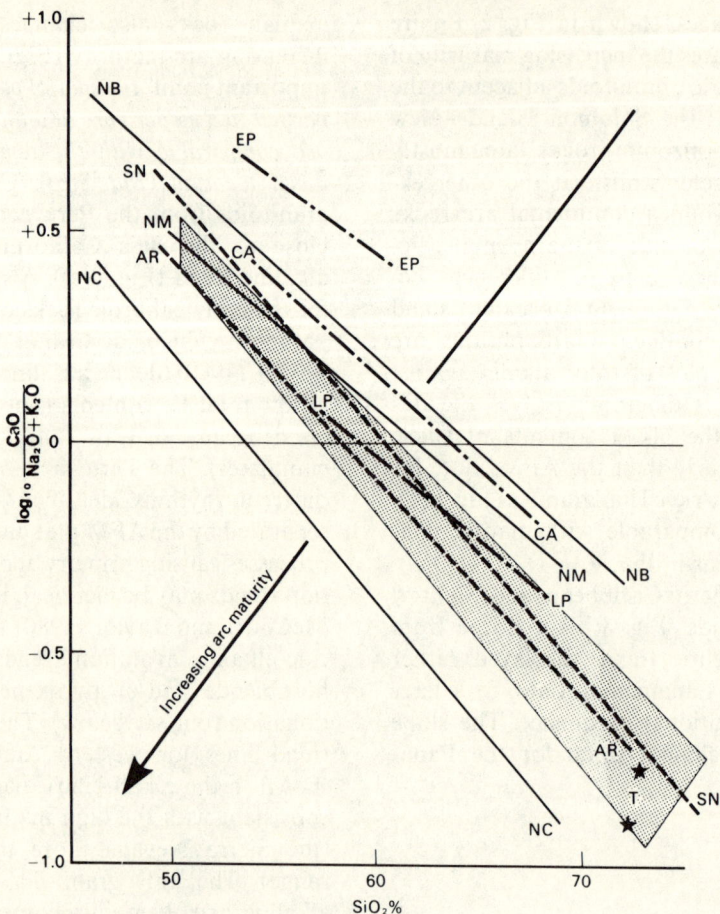

Fig. 10 Calc-alkali ratio–silica trends for intrusive suites from Mesozoic and Tertiary magmatic arcs compared with the range for normal calc-alkaline andesites (shaded, from Fig. 2)

Labelling is as follows: *solid lines* (intrusions in Fig. 8(a)) are NB, New Britain–Solomon Islands arc; NM, New Guinea mobile belt; NC = New Guinea continental arc (data from Hine and Mason, 1978; Måson and McDonald, 1978); *Dot-dash lines* (intrusions in Fig. 8(b)) are EP, Early Panama diorites; CA, North Caribbean arc; LP, Late Panama granodiorites (data from Kesler et al., 1975, 1977); *broken lines* (intrusions in Fig. 8(c)) are SN, Sierra Nevada batholith; AR, Alaska Range batholith (data from Bateman and Dodge, 1970; Richter et al., 1975). The two stars, labelled T, are average Thailand granites quoted by Beckinsale et al. (1979)

mature, late Panamanian, New Guinea mobile belt, and western American batholith granitoids are identified as calc-alkaline, but the New Guinea continental arc magmas are alkali-calcic, ranging to alkaline. The Thai granitoids plot towards the alkali-rich side of the calc-alkaline field, but as these are also homogeneous, silica-rich intrusions, it is not possible to determine their alkali-lime index.

Trace Elements, Isotopes, and Petrogenesis

Perhaps the most widely used indices of magma provenance today are isotopic, but conclusions

TABLE 3 Comparison of certain trace element concentrations for intrusive suites of different tectonic settings in the Papua New Guinea region (all data in p.p.m. and compared at 60 per cent SiO_2; table abbreviated from Mason and McDonald, 1978)

	Island Arc Suites	'Mobile Belt' Marginal Suites	Suites Emplaced in Cratonic Crust	Correlation with Silica
Ba	120–280	100–400	400–800	Positive
Sr	220–900	400–600	700–1100	Variable
Zr	80–110	150–180	150–200	Positive
Nb	<5	8–10	>10	Positive
Ni	5–20	10	<10	Negative
Co	15–25	15–25	<15	Negative
V	150–220	140–160	100–140	Negative

from isotopic data must be shown to be consistent with evidence from other geochemical, geophysical, and geological sources. Figure 11 gives the range of initial $^{87}Sr/^{86}Sr$ ratios for many of the intrusive suites, the locations and major element geochemical characteristics of which are described above. Again, the data are similar to the initial ratios for volcanic rocks (Hawkesworth, this volume, Section VIII) and, in cases where there is a change of initial $^{87}Sr/^{86}Sr$ ratio with time (as in northern Chile; McNutt et al., 1975), parallel changes are found in both intrusive and extrusive suites. The most important deduction from Fig. 11 is that, apart from the Thailand and Nepalese data (discussed later), most areas have a substantial proportion of low values (under 0.706) and this indicates derivation from a Rb-poor source region, probably dominated by the subcontinental lithosphere. This conclusion gains independent support: first, from the fact that the enormous crustal thickening observed in western American continental arcs cannot be explained solely by crustal shortening but requires an input of magma from the mantle; second, from the evidence that crustal temperatures are inadequate to allow spontaneous melting within the crust; and, finally, from the very obvious localization of major magmatic arcs over zones of oceanic plate subduction and potential melting (see Youriker and Vogel, 1976; Brown and Hennessy, 1978, for full discussion of these points).

Against this background of subcontinental magmagenesis, Mason and McDonald (1978) compiled a synthesis of regional trace element data for intrusive rocks of the New Guinea area (Table 3). They noted that, for a given silica concentration, the light incompatible and high field strength cations are enriched with increasing arc maturity whereas transition elements become depleted (see also Tarney and Saunders, 1979). Moreover, their data also show that REE become more strongly fractionated with increasing arc maturity. Again, all these trends are recognized in associated volcanic suites (see elsewhere in this volume) and place constraints on the likely source of magma for both intrusive and extrusive arc activity. For example, according to Wyllie et al. (1976), it is unlikely that such large concentrations of incompatible elements could be incorporated in magmas generated directly by the partial melting of tholeiitic ocean crust subducted beneath magmatic arcs, although this source could supply the early magmas of immature arcs. However, the dehydration of this material and the consequent volatile fluxing of overlying mantle peridotite may produce suitable liquids (see reviews by Green and Mysen, this volume, Section VII), especially if volatiles carry upwards incompatible elements from the subduction zone and enrich the subcontinental lithosphere (Saunders et al., 1980).

The relationship between the trace element and isotopic variations in Andean andesites and their petrogenesis has been reviewed by Thorpe and Francis (1979b, this volume). They showed that the enrichment of certain incompatible (large ion lithophile) trace elements in the An-

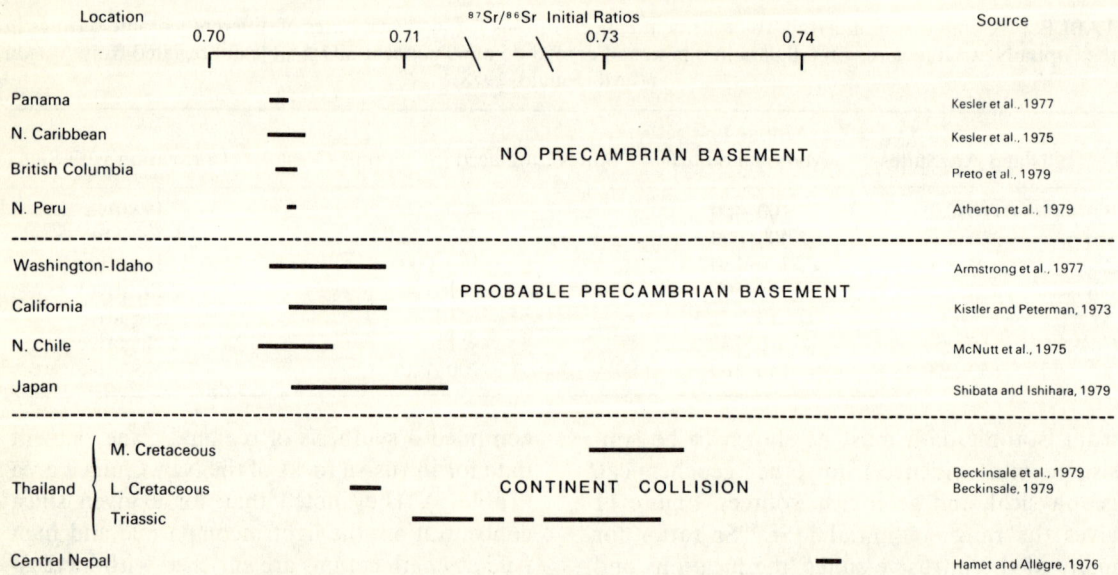

Fig. 11 The ranges of initial Sr isotope ratios for intrusive suites from Mesozoic and Tertiary magmatic arcs. Localities are grouped into those where Precambrian basement is absent (top), those which are thought to have Precambrian basement (middle), and those which came from arc–continent or continent–continent collision zones (bottom)

dean volcanic chain together with silica, K, and initial $^{87}Sr/^{86}Sr$ ratios may be correlated with crustal thickness. The most extreme development occurs over the Chile–Bolivia–Argentina border where the crust has thickened to 70 km (Cummings and Schiller, 1971). During the last 100 Ma, igneous rock initial ratios in northern Chile have increased from 0.7022 to 0.7061 (Fig. 11; McNutt et al., 1975). A petrogenetic model consistent with this evidence was proposed by Brown (1977), involving the addition of mantle-derived diorite (basaltic andesite) magmas to the crust above subduction zones. These magmas crystallize to form low initial ratio intrusions with low concentrations of incompatible elements in the early stages of arc development. Granitoids of the Panamanian, northern Caribbean, and similar immature arcs represent this stage in development. As the input of magma continues, especially in regions already founded on continental crust so the basement rocks, and/or the lower parts of the growing volcaniclastic pile, heat up and partial melting may occur. The first melt will have a composition near to the granite minimum and so will be siliceous and rich in incompatible elements. With time, larger amounts of crustal melt will be added to a continuous supply of mantle-derived magma, thus explaining the observed correlations of chemical and isotopic characteristics with time (Fig. 6), and with crustal thickness for intrusive rocks in magmatic arcs. Another, equally viable explanation of the geochemical trends would be the fractionation of mafic phases from mantle-derived melts and, although there are some difficulties with this argument (Tarney and Saunders, 1979), batholithic rocks might include a cumulate component (Thorpe and Francis, 1979a). Fractional crystallization might therefore accompany crustal melting in modifying the chemical characteristics of mantle-derived batholith magmas (see Pitcher, 1979, for a general review). But the weight of evidence favours the latter process: not only are there isotopic arguments (amplified below) but crustal melting is consistent with the evidence for mixing of crustal and mantle-derived magmas based on phenocryst disequilibrium in andesite melts (Eichelberger, 1978), the constraints of crustal thermal-melting arguments (Younker and Vogel, 1976), and the observation that the western American granitoids are, on average, relatively more acidic than contemporaneous erupted rocks. On the last point, it is well known that the more acidic the composition of a rising magma, the more viscous, low temperature, and hydrous it is likely to be and so the less likely it is to be erupted.

Although the isotope–time trends for northern Chilean batholith rocks were used as a basis for the crust–mantle mixing model explained above, isotopic data for other batholiths may be used to test the model. Despite their wide-ranging petrochemical evolution (e.g. Fig. 6), Sr data for the Lima segment of the Peru coastal batholith (Fig. 11) do not show an analogous increase in isotopic ratio with time (Atherton et al., 1979). If the model is correct, this implies either (1) that any basement was refractory or (2) that remelting occurred only in young rocks without significant radiogenic ^{87}Sr. Both factors are likely to be important, but evidence that the northern limit of the Precambrian basement in Peru may terminate south of the Lima segment (Thorpe et al., 1981) throws the emphasis towards implication (2) (for further discussion see Brown, 1981).

There is also excellent evidence from the northern American batholiths that supports the model of crustal melt addition, and this arises because Precambrian basement rocks do not reach the edge of this plate margin (Fig. 5). Both sets of data for California and the Washington–Idaho region in Fig. 11 show significant east–west variations across the batholith with a change from uniformly low ratios (west) to variable ratios (east) where the host rocks for young intrusives change from recent volcaniclastic rocks to Precambrian basement, either at outcrop or at depth. In the Washington–Idaho region, Armstrong et al. (1977) found no exceptions to the observation that all plutons on the west (trench) side of this line have initial ratios less than 0.703, whereas plutons on the east (continental) side have initial ratios greater than 0.7055. This evidence supports the model that crustal melt is added during the ascent of mantle-derived magma because the isotopic signature of crustal rocks is represented in associated intrusive rocks. Taylor and Silver (1978) have also shown, from O–Sr isotopic correlations in the batholith rocks of California, that a good case can be made for crust–mantle magma interactions.

The evidence from isotopic variations in calc-alkaline rocks is systematically different and has opposite trends to that which would arise from either (1) isotopic exchange of seawater ($^{87}Sr/^{86}Sr = 0.7092$) with ocean crust before it reaches the melt zone (Hawkesworth et al., 1979) or (2) the major involvement in magmagenesis of subducted clastic continental detritus with high $^{87}Sr/^{86}Sr$ (James, 1979; Fyfe, 1979 and this volume, Section X). As pointed out above (see also Taylor and Silver, 1978), the isotopic evidence for a 'crustal' component is controlled mainly by crustal thickness and the nature of the associated basement rocks. For example, initial $^{87}Sr/^{86}Sr$ ratios of the Cretaceous granodiorites of south-central Japan are generally in the range 0.704–0.707 but locally exceed 0.71 (Fig. 11) which, according to Shibata and Ishihara (1979), implies that 'the Japanese island arc basement contains relatively young crustal materials with sporadic old segments'. Many immature continental and island arc granitoids lack the evidence of acidification and isotope trends described above and, presumably, there is no difference in the nature and composition of underthrust ocean crust. Again, this suggests a crustal control rather than one based primarily on subducted radiogenic detritus or major volumes of seawater-affected ocean crust. In the case of the central Andes (southern Peru–northern Chile), the Amazon river discharges Andean sediment into the Atlantic, not the Pacific, where there is only a thin cover (Gibbs, 1967; Meade et al., 1979), so little sediment is available for subduction below this region. Yet the associated calc-alkaline magmas are similar to, and some even have a greater 'crustal' component than, other continental and island arcs where there is more sediment on the adjacent ocean crust (Thorpe et al., 1981). Thus, although there is some evidence that pelagic sediments and perhaps some clastic continental debris is subducted (as proposed by Fyfe, this volume, Section X; *see also* Karig and Kay, 1981), it seems not to be a vital control on magmagenetic processes.

A case has been made that the majority of the calcic and calc-alkaline granitoids from ocean–ocean and ocean–continent convergence zones are mantle-derived. Such a case cannot be made for the more alkali-calcic granitoids of continent collision belts that have much higher initial Sr isotope ratios (e.g. Nepal and Thailand in Fig. 11) reflecting a probable origin by the partial melting of radiogenic crustal rocks. Isotopic data, combined with geothermal arguments based on the exceptional frictional and radiogenic heating of the crust in overthrust terrains, place the origin of these magmas firmly in the crust. The complex tectonic history of Thailand is reflected in the radiogenically immature Lower Cretaceous granitoids which are bounded on both sides by episodes of typical collision zone magmatism. This history has been discussed by Beckinsale *et al.* (1979) and is thought to reflect intermittent subduction from the west beneath the earlier, welded arc–continent suture.

It is clear that the petrochemical indices of maturity in magmatic arcs (Figs. 8–10, Table 3, and earlier discussion) have their parallel in isotopic data because Sr isotope initial ratios increase down Fig. 11, albeit with a marked change from continental margins to continental collision zones. It is also clear that the dominantly mantle-derived intrusive and extrusive suites of most magmatic arcs represent new additions to the continental crust which is growing at a rate close to 0.5 km^3 a^{-1} (Brown, 1979). This is a maximum rate, for it assumes that sediment removal and subduction into the mantle is negligible (a much-debated point; *see* Gass and Wright, 1980), but it implies that at least 10^{10} a would be needed to produce all the continental mass at the present rate. However, it is well known that the rate of crustal processes is likely to have declined in proportion to the Earth's radiogenic heat budget (reviewed by Bickle, 1978) and so the crust can be 'grown' inside the age of the Earth. If crustal processes were much more rapid in the past, this raises the fundamental and also much-debated question of whether modern processes of calc-alkaline intrusive and extrusive magmatism have operated at past times in geological history.

Calc-alkaline Magmas and Crustal History

The majority of the granitoids described in earlier sections are calc-alkaline, varying from calcic in immature island arcs, through true calc-alkaline in mature arcs and continental margins, to alkali-calcic in continental collision zones (*see* Fig. 10). But not all major granitoids are simply related to these examples, in terms of either chemical characteristics or tectonic setting, and certain differences become even more noticeable in older examples. Figure 12 contains data for major granitoid suites of different ages and locations. There are two examples of Phanerozoic alkali-calcic suites in this figure which arise, not from compressional environments, but from abortive or incipient rift systems. These are the Jurassic Nigerian younger granites (Macleod *et al.*, 1971; Wright, 1973) and the Hebridean granites of Skye (Wager *et al.*, 1965), the emplacement of which was related to the opening of the northern Atlantic. These 'extensional' suites differ from the 'compressional' alkali-calcic granitoids (Fig. 10) of collision zones in having a wider compositional diversity resulting from the fractional crystallization of parental basic or intermediate liquids with subalkaline and alkaline affinities. Petro *et al.* (1979), noting that alkali basaltic liquids are produced by melting at greater mantle depths than for tholeiitic or calc-alkaline parental magmas, probably at high P_{CO_2} (Mysen and Boettcher, 1975; Wyllie *et al.*, 1976), suggested that such liquids may undergo fractional crystallization, and become contaminated at crustal levels, to produce alkali-calcic magma series. Clearly, an extensional environment is appropriate for such petrogenesis.

Emslie (1978) has described several extensive Proterozoic gabbro–sodic anorthosite–syenite–

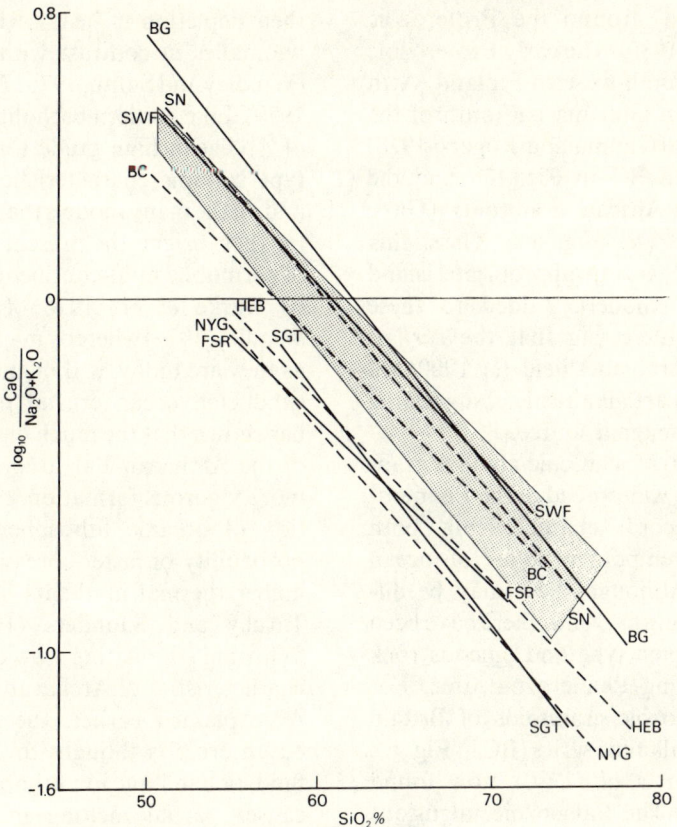

Fig. 12 Calc-alkali ratio–silica plot for granitoid suites of Proterozoic (solid lines) and Phanerozoic (broken lines) ages. Note the separation of calc-alkaline (compressional) and alkali-calcic (extensional) suites. The shaded area is the field of modern calc-alkaline andesites (from Fig. 2). Labelling is as follows: BG, Ben Ghnema, Tibesti (Ghuma and Rogers, 1978); SWF, south-western Finland tonalites (Arth *et al.*, 1978); SN, Sierra Nevada (Bateman and Dodge, 1970); BC, British Caledonides (Brown, 1979; Brown *et al.*, 1980); SGT, Saudi Arabian granite traverse (Gass, 1977, and unpublished data); FSR, Fennoscandian rapakivi granites (Emslie, 1978); HEB, Hebridean Tertiary granites of Britain (Wager *et al.*, 1965); NYG, Nigerian younger granites (Macleod *et al.*, 1971)

rapakivi granite series, for which the type area is Scandinavia and Finland (FSR in Fig. 12), which may well have been generated by this process. Other well known examples occur in the Proterozoic granitoids of Colorado (Barker *et al.*, 1975) and New Mexico (Condie, 1978). The abundance of alkali-calcic and some alkaline magma series, particularly in mid-Proterozoic sequences, may reflect the importance of intraplate, extensional rift magmatism at that time. This is in keeping with contemporary models for the evolution of the Proterozoic crust (reviewed by Windley, 1977) which indicate broad, stable platform areas, that were resistant to fragmentation, but which developed major linear alkaline intrusive suites along the abortive rifts and swells known as mobile belts (for detailed discussion *see* Brown, 1979, 1981). However, calc-alkaline magma series similar to those of the present day, were developed along marginal 'sedimentary basins' around the Proterozoic continents (Brown, 1980). The early Proterozoic tonalite massifs of south-western Finland (Arth et al., 1978), the Ben Ghnema batholith of the Libyan Pan-African (Ghuma and Rogers, 1978) —labelled SWF and BG in Fig. 12—and the Saudi-Arabian Pan-African granitoids (Gass, 1977), are examples (see Fig. 6 of Gass, this volume, Section IX, for an appropriate island arc 'cratonization' model). Added to these conclusions, it is interesting that the *earliest* granitoids of the Arabian shield (c. 1000 Ma old, SGT in Fig. 12) are alkali-calcic, suggesting a deep level parent magma source.

Following the Pan-African, calc-alkaline granitoids are even more widespread in the Phanerozoic geological record, characterizing both fossilized and contemporary zones of ocean plate subduction. Although there may be differences in the ordering of events, there have been few changes in magma type and igneous rock petrochemistry during Phanerozoic time. For example, the Palaeozoic granitoids of Britain form a normal calc-alkaline series (BC in Fig. 12; reviewed by Brown et al., 1980), now found within a continent. The Palaeozoic intrusions now exposed along the coast of central Chile are almost identical in their petrochemistry to adjacent Mesozoic and Tertiary calc-alkaline plutonics (Lopez-Escobar et al., 1979).

So far, this brief discussion has concentrated on a comparison of Proterozoic and Phanerozoic magma series, yet some of the most widespread calc-alkaline magmatic events in the continental crust must have characterized the Archaean (over 2500 Ma ago). Typically, the exposed Archaean cratons comprise high grade calc-alkaline tonalite gneisses surrounded by low grade greenstone assemblages which also contain calc-alkaline volcanics. In many ways, the Archaean high grade suites are reminiscent of modern batholiths. However, the former are often, but not always, poor in K-feldspar and comprise a gabbro–diorite–tonalite–trondhjemite series rather similar to that of modern primitive arcs. Associated geochemical characteristics are their apparently lower K_2O/Na_2O ratios (which could reflect the deeper erosion level), their depletion in heavy REE, and positive Eu anomalies in contrast with modern batholiths (Windley and Smith, 1976; Tarney and Saunders, 1979). Like modern batholiths, the vast majority of Archaean high grade tonalites have mantle-type isotopic characteristics (Moorbath, 1978) and, of the many models that have been proposed for their origin, the present author favours that of permobile mini-continental plates (developed by Burke et al., 1976; Tarney et al., 1976; Windley, 1977) whereby magmas were generated, as they are today, within and above fast moving subducted ocean crustal plates. Bickle (1978) has shown that the much higher heat production of the Archaean Earth could be dissipated by more vigorous formation, cooling, and destruction of oceanic lithosphere. Recognizing the probability of faster spreading rates leading to higher thermal gradients in subduction zones, Tarney and Saunders (1979) proposed the following stimulating view of the petrochemical characteristics of Archaean calc-alkaline suites. As explained earlier, the downgoing slab of ocean crust is thought to contribute a watery fluid, rich in large ion lithophile elements, which causes partial melting in the mantle wedge overlying subduction zones; the rising melts have geochemical patterns that reflect the characteristics of this mantle, the degree of melting, and the subsequent fractional crystallization (Saunders et al., 1980). But in the Archaean, the subducted slab may also have partially melted, leaving residual phases such as garnet or hornblende which retain heavy REE but not Eu (see Arth and Barker, 1976, for partition coefficients). So the liquids generated would directly reflect the low K_2O/Na_2O ratios of ocean crust and would also have the observed REE characteristics.

If these ideas are correct, then the generation of calc-alkaline magmas by shallow partial melting within and above subducted ocean crust has characterized most of geological history. The changes with time are towards a greater involvement of the mantle overlying subduction zones in melt generation, and this may be related to corresponding decreases in thermal gradients and rates of ocean plate subduction. Alkali-calcic suites generated by the fractionation of relatively alkaline parental magmas from deeper mantle source regions complement the calc-alkaline suites. They form in regions of crustal tension and were notably abundant in the mid-Proterozoic, by which time the continental lithosphere may have thickened to the point at which it was able to fracture, fragment, and accumulate in modern plate-like fashion. Sites of fragmentation favour the development of alkali-calcic magmatism in tensional environments whereas accumulation sites are characterized by compressional calc-alkaline magmatism. Although calc-alkaline granitoids have predominated throughout geological history, the ratio between the volumes of both types has varied according to the relative juxtaposition and movements of continental lithospheric fragments.

Summary

(1) The intrusive and volcanic rocks of modern arcs are comparable in their geochemistry, evolution and magmagenesis. They comprise: (a) calcic, K-poor diorite–monzonite stocks of immature island arcs; (b) major calc-alkaline gabbro–diorite–tonalite–granodiorite intrusives of mature continentalized arcs with no ancient sialic basement; (c) calc-alkaline to alkali-calcic suites from ocean–continent arcs intruding mature continental margins with or without Precambrian basement; (d) acid, alkali-calcic, two-mica granites in continent–continent collision zones. Both (b) and (c) show increases in silica and K concentrations, respectively, with both time and distance from the subduction margin. Spatial trends normal to the margin at a given time are similar to the temporal trends in a given location.

(2) At a given location, increasing maturity and thickening of the crust with time, over periods in the range $10^8–10^9$ a, may produce a complete range of magma types from (1)(a) to (1)(d). At the same time, and most markedly between (1)(c) and (1)(d), the amount of mantle-derived melt generally decreases and the amount of crustal melt increases. There is good isotopic and other evidence that the 'crustal' melt component of these granitoids is derived directly by the contamination of rising melts within the crust; concurrent fractional crystallization is also likely to occur. Both contamination and fractionation are favoured by increasingly thick continental crust; therefore, crustal thickness imposes a primary constraint on intrusive rock compositions.

(3) As arcs mature so the volume of intrusive rocks seems to increase whilst that of extrusive rocks decreases. Evidence that the two suites may be genetically related comes from: (a) the erosion of volcanoes to reveal intrusions (e.g. Fig. 4) and centred intrusive complexes (e.g. Fig. 7); (b) their major and trace element similarities, although exposed batholith rocks may be more siliceous; (c) the similarity in chemical and isotopic trends with time for intrusive and extrusive rocks in a single arc; (d) arguments based on crustal thickening and melting conditions which combine to place melt initiation for both suites in the subcrustal lithosphere, except in continent–continent collision zones.

(4) However, it is difficult to prove from field relations that any given magma may be both intrusive and extrusive. The capacity of a magma to erupt is determined by the solidus temperature for both the composition and water content involved; thus, a magma will have a marked tendency either to freeze at depth or to be erupted. Even so the evidence in (3), and the apparently cumulus nature of some zoned granitoids (Pitcher, 1979), indicates that some intrusive bodies should have directly associated volcanic partners.

(5) For a given arc maturity it seems likely that calc-alkaline suites will prove to have become more potassic with time throughout the Earth's history. This trend needs further corroborative field and geochemical evidence, particularly in relation to geochemical variations due to erosion level within batholiths. But such a trend would be consistent with the concept of a decreasing involvement with time of subducted ocean crust, and an increasing involvement of

subcontinental lithosphere in magmagenesis at magmatic arcs.

(6) The calc-alkali ratio–silica plot provides a useful discriminant of compressional (dominantly calc-alkaline) and tensional (dominantly alkali-calcic) magmas series (compare Figs. 2, 10, and 12). The latter are produced by deeper mantle melting than for calc-alkaline magmas and this is followed by substantial fractionation of an alkali basalt parent. Alkali-calcic magma series are distinct from post-tectonic collision zone granitoids, which often have alkaline tendencies, because the latter are usually extremely homogeneous and occupy a very restricted and silicic compositional range. In other words, the evolution of magmas from continental fragmentation through subduction to suturing may follow a sequence from an early alkali-calcic suite (usually small volumes) through the cycle (1)(a)–(1)(c) of major calcic and calc-alkaline suites to alkali-rich silicic plutons during the re-suturing event.

(7) The ratio of calc-alkaline (apparently compressional) to alkali-calcic (apparently extensional) magma series has varied, but generally has been high throughout geological history. At a given time, this ratio is likely to reflect the amount of continental crust undergoing marginal subduction to that undergoing rifting (whether abortive or successful). It is appreciated, however, that geochemical similarities between ancient and modern granites do not *necessarily* prove that generative mechanisms have not changed (cf. Brown, 1981).

(8) Calc-alkaline granite batholiths, formed throughout the Earth's history, have been the major contributors to long term continental growth. It is suggested, although not proven, that this process is still continuing but at a rate which is decreasing exponentially in proportion to the Earth's internal heat production.

Acknowledgements

My thanks go to the many friends and colleagues with whom I have shared discussions on the characteristics and petrogenesis of calc-alkaline intrusive rocks. I. G. Gass, W. S. Pitcher, and R. S. Thorpe kindly reviewed this contribution, and made many useful suggestions for improvements. I am also grateful to John Taylor and Jenny Hill for preparing the illustrations and to Sue Hartnett who typed the manuscript.

REFERENCES

Armstrong, R. L., Taubeneck, W. H., and Hales, P. O. (1977). Rb–Sr and K–Ar geochronometry of Mesozoic granitic rocks and their Sr isotopic composition, Oregon, Washington, and Idaho. *Geol. Soc. Am. Bull.* **88**, 397–411.

Arth, J. G. and Barker, F. (1976). Rare-earth partitioning between hornblende and dacitic liquid and implications for the genesis of trondhjemitic-tonalitic magmas. *Geology* **4**, 534–536.

Arth, J. G., Barker, F., Peterman, D. E., and Friedman, I. (1978). Geochemistry of the gabbro–diorite–tonalite–trondhjemite suite of south-west Finland and its implications for the origin of tonalitic and trondhjemitic magmas. *J. Petrol.* **19**, 289–316.

Atherton, M. P., McCourt, W. J., Sanderson, L. M., and Taylor, W. P. (1979). The geochemical character of the segmented Peruvian coastal batholith and associated volcanics. In *Origin of Granite Batholiths: Geochemical Evidence*. (M. P. Atherton and J. Tarney, eds), Shiva Publishing, Orpington, pp. 45–64.

Baker, P. E. (1968). Comparative volcanology and petrology of the Atlantic island arcs. *Bull. Volc.* **32**, 189–206.

Baker, M. C. W. and Francis, P. W. (1978). Upper Cenozoic volcanism in the central Andes—ages and volumes. *Earth planet. Sci. Letters* **41**, 175–187.

Baird, A. K., Baird, K. W., and Welday, E. E. (1974). Chemical trends across Cretaceous batholithic rocks of southern California. *Geology* **2**, 493–495.

Barker, F., Wones, D. R., Sharp, W. N., and Desbrough, G. A. (1975). The Pikes Peak batholith, Colorado Front Range, and a model for the origin of the gabbro-anorthosite–syenite–potassic granite suite. *Precambrian Res.* **2**, 97–160.

Bateman, P. C. and Dodge, F. C. W. (1970). Variations of major chemical constituents across the central Sierra Nevada batholith. *Geol. Soc. Am. Bull.* **81**, 409–420.

Bateman, P. C. and Eaton, J. P. (1967). Sierra Nevada Batholith. *Science, N.Y.* **158**, 1407–1417.

Beckinsale, R. D. (1979). Granite magmatism in the tin-belt of south-east Asia. In *Origin of Granite Batholiths: Geochemical Evidence* (M. P. Atherton and J. Tarney, eds), Shiva Publishing, Orpington, pp. 34–44.

Beckinsale, R. D., Suensilpong, S. Nakapadungrat, S., and Walsh, J. N. (1979). Geochronology and geochemistry of granite magmatism in Thailand in relation to a plate tectonic model. *J. geol. Soc. Lond.* **136**, 529–540.

Bickle, M. J. (1978). Heat loss from the Earth: a constraint on Archaean tectonics from the relation between geothermal gradients and the rate of plate production. *Earth planet. Sci. Letters* **40**, 301–315.

Brown, G. C. (1977). Mantle origin of Cordilleran granites. *Nature, Lond.* **265**, 21–24.

Brown, G. C. (1979). The changing pattern of batholith emplacement during Earth History. In *Origin of Granite Batholiths: Geochemical Evidence* (M. P. Atherton and J. Tarney, eds), Shiva Publishing, Orpington, pp. 106–115.

Brown, G. C. (1980). Calc-alkaline magma genesis: the Pan-African contribution to crustal growth? In *Evolution and Mineralization of the Arabian–Nubian Shield* (A. M. S. Al-Shanti, ed.), *IAG Bull.* no. 3, pp. 19–29.

Brown, G. C. (1981). Space and time in granite plutonism. *Phil. Trans. R. Soc. A*, **301**, 321–336.

Brown, G. C. and Fyfe, W. S. (1970). The production of granitic melts during ultrametamorphism. *Contr. Mineral. Petrol.* **28**, 310–318.

Brown, G. C. and Hennessy, J. (1978). The initiation and thermal diversity of granite magmatism. *Phil. Trans. R. Soc. A* **288**, 631–643.

Brown, G. C., Plant, J. and Thorpe, R. S. (1980). Plutonism in the British Caledonides: space, time and geochemistry. In *The Caledonides in the U.S.A.* (D. R. Wones, ed.), *VPI and SU Mem.* no. 2, pp. 157–166.

Burke, K., Dewey, J. F., and Kidd, W. S. F. (1976). Dominance of horizontal movements, arc and microcontinental collisions during the later permobile regime. In *The Early History of the Earth*. (B. F. Windley, ed.), Wiley Interscience, New York, pp. 113–129.

Bussell, M. A., Pitcher, W. S., and Wilson, P. A. (1976). Ring complexes of the Peruvian coastal batholith: a long-standing subvolcanic regime. *Can. J. Earth Sci.* **13**, 1020–1030.

Carmichael, I. S. E., Turner, F. J., and Verhoogen, J. (1974). *Igneous Petrology*, McGraw-Hill, New York.

Case, J. E. (1974). Oceanic crust forms basement of eastern Panama. *Geol. Soc. Am. Bull.* **85**, 645–652.

Chappell, B. W. and White, A. J. R. (1974). Two contrasting granite types. *Pacific Geol.* **8**, 173–174.

Chivas, A. R. (1978). Porphyry copper mineralisation at the Koloula igneous complex, Guadalcanal, Solomon Islands. *Econ. Geol.* **73**, 645–677.

Christiansen, R. L. and Lipman, P. W. (1972). Cenozoic volcanism and plate tectonic evolution of the western United States. II. Late Cenozoic. *Phil. Trans. R. Soc. A* **271**, 249–284.

Cobbing, E. J. and Pitcher, W. S. (1972). The coastal batholith of central Peru. *J. geol. Soc. Lond.* **128**, 421–460.

Condie, K. C. (1978). Geochemistry of Proterozoic granitic plutons from New Mexico, U.S.A. *Chem. Geol.* **21**, 131–149.

Cummings, D. and Schiller, G. I. (1971). Isopach map of the Earth's crust. *Earth Sci. Rev.* **7**, 97–125.

Denholm, L. S. (1967). Geological exploration for gold in the Tarua Basin, Viti Levu, Fiji. *N.Z. J. Geol. Geophys.* **10**, 1185–1186.

Desio, A. (1979). Geologic evolution of the Karakorum. In *Geodynamics of Pakistan* (A. Farah and K. A. DeJong, eds), Geological Survey of Pakistan, Quetta, pp. 111–124.

Eichelberger, J. C. (1978). Andesitic volcanism and crustal evolution. *Nature, Lond.* **275**, 21–27.

Emslie, R. F. (1978). Anorthosite massifs, rapakivi granites, and late Proterozoic rifting of North America. *Precambrian Res.* **7**, 61–98.

Evernden, J. F., Stanislav, J. K., and Cherroni, C. M. (1977). Potassium–argon ages of some Bolivian rocks. *Econ. Geol.* **72**, 1042–1061.

Ewart, A. (1976). Mineralogy and chemistry of modern orogenic lavas—some statistics and implications. *Earth planet. Sci. Letters* **31**, 417–432.

Ewart, A. and Stipp, J. J. (1968). Petrogenesis of the volcanic rocks of the central North Island, New Zealand, as indicated by a study of $^{87}Sr/^{86}Sr$ ratios, and Sr, Rb, K, U and Th abundances. *Geochim. cosmochim. Acta* **32**, 699–735.

Ford, J. H. (1978). A chemical study of alteration of the Panguna porphyry copper deposit, Bougainville, Papua New Guinea. *Econ. Geol.* **73**, 703–720.

Fyfe, W. S. (1979). The geochemical cycle of uranium. *Phil. Trans. R. Soc. A* **291**, 433–445.

Garson, M. S. and Mitchell, A. H. G. (1977). Mineralisation at destructive plate margins: a brief review. In *Volcanic Processes in Ore Genesis*, Geological Society of London Special Publication no. 7, pp. 81–97.

Gass, I. G. (1977). The evolution of the Pan-African crystalline basement in N.E. Africa and Arabia. *J. geol. Soc. Lond.* **134**, 129–138.

Gass, I. G. and Wright, J. B. (1980). Continents old and new. *Nature, Lond.* **284**, 217–218.

Gastil, R. G., Phillips, R. P., and Allison, E. C. (1975). Reconnaissance geology of the State of Baja California. *Geol. Soc. Am. Mem.* no. 140.

Ghuma, M. A. and Rogers, J. J. W. (1978). Geology, geochemistry, and tectonic setting of the Ben Ghnema batholith, Tibesti massif, southern Libya. *Geol. Soc. Am. Bull.* **89**, 1351–1358.

Gibbs, R. J. (1967). The geochemistry of the Amazon river system. Part 1. The factors that control the salinity and composition and concentration of the suspended solids. *Geol. Soc. Am. Bull.* **78**, 1203–1232.

Hamet, J. and Allègre, C. J. (1976). Rb–Sr systematics in granite from central Nepal (Manaslu): significance of the Oligocene age and high $^{87}Sr/^{86}Sr$ ratio in Himalayan orogeny. *Geology* **4**, 470–472.

Hamilton, W. (1969). Mesozoic California and the underflow of Pacific mantle. *Geol. Soc. Am. Bull.* **80**, 2409–2430.

Hamilton, W. and Myers, W. B. (1967). The nature of batholiths. *US geol. Surv. prof. Pap.* no. 554–C.

Hawkesworth, C. J., Norry, M. J., Roddick, J. C., Baker, P. E., Francis, P. W., and Thorpe, R. S. (1979). $^{143}Nd/^{144}Nd$, $^{87}Sr/^{86}Sr$, and incompatible element variations in calc-alkaline andesites and plateau lavas from South America. *Earth planet. Sci. Letters* **42**, 45–57.

Hine, R., and Mason, D. R. (1978). Intrusive rocks associated with porphyry copper mineralization, New Britain, Papua New Guinea. *Econ. Geol.* **73**, 749–760.

Hutchison, C. S. (1977). Granite emplacement and tectonic subdivision of Peninsular Malaysia. *Geol. Soc. Malaysia Bull.* **9**, 187–207.

Ishizaka, K. and Yanagi, T. (1977). K, Rb and Sr abundances and Sr isotopic composition of the Tanzawa granitic and associated gabbroic rocks, Japan: low-potash island arc plutonic complex. *Earth planet. Sci. Letters* **33**, 345–352.

Jakeš, P. and Gill, J. (1970). Rare earth elements and the

island arc tholeiite series. *Earth planet. Sci. Letters* **9**, 17–28.

Jakeš, P. and Smith, I. E. (1970). High potassium calc-alkaline rocks from Cape Nelson, Eastern Papua. *Contr. Mineral. Petrol.* **28**, 259–271.

James, D. E. (1979). On the origin of the calc-alkaline volcanics of the central Andes: a revised interpretation. *Carnegie Instn. Wash. Y. B.* **77**, 562–590.

Karig, D. E. and Kay, R. W. (1981). Fate of sediments on the descending plate at convergent margins. *Phil. Trans. R. Soc. A.* **301**, 233–251.

Kesler, S. E. (1978). Metallogenesis in the Caribbean region. *J. geol. Soc. Lond.* **135**, 429–441.

Kesler, S. E., Jones, L. M., and Walker, R. L. (1975). Intrusive rocks associated with porphyry copper mineralization in island arc areas. *Econ. Geol.* **70**, 515–526.

Kesler, S. E., Sutter, J. F., Issigonis, M. J., Jones, L. M., and Walker, R. L. (1977). Evolution of porphyry copper mineralisation in an oceanic island arc: Panama. *Econ. Geol.* **72**, 1142–1153.

Kistler, R. W. and Peterman, Z. E. (1973). Variations in Sr, Rb, K, Na and initial $^{87}Sr/^{86}Sr$ in Mesozoic granitic rocks and intruded wall rocks in central California. *Geol. Soc. Am. Bull.* **84**, 3489–3512.

Klepper, M. R., Robinson, G. D., and Smedes, H. W. (1971). On the nature of the Boulder batholith in Montana. *Geol. Soc. Am. Bull.* **82**, 1563–1580.

Larsen, E. S. (1948). Batholith of southern California. *Geol. Soc. Am. Mem.* no. 29.

Le Fort, P. (1975). Himalayas: the collided range. Present knowledge of the continental arc. *Am. J. Sci.* **275A**, 1–44.

Lopez-Escobar, L., Frey, F. A., and Oyarzun, J. (1979). Geochemical characteristics of central Chile (33°–34°S) granitoids. *Contr. Mineral. Petrol.* **70**, 439–450.

Macleod, W. N., Turner, D. C., and Wright, E. P. (1971). The geology of the Jos Plateau, Vol. 1, General geology. *Bull. geol. Surv. Nigeria* no. 32.

Mason, D. R. and McDonald, J. A. (1978). Intrusive rocks and porphyry copper occurrences of the Papua New Guinea–Solomon Islands region: a reconnaissance study. *Econ. Geol.* **73**, 857–877.

McCourt, W. J. and Taylor, W. P. (1980). In Cobbing, E. J., Baldock, J. McCourt, W. J., Pitcher, W. S., Taylor, W. P. and Wilson, J. J. The geology of the western cordillera of northern Peru. *Mem. Inst. Geol. Sci. Overseas Div.* in press.

McNutt, R. H., Crocket, J. H., Clark, A. H., Caelles, J. C., Farrar, E., and Haynes, S. J. (1975). Initial $^{87}Sr/^{86}Sr$ ratios of plutonic and volcanic rocks of the central Andes between latitudes 26° and 29° south. *Earth planet. Sci. Letters* **27**, 305–313.

Meade, R. H., Nordin, C. F., Curtis, W. F., Rodrigues, F. M. C., DoVale, C. M., and Edmond, J. M. (1979). Sediment loads in the Amazon river. *Nature, Lond.* **278**, 161–163.

Mitchell, A. H. G. and Garson, M. S. (1976). Mineralisation at plate boundaries. *Minerals Sci. Engng.* **8**, 129–169.

Moorbath, S. (1978). Age and isotopic evidence for the evolution of continental crust. *Phil. Trans. R. Soc. A* **288**, 401–412.

Myers, J. S. (1975). Cauldron subsidence and fluidisation: mechanisms of intrusion of the coastal batholith of Peru into its own volcanic ejecta. *Geol. Soc. Am. Bull.* **86**, 1209–1220.

Mysen, B. O. and Boettcher, A. L. (1975). Melting of hydrous mantle. II. Geochemistry of crystals and liquid formed by anatexis of mantle peridotite at high pressures and high temperatures as a function of controlled activities of water, hydrogen and carbon dioxide. *J. Petrol.* **16**, 549–593.

Oba, N. (1977). Emplacement of granitic rocks in the outer zone of southwest Japan and geological significance. *J. Geol.* **85**, 383–393.

Oxburgh, E. R. (1972). Flake tectonics and continental collision. *Nature, Lond.* **239**, 202–204.

Peacock, M. A. (1931). Classification of igneous rock series. *J. Geol.* **39**, 54–67.

Pearce, J. A. and Gale, G. H. (1977). Identification of ore-deposition environment from trace-element geochemistry of associated igneous host rocks. In *Volcanic Processes in Ore Genesis*, Geological Society of London Special Publication no. 7, pp. 14–24.

Pearce, J. A. and Norry, M. J. (1979). Petrogenetic implications of Ti, Zr, Y and Nb variations in volcanic rocks. *Contr. Mineral. Petrol.* **69**, 33–47.

Petro, W. L., Vogel, T. A. and Wilband, J. T. (1979). Major-element chemistry of plutonic rock suites from compressional and extensional plate boundaries. *Chem. Geol.* **26**, 217–235.

Pichler, H. and Weyl, R. (1975). Magmatism and crustal evolution in Costa Rica (Central America). *Geol. Rundschau* **64**, 457–475.

Pitcher, W. S. (1978). The anatomy of a batholith. *J. geol. Soc. Lond.* **135**, 157–182.

Pitcher, W. S. (1979). The nature, ascent and emplacement of granite magmas. *J. geol. Soc. Lond.* **136**, 627–662.

Plant, J., Brown, G. C., Simpson, P. R., and Smith, R. T. (1980). Signatures of metalliferous granites in the Scottish Caledonides. *Trans. Inst. Min. Metall.* **89**, B198–B210.

Preto, V. A., Osatenko, M. J., McMillan, W. J., and Armstrong, R. L. (1979). Isotopic dates and strontium isotopic ratios for plutonic and volcanic rocks in the Quesnel Trough and Nicola Belt, south-central British Columbia. *Can. J. Earth Sci.* **16**, 1658–1672.

Reed, B. L. and Lanphere, M. A. (1973). The Alaska–Aleutian Range batholith: geochronology, chemistry and relation to circum-Pacific plutonism. *Geol. Soc. Am. Bull.* **84**, 2583–2610.

Richter, D. H., Lanphere, M. A., and Matson, N. A. (1975). Granite plutonism and metamorphism, eastern Alaska Range, Alaska. *Geol. Soc. Am. Bull.* **86**, 819–829.

Ringwood, A. E. (1974). The petrological evolution of island arc systems. *J. geol. Soc. Lond.* **130**, 183–204.

Rutland, R. W. R. (1973). On the interpretation of cordilleran orogenic belts. *Am. J. Sci.* **273**, 811–849.

Saunders, A. D., Tarney, J., and Weaver, S. D. (1980). Transverse geochemical variations across the Antarctic Peninsula: implications for the genesis of calc-alkaline magmas. *Earth planet. Sci. Letters* **46**, 344–360.

Scheidegger, K. F. and Kulm, L. D. (1975). Late Cenozoic volcanism in the Aleutian arc: information from ash layers in the northeastern Gulf of Alaska. *Geol. Soc. Am. Bull.* **86**, 1407–1412.

Shackleton, R. M., Ries, A. C., Coward, M. P., and Cobbold, P. R. (1979). Structure, metamorphism and geochronology of the Arequipa massif of coastal Peru. *J. geol. Soc. Lond.* **136**, 195–214.

Shibata, K. and Ishihara, S. (1979). Initial $^{87}Sr/^{86}Sr$ ratios of plutonic rocks from Japan. *Contr. Mineral. Petrol.* **70**, 381–390.

Smith, I. E. (1972). High-potassium intrusives from south-eastern Papua. *Contr. Mineral. Petrol.* **34**, 167–176.

Stern, R. J. (1979). On the origin of andesite in the northern Mariana island arc. *Contr. Mineral. Petrol.* **68**, 207–219.

Sugimura, A. and Uyeda, S. (1973). *Island Arcs: Japan and its Environs*, Developments in Geotectonics, Vol. 3, Elsevier, Amsterdam.

Tarney, J., Dalziel, I. W. D., and De Wit, M. J. (1976). Marginal basin 'Rocas Verdes' complex from S. Chile: a model for Archaean greenstone belt formation. In *The Early History of the Earth* (B. F. Windley, ed.), Wiley-Interscience, New York, pp. 131–146.

Tarney, J. and Saunders, A. D. (1979). Trace element constraints on the origin of cordilleran batholiths. In *Origin of Granite Batholiths: Geochemical Evidence* (M. P. Atherton and J. Tarney, eds), Shiva Publishing, Orpington, pp. 90–105.

Taylor, H. P. and Silver, L. T. (1978). Oxygen isotope relationships in plutonic igneous rocks of the Peninsular Ranges batholith, Southern and Baja California. *USGS Open-file report* no. 78-701, A23–A25.

Taylor, S. R. (1969). Trace element chemistry of andesites and associated calc-alkaline rocks. In *Bull. Oregon St. Dept. Geol. Mineral. Ind.* no. **65**, 43–64.

Thorpe, R. S. and Francis, P. W. (1975). Volcan Ceboruco: a major composite volcano of the Mexican volcanic belt. *Bull. Volc.* **39**, 1–13.

Thorpe, R. S. and Francis, P. W. (1979a). Petrogenetic relationship of volcanic and intrusive rocks of the Andes. In *Origin of Granite Batholiths: Geochemical Evidence* (M. P. Atherton and J. Tarney, eds), Shiva Publishing, Orpington, pp. 65–75.

Thorpe, R. S. and Francis, P. W. (1979b). Variations in Andean andesite compositions and their petrogenetic significance. *Tectonophysics* **57**, 53–70.

Thorpe, R. S., Francis, P. W., and Harmon, R. S. (1981). Andesites in crustal evolution. *Phil. Trans. R. Soc. A* **301**, 305–320.

Wager, L. R., Vincent, E. A., Brown, G. M., and Bell, J. D. (1965). Marscoite and related rocks of the Western Red Hills complex, Isle of Skye. *Phil. Trans. R. Soc. A* **257**, 273–308.

Wilson, P. A. (1975). K–Ar age studies in Peru with special reference to the emplacement of the coastal batholith. Ph.D. thesis, University of Liverpool.

Windley, B. F. (1977). *The Evolving Continents*, Wiley, New York.

Windley, B. F. and Smith, J. V. (1976). Archaean high-grade complexes and modern continental margins. *Nature, Lond.* **260**, 671–675.

Wolfe, J. A., Manuzon, M. S., and Divis, A. F. (1978). The Taysan porphyry copper deposit, southern Luzon Island, Philippines. *Econ. Geol.* **73**, 608–617.

Wright, J. B. (1973). Continental drift, magmatic provinces and mantle plumes. *Nature, Lond.* **244**, 565–567.

Wyllie, P. J., Huang, W. L., Stern, C. R., and Maaloe, S. (1976). Granitic magmas: possible and impossible sources, water contents, and crystallisation sequences. *Can. J. Earth Sci.* **13**, 1007–1019.

Younker, L. W. and Vogel, T. A. (1976). Plutonism and plate dynamics: the origin of circum-Pacific batholiths. *Can. Mineral.* **14**, 238–244.

VII Experimental studies relevant to petrogenesis of andesites

Since the late 1960s, experimental petrology has played an increasingly important role in defining and placing constraints on models for the petrogenesis of andesitic magmas. Prior to this time, the origin of calc-alkaline magmas was discussed in terms of a variety of hypotheses based upon experimental studies on simple mineral systems. These inferences support a number of models for the petrogenesis of calc-alkaline magmas, ranging from fractional crystallization (with or without sialic contamination) to direct partial melting of mantle peridotite. During the late 1960s, a range of geological and geochemical, particularly isotopic, arguments convinced many workers that calc-alkaline andesites were generated at high pressure from mafic or ultramafic sources and that sialic contributions were not of widespread importance. Within this framework two important advances were published during 1968. Green and Ringwood (1968) carried out an experimental investigation of a series of natural calc-alkaline rocks ranging in composition from high alumina basalt to rhyodacite, at pressures of 10–30 kb. The results were interpreted as favouring formation of andesitic magma from fractional crystallization of basaltic magma or from partial melting of solid basalt. A second advance at this time was the demonstration from study of simple model systems by Kushiro et al. (1968) that, in the presence of excess H_2O, enstatite melts incongruently to olivine and a quartz-normative liquid to at least 35 kb (c. 120 km). Experimental data therefore provided a basis for two contrasted models for generation of andesitic magma from mafic and ultramafic source materials. Within the plate tectonic context, Green and Ringwood's model provided an attractive mechanism for production of calc-alkaline magmas by partial melting of the descending oceanic slab. However, in the same context, dehydration of the descending slab might cause hydrous melting and production of andesite magmas within the overlying mantle wedge.

In the context described above, models for petrogenesis of andesite are clearly linked with the subduction of oceanic lithosphere, from the assumption that melting of the descending oceanic crust and/or mantle wedge yields hydrous mafic (basaltic) magmas. In the first of the two contributions in this section, T. H. Green reviews experimental evidence for formation of andesite by partial melting of hydrated mafic crust or fractional crystallization of hydrous mafic magmas at the base of the crust. Based on experiments performed upon natural rock compositions, Green argues that two main models might be invoked for such partial melting and fractional crystallization processes—a model involving amphiboles at 8–12 kb (25–40 km) and a model involving eclogite at over 25 kb (more than 80 km). A transitional garnet-amphibolite model might be applicable to the 12–25 kb (40–80 km) range.

In contrast to models for petrogenesis of andesite based upon partial melting or fractional crystallization of mafic sources, andesitic magma might be derived by hydrous partial melting of ultramafic rock. This model is explained by B. O. Mysen in the second contribution in this section: Mysen argues that the peridotite wedge overlying subducted and dehydrating oceanic lithosphere will become metasomatically altered, such that it can yield primary andesitic melts to pressures of 25 kb (90 km). The compositions of melts deduced from simple systems, or by direct chemical analysis of experimentally formed partial melts are known less accurate than those determined for natural rock samples, and the model is applicable to a generalized andesite composition.

There has been much discussion about models for andesite petrogenesis by fractional crystallization or partial melting of mafic sources or by direct hydrous partial melting of peridotite. This has centred around research at the Australian National University by T. H. Green, D. H. Green, Ringwood, and their co-workers (the GR group) and at the Geophysical Laboratory of the Carnegie Institution of Washington by Kushiro, Mysen, and their co-workers (the KO group). The GR group believes in the use of experiments on natural rock compositions as outlined by T. H. Green in this volume. By contrast, the KO group prefers to apply results derived from simple model systems to make predictions that can be tested by experiments on selected natural rocks. Further controversy between these two groups has resulted from disagreement about several aspects of experimental procedure, including the problem of the grain-size of starting materials, the extent of Fe loss from the sample, the control of the Fe oxidation state, and problems of chemical analysis of the quenched melt. These disagreements are summarized by Mysen *et al.* (1974), Green (1976), and Mysen (this volume).

In spite of these disagreements the phase relationships upon which these models are based can be used quantitatively to evaluate different models for petrogenesis of andesitic magma using petrological and geochemical data. However, rigorous testing of such models is still difficult because of lack of many data such as values of partition coefficients as functions of P, T, and composition, the composition of the fluid phase, and the role of accessory minerals in controlling trace element composition. Both Green and Mysen refer to important ways in which these might affect the petrogenesis of andesite magmas, within the petrogenetic schemes proposed.

REFERENCES

Green, D. H. (1976). Experimental testing of 'equilibrium' partial melting of peridotite under water-saturated, high-pressure conditions. *Can. Mineral.* **14**, 255–268.

Green, T. H. and Ringwood, A. E. (1968). Genesis of the calc-alkaline igneous rock suite. *Contr. Mineral. Petrol.* **18**, 105–162.

Kushiro, I., Yoder, H. S., and Nishikawa, M. (1968). Effect of water on the melting of enstatite. *Geol. Soc. Am. Bull.* **79**, 1685–1692.

Mysen, B. O., Kushiro, I., Nicholls, I. A., and Ringwood, A. E. (1974). A possible mantle origin for andesitic magmas: discussion of a paper by Nicholls and Ringwood. *Earth planet. Sci. Letters* **21**, 221–230.

# Anatexis of mafic crust and high pressure crystallization of andesite

T. H. Green

School of Earth Sciences,
Macquarie University, North Ryde, New South Wales 2113, Australia

ABSTRACT

The major development of calc-alkaline andesite occurs in evolved island arcs with crustal thickness greater than 30 km, and in continental marginal regions. In both cases, the andesite may be derived either by partial melting of hydrated, mafic lower crust or by fractional crystallization of underplated hydrous mafic magmas, directly or indirectly linked to melting processes in the subducted oceanic crust and the mantle wedge overlying the Benioff zone. However, andesites, unlike many basalts, rarely show any evidence of deep level history in the form of high pressure phenocrysts or inclusions, and are probably greatly affected by shallow level processes.

High pressure experimental work is able to delineate broadly possible source compositions and fractional crystallization processes operative in the deep crust and upper mantle for the formation of magmas parental to calc-alkaline andesite. The results for a mafic source composition point to two main models: (1) an amphibole model at 8–12 kb; (2) an eclogite model at over 25 kb, and a transitional garnet-amphibole model in the 12–25 kb range.

Thus at 8–12 kb amphibole is the dominant crystallizing or residual phase from a hydrous mafic composition. Derived liquids have major element characteristics of calc-alkaline andesite and related rocks. Also at over 25 kb, for dry conditions and c. 40–50 per cent melting of a quartz eclogite, an andesite-like liquid may be obtained. In the presence of water, dacite and rhyolite melts may result from lower degrees of melting, but for lower degrees of melting under dry conditions trachytic rather than andesitic liquids form. Trace element patterns, particularly REE, may bear the imprint of these models, but crystallization processes at shallower levels and a residual role for trace element-enriched accessory phases (sphene, apatite, etc.) could significantly modify the trace element concentrations.

The two major controlling factors governing the formation of calc-alkaline andesites are (1) the addition of water to the mantle and deep crust via dehydration and melting reactions in the hydrated, subducted oceanic crust and (2) the presence of a thickened crust in evolved island arc and continental marginal regions forming a density contrast at c. 8–12 kb pressure and encouraging fractional crystallization of upwelling magma at these levels

Introduction

This contribution deals with experimental work related to the petrogenesis of calc-alkaline andesite and, in particular, discusses fractional crystallization of hydrous basaltic magma and the role of melting processes in the mafic parts of subducted lithosphere and of thickened (underplated) continental crust. It is not directly concerned with the origin of orogenic magmas with tholeiitic affinites (e.g. the island arc tholeiitic series; Jakeš and Gill, 1970) which may be related to models involving hydrous anatexis of mantle peridotite and are discussed in the contribution by Mysen in this Section. It is appropriate, however, to note that recent petrological and mineralogical studies question the validity of models involving over 2 per cent H_2O present, and favour essentially dry melting processes (Ewart, 1976a), which some workers regard as occurring in the mafic part of the subducted slab (Marsh and Carmichael, 1974). Certainly, water-saturated melting is eliminated as a viable model for the origin of the voluminous basaltic andesites of the island arc tholeiitic series, but water-undersaturated mantle melting, and subsequent olivine-dominated fractionation, is still a possible mechanism (Nicholls and Ringwood, 1973; Ringwood, 1974, 1975; Wyllie, 1977b; Cawthorn and O'Hara, 1976).

The present chapter reviews evidence for high pressure origin of calc-alkaline andesite, and then discusses experimental work relevant to phase relationships and fractional crystallization trends in melts from a number of mafic compositions at high pressure, together with determination of high pressure phase relationships in andesite and other members of the calc-alkaline series. This has the aim of testing possible connections between andesite and a mafic parent at high pressure either using representative natural rocks or using model compositions. Experimental work falls into two complementary categories: (1) study of complex natural compositions and (2) study of simplified systems in which specific relations not determinable in the complex system may be identified and applied to the natural situation. Because of this, a brief reference to recent experimental work on relevant simple systems will be made before detailed presentation of results for the complex systems.

Geological Evidence for High Pressure Origin and Crystallization on Andesite

It is instructive to apply to andesites the arguments used to support the conclusion that basaltic magmas are generated in the Earth's mantle. In the case of calc-alkaline andesites (1) geochemical data point to a source in either the upper mantle or lower crust (e.g. McNutt et al., 1975; Noble et al., 1975; James et al., 1976; Francis et al., 1977; see also Hawkesworth, this volume, Section VIII; Pearce, this volume, Section VIII); (2) estimated temperatures (c. 950–1100 °C) for calc-alkaline magmas based on both experimental petrology (e.g. Eggler, 1972; Wyllie et al., 1976) and phenocryst geothermometry (e.g. Ewart et al., 1977; Heming, 1977) indicate generation of the magmas in the upper mantle or deepest crust, for generally accepted models of geothermal gradients (Clark and Ringwood, 1964; Oxburgh and Turcotte, 1970). These arguments are similar to those applied to basalts (except for the possible involvement of the lower crust). However, basalts frequently contain direct evidence of mantle origin, in the form of ultramafic xenoliths and megacrysts which could only be stable at high pressure, equivalent to upper mantle depths. Andesites, in contrast, rarely contain xenoliths of high pressure origin. This evidence is restricted to hornblende-rich gabbroic xenoliths in Japan and the West Indies, which indicate a depth of crystallization of 15–40 km (Yamazaki et al., 1966; Green and Ringwood, 1969; Lewis, 1973; Powell, 1978), i.e. the deep crust or uppermost mantle in these areas, and to a suite of mantle-type ultramafic xenoliths (including spinel lherzolites) and Cr-diopside xenocrysts from Japan which point to the passage of the host magma through mantle material (Aoki and Oji, 1966; Shimazu et al., 1974; Katsui et al., 1978).

The ability of basalts to rise rapidly from mantle depths, with concomitant sampling of mantle material, is attributed to the volatile content and subsequent depth at which the magma may become explosive, and erupt rapidly to the surface through tensional zones (Green and Ringwood, 1967b). This is consistent with the common occurrence of ultramafic xenoliths and megacrysts in relatively volatile-rich alkali basalts, basanites, and nephelinites, and their rarity in the generally volatile-poorer tholeiitic magmas. Many workers have suggested that andesites have a significant volatile content (e.g. 0.5–5 per cent H_2O, subordinate CO_2, Cl, F; Eggler, 1972; Anderson, 1973, 1978; Marsh and Carmichael, 1974) and yet xenoliths from deep levels are rare. This contrast with basalts may be attributed to the following:

(1) The tectonic regime for the generation of andesites is dominated by convergent plate tectonics, and the magma rises through the deep crust diapirically, rather than violently, and crystallizes as it moves, leaving behind any denser components, such as ultramafic xenoliths.

(2) The main site of generation of calc-alkaline andesites is the lower crust where melting is possibly initiated by mantle magmatic processes, but where no sampling of actual solid mantle material occurs.

(3) Andesites, as erupted at the surface, are the result of extensive crystal fractionation of mantle-derived initially water-undersaturated magmas, under conditions where crossing of water-saturated liquidus curves occurs at shallow levels. In this situation the magma *cannot* rise without concomitant crystallization (cf. Nicholls and Ringwood, 1973) and derivative rather than primary magmas from the mantle are the result. Any dense, high pressure xenoliths will be deposited along with early-formed crystals during the fractional crystallization process producing the derivative magmas.

(4) Andesites rise through the mantle in a superheated condition (cf. Marsh, 1976).

Further indirect evidence is provided by application of plate tectonic concepts (especially subduction of oceanic crust) and the apparently invariable link between calc-alkaline andesite eruptions and currently or recently active Benioff zones. The correlation of a hiatus in the Benioff zone with possible melting of a subducted slab immediately beneath currently active calc-alkaline volcanoes is of particular relevance since this suggests depths of melting of 80–240 km (Hanus and Vanek, 1978). As emphasized by Ringwood (1974, 1975), a major result of subduction is that it provides a mechanism for transfer of water from the hydrosphere to the mantle via hydrated phases in the subducted slab. These eventually dehydrate, releasing water, and ultimately instigate melting in the mantle overlying the Benioff zone or in the subducted oceanic plate (Ringwood, 1974, 1975).

Typical oceanic tholeiite, when buried to deep levels in subduction zones, will eventually transform to near-anhydrous quartz eclogite (Green and Ringwood, 1969). Thus, in the plate tectonic model, comparatively large volumes of quartz eclogite are being added to the mantle, and yet when natural geological processes such as kimberlitic eruptions sample the mantle, bringing ultramafic and eclogitic xenoliths to the surface, quartz eclogites are found to be extremely rare (O'Hara, 1963; Mathias *et al.*, 1970). One possible explanation is that the quartz eclogite undergoes partial melting, producing silica-rich melts and a low silica eclogitic residuum (Green, 1967; Ringwood, 1975). Such melts may be the precursors of calc-alkaline volcanism.

Finally, it is significant that calc-alkaline andesites are very rare in the earliest stage of development of oceanic island arcs, and characteristically occur after a crust of more than 30 km is formed in island arc areas, or are erupted through a thickened crust in continental marginal regions. The density contrast between the thickened crust ($\rho \sim 2.7$–3.0 g cm^{-3}) and the upper mantle ($\rho \sim 3.3$ g cm^{-3}) may result in a break in the rate of progress of mantle-derived magmas to the surface, with a tendency for the magmas to underplate or interleave with the crust, and fractionally crystallize at pressures of 8–15 kb, corresponding to the crust/mantle interface. This fractional crystallization may be critical for producing, in destructive plate margins, a different magma series (the calc-alkaline series) from that obtained when mantle-derived magmas are not obstructed by a thickened crust in their upward rise to the surface (in which case the island arc tholeiitic series may result). Obviously magma series intermediate between these two extremes could occur where the crust is of intermediate thickness (e.g. between 10 and 30 km). The Talasea peninsula, New Britain may be such an example (Lowder and Carmichael, 1970). Similarly the island arc volcanics of the New Georgia group, Solomon Islands, occur on relatively thin crust. In this case the suite is dominated by basalt to basaltic andesite, not andesite, and the rocks are postulated as resulting from shallow to moderate level fractional crystallization of a hydrous parental basalt (Cox and Bell, 1972).

However, at least one island arc (the Banda arc: Whitford and Jezek, 1977; *see also* Hutchison, this volume, Section III) does not fall into this pattern and has abundant calc-alkaline andesite occurring on a crust *c*. 15 km thick. This arc has unusual tectonic, chemical, and mineralogical characteristics (e.g. the presence of phenocrysts of cordierite and almandine garnet) and the proposed origin involves melting of subducted sediments (Whitford and Jezek, 1977). Such an hypothesis cannot be applied generally to calc-alkaline andesites, and this case remains an unusual exception to the general observation made earlier that voluminous calc-alkaline andesite eruptions only occur in island arc or continental marginal regions where the crust is more than 30 km thick.

Much of the foregoing discussion is indirect evidence concerning the high pressure origin of calc-alkaline andesite, but nevertheless these points need to be considered in petrogenetic models. Thus, in this contribution the nature of melts produced in subducted hydrous oceanic crust, or in thickened and underplated continental crust in continental marginal regions, will be examined, together with the 'intermediate depth' hydrous fractional crystallization of mantle-derived magmas (i.e. crystal fractionation of basalt and andesite at 8–15 kb). Oceanic crust, prior to subduction, may undergo significant hydration and chemical alteration which could affect phase relationships and the nature of melt products obtained (cf. Fyfe, 1975, and this volume, Section X). Accordingly, high pressure experimental work on both a spilite and a K-rich oceanic 'tholeiite' will be dealt with briefly.

Review of High Pressure Experimental Work on Relevant Simple Systems

Plagioclase and pyroxene solid solution series constitute two important mineral groups in calc-alkaline andesites. Yoder (1969) reviewed simple system experimental work on the albite–anorthite and diopside–anorthite–H_2O systems. He emphasized the role of fluctuating water pressure at constant temperature in explaining the complex zoning patterns often observed in plagioclase phenocrysts in andesites. Also, increased water pressure shifts the eutectic in the diopside–anorthite system towards anorthite. This may contribute to the high normative plagioclase content of calc-alkaline andesites. Kushiro (1973) revised the position of the cotectic boundary in the system diopside–albite–anorthite at 1 atm, documented, via use of the electron microprobe, the complexity of the solid solutions (especially pyroxenes) to be expected in the natural systems, and noted that the composition–temperature relation depicted by the plagioclase 'binary' loop is similar to the observed behaviour of plagioclase in basalts and andesites.

Although detailed work has been completed on these systems at atmospheric pressure, only exploratory work has been conducted at elevated pressure (dry or hydrous) and only general predictions can be made, such as the shift of the eutectic in the Di–Ab–An system towards Ab–An with increasing pressure (Lindsley and Emslie, 1968). Similarly, no results are available on the shape of the albite–anorthite binary loop at elevated water pressure, and this is critical in evaluating compositions of *liquidus* plagioclase crystallizing from hydrous calc-alkaline magmas. Lindsley (1968) depicts the shape of the binary loop in the Ab–An system at 10 and 20 kb pressure and demonstrates significant changes in phase relations but largely parallel positions for the binary loop to 10 kb (*see* Fig. 1).

Another major thrust of simple system studies has been to incorporate FeO as one of the components and concentrate on the effect of oxygen fugacity in controlling the appearance of magnetite as a near-liquidus phase in the system $MgO-FeO-Fe_2O_3-SiO_2$. This work was pioneered and then reviewed by Osborn (1969). It demonstrates a mechanism, via fractional crystallization of magnetite from basalt, for obtaining successive liquids having near constant 100 Mg/(Mg + Fe) and increasing silica. This lack of Fe enrichment, especially for intermediate compositions, has long been recognized

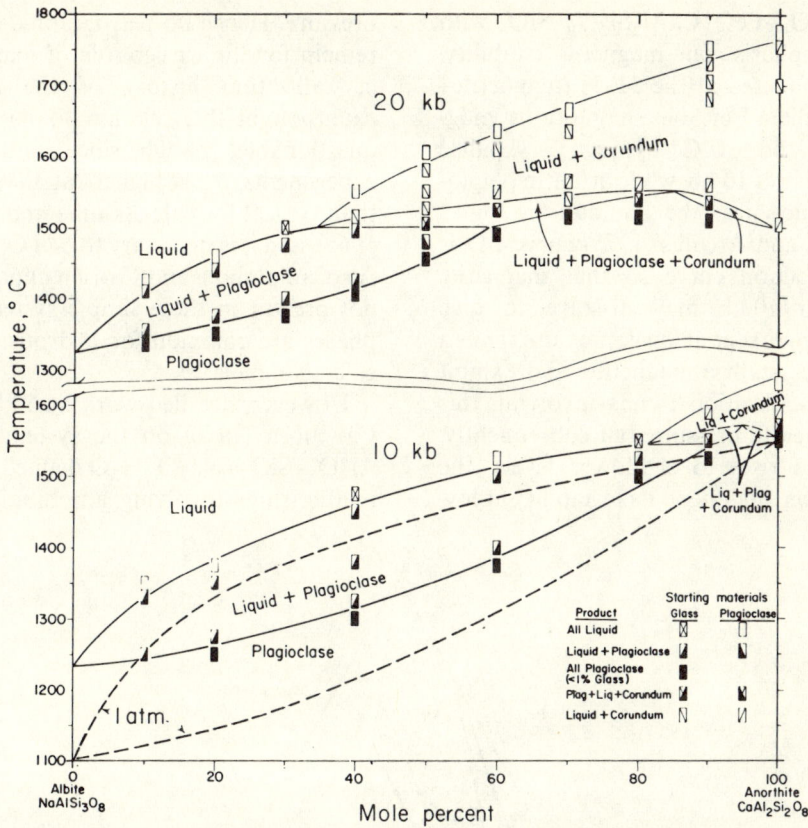

Fig. 1 Melting relations of plagioclase feldspars at 1 atm, 10 kb, and 20 kb. Heavy lines show binary equilibrium in 10 and 20 kb sections; lighter lines show ternary equilibrium. (From Lindsley, 1968; reproduced by permission of the New York State Museum and Science Service)

as a characteristic feature of calc-alkaline rock sequences, clearly distinguishing them from tholeiitic sequences (e.g. via use of an AFM diagram; Poldervaart, 1949; Tilley, 1950). Osborn (1959, 1969) concluded that for the magnetite fractional crystallization mechanism to be valid, constant oxygen fugacity (f_{O_2}) must be maintained, with addition of oxygen to the system. He suggested that these conditions might be attained during hydrous fractional crystallization of basaltic magma, but did not consider the role of hydrous phases, such as amphibole.

Eggler (1974) extended this approach to a portion of the system $CaAl_2Si_2O_8$–$NaAlSi_3O_8$–SiO_2–MgO–Fe–O_2–H_2O–CO_2 and showed that magnetite (or ilmenite) is a near-liquidus phase in silicic liquids at 'normal' f_{O_2} values, such as the NNO buffer, and does not require any increase in oxygen in the system. Total pressure between 1 and 5.5 kb has little effect on the magnetite phase relationships. This suggests that magnetite may crystallize from andesitic-rhyodacitic magmas, but from other high pressure experimental work (on natural systems) (Eggler, 1972; Eggler and Burnham, 1973) it appears unlikely that magnetite is an important near-liquidus phase up to 10 kb pressure for compositions between basalt and andesite at f_{O_2} values measured in these bulk compositions (f_{O_2} approximates the NNO (nickel–nickel oxide) to QFM (quartz–fayalite–magnetite) buffer; Fudali, 1965).

Osborn and Arculus (1975) documented the effect of increased pressure (to 10 kb) on the system Mg_2SiO_4–FeO–$CaAl_2Si_2O_8$–SiO_2 with particular emphasis on magnetite stability. Oxygen was buffered at the M–H (magnetite–haematite) buffer. For these conditions magnetite, olivine, and low Ca pyroxene crystallize near the liquidus at 10 kb, while at 1 atm plagioclase precipitates near the liquidus. In Fig. 2 (from Osborn and Arculus, 1975) curve bD is possibly a reaction curve so that magnetite crystallized at 10 kb may dissolve to give anorthite and pyroxene at lower pressure. Thus a magma may crystallize magnetite, and exhibit fractional crystallization trends involving this phase, at moderate pressure, but subsequently, as the magma rises to shallower levels, the magnetite reacts out, due to its instability at low pressure. Hence no petrographic evidence may remain for the earlier role of magnetite in the crystallization history of the magma. The extension of this relation to natural rocks is questionable, though, since in the laboratory experiments f_{O_2} is higher (at the M–H buffer) than typical for calc-alkaline andesites (Fudali, 1965), and it is necessary to *add* O to the system. Also, amphibole and Ca-rich clinopyroxene are not present in these simple systems, and both phases are common for hydrous conditions in calc-alkaline lavas.

However, detailed work at 5 kb pressure by Cawthorn (1976) on the system CaO–MgO–Al_2O_3–SiO_2–Na_2O–H_2O defined critical phase relationships involving amphibole which may

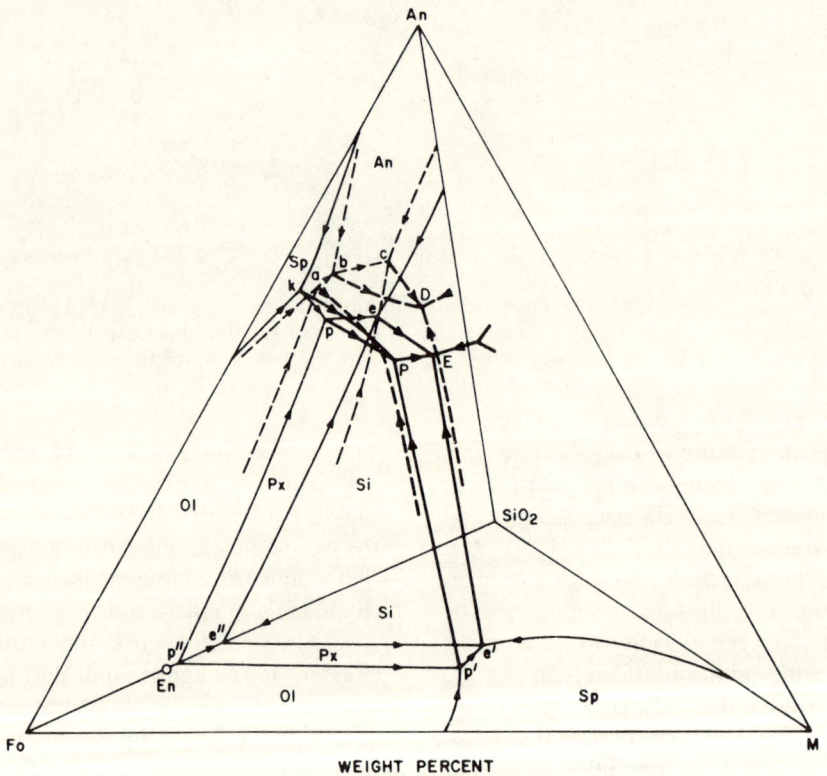

Fig. 2 Tetrahedron representing the system Fo–M–An–SiO_2. Solid lines are boundary curves at 1 atm pressure in air. Broken lines are estimated positions of boundary curves at 10 kb pressure, using a magnetite–haematite buffer. (From Osborn and Arculus, 1975; reproduced by permission of the Carnegie Institution of Washington.) Abbreviations: Fo = Mg_2SiO_4; M = Fe_3O_4; An = $CaAl_2Si_2O_8$; En = enstatite; Px = orthopyroxene; Ol = olivine; Sp = spinel (magnetite); Si = tridymite or cristobalite

have application to amphiboles crystallizing from calc-alkaline magmas (Cawthorn and O'Hara, 1976). Important conclusions are as follows:

(1) Amphibole composition is governed by the bulk composition. Thus amphiboles crystallizing from SiO_2-undersaturated liquids (up to 11 per cent normative nepheline) are themselves SiO_2-undersaturated pargasite, and fractional crystallization may result in silica-oversaturated derivative liquids. Amphiboles crystallizing from silica-saturated liquids are hypersthene-normative.

(2) The amphibole stability field increases markedly with increasing Na_2O content, and amphibole does not crystallize from liquid with less than 3 per cent Na_2O. Thus in natural systems early crystallization of Na-poor phases may cause relative enrichment of Na_2O in the remaining liquid, and eventually when the Na_2O content exceeds the critical value (more than 3 per cent according to Cawthorn's results) amphibole appears in the crystallization sequence.

(3) A reaction relation between olivine plus calcic clinopyroxene and liquid occurs at temperatures where amphibole becomes stable, although in some instances co-precipitation of amphibole and clinopyroxene is possible. This reaction relation suggests that hydrous basic-intermediate magmas in the calc-alkaline series have a large field of crystallization of amphibole (and plagioclase) after a small, near-liquidus temperature interval where olivine and/or calcic clinopyroxene crystallize. This is illustrated schematically in Fig. 3, from Cawthorn and O'Hara (1976), together with the effect of increasing Na_2O enhancing the field of crystallization of amphibole.

In summary, Cawthorn (1976) and Cawthorn and O'Hara (1976) demonstrate for model systems the possible link between a basic parent magma and an intermediate or andesitic derivative via amphibole-dominated equilibrium or fractional crystallization processes. However, the absence of FeO in their study means that the important lack of Fe-enrichment trend in the

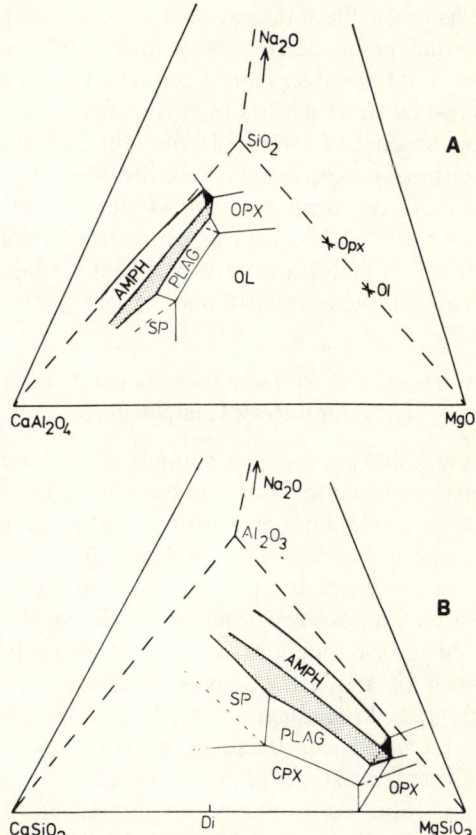

Fig. 3 (A) Phase relations within the volume $CaAl_2O_4$-MgO-SiO_2-Na_2O projected from diopside under 5 kb with excess vapour. SiO_2 lies at the back of the tetrahedron in this view. The amphibole volume (shaded for emphasis) does not intersect the base plane ($CaAl_2O_4$-MgO-SiO_2). The lowest Na_2O level at which it occurs is 3 per cent. The range of amphibole stability widens with increasing Na_2O content. (B) Phase relations within the volume $CaSiO_3$-$MgSiO_3$-Al_2O_3-Na_2O projected from olivine under 5 kb with excess vapour. Al_2O_3 lies at the back of the tetrahedron. Abbreviations: Ol = olivine; Opx = orthopyroxene; Cpx = calcium-rich pyroxene; Plag = plagioclase; Sp = spinel; Amph = amphibole. (From Cawthorn and O'Hara, 1976; reproduced by permission of *The American Journal of Science*)

calc-alkaline series cannot be evaluated. To conclude this discussion on relevant simple system studies it is clear that while both Osborn and coworkers' and Cawthorn's experiments illustrate important points which may have application to the natural system, close modelling of actual liquid derivatives with natural rock types is not possible. Thus it is difficult to relate the two separate models to each other because of the absence of amphibole and clinopyroxene in Osborn's experiments and the absence of magnetite (or ilmenite) in Cawthorn's study. These shortcomings lead us ultimately to consider direct high pressure experimental studies on natural rock compositions.

Review of High Pressure Experimental Work on Natural Rock Compositions

In the following sections composite $P–T$ diagrams are given for tholeiitic basaltic and calc-alkaline andesitic compositions. The actual analyses of the rocks used are listed in Table 1. In general, there is little variation in the sequence of appearance of phases or the position of the phase boundaries for the respective basaltic or andesitic compositional groups, so that generalized diagrams such as Figs. 4–9 can be presented. In some instances, where significant variations in phase boundaries with composition occur, these are plotted in the figures or discussed briefly in the text. The emphasis in this section is on the near-liquidus phases controlling any fractional crystallization trends in the basaltic composition, and evaluation of possible links between andesite and basalt either via fractional crystallization or partial melting of the latter. The diagrams presented and discussed are for the extremes of dry and water-saturated conditions, and for intermediate, water-undersaturated conditions with 5 per cent by weight of H_2O present. As indicated earlier, several lines of evidence suggest between 0.5 and 5 per cent H_2O present in andesite, and the largest body of high pressure experimental data is available for compositions with 5 per cent H_2O added.

Dry conditions (Figs 4, 5)

(1) Andesite and basalt have generally similar liquidus temperatures in the 0–10 kb pressure range, although it is noteworthy that some andesites may have a higher liquidus temperature than associated basalts, for dry conditions (Brown and Schairer, 1971). This is attributed to the high temperature of crystallization of plagioclase and results in a thermal barrier ('plagioclase barrier') between basaltic compositions and andesitic or more evolved compositions (Ringwood, 1975).

(2) From 0 to 10 kb plagioclase is the major liquidus or near-liquidus phase in both compositions. It is joined by olivine and then pyroxene in the basalt, but olivine is not present in the andesite, and instead plagioclase is joined by pyroxene and then quartz, with falling temperature.

(3) The role of a spinel phase is difficult to evaluate for the data available. It is recorded in an olivine tholeiite, quartz tholeiite, and andesite

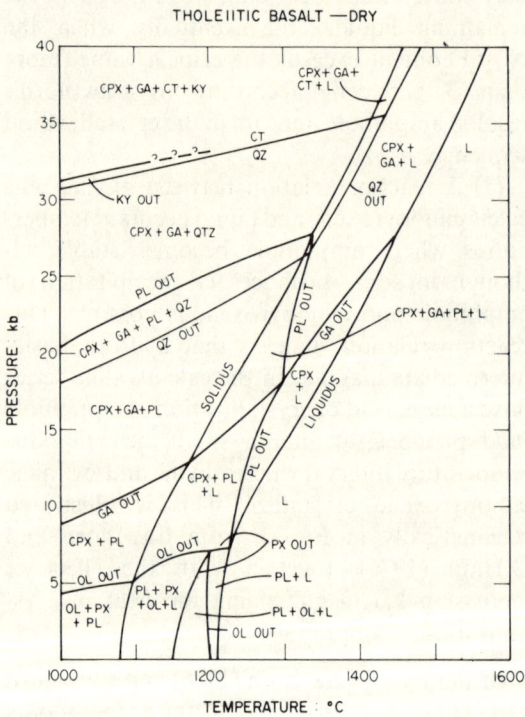

Fig. 4 Composite $P–T$ diagram for anhydrous tholeiitic basaltic compositions. Abbreviations used in this and subsequent figures (Figs. 4–11) are: CPX = clinopyroxene, GA = garnet, CT = coesite, KY = kyanite, L = liquid, PL = plagioclase, QZ = quartz, OL = olivine, PX = pyroxene, AM = amphibole, ZO = zoisite, BI = biotite, MT = magnetite

TABLE 1 Tholeiitic and andesitic compositions (anhydrous) used in high pressure and atmospheric experimental studies

	1	2	3	4	5	6	7	8	9	10	11	12	13	14	15	16	17
SiO_2	47.28	49.11	50.71	48.15	49.93	50.3	52.9	49.8	50.3	49.0	59.79	60.62	62.08	62.2	60.24	60.50	59.10
TiO_2	0.97	2.51	1.70	0.89	1.34	1.7	1.5	1.6	1.6	1.7	0.80	0.56	0.56	1.1	0.69	0.91	0.94
Al_2O_3	17.71	12.74	14.48	18.23	16.75	17.0	16.9	16.5	14.2	16.3	18.43	17.67	17.28	17.3	16.89	17.30	17.80
Fe_2O_3	2.40	0.80	2.16	1.04		1.5	0.3	3.0	3.7		2.34	2.64	0.70	0.3	0.88		
FeO	7.90	10.52	11.53	8.29	11.40	7.6	7.9	5.7	8.0	8.3	3.66	4.40	4.18	5.9	5.35	4.30	6.43
MnO	0.23	0.17	0.22	0.17	0.18	0.16	0.2	0.2	0.1	0.2	0.11	0.18	0.10	0.1	0.13		0.10
MgO	7.70	10.31	4.68	8.94	7.59	7.8	7.0	9.8	9.4	9.8	2.53	2.72	2.86	2.4	3.14	3.80	3.05
CaO	13.95	10.73	8.83	11.29	9.33	11.4	10.0	11.1	8.4	11.0	5.98	6.39	7.71	5.2	7.22	6.30	6.85
Na_2O	1.68	1.97	3.16	2.79	2.92	2.8	2.7	2.6	4.5	2.5	3.85	4.05	2.83	3.3	3.91	4.30	4.27
K_2O	0.14	0.49	0.77	0.14	0.37	0.18	0.6	0.2	0.2	1.4	2.21	0.67	1.58	2.3	1.26	1.69	1.08
P_2O_5	0.04	0.27	0.36	0.07	0.19						0.30	0.10	0.12		0.20		0.22
CIPW norms																	
Q	0.83	2.90	0.76	0.83	2.19	1.06	1.3	1.2	1.2	8.3	11.81	15.80	17.91	15.5	11.47	8.77	8.16
or	14.21	16.66	4.55			23.7	3.5	22.0	35.6	17.0	13.06	3.90	9.33	13.6	7.44	9.98	6.38
ab			26.73	23.09	24.70		22.8		1.4	2.3	32.56	34.20	23.94	27.9	33.07	36.37	36.11
ne				0.27													
an	40.37	24.47	23.06	36.81	31.51	33.3	32.2	32.8	18.0	29.1	26.49	28.02	29.78	25.6	24.82	22.92	26.22
di	23.23	21.94	15.41	15.24	11.19	18.9	14.2	17.7	19.1	20.6	0.99	2.35	6.32	0.2	8.02	6.80	5.28
hy	7.25	19.86	20.90		13.53	12.0	22.6	13.8			9.49	10.72	10.36	14.9	12.03	12.53	15.38
ol	8.69	7.24		20.40	13.90	6.1		5.6	16.9	19.7							
mt	3.48	1.16	3.13	1.51		2.2	0.4	4.4	5.4		3.39	3.77	1.01	0.4	1.28		1.79
il	1.84	4.77	3.23	1.69	2.54	3.2	2.8	3.0	3.0	3.2	1.52	1.04	1.06	2.1	1.31	1.73	
ap	0.09	0.63	0.83	0.16	0.44						0.70	0.23	0.28		0.46		0.51

(1) Gabbro (Stern and Wyllie, 1973, 1978; Lambert and Wyllie, 1972). (2) Olivine tholeiite; 1921 Kilauea Iki (Yoder and Tilley, 1962; Holloway and Burnham, 1972). (3) Quartz tholeiite (Picture Gorge) (Helz, 1976). (4) High alumina basalt (Warner) (Yoder and Tilley, 1962). (5) High alumina olivine tholeiite (Cohen *et al.*, 1967). (6) High alumina olivine tholeiite (Green and Ringwood, 1968). (7) High alumina quartz tholeiite (Green and Ringwood, 1968). (8) Olivine tholeiite (Hellman, 1979). (9) Spilitized 'tholeiite' (Hellman, 1979). (10) K-rich 'tholeiite' (Hellman, 1979). (11) Quartz diorite (Stern and Wyllie, 1973; Lambert and Wyllie, 1974). (12) Andesite (Caribbean 19K-A) (Brown and Schairer, 1971). (13) Andesite (Caribbean 21L) (Brown and Schairer, 1971). (14) Andesite (Green and Ringwood, 1968). (15) Andesite (Fiji) (Green, 1972). (16) Andesite (Paricutin) (Eggler, 1972). (17) Andesite (Mt Hood) (Eggler and Burnham, 1973).

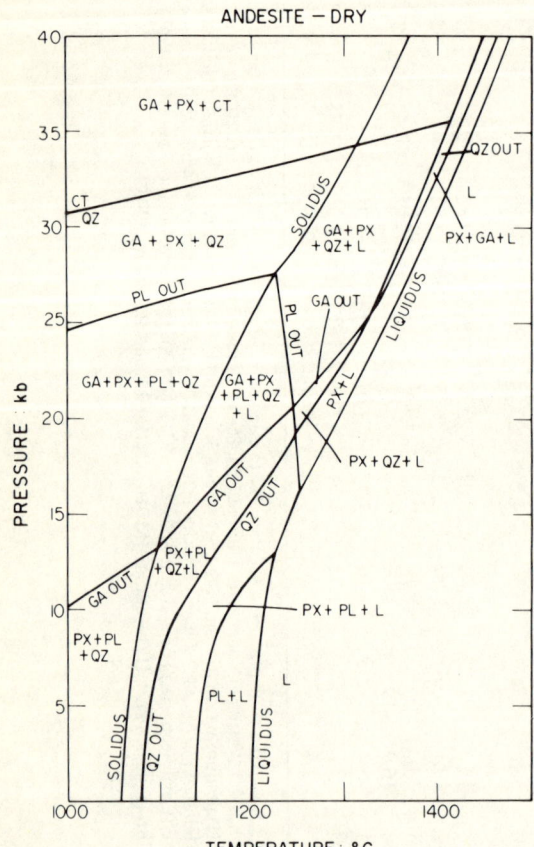

Fig. 5 Composite P–T diagram for anhydrous andesitic compositions

(D. H. Green and Ringwood, 1967*a,b*; T. H. Green and Ringwood, 1968) and appears to have very restricted stability near the solidus in these compositions at pressures between 0 and 13.5 kb. (The f_{O_2} for these experiments was not controlled, but was probably between the NNO and QFM buffers.)

(4) Above *c*. 7 kb in the basaltic compositions and above *c*. 10 kb in the andesitic compositions pyroxene replaces plagioclase on the liquidus, and olivine is eliminated from the basaltic compositions.

(5) Garnet appears above the solidus in both compositions at pressures over 14 kb, and is a near-liquidus phase with clinopyroxene in the basalt at pressures above 25 kb, while it occurs at *c*. 20 °C below the liquidus for most andesites studied up to 40 kb. However, in slightly more silicic compositions with lower 100 Mg/(Mg + Fe) (e.g. analysis 14, Table 1) garnet becomes the liquidus phase at pressures above 25 kb, joined by clinopyroxene at 20 °C below the liquidus.

(6) The plagioclase field contracts with increasing pressure for both compositions. Thus it only appears in the partial melting field near the solidus at above 20 kb, and is not present at all above *c*. 27 kb.

(7) Quartz (or coesite) occurs near the solidus of the basalt at above 20 kb, while at the same pressures in the andesite it is present only 30 °C below the liquidus.

Five per cent H_2O added (Figs 6, 7)

(1) Compared with dry crystallization, the liquidus and solidus are depressed for both compositions, and the field of partial melting greatly expanded to a temperature interval of *c*. 500 °C, compared with *c*. 150 °C for dry conditions.

(2) At low pressure the appearance of plagioclase in both compositions is markedly depressed and for andesites the 'plagioclase barrier' (Ringwood, 1975) is eliminated, with plagioclase no longer the liquidus phase, and the andesite liquidus temperature depressed below that of basalt, for similar water contents.

(3) Amphibole appears in both compositions, reaching a maximum temperature of stability of 1050–1080 °C in the tholeiite at 5–18 kb, and 950–980 °C for the same pressure range in the andesite. The maximum pressure of stability is *c*. 30 kb in the basalt and *c*. 25 kb in the andesite.

(4) Pyroxene together with olivine, and not plagioclase, are important near-liquidus phases in the basalt from 2 to *c*. 14 kb, but olivine reacts out with the incoming of amphibole at *c*. 100 °C below the liquidus. Olivine does not appear in the andesite and pyroxene is the important near-liquidus phase, joined by amphibole 30–100 °C below the liquidus. The phase relations in this P–T region are very dependent on composition, and the relative roles of ortho- and clinopyroxenes and the approach of amphibole to the

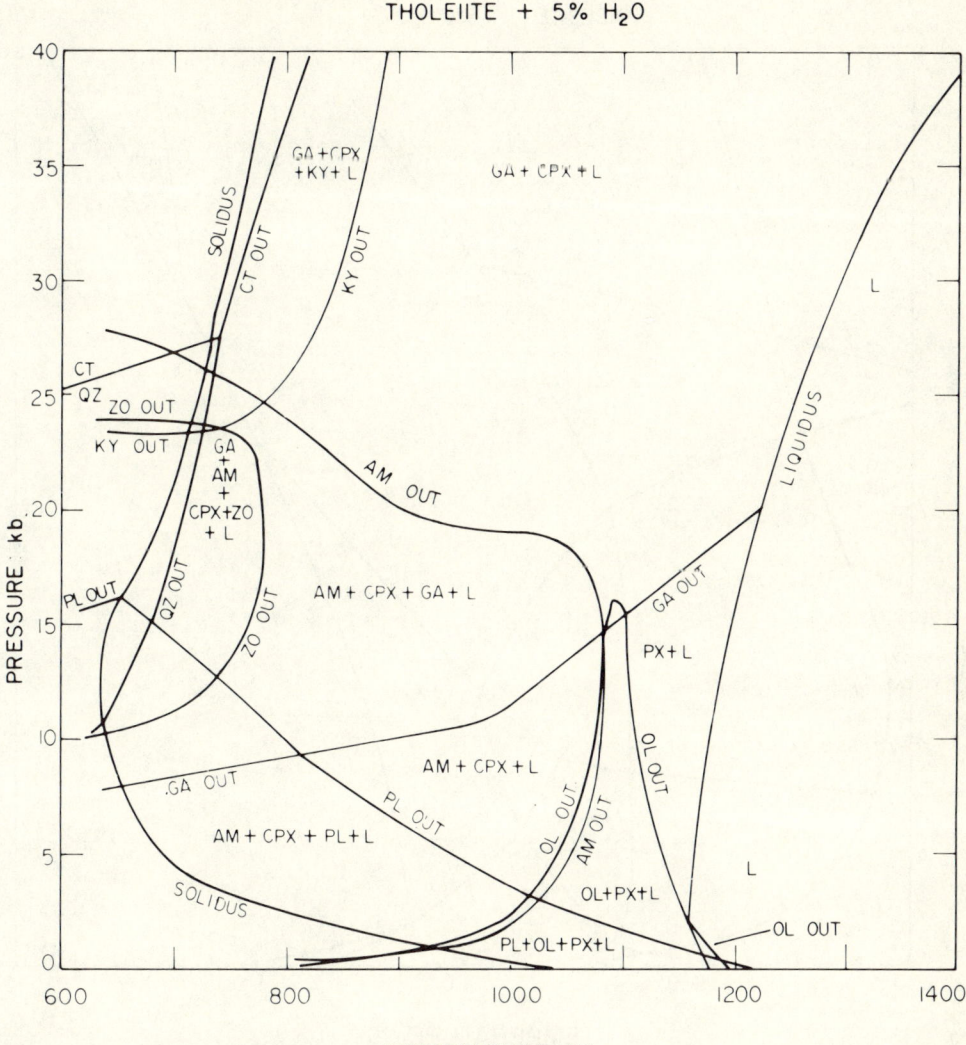

Fig. 6 Composite *P–T* diagram for tholeiitic compositions with 5 per cent by weight of added water

liquidus shows great variation, both as a function of composition and of f_{O_2} (Allen *et al.*, 1975).

(5) Bulk composition and f_{O_2} also are critical in governing the role of magnetite. Thus data from Holloway and Burnham (1972) using H_2O-CO_2 fluid mixes point to the appearance of magnetite in a tholeiite at 0–8 kb and *c.* 1080 °C for f_{O_2} corresponding to the NNO buffer. This temperature is slightly above that of the appearance of amphibole. In contrast, Eggler and Burnham (1973) noted that magnetite disappears at a much lower temperature than the incoming of amphibole in an andesite at pressures above 5 kb for H_2O-CO_2 fluid mixes and the QFM oxygen buffer. However, the role of magnetite is not known at higher pressure and it has not been included in Figs 6, 7.

(6) Garnet appears below the liquidus in both compositions at a pressure of 7–10 kb and by 15 kb is a prominent near-liquidus phase, along with clinopyroxene. Garnet and clinopyroxene are liquidus phases from 20 to at least 40 kb in the basalt, and from *c.* 22 to 30 kb in the andesite. Above 30 kb, garnet is the sole liquidus phase in

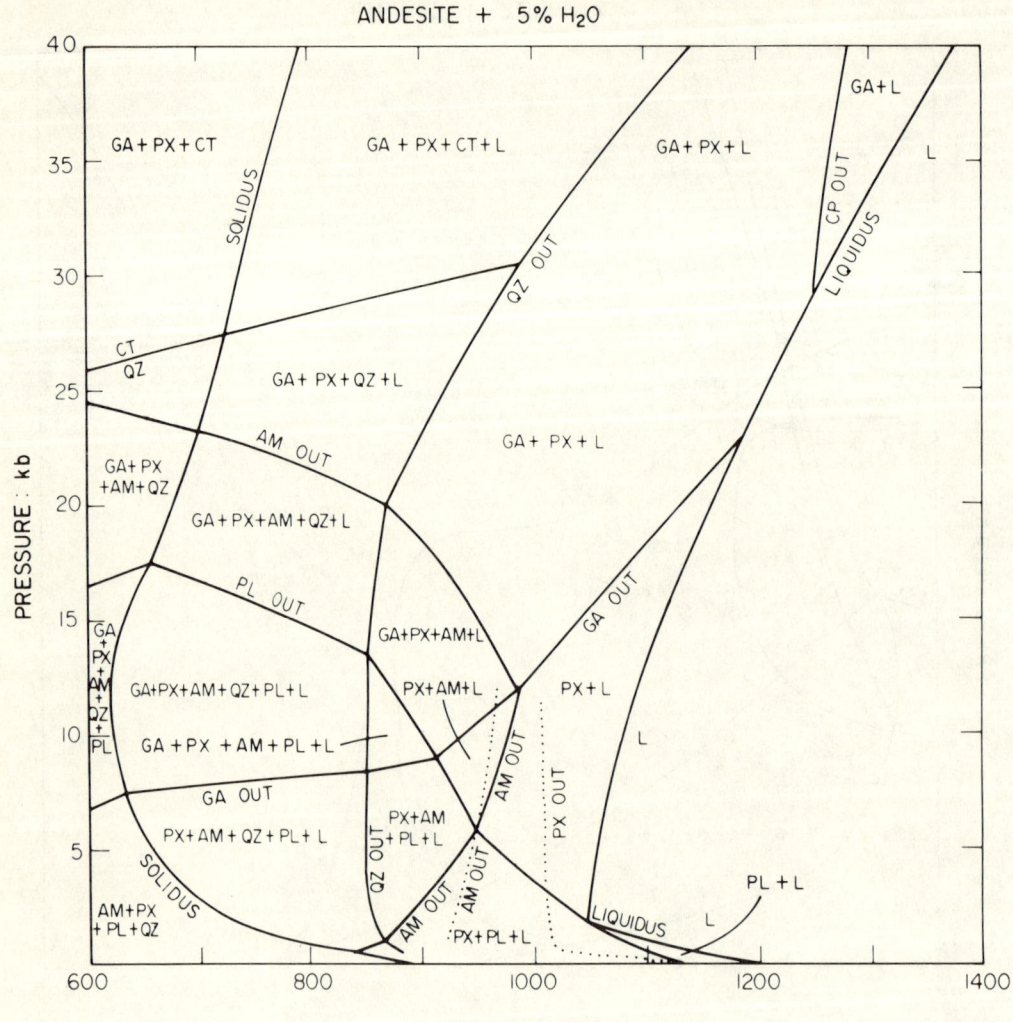

Fig. 7 Composite *P–T* diagram for andesitic compositions with 5 per cent by weight of added water. Dotted lines refer to curves for Mt Hood andesite (Eggler and Burnham, 1973)

the andesite. There is a marked expansion of the field of crystallization of garnet and clinopyroxene in *both* compositions, compared with the phase relations for dry conditions.

(7) Quartz (or coesite) appears within 30–40 °C above the solidus in the tholeiite at pressures above 10 kb, and at c. 200 °C above the solidus in the andesite.

(8) Zoisite is an important near-solidus phase in the tholeiite between 10 and 25 kb at temperatures below 780 °C, while the field of kyanite is expanded in the hydrous tholeiite, compared with the dry tholeiite and occurs at temperatures below 820 °C and pressures above 22 kb. Neither of these phases has been reported from andesite + 5 per cent H_2O, due probably to lack of combined scanning electron microscope/electron microprobe study of definitive runs (this technique has proved invaluable in identifying phases otherwise undetected in near-solidus runs, e.g. Hellman and Green, 1979*b*).

(9) Although not recorded on Fig 6 and 7, sphene has been identified in both andesitic and basaltic compositions. It occurs from near-

solidus temperatures to upwards of 50 per cent melting at a temperature of 1020 °C and at pressures up to 10 kb in a tholeiite, and in model altered tholeiitic compositions (spilite and K-rich 'tholeiite'). Staurolite also is present in the tholeiite at 24–26 kb and 740–760 °C (Hellman and Green, 1979a,b). The altered 'tholeiitic' compositions exhibit an enlarged field of stability of amphibole compared with the unaltered tholeiite, demonstrating a similar effect of alkalis on amphibole stability in the natural system as Cawthorn (1976) observed in simplified systems. Finally, phengitic (at below 800 °C) to phlogopitic mica (800–1100 °C) is a conspicuous phase in the K-rich 'tholeiite' at pressures greater than or of the order of 20 kb.

Water-saturated conditions (Figs 8, 9)

Only major points of difference from the 5 per cent H_2O results are listed.

(1) The temperature of the liquidus is depressed and the field of partial melting decreases to c. 300 °C, compared with the 5 per cent H_2O runs. Additionally, the liquidus has a negative slope at pressures from 0 to 15 kb.

(2) Amphibole is stable to c. 1050 °C and to within a few degrees of the tholeiite liquidus, while, together with clinopyroxene, it is the liquidus phase in the andesite between c. 10 and 15 kb.

(3) The maximum temperature stability of plagioclase is further depressed in both compositions.

(4) The olivine field in the tholeiite expands to the liquidus at pressures up to 10 kb, but it is eliminated with the incoming of amphibole. Olivine does not occur in the andesite.

(5) Pyroxene is a major phase through most of the partial melting field for the tholeiite, but in the andesite it is eliminated at temperatures below 900 °C and pressures below 15 kb in the amphibole field and at temperatures above 900 °C and pressures above 20 kb.

(6) Garnet does not appear in the partial melting field until pressures above 12 kb and is a liquidus phase at pressures above 20 kb in both compositions.

(7) Kyanite appears as a near-liquidus phase in the andesite at pressures above 20 kb.

(8) Biotite is present in the andesite at temperatures below 850 °C and pressures below 17 kb.

(9) Quartz (or coesite) only appears within 50–60 °C of the solidus in the andesite.

Summary of Important Compositional Features

The garnets have relatively low $100 Mg/(Mg + Fe)$ (lower than the values in the co-existing liquid at $T < 900$–950 °C) and show increasing CaO with increasing pressure for near-liquidus runs (Green, 1977; Hellman, 1979; Stern and Wyllie, 1978) but for nearer-solidus runs (e.g. at 820–900 °C) widely diverging CaO trends occur with falling temperature at different pressures (e.g. increasing CaO at 15–20 kb but decreasing CaO at 25–30 kb (Hellman, 1979)). The clinopyroxenes have moderately high $100 Mg/(Mg + Fe)$ and increasing Na_2O with increasing pressure or falling temperature. Al_2O_3 content is high, reflecting high Ca-Tschermak's molecule from 5 to 15 kb and, together with high Na_2O content, high jadeite component at pressures above 15 kb. Amphiboles from the tholeiite are low SiO_2, high Al_2O_3, tschermakitic hornblendes and have relatively low $100 Mg/(Mg + Fe)$, slightly above the co-existing liquid. They have Na/K ratio of 16–33 where crystallizing from a typical oceanic tholeiite composition, changing to 38–68 for a model spilitized oceanic 'tholeiite' and to 2–6 for a model K-rich oceanic 'tholeiite' (Hellman, 1979). The Na/K ratio increases with increasing temperature in the oceanic tholeiite but decreases with increasing temperature for the spilite and K-rich 'tholeiite'. This applies for the pressure range 15–25 kb in all compositions. Near the upper pressure limit of stability, the amphibole composition changes from a tschermakitic hornblende at higher temperature (above 800 °C), to aluminous anthophyllite at lower temperature (below 750 °C), and slightly higher pressure, as determined in the K-rich 'tholeiite'.

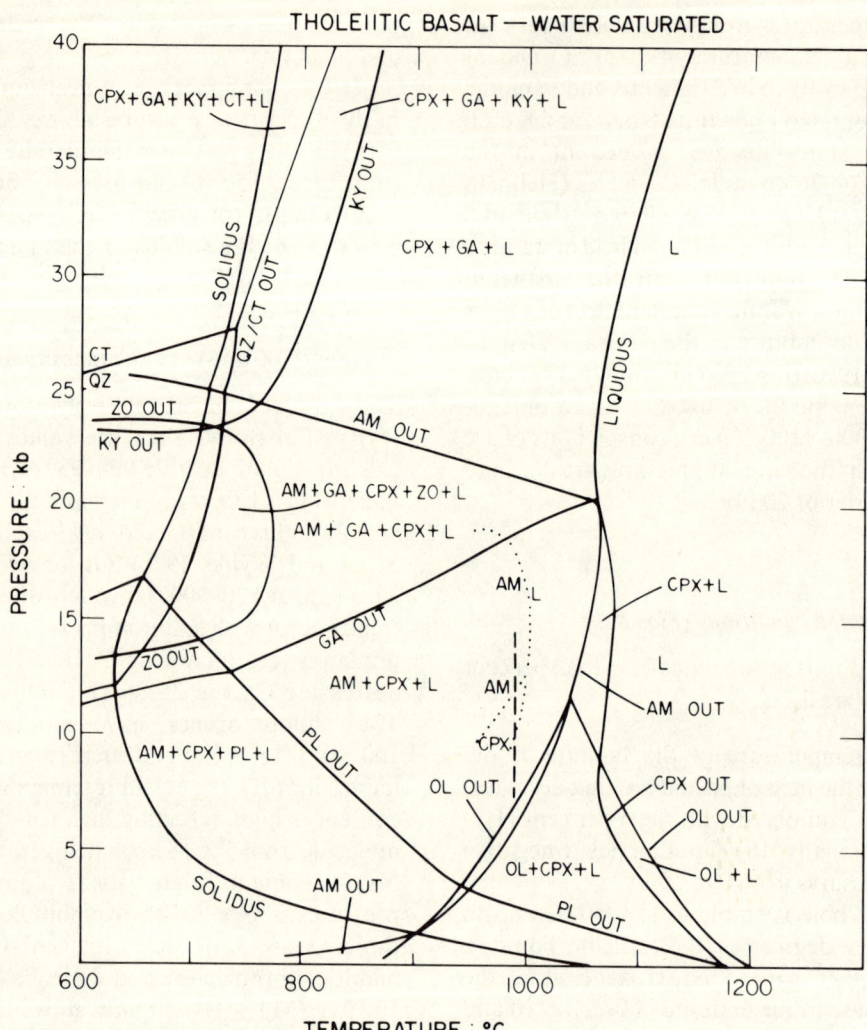

Fig. 8 Composite *P–T* diagram for water-saturated tholeiitic compositions. Dotted lines refer to stability curve for the Picture Gorge tholeiite using a NNO buffer (Allen *et al.*, 1975) while the broken line marks the amphibole-out curve for the Kilauea Iki tholeiite using a M–H buffer (contrasting with the amphibole-out curve for the same rock using the NNO buffer which is represented by the unbroken line—Allen *et al.*, 1975)

Synthesis of Experimental Work and Application to Petrogenetic Models

Basalt and andesite which bear a parent/daughter relationship to one another at some pressure, via either partial melting or fractional crystallization, must have similar liquidus or near-liquidus phases at the necessary degree of melting or crystallization of the basalt to produce the required andesitic liquid composition. However, this statement does not apply where reaction relations occur between early crystallizing phases and liquid or where andesites do not represent initial liquid compositions, and contain a significant concentration of a phenocryst component (cf. Ewart, 1976b, and this volume, Section II). As well as satisfying these broad phase relationships, any model relating

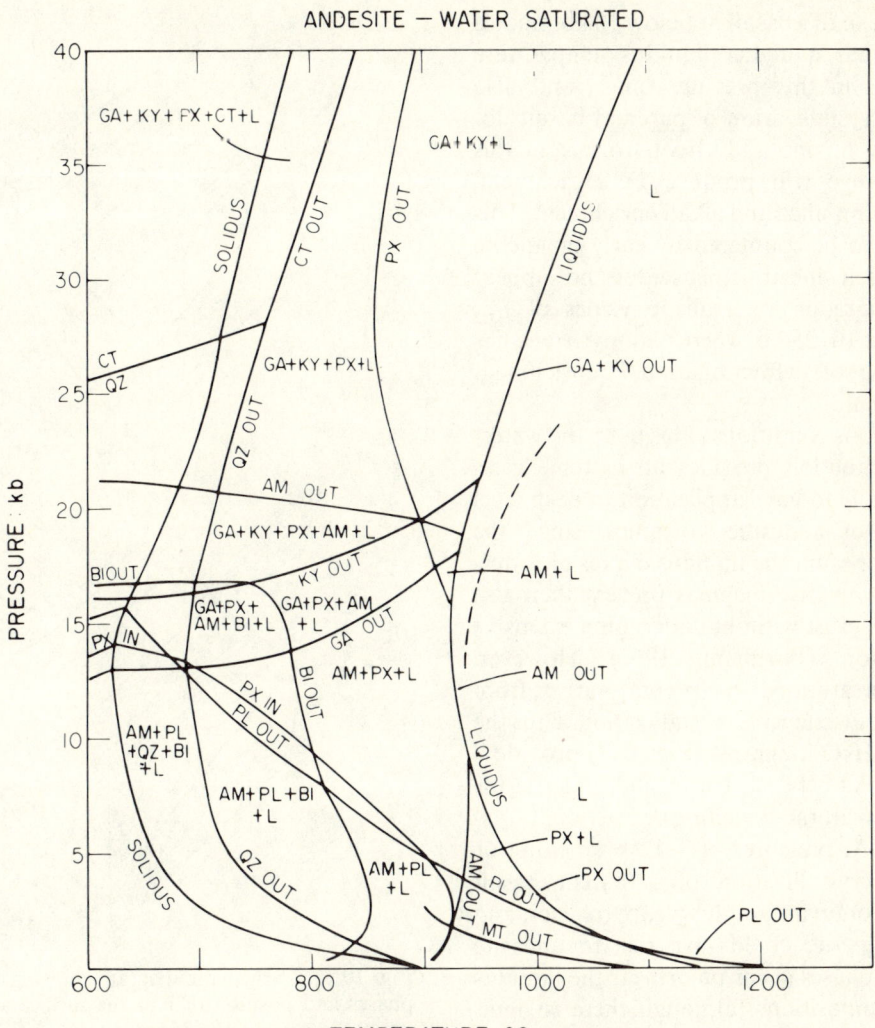

Fig. 9 Composite P–T diagram for water-saturated andesitic compositions. The broken line refers to the liquidus for Mt Hood andesite (Allen et al., 1975)

andesite to a basalt parent must also explain the detailed chemical characteristics. In particular, the lack of Fe enrichment paralleling silica and alkali enrichment in the calc-alkaline suite must be accounted for. The presence of relatively Fe-rich garnet, amphibole, and minor magnetite as important phases in any fractional crystallization process will strongly counteract any Fe-enrichment trend in derivative liquids. Such Fe enrichment occurs for pressures and water contents where crystallization is dominated by olivine and clinopyroxene.

Consideration of these points and examination of the dry phase diagrams (Figs 4, 5) indicates that andesite and basalt may be related at pressures above 25 kb where garnet and clinopyroxene are the near-liquidus phases in both compositions. Thus c. 40–50 per cent melting of quartz eclogite may produce andesite. The presence of quartz only 30 °C below the liquidus for the andesite restricts the silica content of liquids obtained for high degrees of melting under dry conditions to values only slightly greater than andesite. Andesite is unlikely to be a derivative of dry basalt at below 10 kb, since it frequently has a higher liquidus temperature than basalt in this pressure range, and also fractional crystallization of parental basalt dominated by olivine and clinopyroxene in this pressure range will produce Fe enrichment, accompanying silica and alkali enrichment. This is unlikely to be countered by early magnetite crystallization since this phase does not appear near the liquidus for realistic values of f_{O_2}. Similarly, at 10–25 kb, where clinopyroxene has a controlling role in fractionation, Fe enrichment will also result.

For hydrous conditions (Figs 6–9), the water-saturated situation provides an extreme case, but is unlikely to have application to deep level derivation of andesitic volcanics, since the negative slopes on the liquidus curves of water-saturated derivative magmas prevent their rise through the crust without undergoing extensive crystallization (Burnham, 1967). However, water-undersaturated melts may arise from depth without extensive crystallization. Thus the 5 per cent H_2O diagrams (Figs 6–7) provide a guide to links between basalt and andesite for water-undersaturated melting or fractional crystallization. At pressures of 8–12 kb melting or fractional crystallization of a parent basalt involving dominant amphibole, pyroxenes, and minor magnetite could give rise to andesite since these phases occur on or near the liquidus in both compositions (although there is significant variation in sequence of appearance of phases as a function of water content, f_{O_2}, f_{CO_2}, and bulk composition—(Eggler, 1972; Eggler and Burnham, 1973; Allen et al., 1975)). At slightly higher pressure (e.g. 12–25 kb) garnet becomes increasingly important and may be involved. At above 25 kb amphibole is not stable and eclogite melting occurs, but in the hydrous situation quartz is markedly depressed below the andesite liquidus and more silicic liquids may be derived than possible under anhydrous conditions.

Summary diagrams at 15 kb (Fig. 10) and 30 kb (Fig. 11) illustrate the critical phase relations for the range of rock compositions in the calc-alkaline series, as a function of water con-

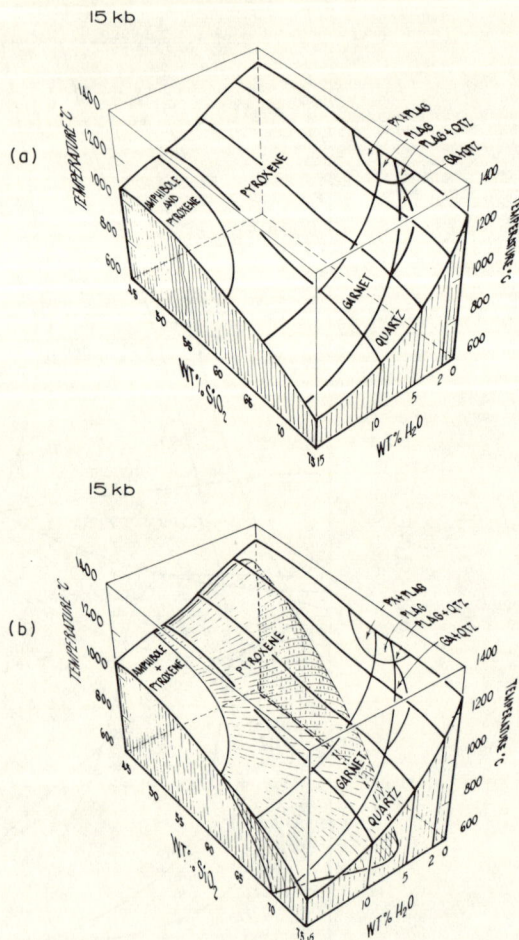

Fig. 10 (a) Schematic diagram showing the liquidus phases and position of liquidus surface at 15 kb for various compositions (simplified to weight per cent SiO_2) and various water contents (additional data in Figs. 10–11 obtained from Green and Ringwood, 1972; Stern and Wyllie, 1973; Stern et al., 1975; Wyllie et al., 1976; Wyllie, 1977a,b). (b) Schematic diagram as in (a) except that the subliquidus amphibole stability field is added

tent. At 15 kb there is a regular fall in liquidus temperature towards silicic compositions. Also Fig. 10 illustrates the suppression of plagioclase and quartz with increasing water content, the appearance of garnet in intermediate to silicic compositions, and the role of amphibole, present only on the liquidus for near water-saturated conditions in basic compositions, but very close to the liquidus in basic-intermediate compositions for water-undersaturated conditions.

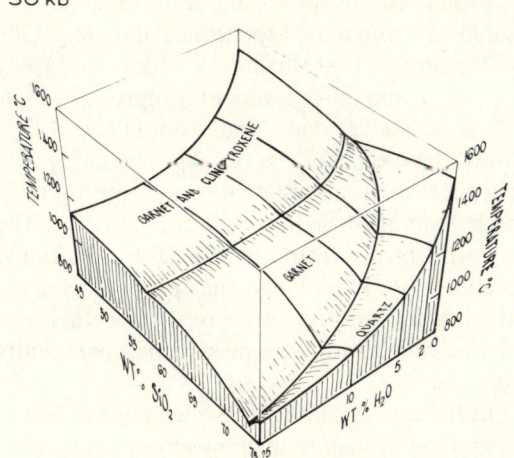

Fig. 11 Schematic diagram showing the liquidus surface at 30 kb for various compositions and water contents (as in Fig. 10)

In contrast, at 30 kb a pronounced thermal trough occurs in liquidus temperatures for silica values slightly higher than andesite for dry conditions, and moving to more silicic compositions with increasing water content. Linked with this, the garnet liquidus field expands, at the expense of quartz, as the water content increases. This trough reflects a series of piercing points, as emphasized by Stern and Wyllie (1978). Thus Fig. 11 does not define the composition of first formed melts in basic compositions at high pressures, but it does indicate that, with increasing temperature and degrees of melting for dry conditions, a liquid slightly more silicic than andesite is the most silica-enriched liquid that can be attained. For near-solidus dry conditions liquids are likely to be alkali enriched (at the expense of the Ca, Mg, and Fe components) for similar absolute silica values (Stern and Wyllie, 1978). Certainly, for dry conditions and low degrees of melting, liquids with higher normative quartz than andesite cannot be obtained, since estimated field boundaries in the system Qz–Ab–Or at 30 kb, anhydrous, show that initial liquids would only have a few per cent normative quartz, and would be syenitic rather than granitic or rhyolitic (Huang and Wyllie, 1975). In the presence of water, liquids with higher silica may be produced, because the primary quartz volume markedly decreases for hydrous conditions (see Fig. 11).

In conclusion, the $P-T$ diagrams and compositional data presented demonstrate two main depth regions (pressure ranges) characterized by specific near-liquidus mineral assemblages, where basalt and andesite may have a parent/daughter relationship: viz. (1) at c. 25–40 km (c. 8–12 kb) amphibole-dominated with subordinate aluminous pyroxenes and minor magnetite—*amphibole model*; (2) at above 80 km (above c. 25 kb) garnet-clinopyroxene-dominated—*eclogite model*. At intermediate depths (40–80 km) both garnet and amphibole may be involved with pyroxene in the parent/daughter linking fractional crystallization mechanism. The phases which are possibly exceptional to the above generalized zones are olivine and clinopyroxene in the amphibole model, because these phases are in known, or predicted, reaction relationship with co-existing liquid. Thus olivine and clinopyroxene may be present in a parental basalt, but not occur in a derivative andesite (cf. Cawthorn and O'Hara, 1976).

Further and more detailed chemical evaluation of the two models has been conducted experimentally by Stern and Wyllie (1978) at 30 kb and Hellman (1979) at 15–25 kb, and by trace element geochemical modelling and close comparison with natural examples (e.g. Gill, 1974, 1978; Lopez-Escobar *et al.*, 1977; Kay, 1977). Stern and Wyllie (1978) demonstrated that the simple eclogite model cannot explain the Ca/(Mg + Fe) values for calc-alkaline suites. Multistage fractional crystallization with removal of crystals at shallower levels is required if products such as calc-alkaline andesites are to be obtained. Since some data suggests that the grossular content of the garnet and the garnet to clinopyroxene ratio increase with increasing pressure (Green and Ringwood, 1968; Green, 1977) different trends for the Ca/(Mg + Fe) values may be obtained for the eclogite model at above 40 kb pressure. This requires further experimental study, but on present evidence it is unlikely to cause an appropriate shift in the Ca/(Mg + Fe) trend. Preliminary calculations by Hellman (1979) for melts in an oceanic tholeiite + 5 per cent H_2O at $c.$ 990 °C and 15 kb in equilibrium with amphibole, pyroxene, and subordinate garnet indicate that while broadly andesitic compositions are obtained, 100 Mg/(Mg + Fe) and Ca/(Mg + Fe), and especially Al_2O_3 values, are higher than found in typical calc-alkaline andesites, and fractional crystallization at shallower levels is required to produce an appropriate andesite composition (cf. Stern and Wyllie, 1978, results for 30 kb). It is noteworthy, however, that these melt compositions at $c.$ 1000 °C and 15 kb are similar to glass compositions determined by Anderson for glass inclusions in phenocrysts from andesites (Anderson, 1978).

Recent trace element geochemical studies (REE in particular) point to either the amphibole model (Gill, 1974; Nicholls and Whitford, 1976), the eclogite model (Condie and Hayslip, 1975; Peccerillo and Taylor, 1976; Lopez-Escobar et al., 1977; Cole, 1978; Kay, 1978) or to complex, multistage models of the type described by Ringwood (1974, 1975) (e.g. Thorpe et al., 1976; Kay, 1977; Dostal et al., 1977a, b; Deruélle, 1978) for different occurrences of calc-alkaline rocks. However, as emphasized by Gill (1978), our knowledge of partition coefficients, proportions of phases involved, roles of CO_2, F, Cl in a fluid phase and distribution of trace elements in this phase, role of accessory phases rich in REE and other trace elements (such as apatite, sphene, etc.) (Gill, 1974; Lopez-Escobar et al., 1977; Hellman and Green, 1979a), and role of altered oceanic crustal material with markedly different trace element distribution from the typical oceanic tholeiite traditionally used in the geochemical modelling, all combine to place severe limits on rigorous attempts to constrain petrogenetic models for the calc-alkaline series through application of trace element geochemistry.

As noted earlier, Ewart (1976b and this volume, Section II) has argued that many andesitic bulk compositions do not represent completely liquid magmas, but rather a combination of liquid and concentrated phenocrysts. In such cases experimental work cannot be applied to prove a parent/daughter relationship between the andesite bulk composition and some basaltic precursor. Fortunately, geochemical data (e.g. Gill, 1978) suggest that this problem does not apply generally, and the experimental approach adopted is a viable one. Anderson (1973, 1978), from detailed studies of glass inclusions in phenocrysts in andesite, argues forcefully for a high alumina basalt parent for andesite. The andesite forms from the basalt by fractional crystallization similar to that proposed for the amphibole model in the deep crust (together with olivine and clinopyroxene which subsequently react out).

In the context of the thickened crust acting as a controlling factor in imposing a depth constraint on the fractional crystallization of hydrous basalt magma, it is significant that Kushiro (1976) has noted the largest viscosity and density change in liquids of $NaAlSi_2O_6$ composition at pressures of between 7.5 and 10 kb, due to a change in the Al co-ordination. If this also occurs in complex silicate liquids then it may have a contributory role towards promoting amphibole, pyroxene, etc., fractional crystallization at depths corresponding to these pressures, and to the crust/mantle interface.

Finally, although in this section the role of water in calc-alkaline rock genesis has been strongly emphasized, other volatile components such as CO_2, Cl, F may also be important, although CO_2 is relatively insoluble in andesite at pressures up to 30 kb (Mysen et al., 1975). Anderson (1973, 1978) noted the presence of up to 0.15 per cent by weight of Cl in silicate glass inclusions in phenocrysts from andesites, and Holloway and Ford (1975) demonstrated the considerable effect of F on increasing the stability of amphibole and hence expanding its possible fractional crystallization regime. The effects of these three components in calc-alkaline genesis merit experimental evaluation.

Conclusion

Any general model for the origin of the calc-alkaline series must attempt to maintain flexibility in the roles of the major phases controlling fractional crystallization, and so allow explanation of the derivation of a variety of calc-alkaline provinces with similar overall major element chemistry, but significantly different trace element geochemistry. This general model rests on two major geological observations that calc-alkaline rocks only occur in (1) convergent plate boundary regions and (2) island arc or continental marginal regions where the crustal thickness is greater than 30–40 km. Point (1) is critical since subduction of hydrated lithosphere allows transport of significant quantities of water into the mantle and thence derivation of voluminous, silica-oversaturated magmas of the island arc tholeiitic series from mantle melting (possibly of depleted and partly residual mantle; Green, 1976). Point (2) is critical since at the density discontinuity of the mantle and crust, the island arc tholeiitic magmas may be forced to undergo fractional crystallization at $c.$ 8–15 kb pressure before final eruption through the crust to the surface. This fractional crystallization, involving dominant subsilicic amphibole and subordinate pyroxenes, magnetite, and sphene, produces derivative calc-alkaline magmas. These

TABLE 2 Synpotic chart showing, in summary form, current petrogenetic alternatives which may explain the gross chemical and petrologic features of calc-alkaline andesites, and their detailed, and variable, trace element distributions (cf. Ringwood, 1975; Marsh, 1976; Ewart, 1976b).

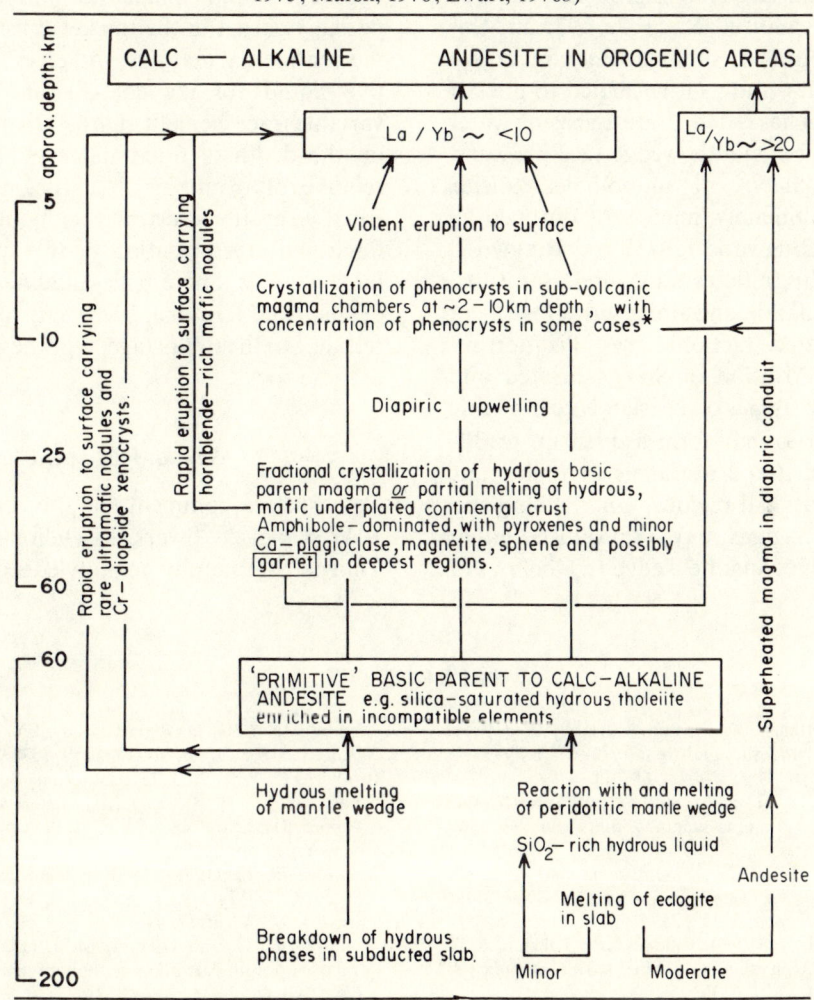

* Ewart, 1976b.

magmas show little or no Fe enrichment, and so contrast with the island arc tholeiitic series which, in its relatively unimpeded rise from the mantle, crystallizes dominant olivine with subordinate pyroxene, and shows a typical tholeiitic Fe enrichment trend. Alternatively, or additionally, in continental marginal regions, significant heating of the thickened crust caused by underplating and intrusion of the mantle-derived hydrous tholeiitic magmas may result in melting of the lower crust and derivation of voluminous intermediate to silicic magmas of the calc-alkaline series.

Thus in the general model (summarized in Table 2), oceanic crust generated in mid-oceanic ridge regions, and frequently significantly altered (e.g. hydrated, spilitized, etc.; Fyfe, 1975 and this volume, Section X), is subducted at convergent plate boundaries, and as it is carried to greater depth hydrous phases dehydrate, releasing water into the overlying mantle wedge (e.g. pressure-induced breakdown of amphibole, zoisite, sphene, and, ultimately, micas). As outlined by Nicholls and Ringwood (1973) and Ringwood (1974, 1975), this initiates water-undersaturated melting of the mantle, and along with subsequent olivine-dominated fractional crystallization results in the derivation of SiO_2-saturated and oversaturated magmas of the island arc tholeiitic series. This series may form a crust up to 30–40 km thick in island arc areas or underplate continental marginal regions. On reaching this stage further magmas may be produced either from the modified mantle wedge (see below) or possibly from hydrous melting of the eclogitic oceanic lithosphere at depths over 80 km (probably 150–200 km). This latter melting results in either SiO_2-rich and alkali-rich magmas for low degrees of melting or andesite-like magmas for high degrees of melting. These magmas generally do not reach the surface without interaction with the mantle wedge (cf. Ringwood, 1974, 1975; Mysen, this volume, Section VII) or extensive fractional crystallization, especially at the crust/mantle interface in island arc and continental marginal areas. The fractional crystallization of the hydrous basic to basaltic andesite magmas at this level (corresponding to c. 8–15 kb) is dominated by amphibole with subordinate pyroxenes, and magnetite and sphene, and possibly garnet in the thicker crustal regions of the continental margins, and gives rise to derivative liquids of the calc-alkaline series, with variable trace element characteristics dependent on the depth of fractional crystallization and relative role of amphibole, garnet, magnetite, etc. Finally, at the crust/mantle boundary, batch fractional crystallization of the type modelled for some Japanese calc-alkaline rock series (Yanagi and Ishizaka, 1978) may be particularly relevant to the multistage hypotheses for genesis of these rocks.

Acknowledgements

Constructive comment on the manuscript by J. W. Cole, A. Ewart, P. Hellman, and R. S. Thorpe is gratefully acknowledged.

REFERENCES

Allen, J. C., Boettcher, A. L., and Marland, G. (1975). Amphiboles in andesite and basalt. I. Stability as a function of $P-T-f_{O_2}$. *Am. Mineral.* **60**, 1069–1085.

Anderson, A. T., Jr (1973). The before-eruption water content of some high-alumina magmas. *Bull. Volc.* **37**, 530–552.

Anderson, A. T., Jr. (1978). Parental magmas of some andesite magma series. *Geol. Assoc. Can., Geol. Soc. Am., Abst. Toronto Meeting*, **3**, 358–359.

Aoki, K., and Oji, Y. (1966). Calc-alkaline volcanic rock series derived from alkaline-olivine basalt magma. *J. geophys. Res.* **71**, 6127–6135.

Brown, G. M. and Schairer, J. F. (1971). Chemical and melting relations of some calc-alkaline volcanic rocks. *Geol. Soc. Am. Mem.* no. 130, 139–157.

Burnham, C. W. (1967). Hydrothermal fluids at the magmatic stage. In *Geochemistry of Hydrothermal Ore Deposits* (H. L. Barnes, ed.), Holt, Rinehart and Winston, New York, pp. 34–76.

Cawthorn, R. G. (1976). Melting relations in part of the system $CaO-MgO-Al_2O_3-SiO_2-Na_2O-H_2O$ under 5 kb pressure. *J. Petrol.* **17**, 44–72.

Cawthorn, R. G. and O'Hara, M. J. (1976). Amphibole-fractionation in calc-alkaline magma genesis. *Am. J. Sci.* **276**, 309–329.

Clark, S. P., Jr. and Ringwood, A. E. (1964). Density distribution and constitution of the mantle. *Rev. Geophys.* **2**, 35–88.

Cohen, L. H., Ito, K., and Kennedy, G. C. (1967). Melting and phase relations in an anhydrous basalt to 40 kilobars. *Am. J. Sci.* **265**, 475–518.

Cole, J. W. (1978). Andesites of the Tongariro volcanic centre, North Island, New Zealand. *J. Volc. geothermal Res.* **3**, 121–153.

Condie, K. C. and Hayslip, D. L. (1975). Young bimodal volcanism at Medicine Lake volcanic centre, Northern California. *Geochim. cosmochim. Acta* **39**, 1165–1178.

Cox, K. G. and Bell, J. D. (1972). A crystal fractionation model for the basaltic rocks of the New Georgia Group, British Solomon Islands. *Contr. Mineral. Petrol* **37**, 1–13.

Deruélle, B. (1978). Calc-alkaline and shoshonitic lavas from five Andean volcanoes (between latitudes 21°45′ and 24°30′S) and the distribution of the Plio-Quaternary volcanism of the south-central and southern Andes. *J. Volc. geothermal. Res.* **3**, 281–298.

Dostal, J., Dupuy, C., and Lefévre, C. (1977a). Rare earth element distributions in Plio-Quaternary volcanic rocks from southern Peru. *Lithos* **10**, 173–183.

Dostal, J., Zentilli, M., Caelles, J. C., and Clark, A. H. 1977b). Geochemistry and origin of volcanic rocks from the Andes (26°–28°S). *Contr. Mineral. Petrol.* **63**, 113–128.

Eggler, D. H. (1972). Water-saturated and undersaturated melting relations in a Paricutin andesite and an estimate of water content in the natural magma. *Contr. Mineral. Petrol.* **34**, 261–271.

Eggler, D. H. (1974). Application of a portion of the system $CaAl_2Si_2O_8$–$NaAlSi_3O_8$–SiO_2–MgO–Fe–O_2–H_2O–CO_2 to genesis of the calc-alkaline suite. *Am. J. Sci.* **274**, 297–315.

Eggler, D. H. and Burnham, C. W. (1973). Crystallization and fractionation trends in the system andesite–H_2O–CO_2–O_2 at pressures to 10 kb. *Geol. Soc. Am. Bull* **84**, 2517–2532.

Ewart, A. (1976a). A petrological study of the younger Tongan andesites and dacites, and the olivine tholeiites of Niua Fo'ou Island, S.W. Pacific. *Contr. Mineral. Petrol.* **58**, 1–21.

Ewart, A. (1976b). Mineralogy and chemistry of modern orogenic lavas—some statistics and implications. *Earth planet. Sci. Letters* **31**, 417–432.

Ewart, A., Brothers, R. N., and Mateen, A. (1977). An outline of the geology and geochemistry and the possible petrogenetic evolution of the volcanic rocks of the Tonga–Kermadec–New Zealand island arc. *J. Volc. geothermal. Res.* **2**, 205–250.

Francis, P. W., Moorbath, S., and Thorpe, R. S. (1977). Strontium isotope data for recent andesites in Ecuador and North Chile. *Earth planet. Sci. Letters* **37**, 197–202.

Fudali, R. F. (1965). Oxygen fugacites of basaltic and andesitic magma. *Geochim. cosmochim. Acta* **29**, 1063–1075.

Fyfe, W. S. (1975). Magma production: influence of the hydrosphere. In *Geodynamics Today: A Review of the Earth's Dynamic Processes*, The Royal Society, London, pp. 29–32.

Gill, J. B. (1974). Role of underthrust oceanic crust in the genesis of a Fijian calc-alkaline suite. *Contr. Mineral. Petrol.* **43**, 29–45.

Gill, J. B. (1978). Role of trace element partition coefficients in models of andesite genesis. *Geochim. cosmochim. Acta* **42**, 709–724.

Green, D. H. (1976). Experimental testing of 'equilibrium' partial melting of peridotite under water-saturated, high-pressure conditions. *Can. Mineral.* **14**, 255–268.

Green, D. H. and Ringwood, A. E. (1967a). An experimental investigation of the gabbro to eclogite transformation and its petrological implications. *Geochim. cosmochim Acta* **31**, 767–833.

Green, D. H. and Ringwood, A. E. (1967b). The genesis of basaltic magmas. *Contr. Mineral. Petrol.* **15**, 103–190.

Green, T. H. (1967). High-pressure experimental investigations on the origin of high-alumina basalt, andesite and anorthosite. Ph.D. thesis, Australian National University, Canberra, Australia.

Green, T. H. (1972). Crystallization of calc-alkaline andesite under controlled high-pressure, hydrous conditions. *Contr. Mineral. Petrol.* **34**, 150–166.

Green, T. H. (1977). Garnet in silicic liquids and its possible use as a P–T indicator. *Contr. Mineral. Petrol.* **65**, 59–67.

Green, T. H. and Ringwood, A. E. (1968). Genesis of the calc-alkaline igneous rock suite. *Contr. Mineral. Petrol.* **18**, 105–162.

Green, T. H. and Ringwood, A. E. (1969). High pressure experimental studies on the origin of andesites. *Oregon St. Dept. Geol. Mineral. Industries, Bull.* no. 65, pp. 21–32.

Green, T. H. and Ringwood, A. E. (1972). Crystallization of garnet-bearing rhyodacite under high-pressure, hydrous conditions. *J. geol. Soc. Aust.* **19**, 203–212.

Hanus, V., and Vanek, J. (1978). Morphology of the Andean Wadati–Benioff zone, andesitic volcanism and tectonic features of the Nazca Plate. *Tectonophysics* **44**, 65–77.

Hellman, P. L. (1979). Geochemical and experimental investigations of subduction-related processes. Ph.D. thesis, Macquarie University, Sydney, Australia.

Hellman, P. L. and Green, T. H. (1979a). The role of sphene as an accessory phase in the high-pressure partial melting of hydrous mafic compositions. *Earth planet. Sci. Letters* **42**, 191–201.

Hellman, P. L. and Green, T. H. (1979b). The high pressure experimental crystallization of staurolite in hydrous mafic compositions. *Contr. Mineral. Petrol.* **68**, 369–372.

Helz, R. T. (1976). Phase relations of basalts in their melting ranges at P_{H_2O} = 5 kb. Part II. Melt compositions. *J. Petrol.* **17**, 139–193.

Heming, R. F. (1977). Mineralogy and proposed P–T paths of basaltic lavas from Rabaul Caldera, Papua New Guinea. *Contr. Mineral. Petrol.* **61**, 15–33.

Holloway, J. R. and Burnham, C. W. (1972). Melting relations of basalt with equilibrium water pressure less than total pressure. *J. Petrol.* **14**, 1–29.

Holloway, J. R. and Ford, C. E. (1975). Fluid-absent melting of the fluoro-hydroxy amphibole pargasite to 35 kb *Earth planet. Sci. Letters* **25**, 44–48.

Huang, W. L. and Wyllie, P. J. (1975). Melting reactions in the system $NaAlSi_3O_8$–$KAlSi_3O_8$–SiO_2 to 35 kilobars, dry and with excess water. *J. Geol.* **83**, 737–748.

Jakeš, P. and Gill, J. (1970). Rare earth elements and the island arc tholeiitic series. *Earth planet. Sci. Letters* **9**, 17–28.

James, D. E., Brooks, C., and Cuyubamba, A. (1976). Andean Cenozoic volcanism: magma genesis in the light of strontium isotopic composition and trace element geochemistry. *Geol. Soc. Am. Bull.* **87**, 592–600.

Katsui, Y., Niida, K., Yamamoto, M., and Nemoto, S. (1978). Genesis of calc-alkaline andesite from Oshima-Oshima and Ichinomegata volcanoes, North Japan (abstract). In *Proceedings of the International Geodynamic Conference on the 'Western Pacific' and 'Magma Genesis'*, Tokyo, Japan, p. 266.

Kay, R. W. (1977). Geochemical constraints on the origin of Aleutian magmas. In *Island Arcs, Deep Sea Trenches, and Back-arc Basins* (M. Talwani and W. C. Pitman, III, eds), Maurice Ewing Series no. 1, American Geophysical Union, Washington, DC, pp. 229–242.

Kay, R. W. (1978). Aleutian magnesian andesites: melts from subducted Pacific Ocean crust. *J. Volc. geothermal Res.* **4**, 117–132.

Kushiro, I. (1973). The system diopside–anorthite–albite: determination of compositions of coexisting phases. *Carnegie Instn. Wash. Yb.* **73**, 502–507.

Kushiro, I. (1976). Changes in viscosity and structure of melt of $NaAlSi_2O_6$ composition at high pressures. *J. geophys. Res.* **81**, 6347–6350.

Lambert, I. B. and Wyllie, P. J. (1972). Melting of gabbro (quartz eclogite) with excess water to 35 kilobars, with geological applications. *J. Geol.* **80**, 693–708.

Lambert, I. B. and Wyllie, P. J. (1974). Melting of tonalite and crystallization of andesite liquid with excess water to 30 kilobars. *J. Geol.* **82**, 88–97.

Lewis, J. F. (1973). Petrology of the ejected plutonic blocks of the Soufrière Volcano, St Vincent, West Indies. *J. Petrol.* **14**, 81–112.

Lindsley, D. H. (1968). Melting relations of plagioclase at high pressures. In *Origin of Anorthosite and Related Rocks* (Y. W. Isachsen, ed.), New York State Museum and Scientific Service, Albury, N.Y., Memoir 18, pp. 39–46.

Lindsley, D. H. and Emslie, R. F. (1968). Effect of pressure on the boundary curve in the system diopside–albite–anorthite. *Carnegie Instn. Wash. Yb.* **66**, 479–480.

Lopez-Escobar, L., Frey, F. A., and Vergara, M. (1977). Andesites and high-alumina basalts from the central-south Chile High Andes: geochemical evidence bearing on their petrogenesis. *Contr. Mineral. Petrol.* **63**, 199–228.

Lowder, G. G. and Carmichael, I. S. E. (1970). Volcanoes and caldera of Talasea, New Britain. *Geol. Soc. Am. Bull.* **81**, 17–38.

McNutt, R. H., Crockett, J. H., Clark, A. H., Caelles, J. C., Farrar, E., Haynes, S. J., and Zentilli, M. (1975). Initial $^{87}Sr/^{86}Sr$ ratios of plutonic and volcanic rocks of the central Andes between latitudes 26° and 29° south. *Earth planet. Sci. Letters* **27**, 305–313.

Marsh, B. D. (1976). Mechanics of Benioff Zone magmatism. In *Geophysics of the Pacific Ocean Basin and Its Margin* (G. H. Sutton, M. H. Manghnani, R. Moberly, and E. U. McAfee, eds), Geophysics Monograph no. 19, American Geophysical Union, Washington, D.C., pp. 337–350.

Marsh, B. D. and Carmichael, I. S. E. (1974). Benioff Zone magmatism. *J. geophys. Res.* **79**, 1196–1206.

Mathias, M., Siebert, J. C., and Rickwood, P. C. (1970). Some aspects of the mineralogy and petrology of ultramafic xenoliths in kimberlite. *Contr. Mineral. Petrol.* **26**, 75–123.

Mysen, B. O., Arculus, R. J., and Eggler, D. H. (1975). Solubility of CO_2 in melts of andesite, tholeiite, and olivine nephelinite composition to 30 kilobars pressure. *Contr. Mineral. Petrol.* **53**, 227–240.

Nicholls, I. A. and Ringwood, A. E. (1973). Effect of water on olivine stability in tholeiites and the production of silica saturated magmas in the island arc environment. *J. Geol.* **81**, 285–300.

Nicholls, I. A. and Whitford, D. J. (1976). Primary magmas associated with Quaternary volcanism in the western Sunda arc, Indonesia. In *Volcanism in Australasia* (R. W. Johnson, ed.), Elsevier, Amsterdam, pp. 77–90.

Noble, D. C., Bowman, H. R., Herbert, A. J., Silberman, M. L., Heropoulos, C. E., Fabbi, B. P., and Hedge, C. E. (1975). Chemical and isotopic constraints on the origin of low-silica latite and andesite from the Andes of Central Peru. *Geology* **3**, 501–504.

O'Hara, M. J. (1963). Melting of bimineralic eclogite at 30 kilobars. *Carnegie Instn. Wash. Yb.* **62**, 76–77.

Osborn, E. F. (1959). Role of oxygen pressure in the crystallization and differentiation of basaltic magma. *Am. J. Sci.* **257**, 609–647.

Osborn, E. F. (1969). Experimental aspects of calc-alkaline differentiation. *Oregon St. Dept Geol. Mineral. Industries, Bull.* no. 65, 33–42.

Osborn, E. F. and Arculus, R. J. (1975). Phase relations in the system Mg_2SiO_4–iron oxide–$CaAl_2SiO_8$–SiO_2 at 10 kbar and their bearing on the origin of andesite. *Carnegie Instn. Wash. Yb.* **74**, 504–506.

Oxburgh, E. R. and Turcotte, D. L. (1970). Thermal structure of island arcs. *Geol. Soc. Am. Bull.* **81**, 1665–1688.

Peccerillo, A. and Taylor, S. R. (1976). Rare earth elements in East Carpathian volcanic rocks. *Earth planet. Sci. Letters* **32**, 121–126.

Poldervaart, A. (1949). Three methods of graphic representation of chemical analyses of igneous rocks. *R. Soc. S. Afr. Trans.* **32**, 177–188.

Powell, M. J. (1978). Crystallization conditions of low-pressure cumulate nodules from the Lesser Antilles island arc. *Earth planet. Sci. Letters* **39**, 162–172.

Ringwood, A. E. (1974). The petrological evolution of island arc systems. *J. geol. Soc. Lond.* **130**, 183–204.

Ringwood, A. E. (1975). *Composition and Petrology of the Earth's Mantle*, McGraw-Hill, New York.

Shimazu, M., Komatsu, M., and Yamada, M. (1974). Chrome-diopsides from the Miocene basalts and andesites in the Misaka Mountainland, Yamanashi Prefecture, Central Japan. *Geol. Soc. Jap. Mem.* no. 11, 59–67.

Stern, C. R. and Wyllie, P. J. (1973). Melting relations of basalt–andesite–rhyolite–H_2O and a pelagic-red clay at 30 kb. *Contr. Mineral. Petrol.* **42**, 313–323.

Stern, C. R. and Wyllie, P. J. (1978). Phase compositions through crystallization intervals in basalt–andesite–H_2O at 30 kbar with implications for subduction zone magmas. *Am. Mineral* **63**, 641–663.

Stern, C. R., Huang, W-L., and Wyllie, P. J. (1975). Basalt–andesite–rhyolite–H_2O: crystallization intervals with excess H_2O and H_2O–undersaturated liquidus surfaces to 35 kilobars, with implications for magma genesis. *Earth planet. Sci. Letters* **28**, 189–196.

Thorpe, R. S., Francis, P. W., and Potts, P. J. (1976). Rare earth data and petrogenesis of andesites from the N. Chilean Andes. *Contr. Mineral. Petrol* **54**, 65–78.

Tilley, C. E. (1950). Some aspects of magmatic evolution. *J. geol. Soc. Lond.* **106**, 37–61.

Whitford, D. J. and Jezek, P. A. (1977). Geochemistry of

Cenozoic and Recent lavas from the Banda Arc, Indonesia. *Carnegie Instn. Wash. Yb.* **76**, 845–855.

Wyllie, P. J. (1977a). Crustal anatexis: an experimental review. *Tectonophysics* **43**, 41–71.

Wyllie, P. J. (1977b). From crucibles through subduction to batholiths. In *Energetics of Geological Processes* (S. K. Saxena and S. Bhattachari, eds), Springer-Verlag, Berlin, pp. 389–433.

Wyllie, P. J., Huang, W-L., Stern, C. R., and Maaloe, S. (1976). Granitic magmas; possible and impossible sources, water contents, and crystallization sequences. *Can. J. Earth Sci.* **13**, 1007–1019.

Yamazaki, T., Onuki, H., and Tiba, T. (1966). Significance of hornblende gabbroic inclusions in calc-alkaline rocks. *J. Jap. Assoc. Mineral. Petrol. econ. Geol.* **55**, 87–103.

Yanagi, T. and Ishizaka, K. (1978). Batch fractionation model for the evolution of volcanic rocks in an island arc: an example from Central Japan. *Earth planet. Sci. Letters* **40**, 252–262.

Yoder, H. S. (1969). Calc-alkaline andesites: experimental data bearing on the origin of their assumed characteristics. *Oregon St. Dept geol. Mineral. Industries, Bull.* no. 65, 77–89.

Yoder, H. S. and Tilley, C. E. (1962). Origin of basalt magmas: an experimental study of natural and synthetic rock systems. *J. Petrol.* **3**, 342–532.

The role of mantle anatexis

B. O. Mysen

Carnegie Institution of Washington, Geophysical Laboratory,
2801 Upton Street, Washington, District of Columbia 20008, USA

ABSTRACT

Published experimental phase equilibrium data indicate that four petrogenetic models for orogenic andesite can reproduce their appropriate major element bulk compositions. These models are: (1) partial melting of hydrous peridotite; (2) partial melting of amphibolite; (3) fractional crystallization of amphibole from tholeiitic melt; and (4) fractional crystallization of an Fe oxide and olivine from a tholeiitic melt. Petrogenetic models of olivine fractionation from tholeiite and partial melting of quartz eclogite are rejected on the basis of phase equilibria. It is also shown that the latter two models are incompatible with trace element patterns of orogenic andesites.

The four accepted models are evaluated using recent trace element partitioning data at high and low pressure, S solubility in silicate melts at high pressures and temperatures, and experimental data relevant to metasomatic alteration of the source rock of andesite in island arcs. None of the models are compatible with all the geochemical and geophysical data from orogenic andesite if the source rock of the primary partial melt is either the decending slab of tholeiitic composition or an unaltered peridotite overlying the descending slab.

Experimental data on crystal–liquid and crystal–vapour trace element partitioning combined with experimentally determined infiltration rates of H_2O in crystalline upper mantle suggest that the peridotite wedge overlying a dehydrating slab of garnet amphibolite will be metasomatically altered. Thermodynamic data on suitable dehydration reactions in combination with geophysical data from island arcs indicate that dehydration will take place, but that the slab material itself will not melt to at least 125 km depth in the mantle. The metasomatically altered peridotite has a major and incompatible trace element composition that yields andesitic primary melts with incompatible trace element patterns similar to those found in orogenic andesite. The transition metal content of such metals can be made compatible with those of orogenic andesite only if 0.5–1 per cent by weight of immiscible Fe-rich sulphide melt separated from the silicate melt during its ascent. Experimentally determined S solubilities in relevant silicate melt compositions at high pressure indicate that such a process is likely to take place. This is, therefore, the favoured petrogenic model for andesite formation in island arcs.

Introduction

In this chapter, dealing with orogenic andesite, the definition given by Taylor (1969) is followed. Gill (1978) pointed out that this definition includes c. 70 per cent of all rocks described as andesite and excludes rocks like icelandites, shoshonites, trachytes, and latites. Further subdivision may be employed (e.g. Miyashiro and Shido, 1976), but will not be attempted here.

Experimental petrologists studying mechanisms of formation and evolution of andesitic rocks often base their models on their own experimental results and exclude other natural and experimental data pertinent to the problem. Furthermore, it is not always recognized that the petrogenesis of andesitic rocks may differ for different regions. It has been shown, for example, that in some areas Fe oxide fractional crystallization may be important in controlling the magmatic evolution (Osborn, 1977), whereas in other regions the andesite magmas may be related to partial melting of peridotite (DeLong, 1974).

Models based on phase equilibrium data frequently include only major element data. In addition to major element composition and petrographic features, any model pretending to explain the genesis of andesitic rock suites must accommodate typical trace element and tectonic constraints. Gill (1978) summarized important trace element data which will also be considered here. He noted that andesites generally show significant light REE enrichment, whereas intermediate REE are not greatly enriched over heavy REE. In fact, a minimum near Nd is sometimes observed (Gill, 1978). Gill (1978) also found that the degree of light REE enrichment does not correlate with silica content of a given suite of andesites. A summary of Ni contents (Gill, 1978) revealed no correlation with silica content of a given andesite suite. Ni concentrations correlate with MgO contents, however. Presumably, other transition metals associated with Ni will show similar behaviour (Taylor, 1969).

Isotope data (Nd, Pb, and Sr isotopes) have been used to deduce a primitive character of the source of andesite. These data appear to rule out significant continental crustal contamination of primary basic magma to generate andesitic liquids (see Boettcher, 1973, and T. H. Green, this volume, Section VII, for reviews of data). Isotopic data often cannot be used to discriminate between a source of abyssal tholeiite composition (e.g. melting of subducting oceanic slab) and peridotite, however. In the present summary, an attempt to conduct such discrimination based on a broad range of experimental data will be discussed.

The average depth of earthquake foci beneath active volcanoes of various convergent plate boundaries has been calculated as 136 ± 4 km (Engdahl, 1973). The magmatic activity has been related to this seismic activity (e.g. Dickinson and Hatherton, 1967; Marsh, 1976). Other investigators have suggested that dehydration–hydration reactions with accompanying volume decrease may be the cause of earthquake activity (Fyfe and McBirney, 1975; Delaney and Helgeson, 1978), however. The latter suggestion may be particularly viable in view of the fact that rocks in or near subduction zones in general and andesitic rocks in particular have high water contents (Delaney et al., 1978; Anderson, 1975) compared with oceanic basalts, for example. Despite these disagreements it is clear that magmatic activity in orogenic zones is related to the existence of an underlying subduction zone. Successful models should accommodate all these considerations, however.

Experimental data have led to petrogenetic models of andesite that can be subdivided into four major groups. These are: (1) direct partial melting of hydrous peridotite (e.g. O'Hara, 1965; Yoder, 1969; Kushiro, 1969a, 1972, 1974; Mysen and Boettcher, 1975a); (2) partial melting of peridotite at depth with continuous re-equilibration with surrounding peridotite during ascent—the last equilibration is presumed to occur in the uppermost portion of the upper mantle (Nicholls and Ringwood, 1973; Ringwood, 1975); (3) partial melting of metamorphosed abyssal tholeiite near the top of a descending slab (Green and Ringwood, 1968; Holloway and Burnham, 1972); and (4) fractional crystallization of a parental liquid of basaltic composition (Green, 1970; Osborn, 1959, 1962; Allen et al., 1975). Each of these groups can be subdivided into subgroups as a function of mineral assemblages, pressures, temperatures, and activity of volatile components.

Models based on field evidence include the same four groups. In addition, mixing of mantle-derived basaltic magma with overlying, acidic crustal rocks has been proposed (e.g. Tilley, 1950). More recently, metasomatic alteration of source regions by a fluid phase derived from a dehydrating slab has been suggested as a precursor to partial melting (Fyfe and McBirney, 1975; Frey and Green, 1974; Mysen, 1979).

Ideally, combinations of all available evidence may reduce the number of models for a given suite of andesite. Recent experimental evidence in combination with field data will be combined in this chapter in an attempt to accomplish this goal.

Volatiles and Silicate Melts

It has become evident from the wealth of data presently available (see below) that volatiles (H_2O and CO_2) play a critical role in controlling formation and evolution of magma in the Earth's upper mantle. Before discussing these data, a brief review of effects of volatiles on silicate melt structure is necessary.

Wasserburg (1957) first suggested that the principle effects of water solution in silicate melts is to break bridging oxygens and replace them with OH^- groups. This theory has subsequently been modified by Burnham (1974) and Eggler and Rosenhauer (1978). In the latter models the OH^- group replacement has been retained in addition to exchange of H^+ from H_2O with exchangeable cations (modifying cations). Finally, with large amounts of dissolved water, additional hydroxylated Al–silicate may be formed. Burnham (1974) used the $NaAlSi_3O_8$ molecule to illustrate the principles of this model:

$NaAlSi_3O_8(melt) + H_2O(vapour) \rightleftharpoons$
$AlSi_3O_7(OH)(melt) + Na(OH)(melt).$ (1)

Additional H_2O may dissolve according to the following mechanism:

$AlSi_3O_7(OH)(melt) + nH_2O(vapour) \rightleftharpoons$
$AlSi_3O_{7-n}(OH)_{2n+1}(melt).$ (2)

Another way of discussing solution mechanisms in silicate melts has been proposed by Hess (1977) using polymer theory (Toop and Samis, 1962a,b). In that model, the modifying cations of silicate melts are related to the silicate network through an equilibrium expression that involves bridging (O^0), non-bridging (O^-) and free oxygen (O^{2-}). Bridging O atoms occur between two tetrahedrally coordinated cations (e.g. Si—O—Si). Non-bridging O atoms link up on a tetrahedrally coordinated cation and a cation of higher O coordination (e.g. M—O—Si). The so-called free O^{2-} anion is bonded only to non-tetrahedrally co-ordinated oxygen (M—O—M).

The activity of silica is positively correlated with the degree of polymerization of silica tetrahedra in the melts. The degree of polymerization of the melt can be expressed (Hess, 1977) as

$$O^0 + O^{2-} \rightleftharpoons 2O^-$$ (3)

with the equilibrium constant

$$K = \frac{(\alpha_{O^-})^2}{\alpha_{O^0}\alpha_{O^{2-}}}.$$ (4)

Solution of exchangeable (modifying) cations such as alkali metals, alkaline earths and water in melts results in significant polarization of the Si—O bond, causing the strength of the Si—O—Si bonds to decrease (Hess, 1977). The weakening of the Si—O—Si bonds become more pronounced with decreasing ionization potential of the modifier (Hess, 1977). H^+ in the form of H_2O is such a modifier. As the result, the activity of silica in a melt decreases with solution of H_2O. Similar, but less pronounced, results are observed for K^+, Na^+, and Ca^{2+} (Kushiro, 1975). In terms of igneous rock-forming minerals, solution of strong melt modifiers such as H_2O results in enhanced stability of olivine relative to orthopyroxene (Kushiro, 1969b) and pyroxene relative to plagioclase (e.g. Eggler, 1972).

Carbon dioxide solution in silicate melts involves formation of new O^0, as the result of O^- being stripped away from the M—O—Si bond, causing formation of CO_3^{2-} and new bridging oxygens. In a formalized way, the solution mechanism of CO_2 may be expressed as follows (Mysen et al., 1976):

$2(SiO_4)^{4-}(melt) + CO_2(vapour) \rightleftharpoons$
$(Si_2O_7)^{6-}(melt) + CO_3^{2-}(melt)$ (5)

when $(SiO_4)^{4-}$ and $(Si_2O_7)^{6-}$ represent the depolymerized and polymerized melt, respectively. Solution of CO_2 results in increased activity of silica in the melt, thus enhancing the stabilization of more polymerized minerals on the liquidus than in the absence of CO_2 (Eggler, 1975). In terms of the polymer theory of Toop and Samis (1962a,b), the solution of CO_2 in the silicate melt results in the equilibrium of equation (3) being shifted to the left.

The discussion illustrated by equations (1)–(4) should be kept in mind when considering the phase equilibria involving anatexis of peridotite in the upper mantle. Very briefly, the data on solubility mechanisms of H_2O indicate that for a peridotite of given mineralogy, a hydrous melt will be more silica-rich than one formed where water is absent. Herein lies the clue to the explanation of generating andesitic melts from peridotite. At given pressure and temperature, the activity of silica or silicate components in melts and crystals in the peridotite + water and dry peridotite system is similar. Because the activity coefficient of silica in melts is lowered with solution of H_2O, higher silica concentrations in the melt are necessary to equilibrate a hydrous melt with a given mineral assemblage than is observed for anhydrous melt (*see also* Carmichael *et al.*, 1974, for discussion).

Phase Equilibrium Data

O'Hara (1965, p. 35) first suggested that melting of a spinel peridotite assemblage in the presence of H_2O will result in a liquid with andesitic affinities. The experimental basis for this prediction was found in the work by Yoder and Chinner (1960). In their experiments with a composition

$$(Ca_3Al_2Si_3O_{12})_{16}(Mg_3Al_2Si_3O_{12})_{84}$$

and *c*. 16 per cent H_2O by weight, Yoder and Chinner (1960) observed two pyroxenes, spinel, and forsterite co-existing with liquid at *c*. 980 °C at 10 kb pressure (Fig. 1). O'Hara (1965) concluded that the liquid must lie in the composition plane $py_{84}gross_{16}$–sp–fo. In terms of a 1-atm normative mineralogy the liquid would lie in the quartz–two pyroxenes–anorthtite volume of the basalt tetrahedron (Yoder and Tilley, 1962). As the result this liquid 'must have a fairly high

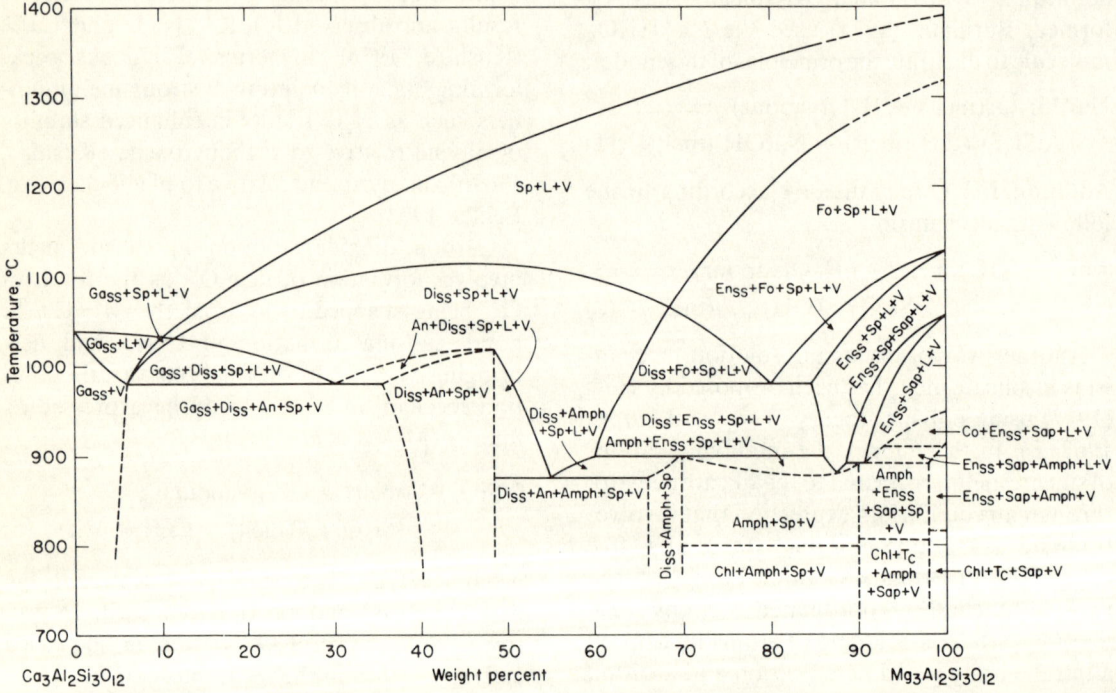

Fig. 1 Phase relations in the system $Mg_3Al_2Si_3O_{12}$–$Ca_3Al_2Si_3O_{12}$–H_2O at 10 kb pressure. Symbols: Fo = forsterite; Ga_{ss} = pyrope–grossular solid solution; sp = spinel; Di_{ss} = diopside–enstatite solid solution; En_{ss} = enstatite–diopside solid solution; An = anorthite; Amph = amphibole; sap = sapphirine, chl = chlorite; Tc = talc; L = liquid; V = vapour. (From Yoder and Chinner, 1960; reproduced by permission of Carnegie Institution of Washington)

content of normative quartz, and probably has a fairly high ratio of hypersthene to diopside in the CIPW norm' (O'Hara, 1965, p. 35).

The first direct demonstration of a significant influence of H_2O on phase equilibria involving upper mantle minerals was provided by Kushiro et al. (1968a). They showed that in the presence of excess H_2O enstatite melts incongruently to olivine and a quartz-normative liquid to at least 35 kb pressure, whereas in the absence of H_2O, enstatite melts congruently at $P \geq 3$ kb (Boyd et al., 1964). Both H_2O and CO_2 occur together in the mantle. Therefore, Eggler (1975) studied the system $MgSiO_3$–CO_2–H_2O at 20 kb and found that enstatite melts congruently with less than c. 60 mol per cent H_2O in the vapour co-existing with crystals and melts. This percentage is likely to increase with increasing pressure.

Despite the observation by Kushiro et al. (1968a), Green (1969) reported that solution of H_2O in silicate melts under pressures corresponding to those of the upper mantle results in enhanced stability of orthopyroxene relative to olivine. Based on this result, Green (1969) suggested that partial melting of hydrous mantle would tend to result in nephelinitic melts. This conclusion was opposed by Kushiro (1969a,b). Kushiro (1969a) studied the phase relations in the system $CaMgSi_2O_6$–Mg_2SiO_4–SiO_2 with and without H_2O (Fig. 2). He noted that addition of H_2O to the system results in a large expansion of the Fo + L field into the quartz-normative field relative to its volume under anhydrous conditions because of the incongruent melting of enstatite. The reaction point P is on the olivine-rich side of the $MgSiO_3$–SiO_2

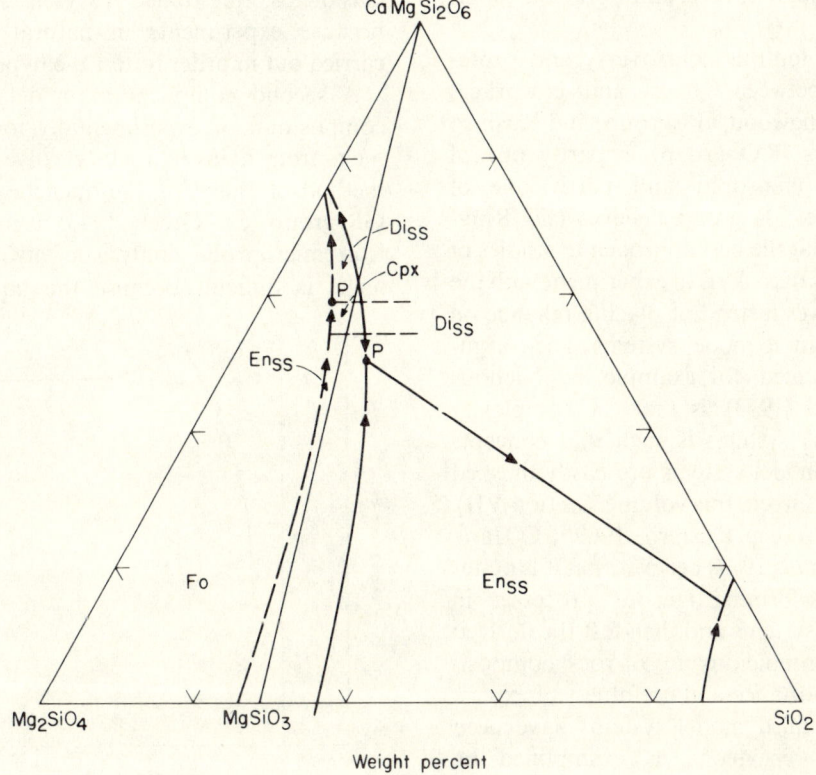

Fig. 2 Phase relations in the system Mg_2SiO_4–$CaMgSi_2O_6$–SiO_2 with (solid lines) and without (broken lines) H_2O at 20 kb. Symbols: Di_{ss} = diopside solid solution; En_{ss} = enstatite solid solution; Fo = forsterite; Cpx = clinopyroxene. (Data from Kushiro, 1969a)

join at $P > 7$ kb in the absence of H_2O because of the congruent melting behaviour of enstatite.

The system Mg_2SiO_4–SiO_2–$CaMgSi_2O_6$–H_2O is, of course, much simplified compared with natural magmas. The absence of alumina and Fe is most notable. The results represent the first illustration, however, of how H_2O affects the melting behaviour of multicomponent systems.

Based on the data of Kushiro (1969a) and those of Kushiro et al. (1968a), Kushiro (1969b) published a discussion of the results of Green (1969), who had noted an expansion of the orthopyroxene field relative to that of olivine in the presence of H_2O in apparent contrast to the results of Kushiro and coworkers. That particular discussion did not resolve any issues, but Green (1973) noted that hydrous melting of 'pyrolite' would result in quartz-normative melts, in an apparent reversal of previous opinions.

The reason for this controversy and subsequent ones between Green and coworkers (Green and Ringwood, GR group) and Kushiro and coworkers (KO group) is partly one of experimental philosophy and partly one of technique. The GR group believes (see Ringwood, 1975) that the best approach to studies of phase relations of rocks is to experiment with the rocks themselves instead of placing reliance on results from simple model systems. Their argument, as presented, for example, by Nicholls and Ringwood (1973), is that the complexity of natural rock systems is such that concepts derived from model systems are easily masked (see also T. H. Green, this volume, Section VII). The KO group (e.g. Kushiro, 1969b; O'Hara, 1971; Mysen et al., 1974) believes that it is better to study rock-forming igneous processes in simple, model systems and then test the derived principles with judicious use of rock compositions. The importance and usefulness of experiments in simplified model systems have been demonstrated repeatedly, as exemplified by classical treatises such as Bowen (1928) and Tuttle and Bowen (1958). The latter approach is favoured because principles are better derived in simple systems where conclusions are not based on what are often subjective interpretations of experimental run products. Such interpretations are necessary when studying results of experiments with natural rock compositions. Furthermore, the influences of extensive parameters (e.g. bulk composition) on experimental results become more difficult to interpret the more components there are in a system. A study of the results of Kushiro (1975) on the effect of major element components on critical liquidus boundaries serves as an illustration of this point. Another example of the success of this approach may be found in the melting studies of natural peridotite by Mysen and Kushiro (1977). The GR group counters this argument (e.g. Nicholls and Ringwood, 1973) by suggesting that the processes in complex natural systems are too involved for studies in three- or four-component systems. Their conclusion is questioned (Mysen et al., 1974) because experiments in natural systems are carried out in order to test the hypothesis.

A second complication in determining the composition of experimentally formed partial melts from hydrous rocks revolves around the method of analysing the quenched liquid. The GR group (e.g. Green, 1973) argues that electron microprobe analysis of hydrous silicate melts is difficult because the liquid may be

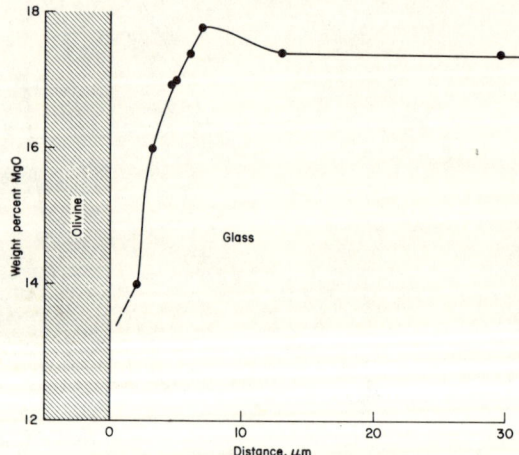

Fig. 3 Variation of MgO content of quenched glass near olivine (Fo_{94}) from anhydrous peridotite at 10 kb and 1375 °C. (From Kushiro, 1974; reproduced by permission of Elsevier Scientific Publishing Co.)

altered by precipitation of minerals during quenching. The alternative to direct analysis is mass-balancing based on composition and modal proportions of the minerals. Mysen and Boettcher (1975a) concluded that the calculated oxide composition of such melts based on the latter technique have more than 30 per cent relative uncertainty because of the difficulty involved in accurate determination of modal composition of experimental run products (see Mysen and Boettcher, 1975a, pp. 576–579, Table 8). Mysen (1978a) and Mysen et al. (1975) estimated the liquid volume from which elements were drawn to form quench minerals. This estimation involved cation diffusion data for silicate melts (e.g. Magaritz and Hofmann, 1978; Hofmann and Magaritz, 1977), quenching rates, and temperatures. It was concluded that elements migrated from liquid volumes that are no more than 10 μm removed from the edge of quench minerals. This conclusion is supported by experimental data by Kushiro (1974) as reproduced in Figs. 3 and 4. As the result of these considerations, Kushiro and coworkers have relied on electron microprobe analysis of quenched glasses when assessing the validity of principles derived from simple systems in experiments in the natural rocks. The reader is referred to Mysen et al. (1974) for such discussions pertaining to andesite genesis.

Kushiro (1972) expanded the original simple system to include Na_2O and Al_2O_3. Results

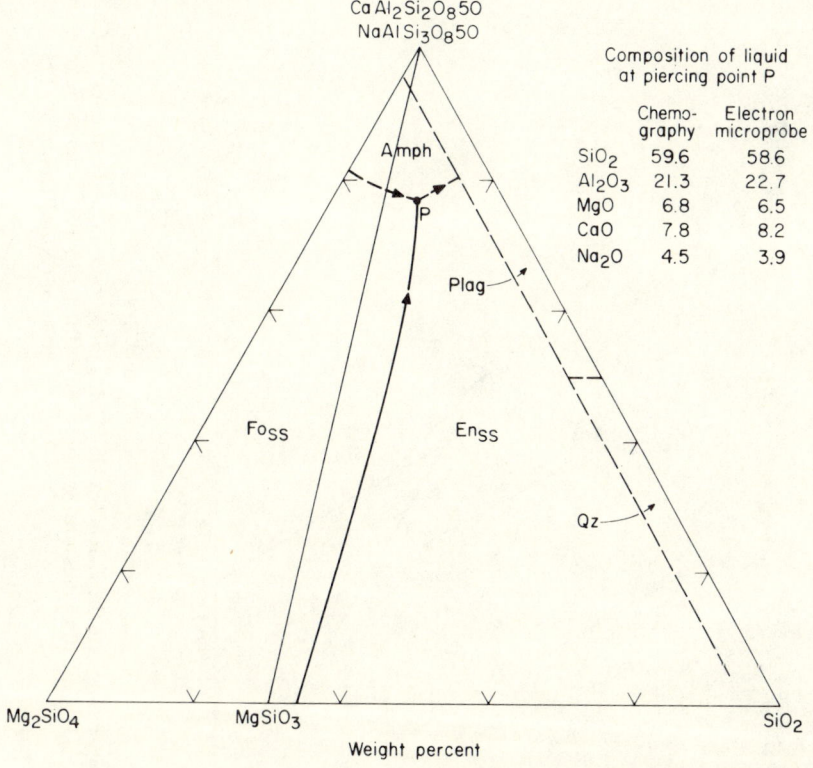

Fig. 4 Phase relations in the system Mg_2SiO_4–$An_{50}Ab_{50}$ (by weight)–SiO_2 with excess H_2O at 15 kb pressure. Point P corresponds to 1000 ± 20 °C with the clinopyroxene volume encountered at 980 ± 20 °C. Table represents composition of liquid at P with chemographic and electron microprobe analysis (recalculated to an anhydrous basis). Symbols: Fo = forsterite; En_{ss} = enstatite solid solution; amph = amphibole; Plag = plagioclase; Qz = quartz. (From Kushiro, 1974; reproduced by permission of Elsevier Scientific Publishing Co.)

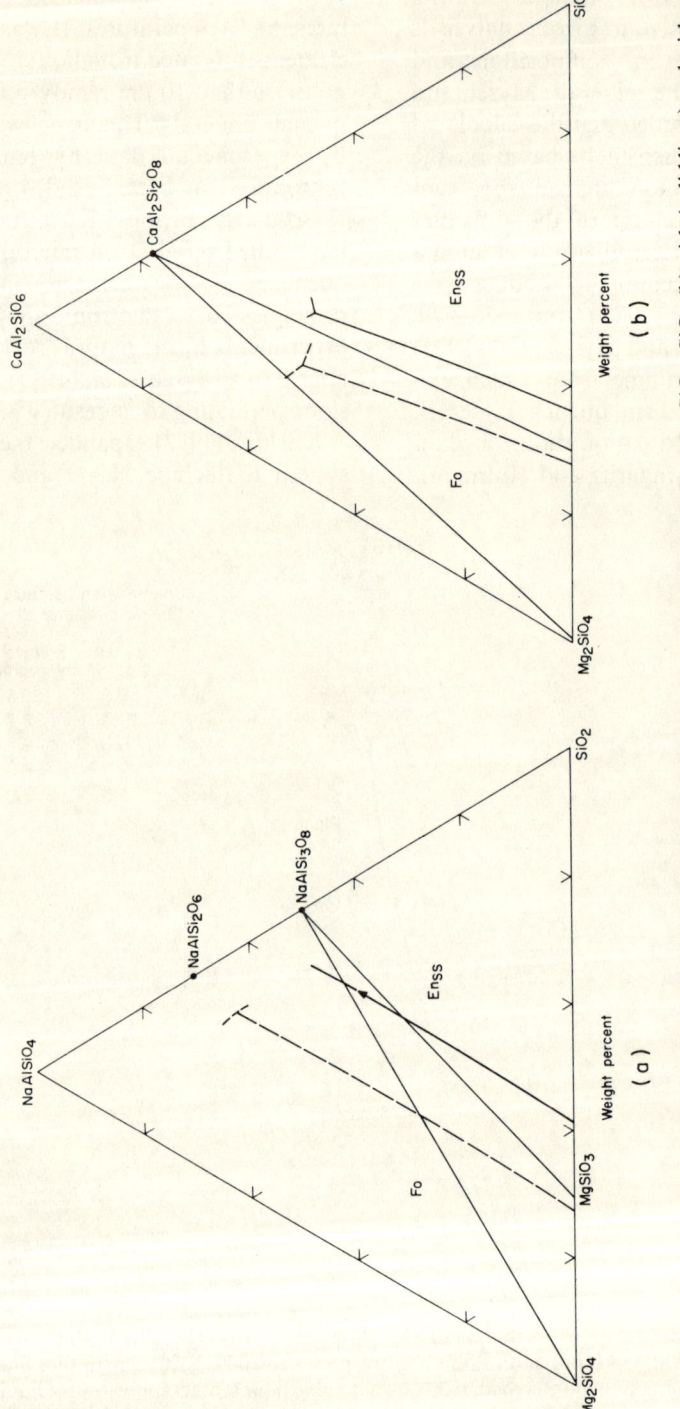

Fig. 5 Phase relations in the system $NaAlSiO_4$–Mg_2SiO_4–SiO_2 (a) and in the system $CaAl_2SiO_6$–Mg_2SiO_4–SiO_2 (b) with (solid lines) and without (broken lines) H_2O at 20 kb pressure. Symbols: Fo = forsterite; En_{ss} = enstatite solid solution. (From Kushiro, 1972; reproduced by permission of Oxford University Press)

from selected joins are shown in Fig. 5 at 20 kb pressure both with and without H_2O. It can be seen from the data in these figures that the critical olivine–orthopyroxene join shifts toward the silica-rich side of all the systems with the addition of H_2O. Based on these data and those of Kushiro (1969a) and Yoder (1969), Kushiro (1972) concluded that partial melts from both spinel and garnet peridotite in the presence of H_2O would be quartz normative to at least 25 kb pressure. In fact, microprobe analysis by Kushiro (1972) of melts from appropriate starting compositions led him to conclude that the partial melts closely resemble those of calc-alkaline andesites or dacites (Kushiro, 1972, p. 330).

Further substantiation of Kushiro's (1972) hypothesis was provided by Kushiro (1974). In this case the system was Mg_2SiO_4–$An_{50}Ab_{50}$–SiO_2–H_2O at 15 kb (Fig. 4). In this diagram the clinopyroxene volume lies 20 °C beneath the piercing point, P, at 15 kb. Consequently, for all practical purposes, the composition of melt P co-exists with olivine, orthopyroxene, clinopyroxene, and pargasitic amphibole at 980 °C at 15 kb. Addition of 10 per cent by weight of $KAlSi_3O_8$ component shifts the location of the piercing point a few per cent toward the plagioclase–quartz join. The normative composition of this melt is compared with an average Cainozoic andesite (Chayes, 1969) in Table 1. The similarity of the two melt compositions is striking. Because the mineral assemblage of the upper mantle at 1000 °C at 15 kb is olivine + orthopyroxene + clinopyroxene + amphibole (pargasite), these observations were taken as evidence for the partial melt of wet peridotite in the upper mantle being andesite (Kushiro, 1974).

In summary, experimental results from simple model systems indicate that the bulk composition of partial melts of wet peridotite resemble calc-alkaline andesite to at least 25 kb and at temperatures c. 1000 °C.

A requirement of wet peridotite being the parent of andesite is that the liquidus temperature of the andesite is similar to the solidus temperature at the pressure of equilibration of the andesite with its parent. Kushiro et al. (1968b)

TABLE 1 Comparison of bulk composition of melt at piercing point P (Fig. 4) with addition of 10 per cent $KAlSi_3O_8$ by weight to the starting material (column 1) and average Cainozoic andesite (column 2) (Chayes, 1969) after Kushiro (1974)

	1	2
SiO_2	60.8	58.17
TiO_2	—	0.80
Al_2O_3	20.8	17.26
Fe_2O_3	—	3.07
FeO	—	4.18
MgO	5.9	3.24
CaO	6.8	6.93
Na_2O	4.0	3.21
K_2O	1.7	1.61
P_2O_5	—	0.21
Total	100.0	98.68

conducted the first experiments with a spinel peridotite nodule from Salt Lake Water, Hawaii, (Table 2) with excess H_2O. The wet solidus in the pressure range corresponding to magma formation in the upper mantle was located near 1000 °C (Fig. 4). Kushiro (1972, 1974) noted that the liquidus of haploandesite in the system Na_2O–CaO–MgO–Al_2O_3–SiO_2–H_2O is near 1025 °C, and that the haploandesitic liquidus in the system Mg_2SiO_4–$An_{50}Ab_{50}$–SiO_2–H_2O is slightly below 1000 °C. The latter temperature is near the solidus temperature of the parent. Nehru and Wyllie (1975) noted that 10 per cent melting of St Paul's peridotite (Table 2) is attained at 1000 °C at 20 kb (25 °C above the wet solidus) in support of the above conclusion.

It should be noted, of course, that natural peridotite includes solid solutions that are not accounted for in the model systems. In fact, potential peridotite parents of partial melts in the mantle also show a spread in bulk composition (Table 2). The influence of such compositional variations on the wet solidus of peridotite was investigated by Mysen and Boettcher (1975b). They found that the solidus temperature is lowered by decreasing Mg/(Mg + Fe) and Ca/Al (Fig. 6), effects that have previously been seen for liquidus temperatures of basalt, for

TABLE 2 Bulk compositions of potential peridotite parents of partial melts

	1	2	3	4	5	6	7
SiO_2	48.02	42.22	43.70	44.82	45.7	43.7	45.10
TiO_2	0.22	0.30	0.25	0.52	0.05	0.20	0.13
Al_2O_3	4.88	4.42	2.75	8.21	1.6	4.0	3.92
Fe_2O_3	1.94	2.86	1.38	2.07	0.77	0.89	1.00
FeO	8.15	4.45	8.81	7.91	5.21	8.09	7.29
MnO	0.14	0.13	0.13	0.19	0.09	0.12	0.14
MgO	32.35	34.61	37.22	26.53	42.8	37.4	38.81
CaO	2.97	3.92	3.26	8.12	0.70	3.50	2.66
Na_2O	0.66	0.43	0.33	0.89	0.09	0.38	0.27
K_2O	0.07	0.11	0.14	0.03	0.04	0.01	0.02
P_2O_5	0.07	—	—	0.04	0.01	<0.01	0.01
Cr_2O_3	0.25	0.50	0.28	0.20	0.41	0.40	0.31
Total	99.72	93.95†	98.25	99.53	97.47	98.69	99.61

(1) Salt Lake spinel peridotite (Kushiro et al., 1968b). (2) St Paul's peridotite (Nehru and Wyllie, 1975). (3) PHN 1611 garnet lherzolite (Nixon and Boyd, 1973). (4) 66SAL-1 garnet lherzolite (Jackson and Wright, 1970). (5) Ga-p(1) garnet lherzolite (Mysen and Boettcher, 1975b). (6) 618-138b.1 spinel lherzolite (White, 1966). (7) 66PAL-3 (Jackson and Wright, 1970) spinel lherzolite (Mysen and Boettcher, 1975b).

† Sample contains c. 6 per cent H_2O by weight.

example (Yoder and Tilley, 1962). The data of Mysen and Boettcher (1975b) are also notable in that they indicate that the solidus determined by Kushiro et al. (1968b) may be as much as 100 °C too high. The solidus of rocks is determined conventionally by optically observing the first appearance of quenched glass in the experimental charges. It has been suggested (Mysen and Boettcher, 1975b) that this technique is open to subjective judgment when there is less than 5 per cent melt in an experimental run product. Mysen and Kushiro (1977) devised a technique using tritiated H_2O to circumvent this potential problem. Upon melting, the H_2O is enriched in the melt over any other condensed silicate phase. The appearance of melt with such H_2O can be easily detected with autoradiography (Mysen and Seitz, 1975) and is, therefore, a tool to determine wet solidus temperatures of rocks. This technique does not rely on optical observation of glass. With this technique, Mysen et al. (1978) found that the wet solidus of garnet peridotite nodule PHN 1611 (Nixon and Boyd, 1973; see Table 2) is between 900 and 915 °C at 20 kb. This observation lends credence to the suggestion (Mysen and Boettcher, 1975b) that the solidus curve of Kushiro et al. (1968b) may be as much as 100 °C too high.

As indicated in the introduction to this section, analysis of partial melts from natural peridotite is done with the electron microprobe. It has been suggested (Green, 1973) that the quality of such analyses may be assessed by computing the Mg–Fe^{2+} partitioning between melt and coexisting olivine. That test relies on data for the expected partition coefficient of Mg and Fe^{2+} between melt and crystals. The test may give a rough indication for dry melts since 1 atm data are available for a variety of melt and olivine compositions (Roeder and Emslie, 1970). No high pressure data exist; furthermore, no information on the effect of H_2O on activity coefficients of Fe^{2+} and Mg^{2+} in hydrous silicate melts is available. It is known that both pressure and water in solution dramatically affect melt structure (Burnham, 1975; Mysen and Kushiro, 1978; Sharma et al., 1978). For example, the activity coefficient of Ni^{2+} in $NaAlSi_2O_6$-rich melt is lowered by 50 per cent between 10 and 20 kb pressure. Moreover, electron microprobe analysis of Fe does not distinguish between Fe^{2+} and Fe^{3+}. The total Fe content of

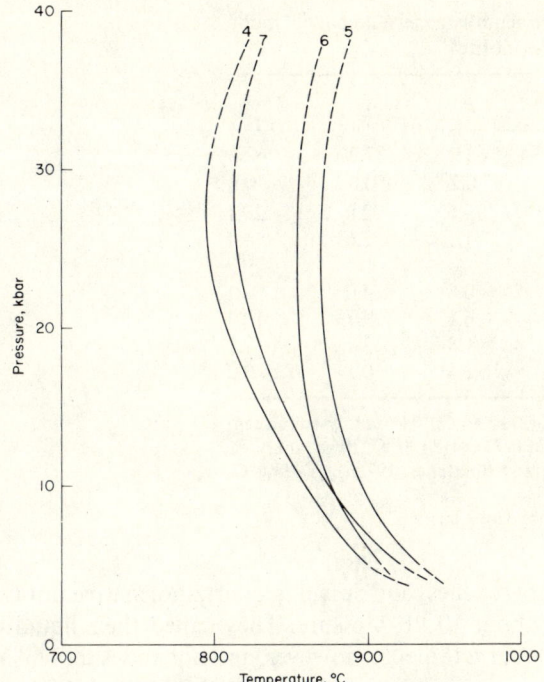

Fig. 6 Solidi of peridotite as a function of peridotite composition in the presence of excess H_2O. Numbers refer to column numbers in Table 2. (From Mysen and Boettcher, 1975b; reproduced by permission of Oxford University Press)

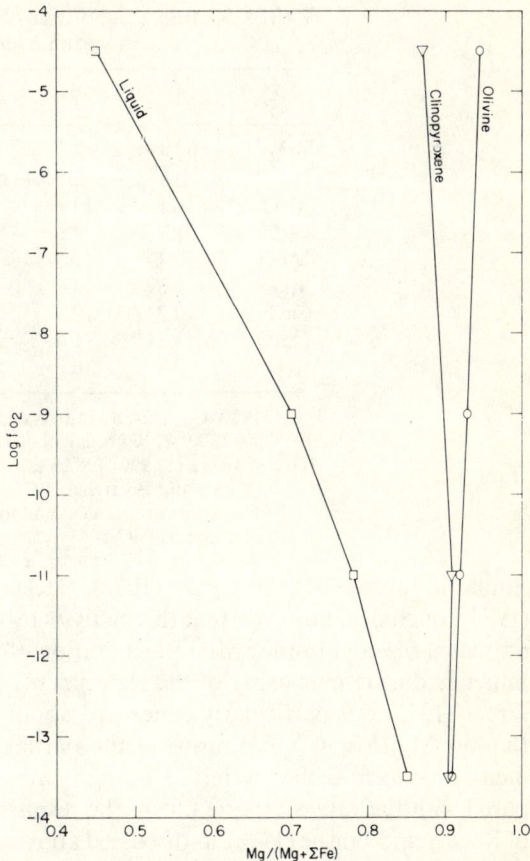

Fig. 7 $Mg/(Mg + \sum Fe)$ of coexisting liquid, clinopyroxene, and olivine during partial melting of peridotite + H_2O as a function of f_{O_2} at 1050°C and 15 kb. (From Mysen, 1975; reproduced by permission of Springer Verlag)

melt and crystals is affected by f_{O_2}. The $Fe^{3+}/\sum Fe$ in these phases shows different functional relationships with f_{O_2}, however. Mysen (1975) showed that the $Mg/(Mg + \sum Fe)$ of hydrous melt at 15 kb and 1050 °C changes from c. 0.5 to 0.85 with an f_{O_2} between that of the magnetite–haematite and that of the iron-wustite buffer (Fig. 7). As the result of these observations, it must be concluded that the $Mg/(Mg + \sum Fe)$ ratio of hydrous melts from peridotite at high pressure is an inappropriate indicator of the $Mg/(Mg + \sum Fe)$ of the coexisting crystalline residue.

The first analysis of partial melts from wet peridotite were provided by Kushiro et al. (1972) at 26 kb and 1190 °C. The starting material was a lherzolite from Salt Lake Crater (Table 2). According to Kushiro et al. (1972), c. 20 per cent of the melt co-existing with clinopyroxene, orthopyroxene, and olivine was formed at this pressure and temperature. The liquid composition is given in Table 3. The $Mg/(Mg + \sum Fe)$ ratio of this glass is 0.49. Kushiro (1972) commented that this liquid composition is similar to that obtained in the system $Na_2O-CaO-Al_2O_3-MgO-SiO_2-H_2O$ under similar pressure and temperature conditions. He concluded, therefore, that the liquid composition obtained in the natural system represented the equilibrium liquid composition. Kushiro et al. (1972) suggested that their data indicate that partial melts from hydrous peridotite are of andesitic to dacitic compositions. Green (1973) conducted experiments with hydrous 'pyrolite'. Electron microprobe analysis of the glass gave results

TABLE 3 Bulk composition of experimentally generated partial melts from hydrous peridotite†

	1	2	3	4	5	6
SiO_2	67.8	59.7	54.6	68.9	57.2	62.2
TiO_2	0.6	2.5	2.2	0.2	0.6	0.8
Al_2O_3	16.1	13.7	10.7	19.6	22.8	20.1
FeO‡	1.1	4.6	8.3	1.1	2.6	1.7
MnO	0.1	0.2	0.2	—	0.1	0.2
MgO	0.6	4.6	11.9	0.8	4.0	2.1
CaO	10.2	12.7	9.4	6.1	9.7	10.4
Na_2O	3.1	≥1.4	1.9	≥1.8	2.7	2.1
K_2O	0.3	0.6	0.4	1.4	0.1	0.2

(1) 1190 °C, 25 kb (Kushiro *et al.*, 1972). (2) 1200 °C, 10 kb, analysed (Green, 1973). (3) 1200 °C, 10 kb, calculated (Green, 1973). (4) 1100 °C, 20 kb, analysed (Green, 1973). (5) 990 °C, 7.5 kb (Mysen and Boettcher, 1975a). (6) 950 °C, 15 kb (Mysen and Boettcher, 1975a).

† All analyses are recalculated to an anhydrous basis.

‡ Total iron as FeO

similar to those of Kushiro *et al.* (1972). Green (1973) concluded, however, that this analysis represented the composition after precipitation of minerals during quenching of the experiments. Green (1973) was particularly concerned about the low $Mg/(Mg + \sum Fe)$ ratios of the partial melts inasmuch as they were too Fe-rich compared with the values expected from the results of Roeder and Emslie (1970), as discussed above. Mysen *et al.* (1974) discussed the $Mg/(Mg + \sum Fe)$ values of these quenched glasses in the light of the inherent f_{O_2} of furnace assemblies used. They concluded that with the f_{O_2} values reported for the furnace assemblies used in these experiments, the $Mg/(Mg + \sum Fe)$ values were at the expected level (*see also* Fig. 7). Green (1973) also reported an electron mictroprobe analysis of quenched glass at 10 kb and 1200 °C (Table 3) and again concluded that the analysis reflected quench problems. He then calculated the liquid composition based on estimated proportions of phases. The composition thus obtained resembles quartz tholeiite (Table 3). Both analyses are probably incorrect because the only co-existing phase is olivine. According to Mysen and Kushiro (1977) such liquids will be olivine-normative.

Nicholls and Ringwood (1973) studied partial melts from hydrous pyrolite and concluded that the liquid in equilibrium with olivine, two pyroxenes, and spinel is quartz-normative until about 10 kb pressure. They called their liquid quartz tholeiite, however, and not andesite. This conclusion was challenged by Mysen and Kushiro (in Mysen *et al.*, 1974), who quoted results by Mysen (1973) and Kushiro *et al.* (1972). Mysen (1973) had found that partial melts from a natural spinel peridotite from Hawaii (White, 1966) resemble andesite to at least 20 kb pressure. Details of the experiments reported by Mysen (1973) were published by Mysen and Boettcher (1975a,b) (*see* Table 3). Nicholls and Ringwood (in Mysen *et al.*, 1974) commented that first, the $Mg/(Mg + \sum Fe)$ of the liquid compositions of Kushiro *et al.* (1972) and Mysen (1973) was too low for a melt in equilibrium with peridotite, and second, that olivine was not a liquidus phase from the melts quoted as equilibrium partial melts by Mysen (1973). Mysen and Kushiro (in Mysen *et al.*, 1974) noted that the $Mg/(Mg + \sum Fe)$ was appropriate in view of the f_{O_2} of the experiments (*see* Mysen, 1975, and above). The lack of observed olivine, even with 10 per cent olivine added (as reported by Nicholls and Ringwood, 1973), is understandable in view of the reaction relation (Kushiro, 1974)

$$\text{Amph} + \text{Opx} \rightleftharpoons \text{Ol} + \text{liq.} \quad (6)$$

It may be argued that the activity of H_2O in the source region of andesitic melts is less than unity. In fact, Marsh (1976) suggested that andesitic melts are essentially anhydrous despite petrological evidence to the contrary (Delaney et al., 1978; Anderson, 1975). Experiments with water pressure less than total pressure can be conducted with sufficiently small amounts of H_2O so that the partial melts are undersaturated with respect to H_2O. From an experimental viewpoint, this technique poses problems. First, silica-rich, water-undersaturated melts notoriously tend not to equilibrate (e.g. Hill and Boettcher, 1970). Second, the activity of H_2O in such melts is unknown and also varies in an unknown fashion with a degree of melting and with intensive variables. With a fixed amount of H_2O, the concentration of H_2O in the melt decreases with increasing degree of melting. The activity coefficient of H_2O in the melt also changes with degree of melting as the result of changing melt bulk composition and temperature. Changing pressure also results in structural changes of hydrous silicate melts, thus resulting in changing activity of water in the melt. The values of these variables are unknown. Furthermore, technical difficulties in determining the degree of melting have in the past made it impossible even to determine water contents of such partial melts. As the result, experiments conducted with this technique have limited applicability.

An alternative technique is to conduct all experiments with a fluid present, but with the activity of H_2O controlled with a suitable diluent (Boettcher et al., 1973). For experiments with silica-rich melts such as andesite, carbon dioxide is a suitable diluent because of the low solubility of CO_2 in such melts (Mysen, 1976; Mysen et al., 1975) even in the presence of H_2O.

The temperatures of the solidus of a peridotite depends on the activity of H_2O and to a much lesser extent on the activity of CO_2 (Mysen and Boettcher, 1975b), as shown in Fig. 8. Mysen and Boettcher (1975a) attempted to assess the roles of H_2O and CO_2 in partial melting of peridotite in the upper mantle. In all these experiments, the f_{O_2} was buffered at the level of the magnetite–haematite buffer to ensure that no reduced carbon species were present in the

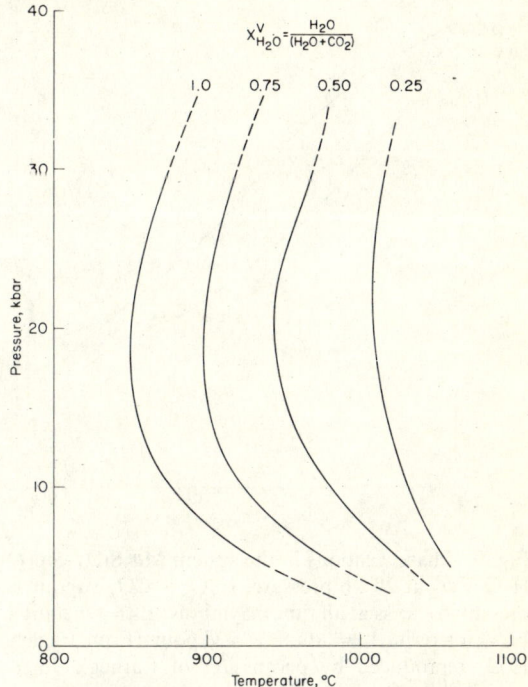

Fig. 8 Solidi of peridotite + H_2O + CO_2 as a function of mole fraction $X^v_{H_2O} = H_2O/(H_2O + CO_2)$ in coexisting vapour. Peridotite bulk composition corresponds to no. 4, Table 2. (From Mysen and Boettcher, 1975b; reproduced by permission of Oxford University Press)

fluids (Boettcher et al., 1973). Because of this high f_{O_2}, the Mg/(Mg + $\sum$ Fe) of all partial melts was around 0.5. The starting material in those experiments was a Hawaiian spinel lherzolite (618–138b. 1; White, 1966) from the Honolulu volcanic series (Table 2). In a multicomponent system such as peridotite–H_2O–CO_2, the bulk composition of the partial melt depends on pressure, temperature, composition of the (H_2O + CO_2) fluid, and bulk composition of the starting material. In Fig. 10, the normative compositions of partial melts are shown as a function of $X^v_{H_2O}(= H_2O/(H_2O + CO_2))$ at constant pressure and temperature). Notably, the bulk composition of the partial melt shifts from olivine- to quartz-normative with $X^v_{H_2O}$ slightly less than 0.6. This number is nearly identical to that observed by Eggler (1973) for the shift of melting behaviour of enstatite (see Fig. 9). A summary of the experimental results of Mysen

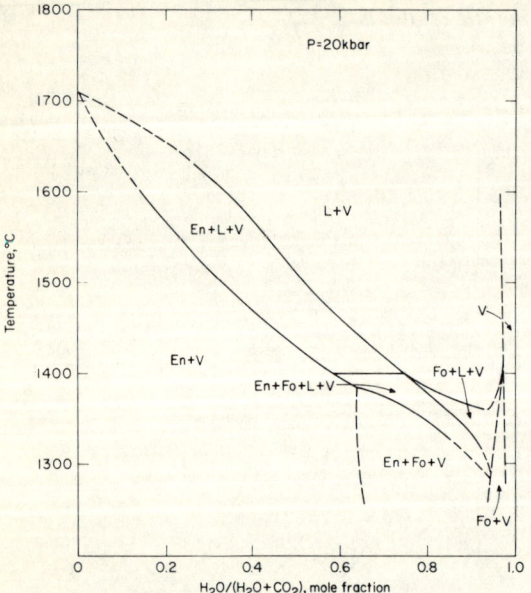

Fig. 9 Phase relations in the system Mg_2SiO_4–SiO_2–H_2O–CO_2 at 20 kb pressure. $H_2O + CO_2$ vapour is present in excess at all times. Symbols: En = enstatite, Fo = forsterite, L = liquid, V = vapour. (From Eggler, 1973; reproduced by permission of Carnegie Institution of Washington)

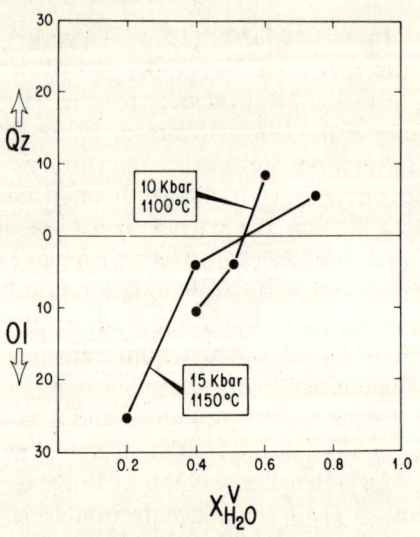

Fig. 10 Degree of silica saturation (expressed as normative quartz and olivine) of partial melts from peridotite as a function of $X^V_{H_2O}$. Starting material as in Fig. 8. (From Mysen and Boettcher, 1975a; reproduced by permission of Oxford University Press)

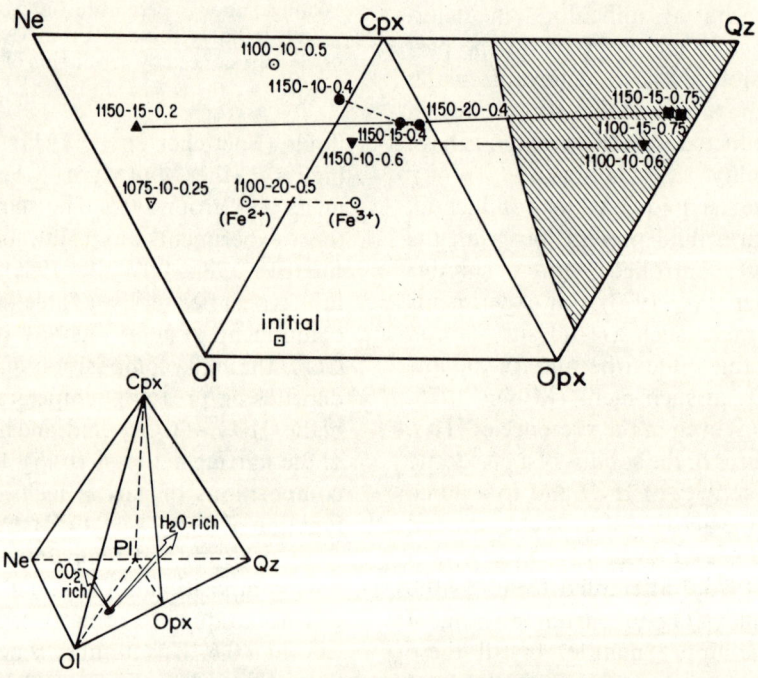

Fig. 11 Summary of analytical data of partial melts from peridotite + H_2O + CO_2 as a function of $X^V_{H_2O}$. The peridotite composition is the same as in Figs 8 and 10 (column 4; Table 2). (From Mysen and Boettcher, 1975a; reproduced by permission of Oxford University Press)

and Boettcher (1975a) in terms of normative mineralogy is given in Fig. 11. It can be seen from the data in this figure that andesitic liquids may be produced by partial melting of peridotite + H_2O + CO_2 with significant amounts of CO_2 in the fluid. In view of the nearly ideal mixing behaviour of CO_2 and H_2O in the mixing range in question (Mysen and Boettcher, 1975b), these data indicate that the activity of H_2O may be lowered by as much as 40 per cent below that of pure H_2O without affecting the phase relations of peridotite significantly, at least not to 20 kb pressure. Data at higher pressure are not available because of analytical problems in acquiring a reliable liquid bulk composition.

Source of Volatiles

Although perhaps only 1 per cent by weight or less H_2O is needed in a peridotite source region of andesitic partial melts, this amount of water exceeds the amount commonly expected in the upper mantle. A source of H_2O is, therefore, needed. Wyllie (1971) first suggested that this source could be the slab of hydrated basaltic material underlying the peridotite. It is necessary that the temperature of dehydration of such a slab is lower than the solidus of wet basalt, however. If not, the slab material itself would begin to melt once H_2O was released. In fact, several thermal models of subduction zones have used melting on the slab–peridotite interface as a boundary condition (e.g. Turcotte and Schubert, 1973). Delaney and Helgeson (1978) calculated temperatures of dehydration equilibria in hydrous basaltic material and concluded that a descending slab could release H_2O to depths of perhaps as much as 125 km without causing melting of the slab itself. If this conclusion is correct, it appears that the H_2O needed for wet partial melting of the peridotite wedge underlying island arcs may be found to be the descending slab. It is necessary, however, for this H_2O to migrate from the slab into the overlying peridotite. It has been suggested that the peridotite is essentially impermeable to H_2O (Marsh, 1976). In an attempt to determine whether this is so, Mysen et al. (1978) conducted experiments to determine infiltration rates of H_2O under pressure conditions of the upper mantle. The peridotite starting material in these experiments was PHN 1611 (Nixon and Boyd, 1973) that was dried thoroughly before the experiment. Two grain sizes, 50 μm and 100 μm, were employed. The source of H_2O was synthetic serpentine that was placed beneath the peridotite layer in the experimental sample container. Mysen et al. (1978) attempted to eliminate all open pore space in the peridotite powder prior to the experiment by subjecting the sample to the experimental pressure and a temperature less than that of the dehydration temperature of serpentine for several days. The infiltration rate of H_2O was then monitored as a function of this

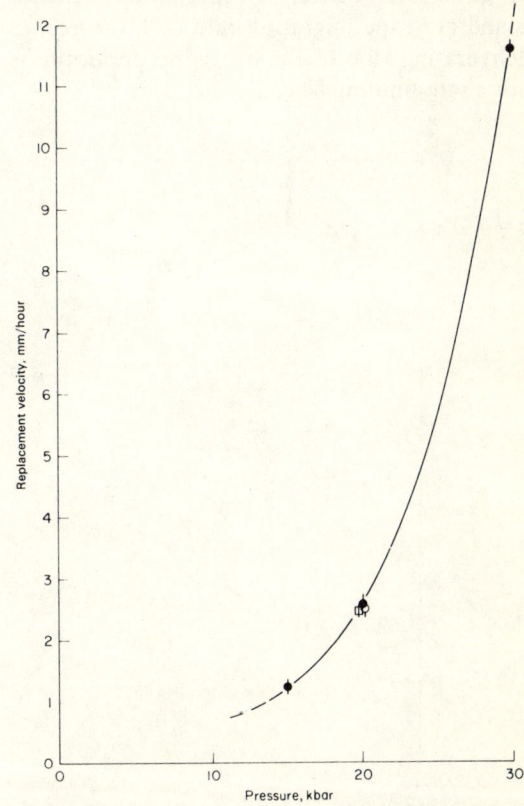

Fig. 12 Replacement velocity (amphibole formation in dry peridotite) as a function of pressure and temperature. Closed symbols, 850°C; open symbols, 650°C. Error bars ±1σ (experimental). Starting material, garnet peridotite nodule PHN 1611 (see Table 2 and Nixon and Boyd, 1973). (From Mysen et al., 1978; reproduced by permission of Carnegie Institution of Washington)

initial annealing period. Once the infiltration rate became independent of annealing time, it was concluded that all pore space in the peridotite was eliminated. The experiments were concluded at pressures up to 30 kb.

The results of the experiments by Mysen *et al.* (1978) are shown in Fig. 12. It can be seen from those data that the infiltration rate of H_2O in peridotite is of the order of millimetres per hour. Furthermore, the rate is strongly dependent on pressure. The data indicate that H_2O may migrate through peridotite upper mantle at the rate of *c.* 1 km a^{-1}. Under normal thermal conditions *c.* 5 Ma is required to melt 1 per cent peridotite. It is concluded, therefore, that if a supply of H_2O from an external source is needed to cause wet melting of peridotite beneath island arcs, the migration rate of H_2O from a dehydrating slab to the overlying peridotite is not a rate-limiting factor.

Trace Element Partitioning

Orogenic andesites often show significant light REE enrichment but without any distinctive correlation between degree of light REE enrichment and major elements that may indicate that this feature reflects a fractionation process (Gill, 1978). One must conclude, therefore, that the REE patterns of andesite reflect the process of partial melting.

Transition metals, likewise, show no correlation with major elements with the exception of Ni and MgO (Gill, 1978). Even the latter correlation is not easily explained in terms of fractional crystallization because such mechanisms will also affect other elements such as Si; the SiO_2 and Ni contents of orogenic andesites show no correlation. Presumably, an element such as Co will show similar relations as that of Ni.

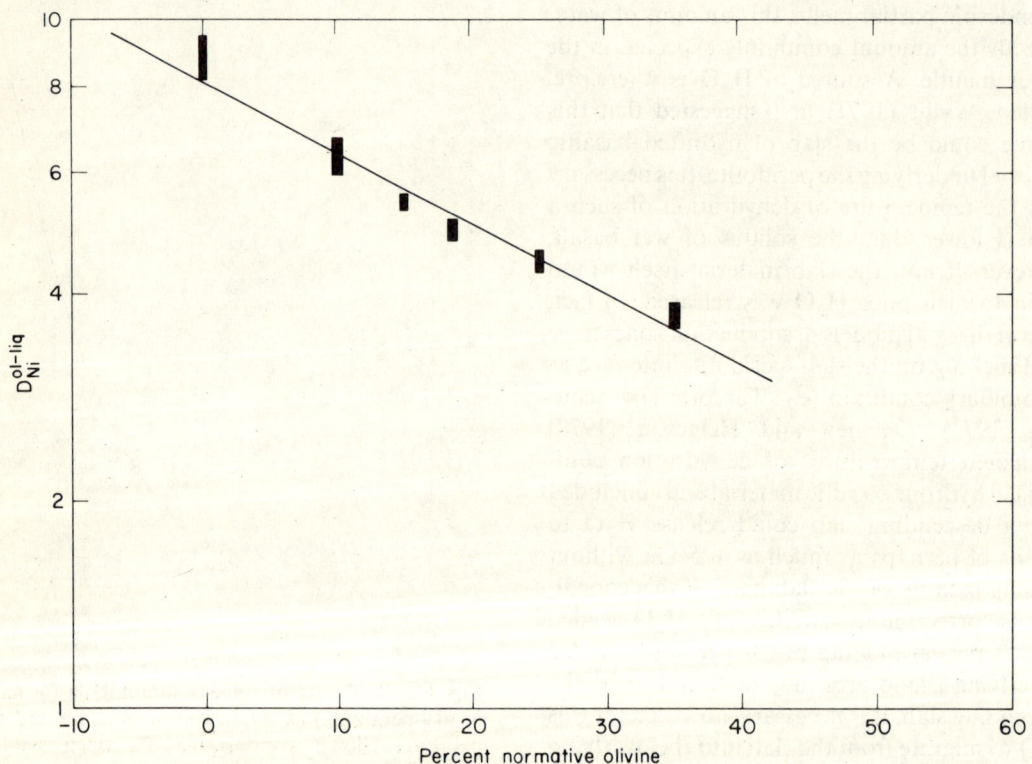

Fig. 13 Ni partition coefficients between olivine and natural silicate melts of basaltic komatiite to peridotitic komatiite composition expressed as normative olivine in the melt. (From Arndt, 1977; reproduced by permission of Carnegie Institution of Washington)

Experimental determinations of crystal–liquid partition coefficients have only recently been initiated. Ni has often been used as an example of a transition metal (e.g. Arndt, 1977; Hart and Davis, 1978; Mysen, 1978a; Mysen and Kushiro, 1978). There is general agreement that nickel and other transition metal partition coefficients depend on the liquid composition (Arndt, 1977; Irvine and Kushiro, 1976; Watson, 1976, 1977; Takahashi, 1978). This bulk compositional dependence presumably reflects the structural variations of silicate melts. Arndt (1977) showed that D_{Ni}^{ol-liq} (= concentration of Ni in olivine/concentration of Ni in liquid) decreases with decreasing silica content of the melt, for example (Fig. 13). This observation is in agreement with that of Watson (1976, 1977), who noted that with two co-existing immiscible melts, the transition metals are favoured by the more basic melt. Hart and Davis (1978) attempted to correlate D_{Ni}^{ol-liq} with MgO content of the melt (Fig. 14) using the system Mg_2SiO_4–Ab–An. It should be noted, however, that in that system, a dependence of D_{Ni}^{ol-liq} on silica content is also observed. Therefore, in the experiments of Hart and Davis (1978) the Si/O ratio of the different melts was kept constant to study the effect of MgO on D_{Ni}^{ol-liq}. The Si/O ratio was assumed to express the degree of silica polymerization of the melt. Some of their data relating the Si/O ratio to the

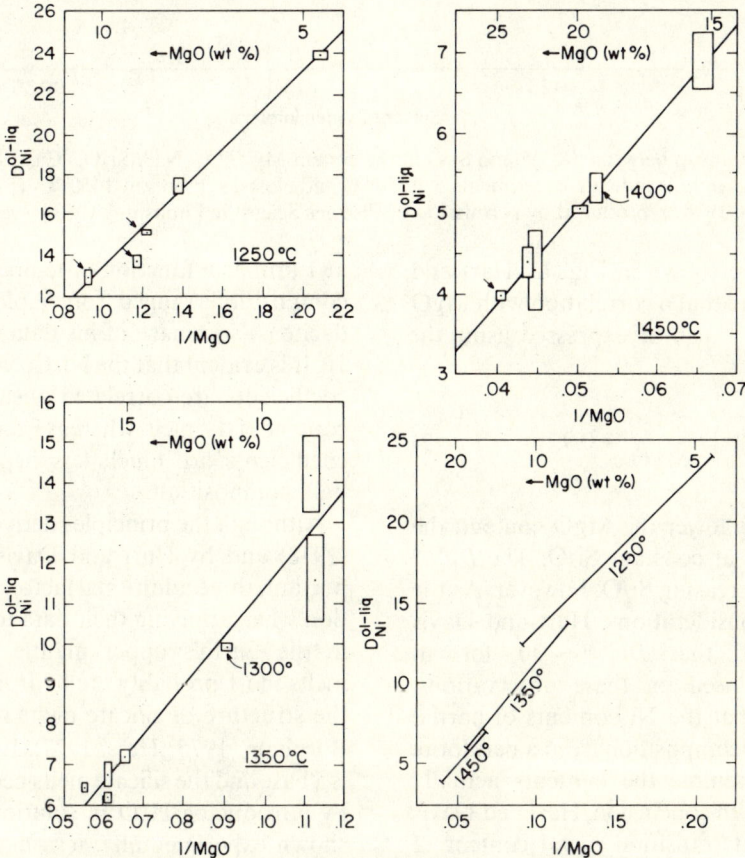

Fig. 14 D_{Ni}^{ol-liq} in the system Mg_2SiO_4–$NaAlSi_3O_8$–$CaAlSi_2O_8$ as a function of MgO content of the melt (in weight per cent) and temperature at 1 atm pressure. The size of the symbols reflects experimental uncertainties. Small arrows indicate experiments on Pt-loop. (From Hart and Davis, 1978; reproduced by permission of Elsevier Scientific Publishing Co.)

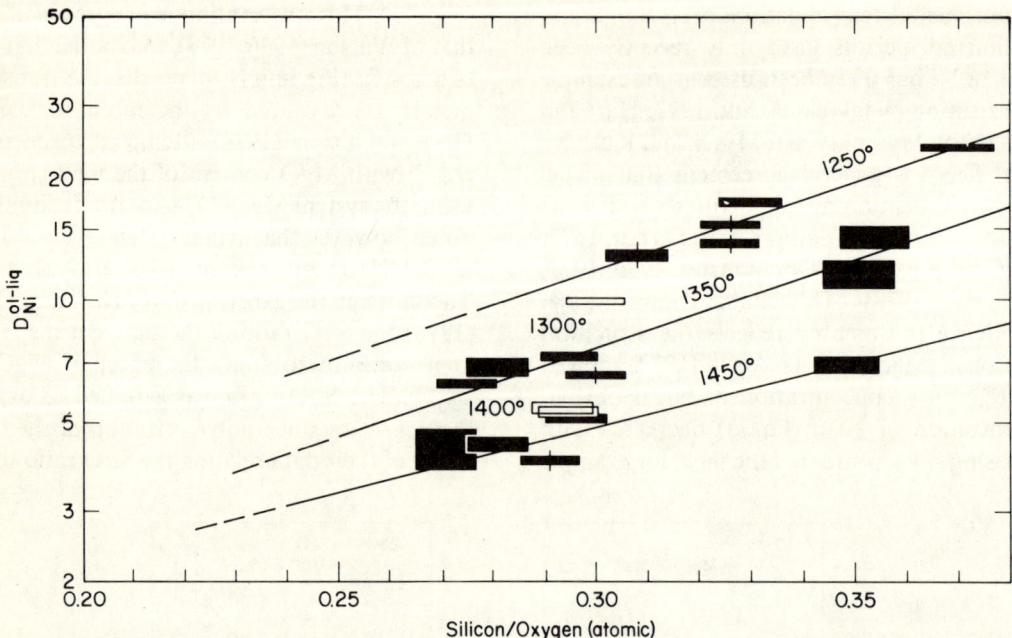

Fig. 15 Relationship between D_{Ni}^{ol-liq} and Si/O in the system Mg_2SiO_4–$NaAlSi_3O_8$–$CaAlSi_2O_8$ at 1 atm pressure. Open symbols indicate experiments at 1350 °C and closed symbols at 1450 °C. (From Hart and Davis, 1978; reproduced by permission of Elsevier Scientific Publishing Co.)

value of D_{Ni}^{ol-liq} are shown in Fig. 15. Hart and Davis (1978) found that a correlation with MgO content and D_{Ni}^{ol-liq} can be expressed using the relation

$$D_{Ni}^{ol-liq} = \frac{24.9}{MgO} - 0.9. \qquad (7)$$

Consequently, the lower the MgO content, the higher the D_{Ni}^{ol-liq} at constant Si/O. The D_{Ni}^{ol-liq} increases with decreasing Si/O, however. As the result of these considerations, Hart and Davis (1978) concluded that $D_{Ni}^{ol-liq} \sim 30$, for an andesitic melt. Based on these observations, they concluded that the Ni contents of partial melts of andesitic composition from a peridotite parent would resemble the contents actually found in andesite. In conclusion, Hart and Davis (1978) stated that transition metal content of andesite indicates that these rocks were formed by direct partial melting of peridotite.

Takahashi (1978) studied Ni, Co, Mn, Mg, and Fe partitioning between olivine and liquid in the system Mg_2SiO_4–Fe_2SiO_4–$K_2O \cdot 4SiO_2$ at 1 atm as a function of temperature. His melt compositions ranged from haplobasalt to haplodacite. A summary of his data is shown in Fig. 16. It is evident that the Ni, Co, and Mg partition coefficients are correlated positively with silica content of the melt, whereas Fe and Mn partition coefficients are much less dependent on melt bulk composition.

Although the principles derived by Takahashi (1978) and by Hart and Davis (1978) are important, three additional factors need consideration when applying their data to partial melting in the Earth's upper mantle. First, andesitic melts most probably are hydrous. Water affects the structure of silicate melts significantly (e.g. Burnham, 1975; Hess, 1977). In principle, H_2O is a base and the silicate melts become more basic by introducing H_2O in solution. Although not shown experimentally, it is likely that crystal–liquid partition coefficients for transition metals are lowered by addition of H_2O to the melt. The extent of lowering is open to speculation until experimental data become available. Second, the structures of silicate melts depend on pressure

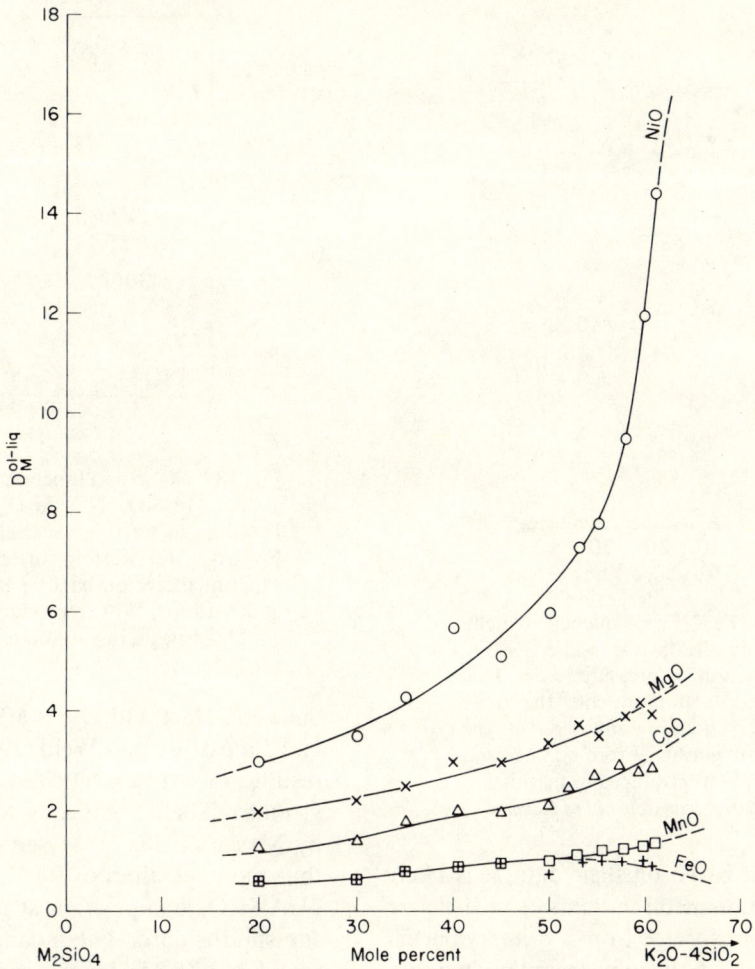

Fig. 16 Relations between transition metal partition coefficients of olivine and liquid and liquid bulk composition at 1atm pressure. (From Takahashi, 1978; reproduced by permission of Pergamon Press)

(Sharma et al., 1978; Kushiro, 1976; Mysen and Virgo, 1978). As the result, activity coefficients of elements in solution depend on the pressure. Mysen and Virgo (1978) showed, for example, that $\gamma_{Fe^{2+}}$ increases and $\gamma_{Fe^{3+}}$ decreases with increasing pressure in a melt of $NaAlSi_2O_6$ composition until a certain pressure maximum is reached. Further pressure increase reverses this relation (Fig. 17). These data indicate that Fe^{2+}/Fe^{3+} partitioning between melts and crystals would be pressure dependent. The data also indicate that other transition metal partition coefficients may be pressure dependent. Mysen and Kushiro (1978) determined D_{Ni}^{ol-liq} for forsterite and a melt of nearly pure $NaAlSi_2O_6$ composition (Fig. 18). As can be seen from their results, the olivine–liquid partition coefficient for Ni increases to a maximum between 5 and 10 kb pressure. An additional pressure increase results in a rapid reduction of D_{Ni}^{ol-liq}. Based on these data, Mysen and Kushiro (1978) warned against using experimental data obtained at low pressure to crystal–liquid equilibria at high pressures. They also concluded that their observations involving Ni may also apply to other transition metals. Mysen and Kushiro (1978)

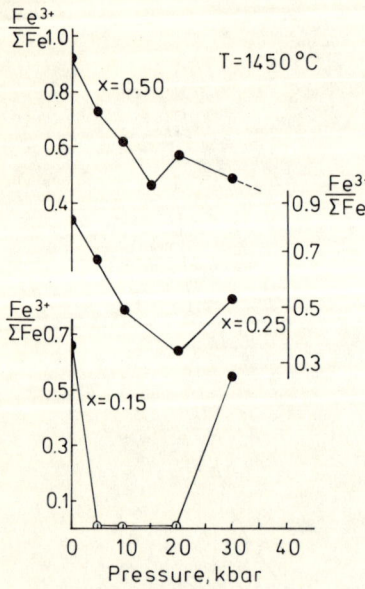

Fig. 17 Fe^{3+}/Fe of quenched melts on the join $NaAlSi_2O_6$–$NaFe^{3+}Si_2O_6$ as a function of pressure. $x = 0.15$, 0.25, and 0.50 represents the mole fraction of acmite component in the starting material. (From Mysen and Virgo, 1978; reproduced by permission of the *American Journal of Science*)

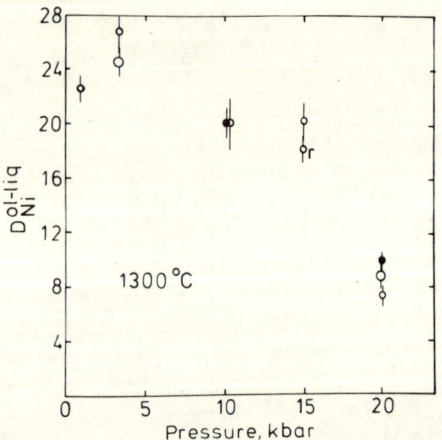

Fig. 18 D_{Ni}^{ol-liq} as a function of pressure in the system Mg_2SiO_4–$NaAlSi_2O_6$. Open symbols, $Fo_{20}Jd_{80}$ as starting material, closed symbols, $Fo_{40}Jd_{60}$ as starting material; r, reversal experiment; error bars, $\pm 1\sigma$. (From Mysen and Kushiro, 1978; reproduced by permission of Carnegie Institution of Washington)

estimated that, based on their data, it is likely that the transition metal contents of andesite are lower than that expected from a primary partial melt of peridotite in the upper mantle. This conclusion is further substantiated by data obtained by Mysen (1978a and unpublished data), who determined Ni partition coefficients as a function of Ni content of the crystalline phases. Mysen (1978a) used a water-saturated melt and temperatures up to 1075 °C (Fig. 19(a)), whereas Mysen (unpublished data) conducted experiments in an anhydrous system at 1300 °C (Fig. 19(b)). These data bring up the third point, for in both cases it was found that D_{Ni}^{ol-liq} depends on Ni content of the crystalline phase in the Ni concentration range corresponding to its abundance range in peridotite in the upper mantle. Mysen (1977) concluded, in fact, that a primary melt of andesitic composition may contain as much as 700 p.p.m. Ni. This concentration far exceeds the maximum abundance of Ni in andesite. Hart and Davis (1978), Shaw (1977), and Lindstrom and Weill (1978) challenged the results of Mysen (1978a) on experimental grounds. Their arguments were, however, met by Mysen (1978a,b). Mysen (1978a) concluded that the low values of D_{Ni}^{ol-liq} (6–8) for melts of $NaAlSi_2O_6$ composition at upper mantle pressures in the natural abundance range of Ni was real (Fig. 19(b)) and not an experimental artefact. The D_{Ni}^{ol-liq} value at 1 atm and within the concentration range of Henry's law was c. 17.

It must be concluded, therefore, that the transition metal contents of andesite are most likely to reflect a fractionation mechanism that was operative subsequent to initial melting. The summary of geochemical data by Gill (1978) indicates that silicate minerals cannot be the cause of transition metal depletion, however. Similar conclusions were reached by Mysen (1977) on phase equilibrium grounds. Gill (1978) agreed with Osborn (1962, 1977) that this phase most probably is an oxide. Mysen (1977) suggested that it is either an oxide or an immiscible sulphide phase or both.

Shima and Naldrett (1975) pointed out that separation of a small amount (less than 1 per cent by weight) of an immiscible Fe-rich sulphide melt would lower transition metal contents of silicate melts without affecting the major element contents of the silicate melt appreciably. This conclusion was substantiated by Mysen and Kushiro (1976) who found that $D_{Ni}^{\text{sulphide melt–basalt melt}}$ has a value c. 200. Mysen and Popp (1978) calculated that $D_{Ni}^{\text{sulphide melt–andesite melt}}$ must be near 500. Experimental data by Rajamani and Naldrett (1978) also indicate that this partition coefficient is near 500. The last-mentioned authors also presented experimental data on Cu partition coefficients. Co partition coefficients were given for basaltic melt (c. 80 at 1255 °C, and 1 atm). Based on the change of Ni partition coefficient from basalt to andesite (274 to 460), the Co partition coefficient is likely to be near 130.

The model involving sulphide melt separation requires a source of S, however. Peridotite and eclogite nodules in kimberlite and alkali basalt contain c. 1 per cent (Cu, Fe, Ni) monosulphide (Meyer and Boctor, 1975; Bishop et al., 1975). It is possible to unite sulphurization–oxidation reactions involving sulphides and silicates, which define f_{S_2} and f_{O_2} at given temperature and pressure. During partial melting of peridotite material such as found in these nodules, the f_{S_2}, f_{O_2}, P, T, and bulk compositions of the silicate melt define the sulphur solubility in the melt. Mysen and Popp (1978) suggested that the reaction

$$S_2 + 4Fe_2SiO_4 \rightleftharpoons 2FeS + 4SiO_2 + 2Fe_3O_4, \quad (8)$$

when accounting for $Fe \rightleftharpoons Mg$ solid solutions, may describe the f_{S_2} and f_{O_2} conditions of the upper mantle. Present experimental technique does not allow independent control of f_{S_2} and f_{O_2} in Fe-bearing systems. The f_{S_2} and f_{O_2} defined by equation (8) may be approximated with f_{S_2} controlled by the PtS–Pt buffer and f_{O_2} with the C–O–CO$_2$ buffer (Mysen and Popp, 1978). The experiments by Mysen and Popp (1978) were conducted with melts of NaAlSi$_3$O$_8$ and CaMgSi$_2$O$_6$ compositions up to 30 kb and 1650 °C. Pertinent experimental results are shown in Fig. 20. Mysen and Popp (1978) noted that despite the absence of Fe in the system, the solubility of S in these melts is five to ten times greater at pressures corresponding to those of the upper mantle than at 1 atm (Haughton et al., 1974; Shima and Naldrett, 1975).

Mysen and Popp (1978) suggested that the S solubility in NaAlSi$_3$O$_8$ melt represented the minimum value for andesitic magmas in the upper mantle and lower crust. Even with this solubility, the S content must be lowered from more than 0.5 per cent by weight in melts in the upper mantle to less than 0.1 per cent by weight by the time the magma reaches the Earth's surface. If this S is exsolved as an immiscible Fe-rich sulphide melt, c. 1.3–1.4 per cent by weight (Fe, Ni)S must separate from the magma during ascent.

Mysen and Kushiro (1976) found that the partition coefficient for Ni between Fe-rich sulphide melt and ferrobasaltic liquid at 1 atm is slightly below 300. This value is most probably a minimum for andesitic melts because the activity of Ni in silicate melts increases with increasing degree of polymerization of the melt (Arndt, 1977; Irvine and Kushiro, 1976; Hart and Davis, 1978), as also indicated by the experimental results of Rajamani and Naldrett (1978). Evolution of Ni, Co, and Cu contents of andesitic magmas as a function of percentage of sulphide melt separation is shown in Fig. 21. The partition coefficients are likely to change as a function of silicate and sulphide bulk composition, pressure, temperature, and element contents (e.g. Arndt, 1977; Rajamani and Naldrett, 1978; Mysen, 1978a,b; Mysen and Kushiro, 1978). Experimental data applicable to such functional relationships are not available, however. In the absence of this information, the partition coefficient is a constant in these calculations. The results are, therefore, only semiquantitative. It should be noted, however, that even if the values of the partition coefficients are too high by a factor of two or three, S-saturated andesite magma contain sufficient S to lower the transition metal contents from those of primary magma in the upper mantle to those found in andesite by separation of immiscible sulphide melt. It is anticipated that effective

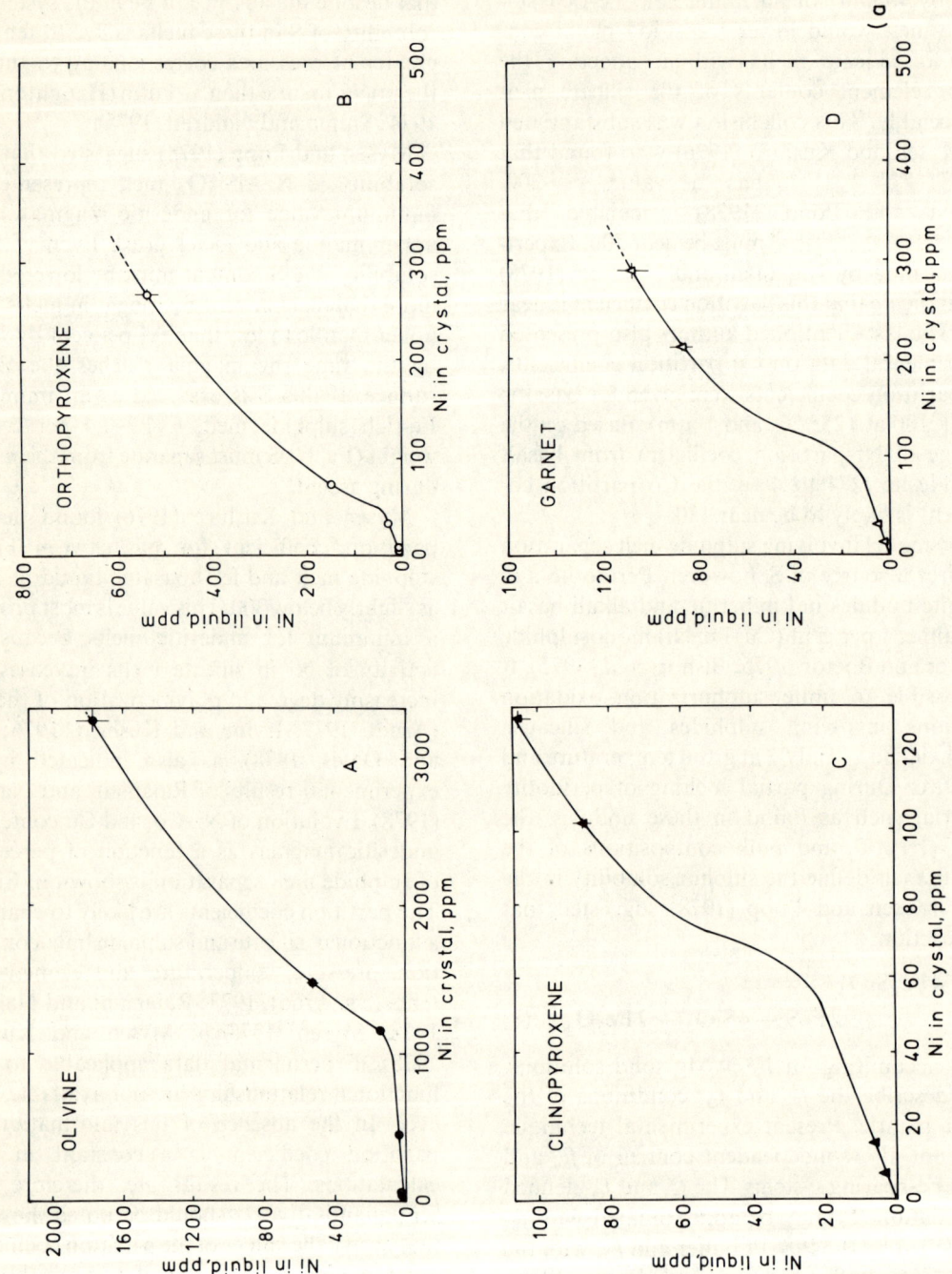

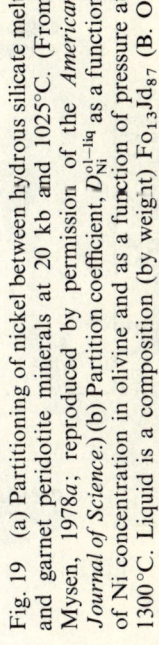

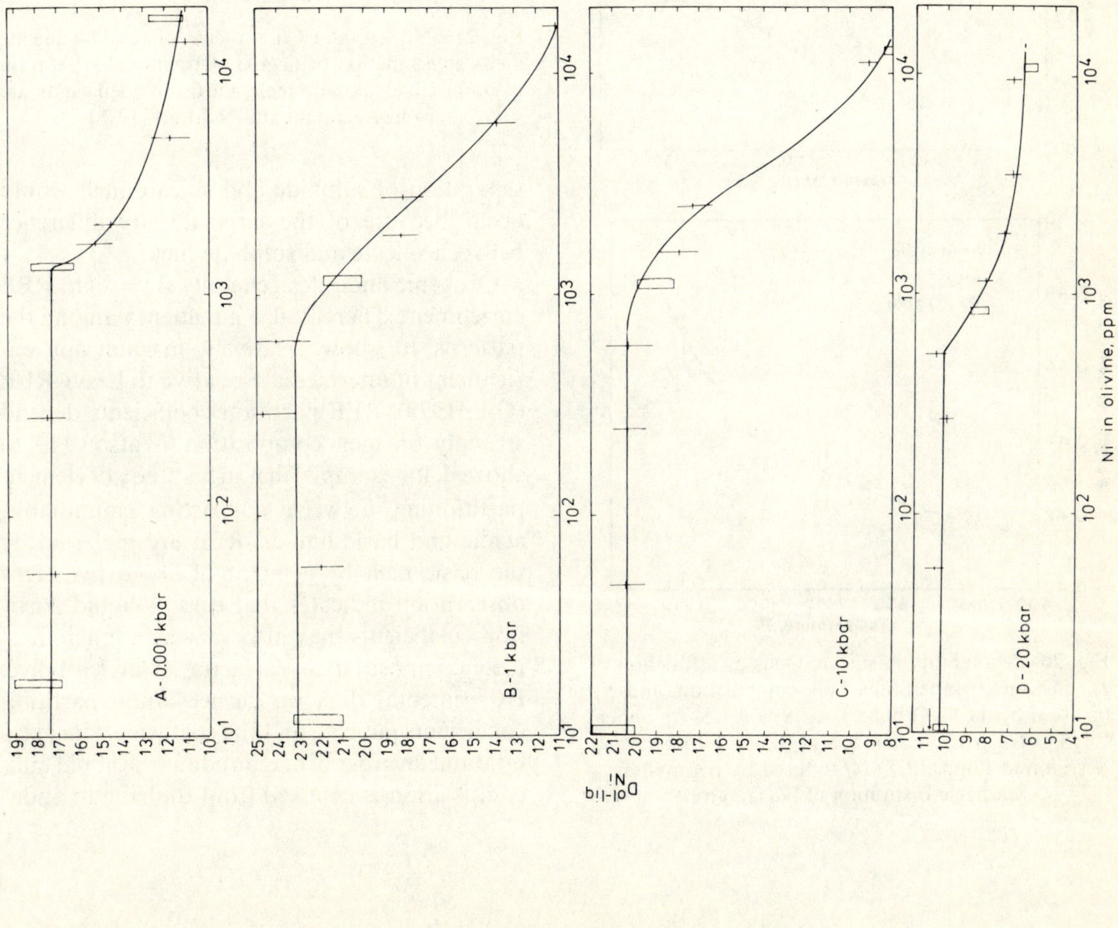

Fig. 19 (a) Partitioning of nickel between hydrous silicate melt and garnet peridotite minerals at 20 kb and 1025°C. (From Mysen, 1978a; reproduced by permission of the *American Journal of Science*.) (b) Partition coefficient, D_{Ni}^{ol-liq} as a function of Ni concentration in olivine and as a function of pressure at 1300°C. Liquid is a composition (by weight) $Fo_{13}Jd_{87}$ (B. O. Mysen, unpublished data)

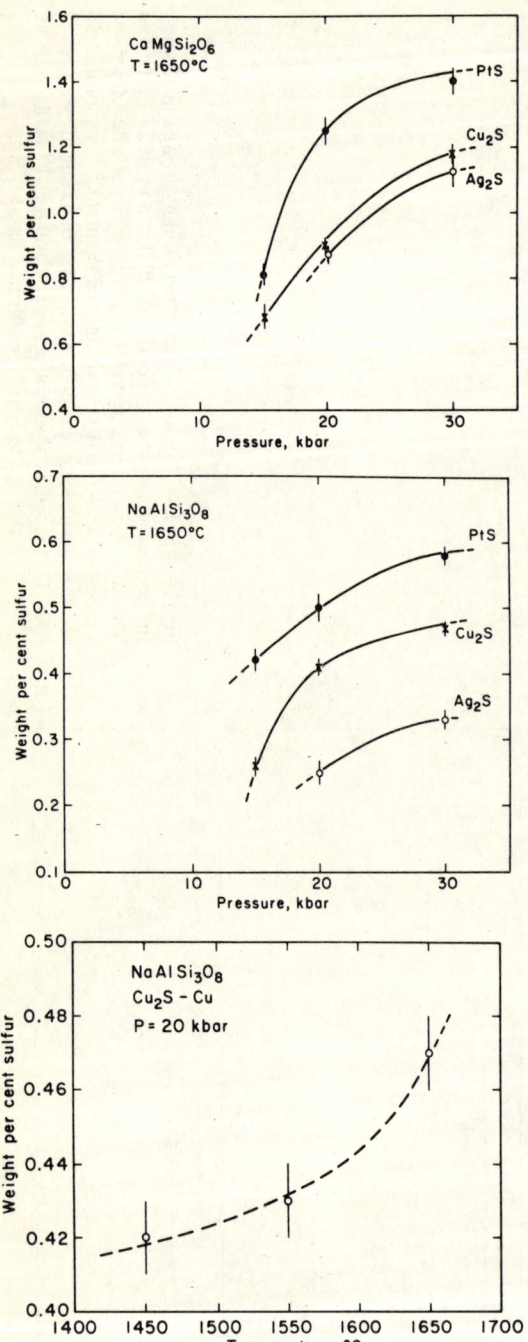

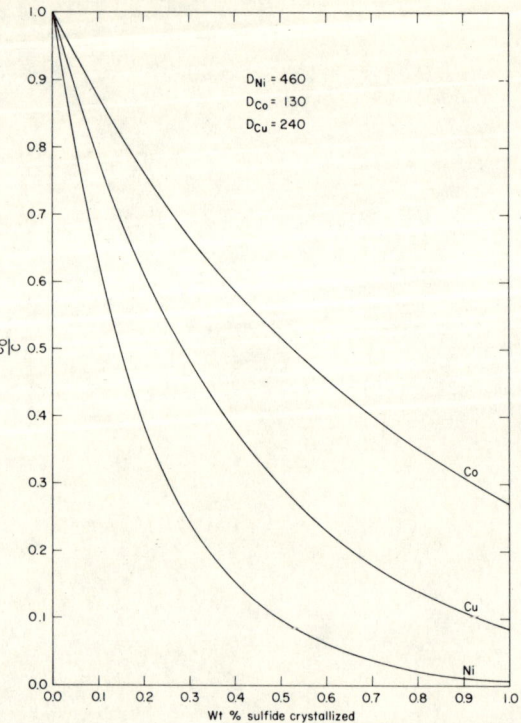

Fig. 21 Ni, Co, and Cu contents of residual silicate melts as a function of degree of fractional separation of immiscible sulphide melt. Partition coefficients are from Rajamani and Naldrett (1978)

Fig. 20 S solubility in silicate melts as a function of f_{S_2}, pressure, temperature, bulk composition, and f_{O_2} (as fixed by the C–O buffer). $Ag_2S = Ag_2S$–Ag buffer; $Cu_2S = Cu_2S$–Cu buffer; PtS = PtS–Pt buffer. (From Mysen and Popp, 1978; reproduced by permission of Carnegie Institution of Washington)

separation of sulphide and silicate melt would occur because of the large density difference between silicate and sulphide melt.

Orogenic andesites generally show light REE enrichment. There is also a tendency among the patterns to show relatively insignificant enrichment of intermediate relative to heavy REE (Gill, 1978). REE partition coefficients depend strongly on melt composition. Watson (1976) showed, for example, that in his study of element partitioning between co-existing, immiscible, acidic and basic liquids, REE are preferred by the basic melt by a factor of about five. This observation indicates that crystal–liquid partition coefficients may also vary as a function of melt composition by a factor of at least five. Experimental data on garnet–liquid partition coefficients indicate that the coefficients increase by about an order of magnitude as the liquid bulk composition is changed from tholeiite to andesite or dacite (*see* Irving, 1978, for review). The fractionation of individual REE partition coefficients is not affected significantly by bulk liquid composition, however. Irving and Frey (1978) summarized available garnet phenocryst/matrix data for matrix compositions ranging from kimberlite to rhyolite. These coefficients varied by nearly two orders of magnitude.

Calculation of REE patterns of andesitic melts requires a choice of partition coefficients. In the present context, coefficients involving acidic liquids should be chosen. It is also necessary to choose an appropriate bulk rock composition of the source, REE contents of the source, and a model of melting to provide a meaningful model for magma genesis in the upper mantle (Shaw, 1970). The bulk rock composition of the source rock controls its modal composition at given pressure and temperature.

In the present model, modal proportions of minerals of garnet- and spinel-peridotite with bulk composition similar to nodule PHN 1611 from Lesotho (Nixon and Boyd, 1973) is chosen (Table 2). This peridotite nodule appears to be representative of an undepleted upper mantle (Boyd and McCallister, 1976; Shimizu, 1975; Mysen and Kushiro, 1977). This conclusion is not meant to imply that all undepleted peridotite upper mantle is similar to nodule PHN 1611, but simply that it is a representative composition. The calculated modal composition of nodule PHN 1611 if equilibrated in the spinel peridotite field is similar to model mantle SP 1 as shown in Table 4. This modal composition is compared with that expected from the same bulk composition in the garnet stability field (GP 1) and also compared with that of nodule PHN 1611 (Boyd and McCallister, 1976).

The model of melting involves the stoicheiometry of the expression describing melting and the extent to which the melt remains in equilibrium with crystalline phases during the melting. Mysen (1977), based on phase equilibrium work in the system $CaO-MgO-Al_2O_3-SiO_2$, suggested that andesitic melt from garnet peridotite could be derived by melting according to the expression

$$Ga + 0.67\ Cpx + 0.38\ Opx \rightleftharpoons 0.45\ Ol + 1.61\ liq, \quad (9)$$

where the coefficients are mass proportions. Melting of garnet-free peridotite may be described with the equation

$$0.54\ Cpx + 0.54\ Opx \rightleftharpoons 0.08\ Ol + liq \quad (10)$$

(Kushiro, 1969*a*).

It is impossible, at present, to decide whether batch or fractional melting most appropriately describe the melting. However, for small degrees of melting (less than 10 per cent melt), this aspect of the melting model does not affect the results appreciably.

REE contents and fractionation patterns of undepleted upper mantle generally have been considered to be between one and three times that of average chondrite (Schmitt *et al.*, 1963, 1964). REE contents at 10 per cent partial melts derived according to equations (9) and (10) are compared with REE patterns of orogenic andesites in Fig. 22. A source with REE contents equal to three times that of average chondrite as well as equal to unity is used. It is clear from the results of these calculations that neither unaltered garnet nor spinel peridotite can be partially melted to provide the appropriate REE patterns. Green and Ringwood (1968) proposed partial melting of quartz eclogite to produce andesite. Marsh and Carmichael (1974) and Marsh (1976) supported such a model on geophysical grounds. Green and Ringwood (1968) did not provide a basis for calculating the stoicheiometry of a melting reaction to produce a melt

TABLE 4 Comparison of modal composition of model spinel peridotite (SP 1) and model garnet peridotite (GP 1) of similar bulk composition (per cent) with garnet peridotite (PHN 1611)

	SP 1	GP 1	PHN 1611
Garnet	—	10.0	10.4
Clinopyroxene	10.0	10.0	11.1
Orthopyroxene	30.1	20.0	19.7
Olivine	56.4	60.0	58.8
Spinel	3.5	—	

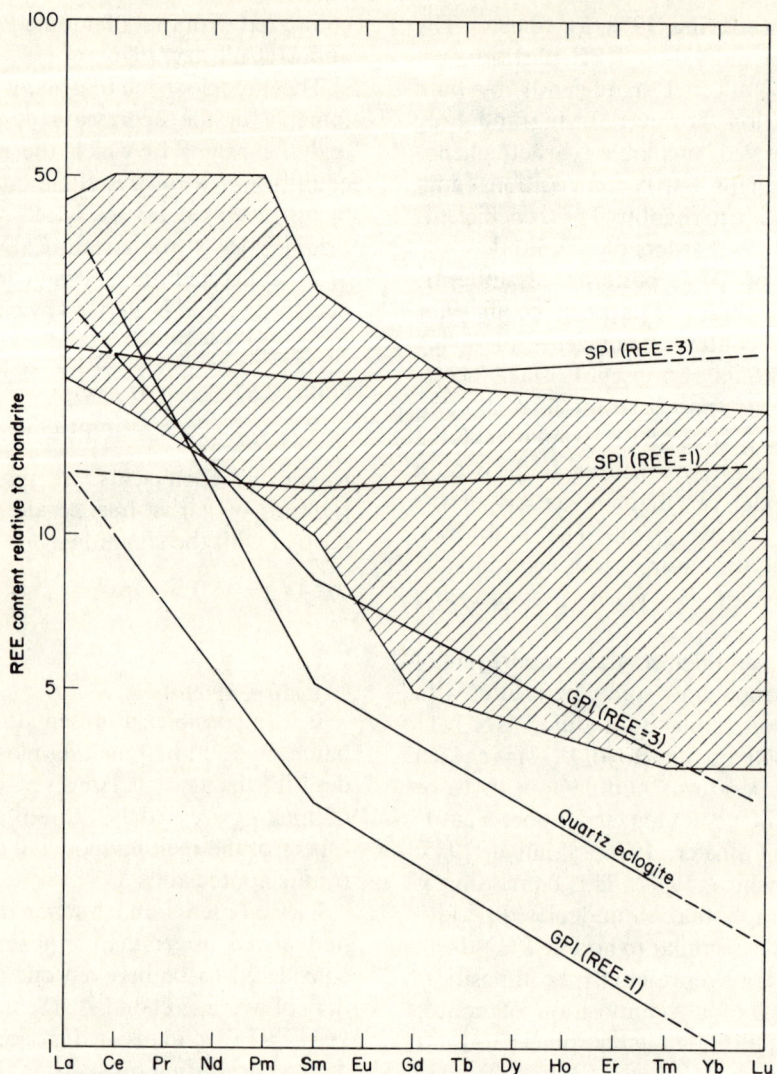

Fig. 22 REE contents of 5 per cent fractional melt from garnet and spinel peridotites GP 1 and SP 1, respectively, compared with REE patterns from orogenic andesites (Gill, 1974; Thorpe *et al.* 1976; Yajima *et al.*, 1972) and with the REE pattern from a 5 per cent fractional melt from quartz eclogite with 10 times average chondrite content. Symbols: GP 1 (REE = 3) and GP 1 (REE = 1), garnet peridotite source rock with 3 and 1 times average chondrite REE abundance; SP 1 (REE = 3) and SP 1 (REE = 1), same as GP 1, but with spinel peridotite source. Partition coefficients are from Mysen (1978c)

of andesitic composition. The REE pattern of the proposed melt can be estimated, however, by first calculating the eclogite norm of abyssal tholeiite. Such an eclogite would have approximately equal proportions of garnet and clinopyroxene and less than 10 per cent quartz. It appears that Green and Ringwood (1968) prefer to melt high Al basalt with clinopyroxene and garnet in the proportions c. 7:3. The REE patterns of partial melts of andesitic compositions derived by this process are also shown in Fig. 22. The metamorphosed abyssal tholeiite considered the source of this magma contains 10 times the REE content of average chondrite with unfractionated REE patterns (Frey et al., 1974). The REE patterns of such melts may resemble those of partial melts derived from garnet peridotite, although the absolute REE abundance is lower in the latter case. However, these patterns do not mimic those of andesite. Based on these observations it is unlikely, therefore, that orogenic andesite is a direct partial melting product of anhydrous quartz eclogite. Gill (1978) concluded similarly, but his conclusion was partially based on too high values of light REE and because he used values of crystal–liquid partition coefficients pertinent to basaltic liquids rather than andesite. Gill (1978) also commented that alkali metals and alkaline earths in orogenic andesites are incompatible with a quartz eclogitic parental rock. Lack of appropriate experimental data on the crystal–liquid partition coefficients makes testing of the various models of magma genesis based on alkali metal and alkaline earth contents difficult.

A role for hydrous minerals in forming orogenic andesite has frequently been suggested (see Boettcher, 1975, for review; see also Allen et al., 1975). Mysen and Boettcher (1975a,b) ruled out an importance of amphibole in peridotite source rock on phase equilibrium grounds.

Amphibole fractionation of a tholeiitic liquid to form andesite has been proposed. Amphibole does not fractionate the individual REE significantly, however (Mysen 1978c). Consequently, the light REE content found in orogenic andesite must already have been present in the original basaltic rock. Basalts in orogenic regions with such patterns have not been found, however. Gill (1978) made further suggestions that other element ratios may be incompatible with amphibole fractionation. Those suggestions, however, must await experimental verification.

Partial melting of amphibolite results in melts with major element compositions that resemble orogenic andesite (Holloway and Burnham, 1972; Allen et al., 1975; Green, 1972). This amphibolite presumably is metamorphosed oceanic crust. Abyssal tholeiites do not show significant light REE enrichment (e.g. Frey et al., 1974). Because no mineral in amphibolite can significantly fractionate individual REE (Mysen, 1978c), a partial melt from amphibolitized oceanic tholeiite will not show significant light REE enrichment. It must be concluded, therefore, that orogenic andesites are not generally formed by melting of amphibolitized oceanic crust in subduction zones.

In summary, the major problem in all models for the formation of orogenic andesites is the content of incompatible elements. None of the simple models summarized above can account for these elements.

Metasomatic Alteration of the Source Rock

Incompatible elements presumably are the most soluble in hydrous fluids (vapour) under the pressure and temperature conditions of the upper mantle. Burnham (1967) first showed, for example, that albite dissolves incongruently in a hydrous fluid under crustal pressure conditions leaving behind a residue that is enriched in Al and Si. Fyfe and McBirney (1975) suggested that fluids derived from a hydrated descending slab of basaltic compositions were enriched in alkalis. As this fluid migrated into overlying peridotite this rock would be metasomatically altered. This alteration would be reflected in the partial melts derived from such a rock.

Before considering metasomatic alteration of element distribution, it is necessary to discuss whether dehydration of the descending slab can be accomplished without melting of the material itself. In summarizing data on depth to the subducting plate, Gill (1978) gave 136 ± 41 km ($\pm 1\sigma$) as the average depth beneath active volcanoes in island arcs. It has been suggested that the earthquake data leading to this conclusion were acquired because of seismic activity resulted from the formation of magma on or near the plate (Dickinson and Hatherton, 1967; Marsh, 1976). Delaney and Helgeson (1978) pointed out, however, that dehydration of hydrous minerals results in a negative volume change. In their view such dehydration may also cause earthquake activity. Delaney and Helgeson (1978) supported their conclusion with thermodynamic data on a variety of probable dehydration reactions and selection of suitable thermal models for the subduction zone. It was also pointed out that, because of the endothermic nature of the dehydration reactions, the descending oceanic plate would act as a heat sink, thus depressing its temperature below the values computed by, for example, Oxburgh and Turcotte (1970) and Toksöz et al. (1971). In the latter two models, melting of the slab material itself would be expected upon dehydration. Delaney and Helgeson (1978), suggested that thermal models of Hasebe et al. (1970), based on heat flow measurements in the Japanese arc, were more appropriate. With these models, Delaney and Helgeson (1978) suggested that the oceanic plate could descend to perhaps as deep as 150 km and dehydrate without melting of the slab material itself. Similar arguments were presented by Anderson et al. (1976).

It appears that a source of hydrous fluid is available for migration into the overlying peridotite wedge with attendant metasomatic alteration (*see* above). Marsh (1976) suggested, however, that even if such fluids were available, they would be essentially immobile in the overlying mantle. Direct experimental measurements on infiltration rates of H_2O in crystalline peridotite under upper mantle pressure and temperature conditions were attempted by Mysen et al. (1978) to 30 kb pressure and 850 °C (Fig. 12). Although the uncertainty of the data in Fig. 12 is difficult to assess, it is clear from these results that the velocity of a hydrous front moving through dry peridotite is so high that the rate of supply of hydrous fluid to the peridotite from a dehydrating slab is not a time-limiting factor of the melting process of the peridotite.

Experimental determination of relevant crystal–vapour partition coefficients has only recently been attempted. Cullers et al. (1973) and Zielinski and Frey (1974) made such determinations for selected REE to c. 5 kb pressure. These authors concluded that insignificant amounts of REE would dissolve in hydrous vapour in equilibrium with olivine, orthopyroxene, or clinopyroxene. It is known, however (Eggler and Rosenhauer, 1978), that the solubility of silicate components in hydrous fluids increases with increasing pressure. This information, coupled with data from peridotite nodules in kimberlite showing evidence for metasomatism in the mantle prior to ascent to the surface (Shimizu, 1975; Erlank and Shimizu, 1977), led Mysen (1979) to determine REE partitioning between hydrous fluids and garnet peridotite minerals to 30 kb pressure. The REE concentrations were in the natural abundance range of REE in garnet peridotite in the upper mantle. These major observations were made: (1) crystal–vapour partition coefficients varied

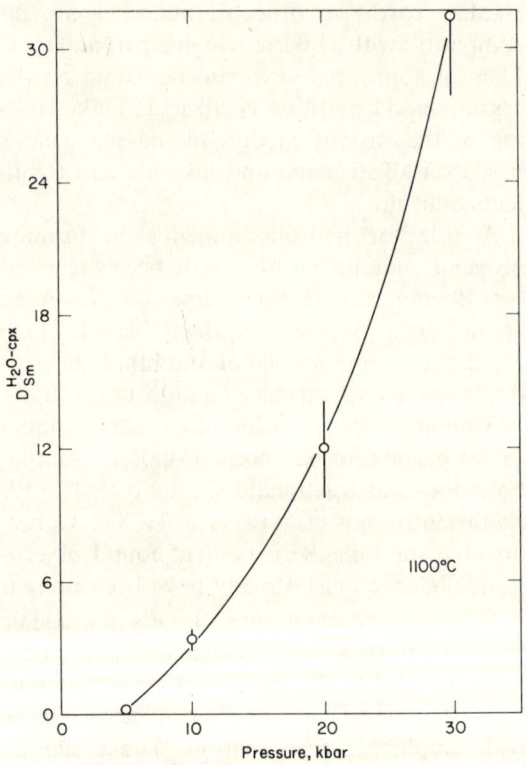

Fig. 23 Partition coefficient for Sm between H_2O-rich vapour and clinopyroxene (diopside) as a function of pressure. Error bars represent $\pm 1\sigma$. (From Mysen, 1979; reproduced by permission of the *American Mineralogist*)

significantly with pressure (Fig. 23) and (2) under upper mantle pressure conditions, REE are generally favoured by the hydrous fluid in equilibrium with a garnet-bearing rock and show strong enrichment in light REE (Fig. 24). Mysen (1979) concluded that because of these observations and because of the mobility of H_2O in crystalline peridotite, metasomatic alteration of the upper mantle is likely. Mysen (1979) developed a model of such metasomatism based on dehydration of an amphibolite to form eclogite (e.g. Essene et al., 1970) where the fluid

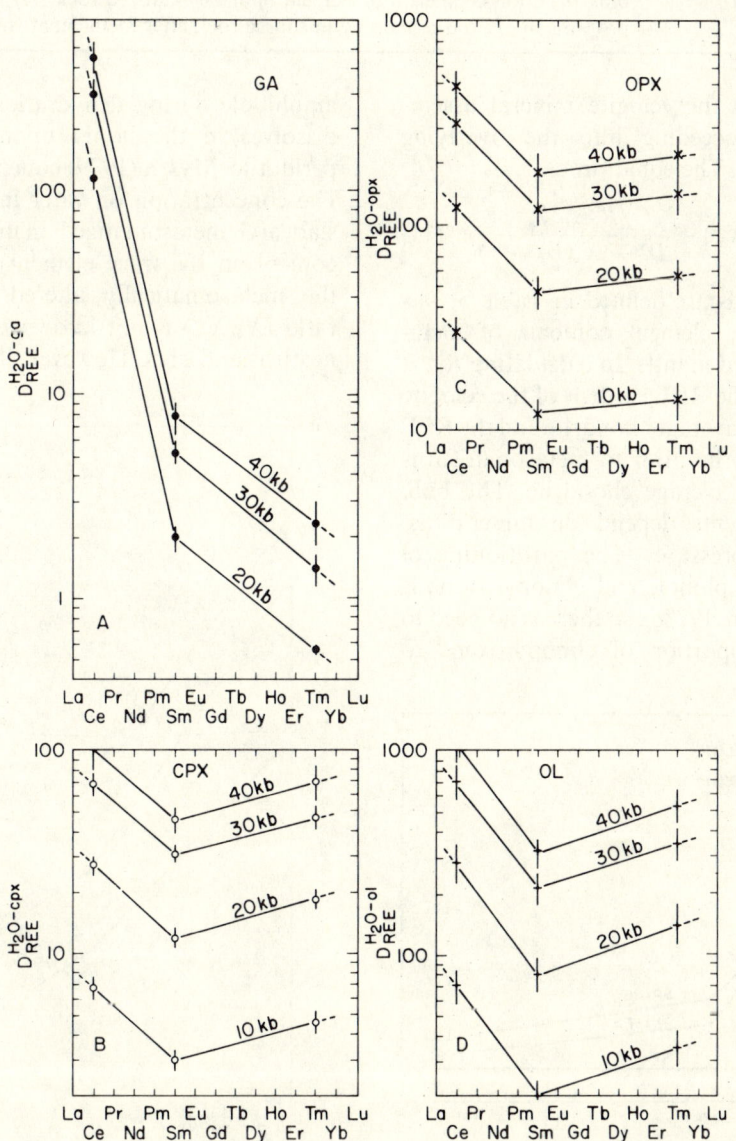

Fig. 24 Partition coefficients for REE between H_2O-rich vapour and garnet peridotite minerals. The curves are based on experimental data at 20 kb, the data shown in Fig. 23, and known crystal–crystal partition coefficients (Mysen, 1978c). (From Mysen, 1979; reproduced by permission of the *American Mineralogist*)

TABLE 5 Definitions of symbols used in equation (11)

R_i	element content of metasomatized rock before alteration
R_i^M	element content of metasomatized rock after alteration
R_i^S	element content of source rock
X_s	proportion of source rock that is converted to fluid
X_m	proportion of fluid in metasomatized rock
D_i^S	bulk partition coefficient (Shaw, 1970) of element in source rock
D_i^M	bulk partition coefficient of element in metasomatized rock
Y_v	proportion of fluid remaining in source rock after metasomatism

equilibrated with the eclogite mineral assemblage before proceeding into the overlying peridotite mantle. The equation

$$R_i^M = R_i(1 - Y_v X_m) + \frac{R_i^S X_m Y_v}{D_i^S - X_s(D_i^S - 1)}, \quad (11)$$

where the symbols are defined in Table 5, was used to calculate element contents of metasomatically altered mantle. In calculating R_i^M, it is assumed that the REE content of the eclogite is 10 times that of average chondrite and the REE content of peridotite prior to metasomatism is equal to that of average chondrite. The bulk partition coefficients depend on mineral assemblages and pressures. The partitioning of REE between amphibole and clinopyroxene is near unity (Mysen, 1978c), so there is no need to consider the proportion of clinopyroxene to amphibole during dehydration. Because water dissolves in the liquid upon melting of the peridotite, Mysen (1979) concluded that $Y_v = 1$. The concentration of water in the dehydrating slab and metasomatized mantle exerts strong control on the trace element concentration in the metasomatically altered peridotite. This ratio (X_s/X_m) is not known in the mantle beneath island arcs. However, if, as indicated by

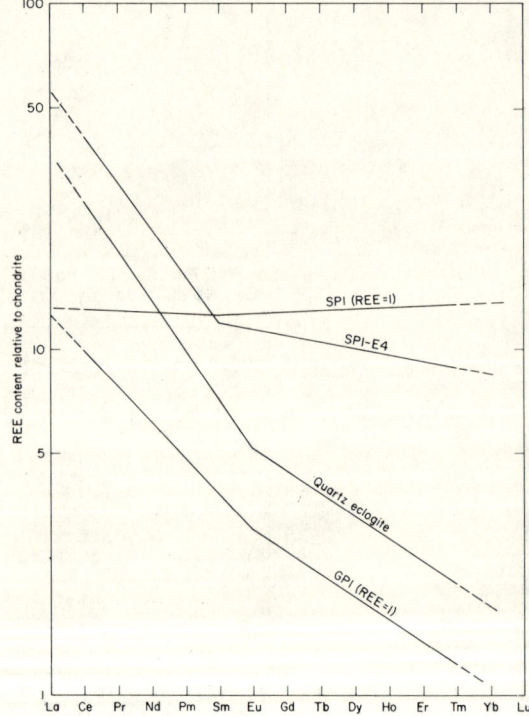

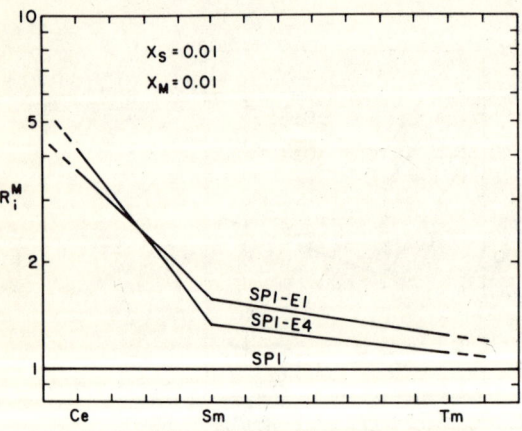

Fig. 25 REE patterns of metasomatized spinel peridotite. R_i = 10 times average chondrite; R_i^D = 1 times average chondrite; E1 = garnet/(garnet + clinopyroxene) = 0.2; E4 = garnet/(garnet + clinopyroxene) = 0.5 (From Mysen, 1979; reproduced by permission of the *American Mineralogist*)

Fig. 26 REE patterns of 5 per cent fractional melt from metasomatized spinel peridotite SP 1–E4 (see text and Fig. 25) compared with patterns for SP 1 (REE = 1), GP 1 (REE = 1) and quartz eclogite (all from Fig. 22). (Data from Mysen, 1978c, 1979)

Anderson (1975), orogenic andesites contain c. 5 per cent H_2O and andesite is formed by 10 per cent partial melting, $X_m \sim 0.005$ may be a reasonable approximation. The value for X_s (proportion of water in eclogite source) depends on the amount and type of hydrous minerals in the source rock. Most probably, the minerals are mainly amphiboles. Amphibolites typically contain c. 25 per cent amphibole, which corresponds to $X_s \sim 0.005$ and, thus, $X_m/X_s \sim 1$. In modelling REE contents of metasomatized peridotite, SP1 and an eclogite with 45 per cent ga, 45 per cent (cpx + amph), and REE contents 10 times that of average chondrite are used. The result of this calculation (Fig. 25) is the REE patterns expected in the peridotite overlying subducting oceanic crust beneath island arcs. The curve SP1-E4 corresponds to the expected pattern of the metasomatized spinel peridotite with modal ga/(ga + cpx) ~ 0.5 of the eclogite. In curve SP 1-E1, the eclogite has ga/(ga + cpx) = 0.2 rather than 0.5 as in SP 1-E4 to indicate the influence of altered modal proportions of minerals in the source rock on the REE patterns of metasomatically altered spinel peridotite. The REE pattern of this peridotite source will be reflected in the partial melt. With melting according to the stoicheiometry of equation (9), the patterns in Fig. 26 result. Notably, these patterns very closely resemble those of orogenic andesite (see also Gill, 1978, for review of data).

In summary, the most appropriate model for the formation of orogenic andesites is partial melting of metasomatically altered spinel peridotite overlying a descending plate with the major and trace element composition of abyssal tholeiite. Once formed, transition metal contents of the partial melts are altered by separation of immiscible Fe-rich sulphide melt. This model is consistent with trace and major element experimental data and the thermal regime of the volcanically active portion of the upper mantle beneath island arcs.

REFERENCES

Allen, J. C., Boettcher, A. L., and Marland, G. (1975). Amphiboles in andesites and basalts. I. Stability as a function of $P-T-f_{O_2}$. *Am. Mineral.* **60**, 1069–1085.

Anderson, A. T. (1975). Some basaltic and andesitic gases. *Rev. Geophys.* **13**, 37–57.

Anderson, R. N., Uyedea, S., and Miyashiro, A. (1976). Geophysical and geochemical constraints at converging plate boundaries. Part I. Dehydration and the downgoing slab. *Geophys. Jl. R. astr. Soc.* **44**, 333–357.

Arndt, N. T. (1977). The partitioning of nickel between olivine and ultrabasic and basaltic komatite liquids. *Carnegie Instn Wash. Yb.* **76**, 553–557.

Bishop, F. C., Smith, J. V., and Dawson, J. B. (1975). Pentlandite–magnetite intergrowth in de Beers spinel lherzolite: review of sulfides in nodules. *Phys. Chem. Earth* **9**, 323–337.

Boettcher, A. L. (1973). Volcanism and orogenic belts—the origin of andesite. *Tectonophysics* **17**, 231–244.

Boettcher, A. L. (1975). Experimental igneous petrology. *Rev. Geophys.* **13**, 75–79.

Boettcher, A. L., Mysen, B. O., and Allen, J. C. (1973). Techniques for the control of water fugacity and oxygen fugacity for experimentation in solid-media, high-pressure apparatus. *J. geophys. Res.* **78**, 5898–5902.

Bowen, N. L. (1928). *The Evolution of Igneous Rocks*, Princeton University Press, Princeton, N.J.

Boyd, F. R. and McCallister, R. H. (1976). Densities of fertile and sterile garnet peridotites. *Geophys. Res. Letters* **3**, 509–512.

Boyd, F. R., England, J. L., and Davis, B. C. T. (1964). Effect of pressure on the melting and polymorphism of enstatite, $MgSiO_3$. *J. geophys. Res.* **79**, 2101–2109.

Burnham, C. W. (1967). Hydrothermal fluids at the magmatic stage. In *Geochemistry of Hydrothermal Ore Deposits* (H. L. Barnes, ed.), Holt, Rinehart and Winston, New York, pp. 34–67.

Burnham, C. W. (1974). $NaAlSi_3O_8-H_2O$ solutions: A thermodynamic model for hydrous magmas. *Bull. Soc. Fr. Mineral. Cristallogr.* **97**, 223–230.

Burnham, C. W. (1975). Thermodynamics of melting in experimental silicate-volatile systems. *Geochim. cosmochim. Acta* **39**, 1077–1084.

Carmichael, I. S. E., Turner, F. J., and Verhoogen, J. (1974). *Igneous Petrology*, McGraw-Hill, New York.

Chayes, F. (1969). The chemical composition of Cenozoic andesites. *Oregon St. Dept Geol. Mineral. Res. Bull.* no. 65, 1–13.

Cullers, R. L., Mederis, L. G., and Haskin, L. A. (1973). Experimental studies on the distribution of rare earths as trace elements among silicate minerals and liquids and water. *Geochim. cosmochim. Acta* **37**, 1499–1513.

Delaney, J. M. and Helgeson, H. C. (1978). Calculation of thermodynamic consequences of dehydration in subducting oceanic crust. *Am. J. Sci.* **278**, 638–686.

Delaney, J. R., Muenow, D. W., and Graham, D. G. (1978). Abundance and distribution of water, carbon and sulfur in glassy rims of submarine pillow basalts. *Geochim. cosmochim. Acta* **42**, 581–594.

DeLong, S. E. (1974). Distribution of Rb, Sr, and Ni in igneous rocks, Central and Western Aleutians, Alaska. *Geochim. cosmochim. Acta* **38**, 245–266.

Dickinson, W. R. and Hatherton, T. (1967). Andesitic volcanism and seismicity around the Pacific. *Science, N.Y.* **157**, 801–803.

Eggler, D. H. (1972). Water-saturated and water-undersaturated melting relations in a Paricutin andesite and an estimate of water content of natural magma. *Contr. Mineral. Petrol.* **34**, 261–271.

Eggler, D. H. (1973). Role of CO_2 in melting processes in the mantle. *Carnegie Instn Wash. Yb.* **72**, 457–467.

Eggler, D. H. (1975). CO_2 as volatile component of the mantle; the system Mg_2SiO_4–SiO_2–H_2O–CO_2. *Phys. Chem. Earth.* **9**, 869–881.

Eggler, D. H. and Rosenhauer, M. (1978). Carbon dioxide in silicate melts. II. Solubilities of CO_2 and H_2O in $CaMgSi_2O_6$ (diopside) liquids and vapors at pressures to 40 kbar. *Am. J. Sci.* **278**, 64–94.

Engdahl, E. R. (1973). Relocation of intermediate depth earthquakes in Central Aleutians by seismic ray tracing. *Nature phys. Sci.* **245**, 23–25.

Erlank, A. J. and Shimizu, N. (1977). Strontium and strontium isotope distributions in some kimberlite nodules and minerals. In *Extended Abstracts.* 2. *International Kimberlite Conference, Santa Fe, New Mexico,* 1977.

Essene, E. J., Hensen, B. J., and Green, D. H. (1970). Experimental study of eclogite and amphibole stability. *Phys. Earth planet. Interiors* **3**, 374–384.

Frey, F. A. and Green, D. H. (1974). The mineralogy, geochemistry and origin of lherzolite inclusions in Victorian basanites. *Geochim. cosmochim. Acta* **38**, 1023–1059.

Frey, F. A., Bryan, W. B., and Thompson, G. (1974). Atlantic ocean floor: geochemistry and petrology of basalt from Legs 2 and 3 of the deep-sea drilling project. *J. geophys. Res.* **79**, 5507–5527.

Fyfe, W. S. and McBirney, A. R. (1975). Subduction and the structure of andesitic volcanic belts. *Am. J. Sci.* **275A**, 284–298.

Gill, J. B. (1974). Role of underthrust oceanic crust in the genesis of Fijian calc-alkalic suite. *Contr. Mineral. Petrol.* **43**, 29–45.

Gill, J. B. (1978). Role of trace element partition coefficients in models of andesite genesis. *Geochim. cosmochim. Acta* **42**, 709–725.

Green, D. H. (1969). The origin of basaltic and nephelinitic magmas in the Earth's mantle. *Tectonophysics* **7**, 409–422.

Green, D. H. (1970). A review of experimental evidence on the origin of basaltic and nephelinitic magmas. *Phys. Earth planet. Interiors* **3**, 221–235.

Green, D. H. (1973). Contrasted melting relations in a pyrolite upper mantle under mid-oceanic ridge, stable crust and island arc environments. *Tectonophysics* **17**, 285–297.

Green, T.H. (1972). Crystallization of calc-alkaline andesite under controlled high-pressure hydrous conditions. *Contr. Mineral. Petrol.* **34**, 150–166.

Green, T. H. and Ringwood, A. E. (1968). Genesis of the calc-alkaline igneous rock suite. *Contr. Mineral. Petrol.* **18**, 105–162.

Hart, S. R. and Davis, K. E. (1978). Nickel partitioning between olivine and silicate melt. *Earth planet. Sci. Letters* in press.

Hasebe, K., Fujii, N., and Uyeda, S. (1970). Thermal processes under island arcs. *Tectonophysics* **10**, 335–355.

Haughton, D. R., Roeder, P. L., and Skinner, B. J. (1974). Solubility of sulfur in mafic magmas. *Ecol. Geol.* **69**, 451–467.

Hess, P. C. (1977). Structure of silicate melts. *Can. Mineral.* **15**, 162–178.

Hill, R. E. T. and Boettcher, A. L. (1970). Water in the Earth's mantle: melting curves of basalt–water and basalt–water–carbon dioxide. *Science, N.Y.* **167**, 980–982.

Hofmann, A. W. and Magaritz, M. (1977). Diffusion of Ca, Sr, Ba and Co in a basalt melt: implications for the geochemistry of the mantle. *J. geophys. Res.* **83**, 5432–5441.

Holloway, J. R. and Burnham, C. W. (1972). Melting relations of basalt with equilibrium water pressure less than total pressure. *J. Petrol.* **13**, 1–29.

Irvine, T. N. and Kushiro, I. (1976). Partitioning of Ni and Mg between olivine and silicate liquids. *Carnegie Instn Wash. Yb.* **75**, 668–675.

Irving, A. J. (1978). A review of experimental crystal/liquid trace element partitioning. *Geochim. cosmochim. Acta* **42**, 743–771.

Irving, A. J. and Frey, A. J. (1978). Distribution of trace elements between garnet megacrysts and host volcanic liquids of kimberlitic to rhyolitic composition. *Geochim. cosmochim. Acta* **42**, 771–789.

Jackson, E. D. and Wright, T. L. (1970). Xenoliths in the Honolulu volcanic series, Hawaii. *J. Petrol.* **11**, 405–430 (1970).

Kay, R. W. (1977). Geochemical constraints on the origin of Aleutian magmas. In *Island Arcs, Deep Sea Trenches and Back-arc Basins* (M. Talwani and W. C. Pittman, III, eds), American Geophysical Union, Washington, D.C., p. 229–242.

Kushiro, I. (1969a). The system forsterite–diopside–silica with and without water at high pressures. *Am. J. Sci* **267A**, 269–294.

Kushiro, I. (1969b). Discussion of the paper. 'The origin of basaltic and nephelinitic magmas in the Earth's mantle' by D. H. Green. *Tectonophysics* **7**, 427–436.

Kushiro, I. (1972). Effect of water on the composition of magmas formed at high pressures. *J. Petrol.* **13**, 311–334.

Kushiro, I. (1974). Melting anhydrous mantle and possible generation of andesitic magma: an approach from synthetic systems. *Earth planet. Sci Letters* **22**, 294–299.

Kushiro, I. (1975). On the nature of silicate melts and its significance in magmagenesis: regularities in the shift of liquidus boundaries involving olivine, pyroxene and silica minerals. *Am. J. Sci.* **275**, 411–431.

Kushiro, I. (1976). Changes in viscosity and structure of melt of $NaAlSi_2O_6$ composition at high pressures. *J. geophys. Res.* **81**, 6347–6350.

Kushiro, I., Yoder, H. S., and Nishikawa, M. (1968a). Effect of water on the melting of enstatite. *Geol. Soc. Am. Bull.* **79**, 1685–1692.

Kushiro, I., Syono, Y., and Akimoto, S.-I. (1968b). Melting of a peridotite nodule at high pressure and high water pressure. *J. geophys. Res.* **73**, 6023–6029.

Kushiro, I., Shimizu, N., Nakamura, Y., and Akimoto, S. (1972). Compositions of co-existing liquids and solid phases formed upon melting of natural garnet and spinel lherzolites at high pressures: a preliminary report. *Earth planet. Sci. Letters* **14**, 19–25.

Lindstrom, D. and Weill, D. F. (1978). Partitioning of

transition metals between diopside and coexisting silicate liquids. I. Nickel, cobalt and manganese. *Geochim. cosmochim. Acta* **42**, 817–833 (1978).

Magaritz, M. and Hofmann, A. W. (1978). Diffusion of Sr, Ba and Na in obsidian. *Geochim. cosmochim. Acta* **42**, 595–605.

Marsh, B. D. (1976). Some Aleutian andesites: their nature and source. *J. Geol.* **84**, 27–45.

Marsh, B. D. and Carmichael, I. S. E. (1974). Benioff zone magmatism. *J. geophys. Res.* **79**, 1196–1206.

Meyer, H. O. A. and Boctor, N. Z. (1975). Sulfide–oxide minerals in eclogite from Stockdale kimberlite, Kansas. *Contr. Mineral. Petrol.* **52**, 57–68.

Miyashiro, A. and Shido, F. (1976). Behavior of nickel in volcanic rocks. In *Volcanoes and Tectosphere* (H. Aoki, ed.), Tokai University Press, Tokyo, pp. 115–121.

Mysen, B. O. (1973). Melting of a hydrous mantle: phase relations of mantle peridotite with controlled water and oxygen fugacities. *Carnegie Instn Wash. Yb.* **72**, 467–478.

Mysen, B. O. (1975). Partitioning of iron and magnesium between crystals and partial melts in peridotite upper mantle. *Contr. Mineral. Petrol.* **52**, 69–76 (1975).

Mysen, B. O. (1976). The role of volatiles in silicate melts: solubility of carbon dioxide and water in feldspar, pyroxene and feldspathoid melts to 30 kb and 1625°C. *Am. J. Sci.* **276**, 969–996.

Mysen, B. M. (1977). Magma genesis in peridotite upper mantle in light of experimental data on partitioning of trace elements between garnet peridotite minerals and partial melts. *Carnegie Instn Wash. Yb.* **76**, 545–550.

Mysen, B. O. (1978a). Experimental determination of nickel partition coefficients between liquid, garnet peridotite minerals and pargasite and concentration limits of behaviour according to Henry's law at high pressure and temperature. *Am. J. Sci.* **278**, 217–243.

Mysen, B. O. (1978b). Limits of solution of trace elements in minerals according to Henry's law: review of experimental data. *Geochim. cosmochim. Acta* **42**, 871–885.

Mysen, B. O. (1978c). Experimental determination of rare earth element partition coefficients between hydrous silicate melt, amphibole and garnet peridotite minerals at upper mantle pressures and temperatures. *Geochim. cosmochim. Acta* **42**, 1253–1263.

Mysen, B. O. (1979). Trace element partitioning between garnet peridotite minerals and water-rich vapor: Experimental data from 5 to 30 kbar. *Am. Mineral.* **64**, 274–289.

Mysen, B. O. and Boettcher, A. L. (1975a). Melting of a hydrous mantle. II. Geochemistry of crystals and liquids formed by anatexis of mantle peridotite at high pressures and high temperatures as a function of controlled activities of water, carbon dioxide and hydrogen. *J. Petrol.* **16**, 549–580.

Mysen, B. O. and Boettcher, A. L. (1975b). Melting of a hydrous mantle. I. Phase relations of natural peridotite at high pressures and high temperatures with controlled activities of water, carbon dioxide and hydrogen. *J. Petrol.* **16**, 520–548 (1975b).

Mysen, B. O. and Kushiro, I. (1976). Partitioning of iron, nickel and magnesium between metal, oxide and silicates in Alle de meteorite as a function of f_{O_2}. *Carnegie Instn Wash. Yb.* **75**, 678–684 (1976).

Mysen, B. O. and Kushiro, I. (1977). Compositional variations of coexisting phases with degree of melting of peridotite in the upper mantle. *Am. Mineral.* **62**, 843–865.

Mysen, B. O. and Kushiro, I. (1978). The effect of pressure on the partitioning of nickel between olivine and aluminous silicate melt. *Carnegie Instn Wash. Yb.* **77**, in press.

Mysen, B. O. and Popp, R. K. (1978). Solubility of sulfur in silicate melts as a function of f_{O_2} and silicate melt composition. *Carnegie Instn Wash.* **77**, in press.

Mysen, B. O., Kushiro, I., Nicholls, I. A., and Ringwood, A. E. (1974). A possible mantle origin for andesitic magmas: discussion of a paper by Nicholls and Ringwood. *Earth planet. Sci. Letters* **21**, 221–230.

Mysen, B. O. and Seitz, M. G. (1975). Trace element partitioning determined by beta-track mapping—an experimental study using carbon and samarium as examples. *J. geophys. Res.* **80**, 2627–2635.

Mysen, B. O. and Virgo, D. (1978). Influence of pressure, temperature and bulk composition on melt structures in the system $NaAlSi_2O_6$–$NaFe^{3+}Si_2O_6$. *Am. J. Sci.* **278**, 1307–1322.

Mysen, B. O., Arculus, R. J., and Eggler, D. H. (1975). Solubility of carbon dioxide in natural nephelinite, tholeiite and andesite melts to 30 kbar pressure. *Contr. Mineral. Petrol.* **53**, 227–239.

Mysen, B. O., Eggler, D. H., Seitz, M. G., and Holloway, J. R. (1976). Carbon dioxide in silicate melts and crystals. I. Solubility measurements. *Am. J. Sci.* **276**, 455–479.

Mysen, B. O., Kushiro, I., and Fujii, T. (1978). Preliminary experimental data bearing on the mobility of H_2O in crystalline upper mantle. *Carnegie Instn. Wash. Yb.* **77**, 793–797.

Nehru, C. E. and Wyllie, P. J. (1975). Composition of glasses from St Pauls peridotite partially melted. *J. Geol.* **83**, 455–471.

Nicholls, I. A. and Ringwood, A. E. (1973). Production of silica-saturated tholeiitic magmas in island arcs. *Earth planet. Sci. Letters* **17**, 243–246.

Nixon, P. H. and Boyd, F. R. (1973). Petrogenesis of the sheared and granular ultrabasic nodule suite in kimberlites. In *Lesotho Kimberlites* (P. H. Nixon, ed.), Cape and Transvaal Printers Ltd., Cape Town, pp. 46–58.

O'Hara, M. J. (1965). Primary magmas and the origin of basalts. *Scottish J. Geol.* **1**, 19–40.

O'Hara, M. J. (1971). Discussion of paper by D. H. Green: 'Composition of basaltic magmas as indicators of conditions of origin': application to oceanic volcanism. *Phil. Trans. R. Soc. A* **268**, 707–725.

Osborn, E. F. (1959). Role of oxygen pressure in the crystallization and differentiation of basaltic magmas. *Am. J. Sci.* **257**, 609–647.

Osborn, E. F. (1962). Reaction series for subalkaline igneous rock series based on different oxygen pressure conditions. *Am. Mineral.* **47**, 211–226.

Osborn, E. F. (1977). Origin of calc-alkalic magma series of Santorini volcano type in light of recent experimental studies. In *Proceedings of the International Congress on Thermal Water, Geothermal Energy, and Vulcanism in the Mediterranean Area, Athens, October 1976*, Vol. 3, pp. 154–167.

Oxburgh, E. R. and Turcotte, D. L. (1970). Thermal structure of island arcs. *Geol. Soc. Am. Bull.* **81**, 1665–1688.

Rajamani, V. and Naldrett, A. J. (1978). Partitioning of Fe, Co, Ni and Cu between sulfide liquid and basaltic melts and the composition of Ni–Cu sulfide deposits. *Econ. Geol.* **73**, 82–94.

Ringwood, A. E. (1975). *Composition and Petrology of the Earth's Mantle*, McGraw-Hill, New York.

Roeder, P. L. and Emslie, R. F. (1970). Olivine–liquid equilibrium. *Contr. Mineral. Petrol.* **29**, 275–289.

Schmitt, R. A., Smith, R. H., Lasch, J. E., Olehy, D. A., and Vasilevskis, J. (1963). Abundances of fourteen rare-earth elements, scandium and yttrium in terrestrial and meteoritic matter. *Geochim. cosmochim. Acta* **27**, 577–622.

Schmitt, R. A., Smith, R. H., and Olehy, D. A. (1964). Rare-earth, yttrium and scandium abundances in meteoritic and terrestrial matter-II. *Geochim. cosmochim. Acta* **28**, 67–86.

Sharma, S. K., Virgo, D., and Mysen, B. O. (1978). Raman study of structure and coordination of Al in $NaAlSi_2O_6$ glasses synthesized at high pressure. *Carnegie Instn Wash. Yb.* **77**, in press.

Shaw, D. M. (1970). Trace element fractionation during anatexis. *Geochim. cosmochim. Acta* **34**, 237–243.

Shaw, D. (1977). Trace element behaviour during anatexis. In *Proceedings of the Chapman Conference on Partial Melting in the Earth's Upper Mantle* (H. J. B. Dick, ed.), American Geophysical Union, Washington, D.C., pp. 189–215.

Shima, H. and Naldrett, A. J. (1975). Solubility of sulfur in ultramafic melts and the relevance of the system Fe–S–O. *Econ. Geol.* **70**, 960–967.

Shimizu, N. (1975). Rare earth elements in garnets and clinopyroxenes from garnet lherzolite nodules in kimberlites. *Earth planet. Sci. Letters* **25**, 26–32.

Takahashi, E. (1978). Partitioning of Ni^{2+}, Co^{2+}, Fe^{2+}, Mn^{2+} and Mg^{2+} between olivine and silicate melts: compositional dependence of partition coefficients. *Geochim. cosmochim. Acta* **42**, 1829–1845.

Taylor, S. R. (1969). Trace element geochemistry of andesites and associated calc-alkaline rocks. *Oregon St. Dept Geol. Mineral. Res. Bull.* no. 65, 43–65.

Tilley, C. E. (1950). Some spects of magmatic evolution. *Q. Jl geol. Soc. Lond.* **106**, 37–61.

Thorpe, R. S., Potts, P. J., and Francis, P. W. (1976). Rare earth data and the petrogenesis of andesite from North Chilean Andes. *Contr. Mineral. Petrol.* **54**, 65–78.

Toksöz, M. N., Minear, J. W., and Julian, B. R. (1971). Temperature fields and geophysical effects of a downgoing slab. *J. geophys. Res.* **76**, 1113–1138.

Toop, G. W. and Samis, C. S. (1962a). Activities of ions in silicate melts. *Trans. Met. Soc. Am. Instn. mech. Engrs.* **224**, 878–887.

Toop, G. W. and Samis, C. S. (1962b). Some new ionic concepts of silicate slags. *Can. Met. Qtly* **1**, 129–152.

Turcotte, D. L. and Schubert, G. (1973). Frictional heating and the descending lithosphere. *J. geophys. Res.* **78**, 5876–5886.

Tuttle, O. F. and Bowen, N. L. (1958). Origin of granite in light of experimental studies in the system $NaAlSi_3O_8$–$KAlSi_3O_8$–SiO_2–H_2O. *Geol. Soc. Am. Mem.* no. 74.

Wasserburgh, G. J. (1957). The effect of H_2O in silicate systems. *J. Geol.* **65**, 15–23.

Watson, E. B. (1976). Two-liquid partition coefficients: experimental data and geochemical implications. *Contr. Mineral. Petrol.* **56**, 119–134.

Watson, E. B. (1977). Partitioning of manganese between forsterite and silicate liquid. *Geochim. cosmochim. Acta* **41**, 1363–1374.

White, R. W. (1966). Ultramafic inclusions in basaltic rocks from Hawaii. *Contr. Mineral. Petrol.* **12**, 245–314.

Wyllie, P. J. (1971). Role of water in magma generation and initiation of diapiric uprise in the mantle. *J. geophys. Res.* **76**, 1328–1338.

Yajima, T., Higuchi, H., and Nagasawa, H. (1972). Variations of rare earth concentrations in pigeonitic and hyperstenitic rock series from Izu–Hakone region, Japan. *Contr. Mineral. Petrol.* **35**, 234–245.

Yoder, H. S. (1969). Calcalkalic andesites: experimental data bearing on the origin of their assumed characteristics. *Oregon St. Dept Geol. Mineral. Bull.* no. 65, 77–89.

Yoder, H. S. and Chinner, G. A. (1960). Grossularite–pyrope system at 10,000 bars. *Carnegie Instn Wash. Yb.* **59**, 78–81.

Yoder, H. S. and Tilley, C. E. (1962). Origin of basalt magmas: an experimental study of natural and synthetic rock systems. *J. Petrol.* **3**, 342–532.

Zielinski, R. A. and Frey, F. A. (1974). An experimental study of the partitioning of a rare earth element (Gd) in the system diopside–vapor. *Geochim. cosmochim. Acta* **38**, 545–565.

Andesites
Edited by R. S. Thorpe
© 1982 John Wiley & Sons

VIII Chemical and isotope characteristics of destructive margin magmas

Chemical and isotopic data provide powerful constraints on models proposed for the petrogenesis of orogenic andesite association (cf. Section VII). In the first contribution in this section J. A. Pearce reviews the chemical characteristics of lavas from destructive plate margins, concentrating on basaltic lavas, since these are the least evolved and might be expected to provide the clearest evidence regarding petrogenetic processes at destructive plate margins. Volcanic arc basalts have distinctive chemical characteristics that enable them to be distinguished from basalts erupted in other tectonic settings. These chemical characteristics are consistent with derivation by partial melting of mantle which has been selectively enriched in incompatible elements of low ionic potential (including K, Rb, and Sr) as a result of input of hydrous fluids from subducted oceanic crust into their mantle source region. Within this framework, island arc basalts are characterized by depletion of incompatible elements with high ionic potential (including Ti, Zr, and Y), indicating high degrees of partial melting, presence of residual phases containing these elements, or melting of depleted mantle. In addition calc-alkaline basalts show enrichment of Th, P, and light REE, posssibly indicating that they contain a sedimentary component derived from subducted oceanic crust.

The petrogenetic arguments based on the chemical characteristics outlined above are complemented by isotopic data reviewed by C. J. Hawkesworth. Therefore, lavas erupted at destructive margins have higher ratios of $^{87}Sr/^{86}Sr$ and $^{207}Pb/^{206}Pb$ in relation to lavas erupted in other tectonic settings. These are assumed to reflect melting of mantle which has been enriched in these elements by material released from the subducted oceanic slab. Island arc lavas are less enriched in these isotopes, suggesting that their mantle source region contains a lower proportion of elements from the subducted slab. By contrast, the enrichment of calc-alkaline lavas is Th, P, and REE (cf. above) allied with low $^{143}Nd/^{144}Nd$ and high $^{207}Pb/^{206}Pb$ might indicate that the mantle source region contains material added by partial melting of the subducted slab and/or a subducted sedimentary component. The presence of continental crust provides an additional degree of uncertainty in interpreting the chemical and isotopic characteristics of destructive margin magmas. In continental areas magmas derived from mantle source regions described above might experience contamination with crustal melts, and some erupted lavas might be derived by partial melting of the lower crust.

Andesites
Edited by R. S. Thorpe
© 1982 John Wiley & Sons

Trace element characteristics of lavas from destructive plate boundaries

J. A. Pearce

Department of Earth Sciences,
The Open University, Milton Keynes MK7 6AA, UK

ABSTRACT

Volcanic arc basalts are all characterized by a selective enrichment in incompatible elements of low ionic potential, a feature thought to be due to the input of aqueous fluids from subducted oceanic crust into their mantle source regions. Island arc basalts are additionally characterized by low abundances (for a given degree of fractional crystallization) of incompatible elements of high ionic potential, a feature for which high degrees of melting, stability of minor residual oxide phases, and remelting of depleted mantle are all possible explanations. Calc-alkaline basalts and shoshonites are additionally characterized by enrichment of Th, P, and the light REE in addition to elements of low ionic potential, a feature for which one popular explanation is the contamination of their mantle source regions by a melt derived from subducted sediment.

By careful selection of variables, discrimination diagrams can be drawn which highlight these various characteristics and therefore enable volcanic arc basalts to be recognized in cases where geological evidence is ambiguous. Plots of Y against Cr, K/Yb, Ce/Yb, or Th/Yb against Ta/Yb, and Ce/Sr against Cr are all particularly successful and can be modelled in terms of vectors representing different petrogenetic processes. An additional plot of Ti/Y against Nb/Y is useful for identifying 'anomalous' volcanic arc settings such as Grenada and parts of the Aleutian arc. Intermediate and acid rocks from volcanic arc settings can also be recognized using a simple plot of Ti against Zr.

The lavas from the Oman ophiolite complex provide a good test of the application of these techniques. The results indicate that the complex was made up of back-arc oceanic crust intruded by the products of volcanic arc magmatism.

Introduction

Many geologists rely heavily on volcanic rocks for providing the evidence by which past destructive plate boundaries can be located. It is therefore important to be able to recognize whether or not a lava sequence was erupted in a volcanic arc. In some cases, this is easily achieved from field relations, petrography, and conventional major element diagrams. Often, however, deformation, metamorphism, or simply erosion have masked these lines of evidence so that more sophisticated means of geochemical characterization must be applied. This is particularly true in many orogenic belts, where allochthonous fragments of island arcs and back-arc basin complexes are thought to exist. The aim of this chapter is to identify the main trace element features by which a lava may be assigned to a volcanic arc magma type; and to show how petrogenetic modelling of these features can lead to a better understanding of the source region, and hence tectonic setting, of the lava concerned. The first part of the chapter deals with basic rocks; the second part extends the results to intermediate and acid rocks; and the final part presents an example of Cretaceous arc volcanism in the Semail thrust sheet of Oman.

Trace Element Characteristics of Destructive Margin Basalts

Average abundances

Table 1 shows the contrast in composition between basalts erupted in volcanic arcs, and basalts erupted at ocean ridges and within plates (away from the influence of subduction zones). Following current convention, the basalts erupted in volcanic arc settings have been subdivided into island arc tholeiites, calc-alkaline basalts, and shoshonites according to the K_2O–SiO_2 discriminant of Peccerillo and Taylor (1976). Basalts erupted in back-arc extensional zones and in volcanic arc settings near lateral edges of subduction zones and where fracture zones are being subducted (DeLong *et al.*, 1975), have been treated as 'anomalous' and omitted from this compilation; they are, however, discussed separately below.

A convenient means of comparing analyses for these various magma types is to plot the data as geochemical patterns. The representative patterns presented in Fig. 1 use a typical mid-ocean ridge basalt as a normalizing composition. For simplicity, the elements used to construct the patterns have been restricted to those with recognizable petrogenetic significance which can be readily analysed by X-ray fluorescence and neutron activation techniques; Ce, Sm, and Yb have been chosen to represent the lanthanide series. The main features of the patterns are summarized below.

Patterns in basalts erupted away from destructive plate margins

As a prerequisite to studying volcanic arc patterns it is necessary to establish the range of patterns in basalts erupted away from any possible influence of subducted oceanic crust. Typical patterns chosen for this purpose, from mid-ocean ridge and within plate settings are given in Fig. 1(a) and (b). Since a tholeiitic

TABLE 1 Chemical characteristics of mid-ocean ridge (MORB), within-plate and volcanic arc basalts

	Mid-Ocean Ridge		Within-Plate		Volcanic Arc		
	Tholeiitic	Transitional	Tholeiitic	Alkaline	Tholeiitic	Calc-Alkaline	Shoshonitic
K_2O (%)†	0.20	(0.51)	(0.5)	(1.5)	0.43	0.94	2.51
Rb (p.p.m.)†	(2)	(6)	(7.5)	(40)	4.7	23	51
Ba (p.p.m.)†	(20)	(60)	(100)	(600)	60	260	609
TiO_2 (%)	1.40	1.39	2.23	2.90	0.84	0.98	0.94
Zr (p.p.m.)	90	96	149	213	40	71	87
Hf (p.p.m.)	2.44	2.93	3.44	6.36	1.17	2.23	2.24
Sm (p.p.m.)	3.26	3.83	5.35	8.87	1.89	3.78	4.88
P_2O_5 (%)	0.12	0.18	0.25	0.64	0.08	0.19	0.44
Ce (p.p.m.)	11.0	23.3	31.3	96.8	6.94	29.3	50.2
Ta (p.p.m.)	0.29	0.85	0.73	5.9	0.10	0.18	0.33
Nb (p.p.m.)	4.6	16	13	84	1.7	2.7	8.4
Th (p.p.m.)	0.26	0.80	0.77	4.5	0.37	1.26	3.6
Sr (p.p.m.)†‡	121	196	290	842	231	428	934
Ni (p.p.m.)†‡	(90)	(130)	(70)	(90)	18	50	14
Y (p.p.m.)‡	33	25	26	25	17	22	22
Yb (p.p.m.)‡	3.22	2.63	2.12	0.89	1.95	2.31	2.55
Sc (p.p.m.)‡	40.6	36.6	32.6	26.2	40.0	32.0	28.3
Cr (p.p.m.)‡	251	411	352	536	111	160	100

† Low ionic potential ($Z/r < 3$).
‡ Compatible with respect to one or more major mantle phases ($D > 1$, for *pl, cpx, opx, ol, gt* or *sp*)
( ) Estimated values (no compilation available).

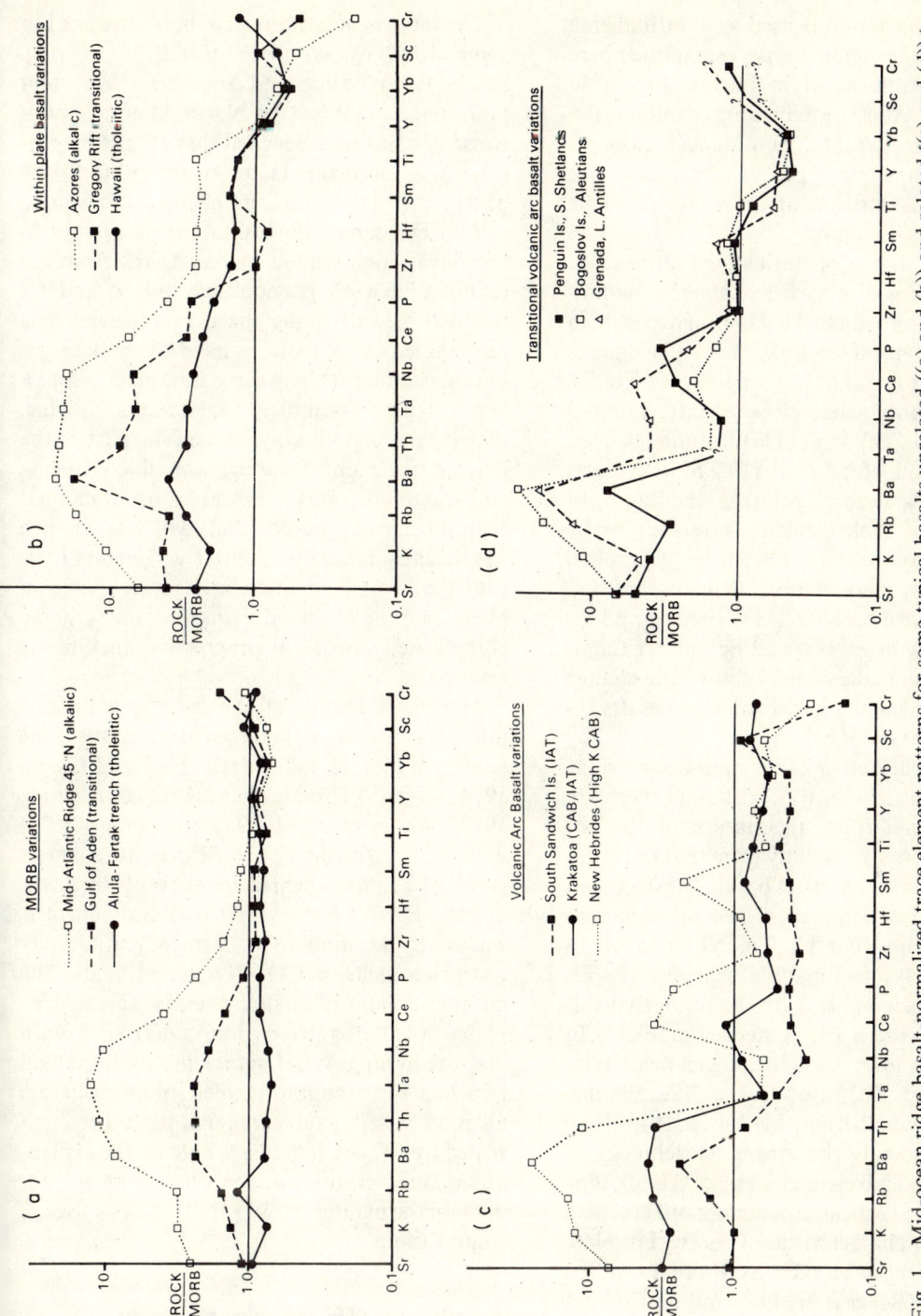

Fig. 1 Mid-ocean ridge basalt-normalized trace element patterns for some typical basalts unrelated ((a) and (b)) and related ((c) and (d)) to subduction. Normalizing values: Sr = 120 p.p.m.; K_2O = 0.15 per cent; Rb = 2.0 p.p.m.; Ba = 20 p.p.m.; Th = 0.2 p.p.m.; Ta = 0.18 p.p.m.; Nb = 3.5 p.p.m.; Ce = 10.0 p.p.m.; P_2O_5 = 0.12 per cent; Zr = 90 p.p.m.; Hf = 2.4 p.p.m.; Sm = 3.3 p.p.m.; TiO_2 = 1.5 per cent; Y = 30 p.p.m.; Yb = 3.4 p.p.m.; Sc = 40 p.p.m.; Cr = 250 p.p.m. Analyses by the author except Mid-Atlantic Ridge 45°N (Wood et al., 1979b), New Hebrides (Gorton, 1977), Penguin Island (Weaver et al., 1979), Bogoslov Island (Kay, 1977), and Grenada (Shimizu and Arculus, 1975)

mid-ocean ridge basalt is used as a normalizing composition, this magma type, exemplified here by a sample from the Alula–Fartak 'trench' in the Gulf of Aden, predictably exhibits flat patterns with elemental abundances close to unity. Other magma types exhibit a selective and systematic enrichment of certain elements relative to this base level.

Two main types of enrichment appear to exist. The first is displayed by tholeiitic within-plate basalts, of which the Hawaiian basalt in Fig. 1(b) is a typical example. Here, all elements from Sr to Ti are enriched whereas Y, Yb, Sc, and Cr retain abundances close to unity. As noted by White and Schilling (1978) in samples from the Azores, and Wood et al. (1979b) in samples from Iceland and adjacent areas, the degree of enrichment of each element is related to its incompatibility relative to garnet lherzolite. Thus the most incompatible elements, Th, Ba, Ta, and Nb, are most enriched whereas Y and Yb (compatible with garnet) and Sc and Cr (compatible with all mafic phases) show little change relative to tholeiitic mid-ocean ridge basalts.

The second type of enrichment is displayed by the transitional and alkaline mid-ocean ridge basalt patterns in Fig. 1(a). Although there are superficial similarities, the nature of the enrichment differs significantly from that observed in the tholeiitic within-plate basalt. In particular, the degree of enrichment of the most incompatible elements (Ba, Th, Ta, Nb) relative to the moderately incompatible elements (P, Zr, Hf, Sm) is much higher, with the result that this part of the pattern has a steeper gradient. In addition, Ti shows negligible enrichment relative to Y and Yb. Inspection of the alkaline within-plate basalt from the Azores (Fig. 1(b)) predictably reveals the strong enrichment of Ba, Th, Ta, and Nb characteristic of this alkaline pattern coupled with an enrichment of Ti relative to Y and Yb characteristic of the within-plate pattern. Other lavas such as those from the Gregory Rift (Baker et al., 1977) in Fig. 1(b) and certain transitional settings such as Afar (Treuil and Varet, 1973) show more complicated patterns, but the overall style of enrichment is similar.

These types of pattern have been discussed in some detail by Wood et al. (1979b), Tarney et al. (1979), White and Schilling (1978), and Sun et al. (1979) for the North Atlantic, and a consensus has been reached that mantle heterogeneity is the major factor contributing to the observed variations in trace element and isotope ratios. The same conclusion clearly applies to the lavas represented here. Apart from Sr (compatible with plagioclase) and Sc and Cr (compatible with mafic phases), all the elements represented in the patterns have low bulk distribution coefficients between basalt magma and a crystallizing assemblage containing olivine, plagioclase, and clinopyroxene. The part of the pattern made up of the incompatible elements will therefore move upwards as fractional crystallization proceeds but will not change significantly in shape. Similar arguments indicate that partial melting is also incapable of explaining the observed variations, unless some complicated form of progressive melting is involved.

The exact nature of the heterogenetities is more controversial. Trace element and isotope studies of mantle nodules (e.g. Frey and Green, 1974; Green, 1973) and basalts (O'Nions et al., 1977; Wood et al., 1979b; Sun et al., 1979) indicate that the upper mantle may have undergone depletion and enrichment events throughout the Earth's history and that enrichment is caused by accumulation of incompatible element-rich melts or CO_2-dominated fluids. The author favours a model whereby the within-plate tholeiite source region is derived from a mid-ocean range basaltic tholeiite source region enriched by a migrating melt phase, whereas alkaline basalt source regions have been enriched by CO_2-rich fluids. Whatever the explanation, however, it is apparent that prior to subduction the mantle shows a well defined range of compositions.

Basalts erupted in volcanic arc settings

Representative patterns for the volcanic arc basalt magma type are presented in Fig. 1(c). It is apparent that these patterns differ in several important respects from the mid-ocean ridge basalt and within-plate basalt patterns of Fig. 1(a) and (b). Three features stand out.

First, all patterns show an enrichment in the elements Sr, K, Rb, and Ba (and sometimes Th) relative to the elements Ta to Cr. The distinguishing feature of the enriched elements is their low ionic potential (charge/radius) and hence their greater tendency to be mobilized by aqueous fluids (Loughnan, 1969). For this reason, many authors (e.g. Best, 1975; Saunders and Tarney, 1979) consider that this component of variation results from enrichment of a mantle source region by aqueous fluids driven off subducted oceanic crust. These fluids would contain the alkali and alkaline earth elements, which have been added to the upper part of the oceanic crust during sea-floor weathering and which are released into an aqueous phase generated by dehydration of the secondary minerals in which they are accommodated. By contrast, the remaining elements, which have high ionic potentials and are therefore relatively immobile in aqueous fluids, remain in the residual phases of the subducted lithosphere. Th is transitional in this respect and may behave as a mobile or immobile element depending on pressure, temperature, and fluid composition (Wood et al., 1979a). This component is therefore related to the shift to higher $^{87}Sr/^{86}Sr$ ratios in volcanic arc basalts as reported by Hawkesworth et al. (1977) and described by Hawkesworth (this volume, Section VIII).

A second feature, displayed by island arc tholeiites, is the low absolute abundance of the elements of high ionic potential relative to the mid-ocean ridge basaltic tholeiite composition. Significantly, both the incompatible elements, Th to Yb (which are enriched in the melt during fractional crystallization), and the compatible element, Cr (which is depleted during fractional crystallization), exhibit low abundances in these patterns, thus precluding the simple explanation that island arc tholeiites are, on average, more primitive than mid-ocean ridge basalt. There are, however, several ways to explain the incompatible element depletion. Proposals include: presence of stable incompatible element-rich minor phases such as rutile, sphene, and zircon in the melt residue (e.g. Dixon and Batiza, 1979; Saunders et al., 1980); a higher degree of partial melting (e.g. Pearce and Norry, 1979); and remelting of already depleted mantle (Green, 1973). Each of these models can be explained in terms of the presence of subduction zone-derived water in the mantle source region, in the first case to raise the O_2 fugacity and thus stabilize residual oxides; and in the second and third cases to depress the mantle solidus. There is no reason why all three mechanisms should not operate simultaneously.

The third feature is peculiar to calc-alkaline basalts and shoshonites and involves enrichment, not only of the alkali and alkaline earth elements, but also Th, Ce, P, and Sm. The other incompatible elements, Ta, Nb, Zr, Hf, Ti, Y, and Yb, remain at low abundances. Gill's (1974) trace element modelling of calc-alkaline basalts from Fiji has shown that, like the island arc tholeiite magma type, the main source of magma is likely to be the upper mantle rather than subducted oceanic crust. However, trace element criteria are not yet sufficient to pinpoint the cause of the selective enrichment of elements such as Ce and P.

Partial melting of subducted sediment of oceanic or continental derivation could prove a component of the required composition but so could hydrous melting of a previously enriched mantle source. Pb isotope studies (Armstrong, 1971; Church, 1976; Kay, 1977) have demonstrated an increase in the $^{207}Pb/^{204}Pb$ ratios of many volcanic arc lavas relative to mid-ocean ridge basalts. This supports a sedimentary origin for some of the Pb, and, by implication (Kay, 1977), for other elements such as Ce with which the ratio is positively correlated (see also Hawkesworth, this volume, Section VIII).

Lavas erupted in 'anomalous' volcanic arc settings

It has been realized for some time that certain volcanoes above subduction zones erupt lavas with the petrographic and geochemical characteristics of alkaline basalt series. Examples include Grenada (Sigurdsson *et al.*, 1973) in the Lesser Antilles arc, New Georgia in the Solomon arc (Stanton and Bell, 1969), Aoba and Ambrym in the New Hebrides (Warden, 1970; Gorton, 1977), and Kanaga in the Aleutian arc (Fraser and Barnett, 1959; Kay, 1977). DeLong *et al.* (1975), in synthesising data from these and other areas, noted that sodic-alkaline lavas were restricted to certain peculiar sites within the world's arc system, namely: (1) along or near lateral edges of subduction zones where hinge faulting is occurring (e.g. Grenada); (2) where a fracture zone or other linear fracture approximately perpendicular to the trench is being subducted (e.g. Kanaga, Aoba, Ambrym, New Georgia). Anomalous magmas have also been found in or adjacent to continental areas during the initial stages of back-arc spreading or back-arc rifting. Examples include: Penguin Island in the Bransfield Strait (Weaver *et al.*, 1979; Tarney *et al.*, this volume, Section IV), Mexico (Thorpe, 1977), and the Jurassic Sarmiento complex in southern Chile (Saunders *et al.*, 1979; Bruhn *et al.*, 1978).

Geochemical patterns from Grenada (Shimizu and Arculus, 1975; Sigurdsson *et al.*, 1973), Penguin Island (Weaver *et al.*, 1979), and Bogoslov Island in the Aleutians (Kay, 1977) are given in Fig. 1(d). It is apparent that all these patterns differ significantly from the patterns for 'normal' volcanic arcs in Fig. 1(c). They can best be interpreted in terms of combinations of volcanic arc and alkaline or within-plate components. Thus Grenada exhibits *both* an enrichment in elements of low ionic potential characteristic of island arc tholeiites *and* an enrichment in all elements from Sr to Sm characteristic of alkaline basalts. The Penguin Island pattern exhibits *both* an enrichment in elements of low ionic potential and in Th, Ce, P, and Sm characteristic of calc-alkaline basalts *and* an enrichment in all elements from Sr to Ti characteristic of within-plate tholeiites. The Bogoslov Island pattern similarly contains island arc tholeiite and within-plate tholeiite components

The most obvious explanations for these patterns are that *either* the mantle source regions have suffered at least two episodes of enrichment, one related and the other unrelated to subduction, *or* that mixing of magmas from two distinct sources has taken place. In the case of Penguin Island, it is probable that the mantle was already enriched in incompatible elements prior to subduction. In the case of Grenada and Bogoslov Island, it is difficult to dispute the explanation of DeLong *et al.* (1975), that these settings had access to sublithospheric regions not normally tapped in island arcs and which contributed the alkaline or within-plate components to the geochemical patterns.

Discrimination and Modelling of the Trace Element Characteristics

Cr–immobile element covariations

One of the principal characteristics of island arc tholeiites to be revealed by the geochemical patterns was the low content of immobile incompatible elements in basalts of a given degree of fractionation. Discriminant diagrams based on Ti as the immobile incompatible element and Cr or Ni as the fractionation index (e.g. Pearce, 1975; Garcia, 1978) have already been published, and appear to discriminate successfully between island arc tholeiite and mid-ocean ridge basalt magma types (e.g. Sharaskin *et al.*, 1981). Here Y is used instead of Ti for the petrogenetic reasons explained below and the resulting discriminant diagram is presented in Fig. 2. Although there is a small overlap between calc-alkaline basalts and mid-ocean ridge basalts, island arc tholeiites are well separated from mid-ocean ridge basalts, being displaced towards lower Y values at a given Cr concentration.

Petrogenetic modelling of the variations in these two elements is greatly assisted by the likelihood, shown in Fig. 1(a) and (b), that neither Y or Cr participate in the processes that cause mantle heterogeneity, possibly because mantle enrichment and depletion events involve transfer only of elements incompatible with garnet lherzolite assemblages. Thus the source regions for most magmas probably contained abundances of Y and Cr that were similar to

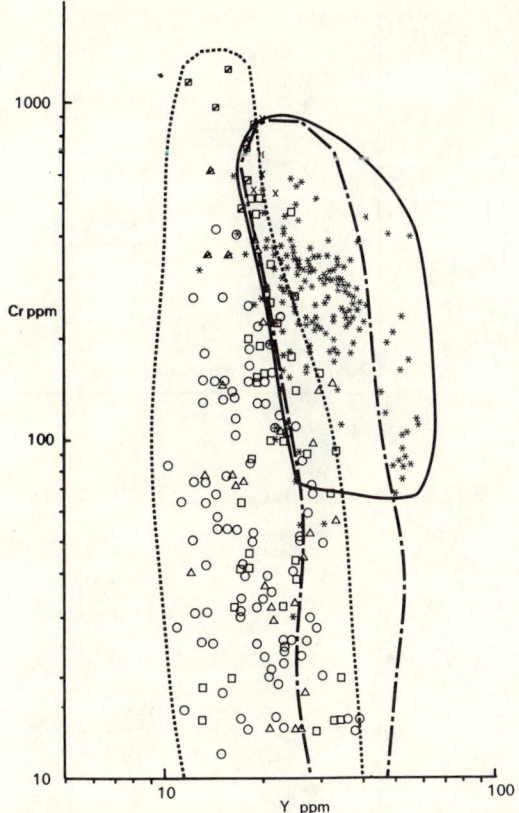

Fig. 2 Discrimination diagram for characterizing island-arc tholeiites on the basis of Y–Cr covariations. Key: *, mid-ocean ridge basalts; ○, island-arc tholeiites; □, calc-alkaline basalts; △, shoshonites (all compositions are basalts except those marked with diagonal lines which are picrites). ———, mid-ocean ridge basalt compositional field; ·—·—·, within-plate basalt compositional field; ······, volcanic arc basalt compositional field

primordial mantle abundances (Wood, 1979), and the variations between basalts can be attributed for the most part to differences in their partial melting and fractional crystallization histories.

Figure 3 shows how petrogenetic pathways can be drawn to link a primordial mantle composition (Wood, 1979) with the composition of mid-ocean ridge basalt and island arc tholeiite magmas. Each of the four pathways highlights one of the hypotheses presented earlier as possible explanations of the difference between these two magma types: first, the degree of partial melting (Fig. 3(a)); second, the presence of residual minor phases (Fig. 3(b)); third, the source composition (Fig. 3(c)); and finally, mixing with melt derived from subducted oceanic crust (Fig. 3(d)).

Figure 3(a) investigates the degree of partial melting as a possible explanation. The partial melting trend drawn was calculated assuming a plagioclase lherzolite source composition of $ol_{0.6}opx_{0.2}cpx_{0.1}pl_{0.1}$ in which the phases melt in the proportions $3:1:4:4$. However, the principle holds for all likely mineral proportions. Because Cr is concentrated in residual phases during mantle melting, the Cr content of the primary melt is buffered for a wide range of melt proportions. Thus, for degrees of melting of less than 50 per cent, the partial melting trend lies subparallel to the Y axis. However, because Cr is strongly partitioned in early mafic crystallizing phases, namely olivine, Cr-spinel, and pyroxene, the fractional crystallization trend is initially nearly vertical. A slightly flatter trend only results once plagioclase, which contains no Cr, starts to crystallize. If, therefore, a typical mid-ocean ridge basalt composition is extrapolated back to the partial melting trend along the fractional crystallization vectors, it intersects the trend at $c.$ 15 per cent melting. By contrast an island arc tholeiite intersects the trend at lower Y values, at between 20 and 40 per cent melting. Although the exact values will vary according to the parameters chosen, it is apparent that the discrimination seen in Fig. 2 *could* be explained in terms of a greater degree of partial melting of the island arc tholeiite magma type.

Figure 3(b) investigates residual minor phases as a possible explanation. Here, the mid-ocean ridge basalt partial melting trend remains the same as before. However, the island arc partial melting trend has been modelled assuming the presence of one or more minor phases which retain Y, and it is therefore displaced to lower Y values. As a consequence, both mid-ocean ridge basalt and island arc tholeiite compositions, when extrapolated backwards along fractional crystallization vectors, will intersect their respective partial melting trends at 15 per cent

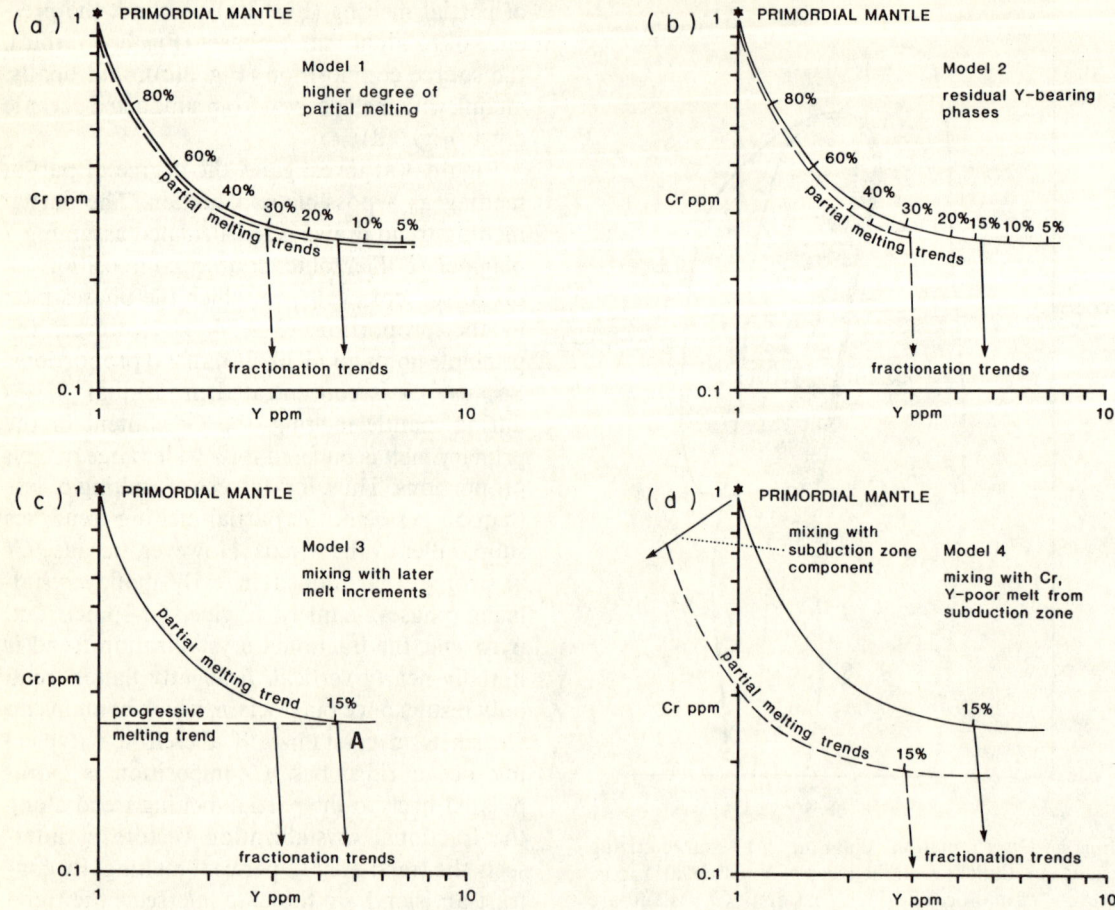

Fig. 3 Possible explanations for the lower Y contents of island-arc tholeiites relative to mid-ocean ridge tholeiites. The diagram shows petrogenetic pathways for mid-ocean ridge (continuous lines) and island-arc tholeiites (broken lines) according to the various models proposed. Scales are all relative to a primordial mantle value of unity. See text for further discussion

melting. Theoretically, therefore, the presence of minor phases in the source region for island arc tholeiites *could* by itself explain the discrimination recorded in Fig. 2.

Figure 3(c) investigates remelting of residual mantle as a mechanism for explaining the low Y values in island arc tholeiites. If the model mantle of Fig. 3(a) undergoes 15 per cent partial melting, it will lie on the partial melting trend at the point A. If the residue from this melting episode is subsequently remelted, the melt produced will lie close to the broken line but at very low Y values. Mixing of this remelt with the initial melt in an island arc setting *could* therefore yield a magma significantly depleted in Y and explain the discrimination seen in Fig. 2.

Finally, Fig. 3(d) investigates the possibility that the low Y values in island arc tholeiites result from mixing of a mantle-derived magma and a magma derived from melting of the underthrust oceanic lithosphere. At present, the product of melting of subducted oceanic crust cannot be modelled accurately because it is impossible to predict either the major or minor phase composition of the residue (Gill, 1974). However, Fig. 3(d) does show that the magma derived from the subducted lithosphere must have low Y. This is possible provided that the residue includes phases containing Y, such as zircon, sphene, or apatite. In this case the mixing process *could* yield the variation required.

Since these four models are not mutually exclusive, each might contribute to the actual explanation. In Fig. 4, the simple equilibrium partial melting model has been adopted to explain differences between some typical volcanic arc suites. It is apparent that variation within any one suite tends to be subparallel with the Cr axis, consistent with a simple variation in the extent of fractional crystallization. The variation between suites is subparallel to the Y axis. Extrapolation of the individual trends back to the partial melting trend indicates that a range of degrees of melting from 40 to 15 per cent would explain the overall variations, with the most primitive island arcs (South Sandwich Islands, Tonga) requiring the greatest degree of partial melting.

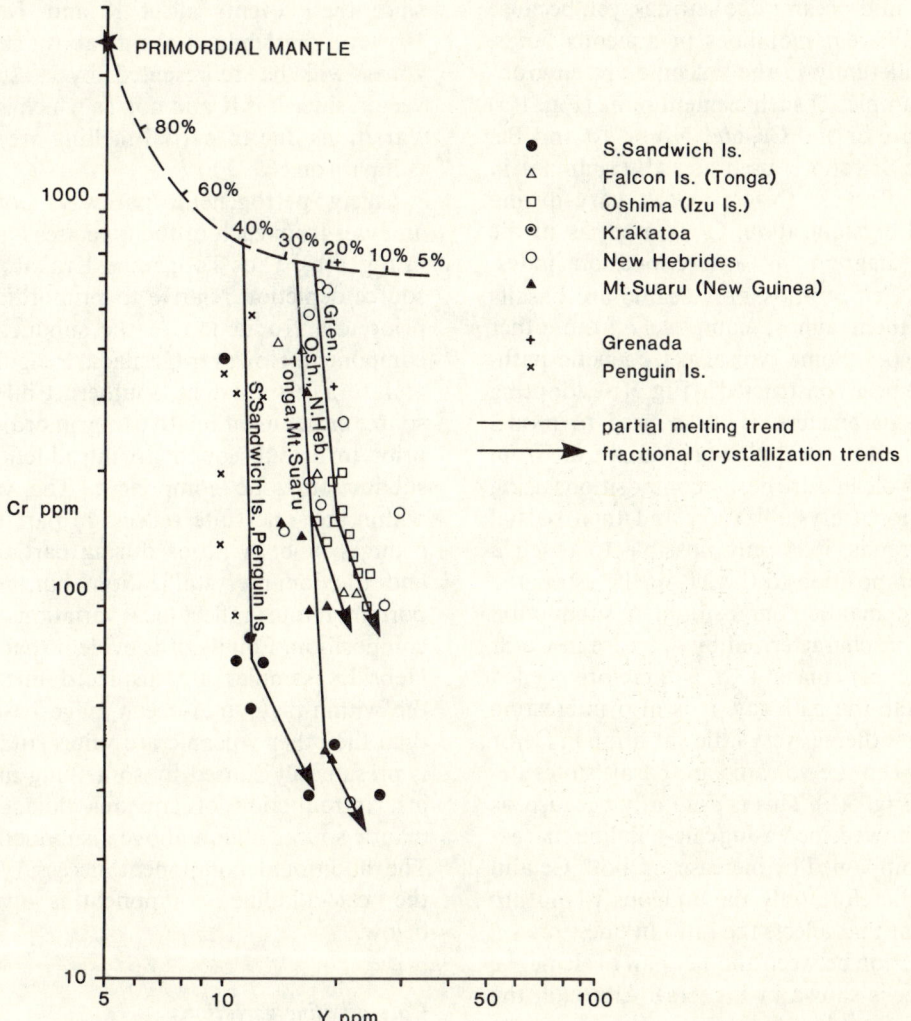

Fig. 4 Variations within some typical volcanic arc suites and the petrogenetic pathways that connect them to primordial mantle composition. Modelling is based on Fig. 3(a). Data by the author, except: New Hebrides (Gorton, 1977); New Guinea (Mackenzie and Chappell, 1972); Penguin Island (Weaver *et al.*, 1979); Grenada, (Sigurdsson *et al.*, 1973; Shimizu and Arculus, 1975)

Mobile element–immobile element covariations

The elements in the geochemical patterns with ionic potentials indicative of mobility in aqueous fluids were K, Rb, Ba, and Sr. Before evaluating the enrichment in volcanic arc basalts of these mobile elements relative to the immobile elements, it is necessary to filter out the effects of other types of enrichment process. The easiest way to do this is to choose a pair of elements which are enriched to a similar extent in within-plate and mid-ocean ridge settings, yet, because of their different mobilities in aqueous fluids, behave differently in the volcanic arc environment. Examples of such element pairs (Fig. 1(a) and (b)) are Sr and Ce, and Nb or Ta and Ba.

The Ce/Sr ratio is used as a discriminant in Fig. 5(a). Because this ratio can vary during fractional crystallization, Cr is used, as in the previous diagram, as a fractionation index. The low Ce/Sr ratios of volcanic arc basalts separate them almost completely from other magma types. Some typical petrogenetic pathways have been constructed in Fig. 5(b), adopting the same parameters as those used to model Fig. 3(a). On extrapolating backwards from observed volcanic arc basalt compositions using first fractional crystallization and then partial melting trends, it is only possible to reach a source composition to the left of the estimated primordial mantle composition. A subduction zone vector, characterized by decrease in Ce/Sr ratio at almost constant Cr, is therefore needed to complete the pathway. It is also interesting to note that there is very little variation in Ce/Sr ratio between the volcanic arc basalt suites depicted in Fig. 5(b). This is primarily because, as Fig. 1(c) showed, increasing calc-alkaline character is accompanied by increases in both Ce and Sr; it is therefore only the aqueous island arc component that affects the ratio in question.

Covariation between another pair of elements, K and Ta, is shown in Fig. 6(a). Although the K/Ta ratio would act as a discriminant on its own, normalization by Yb fulfills a useful function by enabling source variations prior to subduction to be identified. Discrimination of volcanic arc magma types is extremely effective on this diagram; all basalts from this setting exhibit a shift to higher K/Yb ratios relative to the restricted band that is characteristic of mid-ocean ridge and within-plate basalts.

Petrogenetic pathways can be drawn to link the compositions of some typical volcanic arc lava suites with the primordial mantle composition estimated by Wood (1979). Mantle enrichment or depletion events outside the influence of subduction zones will be represented by vectors with a slope of unity on this diagram, since these events affect K and Ta equally. However, enrichment events above subduction zones will be represented by a subvertical vector, since it is K and not Ta which is affected. Variations due to partial melting are minor in comparison.

Likely petrogenetic pathways for typical analyses from each of the type areas are shown in Fig. 6(b). Thus Tonga and Krakatoa require source depletion relative to primordial mantle prior to introduction of the subduction zone component. However Grenada, Penguin Island, and, to a lesser extent, southern Chile, require source enrichment relative to primordial mantle prior to or subsequent to introduction of the subduction zone component. The variability within any one suite reflects in part the small changes in both ratios during partial melting and fractional crystallization, but, more important, it must reflect local variations in source composition. Finally, it is evident that the New Hebrides samples are displaced further from the within-plate–mid-ocean ridge basalt trend than the other volcanic arc suites studied. This is presumably caused by something more than just introduction of aqueous fluids into the mantle source region above a subduction zone. The additional component necessary, termed the 'calc-alkaline' component is investigated below.

Calc-alkaline variations

To isolate the calc-alkaline component, the K/Yb ratio can be replaced by ratios such as Ce/Yb. These ratios do not increase significantly between mid-ocean ridge basaltic tholeiites and

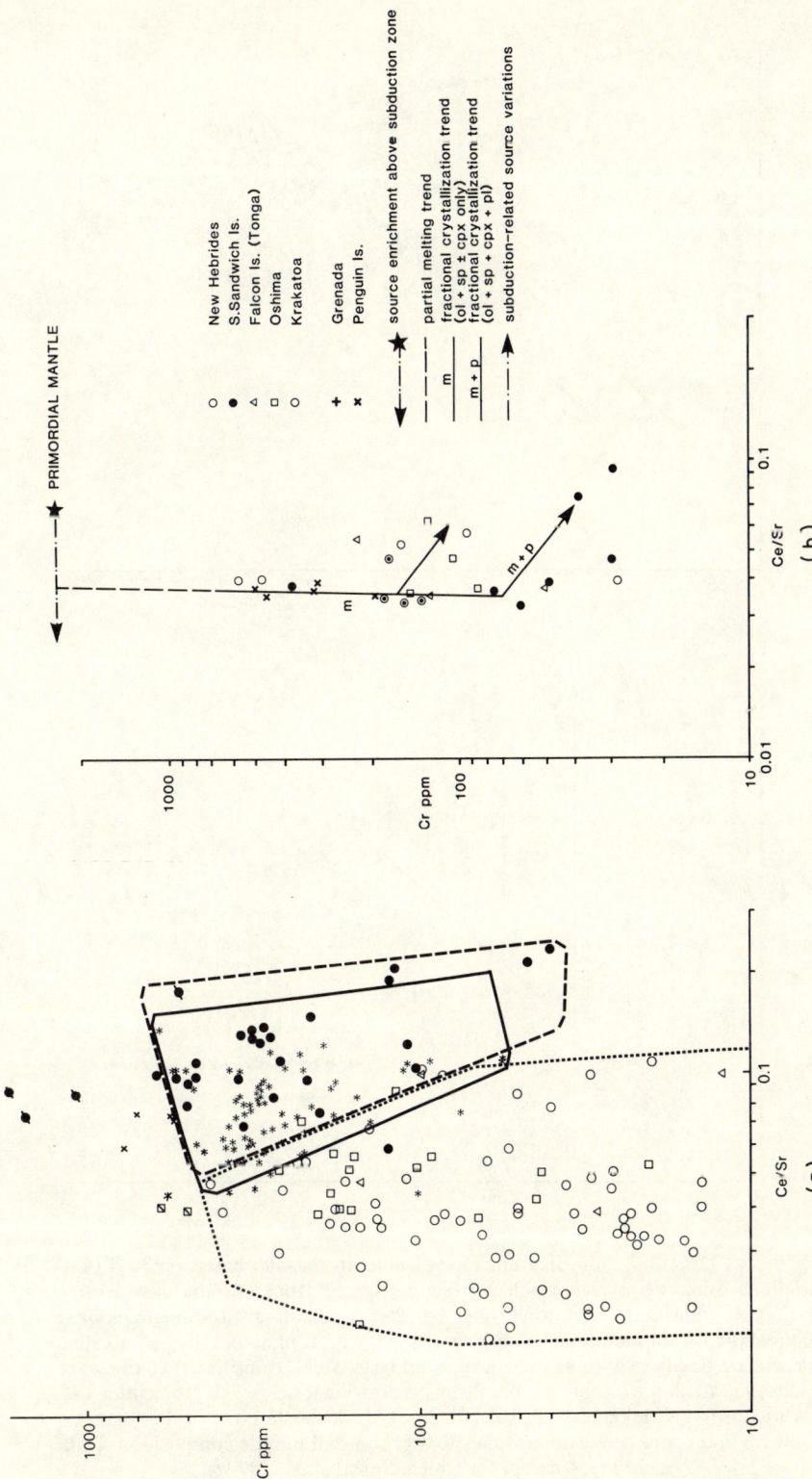

Fig. 5 (a) Discrimination diagram for characterizing volcanic arc basalts on the basis of Ce/Sr ratios (see Fig. 2 for key to symbols). (b) Petrogenetic pathways that connect individual volcanic arc suites to the primordial mantle composition

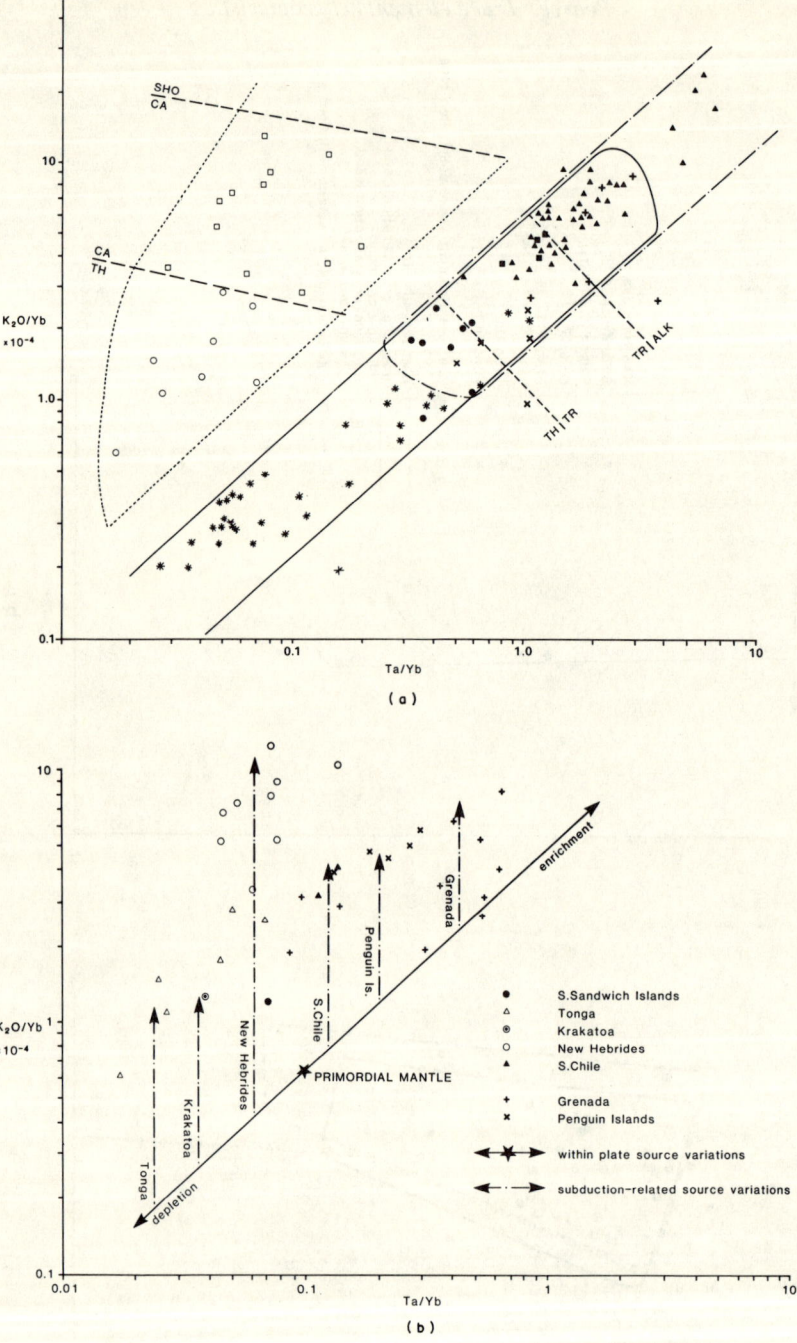

Fig. 6 (a) Discrimination diagram for volcanic arc basalts based on K–Ta co-variations using Yb as a normalizing factor. Key: *, tholeiitic mid-ocean ridge basalts; ×, transitional mid-ocean ridge basalts; +, alkaline within-plate basalts; ●, tholeiitic within-plate basalts; ■, transitional within-plate basalts; ▲, alkaline volcanic arc basalts; ○, tholeiitic volcanic arc basalts; □, transitional volcanic arc basalts; △, alkaline volcanic arc basalts. Compositional fields: TH, tholeiitic; TR, transitional; ALK, alkaline; CA, calc-alkaline; SHO, shoshonitic. (b) Petrogenetic pathways that connect individual suites to a primordial mantle composition. Data as in Fig. 4 except for Chile (Dostal *et al.*, 1977a)

island arc tholeiites but do increase from island arc tholeiites through calc-alkaline basalts to shoshonites. The resulting discriminant diagram is shown in Fig. 7(a). Here the within-plate and mid-ocean ridge basalts occupy a well defined field in which the P/Yb or Ce/Yb and Ta/Yb ratios are positively correlated.

Of the volcanic arc basalts a shift to higher Ce/Yb ratios can be seen in all calc-alkaline and shoshonites, but only some island arc tholeiites. It is the absence of a shift in these rocks, when compared with the large K/Yb shift observed on the previous diagram, that indicates that two processes may be responsible for volcanic arc basalt variation as a whole: one which simply adds elements such as K which are mobile in aqueous fluids; and another which adds elements such as Ce and P as well as K.

Petrogenetic pathways for the individual volcanic arc suites (Fig. 7(b)) can be modelled along the same lines as the previous diagram. Thus the enrichment and depletion vectors with slopes close to unity account for mantle variations unrelated to subduction, whereas subvertical vectors account for subduction-related variations. Pathways can therefore be drawn connecting each of the lava suites to the primordial mantle composition. As before, Grenada, Penguin Island, and southern Chile require superimposition of the subduction-related component on an enriched mantle source; others, e.g. Tonga, Krakatoa, and the New Hebrides, require a depleted mantle source. The most primitive island arc tholeiites, from the South Sandwich Islands, show no calc-alkaline component at all, plotting close to the mantle evolution line.

As far as discrimination is concerned, the suggestion of Wood *et al.* (1979*a*) that Th may be an effective discriminant is supported by the diagram of Th/Yb against Ta/Yb plotted as Fig. 8. There is an overlap on this diagram between island arc tholeiites and mid-ocean ridge basalts. Nonetheless, although K gives a better discrimination than Th, the latter is much less mobile in aqueous fluids and this makes it a particularly valuable element for study of altered rocks.

Variations unrelated to subduction

Figures 5–7 have shown that volcanoes from 'anomalous' arc settings were derived from mantle that had undergone enrichment relative to primordial mantle by a process that was independent of subduction. However, the nature of the enrichment event could not be assessed because the within-plate and alkalinity vectors were parallel or subparallel on these diagrams. It is therefore useful to plot analyses of this type on a diagram which distinguishes between these two types of heterogeneity. One such diagram is Ti/Y against Nb/Y (Fig. 9(a)). Y is used here in the same way that Yb was used in the previous diagrams—as an element which does not participate in the processes that cause mantle heterogeneity but which behaves as an incompatible element during most partial melting and fractional crystallization events. The ratios Ti/Y and Nb/Y therefore highlight mantle heterogeneity but are only slightly affected by subsequent differences in melting and fractional crystallization histories.

As Fig. 1(a) and (b) showed, Nb and Ti are both enriched in within-plate tholeiites relative to mid-ocean ridge basalts; by contrast, Ti is only very slightly enriched and Nb is very strongly enriched in alkaline basalts relative to tholeiitic basalts from either setting. Thus, within-plate character affects both ratios in Fig. 9(a), whereas alkalinity affects the Nb/Y ratio only. Moreover, neither ratio is significantly affected by variations due to subduction. The result is a diagram which is extremely effective in characterizing within-plate basalts and their petrologic character.

The results of plotting the various volcanic arc basalt suites on this diagram can be seen in Fig. 9(b). Whereas 'normal' volcanic arc basalts are slightly depleted relative to primordial mantle in both ratios, the Penguin Island basalts are enriched along the within-plate vector, whereas Grenada basalts are enriched along the alkalinity vector. Clearly, this type of diagram can be used in conjunction with the other diagrams to recognize 'anomalous' volcanic arc basalts in the geological record.

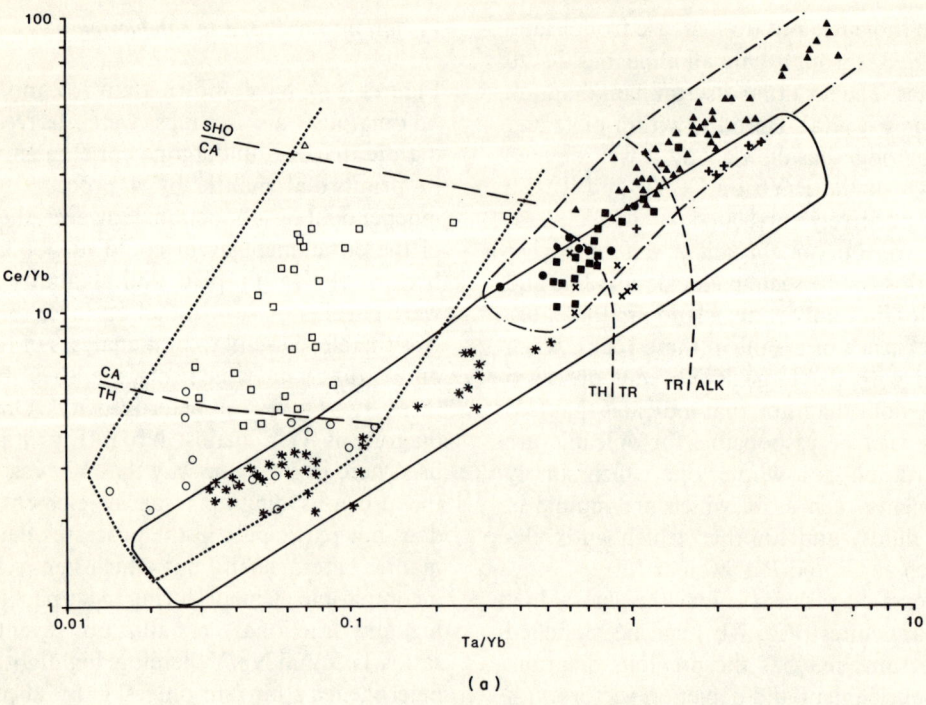

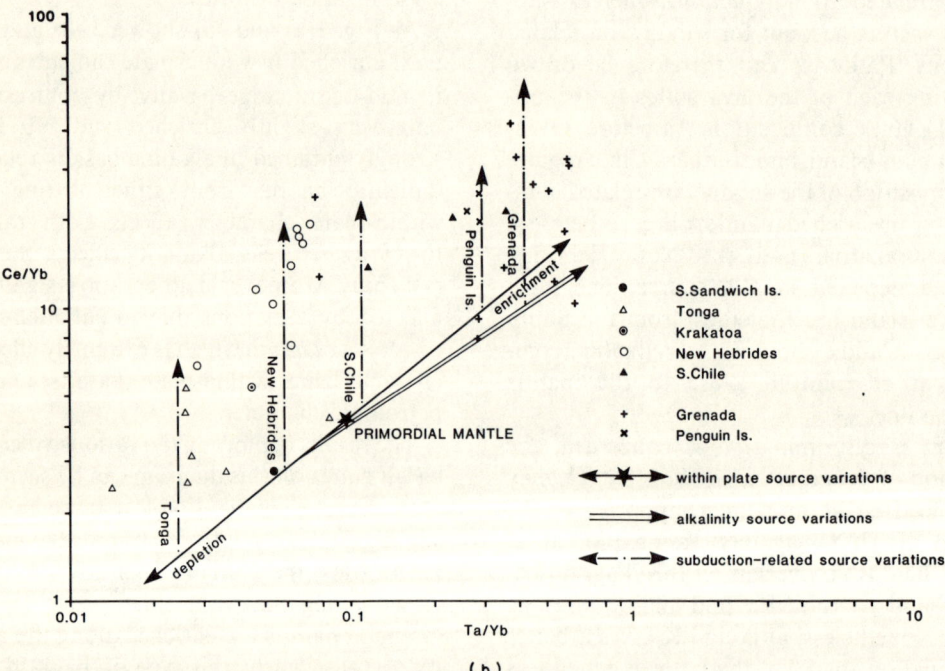

Fig. 7 (a) Discrimination diagrams for calc-alkaline and shoshonitic basalts based on Ce–Ta covariations with Yb as normalizing factor (see Fig. 6(a) for key). (b) Petrogenetic pathways for some typical volcanic arc suites

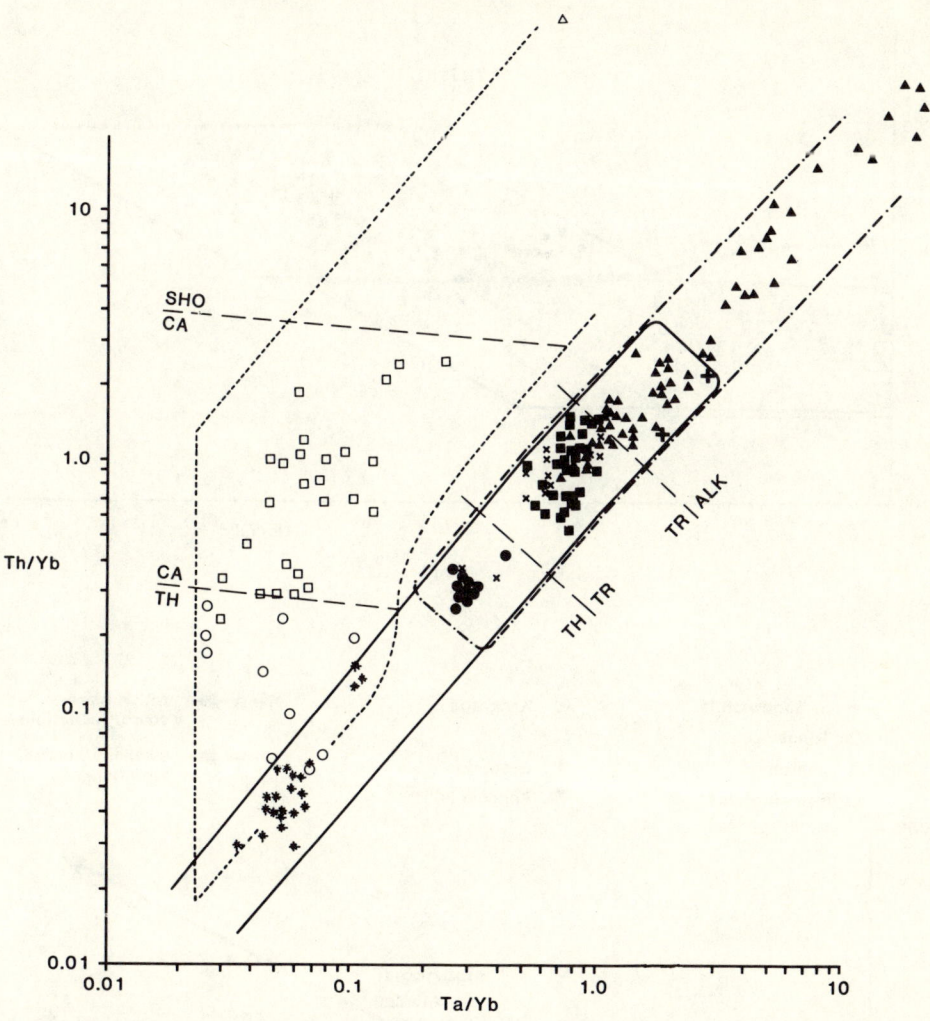

Fig. 8 Discrimination diagram for volcanic arc basalts based on Th–Ta covariations using Yb as a normalizing factor (see Fig. 6(a) for key)

Characteristics of Intermediate and Acid Lavas

Of the geochemical features that characterize basic lavas from destructive plate margins, some can be extended to more evolved compositions, whereas others are masked by fractional crystallization and, to a lesser extent, processes such as volatile loss and crustal contamination.

The effects of fractional crystallization within tholeiitic, calc-alkaline, and shoshonitic suites are summarized in Fig. 10. Of course no two volcanoes are likely to have identical trends, but the diagram serves to illustrate the most important variations. Thus, Rb and Th generally behave as incompatible elements throughout all three series, but may decrease in the extremely acid rocks that are not included in this compilation. Zr, Nb, Ba, and the light REE behave as incompatible elements throughout tholeiitic and calc-alkaline sequences, but can decrease during fractionation from intermediate to acid compositions in high K calc-alkaline and shoshonitic series. This decrease can be explained by crystallization of biotite and zircon at

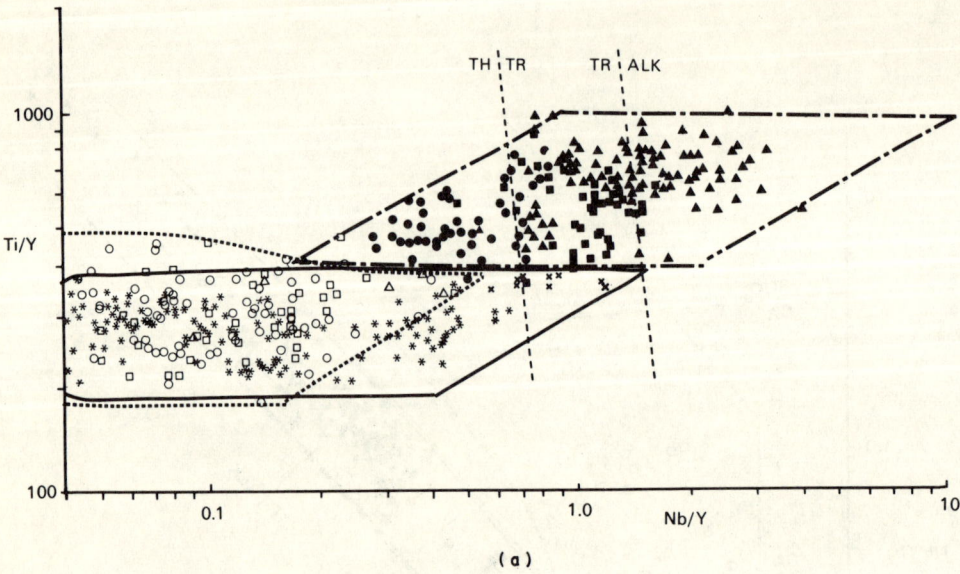

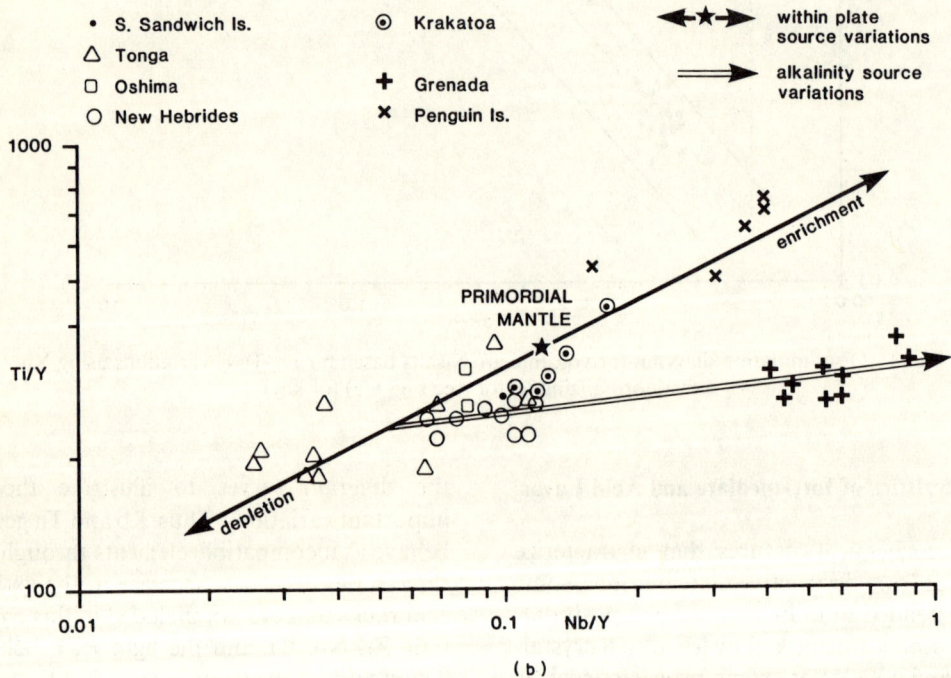

Fig. 9 (a) Discrimination diagram for within-plate and alkaline basalts using Ti/Y and Nb/Y ratios (see Fig. 6(a) for key). (b) Petrogenetic pathways for some typical volcanic arc suites

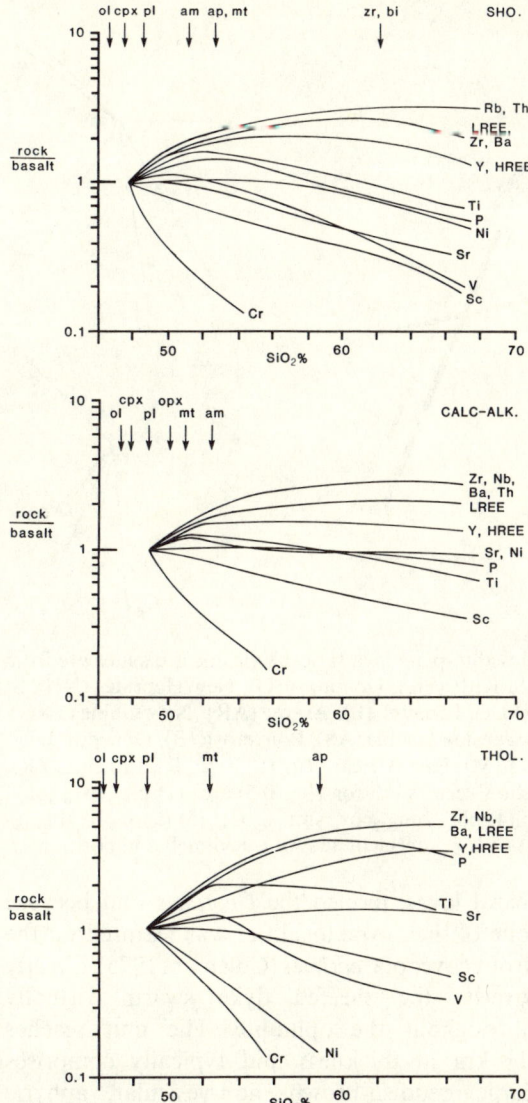

Fig. 10 Variations of some trace elements with SiO_2 content in typical tholeiitic, calc-alkaline and shoshonitic suites. Arrows mark the approximate points at which indicated minerals become crystallizing phases

intermediate compositions. Y and the heavy REE behave as incompatible elements throughout the tholeiitic series, but remain constant or decrease with fractionation during much of the calc-alkaline and shoshonitic series where they can be accommodated within amphibole, biotite, and zircon. Of the other, compatible elements, Cr, Ni, and Sc are accommodated in early crystallizing olivine, pyroxene, and spinel, and all therefore tend to decrease during fractionation—although Ni remains approximately constant within the calc-alkaline series in which olivine is less important as a crystallizing phase. Ti and V both start to decrease once magnetite starts to crystallize and P starts to decrease once apatite starts to crystallize. The point at which magnetite and apatite become crystallizing phases depends on a number of parameters, of which f_{O_2} is one of the most important. Probably, this happens later in the tholeiitic series than in the calc-alkaline series, in keeping with the Fe-enrichment character of the tholeiitic trend.

These variations mean that the basalt discrimination diagrams will have to be modified if they are to be used to study intermediate and acid lavas. For example, a diagram such as Y against Cr cannot be extrapolated to evolved compositions because Cr soon falls below the detection limit and because Y behaves in different ways in the different fractionation series. However, it is possible, by careful choice of axes, to utilize the fact that volcanic arc lavas contain lower abundances of immobile elements than other magma types. A plot of Ti against Zr is particularly effective.

Figure 11 (from Pearce and Norry, 1979) shows the fractionation trends followed by some typical lava sequences from both volcanic arc and within-plate settings. Mid-ocean ridge lavas are not included because there are too few intermediate and acid members for trends to be drawn. The fractionation vectors confirm that magnetite crystallization is most important in causing Ti concentration to decrease relative to Zr during fractionation. Because of the early crystallization of magnetite, volcanic arc lavas never reach the higher Ti contents of within-plate lavas, for any given Zr concentration. The final discriminant boundaries are given in Fig. 13.

Identification of Past Volcanic Arc Sequences: An Example from the Oman Ophiolite Complex

Viewed on a broad scale, the Upper Cretaceous Semail thrust sheet of Oman (Reinhardt, 1969;

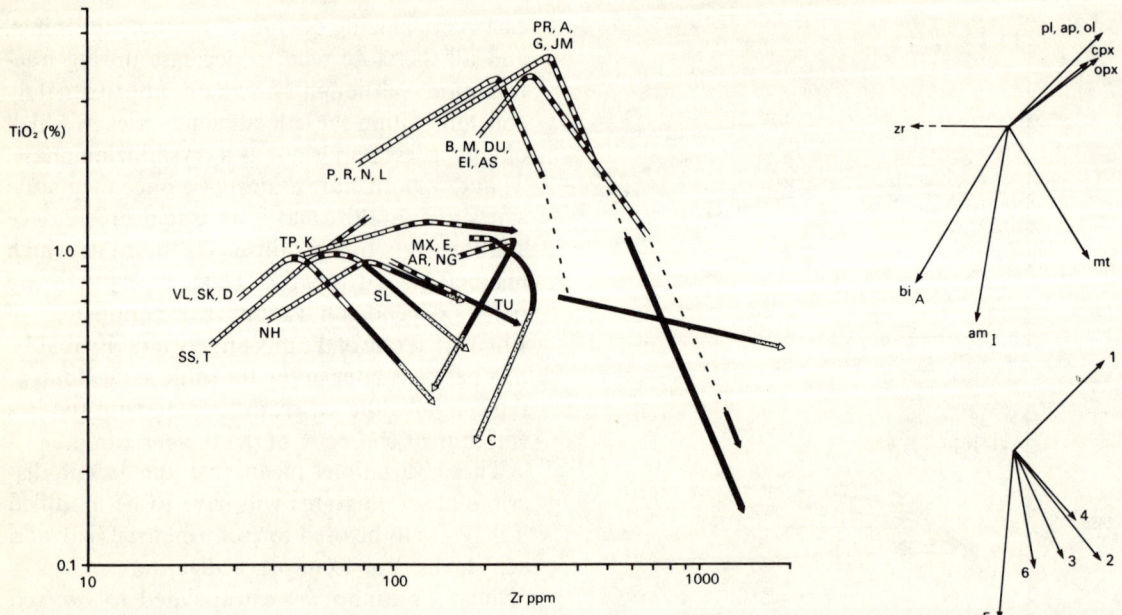

Fig. 11 Ti–Zr fractionation trends in some volcanic arc and within-plate lava series. Volcanic arc suites are from South Sandwich Islands (SS), Tonga (T), Vanua Levu (VL), St Kitts (SK), Dominica (D), New Hebrides (NH), St Lucia (SL), Tangkuban Prahu (TP), Krakatoa (K), Mexico (MX), Ecuador (E), Ararat (AR), New Guinea (NG), Turkey (TU), North Chile (C). Within-plate suites are from Ascension Islands (AS), Bouvetoya (B), Dunedin (DU), Easter Island (EI), Galapagos (G), Jebel Marra (JM), Madeira (M), East African Rift (L, N, P, R), Principe (PR). For details see Pearce and Norry (1979). Modelled fractionation vectors (all for $F = 0.5$) are: (1) $pl_{0.5}cpx_{0.3}ol_{0.2}$ (B); (2) $pl_{0.55}cpx_{0.3}ol_{0.1}mt_{0.05}$ (I); (3) $pl_{0.6}am_{0.35}mt_{0.05}$ (I); (4) $pl_{0.55}am_{0.2}cpx_{0.2}mt_{0.05}$ (I); (5) $(ksp, pl)_{0.6}bi_{0.15}am_{0.2}mt_{0.05}$ (A); (6) $pl_{0.6}cpx_{0.2}am_{0.15}mt_{0.05}$ (A). B = basic; I = intermediate; A = acid melt composition

Glennie et al., 1974) is a stratified igneous sequence from tectonized harzburgite at its base, through layered gabbro, homogeneous gabbro, sheeted dyke swarm, pillow lavas, and, at the top, pelagic sediments. It thus fits perfectly the Penrose Conference (Penrose field conference participants, 1972) definition of an ophiolite complex. However, recent detailed mapping and geochemical analysis of the lava sequence by Alabaster et al. (1980), Alabaster and Pearce (1979), and Pearce et al. (1981) have revealed that the view of Oman as a slice of normal oceanic crust is too simplistic and that the ophiolite appears to be made up of the oceanic crust of a back-arc basin cut by the products of arc magmatism. The geochemical evidence for this interpretation is summarized below.

The volcanic stratigraphy can be subdivided in the field into three distinct units, all erupted within a geologically short space of time. The basal lavas, termed the *Geotimes* unit because one of their type localities was featured on the front cover of *Geotimes* (Coleman, 1975), directly overlie the sheeted dyke swarm virtually throughout the ophiolite. The unit reaches 1½ km in thickness and typically comprises large, reddish-brown, non-vesicular, aphyric pillow lavas interbedded with occasional non-pillowed flows. These lavas form the volcanic 'basement' on which successive lava units were erupted.

The overlying lavas can be subdivided into two groups, named the *Lasail* and *Alley* units after their type localities. The *Lasail* unit comprises a fractionation sequence from basalt through andesite to rhyolite and may reach 1 km in total thickness. The basic members are typically small, grey-green, non-vesicular pillow lavas which directly overlie (with no intervening sediment) the *Geotimes* pillow lavas.

The intermediate and acid members include massive andesite lava flows and swarms of subvolcanic inclined sheets. The *Lasail* unit does not occur throughout the ophiolite but is restricted to centres spaced at intervals along the north–south strike of the complex (Fig. 12). It is therefore thought to represent seamounts erupted on oceanic crust.

The *Alley* unit comprises a second basalt–rhyolite fractional crystallization sequence and reaches a maximum thickness of 0.5 km. It may overlie the extrusive members of the *Geotimes* unit, in which case a thin layer of ferromaganoan sediment separates the units. Alternatively, where no *Lasail* lavas were deposited, the *Alley* lavas overlie the *Geotimes* lavas, in which case a much thicker sedimentary horizon marks the volcanic non-conformity. The basaltic rocks form a series of large, brownish-green, vesicular pillowed flows. The intermediate members

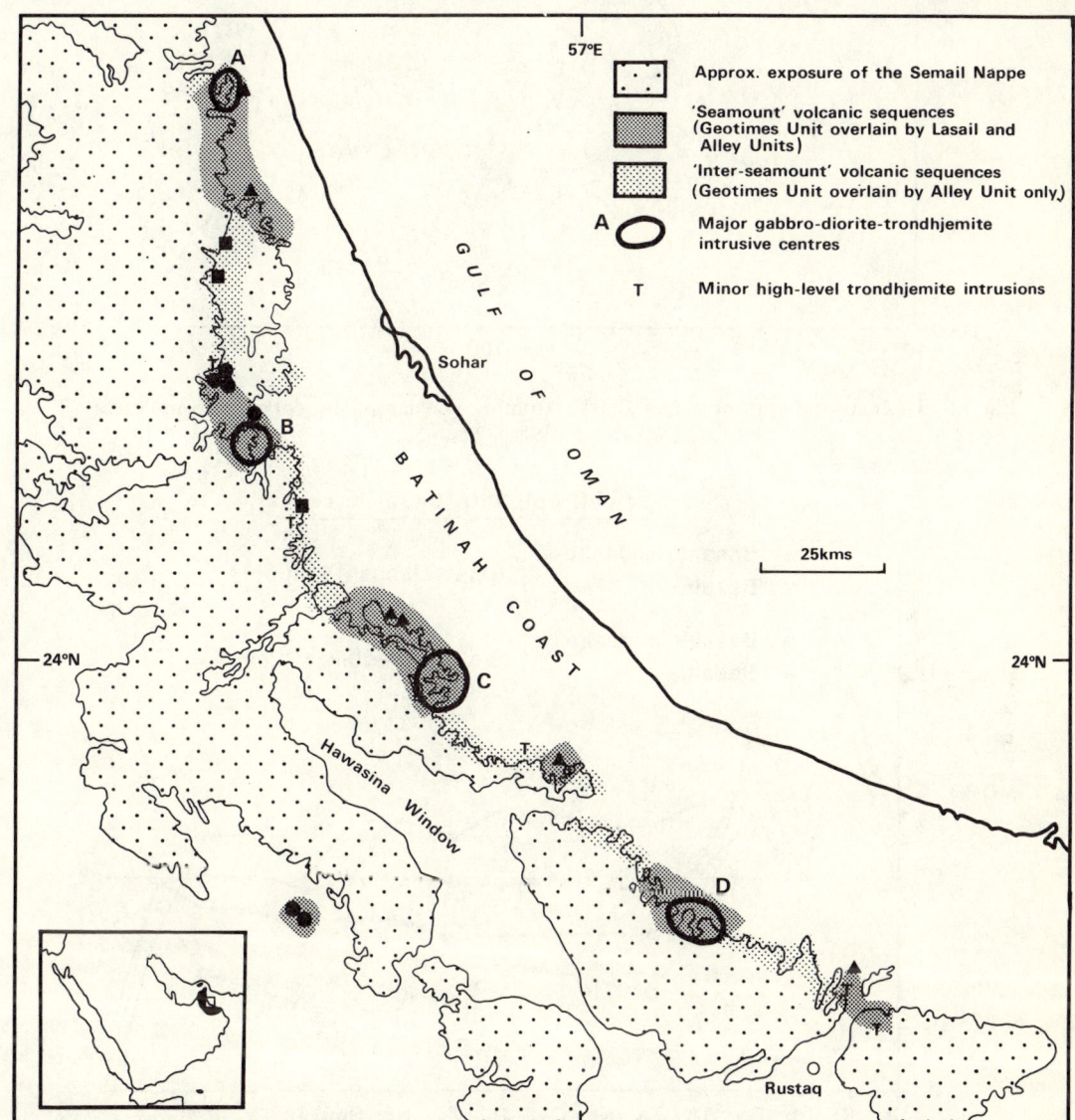

Fig. 12 Distribution of the Lasail volcanic unit along the Oman ophiolite and its relationship to gabbro-tonalite–trondhjemite intrusive centres (from Alabaster *et al.*, 1980). (Symbols show sites of mineralization)

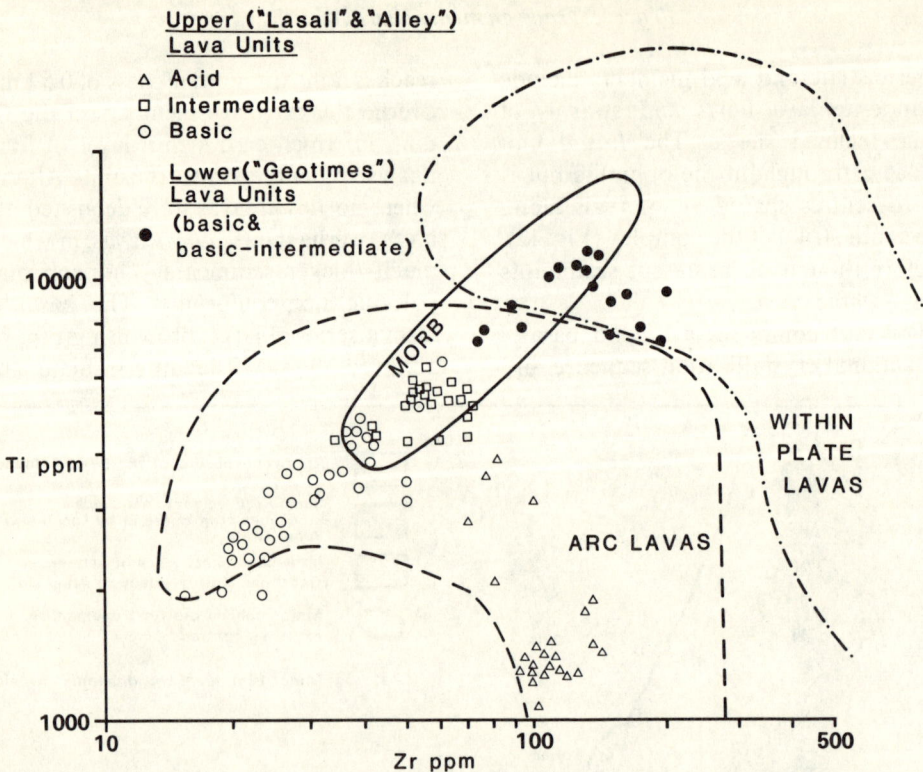

Fig. 13 Ti–Zr covariations for lavas and dykes from the Oman ophiolite complex (from Pearce *et al.*, 1981)

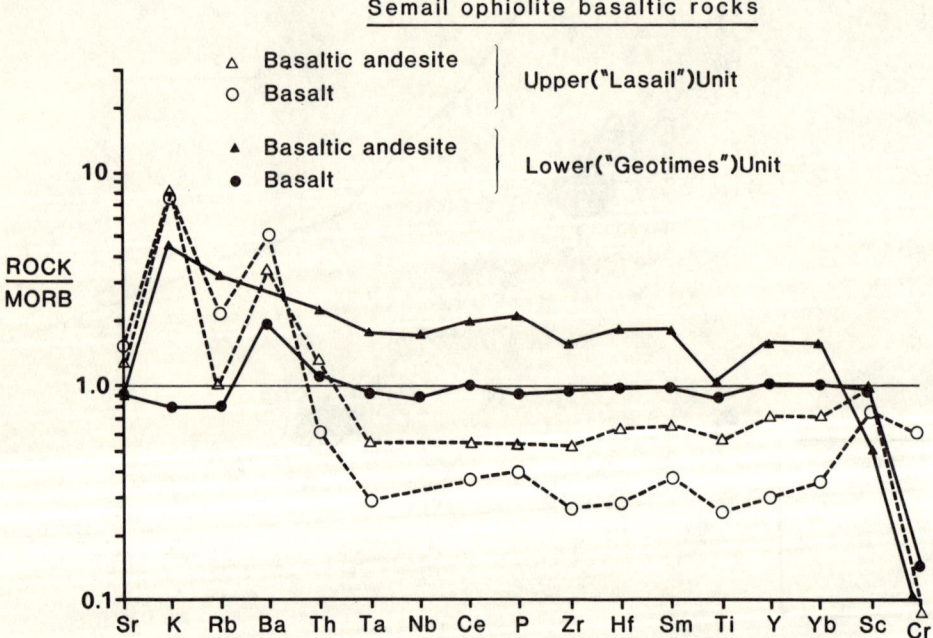

Fig. 14 Geochemical patterns for the lower (Geotimes) lava series and an upper (Lasail) lava series on Oman (see Fig. 1 for normalizing factors) (from Pearce *et al.*, 1981)

include massive flows of andesite and pillow lavas and massive flows of rhyolite and obsidian. The *Alley* lavas are found throughout the ophiolite, and are best developed in faulted depressions of which the 'Alley' itself is the best example (Smewing *et al.*, 1977). If the *Lasail* lavas were erupted in a 'seamount rifting' event, the *Alley* lavas must have been erupted during a 'post-seamount rifting' event.

The first stage in the geochemical interpretation is to plot a diagram of Ti against Zr (Fig. 13) for the three lava units and for the sheeted dykes. The distribution of data points shows: (1) that the *Geotimes* lavas and sheeted dykes have identical compositions; (2) that the *Lasail* and *Alley* lavas have a completely different compositions from the *Geotimes* lavas and follow a typical island arc fractional crystallization trend (cf. Fig. 11). These lines of evidence therefore indicate that the *Lasail* and *Alley* lavas represent a volcanic arc structure developed directly on oceanic crust.

The geochemical patterns (Fig. 14) for some typical basalts from these sequences confirm this interpretation. The *Lasail* and *Alley* lavas show an absolute depletion in the immobile elements, Ta to Cr, and an enrichment of mobile relative to immobile elements. Because of the intense alteration undergone by these lavas, this latter characteristic cannot be regarded as completely diagnostic. However, the degree of enrichment observed is greater than that normally associated with greenschist facies alteration and a subduction zone component in the mantle source region is therefore probable. The *Geotimes* lava has a flat pattern typical of a mid-ocean ridge basaltic tholeiite with the notable exception of an extremely low Cr value.

These two characteristics may be examined in more detail, using the Y–Cr and Th/Yb–Ta/Yb diagrams in Figs. 15 and 16. The *Lasail* and *Alley* lavas exhibit a shift to lower Y values

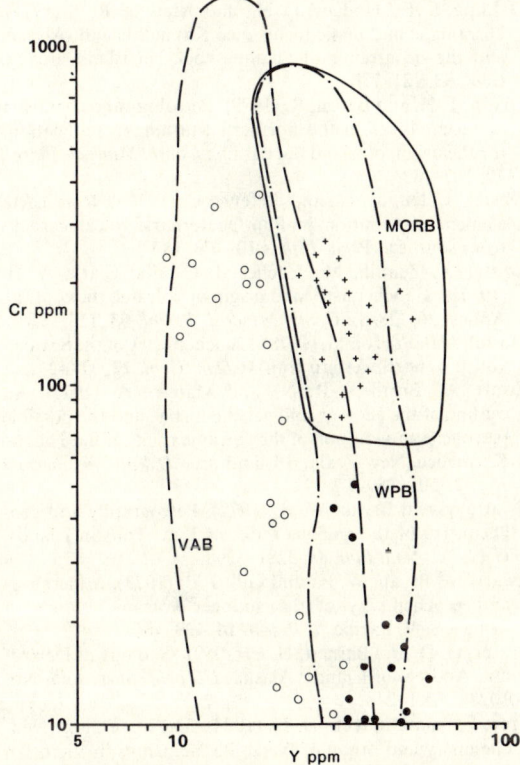

Fig. 15 Plot of lavas from Oman on the Y–Cr discrimination diagram (from Pearce *et al.*, 1981)

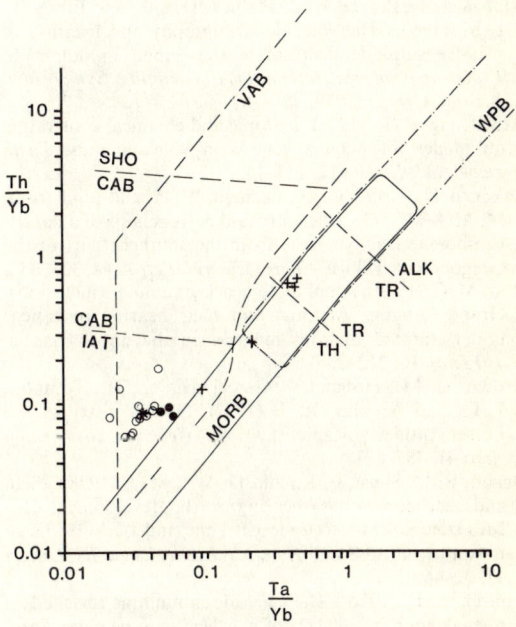

Fig. 16 Plot of lavas from Oman on the Th/Yb–Ta/Yb discrimination diagram (from Pearce *et al.*, 1981) (see Fig. 15 for key)

on the first diagram and to higher Th/Yb values on the second diagram, both diagnostic features of the island arc tholeiite magma type. Interestingly the *Geotimes* lavas occupy a somewhat ambiguous position on both diagrams, but very different from lavas which are thought to belong to 'normal' Tethyan oceanic crust (the 'amphibolites' in Fig. 15). This was cited as evidence that the oceanic crust upon which the arc was built was of back-arc character, containing a small subduction zone component. Lavas of this type have been detected, albeit rarely, in more recent basins (Saunders and Tarney, 1979), although it is possible that they would be more common if it were possible to sample the oceanic crust directly underlying island arc systems.

This example illustrates, therefore, that the geochemical features that characterize recent island arcs can be used to identify past volcanic arcs and thus aid in geological reconstructions. In the case of Oman, the evidence for an Upper Cretaceous arc–basin system has important implications for the reconstruction of the Tethyan ocean.

Acknowledgements

I am grateful to the editor for his helpful comments on the manuscript, to Chris Hawkesworth for petrogenetic discussions, to Dave Wright for help in compiling the discrimination diagrams (to be published in full elsewhere), to Tony Alabaster for his work on the Oman Ophiolite, and to Jenny Hill and John Taylor for drawing the diagrams.

REFERENCES

References containing data from volcanic arcs used in compilation of the discriminant diagrams are marked with an asterisk. Note that only some of these are referred to in the text itself.

Alabaster, T. and Pearce, J. A. (1979). Volcanic stratigraphy of the Oman ophiolite. *EOS* **60**, 62.
Alabaster, T., Pearce, J. A., Mallick, D. I. J., and Elboushi, I. M. (1980). The volcanic stratigraphy and location of massive sulphide deposits in the Oman ophiolite. In *Proceedings of the International Ophiolite Symposium, Nicosia, Cyprus*, 1979, 751–757.
Armstrong, R. L. (1971). Isotopic and chemical constraints on models of magma genesis in volcanic arcs. *Earth planet. Sci. Letters* **12**, 137–142.
Baker, B. H., Goles, G. G., Leeman, W. P., and Lindstrom, M. M. (1977). Geochemistry and petrogenesis of a basalt-benmoreite trachyte suite from the southern part of the Gregory Rift, Kenya. *Contr. Mineral. Petrol.* **64**, 303–332.
Best, M. G. (1975). Amphibole-bearing cumulate inclusions, Grand Canyon, Arizona, and their bearing on silica-undersaturated hydrous magmas in the upper mantle. *J. Petrol.* **16**, 212–236.
*Brown, G. M., Holland, J. G., Sigurdsson, H., Tomblin, J. F., and Arculus, R. J. (1977). Geochemistry of the Lesser Antilles volcanic island arc. *Geochim. cosmochim. Acta* **41**, 785–801.
Bruhn, R. L., Stern, C. R., and De Wit, M. J. (1978). Field and geochemical data bearing on the development of a Mesozoic volcano-tectonic rift zone and back-arc basin in southernmost South America. *Earth planet. Sci. Letters* **41**, 32–46.
Church, S. E. (1976). The Cascade mountains revisited: a re-evaluation in the light of new lead isotope data. *Earth planet. Sci. Letters* **29**, 175–188.
Coleman, R. (1975). *Geotimes* **20** (8), front cover.
*Condie, K. C. and Swenson, D. H. (1973). Compositional variation in three Cascade stratovolcanoes: Jefferson, Rainier and Shasta. *Bull. Volc.* **32**, 205–230.
DeLong, S. E., Hodges, F. N., and Arculus, R. J. (1975). Ultramafic and mafic inclusions, Kanaga Island, Alaska, and the occurrence of alkaline rocks in island arcs. *J. Geol.* **83**, 721–736.
*Dixon, T. H. and Batiza, R. (1979). Petrology and chemistry of recent lavas in the northern Marianas: implications for the origin of island arc basalts. *Contr. Mineral. Petrol.* **70**, 167–182.
*Dostal, J., Dupuy, C., and Lefevre, C. (1977a). Rare earth element distribution in Plio-Quaternary volcanic rocks from southern Peru. *Lithos* **10**, 173–183.
*Dostal, J., Zentilli, M., Caelles, J. C., and Clark, A. H. (1977b). Geochemistry and origin of volcanic rocks of the Andes (26°–28°S). *Contr. Mineral. Petrol.* **63**, 113–128.
*Dostal, J. and Zerbi, M. (1978). Geochemistry of the Savalan volcano (northwestern Iran). *Chem. Geol.* **22**, 31–42.
*Ewart, A., Brothers, R. N., and Mateen, A. (1977). An outline of the geology and geochemistry, and the possible petrogenetic evolution of the volcanic rocks of the Tonga–Kermadec–New Zealand island arc. *J. Volc. geothermal Res.* **2**, 205–250.
*Ewart, A. and Bryan, W. B. (1972). Petrography and geochemistry of the igneous rocks of Eua, Tongan Islands. *Geol. Soc. Am. Bull.* **83**, 3281–3298.
*Ewart, A., Bryan, W. B., and Gill, J. B. (1973). Mineralogy and geochemistry of the younger volcanic islands of Tonga, S.W. Pacific. *J. Petrol.* **14**, 429–465.
Fraser, G. D. and Barnett, H. F. (1959). Geology of Delarof and Andreanof Islands, Alaska. *US geol. Surv. Bull.* No. 1028-I, 211–248.
Frey, F. A. and Green, D. H. (1974). The mineralogy, geochemistry and origin of lherzolite inclusions in Victorian basanites. *Geochim. cosmochim. Acta* **38**, 1023–1059.
Garcia, M. O. (1978). Criteria for the identification of ancient volcanic arcs. *Earth Sci. Rev.* **14**, 147–165.

*Gill, J. B. (1970). Geochemistry of Viti Levu, Fiji, and its evolution as an island arc. *Contr. Mineral. Petrol.* **27**, 179–203.

Gill, J. B. (1974). Role of underthrust oceanic crust in the genesis of a Fijian calc-alkaline suite. *Contr. Mineral. Petrol.* **43**, 29–45.

*Gill, J. B. and Gorton, M. (1973). A proposed geological and geochemical history of Eastern Melanesia. In *The Western Pacific: Island Arcs, Marginal Seas, Geochemistry* (P. J. Coleman, ed.), University of Western Australia Press, Perth, pp. 543–566.

Glennie, K. W., Boeuf, M. G. A., Hughes-Clark, M. W. H., Moody-Stuart, M., Polaar, W. F. H., and Reinhardt, B. M. (1974). *Kon. Nederlands Geol. Mijn. Genoot. Verh.* **31**, 1–423.

*Gorton, M. P. (1977). The geochemistry and origin of Quaternary volcanism in the New Hebrides. *Geochim. cosmochim. Acta.* **41**, 1257–1270.

Green, D. H. (1973). Experimental melting studies on a model upper mantle composition at high pressure under water-saturated and water-undersaturated conditions. *Earth planet. Sci. Letters* **19**, 37–53.

Hawkesworth, C. J., O'Nions, R. K., Pankhurst, R. J., Hamilton, P. J., and Evensen, N. M. (1977). A geochemical study of island-arc and back-arc tholeiites from the Scotia Sea. *Earth planet. Sci. Letters* **36**, 253–262.

*Heming, R. F. (1974). Geology and petrology of the Rabaul caldera, Papua New Guinea. *Geol. Soc. Am. Bull.* **85**, 1253–1264.

*Jakeš, P. and Gill, J. (1970). Rare earth elements and the island arc tholeiite series. *Earth planet. Sci. Letters* **9**, 17–28.

*Jakeš, P. and White, A. J. R. (1972). Major and trace element abundances in volcanic rocks of orogenic areas. *Geol. Soc. Am. Bull.* **83**, 29–40.

*Katsui, Y., Oba, Y., Ando, S., Nishimura, S., Masuda, Y., Kurasawa, H. and Fujimaki, H. (1978). Petrochemistry of the Quaternary volcanic rocks of Hokkaido, North Japan. *J. Fs. Hokkaido V.* **18**, 449–484.

*Kay, R. W. (1977). Geochemical constraints on the origin of Aleutian magmas. In *Island Arcs, Deep-sea Trenches and Back-arc Basins* (M. Talwani and W. C. Pitman, eds), American Geophysical Union, Washington, D.C., pp. 229–242.

*Keller, J. (1974). Petrology of some volcanic rock series of the Aeolian arc, Southern Tyrrhenian Sea: calc-alkaline and shoshonitic associations. *Contr. Mineral. Petrol.* **46**, 29–47.

*Lopez-Escobar, L., Frey, F. A., and Vergara, M. (1976). Andesites from Central South Chile: trace element abundances and petrogenesis. In *IAVCEI International Symposium on Volcanology, Santiago, 1974*, (O. González-Ferran, ed.), pp. 725–761.

*Lopez-Escobar, L., Frey, F. A., and Vergara, M. (1977). Andesites and high alumina basalts from the central-south Chile High Andes: geochemical evidence bearing on their petrogenesis. *Contr. Mineral. Petrol.* **63**, 199–228.

Loughnan, F. C. (1969). *Chemical Weathering of the Silicate Minerals*, American Elsevier, New York.

*Mackenzie, D. E. and Chappell, B. W. (1972). Shoshonitic and calc-alkaline lavas from the highlands of Papua, New Guinea. *Contr. Mineral. Petrol.* **35**, 50–62.

O'Nions, R. K., Hamilton, P. J., and Evensen, N. M. (1977). Variations in $^{143}Nd/^{144}Nd$ and $^{87}Sr/^{86}Sr$ ratios in oceanic basalts. *Earth planet. Sci. Letters* **34**, 13–22.

Pearce, J. A. (1975). Basalt geochemistry used to investigate past tectonic settings on Cyprus. *Tectonophysics* **25**, 41–67.

Pearce, J. A. and Cann, J. R. (1973). Tectonic setting of basic volcanic rocks determined using trace element analyses. *Earth planet. Sci. Letters* **19**, 290–300.

Pearce, J. A. and Norry, M. J. (1979). Petrogenetic implications of Ti, Zr, Y, and Nb variations in volcanic rocks. *Contr. Mineral. Petrol.* **69**, 33–47.

Pearce, J. A., Alabaster, T., Shelton, A. W., and Searle, M. P. (1981). The Oman ophiolite as a Cretaceous arc-basin complex: evidence and implications. *Phil. Trans. R. Soc. A* **300**, 299–317.

*Peccerillo, A. and Taylor, S. R. (1976). Geochemistry of Eocene calc-alkaline volcanic rocks from the Kastamonu area, Northern Turkey. *Contr. Mineral. Petrol.* **58**, 63–81.

Penrose field conference participants (1972). Ophiolites. *Geotimes* **17**, 24–25.

*Philpotts, J. A., Martin, W., and Schnetzler, C. C. (1971). Geochemical aspects of some Japanese lavas. *Earth planet. Sci. Letters* **12**, 89–96.

Reinhardt, B. M. (1969). On the genesis and emplacement of ophiolites in the Oman mountains geosyncline. *Schweiz. Mineral. Petrog. Mitt.* **49**, 1–30.

*Richter, P. and Negendank, J. (1975). Spurenelementersuchungen an Vulkaniten des Tales von Mexiko. *Münster Forsch. Geol. Paläont.* **38/39**, 179–200.

Saunders, A. D. and Tarney, J. (1979). The geochemistry of basalts from a back-arc spreading centre in the East Scotia Sea. *Geochim. cosmochim. Acta* **43**, 555–572.

Saunders, A. D., Tarney, J., Stern, C. R., and Dalziel, I. W. D. (1979). Geochemistry of Mesozoic marginal basin floor igneous rocks from southern Chile. *Geol. Soc. Am. Bull.* **90**, 237–258.

Saunders, A. D., Tarney, J., and Weaver, S. D. (1980). Transverse geochemical variations across the Antarctic peninsula: implications for the genesis of calc-alkaline magmas. *Earth planet. Sci. Letters* **46**, 344–360.

Sharaskin, A. Ya., Dobretsov, N. L., and Sobolev, N. V. (1980). Geochemistry and timing of the marginal basin and arc-magmatism of the Philippine Sea. *Phil. Trans. R. Soc. Lond. A* **300**, 287–297.

*Shimizu, N. and Arculus, R. J. (1975). Rare earth concentrations in a suite of basanitoids and alkalie olivine basalts from Grenada, Lesser Antilles. *Contr. Mineral. Petrol.* **50**, 231–240.

*Sigurdsson, H., Tomblin, J. F., Brown, G. M., Holland, J. G., and Arculus, R. J. (1973). Strongly undersaturated magmas in the Lesser Antilles island arc. *Earth planet. Sci. Letters* **18**, 285–295.

Smewing, J. D., Simonian, K. O., Elboushi, I. M., and Gass I. G. (1977). Mineralized fault zone parallel to the Oman ophiolite spreading axis. *Geology* **5**, 534–538.

Stanton, R. L. and Bell, J. D. (1969). Volcanic and associated rocks of the New Georgia groups, British Soloman Islands Protectorate. *Overseas Geol. Mineral Resources* **10**, 113–145.

*Stern, C., Skewes, M. A., and Duran, M. (1976). Volcanismo orogenico en Chile Austral. In *Primer Congreso Geologico Chileno*, pp. 195–212.

Sun, S.-S., Nesbitt, R. W., and Sharaskin, A. Ya. (1979). Chemical characteristics of mid-ocean ridge basalts. *Earth planet. Sci. Letters* **44**, 119–138.

Tarney, J., Wood, D. A., Saunders, A. D., Cann, J. R., and Varet, J. (1979). Nature of mantle heterogeneity in the North Atlantic: evidence from deep-sea drilling. *Phil. Trans. R. Soc. A* **297**, 179–202.

*Taylor, S. R., Capp, A. C., Graham, A. L., and Blake, D. H. (1969). Trace element abundances in Saipan, Bougainville and Fiji. *Contr. Mineral. Petrol.* **23**, 1–26.

*Taylor, S. R. and White, A. J. R. (1966). Trace element abundances in andesites. *Bull. Volc.* **29**, 177–194.

Thorpe, R. S. (1977). Tectonic significance of alkaline volcanism in eastern Mexico. *Tectonophysics* **40**, 19–26.

*Thorpe, R. S., Potts, P. J., and Francis, P. W. (1967). Rare earth data and petrogenesis of andesite from the North Chilean Andes. *Contr. Mineral. Petrol.* **54**, 65–78.

Treuil, M. and Varet, J. (1973). Critères volcanologiques, pétrologiques et géochimiques de la genése et de la differentiation des magmas basaltiques. Exemple de l'Afar. *Bull. Soc. géol. fr.* **15**, 506–540.

Warden, A. J. (1970). Evolution of Aoba caldera volcano, New Hebrides. *Bull. Volc.* **34**, 107–140.

*Weaver, S. D., Saunders, A. D., Pankhurst, R. J., and Tarney, J. (1979). A geochemical study of magmatism associated with the initial stages of back-arc spreading. *Contr. Mineral. Petrol.* **68**, 151–169.

White, W. M. and Schilling, J.-G. (1978). Nature and origin of geochemical variation in Mid-Atlantic Ridge basalts from Central North-Atlantic, *Geochim. cosmochim. Acta* **42**, 1501–1516.

*Whitford, D. J. and Bloomfield, K. (1975). Geochemistry of Late Cenozoic volcanic rocks from the Nevado de Toluca area, Mexico. *Carnegie Inst. Wash. Yb.* **76**, 207–213.

Wood, D. A. (1979). A variably veined suboceanic upper mantle-genetic significance for mid-ocean ridge basalts from geochemical evidence. *Geology* **7**, 499–503.

Wood, D. A., Joron, J.-L., and Treuil, M. (1979a). A reappraisal of the use of trace elements to classify and discriminate between magma series erupted in different tectonic settings. *Earth planet. Sci. Letters* **45**, 326–336.

Wood, D. A., Tarney, J., Varet, J., Saunders, A. D., Bougault, H., Joron, J.-L., Treuil, M. and Cann, J. R. (1979b). Geochemistry of basalts drilled in the north Atlantic by IPOD leg 49. Implications for mantle heterogeneity. *Earth. planet. Sci. Letters* **42**, 77–79.

Andesites
Edited by R. S. Thorpe
© 1982 John Wiley & Sons

Isotope characteristics of magmas erupted along destructive plate margins

C. J. Hawkesworth

Department of Earth Sciences,
The Open University, Milton Keynes MK7 6AA, UK

ABSTRACT

Combined isotope and trace element data demonstrate that compared with magmas erupted at mid-ocean ridges and in intraplate environments, those generated along destructive plate margins are often *relatively* enriched in ^{87}Sr, ^{207}Pb, and more mobile elements such as K, Sr, Rb, and Pb. These features are attributed to the release of material from the subducted oceanic crust—although the similarity of δ^{18}O values in island arc rocks and uncontaminated basalts suggests that this contribution must be less than 5–10%. The behaviour of the REE is more controversial. Island arc tholeiites tend to be depleted in light REE, suggesting that significant quantities were not introduced from the subducted slab, whereas many calc-alkaline rocks are relatively enriched in both P and light REE. Moreover, the latter appears to be accompanied by relatively low ^{143}Nd/^{144}Nd and radiogenic Pb isotope ratios—both indicative of a sedimentary origin. It is suggested that such differences between the postulated ocean crust components in island arc tholeiites and particularly 'enriched' calc-alkaline rocks may be due to the mobilization of subducted material by dehydration beneath the former, and by partial melting (due perhaps to the presence of a larger sedimentary component) beneath the latter.

Finally, the role of continental crust in andesite magma genesis is briefly reviewed. It is argued that crustal contamination should be accompanied by crystallization, and that particularly for Sr, the concentration of which will be reduced in the liquid by plagioclase fractionation, the effects of contamination should be most marked in the more evolved rock types. Such arguments suggest that parental magmas to some volcanic suites in the central Andes have ^{87}Sr/^{86}Sr ratios in the range 0.706–0.707. Moreover, since these are accompanied by consistent differences in other isotope and trace element ratios, it may be that in areas of unusual crustal thickness high Si–O$_2$ andesites and dacites derived by crustal anatexis.

Introduction

The evolution of the Earth's crust and mantle has fascinated geoscientists for many years, and much has been learnt from the study of the varied isotope composition of elements such as Sr, Nd, and Pb in rocks of different ages. At the present time crustal material appears to be both created and destroyed along destructive plate margins and the net effect of such processes determines recent trends in the evolution of the crust and hence the upper mantle. This review attempts to assess the available Sr, Nd, Pb, and O isotope results on destructive plate margin magmas and to investigate whether the origin of such magmas may be usefully described in terms of a single general model. In pursuit of this aim emphasis has been placed on the more detailed case studies available in the literature and the author is conscious of having omitted many published isotope analyses on andesites and related rock types.

Along a destructive plate margin magmas may include material from both the subducted ocean crust (fresh and altered basalt and continental and oceanic sediment) and the overriding mantle wedge, with the added complication that, particularly in continental areas, the magmas may also suffer contamination *en route* to the surface. Potentially, the study of different isotopes can distinguish which components are present in any magmatic suite, but as more and more data become available two points are increasingly apparent. First, it is most important not to consider any one system in isolation: $^{87}Sr/^{86}Sr$ ratios are arguably more ambiguous than was once thought, whereas others such as Pb tend to be ultrasensitive to a particular component (i.e. subducted sediments). Second, while isotopes may indicate which components are present in a magma, they cannot resolve how they have been introduced. For example, material could be released from the subducted ocean crust either by dehydration or by melting. However, since the relative behaviour of trace elements with different chemical affinities will vary depending on whether the fluid medium is a silicate melt or an aqueous fluid, it may be possible to distinguish such processes by considering the trace element geochemistry of the resultant magmas.

Most of the published isotope work on recent magmatic rocks involves Sr isotopes, but much progress has been made over the past 3–4 years by the application of combined Nd and Sr isotope results. These two systems will therefore be considered together before discussing Pb and O isotopes.

Nd and Sr Isotopes

Mid-ocean ridge and intraplate basalts

Magmas erupted along mid-ocean ridges and in intraplate environments tend to exhibit significant differences in both their isotope and trace element characteristics. Mid-ocean ridge basalts are typically depleted in large ion lithophile elements and have lower $^{87}Sr/^{86}Sr$ ratios than the majority of intraplate magmas. Faure (1977) calculated mean $^{87}Sr/^{86}Sr$ ratios of 0.7028 and 0.7039 respectively for the available results on mid-ocean ridge and ocean island magmatic rocks.

It is now known that such variations in Sr isotopes are usually accompanied by complementary changes in the isotope composition of Nd. Thus, on a diagram of $^{143}Nd/^{144}Nd$ against $^{87}Sr/^{86}Sr$, the majority of recent mantle-derived rocks from mid-ocean ridges and both continental and oceanic intraplate environments plot on a single broad negative trend (Group I, see Fig. 1).

De Paolo and Wasserburg (1976a, 1979) and O'Nions *et al.* (1977) used this broad trend to estimate the present-day $^{87}Sr/^{86}Sr$ ratio of the model bulk Earth. If, as is generally agreed, the Earth has chondritic Sm/Nd and initial $^{143}Nd/^{144}Nd$ ratios, then its present-day $^{143}Nd/^{144}Nd$ ratio is 0.51262 (normalized to $^{146}Nd/^{144}Nd$ = 0.7219). Assuming that Rb/Sr and Sm/Nd have behaved coherently throughout the Earth's history and that the present-day isotope composition of the bulk Earth therefore lies on the main correlation between $^{143}Nd/^{144}Nd$ and $^{87}Sr/^{86}Sr$, its present-day $^{87}Sr/^{86}Sr$ ratio is 0.7045–0.7050 (Fig. 1). Furthermore, most recent mantle-derived volcanic rocks have *higher* $^{143}Nd/^{144}Nd$ and *lower* $^{87}Sr/^{86}Sr$ ratios than the present-day composition of the bulk Earth, which implies that their source regions in the upper mantle have been relatively depleted in Rb and Nd for much of their history. However, many volcanic rocks from both continental and oceanic areas are characterized by high concentrations of large ion lithophile elements and they have clearly not been derived from a large ion lithophile element-depleted source. Thus we may conclude that the enrichment of large ion lithophile elements in their source regions took place comparatively recently, so that insufficient time has elapsed to affect their isotope ratios significantly (Hawkesworth *et al.*, 1979b).

On Fig. 1 samples from Sao Miguel in the Azores, and Roccamonfina and Vesuvius in central Italy, plot to the right of the main correlation. Hawkesworth *et al.* (1979b) empha-

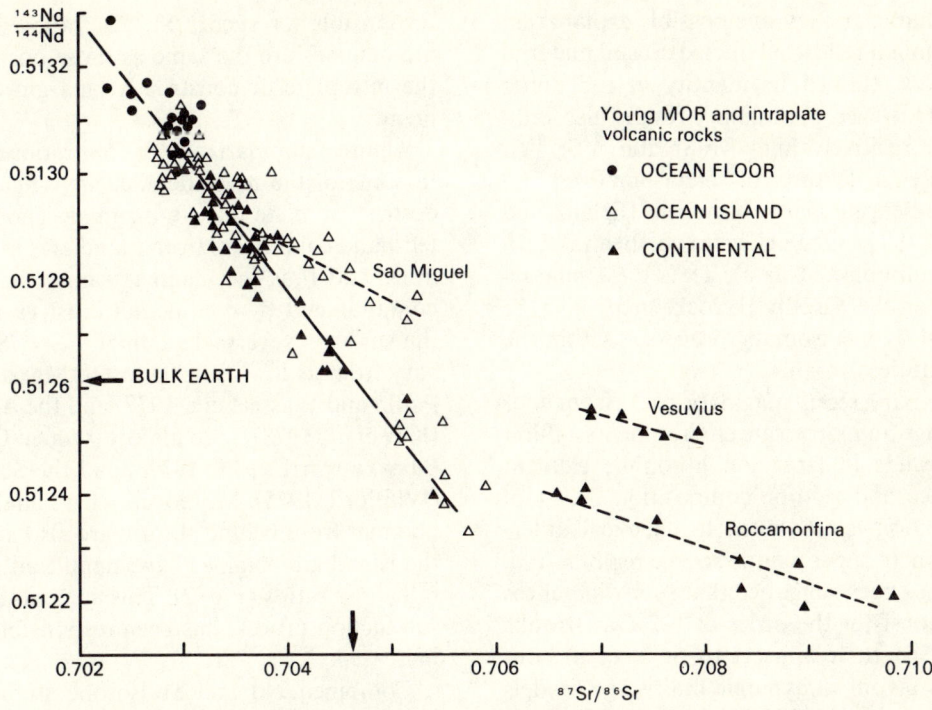

Fig. 1 Initial ^{143}Nd/^{144}Nd and ^{87}Sr/^{86}Sr ratios of young volcanic rocks from mid-ocean ridge and intraplate environments. All the Nd data are normalized to ^{146}Nd/^{144}Nd = 0.7219 and are relative to a BCR–1 value of ^{143}Nd/^{144}Nd = 0.51262. Coarse dashed line = main correlation for most mantle-derived rocks (Group I): fine dashed lines = Group II rock types (after Hawkesworth et al., 1979b). Data from Carter et al. (1978a, 1979), De Paolo and Wasserburg (1976a,b), Dosso and Murthy (1980), Hawkesworth and Vollmer (1979), Hawkesworth et al. (1979a,b), Norry et al. (1980), O'Nions et al. (1977), Richard et al. (1976), White (1979), and White and Hofmann (1978)

sized this point by suggesting that most of the available Nd and Sr isotope results could be usefully considered in two groups. Group I is the main trend of most mantle-derived magmatic rocks and is characterized by a relatively steep slope on the diagram of ^{143}Nd/^{144}Nd against ^{87}Sr/^{86}Sr. The source of large ion lithophile element-enriched magmas in Group I must have been only recently enriched in large ion lithophile elements and those elements appear to have migrated from source regions which had Nd and Sr isotopes that also plotted on the trend of the Group I samples. Moreover, this type of enrichment results in high concentrations of incompatible elements without significantly increasing their Rb/Sr ratios, which in primitive rocks tend to be less than 0.035.

Group II rocks, by contrast, have relatively high Rb/Sr and ^{87}Sr/^{86}Sr ratios and hence plot on more flat-lying trends to the right of the main correlation between Nd and Sr isotopes in Fig. 1. They have only been recognized in two areas so far (Azores and central Italy, Fig. 1) and they clearly represent a minute fraction of the mantle sampled by recent magmatism. However, their interest lies in the fact that they involve a different component to those in Group I. Hawkesworth and Vollmer (1979) and Hawkesworth et al. (1979b) have argued that large ion lithophile element enrichment also took place comparatively recently, but that in this case the elements were derived from material characterized by relatively high Rb/Sr and ^{87}Sr/^{86}Sr ratios. Since these are often taken to be crustal characteristics, one possible explanation is that it might reflect subducted crustal material which had retained its identity in the upper mantle. However, preliminary results also indicate that relatively high Rb/Sr and $^{87}Sr/^{86}Sr$ ratios are a feature of metasomatized K-richterite-bearing mantle xenoliths (Erlank and Shimizu, 1977; C. J. Hawkesworth and A. J. Erlank, unpublished data). Thus these characteristics of the Group II rocks may also be generated by metasomatic processes within the upper mantle.

In summary, recent magmatic rocks from mid-ocean ridge and intraplate environments exhibit a wide range in large ion lithophile element abundance and isotope composition. Most of this variation is attributed to chemical differences in their upper mantle source regions, and, while it has been suggested that such differences might persist for the order of 1–2 Ga (Brooks et al., 1976), there appears to be a recent consensus in favour of dynamic multistage models (e.g. O'Nions et al., 1979; De Paolo and Wasserburg, 1979; Hawkesworth et al., 1979b) in which the present-day chemical heterogeneities are believed to reflect mixing within the upper mantle through geological time. The form of these variations is important in the study of destructive margin magmas in that it provides an indication of the chemical variations which may exist both in the subducted ocean crust and in the overriding mantle wedge. Thus, on Fig. 1 we may predict that before subduction commences the $^{143}Nd/^{144}Nd$ and $^{87}Sr/^{86}Sr$ ratios of what is to become the overriding mantle wedge are most likely to plot on the trend of the Group I samples. In a very small number of cases it might plot on a flat-lying trend similar to the Group II rocks.

Island arcs

The Sr isotope composition of most island arc magmas tends to be more radiogenic than that of mid-ocean ridge basalts, but similar to that of many ocean island magmas (e.g. Faure, 1977). However, for such comparisons to be meaningful we clearly need to ask whether the processes responsible for such $^{87}Sr/^{86}Sr$ ratios in island arc magmas are the same as those operating in the intraplate or constructive margin environment.

Figure 2 summarizes $^{87}Sr/^{86}Sr$ ratios reported in some of the more detailed investigations of destructive plate margin magmas. Those in the left-hand column are from island arcs in oceanic areas and thus the magmas cannot have been contaminated by continental crust *en route* to the surface. Nevertheless, their $^{87}Sr/^{86}Sr$ ratios vary from as low as 0.703 in the Marianas (De Paolo and Wasserburg, 1977) and the Aleutians (Kay et al., 1978) up to almost 0.706 in Grenada (Hawkesworth et al., 1979c) and the Sunda arc (Whitford, 1975). Moreover, where analyses on magmas from behind the arc are also available, the island arc magmas have significantly *higher* $^{87}Sr/^{86}Sr$ ratios (Fig. 2). This suggests that the subduction process has been responsible for an increase in $^{87}Sr/^{86}Sr$.

Combined Nd and Sr isotope studies have now been reported on island arc volcanic rocks from the South Sandwich Islands (Hawkesworth et al., 1977), the Marianas (De Paolo and Wasserburg, 1977), New Britain (De Paolo and Johnson, 1979), and the Lesser Antilles (Hawkesworth et al., 1979c; Hawkesworth and Powell, 1980). The results are summarized on Fig. 3 and the majority are displaced to high $^{87}Sr/^{86}Sr$ ratios relative to the main trend of the Group I rocks (Fig. 1). The Marianas may be the most obvious exception, but since the available $^{87}Sr/^{86}Sr$ ratio on a back-arc basalt from that area is as low as 0.7026, further work is needed to assess whether the isotope composition of these arc basalts still reflects some displacement to higher $^{87}Sr/^{86}Sr$ ratios (*see* Fig. 2).

Continental destructive margins

The Sr isotope ratios of many volcanic rocks erupted along continental margins are similar to those in oceanic island arcs (Fig. 2). However, in some areas, such as Peru (James et al., 1976) and northern Chile (Francis et al., 1977), they are characterized by much higher $^{87}Sr/^{86}Sr$ ratios which might reflect contamination with

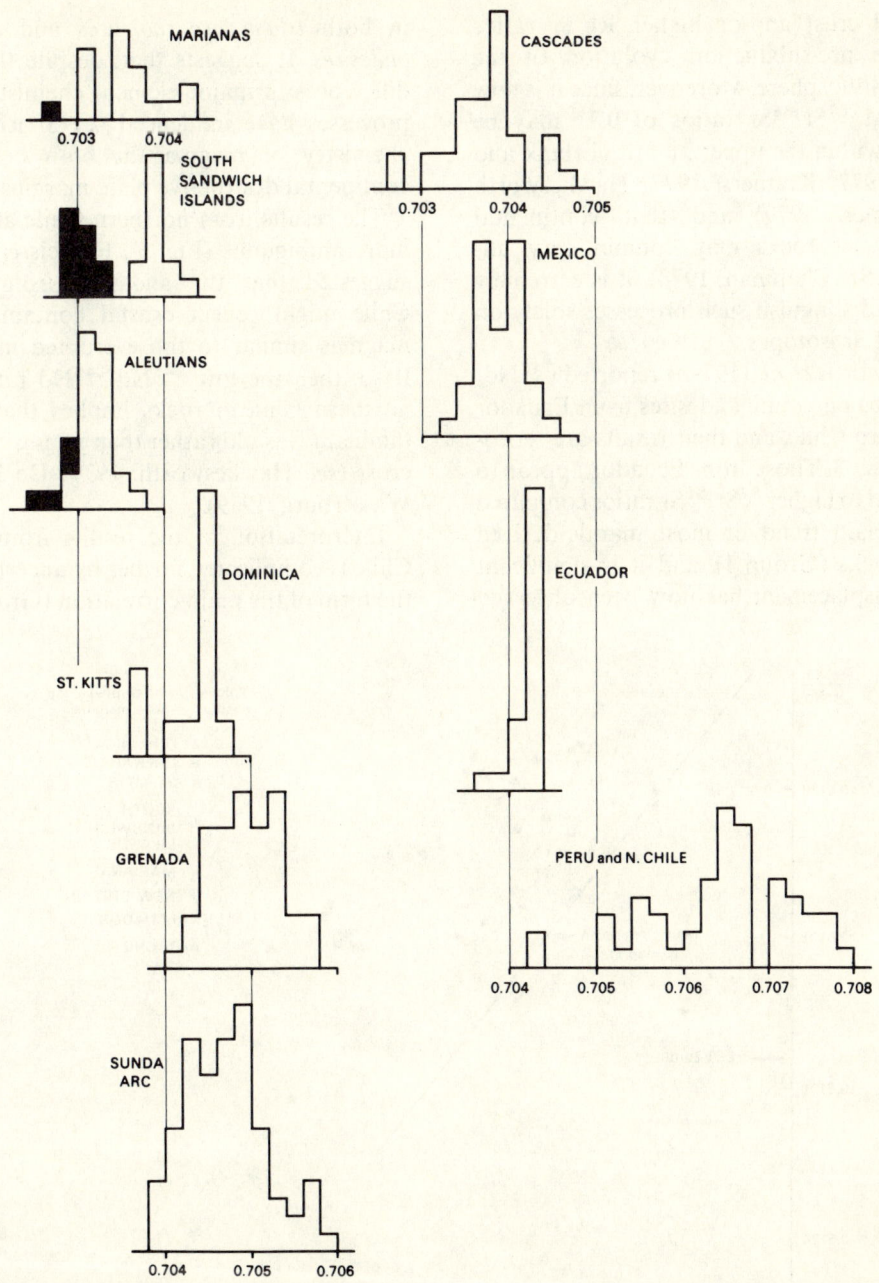

Fig. 2 $^{87}Sr/^{86}Sr$ ratios from volcanic rocks erupted along destructive plate margins: those in black are from basalts in the 'back-arc' environment. Data from Church and Tilton (1973), De Paolo and Wasserburg (1977), Francis *et al.* (1977), Hawkesworth *et al.* (1977, 1979*a,c*), Hawkesworth and Powell (1980), James *et al.* (1976), Kay *et al.* (1978), Meijer (1976), Moorbath *et al.* (1978), Noble *et al.* (1975), Whitford and Bloomfield (1975), and Whitford (1975)

continental crust and/or higher Rb/Sr ratios during the pre-subduction evolution of the overriding lithosphere. Moreover, since it is now known that $^{87}Sr/^{86}Sr$ ratios of 0.71 may be generated within the upper mantle (Erlank and Shimizu, 1977; Kramers, 1977; Hawkesworth and Vollmer, 1979) and that continental granulite-facies rocks may contain very unradiogenic Sr (Chapman, 1978), it is extremely difficult to distinguish such processes solely on the basis of Sr isotopes.

Hawkesworth et al. (1979a) reported $^{143}Nd/^{144}Nd$ ratios on young andesites from Ecuador and northern Chile, and their results are reproduced in Fig. 3. Those from Ecuador appear to be displaced to higher $^{87}Sr/^{86}Sr$ ratios compared with the main trend of most mantle-derived volcanic rocks (Group I), and it is significant that this displacement has now been observed in both *island arc tholeiites* and *continental andesites*. It suggests that, despite the marked differences in major element chemistry, similar processes have influenced the Sr isotope geochemistry of magmas at both oceanic and continental destructive plate margins.

The results from northern Chile are perhaps more ambiguous (Fig. 3). Francis et al. (1977) suggested that the andesites from northern Chile might reflect crustal contamination of magmas similar to those erupted in Ecuador. If so, then the low $^{143}Nd/^{144}Nd$ ratios of the northern Chilean rocks implies that the contaminant was old rather than young continental crust (*see* Hawkesworth, 1979; De Paolo and Wasserburg, 1979).

Interpretation of the results from northern Chile is complicated further by uncertainty over the form of the main correlation (Group I). The

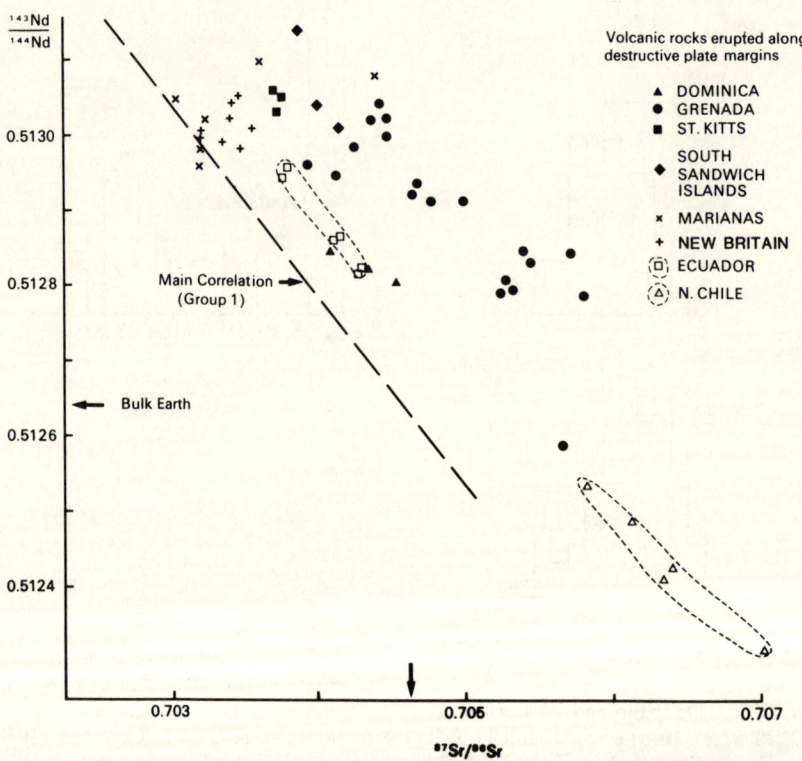

Fig. 3 Nd and Sr isotope ratios in volcanic rocks erupted along destructive plate margins compared with the main trend (Group I, see Fig. 1) of most mantle-derived volcanic rocks. Data from De Paolo and Wasserburg (1977), De Paolo and Johnson (1979), Hawkesworth et al. (1977, 1979a,c), Hawkesworth and Powell (1980)

data may be fitted to a straight line (as in Fig. 1) to a curve which tends to be flatter-lying at high $^{87}Sr/^{86}Sr$ ratios. Estimates of the amount of displacement to higher $^{87}Sr/^{86}Sr$ reflected in the results from northern Chile clearly depend on the model adopted for the Group I samples.

Subducted ocean crust

An increasing volume of Nd and Sr isotope results are now available on the various components which comprise the ocean crust. The majority of unaltered mid-ocean ridge basalts plot at the low $^{87}Sr/^{86}Sr$, high $^{143}Nd/^{144}Nd$ end of the field of Group I rocks on Fig. 1. Ocean island, or 'hot spot', magmas are volumetrically less significant and they tend to have higher $^{87}Sr/^{86}Sr$ and lower $^{143}Nd/^{144}Nd$ ratios.

The Sr isotope composition of present-day ocean water is 0.7092 (Peterman et al., 1970; Spooner, 1976; Hawkesworth and Elderfield, 1978) and it is well documented that hydrothermal alteration of ocean-floor rocks near mid-ocean ridges results in an increase in their $^{87}Sr/^{86}Sr$ ratios (Hart et al., 1974; Spooner et al., 1977). O'Nions et al. (1978) reported Nd isotope analyses on highly altered ocean-floor basalts and it appears that their $^{143}Nd/^{144}Nd$ ratios are not altered significantly by hydrothermal interaction with seawater. Thus basalt/seawater interaction causes marked fractionation between Nd and Sr isotopes with the altered rocks being displaced to high $^{87}Sr/^{86}Sr$ ratios compared with the main trend of the Group I samples (Fig. 4). This reflects both the greater mobility of Sr and the extremely small quantities of Nd that are present in seawater (Høgdahl et al., 1968). Moreover, since material which has been altered hydrothermally is likely to be more easily mobilized during subduction it will presumably exert most influence on the chemical composition of the destructive plate margin magmas.

Results on deep sea sediments and authigenic material such as Mn nodules indicate that the Nd composition of ocean water varies from ocean to ocean (Piepgras et al., 1979; O'Nions et al., 1978; Elderfield et al., 1981). In general, however, authigenic material has Nd and Sr isotope ratios of 0.5125–0.5120 and 0.709 respectively, while sediments with more continental detritus tend to have higher $^{87}Sr/^{86}Sr$ and lower $^{143}Nd/^{144}Nd$ ratios (McCulloch and Wasserburg, 1978).

In summary, the majority of Nd and Sr isotope results from volcanic rocks erupted along destructive plate margins are displaced to high $^{87}Sr/^{86}Sr$ ratios compared with the main trend of mid-ocean ridge and intraplate basalts. Since similar relatively high $^{87}Sr/^{86}Sr$ ratios are observed in both altered ocean-floor basalts and deep sea sediments (Fig. 4) it is argued that the destructive plate margin magmas contain a contribution from the subducted ocean crust. However, it must be emphasized that the isotope results cannot resolve whether that contribution is derived by melting or merely dehydration of the downgoing slab.

Pb Isotopes

For an introduction to the systematics of Pb isotopes, particularly in young volcanic rocks, the reader is referred to Doe (1970) or Faure (1977). Mid-ocean ridge basalts, ocean island basalts, and oceanic and continental sediments have broadly different Pb isotope compositions and this section reviews briefly how this has been exploited in the study of magmatic processes along destructive plate margins.

Figure 5 summarizes (after Sun, 1980) most of the available Pb data from mid-ocean ridge basalts and ocean island volcanics. They are characterized by a relatively flat-lying trend on graphs of $^{207}Pb/^{204}Pb$ against $^{206}Pb/^{204}Pb$ and the ocean island results tend to be more radiogenic than those from mid-ocean ridge basalts. Islands such as Gough, Kerguelen, Reunion, and Bouvet (Fig. 5) which plot above the main trend represent an extremely small fraction of the mantle sampled by recent magmatism. In addition, seawater contains very small amounts of Pb and thus hydrothermal alteration of the oceanic crust appears to have little effect on the Pb isotope ratios (Bass et al.,

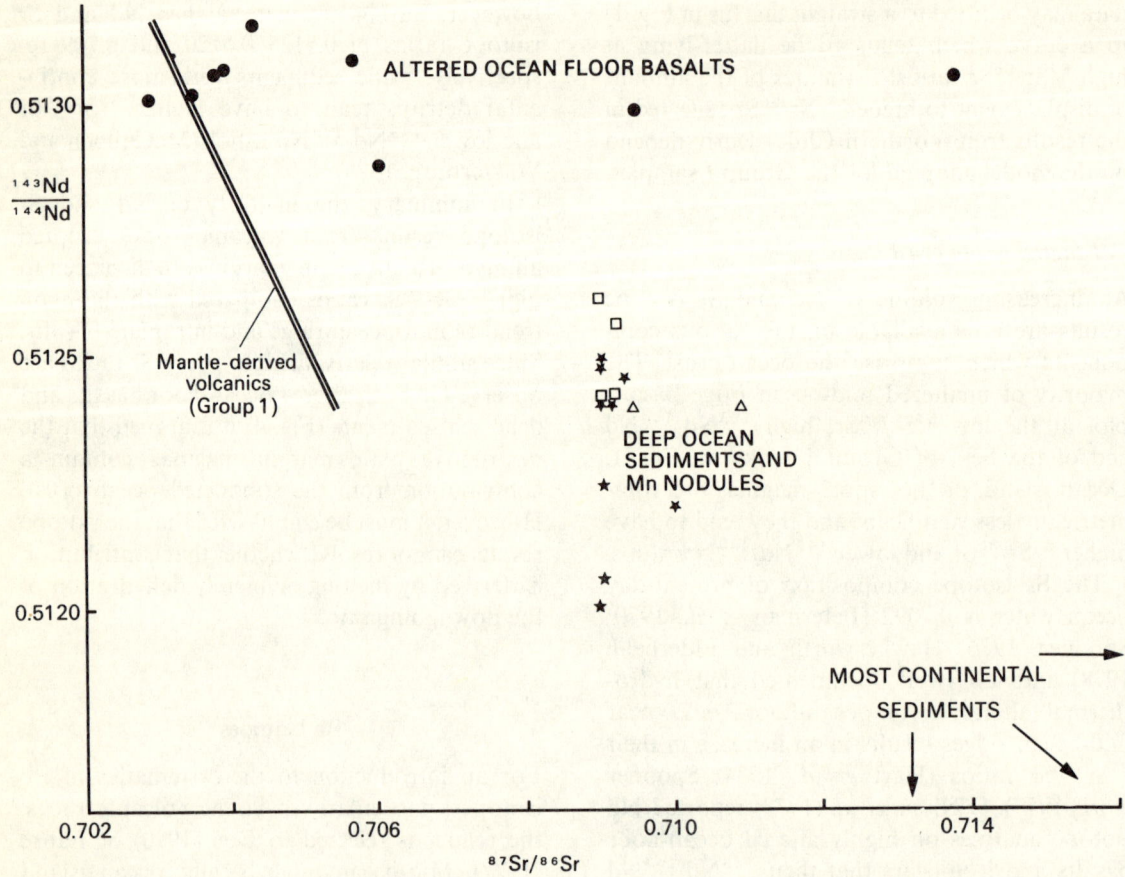

Fig. 4 ^{143}Nd/^{144}Nd and ^{87}Sr/^{86}Sr variations in likely components of a subducted oceanic plate: hydrothermally altered ocean-floor basalt (fresh mid-ocean ridge basalts plot on the low ^{87}Sr/^{86}Sr end of the Group I trend), deep ocean sediments and manganese nodules, and sediments of continental derivation. Key: open squares, metalliferous sediments; stars, Mn-nodules; open triangles, deep sea sediments. Data from O'Nions et al. (1978), McCulloch and Wasserburg (1978), Piepgras et al. (1979), Elderfield et al. (1981)

1973; Meijer, 1976; Church and Tatsumoto, 1975).

Deep sea sediments, Mn nodules, and clay separates have Pb isotope compositions close to the present-day composition of the Pb ore growth curve (Fig. 6). Sediments dominated by continental detritus tend to have higher ^{206}Pb/^{204}Pb ratios, but sediments from both environments have higher ^{207}Pb/^{204}Pb ratios than the main trend of oceanic volcanic rocks.

The majority of volcanic rocks from destructive plate margins have Pb compositions which are more radiogenic than mid-ocean range basalts and typically they form steeper slopes on graphs of ^{207}Pb/^{204}Pb against ^{206}Pb/^{204}Pb. This was demonstrated by early work on rocks from Japan (Kurasawa, 1968; Hedge and Knight, 1969; Tatsumoto and Knight, 1969) and the Lesser Antilles (Armstrong and Cooper; 1971; Donnelly et al., 1971) and has been re-emphasized by recent detailed studies on the Cascades (Church and Tilton, 1973; Church, 1973, 1976), the Aleutian Islands (Kay et al., 1978), and Taiwan (Sun, 1980) (see Fig. 6). The steep trend of these data and the fact that they lie between the fields for mid-ocean ridge basalts and recent sediments has encouraged most authors to interpret them in terms of mixing between

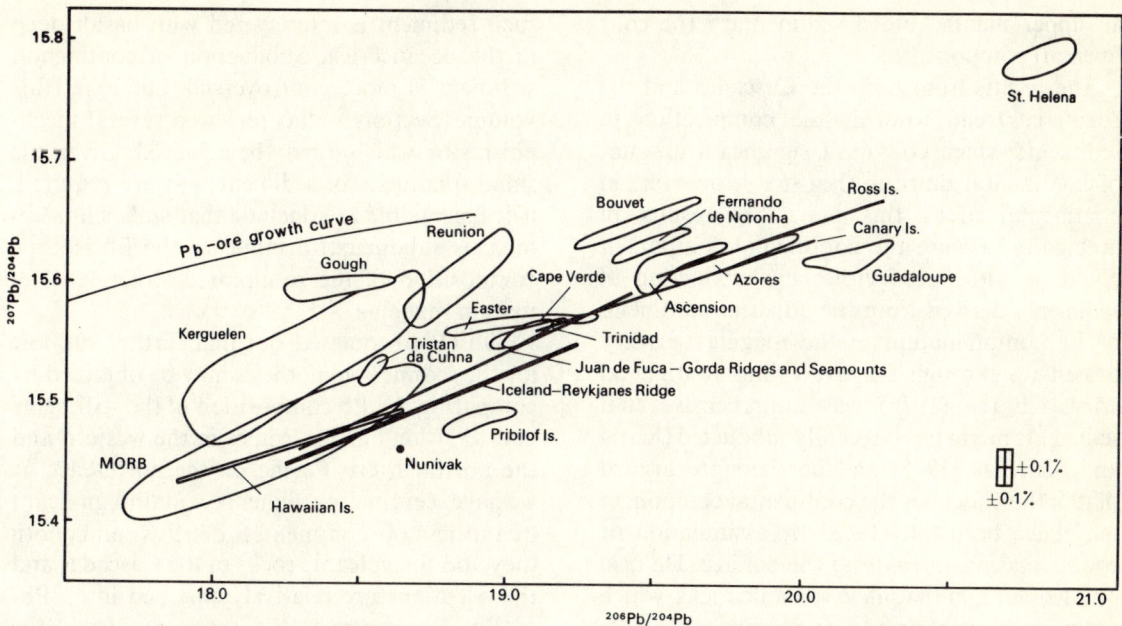

Fig. 5 Graph of $^{207}Pb/^{204}Pb$ against $^{206}Pb/^{204}Pb$ for volcanic rocks from mid-ocean ridges and oceanic islands (after Sun (1980) and references therein). Pb ore growth curve from Cumming and Richards (1975)

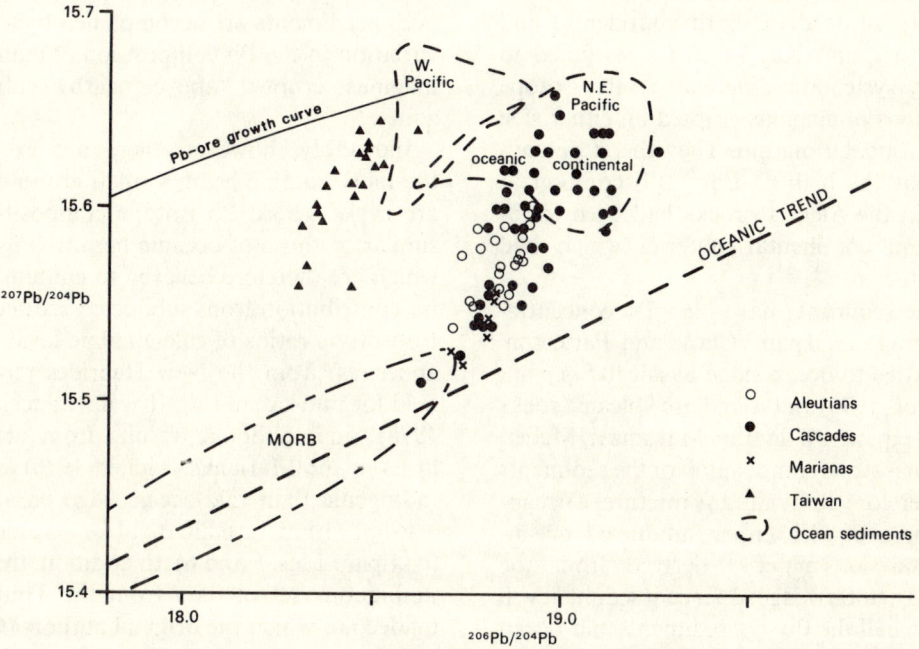

Fig. 6 Graph of $^{207}Pb/^{204}Pb$ against $^{206}Pb/^{204}Pb$ for selected destructive plate margin volcanic rocks compared with the main trend for oceanic volcanics (see Fig. 5) and data from Pacific sediments. Data from Church (1973, 1976), Church and Tilton (1973), Kay *et al.* (1978), Meijer (1976), and Sun (1980). Pb ore growth curve from Cumming and Richards (1975)

an upper mantle and a sedimentary (or continental) component.

The results from both the Cascades and the Aleutians trend towards the composition of sediments which contain a significant amount of continental detritus (Fig. 6). Moreover, in continental areas this inevitably results in ambiguity because the continental component could be introduced either by subduction of sediments derived from the adjacent continent, or by contamination of the magmas as they passed up through the overriding continental crust. Church (1976) was apprehensive that sediments might not be readily subducted (Karig and Sharman, 1975) and he therefore argued that in the Cascades the continental component may have been introduced by assimilation or contamination *en route* to the surface. He also pointed out that the silicic volcanic rocks, which on major element and trace element arguments were probably derived by crustal anatexis (Condie and Hayslip, 1975), have Pb isotope compositions which plot in the field for continental sediments (Fig. 6). In contrast, the Aleutian Islands are situated on both continental and oceanic crust, and Kay *et al.* (1978) failed to detect any systematic difference in the isotope compositions of magmas erupted on either side of the continental margin. They therefore concluded that the high $^{207}Pb/^{204}Pb$ component detected in the Aleutian rocks had been introduced from continental material which had been subducted.

Oceanic sediments have high Pb concentrations (average 25 p.p.m.; Chow and Patterson, 1962) relative to ocean ridge basalt (0.5 p.p.m.; in Kay *et al.*, 1978) and island arc volcanic rocks (e.g. 1–3 p.p.m. Pb in the Marianas; Meijer, 1976). The isotope composition of the sediments should therefore dominate any mixture of ocean-floor sediments with either subducted ocean-floor basalt or material derived from the overriding mantle wedge: 2 per cent sediment will contribute half the Pb in a sediment–mid-ocean ridge basalt mixture. Some oceanic sediment must be subducted because the results of the Deep Sea Drilling Projects have confirmed that such sediment is interlayered with basalt deep in the ocean crust. Subduction of continental sediment is more controversial, but Fyfe (this volume, Section X) has reviewed several mechanisms by which it may be achieved. Given the small quantities of sediment that are required, it is reasonable to conclude that sufficient sediment is subducted to influence the Pb isotope composition of the resultant destructive plate margin magmas.

Sun (1980) pointed out that further support for the sediment hypothesis may be obtained by comparing the Pb composition of the sediments and the island arc volcanics in the western and the north-eastern Pacific. In the north-east, as we have seen, many sediments contain significant quantities of continental detritus and both they and the volcanic rocks of the Cascades and the Aleutians are relatively enriched in $^{206}Pb/^{204}Pb$. By contrast, the sediments from the western Pacific tend to have lower $^{206}Pb/^{204}Pb$; as do the island arc volcanic rocks from Taiwan (Fig. 6), and also Japan. Thus it would appear that variations in the Pb isotope composition of ocean sediments are accompanied by a parallel variation in the Pb composition of many of the magmas erupted above nearby subduction zones.

Inevitably, however, there are exceptions; the most notable being a small group of island arc rocks whose Pb isotope compositions are similar to those of oceanic basalts (Fig. 5) and which are therefore believed to contain little or no contribution from subducted sediment. The Pb isotope ratios of calc-alkaline lavas (mostly andesites) from the New Hebrides plot in the field for mid-ocean ridge basalt (Lancelot *et al.*, 1978), and, although results from active arc lavas in the Marianas (Meijer, 1976) are more radiogenic than mid-ocean ridge basalts, they also fall within the main trend for oceanic basalts (compare Figs. 5 and 6). In addition, there is an ambiguous set of data from the Tonga–Kermadec arc which the original authors (Oversby and Ewart, 1972) interpreted in terms of no sediment involvement, but which Sun (1980) suggested might reflect mixing between mid-ocean ridge basalts and sediments which have been sampled locally in the Tonga trench and the New Zealand Mesozoic geosyncline.

In summary, Pb isotope results from most destructive plate margin volcanics plot on steeper trends on Pb isotope diagrams than do results from mid-ocean ridge and ocean island basalts. They are interpreted in terms of mixing between a sedimentary and an upper mantle component: the latter might be subducted ocean-floor basalt or material derived from the over-riding mantle wedge, while the contribution from the sediments is usually small (below 5 per cent) because of their relatively high Pb contents. Moreover, since most sediments contain a component of Pb derived from continental crust, it may not always be possible in continental areas such as the Cascades (Church, 1976) or the Andes (McNutt et al., 1979) to resolve whether the continental Pb has been introduced from subducted sediment or by crustal contamination en route to the surface. However, the parallel variation in the Pb isotope composition of sediments and island arc volcanic rocks on either side of the Pacific certainly suggests that some Pb is being recycled via the sediments in the subducted ocean crust. Finally, there are two island arc suites (the Marianas and the New Hebrides) the Pb isotope compositions of which suggest that little or no Pb has been introduced from a sedimentary source (Meijer, 1976; Lancelot et al., 1978).

O Isotopes

There is a wide variation in the isotope composition of O in fresh and altered igneous and sedimentary rocks and this has encouraged several scientists to use O isotope ratios in the study of island arc and continental margin volcanic rocks. However, it must be remembered that in sharp contrast to Sr, Nd, and Pb, O is the *most* abundant element in igneous rocks and its isotope composition is therefore much less sensitive to the addition of small amounts of other material: e.g. the c. 2 per cent of sediment suggested by the Pb isotope ratios of the Aleutian magmas (Kay et al., 1978).

Figure 7 summarizes the $\delta^{18}O$ and $^{87}Sr/^{86}Sr$ values of common rock types which might contribute to the composition of magmas generated above subduction zones. These isotope ratios are plotted on the same diagram because they reflect different, and usually unrelated, processes. ($\delta^{18}O$ values are fractionated by phase changes and water/rock interaction at low temperatures (e.g. Savin and Epstein, 1970), while $^{87}Sr/^{86}Sr$ depends on the ratio of Rb to Sr and the decay of ^{87}Rb over long periods of time.) Thus a correlation between $\delta^{18}O$ and $^{87}Sr/^{86}Sr$ may be interpreted as evidence for some comparatively recent mixing process (*see* Turi and Taylor, 1976).

Unaltered basalts from oceanic islands and mid-ocean ridges have $\delta^{18}O$ values generally within 6.0 ± 0.5 per mil (Taylor, 1968; Muelenbachs and Clayton, 1974), while most of their $^{87}Sr/^{86}Sr$ ratios are in the range 0.7025–0.7040 (Hofmann and Hart, 1978). In continental areas the unambiguous recognition of mantle isotope ratios is much more difficult, but it does appear that large ion lithophile element-enriched continental mantle is capable of supporting $^{87}Sr/^{86}Sr$ ratios up to 0.710 (e.g. Erlank and Shimizu, 1977), and perhaps even $\delta^{18}O$ values slightly higher than mid-ocean ridge basalt (*see* Hawkesworth and Vollmer, 1979, for discussion). Fractional crystallization should not alter $^{87}Sr/^{86}Sr$ and it has only a small effect on $\delta^{18}O$. Matsuhisa (1979) reported that within a fractionated sequence of lavas from Japan, $\delta^{18}O$ increased by 1 per mil from basalt to dacite (Group I in Fig. 8).

Seawater interaction with oceanic basalts raises their $\delta^{18}O$ values if the temperature is below 200–300°C and reduces them if the temperature is higher (Muehlenbachs and Clayton, 1976). Thus hydrothermally metamorphosed basic rocks from ophiolite complexes have $\delta^{18}O$ values of 9–13 per mil whereas the metadolerites from the deeper portion of the sheeted complexes have $\delta^{18}O$ values of 3.3–5.3 per mil (Spooner et al., 1974). The $^{87}Sr/^{86}Sr$ ratios of ocean basalts

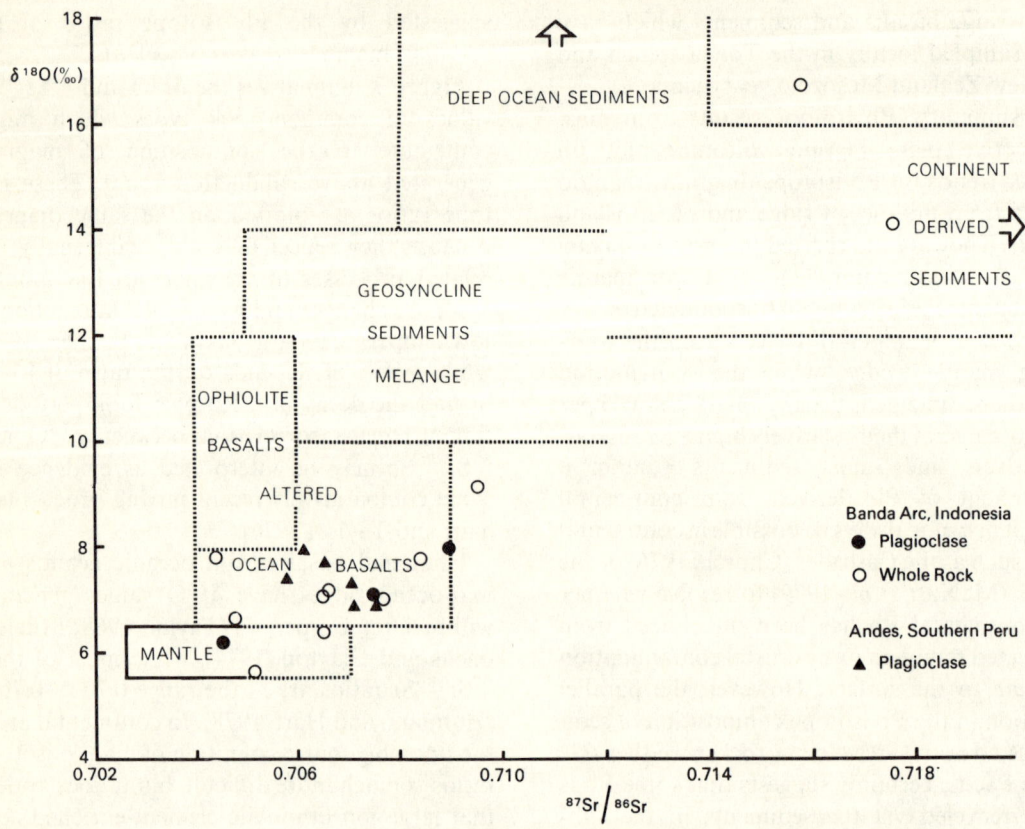

Fig. 7 Ranges of $\delta^{18}O$ relative to SMOW (standard mean ocean water) and $^{87}Sr/^{86}Sr$ in igneous and sedimentary rocks of possible importance in the genesis of andesites compared with results from southern Peru, and the Banda arc (after Magaritz *et al.* (1978) and references therein)

are increased by interaction with seawater (0.709, see Fig. 4) and altered basalts commonly have Sr isotope ratios of 0.704–0.706.

Sedimentary rocks exhibit a much broader range of Sr and O isotope compositions and they are typically enriched in both ^{18}O and ^{87}Sr relative to the mantle (Fig. 7). The $^{87}Sr/^{86}Sr$ ratios of deep ocean sediments tend to be close to or higher than that of seawater; while the Sr in continental sediments is usually much more radiogenic, reflecting the high Rb/Sr ratios of their crustal source regions. $\delta^{18}O$ values range between 16 and 20 per mil in argillaceous sediments and can be as high as 30 per mil in deep ocean sediments (Savin and Epstein, 1970; Knouth and Epstein, 1975; Magaritz and Taylor, 1976).

Many of the destructive margin basalts and andesites which have been analysed for $\delta^{18}O$ have low values very close to those obtained from fresh oceanic basalts (Taylor, 1968; Matsuhisa, 1979). Moreover, although the bulk O isotope composition of the oceanic crust is difficult to assess ($\delta^{18}O$ may be reduced by hydrothermal alteration and increased by weathering), it is most unlikely to be the same as that of fresh mantle-derived volcanics and is probably significantly higher (Fig. 7). Thus, most authors have concluded that such 'normal' $\delta^{18}O$ basalts and andesites cannot be produced by simple melting of subducted oceanic crust, as suggested by Green and Ringwood (1968)—unless their isotope composition was subsequently modified by isotopic exchange with

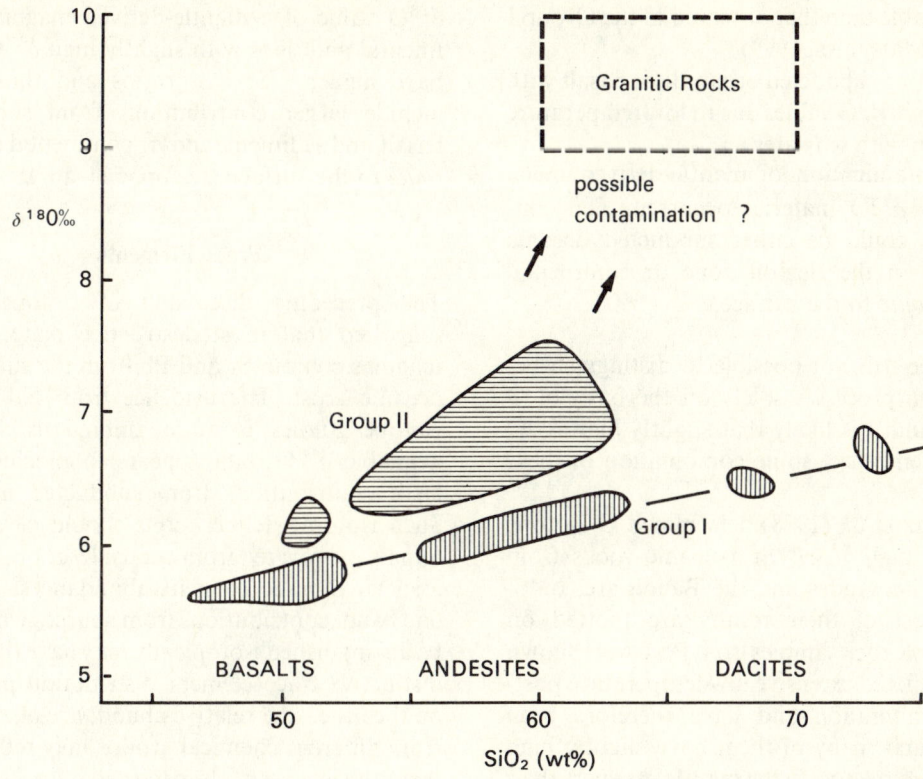

Fig. 8 Variations of $\delta^{18}O$ and SiO_2 in volcanic rocks from East Japan (Matsuhisa, 1979). Group I samples are from Hachijo-jima and are thought to be related by fractional crystallization. Group II rocks by contrast may have been contaminated as they passed through the underlying granitic crust—see text for discussion

the surrounding upper mantle (e.g. Magaritz and Taylor, 1976; Matsuhisa, 1979). However, it is possible that such magmas contain small contributions from, say, subducted sediment since that would only result in very small changes in $\delta^{18}O$: e.g. 5 per cent of sediment ($\delta^{18}O = 16$ per mil) increases the $\delta^{18}O$ value of a mantle-derived magma by just 0.5 per mil.

In detail, many O isotope results are often more ambigous than might be implied by Fig. 7 and the discussion so far. Matsuhisa (1979), in his review of available $\delta^{18}O$ data on volcanic rocks from eastern Japan, recognized two trends on the basis of O isotopes (Fig. 8). Goup I contains a range of rock types from basalt through to dacite in which $\delta^{18}O$ increases from 5.7 to 6.7 per mil as SiO_2 increases. They occur on the oceanic side of Japan and are believed to have been generated by crystal fractionation of mantle-derived basaltic magmas. Group II basalts and andesites, by contrast, are all erupted on to continental crust and are characterized by slightly higher and more variable $\delta^{18}O$ values. Unfortunately, however, this relative enrichment in $\delta^{18}O$ in the Group II rocks can be explained by a variety of mechanisms:

(1) Most of the Group II basalts have $\delta^{18}O$ values of less than 6.5 per mil (Fig. 8), and since this is well within the range obtained from fresh mantle-derived volcanic rocks (Fig. 7), the observed range of $\delta^{18}O$ in the Group I and Group II basalts might merely reflect variations in their probable upper mantle source regions (see also Kyser and O'Neil, 1978).

(2) Group II andesites could be generated by magmatic differentiation of Group II basalts if the isotopic fractionation was slightly larger and

more variable than that observed in the Group I samples (Matsuhisa, 1979).

(3) Melting subducted ocean-floor basalt with slightly high $\delta^{18}O$ values due to low temperature interaction with seawater.

(4) Contamination of mantle-derived melts with high $\delta^{18}O$ material occurred. The contaminants could be either subducted oceanic sediments in the Benioff zone or continental crust *en route* to the surface.

In practice it is not possible to distinguish between such processes solely on the basis of O isotopes and it is likely that slightly high $\delta^{18}O$ values often reflect some combination of these processes.

Magaritz *et al.* (1978) determined O isotope ratios on high $^{87}Sr/^{86}Sr$ volcanic rocks from the Peruvian Andes and the Banda arc, Indonesia. Most of their results are plotted on Fig. 7 (whole rock samples from Peru were shown to have suffered extensive low temperature post-eruption alteration and have therefore been omitted) and many of them have slightly high $\delta^{18}O$ values relative to the mantle. As such, their results are prone to the same ambiguity as the Japanese data (compare with Fig. 8), but as $\delta^{18}O$ and $^{87}Sr/^{86}Sr$ increase, as for example in some of the Banda arc samples, then clearly the case for contamination becomes increasingly convincing. Thus, Magaritz *et al.* (1978) concluded that the Sr and O isotope compositions of the Banda arc samples reflect mixing between mantle-derived and sialic material. Moreover, since the Banda arc is apparently situated on oceanic crust, the sialic contaminant was presumably introduced from the subduction zone.

In summary, many basalts and andesites have O isotope compositions similar to fresh mantle-derived magmas from mid-ocean ridge and intraplate environments and it is, therefore, unlikely that they were derived directly from the subducted oceanic crust. However, because oxygen is so abundant in magmatic rocks its isotope composition is very insensitive to additions of small amounts of high $\delta^{18}O$ material— 5 per cent of oceanic sediment or 10 per cent of altered basalt will not change significantly the $\delta^{18}O$ value of a mantle-derived magma. Continental andesites with slightly high $\delta^{18}O$ often have higher $^{87}Sr/^{86}Sr$ ratios and thus could include larger contributions from subducted basalt and sediment, and/or continental crust *en route* to the surface (*see* pp. 564–567).

Trace Elements

The preceding discussion of isotope ratios suggested that most destructive plate margin magmas contain Sr and Pb from the subducted oceanic crust. The evidence from Nd and O isotope studies is more ambiguous, but the available $\delta^{18}O$ data appears to preclude very large contributions from subducted material. Such isotopic deliberations should clearly not remain separate from consideration of the associated trace element abundances. On the one hand, contributions from sources which can be distinguished isotopically may have their own distinctive trace element distribution patterns; on the other, the relative abundance of elements from different chemical groups may reflect *how* they have been transported—for example whether excess ^{87}Sr from the subducted ocean crust is released during dehydration or partial melting. It is beyond the scope of this contribution to discuss the broader aspects of trace element variations in destructive plate margin magmas (*see* instead, Pearce, this volume, Section VIII) and thus this section is restricted to a few points which can perhaps be related to variations in isotope ratios.

In oceanic areas, where magmas clearly cannot interact with continental crust *en route* to the surface, the observed variations in isotope and trace element abundances may be considered in terms of possible contributions from the overriding mantle wedge, and the sediments and basalts of the subducted ocean crust. The chemical characteristics of the former will reflect the pre-subduction evolution of that segment of the upper mantle, and it seems reasonable to assume that the processes responsible for those characteristics were similar to those responsible for the isotope and trace element variations in mid-ocean ridge and intraplate basalts. That is why, for example, the ^{143}Nd/^{144}Nd and ^{87}Sr/^{86}Sr ratios in island arc rocks are contrasted with the main trend of Group I samples on Fig. 3.

It is well documented that, compared with volcanic rocks from mid-ocean ridge and intraplate environments, those from island arcs are preferentially enriched in alkaline elements (particularly Ba, Rb, K, and Sr) relative to less mobile elements such as the REE, Zr, and Y (Gill, 1974; Hawkesworth et al., 1977; Kay, 1977; Pearce, this volume, Section VIII). Hawkesworth and Powell (1980) used a diagram of Sm/Ce against Sr/Ce to emphasize this point in their discussion of selected basalts and andesites from the Lesser Antilles, and a similar diagram is reproduced in Fig. 9. Relative to the trend of large ion lithophile element-enriched

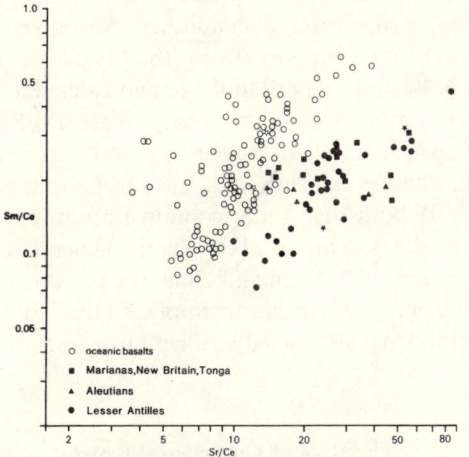

Fig. 9 Graph of Sm/Ce against Sr/Ce for island-arc volcanic rocks (less than 54 per cent SiO$_2$) compared with results from unaltered large ion lithophile element enriched and -depleted oceanic basalts (after Hawkesworth and Powell, 1980). Open stars represent typical enriched and depleted MORB, and filled stars are typical arc tholeiite and calc-alkaline basalt from Sun (1980). Data from De Paolo and Johnson (1979), Dixon and Batiza (1979), Frey et al. (1974), Hawkesworth and Powell (1980 and unpublished), Hawkesworth et al. (1979a,b,c), Kay (1977), Lambert and Holland (1977), O'Nions et al. (1976, 1977), J. A. Pearce (unpublished), Rhodes et al. (1976), Shibita et al. (1979), Sun et al. (1979), and Wood et al. (1979)

and -depleted basalts from mid-ocean ridges and oceanic islands, the majority of those from island arcs are displaced to higher Sr/Ce ratios. This is believed to be the trace element analogue to the relatively high ^{87}Sr/^{86}Sr ratios revealed by combined Nd and Sr isotope studies (Fig. 3), and both are thought to reflect the addition of material from the subducted ocean crust.

Most models for the genesis of island arc tholeiites attribute them to the partial melting in the upper mantle wedge above the subduction zone (Ringwood, 1974). Typically they are relatively *enriched* in alkaline elements and ^{87}Sr, and yet *depleted* in, for example, the light REE (Hawkesworth et al., 1977; Kay, 1977; Dixon and Batiza, 1979; Hawkesworth and Powell, 1980). This requires the addition of a component containing alkaline elements (and ^{87}Sr), but not less mobile elements such as the REE: for reasons outlined earlier, it is probably derived from the subducted ocean crust, but why significant quantities of the REE are not mobilized is not clear. A plausible explanation seems to be that this component is released by dehydration (rather than melting) of the downgoing slab, and that this may not mobilize elements such as the REE, which are not drastically affected by hydrothermal alteration of oceanic basalts (O'Nions et al., 1978) and most of which may tend to be locked in accessory metamorphic minerals derived from authigenic phosphatic and ferruginous phases (Elderfield et al., 1981).

The available results on calc-alkaline rocks are more ambiguous. Major and trace element arguments indicate that in most cases they cannot be derived by fractional crystallization of tholeiitic magmas (e.g. Kushiro and Sato, 1978), or by simple partial melting of the subducted ocean crust (Gill, 1974; Thorpe et al., 1976; Lopez-Escobar et al., 1977; Stern and Wyllie, 1978). Instead, more complex models are needed and most suggest that calc-alkaline magmas contain material from both the subducted crust and the overriding lithosphere—often modified by extensive crystallization (Kay, 1977; Brown et al., 1977; Hawkesworth and Powell, 1980). The major ambiguity arises because, although calc-alkaline rocks are still relatively enriched in alkaline elements and ^{87}Sr (Figs. 9 and 3), they also contain higher concentrations of less mobile elements such as the light REE, Zr, and Y than arc tholeiites. In addition they tend to have lower ^{143}Nd/^{144}Nd (Fig. 10) and more radiogenic Pb isotope ratios than tholeiitic rocks.

Such differences in isotope and trace element abundances might primarily reflect chemical variations which existed in the overriding mantle wedge before subduction commenced. However, such a model implies that coincidently calc-alkaline rocks are generated when subduction occurs beneath 'enriched' mantle, and tholeiites result if it takes place beneath 'depleted' mantle. More plausible perhaps is an alternative model whereby, for some reason, the component from the subducted ocean crust is different in calc-alkaline rocks: in particular it must contain REE (with comparatively low ^{143}Nd/^{144}Nd), radiogenic Pb, and P (Pearce, this volume, Section VIII)—in addition to the more mobile alkaline elements.

Such Nd and Pb isotope ratios are most unlikely to have been derived from either fresh or hydrothermally altered oceanic igneous rocks. Rather they indicate that the ocean crust component in such calc-alkaline rocks must include significant amounts of sedimentary material (see also Armstrong and Cooper, 1971; Kay et al., 1978), and it may be that therein lies the explanation of why this ocean crust component has such different trace element, as well as isotope, characteristics to that in island arc tholeiites. Subduction of significant quantities of sediment would encourage melting to occur at lower temperatures, and such a melt could be expected to contain less mobile elements such as the REE and P. Thus the differences in the ocean crust components may reflect the mobilization of subducted material by dehydration beneath arc tholeiite centres, and by partial melting (due to a larger sedimentary presence) beneath 'enriched' calc-alkaline provinces.

Hawkesworth and Powell (1980) suggested that such a model might explain why, in the Lesser Antilles, island arc tholeiites and calc-alkaline rocks were erupted contemporaneously along strike from one another. Moreover, it gains some support from the available Pb data. Both island arc tholeiite and calc-alkaline rock types contain relatively high Pb/REE ratios (Sun, 1980), suggesting that, as with the alkaline elements (e.g. high Sr/Ce ratios— Fig. 9), both rock suites contain Pb introduced from the subducted slab. Yet in general it is only the Pb in calc-alkaline rocks which is sufficiently radiogenic to imply a large contribution from subducted sediment (e.g. Fig. 6).

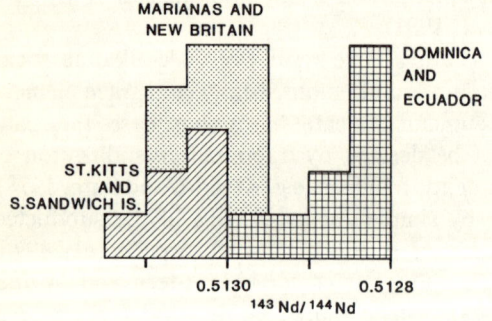

Fig. 10 ^{143}Nd/^{144}Nd variations in island-arc tholeiites (St Kitts and South Sandwich Islands) and mature calc-alkaline provinces (Dominica and Ecuador). The rocks from New Britain and the Marianas are transitional in that although many are calc-alkaline, few are significantly enriched in light REE. Data as for Fig. 3; see text for discussion

The Role of Continental Crust

It has been argued in previous sections that some of the Sr, Pb, and perhaps Nd in destructive plate margin magmas is derived not just from the subducted ocean crust, but specifically from material which has either interacted with or been formed from seawater. Since at a conservative estimate more than 60–70 per cent of the Sr, Pb, and Nd in seawater is derived from the continents, it follows that even in oceanic areas, where there is no continental crust and no continental sediment, island arc magmas contain a component from continental crust. Volumetrically it is clearly trivial, and it is only detected because of the sensitivity of these particular isotope ratios.

This 'seawater' continental component also appears to be present in at least some continental andesites, e.g. Ecuador in Fig. 3 (Hawkesworth et al., 1979a). However, as mentioned earlier, the picture in these areas is further complicated by the possibility of introducing *additional* continental material either by subduction of young continental sediments eroded from the nearby landmass, and/or contamination of the subsequent magmas as they pass through continental crust *en route* to the surface. This latter subject of 'crustal contamination' is extremely controversial; it tends to be postulated largely on the basis of isotope ratios, and only very rately is there any supporting petrographic evidence.

Crustal contamination implies that as a mantle-derived magma passes through the crust its composition is modified by the *addition* of continental material—and in the interests of clarity it will be used only in that context. Thus the term crustal contamination will not be used in this chapter to describe: (1) the introduction of a continental component from the subducted ocean crust; (2) magmas from source rocks which are themselves a mixture of mantle and crustal material, since in that case it is not the magma which becomes contaminated; and (3) magma mixing which is a separate process, clearly recognizable on textural and mineralogical grounds (e.g. Eichelberger, 1975, 1978). The last-mentioned may involve mantle- and crust-derived magmas, but it differs conceptually from crustal contamination by assimilation or isotopic exchange.

Crustal contamination, as defined above, may take place either by bulk assimilation, and/or selectively by crustal dehydration (e.g. Francis et al., 1980). In either case the magma must provide heat to the surrounding crustal rocks, which in turn encourages crystallization. For Sr, such crystallization tends not only to reduce the volume of the residual liquid but also its Sr content, so that changes in $^{87}Sr/^{86}Sr$ due to crustal contamination should be greater in more evolved rock types. This was demonstrated by Taylor et al. (1979) who, in a detailed study of rocks from the Roccamonfina volcanic centre in central Italy, showed that both $\delta^{18}O$ and $^{87}Sr/^{86}Sr$ increased in rocks of increasing SiO_2 content and concluded that fractional crystallization had been accompanied by chemical interaction with the surrounding continental crust. Moreover, such arguments imply that for suites of apparently related magmatic rocks, plots of $^{87}Sr/^{86}Sr$ (or $\delta^{18}O$) against some index of differentiation provide a reasonable method of assessing whether significant contamination with crustal material has taken place.

Figure 11 illustrates the variation of $^{87}Sr/^{86}Sr$ against SiO_2 for three suites from the South American Andes. Those from Ecuador and northern Chile show no tendency for $^{87}Sr/^{76}Sr$ to increase in the more evolved rock types, suggesting that fractional crystallization (Thorpe et al., 1976) occurred without significant crustal contamination. The last statement requires slight amplification: it does not imply that crustal contamination has had *no* effect—it might for example be responsible for some of

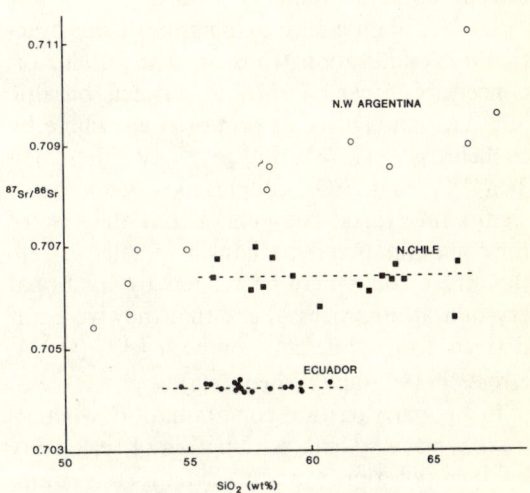

Fig. 11 Graph of $^{87}Sr/^{86}Sr$ against SiO_2 for three Recent South American volcanic suites: Ecuador, northern Chile (San Pablo–San Pedro), and northwestern Argentina (Cerro Galan) (Francis et al., 1977, 1980, unpublished data; G. A. Kretzschmar, unpublished data)

the scatter in the northern Chile results. However, the important point is that the data imply that their parental liquids had $^{87}Sr/^{86}Sr$ ratios in the range 0.706–0.707, and it is that which is difficult to attribute to contamination *during* fractional crystallization. The alternative contamination hypothesis is that the parental liquids were themselves previously contaminated at deeper, hotter levels in the crust. Yet this encounters difficulties hinted at earlier: the volume, and Sr content of the magma will be greater, the expected cumulates produced during contamination have not been found, and the lower crustal contaminant will presumably have lower $^{87}Sr/^{86}Sr$ ratios. On balance, therefore, there is nothing in the available data to suggest that the parental liquids to the northern Chilean and Ecuadorian rocks on Fig. 11 do not have similar Sr isotope ratios to those in their source rocks. For the northern Chilean samples, the source rocks might be *in* the lower crust—but that is a different argument, and one which will be returned to later.

The Argentinian samples on Fig. 11 by contrast show an excellent positive correlation between $^{87}Sr/^{86}Sr$ and SiO_2. It might reflect crustal contamination of a low SiO_2, low $^{87}Sr/^{86}Sr$, mantle-derived magma during fractional crystallization (Francis *et al.*, 1980) or, conceivably, magma mixing between basaltic and dacitic magmas, as proposed elsewhere by Eichelberger (1978). In any case the high $^{87}Sr/^{86}Sr$, high SiO_2 samples clearly contain a significant crustal component, but the test of the contamination hypothesis is to demonstrate that these rocks may be related by fractional crystallization processes and that they were not derived from different source rocks (G. A. Kretzschmar, unpublished data).

In summary, crustal contamination, whether it takes place by bulk assimilation or selectively by incorporation (and/or exchange) of more mobile elements such as K, Rb, Sr, Pb, and O, is likely to be accompanied by crystallization (*see also* Taylor, 1980). Hence the affects of crustal contamination should be most marked in the more evolved rock types (e.g. Fig. 11). If these arguments are correct then neither the Ecuadorian nor the northern Chilean volcanic suites illustrated on Fig. 11 appear to have undergone significant contamination during crystallization. For the Ecuadorian rocks their low $^{87}Sr/^{86}Sr$ ratios (Fig. 2) are consistent with an upper mantle origin (Francis *et al.*, 1977; Hawkesworth *et al.*, 1979*a*), but the higher $^{87}Sr/^{86}Sr$ ratios of 0.706–0.707 suggested for the parental liquids of the northern Chilean samples are more ambiguous. To end this section it is therefore proposed to query very briefly the present conventional wisdom that even in areas of great crustal thickness, such as the northern Chilean Andes, calc-alkaline magmas are derived from the upper mantle—perhaps some are from the lower crust.

A problem in assessing the role of the lower crust in magma genesis is that many of its characteristics, both geochemical and geophysical, are simply transitional between those of the upper crust and the mantle. Arguably, this is most often seen with isotope diagrams (e.g. $\delta^{18}O$ against $^{87}Sr/^{86}Sr$, Fig. 7) on which magmas with transitional characteristics plot between boxes labelled 'crust' and 'mantle', and thus might be generated either by crustal contamination of mantle-derived magmas (Harmon and Halliday, 1980; De Paolo, 1980) or by melting source rocks with intermediate isotope ratios. Such ambiguities may be alleviated if the lower crust experiences granulite facies metamorphism and a reduction in its U/Pb and Rb/Sr ratios relative to Nd/Sm (Tarney and Windley, 1977; Hamilton *et al.*, 1979), since then with time it will develop non-radiogenic Pb (e.g. Chapman and Moorbath, 1977) and comparatively low $^{87}Sr/^{86}Sr$ ratios (Carter *et al.*, 1978*b*). However, it is not clear if such depleted granulite facies rocks are widespread in the lower crust, nor whether in areas of such recent crustal growth as the central Andes there would be sufficient time to generate the distinctive non-radiogenic isotope ratios.

Continental crust is generated along convergent plate boundaries, and yet in upper crustal rocks the distribution of the REE in particular is significantly different from that in island arc magmas. The former appear to be characterized by a marked negative Eu anomaly, and this encouraged S. R. Taylor and his colleagues to argue that they are derived by intracrustal melting of island arc rocks in the presence of residual plagioclase (e.g. Taylor, 1977). Without wishing to become involved in the wider aspects of this 'andesite' model for the evolution of continental crust (see instead Tarney and Weaver this volume, Section X), it may have particular relevance in areas of unusual crustal thickness such as the central Andes.

It is well documented that magmatic rocks in the central Andes (northern Chile and southern Peru) tend to have higher $^{87}Sr/^{86}Sr$ ratios (see Fig. 2) and average SiO_2 contents (e.g. Eichelberger, 1978) than those in island arcs or in the northern Andes (Ecuador). Moreover, there is an increasing quantity of data (e.g. Hammill, 1981) which suggests that compared with rocks in many island arcs, those in the central Andes may also be characterized by low Sr/Nd and high Rb/Nd ratios—together with more pronounced negative Eu anomalies. The $\delta^{18}O$ values on unaltered rocks are in the range 7.5–9.0 per mil (Magaritz et al. 1978; Thorpe et al., 1981), similar to the results of 8.8 per mil reported by James (1979) on a lower crustal xenolith from Kilbourne Hole, New Mexico, and whereas most Pb isotope data suggest a crustal prehistory (McNutt et al., 1979), some even have low $^{206}Pb/^{204}Pb$ ratios indicative of a lower crustal origin (Tilton, 1979). All these features have, in their time, been attributed to crustal contamination processes, but clearly when considered together they might be more simply due to intracrustal melting in the presence of residual plagioclase. Detailed studies of individual volcanic centres in the central Andes are still surprisingly scarce, but as more information becomes available about the trace element ratios in parental liquids of those suites which appear to have been little affected by contamination during fractional crystallization (as, for example, on Fig. 11), it may be necessary to look again at the lower crust as a potential source for intermediate to acid magmas in areas of unusually thick continental crust.

Conclusions

Compared with magmas erupted at mid-ocean ridges and in intraplate environments, those generated along destructive plate margins tend to be enriched in alkaline elements (and Pb) relative to less mobile elements such as the REE Zr, Ti, and Y. This is often accompanied by comparatively radiogenic Sr and Pb isotope ratios and, since both are characteristic of ocean-floor material which has either interacted with or been deposited from seawater, they, and the enrichment in alkaline elements, are attributed to the release of material from the subducted oceanic plate. The similarity of $\delta^{18}O$ values in island arc rocks and uncontaminated basalts indicates that this contribution from the ocean crust is probably less than c. 5–10 per cent.

The behaviour of the REE is less clear cut. Island arc tholeiites tend to be depleted in light REE and yet enriched in alkaline elements (K, Rb, Sr and Ba), suggesting that alkaline elements, but not REE, have been introduced from the subducted slab (e.g. Hawkesworth et al., 1977). Calc-alkaline rocks, however, are also enriched in light REE and P (Pearce, this volume, Section VIII) and, because particularly in the more mature destructive margins such as the Lesser Antilles and Ecuador this is accompanied by lower $^{143}Nd/^{144}Nd$ ratios (Fig. 10), it is probable that at least some of the REE were derived from sediments (see Fig. 4). Moreover, this is consistent with the available Pb isotope data, since many island arc calc-alkaline suites have radiogenic Pb isotope ratios also indicative of a sedimentary origin (Fig. 6). Putting these arguments together, the differences between the ocean crust component in island arc tholeiites and in particularly 'enriched' calc-alkaline rocks may be due to the mobilization of subducted material by dehydration beneath the former, and by partial melting (due to a larger sedimentary presence) beneath the latter.

Much of this ocean crust component in island arc magmas was itself derived from pre-existing continental crust—usually via seawater. It is clear evidence that some continental material is being recycled at convergent plate boundaries and, although O isotope results indicate that volumetrically the amount is small, it is important because it has been demonstrated for those elements which are the basis of most models for the evolution of the crust–mantle system (i.e. elements with isotopes involved in long-lived radioactive decay schemes). In continental areas magmas may also incorporate additional continental material as they pass through the crust. This problem of crustal contamination is extremely controversial at present, not least because the mechanisms are poorly understood and there is little consensus as to the chemical criteria by which it should be evaluated. Most forms of contamination will probably be accompanied by the transfer of heat out of the magma and this should in turn encourage crystallization. It follows that, particularly for Sr, the concentration of which will be reduced by plagioclase crystallization, the effects of contamination should be most marked in more evolved rock types; they not only represent smaller volumes of magma but also have presumably been resident in the crust for longer. Applying these arguments to Andean calc-alkaline rock suites suggests that in some areas of the central Andes parental magmas may have had $^{87}Sr/^{86}Sr$ ratios as high as 0.706–0.707. Moreover, since volcanic rocks in this region tend to have higher SiO_2 contents, $^{87}Sr/^{86}Sr$ ratios, and $\delta^{18}O$ values, together with more radiogenic Pb isotopes and lower Sr/Nd and $^{143}Nd/^{144}Nd$ ratios compared with many island arc calc-alkaline rocks, it may be that some are generated by remelting material in the lower crust. There are as yet insufficient data to test adequately this hypothesis, but it would come as no surprise if in areas of unusual crustal thickness, such as the central Andes, mantle-derived magmas found it more difficult to reach the surface, and having crystallized at depth within the crust were prone to remobilization during subsequent thermal pulses.

Acknowledgements

I thank Dr J. A. Pearce for many entertaining discussions on the problems of magma genesis in this tectonic setting, and Drs J. Marsh and P. van Calsteren for reviewing the manuscript. G. A. Kretzschmar kindly made his unpublished SiO_2 analyses available for Fig. 11.

REFERENCES

Armstrong, R. L. and Cooper, J. A. (1971). Lead isotopes in island arcs. *Bull. Volc.* **35**, 27–73.

Bass, M. N., Moberly, R., Rhodes, J. M., Shih, C., and Church, S. E. (1973). Volcanic rocks cored in the central Pacific, Leg 17, Deep Sea Drilling Project. In *Initial Reports of the Deep Sea Drilling Project*, Vol. 17, US Government Printing Office, Washington, DC, pp. 429–503.

Brooks, C., James, D. E., and Hart, S. R. (1976). Ancient lithosphere: its role in young continental volcanism. *Science, N.Y.* **193**, 1086–1094.

Brown, G. M., Holland, J. G., Sigurdsson, H., Tomblin, J. F., and Arculus, R. J. (1977). Geochemistry of the Lesser Antilles volcanic island arc. *Geochim. cosmochim. Acta* **41**, 785–801.

Carter, S. R., Evenson, N. M., Hamilton, P. J., and O'Nions, R. K. (1978a). Continental volcanics from enriched and depleted source regions: Nd and Sr isotope evidence. *Earth planet. Sci. Letters* **37**, 401–408.

Carter, S. R., Evensen, N. M., Hamilton, P. J., and O'Nions, R. K. (1978b). Neodymium and strontium isotope evidence for crustal contamination of continental volcanics. *Science, N.Y.* **202**, 743–747.

Carter, S. R., Evensen, N. M., Hamilton, P. J., and O'Nions, R. K. (1979). Basalt magma sources during the opening of the North Atlantic. *Nature, Lond.* **281**, 28–30.

Chapman, H. J. (1978). Geochronology and isotope geochemistry of Precambrian rocks from north-west Scotland. D. Phil. thesis, University of Oxford.

Chapman, J. and Moorbath, S. (1977). Lead isotope measurements from the oldest recognized Lewisian geneisses of north-west Scotland. *Nature, Lond.* **268**, 41–42.

Chow, J. J. and Patterson, C. C. (1962). The occurrence and significance of lead isotopes in pelagic sediments. *Geochim. cosmochim. Acta* **26**, 263–308.

Church, S. E. (1973). Limits of sediment involvement in the genesis of orogenic volcanic rocks. *Contr. Mineral. Petrol.* **39**, 17–32.

Church, S. E. (1976). The Cascade Mountains revisited: a re-evaluation in light of the new lead isotopic data. *Earth planet. Sci. Letters* **29**, 175–188.

Church, S. E. and Tatsumoto, M. (1975). Lead isotope relations in oceanic ridge basalts from the Juan de Fuca–Gorda Ridge area, N.E. Pacific. *Contr. Mineral. Petrol.* **53**, 253–279.

Church, S. E. and Tilton, G. R. (1973). Lead and strontium isotopic studies in the Cascade Mountains: bearing on andesite genesis. *Bull. Geol. Soc. Am.* **84**, 431–454.

Condie, K. C. and Hayslip, D. L. (1975). Young bimodal volcanism at Medicine Lake volcanic center, northern California. *Geochim. cosmochim. Acta* **39**, 1165–1178.

Cumming, G. L. and Richards, J. R. (1975). Ore lead isotope ratios in a continuously changing Earth. *Earth planet. Sci. Letters* **28**, 155–171.

De Paolo, D. J. (1980). Sources of continental crust: neodymium isotope evidence from the Sierra Nevada and Peninsular Ranges. *Science, N.Y.* **209**, 684–687.

De Paolo, D. J. and Johnson, R. W. (1979). Magma genesis in the New Britain island arc: constraints from Nd and Sr isotopes and trace-element patterns. *Contr. Mineral. Petrol.* **70**, 367–379.

De Paolo, D. J. and Wasserburg, G. J. (1976a). Nd isotopic variations and petrogenic models. *Geophys. Res. Letters* **3**, 249–252.

De Paolo, D. J. and Wasserburg, G. J. (1976b). Inferences about magma sources and mantle structure from variations of $^{133}Nd/^{144}Nd$. *Geophys. Res. Letters* **3**, 743–746.

De Paolo, D. J. and Wasserburg, G. J. (1977). The sources of island arcs as indicated by Nd and Sr isotopic studies. *Geophys. Res. Letters* **4**, 465–468.

De Paolo, D. J. and Wasserburg, G. J. (1979). Petrogenetic mixing models and Nd–Sr isotopic patterns. *Geochim. cosmochim. Acta* **43**, 615–627.

Dixon, T. H. and Batiza, R. (1979). Petrology and chemistry of recent lavas in the Northern Marianas: implications for the origin of island arc basalts. *Contr. Mineral. Petrol.* **70**, 167–181.

Doe, B. R. (1970). *Lead Isotopes*, Springer-Verlag, New York, Heidelberg, and Berlin.

Donnelly, T., Rogers, J. J. W., Pushkar, P., and Armstrong, R. L. (1971). Chemical evolution of igneous rocks of the eastern West Indies: an investigation of thorium, uranium and potassium distributions, and lead and strontium isotope ratios. *Geol. Soc. Am. Mem.* no. 130, 181–224.

Dosso, L. and Murthy, V. R. (1980). An Nd isotopic study of the Kerguelen Islands: inferences on enriched oceanic mantle sources. *Earth planet. Sci. Letters* **48**, 268–276.

Eichelberger, J. C. (1975). Origin of andesite and dacite: evidence of mixing at Glass Mountain in California and at other circum-Pacific volcanoes. *Geol. Soc. Amer. Bull.* **86**, 1381–1391.

Eichelberger, J. C. (1978). Andesites in island arcs and continetal margins: relationships to crustal evolution. *Bull. Volc.* **41**, 480–500.

Elderfield, H., Hawkesworth, C. J., Greaves, M., and Calvert, S. E. (1981). Rare earth element geochemistry of oceanic ferromanganese nodules and associated sediments. *Geochim. cosmochim. Acta*, **45**, 513–528.

Erlank, A. J. and Shimizu, N. (1977). Strontium and Sr-isotope distributions in some kimberlite nodules and minerals. Paper presented at the Second International Kimberlite Conference, New Mexico.

Faure, G. (1977). *Principles of Isotope Geology*, Wiley, New York.

Francis, P. W., Moorbath, S., and Thorpe, R. S. (1977). Strontium isotope data for recent andesites in Ecuador and North Chile. *Earth planet. Sci. Letters* **37**, 197–202.

Francis, P. W., Thorpe, R. S., Moorbath, S., Kretzschmar, G. A., and Hammill, M. (1980). Strontium isotope evidence for crustal contamination of calc-alkaline rocks from Cerro Galan, north west Argentina. *Earth. planet. Sci. Letters* **48**, 257–267.

Frey, F. A., Bryan, W. B., and Thompson, G. (1974). Atlantic ocean floor: geochemistry and petrology of basalts from Legs 2 and 3 of the Deep Sea Drilling Project. *J. geophys. Res.* **79**, 5507.

Gill, J. B. (1974). Role of underthrust oceanic crust in the genesis of a Fijian calc-alkaline suite. *Contr. Mineral. Petrol.* **43**, 29–45.

Green, T. H. and Ringwood, A. E. (1968). Genesis of the calc-alkaline rock suite. *Contr. Mineral. Petrol.* **18**, 105–162.

Hammill, M. (1981). The Cerro Purico Ignimbrite shield complex, Northern Chile, in press.

Hamilton, P. J., Evensen, N. M., O'Nions, R. K., and Tarney, J. (1979). Sm–Nd systematics of Lewisian gneisses: implications for the origin of granulites. *Nature, Lond.* **177**, 25–28.

Harmon, R. S. and Halliday, A. N. (1980). Oxygen and strontium isotope relationships in the British late Caledonian granites. *Nature, Lond.* **283**, 21–25.

Hart, S. R., Erlank, A. J., and Kable, E. J. D. (1974). Sea-floor basalt alteration: some chemical and Sr-isotopic effects. *Contr. Mineral. Petrol.* **44**, 219–230.

Hawkesworth, C. J. (1979). $^{143}Nd/^{144}Nd$, $^{87}Sr/^{86}Sr$ and trace element characteristics of magmas along destructive plate margins. In *Origin of Granite Batholiths: Geochemical Evidence* (M. P. Atherton and J. Tarney, eds), Shiva Publishing, Orpington, pp. 76–89.

Hawkesworth, C. J. and Elderfield, H. (1978). The strontium composition of interstitial waters from sites 245 and 336 of the Deep Sea Drilling Project. *Earth planet. Sci. Letters* **40**, 423–432.

Hawkesworth, C. J. and Powell, B. M. (1980). Magma genesis in the Lesser Antilles island arc. *Earth planet. Sci. Letters* in press.

Hawkesworth, C. J. and Vollmer, R. (1979). Crustal contamination versus enriched mantle: $^{143}Nd/^{144}Nd$ and $^{87}Sr/^{86}Sr$ evidence from the Italian volcanics. *Contr. Mineral. Petrol.* **69**, 151–169.

Hawkesworth, C. J., O'Nions, R. K., Pankhurst, R. J., Hamilton, P. J., and Evenson, N. M. (1977). A geochemical study of island-arc and back-arc tholeiites from the Scotia Sea. *Earth planet. Sci. Letters* **36**, 253–262.

Hawkesworth, C. J., Norry, M. J., Roddick, J. C., Baker, P. E., Francis, P. W., and Thorpe, R. S. (1979a). $^{143}Nd/^{144}Nd$, $^{87}Sr/^{86}Sr$ and incompatible element variations in calc-alkaline andesites and plateau lavas from South America. *Earth planet. Sci. Letters* **42**, 45–57.

Hawkesworth, C. J., Norry, M. J., Roddick, J. C., and Vollmer, R. (1979b). $^{143}Nd/^{144}Nd$ and $^{87}Sr/^{86}Sr$ ratios from the Azores and their significance in LIL-element enriched mantle. *Nature, Lond.* **280**, 28–31.

Hawkesworth, C. J., O'Nions, R. K., and Arculus, R. J. (1979c). Nd- and Sr-isotope geochemistry of island arc volcanics, Grenada, Lesser Antilles. *Earth planet. Sci. Letters* **45**, 237–248.

Hedge, C. E. and Knight, R. J. (1969). Lead and strontium isotopes in volcanic rocks from northern Honshu, Japan. *Geochem. J.* **3**, 15–24.

Hofmann, A. W. and Hart, S. R. (1978). An assessment of local and regional isotopic equilibrium in a partially molten mantle. *Earth planet. Sci. Letters* **38**, 44–62.

Høgdahl, O. T., Bowen, B. T., and Melson, S. (1968). Neutron activation analysis of lanthanide elements in sea water. *Adv. Chem. Ser.* **73**, 308–325.

James, D. E. (1979). Oxygen isotope research. *Carnegie Instn. Wash. Yb.* **78**, 308–311.

James, D. E., Brooks, C., and Cuyubamba, A. (1976). Andean Cenozoic volcanism: magma genesis in the light of strontium isotopic composition and trace element geochemistry. *Geol. Soc. Am. Bull.* **87**, 592–600.

Karig, D. E. and Sharman, G. F. (1975). Subduction and accretion in trenches. *Geol. Soc. Am. Bull.* **86**, 377–389.

Kay, R. W. (1977). Geochemical constraints on the origin of Aleutian magmas. In *Island Arcs, Deep Sea Trenches and Back-arc Basins* (M. Talwani and W. Pitman, eds), Maurice Ewing Series no. 1, American Geophysical Union, Washington, DC, pp. 229–242.

Kay, R. W., Sun, S.-S., and Lee-Hu, C. N. (1978). Pb and Sr isotopes in volcanic rocks from the Aleutian Islands and Pribilof Islands, Alaska. *Geochim. cosmochim. Acta* **42**, 263–273.

Knouth, L. P. and Epstein, S. (1975). Hydrogen and oxygen isotope ratios in silica from the JOIDES Deep Sea Drilling Project. *Earth planet. Sci. Letters* **25**, 1–10.

Kramers, J. D. (1977). Lead and strontium isotopes in Cretaceous kimberlites and mantle-derived xenoliths from Southern Africa. *Earth planet. Sci. Letters* **34**, 419–431.

Kurasawa, H. (1968). Isotopic composition of lead and concentrations of uranium, thorium and lead in volcanic rocks from Dogo of Oki Island, Japan. *Geochem. J.* **2**, 11–28.

Kushiro, I. and Sato, H. (1978). Origin of some calc-alkalic andesites in the Japanese Islands. *Bull. Volc.* **41**, 576–585.

Kyser, K. T. and O'Neil, J. R. (1978). Oxygen isotope relations among oceanic tholeiites, alkali basalts, and ultramafic nodules. *US geol. Surv. Open-File Report* no. 78-101, 237–240.

Lambert, R. St J. and Holland, J. G. (1977). Trace elements and petrogenesis of DSDP 37 basalts. *Can. J. Earth Sci.* **14**, 809.

Lancelot, J. R., Briueu, L., Westphal, B., and Tatsumoto, M. (1978). Sr and Pb isotopic data bearing on the origin of calc-alkaline lavas of Pliocene and Quaternary age from Peru and New Hebrides active margins. *US geol. Surv. Open-File Report* no. 78-701, 240–241.

Lopez-Escobar, L., Frey, M. A., and Vergara, M. (1977). Andesites and high-alumina basalts from central-south Chile High Andes; geochemical evidence bearing on their petrogenesis. *Contr. Mineral. Petrol.* **63**, 199–228.

Magaritz, M. and Taylor, H. P. (1976). Oxygen, hydrogen and carbon isotope studies of the Franciscan formation, Coast Ranges, California. *Geochim. cosmochim. Acta* **40**, 215–234.

Magaritz, M., Whitford, D. J., and James, D. E. (1978). Oxygen isotopes and the origin of high-$^{87}Sr/^{86}Sr$ andesites. *Earth planet. Sci. Letters* **40**, 220–230.

Matsuhisa, Y. (1979). Oxygen isotopic compositions of volcanic rocks from the East Japan island arcs and their bearing on petrogenesis. *J. Volc. geothermal Res.* **5**, 271–296.

McCulloch, M. T. and Wasserburg, G. J. (1978). Sm–Nd and Rb–Sr chronology of continental crust formation. *Science, N.Y.* **200**, 1003–1011.

McNutt, R. H., Clarke, A. H., and Zentilli, M. (1979). Lead isotopic compositions of Andean igneous rocks, latitudes 26° to 29°S: petrologic and metallogenic implications. *Econ. Geol.* **74**, 827–837.

Meijer, A. (1976). Pb and Sr isotopic data bearing on the origin of volcanic rocks from the Mariana island arc system. *Geol. Soc. Am. Bull.* **87**, 1358–1369.

Moorbath, S., Thorpe, R. S., and Gibson, I. L. (1978). Strontium isotope evidence for petrogenesis of Mexican andesites. *Nature, Lond.* **271**, 437–439.

Muelenbachs, K. and Clayton, R. N. (1972). Oxygen isotope studies of fresh and weathered submarine basalts. *Can. J. Earth Sci.* **9**, 172–184.

Muelenbachs, K. and Clayton, R. N. (1976). Oxygen isotope composition of the oceanic crust and its bearing on sea water. *J. geophys. Res.* **81**, 4365–4369.

Noble, D., Bowman, H., Herbert, A. Silberman, M., Heropoulous, C., Fabbi, B., and Hedge, C. (1975). Chemical and isotopic constraints on the origin of low-silica latite and andesite from the Andes of Central Peru. *Geology* **3**, 501–520.

Norry, M. J., Truckle, P. H., Lippard, S. J., Hawkesworth, C. J., Weaver, S. D., and Marriner, G. F. (1980). Isotopic and trace element evidence from lavas, bearing on mantle heterogeneity beneath Kenya. *Phil. Trans. R. Soc. A* **297**, 259–272.

O'Nions, R. K., Pankhurst, R. J., and Gronvold, K. (1976). Nature and development of basalt magma sources beneath Iceland and the Reykjanes Ridge. *J. Petrol.* **17**, 315–338.

O'Nions, R. K., Hamilton, P. K., and Evensen, N. M. (1977). Variations in $^{143}Nd/^{144}Nd$ and $^{87}Sr/^{86}Sr$ ratios in oceanic basalts. *Earth planet. Sci. Letters* **34**, 13–22.

O'Nions, R. K., Carter, S. R., Cohen, R. S., Evensen, N. M., and Hamilton, P. J. (1978). Pb, Nd and Sr isotopes in oceanic ferromanganese deposits and ocean floor basalts. *Nature, Lond.* **273**, 435–438.

O'Nions, R. K., Evensen, N. M., and Hamilton, P. J. (1979). Geochemical modelling of mantle differentiation and crustal growth. *J. geophys. Res.* **84**, 6091–6101.

Oversby, V. M. and Ewart, A. (1972). Lead isotopic compositions of Tonga–Kermadec volcanics and their petrogenetic significance. *Contr. Mineral. Petrol.* **37**, 181–210.

Peterman, Z., Hedge, C., and Tourtelot, H. (1970). Isotopic composition of strontium in sea water throughout Phanerozoic time. *Geochim. cosmochim. Acta* **34**, 105–120.

Piepgras, D. J., Wasserburg, G. J., and Dasch, E. J. (1979). The isotopic composition of Nd in different ocean masses. *Earth planet. Sci. Letters* **45**, 223–236.

Rhodes, J. M., Blanchard, D. P., Rodgers, K. U., Jacobs, J. W. and Brannon, J. C. (1976). Petrology and chemistry of basalts from the Nazca Plate. Part 2. Major and trace element chemistry. *Initial Reports DSDP* **34**, 239–244.

Richard, P., Shimizu, N., and Allègre, C. J. (1976). $^{143}Nd/^{144}Nd$, a natural tracer: an application to oceanic basalts. *Earth planet. Sci. Letters* **31**, 269–278.

Ringwood, A. E. (1974). The petrological evolution of island arc systems. *J. geol. Soc. Lond.* **130**, 183–204.

Savin, S. M. and Epstein, S. (1970). The oxygen and hydrogen isotope geochemistry of ocean sediments and shales. *Geochim. cosmochim. Acta* **34**, 43–63.

Shibita, T., Thompson, G., and Frey, F. A. (1979). Tholeiitic and alkali basalts from the mid-Atlantic ridge at 43°N. *Contr. Mineral. Petrol.* **70**, 127–141.

Spooner, E. T. C. (1976). The strontium isotopic composition of seawater and seawater–oceanic crust interaction. *Earth planet. Sci. Letters* **31**, 167–174.

Spooner, E. T. C., Beckinsale, R. D., Fyfe, W. S., and Smewing, J. D. (1974). ^{18}O enriched ophiolite metabasic rocks from E. Liguria (Italy), Pindos (Greece) and Troodos (Cyprus). *Contr. Mineral. Petrol.* **47**, 41–62.

Spooner, E. T. C., Chapman, H. J., and Smewing, J. D. (1977). Strontium isotopic contamination and oxidation during ocean floor hydrothermal metamorphism of the ophiolite rocks of the Troodos Massif, Cyprus. *Geochim. cosmochim. Acta* **41**, 873–890.

Stern, C. R. and Wyllie, P. J. (1978). Chemical compositions of phases through the melting intervals of hydrous basalt and andesite compositions at 30 kb: their implications for magma genesis along convergent plate boundaries. *Am. Mineral.* **63**, 641–663.

Sun, S.-S. (1980). Lead isotopic study of young volcanic rocks from mid-ocean ridges, ocean islands and island arcs. *Phil. Trans. R. Soc. A* **297**, 409–446.

Sun, S.-S., Nesbitt, R. W., and Sharaskin, A. Y. (1979). Geochemical characteristics of mid-ocean ridge basalts. *Earth planet. Sci. Letters* **44**, 119–138.

Tarney, J. and Windley, B. F. (1977). Chemistry, thermal gradients and evolution of the lower continental crust. *J. geol. Soc. Lond.* **134**, 153–172.

Tatsumoto, M. and Knight, R. J. (1969). Isotopic composition of lead in volcanic rocks from central Houshu with regard to basalt genesis. *Geochem. J.* **3**, 53–86.

Taylor, H. P. (1968). The oxygen isotope geochemistry of igneous rocks. *Contr. Mineral. Petrol.* **19**, 1–71.

Taylor, H. P. (1980). The effects of assimilation of country rocks by magmas on $^{18}O/^{16}O$ and $^{87}Sr/^{86}Sr$ systematics in igneous rocks. *Earth planet. Sci. Letters* **47**, 243–254.

Taylor, H. P., Giannetti, B., and Turi, B. (1979). Oxygen isotope geochemistry of potassic igneous rocks from the Roccamonfina volcano, Roman comagmatic region, Italy. *Earth planet. Sci. Letters* **46**, 81–106.

Taylor, S. R. (1977). Island arc models and the composition of the continental crust. In *Island Arcs, Deep Sea Trenches and Back-arc Basins* (M. Talwani and W. C. Pitman, eds), Maurice Ewing Series no. 1, American Geophysical Union, Washington, DC, pp. 325–335.

Tilton, G. R. (1979). Isotopic studies of Cenozoic Andean calc-alkaline rocks. *Carnegie Instn. Wash. Yb.* **78**, 298–304.

Thorpe, R. S., Potts, P. J., and Francis, P. W. (1976). Rare-earth data and petrogenesis of andesites from the North Chilean Andes. *Contr. Mineral. Petrol.* **54**, 65–78.

Thorpe, R. S., Francis, P. W., and Harmon, R. S. (1981). Andean andesites and crustal growth. *Phil. Trans. R. Soc. A* **301**, 305–320.

Turi, B. and Taylor, H. P. (1976). Oxygen isotope studies of potassic volcanic rocks of the Roman province, Central Italy. *Contr. Mineral. Petrol.* **55**, 1–31.

White, W. M. (1979). Geochemistry of basalts from the Famous area: a re-examination. *Carnegie Instn. Wash. Yb.* **78**, 325–331.

White, W. M. and Hofmann, A. W. (1978). Geochemistry of the Galapagos Islands: implications for mantle dynamics and evolution. *Carnegie Instn. Wash. Yb.* **77**, 596–606.

Whitford, D. J. (1975). Strontium isotope studies of the volcanic rocks of the Sunda Arc, Indonesia and their petrogenetic significance. *Geochim. cosmochim. Acta* **39**, 1287–1302.

Whitford, D. J. and Bloomfield, K. (1975). Geochemistry of Late Cenozoic volcanic rocks from the Nevado de Toluca area, Mexico. *Carnegie Instn. Wash. Yb.* **74**, 207–213.

Wood, D. A., Tarney, J., Varet, J. Saunders, A. D., Bougault, H., Joron, J. L., Treuil, M., and Cann, J. R. (1979). Geochemistry of basalts drilled in the North Atlantic by IPOD Leg 49; implications for mantle heterogeneity. *Earth planet. Sci. Letters* **42**, 77–97.

Andesites
Edited by R. S. Thorpe
© 1982 John Wiley & Sons

IX Andesite volcanism throughout geological time

At the present day, calc-alkaline volcanic rocks formed above destructive plate margins add, in a major way, to continental growth. An important question is therefore whether these rocks and their intrusive equivalents have played a major part in continental growth throughout geological time. In this chapter three contributors examine the composition and tectonic setting of ancient calc-alkaline rocks and how they may have changed throughout geological time.

Geological time has been divided into three main periods, the Archaean (older than 2500 Ma), the Proterozoic (2500–600 Ma), and the Phanerozoic (younger than 600 Ma), each of which has distinctive rock groups and tectonic characteristics. Archaean rocks, formed between 3800 and 2500 Ma ago, are characterized by tonalite–granodiorite gneiss, and granite–greenstone terranes. The volcanic rocks of such greenstone belts belong to distinctive tholeiitic, calc-alkaline, and komatiitic associations. During this 'permobile' episode, the overall high heat productivity (calculated as being between three and four times higher than present values) was reflected by rapid growth and thickening of continental crust primarily by calc-alkaline igneous activity.

K. C. Condie outlines the characteristics of the various types of andesites in Archaean greenstone belts that seemingly can be distinguished on the basis of trace element contents. In comparison with their modern equivalents, these Archaean andesites have distinctive geochemical characteristics such as high transition trace element concentrations which might indicate higher degrees of mantle partial melting, as might be expected from higher Archaean heat production. From possible modern analogues, such Archaean andesites might have formed in a setting analogous to a modern island arc–marginal basin system. Alternatively, they might have formed in a continental rift environment. Condie argues that, in contrast to modern andesites, Archaean andesites might have formed in both continental rift and continental margin arc systems. Whichever setting was dominant, by the close of the Archaean zone 2500 Ma ago, a variety of geological features suggest that as a result of diminished heat production the lithosphere had tectonic characteristics approaching that at the present day.

This stability is the prime geological characteristic of the Proterozoic. For the first time platform sedimentary rocks were deposited and there is the earliest record of the formation of continental margin geosynclinal sedimentation. Although early and middle Proterozoic igneous rocks are common, and include distinctive rocks such as massive anorthosites and granitic complexes which are similar geochemically to Phanerozoic calc-alkaline plutonic complexes, volcanic associations seem relatively uncommon during this period. So, although these Proterozoic plutonic rocks differ from those of the Archaean, the absence of volcanic products makes it difficult to interpret this period in terms of modern plate tectonics. One suggestion has been that these rocks formed within a 'Proterozoic super-continent' which existed at 2200 Ma ago and was disrupted c. 1000 Ma ago and which, in turn, initiated a new episode of continental fragmentation, plate movement, and collision. Noting that heat production at 1000 Ma ago was only c. 1.8 times that at the present day, it seems likely that magmatic and tectonic processes from this time would be much more analogous to those of the present day than to those of the middle and early Proterozoic and the Archaean. On geothermal and theoretical criteria, it seems plausible to invoke plate tectonic processes to explain the late Proterozoic magmatism. Much discussion of the role played by plate tectonic processes in the formation of Proterozoic and studied continental crust has ensued.

One of the best preserved Upper Proterozoic orogenic belts is the 1200–500 Ma 'Pan-African' of northern Africa and Arabia. Currently, two contrasted hypotheses are invoked to explain the evolution of the Pan-African crust. The first proposes that the Pan-African continental crust evolved over a period of c. 700 Ma as a result of oceanic plate subduction and the resultant formation, evolution, and subsequent accretion of intra-oceanic island arcs. An alternative view is that the Pan-African orogenic belts were ensialic and that their evolution involved massive mobilization of older pre-Upper Proterozoic sialic basement as a result of heat derived from the mantle—a process accompanied by little or no horizontal movements of bordering cratonic areas. As a contribution to resolving these models for continental formation and evolution, I. G. Gass outlines the geological setting and characteristics of calc-alkaline magmatic activity associated with the Pan-African orogenic belts. Gass concludes that the Pan-African crust formed as a result of development of island arc systems and their progressive 'cratonization' by repeated plutonism, volcanism, and attendant sedimentation, and that, for the Pan-African, calc-alkaline magmatism has played a major role in the formation and evolution of Upper Proterozoic continental crust.

As noted above, during the period between the late Proterozoic (1000 Ma ago) and the present day, magmatic and tectonic processes have evolved gradually towards the present pattern. The episode of late Proterozoic continental fragmentation in the North Atlantic continents culminated during the Lower Palaeozoic with the formation of the Appalachian–Caledonian belt. Orogenic activity was accompanied by widespread calc-alkaline magmatism, particularly between 650 and 450 Ma. This orogenic belt is one of the most intensively studied, and during the last decade our understanding of the Appalachian–Caledonian orogenic belt has been enormously advanced by application of plate tectonic concepts. In the final contribution to this chapter, J. G. Fitton *et al.* review the Caledonian orogenic belt of Britain, and outline how an interpretation of these volcanic rocks in plate tectonic terms provides a new insight into Phanerozoic orogenic belts.

Archaean andesites

Kent C. Condie

Department of Geoscience,
New Mexico Institute of Mining and Technology, Socorro, New Mexico 87891, USA

ABSTRACT

Andesites occur in many Archaean greenstone belts and become more abundant with increasing stratigraphic level and in the upper parts of volcanic cycles. In order of decreasing abundance, Archaean andesites occur as tuffs, breccias and agglomerates, flows, and shallow intrusions. Archaean andesites may contain phenocrysts of plagioclase, hornblende, augite, and, less commonly, quartz, orthopyroxene, and magnetite. Major and many trace elements may be redistributed in andesites during secondary processes. REE, Hf, Th, and transition metals are least susceptible to change during such processes.

Archaean andesites can be classified into three types based principally on REE patterns. Type I exhibits slight light REE enrichment, Type II large light REE enrichment, and Type III flat REE patterns with negative Eu anomalies. Archaean andesites differ from modern andesites by their low Al_2O_3 and high FeO, MgO, Y, Ni, Cr, Co, Zn, FeO/Fe_2O_3, and Ni/Co. Geochemical model studies are consistent with one of two modes of production for Archaean andesitic magmas: (1) partial melting of eclogite or amphibolite; (2) fractional crystallization of tholeiite magma in which garnet and/or amphibole are important liquidus phases.

Any model for Archaean andesite production must explain the overall change in magma composition from ultramafic–mafic to mafic and calc-alkaline with stratigraphic height in greenstone successions; the close association of rocks of the tholeiite and calc-alkaline series; the existence of volcanic cycles; the absence of andesite in some belts; and the infrequent production of Type III andesites. Tectonic models involving either or both a convergent plate boundary or a continental rift seem capable of accounting for these features.

Introduction

Andesites are an important volcanic rock type in some Archaean greenstone belts. Two basic types of Archaean greenstone successions are generally recognized, the calc-alkaline and bimodal (Condie, 1976a). The calc-alkaline type contains a range of volcanic compositions from ultramafic through mafic and andesitic to felsic while the bimodal type contains ultramafic–mafic and felsic components and a sparsity or absence of volcanic rocks of intermediate composition. Existing data suggest a geographic provinciality (Condie, 1976b) of these greenstone belts as summarized by the volcanic rock abundances in Table 1. Many of the Archaean greenstone belts in North America are similar to the four listed in Table 1. Noteworthy in all North American belts is the low abundance of ultramafic rocks. North American belts average c. 55 per cent mafic volcanic rocks, 30 per cent andesite, 10 per cent felsic volcanic rocks, and less than c. 5 per cent ultramafic rocks (Goodwin, 1977). Some appear to be bimodal, as illustrated by the greenstone belts in northeastern Minnesota and adjacent Ontario (Arth and Hanson, 1975). In contrast to most North American belts, most African and Australian

TABLE 1 Proportions of volcanic rock types in Archaean greenstone belts (in per cent).

	Ultramafic–mafic†	Mafic	Andesite	Felsic
North America				
1. Birch–Uchi	4	54	29	13
2. Wabigoon	4	58	26	12
3. Abitibi	5	50	37	8
4. Yellowknife	≤1	65	20	14
Africa				
1. Midlands–Bulawayo	10	42	40	8
2. Fort Victoria	10	75	5	10
3. Shabani	12	75	10	3
4. Barberton	24	72	—	4
5. Western Kenya	—	10	75	15
Australia				
1. Coolgardie–Norseman	20	62	5	13

† Mafic rocks include basaltic komatiites and tholeiites closely associated with ultramafic volcanic rocks.

belts contain 10 per cent or more ultramafic rocks and range from bimodal to calc-alkaline. Andesites are particularly uncommon in greenstone belts in South Africa and Australia. Glikson (1976) has also indicated that when two or more ages of greenstone belts are present in the same area, the older belts are bimodal in character and the younger ones are often calc-alkaline. Available radiometric ages suggest that bimodal greenstone belts characterize most of the preserved Archaean (2.7–3.8 Ga) and that calc-alkaline belts formed chiefly during the late Archaean (2.6–2.7 Ga).

Andesites and felsic volcanics become increasingly more abundant in most greenstone successions as a function of increasing stratigraphic height, as illustrated in Fig. 1 for two greenstone belts in the Superior Province. Three stratigraphic sections in each belt are grouped into lower and upper subdivisions. Also, shown on the sections is the distribution of igneous rock series as a function of stratigraphic height. General results may be summarized as follows (after Goodwin, 1977):

(1) Tholeiitic rocks and the tholeiitic series dominate in the lower part of all sections.
(2) Calc-alkaline rocks dominate in the upper parts of all but two subdivisions and dominate in the upper subdivisions, in general. Of the calc-alkaline rocks, andesite is most abundant.
(3) Volcanic rocks of the alkaline series have limited stratigraphic and geographic distribution.
(4) The lower parts of the Manitou and Uchi Lake sections are bimodal.
(5) Ultramafic rocks occur only in lower subdivisions.

Volcanic cycles are well known in Archaean greenstone belts (Anhaeusser, 1971; Hubregtse, 1976). Such cycles occur on scales of tens (minor cycles) to hundreds or thousands of metres (major cycles). Some greenstone belts contain as many as 10 major cycles with each cycle ranging up to 4 km in thickness. In general, each successive cycle in calc-alkaline-type greenstone belts tends to become proportionally more enriched in calc-alkaline components, of which andesite is most abundant.

In terms of areal distribution, calc-alkaline volcanic rocks tend to occur as part of volcanic centres which are tens of kilometres across (Goodwin *et al.*, 1972; Hallberg *et al.*, 1976). These centres appear to have been, at least in part, subareal in character. Pyroclastic rocks tend to become more abundant as the centres are approached. At greater distances, tuffs and

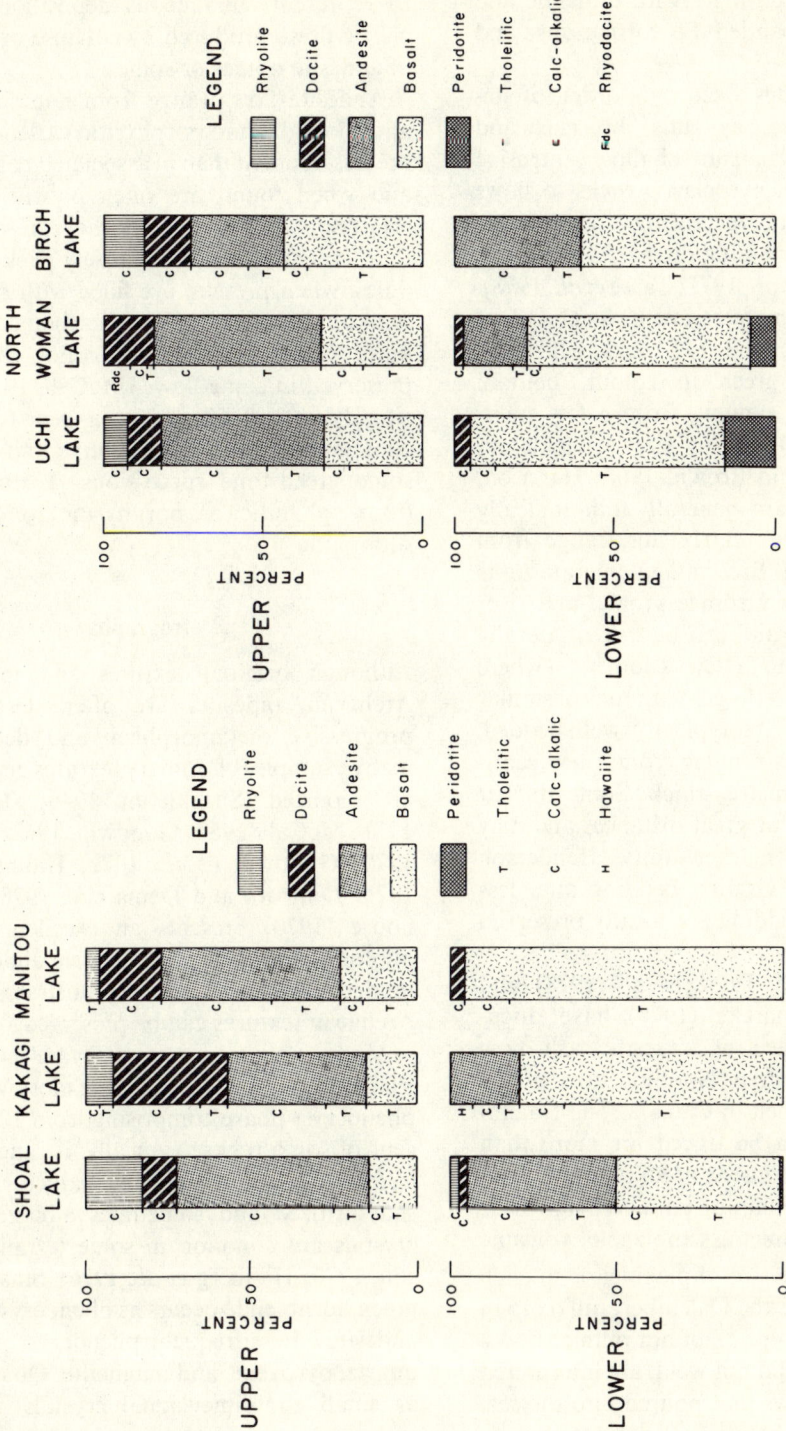

Fig. 1 Weighted mean abundances of volcanic rock types in the lower and upper parts of three volcanic sections from each of two Archaean greenstone belts from the Superior Province in Canada (from Goodwin, 1977; reproduced by permission of the author)

flows merge and interfinger with tholeiitic and ultramafic flows, banded Fe formations, and graywackes.

Archaean andesites occur, in order of decreasing abundance, as tuffs, breccias and agglomerates, flows, and shallow intrusive bodies. The ratio of pyroclastic rocks to flows increases both with increasing stratigraphic height and with decreasing distance to eruptive centres (Goodwin et al., 1972; Tassé et al., 1978). Breccias and agglomerates are most abundant around volcanic vents. These rocks, which are typically gray to green in colour, contain volcanic fragments ranging from a few millimetres to over 30 cm in size (Shackleton, 1946; Henderson and Brown, 1966; Harrison, 1970). Fragments are generally lithologically similar to enclosing matrix and range from angular to rounded. Breccia-agglomerate units are poorly sorted, vary from less than 1 m to over 100 m in thickness, and can be traced laterally over distances up to a few kilometres where they grade into or interfinger with tuffs of similar composition. Tuffs are typically well bedded, with individual beds ranging from a few centimetres to tens of metres thick. Some thicker beds can be traced for great distances and may provide distinctive marker units (Henderson and Brown, 1966). Graded bedding and, less commonly, cross-bedding are locally preserved in tuff units.

Recent studies of Tassé et al. (1978) and Dimroth and Demarcke (1978) have documented the existence of andesitic ash flow eruptions in the Rouyn-Noranda area of the Abitibi greenstone belt in Canada. The Dalembert tuff, which can be traced for more than 15 km in this area, consists of an assemblage of pyroclastic flows containing pumice, plagioclase crystals, and a polymict assemblage of volcanic rock fragments. The size of plagioclase quench crystals varies across the Dalembert tuff (c. 15 m thick), suggesting eruption of hot pumice flows. These flows, which did not weld, are interpreted as subaerial ash flows that poured into the sea. A second facies in the area is characterized by very coarse grained fragmental units that often exhibit reverse grading. These are interpreted to represent subaqueous deposition of pyroclastic flows produced by collapse or explosion of andesite domes or spines.

Andesite flows range from homogeneous to amygdaloidal and porphyritic varieties. Pillows are less frequent than in associated basaltic flows and when found are often poorly developed (Harrison, 1970). Usually they are small (less than 20 cm across) and closely packed. Amygdules, when present, are filled with some combination of quartz, epidote, carbonate, and prehnite. Streaky to lenticular flow banding is preserved in some flows (McCall, 1958). Andesitic dykes and sills, which appear to be penecontemporaneous with eruptive units, occur in some greenstone successions. Textures range from aphanitic or porphyritic to ophitic or subophitic.

Petrography

Although original textures and minerals in Archaean andesites are often destroyed by progressive metamorphism and deformation, many examples of primary textures and minerals are reported (Shackleton, 1946; Huddleston, 1951; McCall, 1958; Goodwin, 1962; Harrison, 1970; Goodwin et al., 1972; Hallberg et al., 1976; Dimroth and Demarcke, 1978). Moorehouse (1970) presents an excellent series of photomicrographs of Archaean andesites and modern counterparts which illustrates how well Archaean textures can be preserved.

Many Archaean andesites are porphyritic. Plagioclase (An_{25}–An_{35}) is the most widespread phenocryst phase comprising from 10 to 40 per cent of some rocks. Typically, it ranges from 1 to 5 mm in length and is partially sericitized or saussuritized and sometimes albitized. Zoned crystals are common in some terrains (Harrison, 1970; Hallberg et al., 1976). Smaller, blue-green hornblende occurs as phenocrysts in some andesites. Less frequent phenocryst phases are quartz, pyroxene, and magnetite. Quartz occurs as small equidimensional crystals sometimes embayed by surrounding matrix; euhedral shapes are preserved in some samples. Both clino- and orthopyroxene are reported from

Archaean andesites. Clinopyroxene (augite) is most widespread and ranges from 1 to 2 mm in length. Remnants of brown orthopyroxene occur in some andesites. Most orthopyroxene is partially or completely replaced by chlorite, Fe–Ti oxides, and actinolite. Small magnetite phenocrysts partially replaced with secondary Fe–Ti oxides, sphene, and leucoxene occur in some andesites.

Aphyric andesites and the groundmass of porphyritic varieties are composed of a fine grained intergrowth of plagioclase microlites, clinopyroxene, and a variety of secondary minerals including some combination of chlorite, actinolite, carbonate, epidote, zoisite, Fe–Ti oxides, sphene, quartz, prehnite, zeolites, and pyrite. Plagioclase is generally similar in composition to plagioclase phenocrysts and may exhibit a pilotaxitic or trachytic texture. Augite is often highly chloritized and orthopyroxene occurs only as pseudomorphs. Pyrite, carbonate, and quartz often occur in veinlets, indicating a post-metamorphic origin. Pseudomorphs of perlitic cracks have been reported in some andesitic rocks that were originally glassy (Harrison, 1970).

Alteration and Classification

Most or all Archaean andesites have undergone some degree of alteration (Condie, 1976b). Such alteration appears to have affected the composition of many rocks, as indicated by the scatter of major and trace element data on conventional variation diagrams. Studies of progressive alteration and low grade metamorphism of modern deep sea tholeiites have been beneficial in enhancing our understanding of the effects of low grade processes on element distributions (Hart, 1969; Christensen et al., 1973; Hart et al., 1974) (Table 2). These studies compare the results of altered and fresh portions of the same sample as well as the results between samples and between rims and centres of pillows. Data indicate that alkaline elements, Fe^{3+}, total Fe, and H_2O are often enriched during alteration and the early stages of metamorphism. Many transition elements (and REE, Nb, Zr, Ta, etc.), however, appear to resist change during such processes. Results of a recent study have shown that major compositional changes can occur in Archaean volcanic terrains during alteration (Condie et al., 1977). This study shows that Au, As, Sb, Sr, Fe^{3+}, Ca, Br, Ga, and U are enriched and H_2O, Na, Mg, Fe^{2+}, K, Rb, Ba, Si, Ti, P, Ni, Cs, Zn, Nb, Cu, Zr, and Co are depleted during epidotization. CO_2, H_2O, Fe^{2+}, Ti, Zn, Y, Nb, Ga, Ta, and light REE are enriched and Na, Sr, Cr, Ba, Fe^{3+}, Ca, Cs, Sb, Au, Mn, and U are depleted during carbonization–chloritization. The elements least affected by epidotization are Hf, Ta, Sc, Cr, Th, and REE, and those least affected by carbonization–chloritization are Hf, Ni, Co, Zr, Th, and heavy REE. Up to c. 10 per cent carbonization and 60 per cent epidotization do not appreciably affect the latter two groups of elements. Recent studies by Beswick and Soucie (1978) suggest that K_2O, Na_2O, CaO, MgO, and FeO (total) have been mobilized in different degrees in volcanic rocks in various parts of the Timaganu greenstone belt in Ontario. These authors propose a graphical procedure for correcting for element redistributions and caution that characterization of Archaean volcanic rocks by major element compositions may be misleading.

TABLE 2 Summary of changes in element concentrations during progressive alteration and metamorphism†

I. Little or no Change	II. Significant Change
TiO_2, Na_2O, Y, REE, Zr, Zn, V, Sc, Hf, Nb, Ta, Co, (Total Fe), (Cr), (Sr), (Ni), (Cu)	Fe^{3+}/Fe^{2+}, K, Cs, Rb, H_2O, SiO_2, CaO, Al_2O_3, MgO, F, Cl, CO_2, Th, U, (Ba)

† ( ) indicates that element sometimes falls in the other category.

Several methods have been proposed for classifying volcanic rocks employing major element concentrations (*see*, for instance, Rittmann, 1952; Taylor *et al.*, 1969; Irvine and Baragar, 1971; Condie and Moore, 1977). These are dependent upon having an accurate representation of the original composition of the rocks. The method of Irvine and Baragar (1971) is based on analyses recalculated on an H_2O- and CO_2-free basis. They also suggest a procedure to correct for anomalous Fe_2O_3/FeO ratios. The classifications of Taylor *et al.* (1969) and Condie and Moore (1977) are based on SiO_2–K_2O distributions only; caution should be used in classifying Archaean volcanic rocks by these methods in that K_2O is a mobile oxide. Perhaps the most useful methods will be methods based chiefly on immobile element distributions (Winchester and Floyd, 1977). The chief disadvantage of such methods is that concentrations of such elements as Zr, Y, Nb, Sc, and Ga must be known.

Composition

Average compositions of andesites from six Archaean greenstone belts and an average for the Superior Province are given in Table 3. Two types of andesite are present in the Midlands belt in Rhodesia, as reflected by averages 1 and 2 (Condie and Harrison, 1976). The variation between the averages ranges by a factor of two to three for most elements. Light REE and especially the La/Yb ratio are more variable. Although the Na_2O/K_2O ratio ranges from 2 to 7, most values are between 3 and 4. SiO_2, Al_2O_3, Zn, Cu, and Co are similar in all averages. Employing REE, which as previously discussed are examples of elements least susceptible to mobilization during secondary process, it is possible to classify Archaean andesites into three types, I, II, and III (Table 4), originally referred to as DAA, LAA, and HAA, respectively (Condie, 1976*b*). Envelopes of variation of REE patterns for each type are given in Fig. 2. Type I shows slightly enriched light REE (*c.* 50 times chondritic) and negligible Eu anomalies. It also has higher FeO, MgO, Ni, Cr, and Zn and lower K_2O, Rb, and Ba than the other types. Type II andesites are notably enriched in light REE (*c.* 200 times chondritic) and also exhibit negligible Eu anomalies. Some greenstone belts, such as the Yellowknife belt in Canada and the Marda Complex in Western Australia, contain only one type of andesite while others, such as the Midlands in Rhodesia and the Nyanzian belts in Kenya, contain both Type I and Type II. In Kenya these types appear to be mixed stratigraphically, although the stratigraphy is not well known in this area (Davis and Condie, 1976). In the Midlands belt, on the other hand, Type I andesites occur only in the Maliyami Formation and Type II only in the overlying Felsic Formation. Type III andesite, which thus far has been described only from the Abitibi belt in Canada (Condie and Baragar, 1974), is characterized by flat REE patterns (30–40 times chondritic) and negative Eu anomalies. These andesites are closely associated with tholeiites with similar REE patterns, although lower REE concentrations. Compared to Types I and II, Type III is also low in Sr and high in Y.

Modern andesites can also be divided into three categories based on composition and tectonic setting (Jakeš and White, 1972; Condie, 1976*a*; Table 4). Arc andesites (AA) occur in immature, oceanic island arcs (such as the Marianas) and near the trench side of mature arcs. Calc-alkaline andesites (CA) are most widespread in modern arc systems, and high K calc-alkaline andesites (HKA) occur in some continental margin arc systems (such as the Andes) which are underlain by thick lithosphere. Although, in terms of many major elements, it is tempting to equate each of the Archaean andesite Types I, II, and III with modern andesites CA, HKA, and AA, respectively, several important differences make such correlations improbable. First, all Archaean andesites differ from modern andesites in terms of their low Al_2O_3 contents and their high FeO, MgO, Y, and FeO/Fe_2O_3, and Ni/Co ratios. Among the transition trace elements Ni, Cr, Co, and Zn are also enriched in Archaean andesites

(Fig. 3). Such enrichments in transition metals also occur in all other Archaean volcanic rocks in comparison with modern equivalents (Condie, 1976b). In addition to the overall differences, many arc andesites (AA) differ from Type III andesites in having lower concentrations of REE and no Eu anomalies (Fig. 2). Some arc andesites, such as those from the South Sandwich Islands

TABLE 3 Average compositions of Archaean andesites

	1	2	3	4	5	6	7	8
SiO_2	56.6	59.7	57.9	56.9	60.8	55.1	56.7	55.9
TiO_2	0.58	0.69	1.2	0.93	0.77	0.96	1.40	0.91
Al_2O_3	15.1	15.8	15.6	14.1	14.7	15.9	15.5	15.6
Fe_2O_3	1.80	1.57	1.42	2.01	1.7	1.99	1.67	2.01
FeO	5.41	5.18	5.00	6.64	5.3	5.86	6.80	5.90
MgO	6.15	3.42	3.01	5.45	3.5	4.30	4.32	4.07
CaO	6.96	3.61	6.92	6.59	5.5	5.90	6.05	5.88
Na_2O	2.86	5.46	4.23	3.37	3.6	3.85	4.00	3.67
K_2O	0.85	1.62	1.35	0.79	1.7	1.14	0.59	1.11
H_2O	3.02	2.90	2.40		1.8	2.84		2.54
FeO/Fe_2O_3	3.0	3.4	3.5	3.3	3.1	2.9	4.1	2.9
Na_2O/K_2O	3.4	3.4	3.1	4.3	2.1	3.4	6.8	3.3
Cr	175	127	100	154	91	105	131	111
Zn				97	61	77	84	80
Cu			47	50	64	63	64	
Ni	95	67	98	73	67	55	75	61
Co	29	22	32	24		29	25	28
Sr	241	291	408	358	319	210	220	242
Rb	24	48	30	24	48	30	25	30
Ba	228	310	420	237	740	361	314	358
Zr	75	65	251	177	200	104		145
La	10	26	37	15	47	12	38	26
Ce	24	50	72	34	78	30	61	46
Sm	2.7	5.1	7.0	4.4	5.3	7.3	5.1	6.2
Eu	0.92	1.5	1.8	1.2	1.5	2.0	1.3	1.7
Tb	0.53	0.57	1.1	0.62	0.77	1.7	0.68	1.2
Yb	1.3	1.6	2.7	1.9	2.13	6.1	1.7	4.0
Lu	0.20	0.24	0.62	0.34	0.31	1.1	0.25	0.63
Y	15	13		26	22	40		14
K/Rb	238	268	325	273	286	315	272	307
Ni/Co	3.2	3.0	3.1	3.0		1.9	3.0	2.2
La/Yb	7.7	16	14	7.6	22	2.0	22	6.5
Eu/Eu*	1.0	1.0	0.82	0.88	0.93	0.75	0.84	0.82
n†	5(5)	3(3)	6(3)	8(8)	15(5)	393(5)	13(3)	582(10)
Classification	II	III	III	I	III	I	III	II–III

† n = number of samples averaged; value in parentheses refers to REE.
(1) Maliyami Formation, Midlands belt, Rhodesia (Condie and Harrison, 1976; Hawkesworth and O'Nions, 1977). (2) Felsic Formation, Midlands belt, Rhodesia (Condie and Harrison, 1976). (3) South Pass greenstone belt, Wyoming (Condie and Baragar, 1974; Bayley et al., 1973). (4) Nyanzian System, Western Kenya (Andesite I; Davis and Condie, 1976). (5) Marda Complex, Western Australia (Hallberg et al., 1976; Taylor and Hallberg, 1977). (6) Abitibi greenstone belt, Ontario (Goodwin, 1977; Condie and Baragar, 1974). (7) Yellowknife greenstone belt, North-West Territories, Canada (Baragar and Goodwin, 1969; Condie and Baragar, 1974). (8) Superior Province average (Goodwin, 1977; Condie and Baragar, 1974).

TABLE 4 Average composition of Archaean andesite groups compared to modern andesites

	Archaean			Modern		
	I	II	III	Arc	Calc-Alkaline	High-K Calc-Alkaline
SiO_2	56.7	58.9	55.1	57.3	59.5	60.2
TiO_2	0.92	0.65	0.96	0.58	0.70	0.95
Al_2O_3	14.0	15.5	15.9	17.4	17.2	16.9
Fe_2O_3	2.3	1.5	1.99	2.5	2.5	2.6
FeO	7.0	4.5	5.86	2.7	5.0	2.8
MgO	5.4	4.5	4.3	3.5	3.4	2.2
CaO	6.6	5.1	5.9	8.7	7.0	5.5
Na_2O	3.4	4.0	3.9	2.6	3.7	3.7
K_2O	0.67	1.9	1.1	0.7	1.6	2.8
H_2O	3.0	3.0	2.8	1.0	1.0	1.0
FeO/Fe_2O_3	3.0	3.0	2.9	1.1	2.0	1.2
Na_2O/K_2O	5.1	2.1	3.4	3.7	2.3	1.2
Cr	125	88	105	40	90	90
Zn	97	81	77	60	65	
Cu	60	36	64	70	100	40
Ni	70	60	55	20	25	40
Co	25	23	29	20	25	20
Sr	278	580	210	240	475	700
Rb	22	75	30	20	40	80
Ba	230	547	361	150	300	700
Zr	150	190	104	90	110	200
La	13	34	12	3	12	43
Ce	31	70	30	6.8	25	84
Sm	3.6	6.7	7.3	2.3	3.0	5.1
Eu	1.1	1.9	2.0	0.9	1.0	1.4
Tb	0.6	1.0	1.7	0.66	0.7	0.56
Yb	1.8	2.4	6.1	2.3	1.9	1.6
Lu	0.3	0.3	1.1	0.4	0.4	0.27
Y	25	35	40	25	20	10
K/Rb	253	210	315	291	332	208
Ni/Co	2.8	2.6	1.9	1.0	1.0	2.0
La/Yb	7.2	14	2.0	1.1	6.3	2.7
Eu/Eu*	0.96	0.90	0.75	1.0	0.90	1.0

After Condie and Harrison (1976), Condie (1976b), and miscellaneous sources.

(Hawkesworth et al., 1977) and the New Hebrides Islands (Gorton, 1977) overlap with Type III andesites in absolute REE abundances and may have negative Eu anomalies. CA and HKA are also somewhat higher in K_2O, Rb, Sr, and Ba than most Type I or II andesites, respectively. Their REE patterns are, however, strikingly similar to the modern groups (Fig. 2).

In terms of igneous rock series, both calc-alkaline and tholeiitic series occur in non-bimodal greenstone belts. Locally, the alkaline series may be present, but rarely is of importance (Goodwin, 1977). Although members of the tholeiitic and calc-alkaline series are closely associated in greenstone successions, they occur as distinct, often mappable units that are interbedded (Gelinas et al., 1977). An example of a greenstone belt with mixed calc-alkaline and tholeiitic components is the Yellowknife belt which is illustrated on an AFM diagram in Fig. 4.

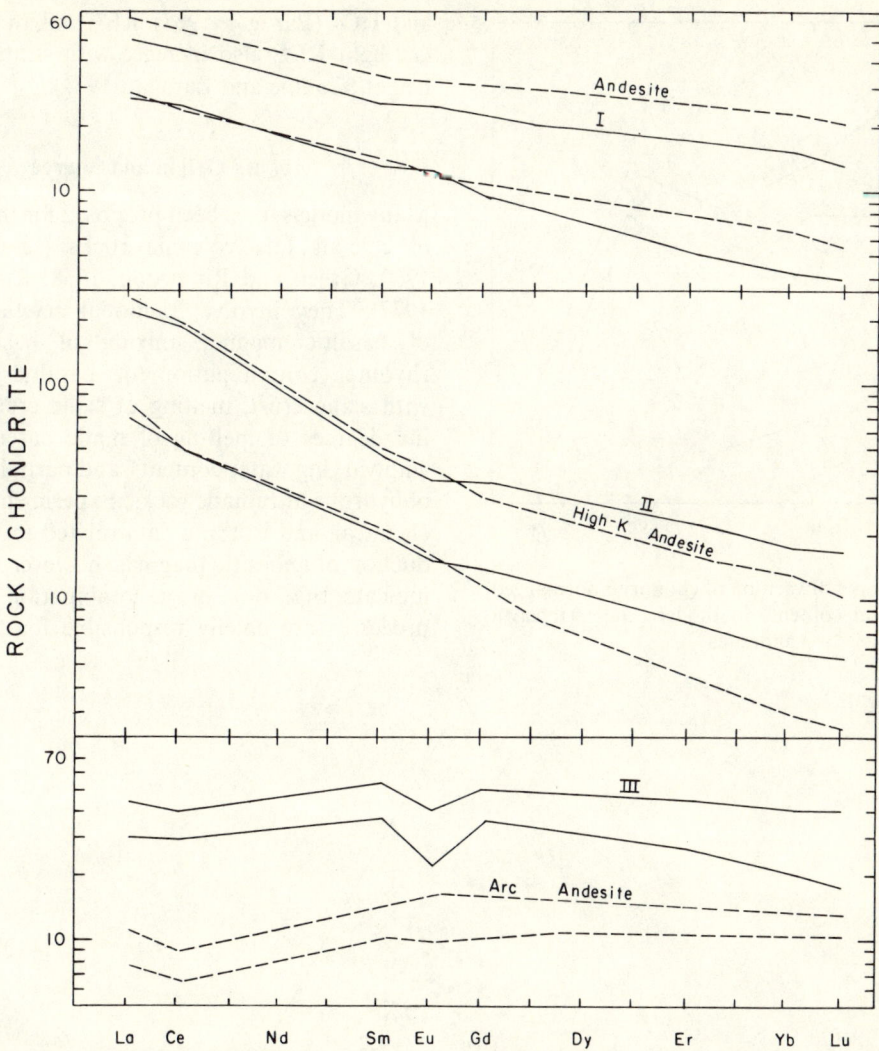

Fig. 2 Envelopes of variation of chondrite-normalized REE patterns for Archaean andesite types I, II, and III (solid lines) compared to modern andesite groups (broken lines, of Table 4)

Members of the calc-alkaline series become increasingly more abundant with increasing stratigraphic level (Baragar, 1966, 1968). In some greenstone belts, a third series, referred to as the magnesian series (or suite) by Jolly (1975) and the komatiitic series by Arndt et al. (1977), has been identified. This series is characterized by rapid changes in MgO with only small changes in FeO (Fig. 5). Jolly (1975) has shown that the Abitibi belt in Ontario can be broadly divided into three subdivisions with the komatiitic series dominating in the lowest part, the tholeiitic series in the middle, and the calc-alkaline series in the uppermost part. Existing data suggest that systematic changes in composition occur in greenstone volcanic successions with stratigraphic height (Baragar, 1966, 1968; Jolly, 1975; Gelinas et al., 1977; Goodwin, 1977). As illustrated by the Abitibi belt in the Noranda area, these changes are characterized, with increasing stratigraphic height, by increases in Al_2O_3 and K_2O and decreases in FeO (total), MgO,

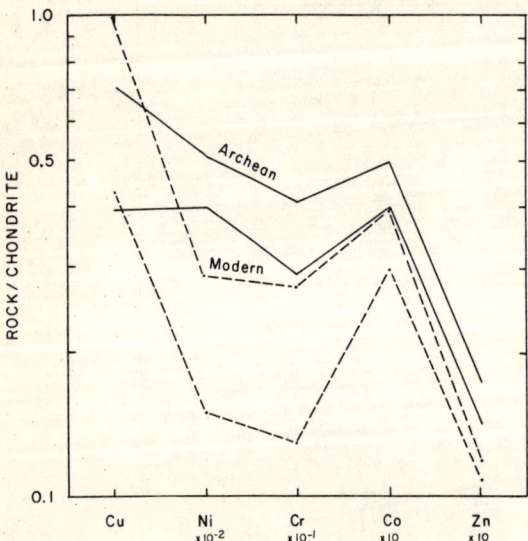

Fig. 3 Envelopes of varition of chondrite-normalized transition metal contents in modern and Archaean andesites

and TiO_2 (Baragar, 1968). REE and, in particular, light REE also increase with stratigraphic height (Condie and Baragar, 1974).

Magma Origin and Source

Many models have been proposed for the origin of calc-alkaline volcanic rocks (*see* Taylor, 1969; Green and Ringwood, 1968; Ringwood, 1977). They involve fractional crystallization of basaltic magma, mixing of basalt and rhyolite, contamination of basaltic magma with sialic crust, melting of sialic crust, varying degrees of melting of mafic parent rocks with varying water contents, and partial melting of hydrous ultramafic rocks. Experimental, geochemical, and isotopic data related to the production of andesitic magmas, however, seem to indicate that one or a combination of two processes are chiefly responsible for the pro-

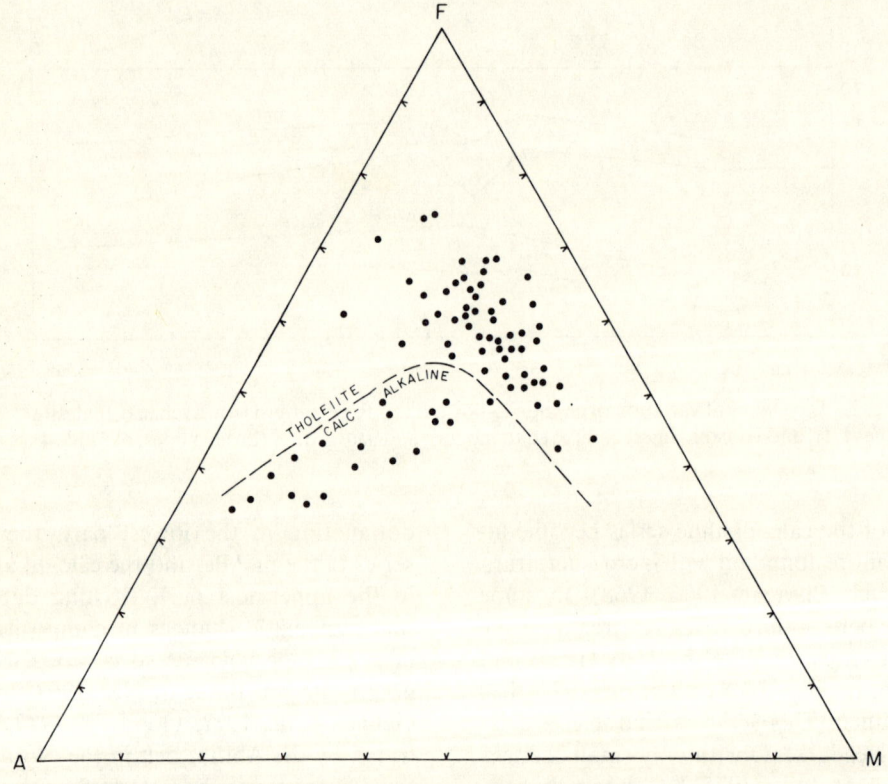

Fig. 4 AFM diagram for volcanic rocks from the Yellowknife greenstone belt, Canada (from Baragar, 1966; reproduced by permission of the National Research Council of Canada from the *Canadian Journal of Earth Sciences*). Tholeiite—calc-alkaline boundary from Irvine and Baragar (1971). $A = Na_2O + K_2O$; $F = FeO$ (total); $M = MgO$

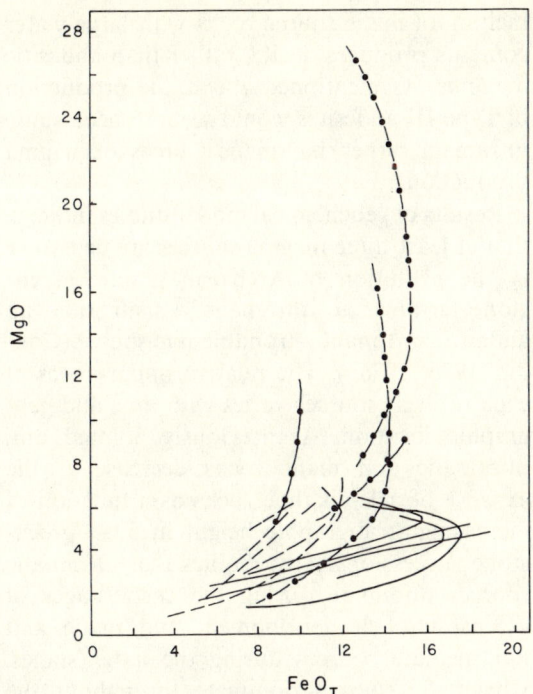

Fig. 5 FeO$_t$–MgO relations of lavas from the Abitibi belt, Canada (from Jolly, 1975; reproduced by permission of Elsevier Scientific Publishing Co.). Each line represents a different stratigraphic traverse. FeO$_t$ = total Fe as FeO. Symbols: —·—·—, komatiitic series; — — — —, calc-alkaline series; ———, tholeiitic series

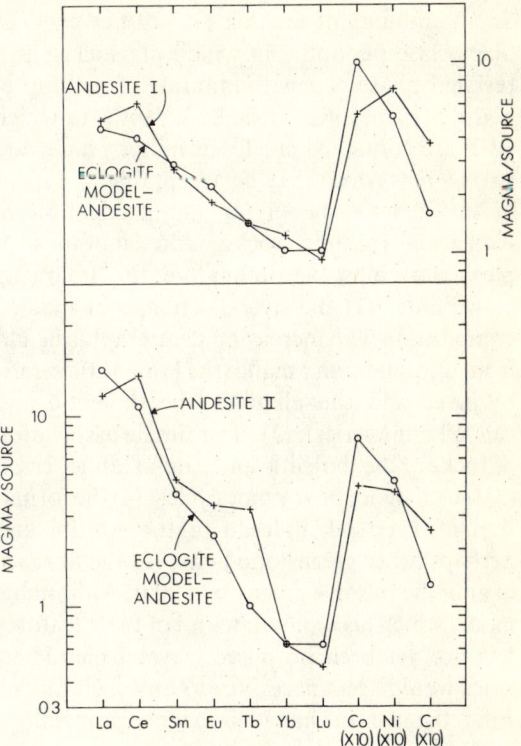

Fig. 6 Source-normalized trace element plot for Type I and Type II Archaean andesites from Kenya compared to model andesite melts (from Davis and Condie, 1976). Upper: 20 per cent modal melting of eclogite with cpx = 0.7, garnet = 0.1, and qtz = 0.2. Lower: 5 per cent modal melting of eclogite with cpx = 0.7, garnet = 0.2, and qtz = 0.1. Source eclogite has composition of average Nyanzian tholeiite

duction of such magmas: (1) varying degrees of melting of mafic mantle sources (amphibolite, eclogite, granulite) with variable water contents; (2) fractional crystallization of tholeiite magmas in which fractionating minerals are high in Al, Fe, and heavy REE (Ringwood, 1977; O'Nions and Pankhurst, 1978). Other mechanisms such as crustal contamination and magma mixing may modify andesite magmas locally after their production in the mantle.

Trace element model studies are particularly valuable in enhancing our knowledge of calc-alkaline magma production and have been applied to both modern and Archaean examples (DeLong, 1974; Condie, 1976b; Kay, 1977). An example of such studies applied to the origin of Type I and Type II Archaean andesites is summarized in Fig. 6. The model andesite composition in each case agrees well with the observed andesite composition and is consistent with an origin for andesites I and II by 20 per cent and 5 per cent partial melting, respectively, of eclogite (Davis and Condie, 1976). Mafic garnet amphibolite and amphibolite source materials also produce acceptable models for production of these andesites; ultramafic sources, however, are not allowed Types I and II. Similar models have been found acceptable for Archaean andesites from other areas (Condie and Harrison, 1976; Condie, 1976b). Hawkesworth and O'Nions (1977) favour a fractional crystallization model in which garnet or amphibole are removed for the production of Type I andesites from Rhodesia. The unique REE distributions in Type III Archaean andesites seem to require a special mechanism for their production (Condie and Baragar, 1974).

Small amounts of melting (10–20 per cent) of plagioclase peridotite in which plagioclase is a residual phase or small amounts of melting of hydrous periodotite at shallow depths in which REE are contained chiefly in minor phases are two models which may be acceptable.

Any general model for the production of Archaean volcanic rocks, and andesites in particular, must accommodate the following constraints: (1) the overall change in magma composition with increasing stratigraphic height from ultramafic and mafic (the komatiitic series) to mafic and calc-alkaline (the tholeiitic and calc-alkaline series); (2) the intimate association of rocks of the tholeiitic and calc-alkaline series; (3) the existence of volcanic cycles; (4) the formation of Type III andesite in the Abitibi and perhaps other greenstone belts; (5) the absence of andesite in some greenstone belts. Although a model which accommodates all of these features has not yet been proposed, several characteristics would seem necessary to any such model. First, the overall change in magma proportions with time in a given greenstone succession is probably controlled directly or indirectly by a falling geothermal gradient (Condie, 1976b). This would result in either or both progressively decreasing amounts of melting of source rocks or greater amounts of fractional crystallization with time. The easiest way to explain the close association of rocks of the calc-alkaline and tholeiitic series is for them to have a common origin. This can be accomplished by fractional crystallization of a tholeiitic parent magma at varying depths. Olivine, pyroxenes, and plagioclase are removed at shallow depths (less than 35 km) and produce the tholeiitic series; clinopyroxene and garnet and/or amphibole (35–80 km) are removed at greater depths, producing the calc-alkaline series. The only way around the volcanic cycle problem seems to be by replenishment of source rocks in the mantle (Condie, 1975). This may be accomplished by the introduction of new material either in mantle plumes or in descending lithospheric slabs. The absence of andesite in some successions may be related to greater water contents in magma source regions (Barker and Peterman, 1974). Partial melting of mafic source rocks with large water contents produces dacitic rather than andesitic magmas. As mentioned above, the production of Type III andesites would seem to necessitate ultramafic rather than mafic sources for magma production.

Results of geochemical model studies indicate that at least three magma sources are necessary in the evolution of Archaean granite–greenstone terrains: an ultramafic, a mafic, and an andesitic or tonalite–trondhjemite source (Condie, 1975, 1976b). The relative importances of each of these sources varies with time and geographic location. As previously pointed out, ultramafic and mafic rocks decrease at the expense of calc-alkaline rocks as a function of increasing stratigraphic height in most greenstone successions. This implies that ultramafic sources dominate during the early stages of greenstone belt development and mafic and intermediate sources during the latter stages. Ultramafic sources dominate throughout the succession in some greenstone terrains such as in Western Australia and mafic sources dominate in others such as in western Kenya. Existing data indicate, however, that all three sources are simultaneously available throughout granite–greenstone belt evolution. Progressive melting and successive tapping of mantle plumes provides a means of obtaining undepleted to depleted mafic and ultramafic magmas from a similar source (Naldrett and Turner, 1977). Because mafic source rocks are required for the production of voluminous Archaean trondhjemite–tonalite magmas (Barker and Arth, 1976), two or more magmatic stages must be involved in the production of sialic crust. It is necessary first to produce large volumes of tholeiitic lavas which sink and are metamorphosed to garnet and/or amphibole-bearing assemblages that serve as sources for trondhjemite–tonalite melts. The general succession from mafic to felsic compositions with time in granite–greenstone terrains appears to reflect a decreasing thermal gradient and hence a shallower depth of melting in the mantle. Volcanic cyclicity necessitates that magma source rocks must be replenished since the same source rock cannot be used more than once for the same kind of magma (Condie, 1975, 1976b).

The origin of the bimodal association in Archaean granitic gneiss complexes (Barker and Peterman, 1974) and the origin of bimodal greenstone belts are important problems related to Archaean magma production. Experimental data indicate that increased amounts of water in mafic parent rocks results in the production of dacitic rather than andesitic melts which are produced in less water-rich systems (Green and Ringwood, 1968; Green, 1972). It has been suggested that the easiest way to explain the absence or sparsity of rocks of andesitic composition in Archaean gneissic complexes is that melting of mafic parent rocks in the mantle occurred under water-rich conditions (Barker and Peterman, 1974). Similar conditions could also apply to the bimodal-type greenstone belts. This would suggest that the calc-alkaline type greenstone belts, which contain andesite, evolved from less water-rich magma source areas. Thus, varying amounts of water liberated by the mantle may have partially controlled the relative abundances of igneous rocks in Archaean granite–greenstone terrains.

It is now clearly established that tholeiites and andesites of Archaean age are enriched in most transition trace metals compared to modern counterparts (Fig. 3; Glikson, 1971; Condie, 1976b). Cr, Ni, and Co are significantly enriched and Zn somewhat enriched in the Archaean rocks compared to modern rocks; Cu, however, shows no enrichment. At least three causes might be considered for the Archaean enrichment (Condie, 1976b; Nesbitt and Sun, 1976): (1) more accumulation of such phases as olivine and sulphides in Archaean magmas; (2) larger degrees of melting in the Archaean than in the Phanerozoic; and (3) the mantle has become depleted in transition trace metals with time. There is no petrographic or chemical evidence to support hypothesis number one. Higher temperatures in the Archaean mantle could result in lowering distribution coefficients for transition metals (which are very temperature sensitive) and also result in larger amounts of melting. The net result would be production of Archaean magmas with higher transition metal contents than Phanerozoic counterparts. Incompatible elements, in turn, should be lower in Archaean mafic and andesitic volcanics. To some extent this is observed for tholeiites, although not for andesites. Naldrett (1973) has suggested a mechanism by which the upper mantle can become depleted in S with time. Experimental data indicate that sulphides in the Archaean upper mantle would be completely melted when the silicate fraction is only partially melted. The sulphides may separate as immiscible liquid droplets and be carried upwards with basaltic magmas, depleting the mantle source in S. Because most transition trace metals are, in part, chalcophyllic, they may also be concentrated in the immiscible liquid droplets and carried upwards, thus depleting the mantle source in these elements. Archaean andesites would therefore inherit enriched metal contents from their mafic sources enriched in metals by this mechanism.

Tectonic Setting

Much has been written regarding the tectonic setting of Archaean greenstone belts (*see* Windley, 1973, 1976, for summaries). Many investigators have pointed out the broad similarity of modern subduction zone-related andesites to Archaean greenstone andesites and, in turn, have suggested a similar tectonic setting for the Archaean varieties (Wilson *et al.*, 1965; White *et al.*, 1971; Jahn *et al.*, 1974; Burke *et al.*, 1976; Condie and Harrison, 1976; Goodwin, 1977). Others have emphasized the differences between modern and Archaean volcanic rocks and have proposed non-subduction zone-type models (Macgregor, 1951; Green, 1972; Condie, 1975).

In the opinion of the author, one, or perhaps both, of two models seem most consistent with available geological and geochemical data from Archaean greenstone belts (Condie, 1979): a convergent plate boundary model or a continental rift model. Tarney *et al.* (1976) propose a model for Archaean greenstone belts involving marginal sea basins and associated arc systems.

The sequence of events in this model is based on the Rocas Verdes Complex of Cretaceous age in Chile. In the model, andesites are produced together with tholeiites by varying degrees of melting in the descending plate. Mixed tholeiites, komatiites, calc-alkaline volcanic rocks, and sediments collect in the marginal basin, which is later deformed during tonalite plutonism. Source area replenishment is provided by the descending slab. Continental rift models for Archaean greenstone belts have been suggested by Windley (1973), Condie (1975), and Condie and Hunter (1976). In common, they have sialic crust which ruptures by downwarping. Such downwarping may be activated by mantle–plume activity, and plumes provide a means for replenishing source rocks for ultramafic and komatiitic lavas and may explain the widespread distribution of calc-alkaline volcanic centres in greenstone belt terrains (Goodwin et al., 1972; Hallberg et al., 1976). It is possible, as suggested by Condie (1979), that both continental rifts and continental margin arc systems existed in the Archaean. Those greenstone belts containing an abundance of ultramafic volcanic rocks may have formed in rift environments and those containing large proportions of andesite, in marginal basin and arc environments.

REFERENCES

Anhaeusser, C. R. (1971). Cyclic volcanicity and sedimentation in the evolutionary development of Archean greenstone belts of shield areas. Geol. Soc. Aust. spec. Publ. 3, 57–70.

Arth, J. G. and Hanson, G. N. (1975). Geochemistry and origin of the early Precambrian crust, northeastern Minnesota. Geochim. cosmochim. Acta 39, 325–362.

Arndt, N. T., Naldrett, A. J., and Pyke, D. R. (1977). Komatiitic and iron-rich tholeiitic lavas of Munro Township, northeast Ontario. J. Petrol. 18, 319–369.

Baragar, W. R. A. (1966). Geochemistry of the Yellowknife volcanic rocks. Can. J. Earth Sci. 3, 9–30.

Baragar, W. R. A. (1968). Major element geochemistry of the Noranda volcanic belt, Quebec–Ontario. Can. J. Earth Sci. 5, 773–790.

Baragar, W. R. A. and Goodwin, A. M. (1969). Andesites and Archean volcanism of the Canadian Shield. Oregon St. Dept. Geol. Mineral. Indust. Bull. 65, 121–142.

Barker, F. and Arth, J. G. (1976). Generation of trondhjemitic–tonalitic liquids and Archean bimodal trondhjemite–basalt suites. Geology 4, 596–600.

Barker, F. and Peterman, Z. E. (1974). Bimodal tholeiitic–dacitic magmatism and the early Precambrian crust. Precambrian Res. 1, 1–12.

Bayley, R. W., Proctor, P. D., and Condie, K. C. (1973). Geology of the South Pass area, Fremont County, Wyoming. US geol. Surv. prof. Pap. no. 793.

Beswick, A. E. and Soucie, G. (1978). A correction procedure for metasomatism in an Archean greenstone belt. Precambrian Res. 6, 235–248.

Burke, K., Dewey, J. F., and Kidd, W. S. F. (1976). Dominance of horizontal movements, arc and microcontinental collisions during the later permobile regime. In The Early History of the Earth (B. F. Windley, ed.), John Wiley, New York, pp. 113–129.

Christensen, N. I., Frey, F., MacDougall, D., Melson, W. G., Peterson, M. N. A., Thompson, G., and Watkins, N. (1973). Deep-sea Drilling Project: properties of igneous and metamorphic rocks of the oceanic crust. Am. geophys. Union Trans. 54, 972–1035.

Condie, K. C. (1975). Mantle-plume model for the origin of Archean greenstone belts based on trace element distributions. Nature, Lond. 258, 413–414.

Condie, K. C. (1976a). Plate Tectonics and Crustal Evolution, Pergamon Press, New York.

Condie, K. C. (1976b). Trace-element geochemistry of Archean greenstone belts, Earth Sci. Rev. 12, 393–417.

Condie, K. C. (1979). Origin and early development of the Earth's crust. Precambrian Res. in press.

Condie, K. C. and Baragar, W. R. A. (1974). Rare-earth element distributions in volcanic rocks from Archean greenstone belts. Contr. Mineral. Petrol. 45, 237–246.

Condie, K. C. and Harrison, N. M. (1976). Geochemistry of the Archean Bulawayan Group, Midlands Greenstone Belt, Rhodesia. Precambrian Res. 3, 253–271.

Condie, K. C. and Hunter, D. R. (1976). Trace-element geochemistry of Archean granitic rocks from the Barberton region, South Africa. Earth planet. Sci. Letters 29, 389–400.

Condie, K. C. and Moore, J. M. (1977). Geochemistry of Proterozoic volcanic rocks from the Grenville Province, eastern Ontario. Geol. Assoc. Can. spec. Pap. 16, 149–168.

Condie, K. C., Viljoen, M. J., and Kable, E. J. D. (1977). Effects of alteration on element distributions in Archean tholeiites from the Barberton greenstone belt, South Africa. Contr. Mineral. Petrol. 64, 75–89.

Davis, P. A., Jr and Condie, K. C. (1976). Trace element model studies of Nyanzian greenstone belts, western Kenya. Geochim. cosmochim. Acta 41, 271–277.

DeLong, S. E. (1974). Distribution of Rb, Sr, and Ni in igneous rocks, central and western Aleutian Islands, Alaska. Geochim. cosmochim. Acta 38, 245–266.

Dimroth, E. and Demarcke, J. (1978). Petrography and mechanism of eruption of the Archean Dalembert tuff, Rouyn-Noranda, Quebec, Canada. Can. J. Earth Sci. 15, 1712–1723.

Gelinas, L., Brooks, C., Perrault, G., Carignan, J., Trudel, P., and Grasso, F. (1977). Chemo-stratigraphic divisions within the Abitibi belt, Rouyn-Noranda District, Quebec. *Geol. Assoc. Can. spec. Pap.* **16**, 265–296.

Glikson, A. Y. (1971). Primitive Archean element distribution patterns: chemical evidence and geotectonic significance. *Earth planet. Sci. Letters* **12**, 309–320.

Glikson, A. Y. (1976). Stratigraphy and evolution of primary and secondary greenstone belts: significance of data from shields of the southern hemisphere. In *The Early History of the Earth* (B. F. Windley, ed.), John Wiley, New York, pp. 257–277.

Goodwin, A. M. (1962). Structure, stratigraphy, and origin of iron formations, Michipicoten area, Algoma District, Ontario, Canada. *Geol. Soc. Am. Bull.* **73**, 561–586.

Goodwin, A. M. (1977). Archean volcanism in Superior Province, Canadian Shield. *Geol. Assoc. Can. spec. Pap.* **16**, 205–241.

Goodwin, A. M., et al. (1972). The Superior Province. *Geol. Assoc. Can. Spec. Pap.* **11**, 527–624.

Gorton, M. P. (1977). The geochemistry and origin of Quaternary volcanism in the New Hebrides. *Geochim. cosmochim. Acta* **41**, 1257–1270.

Green, D. H. (1972). Archean greenstone belts may include terrestrial equivalents of lunar maria? *Earth planet. Sci. Letters* **15**, 263–270.

Green, T. H. and Ringwood, A. E. (1968). Genesis of the calc-alkaline igneous rock suite. *Contr. Mineral. Petrol.* **18**, 105–162.

Hallberg, J. A., Johnston, C., and Bye, S. M. (1976). The Archean Marda Igneous Complex, Western Australia. *Precambrian Res.* **3**, 111–136.

Harrison, N. M. (1970). The geology of the country around Que Que. *Rhod. Geol. Surv. Bull.* no. 67.

Hart, S. R. (1969). K, Rb, Cs contents and K/Rb and K/Cs ratios of fresh and altered submarine basalts. *Earth planet. Sci. Letters* **6**, 295–303.

Hart, S. R., Erlank, A. J., and Kable, E. J. D. (1974). Sea floor basalt alteration: some chemical and isotopic effects. *Contr. Mineral. Petrol.* **44**, 219–230.

Hawkesworth, C. J. and O'Nions, R. K. (1977). The petrogenesis of some Archean volcanic rocks from southern Africa. *J. Petrol.* **18**, 487–520.

Hawkesworth, C. J., O'Nions, R. K., Pankhurst, R. J., Hamilton, P. J., and Evenson, N. M. (1977). A geochemical study of island-arc and back-arc tholeiites from the Scotia Sea. *Earth planet. Sci. Letters* **36**, 253–262.

Henderson, J. F. and Brown, I. C. (1966). Geology and structure of the Yellowknife greenstone belt, District of MacKenzie. *Geol. Surv. Can. Bull.* no. 141.

Hubregtse, J. J. M. (1976). Volcanism in the western Superior Province in Manitoba. In *The Early History of the Earth* (B. F. Windley, ed.), John Wiley, New York, pp. 297–288.

Huddleston, A. (1951). Geology of the Kisii District. *Geol. Surv. Kenya Rep.* no. 18.

Irvine, T. N. and Baragar, W. R. A. (1971). A guide to the classification of the common volcanic rocks. *Can. J. Earth Sci.* **8**, 523–548.

Jakeš, P. and White, A. J. R. (1972). Major and trace element abundances in volcanic rocks of orogenic areas, *Geol. Soc. Am. Bull.* **83**, 29–40.

Jahn, B.-M., Shih, C. Y., and Murthy, V. R. (1974). Trace element geochemistry of Archean volcanic rocks. *Geochim. cosmochim. Acta* **38**, 611–627.

Jolly, W. T. (1975). Subdivision of the Archean lavas of the Abitibi area, Canada, from Fe–Mg–Ni–Cr relations. *Earth planet. Sci. Letters* **27**, 200–210.

Kay, R. W. (1977). Geochemical constraints on the origin of Aleutian magmas. In *Island Arcs, Deep Sea Trenches, and Back-arc Basins* (M. Talwani and W. C. Pitman, eds), American Geophysical Union, Washington, DC, pp. 229–240.

McCall, G. J. H. (1958). Geology of the Gwasi area. *Geol. Surv. Kenya Rep.* no. 45.

Macgregor, A. M. (1951). Some milestones in the Precambrian of Southern Rhodesia. *Trans. geol. Soc. S. Afr.* **54**, 28–74.

Moorehouse, W. W. (1970). A comparative atlas of textures of Archean and younger volcanic rocks. *Geol. Assoc. Can. spec. Pap.* **8**.

Naldrett, A. J. (1973). Nickel sulfide deposits—their classification and genesis, with special emphasis on deposits of volcanic association. *Can. Min. Metal. Bull.* **66**, 45–63.

Naldrett, A. J. and Turner, A. R. (1977). The geology and petrogenesis of a greenstone belt and related nickel sulfide mineralization at Yakabindie, Western Australia. *Precambrian Res.* **5**, 43–103.

Nesbitt, R. W. and Sun, S.-S. (1976). Geochemistry of Archean spinifex-textured peridotites and magnesian and low-magnesian tholeiites. *Earth planet. Sci. Letters* **31**, 433–453.

O'Nions, R. K. and Pankhurst, R. J. (1978). Early Archean rocks and geochemical evolution of the Earth's crust. *Earth planet. Sci. Letters* **38**, 211–236.

Ringwood, A. E. (1977). Petrogenesis in island arc systems. In *Island Arcs, Deep Sea Trenches, and Back-arc Basins* (M. Talwani and W. C. Pitman, eds), American Geophysical Union, Washington, DC, pp. 311–324.

Rittman, A. (1952). Nomenclature of volcanic rocks. *Bull. Volc.* **12**, 75–102.

Shackleton, R. M. (1946). Geology of the Migori gold belt. *Geol. Surv. Kenya Rep.* no. 10.

Tarney, J., Dalziel, I. W. D., and deWit, M. J. (1976). Marginal basin 'Rocas Verdes' Complex from S. Chile: a model for Archean greenstone belt formation. In *The Early History of the Earth* (B. F. Windley, ed.) John Wiley, New York, pp. 131–146.

Tassé, N., Lajoie, J., and Dimroth, E. (1978). The anatomy and interpretation of an Archean volcaniclastic sequence, Noranda region, Quebec. *Can. J. Earth Sci.* **15**, 874–888.

Taylor, S. R. (1969). Trace element chemistry of andesites and associated calc-alkaline rocks, *Oregon St. Dept Geol. Mineral. Res. Bull.* **65**, 43–63.

Taylor, S. R. and Hallberg, J. A. (1977). Rare-earth elements in the Marda calc-alkaline suite: an Archean geochemical analogue of Andean-type volcanism. *Geochim. cosmochim.* **41**, 1125–1129.

Taylor, S. R., Capp, A. C., Graham, A. L., and Blake, D. H. (1969). Trace element abundances in andesites. *Contr. Mineral. Petrol.* **23**, 1–26.

Viljoen, M. J. and Viljoen, R. P. (1969). An introduction to the geology of the Barberton granite-greenstone terrane. *Geol. Soc. S. Afr. spec. Publ.* **2**, 9–23.

White, A. J. R., Jakeš, P., and Christie, D. M. (1971). Composition of greenstones and the hypothesis of sea-floor spreading. *Geol. Soc. Aust. spec. Publ.* **3**, 47–56.

Wilson, H. D. B., Andrews, G., Moxham, R. L., and Ramlal, K. (1965). Archean volcanism in the Canadian Shield. *Can. J. Earth Sci.* **2**, 161–175.

Winchester, J. A. and Floyd, P. A. (1977). Geochemical discrimination of different magma series and their differentiation products using immobile elements. *Chem. Geol.* **20**, 325–343.

Windley, B. F. (1973). Crustal development in the Precambrian. *Phil. Trans. R. Soc. A* **273**, 321–341.

Windley, B. F. (1976). New tectonic models for the evolution of Archean continents and oceans. In *The Early History of the Earth* (B. F. Windley, ed,), John Wiley, New York, pp. 105–111.

Upper Proterozoic (Pan-African) calc-alkaline magmatism in north-eastern Africa and Arabia

I. G. Gass

Department of Earth Sciences,
The Open University, Milton Keynes MK7 6AA, UK

ABSTRACT

Evidence concerning the evolution of the Upper Proterozoic (Pan-African: 1200–500 Ma old) continental crust of north-eastern Africa and Arabia is reviewed. This now typically continental crust, took over 700 Ma to develop. Field evidence suggests that the process started with numerous intra-oceanic island arc systems that evolved by an episodic continuum of magmatic, metamorphic, and sedimentary processes above destructive plate margins, so that, with time, the arcs matured and coalesced into proto-continents. Incomplete, approximately north–south-trending zones of mafic–ultramafic complexes, many identified as 'Penrose-type' ophiolites, are thought to mark the approximate sutures between arc systems that collided c. 1000, 800, and 600 Ma ago. Linear zones containing abundant granite plutons may indicate arc axes. Between 600 and 500 Ma ago the composition of magmatic products changed from calc-alkaline to peralkaline affinity with an order of magnitude increase in Zr, Y, Nb, U, and Th. This change is interpreted as indicating the cessation of subduction and the end of the Pan-African. Global heat productivity calculations are consistent with field and geochemical data in indicating that Pan-African arc systems had equivalent dimensions, and magmatic products similar in composition, to active volcanic arcs. The cratonized island arc model seems to be applicable in Africa as far south as Ethiopia in the east and across northern Africa as far as the West African Archaean craton. If this is the case then $c.\ 5.23 \times 10^8\ km^3$ of new African continental crust was produced during the Upper Proterozoic by calc-alkaline magmatism at destructive plate margins.

Introduction

The portion of the Earth's continental crust described and discussed here is of Upper Proterozoic age (c. 1200–500 Ma old) and forms the crystalline basement to Phanerozoic sedimentary sequences in north-eastern Africa and Arabia. The regional extent of the area covered is shown in Fig. 1 and the rocks described crop out as extensive tracts of basement in western Saudi Arabia, the Egyptian Eastern Desert and the north-eastern Sudan.

The African Precambrian consists of three major and several minor Archaean cratons surrounded and separated by non-cratonic (non-Archaean) Precambrian terrain (Clifford, 1970).

In 1964 Kennedy identified the dominance of K–Ar ages of 450–650 Ma for the non-cratonic Precambrian and proposed the term 'Pan-African tectono-thermal event' to identify this major episode in African geological evolution. Since then, the term 'Pan-African' has been widely used to describe non-Archaean African basement, although the time span has been extended by Rb–Sr and zircon dating to 1200–450 Ma (e.g. Fleck et al., 1976; Hashad, 1980). There are therefore two views on the age range of the Pan-African. Some restrict the term to rocks of 450–650 Ma old. Others, of whom the present author is one, feel that as the longer 1200–450 Ma period is based on isotopic data not available in 1964; its usage for this part of

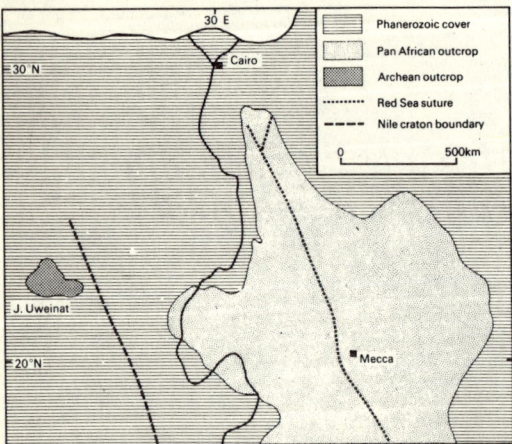

Fig. 1 Sketch map indicating the area discussed in the text. The existence of an Archaean craton to the west of the Nile seems likely on the basis of radiometric dates (Vail, 1976) but it is unusual in that Pan-African granites and Cainozoic volcanic rocks have been emplaced into and erupted on to it in contrast to those other African cratons. The eastern boundary of the craton is very approximate and the Red Sea has been closed to its pre-drift position

Africa and Arabia is more in keeping with Kennedy's original concept. The extended 1200–450 Ma Pan-African age range is used here.

Although the Pan-African basement has all the geological and geophysical characteristics of continental crust, Kennedy (1964), in the pre-plate tectonic early 1960s, could see few indications of classic orogenesis and hence used the term 'tectono-thermal event' to keep the interpretive options open. Since then, and largely in the last five years, two schools of thought concerning the origin of the Pan-African in north-eastern Africa and Arabia have developed. The first maintains that this Pan-African is essentially Archaean crust that was thermally and tectonically reworked during late Proterozoic Pan-African events (1200–500 Ma ago). Interestingly, members of this school have usually worked on Archaean rocks and their conviction seems to be primarily based on the community of character of the two terrains, such as the abundance of andesites and granites (e.g. Kröner, 1977, 1979; Hepworth, 1979) and the occurrence of rocks with Archaean geochemical features within Pan-African terrains (Kemp, 1978; El-Shazly and Engel, 1978).

In the view of the present author, this reasoning takes the inherently unrealistic concept of 'reversed uniformitarianism' to the extreme by invoking that what happened in the Archaean is the key to processes 2000 Ma later in the markedly different heat-flow environment of the Upper Proterozoic. In the past few years, and particularly as a result of a research symposium on the Arabian–Nubian Shield held in Jeddah in 1978, most members of the 'reworked Archaean' school now accept that unmetamorphosed uppermost Proterozoic volcanic rocks and associated cannibalistic sediments of calc-alkaline affinity overlie older and more highly metamorphosed and deformed formations. To explain this situation they (e.g. Kröner, 1977, 1979; Hepworth, 1979) propose that an Archaean basement overlay an Upper Proterozoic Andean-type subduction zone and, as a result, was invaded and blanketed by the products of late Proterozoic destructive margin magmatism.

This paper is based on the alternative hypothesis that this part of the continental crust evolved entirely within the Upper Proterozoic. The envisaged processes started some 1200 Ma ago with numerous immature intra-oceanic island arcs forming as a result of subduction between converging plates of oceanic lithosphere. The island arcs, the proto-continents, evolved by repeated magmatic, metamorphic, and tectonic events associated with concomitant erosion and sedimentation and, as arcs collided, larger 'continental' masses were produced. Finally, when subduction finally ceased, c. 500 Ma ago, the whole region had developed a continental character. This model, based on field, geochemical, and isotopic data, envisages an episodic continuum of magmatic, metamorphic, and sedimentary processes while subduction continued, with tectonic activity occurring primarily when arcs collided. During the 700 Ma span of the extended Pan-African, it is evident that there were numerous subduction zones widely distributed in space and time. Repeatedly, the products of earlier phases must have been invaded and/or blanketed by those of later episodes. In this way similar rock types and associations would be produced at about the same time in arc systems that were then widely separated. Conversely, similar rock types and associations would be produced in the same arc system at different times. With this temporal and spatial community of character, long-range stratigraphic correlations are hazardous and radiometric data on magmatic events cannot be given regional connotations. No detailed evolutionary picture of an individual arc has emerged, although work, primarily in Saudi Arabia, has allowed a timetable of stratigraphic, tectonic, and magmatic events to be erected (Fitch, 1978, Tables 1 and 2) that can also be applied in Egypt and the north-eastern Sudan. In Table 1 a grossly simplified version of this stratigraphic table is presented to provide a time framework for field and geochemical data.

It can be seen from Table 1 that the Pan-African can be divided stratigraphically into three major divisions; these are here termed the Lower, Middle, and Upper Pan-African, respectively. The oldest rocks in the Lower Pan-African age range (1000–1200 Ma ago) are highly deformed and metamorphosed but are comparable in composition to those from present-day, immature, intra-oceanic island arcs. Then, following a distinct stratigraphic, tectonic, and compositional break at c. 1000 Ma ago, the period of 400 Ma between 1000 and 600 Ma ago saw the development during the Middle Pan-African of numerous maturing intra-oceanic island arcs. That these arcs collided at various times is indicated by the emplacement of ophiolite complexes at 1000 and 800 Ma ago and also by several phases of compressional deformation. By 600 Ma ago it seems that most of the Middle Pan-African arcs had coalesced into a unified continental mass. But the continuation of calc-alkaline magmatism during the Upper Pan-African (600–500 Ma ago) indicates the presence of a subduction zone beneath at least part of the region. Finally, at or c. 500 Ma ago, peralkaline magmatism with clear within-plate geochemical characteristics had replaced the previous calc-alkaline associations throughout the entire region. This is taken as indicating the attainment of true continental character and the end of the Pan-African phase of continental evolution.

The field, petrographic, and geochemical characteristics of the three divisions of the Pan-African and the post-Pan-African magmatic products will be reviewed before a concluding tectonic and petrogenetic appraisal is attempted. As geochemical data are still sparse and most geological mapping has been of a reconnaissance variety, the interpretation can only be generalized and semi-quantitative. However, the classification of 'orogenic' (destructive margin-subduction zone) volcanic rocks as reviewed and outlined by Baker (this volume, Section II) is used and the compositional relationships between orogenic volcanic and intrusive rocks as discussed by Brown (this volume, Section VI) are accepted. In this context it is particularly relevant that orogenic associations are characterized by volcanic rocks varying in chemical composition from basic to acid. The *island arc tholeiitic association* of immature island arcs is dominated by basaltic rocks with tholeiitic characteristics (e.g. Fe enrichment in an AFM diagram), whereas in the *calc-alkaline association* of more mature island arcs and continental margins andesites, dacites, and rhyolites are the major rock types. Calc-alkaline andesites are oversaturated intermediate rocks with 55–63 per cent SiO_2, relatively high Al_2O_3 (c. 15–19 per cent), moderate alkalis (Na_2O + K_2O = 4–7 per cent) and Ca contents (CaO = c. 5–6 per cent) and showing no Fe-enrichment trend in an AFM diagram (cf. Baker, this volume, Section II). In addition to these major element characteristics, orogenic associations have distinctive trace element abundances such as low Nb, Y, Zr, U, and Th compared with other volcanic associations (Pearce and Gale, 1977; Pearce and Norry, 1979). The plutonic calc-alkaline association is of gabbro–tonalite–granodiorite–granite in which tonalites and granodiorites are dominant (Brown, this volume, Section VI). Such intermediate calc-alkaline associations can be readily distinguished geochemically from anorogenic peralkaline granitic associations. All magmatic rock types, except

TABLE 1 Simplified subdivision of rock types in the Arabian–Nubian shield. (Largely after Fitch, 1978, Tables 1 and 2, and Brown and Jackson, 1979)

Age (Ma)	Rock Types			Inferred Tectonic Setting	Comments and Other Data
	Plutonic	Volcanic	Sedimentary		
Post-Pan-African	Peralkaline granites characterized by high Ti, Zr, Nb, U, Th	Peralkaline trachytes and rhyolites	Terriginous arkoses and shallow water shales	Continental	Continental character of region established; all magmatism of 'within-plate' variety
500–600	Diachronous end of destructive margin processes				
Upper Pan-African	Calc-alkaline granites and granodiorites with low Ti, Zr, U, Th, and very low Nb	Rhyolites, dacites, trachytes, and andesites (in diminishing order of abundance). Lavas, tuffs, breccias, and agglomerates	Conglomeratic and arenaceous units with granitic and rhyolitic clasts. Stromatolitic limestones	Continental with margins of Andean type	Regionally extensive unmetamorphosed and structurally undeformed silicic volcanic and plutonic rocks
600–670	Major break, regional unconformity, change in composition of magmatism				
Middle Pan-African	Calc-alkaline diorites and granodiorites	Andesites and basaltic andesites with minor rhyolitic and dacitic units	Greywackes and minor arkoses. Stromatolitic limestones and argillaceous shallow water shales.	Numerous mature intra-oceanic island arcs. Major stratigraphic and regional breaks suggest complex evolution of several arcs	c. 600 Ma emplacement of ophiolitic complexes
					c. 800 Ma emplacement of ophiolitic complexes
					Complex deform and matamorphism to green-schist facies
					c. 1000 Ma emplacement of ophiolitic complexes
c. 1000	Distinct structural, compositional, and metamorphic break (orogenesis at c. 960 Ma)				
Lower Pan-African	Diorites, granodiorites, and mafic and ultramafic layered complexes	Basalts, basaltic andesites	Immature greywackes, cherts, shales, occasional limestones	Numerous immature intra-oceanic island arcs	Sparse and highly deformed outcrops. Metamorphosed mainly to amphibolite facies
1200?					

those produced by post-Pan-African within-plate activity, are orogenic and fall within the widest interpretation of the title of this volume—*andesites and related rocks*. Also, it is assumed, because no detailed studies have proved otherwise, that the Pan-African volcanic rocks are the surface expression of the same processes that produced the plutonic masses. So, both plutonic and volcanic rocks are plotted on variation diagrams, although chemical variations due to crystallization at depth are taken into account in interpretation. Lastly, as many of the rocks concerned have been metamorphosed to greenschist facies, the discriminatory but metamorphically immobile trace elements (Ti, Zr, Y, and Nb) are principally used in identifying petrogenetic characteristics (Pearce and Cann, 1973; Pearce and Gale, 1977; Pearce and Norry, 1979). Chemical analyses of representative Pan-African igneous rocks are given in Table 2.

Field, Petrographic, and Geochemical Data

This review starts at the base and works upwards through progressively younger Pan-African formations. The wide variety of local formation and group names are avoided as far as possible and attention is concentrated on describing the main rock types and associations, their compositions, ages, and tectonic settings. In starting at the base a fundamental problem is immediately encountered—just what is the base of the Pan-African succession? In south-western Saudi Arabia, Egypt, and the Sudan, thick sequences of siliceous gneisses derived from sedimentary quartzites occur beneath metavolcanics with 'arc' affinities. These quartzites are thought to represent a passive margin sedimentary wedge flanking the Archaean Nile Craton (*see* Fig. 1). So far, no older basement has been positively identified to the north-east of this zone and all rock types formed in the following 600–700 Ma have arc affinities and are believed to have formed over Proterozoic ocean crust.

Lower Pan-African

There is no direct geochronological evidence to support the proposal (*see* Table 1) that the Lower Pan-African has a time span of 200 Ma between 1000 and 1200 Ma ago. Indeed, opinions differ as to whether these sequences represent the oldest formations or are facies equivalents of younger ones that have been more intensely deformed and metamorphosed than elsewhere (Brown and Jackson, 1979; Greenwood *et al.*, 1976; Fitch, 1978; Schmidt *et al.*, 1973). There are few radiometric dates that fall unequivocally in this time span and, inexplicably, those that do come preferentially from the Egyptian Eastern Desert (Hashad, 1980). Furthermore, recent attempts to date some of the supposedly oldest rocks (e.g. Kröner *et al.*, 1979) give Rb–Sr ages of 800–600 Ma, so there is no precise control over the proposed age range. That this Lower Pan-African exists as a separate entity is best deduced from field and compositional evidence. Brown and Jackson (1979), summarizing field evidence from the southern part of the Arabian Shield, state categorically that there is a distinct structural, compositional, and metamorphic break between overlying more siliceous, less metamorphosed, and relatively undeformed formations and the oldest sequences here allotted to the Lower Pan-African.

Thick sequences (over 12 km) of basalts and basaltic andesites with intra-formational basic greywackes and subordinate carbonate and cherts form the main surface rock types in this division. These were invaded c. 1000 Ma ago by composite gabbroic-dioritic batholiths with initial $^{87}Sr/^{86}Sr$ ratios of 0.7029. In the sediments, as there are no clasts of alkali feldspar and no sign of terrigenous input, Greenwood *et al.* (1976) envisage immature oceanic island arcs with adjacent depositional basins as the preferred tectonic setting. Higher up the sequence, sediments replace volcanic rocks as the dominant rock type and, as there are few compositional differences between them, it is likely that the sediments were derived by the rapid erosion of an adjacent, now underlying, volcanic arc.

In support of the immature arc setting Greenwood *et al.* (1976) and Greenwood and Brown (1973) quoted the limited amount of analytical

TABLE 2 Chemical analyses of representative Pan-African calc-alkaline igneous rocks and post-Pan-African alkaline igneous rocks

Analysis no.:	Lower Pan-African				Middle Pan-African				Upper Pan-African			Post-Pan-African	
	1	2	3	4	5	6	7	8	9	10	11	12
SiO_2	47.56	52.67	58.51	68.37	57.00	61.35	67.80	74.20	67.50	73.65	59.74	73.10
TiO_2	0.62	0.65	1.26	0.35	1.55	0.66	0.40	0.22	0.56	0.11	0.09	0.23
Al_2O_3	15.58	15.42	16.49	16.11	17.30	16.45	15.50	13.90	14.30	13.21	20.42	13.15
Fe_2O_3	3.14	4.54	2.59	1.31	2.40	1.55	1.19	0.57	1.47	1.70	2.28	1.30
FeO	5.77	5.39	6.40	2.33	4.63	5.23	2.82	0.96	2.45	0.21	1.75	2.38
MnO	0.18	0.15	0.15	0.08	0.11	0.13	0.09	0.04	0.07	0.02	0.13	0.12
MgO	9.32	4.81	1.61	0.58	2.92	2.29	1.35	0.22	0.97	0.07	tr.	0.10
CaO	13.32	11.79	3.61	1.61	5.40	5.60	3.78	0.69	2.96	0.56	1.06	0.43
Na_2O	1.49	0.19	4.79	4.06	4.53	3.98	4.41	4.49	3.40	3.85	8.12	5.10
K_2O	1.41	0.07	2.12	4.71	2.38	1.32	1.81	3.71	3.56	5.58	5.29	4.70
H_2O	1.10	2.87	1.51	0.21	0.77	0.77	0.49	0.65	0.86	0.65	1.07	0.24
P_2O_5	0.25	0.09	0.49	0.11	0.54	0.14	0.11	0.01	0.25	0.02	0.04	0.02
CO_2	1.22	2.03	0.96	1.03	0.33	—	—	0.33	—	—	—	—
Total	100.96	100.67	100.49	100.86	99.86	99.47	99.75	99.99	98.35	99.63	99.99	100.87
Rb	—	—	—	—	43	26	37	93	57	123	140	92
Sr	—	—	—	—	601	288	311	128	340	53	55	10
Nb	—	—	—	—	5	2	2	5	3	10	57	77
Zr	—	—	—	—	131	98	99	196	286	116	541	877
Y	—	—	—	—	24	24	13	12	46	37	12	66

(1) Pyroxene basalt, Wadi Bidah area, south-western Saudi Arabia (20°30′N, 41°23.5′E) (Jackaman, 1972, Table 6-1, sample JA 30). (2) Basaltic andesite, Wadi Bidah area, south-western Saudi Arabia (20°34′N, 41°25′E) (Jackaman, 1972, Table 6-2, sample JA 452). (3) Andesite, Wadi Qatan area, south-western Saudi Arabia (18°08′N, 44°06.5′E) (Jackaman, 1972, Table 11-1, sample JM 66). (4) Rhyodacite, Wadi Wassat area, south-western Saudi Arabia (18°22.5′N, 44°13.5′E) (Jackaman, 1972, Table 11-1, sample JM 56). (5) Quartz diorite, 'batholithic association' of north-eastern Sudan (21°12′N, 36°08′E) (Neary et al., 1976, Table 2, sample 15062). (6) Granodiorite, 'batholithic association' of north-eastern Sudan (21°53′N, 36°15′E) (Neary et al., 1976, Table 2, sample 92). (7) Granite 'batholithic association' of north-eastern Sudan (21°52.5′N, 36°09′E) (Neary et al., 1976, Table 2, sample 14406). (8) Rhyolitic tuff, southwest flank of Jebel Asoteriba (21°46.5′N, 36°28′E) (Neary et al., 1976, Table 5, sample 85). (9) Granite, G1 phase from Taif area, south-western Saudi Arabia (21°20′N, 40°27.5′E) (Nasseef and Gass, 1977, Table 7, sample 160). (10) Granite, G2 phase from Taif area, south-western Saudi Arabia (21°19′N, 40°26′E) (Nasseef and Gass, 1977, Table 7, sample 26). (11) Syenite, Salala ring complex, north-eastern Sudan (21°18.5′N, 36°12′E) (Neary et al., 1976, Table 3, sample 14615). (12) Granite (peralkaline), eastern margin of Jebel Shendib (21°58′N, 36°11′E) (Neary et al., 1976, Table 3, sample 95).

data then available (e.g. Jackaman, 1972) which indicated that the basic volcanic rocks of this lowest group are compositionally similar to the present-day immature island arc tholeiites of Jakeš and Gill (1970; Table 2, analyses 1–4).

Since that time other samples of Lower Pan-African diorites–granodiorites have been analysed for major, trace, and rare earth elements (Nasseef and Gass, 1977; Gass, 1977). In Fig. 2, a graph of TiO_2 against Zr regarded by Pearce

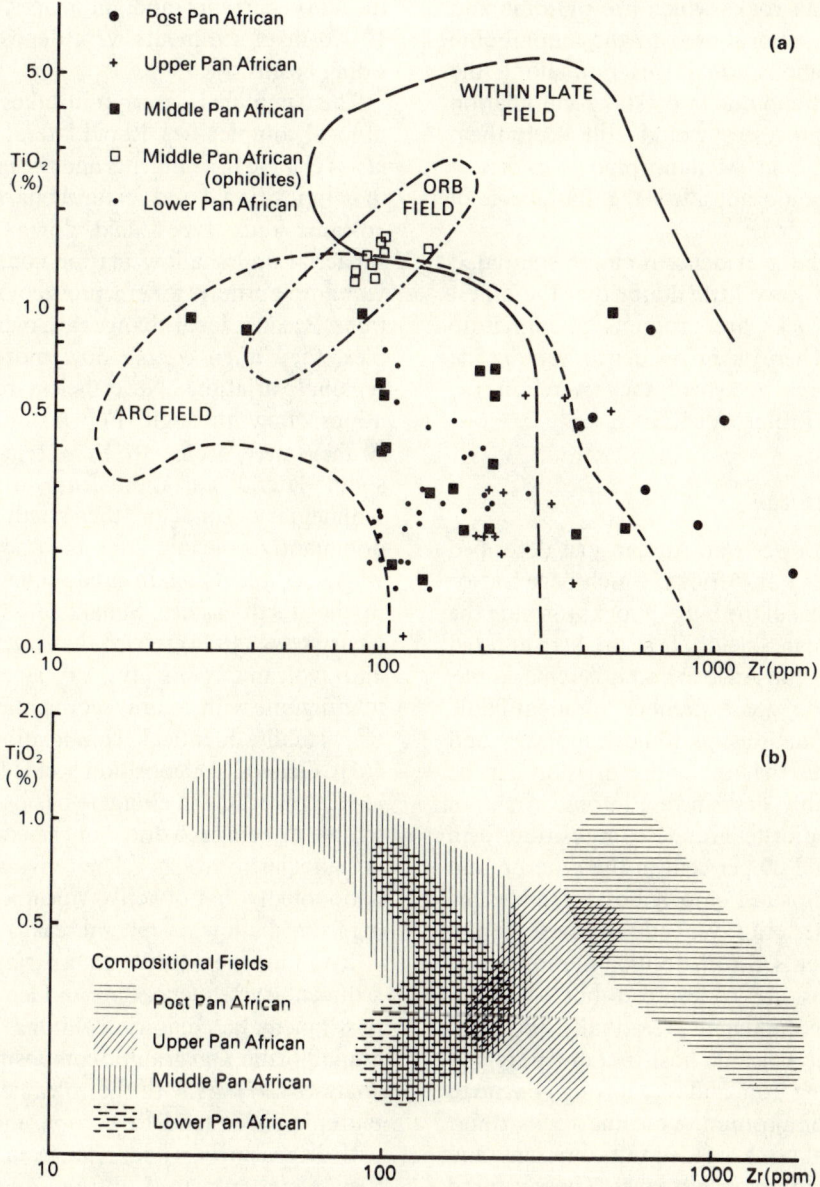

Fig. 2 (a) Plot of TiO_2 against Zr for Pan-African igneous rocks. Composition fields for present-day volcanic rocks from island arc, ocean ridge basalts (ORB) and within-plate settings are after Pearce (1980). (b) Summary of the variation of Ti and Zr abundances with time. Note the progressive increase in Zr whereas Ti remains relatively constant

and Norry (1979) and Pearce (1980) as the most realistic geochemical discriminator between arc and within-plate magmatic products, all specimens so far analysed plot in the arc field. Although the fields on Fig. 2 have been defined by young extrusive rocks, in contrast to many of the Pan-African rocks which are plutonic and have been metamorphosed to the amphibolite facies, the compositional variation along the fractionation trend due to crystal accumulation and related processes would still keep them within the arc field. Metamorphic processes at these facies should not affect the abundance of either TiO_2 or Zr.

So, although the evidence is circumstantial, it is sufficient to leave little doubt that the oldest Pan-African rocks are products of immature island arcs. There is no evidence yet on the number of arcs or where they were on the surface of the Proterozoic Earth.

Middle Pan-African

Between the Lower Pan-African just described and the Upper Pan-African, which began at or c. 600–670 Ma, fall the bulk of rocks forming the Arabian–Nubian Shield. These are here allotted to the Middle Pan-African and it is remarkable, in view of the great number of identifiable individual formations, plutonic complexes, and tectonic events, that the entire division can be relatively simply described. Plutonic rocks of diorite–granodiorite–granite composition form between 50 and 60 per cent of the outcrop and have been emplaced into a host of andesite–dacite–rhyolite eruptives and derived cannibalistic sediments. Stromatolitic limestones and chert horizons are widespread but quantitatively minor components. Gradually, the composition of the volcanic host became more siliceous, changing from andesite through dacite to rhyolite in composition. Volcanic rocks range from subaerial lavas, welded tuffs, breccias, and agglomerates to water-lain ashes. There is rapid lateral and vertical variation in rock type which is characteristic of composite volcanoes. Delfour (1975) and Greenwood et al. (1976) have suggested that these rocks were formed on the flanks of emergent volcanic island arcs. The sediments derived from these volcanic edifices range from fluviatile conglomerates to finer grained arenaceous and argillaceous deposits laid down in shallow marine conditions. Commonly the original sediments were reworked by turbidity currents and, in places, more than 10 000 m of sediments were deposited in subsiding basins.

This simplified description takes no account of local complexities. Rapid lateral and vertical facies change in sediments and the impersistence of compositional and textural characteristics of volcanic rock types, laid down under both subaerial and shallow marine conditions, provide ample criteria for erecting geological formations. Because local changes are many and complex, they often overshadow more significant regional variations. Nevertheless, regional variations show through. For instance in Saudi Arabia there seems to be a regional north–south lateral variation from a dominantly sedimentary zone in the north through a dominantly volcanic zone to a southern zone where sediments again predominate. Similarly, in the north-eastern Sudan, Gass (1955) and Neary et al. (1976) record that tracts of dominantly volcanic rocks give way laterally to, and interdigitate with, mainly sedimentary sequences of virtually identical composition (see Fig. 4(d)). Generally, deposition seems to have been in shallow water in elongate basins, most of the sediments being reworked intermediate and acid volcanoclastic rocks. The environment was undoubtedly that of active volcanic arcs emerging from shallow peripheral seas.

Invading the Middle Pan-African volcanic–sedimentary sequences are numerous syn- and post-kinematic diapiric plutons of dioritic, granodioritic, and granitic composition (Table 2, analyses 5–7). As with the volcanic rocks, these plutonic rocks tend to become more siliceous with decreasing age and it had been thought that they were emplaced during and after well defined orogenic phases.

Recently, however, it has been shown that in the southern Arabian Shield (Cooper et al., 1979) there was essentially an episodic continuum of plutonism. In Fig. 3, compiled largely from Cooper *et al.* (1979), Hashad (1980), and Fleck *et al.* (1976), the ages of various plutonic masses are plotted. The regularity of events throughout the period, from 660 to 820 Ma ago, is evident. With relatively few reliable radiometric dates available, it seems more likely that the gaps in Fig. 3 are due to a lack of isotopic data rather than significant pauses in plutonic activity. Undoubtedly, however, there were maxima in this activity, for in the north-eastern Sudan Neary *et al.* (1976) were unable to subdivide radiometrically numerous granitic-granodioritic masses, all of which gave ages close to 700 Ma.

Another feature of Middle Pan-African magmatism is the spatial distribution of the plutons. The initial impression on studying regional maps is of a random distribution. However, closer inspection suggests that many plutonic bodies lie within linear zones. Particularly clear examples of these granitic zones are shown in Fig. 4(a) and (b). Here, in the southern Arabian Shield (Ramsay *et al.*, 1979) and in the north-eastern Sudan (Neary *et al.*, 1976), there are, respectively, north–south- and NE–SW-trending zones *c.* 50 km wide where granitic masses are markedly more abundant than elsewhere. These linear distributions of plutons contrast with the seemingly random occurrence of many post-Pan-African bodies (e.g. Almond, 1979; Fig. 4(c)).

Ramsay *et al.* (1979), in discussing the granitic zones of the Arabian Shield and other zones with characteristic metamorphic grade or sedimentary/volcanic type, emphasize that, although boundaries are approximate and arbitrary, the zones of granitic plutons do represent discrete identifiable geological entities. Just what these zones or provinces represent in terms of geological processes is less clear—rifted sedimentary basins separated by basement highs, ensialic volcanic/sedimentary basins separated by Andean-type magmatic zones, and eroded ensimatic island arcs with granitic plutons flanked sequentially by volcanic rocks and reworked volcanoclastic sediments have been proposed. Here, the last explanation is preferred and an example of this relationship occurs in the north-eastern Sudan (*see* Fig. 4(b) and (d)).

The model proposed here for the Middle Pan-African is of numerous maturing ensimatic arc systems subsequently swept together by plate tectonic processes to form, by the end of the division, larger 'proto-continental' masses. In this model the original arcs were preserved whilst the intervening oceanic lithosphere was subducted. Occasionally, however, it seems that fragments of oceanic lithosphere were caught up between colliding arcs, for ultramafic–mafic complexes of ophiolitic character (as defined by the 1972 Penrose ophiolite conference) occur in zones across the region (Bakor *et al.*, 1976; Gass, 1977; Frisch and Al Shanti, 1977; Rehaile and Warden, 1978; Shanti and Roobol, 1979). So far *c.* 10 such ophiolite masses have been identified in Saudi Arabia, a further six to eight

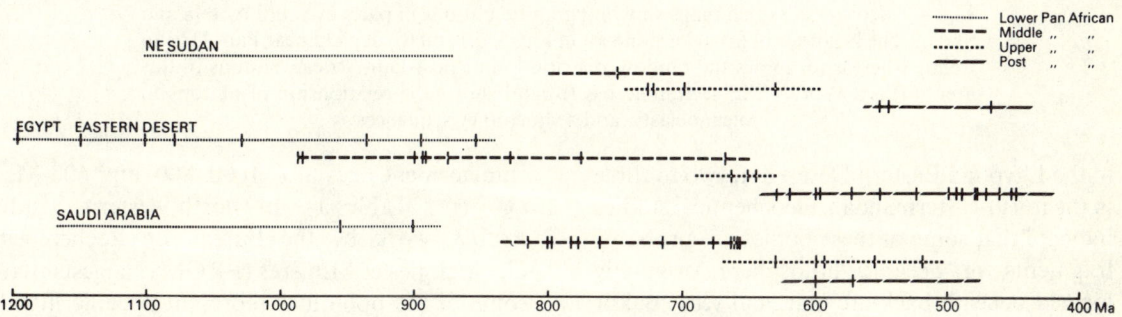

Fig. 3 Schematic representation of the age ranges of the Lower, Middle, and Upper Pan-African, and post-Pan-African activity in the Sudan, Egypt, and Saudi Arabia. Vertical ticks are Rb–Sr whole-rock isochron dates determined mainly for plutonic bodies. Note the continuum of activity and the seemingly diachronous change from one division to another

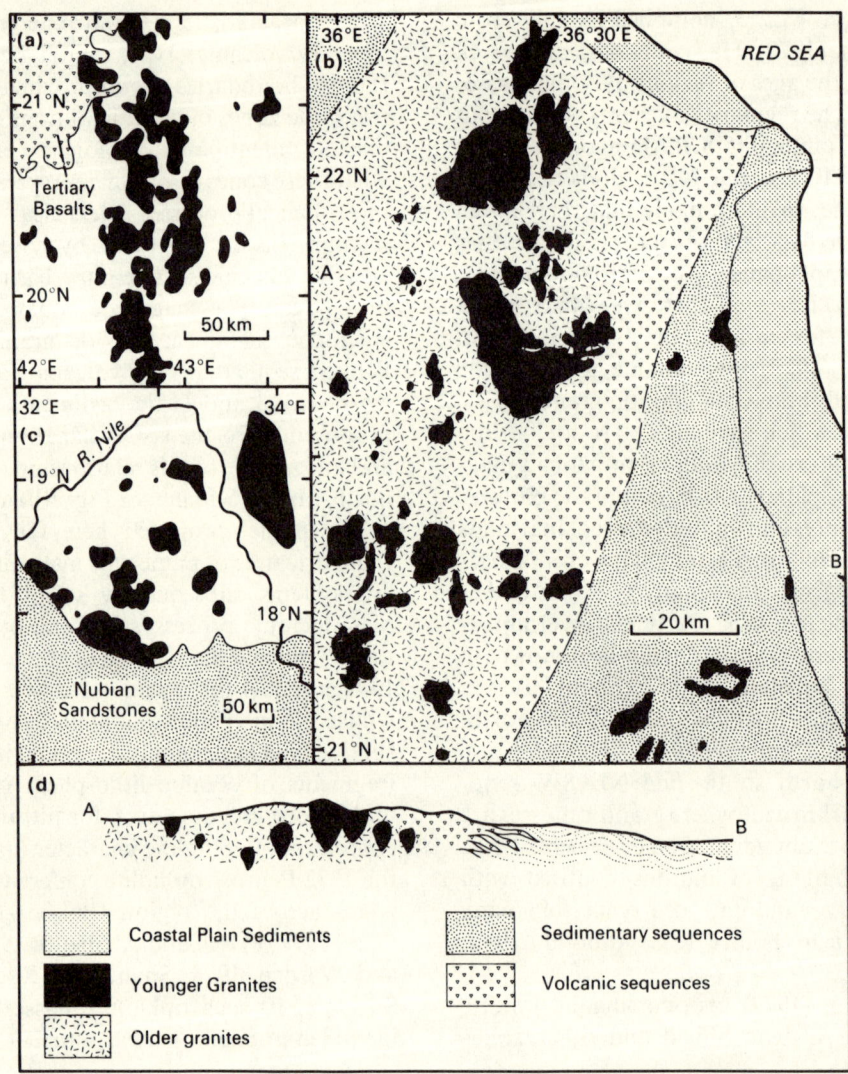

Fig. 4 Geological sketch maps showing granitic plutons in parts of Saudi Arabia and Sudan. The locations of (a)–(c) are shown in Fig. 5. (a) and (b) depict linear Pan-African trends whereas (c) shows the random distribution of post-Pan-African plutons in this area. (d) is a WNW–ESE section across (b) and shows the relationship of plutons to volcanoclastic and sedimentary sequences

in the Egyptian Eastern Desert and two to three in the north-eastern Sudan. Geochemical studies indicate that some of these bodies are probably fragments of oceanic lithosphere originally formed beneath back-arc marginal seas (Bakor et al., 1976). Although no reliable radiometric dates have yet been published for these ophiolite masses, structural relationships with dated formations suggest that there are at least three emplacement ages at c. 1000, 800, and 600 Ma ago (see Table 1). In north-western Saudi Arabia, work by the Bureau de Recherches Géologiques et Minières (BRGM) suggests that some of the ophiolites represent oceanic lithosphere on which all the other rocks were emplaced (Delfour, 1976). In the southern part of the Arabian Shield an 800 Ma emplacement age has been given for the dominantly north–south zone (Brown and Jackson, 1979) whereas other ophiolitic masses such as Jebal al Wask (Bakor et al., 1976) and Jebel Ess (Shanti and Roobol, 1979) seem to have been emplaced at or c. 600 Ma ago.

It is suggested that during the Middle Pan-African there were numerous intra-oceanic arc systems separated by back-arc basins. As arcs matured through geological time the magmatic products became progressively more siliceous and initial $^{87}Sr/^{86}Sr$ ratios increased from 0.7028 to 0.7035. Arcs were accreted, and new subduction zones developed under the composite arc systems. At least three phases of arc collision at c. 1000, 800, and 600 Ma seem to have occurred on the basis of tectonic events and ophiolite emplacement. The magmatic products throughout the period are calc-alkaline and on graphs of Nb against SiO_2 (Gass, 1979) and TiO_2 against Zr (Fig. 2) clearly fall in the volcanic arc magma field of Pearce and Gale (1977) and Pearce (1980), the trend becoming more calc-alkaline and less tholeiitic with time (Fig. 2(b)).

Upper Pan-African

On Table 1 the time span of the Upper Pan-African is given as 670–600 to 600–500 Ma ago, a diachronous period up to 170 Ma long. The base of the division is readily recognizable as unmetamorphosed and relatively undeformed silicic volcanic rocks and related sediments overlying older formations with marked angular unconformity. A conglomerate of regional extent is usually the basal unit and stromatolitic limestones occur as minor components with interbedded high energy fluviatile and shallow water clastic sediments. Both syn- and post-kinematic calc-alkaline granites and granodiorites have been emplaced into the volcanic-sedimentary host.

Extensive tracts of these volcanic rocks occur in the northern part of the Arabian Shield, the Eastern Desert of Egypt, and the north-eastern Sudan, where they are commonly preserved in down-faulted blocks. They seem to be absent in the south, where the level of erosion is deeper. Rock types are mainly pyroclastic in character with ash-fall and flow units much more abundant than lavas. Extensive welded tuff units are particularly conspicuous and their unmetamorphosed character enables multiple eruptive components of a single cooling unit to be identified. These and other features of subaerial volcanism (H. Duyverman, personal communication) indicate the dominantly terrestrial setting of this volcanic activity. The rock types present are mainly rhyolites and rhyodacites with subordinate andesitic flows. Community of composition and structural position allows correlation of these rocks across the region. In Saudi Arabia they belong to the Shammar (557–572 Ma ago; Hadley, 1973) and Murdama (561 and 633 Ma ago; Aldrich et al., 1978) groups. In Egypt they seem to correlate in composition and structural position with the Dokhan volcanic rocks (665 and 654 Ma ago; El-Shazly et al., 1973) and in the north-eastern Sudan with the Asoteriba volcanic rocks (669 ± 24 Ma ago; Neary et al., 1976; Table 2, analysis 8). So, although volcanic activity was diachronous, it had a community of composition and character that identify it as belonging to the Upper Pan-African.

Plutonic rocks of the Upper Pan-African are mainly calc-alkaline granites or granodiorites that occur as both syn- and post-kinematic phases. In a detailed study of a small area in western Saudi Arabia, Nasseef (1971) and Nasseef and Gass (1977) were able to identify three syn-kinematic phases and one post-kinematic phase falling in the age range 610–525 Ma. Of the four magmatic episodes the last, post-kinematic, phase, dated at 525 ± 20 Ma ago, was the most voluminous. These findings are in general agreement with earlier and more extensive investigations by Fleck et al. (1976) based on $^{40}K-^{40}Ar$ mineral studies. In the southern part of the Arabian Shield, Fleck and his co-workers identified a major magmatic phase between 610 and 510 Ma ago with a major pulse between 610 and 540 Ma ago and minor episodes at 535 and 510 Ma ago. For the Egyptian Eastern Desert, Hashad (1980) reported an age range of 675–450 Ma for the Younger or Red granites with maximum activity between 675 and 500 Ma ago, peaking at 600 Ma ago. A similar 600 Ma maximum is reported for the north-eastern Sudan (Neary et al., 1976). Within the Upper Pan-African initial $^{87}Sr/^{86}Sr$ ratios range from 0.7032 to 0.7093.

When plotted on a lime/alkaline index most of the Upper Pan-African rocks are calc-alkaline, although only marginally so; some, on the basis of molar $Al_2O_3/(K_2O + Na_2O)$ ratios are marginally peralkaline. However the trace element abundances (e.g. $Nb-SiO_2$ and Ti–Zr plots) have the distinct characteristics of arc magmatism. This indicates that one or more subduction zones were active under the region, although the lack of metamorphism and deformation, other than block faulting and gentle folding, suggests that a well developed, coherent, and rigid sialic crust had been developed by this time. This contention is supported by the abundance of high energy terrestrial and shallow water arkosic sediments, indicating a high standing continental mass. Just when the Upper Pan-African division ends presents problems for, in many respects, the region already had a continental character, although with active subjacent destructive margin or margins.

What can be deduced geochemically is when the subduction stopped, and this is taken here as the end of Pan-African events. It is marked by the diachronous but well defined, order of magnitude increase of Nb, Y, Zr, U, and Th in magmatic products. Although not strictly accurate, it is convenient to use the terms calc-alkaline and peralkaline in identifying this change in chemical composition. Pearce and Gale (1977) suggest that the Nb content of peralkaline granites is high because the mantle source in such within-plate settings has been enriched in this element by migrating CO_2-rich fluids or interstitial melts. They argue that since Nb is not a hydrophilic element and since it is likely to be compatible with residual mineral assemblages in subducted oceanic crust, the enrichment of Nb is unlikely to occur in normal arc magmas. The incoming of peralkaline magmas covers a wide time span. In the extreme north of the Arabian shield, adjacent to the Gulf of Aqaba, two peralkaline granites have been dated at 600 ± 24 Ma old (Stoesser and Elliot, 1979) and 571 ± 8 Ma old (Baubron et al., 1976), respectively, whereas further south, in Saudi Arabia, east of Jeddah, the calc-alkaline Taif granites are dated at 525 ± 20 Ma old (Nasseef and Gass, 1977; Table 2, analyses 9 and 10). In the north-eastern Sudan two markedly peralkaline masses date at 500 Ma old (Neary et al., 1976), whereas further north in Egypt Hashad (1980) records that calc-alkaline activity continued until 500 Ma ago but that by 450 Ma ago peralkaline magmatism was well established. The obvious interpretation of these data is that subduction beneath the region stopped earlier in some areas than in others. But whether this was along one zone or whether there were several zones that stopped at various times is as yet unknown. It is, however, pertinent to note that, once established, peralkaline magmatism occurred at widely spaced intervals throughout the Phanerozoic.

Petrogenetic and Tectonic Appraisal

Earlier, it was made clear that this article is based on the premise that the Proterozoic Shield of Arabia and north-eastern Africa was formed by progressive and episodic cratonization of originally intra-oceanic island arcs over a period of c. 700 Ma (1200–500 Ma ago). Now that a synopsis of the evidence has been presented it is pertinent to enquire whether any alternative tectonic model is possible. The only model of which the present author is aware that could realistically be invoked is the so-called 'millipede model of ductile plate tectonics' of Wynne-Edwards (1976) who proposed that Proterozoic magmatic and tectonic activity was the in situ response of ensialic lithosphere to large scale increases in the thermal flux of the subjacent mantle. In such a model no rigid plate movements or plate margin magmatism would occur. Referring to the Proterozoic in a global context, Wynne-Edwards (1976) states (p. 928): 'Eugeosynclinal volcanic-clastic sequences are rarities and recognisable island arc sequences, blueschist belts and ophiolites are very scarce. The dominant part of the Proterozoic record in every continent is represented by vast tracts of gneissic metamorphic terrain which are the result of large scale orogenesis'. So far as the Pan-African of northeastern Africa and Arabia is concerned this statement is unacceptable. Any model such as that illustrated in Fig. 5 of Wynne-Edwards' article that involves ductile stretching of sialic crust over mantle thermal highs cannot be applied to this region. Indeed, the absence of older radiometric dates (over 1200 Ma ago), the generally low initial $^{87}Sr/^{86}Sr$ ratios (0.702–0.706), the abundance of volcanoclastic sequences and related cannibalistic sediments, the ubiquitous presence of calc-alkaline magmatic products, and the identification of numerous ophiolites in linear zones of mafic-ultramafic complexes leaves little doubt that the 'cratonized island arc' hypothesis, originally proposed by Greenwood et al. (1976), is substantially correct. As demonstrated earlier, the single arc concept of Greenwood et al. (1976) needs to be modified and a complex of arcs is now thought to be more reasonable.

However, two problems remain: (1) there are no blueschists and (2) some of the mafic-ultramafic rocks have komatiitic characteristics. Fyfe (in discussion after Gass, 1979) asked why low temperature–high pressure metamorphic belts of zeolite – prehnite – pumpellyite – blueschist–eclogite facies are absent in the Pan-African of Africa and Arabia. He went on to suggest that before subduction zones change position and thereby cause up to 30 km uplift of blueschist assemblages, some 700 km of oceanic plate subduction must have occurred. The absence of blueschists in Arabia could indicate small plate motions of less than 700 km—a proposition that fits with the multi-arc models of Gass (1977) and Marzouki and Fyfe (1977). Dixon (1978) has noted the komatiitic character of some ultramafic rocks from the Egyptian Eastern Desert and, comparing these in composition to Archaean komatiites, seemed to prefer an ensialic model of reworked Archaean similar to that of Wynne-Edwards (1976). However, more recently he has affirmed the arc setting of these rocks (Dixon, 1979); this is particularly significant as komatiitic rocks (boninites) occur in modern oceanic trench systems and back-arc basins as well as in Tethyan ophiolites (Cameron et al., 1979).

In the opinion of the present author (Gass, 1979) no significant reason remains to preclude acceptance of the arc model for the Pan-African of the Arabian–Nubian Shield. What does remain, however, is to understand more fully the multitude of complex processes that accompanied and/or caused the conversion of oceanic lithosphere into continental crust; some of these are now discussed.

The features that seem to distinguish the Arabian–Nubian Shield from Proterozoic basement elsewhere are the abundance of unmetamorphosed volcanoclastic rocks and related sediments, and the absence of intense deformation, particularly in the upper part of the sequence. It is also quite evident that the products of surface eruptive and erosional products are most abundant in the north and become progressively more scarce southwards. Seemingly, there has been differential erosion whereby in the north only 1–2 km of basement have been removed in contrast with 5–7 km of basement erosion in the south. This erosional bevel is in response to the repeated uplift of the southern part of the region, now occupied by the Afro-Arabian dome, throughout the Phanerozoic, whilst the northern part was a region of sedimentary deposition for much of that time. The lack of volcanoclastic rocks, arc sequences, and ophiolite complexes that led Wynne-Edwards (1976) to propose his 'millepede' model could simply have been due to deeper erosion, which seems to have affected many Proterozoic terrains.

It was suggested earlier that the reversed uniformitarianism of comparing Upper Proterozoic processes with those of the Archaean is unrealistic. This is substantiated when the relative heat production of the two periods is compared to that of the present day. Brown (1980), taking present-day terrestrial heat production as unity, calculated that in the Archaean it was 5.0–7.5 times greater, whereas in the Pan-African it was between 1.4 and 1.7 times that of present values. These figures suggest that Upper Proterozoic arc systems would have had similar dimensions to those of the present day. And, although slightly steeper thermal gradients and lower upper mantle viscosities may have produced thinner lithospheric plates, steeper subduction zones, and narrower arc systems, the dimensions would, theoretically, be in the order of 10 per cent of present-day values.

Present-day arc systems are narrow, rarely exceeding 100–150 km from trench to back-arc basin, with active volcanism (and presumably plutonism) usually confined to axial zones less than 50 km wide. It would therefore be quite possible to fit 10 or more collided arc systems into the 1500 km north–south or east–west extent of the Arabian–Nubian Shield. Several attempts have been made to identify individual Pan-African arc systems on criteria such as zones of mafic–ultramafic (ophiolite) masses (Bakor et al., 1976; Gass, 1977, 1979; Frisch and Al-Shanti, 1977), regional variations in the geochemical characteristics of magmatic rocks (Greenwood and Brown, 1973; Gass, 1977; Gass and Nasseef, 1980), zones of distinctive metamorphic and/or lithological character (Ramsay et al., 1979), and zones of characteristic mineralization (Al-Shanti and Roobol, 1979). The identification of arcs by these criteria is of dubious validity. For instance, although several mafic–ultramafic masses have been confidently identified as ophiolites as defined by the Penrose Conference (1972) (Bakor et al., 1976; Shanti and Roobol, 1979) and display back-arc geochemical characteristics, it is by no means certain that these 'ophiolite' zones accurately mark the sutures between arc systems.

In the present south-western Pacific arcs, ophiolites are preferentially emplaced along the back-arc margin, but most Tethyan ophiolites are markedly allochthonous and have been moved tens, if not hundreds of kilometres during obduction. All Pan-African ophiolites so far identified have tectonic contacts and Shackleton et al. (1980) believe that the numerous mafic–ultramafic masses of the Egyptian Eastern Desert have been moved so far that the identification of linear ophiolite zones or sutures in that area is unjustified. Nevertheless, in Arabia they may be more trustworthy and their position, tentatively marking arc margins, is shown on Fig. 5. Similarly, recent work on modern arc systems reveals that the supposedly simple correlation between increasing K and Rb and decreasing Sr with depth to the subduction zone (cf. Hutchinson, 1976) is becoming complicated, the variations along the arcs proving to be as great as, or greater than, those across them. So, attempts to identify Pan-African arc systems from compositional variation must be distinctly suspect. Indeed, the variation of subduction zone models proposed, i.e. single easterly-dipping (Greenwood and Brown, 1973), multiple easterly-dipping (Gass, 1977), and single westerly-dipping (Schmidt et al., 1978), proves this point and emphasizes that at present there are insufficient data to justify this approach.

However, a simple arc system that has been eroded to near sea-level should have a plutonic core invading a zone of eruptive products that are flanked by cannibalistic and shallow water sediments. Of these features, the plutonic core is the most likely to be preserved. Study of existing geological maps suggests that several linear granitic zones of appropriate dimensions exist in both Africa and Arabia. Traces of these zones, thought to mark arc axes, are plotted on Fig. 5, the inset rectangle in this figure identifies two granitic zones shown in more detail in Fig. 4(a) and (b). These granite zones seem to be best developed in the north-eastern Sudan, but elsewhere the extensive cover of Recent sediments and Tertiary volcanic rocks leave major gaps, and subsequent deformation, such as that due to the Nadj Fault System in Saudi Arabia, presents complications. Also, geochronological control is poor and the age of the granite zones is not known. It could be anticipated that such zones of magmatic activity would, once established, focus subsequent plutonism. So, plutons of varying ages could well be channelled along the same 'hot' axial zones. Although it is early to place too much reliance on these granitic zones, their abundance does reflect the complexity of the Pan-African arc system, and their dominantly north–south orientation coincides with

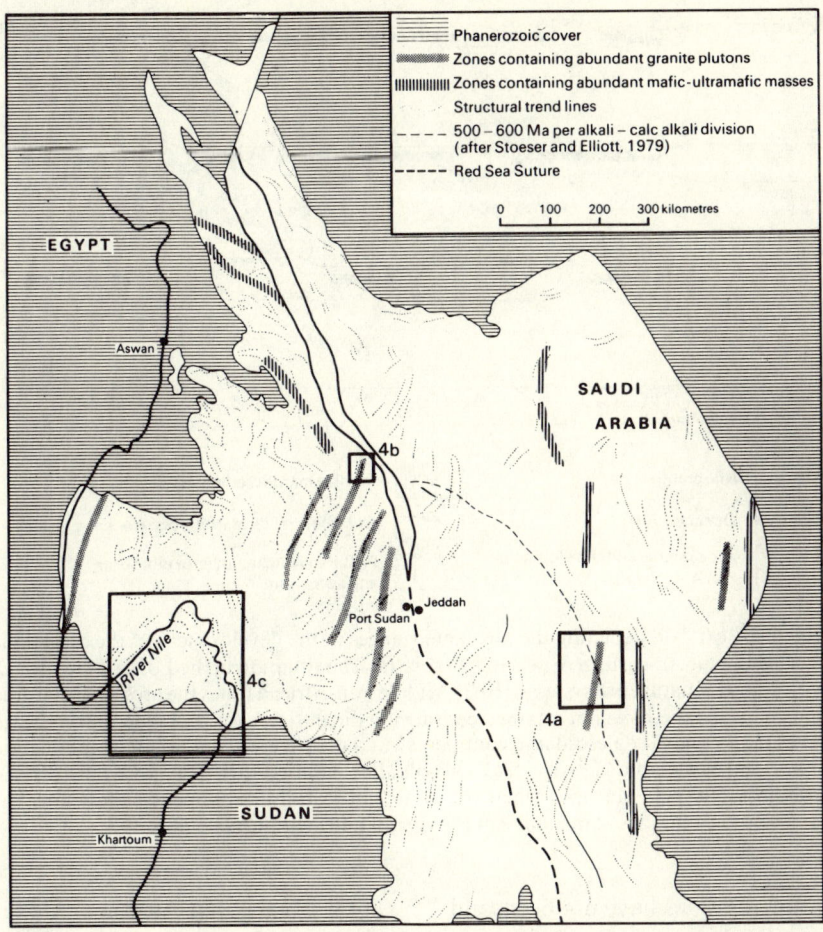

Fig. 5 Regional sketch map showing the disposition of mafic–ultramafic masses (marking the approximate position of arc sutures) linear granitic trends (possible arc axes) and basement structural trends. The Red Sea has been closed to a pre-Miocene position

that of many ophiolite zones and, broadly, with the structural trend lines also shown on Fig. 5.

In Fig. 6 the evolution through time of the arc systems from immature to mature arcs and finally to continental conditions is depicted in cartoon form. The broad stages in evolution are well documented but the passage from Lower to Middle Pan-African is temporally vague and that between Middle and Upper Pan-African is diachronous. Similarly, the final stages in cratonization, marked by the switch from calc-alkaline to peralkaline magmatism, occurred earlier in some areas than others. The line drawn on Fig. 5 is that of Stoesser and Elliot (1979)

and separates 500–600 Ma old Arabian peralkaline and calc-alkaline products. Seemingly, subduction was active south and west of this line after it had ceased to the north and east. There is no evidence as to which way this last subduction zone was inclined.

The case has been presented for the formation of Pan-African continental crust in northeastern Africa and Arabia by the progressive cratonization over 700 Ma of multiple intraoceanic arc systems. Although the broad temporal and spatial parameters of the processes are understood, there is insufficient evidence concerning the many and complex details. Indeed,

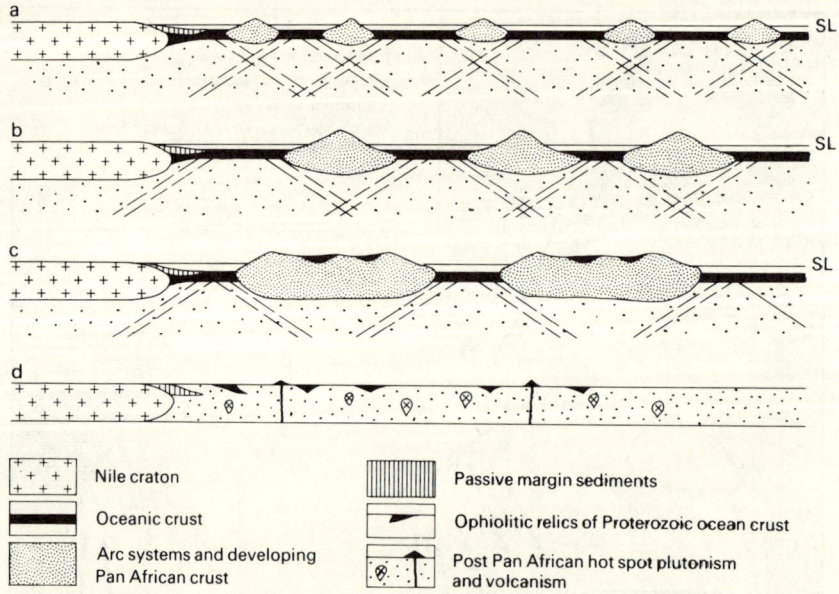

Fig. 6 A cartoon depicting the postulated stages in the development of the crystalline basement in north-eastern Africa and Arabia. (a) The situation in the Lower Pan-African with many immature arc systems. (b) By Middle Pan-African times the arcs have matured and coalesced but have not attained continental dimensions. (c) By Upper Pan-African times the arcs have coalesced into continents but these were still underlain by subduction zones and magmatic activity had calc-alkaline affinity. (d) The post-Pan-African situation (500–600 Ma ago) when the continent is fully developed; subduction has ceased and magmatism is of peralkaline within-plate affinity

some fundamental questions have not been, and seemingly cannot be, answered. One is whether the Pan-African arcs contain fragments of older (Archaean) crust. The problem of older cores within arc systems exists in present-day island arcs of the south-western Pacific, where opinion is divided. Tensional stresses set up above subduction zones can split off slivers of older crust at continental margins and these can be incorporated into arc systems and flanked by destructive margin volcanic products as in the Banda arc of Indonesia (M. G. Audley-Charles, personal communication). Such slivers of continental crust would, from isostatic considerations, tend to be preserved as surface features. In the Pan-African no such microcontinents have been positively identified.

So far, the data presented have come from north-eastern Africa and Arabia. The question arises as to how far the arc cratonization model applies to the Pan-African or Upper Proterozoic elsewhere. It is evident from the literature that in West Africa the Proterozoic around the West African Archaean craton has arc characteristics and genetic models presented by others for the Hoggar (Caby, 1970; Bertrand and Caby, 1978; Caby and Le Blanc, 1973), Mali (Black et al., 1979), Morocco (Le Blanc, 1976), Tibesti (Ghuma and Rogers, 1978; Pegram et al., 1976), and Nigeria (McCurry and Wright, 1977; McCurry, 1976) have been, or can be, interpreted on an island arc cratonization–accretion model such as that presented here. Indeed, the indications are that this model is also applicable to Ethiopia (Chater, 1971; Gilboy, 1970; Kazmin, 1976; De Wit and Aguma, 1977; De Wit and Chewaka, 1978). The regional picture for North Africa is shown in Fig. 7. Elsewhere in Africa the situation is less clear. The Mozambique and Damaran belts could be ensialic

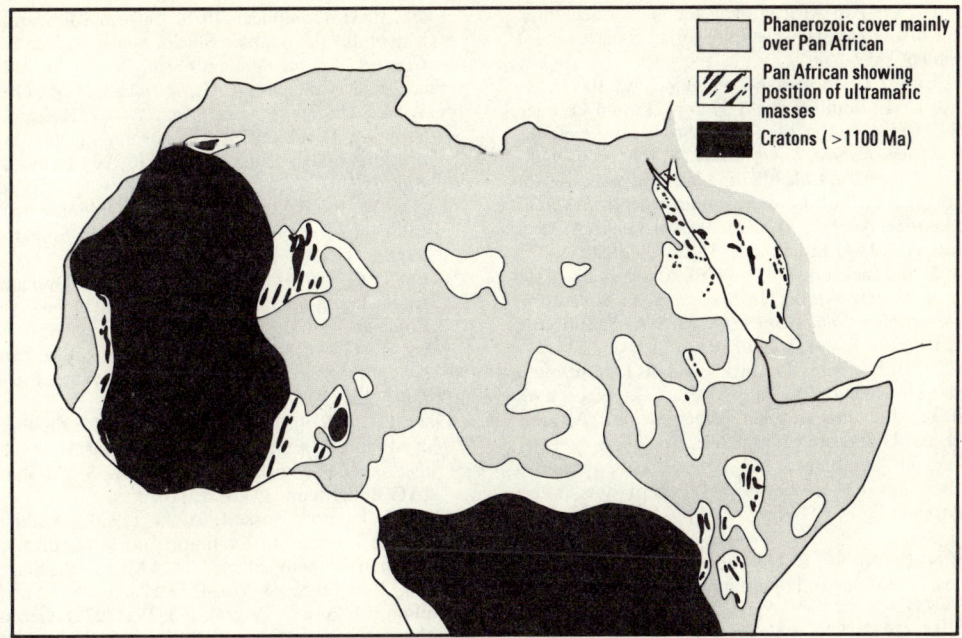

Fig. 7 Regional sketch map of northern Africa depicting cratonic masses, Phanerozoic cover, and Pan-African arc margins as indicated by ophiolitic zones

features or, for the Mozambique belt, Proterozoic continent–continent plate collisions seem more likely (Shackleton, 1979; Kröner, 1979). Nevertheless, despite these uncertainties, the inescapable conclusion, based on Nubian–Arabian Shield data, is that plate tectonics and destructive margin calc-alkaline magmatism, essentially similar to that of the present day, operated between 1200 and 500 Ma ago in northern Africa–Arabia and were responsible for the formation of $c.\ 5.25 \times 10^8$ km^3 of continental crust in the Upper Proterozoic.

Acknowledgements

I thank G. C. Brown, T. Dixon, N. B. W. Harris, A. C. Ries, R. M. Shackleton, and R. S. Thorpe for their constructive comments on the text and John Taylor and Pam Owen for cartographic and clerical services. This is a contribution to IGCP Project 164 'Evolution of the Pan-African in North-eastern Africa and Arabia'. Financial support from NERC and logistic assistance in Egypt from the South Carolina project are gratefully acknowledged.

REFERENCES

Aldrich, L. T., Brown, G. F., Hedge, C., and Marvin, R. F. (1978). Geochronologic data for the Arabian Shield. *US geol. Surv. Open-file Rep.* no. 78–75.

Almond, D. C. (1979). Younger granite complexes of Sudan. In *Evolution and Mineralization of the Arabian–Nubian Shield* (A. M. S. Al-Shanti, ed.), IAG Bulletin no. 3, Vol. 1, 151–164.

Al-Shanti, A. M. S. and Roobol, M. J. (1979). Some thoughts on metallogenesis and evolution of the Arabian–Nubian Shield. In *Evolution and Mineralization of the Arabian–Nubian Shield* (A. M. S. Al-Shanti, ed.), IAG Bulletin no. 3, Vol. 1, 87–96.

Bakor, A. R., Gass, I. G., and Neary, C. R. (1976). Jabal al Wask: an Eocambrian, back-arc ophiolite in NW Saudi Arabia. *Earth planet. Sci. Letters* **30**, 1–9.

Baubron, J. C., Delfour, J., and Violette, Y. (1976). Geochronological measurements on rocks of the Arabian Shield, Kingdom of Saudi Arabia. *BRGM Mineral. Rep.* 76-JED-17.

Bertrand, J. M. L. and Caby, R. (1978). Geodynamic

evolution of the Pan-African orogenic belt: a new interpretation of the Hoggar shield (Algerian Sahara). *Geol. Rundschau* **67**, 357–388.

Black, R., Caby, R., Moussine-Pouchkine, A., Bayer, R., Bertrand, J. M., Boullier, A. M., Fabre, J., and Lesquer, A. (1979). Evidence for late Precambrian plate tectonics in West Africa. *Nature, Lond.* **278**, 223–227.

Brown, G. C. (1980). Calc-alkaline magma genesis: the Pan-African contribution to crustal growth. In *Evolution and Mineralization of the Arabian–Nubian Shield* (A. M. S. Al-Shanti, ed.), IAG Bulletin no. 3, Vol. 3, 19–29.

Brown, G. F. and Jackson, R. D. (1979). An overview of the geology of Western Arabia. In *Evolution and Mineralization of the Arabian–Nubian Shield* (A. M. S. Al-Shanti, ed.), IAG Bulletin no. 3, Vol. 1, 3–10.

Caby, R. (1970). La chaîne Pharusienne dans le nord-ouest de l'Ahaggar (Sahara central, Algérie): sa place dans l'orogénèse due Précambrian supérieur en Afrique. Ph.D. thesis, Université Montpellier, France.

Caby, R. and Le Blanc, M. (1973). Les ophiolites précambriennes sur les bords est et nord due craton ouest Africain. *C.r. 1ere Réunion Ann. Sci. Terre Paris* 19–22.

Cameron, N. E., Nisbet, E. G., and Dietrich, J. J. (1979). Boninites, komatiites and ophiolite basalts. *Nature, Lond.* **280**, 550–553.

Chater, A. M. (1971). The geology of the Megado region of southern Ethiopia. Ph.D. thesis, University of Leeds, England.

Clifford, T. N. (1970). The structural framework of Africa. In *African Magmatism and Tectonics* (T. N. Clifford and I. G. Gass, eds), Oliver & Boyd, Edinburgh, 1–26.

Cooper, J. A., Stacey, J. C., Stoesser, D. G., and Fleck, R. J. (1979). An evaluation of the zircon method of isotopic dating in the southern Arabian craton. *Contr. Mineral. Petrol.* **68**, 429–439.

Delfour, J. (1975). Mineral occurrence documentation system—MODS—revised input manual. *BRGM Rep.* no. 75-JED-1.

Delfour, J. (1976). Volcanisme et gîtes minéraux du bouclier Arabo-Nubien. *Mem. L. Sér. Soc. géol. Fr.* **7**, 137–142.

De Wit, M. J. and Aguma, A. (1977). Geology of the ultramafic and associated rocks of Tulu Dimtu, Wollega. *Ethiopian Inst. Geol. Surv. Note* no. 57.

De Wit, M. J. and Chewaka, S. (1978). Plate tectonic evolution of Ethiopia and the origin of its mineral deposits: an overview. In *Plate Tectonics and Metallogenesis: Some Guidelines to Ethiopian Mineral Deposits*. (S. Chewaka and M. J. De Wit, eds), Ethiopian Institute of Geology, Survey Department, Addis Abbaba.

Dixon, T. H. (1978). Late Precambrian ultramafic magmas in the Egyptian Shield. *Geol. Soc. Am. Abst. Programs.*

Dixon, T. H. (1979). U–Pb ages and chemical characteristics of some pre-Pan-African plutonic rocks in the Egyptian Shield. *5th Conference of the Geological Society of Africa* (*Abstracts*), 75–76.

El-Shazly, E. M. and Engle, A. E. J. (1978). Proterozoic rifting and refractionation of northwestern Africa, *IAG Res. Ser.* **4**, 54–56.

El-Shazly, E. M., Hashad, A. H., Sayyah, T. A., and Bussyuni, F. A. (1973). Geochronology of the Abu Swayel area, South Eastern Desert. *J. Egypt. Geol.* **17**, 1–18.

Fitch, F. H. (compiler) (1978). Informal lithostratigraphic lexicon for the Arabian Shield. *Saudi Arabian Directorale General Mineral Resources tech. Record* TR 1978-1.

Fleck, R. J., Coleman, R. G., Cornwall, H. R., Greenwood, W. R., Hadley, D. G., Prinz, W. C., Ratte, J. C., and Schmidt, D. L. (1976). Geochronology of the Arabian Shield, western Saudi Arabia: K–Ar results. *Geol. Soc. Am. Bull.* **87**, 9–21.

Frisch, W. and Al-Shanti, A. (1977). Ophiolite belts and the collision of island arcs in the Arabian Shield. *Tectonophysics* **43**, 293–306.

Gass, I. G. (1955). The geology of the Dunganab area, Anglo-Egyptian Sudan. M.Sc. thesis, University of Leeds, England.

Gass, I. G. (1977). The evolution of the Pan-African crystalline basement in NE Africa and Arabia. *J. geol. Soc. Lond.* **134**, 129–138.

Gass, I. G. (1979). Evolutionary model for the Pan-African crystalline basement. In *Evolution and Mineralization of the Arabian–Nubian Shield* (A. M. S. Al-Shanti, ed.), IAG Bulletin no. 3, Vol. 1, 11–20.

Gass, I. G. and Nasseef, A. O. (1980). Arabian Shield granite traverse. In *Evolution and Mineralization of the Arabian–Nubian Shield*. (A. M. S. Al-Shanti, ed.), IAG Bulletin No. 3, Vol. 4, 77–82.

Ghuma, M. A. and Rogers, J. J. W. (1978). Geology, geochemistry and tectonic setting of the Ben Ghnema batholith, Tibesti massif, southern Libya. *Geol. Soc. Am. Bull.* **89**, 1351–1358.

Gilboy, C. F. (1970). The geology of the Gariboro region of southern Ethiopia. Ph.D. thesis, University of Leeds, England.

Greenwood, W. R. and Brown, G. F. (1973). Petrology and chemical analysis of selected plutonic rocks from the Arabian Shield, Kingdom of Saudi Arabia. *Saudi Arabian Directorate General Mineral Resources Bull.* no. 9.

Greenwood, W. R., Hadley, D. G., Anderson, R. E., Fleck, R. J., and Schmidt, D. L. (1976). Proterozoic cratonization in southwestern Saudi Arabia. *Phil. Trans. R. Soc. A* **280**, 517–527.

Hadley, D. G. (1973). Geology of the Sahl al Matran quadrangle, northwestern Hijaz Kingdom of Saudi Arabia, *Saudi Arabian Directorale General Mineral Resources Geol. Map* GM-6.

Hashad, A. H. (1980). Present status of geochronological data on the Egyptian basement complex. In *Evolution and Mineralization of the Arabian–Nubian Shield* (A. M. S. Al-Shanti, ed.), IAG Bulletin no. 3, Vol. 2.

Hepworth, J. V. (1979). Does the Mozambique orogenic belt continue into Saudi Arabia? In *Evolution and Mineralization of the Arabian–Nubian Shield* (A. M. S. Al-Shanti, ed.), IAG Bulletin no. 3, Vol. 1, 39–51.

Hutchison, C. S. (1975). Ophiolites in southeast Asia. *Geol. Soc. Am. Bull.* **86**, 797–801.

Hutchison, C. S. (1976). Indonesian active volcanic arc: K, Sr and Rb variations with depth to the Benioff zone. *Geology* **4**, 407–408.

Jackaman, B. (1972). Genetic and environmental factors controlling the formation of massive sulphide deposits of Wadi Bidah and Wadi Wassat. *Saudi Arabian Directorate General Mineral Resources, tech. Record* TR-1972-1.

Jakeš, P. and Gill, J. (1970). Rare earth elements and the island arc tholeiite series: *Earth planet. Sci. Letters* **9**, 17–28.

Kazmin, V. (1976). Ophiolites in the Ethiopian basement. *Ethiopian Inst. Geol. Surv. Note* no. 35.

Kemp, J. (1978). A Middle Proterozoic volcanic-plutonic cycle in Saudi Arabia. *J. Precambrian Res.* **6**, A28.

Kennedy, W. Q. (1964). The structural differentiation of African in the Pan-African (± 500 m.y.) tectonic episode. *Leeds Univ. Res. Inst. Afr. Geol. 8th a. Rep.* 48–49.

Kröner, A. (1977). The Precambrian geotectonic evolution of Africa–plate accretion vs. plate destruction. *Precambrian Res.* **4**, 163–213.

Kröner, A. (1979). Pan-African mobile belts as evidence for a transitional tectonic regime from intraplate orogeny to plate margin orogeny. In *Evolution and Mineralization of the Arabian–Nubian Shield* (A. M. S. Al-Shanti, ed.), IAG Bulletin no. 3, Vol. 1, 21–37.

Kröner, A., Roobol, M. J., Ramsay, C. R., and Jackson, N. J. (1979). Pan-African ages of some gneissic rocks in the Arabian Shield. *J. geol. Soc. Lond.* **136**, 455–462.

Le Blanc, M. (1976). Proterozoic oceanic crust at Bou Azzer. *Nature, Lond.* **261**, 34–35.

Marzouki, F. and Fyfe, W. S. (1977). Pan-African plates additional evidence from igneous events in Saudi Arabia. *Contr. Mineral. Petrol.* **60**, 219–224.

McCurry, P. (1976). The geology of the Precambrian to Lower Palaeozoic rocks of northern Nigeria—a review. In *Geology of Nigeria* (C. A. Kogbe, ed.), Elizabethan Press, Lagos, 15–39.

McCurry, P. and Wright, J. B. (1977). Geochemistry of calc-alkaline volcanics in northwestern Nigeria, and a possible Pan-African suture zone. *Earth planet. Sci. Letters* **37**, 90–96.

Nasseef, A. O. (1971). The geology of the northeastern At-Taif area, Saudi Arabia. Ph.D. thesis, University of Leeds, England.

Nasseef, A. O. and Gass, I. G. (1977). Granitic and metamorphic rocks of the Taif area, western Saudi Arabia. *Geol. Soc. Am. Bull.* **88**, 1721–1730.

Neary, C. R., Gass, I. G., and Cavanagh, B. J. (1976). Granitic association of northeastern Sudan. *Geol. Soc. Am. Bull.* **87**, 1501–1512.

Pearce, J. A. (1980). Geochemical evidence for the genesis and eruptive setting of lavas from Tethyan ophiolites. *Proc. Internat. Ophiolite Symposium, Nicosia, Cyprus.* 261–272.

Pearce, J. A. and Cann, J. R. (1973). Tectonic setting of basic volcanic rocks determined using trace element analyses. *Earth planet. Sci. Letters* **19**, 290–300.

Pearce, J. A. and Gale, G. H. (1977). Identification of ore deposition environment from trace element geochemistry. *Spec. Publ. geol. Soc. Lond.* **7**, 14–24.

Pearce, J. A. and Norry, M. J. (1979). Petrogenetic implications of Ti, Zr, Y and Nb variation in volcanic rocks. *Contr. Mineral. Petrol.* **69**, 33–47.

Pegram, W. J., Register, J. K., Ji, Fullagar, P. D., Ghuma, M. A., and Rogers, J. J. W. (1976). Pan-African ages from a Tibesti massif batholith, southern Libya. *Earth planet. Sci. Letters* **30**, 123–128.

Ramsay, C. R., Jackson, N. J., and Roobol, M. J. (1979). Structural/lithological provinces in a Saudi Arabian Shield geotraverse. In *Evolution and Mineralization of the Arabian–Nubian Shield* (A. M. S. Al-Shanti, ed.), IAG Bulletin no. 3, Vol. 1, 64–84.

Rehaile, M. and Warden, A. J. (1978). Comparison of the Bir Umq and Hamdah ultramafic complexes, Saudi Arabia. *J. Precambrian Res.* **6**, A32.

Schmidt, D. L., Hadley, D. G., Greenwood, W. R., Gonzales, L., Coleman, R. G., and Brown, G. R. (1973). Stratigraphy and tectonism of the southern part of the Precambrian Shield of Saudi Arabia. *Saudi Arabian Directorate General Mineral Resources Bull.* no. 8.

Schmidt, D. L., Hadley, D. G., and Stoesser, D. B. (1978). Late Proterozoic crustal history of the Arabian Shield, southern Nadj Province, Kingdom of Saudi Arabia, *US geol. Surv. Saudi Arabian Project Rep.* SA(IR)-251.

Shackleton, R. M. (1979). Precambrian tectonics in northeast Africa. In *Evolution and Mineralization of the Arabian–Nubian Shield* (A. M. S. Al-Shanti, ed.). IAG Bulletin no. 3, Vol. 2, 1–6.

Shackleton, R. M., Ries, A. C., Graham, R. H., and Fitches, W. R. (1980). Pan-African ophiolitic mélanges in the Egyptian Eastern Desert. *Nature, Lond.* **285**, 472–474.

Shanti, M. and Roobol, M. J. (1979). A late Proterozoic ophiolite complex at Jabal Ess in northern Saudi Arabia. *Nature, Lond.* **279**, 488–491.

Stoesser, D. B. and Elliot, J. E. 1979. Post-orogenic peralkaline and calc-alkaline granites and associated mineralization of the Arabian Shield, Kingdom of Saudi Arabia. *US geol. Surv. Saudi Arabian Project Rep.* SA(IR)–265 1.42.

Vail, J. R. (1976). Outline of the geochronology and tectonic units of the basement complex of the northeast Africa. *Proc. R. Soc. A* **350**, 127–141.

Wynne-Edwards, H. R. (1976). Proterozoic ensialic orogenesis: the millipede model of ductile plate tectonics. *Am. J. Sci.* **276**, 927–953.

Volcanism in the Caledonian orogenic belt of Britain

J. G. Fitton, M. F. Thirlwall and D. J. Hughes

Grant Institute of Geology,
University of Edinburgh, West Mains Road, Edinburgh EH9 3JW, UK,

Department of Earth Sciences,
University of Leeds, Leeds LS2 9JT, UK,

and

Department of Geology,
Portsmouth Polytechnic, Burnaby Road, Portsmouth PO1 3QL, UK

ABSTRACT

The Caledonian orogen represents the site of closure of an ocean (Iapetus) during the late Precambrian and Lower Palaeozoic. The British sector of this belt includes strata originally deposited on both of the opposing continental margins. Volcanic rocks, mostly of calc-alkaline affinities, are abundant within the British Caledonides in two clearly defined provinces:

(1) Ordovician–Lower Silurian (early orogenic) volcanic rocks occur to the south-east of the Iapetus suture in the Lake District, eastern and south-eastern Ireland, and the Welsh basin. The Lake District and Irish volcanic rocks show a variation southwards and with time from transitional tholeiitic/calc-alkaline to truly calc-alkaline in character and probably represent an Ordovician island arc which developed along the south-eastern margin of Iapetus. The Welsh basin evolved on a subsiding continental margin, underlain by Precambrian basement, during the Ordovician and shows a greater diversity of rock types. A *general* variation with time can be recognized from the earliest tholeiitic lavas, through mostly calc-alkaline volcanic rocks of the main group to alkaline rocks erupted at the end of the volcanic episode.

(2) Lower Silurian to Upper Devonian (late orogenic) volcanic rocks associated with the continental Old Red Sandstone facies are restricted to those areas (Scotland, north-eastern England, and northern Ireland) to the north-west of the suture. These volcanic rocks have not been subjected to major Caledonian deformation. The lavas are predominantly calc-alkaline, although shoshonitic types locally occur and the youngest lavas are alkaline. The volcanic rocks of the late orogenic province show a well defined spatial variation with light REE, K, Sr, and Ba all increasing in a west-north-westerly direction. This implies that the final stages of closure of Iapetus were accomplished by oblique subduction beneath the British sector.

Introduction

The Caledonian orogen is well displayed in Britain. It is bounded in north-western Scotland by the Moine Thrust where late Precambrian Moinian metasediments are brought into contact with the Archaean Lewisian gneisses of the stable north-western foreland. The south-eastern margin of the orogen is less well defined since it is largely covered by post-Caledonian sediments. It is generally taken to coincide with the Church Stretton and Pontesford fault systems which

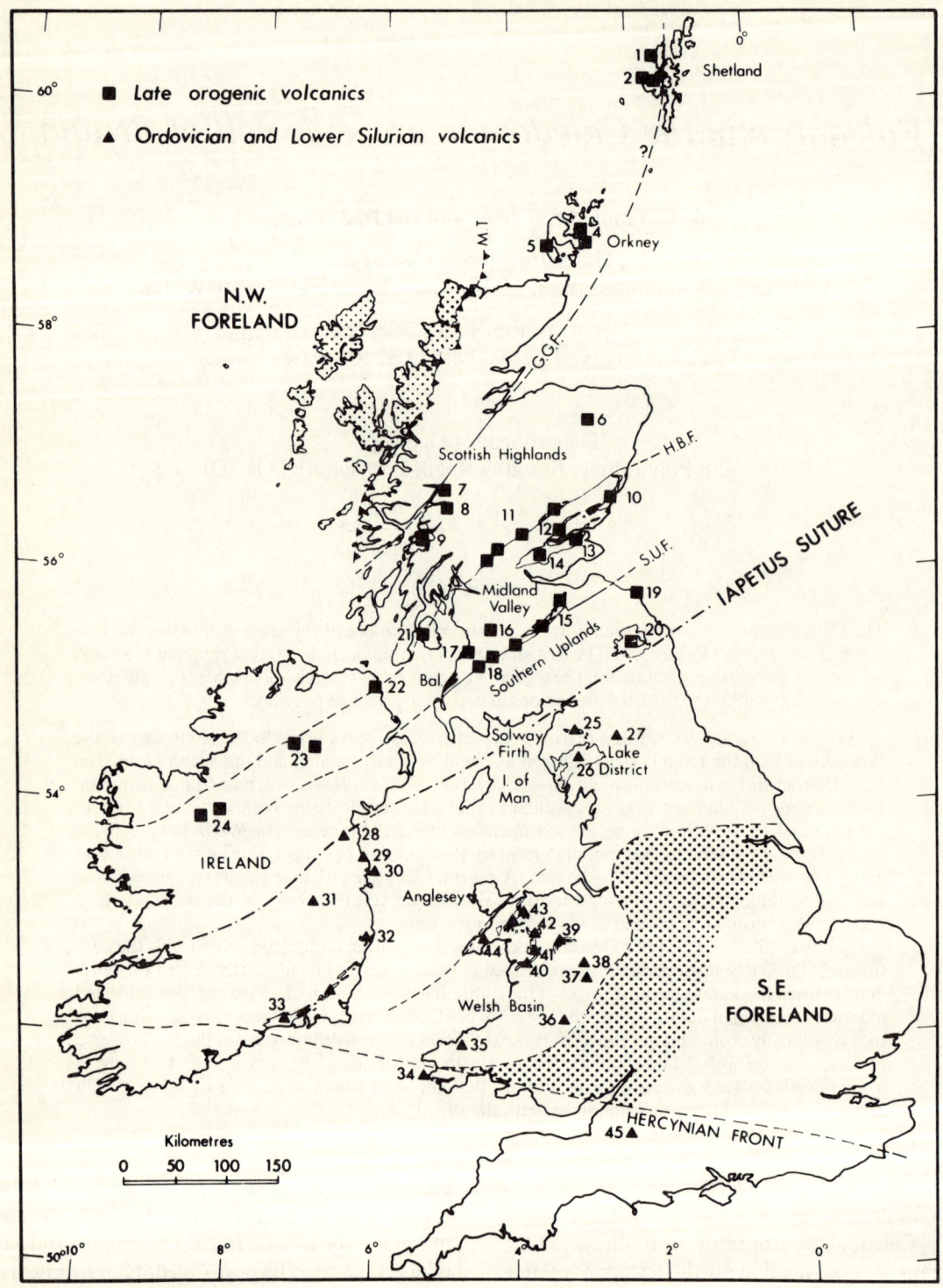

separate the Welsh basin from the stable Precambrian platform beneath the English Midlands.

The main structural elements of the British Calcdonides are shown in Fig. 1. The orogen can be divided into two zones. A metamorphic 'orthotectonic' zone which includes the Scottish Highlands (between the Moine Thrust and the Highland Boundary Fault) with part of north-west Ireland is separated by the Scottish Midland Valley from a non-metamorphic 'paratectonic' zone. The paratectonic zone includes all the Lower Palaeozoic rocks between the Southern Uplands Fault and the south-eastern margin of the orogen. The Midland Valley is a graben filled with Upper Palaeozoic (post-Caledonian) sediments and volcanic rocks and so its position within the Caledonides is unclear. Folded but unmetamorphosed Lower Palaeozoic rocks are found in scattered inliers in its southern part but the occurrence of high grade gneiss xenoliths in Carboniferous diatremes throughout the Midland Valley (Graham and Upton, 1978) indicates the existence of a metamorphic basement.

Wilson's (1966) suggestion that the Caledonian–Appalachian orogen represents the remains of a proto-Atlantic Ocean which closed in the Lower Palaeozoic was followed by attempts by many authors to reconstruct this ocean. These attempts (e.g. Dewey, 1969; Fitton and Hughes, 1970; Church and Gayer, 1973; Phillips et al., 1976) differ in detail but all endorse Wilson's (1966) hypothesis. Most of these models have been reviewed by Moseley (1977). The current consensus is that the two stable forelands (Fig. 1) were separated by an ocean (now referred to as 'Iapetus', following Harland and Gayer, 1972) in the late Precambrian. This ocean closed during the Lower Palaeozoic by subduction of oceanic lithosphere along both of its margins until continental collision occurred in late Silurian to early Devonian times resulting in a major phase of Caledonian deformation.

The Southern Uplands of Scotland comprise Upper Ordovician to Silurian greywackes and shales interpreted by McKerrow et al. (1977) as an accretionary sedimentary wedge which developed in an oceanic trench on the north-western margin of the Iapetus Ocean. Where the base of these sedimentary rocks can be seen they rest on Lower Ordovician pillow lavas often associated with graptolitic black shales, cherts, and serpentinites. These are best developed

---

Fig. 1 Map of the British Isles showing the distribution of andesitic volcanic rocks and the main structural units in the Caledonides. Based on the Institute of Geological Sciences 1:1 584 000 geological and tectonic maps. The location of volcanic outcrops is shown by symbols and the larger outcrops are outlined.

1. Esha Ness
2. Papa Stour
3. Sandness Formation
4. Deerness
5. Hoy
6. Huntly
7. Ben Nevis
8. Glencoe
9. Lorne Plateau
10. Montrose
11. Highland Border
12. Sidlaw Hills
13. North Fife Hills
14. Ochil Hills
15. Pentland Hills and Lanarkshire
16. Distinkhorn
17. Ayrshire Coast
18. Straiton
19. St Abbs Head and Eyemouth
20. Cheviot Hills
21. Arran
22. Cushendall
23. Fintona
24. Curlew Mts
25. Eycott Volcanics
26. Borrowdale Volcanics
27. Cross Fell Inlier
28. Collon and Grangegeeth
29. Balbriggan
30. Lambay Island and Portraine
31. Hill of Allen, Kildare
32. Wicklow
33. Waterford
34. Skomer Island
35. North Pembrokeshire (now part of Dyfedd)
36. Builth Wells
37. Shelve
38. Breidden Hills
39. Berwyn Hills
40. Cader Idris
41. Rhobell Fawr
42. Arenig
43. Snowdonia
44. Lleyn Peninsula
45. Mendip Hills

M.T., Moine Thrust; G.G.F., Great Glen Fault; H.B.F., Highland Boundary Fault; S.U.F., Southern Uplands Fault; Bal., Ballantrae Complex

around Ballantrae in south-western Scotland (Fig. 1), where they have been interpreted as an ophiolite sequence representing an obducted fragment of Iapetus oceanic crust (Fitton and Hughes, 1970; Dewey, 1971; Church and Gayer, 1973). Other occurrences of pillow lavas and serpentinites along the Highland Boundary Fault and as isolated inliers in the Southern Uplands may have a similar origin.

Deep crustal seismic studies (Bamford et al., 1978) have shown that the whole of Britain is now underlain by continental crust with an average thickness of 35 km. It seems certain, therefore, that no large segments of Iapetus Ocean floor now remain beneath Britain. Lewisian-type basement rocks appear to persist as far south as the Southern Uplands Fault whilst the seismically distinct basement rocks beneath central England can be traced northwards as far as the northern part of the Lake District. Between these two major crustal units is a transition zone which may represent the suture between the two contrasted continental masses. The most likely position of the Iapetus suture (after Phillips et al., 1976) is shown on Fig. 1.

Volcanic rocks of calc-alkaline affinities are abundant within the British Caledonides and have been used to define palaeo-subduction zones around the Iapetus Ocean (Fitton and Hughes, 1970). They are also found in late Precambrian inliers in Wales and the Welsh Borders where they may provide evidence for subduction at a very early stage in the evolution of the Caledonides (Thorpe, 1974). The most extensive development of andesitic and related volcanic rocks, however, is to be found among strata of Ordovician to Devonian age, within which two quite distinct volcanic provinces can be recognized.

Considerable thicknesses of volcanic rocks among the Ordovician and Lower Silurian strata of the Lake District, the Welsh basin, and eastern and south-eastern Ireland (Fig. 1) form the first of these provinces. The second province is represented by Silurian and Devonian volcanic rocks associated with the Old Red Sandstone in Scotland and northern England. These volcanic rocks have not been subjected to major Caledonian deformation and will be referred to here as the 'late orogenic volcanic rocks'. The two volcanic provinces are separated completely by the Iapetus suture (Fig. 1).

Calc-alkaline volcanic rocks have not been recorded from the deformed strata north of the suture. The abundance of volcanic debris in the Upper Ordovician greywackes of the Southern Uplands, however, suggests that they may have been originally present here but have subsequently been removed by erosion. Basic pillow lavas (the Tayvallich volcanic rocks) and associated sills in the Cambrian Dalradian metasediments of the south-western Highlands have been interpreted by Graham (1976) as tholeiitic magmas associated with continental rifting along the north-western margin of the Iapetus Ocean.

Ordovician and Lower Silurian Volcanic Rocks

Calc-alkaline volcanic rocks of Ordovician and Lower Silurian age are confined to those parts of the British Caledonides to the south of the Iapetus suture (Fig. 1). They represent a period of intense volcanic activity around the south-eastern margin of the Iapetus Ocean.

The Lake District

The Lower Palaeozoic rocks of the Lake District form an inlier of generally northwards-younging Lower Ordovician (Arenig) greywackes and shales of the Skiddaw Group overlain by thick volcanic successions outcropping to the north and south. To the north the Skiddaw Group sediments pass conformably upwards into the Eycott volcanic rocks (25). (All such numbered references in parentheses refer to localities on Fig. 1.)

Sediments interbedded with the lowermost volcanic units have yielded microfossils suggesting a Lower Llanvirn age (Downie and Soper, 1972). To the south the Skiddaw Group is overlain unconformably by the Borrowdale Volcanic Group (26). The age of the Borrowdale volcanic rocks is not known precisely but must lie between Upper Llanvirn and Middle Caradoc (Wadge, 1978). They are certainly younger than the Eycott volcanic rocks. Volcanic rocks equivalent in age to the Eycott volcanic rocks are not found in the south of the Lake District, although Turner and Wadge (1979) have reported a mid-Arenig volcanic sequence around Millom Park in the south-west. Small outcrops of volcanic rocks correlated with the Eycott and Borrowdale volcanic rocks also occur to the east of the Lake District in the Cross Fell inlier (27) (Wadge, 1978). The last volcanic episode in the Lake District is represented by the Ashgillian Stockdale Rhyolite (Gale et al., 1979). A generalized Ordovician stratigraphy of the Lake District is given in Fig. 2.

The Lower Palaeozoic rocks of the Lake District have been intruded by a host of hypabyssal and plutonic igneous bodies ranging in size from small dykes to large granite stocks. A recent Rb–Sr isochron and K–Ar study of many of the larger intrusions by Rundle (1979) has shown that they range in age from Llanvirnian to Lower Devonian. Thus the gabbros of the Carrock Fell Complex (468 ± 10 Ma old) may be co-eval with the Eycott volcanic rocks while the Threlkeld microgranite and a number of minor bodies have ages (458 ± 9 Ma) similar to that inferred for the Borrowdale volcanic rocks. The two largest granite bodies, the Eskdale Granite and the Ennerdale Granophyre, have Caradoc (429 ± 9 Ma) and Ashgill (421 ± 8 Ma) ages respectively and were therefore emplaced soon after the Borrowdale volcanic episode. The other major granite bodies (Shap and Skiddaw) give ages c. 395 Ma.

The Eycott volcanic rocks are mostly basalt and basaltic andesite lava flows. Pyroclastic rocks account for only a small proportion of the volcanic pile and are concentrated towards the base of the succession. Acid lavas and ignimbrites ranging in composition from dacite to rhyolite also occur but are volumetrically insignificant. Intermediate rocks (andesites) are absent. In contrast the Borrowdale volcanic rocks include a very high proportion of pyroclastic rocks of both ash-fall and ignimbrite types. Andesite is the most common lava type, although basalts and basaltic andesites are also represented. The distribution of rock types (lavas and ignimbrites) in the Eycott and Borrowdale volcanic rocks is shown in Fig. 3. This histogram is *not* based on a statistically sound sampling procedure and so it can only be taken as a rough guide. The phenocryst mineralogy of the two sets of volcanic rocks is summarized in Fig. 4.

All the volcanic rocks in the Lake District, as elsewhere in the Caledonides, are altered to some degree. Olivine and orthopyroxene are always altered to serpentine and chlorite respectively and plagioclase is frequently albitized. Clinopyroxene, however, is usually fresh. Much of this alteration can be ascribed to incipient low-grade regional metamorphism of the volcanic pile, although Millward et al. (1978) have shown that the Borrowdale ignimbrites suffered chemical changes soon after eruption. Because of these alteration effects, great caution must be exercised when interpreting the chemical compositions of these and other volcanic rocks in the Caledonides.

The Borrowdale volcanic rocks have many chemical features in common with modern calc-alkaline volcanic suites. They are mostly quartz- and hypersthene-normative and, in the more acid rocks, corundum-normative. The intermediate members show no trends towards Fe

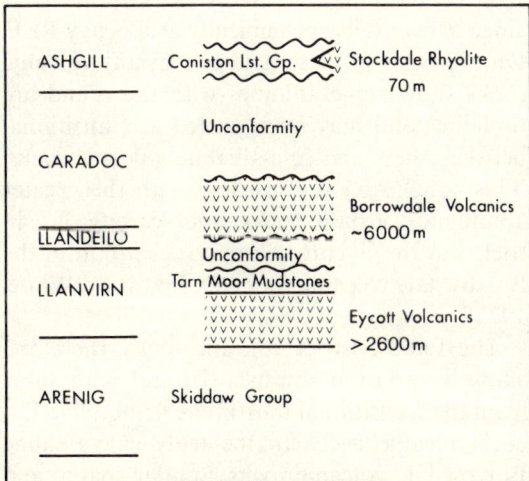

Fig. 2 Generalized Ordovician stratigraphy of the Lake District (after Wadge, 1978)

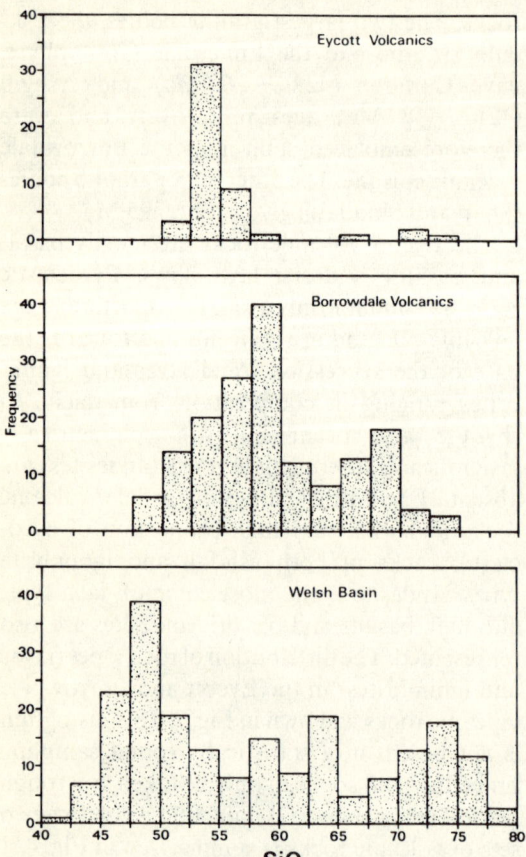

Fig. 3 The frequency of occurrence (number of samples) as a function of SiO_2 content for lavas and ignimbrites from the Eycott and Borrowdale volcanic rocks (data from Fitton, 1971) and the Welsh basin (data from Hughes, 1977)

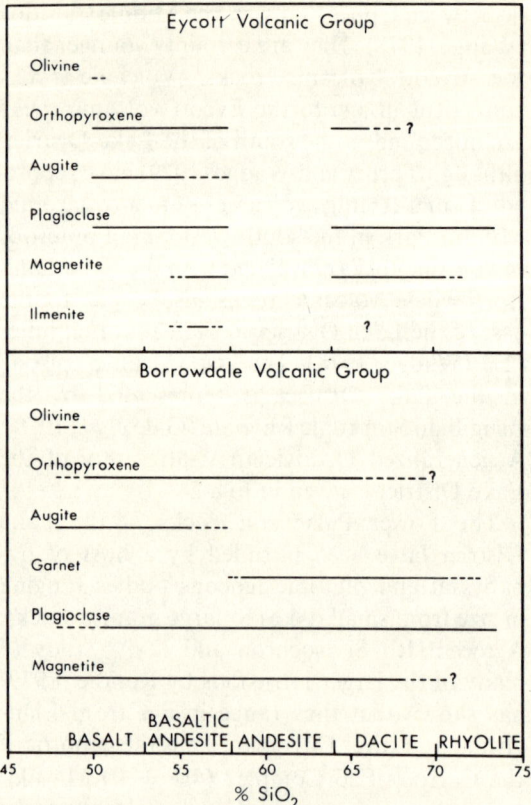

Fig. 4 The distribution of phenocryst phases in the Eycott and Borrowdale volcanic rocks. Broken lines indicate that the phase is seen only rarely

enrichment. Most of the rocks have K_2O/Na_2O ratios of more than 1, which may suggest affinities with modern continental margin calc-alkaline suites (Jakeš and White, 1972). Average compositions of rocks from the Borrowdale volcanic rocks are given in Table 1.

The Eycott volcanic rocks differ chemically from the Borrowdale volcanic rocks in a number of important respects. They are richer in Fe, Ti, and V and have K_2O/Na_2O ratios of less than 1 in all but the most acid members. The Eycott volcanic rocks have consistently lower La/Y ratios than do the Borrowdale volcanic rocks (Fig. 5). This suggests that the Borrowdale volcanic rocks are more enriched in light relative to heavy REE than are the Eycott volcanic rocks since Y behaves geochemically as a heavy REE. On the basis of these data the Eycott volcanic rocks show some affinities with the island arc tholeiites and may be regarded as transitional between these and calc-alkaline volcanic rocks. This conclusion is consistent with the greater abundance of basic rocks (and scarcity of acid rocks) in the Eycott volcanic rocks than in the Borrowdale volcanic rocks (cf. Jakeš and White, 1972).

The Lake District volcanic rocks, therefore, show a variation southwards, and with time, from the transitional tholeiitic/calc-alkaline Eycott volcanic rocks to the truly calc-alkaline Borrowdale volcanic rocks. Similar spatial and temporal variations are now widely recognized in volcanic rocks from modern island arcs (this

TABLE 1 Average compositions of Borrowdale volcanic rocks

	Basalt ($n = 18$)†	Basaltic Andesite ($n = 46$)	Andesite ($n = 46$)	Dacite ($n = 32$)	Rhyolite ($n = 7$)
Percentages‡					
SiO_2	51.04	55.74	59.98	67.38	72.21
Al_2O_3	16.26	17.81	17.76	16.55	14.72
Fe_2O_3 §	11.11	8.84	7.54	4.66	3.17
MgO	7.73	4.63	2.83	0.69	0.75
CaO	8.68	6.53	4.75	1.96	0.58
Na_2O	1.88	2.32	2.67	3.09	2.36
K_2O	1.33	2.42	2.97	4.71	5.55
TiO_2	1.29	1.09	0.91	0.46	0.31
MnO	0.24	0.24	0.20	0.15	0.06
S	0.20	0.16	0.16	0.10	0.08
P_2O_5	0.21	0.19	0.19	0.23	0.21
p.p.m.					
Ba	349	586	747	954	1096
Nb	11	12	14	22	17
Zr	141	179	227	357	282
Y	27	34	40	65	65
Sr	346	310	270	142	85
Rb	49	97	124	211	237
Zn	104	101	93	69	43
Cu	48	26	23	4	5
Ni	145	38	22	2	2
Cr	410	123	62	6	3
V	231	171	119	13	8
La	19	22	35	54	46
CIPW norms‖					
Quartz	5.4	11.0	17.8	25.9	34.9
Corundum	—	—	2.0	3.4	4.3
Orthoclase	7.9	14.4	17.6	27.9	32.9
Albite	16.0	19.7	22.7	26.2	20.0
Anorthite	32.2	31.2	22.4	8.2	1.5
Diopside	7.8	0.1	—	—	—
Hypersthene	20.5	15.3	10.3	3.9	3.3
Magnetite	6.8	5.4	4.6	2.9	1.9
Ilmenite	2.5	2.1	1.7	0.9	0.6
Apatite	0.5	0.5	0.5	0.5	0.5
Pyrite	0.4	0.3	0.3	0.2	0.2

† n = number of analyses.
‡ Recalculated on a volatile-free basis.
§ Total Fe expressed as Fe_2O_3.
‖ Fe_2O_3/FeO taken as 0.808 by weight (Chayes, 1969).

volume, Section II). This chemical variation led Fitton and Hughes (1970) to propose that the Lake District volcanic rocks were erupted in an Ordovician island arc which developed along the south-eastern margin of the Iapetus Ocean above a south-eastward-dipping subduction zone.

The Borrowdale volcanic rocks are unusual among volcanic suites in the abundance of primary phenocryst garnet they contain (Oliver,

1956; Fitton, 1972). They occur in andesites, dacites, and rhyolites and are generally confined to the lower half of the volcanic sequence (the first magmatic cycle of Millward et al., 1978). The garnets are of almandine–pyrope composition with pyrope contents ranging from 31 per cent (molecular) in the andesite garnets to 9 per cent in those from dacites. Spessartine and grossular are minor components with respective ranges of 2.2 and 8.8 per cent (andesites) to 8.8 and 3 per cent (dacites) (Fitton, 1972). Garnets in the andesites are invariably corroded and resorbed and are often surrounded by reaction rims of chlorite and magnetite. Those in the dacites (mostly ignimbrites) are seldom corroded to the same extent and are often euhedral.

The petrogenetic implications of these garnets have been discussed by Fitton (1972) who showed that garnet could not have been a fractionating phase in the evolution of the Borrowdale magmas. The garnets contain high concentrations of heavy REE and Y but very low concentrations of the light REE. Consequently, any fractional crystallization process involving the garnet phenocrysts will result in a large increase in the La/Y ratio of the residual liquids. The Borrowdale volcanic rocks have constant La/Y ratios over the whole compositional range (Fig. 5), which implies that they have not evolved by fractional crystallization of the *observed* phenocryst phases.

The appearance of garnet in the andesites closely follows the disappearance of phenocryst augite, which is abundant in the more basic rocks (Fig. 4). The transition from augite- to garnet-phyric rocks occurs at c. 58 per cent SiO_2

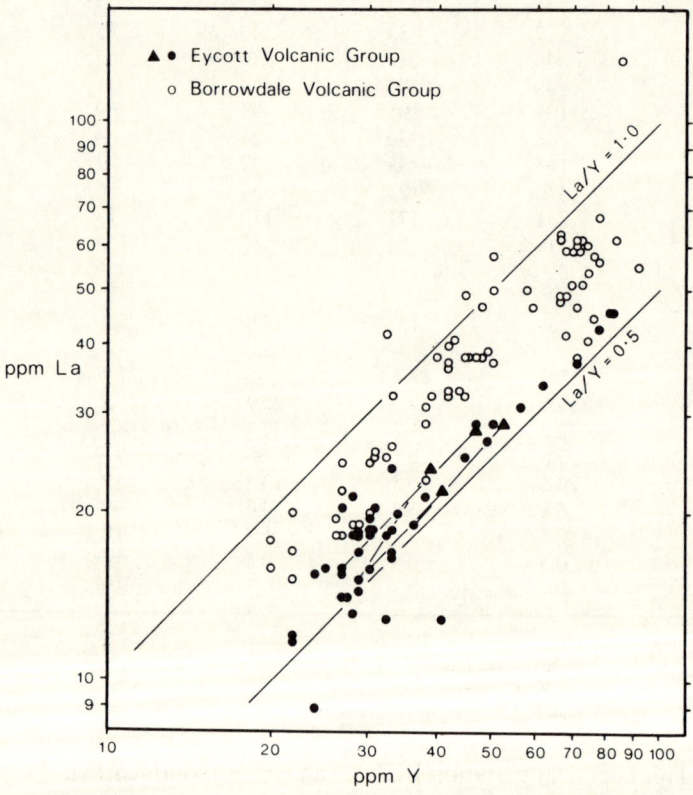

Fig. 5 Plot of La against Y for the Eycott (solid symbols) and Borrowdale (open symbols) volcanic rocks. The solid triangles represent groundmasses separated from highly porphyritic lavas of the Eycott volcanic rocks

and coincides with a change in character from diopside- to corundum-normative (Table 1).

Green (1977) has shown that garnets of similar composition to those in the Borrowdale volcanic rocks can be crystallized from silicic peraluminous liquids at pressures of 5–7 kb. The Borrowdale magmas could, therefore, have crystallized garnet during their ascent through the underlying thick sedimentary pile (Skiddaw Group) which had been deformed and presumably thickened before the Borrowdale volcanic episode. Since the fractional crystallization of garnet is precluded by the REE data, a suite of magmas ranging from basalt to rhyolite must have been in existence by the time garnet began to crystallize. It is possible that this crystallization of garnet was stimulated by the assimilation of Skiddaw Group pelites at 15–20 km depth. The preservation of garnet in the Borrowdale volcanic rocks implies rapid transfer of the magmas to the surface without any intervening low pressure fractional crystallization.

Eastern and south-eastern Ireland

Ordovician volcanic rocks outcrop over a large area of eastern Ireland in two distinct belts (Fig. 1). A northern belt comprises small volcanic inliers at Collon and Grangegeeth (28), Balbriggan (29), Lambay Island and Portraine (30), and the Hill of Allen near Kildare (31). Further south a belt of volcanic rocks extending from Wicklow (32) to Waterford (33) forms the south-eastern belt. The occurrence, tectonic setting, and geochemical compositions of all these volcanic rocks have been reviewed by Stillman and Williams (1979).

Volcanism commenced during the Llanvirn in the northern belt where activity reached a peak in the Caradoc and ended in the Ashgill. In the south-eastern belt volcanism started in the Llandeilo and also reached its maximum activity in the Caradoc. The last magmatism recorded in the south-eastern belt is represented by a series of alkaline dykes. The volcanic rocks in both belts consist largely of subaqueous volcaniclastic deposits and pillow lavas, although the volcanoes sometimes emerged as islands.

The subaqueous eruption of the volcanic rocks has led to spilitic alteration which makes assessment of their original chemical characteristics difficult. Despite this, Stillman and Williams (1979) have shown that the volcanics in both belts are of calc-alkaline to mildly tholeiitic affinities. These authors suggest that the northern belt magmas were rather more tholeiitic than were those in the south-eastern belt. The lavas of the northern belt are predominantly basaltic with no rhyolites whereas rhyolites abound in the south-eastern belt where basalts, andesites, and dacites are relatively uncommon.

The magmatic similarity between the Irish and Lake District volcanic rocks is striking. Volcanism in both areas covers the same time span and the magmas display parallel changes southwards. It seems likely that the Lake District volcanoes and those in Ireland formed part of the same Ordovician island arc system.

The Welsh basin

The Lower Palaeozoic Welsh basin lies some 200 km south-south-west of the Lake District and, unlike the Lake District, the rocks are demonstrably underlain by a complex Precambrian basement (Shackleton, 1954). The tectonic cycle of which the igneous activity was a part began with late Precambrian sedimentation and ended with the complete infilling of the basin and the development of fold mountain systems in the Lower and Middle Devonian.

Volcanicity in the Welsh basin was extensive both in time and space and very varied in character. The bulk of the activity was Ordovician, the oldest volcanics occurring at the base of the Arenig at Rhobell Fawr (41) and the most recent in the Llandovery at Skomer (34) and in several scattered inliers south of the Bristol Channel in the Mendip Hills and Somerset (45). Tholeiitic, transitional, and mildly alkaline basaltic rocks occur with calc-alkaline andesites, dacites, and very extensive rhyolitic rocks. Frequent migrations of the centres of activity occurred throughout the Ordovician. The distribution of the volcanic rocks is shown in Fig. 1, a simplified stratigraphic correlation is presented in Fig. 6,

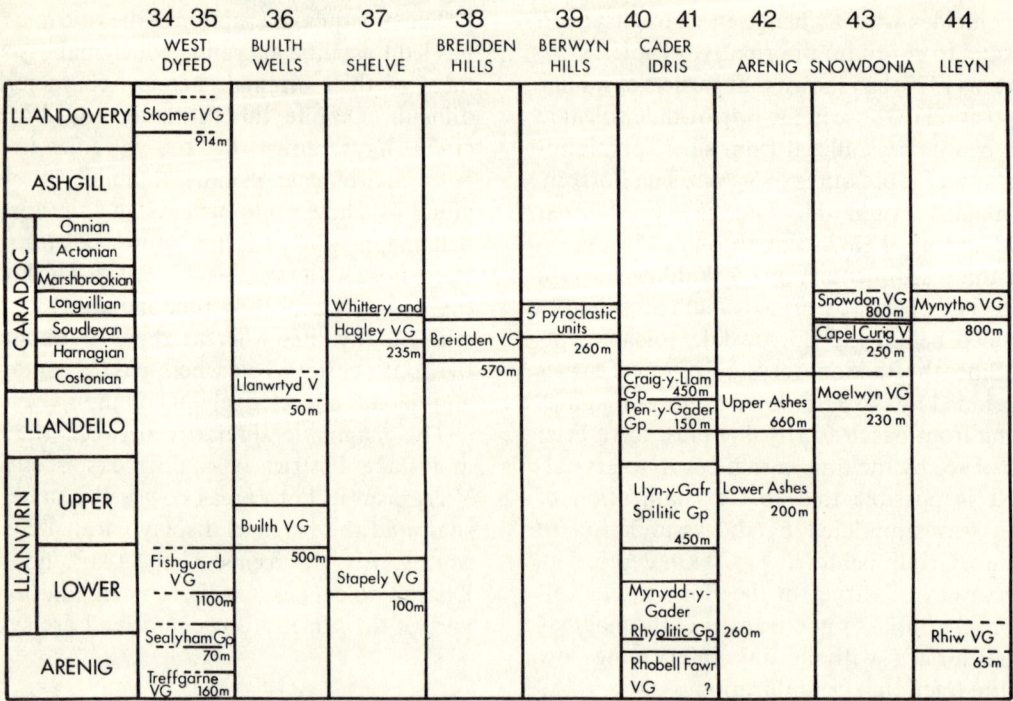

Fig. 6 Stratigraphical correlation diagram for the volcanic rocks of the Welsh basin. V, volcanic rocks; VG, Volcanic Group

and the phenocryst mineralogy of the Welsh volcanic rocks is summarized in Fig. 7.

For the purposes of this discussion three distinct geochemical groups will be considered. First, rocks from the oldest part of the sequence, from the Rhobell Fawr Volcanic Group (41), which show some affinities to the island arc tholeiitic suite. Second, the main group of igneous rocks in the Welsh basin which comprises a variety of basalts, rocks of calc-alkaline affinities, and abundant rhyolitic rocks. Finally, the youngest rocks in the cycle, from the Skomer Volcanic Group (34), which are distinctly alkaline in character.

Rhobell Fawr Volcanic Group (41)

The complex pile of pyroclastic rocks, lavas, and intrusions which comprise the Rhobell Fawr Volcanic Group lie on a post-Tremadocian erosion surface and below the grits which mark the local base of the Arenig (Wells, 1925). The lavas are predominantly basaltic, and are olivine tholeiite in normative composition. The average silica content is high (49.5 per cent); Ti and P are both low. There is no pronounced Fe-enrichment trend, although all of the analysed samples have high Fe contents relative to those in normal calc-alkaline suites. The rocks are not now particularly aluminous (15.4 per cent on average), but analysed clinopyroxene phenocrysts are Al-rich and Ti-poor (Fig. 8). Relatively low Ni and Cr contents are a distinctive feature of the basalts (averaging 39 and 73 p.p.m. respectively), and they also have very low abundances of Ce, Nd, and Y which suggest a flat chondrite-normalized REE pattern.

The rarer, more evolved rocks appear petrographically to be fairly typical basaltic andesites and andesites. Quartz appears in the norm, Ni and Cr drop to very low levels, and the V/Ni ratio is high (25 on average in the andesites), which suggests true calc-alkaline affinities (Taylor *et al.*, 1969). However, Fe remains at relatively high levels (average 7.2 per cent in the

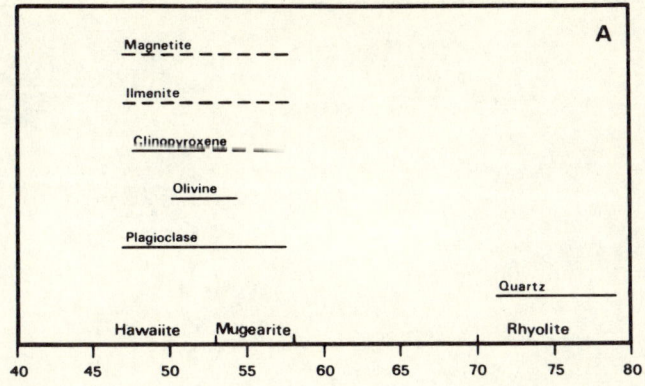

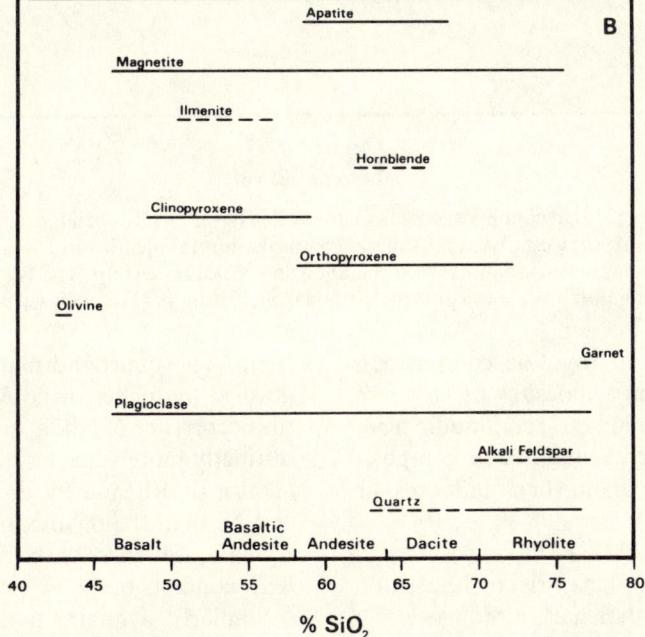

Fig. 7 The distribution of phenocryst phases in the Skomer volcanic rocks (A) and in volcanic rocks from the remainder of the Welsh basin (B). Broken lines indicate that the phase is seen only rarely

andesites), which is more typical of tholeiitic derivatives.

Thus the suite appears to have varied affinities. The basalts are tholeiitic but are not typical tholeiites. They are clearly similar to island arc tholeiites as defined by Jakeš and Gill (1970). However, their K/Rb ratios are not high (average 376 compared with c. 1000 for island arc tholeiites). This may be because the rocks are transitional with a true calc-alkaline suite which would explain the occurrence of some andesites, or possibly it is related to the relative mobility of K and Rb in these rather altered rocks.

The main group

In the majority of exposed sequences, basaltic rocks (either lavas or high level intrusions),

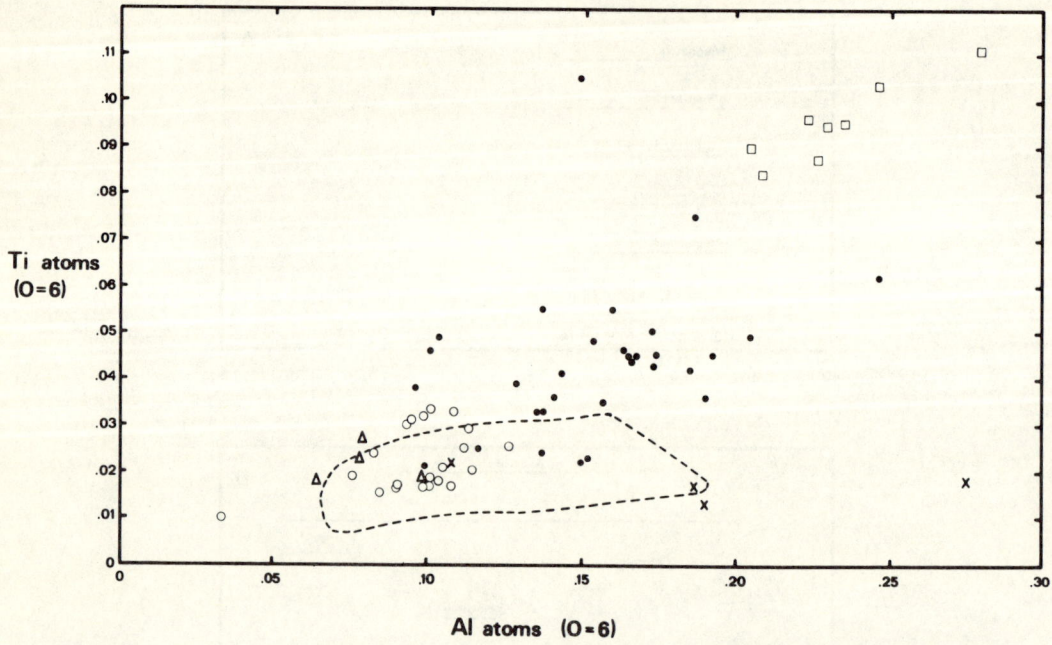

Fig. 8 Ti–Al relationships in clinopyroxenes from the Welsh basin. ×, Rhobell Fawr; △, basaltic andesite (Garn Bodfean, Lleyn); ○, west Dyfedd (formerly Pembrokeshire) tholeiitic intrusions; ●, main group of Welsh dolerites and basalts; □, basalts from the Snowdon Volcanic Group. The field of clinopyroxenes from the Eycott and Borrowdale volcanic rocks (data from Fitton, 1971) is enclosed by the broken line

andesites, and rocks of rhyolitic composition are all abundant. Basaltic andesites are very rare and there is no clear volumetric continuum from andesites through to rhyolites, dacitic compositions being less abundant than andesites or rhyolites.

Basaltic rocks The great majority of the basaltic rocks are normative tholeiites, although a few are slightly nepheline-normative. In mid and north Wales normative olivine tholeiites are predominant whereas in the Builth Volcanic Group (36) and in western Dyfedd (formerly Pembrokeshire, 35) several of the basic intrusions are quartz-normative tholeiites.

Both the quartz-normative and the olivine-normative rocks which occur in western Dyfedd have a distinctly tholeiitic mineralogical character which is unique amongst Welsh basic rocks. Orthopyroxene occurs in many of the rocks, the clinopyroxenes are colourless, Al- and Ti-poor (Fig. 8), and with Fe-enrichment trends very similar to the early part of the Skaergaard trend. The quartz-normative rocks are distinguished by higher orthopyroxene contents and the occurrence of modal quartz. As well as being distinctly more siliceous, they are also richer in alkalis, Zr, Rb, and Ba, and poorer in Sc, V, Cr, and Ni than the olivine tholeiites from western Dyfedd (35). There is very little difference in the REE contents of the two suites.

Similarly, a quartz tholeiitic and an olivine tholeiitic suite occur in the Builth Volcanic Group (36), and although the quartz tholeiites are rather less siliceous than their Dyfedd counterparts the differences between the two suites follow the same pattern and, again, two discrete magma pulses seem likely.

The majority of olivine tholeiites from the main group have remarkably similar bulk compositions, despite the variation among the clinopyroxenes (Fig. 8). They are typically low alumina, low alkalis, and low P tholeiites with a slight tendency towards Fe enrichment. Ti contents vary considerably, ranging from an average of 2.5 per cent TiO_2 in the basalts from the Snowdon Volcanic Group (43) (cf. the Ti contents of the clinopyroxenes from this group; Fig. 8) to less than 1 per cent in some dolerites.

There is no evidence of any similarity to ocean-floor tholeiites or island arc tholeiites. K is not particularly low; K/Rb ratios are typical of Shaw's (1968) 'main trend' for igneous rocks and the REE distributions suggest an enrichment pattern rather steeper than that of the average Hawaiian tholeiite.

Apart from local *in situ* fractional crystallization, such as that described by Cattermole (1969, 1976) and Hawkins (1970) from the Rhiw–Penarfynydd intrusion on the Lleyn Peninsula (44), none of the basic rocks shows any evidence of being involved in a low pressure fractional crystallization sequence. Combined with the rarity of basic–intermediate rocks (Fig. 3), it is therefore probable that the injection of fairly uniform olivine tholeiitic magma was a discrete event not directly genetically related to other igneous activity in the Welsh basin.

Calc-alkaline suite Andesites and related rocks are widely distributed in the Welsh basin. They predominate in the Arenig Treffgarne Volcanic Group in Dyfedd (35), in the Llanvirn–Llandeilo volcanic rocks around Arenig (42), in the Caradoc of parts of Snowdonia (43) and the Lleyn Peninsula (44), and in the Llandovery volcanic rocks of the Mendip Hills in Somerset (45).

Typical calc-alkaline andesites are dominant with fewer dacites and very rare basaltic andesites. Despite considerable variety all of the rocks show clear petrographic affinities with the calc-alkaline suite (Fig. 7). They are all plagioclase-phyric, the phenocrysts are often faintly zoned and are usually andesine-oligoclase in composition, although clear secondary albite is frequently abundant. Colourless clinopyroxene occurs in the groundmass and as phenocrysts in the most basic andesites; these are typically low Ti, low Al augites (e.g. clinopyroxenes from Garn Bodfean; Fig. 8). Chlorite pseudomorphs after orthopyroxene occur infrequently but throughout the range, and hornblende phenocrysts occur rarely. No garnets were found in any of the Welsh andesites or dacites, contrasting with their frequent occurrence in these rocks in the Borrowdale Volcanic Group.

All of the analysed samples are strongly quartz-normative and usually have normative corundum. The average andesite and dacite compositions from the various centres compare very closely with the average andesite and dacite data of Taylor (1969) for major elements, most of the transition elements, and Sr. A significant exception is that V/Ni ratios are much lower in the Welsh andesites (4 compared with 10). Incompatible elements such as Zr, Rb, and REE are more abundant in the intermediate rocks from the Welsh basin. As a consequence of the higher than average Rb contents, the K/Rb ratios are comparatively low (241).

Rhyolitic rocks Acid igneous rocks are ubiquitous throughout the main group, occurring as rhyolitic pyroclastic rocks, flow breccias, intrusive rhyolites, and microgranites. Because of their alkali-rich nature these rocks frequently show the greatest effects of secondary alteration and as alkalis are the elements which, in terms of feldspar composition, characterize the acid rocks, geochemical interpretations drawn from them are often open to some doubt. The majority of the rocks are rhyodacitic or rhyolitic in composition, typical of the acid end of the calc-alkaline suite and, in terms of chemical variation, they apparently follow directly from the dacites. However, acid rocks are significantly more abundant than intermediate rocks in the Welsh basin and it is therefore unlikely that a simple evolutionary sequence from andesite to rhyolite can account for the distribution. The underlying continental crust possibly provides the answer as a source of local melting to produce at least some of the acid rocks. This might also explain the occurrence of the rare, highly alkaline acid rocks, such as the riebeckite-acmite-microgranite intrusion at Mynydd Mawr in Snowdonia (43).

Skomer Volcanic Group (34)

Petrographically, the Skomer Volcanic Group comprises an alkaline suite of hawaiites and rarer mugearites with apparently unrelated acid ignimbrites. Their chemical characteristics clearly confirm the alkaline nature of the basic rocks, with high Na, K, Ti, and P in both hawaiites and mugearites. Total alkalis average 4.5 per cent in the hawaiites and rise to 7.5 per cent in the mugearites. Rb is not abundant (average 24 p.p.m.) and as a consequence K/Rb ratios are typically high. Mg, Ca, Fe, Ti, and Mn fall steadily throughout the hawaiites and in the mugearites. Mg falls more rapidly than Fe and thus there is a tendency towards limited Fe enrichment. Nepheline only occurs rarely in the norms, a typical feature in evolved members of alkali olivine basaltic suites.

The trace element distributions clearly reveal the evolved nature of the sequence. V shows an exponential fall from 400 p.p.m. in the basic hawaiites to 60 p.p.m. in the more evolved mugearites. However, both Cr and Ni are at low levels throughout; the highest Cr content is 70 p.p.m. and the highest Ni content is 72 p.p.m., whereas Zr and REE are at relatively high levels, even in the most basic rocks. This suggests that no true alkali olivine basalts were collected. The whole outcrop was examined and most reasonably fresh flows were sampled, which implies that alkali olivine basalts are, at best, rare rocks in the suite, either because they are unexposed or because they were never erupted in quantity.

The acidic, K-rich ignimbrites which occur in the Skomer Volcanic Group are chemically dissimilar to the sodic, basic rocks, and it is possible that they are the products of minor crustal melting associated with the alkaline volcanic event.

Late Orogenic Volcanism

Outcrops of volcanic rocks associated with the Caledonian Orogeny in Britain display a marked difference to the north and south of the Iapetus suture (Fig. 1). Andesites assigned to the Lower Palaeozoic crop out only to the south, while those assigned to the Old Red Sandstone, regarded as Devonian in age, crop out only to the north of the suture. The latter rocks are the subject of this section.

The Old Red Sandstone sediments associated with the volcanic rocks are of continental facies and are now mainly preserved in the Midland Valley of Scotland and other fault-bounded basins, where their conglomeratic nature indicates the former existence of considerable relief in the Scottish Highlands. Present outcrops of the volcanic rocks are largely restricted to these fault-bounded basins; it is thought that this is not due to a genetic relationship to the faults, but rather to the erosion of the volcanic rocks from the uplifted areas between the basins. This view is substantiated by the occurrence of chemically closely related rocks inside subsided calderas at Ben Nevis (7) and Glencoe (8) and as abundant clasts in Highland-derived Old Red Sandstone conglomerates south of the Highland Boundary Fault. With the exception of Orkney and Shetland, volcanic rocks are absent from Old Red Sandstone sequences north of the Great Glen Fault; it is possible that this apparently abrupt cessation of activity to the north is due to the lateral displacement by faulting of the north-westward continuation of the province.

Unlike the Lower Palaeozoic volcanics to the south, the age ranges of the individual outcrop areas are poorly constrained by fossils. The associated sediments yield plant and fish remains which have been used to suggest a Lower Devonian age (Waterston, 1965) for all the rocks with the exception of those of Orkney and Shetland, where areas 1–4 (Fig. 1) are regarded as Middle Devonian and area 5 as Upper Devonian. Such correlations are cast in doubt by an Rb–Sr isochron age of 415 ± 7 Ma on the Lorne lavas (9) (Brown, 1975), probably representing a Lower Silurian age (cf. the Ashgillian Stockdale Rhyolite; 421 ± 3 Ma in Gale et al. 1979).

The rocks are petrographically very varied, ranging from olivine basalts to rhyolitic ignimbrites. Phenocryst minerals are one or more of olivine, orthopyroxene, calcic clinopyroxene, plagioclase, hornblende, magnetite, ilmenite, apatite, biotite, alkali feldspar, and quartz, the first four being dominant. Pyroclastic rocks are much less common than in modern andesitic provinces, and are mainly represented by laharic deposits at Lorne (9) and the Ochil Hills (13 and 14). Xenoliths are rare in the lavas, but two flows of a highly altered andesite in the Ochil Hills near Auchtermuchty contain coarse gabbroic xenoliths dominated by amphibole pseudomorphs. The very low Ni content of the xenoliths restricts their significance to the genesis of some of the more evolved rocks.

Chemical affinities with modern andesitic rocks include the generally quartz-normative nature of the mafic members, the very minor Fe enrichment relative to Mg in intermediate members, the low content of Nb (less than 25 p.p.m.), the high alumina nature of mafic rocks, and a peraluminous tendency in siliceous rocks (more than 62 per cent SiO_2). Pyroxenes show only slight Fe enrichment with increasing host-rock silica, and are poor in Ti, Na, and tetrahedral Al. Despite these general characteristics, the province shows a high degree of petrographic and chemical variability, often between successive flows. The rocks are richer in P, Ti, Ni, K, Ba, Sr, Rb, and the light REE than most modern andesites.

Discussion of petrography and chemistry is facilitated by a fivefold geographic division of the province.

Southern Uplands

Siluro-Devonian volcanic rocks in the Southern Uplands crop out in two areas: at St Abbs Head and Eyemouth (19) and in the Cheviot Hills (20). In both these areas they rest unconformably on folded Silurian and Ordovician greywackes and shales, the youngest of Wenlock age.

Volcanism in the Cheviot Hills was dominated by the eruption of monotonous acid andesites and dacites (61–67 per cent SiO_2) with phenocrysts of bronzite–hypersthene, augite, andesine–labradorite, apatite, ilmenite, and infrequent biotite. The rocks have high contents of Th, K, Rb, and light REE relative to other acid andesites in the province (Table 2, analysis 1). Harker-type variation diagrams display generally linear variation of element concentrations with silica, and major element variation may be interpreted as resulting from fractionation of the phenocryst phases. Certain trace elements do not support this hypothesis, however; in particular Zr, light REE, Y, and Nb show depletion with increasing silica and Th. Phenocryst ilmenite contains up to 2000 p.p.m. Zr and apatite up to 4000 p.p.m. Ce, but concentrations of $c.$ 12 000 and 9000 p.p.m. respectively would be required to produce the observed depletions. Average Zr in terrestrial ilmenites has been quoted as 330 p.p.m. (Arrhenius et al., 1971); it is therefore thought unlikely that Zr distribution could be controlled by fractional crystallization of a more zirconian ilmenite at depth, and it seems probable that some form of contamination or mixing hypothesis is needed to account for the chemical variation.

The volcanic rocks of the St Abbs Head–Eyemouth district are severely altered but are typically olivine-orthopyroxene-phyric in the mafic members, developing amphibole and plagioclase and losing olivine in more siliceous rocks. The less mobile elements, P, light REE, Zr, Th, Y, and Nb, show a striking distinction between a suite of rocks showing rapid enrichment in these elements with increase in SiO_2 content, with final depletion in all but Th in the most siliceous member (60 per cent SiO_2), and a suite showing much slower enrichment or minor depletion and far more scatter. K and Rb can be shown to follow the same pattern of two suites, despite their much greater mobility during alteration, leading to two separate values for K at constant silica. It is possible that both suites may originate from the same parent. The rapid enrichment trend could lead to the high incompatible element concentrations characteristic of the Cheviot Hills lavas.

Midland Valley

The Scottish Midland Valley has the greatest development of the Siluro-Devonian andesitic rocks, with a thickness of some 2400 m (Francis et al., 1970) in the Ochil Hills (14). The base of the Old Red Sandstone sequence is only seen at the margins of the area, where the rocks rest unconformably on Ordovician to Upper Llandovery greywackes and shales in the south, and

IX Andesite volcanism throughout geological time

TABLE 2 Means, ranges, and selected analyses of volcanic rocks from the late orogenic province

	1	2	3	4	5	6	7	8	9
Per cent ‡									
SiO_2	63.26	54–56	56.99	52.34	53.15	58.52	64.53	64.03	51.44
Al_2O_3	15.97	15.70–18.71	15.34	16.49	16.58	16.71	16.53	16.54	19.86
Fe_2O_3†	4.89	6.72–10.98	7.00	9.14	8.34	6.77	4.42	4.56	7.15
MgO	3.00	2.46–7.41	7.25	6.90	5.87	4.20	2.37	2.29	4.45
CaO	3.79	3.31–8.86	6.24	8.63	7.84	5.28	3.78	3.69	7.49
Na_2O	3.59	3.18–5.09	3.73	3.03	3.88	4.13	4.38	4.30	5.37
K_2O	3.96	0.91–3.00	1.85	1.04	2.12	2.55	3.01	3.23	1.56
TiO_2	0.90	0.88–1.74	1.15	1.40	1.39	1.10	0.61	0.71	1.24
MnO	0.07	0.05–0.19	0.13	0.14	0.12	0.08	0.07	0.07	0.13
P_2O_5	0.31	0.18–0.60	0.24	0.21	0.47	0.38	0.22	0.24	0.73
Total	99.74	—	99.92	99.32	99.76	99.72	99.92	99.66	99.42
LOI	2.24	—	0.75	1.56	3.32	2.87	0.62	1.51	3.42
p.p.m.§									
Ni	53	5–208	176	130	157	106	35	36	32
Cr	90	3–449	367	351	302	207	60	69	17
V	90	108–219	119	157	169	129	79	78	167
Sc	13	12–38	18	24	22	18	10	10	20
Cu	21	12–90	32	25	29	29	18	17	45
Zn	90	47–550	68	76	82	75	62	65	50
Sr	488	277–817	559	429	1151	1063	1048	963	1266
Rb	143	8–69	56	25	34	46	59	70	20
Zr	287	144–351	201	160	197	198	139	222	196
Nb	15	6–18	11	8	13	12	8	10	42
Ba	827	311–805	468	247	1006	1100	1386	1368	2400
Th	23	<1.5–12	11	<1.5	3	5	4	5	<1.5
La	58	16–46	27	17	35	35	28	35	56
Ce	118	33–100	55	34	78	75	57	72	102
Nd	45	18–48	26	18	38	34	26	31	45
Sm	7	2–10	4	6	7	6	4	5	6
Y	20	21–38	21	26	21	18	13	17	24

(1) Mean Cheviot acid andesite–dacite with SiO_2 range 60–67 per cent (44 samples). (2) Ranges for 69 Midland Valley samples with 54–56 per cent SiO_2. (3) Bronzite-andesite from the Sidlaw Hills. (4) Olivine-plagioclase-phyric basalt from the Sidlaw Hills. (5) Mean Lorne Plateau basalt–basic andesite with SiO_2 less than 55 per cent (23 samples). (6) Mean Lorne Plateau andesite with SiO_2 range 55–62 per cent (32 samples). (7) Mean Ben Nevis acid andesite–dacite with SiO_2 range 62–67 per cent (11 samples). (8) Mean Glencoe acid andesite–dacite with SiO_2 range 62–67 per cent (9 samples). (9) Hawaiite from Hoy.
   † Total iron expressed as Fe_2O_3.
   ‡ All samples were ignited at 1100 °C before analysis.
   § Recalculated on a volatile-free basis.

on probably Cambro-Ordovician Dalradian metasediments in the north. The volcanic rocks and the associated conglomerates and sandstones are overlain unconformably by a less conglomeratic sequence of Upper Devonian to Carboniferous sediments, including in the early Carboniferous a major development of alkaline basalt lavas (Macdonald, 1975).

The phenocryst mineralogy of the Siluro-Devonian volcanic rocks is extremely varied, with the appearance of all the phases noted previously. Phenocryst plagioclase is present in more than 70 per cent of basalts and basic andesites, and olivine and pyroxene seldom co-exist without plagioclase. The bulk of the rocks fall in the range 52–63 per cent SiO_2 with displacement of the mode to more basic compositions in the Sidlaw Hills (12) and in the south of the region. The rocks also show great variability in chemical characteristics, and there is little correlation between chemical composition and phenocryst mineralogy. Rocks similar to the St Abbs rapid enrichment trend are present in the Pentland Hills (15), displaced towards less siliceous compositions, but most of the rocks show relatively slow incompatible element enrichment with increasing silica content, with depletion of all but K, Th, and Rb in the most acid members (over 65 per cent SiO_2). However, wide variation in chemical composition is present at constant silica, as indicated by the ranges in composition for a small silica range in the basic andesites (Table 2, analysis 2). A pattern is present in this wide variation, for each rock tends to be enriched to the same degree for every incompatible element, so that Ce/Zr, for example, remains approximately constant.

The existence of andesites with over 150 p.p.m. Ni is noteworthy in these ranges: such rocks characteristically have the phenocryst assemblage bronzite–plagioclase and also have high MgO and Cr contents. Analysis 3 in Table 2 is of a bronzite–andesite from the Sidlaw Hills where it is one of the most siliceous and yet also one of the most MgO-, Ni-, and Cr-rich rocks in the area. Variation diagrams for the Sidlaw Hills using MgO as abscissa demonstrate the existence of two subparallel trends, particularly well defined for light REE and Zr, the most incompatible element-rich of which originates in the bronzite–andesite of Table 2. Rocks so defined as related to the bronzite–andesite are also richer in Sr, precluding derivation of the bronzite–andesite from a less siliceous but equally magnesian parent (analysis 4) by fractionation of calcic plagioclase. When samples from a wider geographic area are considered the differentiation between suites is no longer clear-cut, and it seems probable that several source materials, also indicated by variation in Zr/Nb, and several evolutionary courses are required to produce the chemical variation. The problem of recognizing evolutionary courses is compounded by the restriction in reliability of MgO contents to the least altered rocks.

Marked geographical variation is noticeable in Midland Valley rocks. Fe enrichment on an AFM diagram is more pronounced in the south, while La/Y, K/Th, and Sr increase in a north-westerly direction. In outcrops near the Highland Boundary Fault (11) rocks transitional to some of the Lorne lavas appear: quartz 'xenocrysts', the paucity of phenocryst plagioclase in mafic rocks, and the occurrence of an ignimbrite are all 'Highland' features.

Grampian Highlands

In the four outcrops in the Grampian Highlands (6–9), Siluro-Devonian volcanic rocks rest unconformably on Dalradian metasediments (Precambrian to Ordovician), and in the western outcrops are intruded by the youngest Caledonian granites. Phenocryst plagioclase is rare in rocks with less than 60 per cent SiO_2, but again the rocks are mineralogically very varied, and c. 30 different phenocryst assemblages have been recognized in the Lorne area (9). Single quartz crystals occur frequently in basic andesites, often rounded and embayed and always surrounded by an augite reaction rim. Quartzite xenoliths are not surrounded by a reaction rim, and it is thought that the single-crystal quartz is a high pressure phenocryst phase.

The Lorne volcanic rocks vary from basalts to rhyolitic ignimbrites, with all silica contents well represented in the range 52–64 per cent SiO_2, but the Glencoe volcanic rocks (8) are more siliceous, with a mode at c. 60 per cent SiO_2 and with the development of thick rhyolite lavas and ignimbrites. The Ben Nevis lavas (7) are monotonous hornblende–two pyroxenes–plagioclase acid andesites and dacites, with 62–66 per cent SiO_2. All three areas have distinct chemical characteristics, but there is overlap. Harker variation trends for the major elements are very tight straight lines, except where affected by alteration, but as with the Cheviot lavas this simple appearance is not confirmed by the 'incompatible' trace elements. Major element variation may not be explained by fractional crystallization of the phenocryst phases; for example, Al_2O_3 and Sr do not behave incompatibly, despite the absence of aluminous phenocryst phases in rocks more basic than 60 per cent SiO_2; and apatite phenocrysts are very rare, despite the compatible behaviour of P_2O_5. As with the Cheviot lavas, Nb, Y, and light REE show depletion with increasing silica content (Table 2, analyses 5 and 6), but the respective variation diagrams are very scattered. Zr is scattered between 140 and 250 p.p.m. for all silica contents. Again, as with the Cheviot lavas, it is difficult to produce these variations by fractionation of ilmenite and apatite at depth. Microprobe studies have not confirmed the high concentration of Zr reported by Haslam (1968) from a Ben Nevis hornblende (4000 p.p.m.): Zr contents of hornblende are uniformly below the detection limit of c. 150 p.p.m. The Ben Nevis lavas have the same compatible element compositions as those from Glencoe, but have much lower 'incompatible' element concentrations (Table 2, analyses 7 and 8: the contrast is greater than shown by these averages, for two of the oldest Ben Nevis rocks are identical to Glencoe rocks). Again, it seems likely that a dual source petrogenetic model is required, and, in Lorne, the possibility of crustal contribution is restricted by an initial Sr isotope ratio of 0.7044 (Brown, 1975).

Further chemical variety in Lorne is provided by a small group of apatite-biotite-phyric lavas and related minor intrusions. These rocks have K_2O/Na_2O ratios greater than 1, Sr contents of more than 2000 p.p.m., and La/Y ratios of over 6, and are considered to have shoshonitic affinities. Despite the great variability in concentrations of 'incompatible' trace elements, there is much less chemical variability than in the Midland Valley (Table 2). For example, no rocks are more aluminous than 17.8 per cent Al_2O_3 occur, and there are no basalts or basic andesites (under 57 per cent SiO_2) with less than 70 p.p.m. Ni.

Shetland and Orkney

Siluro-Devonian volcanic rocks of the Orcadian basin are regarded as younger than the rest of the province on palaeontological grounds. In Shetland a wide range of rocks crop out in three isolated areas, one of which (3) has suffered intense polyphase deformation since volcanism. The relationships between these areas and Orkney are difficult to elucidate because of the possibility of offset on splays of the Great Glen Fault system. Petrographically, the rocks are comparable to those of other parts of the province, although there is apparently a basalt–rhyolite bimodal distribution at Sandness (3) and Papa Stour (2). Chemically the rocks resemble modern andesitic suites more closely than do the rest of the rocks of the province, in that Y, Zr, Nb, and light REE show continuous enrichment with silica. An unusual feature shown by some of the rocks is very high concentrations of MnO and Zn, up to 0.6 per cent and 1200 p.p.m. respectively. P, Ti, and K are high, in common with the rest of the province. Intermediate lavas show a greater degree of Fe enrichment relative to Mg than is common elsewhere in the province.

The Orkney lavas of Deerness (4) have been described as comparable to Carboniferous alkaline types by Kellock (1969). The rocks are uniformly basaltic at both Deerness (4) and Hoy (5), with phenocrysts from olivine, plagioclase, and calcic clinopyroxene; those of Deerness show secondary development of analcite and zeolites (E. Kellock, personal communication), while those of Hoy may have primary groundmass analcite. All four samples analysed from Hoy show normative nepheline. This is considered to be primary, for unaltered olivine is present in two of the rocks, and analyses show very high concentrations of P, Nb, and light REE relative to basalts in the rest of the province (Table 2, analysis 9). These are clearly alkaline olivine basalts and hawaiites, but their TiO_2 content distinguishes them from Carboniferous alkaline basalts which have TiO_2 from 2.5 to 4 per cent (Macdonald, 1975). Whitford (1975) considers that alkaline basalts related to island arcs may be distinguished from typical continental alkaline basalts by their lower TiO_2 content. Two of the four samples analysed from Deerness show normative nepheline, but as the rocks are greatly altered and show contents of Nb, TiO_2, and light REE very similar to the lavas of Esha Ness (1) it is thought that they are not alkaline basalts.

Ireland

Volcanic rocks of the province are preserved *in situ* in Ireland at only two localities (23 and 24), in both of which only acid andesites and dacites are present as a few lavas and pyroclastic deposits interbedded with Old Red Sandstone sediments. Lava clasts in conglomerates at Cushendall (22) strongly resemble those from Arran (21), but again are very siliceous. Comparison with Scottish rocks is rendered difficult by the small number of outcrops and their siliceous nature, but the Cushendall and Arran rocks resemble those of Lorne (9) most closely, while those furthest west in Ireland resemble some of the northern Midland Valley rocks. Concentrations of all elements are well within the range of most Scottish rocks, and those of Ireland are clearly part of the same province.

Spatial chemical variation in the late orogenic volcanic province

The Siluro-Devonian volcanic rocks north of the suture of Phillips *et al.* (1976) have many chemical and petrographic features in common with modern andesitic volcanic suites, and may be described as calc-alkaline. Most modern andesitic volcanoes are located above Benioff zones of deep focus earthquakes, generated within subducting oceanic lithospheric plates. Regular spatial variation in andesite composition with depth to the Benioff zone has been described by a number of authors: Arculus and Johnson (1978) summarize these published variations, and conclude that it is generally held that the degree of 'Fe enrichment' decreases, and, at constant silica, K, Sr, Ba, Rb, Rb/K, Th, Ti, and P all rise with increasing depth to the Benioff zone, and thus with increasing distance from the oceanic trench.

It is clear from the preceding discussion of the chemical characteristics that comparison of K_2O content at constant silica as in Dickinson (1975) is impractical for the Siluro-Devonian rocks. This is primarily because of the great variation in degree of evolution, measured by incompatible element concentration, at constant silica content. While this variation must in some cases (e.g. Glencoe and Ben Nevis) reflect different concentrations of these elements in the source materials, in others (e.g. St Abbs) it is probably due to a variety of evolutionary pathways. Comparisons of the most mafic rocks of each area are not too meaningful because of the small number of such samples and the ease with which K may be mobilized by alteration.

In most areas, Sr remains at a nearly constant level with increasing silica up to *c*. 63 per cent SiO_2. If fractional crystallization is of importance in controlling evolution, then the bulk fractionated composition must have a Sr content comparable to that of the lavas, and variation in its silica content between areas could not produce an inter-area variation in Sr content. Figure 9(a) demonstrates that there is a considerable difference in Sr content between areas. The ranges in this diagram do not include the Lorne 'shoshonites', rocks with over 63 per cent SiO_2, or a number of rocks identified petrographically as severely altered. That none of these omissions affects the conclusion of regional variation in Sr content is illustrated in Fig. 10, in which the $Sr-SiO_2$ relation is presented for the 460 samples analysed from areas 7–18 of Fig. 1, out of a total of 580 samples.

While comparisons of absolute concentrations of incompatible elements are not valid, it is clear that if an incompatible element could be found which was not susceptible to regional variation then it could be used as an index of evolution instead of silica, and concentrations of other incompatible elements could be normalized by ratio. Unfortunately, of the 27 elements analysed, only Th, K, and Rb behave

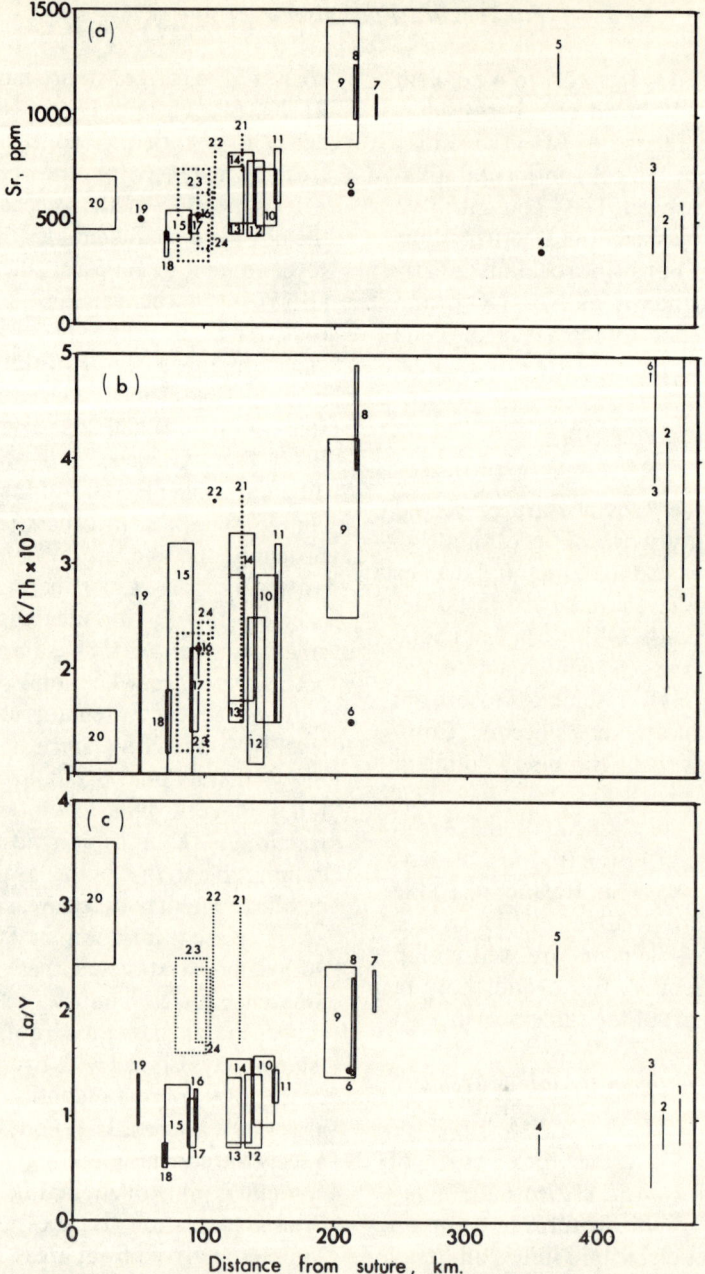

Fig. 9 Ranges of Sr, K/Th, and La/Y plotted against distance from the suture for volcanic rocks of the late orogenic province. Numbers refer to the outcrops marked on Fig. 1. Individual outcrops are shown as rectangles if they have substantial width normal to the suture, and as single dots (e.g. area 6) if the values represent a very small number of samples. Data from the Irish and Arran samples (areas 21–24) are shown dotted. The Lorne Plateau shoshonites and the more altered rocks are not included in the diagram. Ranges for K/Th are not quoted for Th values of less than 5 p.p.m. Areas 1–6, 16, 18, and 21–24: fewer than 10 samples per area. Areas 7, 8, 11, 17, and 19: 11–25 samples per area. Areas 9, 10, 12–15, and 20: more than 40 samples per area

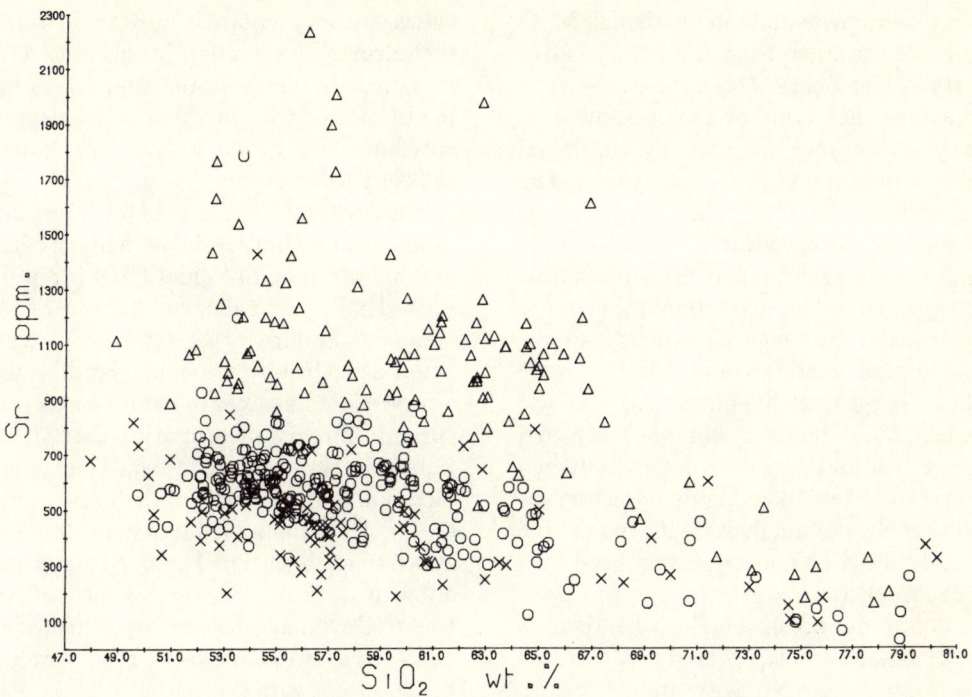

Fig. 10 Sr–SiO$_2$ variation diagram for all rocks of areas 7–18 (Fig. 1) of the late orogenic province. △, Highland lavas (7–9); ○, North Midland Valley lavas (11–14); ×, South Midland Valley lavas (15–18)

consistently incompatibly in the basalt-andesite range. K/Rb shows little geographic variation: it ranges from c. 250 to 500, with some tendency to lower values in the south, and shows a slight decrease with evolution in each area. K/Th (Fig. 9(b)), Rb/Th, Ba/Th, and Ba/K all show a geographic distribution similar to that of Sr, although the low levels (mainly below 20 p.p.m.) and the poor analytical precision of Th (± 1.5 p.p.m., 2σ) give rise to a wide scatter. La/Y also shows a similar pattern to Sr (Fig. 9(c)), and remains nearly constant with evolution in each area.

It is clear from Fig. 9 that rocks from areas 7–18 show a regular increase in Sr, K/Th, and La/Y with distance north-westward normal to the suture of Phillips et al. (1976). Rocks in these areas represent the bulk of the volcanism of the province. In addition, Ba/K, Ba/Th, Rb/Th, Ce/Sm, and, less clearly, Ti and P, all increase in the same direction, while the degree of 'Fe enrichment' shows a slight decrease, as possibly does Th. No simple mineral control could cause these spatial variations, and their resemblance to those of modern andesitic suites strongly suggests the former presence of a subduction zone dipping north-westward beneath Scotland.

McKerrow et al. (1977) have interpreted the Siluro-Ordovician sediments of the Southern Uplands as an accretionary wedge related to northward subduction; the associated oceanic trench migrating southward from the Southern Uplands Fault to the region of the Solway Firth during this period. It is generally considered that plate collision occurred and subduction ceased prior to the bulk of the volcanism considered here, although the age range of the volcanism is ill-defined. It is possible that the volcanism may have been related to the stationary remains of subducted Iapetus Ocean lithosphere, detached beneath Scotland from the collided continents above. Subduction is not thought to be currently active beneath the volcanoes of the Californian Cascades. The occurrence of bronzite-andesites with high MgO and Cr has been noted from the latest vent of Mt Shasta (Anderson, 1977) and these have been interpreted as the result of the melting of a refractory source rock initiated by continued volatile loss from the now stationary slab. The bronzite-andesites of the Midland Valley are, however, interbedded with more normal andesites, and their higher content of incompatible elements precludes their derivation from a more refractory source. The high Ni and Cr contents of many of the andesites and basalts create difficulties for general hypotheses of andesite magma generation by the melting of subducted lithosphere. The lower portions of the subducted lithosphere are likely to be highly refractory, as will be the gabbroic cumulates at the base of the crustal layer. If basaltic material produced by a hypothetical Iapetus mid-ocean ridge was similar to modern mid-ocean ridge basalts, then its content of Ni and Cr was probably low (often less than 150 p.p.m. Ni; Kay and Hubbard, 1978): it is very difficult to generate andesites and basalts with Ni contents of more than 100 p.p.m. by melting this. Accordingly, a mixture of two primary sources (e.g. mantle and subducted material) is required to explain both the high Ni and Cr concentrations and the high lithophile element content.

Figure 9 demonstrates a number of anomalies in magma composition relative to distance from the suture. The parameters Sr, K/Th, and La/Y are relatively low in the Shetland and Orkney (1–5) and Huntly (6) lavas and relatively high in the Irish rocks, while the Cheviot and St Abbs rocks are also anomalous. The last two areas are unusual in their close proximity to the suture, and the observed increases in Sr and La/Y from St Abbs to Cheviot may suggest a relation to a different subduction system, or not to a subduction system at all. In their lack of inherited zircons (Pidgeon and Aftalion, 1978), the Devonian Southern Upland granites resemble more closely those of the Shap and Weardale plutons (northern England) than those of the more northerly Scottish Devonian plutons.

Conclusions drawn from the St Abbs Sr values are very tentative, however, owing to the high degree of alteration in this area. It should be noted that, despite the similarities between the Lorne and Cheviot lavas in low degree of Fe enrichment, high La/Y, etc., the latter have radically lower Ba and Sr contents.

The apparent anomalies of the Irish, Shetland, Orkney and Huntley lavas suggest that geographic variation of Sr and La/Y is not best described by perpendicular distance from the suture of Phillips et al. (1976), but would be better described by perpendicular distance from a curved line convex to the south-east, eventually curving into parallelism with the Great Glen Fault. Futher evidence for this change in strike direction may be found in the northward strike of the Dalradian metasediments of Shetland and the north-eastern Highlands, and it would allow the suture to pass to the north of Norway, where Devonian nappes are directed south-eastward over the Swedish craton. It would also be consistent with the existence of the tholeiitic members of a magmatic arc on the convex side of the arc. It should be remembered however, that volcanism is only sparsely preserved in Ireland, and the position of the Shetland outcrops relative to each other and to the mainland may have been greatly modified by displacement on the Great Glen Fault since eruption. The possibly more recent age of the Shetland rocks also provides scope for other hypotheses on the relation between Shetland and mainland lavas.

Discussion

The volcanic rocks described here were erupted over a period of 150 Ma from early Ordovician to late Devonian times. They show a wide diversity in magmatic character, from rocks with tholeiitic affinities such as the Eycott and Rhobell Fawr volcanic rocks, through truly calc-alkaline types represented by the Borrowdale volcanic rocks and most of the late orogenic and main group Welsh volcanic rocks, to the alkaline rocks of Skomer and Hoy and the shoshonitic rocks of the Lorne Plateau. However, these rocks probably represent only a small proportion of the magmatism associated with the closure of the Iapetus Ocean. Wright (1976) has suggested that the ocean was in existence more than 1000 Ma ago and closed by successive subduction and marginal basin extension episodes along its north-western and south-eastern margins. Occasional glimpses of volcanism associated with these early episodes are provided by the late Precambrian volcanic rocks of Wales and the Welsh Borders (Thorpe, 1974) and the Tayvallich volcanic rocks of the Scottish Dalradian (Graham, 1976). Most of the volcanic rocks erupted during these early phases, however, have been eroded away to provide the sediments and metasediments which now form the Caledonides. Those volcanic rocks which remain were mostly erupted during the later stages of closure.

The earliest andesitic volcanic rocks in the British sector of the Caledonian mobile belt are found at the base of the Ordovician successions of the Welsh basin, at Rhobell Fawr (41) and Treffgarne (35). While these early Welsh volcanoes were erupting, thick sequences of shale and greywacke were accumulating to the north to form the Manx Slates and Skiddaw Group of the Isle of Man and the Lake District respectively. It is tempting to draw an analogy between these sediments and those of the Southern Uplands which McKerrow et al. (1977) regard as an accretionary prism associated with late Ordovician and Silurian subduction on the other side of the Iapetus Ocean.

By Llanvirn times, volcanism had spread out of the Welsh basin north-westwards to the Lake District and south-eastern Ireland. The Eycott volcanic rocks (25) were erupted conformably on to the Skiddaw Group of the northern Lake District and the whole sequence tilted northwards and eroded before the eruption of the Borrowdale volcanic rocks (26) in the Llandeilo/Caradoc. These latter volcanic rocks were erupted contemporaneously with those in south-eastern Ireland and the main group of the Welsh basin, and represent the climax of volcanic activity in the Ordovician–Silurian province on the south-eastern margin of Iapetus. Activity in this province declined rapidly after the Caradoc, with the alkaline Skomer volcanic rocks representing the last phase of activity.

The analogy between the volcanic rocks of the Ordovician–Silurian province and those of a modern destructive plate margin is clear. The general magmatic trend from tholeiitic through calc-alkaline to alkaline both with time and with distance from the proposed suture can be matched closely with similar trends observed in modern island arc and continental margin environments. Whether the Lake District–south-eastern Ireland island arc was built on continental or oceanic crust is debatable, but it is now underlain by continental crust (Bamford et al., 1978). The ocean trench associated with this arc in mid-Ordovician times must have been sited to the north of the present Lake District, possibly in the region of the Solway Firth, along the line of the Iapetus suture.

If we interpret the Manx Slates and Skiddaw Group sediments as an accretionary prism analogous with the Southern Uplands, then the location of the Borrowdale volcanic rocks unconformably on top of this prism sequence implies that the volcanic front migrated north-westwards during the Ordovician. During the same period the site of subduction must have moved from the Lake District north-westwards towards the suture. It is well known from modern examples (Karig and Sharman, 1975) that the arc–trench gap widens with time as the accreting prism grows, and that the volcanic front remains stationary relative to the prism. The rare occurrence of magmatism within the arc–trench gap is regarded as anomalous and warrants special explanation (Marshak and Karig, 1977). The occurrence of the Borrowdale volcanic rocks on what may be an accretionary prism (Skiddaw Group) can best be explained by two phases of subduction during the Ordovician–Silurian volcanic episode. Such a model is illustrated in Fig. 11. This must be regarded as tentative as it is dependent upon the interpretation of the Skiddaw Group and Manx Slates. The recent discovery of an Arenig volcanic sequence within the Skiddaw Group (Turner and Wadge, 1979) may require some modification to the model.

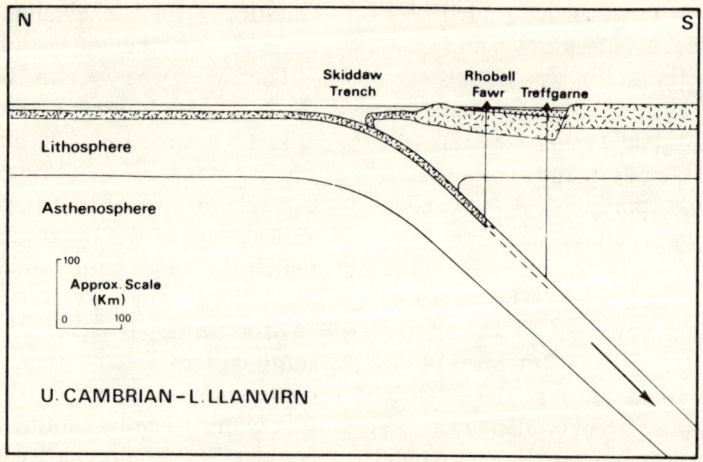

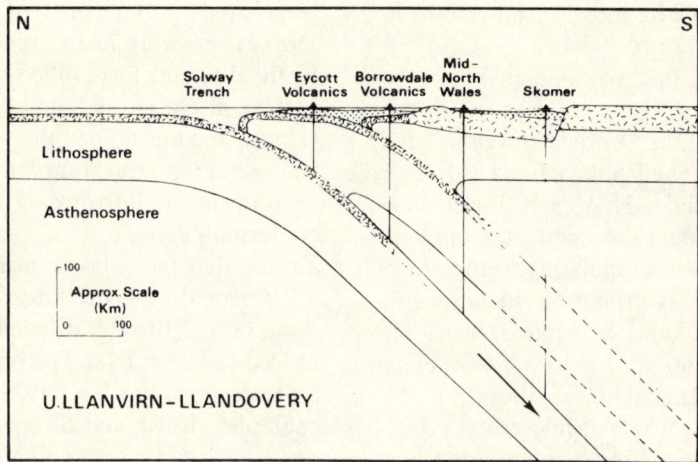

Fig. 11 Plate tectonic model involving two stages of subduction to account for volcanicity along the south-eastern margin of the Iapetus Ocean from the Upper Cambrian to the Lower Silurian

The decline in volcanic activity in England, Wales, and south-eastern Ireland towards the end of the Ordovician implies that south-eastwards subduction ceased at around this time. The final stage in the closure of the Iapetus Ocean appears to have been accomplished by subduction beneath its north-western margins in what is now Scotland and northern Ireland. This stage began in the Caradoc with the formation of the earliest segments of the Southern Uplands accretionary prism and continued at least into the Wenlock (McKerrow et al., 1977). It is possible that volcanism in the late orogenic province was initiated by this subduction episode. The volcanic rocks associated with the Old Red Sandstone have traditionally been assigned to the Devonian period but Rb–Sr isochron dating by Brown (1975) has given a Lower Silurian age (415 ± 7 Ma) for the Lorne Plateau lavas. None of the scant palaeontological evidence for the age of the Lower Old Red Sandstone can rule out such an age for the associated volcanic rocks.

In the absence of reliable stratigraphical control it is not possible to discuss temporal variation in the late orogenic volcanic rocks except in the most general terms. The youngest lavas in the province, the Hoy lavas (5) of Orkney, are alkaline in character whereas the other lavas are generally calc-alkaline. A much more detailed picture, however, emerges when spatial variation is considered. The light REE, K, Sr, and Ba all increase in abundance in a west-north-westerly direction and may reflect the direction of subduction beneath the province. The strike of the implied subduction zone is oblique to the Iapetus suture in the British sector of the Caledonides.

Volcanism in the late orogenic province accompanied the final closure of Iapetus and may have persisted for some time after active subduction had ceased. This closure may have been responsible for the end-Silurian deformation of the Lower Palaeozoic strata south of the suture and for the much more intense deformation of the Silurian rocks of Scandinavia. There is no clear evidence, however, for strong end-Silurian deformation in the rocks north of the Southern Uplands where most of the late orogenic volcanics occur. The absence of intense deformation from the British sector may be explained by the oblique closure implied by the chemical variation in the late orogenic volcanics. If the suture bends northwards to the north-east of Britain then closure of the Scandinavian sector of Iapetus would have resulted in collision and would account for the much more intense deformation which occurred at this time in the Norwegian Caledonides.

In conclusion, we have in the British Isles a well preserved record of volcanism spanning the late stages of closure and final extinction of a major ocean. In this chapter we have attempted to show how geochemical studies on ancient andesitic rocks may be used together with modern analogues to build up a picture, albeit incomplete, of the waning phases of this ocean.

REFERENCES

Anderson, A. T. (1977). Bronzite andesite: indicator of dead subduction? (Abstract). IASPEI/IAVCEI Assembly, Durham, p. 48.

Arculus, R. J. and Johnson, R. W. (1978). Criticism of generalised models for the magmatic evolution of arc-trench systems. *Earth planet. Sci. Letters* **39**, 118–126.

Arrhenius, G., Everson, J. E., Fitzgerald, R. W., and Fujita, H. (1971). Zirconium fractionation in Apollo 11 and Apollo 12 rocks. In *Proceedings of the 2nd Lunar Science Conference*, Vol. 2, M.I.T. Press, Cambridge, Mass., pp. 169–176.

Bamford, D., Nunn, K., Prodehl, C., and Jacob, B. (1978). LISPB-IV. Crustal structure of Northern Britain. *Geophys. Jl R. astr. Soc.* **54**, 43–60.

Brown, J. F. (1975). Rb–Sr studies and related geochemistry on the Caledonian calc-alkaline igneous rocks of N.W. Argyllshire. D.Phil. thesis, University of Oxford.

Cattermole, P. J. (1969). A preliminary geochemical study of the Mynydd Penarfynydd intrusion, Rhiw igneous complex, southwest Lleyn. In *The Pre-Cambrian and Lower Palaeozoic Rocks of Wales* (A. Wood, ed.), University of Wales Press, Cardiff, pp. 435–446.

Cattermole, P. J. (1976). The crystallisation and differentiation of a layered intrusion of hydrated alkali olivine basalt parentage at Rhiw, North Wales. *Geol. J.* **11**, 45–70.

Chayes, F. (1969). The chemical composition of Cenozoic andesite. *Oregon St. Dept Geol. Mineral Industries Bull.* no. 65, 1–11.

Church, W. R. and Gayer, R. A. (1973). The Ballantrae ophiolite. *Geol. Mag.* **110**, 497–510.

Dewey, J. F. (1969). Evolution of the Appalachian/Caledonian orogen. *Nature, Lond.* **222**, 124–129.

Dewey, J. F. (1971). A model for the Lower Palaeozoic evolution of the southern margin of the early Caledonides of Scotland and Ireland. *Scott. J. Geol.* **7**, 219–240.

Dickinson, W. R. (1975). Potash–depth (K–h) relations in continental margin and intra-oceanic magmatic arcs. *Geology* **3**, 53–56.

Downie, C. and Soper, N. J. (1972). Age of the Eycott Volcanic Group and its conformable relationship to the Skiddaw Slates in the English Lake District. *Geol. Mag.* **109**, 259–268.

Fitton, J. G. (1971). The petrogenesis of the calc-alkaline Borrowdale Volcanic Group, Northern England. Ph.D. thesis, University of Durham.

Fitton, J. G. (1972). The genetic significance of almandine-pyrope phenocrysts in the calc-alkaline Borrowdale Volcanic Group, Northern England. *Contr. Mineral. Petrol.* **36**, 231–248.

Fitton, J. G. and Hughes, D. J. (1970). Volcanism and plate tectonics in the British Ordovician. *Earth planet. Sci. Letters* **8**, 223–228.

Francis, E. H., Forsyth, I. H., Read, W. A., and Armstrong, M. (1970). The Geology of the Stirling District (explanation of one-inch Geological Sheet 39). *Mem. Geol. Surv. Gt Britain (Scotland)*, HMSO, Edinburgh.

Gale, N. H., Beckinsale, R. D., and Wadge, A. J. (1979). A Rb-Sr whole rock isochron for the Stockdale Rhyolite of the English Lake District and a revised mid-Palaeozoic time scale. *J. geol. Soc. Lond.* **136**, 235–242.

Graham, A. M. and Upton, B. G. J. (1978). Gneisses in diatremes, Scottish Midland Valley: petrology and tectonic implications. *J. geol. Soc. Lond.* **135**, 219–228.

Graham, C. M. (1976). Petrochemistry and tectonic

significance of Dalradian metabasaltic rocks of the SW Scottish Highlands. *J. geol. Soc. Lond.* **132**, 61–84.

Green, T. H. (1977). Garnet in silicic liquids and its possible use as a *P–T* indicator. *Contr. Mineral. Petrol.* **65**, 59–67.

Harland, W. B. and Gayer, R. A. (1972). The Arctic Caledonides and earlier oceans. *Geol. Mag.* **109**, 289–314.

Haslam, H. W. (1968). The crystallisation of intermediate and acid magmas at Ben Nevis, Scotland. *J. Petrol.* **9**, 84–104.

Hawkins, T. R. W. (1970). Hornblende gabbros and picrites at Rhiw, Caernarvonshire. *Geol. J.* **7**, 1–24.

Hughes, D. J. (1977). The petrochemistry of the Ordovician igneous rocks of the Welsh Basin. Ph.D. thesis, University of Manchester.

Jakeš, P. and Gill, J. (1970). Rare earth elements and the island arc tholeiitic series. *Earth planet. Sci. Letters* **9**, 17–28.

Jakeš, P. and White, A. J. R. (1972). Major and trace element abundances in volcanic rocks of orogenic areas. *Geol. Soc. Am. Bull.* **83**, 29–40.

Karig, D. E. and Sharman, G. F., III (1975). Subduction and accretion in trenches. *Geol. Soc. Am. Bull.* **86**, 377–389.

Kay, R. W. and Hubbard, N. J. (1978). Trace elements in ocean ridge basalts. *Earth planet. Sci. Letters* **38**, 95–116.

Kellock, E., (1969). Alkaline basic igneous rocks in the Orkneys. *Scott. J. Geol.* **5**, 140–153.

Macdonald, R. (1975). Petrochemistry of the early Carboniferous (Dinantian) lavas of Scotland. *Scott. J. Geol.* **11**, 269–314.

McKerrow, W. S., Leggett, J. K., and Eales, M. H. (1977). Imbricate thrust model of the Southern Uplands of Scotland. *Nature, Lond.* **267**, 237–239.

Marshak, R. S. and Karig, D. E. (1977). Triple junctions as a cause for anomalously near-trench igneous activity between the trench and volcanic arc. *Geology* **5**, 233–236.

Millward, D., Moseley, F., and Soper, N. J. (1978). The Eycott and Borrowdale Volcanic Rocks. In *The Geology of the Lake District* (F. Moseley, ed.), Yorkshire Geological Society, Leeds, pp. 99–120.

Moseley, F. (1977). Caledonian plate tectonics and the place of the English Lake District. *Geol. Soc. Am. Bull.* **88**, 764–768.

Oliver, R. L. (1956). The origin of garnets in the Borrowdale Volcanic Series and associated rocks, English Lake District. *Geol. Mag.* **93**, 121–139.

Phillips, W. E. A., Stillman, C. J., and Murphy, T. (1976). A Caledonian plate tectonic model. *J. geol. Soc. Lond.* **132**, 579–609.

Pidgeon, R. T. and Aftalion, M. (1978). Cogenetic and inherited zircon U–Pb systems in granites: Palaeozoic granites of Scotland and England. In *Crustal Evolution in Northwestern Britain and Adjacent Regions* (D. R. Bowes and B. E. Leake, eds), *Geological Journal* Special Issue no. 10, pp. 183–220.

Rundle, C. C. (1979). Ordovician intrusions in the English Lake District. *J. geol. Soc. Lond.* **136**, 29–38.

Shackleton, R. M. (1954). The structural evolution of North Wales. *Liverpool Manchester Geol. J.* **1**, 261–297.

Shaw, D. M. (1968). A review of K–Rb fractionation trends by covariance analysis. *Geochim. cosmochim. Acta* **32**, 573–601.

Stillman, C. J. and Williams, C. T. (1979). Geochemistry and tectonic setting of some Upper Ordovician volcanic rocks in east and southeast Ireland. *Earth planet. Sci. Letters* **42**, 288–310.

Taylor, S. R. (1969). Trace element chemistry of andesites and associated calc-alkaline rocks. *Oregon St. Dept Geol. Mineral Industries Bull.* no. **65**, 43–63.

Taylor, S. R., Kaye, M., White, A. J. R., Duncan, A. R., and Ewart, A. (1969). Genetic significance of Co, Cr, Ni, Sc and V content of andesites. *Geochim. cosmochim. Acta* **33**, 275–286.

Thorpe, R. S. (1974). Aspects of magmatism and plate tectonics in the Precambrian of England and Wales. *Geol. J.* **9**, 115–136.

Turner, R. E. and Wedge, A. J. (1979). Acritarch dating of Arenig volcanism in the Lake District. *Proc. Yorks. geol. Soc.* **42**, 405–414.

Wadge, A. J. (1978). Classification and stratigraphical relationships of the Lower Ordovician rocks. In *The Geology of the Lake District* (F. Moseley, ed.), Yorkshire Geological Society, Leeds, pp. 68–78.

Waterston, C. D. (1965). Old Red Sandstone. In *The Geology of Scotland* (G. Y. Craig, ed.), Oliver and Boyd, Edinburgh, pp. 269–308.

Wells, A. K. (1925). The geology of the Rhobell Fawr district (Merioneth). *Q. Jl geol. Soc. Lond.* **81**, 463–538.

Whitford, D. J. (1975). Strontium isotopic studies of the volcanic rocks of the Sunda arc, Indonesia, and their petrogenetic implications. *Geochim. cosmochim. Acta* **39**, 1287–1302.

Wilson, J. T. (1966). Did the Atlantic close and then re-open? *Nature, Lond.* **211**, 676–681.

Wright, A. E. (1976). Alternating subduction direction and the evolution of the Atlantic Caledonides. *Nature, Lond.* **264**, 156–160.

Andesites
Edited by R. S. Thorpe
© 1982 John Wiley & Sons

X Implications of andesite volcanism

Considering the composition, distribution, and petrogenesis of andesites as outlined in the preceding nine sections, the widespread occurrence of volcanic arcs and continental margins characterized by andesite volcanism has important geological implications. These include the processes of continental growth, the circulation of elements through the plate tectonic cycle, and the evolution of the oceans and atmosphere. Finally, and arguably most important, many important mineral deposits are associated with andesites and their intrusive equivalents.

As the continental crust is of broadly intermediate chemical composition, and since andesites are the most abundant volcanic rocks, a popular model for the origin and growth of continental crust involves formation of island arcs which become accreted laterally to form continental masses. This is the andesite crustal growth model. B. L. Weaver and J. Tarney outline the major physical and chemical characteristics of the continental crust and use these to examine the andesite model in detail. They argue that, as the continental crust as a whole is more silicic than most island arcs, vertical accretion of intrusive rocks at continental margins is more important than lateral accretion. However, since probably over 70 per cent of the continental crust was formed during the Archaean, differences between Archaean and younger crustal rocks may reflect differences in the composition of the accreting mantle-derived magma according to this model. Such differences in chemical composition between Archaean and younger rocks are interpreted in terms of an increased component derived by partial melting of subducted oceanic crust during the Archaean, in comparison with younger volcanic arc magmas which might be derived dominantly from the underlying mantle wedge.

The complexity of interactions between subducted oceanic crust, the mantle wedge, and continental crust involved in the plate tectonic cycle is a theme emphasized by W. S. Fyfe, who argues that andesites are a product of mixing of continental, oceanic crust, and mantle components ('geosphere mixing'). The crustal components include elements fixed in subducted altered oceanic crust, subducted sediments, and possibly subducted continental crust. In such a complex situation it is difficult to determine whether the continental crust is presently growing in volume, whether it has reached a steady state, or whether it is even diminishing in volume. The abundance of andesite volcanoes in continental and oceanic areas means that such volcanoes make important contributions to hydrosphere and atmosphere compositions. Further, as emphasized by Walker (this volume, Section V), gas content exerts a major control on the style of andesite eruptions. W. I. Rose *et al.* therefore explain how estimates of gas composition can be made for andesite magmas, and, using S and Cl as examples, emphasize the large scale of andesite gas emission and show how the amount and composition of emitted gases are related to the style of volcanic eruption.

In addition to hazards of volcanic eruption (Section V), the major practical aspect of study of andesite volcanoes lies in the mineral deposits associated with andesites and intrusive equivalents at convergent plate margins. The features of such mineral deposits are reviewed by A. H. G. Mitchell and R. D. Beckinsale, who consider deposits formed in four plate tectonic settings. Volcanic arc settings include 'magmatic arc' mineralization (which includes porphyry Cu and Kuroko-type sulphide deposits) and 'back-arc' magmatic belts (which include Sn–Ag and Sn–W deposits). Further settings are 'outer-arc' and collision belt mineralization (which include Sn and magmatic U deposits). Although the settings of these mineral deposits can be explained elegantly in terms of plate tectonic processes, Mitchell and Beckinsale conclude that, for most mineral deposits, it is the nature of the associated igneous rocks and style of mineralization that still provide the best guides for mineral exploration.

Andesites
Edited by R. S. Thorpe
© 1982 John Wiley & Sons

Andesitic magmatism and continental growth

B. L. Weaver and J. Tarney

Department of Geology,
University of Leicester, University Road, Leicester LE1 7RH, UK

ABSTRACT

The role of andesitic magmatism in crustal growth is examined in relation to the known geophysical, geochemical, and isotopic characteristics of the continental crust, the compositions of volcanic island arcs and continental margin Cordilleran volcanic and plutonic suites, and petrogenetic models for the generation of calc-alkaline magmas. Models for crustal formation and growth range from those which demand considerable intracrustal fractionation, such as with the lateral accretion of andesitic island arcs, to those which require only limited intracrustal element transfer and consider that much of the continental crust represents essentially liquid compositions. The nature of the lower crust is a major uncertainty in these crustal growth models. Available evidence on Archaean deep crustal granulites argues against the Archaean lower crust being predominantly cumulate or composed of the refractory residues of intracrustal melting.

Calc-alkaline plutonism appears to be dominant over andesitic volcanism as the agent for crustal growth. The compositions of intra-oceanic island arcs are too basic to yield average crustal compositions. Vertical accretion of calc-alkaline plutonic magmas at continental margins is therefore more important than lateral accretion of island arcs. In zones of active crustal-growth, magmas rise to a level consistent with their density and produce a density-stratified crust, possibly enhanced by fractional crystallization.

Isotopic evidence suggests that a significant proportion of the continental crust grew during the Archaean, and subsequently at a decreasing rate until the present day. Archaean crustal compositions show many similarities, but also some differences, to those in modern Cordilleran belts. These differences probably reflect an increased component derived from hydrous partial melting of the subducted ocean crust in the Archaean, consistent with the higher heat flow; at the present day this component may be limited to fluids carrying silica and large ion lithophile elements from the subducting ocean crust as it dehydrates.

Extraction of crustal material from the mantle with time is, to a first approximation, responsible for the apparent depletion of incompatible elements in the upper mantle (as represented by the normal (N-type) ocean basalt mantle source). Some recycling of continental material back into the mantle as subducted sediment may have occurred. However accretion of sediment at subduction zones is a contributing factor in crustal growth, and may be an important source for S-type granites.

Introduction

Whereas it is generally agreed that continental material ultimately originates from within the mantle, perhaps through multistage magmatic processes, there is much debate as to the overall bulk composition of the continental crust, the meaning of its vertical petrological and chemical zonation, and the nature of the crust-generating processes. There is also argument as to whether the crust has grown continuously or episodically with time, whether it is growing at a decreasing rate, or whether the crustal volume is relatively constant such that continental material is being recycled back into the mantle as new crustal material is being generated. Some of these general points have been discussed by Brown (1979).

Andesites have always played a prominent part in models of crustal growth. This arises from the widely held view that the continental crust is broadly andesitic in composition, from the fact that andesite is perhaps the most voluminous magma type in continental areas, and from the undoubted success of the andesite model (e.g. Taylor, 1977) in explaining many geochemical features of the continental crust. The aim of this contribution is to review the role of andesite in models for crustal growth—through examining the available information on the composition of the upper and lower crust, on the composition of island arcs, the nature of crust-generating processes, the isotopic composition of crust and mantle, and the overall geochemical balance between crust and mantle.

The Composition of the Continental Crust

Many attempts have been made at deriving the chemical composition of the continental crust, from the early estimate of Goldschmidt (1933) to the more recent estimate of Taylor (1977). Most such estimates are based on the assumption that the composition of the total continental crust can be approximated by appropriate proportions of 'typical' basalt, granite, and sediment. The need to make such assumptions arises due to the largely unknown chemical nature of the deeper levels of the continental crust, which are not easily amenable to sampling. Because of the relatively well known composition of the upper continental crust (representing perhaps the upper 15 km of the crust which is well sampled by sediments), models for the composition of the lower continental crust are rigidly constrained by the assumed bulk crust composition. An alternative, and preferable, approach is to estimate lower crustal compositions by analysis of either deep crustal xenoliths sampled by volcanic activity, or of surface exposures of rocks presumed to have originated at deep crustal levels. Typical lower crust compositions

TABLE 1 Estimates of the composition of the continental crust

	A	B	C	D	E
SiO_2	61.9	63.1	60.2	61.9	62.5
TiO_2	1.1	0.8	1.0	0.8	0.7
Al_2O_3	16.7	15.2	15.6	15.6	15.6
Fe_2O_3	7.7	6.7	7.9	6.9	6.1
MnO	—	—	0.1	0.1	0.1
MgO	3.5	3.1	3.9	3.1	3.2
CaO	3.4	4.1	5.8	5.7	6.0
Na_2O	2.2	3.4	3.2	3.1	3.4
K_2O	4.2	3.0	2.5	2.9	2.3
P_2O_5	—	—	0.2	0.3	0.2

Average continental crust estimates are from the following: (A) Goldschmidt (1933); (B) Vinogradov (1962); (C) Taylor (1964); (D) Ronov and Yaroshevski (1969); (E) Holland and Lambert, (1972).

derived in this way, coupled with upper crustal averages, should provide a more reliable estimate of the composition of the continental crust.

Goldschmidt's (1933) average continental crust estimate was based upon the composition of glacial clays thought to have sampled large areas of the Scandinavian Shield. A number of later estimates for the composition of the continental crust combined various proportions of basalt and granitic components (Vinogradov, 1962; Taylor, 1964; Ronov and Yaroshevski, 1969). On the other hand, Holland and Lambert (1972) derived a total crust composition based upon a mixture of upper crust (as represented by shield averages) and average granulite facies rocks (those presumably underlying the major shields) which is significantly different to averages derived from simple igneous rock compositions. An intermediate composition, with 60–63 per cent SiO_2, for the continental crust is indicated by the above estimates. Some of these average compositions are given in Table 1.

The Andesite Model of Crustal Composition and Crustal Growth

A rather different approach to the problem of continental crust composition was proposed by Taylor (1967). He noted that andesitic island arcs, which are mostly quite young features, represent a volumetrically abundant new addition of primitive continental material from the mantle. Because such arcs are formed at destructive plate margins, they can be regarded as essentially transient features on a geological time scale, and must eventually be accreted laterally during continental collision and orogenesis. Assuming that such arcs have formed regularly throughout geological history, they potentially represent a major mechanism of crustal growth by lateral accretion. If so, then the compositions of island arcs can furnish a reasonable estimate of the bulk continental crust composition. This is the essence of the popular 'andesite model' for crustal growth developed over a number of years by S. R. Taylor and co-workers (Taylor, 1967, 1977, 1979; Jakeš and Taylor, 1974; Taylor and McLennan, 1981a,b). In effect, the model proposes that the mean composition of the continental crust can be approximated by the composition of the average island arc andesite. Because a wealth of geochemical data now exists for andesites the model thus provided reasonable estimates of crustal abundances for a wide range of elements.

The andesite model can also be used to estimate the composition of the lower crust. A basic assumption is that, as island arcs are accreted laterally during orogenesis, they suffer intracrustal melting to produce a granodioritic upper crust and a more refractory, residual, lower crust. For simplicity the model assumes one-third upper crust and two-thirds lower crust. Taylor (1977, 1979) and Nance and Taylor (1976) have shown that post-Archaean sediment REE patterns are surprisingly consistent, which probably reflects efficient sampling of the debris from a wide range of upper crustal sources. The average clastic sediment is assumed to have a REE pattern representative of that of the average upper crust. Together with the REE pattern of the average andesite, representing the bulk crust, this provides a means for estimating the average REE distribution in the lower continental crust. The respective chondrite-normalized REE patterns are shown in Fig. 1. It is possible to make estimates for the abundances

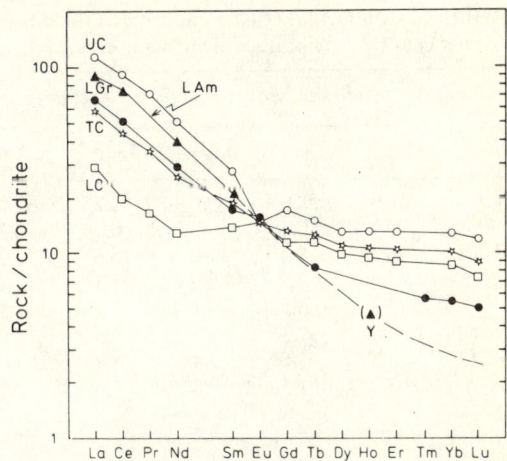

Fig. 1 REE patterns of some suggested continental crustal components. UC, TC, and LC are the average upper crust, total crust, and (calculated) lower crust compositions based on the andesite model for crustal growth (Taylor, 1977). LGr is the mean REE pattern for Lewisian lower crustal granulites, and LAm the pattern for Lewisian amphibolite facies (intermediate crustal level) gneisses (Weaver and Tarney, 1980a,b). Data given in Table 2

of other major and trace elements in the lower crust based on the REE data, as has been done by Taylor (1977). The compositions of the bulk crust, upper crust, and lower crust estimated in this way are listed in Table 2. Mass balance of major elements yields a rather mafic composition for the lower crust (Table 2).

Relative to the bulk crust of assumed average andesite composition, REE abundances in the upper crust are higher by a factor of c. 1.6 for the light REE, but heavy REE abundances are similar. The calculated chondrite-normalized REE pattern for the lower crust is essentially flat; however, if the assumed proportion of upper crust is greater than one-third the calculated lower crust, then the REE pattern would show light REE depletion. The average upper crustal sediment has a prominent negative Eu anomaly which requires a compensating positive Eu anomaly in the lower crust.

Although there are several assumptions inherent in the andesite model, the results, at least superficially, are entirely reasonable. It provides for a more mafic lower crust, which is

TABLE 2 Continental crust compositions (oxides in per cent by weight, trace elements in p.p.m.)

	Andesite Model			Archaean	
	UC	TC	LC	LGr	LAm
SiO_2	66.0	58.0	54.0	61.2	67.3
TiO_2	0.6	0.8	0.9	0.5	0.3
Al_2O_3	16.0	18.0	19.0	15.6	15.9
FeO*	4.5	7.5	9.0	5.3	3.4
MgO	2.3	3.5	4.1	3.4	1.2
CaO	3.5	7.5	9.5	5.6	3.4
Na_2O	3.8	3.5	3.4	4.4	4.6
K_2O	3.3	1.5	0.6	1.0	2.2
Cr	35	55	65	88	27
Ni	15	20	23	58	16
Zr	240	100	30	202	199
Rb	110	50	20	12	62
Sr	350	400	425	569	541
Ba	700	350	175	757	838
Pb	15	7	3	13	18
Th	10.5	2.5	—	0.42	9
U	2.5	1	0.25	0.05	—
La	38	19	9.5	22	29
Ce	80	38	17	44	64
Pr	8.9	4.3	2.0	—	—
Nd	32	16	8	18.5	(25)
Sm	5.6	3.7	2.75	3.3	4.3
Eu	1.1	1.1	1.1	1.18	—
Gd	4.7	4.2	3.95	—	—
Tb	0.77	0.64	0.58	0.43	—
Dy	4.4	3.7	3.35	—	—
Ho	1.0	0.82	0.73	—	—
Er	2.9	2.3	2.0	—	—
Tm	0.5	0.4	0.35	0.19	—
Yb	2.8	2.3	1.9	1.2	—
Lu	0.4	0.3	0.25	0.17	—
Y	27	22	20	9	9
Ce_N/Yb_N	7.3	4.4	2.3	9.3	—

UC, TC, and LC are upper crust, total crust, and (calculated) lower crust compositions respectively, based on the andesite model for crustal growth (Taylor, 1977). LGr is mean composition of Lewisian granulites (= Archaean lower crust); LAm is mean composition of amphibolite-facies gneisses from Scotland and Greenland (= Archaean intermediate crustal level). Figures in brackets are interpolated values. From Weaver and Tarney (1980a).

broadly consistent with seismic data. The calculated positive Eu anomaly in the lower crust would agree with a refractory residue dominated by residual plagioclase. The low contents of radioactive heat-producing elements and some other lithophile elements in many lower crustal granulites (Heier, 1973; Tarney and Windley, 1977) could be interpreted as indicating the removal of these elements in a melt phase from the lower crust (e.g. Fyfe, 1973). It is necessary, of course, to examine various aspects of the andesite model more closely as well as to consider alternative models. The model is strongly dependent on the estimated composition of young island arcs and post-Archaean (mainly Phanerozoic) sediments, although a significant fraction of the continental crust appears to have been generated during the Archaean.

Taylor and co-workers (Nance and Taylor, 1976, 1977; Taylor, 1979; McLennan et al., 1979; Bavinton and Taylor, 1980; McLennan and Taylor, 1980) have in fact shown that there are compositional differences between Archaean and post-Archaean sediments which probably reflect differences in upper crust composition. In general, relative to post-Archaean sediments, Archaean sedimentary rocks have lower Th and U contents and lower Th/U ratios and have less fractionated REE patterns (La_N/Yb_N = 4.6 as opposed to 11.6), lower REE abundances ($\sum REE$ = 70 p.p.m. compared with 183 p.p.m.) and lack a prominent negative Eu anomaly. To illustrate this latter point the chondrite-normalized REE patterns of the average Archaean and average post-Archaean sediments as estimated by Taylor (1979) and Taylor and McLennan (1981b) are shown in Fig. 2. This difference in estimated upper crustal REE distributions with time has been attributed to a lack of intracrustal melting in the Archaean (Taylor and McLennan, 1981b) such that exposed and unexposed segments of the Archaean crust were similar and that a separate distinctive lower crust composition in the Archaean may not be required.

To account for the Archaean sediment REE abundance patterns, Taylor and McLennan (1981b) suggest that the sediments are sampling a mixture of the bimodal tholeiite–tonalite suite commonly found in Archaean terrains (Barker and Peterman, 1974) rather than sampling a crustal composition equivalent to that of modern

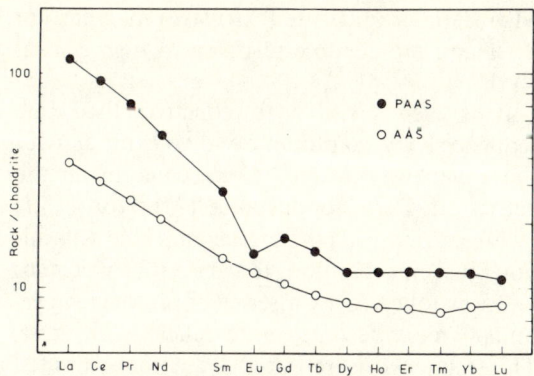

Fig. 2 REE patterns of post-Archaean average sediment (PAAS, ●) (Taylor, 1977) and Archaean average sediment (AAS, ○) (Taylor, 1979)

island arcs; thus they do not imply any specific tectonic environment for the generation of Archaean continental crust. However, according to this model the change in sedimentary composition at the Archaean–Proterozoic boundary (c. 2.5 Ma ago) must mark a period of extensive crustal re-working such that the continental crust accreted by this time (estimated by Taylor and McLennan (1981b) as c. 70 per cent at the present crustal volume) underwent partial fusion at deep crustal levels to yield residual granulite facies assemblages. Such a major event as this of course should be recorded in the isotope systematics of ancient crust. Initial Sr isotope ratios for Proterozoic granitic rocks (as compiled for instance by Glikson (1979)) do in fact scatter to much higher values than their Archaean equivalents.

In summary, there are uncertainties as to whether the andesite model, in spite of its attractive simplicity, can be applied to the Archaean when at least the major proportion of crustal growth took place. Below we examine more critically the petrogenesis of Archaean crustal compositions, particularly those from Archaean granulite terrains which may be representative of the deeper crust. First, however, it needs to be asked whether the andesite model in its basic form can account for crustal growth during Phanerozoic times. There are two main questions here. One concerns the compositions of modern intra-oceanic island arcs, which are the primary material for crustal growth; the second is whether such arcs can be remobilized as they are laterally accreted on to continents.

Although it is a widely held assumption that most island arcs are andesitic, recent studies of intra-oceanic arcs and remnant arcs by deep sea drilling, dredging, and direct subaerial sampling have shown that only a minor proportion of the lavas have the composition of 'average' andesite (which includes, of course, continental andesites). Thus the lavas making up the Kyushu–Palau Ridge, the West Mariana Ridge, and the active Mariana arc, all formed during the last 40 Ma, during the evolution of the Mariana arc system, are mostly island arc tholeiites or calc-alkaline basalts and basaltic andesites (see review by Tarney at al., 1981). Similarly, young volcanic arcs such as Tonga and the South Sandwich are also mainly basaltic. More mature arcs, such as the Lesser Antilles, do erupt andesite in greater volume; but seismic studies on the Lesser Antilles (Boynton et al., 1979) have shown that the bulk composition of this arc is more mafic than andesite, reflecting a dominant component of more tholeiitic material or mafic cumulate at depth.

Remobilizing an island arc during lateral accretion requires considerable thermal energy. Thermal models of present-day subduction zones (Anderson et al., 1978) tend to discount the possibility that the subducted ocean crust actually melts even when taken down within the mantle to depths of c. 80–150 km. It is a curious fact that in collision belts such as the Alpine–Himalayan system and in south-eastern Asia, where there is evidence of island arc accretion, the amount of granite generated during the collision phase is very small. On the other hand, along the central Andean margin of South America there is abundant granite (s.l.) but no evidence of island arc accretion. Brown (this volume, Section VI) also emphasizes that the respective granite compositions differ, those in collision zones being dominantly alkali-calcic, whereas those at Cordilleran margins are dominantly calc-alkaline tonalites.

Thus island arcs, if accreted laterally in the past (assuming that they were as common as at the present day and were of broadly similar composition), do not appear to have contributed significantly to the presently observed crustal compositions of Precambrian terrains.

Continental Andesites and Crustal Growth

An alternative to growth of the continental crust by lateral accretion of volcanic island arcs would be crustal growth at continental margins by essentially vertical accretion of andesitic magmas. This process, which retains the main compositional features of the andesitic model, is favoured by Thorpe and Francis (1979a) and Thorpe et al. (1981) for crustal growth in the Andes. Assessing whether the compositions of the erupted andesitic magmas represent juvenile additions from the mantle system or whether the magmas have suffered extensive fractional crystallization and/or contamination with the continental crust itself is more difficult in this environment. Of course, it is perhaps more likely that the dominant contribution to crustal growth may come from the addition of plutonic magmas, rather than extrusive magmas, to the continental crust. Thorpe et al. (1981) have estimated that the plutonic magmas exceed the volume of erupted lavas by at least a factor of 10. In this case, how are the compositions of extrusive lavas (dominantly andesites) related to the compositions of plutonic batholiths (dominantly slightly more silicic tonalites)?

Thorpe and Francis (1979a) discuss this problem with reference to the Andes, where they find difficulty in establishing direct links between plutonic and extrusive activity, with ring complexes lacking volcanic counterparts, and major volcanic cauldron structures lacking plutonic counterparts. They suggest that plutonic and extrusive magmatism might be separated in time, with early predominantly plutonic acitivity being followed later by dominantly volcanic activity, and propose that this may be governed by the water content of the mantle-derived magmas. Thus the early tonalitic (plutonic) magmas, being rather wet, crystallize at deep crustal levels (cf. Harris et al., 1970) whereas later andesitic (extrusive) magmas are drier and are capable of rising to high crustal levels.

It is also possible to construe this time sequence between plutonic and volcanic activity the other way round. Cretaceous magmatic activity in Peru, for instance (Atherton et al., 1979), started with the extrusion of the voluminous Casma basaltic to andesitic volcanics and was followed by a series of plutonic super-units which become more silicic with time. This sequence is repeated, to a more limited extent, in the Tertiary with the extrusion of the Calipuy volcanic rocks and later plutons. A possible explanation for this is that at the start of magmatic cycle the crust is rigid enough to fracture and permits the rise of magmas to the surface; however, as thermal gradients within the deeper crust increase as a result of this magmatic activity, the crust is no longer able to sustain fractures and magmas rise to a level dependent on their density and water content. The conspicuous lack of pegmatites in Andean batholiths suggests that the magmas are not excessively hydrous. A consequence of either model is that, in broad terms, the crust may be density stratified, with the more basic and intermediate magmas being concentrated at deeper levels in the crust.

From the discussion so far it would appear that the volcanic contribution to crustal growth may be subordinate to the plutonic contribution (cf. Tarney, 1976), and that essentially *vertical* accretion of plutonic magmas may be more important than *lateral* accretion of volcanic island arcs. Lateral accretion of island arcs (without mobilization) may, however, be an important factor in crustal growth in Phanerozoic collision zones. Because volcanic sequences are vulnerable to erosion, they also add significantly to the sedimentary budget; accretionary sedimentary prisms are also, of course, a notable feature of some continental margin subduction zones (Karig and Kay, 1981). These contributions to crustal growth will be considered further below. Continental growth during the Precambrian, however, appears to be dominated by plutonism, a petrogenetic interpretation of which is dependent on knowledge of both upper and lower crust compositions.

Geochemical Nature of the Lower Crust

As we have already stressed, the chemical composition of the lower crust is a vital, yet poorly constrained, component in any crustal model. Geophysical constraints are provided by seismic data coupled with laboratory measurements of potential lower crust samples. Geochemical constraints are provided by studies of high pressure granulite terrains now uplifted and exposed by erosion and by granulitic nodules in volcanic breccia pipes.

Refinement of seismic refraction studies, coupled to laboratory measurements of seismic velocities in various rock types, has led to the development of a coherent picture of crustal structure broadly compatible with the geological data (i.e. Smithson and Brown, 1977; Smithson, 1978). The geophysical implications are of a rather heterogeneous continental crust, both laterally and vertically (Smithson, 1978), and have led to models suggesting an upper crust composed of sedimentary and metamorphic rocks intruded by granite, and underlain by a migmatitic middle crust. The lower crust is considered to consist of both igneous and metamorphic components ranging in composition from gabbro to granite (Smithson and Brown, 1977).

Perhaps of more relevance is an estimate of the bulk composition of the lower continental crust, which can be obtained from the geophysical data. Lower crustal seismic P wave velocities generally range from 6.4 to 6.8 km s^{-1} (Smithson, 1978), which are too low for gabbroic rocks (with a velocity of $c.$ 7 km s^{-1}) but are within the range of anhydrous granitic rocks (charnockites) and dioritic rocks, with respective velocities of 6.5 and 6.8 km s^{-1} (Smithson, 1978). The mean composition of the heterogeneous granulitic lower crust appears to approach that of diorite, although rocks of a dioritic composition may be relatively minor (Smithson, 1978). Hence an intermediate, rather than basic, composition is indicated for the lower crust, and the bulk continental crust must have a composition corresponding to quartz diorite/granodiorite (Smithson, 1978).

That the lower continental crust must have the depletion in K, Rb, U, Th, and Cs characteristic of granulites (e.g. Tarney and Windley, 1977) can be demonstrated from thermal considerations. Knowing the average contents of these elements in the upper crust, the present heat flow through the continental crust requires that the lower half of the continental crust be drastically depleted in radioactive heat-producing elements compared with the levels normally found in intermediate igneous rocks (Heier, 1973).

Archaean granulites commonly record in their mineral assemblages evidence of an origin in the deep crust ($c.$ 30–45 km; Tarney and Windley, 1977; O'Hara, 1977) and have the appropriate seismic characteristics. Given that a substantial proportion of the crust was in existence before the beginning of the Proterozoic ($c.$ 70 per cent according to Windley (1977) and Taylor and McLennan (1981b); perhaps equal to the present volume according to the model of Armstrong (1981)), then it is clearly important to establish the chemical characteristics of the deep Archaean crust.

Many Archaean high grade gneiss–granulite terrains have rather similar characteristics, being essentially intermediate in composition ($c.$ 61 per cent SiO$_2$). However, they commonly have a bimodal composition distribution (Barker and Peterman, 1974; Tarney, 1976) comprising a minor basic (tholeiitic) component and a dominant (tonalite–trondhjemite) 'grey gneiss' quartzo-feldspathic component with broadly calc-alkaline geochemical characteristics. In high grade terrains such as the Lewisian of north-western Scotland the granulite complex differs from the juxtaposed amphibolite–facies complex not only in having much lower concentrations of K, Rb, Th, and U, but also in having a higher proportion of mafic gneisses (Sheraton et al., 1973). An average analysis for Lewisian granulites, based on a large number of X-ray fluorescence analyses (Sheraton et al., 1973) is included in Table 2 for comparison with other proposed crustal compositions. Table 2 also contains REE abundances for the average Lewisian granulite (Weaver and Tarney, 1980a, b) which are very close to those published recently by Muecke et al. (1979). The chondrite-normalized REE pattern for this average granulite is plotted in Fig. 1.

In some respects the REE pattern for the *average* Lewisian granulite resembles that of the calculated lower crust REE pattern based on the andesite model. The total REE abundance is similar and there is a positive Eu anomaly, although the pattern is more fractionated ($Ce_N/Yb_N = 9.3$ compared with 2.3). Whereas the positive Eu anomaly could be taken to indicate the presence of residual feldspar and that the granulites are lower crustal residues (Muecke et al., 1979), the petrogenetic relationships between the individual gneiss components suggest a different interpretation. Chondrite-normalized REE patterns of the separate mafic, intermediate, tonalitic, and trondhjemitic gneiss components are shown in Fig. 3. It is apparent that it is the more silicic trondhjemitic and tonalitic gneisses which have the positive Eu anomaly. The basic gneisses on the other hand,

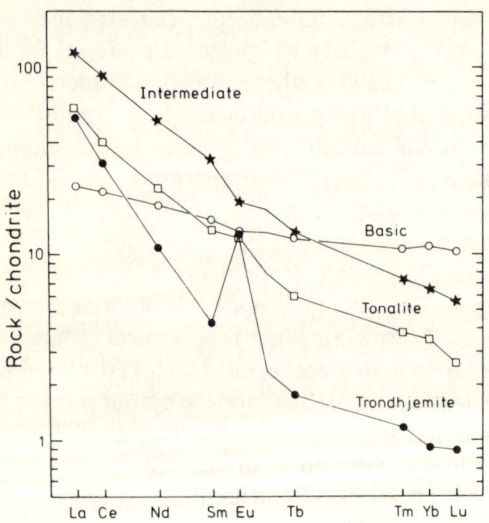

Fig. 3 Average REE patterns for the main rock types comprising the Lewisian lower crustal granulite complex, north-western Scotland. Averages calculated from the data of Weaver and Tarney (1980b)

which should in theory be the plagioclase-rich granulite–facies residua, have slight negative Eu anomalies. The more siliceous samples with positive Eu anomalies tend to have an intrusive relationship with the other rock types in the field and clearly represent residual liquids rather than refractory residues. Fractionation leading to evolved liquids with marked positive Eu anomalies has been described from a Proterozoic gabbro–diorite–tonalite–trondhjemite suite in south-western Finland (Arth et al., 1978), so the feature is by no means unique. Many Archaean silicic plutons have positive rather than negative Eu anomalies (e.g. Arth and Hanson, 1975). REE patterns for the charnockitic Madras granulites (Weaver, 1980) also show positive Eu anomalies whereas the associated mafic granulites have mostly negative Eu anomalies. The majority of the silicic gneisses with positive Eu anomalies also have highly fractionated REE patterns, showing similar light REE enrichment but strong heavy REE depletion relative to modern Andean tonalites (e.g. Lopez-Escobar et al., 1979; Atherton et al., 1979).

Although an earlier study of Lewisian granulites from the island of Coll, Inner Hebrides (Drury, 1978) suggested that the low total REE levels and positive Eu anomalies in Lewisian tonalitic granulites reflected their origin as cumulates from a parental dacitic liquid, we feel that the overwhelming evidence now supports the view that in granulite terrains it is the more siliceous rock types which have the positive Eu anomaly (Tarney et al., 1979). In a more detailed study of the petrogenesis of Lewisian granulites, Weaver and Tarney (1980b), using REE data for a wide range of representative samples, have argued that most of the granulites actually represent liquid compositions rather than cumulates or refractory residues. Their modelling of the REE and other trace element relationships suggested that the compositional range within the tholeiitic mafic gneisses could best be accounted for by low pressure fractional crystallization involving removal of anhydrous mineral phases such as olivine, pyroxenes, and plagioclase. On the other hand, the compositional variation within the intermediate tonalitic and trondhjemitic gneisses could be most successfully interpreted as resulting from high pressure hydrous partial melting of a mafic source involving residual garnet and/or hornblende. Weaver (1980) came to broadly similar conclusions in his study of the Madras granulites. Some rock types within the granulite complex can be recognized as cumulates or remelts of the tonalitic or mafic gneisses themselves, but these make up a relatively minor component of the granulite terrain.

The conclusion that the majority of the rock types in deep crustal granulites appear to represent liquid compositions rather than cumulates or the refractory residues from intracrustal melting has important implications for crustal growth processes. These will be considered in more detail below. The low K, Rb, Cs, Th, and U contents of many granulite terrains (Heier, 1973; Tarney and Windley, 1977), previously considered as being due to removal of lithophile elements during intracrustal melting (e.g. Fyfe, 1973), are instead now regarded as being removed in a CO_2-rich fluid phase (Tarney and Windley, 1977; Collerson and Fryer, 1978; Weaver, 1980) without significant melting. Melting would considerably modify the REE characteristics of residual granulites, but the Nd isotope systematics of Lewisian gneisses are preserved through the granulite facies event (Hamilton et al., 1979), which argues strongly against intracrustal melting. Moreover, other lithophile elements such as Ba, which might be expected to be removed during melting, remain high in granulite facies gneisses (Tarney, 1976).

The above arguments do not, of course, exclude the possibility of cumulates in the lower crust nor that some crustal melting should not take place. Rogers (1977), for instance, has described mafic granulite xenoliths from breccia pipes which have the positive Eu anomalies characteristic of cumulates. Mafic magmas emplaced into the lower crust might be expected to undergo fractional crystallization in the same way as many well known examples emplaced at high crustal levels.

Whereas the discussion so far has centred on Archaean granulites because of the probable pre-eminence of Archaean crust in the total crust volume, younger granulites do occur. Those in the Central European belt (of late Proterozoic or younger age) occur as tectonic slices and in volcanic pipes. They are, however, very different in character from the granulites of Archaean high grade terrains. Those described by Tarney and Windley (1977) and Dupuy et al. (1979) are not significantly lower in K, Rb, Cs, Th, and U than average upper crustal values, their REE patterns are much less fractionated than those of Lewisian gneisses and show neither positive Eu anomalies nor heavy REE depletion. The basic granulites, which form a prominent component of these granulite assemblages, have the geochemical composition of continental tholeiites (Dupuy et al., 1979) and show few characteristics expected of cumulates or refractory residua. REE patterns of the silicic granulites have negative Eu anomalies and are broadly similar to the REE patterns of post-Archaean sediments. Since many are garnetiferous and corundum-normative they could well be metasediments, perhaps underplated to deep levels as part of an accretionary wedge. In summary these younger granulites, like those in the Archaean, do not have the characteristics of cumulates or refractory residua. Basic rocks are abundant, but silicic rocks are always present.

Isotopes and Crustal Growth

Sr, Pb, and Nd isotope ratios, and particularly a combination of these, are powerful tools in deciphering possible sources for the derivation of continental crust-forming magmas, the potential source regions (the mantle and lower and upper continental crust) having different isotopic signatures (see Hawkesworth, this volume, Section VIII). For instance, the present-day upper mantle represents a primitive reservoir for Sr, Pb, and Nd isotopes, relative to which the lower continental crust has comparable $^{87}Sr/^{86}Sr$ ratios and comparable, or lower,

$^{207}Pb/^{204}Pb$ and $^{206}Pb/^{204}Pb$, but very unradiogenic $^{143}Nd/^{144}Nd$, due to the strong depletion in Rb, U, and Th in the lower crust (and hence low Rb/Sr, U/Pb and Th/Pb ratios) but quite fractionated REE patterns with high Nd/Sm ratios. On the other hand, the upper continental crust is conversely enriched in Rb relative to Sr and in U and Th relative to Pb, and therefore typically has rather radiogenic $^{87}Sr/^{86}Sr$, $^{207}Pb/^{204}Pb$, and $^{206}Pb/^{204}Pb$, but, again due to light REE enrichment in the upper crust, unradiogenic $^{143}Nd/^{144}Nd$. Theoretically then, the isotopic composition of crust-forming igneous rocks should enable recognition of mantle or crustal sources for these rocks, plus any contribution from crustal contamination of the magmas, either at mantle depths (from subducted sediment with an upper crustal isotopic signature) or by contamination/assimilation processes operating at crustal depths. However, the role of mantle heterogeneity with regard to trace element levels and isotope ratios, now well demonstrated for both the oceanic and subcontinental mantle (e.g. Tarney et al., 1980; Menzies and Murthy, 1980) makes the interpretation of isotope data rather more complex.

Sr isotopes

The Sr isotope geochemistry of ancient gneisses and their implications to processes of crustal generation have been well reviewed by Moorbath (1976, 1977, 1978). Only the essence of the Sr isotope argument will be outlined here.

Calculated initial $^{87}Sr/^{86}Sr$ ratios for the major rock units of Archaean terrains from all continents group together closely, and have values not appreciably higher than that expected in the upper mantle at the appropriate time, assuming single-stage Sr isotope evolution in the mantle reservoir. This is illustrated diagrammatically in Fig. 4. Moreover, low initial $^{87}Sr/^{86}Sr$ ratios in younger rock units preclude their formation by mobilization of older pre-existing units, which, by the time of derivation of the younger rocks, would have developed high $^{87}Sr/^{86}Sr$ ratios. This point is illustrated

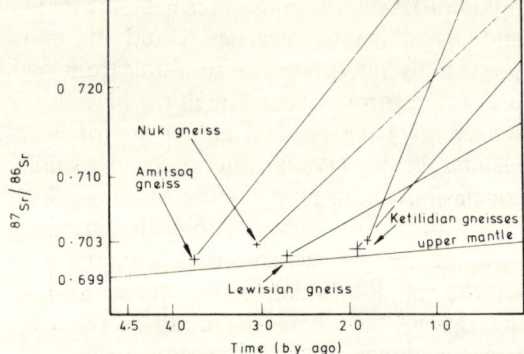

Fig. 4 $^{87}Sr/^{86}Sr$ isotope evolution in Precambrian gneisses from Greenland and Scotland. Adapted from Moorbath (1976)

on Fig. 4, especially in relation to the Archaean gneisses of south-western Greenland. At the time of generation of the Nûk gneisses (at c. 2.9 Ma ago), the Amitsoq gneisses (generated at 3.7 Ma ago) had developed a $^{87}Sr/^{86}Sr$ ratio far in excess of the initial $^{87}Sr/^{86}Sr$ ratio observed in the Nûk gneisses (Fig. 4). On this basis it is impossible to derive the Nûk gneisses by reworking of the older Amitsoq gneisses. Rather, both the Nûk and Amitsoq gneisses must have been generated from a low Rb/Sr reservoir. These relationships suggest that the gneiss precursors must have been derived from an upper mantle source at, or very close to, the measured isochron age, and represent essentially juvenile additions to the continental crust. The time interval between extraction of the juvenile igneous material from the upper mantle, and/or subducted oceanic lithosphere, and formation of the regional metamorphic characteristics of the derived gneiss complex is constrained by the Rb–Sr data as being of the order of 50–100 Ma.

The interpretation of Sr isotope data from recent sites of crustal growth (i.e. continental margins and island arcs) is rather less clear. Many of the Mesozoic and Tertiary granitic (sensu lato) batholiths from the Cordilleran margin of North and South America have low, if somewhat variable, initial $^{87}Sr/^{86}Sr$ ratios (i.e. 0.702–0.703 for Mesozoic plutons from the central Andes; McNutt et al., 1975) indicating a subcrustal source. Volcanic rocks from the destructive continental margins of the Cascades, Mexico, and Ecuador have $^{87}Sr/^{86}Sr$ ratios of 0.7030–0.7045 (Church and Tilton, 1973; Moorbath et al., 1978; Francis et al., 1977), while $^{87}Sr/^{86}Sr$ ratios in Peruvian and northern Chilean volcanic rocks are higher, at 0.705–0.708 (James et al., 1976; Francis et al., 1977; Hawkesworth et al., 1979a). Island arc volcanic rocks, removed from the influence of continental crust, have a mean $^{87}Sr/^{86}Sr$ ratio of 0.7044 (Faure, 1977) but with a range from 0.703 to 0.706 (i.e. Hawkesworth, 1979, Fig. 1). For comparison, ocean-floor basalts have a mean $^{87}Sr/^{86}Sr$ ratio of 0.7028 (Faure, 1977), while back-arc basalts associated with island arcs have lower $^{87}Sr/^{86}Sr$ than do the island arc volcanic rocks (Hawkesworth, 1979). This last point suggests that it is the subduction process which is responsible for the increase in $^{87}Sr/^{86}Sr$ at destructive plate margins. While this would perhaps be compatible with contamination of the mantle source region with subducted sediment (with an upper crustal isotope signature), the general enrichment of large ion lithophile elements over other incompatible elements in destructive margin magmas is often regarded as reflecting melting in the mantle wedge overlying the slab, which has undergone enrichment by large ion lithophile element-bearing hydrous fluids derived from the slab (e.g. Thorpe et al., 1976; Lopez-Escobar et al., 1977). This fluid would also carry radiogenic Sr from the altered ocean crust into the mantle wedge, accounting for the enhanced $^{87}Sr/^{86}Sr$ ratios of these magmas. However, the high $^{87}Sr/^{86}Sr$ ratios in the Peruvian and northern Chilean volcanic rocks are thought to reflect limited interaction with continental crustal material at crustal depths (Thorpe and Francis, 1979b).

Pb isotopes

Moorbath (1977) has reviewed the available Pb isotope data on ancient gneisses. Briefly, Pb–Pb whole-rock isochron ages for Archaean gneisses agree well with corresponding Rb–Sr ages. In this case, Moorbath (1977) considers that the simplicity of the Pb–Pb isotope systematics suggests derivation of the gneiss precursors from a source region with a rather homogeneous U/Pb ratio, approximating to single-stage evolution of Pb isotopes from the time of formation of the Earth to the measured isochron age. An upper mantle (or upper mantle-derived basic lithosphere) source is suggested by the $^{238}U/^{204}Pb$ ratios for the source regions calculated from the appropriate primary growth curve (Moorbath, 1977).

Available Pb isotope data for destructive plate margin magmas appear to be rather conflicting. In the case of island arc volcanics, Kay et al. (1978), for instance, identified a high $^{207}Pb/^{204}Pb$ component (equated with old continental crust) in volcanic rocks from the Aleutian arc, indicating mixture of a few per cent of continent-derived sediment with melt derived from the subduction zone, while Pb isotope data on volcanic rocks from the Marianas arc suggest that subducted sediment is not involved in magma genesis in this arc (Meijer, 1976). A rather more complex situation appears to exist in the western Cordilleran margin of North America (Zartman, 1974). Here the rocks of the western coastal batholith and associated andesites have rather homogeneous Pb isotope ratios comparable to those of island arc volcanics, and represent a juvenile, uncontaminated addition to the continental crust from the mantle–oceanic lithosphere system (Zartman, 1974). On the other hand, the Mesozoic–Cainozoic igneous rocks to the east of the coastal batholith have homogeneous, but rather radiogenic, Pb isotopes, while those of the Eastern Cordillera have a somewhat dispersed, but unradiogenic, Pb isotope signature. These contrasted Pb isotope patterns are thought to indicate contamination by the late Precambrian to Palaeozoic sediments and Precambrian crystalline basement, respectively, through which the magmas were intruded (Zartman, 1974).

Hence, while it appears from both Sr and Pb isotopes that continental crustal material may have some (but highly variable) influence upon the isotopic composition of orogenic magmas, the dominant radiogenic component of these magmas is inherited from the source region, and therefore largely reflects the isotope systematics of the mantle reservoir.

However, contamination of the upper mantle reservoir by recycled crustal Pb (via sediment subduction) plays an important role in Pb isotope models for the evolution of the crust–mantle system. Doe and Zartman (1979) have produced a dynamic model for Pb isotope evolution in the mantle and continental crust systems which recognizes the complementary nature of the upper and lower continental crust, the former being enriched in U, Th, and radiogenic Pb relative to the latter. Continental growth is considered to have commenced 4000 Ma ago, with evenly spaced orogenies (every 400 Ma) contributing equal volumes of magma to the continental crust. However, the crustal addition in each orogeny comprises a component of recycled continental crust as well as a juvenile component, but with the juvenile component dominant, such that the volume of the continents grows with time. The present-day Pb isotope ratios predicted by this model agree well with observed ratios in both crustal and mantle-derived rocks.

By contrast, Armstrong (1968) and Armstrong and Hein (1973) have developed a dynamic model for Sr and Pb isotope evolution in the continental crust plus mantle which is rather more extreme in its basic assumptions. Based upon apparently constant freeboard (relative elevation of the continents with respect to sea level) since the end of the Archaean, Armstrong (1968) and Armstrong and Hein (1973) consider that the growth of the continents was complete by c. 2500 Ma ago. Since that time the continents have been in equilibrium, with the rate of sediment subduction equalling that of magma addition to the crust from the mantle. According to this model the apparent single-stage evolution of Sr and Pb isotope ratios in crustal rocks, believed by Moorbath (1977) to reflect continuing crustal growth, is compatible with rapid crustal recycling, with the mantle system buffering the observed isotope evolution.

Such models must rely on supporting geological evidence for the subduction of considerable volumes of sediment, and this point will be discussed later.

Nd and correlated Nd–Sr isotopes and crustal growth

Initial Nd isotope ratios available for Archaean gneisses are summarized in Fig. 5. All of the data plot on, or very close to, the chondritic growth curve for ^{143}Nd/^{144}Nd, suggesting that the precursor magmas for these gneisses were generated from an undepleted chondritic mantle source region (Hamilton et al., 1979), a conclusion in accord with data from other isotopic systems. For instance McCulloch and Wasserburg (1978) derived model ages (that is, ages obtained assuming derivation from the mantle reservoir) of 2700–2500 Ma for composite rocks from the Canadian shield for both Nd and Sr isotope data, and these model ages agree well with K–Ar ages and Rb–Sr whole-rock isochron ages for these rocks, implying a major period of generation of juvenile continental crust at 2700–2500 Ma (McCulloch and Wasserburg, 1978). Combined Nd, Sr, and Pb

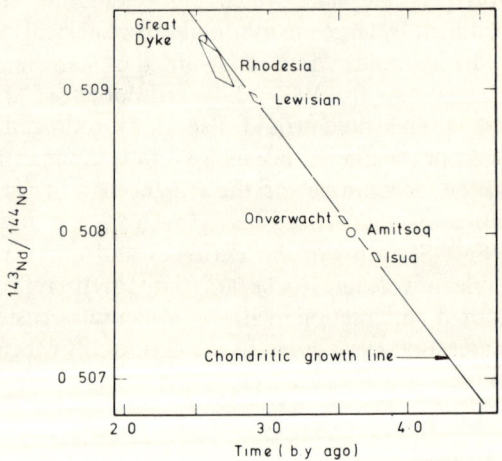

Fig. 5 Graph of initial ^{143}Nd/^{144}Nd ratio against age for Archaean rocks. The diagonal line represents the growth of ^{143}Nd/^{144}Nd in a chondritic reservoir which had a ^{143}Nd/^{144}Nd ratio of 0.50682 at 4550 Ma ago. Adapted from Hamilton et al. (1979)

data for Lewisian granulites (Hamilton et al., 1979) imply that c. 200 Ma elapsed between the removal of the gneiss precursors from the mantle reservoir and stabilization (cessation of granulite facies metamorphism) of this segment of continental crust. This agrees with the Sr and Pb isotope evidence for a very limited crustal prehistory for the gneiss precursors.

There is, as yet, relatively little Nd isotope data for Cordilleran margin plutons or lavas, although a considerable body of data exists for island arc volcanic rocks. In particular, great use of combined Nd and Sr isotope data has been made in the interpretation of recent subduction related magmas. The study of both Nd and Sr isotopes in young oceanic, and some continental, basalts has revealed a strong negative correlation between Sr and Nd isotopes in these rocks (i.e. DePaolo and Wasserburg, 1976; O'Nions et al., 1977), which is commonly termed the mantle 'main trend' and illustrated in Fig. 6. Basalts plotting in the north-western quadrant of Fig. 6 (typically mid-ocean ridge and ocean island basalts) have been derived from a mantle source with a time-integrated depletion in the light REE (i.e. Nd relative to Sm) and in Rb relative to Sr, whereas those

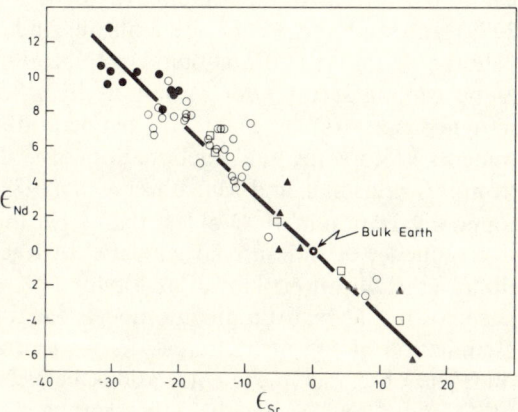

Fig. 6 Nd and Sr isotope compositions of recent basalts expressed as differences (in parts per 10^4) between the observed ratios and those inferred for the present-day bulk Earth: ●, mid-ocean ridge basalts; ○, ocean island basalts; □, continental flood basalts; ▲, other continental basalts. Data from DePaolo and Wasserburg (1976) and O'Nions et al. (1977)

basalts which plot in the south-eastern quadrant of Fig. 6 (some continental basalts) have been derived from a mantle source with a time-integrated enrichment in the light REE and in Rb relative to Sr, both in relation to the bulk Earth. Young destructive margin magmas often display in their Sr and Nd isotopic ratios a divergence from the 'main trend' correlation (e.g. Hawkesworth et al., 1977, 1979a,b), plotting to the right of the 'main trend' of Fig. 6, although not all island arc magmas show this feature (DePaolo and Wasserburg, 1977; DePaolo and Johnson, 1979). Possible explanations for this divergence from the 'main trend' in the case of island arc magmas are: (1) contamination of the magma source region by subducted authigenic sediments, or (2) introduction of radiogenic Sr, but not Nd, into the mantle wedge overlying the subducting slab during dehydration of the slab (i.e. Hawkesworth, 1979; Hawkesworth et al., 1979a,b). The second alternative is generally considered to be the more likely (Hawkesworth, 1979; Hawkesworth et al., 1979b). However, for continental margin magmas a third alternative, namely contamination by assimilation of continental crust, must be considered, and indeed the chemical characteristics of some Andean andesites are consistent with a small component of assimilated crust (Thorpe and Francis, 1979a), but the chemical character of these magmas seems largely inherited from the subduction zone, with dehydration processes in the slab causing the large ion lithophile and $^{87}Sr/^{86}Sr$ enrichment observed in these magmas.

The negative correlation between Sr and Nd isotopes displayed by many mantle-derived rocks (Fig. 6) may yield information regarding the evolution of the continental crust–mantle system and the processes of crustal growth, and various attempts have been made at explaining the origin of the 'main trend' correlation. DePaolo and Wasserburg (1979) consider the slope of the 'main trend' to be incompatible with an origin due to contamination of a uniform mantle reservoir by recycled continental crustal material, while Allègre et al. (1979) have demonstrated that the 'main trend' correlation reflects melt extraction from the mantle rather than, for instance, mixing processes involving separate mantle reservoirs. The only high $^{87}Sr/^{86}Sr$, low $^{143}Nd/^{144}Nd$ reservoir which could realistically complement the 'main trend' correlation is the continental crust, and a number of attempts have been made at modelling the 'main trend' by extraction of continental crust from the mantle reservoir. Jacobsen and Wasserburg (1979) considered two possible models: (1) that the continents form throughout geological time by melt extraction from an undepleted mantle reservoir, the residue of this process forming the depleted mantle reservoir which is today the source for mid-ocean ridge basalts; (2) new magma additions to the continents are derived from a mantle reservoir which becomes increasingly depleted with time by repeated melt extraction. Both of these models suggest that the rate of growth of the continents during the last 500 Ma has been much less than the average growth rate, and that the continents have formed from c. 30 per cent of the mantle reservoir, leaving 70 per cent undepleted. However, only model (1) agrees with the observed Nd isotope data, in which case the 'main trend' can be obtained by mixing of the depleted and undepleted mantle reservoirs. DePaolo (1979) used the Nd–Sr isotopic correlation to model the Sr, Rb, Sm, and Nd contents of the continental crust and derived a rather low Rb/Sr ratio of less than 0.1 for the bulk crust, and suggested from this that the continental crust may be more mafic in composition than is generally thought. Extending this approach, Weaver and Tarney (1980a) have modelled the development of the 'main trend' based upon extraction of melts from the mantle with the composition of observed Archaean crustal components, and found that extraction of continental crust with these compositions from the mantle reservoir to be perfectly compatible with a simple model for the generation of the mantle isotope correlation. Moreover, the modelling suggests that a considerable proportion of the continental crust is required to have the low Rb/Sr ratios but high Nd/Sm ratios typical of Archaean granulites, such that there is a considerable crustal reservoir with unradiogenic Sr, Pb, and Nd isotopes. On the other hand, models for the extraction of continental crustal components with the compositions suggested by the andesite model (i.e. Table 2) are not compatible with a simple model for the development of the mantle 'main trend' (Weaver and Tarney, 1980a).

O isotopes

O isotopes, in particular the isotopic ratio of $^{18}O/^{16}O$, are useful in discriminating between potential magma sources, especially if used in conjunction with Sr isotope data. A number of suites of igneous rocks display a good positive correlation between $\delta^{18}O$ and the initial $^{87}Sr/^{86}Sr$ ratio (i.e. Taylor, 1980).

Such correlations for andesites from the Banda island arc and the Peruvian Andes (Margaritz et al., 1978) have been taken to indicate contamination of the source of the andesites, probably by subducted sea-floor sediment. Alternatively, fractional crystallization accompanied by assimilation of continental crustal material has been invoked to explain the O–Sr isotope correlation in alkaline volcanic rocks from Roccamonfina (Taylor et al., 1979) and andesites from the Chilean Andes (Thorpe et al., 1981). Fractional crystallization alone cannot account for variations in $\delta^{18}O$ in igneous rocks; at magmatic temperatures mineral vectors for the fractionation of ^{18}O from ^{16}O are small, and Matsuhisa et al. (1973) found a variation in $\delta^{18}O$ of less than 1 per mil in a sequence of andesitic lavas related by fractional crystallization. However, Taylor (1980) has shown that realistic mixing models for the assimilation of continental crust do not reproduce the observed linear correlations between $\delta^{18}O$ and $^{87}Sr/^{86}Sr$. Rather, the isotopic relationships must be inherited from their source regions and reflect the patterns that existed in the parent rocks (or magmas) prior to, or during, melting. In this case the $\delta^{18}O-^{87}Sr/^{86}Sr$ correlation in andesites could reflect sediment involvement in calc-alkaline magma genesis (the sedimentary component having high $\delta^{18}O$ and $^{87}Sr/^{86}Sr$ ratios), as suggested by Magaritz et al. (1978), or perhaps represent merely the influence of ^{18}O- and ^{87}Sr-enriched fluids derived from the subducting slab. With respect to the latter point, the upper portions of the oceanic crust (as represented in ophiolite complexes) have, besides high $^{87}Sr/^{86}Sr$ ratios, relatively high $\delta^{18}O$ (up to +13) due to exchange with heated ocean water at mid-ocean ridges (Taylor, 1980). Dehydration of amphibolite to quartz eclogite, at relatively low temperatures, in the downgoing slab could thus generate $\delta^{18}O$-enriched fluids, especially as quartz fractionates ^{18}O strongly from ^{16}O.

Discussion

The central issues in an understanding of continental crust development concern the problem of its bulk chemical composition and the extent to which the crust has been subsequently modified by magmatic, metasomatic, or metamorphic processes. On the one hand the crust can be conceived as being grossly modified by magmatic processes, such that fractional crystallization or partial melting events have produced the chemical differentiation into upper and lower crust, while on the other hand the continental crust can be considered as having an essentially primary chemical composition little modified by secondary igneous processes, but with the radioactive heat-producing elements being redistributed within the crust by metamorphic–metasomatic processes. In the latter case the more basic nature of the lower crust is essentially a density stratification, the more mafic bodies or magmas not attaining such high crustal levels as contemporaneous granitic–tonalitic magmas because of their intrinsically higher density.

If crust-building magmas are ultimately mantle derived, as seems to be suggested by the trace element and isotopic compositions of ancient gneisses and many younger orogenic magmas, then the major element compositions of these magmas are subject to constraints from experimental data. Hydrous melting of mantle pyrolite can yield a liquid no more siliceous than andesite (Ringwood, 1977). Moreover, fractionation of more mafic Benioff zone magmas such as silica-saturated tholeiites are unlikely to produce evolved liquids more siliceous than andesite unless crystal fractionation occurred at depths shallower than 30 km (cf. Ringwood, 1977). It is apparently difficult, therefore, to generate continental crust with a bulk composition more silicic than andesite, by single-stage melting of the mantle.

Experimental data for the partial fusion of a mafic source indicates, however, that more siliceous magmas are possible. Hydrous melting of basalt at 5 kb pressure, for instance, produced a liquid of tonalitic to trondhjemitic composition (Helz, 1976). Such a mechanism may be relevant to the Archaean, where trace element evidence from silicic gneisses, as we have seen, suggests derivation of these magmas from a mafic, garnet-bearing source, commonly thought to be eclogite or garnet amphibolite (O'Nions and Pankhurst, 1974, 1978; Condie and Hunter, 1976; Glikson, 1976; Tarney et al., 1979; Weaver and Tarney, 1980b).

If subduction occurred in the Archaean (Burke et al., 1976; Bickle, 1978), then it is possible that, with the higher geothermal gradient, the processes of magma generation in Benioff zones may have been different. Melting of the mafic crust of the slab itself may have taken place, whereas in modern subduction zones the available geochemical data on calc-alkaline magmas tend to favour melting of the mantle wedge overlying the slab, under the influence of slab-derived fluids. Nevertheless, there are many similarities between ancient and modern calc-alkaline magmas. Both have high concentrations of the large ion lithophile elements (K, Rb, Cs, Ba, Sr, Th, U, light REE) relative to that of the high field strength elements (Nb, Ta, P, Hf, Zr, Ti) in comparison with normal tholeiitic or alkaline magmas (Tarney and Saunders, 1979; Saunders et al., 1980; Perfit et al., 1980). The decoupling in the behaviour of large ion lithophile elements from the high field strength elements in subduction zone magmas is generally attributed on the one hand to large ion lithophile element-bearing hydrous fluids derived from dehydration of the subducting slab entering, and promoting melting in, the source regions of calc-alkaline magmas, and on the other hand to the hydrous melting conditions enhancing the stability of minor mineral phases such as ilmenite, rutile, apatite and zircon, which retain the high field strength elements (Saunders et al., 1980). The only major difference in trace element composition between modern and Archaean calc-alkaline magmas concerns the strong relative heavy REE depletion displayed by the latter. This is most easily accommodated by permitting a melt component, in equilibrium with residual quartz eclogite, to be derived from the slab as well as a fluid component.

Thermal modelling of modern subduction zones (Anderson et al., 1978) suggests that the slab itself does not melt significantly. Thermal modelling for the Archaean suggests that the higher heat flow (two to three times that at present) would result in rather rapid sea-floor spreading, with a higher total ridge length and with short convection cells (Burke et al., 1976; Bickle, 1978; Davies, 1979). The young, warm lithosphere would be subducted at a shallow angle (cf. Molnar and Atwater, 1978), a situation that typifies modern Cordilleran belts such as the Andes which are zones of recent active crustal generation. This would at least be consistent with the rapid and extensive crustal growth which characterizes the Archaean. Moreover, the subduction of warm oceanic lithosphere into the mantle might permit dehydration reaction boundaries to become coincident with the wet solidus for the gabbro/eclogite transition in the subducting lithosphere (Wyllie, 1979), enabling dehydration and fusion of the slab to occur concomitantly, leading to the generation of the dominant tonalite–trondhjemite components of the bimodal suite which characterize Archaean terrains (Barker and Peterman, 1974; Weaver and Tarney, 1980b).

In modern zones of crustal generation, such as the Andes, there appears to be an anomaly between the dominantly tonalitic to granodioritic composition of the crust-building magmas, and the experimental constraints suggesting that mantle-derived magmas may be 'no more siliceous than andesite'. Deep crustal granulite terrains, such as the Lewisian, also have an appreciable amount of modal quartz (Sheraton et al., 1973). It seems then that there is a 'silica balance' problem in that the bulk crust is more siliceous than is permitted by the experimental constraints for mantle-derived melts. A possible solution is that silica, in addition to the large ion lithophile elements, is transported in appreciable amounts from the subducted slab as fluids are released during dehydration (cf. Anderson et al., 1978). This might significantly modify the bulk composition of the overlying mantle wedge and allow the generation of more siliceous melts than would otherwise be possible. Silica is certainly mobile in subduction zones, even at quite low temperatures, as evidenced by the very extensive quartz veins seen in subduction complexes such as the blueschist terrain on the Île de Groix, Brittany, the New Harbour Group in Anglesey, Wales, and the major accretionary prism in southern Chile (unpublished observations of the authors). The silica may not wholly be mobilized from subducted sediments. The subducting basaltic ocean crust will develop mineral assemblages dominated by silica-poor phases such as chlorite, epidote, and hornblende (low pressure) and garnet and jadeiitic pyroxene (high pressure). Many olivine tholeiites transform to quartz eclogite (Green and Ringwood, 1967), producing a source of free silica (5–10 per cent) in the downgoing slab which may be dissolved and removed in hydrous fluids. In this respect it may be relevant that quartz-bearing eclogites are rare amongst the mafic xenoliths of mantle origin in kimberlite pipes (Mathias et al., 1970).

Clearly such processes, although difficult to constrain, could result in the production, from modified mantle pyrolite, of magmas which are more siliceous than andesite. Entry of hydrous fluids into the mantle wedge at greater depth would also promote incipient melting and upward migration and concentration of lithophile element-enriched liquids (Tarney et al., 1981), thus accounting for the generally high large ion lithophile element contents of calc-alkaline magmas without necessarily requiring that all large ion lithophile elements are derived from the downgoing slab through fluid transfer. Differences between Archaean and modern crustal compositions could be related to the increased magma contribution from partial melting of the subducted ocean crust during the Archaean. Such magmas would be more voluminous (i.e. higher rate of crustal generation), less potassic (being derived from relatively K-poor oceanic crust), and have REE patterns showing heavy REE depletion (because residual garnet in the mafic source retains the heavy REE).

Primary magmas from the subduction zone may, however, be modified by fractional crystallization during diapiric uprise, or by contamination with subducted continental material (i.e. sediments), or by assimilation of the continental crust itself. The concentrations of Ni and Cr in subduction zone magmas are generally much lower than expected for melts in equilibrium with mantle mineral assemblages, suggesting that crystallization of mafic phases has occurred during ascent of these magmas (Ringwood, 1977). Very high partition coefficient values for Ni in olivine and Cr in pyroxene require only minor crystallization of these phases, however. Recently Thorpe et al. (1981) have proposed that intermediate orogenic magmas (i.e. continental andesites) have evolved through fractional crystallization, within the lower crust, from a more basic parent magma. In order to conserve the intermediate composition of the lower continental crust it is necessary to remove the ultramafic and mafic cumulates produced by this process. Gravitational settling of these cumulates through the lower crust, and their removal by convection in the mantle wedge, is suggested (Thorpe et al., 1981). This process conflicts to some extent with Jordan's (1978) concept of a thick, refractory, low density, upper mantle tectosphere which, he argues, underlies continents and which is less likely to be involved in convective recycling. Nonetheless, the proposal suggests a convenient method of getting rid of unwanted cumulates; the alternative is that more siliceous primary magmas must somehow be generated in subduction zones.

Any model seeking to explain the growth and development of the continental crust needs to adequately account for the negative Eu anomaly observed in post-Archaean sediments and, by inference, in the post-Archaean upper crust. The upper crustal positive Eu anomaly may be complementary to: (1) a positive Eu anomaly in lower crustal residues of intra-crustal melting; (2) a positive Eu anomaly in plagioclase-rich cumulates resulting from fractional crystallization of magmas ascending through the crust; or (3) a positive Eu anomaly in plagioclase-rich residues (amphibolite–granulite) of partial melting of the subducting slab at mantle depths.

The first alternative is invoked by Taylor (1977) in the andesite model for the continental crust, but requires a substantial proportion of the deeper crust (over half the total crust) to have a net positive Eu anomaly. As we have seen, although deep crustal xenoliths with positive Eu anomalies are known (Rogers, 1977), it still needs to be demonstrated that such xenoliths are typical of a large volume of post-Archaean lower crust. The second alternative is preferred by Thorpe et al. (1981), among others, and necessitates substantial fractionation of plagioclase. Whether fractional crystallization on the scale required is possible in more siliceous tonalitic to granitic magmas, which are highly viscous, is debateable. Direct evidence of cumulates in deeply eroded sections of the Andean batholith is meagre (Tarney and Saunders, 1979), and they must therefore be removed at lower crustal depths. The third alternative provides a convenient mechanism for disposing of the compensating positive Eu anomaly, although the process is fraught with uncertainties as to whether plagioclase could remain stable at the depths of magma generation in subduction zones (80–150 km); moreover, it requires moderately siliceous primary magmas. Clearly further information on the nature of the post-Archaean lower crust is required before the Eu anomaly problem can be solved satisfactorily. The Eu problem is not critical with regard to the petrogenesis of Archaean crust.

Recently, interest has arisen into the possibility of subduction of sediment at Benioff zones (e.g. Karig and Kay, 1981). The apparent lack of sediment in the Mariana trench, as revealed by recent deep sea drilling, and an apparent deficiency in the expected amount of accreted sediment in the Middle America trench and along parts of the Andean margin, certainly lend credibility to suggestions (Armstrong, 1968, 1981; Fyfe, this volume, Section X) that crustal material is being recycled into the mantle. The main point of contention, as we have noted earlier, is the scale of the process. Evidence for wholesale sediment involvement in island arc and continental margin calc-alkaline magma genesis is not compelling (see Thorpe et al., 1981; Karig and Kay, 1981), although isotopic data do permit limited sediment involvement (Kay et al., 1978). Unfortunately, the fact that subcontinental mantle nodules (Menzies and Murthy, 1980) and continental flood basalts (Kyle, 1980) have $^{87}Sr/^{86}Sr$ ratios which exceed those of many calc-alkaline magmas suggests that parts of the subcontinental mantle may have suffered large ion lithophile element metasomatism (cf. Lloyd and Bailey, 1975; Tarney et al., 1980); the involvement of such mantle material in calc-alkaline magma genesis would be difficult to distinguish from crustal involvement.

Sediment accretion at continental margins has been equally convincingly demonstrated (e.g. Karig and Kay, 1981) but is related to climate and sedimentation rate (Ziegler et al. 1981). Some accretionary wedges approach normal crustal thickness, are mainly composed of psammites and pelites, and are highly deformed. They constitute a net contribution to crustal growth, although secondary in nature. Such wedges can potentially be mobilized during continental collision, or even, if over-thickened, during arc–continent collision. They may provide a volumetrically suitable source for S-type granites (Chappell and White, 1974), which are an important, although minor, component of the continental crust. S-type granites characteristically have large negative Eu anomalies, as may be expected if they are derived from a sedimentary source which already has a negative Eu anomaly. Recycling of sediments within the crust in this way would enhance the upper crustal (sediment) negative Eu anomaly. It is unlikely that sedimentary recycling, even if repeated on a major scale since the Archaean, could generate the upper crustal Eu anomaly. This would necessitate, in part, the compensating positive Eu anomaly being transferred to another sedimentary component which would need to be removed from the crustal cycle. Limestones are such a potential component. Calcareous sediments generally occur near the base of ocean-floor sedimentary sequences, and hence are more likely to be subducted then any other sedimentary component. Limestones are often rich in Sr, and, as there is often a broad correlation between Sr^{2+} and Eu^{2+} in crustal rocks, it might be expected that limestones would have a positive Eu anomaly. However, available data on limestones (Haskin et al., 1966) indicates that most have negative Eu anomalies, presumably reflecting the negative Eu anomaly in seawater; the mechanism, then, seems not to be viable.

Finally, it is useful to consider briefly the geochemical effects of crustal extraction on the composition of the mantle. The lithophile element enrichment in the crust, to a first approximation, complements the lithophile element depletion (relative to the estimated primordial mantle composition) of the mantle source for N-type mid-ocean ridge basalts. Most geochemists would consider that there are essentially three elemental reservoirs: the crust, the N-type mid-ocean ridge basalt mantle source, and an undepleted mantle reservoir which provides the material for alkaline basalts and which may be located in the lower mantle (Jacobsen and Wasserburg, 1979; O'Nions et al., 1979; Dupré and Allègre, 1980). However, the distribution patterns of some incompatible elements suggest a slightly more complex

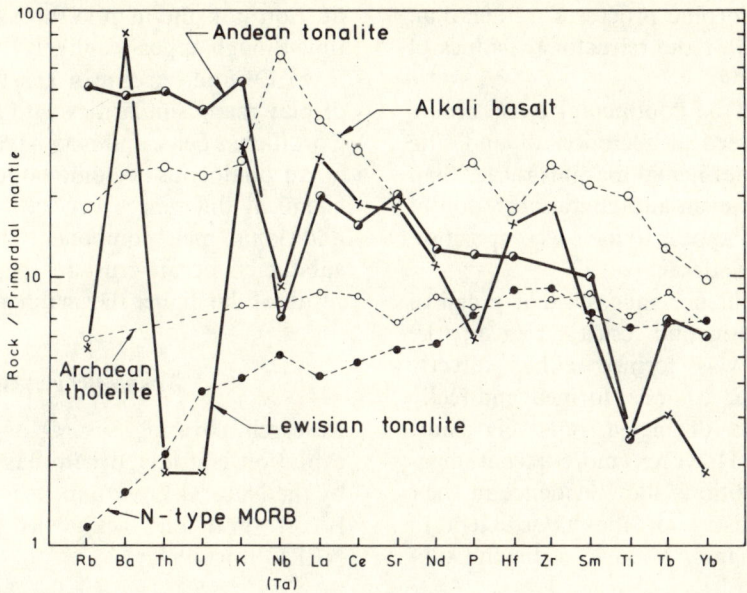

Fig. 7 Primordial mantle normalized (Wood et al., 1979) plot of incompatible element abundances in an Archaean tonalite (average of three tonalitic gneisses from Weaver and Tarney, 1980b), a Tertiary Andean tonalite (Lopez-Escobar et al., 1979), and a number of basalt types. From Weaver and Tarney (1980b)

situation. To illustrate this, Fig. 7 shows a plot of incompatible element abundances in an Archaean (granulite-facies) tonalite and a younger Andean tonalite, as well as a number of basalt types, normalized to the estimated abundances in the primordial mantle (Wood et al., 1979). Such plots emphasize the relative trace element differences between rock types. The granulite facies Archaean tonalite is strongly depleted in Th, U, K, Rb, (and Cs) relative to the modern tonalite, confirming that the Archaean lower crust is impoverished in heat-producing elements and may be a significant reservoir for unradiogenic Pb isotopes (Moorbath, 1977; O'Nions et al., 1979). The heavy REE depletion in the Archaean tonalite is attributable to equilibration with residual garnet during magma generation, as discussed earlier. Both Archaean and modern tonalites exhibit the marked relative depletion in Nb and Ta which characterizes most continental crust material (Saunders et al., 1980). Unfortunately a marked 'positive' Nb–Ta anomaly, which might compensate for the 'negative' Nb–Ta anomaly in the crust, is not present in N-type mid-ocean ridge basalt (although there is a *slight* positive anomaly). Alkaline basalts do, however, display a strong 'positive' Nb–Ta anomaly, which suggests that somehow the alkaline basalt mantle source must be considered in models describing crust-mantle evolution (cf. Tarney et al., 1980).

Conclusions

(1) Although many models for the chemistry of the continental crust propose an andesitic composition for the bulk crust, geophysical and geochemical evidence suggests that the continental crust may have a composition more siliceous than andesite.

(2) If the bulk continental crust is andesitic in composition, then the granodioritic nature of the upper crust requires that lower crustal granulites have a residual, mafic chemistry. However, the chemistry of Archaean lower crustal granulites is best interpreted as representing primary, intermediate (tonalitic) compositions little modified by subsequent

magmatic/metamorphic processes, rather than crystal cumulates or the refractory residues of intracrustal melting.

(3) Growth of the continental crust occurs dominantly by vertical accretion of andesitic (s.l.) plutons at continental margins rather than by andesitic volcanism and lateral accretion of island arcs, which appear to have a composition more mafic than andesite.

(4) Isotopic evidence suggests that much of the present continental crust grew in the Archaean, and was formed either directly from the mantle, or was formed indirectly from the mantle during a relatively short geological time. However, more recent magmatic crustal additions show evidence in their isotope geochemistry for the involvement of crustal contaminants. Massive sediment subduction into the mantle throughout geological time though appears unlikely.

(5) Overall, Archaean crustal compositions display many similarities with those in modern Cordilleran belts, and may have been generated in an analogous tectonic situation. The main chemical differences can be attributed to an additional melt component derived from the subducted ocean crust in the Archaean, as a result of the higher thermal regime.

Acknowledgements

Research covering several aspects of crustal evolution considered here has been supported by the Natural Environment Research Council. B.L.W. gratefully acknowledges receipt of a NERC Studentship.

REFERENCES

Allègre, C. J., Ben Othman, D., Polve, M., and Richard, P. (1979). The Nd–Sr isotopic correlation in mantle materials and geodynamic consequences. *Phys. Earth planet. Inter.* **19**, 293–306.

Anderson, R. N., DeLong, S. E., and Schwarz, W. M. (1978). Thermal model for subduction with dehydration in the downgoing slab. *J. Geol.* **86**, 731–739.

Armstrong, R. L. (1968). A model for the evolution of strontium and lead isotopes in a dynamic Earth. *Rev. Geophys.* **6**, 175–199.

Armstrong, R. L. (1981). Radiogenic isotopes: evidence for continental growth or recycling? The case for a near-steady-state no-continental growth crustal-recycling Earth. *Phil. Trans. R. Soc. A* **301**, 443–472.

Armstrong, R. L. and Hein, S. M. (1973). Computer simulation of Pb and Sr isotope evolution of the Earth's crust and upper mantle. *Geochim. cosmochim. Acta* **37**, 1–18.

Arth, J. G. and Hanson, G. N. (1975). Geochemistry and origin of the early Precambrian crust of northeastern Minnesota. *Geochim. cosmochim. Acta* **39**, 325–362.

Arth, J. G. Barker, F., Peterman, Z. E., and Friedman, I. (1978). Geochemistry of the gabbro–diorite–tonalite–trondhjemite suite of Southwest Finland and its implications for the origin of tonalitic and trondhjemitic magmas. *J. Petrol.* **19**, 289–316.

Atherton, M. P., McCourt, W. J., Sanderson, L. M., and Taylor, W. P. (1979). The geochemical character of the segmented Peruvian coastal batholith and associated volcanics. In *Origin of Granite Batholiths: Geochemical Evidence* (M. P. Atherton and J. Tarney, eds), Shiva Publishing, Orpington, pp. 45–64.

Barker, F. and Peterman, Z. E. (1974). Bimodal tholeiitic-dacitic magmatism and the early Precambrian Crust. *Precambrian Res.* **1**, 1–12.

Bavinton, O. A. and Taylor, S. R. (1980). Rare earth element geochemistry of Archaean metasedimentary rocks from Kambalda, Western Australia. *Geochim. cosmochim. Acta* **44**, 639–548.

Bickle, M. J. (1978). Heat loss from the Earth: a constraint Archaean tectonics from the relation between geothermal gradients and the rate of plate production. *Earth planet. Sci. Letters* **40**, 301–315.

Boynton, C. H., Westbrook, G. K., Bott, M. H. P., and Long, R. E. (1979). A seismic refraction investigation of crustal structure beneath the Lesser Antilles island arc. *Geophys. Jl R. astr. Soc.* **58**, 371–393.

Brown, G. C. (1979). The changing pattern of batholith emplacement during Earth history. In *Origin of Granite Batholiths: Geochemical Evidence* (M. P. Atherton and J. Tarney, eds), Shiva Publishing, Orpington, pp. 106–115.

Burke, K., Dewey, J. F., and Kidd, W. S. F. (1976). Dominance of horizontal movements, arc and microcontinental collisions during the later permobile regime. In *The Early History of the Earth* (B. F. Windley, ed.), Wiley, Chichester, pp. 113–129.

Chappell, B. W. and White, A. J. R. (1974). Two contrasting granite types. *Pacific Geol.* **8**, 173–174.

Church, S. E. and Tilton, G. R. (1973). Lead strontium isotopic studies in the Cascade Mountains: bearing on andesite genesis. *Geol. Soc. Am. Bull.* **84**, 431–454.

Collerson, K. D. and Fryer, B. J. (1978). The role of fluids in the formation and subsequent development of early crust. *Contr. Mineral. Petrol.* **61**, 151–167.

Condie, K. C. and Hunter, D. R. (1976). Trace element geochemistry of Archaean granitic rocks from the Barberton region, South Africa. *Earth planet. Sci. Letters* **29**, 389–400.

Davies, G. F. (1979). Thickness and thermal history of continental crust and root zones. *Earth planet. Sci. Letters* **44**, 231–238.

DePaolo, D. J. (1979). Implications of correlated Nd and Sr

isotopic variations for the chemical evolution of the crust and mantle. *Earth planet. Sci. Letters* **43**, 201–211.
DePaolo, D. J. and Johnson, R. W. (1979). Magma genesis in the New Britain island-arc: constraints from Nd and Sr isotopes and trace element patterns. *Contr. Mineral. Petrol.* **70**, 367–379.
DePaolo, D. J. and Wasserburg, G. J. (1976). Inferences about magma sources and mantle structure from variations of $^{143}Nd/^{144}Nd$. *Geophys. Res. Letters* **3**, 743–746.
DePaolo, D. J. and Wasserburg, G. J. (1977). The sources of island arcs as indicated by Nd and Sr isotopic studies. *Geophys. Res. Letters* **4**, 465–468.
DePaolo, D. J. and Wasserburg, G. J. (1979). Petrogenetic mixing models and Nd–Sr isotopic patterns. *Geochim. cosmochim. Acta* **43**, 615–627.
Doe, B. R. and Zartman, R. E. (1979). Plumbotectonics: the Phanerozoic. In *Geochemistry of Hydrothermal Ore Deposits*, 2nd edn (H. L. Barnes, ed.), Wiley–Interscience, New York, pp. 22–70.
Drury, S. A. (1978). REE distributions in a high-grade Archaean gneiss complex in Scotland: implications for the genesis of ancient sialic crust. *Precambrian Res.* **7**, 237–257.
Dupré, B. and Allègre, C. J. (1980). Pb–Sr–Nd isotopic correlation and the chemistry of the North Atlantic mantle. *Nature, Lond.* **286**, 17–22.
Dupuy, C., Leyreloup, A., and Verniéres, J. (1979). The lower continental crust of the Massif Central (Bournac, France)—with special references to REE, U and Th composition, evolution, heat-flow production. *Phys. Chem. Earth* **11**, 401–415.
Faure, G. (1977). *Principles of Isotope Geology*, Wiley, New York.
Francis, P. W., Moorbath, S., and Thorpe, R. S. (1977). Strontium isotope data for Recent andesites in Ecuador and North Chile. *Earth planet. Sci. Letters* **37**, 197–202.
Fyfe, W. S. (1973). The granulite facies, partial melting and the Archaean crust. *Phil. Trans. R. Soc. A* **273**, 457–462.
Glikson, A. Y. (1976). Trace element geochemistry and origin of early Precambrian acid igneous series, Barberton Mountain Land, Transvaal. *Geochim. cosmochim. Acta* **40**, 2161–1280.
Glikson, A. Y. (1979). Early Precambrian tonalite–trondhjemite sialic nuclei. *Earth Sci. Rev.* **15**, 1–73.
Goldschmidt, V. M. (1933). Grundlagen der quantitativen geochemie. *Fortsch. Min. Krist. Petrog.* **17**, 112.
Green, D. H. and Ringwood, A. E. (1967). An experimental investigation of the gabbro to eclogite transformation and its petrological implications. *Geochim. cosmochim. Acta* **31**, 767–833.
Hamilton, P. J., Evensen, N. M., O'Nions, R. K., and Tarney, J. (1979). Sm–Nd systematics of Lewisian gneisses: implications for the origin of granulites. *Nature, Lond.* **277**, 25–28.
Harris, P. G., Kennedy, W. Q., and Scarfe, C. M. (1970). Volcanism versus plutonism—the effect of chemical composition. In *Mechanism of Igneous Intrusion* (G. Newall and N. Rast, eds), *Geological Journal* Special Issue no. 2, pp. 187–200.
Haskin, L. A., Wildeman, T. R., Frey, F. A., Collins, K. A., Keedy, C. R., and Haskin, M. A. (1966). Rare earths in sediments. *J. geophys. Res.* **71**, 6091–6105.
Hawkesworth, C. J. (1979). $^{143}Nd/^{144}Nd$, $^{87}Sr/^{86}Sr$ and trace element characteristics of magmas along destructive plate margins. In *Origin of Granite Batholiths: Geochemical Evidence* (M. P. Atherton and J. Tarney, eds), Shiva Publishing, Orpington, pp. 76–89.
Hawkesworth, C. J., O'Nions, R. K., Pankhurst, R. J., Hamilton, P. J., and Evensen, N. M. (1977). A geochemical study of island-arc and back-arc tholeiites from the Scotia Sea. *Earth planet. Sci. Letters* **36**, 253–262.
Hawkesworth, C. J., Norry, M. J., Roddick, J. C., Baker, P. E., Francis, P. W., and Thorpe, R. S. (1979a). $^{143}Nd/^{144}Nd$, $^{87}Sr/^{86}Sr$, and incompatible element variations in calc-alkaline andesites and plateau lavas from South America. *Earth planet. Sci. Letters* **42**, 45–57.
Hawkesworth, C. J., O'Nions, R. K., and Arculus, R. J. (1979b). Nd and Sr isotope geochemistry of island arc volcanics, Grenada, Lesser Antilles. *Earth planet. Sci. Letters* **45**, 237–248.
Heier, K. S. (1973). Geochemistry of granulite facies rocks and problems of their origin. *Phil. Trans. R. Soc. A* **273**, 429–442.
Helz, R. T. (1976). Phase relations of basalts in their melting ranges at $P_{H_2O} = 5$ kb. II. Melt compositions. *J. Petrol.* **17**, 139–193.
Holland, J. G. and Lambert, R. St. J. (1972). Major element chemical composition of shields and the continental crust. *Geochim. cosmochim. Acta* **36**, 673–683.
Jacobsen, S. B. and Wasserburg, G. J. (1979). The mean age of mantle and crustal reservoirs. *J. geophys. Res.* **84**, 7411–7427.
Jakeš, P. and Taylor, S. R. (1974). Excess europium content in Precambrian sedimentary rocks and continental evolution. *Geochim. cosmochim. Acta* **38**, 739–745.
James, D. E., Brooks, C., and Cuyubamba, A. (1976). Andean Cenozoic volcanism: magma genesis in the light of strontium isotopic composition and trace element geochemistry. *Geol. Soc. Am. Bull.* **87**, 592–600.
Jordan, T. H. (1978). Composition and development of the continental tectosphere. *Nature, Lond.* **274**, 544–548.
Karig, B. E. and Kay, R. W. (1981). Fate of sediments on the descending plate at convergent margins. *Phil. Trans. R. Soc. A* **301**, 233–251.
Kay, R. W., Sun, S. S., and Lee-Lu, C. N. (1978). Pb and Sr isotopes in volcanic rocks from the Aleutian Islands and Pribilof Islands, Alaska. *Geochim. cosmochim. Acta* **42**, 263–273.
Kyle, P. R. (1980). Development of heterogeneities in the subcontinental mantle: evidence from the Ferrar Group, Antarctica. *Contr. Mineral. Petrol.* **73**, 89–104.
Lloyd, F. E. and Bailey, D. K. (1975). Light element metasomatism of the continental mantle: the evidence and consequences. *Phys. Chem. Earth* **9**, 389–416.
Lopez-Escobar, L., Frey, F. A., and Vergara, M. (1977). Andesites and high-alumina basalts from central-south Chile High Andes: geochemical evidence bearing on their petrogenesis. *Contr. Mineral. Petrol.* **63**, 199–228.
Lopez-Escobar, L., Frey, F. A., and Oyarzun, J. (1979). Geochemical characteristics of Central Chile (33°–34°S) granitoids. *Contr. Mineral. Petrol.* **70**, 437–450.
Margaritz, M., Whitford, D. J., and James, D. E. (1978). Oxygen isotopes and the origin of high $^{87}Sr/^{86}Sr$ andesites. *Earth planet. Sci. Letters* **40**, 220–230.
Matsuhisa, Y., Matsubaya, O., and Sakai, H. (1973). Oxygen isotope variations in magmatic differentiation processes in the volcanic rocks of Japan. *Contr. Mineral. Petrol.* **39**, 277–288.

Mathias, M. J., Siebert, J., and Rickwood, P. (1970). Some aspects of the mineralogy and petrology of ultramafic xenoliths in kimberlite. *Contr. Mineral. Petrol.* **26**, 75–123.

McCulloch, M. T. and Wasserburg, G. J. (1978). Sm–Nd and Rb–Sr chronology of continental crust formation. *Science, N.Y.* **200**, 1003–1011.

McLennan, S. M. and Taylor, S. R. (1980). Th and U in sedimentary rocks: crustal evolution and sedimentary recycling. *Nature, Lond.* **285**, 621–624.

McLennan, S. M., Fryer, B. J., and Young, G. M. (1979). Rare earth elements in Huronian (Lower Proterozoic) sedimentary rocks: composition and evolution of the post-Kenoran upper crust. *Geochim. cosmochim. Acta* **43**, 375–388.

McNutt, R. H., Crockett, J. H., Clark, A. H., Caelles, J. C., Farrar, E., Haynes, S. J., and Zentilli, M. (1975). Initial $^{87}Sr/^{86}Sr$ ratios of the plutonic and volcanic rocks of the central Andes between latitudes 26° and 29° south. *Earth planet Sci. Letters* **27**, 305–323.

Meijer, A. (1976). Pb and Sr isotopic data bearing on the origin of volcanic rocks from the Mariana arc system. *Geol. Soc. Am. Bull.* **87**, 1358–1369.

Menzies, M. and Murthy, V. R. (1980). Nd and Sr isotope geochemistry of hydrous mantle nodules and their host alkali basalts: implications for local heterogeneities in metasomatically veined mantle. *Earth planet. Sci. Letters* **46**, 323–334.

Molnar, P. and Atwater, T. (1978). Interarc spreading and Cordilleran tectonics as alternatives related to the age of subducted oceanic lithosphere. *Earth planet. Sci. Letters* **41**, 330–340.

Moorbath, S. (1976). Age and isotope constraints for the evolution of Archaean crust. In *The Early History of the Earth* (B. F. Windley, ed.), Wiley, Chichester, pp. 351–360.

Moorbath, S. (1977). Ages, isotopes and evolution of Precambrian continental crust. *Chem. Geol.* **20**, 151–187.

Moorbath, S. (1978). Age and isotope evidence for the evolution of continental crust. *Phil. Trans. R. Soc. A* **258**, 401–413.

Moorbath, S., Thorpe, R. S., and Gibson, I. L. (1978). Strontium isotope evidence for petrogenesis of Mexican andesites. *Nature, Lond.* **271**, 437–439.

Muecke, G. K., Pride, C., and Sarkar, P. (1979). Rare-earth element geochemistry of regional metamorphic rocks. *Phys. Chem. Earth* **11**, 449–464.

Nance, W. B. and Taylor, S. R. (1976). Rare earth element patterns and crustal evolution. I. Australian post-Archaean sedimentary rocks. *Geochim. cosmochim. Acta* **40**, 1539–1555.

Nance, W. B. and Taylor, S. R. (1977). Rare earth element patterns and crustal evolution. II. Archaean sedimentary rocks from Kalgoorlie, Australia. *Geochim. cosmochim. Acta* **41**, 225–231.

O'Hara, M. J. (1977). Thermal history of excavation of Archaean gneisses from the base of the continental crust. *J. geol. Soc. Lond.* **134**, 185–200.

O'Nions, R. K. and Pankhurst, R. J. (1974). Rare earth element distribution in Archaean gneisses and anorthosites, Godthåb area, West Greenland. *Earth planet. Sci. Letters* **22**, 328–338.

O'Nions, R. K. and Pankhurst, R. J. (1978). Early Archaean rocks and geochemical evolution of the Earth's crust. *Earth planet. Sci. Letters* **38**, 211–236.

O'Nions, R. K., Hamilton, P. J., and Evensen, N. M. (1977). Variations in $^{143}Nd/^{144}Nd$ and $^{87}Sr/^{86}Sr$ ratios in oceanic basalts. *Earth planet. Sci. Letters* **34**, 13–22.

O'Nions, R. K., Evensen, N. M., and Hamilton, P. J. (1979). Geochemical modeling of mantle differentiation and crustal growth. *J. geophys. Res.* **84**, 6091–6101.

Perfit, M. R., Gust, D. A., Bence, A. E., Arculus, R. J., and Taylor, S. R. (1980). Chemical characteristics of island arc basalts: implications for mantle sources. *Chem. Geol.* **30**, 227–256.

Ringwood, A. E. (1977). Petrogenesis in island arc systems. In *Island Arcs, Deep Sea Trenches and Back-arc Basins* (M. Talwani and W. C. Pitman, eds), Maurice Ewing Series no. 1, American Geophysical Union, Washington, DC, pp. 311–324.

Rogers, N. W. (1977). Granulite xenoliths from Lesotho kimberlites and the lower continental crust. *Nature, Lond.* 681–684.

Ronov, A. B. and Yaroshevsky, A. A. (1969). Chemical composition of the Earth's crust. In *The Earth's Crust and Upper Mantle* (P. J. Hart, ed.), American Geophysical Union Monograph no. 13, American Geophysical Union, Washington, DC, pp. 37–57.

Saunders, A. D., Tarney, J., and Weaver, S. D. (1980). Transverse geochemical variations across the Antarctic Peninsula: implications for genesis of calc-alkaline magmas. *Earth planet. Sci. Letters* **46**, 344–360.

Sheraton, J. W., Skinner, A. C., and Tarney, J. (1973). The geochemistry of the Scourian gneisses of the Assynt district. In *The Early Precambrian of Scotland and Related Rocks of Greenland* (R. G. Park and J. Tarney, eds), University of Keele, Keele, pp. 13–30.

Smithson, S. B. (1978). Modelling continental crust: structural and chemical constraints. *Geophys. Res. Letters* **5**, 749–752.

Smithson, S. B. and Brown, S. K. (1977). A model for lower continental crust. *Earth planet. Sci. Letters* **35**, 134–144.

Tarney, J. (1976). Geochemistry of Archaean high-grade gneisses, with implications as to the origin and evolution of the Precambrian crust. In *The Early History of the Earth* (B. F. Windley, ed.), Wiley, Chichester, pp. 405–417.

Tarney, J. and Saunders, A. D. (1979). Trace element constraints on the origin of Cordilleran batholiths. In *Origin of Granite Batholiths: Geochemical Evidence* (M. P. Atherton and J. Tarney, eds), Shiva Publishing, Orpington, pp. 90–105.

Tarney, J. and Windley, B. F. (1977). Chemistry, thermal gradients and evolution of the lower continental crust. *J. geol. Soc. Lond.* **134**, 153–172.

Tarney, J., Weaver, B. L., and Drury, S. A. (1979). Geochemistry of Archaean tonalitic and trondhjemitic gneisses from Scotland and East Greenland. In *Trondhjemites, Dacites and Related Rocks* (F. Barker, ed.), Elsevier, Amsterdam, pp. 275–299.

Tarney, J., Wood, D. A., Saunders, A. D., Cann, J. R., and Varet, J. (1980). Nature of mantle hetergeneity in the North Atlantic: evidence from deep sea drilling. *Phil. Trans. R. Soc. A* **297**, 179–202.

Tarney, J., Saunders, A. D., Mattey, D. P., Wood, D. A., and Marsh, N. G. (1981). Geochemical aspects of back-arc spreading in the Scotia Sea and west Pacific. *Phil. Trans. R. Soc. A* **300**, 263–285.

Taylor, H. P. (1980). The effects of assimilation of country rocks by magmas on $^{18}O/^{16}O$ and $^{87}Sr/^{86}Sr$ systematics in igneous rocks. *Earth planet. Sci. Lett.* **47**, 243–254.

Taylor, H. P., Giannetti, B., and Turi, B. (1979). Oxygen isotope geochemistry of potassic igneous rocks from the Roccamonfina volcano, Roman comagmatic region, Italy. *Earth planet. Sci. Lett.* **46**, 81–106.

Taylor, S. R. (1964). Abundance of elements in the continental crust: a new table. *Geochim. cosmochim. Acta* **28**, 1273–1285.

Taylor, S. R. (1967). The origin and growth of continents. *Tectonophysics* **4**, 17–34.

Taylor, S. R. (1977). Island arc models and the composition of the continental crust. In *Island Arcs, Deep Sea Trenches and Back-arc Basins* (M. Talwani and W. C. Pitman, eds), Maurice Ewing Series no. 1, American Geophysical Union, Washington, DC, pp. 325–335.

Taylor, S. R. (1979). Chemical composition and evolution of the continental crust: the rare earth element evidence. In *The Earth: Its Origin, Structure and Evolution* (M. W. McElhinny, ed.), Academic Press, London, pp. 353–376.

Taylor, S. R. and McLennan, S. M. (1979). The rare earth element evidence in Precambrian sedimentary rocks: implications for crustal evolution. In *Precambrian Plate Tectonics* (A. Kroner, ed.), Elsevier, Amsterdam, in press.

Taylor, S. R. and McLennan, S. M. (1980). The composition and evolution of the continental crust: rare earth element evidence from sedimentary rocks. *Phil. Trans. R. Soc. A* in press.

Thorpe, R. S. and Francis, P. W. (1979a). Petrogenetic relationships of volcanic and intrusive rocks of the Andes. In *Origin of Granite Batholiths: Geochemical Evidence* (M. P. Atherton and J. Tarney, eds), Shiva Publishing, Orpington, pp. 65–75.

Thorpe, R. S. and Francis, P. W. (1979b). Variations in Andean andesite compositions and their petrogenetic significance. *Tectonophysics* **57**, 53–70.

Thorpe, R. S., Potts, P. J., and Francis, P. W. (1976). Rare earth data and petrogenesis of andesite from the North Chilean Andes. *Contr. Mineral. Petrol.* **54**, 65–78.

Thorpe, R. S., Francis, P. W., and Harmon, R. S. (1981). Andean andesites and crustal growth. *Phil. Trans. R. Soc. A* **301**, 305–320.

Vinogradov, A. P. (1962). Average contents of chemical elements in the principal types of igneous rocks of the Earth's crust. *Geochemistry* **7**, 641–664.

Weaver, B. L. (1980). Rare-earth element geochemistry of Madras granulites. *Contr. Mineral. Petrol.* **71**, 271–279.

Weaver, B. L. and Tarney, J. (1980a). Continental crust composition and the nature of the lower crust: constraints from the mantle Nd–Sr isotope correlation. *Nature, Lond.* **286**, 342–346.

Weaver, B. L. and Tarney, J. (1980b). Rare-earth geochemistry of Lewisian granulite-facies gneisses, N.W. Scotland: implications for the petrogenesis of the Archaean lower continental crust. *Earth planet. Sci. Letters* **51**, 279–296.

Windley, B. F. (1977). Timing of continental growth and emergence. *Nature, Lond.* **270**, 426–428.

Windley, B. F. and Smith, J. V. (1976). Archaean high grade complexes and modern continental margins. *Nature, Lond.* **260**, 671–675.

Wood, D. A., Joron, J.-L., Treuil, M., Norry, M. J. and Tarney, J. (1979). Elemental and Sr isotope variations in basic lavas from Iceland and the surrounding ocean floor. *Contr. Mineral. Petrol.* **70**, 319–339.

Wyllie, P. J. (1979). Magmas and volatile components. *Am. Mineral* **64**, 469–500.

Zartman, R. E. (1974). Lead-isotopic provinces in the Cordillera of the western United States and their geologic significance. *Econ. Geol.* **69**, 792–805.

Ziegler, A. M., Barrett, S. F. and Scotese, C. R. (1981). Palaeoclimate, sedimentation and continental accretion. *Phil. Trans. R. Soc. A* **301**, 253–264.

Andesites product of geosphere mixing

W. S. Fyfe

Department of Geology,
University of Western Ontario, London, Ontario, Canada N6A 5B7

ABSTRACT

There is increasing evidence that andesites are the product of mixing continental components (via the hydrosphere–biosphere–atmosphere bridge) with subducted oceanic lithosphere. When melts from the regions above subduction zones reach continental crust they induce crustal fusion and further magma mixing. There is still great uncertainty concerning the volume of subducted sediments, but more evidence for massive subduction of pelagic sediments is now available and possible mechanisms are becoming apparent. The question as to whether or not continental crust is growing at the present time cannot be answered in its entirety.

Introduction

Before trying to write this paper, the author considered comments made elsewhere in this volume about andesites and their origin. To a student new to this problem, the conclusions would be highly confusing. At the extremes of the ideas, he would find andesites derived by partial melting of subducted oceanic crust (Marsh, this volume, Section III) or subducted sialic crust (Hutchison, this volume, Section III). In between, he would find complex polystage melting and fractional crystallization processes involving these sources together with magmas derived from variably transformed ('enriched' or 'metasomatized') mantle wedge (Mysen, this volume, Section VII).

If one reflects on average chemical analyses of andesites, they are often rather similar to typical ridge basalts, but with varying additions of a 'crustal' component. They are 'contaminated' basalt and rhyolite. Two recent papers illustrate the problem nicely. Thus, Whitford and Jezek (1979), considering the rocks of the Banda arc, conclude that, 'The most likely cause (of high Sr isotope ratios) is subduction and subsequent melting of either sea-floor sediments or continent crust'. In the same journal, Stern (1979), studying the northern Mariana arc, concludes 'that the source region of these parental liquids lies in the mantle, not subducted crust'. Andesites are variable, and the problem of their origin must explain such variation.

At the outset of the discussion, there are certain boundary conditions on which most agree. First, andesites are associated with the process of subduction, the mechanism of which is one of the great remaining questions of Uyeda (1978). Second, ocean-floor crust, produced at ridges, is approximately quantitatively subducted, and both ophiolites and old ocean crust are volumetrically trivial (Hallam, 1976). Third, subducted ocean-floor crust does not have the same chemical composition as new ocean-floor crust formed from the mantle—what comes up is not the same as what goes down (Fyfe, 1978).

What is Subducted?

The ocean-floor crust and the cool oceanic lithosphere are subducted to depths in the order of 700 km. The shape of the subduction zone is defined by modern earthquakes. The upper 10 km or so of this crust has been intensely studied in recent years, and the assemblage of sediments, pillow lavas, flows, dykes, and intrusions comprises the ophiolite suite. Early workers on spilites (see Hyndman, 1972, p. 99) recognized that these surface rocks of the sea-floor crust had highly variable compositions, but on average showed considerable enrichment in H_2O, CO_2, O_2, K_2O, Na_2O relative to fresh oceanic basaltic rocks. Gradually, modern studies are providing more reliable data, but most of the trends originally proposed remain valid.

Data based on ocean-floor heat-flow patterns have shown that the cooling of the new lithosphere involves massive hydrothermal convection processes. A recent account of studies of the Indian Ocean arrives at a typical conclusion. Thus, Anderson et al. (1979) conclude that 'more than one-third of the entire surface of the world's ocean floor contains presently active geothermal convection that is cellular in plan form'. These authors also show that seawater flow through the ocean crust certainly continues in rocks 55 Ma old, far from the active ridges.

By careful analysis of heat flow patterns, Wolery and Sleep (1976) have shown that 30–40 per cent of the ridge thermal energy is involved in driving thermal convection. Given reasonable assumptions of the temperature to which the seawater is heated, they estimate that the average annual flux of seawater through the ocean crust is in the order of 500 km^3 (5×10^{17} g). As the rate of production of new crust is about 10 km^3 (3×10^{16} g), on average, every gram of new basalt may react with 20 g of seawater. If, in fact, convection continues in much older rocks (cf. Anderson et al., 1979), then this ratio will be larger. This ratio sets an absolute limit on possible fixation of seawater components. As such reactions involve exchange over the complex P-T-composition path of a convection cell, chemical complexity in ophiolites must be no surprise.

Recently, Fyfe and Lonsdale (1980) have reviewed some of the available data on direct observations of sea-floor thermal systems and the chemical processes. It is now clear that significant additions of O_2, S, CO_2, H_2O, Na_2O, K_2O, MgO, U, and Rb occur while depletions occur for Ca and most transition metals. Silica is too abundant to be significantly changed, but some is leached. Exchange of some seawater isotopic species ($^{18}O/^{16}O$, $^{87}Sr/^{86}Sr$) may drastically change the primary mantle values.

What Must Come Back

Some of the subducted species must be returned to the surface on a large scale. Here, we shall consider only two species. The average water content of ophiolites (in hydrated minerals) is c. 5 per cent, depending on exactly how much serpentinite is considered. Given a subduction rate of ophiolite of 10 km^3 a^{-1}, 1.5×10^{15} g a^{-1} of water will be subducted to significant depths (controlled by the specific metamorphic processes in the slab; Fyfe and McBirney, 1975; Wyllie, 1979). At this rate, the ocean mass would be subducted in 1 Ga. Plate tectonics has certainly been occurring for more than 1 Ga, and perhaps for much longer (Black et al., 1979). To maintain significant ocean volumes, water must be returned to the surface. Some may come back at the blueschist stage of metamorphism, but there is little problem in finding a source for deeply locked volatiles in minerals of high thermal stability, like phlogopite, which would be stable to depths exceeding 100 km. Phlogopite is the natural product of reaction between ultramafic materials and K-enriched spilites.

Recently, Bloch and Bischoff (1979) have considered K balance. They consider that about half the K delivered to the oceans is fixed in spilites. Given present rate processes, this would subduct all continental K in 10^{10} a. Similar arguments apply to U (Aumento, 1979; Fyfe, 1979). At least some continental components are recycled on a significant scale.

Are Sediments Subducted?

Gilluly (1971) was one of the first workers to suggest that massive volumes of sediment might be subducted. The subject is one of rather intense modern debate, and only a great deal more research, involving detailed seismic studies and drilling into trench walls, will eventually provide the data needed for realistic answers. In many places, for example the Japan Trench (Uyeda, 1978) and the Mariana Trench (Jones et al., 1979), the volume of sediments present appears to be smaller than expected in the absence of sediment subduction.

The conceptual problem involves finding the mechanism by which light crustal materials (clay sediments, greywackes) can be pulled into denser mantle. But, we have all observed convective drag in action, while making certain species of soup! And, once the sediments and greywackes reach the blueschist or eclogite facies, they are denser than crust.

Recently, there have been suggestions as to how this may happen. Jones et al. (1979) have discussed the mechanism of bending and cracking of the lithosphere, while it commences to descend into the mantle. They suggest that fault blocks and cracks may be involved in trapping sediment to be subducted. Lister's (1977) discussion of tilted fault blocks formed at ridges provides another mechanism for trapping wedges of sediment. Hargraves (1978) has suggested that, if convection is sufficiently vigorous, viscous drag may pull continental materials into the mantle. Molnar and Gray (1979) show that it is mechanically possible to subduct microcontinents. The large problems appear to be associated with the detailed mechanism of binding the initially light materials to mafic lithosphere. It is clear that sediment subduction, if it occurs, will depend on many factors, such as spreading rate, subduction angle, and sediment type. But it is significant that not all continental masses become younger toward their edges, as the 2 Ga ages of the Peruvian coast indicate (Dalmayrac et al., 1977). Most of the Brazilian Atlantic margin is also 2 Ga old. And the present author is intrigued that, while clay-rich sediments dominate deep sea pelagic sediments, the anticipated metamorphic product, kyanite, is rare in blueschists of subduction regimes (cf. Uyeda, 1978, pp. 186–188; Fyfe, 1980).

Andesites and Batholiths

Brown (1977) and Brown and Hennessy (1978) have stressed that the bulk of the magmas in Cordilleran situations are mantle-derived. Much of their analysis is based on the significance of the Sr isotope systematics. They have also estimated that the volumes of such materials produced is in the range $0.1-0.5$ km^3 a^{-1} and suggested that crustal thickening in Andean situations is caused by igneous under- and over-plating and not by compressional tectonics.

But the fact is that the Sr isotope data, and the O isotope data, cannot be explained by mantle sources alone (*see* Taylor and Silver, 1978), and many aspects of trace and major element chemistry require a crust–mantle mix, as recognized by Brown (1977) (*see also* Wyllie, 1979).

In many ways, the present author thinks that the recent work of Eichelberger (1975, 1978, 1979a, b) goes a long way toward providing a key to the andesite problem. By careful study of phenocryst reactions, and the silicate melt fluid inclusion in the phenocrysts in 'andesites', and their chemical trends, he has shown that some of these magmas are a product of mixing of rhyolite (crust-derived) with basalt (mantle-produced). He has also suggested an interesting mechanism for effective mixing, when wet mafic magma vesiculates as it enters a cooler acid magma chamber, and floats off into the lighter magma. He presents convincing field evidence for the occurrences of the phenomenon, but diffusional mixing across convecting interfaces may also be a significant mixing process. Based on the mixing concept, he shows how the crustal component of an andesite can be obtained from the bulk composition, and how well this can be correlated with the Sr isotope systematics. Thus, the crustal component may vary from c. 60 per cent in some Andean types, to 20 per cent in the Kermadec lavas. There is a general correlation of composition with crustal thickness (cf. Taylor and Silver, 1978). Thus, the chemical trends are considered to be dominated by mixing processes, and not by all the normal fractional crystallization processes.

The Eichelberger approach has great appeal, due to its simplicity and the large number of field petrographic and geochemical observations it correlates. Given such a concept, it becomes a simple matter to explain the variations in the igneous rocks above subduction zones. The processes involved in subduction melting may include: (1) subduction of ophiolite enriched in $H_2O, CO_2, S, U, K, Na, {}^{87}Sr$, and other elements (seawater–atmosphere components); (2) release of H_2O-rich fluids high in K_2O, Na_2O, U, etc., with or without fusion of the slab; (3) mixing of the fluid from (2), aqueous soup or melt, with hotter mantle, recharging of the hotter, depleted refractory continental tectosphere (Jordan, 1979), and triggering production of 'basalt' magma volumetrically dominated by mantle components; (4) over- and under-plating of the crust by basaltic magmas ($\rho_{melt} \simeq \rho_{crust}$), with crustal heating and eventual partial fusion of crust, producing rhyolite magmas; (5) mixing of rhyolite-basalt magmas to varying degrees, producing the hybrid rocks of the 'andesite' family and their intrusive equivalents.

The model has great simplicity. It seems to provide an explanation of the modern situation above subduction zones. It has fascinating applications to ancient volcanic regimes, where denser and hotter magma types were also present. It demonstrates the types of processes that may be involved in crust and mantle evolution, above descending convective systems. It is also clear that experimental confirmation of the steps (1)–(5) will be very difficult; but, as Eichelberger has observed, Nature has provided many of the clues in the rock types, their mineral disequilibria, and their inclusions.

While there may be argument about details, the present author feels that there can no longer be doubt that the subduction process involves mixing of atmosphere, hydrosphere, and continental components, back into the mantle. The exact scale of the process remains to be quantified.

REFERENCES

Anderson, R. N., Hobart, M. A., and Langseth, M. G. (1979). Geothermal convection through oceanic crust and sediments in the Indian Ocean. *Science, N.Y.* **204**, 828–832.

Aumento, F. (1979). Distribution and evolution of uranium in the oceanic lithosphere. *Phil. Trans. R. Soc. A* **291**, 423–431.

Black, R., Caby, R., Moussine-Pouchkine, A., Bayer, R., Bertrand, J. M., Boullier, A. M., Fabre, J., and Lesgner, A. (1979). Evidence for late Precambrian plate tectonics in West Africa. *Nature, Lond.* **278**, 223–227.

Bloch, S. and Bischoff, J. L. (1979). The effect of low-temperature alteration of basalt on the oceanic budget of potassium. *Geology* **7**, 193–196.

Brown, G. C. (1977). Mantle origin of Cordilleran granites. *Nature, Lond.* **265**, 21–24.

Brown, G. C. and Hennessy, J. (1978). The initiation and thermal diversity of granite magmatism. *Phil. Trans. R. Soc. A* **288**, 631–643.

Dalmayrac, B., Lancelot, J. R., and Leyreloup, A. (1977). Two-billion year granulites in the late Precambrian basement along the Southern Peruvian coast. *Nature, Lond.* **198**, 49–51.

Eichelberger, J. C. (1975). Origin of andesite and dacite: evidence of mixing at Glass Mountain, California and at other circum-Pacific volcanoes. *Geol. Soc. Am. Bull.* **86**, 1381–1391.

Eichelberger, J. C. (1978). Andesitic volcanism and crustal evolution. *Nature, Lond.* **275**, 21–27.

Eichelberger, J. C. (1979a). Andesites in island arcs and continental margins: relationship to crustal evolution. *Bull. Volc.* in press.

Eichelberger, J. C. (1979b). Vapor exsolution and mixing during replenishment of crustal magma reservoirs. *Nature, Lond.* in press.

Fyfe, W. S. (1978). The evolution of the Earth's crust: modern plate tectonics to ancient hot spot tectonics? *Chem. Geol.* **23**, 89–114.

Fyfe, W. S. (1979). The geochemical cycle of uranium. *Phil. Trans. R. Soc. A* **291**, 433–445.

Fyfe, W. S. (1980). Crust formation and destruction. In *The J. T. Wilson Symposium* (D. Strangeway, ed.), Geological Association of Canada Special Paper, in press.

Fyfe, W. S. and McBirney, A. R. (1975). Subduction and the structure of andesitic volcanic belts. *Am. J. Sci.* **275A**, 285–297.

Fyfe, W. S. and Londsdale, P. (1980). Ocean floor hydrothermal activity. In *The Sea*, Vol. 5 (C. Emiliani, ed.), John Wiley, New York, in press.

Gilluly, J. (1971). Plate tectonics and magmatic evolution. *Soc. Am. Bull.* **82**, 2387–2396.

Hallam, A. (1976). How closely did the continents fit together? *Nature, Lond.* **262**, 94–95.

Hargraves, R. B. (1978). Punctuated evolution of tectonic style. *Nature, Lond.* **276**, 393–399.

Hyndman, D. W. (1972). *Petrology of Igneous and Metamorphic Rocks*, McGraw-Hill, New York.

Jones, G. M., Hilde, T. W. C., Sharman, G. F., and Agnew, D. C. (1979). Fault patterns in outer trench walls and their tectonic significance. *J. Phys. Earth* in press.

Jordan, T. H. (1979). The deep structure of the continents. *Scient. Am.* **240**, 92–107.

Lister, C. R. B. (1977). Qualitative models of spreading-centre processes, including hydrothermal penetration. *Tectonophysics* **37**, 203–218.

Molnar, P. and Gray, D. (1979). Subduction of continental lithosphere: some constraints and uncertainties. *Geology* **7**, 58–63.

Stern, R. J. (1979). On the origin of andesite in the Northern Mariana Island Arc: implications from Agrigan. *Contr. Mineral. Petrol.* **68**, 207–220.

Taylor, H. P. and Silver, L. T. (1978). Oxygen isotope relationships in plutonic igneous rocks of the Peninsular Ranges batholith, Southern and Baja California. *US geol. Surv. Open-file Report* no. 78–701, 423–425.

Uyeda, S. (1978). *The New View of the Earth*, W. H. Freeman Co., San Francisco.

Whitford, D. J. and Jezek, P. A. (1979). Origin of late Cenozoic lavas from the Banda Arc, Indonesia: trace element and Sr isotope evidence. *Contr. Mineral. Petrol.* **68**, 141–150.

Wolery, T. J. and Sleep, N. H. (1976). Hydrothermal circulation and geochemical flux at mid-ocean ridges. *J. Geol.* **84**, 249–275.

Wyllie, P. J. (1979). Magmas and volatile components. *Am. Mineral.* **64**, 469–500.

Eruptive gas compositions and fluxes of explosive volcanoes: budget of S and Cl emitted from Fuego volcano, Guatemala

W. I. Rose Jr, R. E. Stoiber, and L. L. Malinconico

Department of Geology and Geological Engineering,
Michigan Technological University, Houghton, Michigan 49931, USA

and

Department of Earth Sciences,
Dartmouth College, Hanover, New Hampshire 03755, USA

ABSTRACT

Because of the force of such eruptions, direct eruptive gas determinations in major explosive eruptions have not been obtained. Nevertheless, the S and Cl budget of explosive eruptions can be constrained by several measurements including: (1) microprobe analyses of glass inclusions in phenocrysts; (2) analyses of scavenged acids *on* fresh ash; (3) analyses of residual S and Cl trapped *in* tephra; (4) remote correlation spectrometry of gases emitted by volcanoes in quiet intervals between eruptions; (5) determination of gas/particle ratios of S and Cl by airborne sampling inside small eruption clouds; and (6) direct measurements of Cl/S in passive emissions from craters.

Data determined by all the above methods, some of which were obtained at a low level of activity, are available for the recent activity at Fuego volcano, Guatemala. The large eruption of October 1974 released 2.2×10^{14} g of ash. Constrained extrapolation allows us to estimate that it also released 1.6×10^{12} g of S and 6.2×10^{10} of Cl. The flux of S and Cl for 3 years of passive emission following October 1974, when added to the eruptive gases, make the emission totals equal 2.0×10^{12} g S and 4.7×10^{11}–11×10^{11} g Cl. Based on the pre-eruption S and Cl content of Fuego magma, estimated from glass inclusions, the mass of magma required is 1×10^{15} g, about five times the amount erupted. Most of the S released and only a small fraction of the Cl was in the short-lived 1974 eruption. S is thus preferentially emitted during explosive eruptions and Cl during low level activity. The data suggest that rates of S emission during major eruptions are more than 100 times the largest measured rates during low level emissions.

Introduction

This paper is an attempt to estimate volcanic gas compositions and volumes from explosive vents and to demonstrate how this information can be used. Because analytical techniques are more diverse and precise for S and Cl, we have focused on these since the principles derived may be applicable to other elements.

Measurement Techniques

Several investigators have tried direct collection and analysis of volcanic gases as they are erupted, with collection techniques ranging from simple vacuum cylinders to liquid traps and adsorption columns (Finlayson, 1970; Sigvaldason and Elisson, 1968; Huntington, 1973; Zettwoog and Tazieff, 1972; Giggenbach, 1976; Naughton et al., 1963; Tazieff et al., 1970). One problem concerning these techniques is contamination. As soon as the gas is released from the magma it begins re-equilibrate and to react with the wall rock, and finally with air. Direct sampling techniques have the additional problem of gases reacting with the sample container (Gerlach, 1978). Perhaps most important, direct collection techniques can be used only at fairly 'quiet' and accessible volcanoes and therefore provide a somewhat unrepresentative sample. This is true particularly for explosive vents. Anderson (1975) recently reviewed data on volcanic gases and listed several analyses that he considered the least contaminated. They were all from non-orogenic basaltic vents (Table 1). As Anderson noted, these data suggest compositions that are too rich in SO_2 and CO_2 and too poor in H_2O when compared with glassy submarine lava of Kilauea (Moore and Fabbi, 1971). Furthermore, they give us little real insight into the magmatic gases of orogenic volcanoes, where magmas are believed to form by different mechanisms.

Direct sampling of gases at explosive vents is usually dangerous or impossible. In such cases other techniques have been employed to try to estimate the gas compositions. Analyses of glass inclusions trapped inside phenocrysts (Anderson, 1974, 1975, 1979; Rose et al., 1978a) have allowed the determinations of S and Cl abundances in the melt prior to eruption. Anderson has show that the interpretation of microprobe glass inclusion data must be made carefully, bearing in mind the possibility of leakage and post-entrapment crystallization. Examples of such data for the 1974 magma of Fuego volcano are given in Fig. 1. The figure shows that Cl is enriched with K_2O during crystallization of the melt, while S shows depletion. This probably reflects the greater solubility of Cl in the magma. As the magma is crystallizing and cooling, Cl is retained in the magma while S is lost. Thus the S and Cl contents of K_2O-poor inclusions may closely reflect the initial magmatic S and Cl contents (see also Anderson, 1979). From this

TABLE 1 Compositions of igneous gases from basaltic vents (after Anderson, 1975)†

Gas	Range (vol. %)	Mean ± S.D. (vol. %)	Mean (wt %)
H_2O	21–94	59.2 ± 27.9	34.9
H_2	0.1–4.7	1.5 ± 1.6	0.1
CO_2	0.5–40.9	17.0 ± 16.9	24.5
CO	0.04–2.4	0.9 ± 0.9	0.8
SO_2	1.3–59	18.8 ± 19.1	39.5
S_2	0.02–0.04		
H_2S	0.01–9		
HCl	0.02–0.4		
N_2	0–8.3		
		Total	97.4

† Data is a summary of eight samples, four from Kilauea, one each from Surtsey, Erta Ale, Nyiragongo, and Etna.

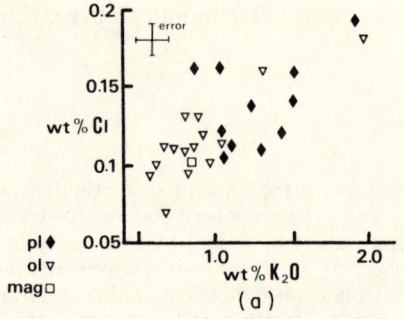

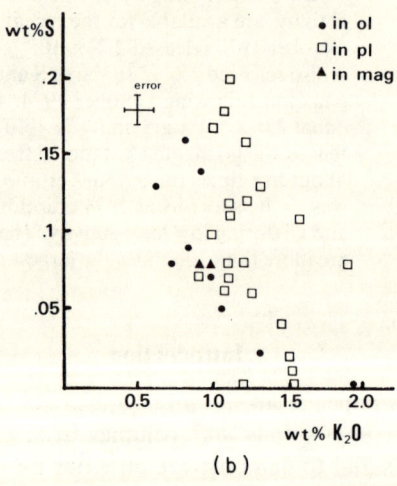

Fig. 1 (a) Plot of Cl against K_2O and (b) plot of S against K_2O for trapped glass inclusions from phenocrysts of 1974 ash from Fuego volcano (after Rose et al., 1978a)

data and the detailed studies of Harris (1979), we infer that parental Fuego magma contained 2800 p.p.m. S and 800 p.p.m. Cl (Table 2). Another indirect technique is the collection of condensates from high temperature fumaroles around the vents (Stoiber and Rose, 1970; Naughton et al., 1976). These condensate analyses suggest that the S/Cl ratio varies directly with the intensity of the volcanic activity (Noguchi and Kamiya, 1963; Stoiber and Rose, 1970; Menyailov, 1975). Other investigators have found that the Cl/F ratio (Naughton et al., 1975a, b), the He content (Thomas and Naughton, 1978), or the H content (Rose et al., 1980) may also vary with intensity of activity. A less direct technique for inferring the gas composition involves the examination of incrustations from around the fumaroles (Stoiber and Rose, 1974; Naughton et al., 1974). In the hottest fumarolic zones, where the gases have not reacted with the wall rock, incrustations may reflect the composition of the gas. However, in both fumarolic gas and incrustation analysis one must question how well the gases studied reflect the actual volcanic gas.

Remote sensing techniques have been employed to detect the gaseous components in volcanic gas (Murata, 1960; Delsemme, 1960; Naughton et al., 1969). Remote sensing by correlation spectrometry is being used to measure the mass flow of SO_2 from volcanic vents (Stoiber and Jepsen, 1973; Okita and Shimoura, 1975; Haulet et al., 1977). The mass flow of SO_2 as measured by spectroscopy shows a correlation with the intensity of volcanic activity similar to the S/Cl ratios and might provide a simple technique for monitoring volcanic activity (Crafford, 1977; Malinconico, 1978, 1979).

Application of Data

Measurement of the flux of a given volcanic gas allows the calculation of minimum volumes of magma needed to account for the gas emitted. An example of such a calculation for sulphur dioxide is shown in Table 3. We have chosen San Cristobal volcano, in northern Nicaragua, a basaltic composite volcano which has recently renewed activity after a repose of several centuries (Hazlett, 1977; Decker, 1973). The activity consisted of intense gaseous emission, but with little eruption of magma. The calculations (Table 3) suggest that since 1971, when degassing began, 0.09 km^3 of magma is required to produce the observed SO_2 emission. Several assumptions have been made to obtain this estimate. First, the initial and final concentrations of S in the magma were taken from analyses of glassy ashes from Fuego volcano. (These analyses are shown in Table 2 and will also be used in other calculations.) The similarity of these values to those for San Cristobal may or may not be close. Second, it is assumed that the magma degasses 97 per cent of its S. If the magma did not lose 97 per cent of its S, then the magma volume required to produce the observed output would be greater. Third, we assumed that the system was closed to gases from deep sources.

While the technique for remotely monitoring the SO_2 flux is routine, it has disadvantages. As was pointed out by Crafford (1977), the technique cannot be used during major pyroclastic

TABLE 2 Concentrations of S and Cl in the 1974 Fuego eruption (from Rose, 1977; Harris, 1979)

	S (p.p.m.)	Cl (p.p.m.)
A. Glass inclusions (original magma)	2800†	800†
B. Ash leachates	530	140
C. Insoluble bulk silicate ash	75	220
D. Proportion of S or Cl degassed (1–C)/A	0.97	0.73

† These values modified from values quoted by Rose (1977) after intensive study by Harris (1979).

TABLE 3 Magma volume calculation at San Cristobal volcano, Nicaragua, from SO_2 mass flow data

From 1971 to 1980, i.e., 9 years, constant degassing occurred at a rate of $c.$ 400 t d^{-1} of SO_2 gas = 200 t d^{-1} S. Therefore in 9 years the total amount of S produced is given by

$$\text{Total S produced} = 6.57 \times 10^5 \text{ t S}$$
$$= 6.57 \times 10^{11} \text{ g S}$$

Thus, using the value of 0.97 from Table 2, row D, for the proportion of S degassed,

$$\text{Total S} = 6.57 \times 10^{11} \times \frac{1}{0.97}$$
$$= 6.77 \times 10^{11} \text{ g S}$$

Amount of basalt needed to produce 6.77×10^{11} g S given that the concentration in the basalt is 2800 p.p.m. is found as follows:

$$\text{Amount of basalt} = 6.77 \times 10^{11} \div \frac{2800}{10^6}$$
$$= 2.42 \times 10^{14} \text{ g}$$

Finally, the volume of 2.42×10^{14} g basalt, assuming a density of 2.8 g cm^3, is given by

$$\text{Volume of basalt} = \frac{2.42 \times 10^{14}}{2.8}$$
$$= 8.6 \times 10^3 \text{ cm}^3$$
$$= 0.09 \text{ km}^3$$

eruptions, since the eruption cloud is heavily laden with ash and does not transmit light. In large eruptions, indirect methods for inferring gas compositions complement the remote sensing data. Initial concentrations of S and Cl in the magma can be inferred (after eruption) by measuring the concentrations in glass inclusions in phenocrysts, and final concentrations obtained by measuring the concentrations in the erupted ash (Anderson, 1974; Rose, 1977). The difference between the initial and final concentrations can be used to determine the amount released from the magma. A portion of the released gas is then scavenged by erupted ash in the eruption cloud and can be studied by leachate analysis (Taylor and Stoiber, 1973). Rose (1977) made measurements of the ashes from the 1974 eruption of Fuego volcano (Table 3) and, based upon the estimated amount of ash erupted, calculated the minimum amounts of S and Cl released to the atmosphere (2.2×10^5 g S and 1.6×10^5 g Cl). These values are minima because intrusive magma may also have contributed to the total amount of volatiles erupted.

This calculation can be carried a step further with information provided by airborne sampling of small eruptive plumes from Fuego (Rose *et al.*, 1978*b*). One of the airborne experiments determined the fractions of total S and Cl that are contained in particles and in gaseous form (Lazrus *et al.*, 1980). Results for S and Cl are summarized in Table 4. Using this information, data on scavenging and ash emission rates

TABLE 4 Percentages of total S and total Cl in particles from Central American volcanoes. The remaining percentages are as gaseous S and Cl (chiefly SO_2 and HCl) (from Lazrus *et al.*, 1980)

	Averages		Maxima	
	S	Cl	S	Cl
Fuego	2.2 ± 1.9	22 ± 14	5.7	47
Pacaya	1.3 ± 1.1	12 ± 11	2.6	26
Santiaguito	4.0 ± 2.8	12 ± 10	7.3	25

TABLE 5(a) S emitted (as SO_2) during the four phases of the 1974 eruption of Fuego volcano

Data

(1) Percentage of S on the ash versus the total S = 7.3 per cent (maximum value from Table 4). Therefore 13.7 times as much S was released from the magma
(2) Amount of S scavenged by the ash = 530 p.p.m. (Table 2)
(3) Amount of ash erupted during the four major events = 2.2×10^{14} g (from Rose et al., 1978a).

Calculations

$$\text{Amount of S released} = \frac{530 \text{ g S}}{10^6 \text{ g ash}} \times 2.2 \times 10^{14} \text{ g ash} \times 13.7$$

$$= 1.6 \times 10^{12} \text{ g S}$$

Averaged over four events,

$$\text{Amount of S released} = 4.0 \times 10^{11} \text{ g S per event}$$

$$= 8.0 \times 10^{11} \text{ g } SO_2 \text{ per event}$$

allow an estimate of the total S actually released during the four major eruptive events of the 1974 Fuego eruption (Table 5(a)). For a complete description of Fuego's activity, *see* Rose et al., (1978a). The value of 8.0×10^{11} g SO_2 per event is remarkable (the largest value recorded by spectrometer from a volcano has been about 1×10^{10} g SO_2 per day (R. E. Stoiber, unpublished data). A similar calculation has been made for Cl, but the result is one and a half orders of magnitude less (Table 5(b)).

If the amount of S and Cl that has been degassing in the interval following the 1974 eruption and ending with the renewed activity of Fuego in September 1977 is added to the amounts produced in the 1974 eruption, an estimate of the magma volumes required to produce the observed emission can be calculated (Table 6). Using both the S and the Cl values, $c\ 10^{15}$ g of basaltic melt would be needed to account for the observed emission. The fact that the calculation requires the same mass of magma for both elements is consistent with the validity of this approach. The mass of basaltic melt required is about five times the mass of material erupted during the 1974 eruption (2.2×10^{14} g).

Figure 2 shows the relative proportions of S and Cl released in explosive activity and passive emission. The majority of the S was released during eruptive activity while only a small fraction of the Cl was emitted during the eruption.

The calculations on passive emissions shown are based on (1) correlation spectrometry of passive emissions from Fuego to estimate SO_2 flux and (2) extensive measurements of Cl/S

TABLE 5(b) Cl emitted during the four phases of the 1974 eruption of Fuego volcano

Data

(1) Percentage of Cl on the ash versus the total chlorine = 47 per cent (maximum from Table 4). Therefore twice as much Cl was released from the magma
(2) Amount of chlorine scavenged by the ash = 140 p.p.m. (Table 3)
(3) Amount of ash erupted during the four major events = 2.2×10^{14} g (from Rose et al., 1978a).

Calculations

$$\text{Amount of Cl released} = \frac{140 \text{ g Cl}}{10^6 \text{ g ash}} \times 2.2 \times 10^{14} \text{ g ash} \times 2$$

$$= 6.2 \times 10^{10} \text{ g Cl}$$

TABLE 6 Calculation of degassing of S and Cl from Fuego volcano, 1974–77

I. *S data*
   A. Degassing of S during 1974 eruption (Table 5(a)) 1.6×10^{12} g
   B. Three years of passive degassing, 1974–77, at 4300 g s^{-1} (Table 7) plus 24 days of stronger activity at 12 000 g s^{-1} 4.3×10^{11} g
   C. Total S released by Fuego, 1974–1977 (A + B) 2.0×10^{12} g
   D. Amount of basalt needed to produce C assuming initial magmatic S of 2800 p.p.m. and that 97 per cent of S is degassed (Table 3) =

$$\frac{C \times 10^6 \text{ g basalt}}{(2800 - 75) \text{ g S}} = \quad = 0.7 \times 10^{15} \text{ g}$$

II. *Cl data*
   A. Degassing of Cl during 1974 eruption (Table 5(b)) 6.2×10^{10} g
   B. Three years of passive degassing, 1974–77, at 4300–10 000 g s^{-1} (Table 7) $\begin{cases} 4.1 \times 10^{11} \text{ g (min)} \\ 1.0 \times 10^{12} \text{ g (max)} \end{cases}$
   C. Total Cl released by Fuego, 1974–1977 (A + B) $4.7 \times 10^{11} – 1.1 \times 10^{12}$ g
   D. Amount of basalt needed to produce C assuming initial magmatic Cl of 800 p.p.m. and degassing of 73 per cent (Table 3) =

$$\frac{C \times 10^6 \text{ g basalt}}{(800 - 220) \text{ g Cl}} = \quad 0.8 \times 10^{15} – 1.9 \times 10^{15} \text{ g}$$

III. *Estimate of participation of intrusive magma*
   A. Mass of magma erupted, 1974 (Rose *et al.*, 1978a) 2.2×10^{14} g
   B. Proportion of estimated mass of degassed magma (I.D., II.D) to mass of magma erupted (III.A) 3.2–8.6

ratios in condensates from volcanic craters, ash leachates, and direct eruption cloud gas sampling (Table 7), which together with the spectrometry controls the Cl flux estimates. The differential degassing of S and Cl shown in Fig. 2 agrees with prior conclusions relating Cl/S ratios to eruptive activity patterns (Noguchi and Kamiya, 1963; Stoiber and Rose, 1970, 1973, 1974; Rose, 1977) and also with the glass inclusion data shown in Fig. 1 and discussed above.

As a postscript to our calculations, it may be useful to point out a significant uncertainty: our lack of actual data on Cl and S gas to particle

TABLE 7 Comparison of Cl/S ratios and S and Cl emission in various volcanic samples

	Cl/S Ratio	Emission Rate (kg s^{-1})
A. Leachates of ash from the 1974 Fuego eruption (Rose, 1977)	0.26	
B. Direct gas analyses inside small eruption clouds of Fuego (Lazrus *et al.*, 1980)	0.40	
C. Condensate analyses of other active craters (Stoiber and Rose, 1970)	2.3	
D. Condensate analyses (average of 78 observations) of fumaroles within active crater of San Cristobal, Nicaragua, 1972–77	2.06†	
E. Range of values assumed for most low level activity (based on above)	1.0–2.3	
F. Flux of S during passive emission (based on correlation spectrometry)		4.3
G. Flux of Cl during passive emission (based on E × F)		4.3–10

† We believe this is the best available data on emissions from a fuming crater during a quiet period. The Cl/S value is log-normal mean of 75 condensates from a 200–300 °C fumarole within the crater collected at 2 week intervals.

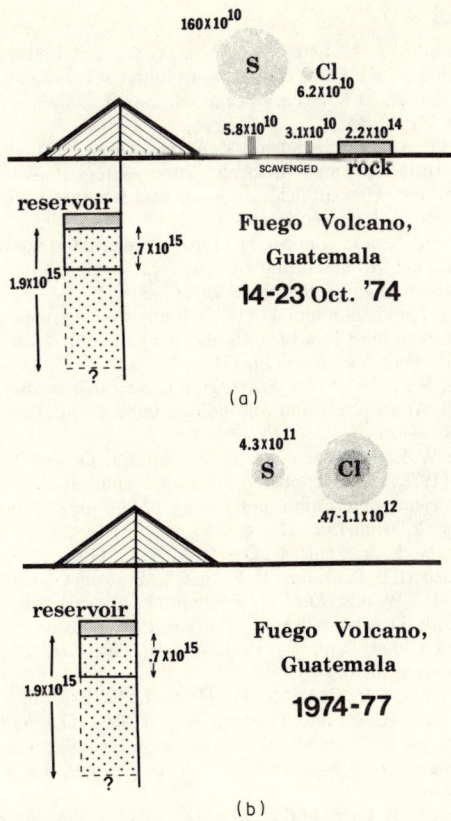

Fig. 2 Schematic diagram to represent the masses of S and Cl degassing from Fuego volcano and the magma masses required to account for them. (a) October 1974 explosive eruption; (b) 1974–77 passive degassing following eruption. All numbers given are in grams

partitioning in a large eruption. The data of Table 4 were collected from small eruptions and, in order to try to estimate how a large eruption might differ, we chose to use the maximum values. The validity of these data must be checked. Another need is to begin to accumulate similarly constrained data for other magmatic gas components such as H_2O, CO_2, and H_2.

Conclusions

Explosive volcanic vents do not allow simple collection and analysis techniques to determine the composition of volcanic gases. Remote sensing techniques can provide useful information but are less effective during major eruptions. Perhaps the best approach is one that combines many sampling methods: (1) analysis of glass inclusions for magma chamber compositions; (2) ash analysis for residual trapped gases; (3) leachate analysis for scavenged gases; (4) direct sampling for gas/particle ratios; (5) spectrometer measurements for flux rates; and (6) fumarolic gas sampling during quiet periods for passive emission gas ratios. Using all these techniques allows calculations of gas budgets for large eruptions like the 1974 eruption of Fuego. Even these inferences are extrapolations as we have yet to obtain direct samples or measurements from a large explosive eruption.

TABLE 8 Sources of data used in this paper. Please refer to the original sources for discussion of uncertainties in these data

1. Original magmatic concentrations of S and Cl (Tables 3,6)	Microprobe analyses of trapped glass inclusions in phenocrysts (Rose *et al.*, 1978a; Harris, 1979).
2. Passive emission of S and Cl (Tables 6,7)	Correlation spectrometry (Crafford, 1977; Stoiber and Bratton, 1978) and condensate geochemistry (Stoiber and Rose, 1970, 1973)
3. Mass of magma erupted (Table 6)	Isopach maps (Rose *et al.*, 1978a).
4. Percentage of S and Cl retained in quenched silicates (Tables 3,6)	Analysis of tephra for Cl and S (Rose, 1977)
5. Partitioning of S and Cl released to atmosphere between liquid aerosol and gas (Tables 4,6)	Analysis of fluoropore and Whatman filters used in direct sampling of eruption clouds (Lazrus *et al.*, 1979).
6. Cl and S concentrations in the liquid aerosol of the eruption (Tales 3,6)	Analysis of ash leachates (Taylor and Stoiber, 1973, Rose *et al.*, 1978a).

REFERENCES

Anderson, A. T. (1974). Chlorine, sulfur and water in magmas and oceans. *Geol. Soc. Am. Bull.* **85**, 1485–1492.

Anderson, A. T. (1975). Some basaltic and andesitic gases. *Rev. Geophys. Space Phys.* **13**, 37–55.

Anderson, A. T. (1979). Water in some hypersthenic magmas. *J. Geol.* **87**, 509–531.

Crafford, T. C. (1977). Remote sensing of SO_2 in the plume of Fuego Volcano, Guatemala. *Bull. Volc.* **40**, 1–21.

Decker, R. W. (1973). State of the art of volcano forecasting: *Bull. Volc.* **37**, 372–393.

Delsemme, A. H. (1960). Volcanic flame spectroscopy. *Acad. R. Sci. d'Outre mer Bull. Sci. Bruxelles* **6**, 507–519.

Finlayson, J. B. (1970). The collection and analysis of volcanic and hydrothermal gases. *Geothermics*, Special Issue, Vol. 2 (2), 1344–1354.

Gerlach, T. (1978). Restoration of the 1970 volcanic gas analyses of Mt Etna, Sicily. *Trans. Am. geophys. Un.* **59**, 1223.

Giggenbach, W. F. (1976). A simple method for the collection and analysis of volcanic gas samples. *Bull. Volc.* **39**, 1–14.

Harris, D. M. (1979). Geobarometry and geothermometry of individual crystals using H_2O, CO_2, S and major element concentrations in silicate melt inclusions. *Geol. Soc. Am. Abstr. Program.* **11**, 439.

Haulet, R., Zettwoog, P., and Sabroux, J. C. (1977). Sulphur dioxide discharge from Mount Etna. *Nature, Lond.* **268**, 715–717.

Hazlett, R. W. (1977). Geology and hazards of San Cristobal Volcanic Complex, Nicaragua. M.A. thesis, Dartmouth College, Hanover, N.H.

Huntingdon, A. T. (1973). The collection and analysis of volcanic gases. *Phil. Trans. R. Soc. A* **274**, 119–128.

Lazrus, A. L., Cadle, R. D., Gandrud, R. W., Greensburg, J. P., Huebert, B. J., and Rose, W. I., Jr (1980). Trace chemistry of the stratosphere and of volcanic eruption plumes. *J. geophys. Res.* in press.

Malinconico, L. (1978). Remote sensing of SO_2 as a tool for predicting volcanic eruptions. *Geol. Soc. Am. Abstr. Program.* **10**, 115.

Malinconico, L. (1979). Fluctuations in SO_2 emission during recent eruptions of Etna. *Nature, Lond.* **278**, 43–45.

Menyailov, I. A. (1975). Prediction of eruptions using changes in composition of volcanic gases. *Bull. Volc.* **34**, 112–215.

Moore, J. G. and Fabbi, B. P. (1971). An estimate of the juvenile sulfur content of basalt. *Contr. Miner. Petrol.* **33**, 118–127.

Murata, K. J. (1960). Occurrence of CuCl emission in volcanic flames. *Am. J. Sci.* **258**, 769–772.

Naughton, J. J., Heald, E. F., and Barnes, J. J. (1963). The chemistry of volcanic gases. 1. Collections and analysis of equilibrium mixtures by gas chromatography. *J. geophys. Res.* **68**, 538–544.

Naughton, J. J., Derby, J. V., and Glover, R. B. (1969). Infrared measurements on volcanic gas and fumes: Kilauea eruption, 1968. *J. geophys. Res.* **74**, 3273–3277.

Naughton, J. J., Lewis, V., Hammond, D., and Nishimoto, D. (1974). The chemistry of sublimates collected directly from lava fountains at Kilauea Volcano, Hawaii. *Geochim. cosmochim. Acta* **38**, 1679–1690.

Naughton, J. J., Finlayson, J. B., and Lewis, V. A. (1975a). Some results from recent chemical studies at Kilauea Volcano, Hawaii. *Bull. Volc.* **39**, 64–69.

Naughton, J. J., Lewis, V., Thomas, D., and Finlayson, J. B. (1975b). Fume compositions found at various stages of activity at Kilauea Volcano, Hawaii. *J. Geophys. Res.* **80**, 2963–2966.

Naughton, J. J., Greenberg, V. A., and Goguel, R. (1976). Incrustations and fumarolic condensates at Kilauea Volcano, Hawaii: field, drill-hole and laboratory observation. *J. Volc. geothermal Res.* **2**, 149–165.

Noguchi, K. and Kamiya, H. (1963). Prediction of volcanic eruption by measuring the chemical composition and amount of gases. *Bull. Volc.* **26**, 367–378.

Okita, T. and Shimoura, D. (1975). Remote sensing measurements of mass flow of sulfur dioxide gas from volcanoes. *Bull. Volc. Soc. Japan* **19**, 151–157.

Rose, W. I., Jr (1977). Scavenging of volcanic aerosol by ash: Atmospheric and volcanologic implications. *Geology* **5**, 621–624.

Rose, W. I., Jr, Anderson, A. T., Woodruff, L. G., and Bonis, S. (1978a). The October 1974 basaltic tephra from Fuego Volcano: description and history of the magma body. *J. Volc. geothermal. Res.* **4**, 3–53.

Rose, W. I., Jr, Cadle, R. D., Heidt, A. L., Gillette, D. A., Huebert, B. J., Stoiber, R. E., Self, S., Bratton, G., Chuan, R. L., Woods, D. C., Friedman, I., Zielinski, R. A., Smith, D., and Wilson, L. (1978b). 1978 Volcanic Eruption Cloud Sampling Project. *Geol. Soc. Am. Abstr. Program* **10**, 480–481.

Rose, W. I., Jr, Cadle, R. D., Heidt, L. E., Friedman, I., Lazrus, A. L., and Huebert, B. J. (1980). Gas and H isotopic analyses of volcanic eruption clouds in Guatemala sampled by aircraft. *J. Volc. geothermal. Res.* **6**, in press.

Sigvaldason, G. E. and Elisson, G. (1968). Collection and analysis of volcanic gas at Surtsey, Iceland. *Geochim. cosmochim. Acta* **32**, 797–805.

Stoiber, R. E. and Bratton, G. (1978). Airborne correlation spectrometer measurements of SO_2 in eruption clouds from Guatemalan volcanoes. *Trans. Am. geophys. Un.* **59**, 1222.

Stoiber, R. E. and Jepsen, A. (1973). Sulfur dioxide contribution to the atmosphere by volcanoes. *Science, N.Y.* **182**, 577–578.

Stoiber, R. E. and Rose, W. I., Jr. (1970). Geochemistry of Central American volcanic gas condensates. *Geol. Soc. Am. Bull.* **81**, 2891–2912.

Stoiber, R. E. and Rose, W. I., Jr. (1973). Cl, F and SO_2 in Central American volcanic gases. *Bull. Volc.* **37**, 454–460.

Stoiber, R. E. and Rose, W. I. J. (1974). Fumarole incrustations at active Central American volcanoes. *Geochim. cosmochim. Acta* **38**, 495–516.

Taylor, P. S. and Stoiber, R. E. (1973). Soluble material on ash from active Central American volcanoes. *Geol. Soc. Am. Bull.* **84**, 1031–1042.

Tazieff, H., LeGuern, F., and Zettwoog, P. (1970). Gas sampling in deep active craters. CEA, Direction de la Physique Section Magnetohydrodynamique MHD/RT 551.

Thomas, B. M. and Naughton, J. J. (1978). Changes in fumarolic gas compositions as precursors to volcanic eruptions: *Geol. Soc. Am. Abstr. Program* **10**, 504.

Zettwoog, P. and Tazieff, H. (1972). Instrumentation for measuring and recording mass and energy transfer from volcanoes to atmosphere. *Bull. Volc.* **36**, 1–19.

Andesites
Edited by R. S. Thorpe
© 1982 John Wiley & Sons

Mineral deposits associated with calc-alkaline rocks

A. H. G. Mitchell and R. D. Beckinsale

United Nations Development Project,
7285 A.DC, Mia Road, Pasay City, Metro Manilla, Philippines

and

Geochemical Division, Institute of Geological Sciences,
64–78 Gray's Inn Road, London WC1X 8NG, UK

ABSTRACT

Mineralization associated with magmatic rocks at convergent plate margins occurs in four characteristic tectonic settings. The four plate tectonic settings are: magmatic arcs, back-arc fold-thrust belts, outer arcs, and continent–continent or continent–magmatic arc collision belts. It is useful to make the distinction between I-type and S-type granites on the basis of geochemical features, although such a simple twofold division is bound to be oversimplified. I-type granites are characteristic of island arcs whereas S-type granites are characteristic of back-arc fold-thrust belts, outer arc belts, and particularly of collision belts. The types of mineralization most characteristic of magmatic arcs are porphyry Cu deposits and Kuroko-type stratiform base metal deposits. Porphyry deposits are associated with uplift, mountain belts, and composite andesite volcanoes, whereas Kuroko deposits are associated with subsidence and are formed in submarine conditions. Certain Sn deposits such as those of eastern Malaysia, stratiform Sb–W–Hg deposits, and magnetite–haematite–apatite extrusive deposits may also be characteristic of magmatic arcs. Back-arc magmatic belts such as that in Bolivia are characterized by Sn–Ag and Sn–W deposits. Outer-arc magmatic belts are not commonly mineralized but in some cases, such as the Sn deposits of the Outer Zone of Japan, reflect the older mineralization exposed in the magmatic arc, which suggests that they may reflect recycling of earlier components. Finally, collision belts such as that of central south-eastern Asia are characterized by Sn granites and perhaps by disseminated magmatic U deposits such as the granites of the Western Massif Central in France or the Rössing deposit of Namibia.

Introduction

The distribution of calc-alkaline volcanism in arcuate belts above Benioff zones was partly explained in 1967 by the hypothesis of subduction of oceanic lithosphere. Several more years elapsed, however, before economic geologists began to discuss the distribution of mineral deposits associated with these igneous rocks in terms of plate boundaries. This delay was perhaps due, first, to a lack of appreciation of the close genetic relationship between many types of sulphide ore body and their host rocks, despite the work of Schneiderhohn (1941), Stanton (1955), and Oftedahl (1958), and second, to the view, still largely justifiable today, that the plate tectonic hypothesis could contribute little to the discovery of the exact location of an ore body.

Nevertheless, by 1972 a number of authors had noted the distribution of late Mesozoic and Cainozoic sulphide deposits above contemporaneous Benioff zones (Sillitoe, 1970; Snelgrove, 1971; Guild, 1971) and attempts were made to relate the source of the metals to subduction-related processes occurring within and above the underthrusting plate. Subsequently, the ore deposits associated with calc-alkaline rocks in many Phanerozoic and older orogens and modern arc systems have been interpreted as related to subduction of oceanic lithosphere.

Mineralization at convergent plate boundaries in which continental rocks are present on the subducting plate (in other words, continent–continent or continent–island arc collision belts) was not subject to plate tectonic interpretations until *c*. 1974, when interest was aroused in the possibility of generating magmas in continental collisions. This early lack of interest in magmas and mineralization in collision belts is largely explained as a consequence of their structural and geological complexity and the relative scarcity of Cainozoic collision zones in which present plate convergence is indicated by seismic activity.

We here consider mineralization associated with magmatic rocks in four tectonic settings, the first three of which lie on overriding plate margins and are contemporaneous with and related to subduction of oceanic lithosphere while the fourth setting is confined to continental forelands in collision belts.

Tectonic Settings for Magmatism and Mineralization at Convergent Plate Boundaries

The four tectonic settings for magma generation at converging plate boundaries are magmatic arcs, back-arc magmatic belts, and outer arcs on the overriding plate above subducting oceanic lithosphere (Fig. 1(A)), and foreland fold-thrust belts on the subducting plate in collision belts

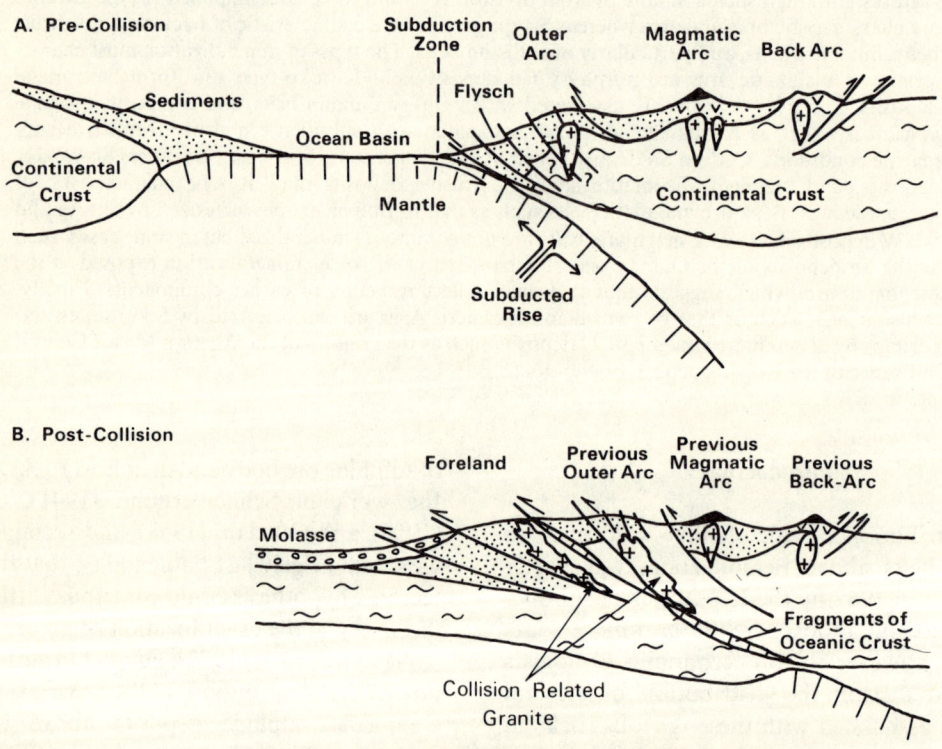

Fig. 1 Tectonic settings for magmatic rock-related mineralization at convergent plate margins. (A) Magmatic arc, back-arc magmatic belt, and outer arc settings; (B) collision-related foreland fold-thrust belts

(Fig. 1(B)). The granitoid rocks of these tectonic settings may be divided on the basis of geochemical features into I-types and S-types (Chappell and White, 1974). I-type granites are characteristic of magmatic arc settings, while S-type granites are characteristic of collision belts, back-arc fold-thrust belts, and perhaps outer arcs. The geochemical features which may be used to distinguish I-type and S-type granites are listed in Table 1, although it should be emphasized that these features are only characteristic tendencies and the classification should not be regarded as rigid. The essential distinction is between S-type magmas produced by fusion of typical continental material or incorporation of a high proportion of continental material into a mantle-derived melt, and I-type magmas derived by partial melting followed by fractional crystallization more or less direct from the mantle. This classification has been discussed in more detail by Beckinsale (1979). For example, in S-type granites such geochemical features as high $^{87}Sr/^{86}Sr$ ratios, high $\delta^{18}O$ values, low Na_2O contents, and low f_{O_2} values reflect the concentration in continental material relative to the mantle of Rb and ^{18}O-enriched phases, the partition of Na into seawater during crustal evolution, and the presence of reducing agents such as black shales in crustal sequences. This bimodal classification into I-type and S-type magmas is rather idealized and, although it does work well for many granites, may be difficult to apply in every case. In particular, for example, remelting of a mantle-derived igneous complex which had become incorporated into continental crust would yield a new parental magma with

TABLE 1 Summary of tectonic settings, geochemical characteristics, and mineralization associated with I-type and S-type granites

	I-Types: Magmatic Arcs	S-Types: Back Arcs, Outer Arcs, Collision Zones
Composition	High Na_2O contents ($>3.2\%$ Na_2O at 5.0% K_2O) Granites are part of a broad compositional spectrum from basic to acid	Low Na_2O contents because source region has already undergone a sedimentary cycle losing Na to seawater. Limited range of silica-rich chemical compositions.
Sr Isotopic composition	Typically low initial $^{87}Sr/^{86}Sr$ ratios $\lesssim 0.7080$	Typically high initial $^{87}Sr/^{86}Sr$ ratios $\gtrsim 0.7080$
Oxidation state of magma	High f_{O_2} resulting in crystallization of Fe_3O_4 as an accessory mineral	Low f_{O_2} (due to presence of reducing agents, e.g. graphitic shales in crustal sequences) resulting in crystallization of ilmenite as an accessory mineral
O isotope composition	Relatively low $\delta^{18}O$ values typically $\lesssim 10\%_{oo}$ SMOW	Relatively high $\delta^{18}O$ values typically $\gtrsim 10\%_{oo}$ SMOW
Characteristic accessory minerals	Hornblende, biotite	Muscovite, cordierite, tourmaline
Associated mineralization	Typically porphyry Cu deposits	Typically Sn deposits
Simplified petrogenesis	Partial melting of subducted oceanic lithosphere $\pm$ mantle overlying Benioff zone followed by fractional crystallization. Possibility of magma contamination with continental material.	Anatexis of continental crust followed by fractional crystallization

I-type features (except possibly ^{143}Nd/^{144}Nd model ages) despite the fact that in geological terms the process leading to magma generation is crustal fusion. Brown (this volume, Section VI) has described the spectrum of compositions of granitic rocks in more detail.

Magmatic arcs, as well as being the geological and topographical feature most characteristic of overriding plate margins, are also the most abundantly mineralized. They lie between 80 and several hundred kilometres from the submarine trench, and from 120 to 180 km above the underlying Benioff zone, forming either Cordilleran belts or island arcs. Volcanic rocks range from high alumina basalt to rhyolite in composition but are predominantly andesitic. The intrusive equivalents are diorite–trondhjemite bodies, and granite plutons are rare. As was recognized long ago, they are largely calc-alkaline, although island arc tholeiites characterize immature volcanic arcs. Isotope data suggest a magma source largely or entirely in the mantle.

Back-arc magmatic belts lie on the landward side of some continental margin magmatic arcs and form elevated belts with plutonic and in some cases volcanic rocks lying up to 700 km from the trench. On their continental side these belts commonly include a fold-thrust belt with thrusts dipping towards the ocean and along which considerable crustal shortening has taken place. Examples of such belts are the Eastern Cordillera of Bolivia and the Cretaceous foreland fold-thrust belt of the North American cordillera (Dickinson, 1976). The magmatic rocks in the back-arc belts are mostly more acidic than those of the magmatic arc and include rhyolites and granites. In some cases, these have higher initial ^{87}Sr/^{86}Sr ratios than analogous rocks in magmatic arcs, thus suggesting some involvement of crustal material in magma genesis.

Outer arcs, on the ocean side of some magmatic arcs, are considered to consist largely of imbricate wedges of ocean-floor and trench-fill flysch-type sediments off-scraped and tectonically emplaced above the subduction zone. In some of these outer arcs, for example the Shimanto Belt of south-western Japan, in westernmost Sumatra, and in southern Alaska, granitic intrusions have been described, of the same age but much closer to the trench than the associated magmatic arc. Marshak and Karig (1977) and DeLong *et al.* (1979) have suggested that these plutons developed above a subducting ocean rise which provided a source of heat in the normally cold descending slab of oceanic lithosphere.

Collision-related foreland fold-thrust belts, resulting from collision either of two continents or of a continent and island arc following ocean closure, have attracted relatively little attention as possible sites of magmatism. However, the occurrence of Tertiary collision-related granites on the Indian plate in the Himalayas has led to suggestions that some older belts of major granite batholiths were also generated in the subducting continental plate during collision with the overriding plate. The Himalayan granites characteristically contain two micas and occur as sheets and stocks. The Manaslu Granite of Nepal has an exceptionally high initial ^{87}Sr/^{86}Sr ratio of 0.74, suggesting an origin by crustal fusion (Hamet and Allègre, 1976).

The tectonic settings described above are defined mostly from late Mesozoic and Cainozoic terrains. Such settings can be confidently recognized throughout the Phanerozoic and probably most of the Proterozoic. However, in the Archaean some tectonic settings and related mineral deposits were significantly different. These include the Ni sulphide deposits of the Archaean greenstone belts and Witwatersrand-type quartz-pebble conglomerate Au–U deposits which may reflect particular stages in the long term evolution of the continental crust as it differentiates from the mantle.

Magmatic Arc Mineralization

Mineralization in magmatic arcs occurs in both intrusive rocks (e.g. porphyry Cu) and extrusive rocks (e.g. Kuroko-type stratiform sulphides), the latter including some less important stratabound types of metal deposit.

Porphyry Cu deposits

Porphyry Cu deposits are large tonnage, low grade deposits of Cu ranging from 10^7 to 10^9 t of ore with a grade of between c. 0.4 and 1.0 per cent Cu disseminated in predominantly igneous host rocks which commonly include a 'porphyry' intrusive phase. Porphyry deposits of Cu–Mo, largely confined to cordilleran arcs, and Cu–Au, predominant in island arcs, are important sources of these other metals, and porphyry Mo deposits lacking Cu are abundant in some arcs. More than 70 per cent of the world's annual Cu production is derived from porphyry deposits and for Mo the figure is close to 100 per cent.

The distribution of known porphyry Cu deposits in association with calc-alkaline magmatism in young mountain belts, especially in the Andes and in western North America (Fig. 2), prompted increased prospecting from

Fig. 2 Distribution of porphyry deposits in North and South America, after Lowell (1974) and Armstrong et al. (1976)

the mid-1960s onwards in the island and continental margin arcs of the Pacific and the Caribbean. This has resulted in many new discoveries, for example in the Solomon Islands, Papua New Guinea, the Philippines, Fiji, Sumatra, and especially in the Caribbean area. The relative abundance of porphyry deposits in these orogens of late Mesozoic and Cainozoic age has led to suggestions that this style of mineralization has been characteristic of only about the last 100 Ma of geological time. However, as noted above, by the early 1970s it was appreciated that ore deposits of this type and their associated igneous host rocks of the magmatic arcs overlie, and are related to, areas experiencing subduction of oceanic lithosphere. The presence of porphyry Cu deposits in areas which do not overlie active subduction zones such as the Appalachians, the Caledonides of Scotland, the Urals, central Thailand, and Taiwan can be explained, to the satisfaction of most geologists outside the USSR, on the basis of a model in which the magmatism and mineralization were related to *contemporaneous* subduction of oceanic crust, but that continent–continent or continent–island arc collision following ocean closure occurred subsequently. The relative scarcity of older porphyry Cu deposits reflects their formation at high structural levels within a magmatic arc and consequent removal by erosion. Strata-bound pyritic ore bodies of Kuroko type, which are also characteristic of island arcs, tend to be preserved in older deposits in sedimentary basins.

Features common to most porphyry deposits include a plutonic or subvolcanic calc-alkaline diorite to granodiorite host rock (Kesler *et al.*, 1975). In and around this host rock, hydrothermal alteration, in many cases affecting an area of several square kilometres, is more pronounced and extensive than in any other type of ore body, with a characteristic concentric zonation of hydrothermal minerals around a central zone of potassic alteration (Guilbert and Lowell, 1974) as indicated in Fig. 3(a).

Primary ore minerals in the zone of hypogene or primary mineralization include chalcopyrite with bornite and abundant pyrite. Surface oxidation and supergene enrichment through leaching and redeposition of copper by downward percolation of ground water are common features of these deposits, resulting in subsurface zones or 'blankets' in which the grade of hypogene 'protore' is increased by a factor of up to about five, forming grades of 1 per cent copper or more. Important ore blankets can occur here in positions related to erosional base levels. Common supergene ore minerals are chalcocite, covellite, and cuprite.

Isotopic and other geochemical studies indicate that the outer alteration zones surrounding porphyry copper deposits are caused by circulating hydrothermal systems of meteoric ground waters and/or formation fluids (Taylor, 1974).

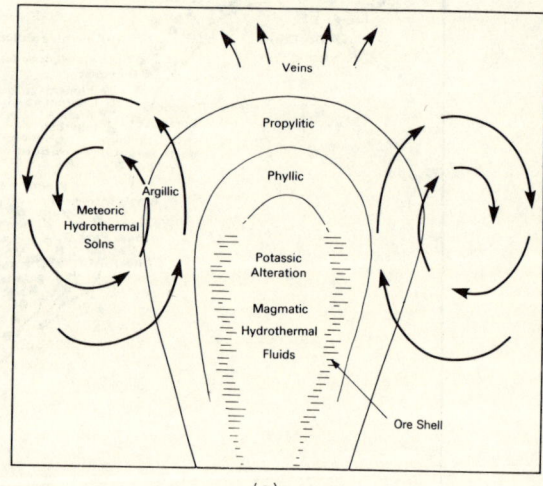

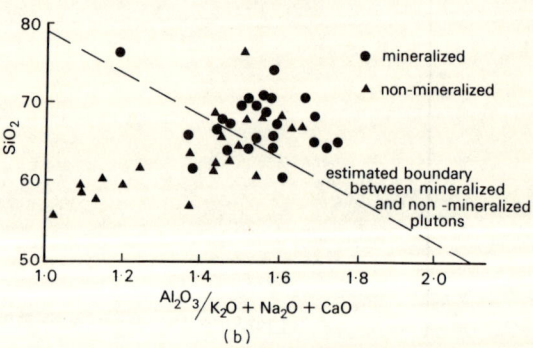

Fig. 3 (a) Hydrothermal alteration zones and water circulation in porphyry Cu systems. (b) Graph of SiO_2 against $Al_2O_3/(K_2O + Na_2O + CaO)$ for porphyry Cu and non-mineralized intrusive rocks, from Feiss (1978)

The circulation is convective and driven by heat from the cooling intrusive. The central potassic alteration, however, is produced by primary magmatic fluids, and although the ore shells may tend to occur where these two different fluid circulation systems are in contact, the porphyry stock itself is the primary source of most of the S, Cu, and other metals; with continued cooling the meteoric hydrothermal system tends to contract on to the central potassic (K-feldspar–biotite) alteration zone, superimposing on it the argillic and propyllitic (sericite–pyrite) zones.

The association of porphyry deposits with calc-alkaline intrusives probably depends on the partitioning of Cu and similar metals between silicate magmas and crystallizing minerals (Feiss, 1978). The Cu^{2+} ion in silicate melts is probably preferentially partitioned into sites with octahedral co-ordination rather than tetrahedral sites. Burns and Fyfe (1964) showed that the proportion of tetrahedral sites in silicate melts increases with alkali and silica and decreases with alumina contents. Thus, in aluminous calc-alkaline magmas with a high proportion of octahedral sites, Cu tends to partition into the residual silicate liquid, and becomes available to enter a late magmatic hydrothermal phase. In contrast, in silicate liquids with alkali/alumina ratios higher than those found in calc-alkaline rocks Cu tends to partition into crystallizing minerals such as biotite rather than into residual liquids. Such concentration into crystallizing minerals may have the effect of 'fixing' the Cu, making it unavailable for sulphide mineralization in the late cooling stages of a magma body. Figure 3(b) shows that in accord with these arguments mineralized and unmineralized intrusives are discriminated quite efficiently using a plot of SiO_2 against $Al_2O_3/(K_2O + Na_2O + CaO)$.

Initial $^{87}Sr/^{86}Sr$ ratios from intrusive rocks associated with porphyry Cu deposits in oceanic island arcs are low, in the range 0.703–0.704 (Kesler et al., 1975) and are comparable to those of unmineralized intrusive and volcanic rocks in island arcs. This indicates that the magmas are essentially mantle derived, that is either from the downgoing subducted slab of oceanic lithosphere or more probably from the overlying mantle wedge (Mysen, this volume, Section VII). In continental settings (e.g. North America) the intrusions tend to have more potassic compositions and initial $^{87}Sr/^{86}Sr$ ratios which range from c. 0.704 up to c. 0.709, probably reflecting involvement, perhaps rather minor contamination, with continental crust. The geochemical characteristics of the intrusive rocks and style of porphyry Cu mineralization in continental settings are so similar to those in oceanic island arcs that the magmas in both are probably largely derived from beneath the continental crust. Gustafson (1979) has provided an informative summary of much recent data for porphyry copper deposits.

Kuroko-type stratiform base metal sulphide deposits

A number of pyritic sulphide ore bodies conformable with the stratification of their predominantly volcanic host rocks, and often referred to as massive pyritic sulphides, were first recognized as syngenetic and volcanogenic in origin by Stanton (1955, 1960). Kuroko-type deposits, named from the mines on Honshu Island in Japan, are a major type of stratiform pyritic sulphide; they are polymetallic ores associated with acidic calc-alkaline rocks of volcanic arcs.

Kuroko-type deposits occur as relatively small lens-like bodies containing up to a few million tonnes of ore lying parallel to the bedding of the associated adjacent dacitic to rhyolitic submarine lavas and volcaniclastic rocks. In most Phanerozoic deposits the metals, in order of abundance, are Fe, Zn, Pb, Cu, with minor and variable amounts of Ag and Au. Although the total production of metals from ore bodies of this type is much less than that from porphyry Cu deposits described above, the pyritic ores are of higher grade (commonly at least 7 per cent combined Pb–Zn) and are usually worked as underground mines. Cainozoic deposits, many of which remain more or less in the broad tectonic setting in which they formed, occur within the volcanic sequences of some island arcs, although rather similar Mesozoic deposits are known from submarine volcanic rocks of Andean or cordilleran-type continental margins, for example the Campo Morado Zn–Pb–Cu–Ag–Au deposits of Mexico (Lorinczi and Miranda, 1978), and the manto-type deposits of Chile (Ruiz et al., 1971).

Older pyritic massive sulphides, mostly now within continents as a result of post-mineralization continental collision, are known from Lower Palaeozoic (Bathurst, New South Wales; Buchans Mine, Newfoundland; Bawdwin Mine, Burma), Upper Palaeozoic (Iberian Pyrite Belt), and Mesozoic (Western Mine, Vancouver Island) rocks; a number of similar deposits, commonly with relatively low Pb content (Sangster, 1976; Lambert, 1977) are also known from Archaean greenstone belts (e.g. Noranda and Kirkland Lake, Canadian Shield) and from Proterozoic successions (Flin Flon, Canada).

The late Miocene Kuroko ores of Japan, within the Green Tuff belt on Honshu and

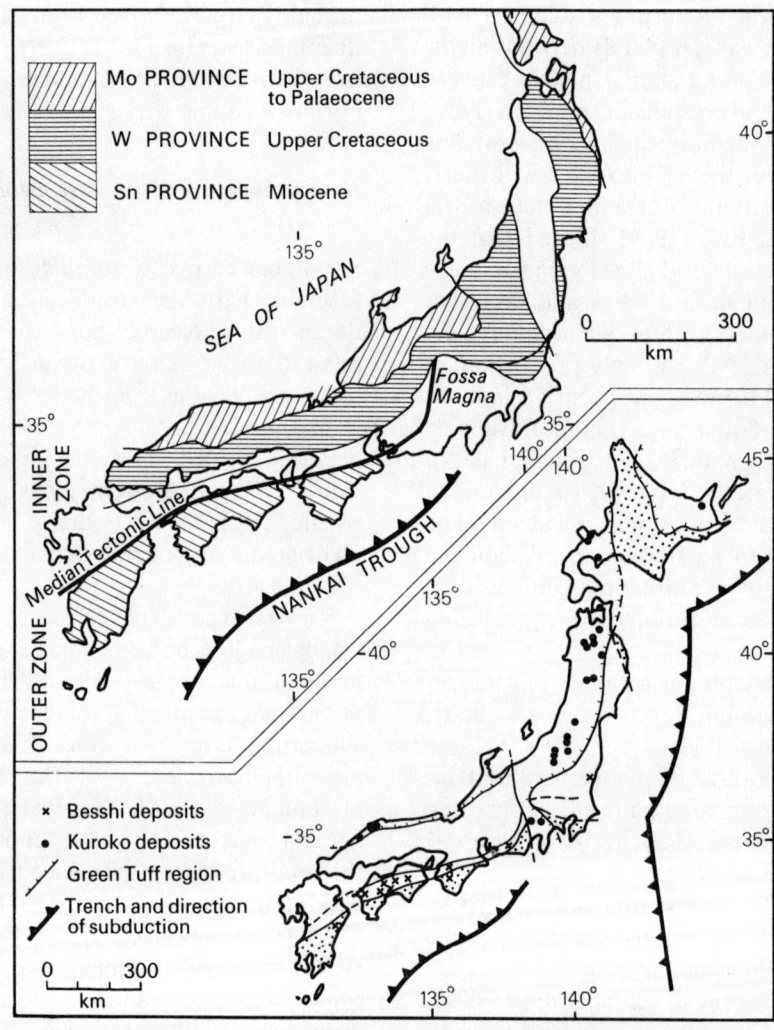

Fig. 4 Distribution of magmatic arc Kuroko deposits in Japan (after Tatsumi et al., 1972), and magmatic arc (W and Mo) and outer arc (Sn) provinces in southwestern Honshu (modified from Ishihara, 1973)

Hokkaido (Fig. 4), are widely quoted as a 'modern' analogue of older stratiform massive sulphides in silicic volcanic rocks (e.g. Sawkins, 1972). The Japanese Kuroko deposits appear to occupy a very narrow stratigraphic interval of not more than 200 ka (Ueno, 1975) within a succession of rhyolitic lava flows, domes, breccias, and tuffs forming the upper part of a andesitic to felsic igneous cycle lasting c. 10 Ma (Urabe and Sato, 1978). More than 50 deposits are known within a belt more than 500 km long and over 60 km wide parallel to and within the Miocene volcanic arc and corresponding to the zone of Quaternary tholeiitic volcanoes. The best described mines are those at Kuroko (Horikoshi, 1969), where the massive stratiform sulphides include a yellow pyritic Cu ore at the base overlain by black polymetallic ores with Pb, Zn, and barite; siliceous stockwork and disseminated ore bodies, and locally gypsum deposits, commonly underlie the massive sulphides.

Most authors, following Horikoshi (1969), relate the formation of the ore bodies to submarine hot spring activity following steam explosions on the flanks of rhyolite domes (Fig. 5). The ore fluid which deposited sulphides on the sea-floor is considered to have been a mixture of magmatic hydrothermal solutions with large dominant volumes of seawater (Urabe and Sato, 1978).

The role of faults bounding the down-faulted green tuff belt in controlling mineralization has also been emphasized (Scheibner and Markham, 1976), although deposits remote from the basin margins are known. Later Cu–Pb–Zn and Ag–Au veins in the Green Tuff belt are considered to have been deposited from fluids of predominantly meteoric origin, implying that the Japanese magmatic arc emerged from the sea by the late Miocene (Hattori and Sakai, 1979).

In Fiji a massive Zn–Cu–Pb sulphide deposit in lower Pliocene dacite and rhyolite lavas and volcaniclastic rocks on Vanua Levu is of particular interest because of the preservation of various lithological ore types, including breccia

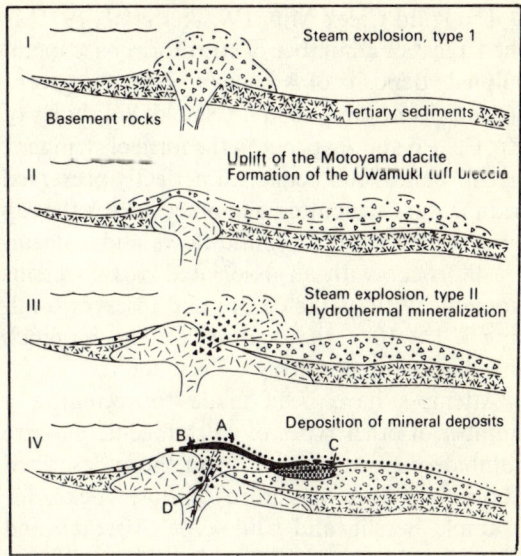

Fig. 5 Schematic cross-sections showing formation of Kuroko deposit at Kosaka volcano, Japan, after Horikoshi (1969). A, haematite, quartz, and gypsum; B, Kuroko ore; C, gypsum ore; D, stockwork and fissure-filling vein deposits

and stockwork mineralization, suggesting to Colley and Rice (1975) a possible genetic relationship to porphyry Cu deposits. However, it seems equally possible that porphyry Cu deposits and Kuroko deposits are spatially mutually exclusive; the former occurring in mountain ranges and the latter in subsiding belts. The absence of porphyry deposits in Japan would then reflect the unusually large scale subsidence accompanying the 'Green Tuff' magmatic arc volcanism.

The Buchans mine in Newfoundland lies within a thick and extensive zone of Ordovician to early Silurian volcanic and volcaniclastic rocks interpreted as a Lower Palaeozoic island arc (Thurlow et al., 1975). The mineralization, mostly massive and consisting of sphalerite, galena, and chalcopyrite, with minor pyrite and tetrahedrite, occurs as lens-like bodies up to 1 km in length at two horizons, each within a sequence of volcaniclastic sediments, siltstones, and dacitic flows and breccias. The mine has produced 16×10^6 t of ore containing Zn, Cu, Pb, and minor Ag and Au.

The Kidd Creek Mine (Walker et al., 1975) is the largest of a number of stratiform base metal sulphide deposits of Archaean age in the greenstone belts of the Canadian Shield. Sulphides of Zn, Cu, Pb and Ag occur in the form of stringers, pyritic bands, and beds with perfectly preserved sedimentary structures, and breccia, near the top of a succession of rhyolitic flows and volcaniclastic rocks with an associated carbonaceous horizon. With extracted ore and reserves totalling 120×10^6 t, the deposit contains as much metal as many porphyry Cu ore bodies.

Attempts have been made to recognize a number of other types of volcanogenic massive sulphide deposit, distinct from those described above in terms of nature of host rocks, for example Besshi- and Chile-type (Mitchell and Garson, 1976), and proportion of ore metals (Gilmore, 1971).

The source of the metal in the Kuroko deposits of Japan has been discussed in some detail. Horikoshi (1976) considered the metals to be genetically related to the subduction-controlled generation of tholeiitic magma, the initiation of which coincided with the late Miocene mineralization. Sato (1976) suggested that the metals were concentrated in the presence of water in a differentiating magma body derived either from subducted ocean floor or from the overlying mantle, and that the volume of magma and degree of differentiation determined whether mineralization occurred.

Separation of ore fluids during the final stages of fractionation of magma, or leaching of older igneous rocks by circulating hydrothermal brines, were the genetic processes favoured by Lambert and Sato (1974), and similar origins for the Archaean Cu, Zn, Ag and Au deposits of the Noranda area were suggested by Lambert (1977) on the basis of Pb isotope data. Although rarely discussed in the literature, the very narrow stratigraphic interval within which the Kuroko ores of Japan were apparently emplaced does suggest a regional tectonic control on mineralization. One possible regional control may have been subsidence and caldera collapse related to the eruption of the rhyolites and pyroclastic rocks comprising the Green Tuff belt.

Other deposits possibly related to magmatic arcs

Sn and associated metals

Most of the western world's annual production of Sn, approaching 200 000 t, is from placer deposits derived from primary mineralization associated with belts of S-type granitic rocks emplaced at convergent plate boundaries. Primary Sn deposits formed at convergent plate boundaries can occur in magmatic arcs, in back-arc magmatic belts, and in collision belts. The Sn is mostly associated with plutons of granite and quartz monzonite, commonly peraluminous and including muscovite which is probably secondary as an abundant mineral phase.

Metals commonly present with the Sn, sometimes in economic quantities, are W, Mo, Bi, Ta, and Nb. Sn deposits related to granites and pegmatites associated with peralkaline rocks typical of intracontinental rifts, for example in Nigeria and Niger, are not described in this chapter.

Sn deposits are less regular in grade and occurrence than, for example, massive stratiform sulphide or porphyry deposits. A number of types of mineralization are known (e.g. Hosking, 1974) including those in pegmatites, skarns, pipes, lodes, and veins. Most primary deposits, other than those in pegmatites, are associated with the apical parts or margins of granitic stocks representing highly evolved crustal melts. In most deposits the granites show appreciable hydrothermal alteration with albitization and introduction of Li, F, and Si in greisen, tourmalinization, and abundant quartz veins.

Sn mineralization in magmatic arcs is restricted to cordilleran-type predominantly granodioritic magmatic arcs in which S-type granites are present locally. Examples in late Cainozoic magmatic arcs, a tectonic setting which remains broadly similar to that at the time of mineralization, are the mineralization in the Inner Zone of south-western Japan, the Sn occurrences in the Miocene granites of the Aleutian arc in Alaska (Reed and Lanphere, 1973), and cassiterite in the Oligocene rhyolitic lavas of the Sierra Madre Occidental, Mexico (Swanson et al., 1978).

In the Inner Zone or Japan Sea side of south-western Japan (Ishihara, 1973) a belt of late Cretaceous to Palaeocene granitic rocks, largely with S-type characteristics, extends through much of Honshu Island roughly parallel to the coast (see Fig. 4); numerous occurrences of W and minor Sn and Cu mineralization are known in this zone, mostly forming vein-type deposits with scheelite and some cassiterite. Ishihara (1977) has shown that the Sn-bearing plutons commonly contain minor ilmenite, as opposed to magnetite and ilmenite in the Sn-free intrusives. The common association of cassiterite and tungsten deposits with S-type granites has been explained (Ishihara, 1978) by the observation that in the I-type granites Sn in the tetravalent state substitutes in sphene, magnetite, and ilmenite, resulting in low concentrations in the residual liquid. Although now an island arc, south-western Japan was attached to the Asian continent during the mineralization forming a cordilleran-type arc underlain by continental crust, a tectonic setting very broadly analogous to that of Alaska and Mexico.

An example of Sn mineralization in an older magmatic arc is provided by the Sn ores of eastern Malaya (Fig. 7), within a late Palaeozoic arc system developed above eastward-subducting ocean floor before Triassic collision with western Malaya to the west.

Sb-W-Hg strata-bound deposits

Many late Mesozoic and Cainozoic W and Sb deposits are clearly epigenetic and associated with granitic plutons, for example those in south-eastern China (Hutchison and Taylor, 1978). However, an important class of ore bodies, which could possibly include some of those previously thought to be epigenetic, comprises the scheelite-stibnite–cinnabar deposits first recognized in the eastern Alps as strata-bound and associated with volcanic rocks (Maucher, 1965).

In the eastern Alps the ore bodies are mostly of three types: W (scheelite) with minor Mo, Cu, Bi; Sb (stibnite) with minor As, W, Cu; and Hg (cinnabar). The W (Mo-Cu-Bi) type of deposit includes the recently discovered large body at Falbertal. Similar types of mineralization are found in Sardinia and Turkey (Maucher, 1976).

Although the three types of ore occur at different localities, they are all associated with broadly similar Lower Palaeozoic submarine volcanic and commonly carbonaceous sedimentary rocks, some of which are metamorphosed. Holl and Maucher (1976) and Holl (1977) consider that the mineralization was initially syngenetic and genetically related to eruption of the adjacent lavas, but that much of the ore was remobilized during subsequent deformation and metamorphism to form veins discordant to the bedding. The volcanic rocks are mostly basic, but rocks of intermediate composition and quartz porphyries are also present. Holl (1977) has interpreted the distribution of the W, Sb, and Hg occurrences in the Alps in terms of their position relative to an inferred underlying Lower Palaeozoic north-dipping Benioff zone, although there is little direct evidence for this subduction other than the volcanics associated with the mineralization itself.

Cainozoic examples of strata-bound scheelite-stibnite mineralization within andesitic and dacitic volcanic rocks have been described from Turkey (Maucher, 1976), and Hg mineralization is not uncommon in young volcanic arcs such as the Philippines.

Magnetite-haematite-apatite deposits

A number of Phanerozoic, Proterozoic, and possibly Archaean Fe deposits associated with silicic volcanic rocks and consisting largely of magnetite with minor haematite, fluorapatite, and actinolite are known, and controversy continues as to whether many of these are extrusive, intrusive, or exhalative-sedimentary in origin (e.g. Parak, 1975). It was suggested by Sillitoe (1972) that some Fe ores of this type form above Benioff zones but closer to the trench than porphyry Cu deposits.

In northern Chile, the Pliocene–Pleistocene El Laco deposits (Park, 1961), in which the tectonic setting and mode of occurrence of the ore in body are well preserved, provide a possible analogue for some of the older ore bodies. The El Laco deposits occur in flows c. 20 m thick overlying ignimbrites and andesite lavas and are more or less co-eval with nearby rhyolites; they are associated with craters around the margins of a caldera. The ore contains 50 per cent Fe and reserves are estimated at 10^9 t. Geochemical evidence suggests that the flows were derived from the underlying Palaeozoic ferruginous sedimentary rocks (Frutos and Oyarzun, 1975), although the mobilization and emplacement of the ore bodies was directly related to intrusion and eruption of the calc-alkaline lavas in the easternmost part of the Cainozoic Andean magmatic arc. To the south, in a belt parallel to the coast, numerous deposits of Mesozoic and Cainozoic age with broadly similar mineralogy are associated with mostly andesitic rocks; most are related to intrusions and are hydrothermal in origin (Bockstrom, 1977).

In the Sierra Madre Occidental of Mexico a major source of Fe ore is provided by haematite, mortite, and magnetite eruptive rocks within a thick succession of rhyolitic rocks in a caldera complex (Swanson et al., 1978).

Au

Au deposits and occurrences occur in a number of settings associated with Cainozoic and Mesozoic igneous rocks of mostly intermediate calc-alkaline composition. They include auriferous quartz veins around dioritic plutons in a number of magmatic arcs. In eruptive rocks, Au occurs as auriferous quartz veins in meta-andesite, as in the Hauraki Peninsula, New Zealand, and Au tellurides as in the intermittently worked Emperor Mine in Fiji. The possible existence of economic porphyry Au deposits has also recently been discussed (Sillitoe, 1979).

Back-Arc Magmatic Belt Mineralization

Sn–Ag and Sn–W deposits

Of the late Mesozoic to Cainozoic Sn provinces situated on overriding plate margins, only two are major producers of the metal: Bolivia, and the Western Belt of south-eastern Asia in Burma and western Thailand.

The best known and most productive of these is in Bolivia (Fig. 6), where production is entirely from primary deposits, mostly associated with volcanic or, less commonly, subvolcanic silicic rocks. However, it has been suggested that at least one of these deposits includes large volumes of low grade, disseminated Sn in an igneous host rock, forming a deposit analogous to that of porphyry Cu (Sillitoe et al., 1975), and in some cases vein systems occur in sedimentary se-

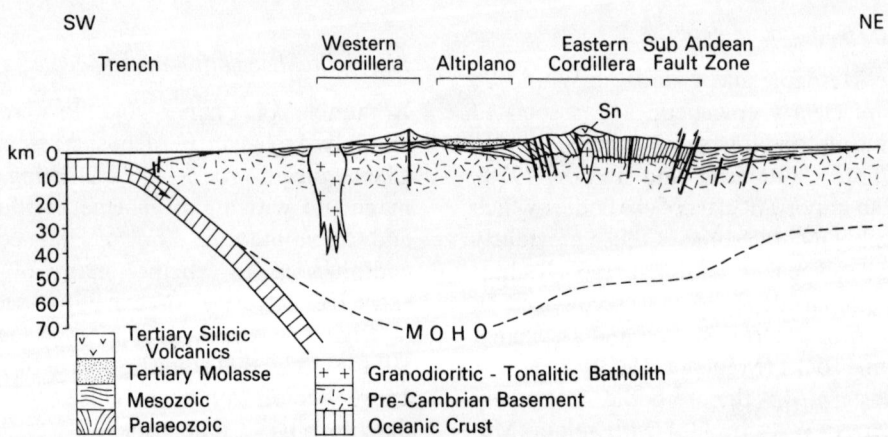

Fig. 6 Schematic cross-section through Andes near northern end of Bolivian Sn belt, after Cobbing and Pitcher (1972). Projected position of Tertiary Sn deposits of southern Bolivia also shown

quences with little or no direct evidence of igneous activity (Grant et al., 1979). Typically the veins are complex, with Sn, Sn-Ag, and base metal assemblages. The volcanic rocks forming the Bolivian Sn belt are usually quartz porphyries ranging from dacite to rhyolite in composition, occurring as lavas, breccias, and pyroclastic rocks resting on Palaeozoic sediments and exhibiting a central-type alteration zone in which the vein systems occur. Locally, central stock-like subvolcanic bodies of mineralized porphyry and breccia intruding the volcanic rocks are exposed. The alteration assemblages tend to fall in the sequence quartz-tourmaline, tourmaline-sericite, and sericite-clay minerals from the centre outwards. The Sn porphyry type of deposit appears to be the earliest style of mineralization, reflecting pervasive high temperature alteration (homogenization temperatures above 450 °C have been recorded by Grant et al., 1979), and subsequent fracturing and establishment of a hydrothermal circulation system, probably involving increasing access to fluids of meteoric origin, led to the formation of the vein systems which cut the earlier patterns of alteration. With the progressive exhaustion of the high grade vein deposits and the rapidly rising price of Sn, the Bolivian porphyry-type deposits may become more and more significant.

There are fewer published descriptions of the Burma–western Thailand belt (Fig. 7). The Sn and W are only associated with plutonic rocks, mostly biotite or two-mica granites and adamellites; the host rocks everywhere comprise argillaceous sedimentary rocks of the Carboniferous Phuket Group (Garson et al., 1975). The plutons range from Cretaceous to possibly Eocene in age and include a late Palaeocene mineralized granite in Burma (Brook and Snelling, 1976). In general the proportion of W in the form of wolframite and less commonly scheelite increases northwards through the belt.

Genesis of the Bolivian and Burma Sn belts has been related to the subduction of ocean floor, with metal either derived from the Benioff zone in the case of Bolivia (Sillitoe, 1972), or extracted from continental crust by F expelled from subducting ocean floor along a migrating Benioff zone in the case of Burma (Mitchell and Garson, 1972). However, the two belts have certain major features in common which may be of genetic significance. Unlike most magmatic arcs, both are concave towards the adjacent ocean, the Bolivian belt markedly so, and, allowing for late Cainozoic dextral movement on the Sagaing Fault in Burma (Mitchell, 1977), both belts formed on the landward side of magmatic arcs. The Bolivian belt is bordered on its landward side by a fold-thrust belt with overthrusts directed towards the continent, and recent mapping in Burma indicates that the northern part of the Burma–western Thailand belt is similarly bordered by a series of post-mid-Cretaceous, *en echelon*, continent-directed thrusts. Although the Bolivian fold-thrust belt extends north and south beyond the Sn belt, and the southern extension of the thrust belt in

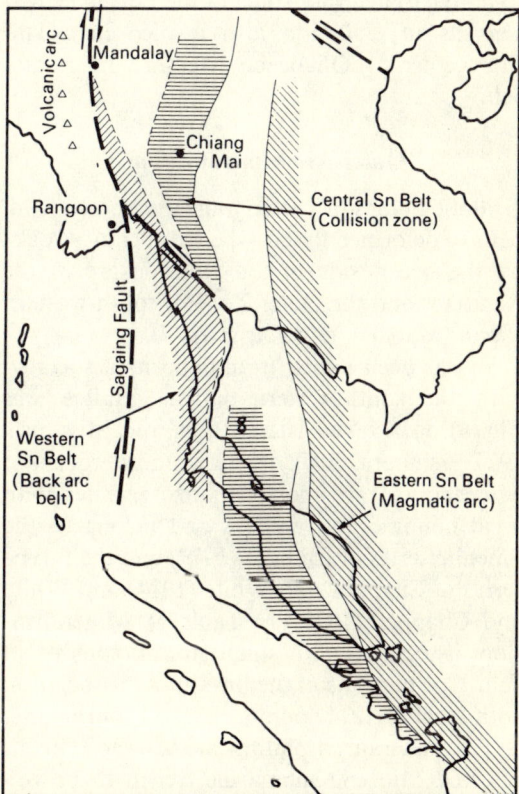

Fig. 7 Eastern, central and western Sn belts of southeastern Asia, after Mitchell (1977)

Burma is obscured by the Gulf of Siam, a genetic relationship is probable between the curvature of the Sn belts, the 'back-arc' thrusts, and the granites (Mitchell, 1979). Possibly major movement along back-arc thrusts, antithetic to subduction, is characteristic of arcs concave to the ocean, and results in generation of intracontinental anatectic granites, in some cases with initial $^{87}Sr/^{86}Sr$ ratios of c. 0.73 (Beckinsale et al., 1979).

Other deposits

While Sn and W are characteristic of the two back-arc belts considered above, there is some evidence that in western North America a number of other types of deposit formed in this type of setting. The porphyry Cu deposits associated with the Boulder batholith were emplaced within the contemporaneous back-arc or 'foreland' fold-thrust belt of Dickinson (1976), and it could be argued that some of the porphyry Mo deposits in Colorado also formed in a late Cretaceous to Oligocene back-arc magmatic belt.

Outer-Arc Mineralization

Granitic bodies of Tertiary age are known from belts of deformed flysch-type sedimentary rocks on the ocean side of the volcanic arc in the Aleutians and the Outer Zone of south-western Japan (Marshak and Karig, 1977).

Sn has been mined from deposits associated with the south-western Japan granites (*see* Fig. 4) on Kyushu Island (Oba and Miyahisa, 1977; Ishihara, 1978). The granitic rocks of Miocene age are mostly monzo-granites with local minor eruptive rocks and belong to the ilmenite series of Ishihara (1977) and the S-type series of Chappell and White (1974) and White and Chappell (1977, *see* Table 1). Mineralization, associated with small stocks, consists of vein-type deposits of the Sn–Cu–As association with local Pb, Zn, and fluorite, and tourmaline.

The mineralized plutons are too close to the inferred Miocene submarine trench to be volcanic arc deposits, and Ishihara (1978) suggested that the magma and metals were derived from underlying continental crust. However, interpretation of the surrounding pre-pluton greywackes and mudstones as trench sediments tectonically accreted in a subduction zone (Marshak and Karig, 1977) would suggest that only late Mesozoic to early Cainozoic sedimentary rocks were available at depth, and that partial melting of these, rather than of older continental crust, must have supplied the magma and Sn. As trench sediments are derived largely from the magmatic arcs, which in Japan contains older Sn deposits described above, it is possible that the trench sediments and consequently the outer arc sedimentary rocks were anomalously rich in Sn.

Collision-belt Mineralization

Sn deposits

Most Sn provinces are associated with belts of S-type granitic plutons (Chappell and White, 1974; White and Chappell, 1977) which differ from the mostly I-type granodioritic belts of cordilleran or island arc subduction-related magmatic arcs in composition, in absence of associated volcanic rocks, and in being late to post-tectonic with respect to deformation and metamorphism of the country rocks (Table 1). They differ from outer arc granites in composition, and in the nature of the pre-pluton host rocks, which in the case of the S-type granites include shallow marine sedimentary rocks. It has been argued that some of these mineralized granitic belts are analogous to the Oligocene granites of the Himalayas, and were emplaced within continental crust of the subducting plate either during or immediately following its collision with a continent or island arc on the overriding plate (Mitchell, 1978).

Sn occurrences in the Himalayas

The Himalayan granites described from Nepal by Hagen (1969), Gansser (1964), Le Fort (1975), and others form major plutons and inclined sheets intruding metamorphic and Palaeozoic sedimentary rocks in the Higher Himalayas (Fig. 8). Most abundant are tourmaline-bearing

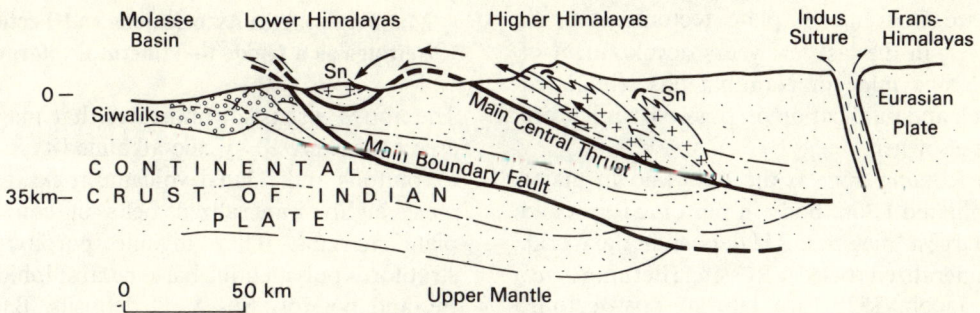

Fig. 8 Tectonic settings of Himalayan collision-related Sn-bearing granites, modified from Andrieux et al. (1976)

two-mica granites and leucogranites, with common pegmatites and aplites; the Manaslu Granite has yielded a late Oligocene Rb/Sr age, with a high initial $^{87}Sr/^{86}Sr$ ratio of 0.74 (Hamet and Allègre, 1976). Similar granites are present in the Lower Himalayas to the south within rocks interpreted as *klippe* of the Higher Himalayan rocks (Hagen, 1969). The age and tectonic setting of the Higher Himalayan granites indicate that they were emplaced in the Indian plate during movement on the Main Central Thrust, following the India–Asia collision. The high initial ratios suggest that the magma was anatectic and resulted from partial melting of continental crust at depth (Andrieux et al., 1976).

Sn mineralization has been described in most detail from the Lower Himalayas (Talalov, 1976), where the Dandeldhura granite is locally extensively greisenized and cassiterite, xenotime, scheelite, fluorite, topaz, and tantalo-columbite are present at the granite margins. The generation of the Lower Himalayan granites, like those of the Higher Himalayas, can be interpreted most easily in terms of the India–Asia collision (Andrieux et al., 1976); however, Rb–Sr whole-rock isochron ages (R. D. Beckinsale, unpublished) indicate that some of them are much older than the collision event.

Sn mineralization in older collision belts

Sn-bearing granites in older orogens, interpreted on the basis of their tectonic setting as analogues to those of the Himalayas, include the Hercynian granites of south-western England and the Erzgebirge (Dewey and Burke, 1973; Mitchell, 1974), the late Triassic granites extending through central Thailand and the Main Range of Malaysia into Indonesia (Mitchell and Garson, 1976; Mitchell, 1977, Beckinsale et al., 1979), and possibly the mineralized granitic plutons of the Okcheung zone in Korea (R. H. Sillitoe, personal communication).

The late Triassic granites of south-eastern Asia have high initial $^{87}Sr/^{86}Sr$ ratios in Thailand (Von Braun et al., 1976, Beckinsale et al., 1979) and Malaysia (Bignell and Snelling, 1977), suggestive of crustal origin, and those in the Bangka and Billiton Islands in Indonesia have yielded both a high initial $^{87}Sr/^{86}Sr$ ratio and Pb isotope data indicating that Pb in galena associated with Sn mineralization originated in continental crust (Jones et al., 1977).

A collision-related origin for these granites is by no means undisputed. For example, it has been suggested that the south-western England granites were emplaced on the overriding plate during thrusting away from the subduction direction (Bromley, 1976), that is in essentially a back-arc magmatic belt setting as described above. However, pending new information, the south-western England granites can still be interpreted most simply as syn-collision plutons emplaced at a high structural level into former outer arc rocks.

Magmatic U deposits

The formation of deposits of U associated with plutons has not, until very recently, been discussed in terms of plate tectonic settings. However, in the last few years development of the Rössing mine in Namibia has stimulated research and publication on possible similar ore bodies elsewhere.

The Rössing mine is the only known major disseminated U ore body in plutonic rocks and is the largest 'magmatic' U deposit in the world. The mineralized rocks at Rössing (Berning et al., 1976; Jacob, 1978) are late to post-tectonic pegmatitic granites of anatectic origin within high grade, regionally metamorphosed, and locally migmatitic rocks of the latest Proterozoic Damara Group. The mineralized pegmatitic rocks are largely confined to the carbonate-bearing Rössing Formation, but granites and leucogranites of similar or slightly greater age occupy synclines in the isoclinally folded overlying sequence. The U, mostly in the form of uraninite, is considered to have been concentrated during differentiation of anatectic melts derived from partial melting of underlying Damara Supergroup rocks.

Other deposits of magmatic U have been compared with Rössing, for example the deposits in Devonian two-mica granites in the western Massif Central of France (Leroy, 1978) and the lower Palaeozoic Lithonia Gneiss in Georgia and uraninite-bearing pegmatites in the northern North Carolina Blue Ridge area (Rogers et al., 1978). In general these deposits are associated with pegmatites and dykes, with high initial $^{87}Sr/^{86}Sr$ ratios, confined to zones of high grade, regional metamorphism with anatectic granites and leucogranites.

The composition of the granites, their high initial $^{87}Sr/^{86}Sr$ ratios, and associated regional metamorphism all suggest similarities with the collision-related Himalayan granites; Leroy (1978) compared the Massif Central granites to the Oligocene Manaslu Granite of northern Nepal, and Moreau (1976) suggested that they were related to a Devonian continental collision. A collision-related origin for the Rossing Granite is not incompatible with the regional geology of the 'Pan-African' orogeny of the Damara Belt.

Magmatic-metal Associations and Tectonic Settings as a Guide to Mineral Exploration

The above discussion indicates that magmatic arcs, characterized by calc-alkaline I-type granitic plutons and related volcanic rocks, are the most highly mineralized belts of converging plate margins. They include porphyry Cu, stratiform polymetallic base metal sulphide, Au, Fe, and possibly Sn–W–U deposits. Back-arc magmatic belts contain S-type granites, with Sn, W, and U deposits and possibly I-type granites with porphyry Cu and Mo. Outer arc granites, predominantly S-type, are rarely mineralized but can contain Sn and associated metals. Collision-related S-type granites, emplaced in the foreland of the subducting continental plate, contain important deposits of Sn, W, and fluorite, and in some cases U.

It is sometimes argued that the plate tectonics concept has been of great value in mineral exploration by indicating the presence of belts favourable for particular types of metallic deposit. However, the observed association of certain types of metallic mineral deposit with magmatic rocks of particular ranges in composition (e.g. porphyry Cu deposits with diorite–granodiorite) was used with considerable success as a guide to regional exploration long before a link with subduction of ocean floor was suggested. That this is an empirical relationship, and not dependent on hypotheses of origin or involving tectonic setting, is indicated by the popularity of the rock–mineral association concept with Russian authors (e.g. Talalov, 1976) who in general have not favoured mineralization models related to processes of ocean-floor spreading, subduction, and continental drift.

It is probable that ideas on the relationship of plate convergence to magma emplacement and metal concentration may be of value to exploration in the future. But at present is appears that the presence of calc-alkaline I-type or S-type granitic rocks can be predicted from analogy with the tectonic settings of better known orogenic belts elsewhere, irrespective of genetic hypotheses involving movement of crustal plates.

Once the magmatic belt has been identified, it is the nature of the igneous rock and presence of mineral occurrences themselves which indicate respectively the metals likely to be present and whether the belt is in fact mineralized.

Acknowledgement

R. D. Beckinsale thanks the Director of the Institute of Geological Sciences for permission to publish.

REFERENCES

Andrieux, J., Brunel, M., and Hamet, J. (1976). Metamorphism, granitisation and relations with the Main Central Thrust in Central Nepal: Rb^{87}/Sr^{87} age determination and discussion. In *Colloque sur l'ecologie et la geologie de l'Himalaya* no. 268, Centre National de la Recherche Scientifique, Paris, pp. 31–40.

Armstrong, R. L., Harakal, J. E., and Hollister, V. F. (1976). Age determination of late Cenozoic porphyry copper deposits of the North American cordillera. *Trans. Instn Min. Metall. B, Appl. Earth Sci.*, **85**, B239–B244.

Beckinsale, R. D. (1979). Granite magmatism in the tin belt of southeast Asia. In *Origin of Granite Batholiths: Geochemical Evidence* (M. P. Atherton and J. Tarney, eds), Shiva Publishing, Orpington, pp. 34–44.

Beckinsale, R. D., Suensilpong, S., Nakapadungrat, S., and Walsh, J. N. (1979). Geochronology and geochemistry of granite magmatism in Thailand in relation to a plate tectonic model. *J. geol. Soc. Lond.* **136**, 529–540.

Berning, J., Cooke, R., Heimstra, S. A., and Hoffman, U. (1976). The Rössing uranium deposit, South West Africa. *Econ. Geol.* **71**, 351–368.

Bignell, J. D. and Snelling, N. J. (1977). Geochronology of Malayan granites. *Overseas Geol. Mineral. Resources* no. 47.

Bockstrom, A. A. (1977). The magnetite deposits of El Romeral, Chile. *Econ. Geol.* **72**, 1101–1130.

Bromley, A. V. (1976). Granites in mobile belts—the tectonic setting of the Cornubian batholith. *Cambourne School Mines* **76**, 40–47.

Brook, M. and Snelling, N. J. (1976). K/Ar and Rb/Sr age determinations on rocks and minerals from Burma. *Inst. ged. Sci. Rep.* no. 76/12.

Burns, R. G. and Fyfe, W. S. (1964). Site preference energy and selective uptake of transition-metal ions during magmatic crystallization. *Science, N.Y.* **144**, 1001–1003.

Chappell, B. W. and White, A. J. R. (1974). Two contrasting granite types. *Pacific Geol.* **8**, 173–174.

Cobbing, E. J. and Pitcher, W. S. (1972). The coastal batholith of Central Peru. *J. geol. Soc. Lond.* **128**, 421–460.

Colley, H. and Rice, C. M. (1975). A Kuroko-type ore deposit in Fiji. *Econ. Geol.* **70**, 1373–1386.

DeLong, S. E., Schwarz, W. M., and Anderson, R. N. (1979). Thermal effects of ridge subduction. *Earth planet. Sci. Letters*, **44**, 239–246.

Dewey, J. F. and Burke, K. C. A. (1973). Tibetan, Variscan and Precambrian basement reactivation: products of continental collision. *J. Geol.* **81**, 683–692.

Dickinson, W. R. (1976). Sedimentary basins developed during evolution of Mesozoic–Cenozoic arc–trench system in Western North America. *Can. J. Earth Sci.* **13**, 1268–1287.

Feiss, P. G. (1978). Magmatic sources of copper in porphyry copper deposits. *Econ. Geol.* **73**, 397–404.

Frutos, J. and Oyarzun, J. (1975). Tectonic and geochemical evidence concerning the genesis of El Laco magnetite lava flow deposits, Chile. *Econ. Geol.* **70**, 988–990.

Gansser, A. (1964). *Geology of the Himalayas*, Wiley-Interscience, London.

Garson, M. S., Young, B., Mitchell, A. H. G., and Tait, B. A. R. (1975). The geology of the tin belt in Peninsular Thailand around Phuket, Phangnga and Takua Pa. *Inst. geol. Sci., Overseas Mem.* no. 1, HMSO, London.

Gilmore, P. A. (1971). Strata-bound massive pyritic sulphide deposits—a review. *Econ. Geol.* **66**, 1239–1244.

Grant, J. N., Halls, C., Salinas, W. A., and Snelling, N. J. (1979). K-Ar ages of igneous rocks and mineralization in part of the Bolivian tin belt. *Econ. Geol.* **74**, 838–51.

Guilbert, J. M. and Lowell, I. D. (1974). Variations in zoning patterns in porphyry ore deposits. *Can. Min. Metall. Bull.* **67**, 99–109.

Guild, P. W. (1971). Massive sulphides versus porphyry deposits in their global tectonic settings. Offprint no. G 13, MMIJ–AIME Joint Meeting, May 1972, Tokyo, pp. 1–12.

Gustafson, L. B. (1979). Porphyry copper deposits and calc-alkaline volcanism. In *The Earth: Its Origin, Structure and Evolution* (M. W. McElhinny, ed.), Academic Press, London, pp. 427–468.

Hagen, T. (1969). Uber den Geologischen Bau des Nepal Himalayas. *St Gall. naturw. Ges.* **76**, 3–48.

Hamet, J. and Allègre, C. J. (1976). Rb-Sr systematics in granite from central Nepal (Manaslu): significance of the Oligocene age and high Sr^{87}/Sr^{86} ratio in Himalayan orogeny. *Geology* **4**, 470–472.

Hattori, J. and Sakai, H. (1979). D/H ratios, origins and evolution of the ore-forming fluids for the Neogene veins and Kuroko deposits of Japan. *Econ. Geol.* **74**, 535–555.

Holl, R. (1977). Early Palaeozoic are deposits of the Sb–W–Hg formation in the Eastern Alps and their genetic interpretation. In *Time and Strata-bound Ore Deposits* (D. D. Klem and H. J. Schneider, eds), Springer-Verlag, Berlin, pp. 169–198.

Holl, R. and Maucher, A (1976). The strata-bound ore deposits in the Eastern Alps. In *Handbook of Strata-bound and Stratiform Ore Deposits*, Vol. 5 (K. H. Wolf, ed.), Elsevier, Amsterdam, pp. 1–36.

Horikoshi, Ei. (1969). Volcanic activity related to the formation of the Kuroko-type deposits in the Kosaka District, Japan. *Mineral. Deposita* **4**, 321–345.

Horikoshi, Ei. (1976). Development of Late Cenozoic petrogenetic provinces and metallogeny in northeast Japan. *Geol. Assoc. Can. spec. Pap.* no. 14, 121–146.

Hosking, K. F. G. (1974). The search for deposits from

which tin can be profitably recovered now and in the foreseeable future. In *Fourth World Conference on Tin*, Vol. 1, *World Tin Resources*, International Tin Council, London, pp. 21–38.

Hutchison, C. S. and Taylor, D. (1978). Metallogenesis in SE Asia. *J. geol. Soc. Lond.* **135**, 407–428.

Ishihara, S. (1973). Molybdenum and tungsten provinces in the Japanese islands and North American cordillera: an example of asymmetrical zoning in Pacific-type orogeny. *Bur. Mineral Resources Geol. Geophys. Bull.* no. 141, 173–189.

Ishihara, S. (1977). The magnetite and ilmenite series granitic rocks. *Mining Geol.* **27**, 293–305.

Ishihara, S. (1978). Metallogenesis in the Japanese island arc system. *J. geol. Soc. Lond.* **135**, 389–406.

Jacob, R. E. (1978). Granite genesis and associated mineralization in parts of the Central Damara Belt. In *Mineralization in Metamorphic Terraces* (W. J. Verwoerd, ed.), van Schaik, Pretoria, pp. 417–432.

Jones, M. T., Reed, B. L., Doe, B. R., and Lanphere, M. A. (1977). Age of tin mineralization and plumbo tectonics, Belitung, Indonesia. *Econ. Geol.* **72**, 745–752.

Kesler, S. E., Jones, L. M., and Walker, R. L. (1975). Intrusive rocks associated with porphyry copper mineralization in island arc areas. *Econ. Geol.* **70**, 515–526.

Le Fort, P. (1975). Himalayas: the collided range, present knowledge of the continental arc. *Am. J. Sci.* **275A**, 1–44.

Leroy, J. (1978). The Morgnac and Fanay uranium deposits of the La Crouzille District (Western Massif Central, France): geologic and fluid inclusion studies. *Econ. Geol.* **73**, 1611–1634.

Lambert, I. B. (1977). Notes on exploration guides for stratiform lead–zinc ores, geochemical and geobiological evolution in the Precambrian, and massive sulphide deposits of the Nevada area. *Mineral. Resources Lab. tech. Comm.* no. 61, CSIRO, Canberra.

Lambert, I. B. and Sato, T. (1974). The Kuroko and associated ore deposits of Japan: a review of their features and metallogenesis. *Econ. Geol.* **69**, 1215–1236.

Lorinczi, G. I. and Miranda, J. C. (1978). Geology of the massive sulphide deposits of Campo Morado, Guerrero, Mexico. *Econ. Geol.* **73**, 180–191.

Lowell, J. D. (1974). Regional characteristics of porphyry copper deposits of the Southwest. *Econ. Geol.* **69**, 601–617.

Marshak, R. S. and Karig, D. E. (1977). Triple junctions as a cause for anomalously near-trench igneous activity between the trench and volcanic arc. *Geology* **5**, 233–236.

Maucher, A. (1965). Die Antimon-Wolfram-Quesckailber Formation und ihre Beziehungern zu magmatismus and Geotektonic. *Freiberger Forschungsh.* no. C186, 173–188.

Maucher, A. (1976). The strata-bound cinnabar–stibnite–scheelite deposits. In *Handbook of Strata-bound and Stratiform Ore Deposits*, Vol. 7 (K. H. Wolf, ed.), Elsevier, Amsterdam, pp. 477–503.

Mitchell, A. H. G. (1974). Southwest England granites: magmatism and tin mineralization in a post-collision tectonic setting. *Instn Min. Metall. Trans. B, Appl. Earth Sci.* **83**, B95–B97.

Mitchell, A. H. G. (1977). Tectonic settings for emplacement of southeast Asian tin granites. *Geol. Soc. Malaysia Bull.* **9**, 123–140.

Mitchell, A. H. G. (1978). Geosynclinal and plate tectonic hypotheses: significance of late orogenic Himalayan tin granites and continental collision. In *Proceedings of the 11th Commonwealth Mining and Metallurgy Congress*, Institution of Mining and Metallurgy, Hong Kong, Paper 37, pp. 1–13.

Mitchell, A. H. G. (1979). Rift, subduction and collision-related tin belts. *Geol. Soc. Malaysia Bull.* **11**, in press.

Mitchell, A. H. G. and Garson, M. S. (1972). Relationship of porphyry copper and circum-Pacific tin deposits to palaeo-Benioff zones. *Instn. Min. Metall. Trans. B, Appl. Earth Sci.* **81**, B10–B25.

Mitchell, A. H. G. and Garson, M. S. (1976). Mineralization at plate boundaries. *Min. Sci. Engng.* **8**, 129–169.

Moreau, M. (1976). L'uranium et les granitoides—essai de interpretation. In *Geology, Mining and Extractive Processes of Uranium*, Institution of Mining and Metallurgy, London, pp. 83–102.

Oba, N. and Miyahisa, M. (1977). Relations between chemical composition of granitic rocks and metallization in the Outer Zone of southwest Japan. *Geol. Soc. Malaysia Bull.* **9**, 67–74.

Oftedahl, C. (1958). A theory of exhalative-sedimentary ores. *Geol. Foren. Stockh. Fork.* **80** (1), 1–19.

Parak, T. (1975). Kiruna iron ores are not 'intrusive-magmatic ores of the Kiruna type'. *Econ. Geol.* **70**, 1242–1258.

Park, C. F. (1961). A magnetite flow in northern Chile. *Econ. Geol.* **56**, 431–436.

Reed, B. L. and Lanphere, M. A. (1973). Alaska–Aleutian Range batholith: geochronology chemistry and relation to circum-Pacific plutonism. *Geol. Soc. Am. Bull.* **84**, 2583–2610.

Rogers, J. W., Ragland, P. C., Nishimori, R. K., Greenberg, J. K., and Hauck, S. A. (1978). Varieties of granitic uranium deposits and favourable exploration areas in the Eastern United States. *Econ. Geol.* **73**, 1539–1555.

Ruiz, F. C., Aguilar, A., Egert, E., Espinosa, W., Peebles, F., Quezada, R., and Serrano, M. (1971). Strata-bound copper sulphide deposits of Chile. *Soc. Min. Geol. Japan spec. Issue* no. 3, 252–260.

Sangster, D. F. (1976). Possible origins of lead in volcanogenic massive sulphide deposits of calc-alkaline affiliation. *Geol. Assoc. Can. spec. Pap.* no. 14, 103–104.

Sato, T. (1976). Origin of the green tuff metal province of Japan. *Geol. Assoc. Can. spec. Pap.* no. 14, 105–120.

Sawkins, J. F. (1972). Sulphide ore deposits in relation to plate tectonics. *J. Geol.* **80**, 377–397.

Scheibner, E. and Markham, N. L. (1976). Tectonic setting of some strata-bound massive sulphide deposits in New South Wales, Australia. In *Handbook of Strata-bound and Stratiform Ore Deposits*, Vol. 6 (K. H. Wolf, ed.), Elsevier, Amsterdam, pp. 55–77.

Schneiderhohn, H. (1941). *Lehrbuch de Lagerstattenkunde 1*, Fischer, Jena.

Sillitoe, R. H. (1970). South American porphyry copper deposits and the new global tectonics. *Resumenes Primer Congr. Latinoamericano Geol., Lima, Peru*, 254–256.

Sillitoe, R. H. (1972). Model for origin of porphyry copper deposits. *Econ. Geol.* **67**, 184–197.

Sillitoe, R. H. (1979). Some thoughts on gold-rich porphyry copper deposits. *Mineral. Deposita* **14**, 161–174.

Sillitoe, R. H., Halls, C., and Grant, N. J. (1975). Porphyry tin deposits in Bolivia. *Econ. Geol.* **70**, 913–927.

Snelgrove, A. K. (1971). Metallogeny and the new global tectonics. *Mineral Resources Explor. Inst. Bull.* no. 76, 130–149.

Stanton, R. L. (1955). Lower Palaeozoic mineralization near Bathurst, New South Wales. *Econ. Geol.* **50**, 681–714.

Stanton, R. L. (1960). General features of the conformable pyritic ore bodies. *Can. Inst. Min. Metall. Trans.* **63**, 22–27.

Swanson, E. R., Keitzer, R. P., and Clabaugh, S. E. (1978). Tertiary volcanism and caldera development near Durango City, Sierra Madre Occidental, Mexico. *Geol. Soc. Am. Bull.* **89**, 1000–1012.

Talalov, V. A. (1976). Main features of magmatism and metallogeny of the Nepalese Himalayas. In *Colloque sur l'ecologie et la geologie de l'Himalaya* no. 268, Centre National de la Recherche Scientifique Paris, pp. 409–430.

Tatsumi, T., Takagi, Y., and Otagaki, T. (1972). Geology of the Kuroko deposits. Preprint no. T1b1, Joint Meeting of MMIJ–AIME and the Society of Petroleum Engineers, New York.

Taylor, H. P. (1974). The application of oxygen and hydrogen isotope studies to problems of hydrothermal alteration and ore deposition. *Econ. Geol.* **69**, 843–883.

Thurlow, J. G., Swanson, E. A., and Strong, D. F. (1975). Geology and lithogeochemistry of the Buchans polymetallic sulphide deposits, Newfoundland. *Econ. Geol.* **70**, 130–144.

Ueno, H. (1975). Duration of the Kuroko of the Kosaka mineralization episode. *Nature, Lond.* **253**, 428–429.

Urabe, T. and Sato, T. (1978). Kuroko deposits of the Kosaka Mine, Northeast Honshu, Japan—products of submarine hot springs on Miocene sea floor. *Econ. Geol.* **73**, 161–179.

Von Braun, E., Besang, C., Eberlie, W., Harre, W., Krenzer, H., Lenz, H., Muller, P., and Wendt, T. (1976). Radiometric age determinations of granites in Northern Thailand. *Geol. Jb. B* **21**, 171–204.

Walker, R. R., Matulich, A., Amos, A. C., Watkins, J. J., and Mannard, G. W. (1975). The geology of the Kidd Creek Mine. *Econ. Geol.* **70**, 80–89.

White, A. J. R. and Chappell, B. W. (1977). Ultrametamorphism and granite genesis. *Tectonophysics* **43**, 7–22.

Geographical Index

'Single page numbers (e.g. 10) refer to individual entries of the listed term, grouped page numbers (e.g. 10–20) indicate sequences of individual entries of the listed term, while emphasised (bold) page numbers (e.g. **10–20**) indicate major reference sections to the listed term'.

Abitibi Lake, Ontario, Canada, 576, 578, 580, 581, 583, 585, 586
Acambay, Mexico, 408
Acatenango, Guatemala, 158, 159, 162
Acatlan, Mexico, 408
Achilleion, Volos-Atalanti Group, Aegean Sea, Europe, 318, 322
Acigol, Turkey, 335, 337, 338, 344
Adagdak Volcano, Aleutian Islands, 105
Adak Island, Aleutian Islands, 100, 103, 105
Adana, Turkey, 330
Adana Basin, Turkey, 328
Adanac, Saskatchewan, Canada, 681
Aden, Gulf of, 346, 527, 528
Admiralty Islands, Papua New Guinea, 226, 230
Aegean Arc (*see also* Hellenic arc), 3, 27, 40, 41, 50, 60, 70, 72, 73, **317–323**, 346
Aegean Microplate, 317, 329, 346
Aegean Sea, 307, 308, 317, 329
Aegina, Aegean Sea, 316–322, 323
Aeolian Arc (*see also* Calabrian arc) Tyrrhenian Sea, 3, 19, 27, 40, 41, 50, 60, 70, 72, 73, **308–316**
Afar, 528
African Plate, 2, 307–309, 317, 318, 328, 329, 347, 353, 373
Afro-Arabian Platform, 328
Afyon, Karahisar, Turkey, 330, 332–334, 344
Agattu Island, Aleutian Islands, 100, 103
Agrigan, Mariana Islands, 294, 297, 299, 300, 302
Agua, Guatemala, 150, 153, 158, 159
Agung, Bali, Indonesia, 215, 218, 411, 419
Aira, Kyushu, Japan, 262, 273
Aird Hills, Papua New Guinea, 226, 241
Ajo, Arizona, USA, 681
Ajoupa Bouillon, Mount Pelée, Martinique, 416
Ajusco-Xitle Complex, Mexico, 138
Akerlundh, North Antarctica, 387
Akita-Komagatake, Honshu, Japan, 262, 285, 288, 410
Akrotiri, Santorini, Aegean Sea, Europe, 320
Akun Volcano, Aleutian Islands, 101, 105
Akutan Volcano, Aleutian Islands, 101, 105

Alamagan, Mariana Islands, 294, 297, 299
Alanya, Turkey, 330
Alaska, USA, 18, 27, 29, 32, 40–42, 44, 46, 48, 50, 60, 70, 72, 73, 90, 99, 100, 103, 105, 419, 442, 444, 449, 450, 680, 686, 687
Alaska Peninsula, Alaska, USA, 101, 102, 104, 106, 107, 109
Alaska, Gulf of, 100
Alborz, Iran, 328, 329
Aleutian Islands, 25, 27, 29, 32, 40, 41, 42, 44, 46, 48, 50, 53, 54, 60, 70, 71–73, 88, 90, 94, **99–114**, 295, 396, 438, 442, 527, 530, 552, 553, 556–559, 690
Aleutian Island Arc, 18, 27, 40, 41, 50, 60, 63, 70, 72, 73, 79, 99, 102, 105, 439, 441, 444, 525, 530, 649, 686
Aleutian Trench, 102, 260
Alexander Island, Bellingshausen Sea, Antarctica, 375, 377
Alicudi, Aeolian Islands, 308, 309, 312, 313, 316
Almeria, Spain, 27, 40, 41, 50, 60, 70, 72, 73
Alor, Indonesia, 208, 209, 215, 219
Alps, 3, 5, 6, 307, 308, 438, 447, 643, 687
Altiplano, Mexico, 138, 139, 140, 141, 142
Aluk Ridge, Antarctica, 371, 372, 373
Alula-Fartak Trench, Gulf of Aden, 527, 528
Amaw, Aleutian Islands, 99, 101, 104, 105, 110
Amatignak Island, Aleutian Islands, 100, 103
Amatitlan, Guatemala, 150
Amazon River, 190, 453
Ambon, Indonesia, 207, 210
Ambon Hitu, Indonesia, 210, 214, 216, 219, 220, 222
Ambon Laitimor, Indonesia, 210, 214, 216, 219, 222
Ambrym, New Hebrides, Pacific Ocean, 530
Amchitka Island, Aleutian Islands, 100, 103, 105
Amitsoq, Greenland, 648, 650
Amlia Island, Aleutian Islands, 100, 105
Amukta Volcano, Aleutian Islands, 100, 105
Anak Krakatau, Indonesia, 211
Anatahan, Mariana Islands, Pacific Ocean, 294, 295, 296, 297
Anatolia, Turkey, 27, 40, 41, 50, 60, 70, 72, 73, 318, **327–349**

Anchorage, Alaska, 99
Andaman Sea, Indian Ocean, 208
Andes, 9, 11, 14, 17, 18, 20, 25, 27, 29, 30, 40, 41, 42, 50, 60, 70, 72, 73, 78, 79, 80, 116, 122, **187–205**, 397, 415, 438, 444, 452, 453, 549, 559, 560, 562, 565–568, 580, 643, 644, 646, 648, 651, 652, 654–657, 665, 681, 688
Andes, Eastern Cordillera, 188, 190–192, 196, 198
Andes, Western Cordillera, 188, 190–193, 198
Andreanof Island, Aleutian Islands, 100, 103
Anglesey, Wales, UK, 612, 654
Anglona, Sardinia, Italy, 354
Anguilla, Lesser Antilles, 169, 172
Aniakchak Crater, Alaska, 101, 105, 419
Ankara, Turkey, 329, 330
Antalya, Turkey, 330
Antamina, Peru, 681
Antarctic Peninsula, Antarctica (Graham Land), 351, **371–400**
Antarctic Plate, 2, 188, 189, 198, 373, 376
Antigua, Lesser Antilles, 169
Antimilos, Milos Group, Aegean Sea, 318
Antiparos, Aegean Sea, 317–320, 322
Antizana, Ecuador, 190
Antofagasta, Chile, 191, 195
Anvers Island, Antarctica, 374, 375
Aoba, New Hebrides, 530
Apennine Mountains, Italy, 308
Api, Wetar, Indonesia, 215, 221
Apoyo, Nicaragua, 150
Appalachian Mountains, USA, 574, 682
Aqaba, Gulf of, Red Sea, 602
Arabia, 346, 347, 356, 574, 591–609
Arabian–Nubian Shield, 592, 594, 598, 603, 604, 607
Arabian Plate, 329, 346, 347
Arabian Platform, 329, 346
Arabian Shield, 595, 598–602
Aragats, USSR, 331, 339
Ararat, Turkey, 331, 339–341, 344, 542
Aravil Volcano, Turkey, 340
Arctic Ocean, 100
Arenal, Costa Rica, 150, 156, 419
Arenig, Gwynnedd, Wales, UK, 613, 620, 623
Arequipa, Peru, 195, 365, 366
Argentina, 19, 188, 194–199, 201, 452, 565, 566
Argentine Islands, Antarctica, 378
Argentinian Puna, 193
Arica, Chile, 191, 194, 195
Armenia, USSR, 338, 339, 340, 346, 347
Arran, Isle of, Strathclyde, Scotland, UK, 613, 629, 630
Aru-Timor Trough, Indonesia, 210
Asama, Honshu, Japan, 262, 274, 288, 406, 410, 419, 420
Ascension Island, Atlantic Ocean, 542, 557
Aso, Kyushu, Japan, 262, 273, 286, 287, 408, 419, 420
Asuncion, Mariana Islands, 294, 297
Aswan, Egypt, 605

Atalanti Channel, Atalanti, Aegean Sea, 318
Atalanti, Aegean Sea, 316, 318–322
Atauro, Indonesia, 210, 215, 219
Athens, Greece, 318
Atitlan, Guatemala, 150, 153, 157–162
Atka Island, Aleutian Islands, 100, 106, 107, 108
Atka, Atka Island, Aleutian Islands, 100
Atlantic Ocean, 3, 167–170, 172, 178, 351, 372, 376, 377, 381, 396, 422, 453, 528, 574, 613, 665
Atlantic Plate, 171
Atotonilco, Mexico, 141, 142
Attu Island, Aleutian Islands, 100, 103
Auchtermuchty, Fife, Scotland, UK, 625
Augustine Volcano, Alaska, 101, 105
Australia, 27, 40, 222, 227, 236, 241, 442, 449, 575, 576
Australian Platform, 207, 208, 209, 222
Aves Island, West Indies, 169
Aves Ridge, Caribbean Sea, 169, 170, 172
Awu, Moluccas, Indonesia, 217, 219
Ayacucho, Peru, South America, 199
Ayarza, Guatemala, 150, 153
Ayrshire, Scotland, UK, 613
Azores, Atlantic Ocean, 527, 528, 550, 551, 557

Bagana, Bougainville Island, Papua New Guinea, 226, 231–234, 237, 238–242
Baghdad, California, USA, 681
Bagiai, Karkar, Papua New Guinea, 237
Bajah, Java, Indonesia, 221
Balbi, Bougainville Island, Papua New Guinea, 226, 240
Balbriggan, Republic of Ireland, 613, 619
Bali, Indonesia, 207, 209, 214, 215, 217, 218, 221
Balikesir, Turkey, 334, 344
Ballantrae, Strathclyde, Scotland, UK, 612–614
Balsamo, El Salvador, 153, 154
Bam, Papua New Guinea, 226, 231, 234, 236, 237, 239
Bamus, New Britain, Papua New Guinea, 226, 234, 238
Banda, Indonesia, 208, 219
Banda Api, Banda Sea, Indonesia, 216, 219
Banda Arc, Indonesia, 207, 210, 216, 219, 222
Banda Basin, Indonesia, 208
Banda Neira, Banda Sea, Indonesia, 216, 219
Banda Sea, Indonesia, 207–210, 216, 219, 222
Banda-San, Honshu, Japan, 12, 262, 273, 285, 411
Bangeang Api, Sumbawa, Indonesia, 219
Bangka, Indonesia, 209, 691
Barba, Costa Rica, 150
Barbados, Lesser Antilles, 169, 170, 172
Barbados Ridge, Atlantic Ocean, 169, 170, 172
Barberton, South Africa, 576
Basiluzzo, Panarea, Aeolian Islands, Europe, 313
Basse Terre, Guadeloupe, Lesser Antilles, 180
Bathurst, New South Wales, Australia, 684
Batinah, Oman, Southwest Asia, 543

Geographical Index

Battle Mountain, Nevada, USA, 681
Batu Tara, Indonesia, 214, 215, 219, 221
Batur, Bali, Indonesia, 214, 215, 219, 221
Bawdwin Mine, Burma, 684
Bay of Plenty, New Zealand, 248, 249, 253–256
Belize, Central America, 443
Ben Ghnema, Libya, 455, 456
Benkulu, Sumatra, Indonesia, 221
Ben Nevis, Highlands, Scotland, UK, 613, 624, 626–629
Bequia, Lesser Antilles, 177
Bergama, Turkey, 330, 334, 344, 345
Bering Sea, 100, 102, 105
Bering Shelf, 99
Berwyn Hills, Powys/Clwyd, UK, 613, 620
Bethlehem, British Columbia, Canada, 681
Betic Cordillera, Spain, 308
Bezymianny, Kamchatka, USSR, 410, 419, 420
Big Ben, Montana, USA, 681
Bijar, Iran, 331, 338–341, 344
Billiton, Indonesia, 209, 691
Bingham Canyon, Utah, USA, 681
Bingol, Turkey, 331, 240
Birch Lake, Ontario, Canada, 576, 577
Bismarck Sea, Papua New Guinea, 227, 230, 231, 233, 236–240
Bismarck Volcanic Arc, Papua New Guinea, 226, 229, 230, 233, 237, 242
Bitlis, Turkey, 329
Black Rocks, St. Kitts, Lesser Antilles, 18
Black Sea, Europe, 329, 330
Black Sea Plate, 329
Blue Lake, Tongariro Massif, New Zealand, 249
Blue Ridge Mountains, USA, 692
Bobrof Volcano, Aleutian Islands, 105
Bodrum, Turkey, 318, 330, 332–334, 344
Bogoslof Island, Aleutian Islands, 79, 99, 101, 103–106, 110, 527, 530
Boisa, Papua New Guinea, 226, 234, 237, 238, 239
Bolivia, 188, 191, 194–199, 201, 447, 449, 452, 677, 680, 688, 689
Bolivian Altiplano, 190, 191, 193, 195, 196
Bonin Islands, 27, 40, 41, 50, 60, 70, 72, 73, 259, 270, 271, 296, 300
Bonin Trench, 261, 296
Boqueron Volcano, El Salvador, 154, 158, 159, 161, 162
Boqueroncito, El Salvador, 158, 159
Borrowdale, Cumbria, UK, 614, 619, 622, 632–634
Bosana, Sardinia, Italy, 354, 355, 357, 362, 365, 367
Bougainville Island, Papua New Guinea, 226–233, 235, 239, 240, 242,.443
Bouvet Island, Atlantic Ocean, 555, 557
Bouvetoya, Atlantic Ocean, 542
Braden, Chile, 681
Bransfield Strait, South Shetland Islands, Antarctica, 371, 372, 374–376, 379, 381, 387, 390, 392, 397, 530

Brazil, 444, 665
Breidden Hills, Wales, UK, 613, 620
Brenda, British Columbia, Canada, 681
Bridgeman Islands, Antarctica, 371, 374–376, 379, 388, 390–393, 395
Bristol Channel, England, UK, 619
British Columbia, Canada, 116, 118, 123, 438, 444, 451, 681
Brittany, France, 654
Brokeoff Volcano, California, USA, 18
Bromo, Java, Indonesia, 213, 218
Buchans Mine, Newfoundland, Canada, 684, 685
Builth Wells, Powys, Wales, UK, 613, 620, 622
Bulawayo, Zimbabwe, 576
Buldir, Aleutian Islands, 99, 100, 103–106
Burdwood Bank, Scotia Sea, Antarctica, 374
Burma, 208, 684, 688–690
Burmese Arc, 208
Burney Volcano, Chile, 188
Buru, Indonesia, 209, 210
Butte, Montana, USA, 681
Butung, Indonesia, 209
Byers Peninsula, Livingstone Island, Antarctica, 376, 377, 385–387, 389, 391

Cader Idris, Gwynedd, Wales, UK, 613, 620
Cairo, Egypt, 592
Calabria, Italy, 308, 309, 310
Calabrian Arc, 27, 40, 41, 50, 60, 70, 72, 73, **307–317**
Calama, Chile, 194
Calapooya Valley, Cascades, USA, 129
Calbuco, Chile, 192
California, USA, 18, 116, 117, 120–124, 127, 139, 428, 438, 444, 445, 448, 449, 453, 632
California, Gulf of, USA, 138, 139
Calipuy, Peru, 446, 644
Campo Morado, Mexico, 684
Canada, 15, 100, 105, 124, 444, 445, 577, 578, 580, 581, 584, 585, 650, 684, 686
Canakkale, Turkey, 330
Cananea, Sonora, Mexico, 681
Canary Islands, Atlantic Ocean, 416, 557
Cap Marargiu, Northwest Sardinia, Italy, 354, 358
Cape Verde, Atlantic Ocean, 557
Caradoc, Gwynnedd, Wales, UK, 623
Caribbean Arc, 28, 449–452
Caribbean Plate, 149, 167, 168, 171, 189, 198
Caribbean Sea, 138, 167–170, 172, 422, 438, 442, 433, 448, 449, 473, 682
Carlisle Volcano, Aleutian Islands, 101, 105
Caroline Plate, 226, 229
Carpathian Mountains, Europe, 308
Carrock Fell, Cumbria, England, UK, 615
Casapalca, Peru, 681
Cascade Province, USA, 25, 27, 29, 32, 40–42, 44, 46, 48, 50, 53, 54, 60, 63, 64, 70–73, 78, 79, 88, 90, **115–135**, 182, 315, 439, 553, 556–559, 632, 649
Casino, Yukon, Canada, 681

Casma, Peru, 644
Caspian Sea, USSR, 329, 331
Catface, British Columbia, Canada, 681
Cayman Trench, Caribbean Sea, 443
Ceboruco, Mexico, 138, 140, 141
Celebes Sea, Indonesia, 209
Central America, 3, 18, 28, 122, 139, **149–166**, 182, 438, 442, 443, 448, 449, 672
Central Iran Range, Iran, 327, 328, 329
Cerro Verde, Peru, 681
Chagulak Island, Aleutian Islands, 100, 105
Chalatenango, El Salvador, 153
Chancay, Peru, 446
Chanka, Chile, 194
Chao, Chile, 194
Chaucha, Ecuador, 681
Cheviot Hills, Northumberland, UK, 613, 625, 626, 628, 632
Chiang Mai, Thailand, 689
Chichon, Mexico, 138
Chiginadak, Aleutian Islands, 101, 105
Chile, 19, 187–189, 191, 192, 194–196, 197–199, 201, 202, 372, 374, 375, 377, 386, 391, 407, 438, 439, 445, 451–453, 456, 530, 534, 536–538, 542, 552–555, 565, 566, 567, 649, 652, 654, 687
Chile Rise (Ridge), Pacific Ocean, 188, 189, 198, 373
China, 270, 281, 687
Chokai, Honshu, Japan, 262, 285
Christiana Islands, Aegean Sea, 318, 319, 321, 322
Church Stretton, Shropshire, UK, 611
Chuquicamata, Chile, 681
Cinder Spur, King George Island, Antarctica, 376, 379
Cixerri, Sardinia, Italy, 354, 360
Clarence Island, Scotia Sea, Antarctica, 376
Coalville Ridge, Pacific Ocean, 246
Coatepeque, El Salvador, 150
Cocos Plate, 137, 139, 149, 189, 198
Cofre De Perote, Mexico, 138, 140
Cold Bay, Alaska, 101, 103, 104
Colima, Mexico, 138, 140, 142, 145
Coll, Inner Hebrides, Scotland, UK, 646
Collon, Louth, Republic of Ireland, 613, 619
Colombia, 187, 188, 443
Colorado, USA, 455, 690
Columbia River, USA, 117, 118
Columbia River Plateau, USA, 27, 40, 41, 50, 60, 70, 72, 73
Concepcion, Nicaragua, 150, 155
Conchagua, El Salvador, 150
Coniston, Cumbria, UK, 615
Cook Island, South Sandwich Islands, 16
Coolgardie, Western Australia, 576
Coppermine Cove, Robert Island, Antarctica, 377
Corinth, Greece, 318
Corsica, France, 353
Cornwallis Island, Antarctica Peninsula, Antarctica, 378

Coseguina (Cosiguina), Nicaragua, 150, 155, 156
Costa Rica, 43, 149, 156, 157, 443, 444
Cotopaxi, Ecuador, 188, 190, 199, 415
Craigmont, British Columbia, Canada, 681
Crater Lake, Oregon, USA, 117, 126, 408, 419, 420, 422
Cross Fell, Cumbria, Scotland, UK, 613, 615
Cuajone, Bolivia, 681
Cuba, Greater Antilles, 168, 443
Curlew Mountains, Republic of Ireland, 613
Curtis Island, Kermadec Islands, 246, 254, 255
Cuscatlán, El Salvador, 153
Cushendall, England, UK, 613, 629
Cyprus Arc, 328

Daisen, Honshu, Japan, 262, 287
Daisetsu, Hokkaido, Japan, 262, 286
Dalembert, Quebec, Canada, 578
Damar, Banda Sea, Indonesia, 208, 215, 216, 219, 220
Dana Volcano, Alaska, 101, 105
Danau Complex, Java, Indonesia, 213, 218
Dandelhura, India, 691
Davidof, Aleutian Islands, 103
Dawson Strait, Papua New Guinea, 226, 227, 230
Dead Sea, Israel/Jordan, 347
Deception Island, South Shetland Islands, Antarctica, 21, 27, 40, 41, 62, 73, 371, 375, 376, 379, 386, 388, 390–393, 395
Deerness, Orkney Islands, UK, 613, 628, 629
Delarof, Aleutian Islands, 103
Dempo, Sumatra, Indonesia, 212, 218
D'Entrecasteaux Islands, Papua New Guinea, 226, 229, 230, 240
Desirade, Lesser Antilles, 169, 172
Dieng Complex, Java, Indonesia, 213, 218
'Discovery Arc', Antarctica, 381, 388, 394
Disputada, Chile, 681
Distinkhorn, Strathclyde, Scotland, UK, 613
Dodecanese Islands, Greece, 427
Doma Peaks, Papua New Guinea, 226, 232, 234, 241, 242
Dominica, Lesser Antilles, 17, 167–170, 176, 177, 179, 442, 443, 542, 553, 554, 564
Drake Passage, Southern Ocean, 351, 371, 372, 375, 397
Drake Plate, Antarctica, 373, 375
Dry Creek, Alaska, USA, 681
Duau, Papua New Guinea, 226, 241
Dukono, Moluccas, Indonesia, 216, 217, 219
Dunedin, New Zealand, 542
Durango, Mexico, 138
Dutch Harbour, Alaska, 101
Dyfed, Wales, UK, 613, 620, 622, 623

Earl, Oregon, USA, 681
East Coast, North Island, New Zealand, 256
East Africa, 416, 542

Geographical Index

East Japan Volcanic Belt, Japan, 260, 263, 265, 267, 271, 288
East Pacific Rise, Pacific Ocean, 138, 139
Eastcape Ridge, New Zealand, Pacific Ocean, 246
Easter Island, Pacific Ocean, 542, 557
Eastern Papua, Papua New Guinea, 225, 226, 229, 231–233, 235, 239, 240–242
Ecuador, South America, 18, 57, 187, 188, 190, 192, 197–199, 202, 386, 438, 445, 542, 553, 554, 564–567, 649
Edgecumbe, New Zealand, 249
Edremit, Turkey, 330, 333, 334, 344, 345
Egmont Volcanic District, New Zealand, 248
Egypt, 595, 599, 601, 602, 605
Egyptian Eastern Desert, 591, 595, 599, 600, 601, 603, 604
Elephant Island, South Shetland Islands, Antarctica, 376
El Laco, Chile, 687, 688
El Salvador, 43, 152–154, 156, 157, 160, 162, 164
El Salvador, Chile, 681
Ely, Nevada, USA, 681
Emmons Volcano, Aleutian Islands, 105
Emperor Mine, Fiji, Pacific Ocean, 688
Endako, British Columbia, Canada, 681
Ennerdale, Cumbria, UK, 615
Eolian Islands, Italy, 20
Erciyes, Turkey, 330, 335, 337, 338, 344, 345
Erevan, USSR, 331
Erta, Ale, Ethiopia, 670
Ersurum, Turkey, 31, 338, 339
Erzgebirge, Czechoslovakia/East Germany, 691
Eshna Ness, Shetland Islands, UK, 613, 629
Eskdale, Cumbria, UK, 615
Ethiopia, 416, 591, 606
Eua, Tonga, 16, 245, 247
Eurasian Plate, 270, 307, 317, 318, 327–329, 345, 691
European Plate, 353
Exotica, Chile, 681
Eyemouth, Scotland, UK, 613, 625
Eycott, Cumbria, UK, 614–616, 618, 622, 632, 634
Ezine, Turkey, 330, 332, 335, 337, 344

Falbertal, Austria, 687
Falcon Bank, Tonga, 247
Falcon Islands, Tonga, 533, 535
Falkland Islands, Atlantic Ocean, 374
Fantale, Ethiopia, 687
Favenc, Papua New Guinea, 226, 241
Feni, Papua New Guinea, 225, 226, 229, 230, 233
Fernando de Noronha, Brazil, 557
Ferro Bamba, Peru, 681
Fiji, 13, 15, 19, 27, 40, 41, 50, 60, 70, 71, 73, 246, 442, 473, 529, 682, 685, 688
Fildes Peninsula, King George Island, Antarctica, 376, 378, 379, 386, 389, 391
Filicudi, Aeolian Islands, Europe, 309, 311–312, 313, 316

Finland, 455, 456, 646
Fintona, Tyrone, UK, 613
Fisher Volcano, Aleutian Islands, 105, 419
Flin Flon, Manitoba, Canada, 684
Flores Island, Indonesia, 207–208, 209, 215
Flores Sea, Indonesia, 211
Foca, Turkey, 330, 332, 334, 344
Fonualei, Tonga, 247, 248, 251, 254, 301
Fonseca (Gulf), El Salvador/Honduras, 150
Fortalenza, Peru, 446
Fort Glenn, Alaska, 101
Fort Victoria, Zimbabwe, 576
Fossa Della Felci, Aeolian Islands, 312
Fossa Di Vulcano, Vulcano, Aeolian Islands, 313
'Fossa Magna', Honshu, Japan, 261, 270, 271, 288, 684
Fourpeaked Mountain, Alaska, 101, 105
Fox Islands, Aleutian Islands, 101
Frailes Plateau, Bolivia, 191, 195, 196
France, 677, 692
Frosty Volcano, Alaska, 101, 103, 105
Fuego, Guatemala, 150, 153–162, 416, 418, 669–676
Fuji, Honshu, Japan, 259, 262, 278, 280, 286
Funagata, Honshu, Japan, 262, 285

Galan (Cerro), Argentina, 195, 196, 565
Galapagos Islands, 189, 542
Galunggung, Java, Indonesia, 213, 218
Gareloi Volcano, Aleutian Islands, 100, 103, 105
Garn Bodfean, Lleyn Peninsula, Wales, UK, 622, 623
Garove, Witu Islands, Papua New Guinea, 226, 234, 238, 239
Gassan, Honshu, Japan, 262, 285
Georgia, USA, 692
Glacier Peak, Washington, USA, 117, 681
Glencoe, Scotland, UK, 613, 624, 626–629
Gondwanaland, 372, 373, 396
Gorda Ridge (Rise), Pacific Ocean, 557
Goropu, (Waiowa), Papua New Guinea, 226, 231, 232, 240
Gough Island, South Atlantic Ocean, 555, 557
Graham Land, Antarctica, 378, 381
Grampian Highlands, Scotland, UK, 627
Grangegeeth, Louth, Republic of Ireland, 613, 619
Gransile, British Columbia, Canada, 681
Greater Antilles, 168, 443
Great Glen, Scotland, UK, 612, 613, 624, 628, 632
Great Sitkin Volcano, Aleutian Islands, 100, 103, 105, 107
Greece, 427
Greenland, 642, 648
Gregory Rift, Kenya, 527, 528
Grenada, Lesser Antilles, 13, 14, 19, 20, 79, 167–182, 525, 530, 533, 538, 540, 552, 553, 554
Grenada Trough, Caribbean Sea, 169, 170
Grenadines (The), Lesser Antilles, 173, 175, 177, 179
Guacha (Cerro), Bolivia, 195, 196
Guadalajara, Mexico, 140, 142, 143

Geographical Index

Guadalcanal, Solomon Islands, Pacific Ocean, 442
Guadeloupe, Lesser Antilles, 168–170, 172–174, 177, 180, 181, 411, 557
Guallatiri, Chile, 191
Guam, Mariana Islands, 294, 295
Guanajuanto, Mexico, 140
Guatemala, 43, 138, 139, 149, 150, 152–157, 160, 162, 164, 407, 669–676
Guayaquil, Ecuador, 190
Guayas River, Ecuador, 190
Guguan, Mariana Islands, 294, 297, 299
Guntur, Java, Indonesia, 213, 218
Gunung Api, Banda Sea, Indonesia, 207, 216, 219
Guyana, 444

Ha'Apai Group, Tonga, 247
Hachigo-Higashi-Yama, Izu Islands, Japan, 262, 285, 286
Hachigo-Jima, Izu Islands, Japan, 561
Hachigo-Nishiyama, Izu Islands, Japan, 262
Hagen, Papua New Guinea, 226, 234, 241, 242
Haiti, Greater Antilles, 443
Hakkoda, Honshu, Japan, 262, 285
Hakone, Huzi, Japan, 13, 16, 275, 283, 419, 420
Halmahera, Indonesia, 207, 209
Hamilton, New Zealand, 249
Hasan, Turkey, 330, 335, 338, 344
Haura, Peru, 446
Hauraki Peninsula, North Island, New Zealand, 688
Hauraki Volcanic Region (Northland Arc), New Zealand, 248, 249
Havre Rock, Kermadec Islands, 246
Havre Trough, 245, 246
Hawaii, USA, 280, 303, 497, 500, 501, 527, 528, 557
Hawasina Window, Oman, 543
Hawkes Bay Basin, North Island, New Zeland, 256
Haybi, Oman, 545
Hebrides, Scotland, UK, 454, 455, 646
Heddleston, Montana, USA, 681
Hekla, Iceland, 403, 407
Hellenic Arc, 27, 40, 41, 50, 60, 70, 72, 73, 317–323
Hellenic Trench, Aegean Sea, Europe, 318
Henderson, Colorado, USA, 681
Herbert Island, Aleutian Islands, 101, 105
Hibok-Hibok, Philippines, 419
Hill of Allen, Kildare, Republic of Ireland, 613, 619
Himalayas, 3, 5, 6, 9, 438, 447, 643, 680, 690–692
Hoggar, Algeria, 606
Hokkaido, Japan, 261–263, 265, 268, 272, 277, 285, 288, 408, 411, 685
Honduras, Central America, 43, 150, 443
Honshu, Japan, 261, 265, 267, 268, 270–274, 285–289, 683, 684, 687
Hope Bay, Antarctic Peninsula, Antarctic, 376, 377
Hoy, Orkney Islands, UK, 613, 626, 628, 632, 635
Huasteca Plain, Mexico, 138
Hudson Bay Mountain, British Columbia, Canada, 681

Hunga Ha'Apai, Tonga, Pacific Ocean, 247, 248, 250, 254
Hunga Tonga, Tonga, Pacific Ocean, 247, 248, 250
Huntly, Grampian, Scotland, UK, 613, 632
Huzi, Japan, 13, 14

Iceland, 407, 416, 528, 557
Ichinomegata, Honshu, Japan, 262, 267, 273
Idaho, USA, 118, 122, 444, 451, 453
Ilak, Aleutian Islands, 103
Ile de Groix, Brittany, France, Europe, 654
Iliamna Volcano, Alaska, 101, 105
Ili Boleng, Indonesia, 215, 219
Ilopango, El Salvador, 150
India, 447, 691
Indian Ocean, 207, 208, 210, 211, 220, 664
Indian plate (Indo-Australian plate), 2, 208, 225, 226, 228, 229, 236, 240, 256, 680, 691
Indonesia, 16, 28, 63, **207–224**, 299, 311, 438, 560, 562, 606, 691
Indonesian Arc, 13, **207–224**
Indus River, India/Pakistan, 691
Ingerbell, British Columbia, Canada, 681
Ionian Sea, Europe, 310
Iquique, Chile, South America, 191, 194
Iran, 3, 6, **327–349**
Iranian Plate, 329, 347
Irazu, Costa Rica, 150, 156, 407
Ireland, (Republic of Ireland and UK), 611, 612, 614, 619, 629, 630, 632–634
Irruptunca, Chile, South America, 191
Irving Point, Visokoi, South Sandwich Islands, 16
Isanotski Volcano, Aleutian Islands, 101, 105
Ishikari Lowland, Japan, 261, 268
Islas Quemadas, E. Salvador, 154, 158, 159
Isle of Man, UK, 612, 633
Isluga, Chile, 191
Isparta, Turkey, 330, 332–334, 344
Istanbul, Turkey, 329, 330
Isthmus of Corinth (Crommyonia), Aegean Sea, 316–319, 321, 322
Isua, Greenland, 650
Italy, 309, 353, 411, 550, 551, 565
Ithaca Peak, Nevada, USA, 681
Ito, Kyushu, Japan, 408, 409, 420
Iwaki, Honshu, Japan, 262, 285
Iwo Jima, Volcano Islands, Pacific Ocean, 296, 298–300, 303, 304
Izalco, El Salvador, 150, 155, 156, 158, 159, 162
Izmir, Turkey, 318, 329, 330, 333, 334, 344, 345
Iztacchautl, Mexico, 138, 140
Izu Arc, Pacific Ocean, 293, 296, 300, 303
Izu-Bonin Arc (*see* Izu-Mariana Arc), 259, 260, 261, 263, 265, 268, 272, 288
Izu Islands, Japan, 14, 27, 40, 41, 50, 60, 70, 72, 73, 261, 271, 275, 278, 280, 283, 284, 288, 297, 533
Izu-Mariana Arc, 260, 262, 269, 272, 275, 276–278, 284, 285, 288, 298

Geographical Index

Izu-Oshima, Izu Islands, Japan, 285
Izu Peninsula, 14, 16, 261, 272, 278, 288, 296
Izu Trench, Pacific Ocean, 261

Jalisco, Mexico, 140
Jamaica, Greater Antilles, 442, 443
James Ross Island, Weddell Sea, Antarctica, 372, 374, 379, 381, 387, 393, 395
Japan, 1, 3, 13, 14, 19, 27, 40, 41, 50, 60, 63, 70, 72, 73, 182, 214, **259–292**, 396, 405, 408, 411, 415, 419, 438, 442–444, 451, 453, 466, 484, 556, 558, 559, 561, 562, 677, 680, 683–687, 690
Japanese Arcs, **259–292**, 439, 441, 453, 516
Japan Sea (Sea of Japan), 13, 260, 261–264, 266, 267, 269, 270, 273, 285, 289, 684, 687
Japan Trench, 261, 665
Jason Peninsula, Antarctica, 377, 379
Jatibarang, Java, Indonesia, 210
Java, Indonesia, 207–211, 213, 214, 217, 218, 220, 221
Java Sea, Indonesia, 208, 209, 211
Jebel Asoteriba, Sudan, Africa, 596, 601
Jebel Ess, Saudi Arabia, 601
Jebel Marra, Sudan, 542
Jebel Shendib, Sudan, 596
Jebel Uweinat, Egypt, 592
Jeddah, Saudi Arabia, 602, 605
Jorullo, Mexico, 138
Jebel Al Wask, Northwest Saudi Arabia, 601
Juan de Fuca Strait, Canada/USA, 557

Kagalaska Island, Aleutian Islands, 100, 103
Kagamil Volcano, Aleutian Islands, 101, 105
Kaimanawa Range, North Island, New Zealand, 256
Kakagi Lake, Superior Province, Ontario, Canada, 577
Kakaramea, New Zealand, 249, 250
Kalimantan, Borneo, Indonesia
Kammena Vourla, volos-Atalanti Group, Aegean Sea, 318, 322
Kamchatka, USSR, 1, 27, 40, 41, 50, 60, 70, 72, 73, 260–263, 270–272, 274, 278, 280, 285, 288
Kampu, Honshu, Japan, 262, 285
Kanaga Volcano, Aleutian Islands, 100, 103, 105, 530
Kanto Plain, Japan, 261, 268
Kao, Tonga, 247, 248, 250, 251
Karaburun, Izmir, Turkey, 345
Karadag, Turkey, 335, 338, 344
Karaman, Turkey, 330, 335
Karangahape, New Zealand, 249, 254
Karapinar, Turkey, 330, 335, 337, 338, 344
Karkar Volcano, Papua New Guinea, 225, 226, 231, 232, 234, 237
Karpathos, Aegean Sea, 319
Kars, Turkey, 331, 339, 340, 341, 344
Kars Plateau, Turkey, 339, 340
Kasatochi Island, Aleutian Islands, 100, 105
Katmai Volcano, Alaska, 101–103, 105–107, 408, 411, 421

Kavalga, Aleutian Islands, 103
Kawah Idjen, Java, Indonesia, 213, 218
Kaweka Range, North Island, New Zealand, 256
Kayseri, Turkey, 330, 335
Kelang, Banda Sea, Indonesia, 210, 214, 216, 219
Kelud, Java, Indonesia, 213, 218
Kenya, 576, 580, 581, 585, 586
Kerguelen Island, Indian Ocean, 555, 557
Kermadec Arc, **245–258**, 558
Kermadec Islands, 27, 40, 41, 50, 60, 70, 72, 73, 79, **245–258**, 368, 665
Kermadec Trench, 246
Khartoum, Sudan, 605
Khavar, Iran, 331, 339
Khyostof Island, Aleutian Islands, 103
Kick-em-Jenny, Grenadines, Lesser Antilles, 173
Kidd Creek, Ontario, Canada, 681
Kii Peninsula, Japan, 261, 268
Kilauea Iki, Hawaii, 473, 478, 670
Kilbourne Hole, New Mexico, USA, 567
Kildare, Republic of Ireland, 613, 619
Kimolos, Milos Group, Aegean Sea, 318, 319
King George Island, South Shetland Islands, Antarctica, 376, 378, 379, 386, 387, 390
Kirkland Lake, Ontario, Canada, 684
Kiska Volcano, Aleutian Islands, 100, 103, 105
Kita Iwo Jima, Volcano Islands, 296, 298, 299, 303
Klamath Mountains, California, USA, 118
Kolmovos Volcano, Santorini, Aegean Sea, 322
Komagatake, Honshu, Japan, 419, 420, 424
Konia Volcano, Aleutian Islands, 100
Koniuji Volcano, Aleutian Islands, 100, 105
Konya, Turkey, 330, 335, 338, 344
Konya Massif, Turkey, 335
Korea, 261, 270, 281, 691
Korovin Volcano, Aleutian Islands, 100, 105
Kos, Aegean Sea, 317–319, 320–322
Kosaka Volcano, Honshu, Japan, 685
Krakatau, Java, Indonesia, 207, 211, 213, 214, 218, 299, 408, 411, 420, 527, 533, 534, 536, 537, 538, 540, 542
Kukak Volcano, Aleutian Islands, 105
Kula, Turkey, 330, 332, 335, 337
Kula Plate, 102
Kulal Volcano, Alaska, 101
Kunashiri, Japan, 261
Kupreanof Island, Alaska, 101
Kurikoma, Honshu, Japan, 262, 285
Kurile Arc, 259–262, 265, 268–270, 272, 274, 276–278, 284, 285, 288
Kurile, Basin, Sea of Okhotsk, 261
Kurile Islands, USSR, 1, 27, 40, 41, 50, 60, 70, 72, 73, 260, 263, 267, 271, 280, 288
Kurile Trench, Pacific Ocean, 261
Kuroko, Honshu, Japan, 685
Kutahya, Turkey, 330
Kutcharo, Hokkaido, Japan, 262, 273, 420
Kyushu, Japan, 261, 270, 272, 287, 288, 690
Kyushu-Palau Ridge, Philippine Sea, 261, 271, 643

La Caridad, Sonora, Mexico, 681
La Paz, Bolivia, S. America, 195
La Soufriere, Guadeloupe, Lesser Antilles, 173, 174
Laguna Jaya Khota, Bolivia, 195, 196
Lake Coatepeque, El Salvador, 158, 159
Lake District, England, UK, 611, 612, 614–619, 633
Lake Sevan, USSR, 331
Lake Taupo, New Zealand, 255
Lake Titicaca, Bolivia–Peru, 195
Lake Toba, Sumatera, Indonesia, 207, 210–212, 218–220, 222
Lake Tuz, Turkey, 330
Lake Urmia, Iran, 331
Lake Van, Turkey, 327, 329, 331, 338–341, 344, 346, 347
Lambay Island, Republic of Ireland, 613, 619
Lamongan, Java, Indonesia, 213, 218
Lanarkshire, Scotland, UK, 613
Langila, New Britain, Papua New Guinea, 226, 229, 231, 232
Lares, Puerto Rica, USA, 681
Larsen Ice Shelf, Antarctic Peninsula, Antarctica, 379
Lasail, Oman, 544, 545
Lascar, Chile, South America, 191
Lassen Peak, California, USA, 27, 40, 41, 50, 60, 70, 72, 73, 116, 117, 132
Late, Tonga, Pacific Ocean, 247–249, 254, 301
Lau Basin, Tonga, Pacific Ocean, 20, 246
Lau Ridge, Pacific Ocean, 246
Laut, Indonesia, 209
Lautaro Volcano, Chile, South America, 188
Lawu, Java, Indonesia, 213, 218
Lentia, Vulcano, Aeolian Islands, 312
Lesotho, Africa, 512
L'Esperance Rock, Kermadec Islands, Pacific Ocean, 246, 254
Lesser Antilles (West Indies), 3, 13, 14, 17–20, 79, **167–185**, 396, 411, 415, 424, 425, 439, 443, 466, 527, 550, 552, 556, 563, 564, 567, 643
Lewotolo, Indonesia, 215, 219
Liard, North West Territories, Canada, 681
Libya, 456
Lihir, Papua New Guinea, 226, 230
Likhades Islands, Volos-Atalanti Group, Aegean Sea, 318, 322
Lima, Peru, 445, 453
Lingga, Indonesia, 209
Lions Rump, King George Island, Antarctica, 376, 379
Lipari, Aeolian Islands, 309–313, 316
Little Sitkin Volcano, Aleutian Islands, 100, 103, 105
Livingstone Island, Drake Passage, Antarctica, 376, 377, 385, 387
Llaima, Chile, South America, 192
Lleyn Peninsula, Gwynnedd, Wales, UK, 613, 620, 623
Llullailaco, Chile, 191

Logudoro, Sardinia, Italy, 354, 355, 357, 365
Lokon Empung, Moluccas, Indonesia, 216, 217, 219
Lolobau, New Britain, Papua New Guinea, 226, 231
Loloru, Bougainville Island, Papua New Guinea, 226, 240
Lombok Island, Indonesia, 207, 209, 213, 215, 219
Long Island, Papua New Guinea, 226, 236, 237, 242
Longquimay, Chile, 192
Lopevi, New Hebrides, 410
Lorne Plateau, Strathclyde, UK, 613, 624–630, 632, 634
Los Chocoyos, Guatemala, 153, 157–160, 161
Los Humeros, Mexico, 140
Louisiade Archipelago, Papua New Guinea, 226, 229, 240
Lumaravi-Archangelo, Santorini, Aegean, 320
Lusancay Island, Papua New Guinea, 226, 240, 241
Luzon, Philippines, 261

Macauley Island, Kermadec Islands, 246, 250, 251, 254
Madeira, 542
Madera, Nicaragua, 150
Madura, Indonesia, 209
Mageik Volcano, Alaska, 101, 105
Maghrebinian Belt, Africa, 308
Makassar Strait, Indonesia, 209
Makian, Moluccas, Indonesia, 217, 219
Makushin Volcano, Aleutian Islands, 101, 105
Malatya, Turkey, 330, 338
Malaysia, 207, 209–211, 447, 687, 691
Malazgirt, Turkey, 331, 340
Mali, Africa, 606
Malinche, Mexico, 138, 140
Maliyami, Zimbabwe, 580, 581
Mamuta (Cerro), Chile, 195, 196
Manam, Papua New Guinea, 226, 231, 232, 237
Manawahe, New Zealand, 249
Manaslu, Nepal, 680, 691, 692
Mandalay, Burma, 689
Mangani, Sumatera, Indonesia, 221
Manitou Lake, Ontario, Canada, 576, 577
Manuk, Banda Sea, Indonesia, 216, 219
Manus Basin, 226–228, 238
Manus Island, Papua New Guinea, 227
Marapi, Sumatera, 212, 218
Marble Peak, Papua New Guinea, 226, 241, 242
Marda, Western Australia, 580, 581
Mare, Moluccas, Indonesia, 217, 219
Mariana Islands (*see also* Izu-Mariana arc), 1, 27, 40, 41, 50, 60, 70, 72, 73, 106, 259, 260, 263, 265, 270, 274, 278, **293–306**, 380, 439, 442, 552–554, 557–559, 564, 580, 643, 649, 663
Mariana Ridge, Pacific Ocean, 643
Mariana Trench, Pacific Ocean, 20, 294, 295, 656, 665
Mariana Trough, Pacific Ocean, 294, 295, 297, 302
Marie Byrd Land, Antarctica, 372, 395

Geographical Index

Marie Galante, Lesser Antilles, 168–170
Marmilla, Sardinia, Italy, 354–356, 358–364
Marmolejo, Chile, 202
Marsili Seamount, Aeolian Islands, 308–310
Martinique, Lesser Antilles, 167–174, 176, 177, 180, 408, 415, 417, 419, 422, 427, 428, 430
Martins Head, King George Island, Antarctica, 376, 379
Masaya, Nicaragua, 150, 155
Mashu, Hokkaido, Japan, 262, 285
Massif Central, France, 677, 692
Maug, Mariana Island 294, 297–299
Maungakakaramea, New Zealand, 249
Maungaongaonga, New Zealand, 249
Maupura, Banda Sea, Indonesia, 216, 219
Mayon, Philippines, 419, 420
Mazatlan, Mexico, 138
Mecca, Saudi Arabia, 592
Medicine Lake, California, USA, 27, 40, 41, 50, 60, 64, 70, 72, 73, 117
Mediterranean Basin, Europe, 307, 308, 351
Mediterranean Island Arcs, **307–325**
Melanesia, 13
Menderes Massif, Turkey, 329, 335
Mendip Hills, Somerset, UK, 613, 619, 623
Merapi, Java, Indonesia, 213, 218, 409, 419
Merbabu, Java, Indonesia, 213, 218
Methana, Aegean Sea, 316–319, 321, 322
Metis Shoal, Tonga, Pacific Ocean, 247, 251, 252, 255
Mexico, 3, 43, **137–147**, 149, 150, 406, 408, 411, 428, 439, 443, 444, 530, 542, 553, 649, 684, 686–688
Mexico City, Mexico, 138, 141, 142, 144
Mexico, Gulf of, 138, 139, 141, 142
Mianeh, Iran, 331, 339
Michiquillay, (Michiquiay), Peru, 681
Microthebe, Volos-Atalanti Group, Aegean Sea, 318, 322
Mid-Atlantic Ridge, 171, 527
Middle America Trench, 137–139, 443, 656
Midland Valley, Scotland, UK, 612, 613, 624–629, 631, 632
Midlands, Zimbabwe, 576, 580, 581
Milas, Turkey, 330
Milos Group, Aegean Sea, 317–322
Millom Park, Cumbria, UK, 615
Minami Iwo Jima, Volcano Islands, 296–298
Minnesota, USA, 575
Minor Caucasus, 328, 329, 339
Miravalles, Costa Rica, 150
Misool, Indonesia, 209
Miyakesima, Izu Islands, Japan, 403
Moluccas, Indonesia, 216, 217, 219
Molucca Sea, Indonesia, 207, 210
Moluccan Sea Collision Complex, Indonesia, 207, 208, 210
Mombacho, Nicaragua, 150
Momotombo, Nicaragua. 150, 155, 156
Montana, USA, 118, 445

Monte Dei Porri, Salina, Aeolian Islands, 312
Monte Rosa, Lipari, Aeolian Islands, 312
Montresta, Sardinia, Italy, 354
Montrose, Tayside, Scotland, UK, 613
Montserrat, Lesser Antilles, 12, 168–170, 173, 174, 175–177, 179–181
Morenci, Arizona, USA, 681
Moriyoshi, Honshu, Japan, 262, 285
Morne Microtrin, Dominica, Lesser Antilles, 173
Morne Rouge, Mount Pelée, Martinique, 416
Morocco, 606
Morococala, Bolivia, 195, 196
Morococha, Peru, 681
Mount Adams, Washington, USA, 117
Mountain, The, Saba, Lesser Antilles, 173
Mount Arcuentu, Sardinia, Italy, 354–356, 359–364, 365, 367
Mount Baker, Washington, USA, 116, 117
Mount Cannisones, Northwest Sardinia, 354, 358, 359, 363–365, 367
Mount Chiginagak, Alaska, USA, 101
Mount Cleveland, Aleutian Islands, 101, 105
Mount Denison, Alaska, USA, 101
Mount Douglas, Alaska, USA, 101, 105
Mount Egmont, New Zealand, 246, 248
Mount Etna, Sicily, Italy, 309, 670
Mount Fromma, Northwest Sardinia, Italy, 354, 358, 359, 367
Mount Garibaldi, British Columbia, Canada, 117
Mount Gisborne, White Island, New Zealand, 251
Mount Hood, Oregon, USA, 116, 117, 132, 404, 473, 476, 479
Mount Jefferson, Oregon, USA, 117, 125, 128, 129, 132
Mount Kupreano Volcano, Alaska, USA, 101
Mount Lamington, Papua New Guinea, 225, 226, 231, 232, 240, 410, 411, 419, 420
Mount Larenta, Sardinia, Italy, 354, 359, 361, 362, 364
Mount Maitland, Grenada, Lesser Antilles, 20
Mount Mazama (Crater Lake), Oregon, USA, 126, 273, 428
Mount McLoughline, Oregon, USA, 117
Mount Misery, St. Kitts, Lesser Antilles, 173, 174, 176, 180, 411
Mount Moffet, Aleutian Islands, 100, 105
Mount Narcao, Sardinia, Italy, Europe, 354, 360–362
Mount Ozzastru, Sardinia, Italy, 354, 358, 359, 363–365
Mount Pelée, Martinique, Lesser Antilles, 167, 172–174, 180, 408, 409, 411, 415–417, 419, 420, 422, 428, 430, 431
Mount Peulik, Alaska, USA, 101, 105
Mount Rainer, Washington, USA, 117, 129, 132
Mount Redoubt, Alaska, USA, 101, 105
Mount Rugiu, Northwest Sardinia, Italy, 354, 358
Mount St. Catherine, Grenada, Lesser Antilles, 20, 173

Mount Saint Helens, Washington, USA, 116, 117, 410, 411, 412
Mount Seda Oro, Northwest Sardinia, Italy, 354–356, 358, 359, 361–367
Mount Shasta, California, USA, 117, 132, 632
Mount Spurr, Alaska, 101, 105
Mount Suaru, Papua New Guinea, 533
Mount Tiloromo, Sardinia, Italy, 354, 358
Mount Tulik, Aleutian Islands, 101
Mount Vsevidof, Aleutian Islands, 101, 105
Mount Wilhelm, Papua New Guinea, 227
Moyuta, Guatemala, 150
Mozambique, 606, 607
Muertos Trench, Caribbean Sea, 443
Mugla, Turkey, 330
Muriah, Java, Indonesia, 213, 214, 218, 220, 221
Mynydd Mawr, Gwynnedd, Wales, UK, 623

Nacozari, Sonora, Mexico, 681
Nadji, Saudi Arabia, 604
Naknek, Alaska, 101
Namibia, 677, 692
Nankai Trough, 260, 684
Nantai, Honshu, Japan, 261, 683
Natuna, Indonesia, 209
Nazca Plate, 2, 187–189, 198, 373
Nazca Ridge, Pacific Ocean, 189, 198
Nea Kameni, Santorini, Aegean Sea, 320
Near Islands, Aleutian Islands, 100, 103
Negro (Cerro), Nicaragua, 150, 155, 156, 158, 159
Nejapa, Nicaragua, 150
Nekoma, Honshu, Japan, 262, 285
Nemrut, Turkey, 331, 338, 340
Nepal, 447, 451, 454, 680, 690, 692
Nevada, USA, 120, 124, 422
Nevado de Toluca, Mexico, 138, 140, 141, 142–143
Nevado Ojos Del Salado, Chile/Argentina, South America, 194
Nevis, Lesser Antilles, 169, 173, 175–177
Nevis Peak, Nevis, Lesser Antilles, 173
Newberry Volcano, Oregon, USA, 27, 40, 41, 50, 60, 70, 72, 117
New Britain, Papua New Guinea, 1, 13, 16, 225–229, 238, 242, 442, 449, 450, 467, 552, 554, 564
New Britain Trench, Solomon Sea, 227, 229, 238
Newfoundland, Canada, 684, 685
New Georgia, Solomon Islands, Pacific Ocean, 530
New Guinea, 13, 19, 209, 226–229, 236, 240, 390, 439, 440, 441, 442, 448–451, 467, 533, 542
New Guinea Highlands, Papua New Guinea, 80, 225, 226, 229, 232, 233, 235, 239–242
New Hebrides, 27, 40, 41, 50, 60, 70, 72, 73, 302, 366, 527, 530, 533–538, 540, 542, 558, 559, 582
New Ireland, Papua New Guinea, 226, 227, 229, 230
New Ireland Arc, Papua New Guinea, 225
New Mexico, USA, 19, 455, 567
New South Wales, Australia, 684

New Zealand, 1, 27, 40, 41, 50, 60, 63, 70, 72, 73, 79, 80, **245–258**, 365, 366, 368, 372, 406, 411, 415, 418, 424, 438–442, 559, 688
Ngatoro Basin, 245, 246, 249
Ngauruhoe, Tongariro Massif, New Zealand, 249, 250, 418, 419, 424
Nias, Indonesia, 209
Nicaragua, 43, 149, 150, 152, 154–157, 163, 164, 438, 443, 444, 671, 672, 674
Niger, 686
Nigeria, 454, 455, 606, 686
Niijima, Izu Islands, Japan, 262, 286
Nikolski, Aleutian Islands, 101
Nila, Banda Sea, Indonesia, 216, 219
Nilahue, Chile, 192
Nile, River, 592, 595, 600, 605
Niragongo, 670
Niseko, Hokkaido, Japan, 262, 285
Nishina Jima, Volcano Islands, 296, 298, 299
Nisyros, Aegean Sea, 316–323
Niuatoputapu, Tonga, 246, 248, 254
Nomuka Group, Tonga, 247, 248
Noranda, Quebec, Canada, 578, 583, 684, 686
Norseman, Australia, 576
North American Plate, 2, 115, 138, 149, 171
North Atlantic Ridge, 396
North Bismarck Plate, 226, 229
North Carolina, USA, 692
North Crater, Tongariro Massif, New Zealand, 249
North-East Honshu (North-East Japan) Arc, Japan, 259–262, 264–272, 276–278, 284, 285, 287, 288
North Fife Hills, Fife, Scotland, UK, 613
North Fork, California, USA, 681
North Island, New Zealand, 245, 256
Northland Arc, New Zealand, 245, 248–250, 253
North Meyer Islands, Kermadec Islands, 254
North-West Territories, Canada, 581
Norway, 632, 635
Novarupta, Alaska, USA, 421
Nunivak Island, Alaska, USA, 557
Nûk, Greenland, 648
Nyanza, Kenya, 580, 581
Nyiragongo, Zaïre, 670

Obi, Indonesia, 209
Ochil Hills, Scotland, UK, 613, 625
Ogasawara Islands, Izu-Bonin arc, 261
Ogliuga, Aleutian Islands, 103
Ohakune, New Zealand, 249
Okcheung, Korea, 691
Oki-Dogo Islands, Japan, 262, 267, 286
Okmok Volcano, Aleutian Islands, 101, 105
Olca, Chile, 191
Ollague, Chile, 191, 199
Oman, 525, 526, 541–546
Oman, Gulf of, 543
Omuroyama, Izu Islands, Japan, 262, 286
Ontario, Canada, 575, 579, 581, 583

Onverwacht, South Africa, 650
Oregon, USA, 116–124, 126, 127, 129, 131, 419, 422
Oregon Coast Range, USA, 116, 118, 119, 120
Oriental Province, Mexico, 138
Orizaba, Mexico, 138, 140, 141, 142–143
Orkney Islands, Scotland, UK, 612, 624, 628, 635
Orosi, Costa Rica, 150
Oshima (Izu-Oshima), Izu Islands, Japan, 274, 296, 297, 533, 535, 540
Oshima-Kojima, Hokkaido, Japan, 262, 285
Oshima-Oshima, Hokkaido, Japan, 262, 285
Osorno, Chile, 192

Pacaya, Guatemala, 150, 155, 156, 158, 159
Pacha (Cerro), Lake Coatepeque, El Salvador, 158, 159
Pachuca (Hidalgo), Mexico, 140
Pacific–Antarctic Ridge, 373
Pacific–Kula Ridge, 270
Pacific Ocean, 1, 3–5, 12, 13, 16, 19, 25–29, 34, 40, 41, 43, 44, 46, 48, 50, 51, 53, 54, 56–58, 60, 68–70, 72–77, 79, 80, 88–95, 100, 105, 116–119, 122, 133, 137, 139, 153, 190, 195, 209, 227, 245, 246, 248, 259–262, 264, 269–271, 285, 295, 351, 371, 372, 373, 375–377, 396, 397, 438, 440, 442, 444, 448, 449, 453, 557–559, 604, 606, 682
Pacific Plate, 2, 102, 104, 105, 138, 225, 226, 229, 239, 256, 270, 295
Pagan Island, Mariana Islands, 293, 294, 296, 297, 299–304
Pago, New Britain, Papua New Guinea, 226, 231
Palaea Kameni, Santorini, Greece, Aegean Sea, Greece, 320
Palinuro Seamount, Aeolian Islands, Italy, 308, 309
Palma Sola, Mexico, 138, 139
Paluqek, Flores, Indonesia, 215, 219
Panama, 438, 443, 444, 448–452
Panarea, Aeolian Islands, Italy, 309, 311–313, 316
Panguna Mine, Bougainville Island, Papua New Guinea, 239
Panizos (Cerro), Bolicia, 195, 196
Papa Stour, Shetland Islands, UK, 613, 628
Papan Dajan, Java, Indonesia, 213, 218
Papua New Guinea, 1, 27, 40, 41, 50, 57, 60, 63, 70, 72, 73, 80, **225–244**, 271, 447, 452, 682
Parece Vela Basin, Mariana Islands, 294
Paricutin, Mexico, 138, 142, 403, 406, 473
Paros, Peru, 446
Paso Pino Hacado, Chile/Argentina, 192
Pastos Grandes, Bolivia, 195, 196
Patagonia, Argentina, 20, 201, 386
Patmos, Aegean Sea, Greece, 318
Paulet Island, Antarctica, 387
Pavlof Volcano, Alaska, USA, 101, 103–105
Pavlof Sister Volcano, Alaska, USA, 101, 105
Payun Matru, Argentina, 198
Peleng, Indonesia, 209

Pembrokeshire (Dyfed), Wales, UK, 613, 622
Penarfynydd, Gwynedd Wales, UK, 623
Penguin Island, Antarctica, 371, 374, 376, 379, 388, 390–393, 395, 527, 530, 533–538, 540
Pentland Hills, Scotland, UK, 613, 627
Peristeria, Santorini, Aegean Sea, 320
Peru, 19, 187, 188, 191, 193–195, 197–199, 202, 365, 366, 368, 377, 411, 438, 444, 445, 448, 449, 451, 453, 552, 553, 560, 562, 567, 644, 649, 652
Peru–Chile Trench, 189, 198
Petaquilla, Panama, 681
Philippines, 27, 40, 41, 50, 60, 70, 72, 73, 208, 220, 412, 442, 682, 687
Philippine Sea, 261, 262, 269
Phuket, Thailand, 689
Phyriplake-Traphores, Milos, Aegean Sea, 321
Pichincha, Ecuador, 190
Pico de Orizaba, Mexico, 139–143
Picture Gorge, Idaho, USA, 478
Pihanga, New Zealand, 249, 250
Pino Hachado, Chile/Argentine, 198, 199, 201
Pliny Trench, Aegean Sea, 318
Poás, Costa Rica, 150
Pocho, Argentina, 198
Pogromni Volcano, Aleutian Islands, 101, 105
Poliegos, Milos Group, Aegean Sea, 318
Pollara, Salina, Aeolian Islands, 312
Pontesford, Shropshire, England, UK, 611
Pontus, Turkey, 328, 329, 339
Popocatepetl, Mexico, 138, 140, 141, 142–143
Poros, Aegean Sea, 316–319, 321–323
Porphyrion, Volos-Atlanti Group, Aegean Sea, 318, 322
Port Lockroy, Wiencke Island, Antarctica, 378
Port Sudan, Sudan, 605
Portraine, Dublin, Republic of Ireland, 613, 619
Potrerillos, Chile, 681
Pozzomaggiore, Sardinia, Italy, 354
Pribilof Islands, Bering Sea, 101, 557
Primavera, Mexico, 140
Primoreye, Japan, 264
Principe, Gulf of Guinea, Africa, 542
Pta Cugputtada, Northwest Sardinia, Italy, 354, 358, 362, 365–367
Puerto Montt, Chile, 192
Puerto Rico, Greater Antilles, 168, 442, 443
Puerto Rico Trench, North Atlantic Ocean, 169, 170, 171, 443
Puhipuhi, New Zealand, 249
Pukeonake, Tongariro Massif, New Zealand, 249
Pura Beser, Indonesia, 215, 219
Pureora, New Zealand, 249
Purico (Cerro) Chile, 195, 196
Purple Volcano, Alaska, 101, 105
Putana, Chile, 191
Puturge, Turkey, 329
Puyehue, Chile, 192
Pyramid, Alaska, USA, 681

Qualibou, St. Lucia, Lesser Antilles, 173
Queensland, Australia, 25, 27, 38, 40, 41, 44, 46, 48, 50, 52, 57, 59, 61, 65, 66, 68, 71, 79, 80
Quellaveco, Peru, 681
Quemado (Cerro), Guatemala, 157–159, 162
Questa, New Mexico, USA, 681
Quezaltenango, Guatemala, 157
Quill The, St. Eustatius, Lesser Antilles, 173
Quilotoa, Ecuador, 190
Quito, Ecuador, 190
Quizapu, Chile, 307

Rabaul, New Britain, Papua New Guinea, 226, 229, 231, 235, 239, 241, 390, 421
Rabaul Province, Papua New Guinea, 231, 233
Rangitoto Range, North Island, New Zealand, 256
Rangoon, Burma, 689
Raoul Island, Kermadec Islands, 246, 248, 250, 251, 254
Rat Island, Aleutian Islands, 103, 105
Raung, Java, Indonesia, 213, 218
Raung, Moluccas, Indonesia, 217, 219
Recheschnoi Volcano, Aleutian Islands, 101, 105
Red Crater, Tongariro Massif, New Zealand, 249, 250
Red Sea, 346, 347, 592, 600, 605
Reef Point, Cook Island, South Sandwich Islands, 16
Reunion Island, Indian Ocean, 555, 557
Reventador, Ecuador, 188
Reykjanes Ridge, Iceland, 557
Rhobell Fawr, Gwynedd, Wales, UK, 613, 619, 620, 622, 632–634
Rhodes, Aegean Sea, 318, 319
Rhodesia (Zimbabwe), 580, 581, 585, 650
Rhiw, Gwynedd, Wales, UK, 623
Rincon De La Vieja, Costa Rica, 150
Rindjani, Lombok, Indonesia, 215, 219
Rinihue, Chile, 192
Rio Blanco, Chile, 681
Rio Caliente, Mexico, 428
Rio Verde, Mexico, 138
Rishiri, Hokkaido, Japan, 262
Ritter, New Britain, Papua New Guinea, 226, 230, 231
Rivière Blanche, Mount Pelée, Martinique, Lesser Antilles, 409, 410
Robert Island, Antarctica, 377
Roccamonfina, Italy, 550, 551, 565, 652
Roman Province, Italy, 25, 27, 29, 36, 40, 44, 46, 48, 53–55, 57, 59, 60, 70, 71, 73, 79
Romang, Banda Sea, Indonesia, 208, 216, 219
Rosanel Island, Antarctica, 387
Rosita, Nicaragua, 681
Ross Island, Antarctica, 387, 395, 396, 557
Rossing Mine, Namibia, 692
Rota, Mariana Islands, 294, 295
Rotorua, New Zealand, 249–251

Rouyn, Quebec, Canada, 578
Ruapehu, New Zealand, 249, 250, 251
Rumble III, Pacific Ocean, 246, 248, 250, 251
Rusteq, Oman, 543
Ryukyu Arc, Japan, 259–261, 265
Ryukyu Islands, Japan, 287
Ryukyu Trench, 260

Saba, Lesser Antilles, 168, 169, 173, 175–177, 180, 182
Sagami Trough, Pacific Ocean, 261
Sabalan, Iran, 331, 338–341, 344, 345
Sahand, Iran, 331, 338–340, 342, 344, 346, 347
St. Abbs Head, Scotland, UK, 613, 625, 627, 629, 632
St. Andrew Strait, Admiralty Island, Papua New Guinea, 226, 230
St. Bartholomew, Lesser Antilles, 169
St. Eustatius, Lesser Antilles, 168, 169, 173, 175, 180
St. George Island, Pribilof Islands, 101
St. Helena, Atlantic Ocean, 557
St. Kitts, Lesser Antilles, 14, 18, 168, 169, 171–174, 175–177, 179, 180, 542, 553, 554, 564
St. Lucia, Lesser Antilles, 17, 18, 168, 169, 171, 173, 176, 177, 181, 542
St. Martin, Lesser Antilles, 169, 443
St. Paul Rocks, Atlantic Ocean, 497, 498
St. Paul Island, Pribilof Islands, 101
St. Pierre, Martinique, Lesser Antilles, 410, 411, 415, 416
St. Vincent, Lesser Antilles, 17, 167–169, 171, 173–177, 179–181, 408–411, 415, 419, 420, 422, 424
Saipan, Mariana Islands, 294, 295
Sakhalin, USSR, 261
Sakurajima, Kyushu, Japan, 262, 274, 288, 404, 420
Salak, Java, Indonesia, 213, 218
Salala, Sudan, 596
Salar de Atacama, Chile, 194
Salar de Uyuni, Bolivia, 194
Salida, Sumatera, Indonesia, 221
Salina, Aeolian Islands, 309, 311–313, 315, 316, 323
Salt Lake Crater, Hawaii, 498, 499
Sambe, Honshu, Japan, 262, 287
Samoa, Pacific Ocean, 246
San Andres, Mexico, 139
San Antioco, Sardinia, Italy, 354
San Cristobal, Nicaragua, 150, 154, 156, 671, 672, 674
San Juan Volcanic Field, USA, 79
San Luis, Mexico, 138
San Martin Tuxtla (San Andres Tuxtla), Mexico, 138, 139
San Miguel, El Salvador, 150, 155, 156, 158, 159
San Pablo, Chile, 191, 194, 199, 565
San Salvador, El Salvador, 154
San Vicente, El Salvador, 150
Sanak Island, Aleutian Islands, 101
Sandness, Shetland Islands, UK, 613, 628

Sandwich Plate, 373
Sanganguey, Mexico, 140
Sangay, Ecuador, 188, 190
Sangeang Api, Sumbawa, Indonesia, 214
Sangenges, Sumbawa, Indonesia, 219, 221
Sangihe, Indonesia, 207, 209, 210, 220
Santa Ana, El Salvador, 150, 162
Santa Maria, Guatemala, 150, 153, 156–160, 162, 407
Santa Rita, New Mexico, USA, 681
Santiaguito, Guatemala, 150, 155–157, 161, 162, 404, 410, 416
Santorini, Greece, Aegean Sea, 307, 314–322, 323, 415
Sao Miguel, Azores, 550, 551
Sardinia, Italy, 308, 331, **353–370**, 687
Sardinian Microplate, 351
Sarichef Volcano, Aleutian Islands, 100
Sarigan Island, Mariana Islands, 293, 294, 296–301, 302
Sarmiento, Chile, 21, 391, 530
Saronic Gulf, Greece, 321
Sarroch, Sardinia, Italy, Europe, 354, 360, 362
Sasak, Sulawezi, Indonesia, 220
Saudi Arabia, 455, 591, 593, 595, 596, 598–602, 604, 605
Scandinavia, 455, 635, 640
Schouten Island, Papua New Guinea, 226, 236, 237
Sciarra Volcano, Stromboli, Aeolian Islands, 313
Scotia Arc, 21, 351, **371–400**
Scotia Plate, Antarctica, 373
Scotia Sea, Antarctica, 351, 372, 376, 377, 381, 397
Seal Nunataks, Antarctica, 374, 379, 388, 393, 395
Sea of Okhotsk, 261, 262
Seguam Volcano, Aleutian Islands, 100, 105, 106
Segula Volcano, Aleutian Islands, 100, 103, 105
Semangko, Sumatera, Indonesia, 208, 209
Semail, Oman, 541, 544
Semeru, Java, Indonesia, 213, 218
Semichi Island, Aleutian Islands, 100
Semisopochnoi Volcano, Aleutian Islands, 100, 103, 105–107
Seraja, Bali, Indonesia, 215, 218
Seram, Indonesia, 208–210, 222
Seram Trough, Indonesia, 222
Serva, Banda Sea, Indonesia, 212, 216, 218, 219, 222
Setouchi, Honshu, Japan, 261, 271, 286
Shabani, Zimbabawe, 576
Shap, Cumbria, UK, 615
Sheep Mountain, Colorado, USA, 681
Shelve, Salop, England, UK, 613, 620
Shemya Island, Aleutian Islands, 100, 103
Shetland Islands, Scotland, UK, 612, 624, 628, 632
Shetland Plate, Antarctica, 373
Shelve, Shropshire, England, UK, 620
Sheveluch Volcano, Kamchatka, 419
Shikoku, Japan, 261, 267, 268, 270
Shikoku Basin, 261, 267, 296

Shikotsu, Hokkaido, Japan, 273, 408, 420
Shiretoko Peninsula, Japan, 261
Shirouma-Oike, Honshu, Japan, 272
Shiskaldin Volcano, Aleutian Islands, 101, 105
Shoal Lake, Ontario, Canada, 577
Showa-Shin-zan, Hokkaido, Japan, 411
Shumagin Islands, Aleutian Islands, 101
Siam, Gulf of, 690
Sibajak, Sumatera, Indonesia, 212, 218
Siberia, USSR, 100, 105, 270
Siberut, Indonesia, 209
Sicily, Italy, 308, 309
Sidlaw Hills, Tayside, Scotland, UK, 613, 626, 627
Sierra De San Carlos, Mexico, 138
Sierra De Tamaulipas, Mexico, 138
Sierra Madre, Mexico, 686, 688
Sierra Negra, Mexico, 138, 140
Sierra Nevada, California, USA, 116, 444, 445, 449, 450, 455
Sikhot Alin, Khrebet, USSR, 261, 270
Simandir, Santorini, Aegean Sea, 320
Simeule, Indonesia, 209
Singkep, Indonesia, 209
Sin Iwo Jima, Volcano Islands, 296, 298
Sirung, Indonesia, 215, 219
Sivas, Turkey, 330, 338, 339
Siwalik Range, Nepal, 691
Skaergaard, Greenland, 283, 622
Skalgul, Aleutian Islands, 103
Skaros-Therasia, Santorini, Aegean Sea, 320
Skiddaw, Cumbria, England, UK, 614, 615, 619, **633**, 634
Skomer Island, Dyfed, Wales, UK, 613, 619, 621, 623, 632–634
Skye, Hebrides, Scotland, UK, 454
Slamet, Java, Indonesia, 213, 218
Smith Island, Antarctica, 376
Snake River Plain, Idaho, USA, 27, 40, 41, 50, 60, 70, 72, 73, 122
Snowdonia, Gwynnedd, Wales, UK, 613, 620, 622, 623
Sohar, Oman, 543
Solomon Islands, 27, 40, 41, 50, 60, 70, 72, 73, 226, 229, 239, 442, 450, 467, 530, 682
Solomon Sea, Papua New Guinea, 225, 227, 241, 448, 449
Solomon Sea Plate, 226, 227, 228, 229, 238, 239, 240
Solor, Indonesia, 215, 219
Solway Firth, UK, 612, 631, 633, 634
Sombrero, Lesser Antilles, 168
Somerset, England, UK, 619, 623
Sorikmarapi, Sumatera, Indonesia, 212, 218
Soromundi, Sumbawa, Indonesia, 219, 221
Soufrière, St. Vincent, Lesser Antilles, 167, 173, 174, 180, 408, 410, 415, 419, 420, 422
Soufrière Hills, Montserrat, Lesser Antilles, 173, 180
South Africa, 576

South America, 3, 25, 27, 29, 30, 40, 41, 42, 44, 46, 48, 50, 53, 54, 60, 70, 72, 73, 78, 94, 170, 187–189, 193, 198, 371–373, 376, 383, 444, 565, 643, 648, 681
South American Plate, 2, 171, 188, 189, 198, 373, 376, 397
South Bismarck Plate, 226–230, 236–238
South Fiji Basin, 246
South Georgia, Falkland Island Dependencies, Antarctica, 376
South Honshu Ridge, (West Mariana Ridge), 294–297
South Meyer Islands, Kermadec Islands, 254
South Pass, Wyoming, USA, 581
South Sandwich Islands, 3, 14–16, 20, 302, 351, 372, 377, 380, 381, 388, 394, 397, 439–442, 527, 533, 535–538, 540, 542, 552–554, 564, 582, 643
South Scotia Ridge, Antarctica, 372, 376, 377, 381
South Shetland Islands, Antarctica, 25, 27, 38, 40, 41, 44, 46, 48, 50, 73, 371–378, 380–391, 395, 396, 397, 527
South Shetland Trench, Antarctica, 375, 376, 395
South Soufriere Hill, Montserrat, Lesser Antilles, 174, 180
South Volcano, Pagan Island, Mariana Islands, 296
South-West Honshu Arc, 259–261, 265, 267, 268, 285
South-West Japan Arc (South-West Honshu and Rykyu Arcs), 259, 262, 272, 275–278, 284, 287–289
Southern Uplands, Scotland, UK, 612–614, 625, 632–635
Spain, 27, 40, 41, 50, 60, 70, 72, 73, 308
Sri Lanka, 211
Statia, Lesser Antilles, 176
Stockdale, Cumbria, England, UK, 615, 624
Strabo Trench, Aegean Sea, 318, 319
Straiton, Strathclyde, Scotland, UK, 613
Strawberry Mountain, Oregon, USA, 120, 121
Stromboli, Aeolian Islands, 308, 309, 311–313, 316
Strombolicchio, Stromboli, Aeolian Islands, 313
Sudan, 591, 593, 595, 596, 598–602, 604, 605
Sula, Indonesia, 209
Sulawezi, Indonesia, 207, 209, 210, 214, 216, 220
Sulawezi Sea, 210
Sumaco, Ecuador, 188, 190, 198, 199
Sumatera, (Sumatra), Indonesia, 207–212, 214, 218, 220, 221, 680, 682
Sumba, Indonesia, 209
Sumbawa Island, Indonesia, 207–209, 211, 214, 215, 221
Sumbing, Java, Indonesia, 213, 218
Sunda Island Arc, Indonesia, 207, 208, 211, 447, 552, 553
Sunda Shelf, Indonesia, 207, 209, 221
Sundoro, Java, Indonesia, 213, 218
Suoh, Sumatera, Indonesia, 211
Superior Province, Ontario, Canada, 576, 581
Suphan Dag, Turkey, 331, 340, 342, 344, 346, 347
Surtsey, Iceland, 670
Sweden, 632

Tabar, Papua New Guinea, 225, 226, 229, 230, 233
Tabriz, Iran, 331
Tacana, Guatemala, 150
Tafahi, Tonga, 246, 248, 254
Taif, Saudi Arabia, 596, 602
Taiwan, 27, 40, 41, 50, 60, 70, 72, 73, 261, 556–558, 682
Tajamulco, Guatemala, 150, 158, 159
Takab, Iran, 331
Talakmau, Sumatera, Indonesia, 212, 218
Talasea, New Britain, 16, 64, 467
Talaud, Indonesia, 209, 217
Talaud-Mayu Ridge, Indonesia, 209, 210
Talawe Volcano, Bismarck Sea, Papua New Guinea, 231
Tanaga Volcano, Aleutian Islands, 100, 103, 105, 106
Tama Lakes, Tongariro Massif, New Zealand, 249, 250
Tambora, Sumbawa, Indonesia, 207, 211, 214, 219, 411
Tandikat, Sumatera, Indonesia, 212, 218
Tanga, Papua New Guinea, 226, 230
Tangkuban Prahu, Java, Indonesia, 213, 218, 542
Tapadaa, Sulawezi, Indonesia, 220
Tarn Moor, Cumbria, England, UK, 615
Tauhara, New Zealand, 249
Taumarunui, New Zealand, 249
Taupo, New Zealand, 249, 254, 255, 257
Taupo Arc, New Zealand, **245–258**
Taupo Volcanic Zone, New Zealand, 79, 80, 245, 246, 248–255, 440
Tauranga, New Zealand, 149, 245
Taurus Mountains, Turkey, 327, 329, 331, 339, 342
Tavurvur, Rabaul, Papua New Guinea, 231, 239
Tayvallich, Argyllshire, Scotland, UK, 614, 633
Tbilisi, USSR, 331
Tecapa, El Salvador, 150
Tecuamburro, Guatemala, 150
Teheran, Iran, 329
Telica, Nicaragua, 150
Te Mari, Tongariro Massif, New Zealand, 249, 250
Temuco, Chile, 192
Tendurek, Turkey, 331, 338, 340
Tequila, Mexico, 140
Ternate, Moluccas, Indonesia, 210, 216, 217, 219
Teun, Banda Sea, Indonesia, 216, 219
Texas, USA, 139
Teziutlan, Mexico, 140
Thailand, 447–451, 454, 682, 688, 689, 691
Thera, Santorini, Aegean Sea, 320
Thiesi, Sardinia, Italy, 354
Thingmuli, Iceland, 315
Three Kings Rise, Indian Ocean, 246
Three Sisters, Oregon, USA, 117, 126, 128
Tibesti Massif, Libya, 455, 606
Tibet, 447
Tidore, Moluccas, Indonesia, 217, 219
Timaganu, Ontario, Canada, 579
Timor Island, Indonesia, 208, 209, 222

Timor Sea, 209
Timor Trough, Indonesia, 208
Tinian, Mariana Islands, 294, 295, 297
Tintaya, Peru, 681
Titiraupenga, New Zealand, 249
Tjerimai, Java, Indonesia, 213, 218
Tjikurai, Java, Indonesia, 213, 218
Tobago, Lesser Antilles, 169
Tobago Trough, 169, 170
Tobelo, Moluccas, Indonesia, 217, 219
Tocopilla, Chile, 194
Tofua, Tonga, 15, 247, 248, 250, 251
Tokachi, Hokkaido, Japan, 262, 286, 420
Tokyo, Honshu, Japan, 296, 297
Toliman, Guatemala, 158, 159, 162
Toluca, Mexico, 141–143
Tondano, Moluccas, Indonesia, 217
Tonga, 1, 9, 14–16, 27, 40, 41, 50, 60, 63, 64, 70, 72, 73, 79, **245–258**, 365, 368, 533, 538, 540, 542
Tonga Arc, **245–258**, 301, 380, 439, 440–442, 533, 558, 643
Tonga-Kermadec—New Zealand Arc, **245–258**, 368
Tonga Trench, 245, 246, 248, 559
Tongariro, Tongariro Volcanic Centre, New Zealand, 250, 256, 406
Tongariro Massif, New Zealand, 249, 250, 251
Tongariro Volcanic Centre, New Zealand, 245, 248–250, 252, 254–256
Tongatapu, Tonga, 247, 248
Toquepala, Chile, 681
Towada, Honshu, Japan, 262, 285, 420
Tower Island, Bransfield Strait, Antarctica, 379
Trachylas, Milos, Aegean Sea, 321
Trafalgar, Papua New Guinea, 226, 240, 241
Trans-Pecos Province, Texas, USA, 139
Treffgarne, Dyfed, Wales, UK, 623, 633, 634
Trinidad Lesser Antilles, 169, 557
Tristan Da Cuhna Island, Atlantic Ocean, 557
Tronador, Chile, 192, 197
Tuluman, Admiralty Island, Papua New Guinea, 226, 230
Tunguraha, Ecuador, 190
Tupungato, Chile, 202
Turkey, 3, 6, 308, 412, 542, 687
Turkish (Anatolian) Plate, 318, 329, 347
Turret Point, King George Island, Antarctica, 376, 379
Tuzla, Turkey, 332
Two Hummock Island, Bransfield Strait, Antarctica, 379
Tyrrhenian Abyssal Plain, 308, 310, 317
Tyrrhenian Sea, 19, 307–309, 316, 366

Uchi Lake, Superior Province, Ontario, Canada, 576, 577
Ulak Island, Aleutian Islands, 102, 103
Ulawun, New Britain, Papua New Guinea, 226, 230–232, 238
Uliagal Island, Aleutian Islands, 101, 105

Umnak Island, Aleutian Islands, 99, 101, 103, 106, 107
Unalaska Island, Alaska, USA, 99, 101, 103, 106
Unalga, Aleutian Islands, 103
Una Una, Moluccas, Indonesia, 217, 219
Unea, Witu Island, Papua New Guinea, 226, 239
Ungaran, Java, Indonesia, 213, 218
Unimak Island, Alaska, 99, 101, 104, 106
Unzen, 411
Uracas (Farallon De Pajaros), Mariana Islands, Pacific Ocean, 294–299
Urad, Colorado, USA, 681
Urals, USSR, 682
Urgup, Turkey, 330, 335
Urgup Basin, Turkey, 335, 338, 344
Urla, Turkey, 333, 335, 337
USA, 3, 25, 27–30, 32, 40–42, 44, 46, 48, 50, 51, 53, 54, 57, 60, 63–65, 68, 70–73, 78–80, 92–95, 115, 117, 138, 412, 438, 444, 445
USSR, 682
Ustica, Aeolian Islands, 309
Usu, Hokkaido, Japan, 262, 285, 286, 404, 411
Utah, USA, 64, 65
Utuado, Puerto Rico, USA, 681
Uturunca Volcano, Bolivia, 199
Uyuni, Bolivia, 194

Valdivia, Chile, 192
Valle De Puebla, Mexico, 142–143
Valle De Mexico, Mexico, 143
Valley, British Columbia, Canada, 681
Valley, of Desolation, Dominica, Lesser Antilles, 173
Valley of Mexico, Mexico, 141
Valley of 10,000 Smokes, Alaska, 419–421, 424
Vancouver Island, British Columbia, Canada, 684
Vanua Levu, Fiji, Pacific Ocean, 542, 685
Varzin, Rabaul, Papua New Guinea, 239
Vava'u Group, Tonga, Pacific Ocean, 247, 248
Venezuela, 168, 443
Venezuela Basin, Caribbean Sea, 169
Veniaminof Volcano, Alaska, 101, 105, 106
Verde (Cerro), Peru, 681
Vesuvius, Naples, Italy, 415, 550, 551
Victoria Land, Antarctica, 395
Victory, Papua New Guinea, 226, 234, 240, 241
Villacollo (Cerro), Chile, 195, 196
Villanova Monte Leone, Sardinia, Italy, 354
Villarica, Chile, 192, 199
Virgin Islands, Lesser Antilles, 169, 442, 443
Visokoi, South Sandwich Islands, 16
Viti Levu, Fiji, 19
Vokeo, Papua New Guinea, 226, 229, 236
Volcan Hudson, Chile, 189, 198, 201
Volcano Islands, Pacific Ocean, 1, **293–306**
Volos, Aegean Sea, 316, 318–322
Vulcan, Rabaul, Papua New Guinea, 231, 239
Vulcanello, Vulcano, Aeolian Islands, 312, 313
Vulcano, Aeolian Islands, 309, 311–313, 315, 316

Wabigoon Lake, Ontario, Canada, 576
Wadi Bidah, Saudi Arabia, 596
Wadi Qatan, Saudi Arabia, 596
Wadi Wassat, Saudi Arabia, 596
Waitakere Volcanic Arc, New Zealand, 246, 248
Wales, UK, 614, 620, 622, 623, 632–634, 654
Washington, USA, 116–120, 123, 445, 451, 453
Waterford, Waterford, Republic of Ireland, 613, 619
Wuatom Island, Rabaul, Papua New Guinea, 239
Weddell Sea, Antarctica, 351, 371–373, 396, 397
Welsh Basin, UK, 611–614, 616, 619–624, 633
Westdahl Volcano, Aleutian Islands, 101, 105
Western Australia, 580, 581, 586
Western Mine, Vancouver Island, British Columbia, Canada, 684
West Indies, see Lesser Antilles
West Japan Volcanic Belt, Japan, 260
Wetar, Banda Sea, Indonesia, 209, 210, 214–216, 219, 221
Whakatane, New Zealand, 249
Whale Island, New Zealand, 249
White Island, New Zealand, 245, 249, 251, 252, 254
Wicklow, Wicklow, Republic of Ireland, 613, 619

Wiencke Island, Antarctica, 378
Willamette Valley, Oregon, USA, 119, 120
Willamette-Puget Sound Depression, Oregon, USA, 118, 120, 124
Willaumez Peninsula, New Britain, Papua New Guinea, 227, 238
Witu Island, Papua New Guinea, 226, 229, 238, 239
Woodlark Basin, Solomon Sea, 226, 228, 230
Woodlark Rise, Solomon Sea, Papua New Guinea, 226, 228, 229
Wyoming, USA, 64, 65, 118, 581

Yali, Aegean Sea, 317–322
Yelia, Papua New Guinea, 226, 232, 241, 242
Yellowknife, North West Territories, Canada, 576, 580, 581, 583, 584
Yellowstone Park, USA, 19, 122
Yerington, Nevada, USA, 681
Yotei, Hokkaido, Japan, 285
Yunaska Volcano, Aleutian Islands, 101, 105

Zealandia Bank, Mariana Islands, Pacific Ocean, 297
Zagros Range, Iran, 328, 331

Subject Index

'Single page numbers (e.g. 10) refer to individual entries of the listed term, grouped page numbers (e.g. 10–20) indicate sequences of individual entries of the listed term, while emphasised (bold) page numbers (e.g. **10–20**) indicate major reference sections to the listed term'.

Absarokite, 40
Abyssal basalt/tholeiite (*see* Mid-ocean ridge basalt)
Accretionary prism/wedge, 4, 222, 256, 613, 631, 633, 634, 639, 644, 647, 654, 656, 680
Actinolite, 579, 687
Active continental margin, 1, 3, 6, 9, 12, 17, 25, 29–33, 41, 115, 338
Adamellite, 443, 623, 689
Aegirine, 282
Ag_2S–Ag buffer, 512
Ag mineral deposits, 221
Agglomerate, 252, 332, 335, 339, 575, 578, 594, 598
Agglutinate, 405
Airborne sampling (small eruptive plumes), 669, 672, 674, 675
Air-fall phase (plinian eruption), 422
Air-fall pumice, 312, 320
Air-fall pyroclastic rocks/air-fall ash, 5, 151, 239, 407, 409–411, 415, 419, 424, 426
Albite–anorthite (plagioclase) system, 468, 469
Albitization, 578, 686
Alkali amphibole, 280, 282
Alkali(ne) basalts, **19, 20**, 41, 79, 107, 111, 140, 157, 167, 173, 177, 178, 181, 187, 196, 200, 201, 230, 240, 241, 267, 270, 284, 286, 333, 340, 380, 387, 393, 395, 396, 454, 458, 466, 509, 526–530, 536, 537, 540, 619, 628, 629, 656, 657
Alkali–calcic igneous rocks, 437, 439, 440, 450, 454–458, 643
Alkali feldspar (*see also* K-feldspar), 160, 235, 280, 335, 428, 595, 621, 624
Alkali–lime index, 12, 440, 441, 450
Alkali(ne) olivine basalt, 279, 281, 287, 379, 393, 624
Alkaline volcanic rocks, 9, 11–14, 97, 119, 137–141, 145, 172, 175, 181, 187, 189, 198, 199, 203, 212, 225, 229, 230, 233, 259, 274, 275, 278, 279–282, 286, 295, 301, 302, 318, 327, 330, 333, 335, 336, 339–341, 345–347, 359, 372, 376, 392, 395, 396, 416, 440, 442–445, 447, 450, 454, 455, 457, 458, 484, 530, 536–540, 576, 582, 596, 611, 619, 620, 623, 624, 626, 628, 633, 635, 652, 653

Allanite, 160
Almandine, 214, 222, 468, 618
Alteration (volcanic rocks), 121, 271, 367, 390, 395, 555, 556, 560, 562, 563, 579, 580, 623
Ambonites, 210, 214, 219, 220, 222
Amphibole, 15–17, 20, 58, **66–68**, 80, 132, 140, 144, 160, 161, 167, 175, 178, 179, 182, 197, 214, 221, 237, 251, 283, 355, 396, 439, 463, 465, 469, 470–472, 474, 475, 477, 479–484, 497, 500, 503, 515, 517, 519, 541, 542, 575, 585, 586, 625
Amphibolite, 14, 256, 267, 515, 519, 545, 546, 575, 585, 594, 598, 653, 655
Analcite, 230, 628
Andalusite, 312
Andesine, 59, 125, 197, 303, 623, 625
Andesite crustal growth model, 637, 639, **640–645**
Andesite line, 12
Andesite, origin of term, 11, 12
Ankaramite, 213
Anorogenic basalt–tholeiitic andesite association, 25, 27, 38–41, 44–52, 57, 59, 61, 65–68, 79–81
Anorthite, 13, 17, 66, 160, 167, 175, 179, 468, 470, 492
Anorthoclase, 57, 59, 230
Anorthosite, 573
Anthophyllite, 477
Apatite, 80, 160, 230, 235, 251, 356, 396, 465, 482, 533, 541, 621, 625, 628, 654
Aplite, 445, 691
Aplogranite, 445
Arc-trench gap (*see also* Trench-arc gap), 633
Archaean (general), 6, 14, 21, 573, **575–590**, 592, 637, 639, 642, 643, 645, 648, 650, 652–655, 658, 680, 684, 686, 687
Archaean andesites, 573, **575–590**
  gneiss–granulite terraine, 645
  granite–greenstone terrains, 586
  granulite, 645, 647, 652
  heat production, 456
  heat flow, 639, 654
  intrusive rocks, 456

Argillic alteration, 682, 683
Aseismic front, 265
Ash/ash layer, 107, 153, 156, 211, 231, 273, 274, 405–407, 669–675
Ash cone, 313
Ash-cloud surges/ash-cloud surge deposits, 418, 423, 424
Ash-fall/ash-fall deposits, 242, 259, 273, 298, 299, 411, 601, 615
Ash-flow/ash-flow, 107, 109, 156, 270, 271, 408, 423, 425, 427, 578, 601
Ash-flow sheet, 422
Ash-flow tuff, 408
Ash horizons (ocean floor), 102–104, 153, 207, 209, 210
Ash hurricane, 426
Ash leachates, 164, 671, 672, 674, 675
Ash/pumice flow, 424
Ash turbidites, 429
Assimilation, 13, 78, 132, 255–257, 394, 558, 566, 619, 648, 651, 652, 655
Atmosphere, 637, 663, 666, 672
Au mineral deposits, 221, 688
Augite, 15, 20, 25, 26, 51, 57, **62–65**, 74–78, 81, 123, 143, 160, 167, 175, 178, 211–214, 230, 251, 252, 280, 282, 303, 340, 355, 575, 579, 618, 623, 625
Authigenic materials/sediment, 555, 563, 650

Bandaite, 12
Back-arc basalt, 552, 553, 649
Back-arc basin, 193, 239, 274, 307, 317, 376, 442, 447, 525, 526, 542, 546, 600, 601, 603, 604
Back-arc magmatic belt, 678, 679, 680
Back-arc magmatic belt mineralization, 637, 677, 678, 686, **688–690**, 691, 692
Back-arc setting/spreading, 11, 20, 260, 263, 286, 308, 372, 373, 376, 377, 380, 381, 392, 397, 530, 553
Ballistic blocks, 406
Banakite, 40
Banded Fe formation, 578
Barite, 685
Basanite, 167, 173, 175, 181, 201, 211, 230, 466
Basanitoids, 19, 167, 177, 178
Base surge deposits, 425
Batch fractional crystallization, 484
Batholith, 1, 79, 102, 116, 118, 190, 270, 271, 437, 438, 444–446, 448–450, 452, 453, 456–458, 595, 596, 644, 648, 649, 680, 688, 690
Benioff zone (Wadati-Benioff zone), 3, 13, 15, 19, 26, 78, 79, 104, 108, 109, 115, 154, 155, 187–189, 207–210, 214–217, 220–222, 228, 229, 237–239, 241, 242, 245, 248, 259, 262–265, 267, 284, 285, 289, 296, 297, 307, 308, 310, 316–318, 397, 465, 467, 562, 629, 653, 656, 677, 679, 680, 687, 689
Benmoreite, 230, 340
Besshi-type mineral deposits, 684, 686

Bimodal tholeiite–tonalite (trondhjemite) suite, 642, 645, 654
Bimodal volcanic association, 27, 41, 51, 80, 118, 151, 180, 197, 207, 211, 213, 214, 216, 271, 575, 576, 587, 628, 642, 645
Bingham plastic, 419
Biosphere, 663
Biotite, 15, 17, 25, 55–60, **66**, 74–78, 160, 161, 175, 197, 207, 211, 213, 214, 221, 235, 260, 272, 280, 282, 283, 286–288, 313, 333, 338–340, 344, 355–357, 477, 539, 541, 542, 624, 625, 628, 679, 683, 689
Biotite–leucite tephrite, 214
Biotite–liparite, 320
Block and ash deposit, 422, 425, 428
Blocky lava flow, 240, 249
Blueschist, 4, 5, 227, 602, 603, 664, 665
Boninites, 11, 21, 271, 603
Bornite, 682
Bouguer gravity anomaly, 210, 263
Breadcrust block (bomb), 406
Breccia, 299, 575, 578, 594, 598, 623, 685, 686, 689
Breccia pipe, 645, 647
Bronzite, 62, 625, 627
Bronzite andesite, 259, 270, 271, 626, 627, 632
Bytownite, 12, 59, 175, 214, 251

$CaAl_2O_4$–MgO–SiO_2–Na_2O system, 471
$CaAl_2SiO_6$–Mg_2SiO_2 system, 496
CaO–MgO–Al_2O_3–SiO_2 system, 513
CaO–MgO–Al_2O_3–Na_2O–H_2O system, 470
$CaSiO_3$–$MgSiO_3$–Al_2O_3–Na_2O system, 471
$CaMgSi_2O_6$–Mg_2SiO_4–SiO_2 system, 493
Calc–alkali ratio, 440, 441, 450, 455, 458
Calcic igneous rocks, 437, 439–441, 449, 454, 457, 458
Caldera, 107, 120, 126, 140, 144, 152, 196, 229, 231, 237, 239, 241, 249–251, 259, 263, 268, 273, 275–277, 299, 313, 319, 320, 338, 379, 408, 421, 446, 624, 686, 688
Caldera, Crater Lake (Krakatau) type, 126, 259, 268, 273
Caldera, Haruna type, 268, 273
Caldera, Kilauea type (Oshima), 268
Caldera, Valles type, 273
Caledonian orogenic belt (Caledonian orogen, British Caledonides), 574, **611–636**,
 orthotectonic zone, 613
 paratectonic zone, 613
Carbonate, 578, 579
Carbon dioxide, 480, 482, 491, 501–503, 528
 role in metamorphism, 647
 solution in silicate melts, 491
Cassiterite, 686, 687, 691
Cation diffusion, 495
Cauldron subsidence, 239
Cauldron structure, 644
Ce anomalies, 395
Central volcano, 331, 335, 338, 339, 340

Chalcocite, 682
Chalcopyrite, 682, 685
Charnockite, 645, 646
Chile-type mineral deposits, 686
Chlorine, budget in volcanic gas, **669–676**
Chlorite, 579, 615, 618, 623, 654
Chrome-spinel (chromite), 314, 322, 393, 531
Ce anomaly, 387, 390
Cinnabar, 687
Clinopyroxene, 17, (see 62–66) 99, 109, 111, 143, 144, 160, 161, 175, 179, 202, 230, 235, 251, 252, 280, 283, 288, 333, 335, 338, 339, 355–357, 364, 366, 380, 393, 394, 470–472, 474, 475, 477, 479, 480–482, 497, 499, 513, 514, 516, 518, 528, 535, 541, 542, 578, 579, 615, 620–624, 628
C–O buffer, 512
C–O–CO$_2$ buffer, 509
Coesite, 474, 476, 477
Co-ignimbrite lag-fall deposit, 412, 426
Cold rock avalanches, 416, 420, 431
Collision zone, 1–6, 9, 97, 207, 210, 227, 228, 236, 317, 327, 328, 345, 347, 435, 437–447, 449, 451, 454, 457, 458, 573, 591–592, 601, 604, 606, 607, 613, 631, 635, 641, 643, 644, 656, 678–680, 682, 684, 687, 690, 692
Collision belt mineralization, 637, 677, 678, 686, 689, **690–692**
Comendite, 41, 230
Composite volcano (cone), 106, 107, 115, 121–123, 126, 174, 190, 192–194, 201, 231, 250, 251, 259, 263, 268, 271–274, 276–278, 293, 312, 313, 320, 379, 598, 677
Compound volcano, 194
Condensates (from fumaroles), 671, 674, 675
Continental crust, composition, **640**, 641, 642
   contamination, 11, 12, 14, 133, 137, 144, 145, 157, 167, 168, 177, 180, 182, 187, 202, 203, 220, 222, 245, 259, 288, 289, 314, 323, 353, 364–368, 394, 457, 463, 490, 523, 550, 539, 549, 552–554, 558, 559, 561, 562, 565–568, 584, 585, 625, 644, 648, 649, 651, 652, 658, 663, 679, 683
   fragmentation, 573, 574
   growth, 437, 458, 566, 573, 591, 607, 637, **639–658**, 663
   role in petrogenesis, **564–567**
Continental flood basalt, 651, 656
Continental margin volcanism (magmatism), 17–18, 25, 27, 78, 97, 115, 133, 307, 308, 327, 353, 363, 415, 416, 435, 437, 438, 447–451, 454, 465, 468, 483, 484, 552, 554, 573, 580, 588, 593, 616, 633, 637, 639, 644, 648, 658, 682
Continental tholeiites, 361, 647, 651
Convection, mantle wedge, 21, 655
   seawater through ocean crust, 664
Cordierite, 207, 210, 214, 219, 222, 310, 312, 468, 679
Cordilleran-type orogeny (orogenic belt), 5, 6, 639, 648, 651, 654, 658, 665, 680, 681, 684, 686, 687, 690

Coulée, 410
Covellite, 682
Cr-diopside xenocrysts, 466, 483
Craton, 80, 444, 452, 456, 574, 591, 592, 606, 607, 632
Cratonized island arc model, 591
Cristobalite, 280, 355, 428
Cryptodome, 403, 410, 411
Crustal contamination (see Continental crust, contamination)
Crustal (ground) deformation, 274
Crustal fusion (anatexis/melting), 1, 5, 6, 79, 80, 109, 203, 207, 220, 255, 257, 322, 333, 368, 435, 437, 451–453, 457, 567, 663, 680
Cu$_2$S–Cu buffer
(Cu, Fe, Ni) monosulphide, 509
Cummingtonite, 160, 161
Cumulates, 175, 230, 566, 639, 643, 646, 647, 655, 658
Cumulodome (see also Dome), 231
Cuprite, 682

Dehydration of oceanic crust (lithosphere, slab), 4, 5, 133, 167, 201, 225, 236, 237, 256, 390, 396, 452, 463, 465, 467, 484, 489, 490, 503, 504, 515–518, 549, 550, 555, 562, 563, 567, 639, 651, 653, 654, 666
Dense andesite surge (pyroclastic flow), 423
Detached slab, 239, 308, 310, 316, 346, 588
Diatreme, 613
Diopside (diopsidic pyroxene), 62, 214, 322, 396, 493
Diopside–albite–anorthite system, 468
Diopside–anorthite–H$_2$O system, 468
Diorite, 239, 437, 442, 443, 445–447, 449, 450, 452, 457, 543, 594, 595, 597, 598, 645, 680, 682, 688, 692
Differentiation index, 232
Discrimination diagram (discriminant), 525, 530, 531, 534, 540
Divergent volcano, 115, 126, 128
Dome, 107, 122, 124, 126, 128, 140, 151, 167, 173–175, 194, 221, 259, 271–273, 276–278, 293, 312, 319–322, 332, 335, 338, 339, 340, 355, 403, 410, 411, 415, 416, 422, 423, 425, 429–431, 578, 685
Double arc system, 259, 260, 267
Double-planed seismic zone, 264, 265
Dunite, 442

Earth tides (solid), 154
Eclogite (see also Quartz eclogite), 14, 17, 99, 111, 112, 164, 220, 463, 465, 480, 483, 484, 514, 517–519, 575, 603, 665
Eclogite model (partial melting/fractional crystallization), **478–484**
Eclogite xenolith (nodule), 467, 509
Edenite, 67
Enstatite, 58, 110, 144, 463, 493, 494, 501
Epidote, 578, 579, 654
Eruptive gas compositions, **669–675**

Eruption rate, 272, 273
Eu anomalies, 157, 202, 314, 344, 385, 386, 389, 391, 393, 394, 456, 580–582, 641, 642, 646, 647, 655
Eutaxitic texture, 428
Experimental petrology, philosophy of, 494
techniques, 494
Extrusion of lava, **410–411**

Fault plane solution, 102, 171
Fayalite, 197
Fe-rich sulphide melt, 489
(Fe, Ni)S, 509
Fe–Ti oxides, 28, 51, 57, **66**, **68**, 141, 143, 211, 213, 230, 235, 251, 333, 338, 339, 355, 356, 363, 421, 579
Fe–wustite buffer, 499
Feldspathoids, 57, 77, 140, 214, 221
Ferromanganoan sediment, 543
Ferropigeonite, 280
Ferruginous arthigenic minerals, 563
Fiamme, 428
Fissure eruption, 124, 250, 263, 421
Fluorapatite, 80, 687
Fluorite, 690, 691
Flood basalts/lavas, 120, 122
Flow differentiation, 9, 51, 76, 78
Fluidisation/fluidized pyroclastic flow, 408, 415, 431
Fold-thrust belt (back-arc), 677, 679, 680, 689, 690
Fold-thrust belt (collision-related), 678, 680
Forsterite (forsteritic olivine), 58, 178, 252, 492, 507
Fossil fumarole pipe, 424, 428
Fracture zone, 526, 630
Freeboard (continental), 650
Frontal arc, 295, 296
Fumaroles/fumarolic activity, 164, 194, 211, 230, 251, 274, 317, 421, 671, 674, 675

Gabbro, 119, 267, 299, 385, 437, 442–446, 448, 449, 457, 473, 542, 543, 593, 595, 625, 645
Gabbroic cumulates, 314, 632
Gabbro–diorite–tonalite–trondhjemite series/suite, 456, 646
Gabbro/eclogite transition, 654
Gabbro–sodic anorthosite–syenite–rapakavi granite series, 454
Galena, 685, 691
Garnet, 17, 111, 112, 137, 141, 145, 160, 164, 182, 201–203, 210, 289, 310, 312, 314, 367, 387, 393, 456, 474, 475, 477, 479, 480–484, 512–514, 517–519, 575, 585, 586, 617–619, 621, 623, 647, 654, 657
Garnet–amphibolite, 463, 465, 489, 585, 653
Garnet–cordierite andesite, 310
Garnet lherzolite, 167, 178, 182, 498, 528, 530,
Garnet peridotite, 201, 367, 393, 497, 498, 511, 513–517
Gas condensates, 164
Gas content of andesite/lava, 401, 403, 404, 405

Geoelectricity, 274
Geosphere mixing, 637, **663–667**
Geothermal gradient (*see also* Thermal gradient), 586, 653
Geothermal anomalies (*see also* Heat flow), 274
Glacio-eustatic terrace, 309
Glass former constituents, 404
Glass inclusions in phenocrysts, 669, 670–672, 674, 675
Graben, 137, 138, 260, 329, 332, 335, 346, 353
Grain flow (pyroclastic flow), 415, 430
Granite/granitic rocks, 437, 444, 446, 449, 450, 586, 593, 594, 596, 598–601, 604, 605, 615, 623, 627, 643, 645, 648, 653, 655, 677, 680, 686, 689–692
Granite–greenstone belt/terrains, 586, 587
Granite gneiss complex, 587
Granite minimum, 452
Granitoid, 435, 437, 438, 441, 442, 444, 447–450, 453, 679
Granodiorite, 271, 378, 384, 435, 443–445, 447, 448, 450, 453, 457, 593, 594, 596–599, 601, 623, 641, 645, 657, 682, 686, 688, 690, 692
Granodiorite gneiss, 573
Granulite, 164, 554, 566, 639–643, 645–647, 651, 655, 657
Granulite xenoliths (nodules), 645, 647
Graphite, 214, 679
Gravity anomalies/regional variations (*see also* Bouguer gravity anomalies), 3, 99, 151, 162, 163, 167–170, 181, 264, 266
Gravitational collapse (dome/eruptive column) 415, 429
Greenstone belt, 14, 15, 456, 573, 575–581, 583, 584, 586–588, 680, 684
Greisen, 447, 686, 691
Green Tuff region (Japan), 271, 684–686
Grey geniss, 645
Grossularite, 618
Ground surge/deposit, 174, 418, 423, 424
Gypsum, 685

Haematite, 499, 685, 687, 688
Harzburgite, 442, 542
Hastingsite, 67, 167, 179
Häuyne, 57, 199, 230
Hawaiite, 12, 201, 279, 282, 299, 333, 379, 393, 577, 623, 624, 726, 628
Heat flow, 3, 4, 263, 264, 266, 267, 270, 307, 308, 317, 592, 664
Heat production, 458, 573, 591, 603
Henry's law, 508
Hercynian front, 612
Hercynian granites, 691
High-alkali tholeiitic series, 259, 278–280, 284
High-alumina basalt, 13, 115, 141, 149, 157, 162, 179, 187, 192, 197, 199, 201, 202, 207, 213, 253–255, 259, 267, 270, 278, 280, 281, 307, 312–314, 320, 322, 353, 363, 463, 473, 482, 514, 680

High-alumina basalt series, 214, 259, **278–281**, 284, 286, 300, 301, 307, 312, 322
High-alumina olivine tholeiite, 473
High-alumina quartz tholeiite, 473
High-K calc-alkaline/andesite series, 25, 26, 28, 29, 40–43, 51–58, 61–65, 70 74, 92, 93, 207, 212, 213, 216, 221, 259, 271, 307, 311, 312, 316, 318, 320, 322, 327, 330, 332–336, 338, 339, 341–346, 539, 580, 582, 583
High pressure phenocryst/inclusion, 465, 466, 627
Hinge faulting, 530
Hornblende, 11, 12, 17, 25, 51, 54–60, **66**, 76–78, 125–175, 207, 213, 214, 221, 230, 235, 252, 260, 272, 282, 286, 313, 333, 338–340, 344, 356, 449, 456, 575, 578, 621–624, 628, 647, 679
Hornblende andesite, 194, 197, 293, 299, 302, 312, 627
Hornblende basalt, 79
Hornblende dacite, 158
Hornblende gabbro, 267, 466, 483
Hot-avalanche/hot-rock avalanche, 194, 416
Hot-spot/magmas, 102, 230, 555, 606
Hot springs, 267, 274, 685
Hyaloclastite, 271, 335, 338
Hybridization, 12
Hydrosphere, 637, 663, 666
Hydrothermal alteration (associated with mineral deposits), 682, 683, 686, 688, 689
Hydrothermal alteration (ocean crust), 555, 556, 560, 563, 564, 664
Hydrothermal brine, 686
Hypersthene, 17, 62, 123, 140, 160, 175, 212, 214, 251, 281, 314, 359, 493, 625
Hypersthenic rock series, 13, 259, 275, **279–284**, 299, 301

I-type granite/granitoid, 438, 677, **679**, 680, 687, 690, 692
Iapetus Ocean, 611, 613, 614, 617, 631–635
Iapetus Suture, 612, 614
Icelandite, 15, 489
Ignimbrite, 5, 14, 18, 122, 140, 151, 187, 189–191, 193–199, 201, 203, 207, 210, 211, 219, 220, 319, 331, 332, 334, 338–340, 353, 355, 364–366, 368, 401, 403, 407, **408–409**, 411, 412, 424, 426–428, 615, 618, 624, 627, 688
Ignimbrite (pyroclastic flow type), 423
Ilmenite, 57, 66, 68, 109, 160, 161, 207, 212, 396, 397, 469, 472, 621, 624, 625, 628, 654, 679, 687, 690
Immiscible melt, 505, 508
Immiscible sulphide melts, 509, 512, 519, 587
Ignimbrite shield volcano, 196
Incompatible elements, 11, 17, 115, 131, 137, 162, 167, 177, 180, 181, 202, 221, 222, 225, 235, 237–242, 253, 256, 259, 289, 302, 316, 317, 384, 390, 391, 393–396, 441, 451, 452, 489, 515, 523, 525, 528–530, 537, 539, 541, 551, 587, 627–629, 631, 632, 639, 649, 656, 657

Incrustations (around fumaroles), 164, 671
Inter-arc basin, 295, 302
Inter-arc trough, 442
Intermountain basin, 328, 329
Intracrustal melting, 639, 641, 642, 647, 655, 658
Intraplate volcanic rocks, 549–552, 555, 562, 563
Intrusion breccia, 446
Inversion curve (orthopyroxene–clinopyroxene), 356
Island arc calc-alkaline series, **16–17**, 144, 323, 353, 363, 567, 568
  cratonization, 456, 574, 591, 602, 603, 605, 606
  tholeiite series (low K tholeiite series), 9, 11, 13, **14–16**, 17, 18, 21, 29, 97, 167, 175, 177, 181, 245, 252, 293, 295, 311, 334, 351, 361, 372, 377, 380, 386, 390, 394, 440, 465, 466, 467, 483, 484, 526, 528, 530–533, 536, 537, 546, 549, 554, 563, 564, 567, 580, 582, 583, 593, 597, 616, 620, 621, 623, 643, 680
Island arc/volcanic rocks, 1, 3, 4, 6, 9, 11, 12, **14–17**, 19, 25, 29, 34–36, 38, 39, 78, 80, 97, **99–112**, 133, 137, 144, **167–182**, 188, 192, 197, **207–323**, 367, 371, 386, 394, 415, 416, 424, 435, 435, 437–440, 442, 443, 447, 449, 452, 453, 465, 468, 483, 484, 489, 503, 504, 516, 518, 519, 523, 525, 546, 549, 552, 554, 558, 559, 563, 564, 566, 567, 573, 574, 580, 587, 588, 591–595, 597–599, 601, 602, 604–606, 611, 617, 619, 633, 637, 639–644, 648, 649, 651, 658, 677, 678, 680–683, 687, 690
Isotopes and continental growth, **647–653**

Jadeitic pyroxene, 654

K-feldspar (*see also* Alkali feldspar), 251, 314, 355, 357, 442, 443, 447, 456, 497, 542, 683
Kaersutite/kaersutitic amphibole, 280, 282, 335
K-*h* correlation/relationship, 110, 144, 187, 202, 203, 307, 316, 322, 371, 384, 386, 394–397
K-rich oceanic 'tholeiite', 477
K-rich series (Roman Volcanic Province), 27, 36, 37, 44–49, 53, 54, 57, 59, 60, 70, 71, 79
K-richterite, 552
Kimberlite, 396, 467, 509, 513, 516, 654
Komataiite/komataiitic series, 21, 573, 583, 585, 586, 588, 603
Komataiite, basaltic, 504, 576
Komataiite, peridotitic, 504
Krakatoan-style eruptions, 151, 152
Kuroko-type sulphide deposits, 637, 677, 680, 682, **683–686**
Kyanite, 476, 477, 665

Labradorite, 59, 125, 197, 214, 252, 335, 625
Lahar, 211, 243, 410, 625
Laminar flow (pyroclastic flows), 415, 419, 430
Lateral accretion (island arcs), 637, 639, 641, 643, 644, 658
Lateral faulting, 5, 6

Latite, 332–334, 342, 343, 345, 489
Lava lake, 271
Lava plug, 250, 252, 377, 379, 389
Lead isotopes, 80, 109, 180, 253, 279, 288, 302, 303, 490, 523, 529, 549, **555–559**, 562, 564, 567, 568, 647, **649–650**, 652, 657, 686, 691
Lead ore growth curve, 556, 557, 649
Least squares (linear) mixing calculation, 237, 255, 364, 365, 367
Leucite, 57, 201, 207, 211, 214, 215, 221, 230, 313, 314, 339
Leucite basanite, 201, 214
Leucite tephrite, 13, 19, 214, 218, 312, 313, 316
Leucite tephritic series, 307, 311, 313
Leucitite, 214, 218, 333
Leucogranite, 445, 447
Leucoxene, 579
Lewisian metamorphic rocks, 611, 614, 641, 642, 645, 647, 648, 650
Lewisian amphibolites, 641, 642, 645
Lewisian granulites, 641, 642, 645, 646, 651, 654
Lherzolite, 396, 499
Lithic tuff breccias, 299
Loveringite, 80
Low alkali tholeiite, 259, 269, 270, 284, 285
Low alkali tholeiitic series, 259, **278–279**, 284, 287
Low K series (see also Island arc tholeiite series), 14, 15, 17, 19, 25, 26, 28, 29, 40–43, 51–54, 56–59, 61–64, 66, 70–74, 76, 80, 88, 89, 108, 192, 212, 313, 334, 386
Low-Si andesite, 310, 311, 320
Lower crust, 566, 567, 639, 640, 641, **645–647**, 655–657
  seismic P wave velocities, 645

Maar, 196, 273, 332, 335, 425
Mafic inclusions, 267
Magma mixing, petrogenetic models, 490, 530, 533, 565, 566, 584, 585, 663
  petrographic/geochemical evidence, 167, 168, 174, 177, 180, 182, 212, 213, 565, 566, 625, 663, 665, 666
Magma reservoir/chamber, 240, 260, 286, 288, 368, 422, 431, 675
Magma segregation/extraction, 13, 111, 112
Magmatic arc mineralization, 637, 677, 678, **680–688**
Magmatic U deposits, 691, 692
Magnetic anomalies, 102, 151, 208, 248, 270, 274, 373, 374, 375, 380
Magnetite, 17, 99, 109, 110, 123, 161, 162, 167, 175, 179, 207, 212, 214, 251, 280, 314, 368, 380, 468–470, 472, 475, 479, 480, 483, 484, 499, 541, 542, 575, 578, 579, 618, 621, 624, 670, 679, 687, 688
Magnetite–haematite (M–H) buffer, 470, 501
Magnetite–haematite–apatite mineral deposits, 677, 687, 688
Main boundary Fault (Himalayas), 691

Main Central Thrust (Himalayas), 691
Manganese nodules, 387, 555, 556
Mantle diapirs, 14, 21, 397
Mantle heterogeneity, 289, 368, 395, 528, 531, 537, 552, 648
Mantle metasomatism, 367, 463, 489, 490, **515–519**, 656
Mantle peridotite (see Peridotite)
  plume, 586, 588
  wedge (asthenospheric mantle wedge), 14, 21, 133, 188, 201, 207, 221, 245, 255, 256, 265, 266, 289, 390, 396, 437, 463, 465, 483, 484, 550, 552, 558, 559, 562, 564, 637, 649, 653, 654, 683, 686
  xenoliths (nodules), 396, 528, 552, 656
Marginal basin (sea), 1, 3, 4, 11, 20, 21, 80, 245, 270, 317, 372, 376, 391, 392, 573, 582, 588, 600, 633
Marginal basin basalt, 13, 20, 21
'Marid' nodule, 396
Median mass, 328, 329
Megacrysts, 466
Mélange, 4, 5, 560
Melilite olivine nephelinite, 287
Metalliferous sediments, 556
Metamorphism/metamorphic rocks, 102–104, 116, 525, 579, 591–595, 598, 599, 601, 603, 604, 606, 611, 613, 615, 627, 632, 633, 640–643, 645, 650, 653, 664, 687, 690, 692
$Mg_3Al_2Si_3O_{12}$–$Ca_3Al_2Si_3O_{12}$–H_2O system, 492
MgO–FeO–Fe_2O_3–SiO_2 system, 468
Mg_2SiO_4–Ab–An system, 505
Mg_2SiO_4–$An_{50}Ab_{50}$–SiO_2 system, 495
Mg_2SiO_4–$An_{50}Ab_{50}SiO_2$–H_2O system, 497
Mg_2SiO_4–$CaMgSi_2O_6$–SiO_2 system, 493
$MgSiO_3$–CO_2–H_2O system, 493
Mg_2SiO_4–FeO–$CaAl_2Si_2O_8$–SiO_2 system, 470
Mg_2SiO_4–Fe_3O_4–$CaAl_2Si_2O_8$–SiO_2 system, 470
Mg_2SiO_4–Fe_2SiO_4–$K_2O.SiO_2$ system, 506
Mg_2SiO_4–$NaAlSi_2O_6$ system, 508
Mg_2SiO_2–SiO_2–$CaMgSi_2O_6$–H_2O system, 494
Mg_2SiO_4–SiO_2–H_2O–CO_2 system, 502
Mica (see also Two-mica granite), 144, 145, 396, 484
Micro-earthquakes, 274
Mid-ocean ridge, 395, 549, 550–552, 555, 557, 653
Mid-ocean ridge basalt/tholeiite (abyssal/deep sea/ocean floor-basalt/tholeiite), 11, 15, 21, 109, 111, 178, 179, 182, 188, 253, 289, 295, 362, 363, 371, 389, 390, 393, 397, 439, 467, 477, 482, 490, 514, 515, 519, 526–532, 534, 536, 537, 541, 544, 545, 550, 555, 556, 558–560, 563, 579, 623, 632, 639, 649, 651, 652, 656, 657, 663
Migmatite, 645, 692
Millepede model of ductile plate tectonics, 602
Minoan eruption, 319, 320
Mineral deposits/mineralization, 7, 121, 221, 239, 442, 447, 543, 604, 637, **677–693**
Mineral exploration, 637, 692, 693
Molasse basin, 329
Monogenetic cones, 122, 140, 142, 146, 151

Monogenetic lava extrusions, 194
Monogenetic volcanoes, 152, 259, 263, 267, 268, 273, 274, 287
Monzogranites, 445, 690
Monzonite, 437, 447, 449, 457
Mortite, 688
Mount Katami tuff flow type (pyroclastic flow), 423
Mud eruptions/flows, 211, 274, 406, 411, 416, 420, 424
Mugearite, 12, 279, 282, 624
Muscovite, 447, 679, 686

$NaAlSi_2O_6$–$NaFe^{3+}Si_2O_6$ system, 508
$NaAlSiO_4$–Mg_2SiO_4–SiO_2 system, 496
Na_2O–CaO–MgO–Al_2O_3–SiO_2–H_2O system, 497
Na_2O–CaO–Al_2O_3–MgO–SiO_2–H_2O system, 499
Nappes, 5, 328, 543, 632
Neodymium isotopes, 187, 197, 201, 202, 238, 394, 490, 523, 549, **550–555**, 556, 559, 562–564, 567, 568, 647, **650–652**, 680
Nepheline, 221, 335
Nepheline normative volcanic rocks, 11, 12, 19, 29, 41, 110, 140, 141, 151, 215, 340, 622, 624, 628, 629
Nepheline syenite, 119
Nephelinite, 466, 493
Newtonian fluid, 404
Nichols' formula, 403
Nickel–nickel oxide (NNO) buffer, 469, 474, 475
Ni sulphide, 178
Ni sulphide mineral deposits, 680
Nuée ardente, 17, 18, 140, 144, 168, 174, 211, 231, 232, 403, 406, 408, 409, 411, 417, **422–427**, 430, 431

Obsidian, 123, 151, 312, 319, 320, 340, 545
Ocean ash layers, 207, 297, 422
Ocean crust, 1, 4, 14, 79, 99, 110–112, 182, 187, 200, 201, 225, 256, 289, 293, 295, 310, 386, 390, 438, 442, 443, 452, 453, 457, 465, 467, 468, 482, 515, 519, 525, 542, 546, 550, 552, 555, 564, 567, 595, 606, 614, 653, 654, 663, 664, 688
Ocean ridge/rise, 1, 2, 3, 189, 237, 238, 270, 373, 375, 654, 680
Ocean island volcanic rocks, 550, 552, 555, 557, 559, 563, 651
Ocean ridge/floor basalt/tholeiite (*see* Mid-ocean ridge basalt/tholeiite)
Ocean sediment, 4, 14, 102, 132, 167, 168, 180–182, 375, 394, 550, 555–558, 560, 562
Oceanic tholeiite/ocean floor basalt (*see* Mid-ocean ridge basalt/tholeiite)
Ocean water (*see* Seawater)
Oligoclase, 160, 623
Olististrome, 295
Olivine, 9, 12, 17, 20, 25, 26, 51, 53, 56–58, **59–62**, 64–66, 73–78, 81, 99, 109, 110, 123, 132, 140, 141, 143, 144, 160–162, 164, 167, 175, 178, 179, 182, 197, 201, 202, 207, 211, 213–215, 218, 230, 235, 242, 245, 251, 252, 280, 282, 283, 288, 289, 303, 307, 313, 314, 333, 335, 338, 339, 355, 356, 357, 364, 366, 380, 393, 394, 439, 463, 466, 470–472, 474, 477, 479, 480–482, 484, 491, 493, 494, 497, 499, 500, 504, 505, 511, 513, 516, 528, 531, 535, 541, 542, 586, 587, 615, 621, 624, 625, 628, 646, 670
Olivine–orthopyroxene join ($NaAlSiO_4$–Mg_2SiO_4–SiO_2 system), 497
Olivine basalt, 41, 158, 173, 255, 624, 626
Olivine basalt–trachyte/phonolite association, 12
Olivine-bearing andesite/low-Si andesite, 245, 248, 249, 252, 254, 255
Olivine gabbro, 267
Olivine nephelinite, 230
Olivine-normative basalt, 29, 151, 233, 500, 501
Olivine tholeiite, 11, 15, 111, 201, 280, 390, 472, 473, 620, 622, 623, 654
Olivine websterite, 267
Ophiolite(complex/ophiolitic rocks), 21, 172, 192, 227, 245, 268, 390, 391, 525, 541–545, 559, 560, 591, 593, 594, 597, 599, 600–607, 614, 653, 663, 664, 666
Ore blanket, 682
Orogenic areas/volcanic rocks, 1, 6, 9, 11, **12–14**, 21, 68–74, 227, 241, 270, 293, 327–329, 338, 363, 435, 445, 490, 574, 592, 593, 641, 650, 653, 678, 682, 691, 692
Orthopyroxene, 15, 17, 25, 51, 53, 56–58, **62–66**, 76–78, 109, 140, 141, 143, 144, 161, 175, 179, 197, 214, 235, 251, 252, 280, 282, 283, 333, 338, 339, 355–357, 474, 491, 493, 497, 499, 500, 513, 516, 541, 575, 578, 579, 615, 621–625
Outer arc, 168, 260, 267, 442, 447, 677–679, 680, 690
Outer arc mineralization, 637, 677, 678, 684, **690–691**, 692
Oxygen fugacity, 110, 302, 314, 315, 322, 355, 356, 468–470, 474, 475, 480, 499–501, 509, 529, 541, 679
Oxygen isotopes, 79, 162, 208, 219, 220, 222, 453, 549, **559–562**, 565–568, **652–653**, 664, 665, 679

Pacific (branch of) igneous rocks, 12
Paired metamorphic belt, 5, 270
Palaeomagnetism, 116, 168
Pan-African, 456, 574, **591–609**, 692
Pargasite, 66, 67, 160, 230, 471, 497
Partial melting/anatexis, amphibole peridotite, 144
   amphibolite, 489, 515, 575, 585, 653
   continental crust/lower continental crust, 78–80, 220, 222, 322, 365, 368, 452, 484, 523, 549, 558, 568, 584, 624, 643, 666, 679, 686, 690, 692
   eclogite/quartz eclogite, 112, 164, 220, 489, 515, 575, 585, 653
   garnet lherzolite/peridotite, 141, 178, 182, 393
   granulite, 164, 585
   mafic rocks, 584, 585, 647, 653

mantle peridotite, 19, 20, 111, 129, 137, 141, 144, 145, 155, 164, 167, 178, 187, 201, 202, 207, 220, 221, 241, 245, 255, 256, 289, 303, 307, 314, 317, 322, 328, 345, 367, 387, 392, 393, 395, 396, 456, 463, 465, 489, 490, 492, 493, 497, 499–503, 506, 523, 528, 531–535, 563, 584–586, 663, 679
  natural peridotite, 498
  oceanic crust (subducted), 11, 112, 164, 178, 179, 182, 201, 220, 221, 256, 322, 328, 390, 452, 456, 463, 465, 467, 472, 490, 515, 523, 550, 555, 562, 563, 567, 588, 639, 653, 655, 679
  pyrolite, 500, 653
  metasomatized spinel peridotite, 515–519
  plagioclase peridotite, 586
  spinel peridotite, 144, 367, 492
  subducted continental crust/sediment, 523, 529, 663, 690
Partition coefficient/trace element partitioning, barium, 364
  chromium, 364
  cobalt, 364, 506, 507, 509, 512
  copper, 509, 512
  crystal–liquid, 131, 141, 144, 146, 212, 301, 361, 363, 364, 367, 395, 456, 489, **504–515**, 528, 587
  crystal–vapour, 516, 517
Partition coefficients, FeO, 506, 507
  trace element partitioning, manganese, 506, 507
  MgO, 506, 507
  trace element partitioning, nickel, 364, **504**, 505–512
    Rare Earth Elements, 512, 513
    rubidium, 364
    strontium, 364
    vanadium, 364
    zinc, 364
Peacock alkali–lime index, 439, 440
Pegmatite, 644, 686, 691, 692
Pelean pyroclastic flows, 422
Pelean-type eruptions, 140, 174, 231, 401, **409–410**
Peralkaline granite, 593, 594, 596, 602
Peralkaline rocks/magmatism, 440, 591, 593, 594, 602, 605, 606, 686
Peralkaline rhyolite (see Comendite), 230, 594
Peralkaline trachyte, 594
Peridotite, 11, 14, 20, 21, 78, 80, 99, 110–112, 137, 141, 144, 145, 164, 178, 187, 201–203, 255, 288, 307, 314, 317, 322, 353, 367, 452, 483, 489, 490, 492, 494, 497, 499, 501, 503, 518
Peridotite nodule, 509, 516
Petrogenetic modelling/petrogenetic pathways, **530–538**, 540
Petrographic provinces, 12
Phase diagram, andesite (anhydrous), 472, 474
Phase diagram, andesite (5 per cent water), 474–476
Phase diagram, andesite (water-saturated), 477, 479
Phase diagram, tholeiitic basalt (anhydrous), 472, 474
Phase diagram, tholeiitic basalt (5 per cent water), 474–476

Phase diagram, tholeiitic basalt (water-saturated), 477, 478
Phengite, 477
Phenocryst contents/occurrence, 9, 25, 26, **51–59**, 74, 75–76, 78, 81, 99, 109, 123, 162, 175, 343, 357, 483
Phenocryst disequilibrium, 452
Phenocryst mineralogy/composition, 9, 17, 25, 26, 28, **58–68**, 78, 109, 287, 343, 355–357, 615, 616, 620, 621
Phlogopite, 66, 68, 80, 133, 144, 230, 477, 664
Phlogopite eclogite, 256
Phonolite, 214, 218
Phonolitic leucitite, 333
Phonolitic tephrite, 230
Phosphatic authigenic minerals, 563
Phreatic eruption/activity, 173, 174, 196, 211, 251, 273, 317, 321, 335, 410
Phreatomagmatic eruption/activity, 251, 335, 410, 425
Phyllic alteration, 682
Picrite, 531
Pigeonite, 15, 51, 57, 63, 65, 66, 77, 78, 81, 140, 160, 207, 213, 251, 280–283, 314, 355, 357
Pigeonitic rock series, 13, 14, 275, **278–284**, 300, 301
Pillow lava, 250, 271, 338, 355, 442, 542, 543, 545, 578, 613, 614, 619, 664
Pinch and swell structure (pyroclastic rocks), 424
Pit crater, 249
Placer deposit, 686
Plagioclase, 9, 11, 12, 15–17, 20, 25, 26, 51, 57, **59–61**, 75–78, 81, 99, 109, 125, 143, 144, 160–162, 175, 197, 201–203, 207, 211–213, 214, 230, 235, 237, 251, 252, 289, 313, 314, 333, 335, 338–340, 355–357, 361, 364, 366, 380, 391, 393, 394, 468, 470–472, 474, 477, 480, 483, 491, 528, 535, 541, 542, 549, 560, 567, 568, 575, 578, 579, 586, 615, 621, 623–628, 642, 646, 655, 670
Plagioclase barrier (thermal barrier), 472, 474
Plagioclase cumulate, 267
Plagioclase lherzolite, 267, 531
Plagioclase–liquid geothermometer, 356
Plagioclase–quartz join (Mg_2SiO_4–$Ab_{50}An_{50}$–SiO_2 system), 497
Plagiogranite, 391
Plateau basalts/lavas, 139, 140, 141, 192, 203, 281, 338, 340, 373
Plinean eruptions, 140, 174, 211, 259, 401, 403, 422, **407–408**, 410, 429, 431
Plug flow, 415, 419, 430
Polymerization of silicate melts, 404, 491, 505, 509
Porphyry copper mineral deposit, 220, 239, 442, 637, 677, 679, 680, **681–683**, 685–688, 690, 692
Potassic alteration, 682, 683
Prediction of volcanic eruptions, 7, 174, 274
Preferred orientation of volcanoes, 263
Prehnite, 578, 603
Primordial mantle composition, 532, 538, 540, 657
Principal components analysis, 76, 77

Subject Index

Propyllitic alteration, 682, 683
Proterozoic rocks, 454–457, 573, 574, **591–607**, 643, 646, 647, 684
Proterozoic super-continent, 573
Pseudo-isochron, 157, 241
PtS–Pt buffer, 509, 512
Pumice and ash deposit, 324, 425
Pumice–ash flow, 423, 424, 431
Pumice and ash deposit (pyroclastic flow type), 423, 425
Pumice eruption/deposit, 250, 273, 279, 312, 317, 319–321, 407–409, 428, 429, 578
Pumice-explosion crater, 312
Pumice-fall eruption, 259, 273, 312, 407, 408
Pumice flow, 18, 173, 174, 410, 421, 423, 426–428
Pumice-flow deposits, 421, 578
Pumice surge, 427
Pumpellyite, 603
Pyrite, 579, 682, 683, 685, 686
Pyroclastic avalanche, 250, 259, 273
Pyroclastic cone, 249, 273, 276, 278
Pyroclastic fall (see air-fall), 17, 140, 151, 167, 173, 174, 180, 190, 259, 263, 276, 277, 405, 407, 424
Pyroclastic flow, 5, 15, 17, 140, 167, 168, 173, 174, 190, 196, 239, 250, 259, 263, 268, 269, 272–278, 401, 406, 408, 409, 410, **415–431**
  ash flow type, 423
  ash–pumice surge, 423
  block and ash, 423
  concealed fissure–orifice type, 423
  degassing structure, 428
  fissure-eruption type (Valley of 10000 Smokes), 423
  Katmaian type, 423
  Krakatoan type, 423
  Merapi lateral disintegration, 423
  Merapi type, 423
  Novarupta tuff flow type, 423
  Nuée ardente (pyroclastic flow type), 423
  Nuée ardente du massif du Katmai, 423
  Nuée ardente peléene d'avalanche, 423
  Nuée ardente peléene d'explosion dirigée, 423
  Nuée ardente d'explosion vulcanienne, 423
  Nuée ardente surge, 427
  Peléan type, 423
  Pelée discharge lateral type, 423
  Pelée type, 423
  Pelée vertical type, 423
  Pelée directed lateral type, 423
  St. Vincent type, 140, 173
  St. Vincent vertical type, 423
  scoria surge, 423, 427
  scoria and ash deposit, 423–425
  Sourfiere type, 423
  Valley of 10000 Smokes type, 423
Pyroclastic flow deposits, composition and grain size analysis, 426–427
Pyroclastic plateaux/sheet, 272, 408

Pyroclastic rocks, 5, 15, 41, 76, 119, 120, 122, 149, 180, 192–194, 250, 251, 263, 270, 272, 273, 276, 277, 293, 298, 299, 321, 331–333, 339, 355, 377, 379, **405–410**, **415–431**, 576, 578, 601, 615, 620, 623, 624, 629, 671, 686, 689
Pyroclastic surge, 335, 409, 410, 415, 429
Pyrolite, 17, 21, 494, 499
Pyrope, 618
Pyroxene, 16, 17, 57, 59, 60, **62–66**, 73, 80, 123, 143, 158, 164, 167, 179, 197, 201, 202, 207, 212, 213, 280, 289, 299, 310, 313, 314, 338, 355, 439, 449, 468, 470, 472, 474, 477, 480–484, 491, 492, 500, 531, 541, 578, 586, 625, 627, 646

Quartz, 17, 56, 57, 77, 140, 160, 175, 182, 197, 207, 213, 214, 251, 272, 286, 287, 333, 338–340, 355, 357, 472, 474, 476, 477, 479–481, 492, 493, 514, 575, 578, 579, 620, 621, 627, 653, 685, 686, 688, 689
Quartz–albite–orthoclase system, 481
Quartz diorite, 121, 271, 442, 444, 445, 448, 473, 596, 645
Quartz eclogite, 111, 112, 465, 467, 479, 513–515, 518, 653, 654
Quartz–fayalite–magnetite (QFM) buffer, 469, 474, 475
Quartzite xenoliths, 627
Quartz latite, 422
Quartz monzonite, 442, 445, 686
Quartz-normative basalt/tholeiite (*see also* Silica-oversaturated basalt), 15, 29, 41, 151, 201, 229, 230, 233, 358, 472, 473, 500, 501, 622, 623, 625
Quartz porphyry, 443, 687, 689

Radiogenic heat/radioactive heat-producing elements, 454, 642, 645, 653, 657
Rapakivi granite, 455
Rare earth elements (REE), 11, 14, 15, 17, 19, 20, 25, 78, 80, 108, 127, 130, 137, 141, 157, 160, 164, 177, 197, 199, 202, 203, 207, 220–222, 225, 233–235, 239–242, 253, 254, 256, 259, 260, 279, 285, 286, 289, 302–304, 314, 344, 345, 356, 361, 362, 365, 366, 383–387, 389, 390–395, 441, 451, 456, 465, 482, 483, 490, 504, **512–519**, 523, **525–541**, 549, **562–564**, 566, 567, 575, 579–586, 597, 611, 616, 618–620, 622–625, 627, 628, 635, 641–643, 646–648, 651, 654, 655, 657
Reaction relationships/relations, 282, 283, 357, 471, 500
Reaction rim, 283, 618, 627
Remote correlation spectrometry, 164, 669, 671–675
Remnant arc, 295, 643
Residual phase (in partial melting), 80, 179, 523, 525, 529, 531–533
Resurgent caldera, 196
Rheology of andesite, 401, 404
Rheomorphic welded tuff, 428
Riebeckite-acmite microgranite, 623

Rifting (rifted environment), 80, 137, 139, 196, 228, 454, 455, 458, 530, 545, 573, 575, 587, 588, 599, 614, 686
Ring complex/structure, 446, 644
Rutile, 80, 111, 396, 397, 529, 654

S-type granite/granitoid, 438, 656, 677, **679**, 686, 687, 690, 692
Salite, 62, 65
Sanidine, 25, 54, 57, 77, 333, 339, 340
Saussuritization/epidotization, 578, 579
Sb–W–Hg mineral deposits, 677, 687
Scavenged acids on volcanic ash, 669
Scheelite, 687, 689, 691
Scoria cone (cinder cone), 106, 123, 151, 152, 249, 259, 299, 332, 379, 405, 406
Scoria flow, 418, 423, 424, 426, 427
Sea-floor weathering, 529
Sea-floor spreading, 1, 133, 228, 230, 270, 295, 346, 347, 372, 373, 375, 376, 456, 654, 692
Seawater, 14, 386, 387, 390, 395, 453, 555, 559, 560, 564, 555, 565, 567, 656, 664, 666, 679, 685
Segmentation of volcanic belt, 6, 99, 104, 105, 149–152, 154, 156
Seismicity/seismic zone, 3, 9, 99, 102, 115, 138, 149–152, 154, 155, 162, 163, 167, 168, 170–173, 187, 188, 207, 208, 210, 220, 229, 245, 262, 264, 267, 274, 307–310, 314, 316–319, 328, 375, 410, 490, 515, 645, 678
Seismic crisis, 173, 174
Sericitization, 578, 683, 689
Serpentine, 503, 615
Serpentine, 14, 17, 613, 614, 664
Sheeted dyke swarm/complex, 542, 545, 559
Shield (*see also* Craton), 640, 650
Shield volcano, 106, 115, 123, 125, 128, 196
Shoshonitic basalt, 312, 313, 315, 333, 334, 342, 343, 538, 539
Shoshonitic association/series, 9, 11, 13, 14, 17, **19**, 20, 25, 26, 28, 29, 40–43, **50–66**, **70–74**, 76, 78, 79, 94, 95, 140, 187, 189, 198, 199, 201–203, 207, 212, 214, 218, 221, 235, 302, 307, 309–314, 316, 320, 327, 330, 332–334, 339, 342–346, 440, 447, 489, 525–527, 529, 536–541, 545, 611, 628, 630
Silica-undersaturated (alkaline) basalt, 19, 29, 152, 177
Sillar, 428
Sillimanite, 214, 312
Sn mineral deposits, 679, 686, 687, 690–692
Sn–Ag mineral deposits, 688
Sn–W mineral deposits, 688, 692
Sodalite, 57, 230
Solfatara/solfataric activity, 211, 230, 232, 241, 251, 274
Spacing of volcanic centres/islands, 104, 106, 296–298
Spatter cone/spatter, 332, 340, 405
Spessartine, 618

Sphalerite, 685
Sphene, 356, 396, 465, 476, 482–484, 529, 533, 579, 687
Spilite/spilitic alteration, 1, 15, 256, 468, 477, 484, 619, 664
Spilitised 'tholeiite', 473, 477
Spinel (*see also* Chrome-spinel), 20, 123, 143, 144, 167, 178, 179, 202, 367, 472, 492, 497, 500, 513, 535, 541
Spinel cumulate, 267
Spinel peridotite/lherzolite, 267, 367, 492, 497, 498, 500, 501, 513, 514, 518, 519
Spinel lherzolite xenoliths, 466
Staurolite, 477
Stibnie, 687
Strombolian volcanic activity, 313, 401, 403, **405–406**, 409
Strontium isotopes, 14–21, 80, 109, 110, 127, 129, 131, 132, 137, 141, 144, 145, 157, 164, 167, 175–177, 181, 187, 197, 199, 201–203, 207, 208, 217, 218–222, 225, 234–241, 253, 254, 260, 288, 289, 302, 303, 314, 316, 323, 333, 344, 345, 365, 386, 389, 392, 395, 397, 451, 454, 490, 529, 549, **550–555**, 556, 559, 560, 562–568, 601–603, 628, 643, 647, **648–649**, 650–653, 656, 663–666, 679, 680, 683, 690–692
Subduction complex, 4, 654
Subduction (process), 11, 13, 14, 21, 27, 99, 102–104, 129, 132, 133, 137, 139, 145, 146, 168, 170, 181, 182, 207, 208, 220, 228, 229, 237–242, 245, 248, 262, 268, 270, 285, 288, 289, 298, 307, 308, 310, 327, 345–347, 351, 353, 372–377, 380, 381, 390, 392, 394, 396, 397, 437, 447, 451, 452, 456, 458, 467k, 483, 484, 530, 531, 537, 552, 555, 565, 591, 592, 602, 603, 605, 613, 614, 629, 631, 633, 635, 649, 651, 653, 654, 663, 665, 666, 678, 684, 686, 687, 690, 691
Subduction, oblique, 611, 635
Subduction of sediment, 222, 390, 453, 454, 468, 525, 550, 552, 558, 559, 561, 562, 564, 637, 639, 648–652, 654, 656, 658, 663, **664–665**
Subduction zone, 2, 9, 27, 79, 127, 137, 139, 167, 168, 171, 172, 178, 181, 187, 201, 203, 210, 226, 227, 236, 245, 248, 255, 270, 271, 308, 328, 345, 351, 353, 372, 394, 395, 443, 452, 456, 467, 490, 515, 516, 526, 529, 534, 535, 545, 546, 558, 559, 562, 587, 592, 593, 601, 602, 604–606, 617, 631, 632, 635, 639, 644, 649, 651, 653, 654, 655, 666, 678, 680, 682, 690
Sulphur, budget in volcanic gas, **669–676**
  depletion in upper mantle, 587
Sulphur dioxide, 164
Sulphur fugacity, 509
Sulphides, in Archaean mantle, 587
Sulphur solubility in silicate melts, 489, 509, 512
Supergene enrichment, 682
Surge/surge deposit (*see also* Pyroclastic surge), 409, 410, 416, 418, 422–425

Suture, 6, 346, 347, 591, 592, 605, 612, 629, 630–633, 635
Syenite, 447, 481, 596
Syenogranites, 445

Tantalo-columbite, 691
Tectosphere, 396, 655, 666
Temporal variations (in volcanic products), 351, 381, 634
Temperature of magma, 110, 141, 396, 466
Temperatures in Archaean mantle, 587
Tephra, 156, 250, 251, 379, 669, 675
Tephrite, 199, 211, 214, 230, 314–316
Tephritic phonolite, 230
Tephritic series, 316
Tetrahedrite, 685
Thermal regime in subduction zone, 237, 368, 456, 516, 643, 654, 658
Thermal gradients (*see also* Geothermal gradients), 604, 644
Tholeiitic basalt, 13, 15, 107, 111, 118, 132, 157, 168, 214, 216, 230, 236, 240, 241, 251, 281, 286, 308, 366, 621, 653
Tholeiitic mafic gneiss, 646
Tholeiitic series, 212–214, 221, 233, 235, 250, 275–279, 281, 285, 286, 301, 315, 340, 358, 359, 363, 365, 367, 368, 396, 416, 439, 440, 442, 443, 447, 454, 466, 469, 472, 475–477, 484, 513, 536, 538–541, 563, 573, 575–578, 582–588, 593, 601, 611, 614, 619, 632, 633, 643, 653, 657, 685, 680
Tholoid, 251
Titanaugite, 137, 140
Titanomagnetite (titaniferous magnetite), 25, 57, 66, 68, 76, 160, 167, 175, 179, 251, 252, 335, 356, 357, 366
Titanpyroxene, 66
Tonalite, 384, 385, 435, 437, 443–447, 456, 457, 543, 586, 588, 593, 643, 644, 646, 653–655, 657, 688
Tonalite gneiss, 456, 573, 645, 646, 647, 657
Topaz, 691
Tourmaline, 447, 679, 686, 689, 690
Trace elements in orogenic volcanic rocks, **68–74**, **526–530, 539–541, 562–564**
Trachyte, 19, 41, 71, 187, 201, 229, 230, 240, 279, 282, 303, 312–315, 318, 319, 320, 333, 334, 342–344, 465, 489, 594
Trachyandesite, 199, 214, 230, 232, 233, 235, 240, 241, 279, 282, 287, 293, 299, 302, 303
Trachybasalt, 140, 142, 214, 230, 232, 233, 235, 240, 241, 287
Transcurrent faulting, 329
Transform fault, 150, 228, 229, 230, 237, 245
Transition elements, 362, 363, 489, 490, 504–509, 519, 573, 575, 579, 580, 582, 584, 587, 623
Trench/ocean trench, 99, 102, 104, 115, 118, 133, 137, 138, 146, 152, 198, 200, 207, 208, 225–230, 232, 233, 236–239, 241, 242, 245, 246, 248, 256, 259, 260–262, 264, 266, 268, 269, 295, 296, 307–310, 317–319, 373, 375, 376, 378, 387, 394, 442–444, 453, 604, 613, 665, 684, 687, 690
Trench-volcano gap (*see also* arc-trench gap), 263, 265, 268, 285
Tridymite, 280, 355, 428
Triple junction, 259, 260
Trondhjemite, 383, 437, 543, 586, 653, 654, 680
Trondhjemite gneiss, 646, 647
Tschermakite/tschermakitic amphibole, 67, 577
Tsunami, 231, 274, 411
Tuff, 118, 119, 120, 124, 210, 211, 248, 250, 271, 299, 332, 335, 429, 575, 576, 578, 594, 596, 685
Tuff-ring, 196, 335
Turbulent flow (pyroclastic flow), 419, 429, 430, 431
Two-mica granite, 447, 457, 680, 689, 691, 692

U mineral deposits, 677, 691, 692
Ultramafic layered complex, 594
Ultrabasic cumulate blocks, 167, 174, 175, 179, 180
Ultramafic lavas (Archaean greenstone belts), 575, 578, 586, 588, 603
Ultramafic xenolith/nodule, 267, 466, 467, 483
Ulvöspinel, 66
Underflow (pyroclastic flow), 415, 416, 418, 423, 424, 430, 431
Underplating, 5, 80, 465, 468, 483, 665, 666
Undersaturated basalt (*see also* Alkaline basalt), 137, 167, 178
Undersaturated leucite-bearing rocks, 214
Undersaturated rock, 335, 340
Uraninite, 692

Valles-type cauldron (*see also* Resurgent caldera), 271
Vapour phase crystallization, 428
Veined mantle model, 396, 397
Vertical accretion (continental crust), 637, 639, 644, 658
Vesicularity of rock/vesicles/vesiculation, 406–408, 421, 422, 424–426, 430, 431
Vertical explosion type (Krakatoan) (pyroclastic flow), 423
Viscosity of lava, 51, 273, 299, 401, 403, 404, 407, 409, 410, 453
Volatiles in lavas/silicate melts, 260, 279, 286, 289, 404, 427, 466, **491–492**
Volatile loss, 540
Volcanic bomb, 405, 426
Volcanic centres, spacing, 261–263
Volcanic centres, evolution of (volcanic evolution), 106, 107, 272, 273, 293, 299
Volcanic cycles, Archaean, 575, 576, 586
Volcanic front, 3, 6, 104, 105, 110, 149, 151, 153, 155, 162, 164, 193, 259, 260–269, 271, 272, 279, 284–286, 288, 327, 346, 633
Volcanic gas, 164, 274, 405, 406, 408, 410, 421, 429, 431, 637, **669–675**
Volcanic gas, direct sampling, 670
Volcanic hazard, **411–412**

Volcano height, 163, 263, 297, 298
Volcano-tectonic depression, 120, 273, 421
Volcanic tremors, 274
Volume of volcanic rocks, 104, 106, 107, 115, 120, 121, 123, 128, 129, 181, 191, 237, 241, 260, 270, 272, 273, 275–278, 342, 420, 440, 445
Vulcanian eruption, 401, 403, **406–407**, 409

Water, content in magmas, 110, 162, 356, 404, 406, 457, 466, 480, 481, 490, 501, 503, 586, 644
Water, infiltration rate in mantle, 503, 504, 516
Water, solution in silicate melts, 491, 506
Websterite, 267
Welded air-fall tuff, 407
Welded ignimbrite, 18, 408, 428
Welded tuff, 406, 428, 598, 601
Welding (pyroclastic flows), 421, 422, 427, 428
Within-plate volcanism/volcanic rocks, 3, 526–530, 534, 536–538, 540–542, 544, 545, 593, 594, 597, 598, 602, 606

Witwatersrand-type Au–U mineral deposit, 680
Wolframite, 689

Xenocrysts, 57, 65, 66, 76, 81, 288, 627
Xenoliths, 182, 214, 220, 222, 256, 288, 310, 314, 322, 333, 640, 654, 655
Xenoliths, cognate, 180
Xenoliths, gneiss, 613
Xenotime, 691

Yield strength of lava, 403, 404, 406, 407, 409, 410, 415, 419, 430, 431

Zeolite, 390, 579, 603, 628
Zircon, 356, 396, 397, 529, 533, 539, 541, 591, 632, 654
Zoisite, 476, 484, 579
Zone refining, 202, 394, 395